Molecular Neurology

Molecular Neurology

Edited by

Stephen G. Waxman

ELSEVIER

AMSTERDAM • BOSTON • HEIDELBERG • LONDON
NEW YORK • OXFORD • PARIS • SAN DIEGO
SAN FRANCISCO • SINGAPORE • SYDNEY • TOKYO

Academic Press is an imprint of Elsevier

Elsevier Academic Press
30 Corporate Drive, Suite 400, Burlington, MA 01803, USA
525 B Street, Suite 1900, San Diego, California 92101-4495, USA
84 Theobald's Road, London WC1X 8RR, UK

This book is printed on acid-free paper. ♾

Library of Congress Cataloging-in-Publication Data
Application submitted

British Library Cataloguing in Publication Data
A catalogue record for this book is available from the British Library

ISBN: 978-0-12-369509-3

For all information on all Elsevier Academic Press publications visit our Web site at www.books.elsevier.com

Printed in China
07 08 09 10 9 8 7 6 5 4 3 2 1

Contents

Contributor's List

Samuel F. Berkovic
Epilepsy Research Centre and Department of Medicine (Neurology), University of Melbourne, Austin Health, Heidelberg, Victoria, Australia.

Kaya Bilguvar
Program of Neurogenetics, Department of Neurosurgery, Yale University School of Medicine, New Haven, CT, USA; Health Sciences Institute, Department of Medical Biology and Genetics, Marmara University, Turkey.

Craig Blackstone
Cellular Neurology Unit, NINDS, National Institutes of Health, Bethesda, MD, USA.

Michael H. Bloch
Child Study Center and the Departments of Pediatrics and Psychology, Yale University School of Medicine, New Haven, CT, USA.

Hal Blumenfeld
Departments of Neurology, Neurobiology, and Neurosurgery, Yale University School of Medicine, New Haven, CT, USA.

D. E. Bredesen
Buck Institute for Age Research, Novato, CA, USA; Department of Neurology, University of California at San Francisco, CA, USA.

Susan B. Bressman
Department of Neurology, Beth Israel Medical Center, New York, NY, USA.

Michelle Brucal
Wayne State University School of Medicine, Center for Molecular Medicine and Genetics, Detroit, MI, USA.

Edward A. Burton
Departments of Neurology, and Molecular Genetics and Biochemistry, University of Pittsburgh School of Medicine, Pittsburgh, PA, USA.

Josep Dalmau
Department of Neurology, Division of Neurooncology, University of Pennsylvania, Philadelphia, PA, USA.

Ted M. Dawson
Institute for Cell Engineering, Departments of Neurology and Neuroscience, Johns Hopkins University School of Medicine, Baltimore, MD, USA.

Valina L. Dawson
Institute for Cell Engineering, Departments of Neurology and Neuroscience, Johns Hopkins University School of Medicine, Baltimore, MD, USA.

Chantal Depondt
Erasme Hospital, Brussels, Belgium.

Michael L. DiLuna
Resident, Department of Neurosurgery, Yale University Center of Medicine, New Haven, CT, USA.

Salvatore DiMauro
Department of Neurology, Columbia University Medical Center, New York, NY, USA.

Michel D. Ferrari
Department of Neurology, Leiden University Medical Center, Leiden, The Netherlands.

David J. Fink
Department of Neurology, University of Michigan; VA Ann Arbor Healthcare System, Ann Arbor, MI, USA.

Alexander Flügel
Max-Planck Institute for Neurobiology, Department of Neuroimmunology, Martinsried, Germany.

Rune R. Frants
Department of Human Genetics, Leiden University Medical Center, Leiden, The Netherlands.

Joseph C. Glorioso
Department of Molecular Genetics and Biochemistry, University of Pittsburgh School of Medicine, Pittsburgh, PA, USA.

Peter J. Goadsby
Queen Square, Institute of Neurology, London, United Kingdom; Department of Neurology, University of California, San Francisco, CA, USA.

Alan L. Goldin
Departments of Microbiology and Molecular Genetics, and Anatomy and Neurobiology, University of California, Irvine, CA, USA.

Murat Gunel
Program of Neurogenetics, Section of Neurovascular Surgery; Neuroscience Intensive Care Unit, Department of Neurosurgery, Yale School of Medicine, New Haven, CT, USA.

Noam Y. Harel
Department of Neurology, Yale University School of Medicine, New Haven, CT, USA.

Ingo Helbig
Epilepsy Research Centre and Department of Medicine (Neurology), University of Melbourne, Austin Health, Heidelberg, Victoria, Australia.

Thomas M. Hemmen
Department of Neuroscience, University of California, La Jolla, CA, USA

Fuki M. Hisama
Division of Genetics, Childrens Hospital Boston, Harvard Medical School, Boston, MA, USA.

Bradley T. Hyman
Harvard Medical School, Massachusetts General Hospital, Boston, MA, USA.

Martin Ingelsson
Uppsala University, Department of Public Health, Molecular Geriatrics, Uppsala, Sweden.

Dennis R. Johnson
Department of Pharmacology, Yale University School of Medicine, New Haven, CT, USA.

John Kamholz
Wayne State University School of Medicine, Center for Molecular Medicine and Genetics; Department of Neurology, Detroit, MI, USA.

Marcus Kaul
Center for Neuroscience and Aging Research, Burnham Institute for Medical Research; Department of Psychiatry, University of California at San Diego, La Jolla, CA, USA.

Jeffery D. Kocsis
Department of Neurology, Yale University School of Medicine, New Haven, CT, USA.

Gert Jan Lammers
Department of Neurology, Leiden University Medical Center, Leiden, The Netherlands.

James F. Leckman
Child Study Center and the Departments of Pediatrics and Psychology, Yale University School of Medicine, New Haven, CT, USA.

Jun Li
Wayne State University School of Medicine, Department of Neurology, Detroit, MI, USA.

Stuart A. Lipton
Center for Neuroscience, Stem Cells, and Aging Research, Burnham Institute for Medical Research; Department of Neurosciences, and Department of Psychiatry, University of California at San Diego; Molecular Neurobiology Laboratory, The Salk Institute for Biological/Studies, Department of Molecular and Integrative Neurosciences and Department of Molecular and Experimental Medicine, The Scripps Research Institute, La Jolla, CA, USA.

Nicholas J. Maragakis
Department of Neurology, Johns Hopkins University, Baltimore, MD, USA.

P. Mehlen
Apoptosis, Cancer, and Development Laboratory, Université de Lyon, Centre Léon Bérard, Lyon, France; Buck Institute for Age Research, University of California, Novato, CA, USA.

Richard I. Morimoto
Department of Biochemistry, Molecular Biology, and Cell Biology, Rice Institute for Biomedical Research, Northwestern University, Evanston, IL, USA.

Kai Orton
Department of Biochemistry, Molecular Biology, and Cell Biology, Northwestern University, Evanston, IL, USA.

Sebastiaan Overeem
Department of Neurology, Radboud University Nijmegen Medical Center, Nijmegan, the Netherlands; Department of Neurology, Leiden University Medical Center, Leiden, the Netherlands.

Laurie Ozelius
Department of Genetics and Genomic Sciences, Mount Sinai School of Medicine, New York, NY, USA.

Massimo Pandolfo
Department of Neurology, Brussels Free University; Erasme Hospital, Brussels, Belgium.

Juan M. Pascual
Departments of Neurology, Physiology, and Pediatrics, University of Texas Southwestern Medical Center; Children's Medical Center; UT Southwestern Hospitals and Clinics, Dallas, TX, USA.

Henry L. Paulson
Department of Neurology, University of Iowa College of Medicine, Iowa City, IA, USA.

Stephen J. Peroutka
Franchise Development Leader, Pain, Johnson & Johnson, Titusville, NJ, USA.

Ognen A. C. Petroff
Department of Neurology, Yale University, School of Medicine, New Haven, CT, USA.

Christopher B. Ransom
Department of Neurology, Yale University School of Medicine, New Haven, CT, USA.

R. V. Rao
Buck Institute for Age Research, Novato, CA, USA.

Neggy Rismanchi
Cellular Neurology Unit, NINDS, National Institutes of Health, Bethesda, MD, USA.

Jeffrey D. Rothstein
Department of Neurology, Johns Hopkins School of Medicine, Baltimore, MD, USA.

Joseph M. Savitt
Institute for Cell Engineering and the Department of Neurology, Johns Hopkins University School of Medicine, Baltimore, MD, USA.

Ingrid E. Scheffer
Department of Medicine and Paediatrics and Royal Children's Hospital, University of Melbourne, Austin Health Hospital, Melbourne, Australia.

Eric A. Schon
Departments of Neurology and Genetics and Development, Columbia University Medical Center, New York, NY, USA.

Michael Shy
Wayne State University School of Medicine, Center for Molecular Medicine and Genetics; Department of Neurology, Detroit, MI, USA.

Stephen M. Strittmatter
Department of Neurology, Yale University School of Medicine, New Haven, CT, USA.

Mehdi Tafti
Center for Integrative Genomics, University of Lausanne, Lausanne, Switzerland.

Gamze Tanriover
Program of Neurogenetics, Department of Neurosurgery, Yale University School of Medicine, New Haven, CT, USA.; Department of Histology, School of Medicine, Akdeniz University, Turkey.

Sokol V. Todi
Department of Neurology, University of Iowa College of Medicine, Iowa City, IA, USA.

Arn M.J.M. van den Maagdenberg
Departments of Neurology and Human Genetics, Leiden University Medical Center, Leiden, The Netherlands.

Jeffery M. Vance
Duke University Medical Center, Center for Human Genetics, Durham, NC, USA.

Angela Vincent
Neurosciences Group, Department of Clinical Neurology, Weatherall Institute of Molecular Medicine, John Radcliffe Hospital, Oxford, United Kingdom.

Cindy Voisine
Department of Biochemistry, Molecular Biology, and Cell Biology, Rice Institute for Biomedical Research, Northwestern University, Evanston, IL, USA.

Stephen G. Waxman
Department of Neurology and Center for Neuroscience and Regeneration Research, Yale University School of Medicine; Rehabilitation Research Center, Veterans Affairs Medical Center, New Haven, CT, USA.

Hartmut Wekerle
Max-Planck Institute for Neurobiology, Department of Neuroimmunology, Martinsried, Germany.

Aislinn J. Williams
Department of Neurology, University of Iowa, Iowa City, IA, USA.

John N. Wood
Molecular Nociception Group, Biology Department, University College, London, United Kingdom.

Yvonne S. Yang
Department of Neurology, Yale University School of Medicine, New Haven, CT, USA.

Justin A. Zivin
Department of Neuroscience, University of California San Diego School of Medicine, La Jolla, CA, USA.

Preface

Around the world from Aarhus to Zurich, medical students learn the essentials of neurology; their senior colleagues in practice are challenged on a daily basis by patients who fall within the realm of neurology. Laboratory scientists interested in the nervous system are, likewise, very aware of diseases that are characterized as neurological, and are thus well acquainted with the boundaries of neurology. What, then, is *molecular neurology*? And why do we need it?

Neurology traditionally has rested upon a systematic and meticulous system of diagnosis and classification of disorders of the nervous system, based first on localization within the nervous system, and second on pathological identity. Thus, for many decades neurologists have approached each patient with an emphasis on answering the questions: Where and what is (are) the lesion(s)? This localizationist approach is one of the bastions of classical neurology and it has served the discipline well. Good neurologists, indeed, are regarded by clinicians in other specialties as superb diagnosticians.

The therapeutic arm of neurology, however, has had a somewhat shorter history. For example, thirty years ago, medical students at some institutions were taught *not* to make the diagnosis of multiple sclerosis early in its course, since there was little that could be done. Strokes were treated with bed rest, and not much else. Patients who sustained spinal cord injuries were told, in no uncertain terms, that there was no hope.

All this is changing. Neurology is truly at the beginning of a revolution, from therapeutically nihilistic to therapeutically active. Effective therapies are now available for some neurological diseases that were previously untreatable. And other new therapies are on the way.

Why, then, a molecular neurology? Medicines work by targeting molecules. The more specific the targeting, the more specific the actions, and the fewer the side effects (indeed, it is the presence of unacceptable side effects that commonly derails the development of new therapeutic candidates). Molecular neurobiology is advancing at a stunningly spectacular rate. And as it does so, it is revealing important clues to the pathogenesis and pathophysiology of neurological diseases, and of the therapeutic targets that they present.

This book highlights—for graduate and MD–PhD students, research fellows and research-oriented clinical fellows, and researchers in the neurosciences and other biomedical sciences—the principles underlying molecular medicine as related to neurology. This book is not meant to be a comprehensive or encyclopedic compendium. Rather, it presents *principles and disease examples* relevant to molecular neurology, and reflects the concepts, excitement, and sense of forward motion of this field. In providing these sketches of progress, the chapters in this book also illustrate the trajectory of neurology, from a descriptive, anatomically-based specialty into a mechanistic, molecularly-based discipline. As neurology becomes more molecular, it will undoubtedly become more therapeutic.

Ten years from now, neurology will almost certainly be a more therapeutic specialty than it is today. This will reflect in large part the evolution of neurology into a molecular discipline. As they use this book, readers therefore are urged to consider the therapeutic implications of a molecular neurology, and are encouraged to think about the molecular underpinnings of neurology not only in terms of what we understand today, but also with respect to what we may accomplish for people with diseases of the nervous system tomorrow.

Stephen G. Waxman, MD, PhD

1

Genetics as a Tool in Neurology

Dennis R. Johnson and Fuki M. Hisama

I. Introduction

Genetics is the study of heredity, and seeks to explain the mechanism of hereditary transmission, as well as the genetic basis of individual variation. Medical genetics is the science of genetically associated biologic variation relevant to human traits and diseases. In industrialized countries, improvements in public health and medical care have resulted in a marked reduction in mortality from infectious causes, or nutritional deficiencies; during this time, there has been a corresponding increase in awareness of the role of genetic factors in human diseases, including neurological disease.

Although genetic disorders sometimes have been perceived as rarely encountered outside of a tertiary medical center, this notion is no longer true. Although a particular genetic disease may be rare, defined as affecting fewer than one in 250,000 individuals, some genetic disorders such as spinal muscular atrophy and cystic fibrosis are common in the population, affecting from one in 500 to one in 1,000 people.

In aggregate, the thousands of single-gene disorders affect millions of people. The proportion of recognized genetic diseases has increased in both the pediatric and adult populations as our ability to identify and understand the role of genes in biology and medicine has increased. New genetic disorders continue to be described. In 1994, McKusick's Mendelian Inheritance in Man listed 6,678 monogenic human traits, and accessing the online version on September 6, 2006 showed 17,033 entries, including over 1,500 Mendelian traits with an unknown molecular basis (www.ncbi.nlm.nih.gov/OMIM). Many chronic disorders affecting the nervous system are genetically determined. In others, genetic risk factors increase the likelihood of developing certain diseases, but do not always result in the disease.

In the last 15 years, hundreds of human disease genes have been mapped and cloned because of the explosion of basic knowledge in genomics and molecular genetics. The identification of these genes has led to novel insights into disease pathogenesis in humans and model systems. A growing number of genetic tests are available on a clinical basis to assist in the diagnosis of a symptomatic patient, to accurately predict risk to family members, and to assess the likelihood of a serious drug side effect. Sophisticated genetic methods are now being applied to the next genetic frontier: the challenging task of identifying genes contributing to common diseases such as Alzheimer's disease and schizophrenia. These genetically complex disorders are thought to arise from the cumulative effects of variants in several genes, in combination with environmental effects. The goal of this chapter is to provide an overview of the genetic principles that have transformed neurology into a molecular specialty.

II. Structure and Function of Genes and Chromosomes

A. DNA Structure and Replication

The information for the development of an organism and the specific functions of cells, tissues, and organs is stored in its DNA. The double helix structure of DNA was described a little more than 50 years ago (Watson and Crick, 1953). DNA is a nucleic acid composed of *nucleotide bases* (adenine (A), thymine (T), guanine (G), and cytosine (C)), a *deoxyribose sugar*, and *phosphate groups*. These components are organized into two sugar-phosphate strands of opposing polarity, and paired nucleotide bases. Each pair consists of a purine (guanine or adenine) and a pyrimidine (thymine or cytosine). Guanine pairs with cytosine via three hydrogen bonds, thymine and adenine pair via two hydrogen bonds. Every time a cell divides, its DNA must be replicated in the daughter cells. During the process of replication, the two strands of DNA separate, and each is copied by DNA polymerase. Thus, DNA replication is semiconservative, with each double helix consisting of a "new" and an "old" strand (Alberts et al., 2002).

B. Gene Expression and Transcription

The genetic information in DNA is transcribed into RNA in the nucleus, or in the case of a few genes, in the mitochondria, by means of a DNA-directed RNA polymerase; the RNA crosses into the cytoplasm where translation of the information into polypeptides occurs on ribosomes. Both post-transcriptional (RNA splicing, capping, and polyadenylation) and post-translational modifications (cleavage into a mature peptide, hydroxylation, phosphorylation, the addition of carbohydrate or lipid groups) may take place. The basic theme of unidirectional transfer of information from DNA®RNA®protein has been termed the *central dogma* of molecular biology (Alberts et al., 2002). Although the term reflects its near universal occurrence, there are exceptions to this important principle. Eukaryotic cells contain sequences that encode reverse transcriptases that enable making cDNA copies of RNA transcripts, which can then be inserted into the genome.

The vast majority of DNA is not transcribed. Only approximately 1.5% of the DNA in mammalian cells codes for proteins (Strachan & Read, 2004). The remaining noncoding DNA (formerly called "junk DNA" contains the intronic sequences within genes, various classes of highly repetitive DNA, nonfunctional copies of genes (pseudogenes) and noncoding RNAs. The potential role of noncoding DNA is the subject of considerable interest (Mattick, 2004).

C. Chromosomes

In eukaryotic cells, DNA is organized into chromosomes (Tyler-Smith & Willard, 1993). The correct diploid number of human chromosomes was determined to be 46 (Tijo & Levan, 1956). This includes 22 paired maternal and paternal autosomes, and a pair of sex chromosomes. During interphase, a chromosome consists of a single, extended DNA molecule and its closely associated histone and nonhistone proteins. The packaging of DNA is a dynamic, reversible, highly organized process that takes place on multiple levels and enables a 10,000 fold compaction of the DNA. The basic unit of packaging is the nucleosome, consisting of a core of eight histone proteins around which the DNA is coiled (Jenuwein & Allis, 2001). A string of adjacent nucleosomes are coiled into a chromatin fiber of 30 nm diameter, visible by electron microscopy. These in turn form a long looped chromosome segment.

With each cell division, the chromosomes become even more condensed, and the DNA content is replicated. The typical textbook image of chromosomes reflects a brief stage of the cell cycle, their most highly condensed state during metaphase, in which the chromosome has been replicated, and exists briefly as a two-chromatid entity.

The development from single-cell zygote to multicellular organism requires millions of cell divisions, which occur through the process of mitosis. The result of mitosis is the formation of two daughter cells containing identical genetic information. During mitosis, the chromosomes condense and become visible (prophase), the nuclear envelope disappears, and the chromosomes migrate to the equatorial plane (prometaphase). Importantly, during mitosis, the maternal and paternal homologous chromosomes do not pair during metaphase, in contrast to meiosis (discussed later). In anaphase, each centromere splits and the two chromatids of each chromosome migrate to opposite poles; in telophase, the chromosomes reach the poles, and the nuclear membrane reforms.

D. Meiosis

Meiosis is the specialized process of cell division by which gametes are formed. Somatic cells possess a diploid DNA content or two copies of the chromosome set. In humans, the diploid number of chromosomes is 46, including the sex-determining X and Y chromosomes. Meiosis involves a single round of DNA replication, but two rounds of cell division; the products (spermatozoa or an oocyte) contain the haploid number of chromosomes (23 in humans). Meiosis differs fundamentally from mitosis in several respects. First, the products of mitosis are diploid, the products of meiosis are haploid. Second, in meiosis, the homologous chromosomes pair up. Third, genetic recombination arises from the exchanges between the homologous chromosomes (crossing over), a process that has proven important in positional cloning. Finally, the products of mitotic division are generally identical to each other, whereas the products of meiotic division differ genetically from one another due to the exchange of genetic information that occurs during crossing over.

Chromosomal abnormalities may affect human reproduction or cause recognizable genetic syndromes. Problems arising from chromosomal aberrations include infertility; fetal loss; gain or loss of complete chromosomes resulting in recognizable phenotypes such as Down syndrome (47,XX, +21), Turner syndrome (45,X), or other structural abnormalities such as deletion, insertion, inversion, duplication, translocation, or formation of a ring chromosome (ISCN, 1995; Shaffer & Tommerup, 2005). Although many chromosomal abnormalities cause a neurological phenotype, particularly mental retardation with or without epilepsy, they are often multisystem diseases and space does not permit detailed discussion. For further information, refer to Gardner and Sutherland, 1996. In the clinical setting, cytogenetic studies of human chromosomes are performed using accessible tissues: lymphocytes from blood, fibroblasts from skin biopsy, or fetal cells obtained by amniocentesis or chorionic villus sampling.

E. The Human Genome Project

The genome is the total sum of genetic material in human cells. The Human Genome Project (see Table 1.1) was based upon the central importance of DNA to understanding the function of genes, their role in human disease, and the potential for medical benefits. The Office of Human Genome Research was established in 1988, and the Human Genome Project originally was envisioned as a 15-year effort to produce high resolution genetic and physical maps leading to the sequence of the human genome as the primary goal. The Human Genome Project became an international project carried out in specialized genome centers with high throughput sequencing capability, and the ability to release the sequence to publicly available databases. Secondary goals included new technological developments in DNA sequencing tools and technology, bioinformatics, genome projects for several widely studied model organisms (*E. coli*, *Saccharomyces cerevisiae*, *Caenorhabditis elegans*, *Drosophila melanogaster*, and mouse), and the ethical, legal, and societal implications (termed ELSI) of human genome studies. Competition from a private company, Celera, and technical advances sped up the timeline of the project. In 2001, a draft of the human genome was released (2001; Lander et al., 2001), and the finished sequence of the human genome and the model organisms was available by 2003.

F. Overview of the Human Genome

The genome comprises 3,000 Mb encoding an estimated 35,000 genes compared with 18,425 genes in *C. elegans* and 13,601 genes in the fruit fly (Claverie, 2001). Organismal complexity, therefore, does not fully reflect genome complexity, and is incompletely understood, but could be explained by increased alternative splicing in complex organisms or perhaps by noncoding DNA.

Human genes vary in size from less than 1 Kb to the 2.4 Mb dystrophin gene. Although most protein coding genes possess introns, a few lack introns, including the dopamine D1 and D5 receptors. Because of the complexity of the human brain, over half of genes are thought to be expressed predominantly or exclusively in the nervous system. It is therefore not surprising that a number of neurological diseases have a genetic basis. Human genes are not evenly distributed on chromosomes. Instead, there are "deserts" devoid of genes (e.g., regions of heterochromatin), and "oases" of gene-rich regions. In general, the distinctive Giemsa staining pattern of light and dark bands observed on a karyotype reflect gene-dense and gene-poor regions, respectively.

G. Genetic Diversity: Normal Variation and Genetic Disease

Although the DNA sequence of any two humans is 99.9 percent identical, nevertheless, the degree of human genetic

Table 1.1 Internet Resources for Genetics

National Human Genome Research Institute	(www.genome.gov/)
Online Mendelian Inheritance in Man	(www.ncbi.nlm.nih.gov/Omim/)
Genetic Testing, Genetic Clinics and Gene Reviews	www.genetests.org
National Organization for Rare Diseases	www.rarediseases.org
National Coalition for Health Professional Education in Genetics	www.nchpeg.org
Human Genome Organization	www.gene.ucl.ac.uk/hugo
Human Gene Mutation Database	www.hgmd.cf.ac.uk
American Society of Human Genetics	www.ashg.org
National Society of Genetic Counselors	www.nsgc.org
American College of Medical Genetics	www.acmg.net
American Board of Medical Genetics	www.abmg.org

variation is remarkable, and readily detected by the casual observer. Genetic variation also influences the most common medical tests such as blood pressure, cholesterol, and glucose, as well as causes the development of diseases such as neurofibromatosis, familial Alzheimer's disease, certain forms of epilepsy, among others. Genetic variation may also affect the effectiveness or the predisposition to side effects of the conventional drugs currently on the market. A number of liver enzymes have polymorphisms that affect the rate of drug metabolism. For example, the CYP2D6 gene in the P450 cytochrome system affects the metabolism of many psychoactive drugs, including tricyclic antidepressants and atypical antipsychotics, so that a dose that causes toxicity in a patient with low enzyme activity may be nontherapeutic in a patient with high enzyme activity (Wolf et al., 2000).

All genetic variation arises from a change in the DNA sequence, termed a mutation. Mutations may occur in a somatic cell and be passed on to its daughter cells (a critical event in the development of many cancers) or a mutation may affect a germline cell (and thus be capable of transmission from one generation to the next). Mutations may be induced by exogenous agents such as ionizing radiation, ultraviolet radiation, various classes of chemicals such as alkylating agents, nitrogen mustard, and formaldehyde. Spontaneous mutations arise during the process of DNA replication that continues throughout an individual's lifetime. Efficient cellular DNA repair systems correct the vast majority of mutations, so that the normal mutation rate is low, but is not zero. The residual mutations may (1) be functionally neutral and thus clinically silent or associated with normal variation, (2) may result in disease, or (3) may provide some selective advantage that will provide a substrate for evolution.

The chromosomal location of a gene is its locus. The different, alternate forms of a gene are referred to as *alleles*. If an individual has the same allele on both the maternal and the paternal chromosomes, the person is homozygous. If the two alleles differ, the individual is heterozygous. The alleles present at a specified locus comprise the genotype. A locus in which two or more alleles have frequencies less than 1 percent of the normal population is termed a polymorphism.

III. Genetic Medicine

Genetics is the study of inheritance and *medical genetics* is the branch of medicine that specializes in inherited diseases (2006; McKusick, 1993). Neurogenetics, by extension, aims to identify and characterize the inherited components of neurological disease, and in so doing, contributes broadly to the elucidation of neurological disease mechanisms (Rosenberg, 2001). Clearly, ever more powerful contemporary molecular biological technologies and the availability of the sequence of the human genome have enhanced the potential yield of applying genetic approaches in the clinic. However, optimizing the outcome from the evaluation of a patient/family requires an understanding of a number of key concepts, strategies or approaches—the tools of clinical genetics. In addition, it may soon be possible to sequence an individual's entire genome at reasonable cost, thus opening the door to *genomic medicine* (Guttmacher & Collins, 2002). Conceivably, one might envision an approach to personalized medicine that includes sequencing an individual's genome at birth, thus generating a personal genome database, which can then be interrogated repeatedly throughout the patient's lifetime to predict and prevent diseases for which the person is at high risk, as well as optimize therapeutic interventions as genome guided therapies evolve.

A. The Importance of the Pedigree

The pedigree is one of the most essential tools in genetics. It depicts which family members are affected with a genetic condition, which family members are unaffected, and the relationships among the family members. A standardized format for pedigree notation is widely accepted (Bennett et al., 1995). The simplest human genetic diseases are those in which the genotype at a single locus is necessary and sufficient to result in the disease. This category of diseases is known as Mendelian diseases, because they follow the principles discovered by the Augustinian monk Gregor Mendel (1822–1884) by crossbreeding garden peas in the monastery gardens. Mendel's two fundamental principles are the *principle of segregation*, which states that only one of a pair of genes is transmitted to the offspring; and the *principle of independent assortment*, which states that genes at different chromosomal loci are transmitted independently of each other. Mendel also defined dominance and recessiveness, recognizing that the effects of one allele may mask those of the second allele. Many neurological diseases follow a Mendelian inheritance pattern (see Table 1.2; Pulst, 2003). They may be recognized because of their characteristic pedigree patterns (see Figure 1.1).

B. Mendelian Patterns of Inheritance

1. Autosomal Dominant Inheritance

An autosomal dominant (AD) trait is observable in the heterozygote state. In a classical AD trait, males and females are equally likely to be affected, and equally likely to transmit the trait to their offspring, there is no skipping of generations (i.e., at least one person in each generation is affected), vertical transmission from parent to child occurs, unaffected persons do not transmit the trait, and an affected individual passes on the trait to half of his or her offspring. Examples of neurological diseases inherited as an autosomal trait include Huntington's disease, myotonic dystrophy 1 and 2, neurofibromatosis 1 and 2, and many forms of spinocerebellar ataxia.

Table 1.2 Selected Neurological Disease Genes[a]

Disease	Inheritance	Locus	Gene	Gene Product	Year Identified
Peripheral Neuropathies					
Charcot-Marie-Tooth (CMT) 1A	AD	17p11.2	PMP22	Peripheral myelin protein 22	1991
X-Linked Charcot-Marie-Tooth	XLD	Xq13.1	GJB1	Connexin 32	1993
CMT 4A	AR	19q13.1	PRX	Periaxin	2001
Muscular Dystrophies					
Duchenne muscular dystrophy	XLR	Xp21.2	DMD	Dystrophin	1987
Myotonic dystrophy 1	AD	19q13.2	DMPK	Dystrophia myotonica protein kinase	1992
Myotonic dystrophy 2	AD	3q13	ZNF9	Zinc finger protein 9	2001
Neurocutaneous Diseases					
Tuberous sclerosis	AD	9q34	TSC1	Hamartin	1997
Tuberous sclerosis	AD	16p13.3	TSC2	Tuberin	1993
Neurofibromatosis 1	AD	17q11.2	NF1	Neurofibromin	1990
Neurofibromatosis 2	AD	22q12.2	NF2	Neurofibromin 2 or Merlin	1993
Mental Retardation Syndromes					
Fragile X syndrome	XLR	Xq27.3	FMR1	Frataxin	1991
Rett syndrome	XLD	Xq28	MECP2	Methyl CpG-binding protein 2	1999
Cortical Development					
Schizencephaly	AD	10q26	EMX2	Homeobox containing "Empty Spiracles" homolog	1996
X-linked lissencephaly and double cortex syndrome	XL	Xq22.3	DCX	Doublecortin	1998
Isolated lissencephaly sequence	AD	17p13.3	LIS1	Platelet activating factor acetylhydrolase, isoform 1B, alpha subunit	1993
Stroke					
CADASIL	AD	19p13.2	NOTCH3	Notch3	1996
Leukodystrophy					
Adrenoleukodystrophy	XLR	Xq28	ABCD1	ATP-binding cassette, subfamily D	1993
Leukoencephalopathy with vanishing white matter or ovarioleukodystrophy	AR	2p23 14q24 12	EIF2B4 EIF2B2 EIF2B1	Subunits of eukaryotic translation initiation factor EIF2B	2002 2002 2002
Neurodegenerative diseases					
Spinocerebellar ataxia 1	AD	6p23	ATXN1	Ataxin 1	1994
Spinocerebellar ataxia 2	AD	12q24	ATXN2	Ataxin 2	1996
Huntington disease	AD	4p16.3	Huntington	Huntington	1993
Parkinson disease	AD	4q21	SNCA	Alpha synuclein	1997
Parkinson disease	AR	6q25	PARK2	Parkin	1998
Parkinson disease	AR	1p36	DJ1	DJ1	2003
Parkinson disease	AD	12q12	LRRK2	Dardarin	2004
Alzheimer disease	AD	21q21	APP	Amyloid precursor protein	1991
Alzheimer disease	AD	14q24.3	PS1	Presenilin 1	1995
Alzheimer disease	AD	1q31-q42	PS2	Presenilin 2	1995
Epilepsies					
Generalized epilepsy febrile seizures +	AD	2q24	SCNA1	Voltage-gated sodium channel	2000
Generalized epilepsy febrile seizures +	AD	2q24	SCN2A1	Voltage-gated sodium channel	2004
Benign familial neonatal convulsions	AD	20q13.3	KCNQ2	Voltage-gated potassium channel	1998
Juvenile myoclonic epilepsy	AD	5q34-q35	GABRA1	Gaba A receptor subunit	2002
Unverricht-Lundborg	AR	21q22.3	EPM1	Cystatin B	1996
Lafora disease	AR	6q24	EPM2A	Laforin	1998
Channelopathies					
Hyperkalemic periodic paralysis	AD	17q23	SCN4A	Voltage-gated sodium channel	1991
Hemiplegic migraine 1	AD	19p13	CACNA1A	Voltage-gated calcium channel	1996
Myotonia congenita	AD/AR	7q35	CLCN1	Voltage gated chloride channel	1992
Erythromelalgia	AD	2q24	SCN9A	Voltage-gated sodium channel	2004

[a]This table is not comprehensive, and illustrates only a few of the wide range of neurological conditions for which the genetic basis is known.

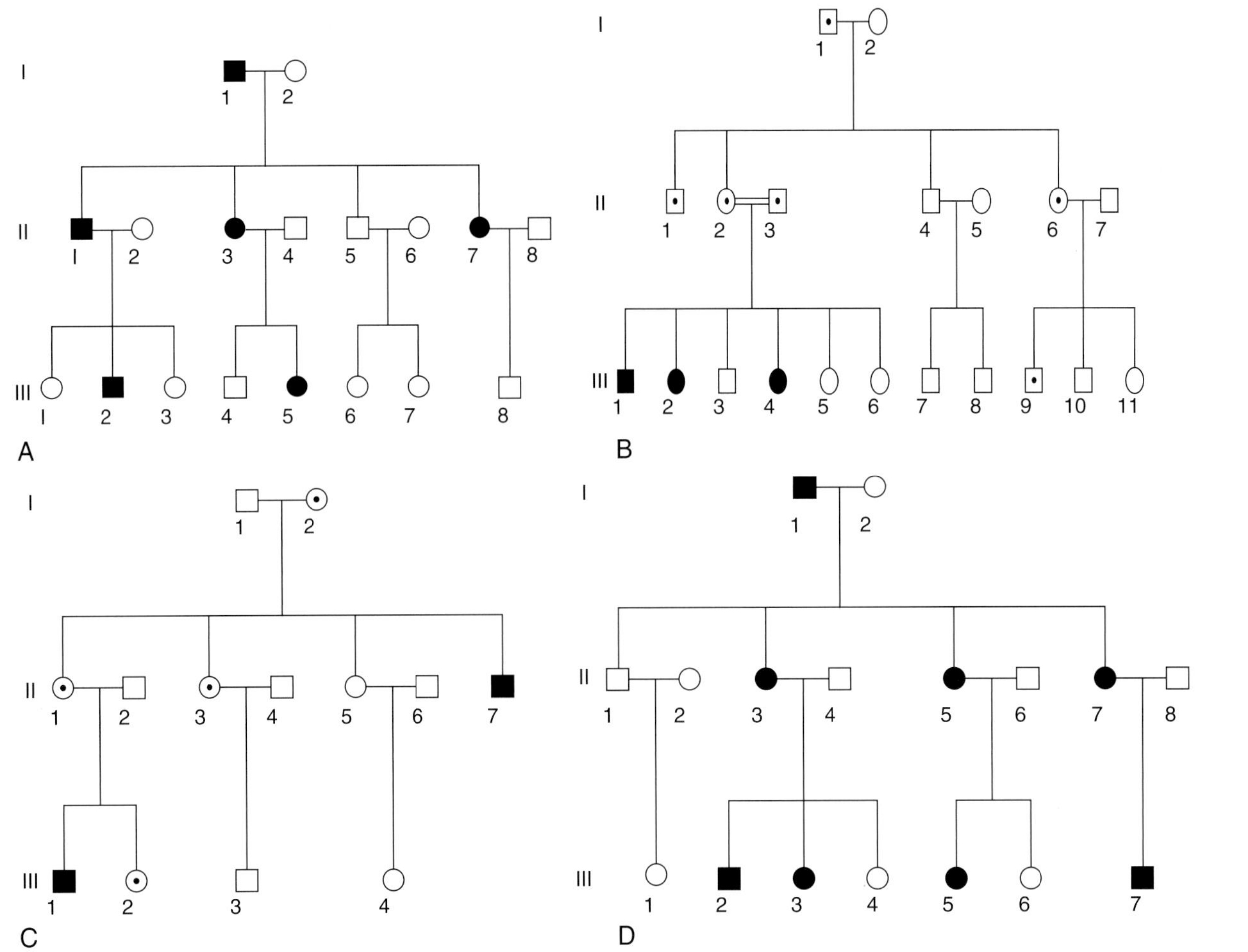

Figure 1.1 Pedigrees illustrating Mendelian patterns of inheritance. (**A**) Autosomal dominant (**B**) Autosomal recessive (**C**) X-linked recessive (**D**) X-linked dominant. Males are shown as squares; females are circles; affected status is indicated by a dark, filled circle or square; unaffected status is unfilled; carrier status is depicted by a dot. Although carriers are shown here for illustrative purposes in order to follow the inheritance of the mutant allele, they are frequently clinically undetectable, and therefore indistinguishable from unaffected individuals. The double horizontal line indicates a consanguineous marriage. Roman numerals indicate the generation, Arabic numbers indicate the individual.

2. Autosomal Recessive Inheritance

An autosomal recessive trait is observable in the homozygous state. In a classical AR trait, both parents are unaffected, heterozygous carriers. Statistically, there is a one in four chance that a child will be homozygous for the mutant allele and affected. Males and females are equally likely to be affected. The transmission pattern is horizontal (one or more siblings may be affected, but successive generations are not affected), and the rate of consanguinity between the parents in increased. Consanguinity (Latin, "with blood") refers to mating between relatives, such as first-cousin marriages, or uncle-niece marriages, which are traditional in some cultures. Examples of neurological diseases inherited as autosomal recessive traits include spinal muscular atrophy, metachromatic leukodystrophy, Gaucher disease.

3. Sex-linked Inheritance

The X and Y chromosomes determine sex. In addition, the human X chromosome contains hundreds of other genes. Because females have two copies of the X chromosome, whereas males have only one (they are hemizygous), diseases caused by genes on the X chromosome, most of which are X-linked recessive, predominantly affect males.

Examples of X-linked recessive diseases include adrenoleukodystrophy, Fragile X syndrome, Duchenne muscular dystrophy, and red-green color blindness.

The pedigree in these disorders is characterized by affected males related through carrier females, and absence of father-to-son transmission. If females are affected, their symptoms are usually milder, and they are termed manifesting heterozygotes. This situation typically arises from

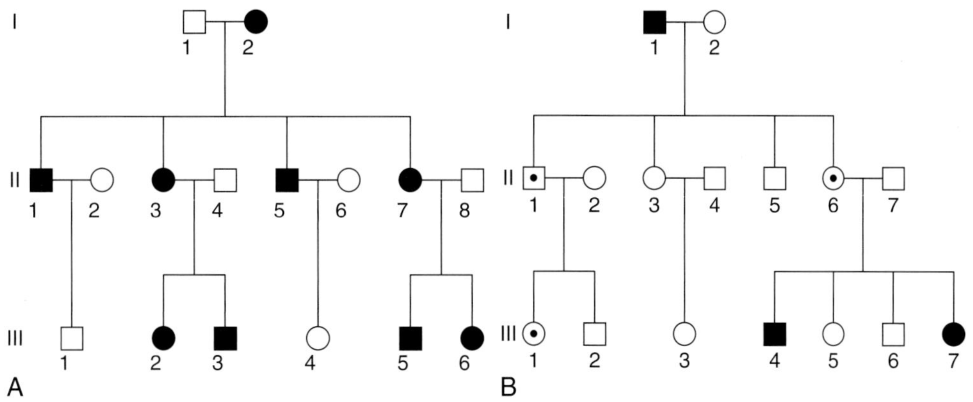

Figure 1.2 Pedigrees illustrating non-Mendelian patterns of inheritance. **A.** Mitochondrial inheritance. Note that an affected male does not pass on his mitochondria, so no offspring are affected. All the offspring inherit their mother's mitochondria. All offspring of an affected female are depicted here as affected, although some may be clinically unaffected because of unique features of mitochondrial inheritance such as heteroplasmy. **B.** An imprinted pedigree. Whether the mutant allele causes a phenotype or not is dependent on the sex of the transmitting parent. Compare individuals III-1 and II-1, who inherited the mutant allele from their father and are carriers, with individuals III-4 and III-7, who inherited the mutant allele from their mother and who are affected. See Figure 1.1 for interpretation of the symbols.

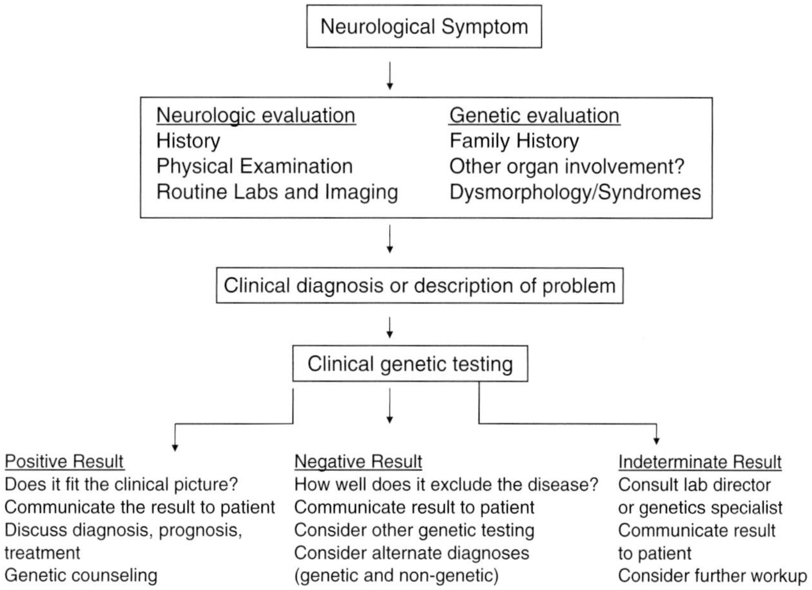

Figure 1.3 The workup of the patient with a suspected neurogenetic condition.

skewed X-inactivation with preferential inactivation of the normal X chromosome in their cells. In the normal situation, X chromosome inactivation is random, with inactivation of a woman's paternal X chromosome in some cells, and inactivation of her maternal X chromosome in others. In other, rare cases, women with only a single copy of the X chromosome (45, X) or with structural abnormalities of the X chromosome may manifest an X-linked recessive condition.

4. X-linked Dominant Inheritance

X-linked dominant disorders are seen more commonly in females than in males, or in the case of some diseases, affect only females. In the latter case, it is thought that the hemizygous males are so severely affected, they do not survive. This may be reflected in the pedigree by multiple miscarriages or male infant deaths. Examples of X-linked dominant disorders include Rett syndrome, the X-linked lissencephaly and double-cortex syndrome, and incontinentia pigmenti type 1, characterized by dermatological, ocular, dental, and neurological abnormalities.

C. Non-Mendelian Patterns of Inheritance

1. Mitochondrial Inheritance

Mitochondria have their own genome, separate from the nuclear genome. Mitochondrial DNA is localized to the cytoplasm, and is inherited strictly from the mother (see Figure 1.2). As a matter of perspective, in considering an individual's genetic makeup, if one goes back five generations, 32 ancestors will have contributed to that individual's nuclear genome, whereas a single woman contributed the individual's mitochondrial DNA. Mitochondrial genetics demonstrates several other special features, including heteroplasmy and a threshold effect, and so is an exception to the laws of Mendel. For further details, see Chapters 3 and 33.

2. Imprinting

Imprinting is an exception to Mendelian principles, and refers to differential "marking" of maternal and paternal alleles of specific genes or chromosome regions during gametogenesis, leading to expression of one allele, and silencing of the other allele, based upon the sex of the parent who contributed the allele (Ferguson-Smith & Surani, 2001). Remarkably, this molecular memory is erased and reset in the germline at each generation (see Figure 1.2). Imprinted genes, then, are unusual in that only the maternal *or* paternal version of the gene (not both) is functional in the offspring. Loss of expression of imprinted genes may occur by several molecular mechanisms, most commonly via deletion of the normally expressed parental allele, or by uniparental disomy, in which both alleles are inherited from a single parent. Aberrant imprinting is known to cause human genetic diseases such as Prader-Willi syndrome or Angelman syndrome (Nicholls et al., 1998).

D. Phenotype

The genotype represents an individual's genetic constitution at a specified locus, but the *phenotype* is what is observed clinically. The clinical geneticist approaches disease by stringently characterizing the phenotype. Describing clinical presentations in terms of phenotype implies the existence of an underlying genotype or gene sequence-based modification for the clinical features at hand. There is great power and clarity in characterizing disease presentations in terms of phenotype. It invites one to unify the seemingly disparate clinical features of a given patient and affected family members as potentially attributable to a single gene defect; for example, premature ovarian failure in a woman, learning disability with autistic features causing Fragile X syndrome in her son, and late onset ataxia in her father may all be caused by a premutation or a full mutation in the FMR1 gene they share (Hagerman & Hagerman, 2004).

Phenotype is often variable within families and among different patients with the same disorder. A given gene mutation may be expressed to varying degrees from one patient to another (variable expression) with some core phenotypic feature being common to most patients. Similarly, genetic heterogeneity, or the notion that mutations in different genes can give rise to a common phenotype, is also possible, considering for example the situation in which mutations of different genes in a single biochemical pathway may ultimately give rise to similar phenotypic manifestations, as in leukoencephalopathy with vanishing white matter disease, which has been associated with mutations in at least five subunits of eukaryotic translation initiation factor 2 (van der Knaap et al., 2006). Thus, the painstaking detail-driven quantification of clinical phenotype is an essential first step toward identifying the underlying genetic defect as it empowers one to view a given patient in the context of other related or unrelated patients.

IV. The Neurogenetic Evaluation

A. Clinical Evaluation

Careful neurological evaluation is an important first step. Once the problem is defined in general terms insofar as age of onset and course of the disease (acute or chronic, episodic or progressive), and the nature of the problem is established (pure ataxia, demyelinating neuropathy, partial epilepsy), appropriate laboratory tests to exclude nongenetic causes of disease, and clinical electrophysiology and diagnostic imaging have been performed to further define the disease, consideration should be given to as to whether the problem is genetic in origin or not (see Figure 1.3).

If there are multiple other affected family members, with the disease segregating in a Mendelian pattern, then the answer is straightforward. In most cases, however, the answer is not readily apparent. The absence of a family history does not exclude a genetic cause of disease; either because of a new dominant mutation in the proband causing sporadic disease, because of a paucity of males in a small family with an X-linked recessive disease, because it may be a first child affected with a recessive disease, or in the setting of adoption with an unknown family history. In the absence of a family history, a genetic etiology may be suspected because of the unusually young onset of a disease (e.g., a teenager with parkinsonism, or a 40-year-old with Alzheimer's disease) or because of distinctive features of the disease itself (e.g., a patient with long, slender fingers, lens dislocation, and stroke may have underlying homocystinuria). Evaluation by a clinical geneticist is helpful in recognizing minor anomalies, dysmorphic features, and developmental anomalies if they are suspected. The input of other specialists is frequently required; for example, an ophthalmologist to evaluate retinal pigmentary changes, or a cardiologist to evaluate cardiomyopathy associated with Friedreich's ataxia or muscular dystrophy (Hanna & Wood, 2002).

A detailed three-generation pedigree should be recorded, depicting all family members, their ages, their status (affected or unaffected), as well as miscarriages, adoptions, consanguinity, and ethnic background. The pedigree is never fully completed, because of the addition and loss of family members, and the availability of new medical information. Whenever possible, the diagnosis of other family members should be verified by obtaining copies of their medical records.

B. Clinical Diagnostic Genetic Testing

If a genetic diagnosis is suspected, the clinician should then determine if there is a clinical genetic test available to confirm the diagnosis. Genetic testing is defined as the analysis of DNA, RNA, chromosomes, proteins, and certain metabolites in order to detect heritable disease-related genotypes, mutations, or karyotypes for clinical purposes (Harper, 1998; Holtzman & Watson, 1999). There are a variety of academic and commercial labs that perform clinical genetic testing (see Table 1.2). Many specialize in an area of biochemical, cytogenetic, or molecular testing, but with the advent of fluorescent *in situ* hybridization and microarray testing, the traditional distinctions between the types of testing is becoming blurred. It can also be possible and useful to employ more than one type of testing in the evaluation of a disorder; for example, measuring very long chain fatty acids in plasma and DNA testing, in the workup of adrenoleukodystrophy.

A clinical genetic test is one in which patient specimens are examined by a lab certified to perform clinical testing, and written results are reported to the clinician for the purpose of diagnosis, prevention, or treatment in the care of individual patients. It is vital to realize that a genetic test result has implications not only for the individual, but also for family members. Thus, in medical genetics clinics, frequently multiple family members are seen either together or sequentially in consultation.

The interpretation of genetic test results can be challenging. In one widely cited study (Giardiello et al., 1997), one third of physicians given results of a genetic test did not interpret it correctly. Because of genetic heterogeneity and the specific nature of genetic testing, a negative test result does not rule out clinically similar diseases (e.g., a negative genetic test for spinocerebellar ataxia type 1 does not rule out spinocerebellar ataxia type 2 or type 3 or type 6, nor does it mean that a patient does not have ataxia). A patient with all the clinical signs of ataxia and perhaps with cerebellar atrophy on magnetic resonance imaging does have ataxia, even if 20 genetic tests are negative. Rather, it means that the correct gene was not tested, or the molecular mechanism was not identified by the testing that was performed. Because of reduced penetrance (having a genotype associated with the disease, but not expressing the phenotype) or age-dependent penetrance, a positive genetic result does not prove the person is or will become affected (Burke, 2004; Burke & Zimmern, 2004). Additional possible outcomes include mutations of unknown significance and indeterminate results. A person at risk for Huntington's disease with a CAG expansion of 37 repeats may or may not be clinically affected in the future, whereas a person with 30 repeats is unaffected, and a person with 50 repeats will develop HD if he or she lives long enough. If a genetic test result is not understood by the ordering clinician, then the assistance of the diagnostic laboratory director or referral to a medical genetics clinic should be sought before sharing the results with the patient.

However, the era of genetic testing for neurologic disease is here; in the right setting and with careful interpretation, genetic testing is one of the most accurate and specific approaches in the diagnostic armamentarium (1996, Practice parameter: Genetic Testing alert. Statement of the Practice Committee Genetics Testing Task Force of the American Academy of Neurology). In many cases, a genetic test result can exclude a particular diagnosis with certainty, or it can pinpoint a specific cause. All further discussion of the prognosis, anticipatory guidance, recurrence risk, ability to perform prenatal diagnosis is critically dependent upon accuracy of diagnosis, and this process is enhanced by the intelligent use and interpretation of genetic testing.

C. Research Genetic Testing

In the case of some patients, no clinical genetic test is available, or the available clinical genetic testing has not yielded a diagnosis. However, disease-causing mutations may have been published in a research journal. In this case, if the clinician and the patient wish to pursue the diagnosis, then research testing may be requested. Research testing should not be regarded as equivalent to clinical genetic testing, nor should it be requested as a cost-free alternative. Research testing primarily contributes to the further investigation and understanding of a disease. Research testing frequently takes longer (months to years) to achieve a result, and compared to clinical genetic testing, achieving no result is significantly more likely. Research labs are not subject to the stringent oversight and regulations of a clinical lab. Like any other research involving human subjects, research genetic testing requires informed, written consent of the participant or the parent or legal guardian. Depending on the local ethics board, results may or may not be allowed to be communicated back to the patient, or only under special circumstances.

D. Genetic Counseling

Genetic counseling goes hand in hand with the practice of medical genetics (Harper, 2001). In 1975, genetic counseling was defined by a committee of the American Society of Human Genetics as follows:

> Genetic counseling is a communications process which deals with the human problems associated with the occurrence

> or risk of occurrence of a genetic disorder in a family. This process involves an attempt by one or more appropriately trained persons to help the individual or family to: (1) comprehend the medical facts including the diagnosis, probable course of the disorder, and the available management, (2) appreciate the way heredity contributes to the disorder and the risk of recurrence in specified relatives, (3) understand the alternatives for dealing with the risk of recurrence, (4) choose a course of action which seems to them appropriate in their view of their risk, their family goals, and their ethical and religious standards, and act in accordance with that decision, and (5) to make the best possible adjustment to the disorder in an affected family member and/or to the risk of recurrence of that disorder (1975, Genetic Counceling. Am J Hum Genet).

Inherent in this definition is not only a more traditional medical model of understanding and communicating the facts about a disease, but also information specific to the hereditary nature of genetic disease such as calculating individual risk of recurrence of a disease, taking into account variable expression, reduced penetrance, and refinement of numerical risk by Bayes theorem, communicating available reproductive options in a neutral manner, and allowing the family to come to a decision about future reproduction, with respect, support, and medically accurate information provided by the counselor (Baker et al., 1998). The principle of allowing the family to determine the decision, rather than the counselor making a recommendation for a course of action, is termed *nondirectiveness*. Skilled genetic counseling is complex and can be time-consuming. Thus, in 1969, the first U.S. graduate program to educate master's level professionals in the art and science of genetic counseling was founded at Sarah Lawrence College. Since then, the field of genetic counseling has grown, with the development of additional training programs, the creation of a professional society, and professional certification available through the American Board of Genetic Counseling.

V. Identification of Human Disease Genes

A. Position Independent Methods

Prior to the advent of positional cloning, a few genes causing human disease were identified on the basis of visible cytogenetic defects or by pinpointing a biochemical pathway, which led to identification of the relevant gene. Each of these efforts required insight into the mechanism of the disease, finding a particularly informative patient or family, and just plain luck.

Position-independent methods include the following:

1. Knowledge or identification of the dysfunctional protein, and working backward toward gene identification, as in the case of late-infantile neuronal ceroid lipofuscinosis (Sleat et al., 1997).
2. Identification of a candidate gene in a mouse model, and isolating and testing the orthologous human gene.
3. Localizing a disease gene by a chromosomal abnormality such as an unbalanced translocation, inversion, or monosomy of part of 17p in Miller-Dieker syndrome (Dobyns et al., 1983), which led to the identification of mutations in LIS1 (Lo Nigro et al., 1997) as the cause of isolated lissencephaly sequence.
4. Expression array experiments comparing the levels of mRNA in a series of patients versus a series of controls.
5. Knowledge of the probable mechanism of the disease.

For example 5, once it became clear that this was a unifying mechanism for many inherited forms of ataxia, genomic DNA from affected patients who were negative for known causes of inherited ataxia could be screened on a research basis for repeat expansions, which were subsequently cloned and characterized. This method was used to identify the molecular basis of spinocerebellar ataxia type 8 (Koob et al., 1999).

B. Positional Cloning

Positional cloning is the identification of a disease gene based purely upon its chromosomal location, in the absence of knowledge of the gene's function or disease pathogenesis (Collins, 1992). The first genes identified by this type of approach were for the X-linked diseases, chronic granulomatous disease and Duchenne muscular dystrophy, partly because of the advantage provided by excluding a locus on any of the 22 autosomes. In 1993, the gene for Huntington's disease on chromosome 4 was identified by a positional cloning approach (1993), and a new era in human molecular genetics had begun. The basic approach of positional cloning consists of linkage analysis to define a genetic interval, obtaining overlapping clones containing the DNA from the region, identifying the dozen or more genes in the region, and screening them individually for mutations in affected people.

In model organisms such as *Drosophila melanogaster*, it is relatively easy to narrow a genetic interval containing the gene of interest by performing crosses. In humans, since this is not possible, linkage analysis depends upon informative meioses, usually in large families. The power and versatility of the positional cloning approach has led to widespread application to human disease and resulted in the successful identification of hundreds of genes for human Mendelian disorders in the last 15 years. With the advent of PCR, the availability of dense maps of markers, large insert clones, polymorphic markers arrayed on a chip, and computerized searches to identify transcripts within a target region, the major limiting factor now in identifying a new Mendelian disease gene is the identification of appropriate families who have undergone meticulous clinical phenotyping.

C. Mapping Complex Genetic Disorders

Most of the progress in neurogenetics has been in the identification of Mendelian disease genes, and the translation of that research into the clinic. It has been estimated that 10 percent of neurology patients have a Mendelian disorder (Hanna & Wood, 2002). This implies then, that the vast majority of patients have neurological disorders in which a few genes each contribute (oligogenic), many genes each contribute a small amount (polygenic), or predisposition is caused by a combination of genetic and environmental factors (Mayeux, 2005a; Wright, 2005). This includes disorders such as late onset Alzheimer's disease, multiple sclerosis, stroke, brain tumors, schizophrenia, and autism. Although the identification of genetic risk factors for complex traits is fraught with difficulties, and many reported genes predisposing to complex traits are not replicated by independent research, nonetheless, the search for genes for complex traits is the focus of many studies. This is partly because of the relative burden of disease caused by rare, Mendelian forms of a disease compared with the common forms of the disease, and therefore, the potential impact of a true finding on many-fold more patients. There are perhaps fewer than 1,000 patients worldwide with the rare, Mendelian forms of early-onset familial Alzheimer's disease, whereas there are over 4 million patients with the common form of Alzheimer's disease in the United States alone.

Since family clustering of a disease may be from shared genes or shared environment, there are a number of ways to estimate the hereditability, defined as the proportion of causation of a trait that is due to genetic causes. Comparing the concordance of monozygotic twins (who are genetically identical) versus dizygotic twins (who share one half of their genes), adoption studies (determining whether adopted children's risk of developing a disease more closely resembles their biological relatives or their adoptive family members) and segregation analysis are methods to ascertain whether familial segregation of a disease is genetic.

The paradigm of linkage analysis followed by positional cloning that has proven so effective in Mendelian disorders has a number of disadvantages for the analysis of complex traits, including the effects of phenotyping uncertainty, locus and clinical heterogeneity on the lod score (Gordon & Finch, 2005).

Many consider genetic association analysis to be the best method for identifying genetic variants related to common, complex diseases (Hirschhorn & Daly, 2005; Mayeux, 2005b; Neale & Sham, 2004). Association analyses are nonparametric, so they do not require specifying the inheritance pattern of a disease, they are better suited to the study of small-to-medium sized families, they may be conducted using patients and controls or as family-based association tests. Finally, for a fixed genotypic relative risk, association studies have greater power requiring smaller sample sizes to detect gene effects than do comparable studies using linkage analysis (Sklar, 2001).

Genetic association studies have historically suffered from lack of replication. Guidelines for submitting an association study have been proposed to increase the likelihood of a true association, and not a false association due to population stratification, multiple comparisons, and other causes (Bird et al., 2001).

With the advent of microarray technology, thousands of single nucleotide polymorphism (SNPs) may be interrogated simultaneously. This has led to genome-wide association studies with 10,000 to 500,000 SNPs, with some impressive results, such as the identification of two complement factor H SNPs associated with an increased risk of age-related macular degeneration in a case-control study performed by Hoh and colleagues (Klein et al., 2005).

VI. Methods for Human Molecular Genetic Analysis

The following is a brief introduction to some of the basic concepts and methods relevant to the molecular diagnostic evaluation of patients with suspected single gene defects in nuclear genes. For a wider perspective see Cooper & Krawczak (1993) and Stenson et al. (2003). Since the vast majority of genetic testing ordered by neurologists is molecular genetic (DNA-based) diagnostic testing, that will be the focus of this discussion. Biochemical genetic testing and cytogenetic testing are important diagnostic tools, but beyond the scope of this chapter.

Molecular genetic testing for the clinical evaluation of Mendelian traits involves identifying known or unknown disease-causing mutations in a gene of interest. These may include several types of mutations, including but not limited to established or novel single nucleotide substitutions, dynamic microsatellite repeat expansion mutations, as well as gene deletions and duplications. Gene rearrangements and epigenetic changes such as altered methylation status can also occur but will not be discussed. In addition, linkage analysis following disease associated markers within an individual family may also be useful for diagnostic risk assessment in certain settings. Identifying a mutation has multiple implications. It confirms the genetic basis of the disease and facilitates ongoing anticipatory medical and educational guidance, since one can now look to the broader cohort of patients having mutations in the same gene for comparison. In addition, from the point of view of inheritance, one can now test other family members as well as provide more complete recurrence risk genetic counseling, including potentially prenatal or predictive testing.

A. Polymerase Chain Reaction in Molecular Diagnostics

The polymerase chain reaction (PCR), a cornerstone of modern molecular biology, has been used extensively in the clinical molecular diagnostic setting (Saiki et al., 1985). Briefly, PCR involves amplifying the endogenous target DNA (template) of interest through repetitive cycles of (1) heat-induced separation (denaturation) of the parent DNA template strands followed by (2) temperature optimized annealing of complementary oligonucleotides primers and (3) subsequently heat stable DNA polymerase driven extension of the primers along the endogenous DNA template. Each cycle doubles the number of molecules present at the start of the cycle so that after 30 or so cycles one can expect to have generated approximately a billion copies of the target DNA from the original target (Cooper & Krawczak, 1993).

A few precautions relevant to diagnostic PCR need to be acknowledged. In the clinical diagnostic setting, quality control requires that PCR be carried out in such a manner so as to eliminate the possibility of cross contamination of samples as well as the inclusion of negative and positive controls. Negative controls (lacking template DNA) are used to monitor for reagent purity or contamination. Positive controls ensure that the signal is specific to the target. It is important to note that enzyme infidelity may generate sequence errors that do not reflect the actual sequence of the patient's gene, and if such errors arise early in the PCR they may represent a significant fraction of the product. Unknown polymorphisms present in the target sequence of the patient's DNA, within PCR primer binding sites, may also inhibit amplification.

B. Types of Gene Mutations

Disease-causing mutations in the context of inherited single gene disorders are changes in the DNA sequence of individual genes that are both necessary and sufficient to give rise to the abnormal phenotype. For the purpose of illustrating the range of mutations associated with neurogenetic disorders with an emphasis of laboratory diagnostics, the following types of mutations will be discussed:

- ▲ Single nucleotide substitutions
- ▲ Microsatellite expansion mutations
- ▲ Deletion/amplification mutations

1. Nucleotide Substitution Mutations

Gene mutations of the single nucleotide substitution type (such as C®T or G®A) occur in either coding or noncoding regions of genes. Nonsense mutations generate premature stop codons, which can lead to truncated protein products. Missense mutations result in an amino acid substitution, which may alter the function of the resulting protein. Frameshift mutations alter the reading frame and typically lead to one or more amino acid substitutions followed by a stop codon. Single nucleotide substitution mutations can also result in abnormal transcription via alterations in splicing or processing at the messenger RNA level. Gene promoter as well as 5′ Cap site and 3′ polyadenylation signal mutations may also occur (Cooper & Krawczak, 1993). Single nucleotide substitutions are found in a wide variety of neurological diseases, including channelopathies (Hanna, 2006), early-onset familial Alzheimer's disease (Goate et al., 1991; Levy-Lahad et al., 1995; Sherrington et al., 1995), CADASIL (Orlandi et al., 2003), neurofibromatosis type 1 and type 2 (Yohay, 2006), and such.

Traditionally, much effort has gone into developing mutation screening methodologies to search for previously unreported mutations such as single-stranded conformational polymorphism analysis, heteroduplex analysis, and denaturing high performance liquid chromatography. These methods narrow the region of the gene to be sequenced and are followed by sequencing the variant fragments from candidate regions of the gene under investigation to confirm a base change. However, the necessity of screening approaches in the diagnostic lab has been reduced substantially by accurate, direct, automated, bidirectional sequencing of PCR fragments from the gene in question as the most cost-effective and high-yield approach of choice for the molecular diagnosis of Mendelian traits characterized by point mutations and small deletions.

Methodologies for mutation detection continue to evolve with advances in sequence-based technologies. More accurate and sensitive analytical tools continue to deliver more information at lower cost, often generating new or unforeseen questions in the process. For instance, complete sequencing of a gene may reveal the absence of a known mutation but the identification of a previously unknown nucleotide variant requiring additional inquiry to determine whether or not this variant represents a newly identified mutation or is a nonpathogenic polymorphism.

Molecular analysis of autosomal dominant traits is conducted with the realization that only one of the two alleles of interest will contain the disease-associated mutation. Thus PCR amplification will generate a product that is in effect a 50/50 mixture of wild type and mutant alleles, assuming both alleles amplify with equal efficiency. Sequencing this PCR product will yield identical sequence peak intensities for all bases except for the position where the mutant base appears, and this position will show the superposition of normal and mutant bases, each at half intensity. In the case of recessive disease with homozygous base substitution, a full intensity new base peak is observed, since both alleles contain the same mutation at the same position. Compound heterozygotes or patients who are affected with a recessive disease by virtue of having a different disease-associated mutation in each allele will yield sequence data showing half wild type and half mutant signal intensity at two different positions in the sequenced alleles.

2. Microsatellite Expansion Mutations

Several neurological diseases are characterized by microsatellite sequence instability including Huntington's disease, myotonic dystrophy types 1 (DM1) and 2 (DM2), Fragile X syndrome, Friedreich's ataxia, as well as spinocerebellar ataxia types 8, 10, 12, to name a few (Richards, 2001). Although repetitive DNA sequences are distributed throughout the genome, most are stably transmitted. However, in disorders caused by trinucleotide repeat microsatellite expansions, such as Huntington's disease and Fragile X or by a tetranucleotide repeat expansion in the case of DM2, the microsatellite sequence is inherently unstable. During meiosis, expansion of the microsatellite sequence may occur, leading to a deficiency in the respective protein. For each disease, once a numerical threshold of a certain number of repeats is passed, the expansion mutation causes disease. This dynamic nature of microsatellite expansion disease mutations creates unique molecular diagnostic challenges.

a. FMR1 The Fragile X syndrome was one of the first microsatellite expansion diseases to be described (Kremer et al., 1991; Yu et al., 1991). It is the most common form of inherited developmental delay in boys, occurring in approximately 1 in 4,000 males (Turner et al., 1996). It is notable for an expanded CGG repeat (greater than 200 repeats) in the 5′ untranslated region of the FMR1 gene leading to epigenetic hypermethlation of FMR1 promoter and loss of function in affected males. However another phenotype manifests in adults over 50 years old with expansions of the FMR1 gene in the 55 to 200 range (premutation range). This is referred to as Fragile X associated tremor ataxia syndrome (FXTAS), and is characterized by gait instability, tremor, cognitive defects, MRI signal abnormalities in the middle cerebellar peduncles and adjacent white matter, and intranuclear neuronal and astrocytic inclusions (Hagerman et al., 2001; Jacquemont et al., 2003). Some one third of older males with premutation alleles will develop symptoms of FXTAS (Jacquemont et al., 2004).

Thus a key objective in FMR1 gene analysis is determining the precise size of the single FMR1 allele in males, and the two FMR1 alleles in females, since the propensity of premutation alleles to expand to a full mutation while being transmitted through a female (mother to child) is proportional to the size of the expansion in the mother. Restriction enzyme digestion followed by Southern blotting to detect expansions in the pre- and full mutation range of FMR1 has served as the diagnostic approach to Fragile X syndrome for some time, though it is limited in its ability to resolve small differences in size between large, normal, and small permutation alleles. More recent approaches are PCR-based, using a combination of methylation and nonmethylation sensitive PCR primers as well as CGG-based expansion primers (Zhou et al., 2004).

b. Myotonic Dystrophy Type 1 and Type 2 The myotonic dystrophies type 1 and 2 (DM1 and 2) are both microsatellite instability diseases with a similar multisystem clinical presentation notable for myotonia, muscular dystrophy, cataracts, diabetes, testicular failure, hypogammaglobulinemia, and cardiac conduction defects. DM2 lacks a severe, congenital form, which is seen with DM1 (Day et al., 2003). DM1 is caused by abnormal CTG expansion in the 3′ untranslated region of *dystrophia myotonica protein kinase gene* and DM2 by a pathologic tetranucleotide (CCTG) expansion of the first intron of the *zinc finger protein 9 gene* (Ranum & Day, 2004).

A unique combined PCR-Southern blot approach to the diagnosis of autosomal dominant myotonic dystrophy type 2 (DM2, proximal myotonic myopathy PROMM) has been described (Day et al., 2003). The DM2 molecular diagnostic signature is an approximately 5,000 CCTG (tetranucleotide) repeat expansion of the first intron of the *zinc finger protein 9 gene*, which can be detected by PCR using a forward primer located outside of the expansion and a reverse primer containing sequence complementary to the CCTG repeat expansion such that a full range of expanded repeats are generated. Myotonic dystrophy type 2 is characterized by somatic instability such that the PCR reaction yields a smear. The resulting PCR smear is then transferred to a membrane and probed with an internal probe as a Southern blot (Day et al., 2003) revealing the extent to which expansion has occurred. Using this approach the diagnostic molecular detection rate was 99 percent compared to 80 percent via Southern blot.

3. Deletion/Amplification Mutations

The diagnosis of gene deletion diseases such Duchenne muscular dystrophy (DMD) and Becker muscular dystrophy (BMD) highlight the unique challenges of demonstrating a diagnostic defect in gene dosage since about two thirds of muscular dystrophy patients have the disease on the basis of a large deletion or duplication, and the remaining one third harbor point mutations (Prior & Bridgeman, 2005). Diagnostic tests, which are based on a negative finding such as an absent or reduced signal in the case of a deletion, always carry the extra burden of proof to show that the loss of the diagnostic signal is due to the true absence of a portion of the gene as opposed to failure of the assay to generate a signal (a false positive) secondary to failure to achieve optimal probe hybridization or PCR failure, for example.

A male affected with X-linked Duchenne muscular dystrophy (DMD) or Becker muscular dystrophy (BMD), for instance, may be deleted in some portion of his only copy of the dystrophin gene. At the molecular level, most DMD cases are due to deletions or duplications that are out of frame, resulting in a truncated nonfunctional protein.

The phenotypically milder BMD patients have in-frame deletions and duplications with the preservation of some dystrophin protein expression.

The multiplex ligation-dependent probe amplification (MLPA) test is a molecular assay that detects deletions and gene amplifications through the hybridization of a set of adjacent probes across the gene of interest followed by ligation and subsequent PCR amplification and quantification of the ligated and amplified probes (Lai et al., 2006). In the case of a male with a deletion the relevant probes cannot anneal, and ligation and PCR are unable to proceed resulting in an absent signal for the deleted region or a reduced signal reflecting reduced gene dosage in the case of a DMD carrier female.

MLPA as well as Southern blot analysis are also amenable to the diagnosis of gene amplifications such as duplications. Here the diagnostic signal is an increased dosage or intensity beyond that expected under normal circumstances. Accordingly, MLPA also has been used to identify duplications in the DMD gene (Lai et al., 2006). Most patients with Charcot-Marie-Tooth neuropathy type 1 exhibit a duplication chromosome 17p11.2-p12 that contains the peripheral myelin protein 22 gene (Park et al., 2006; Patel et al., 1992), and again the MLPA approach also has been used to diagnose gene amplification in the setting of Charcot-Marie-Tooth disease (Slater et al., 2004).

VII. Treatment of Genetic Diseases

In the workup of a patient presenting with a neurological problem, after the diagnosis has been established by clinical evaluation and genetic testing, and genetic counseling has been provided to the patient and family about recurrence risk, then how does the diagnosis of a genetic condition inform treatment now and in the future (Varmus, 2002)?

There are several potential implications. First, the correct diagnosis can assist the patient and physician in avoiding ineffective or even adverse treatments. For example, a patient with stroke due to Ehlers Danlos type IV, hemiplegic migraine, or mitochondrial encephalopathy, lactic acidosis, and stroke (MELAS) is less likely to benefit from conventional stroke treatments such as tissue plasminogen activator, carotid endarterectomy, or hypertension and cholesterol management. Instead, recommendations should be tailored to the individual's disease. Second, there are nongenetic approaches to the treatment of genetic diseases. Third, there are some effective and many other emerging gene-based therapies for genetic diseases.

A. Medical Treatment of Neurogenetic Diseases

The medical treatment of genetic diseases has a distinguished history, particularly in the area of inborn errors of metabolism. Dietary modification to limit intake of foods containing a substrate that builds up in the absence of a necessary enzyme is still the mainstay of treatment for a host of diseases, including amino acidopathies, and urea cycle disorders. Hematopoietic stem cell transplantation or liver transplantation are also used to treat some metabolic diseases, albeit these are complicated medical procedures associated with a significant potential for morbidity or mortality (Mahmood et al., 2005; McBride et al., 2004).

Next, a range of conventional medical and surgical treatments are available to treat the tumors that arise in tuberous sclerosis, severe scoliosis of neurofibromatosis, dystonia of idiopathic torsion dystonia, or the epilepsy from autosomal dominant partial epilepsy with auditory features (ADPEAF). Finally, special education, social services, and patient support groups can improve adaptation to a disease and maximize outcome. Indeed, one of the functions of a genetics clinic or a neurogenetics clinic is to coordinate the disparate subspecialty services that may be required in the care of a single patient. This underscores the notion that the treatment of genetic diseases is not limited to a strictly genetic approach.

B. Enzyme Replacement Therapy

Another approach to the treatment of metabolic, in particular, lysosomal storage diseases, has been replacement of the missing enzyme, not by replacing the gene, but by replacing the defective protein product. Type 1 Gaucher disease is an autosomal recessive lysosomal storage disease that is the most common genetic disorder observed in the Ashkenazi Jewish population. It is caused by glucocerebrosidase deficiency, resulting in the intralysosomal accumulation of glucocerebroside in tissues of the reticuloendothelial system (Charrow et al., 2004). Key clinical features include hepatosplenomegaly, anemia, thrombocytopenia, osteopenia, fractures, acute and chronic bone pain, and osteonecrosis. Type 2 Gaucher disease and Type 3 Gaucher disease (Norrbottnian form) are allelic disorders caused by mutations in the same gene as Type 1, distinguished by clinical presentation. In Type 2, neurological presentation in infancy accompanies the systemic symptoms. Type 3 is characterized by more slowly progressive neurological problems such as ataxia, spastic paraplegia, seizures, dementia, or supranuclear gaze palsy.

Type 1 Gaucher disease is treatable by enzyme replacement therapy with recombinant human macrophage targeted human glucocerebrosidase (imiglucerase, Cerezyme®, Genzyme Corporation, Cambridge, MA). Treatment has been shown to effect breakdown of stored glucocerebroside, resulting in reduction of hepatosplenomegaly, improvement of hematological parameters, and increase in bone mineralization (Weinreb et al., 2002). The effects of enzyme treatment on the CNS effects of the neuronopathic forms has been variable with improvement in some cases, slowing of progression in others, no improvement in still others (Vellodi et al., 2001).

Enzyme replacement therapy is also available for Fabry disease, an X-linked lysosomal storage disease characterized by renal dysfunction, painful distal extremities, and vascular disease (Desnick et al., 2003; Wilcox et al., 2004). In 2006, the Food and Drug Administration approved an alglucosidase alfa (Myozyme®) for the treatment of Pompe disease, a glycogen storage disease affecting muscle.

C. Gene Therapy

Gene therapy involves the modification of a patient's cells by transfer of normal genes in order to treat a disease. Particularly for the subgroup of fatal, progressive neurodegenerative genetic diseases for which treatment is largely symptomatic and supportive at present, such as Huntington's disease, familial Alzheimer's disease, familial amyotrophic lateral sclerosis, and various forms of muscular dystrophy, gene therapy would fulfill a long-held dream. Further details on this topic may be found in Chapter 7.

D. RNA Interference (RNAi)

RNA interference is an evolutionarily conserved process that directs gene silencing of gene expression in a sequence-specific manner (Downward, 2004; Dykxhoorn et al., 2003). Multicellular organisms possess a conserved protein machinery that recognizes double-stranded RNA, which is degraded by the enzyme, dicer, into short segments of approximately 20 nucleotides called small interfering RNAs (siRNAs). A protein complex called RNA-induced silencing complex (RISC) uses the siRNA to recognize and degrade single-stranded RNA (such as mRNA) with the same sequence. RNA interference is a system whereby post-transcriptional gene silencing is used to rid the cell of viruses, to regulate mobile genetic elements such as repetitive sequences, and as a mechanism for fine-tuning normal gene expression during development, apoptosis, and other processes.

RNAi is used routinely in the biomedical research setting to block the expression of a given gene. In *C. elegans*, which feed on bacteria, it is possible to cause RNA interference by feeding them genetically engineered bacteria expressing double-stranded RNA targeting the gene of interest. In mammalian cells, synthetic RNAs are complexed with cationic lipids in order to pass through the cell membrane, or introduced by genetically engineered viral vectors such as retroviruses, adenoviruses, or lentiviruses (Hommel et al., 2003). Once in the cell, the viral construct directs expression of a short hairpin RNA, which is processed to produce a functional siRNA.

Given this background, there is obvious potential to harness RNAi as a therapeutic approach toward blocking the expression of a dominantly inherited neurological disease caused by a toxic gain of function mutation. In fact, several academic research groups and commercial companies are attempting to do just that. RNAi treatment has been reported in a transgenic mouse model of spinocerebellar ataxia type 1 (Xia et al., 2004), one of the group of neurodegenerative diseases caused by polyglutamine expansion, as well as in a mouse model of Huntington's disease (Harper et al., 2005; Machida et al., 2006) and amyotrophic lateral sclerosis (Ralph et al., 2005). In each of these, pathological and clinical improvements were reported. Although these reports are encouraging, more research is needed to avoid knocking down homologous genes (off-target effects), toxicity from the viral vectors, and to optimize the stability, delivery, and long-term safety of this method.

E. MicroRNAs

MicroRNAs (miRNAs) are transcribed from endogenous DNA as hairpin precursors. The RNA-induced silencing complex allows one strand of the microRNA to bind to the 3′ untranslated region of specific mRNAs, thus disrupting the translation of the message and expression of the corresponding protein. MicroRNAs normally regulate such fundamental processes as development, apoptosis, and patterning of the nervous system. Several hundred miRNAs have been discovered experimentally in animals, and computational methods predict many more may exist (Berezikov et al., 2006). Now, microRNAs have emerged as a potential therapeutic target. In mice, exogenous delivery of a single-stranded molecule complementary to miR-122, which is highly expressed in liver, has been shown to result in degradation of endogenous miR-122 and a significant physiological effect with a 44 percent drop in cholesterol levels (Krutzfeldt et al., 2005). RNA-based approaches are a potentially exciting therapeutic approach, but much research needs to be done to answer questions about the delivery to various tissues, stability, secondary effects such as silencing of related mRNAs, and other potential cellular responses.

References

(1975). Genetic counseling. *Am J Hum Genet* **27**, 240–242.

(1993). A novel gene containing a trinucleotide repeat that is expanded and unstable on Huntington's disease chromosomes. The Huntington's Disease Collaborative Research Group. *Cell* **72**, 971–983.

(1996). Practice parameter: Genetic testing alert. Statement of the Practice Committee Genetics Testing Task Force of the American Academy of Neurology. *Neurology* **47**, 1343–1344.

(2001). The human genome. Science genome map. *Science* **291**, 1218.

(2006). *Emery and Rimoin's Principles and Practice of Medical Genetics*. Churchill Livingstone, New York.

Alberts, B., Johnson, A., Lewis, J., Raff, M., Roberts, K., and Walter, P. (2002). *Molecular Biology of the Cell*. Garland Publishing, Inc., New York.

Baker, D. L., Schuette, J. L., and Uhlmann, W. R. (1998). *A Guide to Genetic Counseling*. Wiley-Liss, New York.

Bennett, R. L., Steinhaus, K. A., Uhrich, S. B., O'Sullivan, C. K., Resta, R. G., Lochner-Doyle, D. et al. (1995). Recommendations for standardized human pedigree nomenclature. Pedigree Standardization Task Force of the National Society of Genetic Counselors. *Am J Hum Genet* **56**, 745–752.

Berezikov, E., Cuppen, E., and Plasterk, R. H. (2006). Approaches to microRNA discovery. *Nat Genet* **38 Suppl**, S2–7.

Bird, T. D., Jarvik, G. P., and Wood, N. W. (2001). Genetic association studies: Genes in search of diseases. *Neurology* **57**, 1153–1154.

Burke, W. (2004). Genetic testing in primary care. *Annu Rev Genomics Hum Genet* **5**, 1–14.

Burke, W. and Zimmern, R. L. (2004). Ensuring the appropriate use of genetic tests. *Nat Rev Genet* **5**, 955–959.

Charrow, J., Andersson, H. C., Kaplan, P., Kolodny, E. H., Mistry, P., Pastores, G. et al. (2004). Enzyme replacement therapy and monitoring for children with type 1 Gaucher disease: Consensus recommendations. *J Pediatr* **144**, 112–120.

Claverie, J. M. (2001). Gene number. What if there are only 30,000 human genes? *Science* **291**, 1255–1257.

Collins, F. S. (1992). Positional cloning: Let's not call it reverse anymore. *Nat Genet* **1**, 3–6.

Cooper, D. N. and Krawczak, M. (1993). *Human Gene Mutation*. Bios Scientific Pub. Ltd.

Day, J. W., Ricker, K., Jacobsen, J. F., Rasmussen, L. J., Dick, K. A., Kress, W. et al. (2003). Myotonic dystrophy type 2: molecular, diagnostic and clinical spectrum. *Neurology* **60**, 657–664.

Desnick, R. J., Brady, R., Barranger, J., Collins, A. J., Germain, D. P., Goldman, M. et al. (2003). Fabry disease, an under-recognized multisystemic disorder: Expert recommendations for diagnosis, management, and enzyme replacement therapy. *Ann Intern Med* **138**, 338–346.

Dobyns, W. B., Stratton, R. F., Parke, J. T., Greenberg, F., Nussbaum, R. L., and Ledbetter, D. H. (1983). Miller-Dieker syndrome: Lissencephaly and monosomy 17p. *J Pediatr* **102**, 552–558.

Downward, J. (2004). RNA interference. *BMJ* **328**, 1245–1248.

Dykxhoorn, D. M., Novina, C. D., and Sharp, P. A. (2003). Killing the messenger: Short RNAs that silence gene expression. *Nat Rev Mol Cell Biol* **4**, 457–467.

Ferguson-Smith, A. C. and Surani, M. A. (2001). Imprinting and the epigenetic asymmetry between parental genomes. *Science* **293**, 1086–1089.

Gardner, R. J. M. and Sutherland, G. R. (1996). Chromosome abnormalities and genetic counseling. In A. G. Motulsky, P. S. Harper, M. Bobrow, and C. Scriver, Eds., *Oxford Monographs on Medical Genetics No. 29*. New York.

Giardiello, F. M., Brensinger, J. D., Petersen, G. M., Luce, M. C., Hylind, L. M., Bacon, J. A. et al. (1997). The use and interpretation of commercial APC gene testing for familial adenomatous polyposis. *N Engl J Med* **336**, 823–827.

Goate, A., Chartier-Harlin, M. C., Mullan, M., Brown, J., Crawford, F., Fidani, L. et al. (1991). Segregation of a missense mutation in the amyloid precursor protein gene with familial Alzheimer's disease. *Nature* **349**, 704–706.

Gordon, D. and Finch, S. J. (2005). Factors affecting statistical power in the detection of genetic association. *J Clin Invest* **115**, 1408–1418.

Guttmacher, A. E. and Collins, F. S. (2002). Genomic medicine—A primer. *N Engl J Med* **347**, 1512–1520.

Hagerman, P. J. and Hagerman, R. J. (2004). The fragile-X premutation: A maturing perspective. *Am J Hum Genet* **74**, 805–816.

Hagerman, R. J., Leehey, M., Heinrichs, W., Tassone, F., Wilson, R., Hills, J. et al. (2001). Intention tremor, parkinsonism, and generalized brain atrophy in male carriers of fragile X. *Neurology* **57**, 127–130.

Hanna, M. G. (2006). Genetic neurological channelopathies. *Nat Clin Pract Neurol* **2**, 252–263.

Hanna, M. G. and Wood, N. W. (2002). Running a neurogenetic clinic. *J Neurol Neurosurg Psychiatry* **73 Suppl 2**, II2–4.

Harper, P. S. (1998). Promoting safe and effective genetic testing in the United States: Final Report of the Task Force on Genetic Testing. *Community Genet* **1**, 91–92.

Harper, P. S. (2001). *Practical Genetic Counseling*. Arnold, London.

Harper, S. Q., Staber, P. D., He, X., Eliason, S. L., Martins, I. H., Mao, Q. et al. (2005). RNA interference improves motor and neuropathological abnormalities in a Huntington's disease mouse model. *Proc Natl Acad Sci U S A* **102**, 5820–5825.

Hirschhorn, J. N. and Daly, M. J. (2005). Genome-wide association studies for common diseases and complex traits. *Nat Rev Genet* **6**, 95–108.

Holtzman, N. A. and Watson, M. S. (1999). Promoting safe and effective genetic testing in the United States. Final report of the Task Force on Genetic Testing. *J Child Fam Nurs* **2**, 388–390.

Hommel, J. D., Sears, R. M., Georgescu, D., Simmons, D. L., and DiLeone, R. J. (2003). Local gene knockdown in the brain using viral-mediated RNA interference. *Nat Med* **9**, 1539–1544.

ISCN. (1995). *An International System for Human Cytogenetic Nomenclature*. Karger, Basel.

Jacquemont, S., Hagerman, R. J., Leehey, M., Grigsby, J., Zhang, L., Brunberg, J. A. et al. (2003). Fragile X premutation tremor/ataxia syndrome: Molecular, clinical, and neuroimaging correlates. *Am J Hum Genet* **72**, 869–878.

Jacquemont, S., Hagerman, R. J., Leehey, M. A., Hall, D. A., Levine, R. A., Brunberg, J. A. et al. (2004). Penetrance of the fragile X-associated tremor/ataxia syndrome in a premutation carrier population. *JAMA* **291**, 460–469.

Jenuwein, T. and Allis, C. D. (2001). Translating the histone code. *Science* **293**, 1074–1080.

Klein, R. J., Zeiss, C., Chew, E. Y., Tsai, J. Y., Sackler, R. S., Haynes, C. et al. (2005). Complement factor H polymorphism in age-related macular degeneration. *Science* **308**, 385–389.

Koob, M. D., Moseley, M. L., Schut, L. J., Benzow, K. A., Bird, T. D., Day, J. W., and Ranum, L. P. (1999). An untranslated CTG expansion causes a novel form of spinocerebellar ataxia (SCA8). *Nat Genet* **21**, 379–384.

Kremer, E. J., Pritchard, M., Lynch, M., Yu, S., Holman, K., Baker, E. et al. (1991). Mapping of DNA instability at the fragile X to a trinucleotide repeat sequence p(CCG)n. *Science* **252**, 1711–1714.

Krutzfeldt, J., Rajewsky, N., Braich, R., Rajeev, K. G., Tuschl, T., Manoharan, M., and Stoffel, M. (2005). Silencing of microRNAs in vivo with 'antagomirs'. *Nature* **438**, 685–689.

Lai, K. K., Lo, I. F., Tong, T. M., Cheng, L. Y., and Lam, S. T. (2006). Detecting exon deletions and duplications of the DMD gene using Multiplex Ligation-dependent Probe Amplification (MLPA). *Clin Biochem* **39**, 367–372.

Lander, E. S., Linton, L. M., Birren, B., Nusbaum, C., Zody, M. C., Baldwin, J. et al. (2001). Initial sequencing and analysis of the human genome. *Nature* **409**, 860–921.

Levy-Lahad, E., Wijsman, E. M., Nemens, E., Anderson, L., Goddard, K. A., Weber, J. L. et al. (1995). A familial Alzheimer's disease locus on chromosome 1. *Science* **269**, 970–973.

Lo Nigro, C., Chong, C. S., Smith, A. C., Dobyns, W. B., Carrozzo, R., and Ledbetter, D. H. (1997). Point mutations and an intragenic deletion in LIS1, the lissencephaly causative gene in isolated lissencephaly sequence and Miller-Dieker syndrome. *Hum Mol Genet* **6**, 157–164.

Machida, Y., Okada, T., Kurosawa, M., Oyama, F., Ozawa, K., and Nukina, N. (2006). rAAV-mediated shRNA ameliorated neuropathology in Huntington disease model mouse. *Biochem Biophys Res Commun* **343**, 190–197.

Mahmood, A., Dubey, P., Moser, H. W., and Moser, A. (2005). X-linked adrenoleukodystrophy: Therapeutic approaches to distinct phenotypes. *Pediatr Transplant* **9 Suppl 7**, 55–62.

Mattick, J. S. (2004). RNA regulation: A new genetics? *Nat Rev Genet* **5**, 316–323.

Mayeux, R. (2005a). Mapping the new frontier: Complex genetic disorders. *J Clin Invest* **115**, 1404–1407.

Mayeux, R. (2005b). Mapping the new frontier: Complex genetic disorders. *J Clin Invest* **115**, 1404–1407.

McBride, K. L., Miller, G., Carter, S., Karpen, S., Goss, J., and Lee, B. (2004). Developmental outcomes with early orthotopic liver transplantation for infants with neonatal-onset urea cycle defects and a female patient with late-onset ornithine transcarbamylase deficiency. *Pediatrics* **114**, e523–526.

McKusick, V. A. (1993). Medical genetics. A 40-year perspective on the evolution of a medical specialty from a basic science. *JAMA* **270**, 2351–2356.

Neale, B. M. and Sham, P. C. (2004). The future of association studies: gene-based analysis and replication. *Am J Hum Genet* **75**, 353–362. Epub 2004 Jul 2022.

Nicholls, R. D., Saitoh, S., and Horsthemke, B. (1998). Imprinting in Prader-Willi and Angelman syndromes. *Trends Genet* **14**, 194–200.

Orlandi, G., Inzitari, D., and Frederico, A. (2003). The neurologist and stroke. *Neurol Sci* **23**, 323–325.

Park, H. K., Kim, B. J., Sung, D. H., Ki, C. S., and Kim, J. W. (2006). Mutation analysis of the PMP22, MPZ, EGR2, LITAF, and GJB1 genes in Korean patients with Charcot-Marie-Tooth neuropathy type 1. *Clin Genet* **70**, 253–256.

Patel, P. I., Roa, B. B., Welcher, A. A., Schoener-Scott, R., Trask, B. J., Pentao, L. et al. (1992). The gene for the peripheral myelin protein PMP-22 is a candidate for Charcot-Marie-Tooth disease type 1A. *Nat Genet* **1**, 159–165.

Prior, T. W. and Bridgeman, S. J. (2005). Experience and strategy for the molecular testing of Duchenne muscular dystrophy. *J Mol Diagn* **7**, 317–326.

Pulst, S. M. (2003). Neurogenetics: Single gene disorders. *J Neurol Neurosurg Psychiatry* **74**, 1608–1614.

Ralph, G. S., Radcliffe, P. A., Day, D. M., Carthy, J. M., Leroux, M. A., Lee, D. C. et al. (2005). Silencing mutant SOD1 using RNAi protects against neurodegeneration and extends survival in an ALS model. *Nat Med* **11**, 429–433.

Ranum, L. P. and Day, J. W. (2004). Myotonic dystrophy: RNA pathogenesis comes into focus. *Am J Hum Genet* **74**, 793–804.

Richards, R. I. (2001). Dynamic mutations: A decade of unstable expanded repeats in human genetic disease. *Hum Mol Genet* **10**, 2187–2194.

Rosenberg, R. N. (2001). Genomic neurology: A new beginning. *Arch Neurol* **58**, 1739–1741.

Saiki, R. K., Scharf, S., Faloona, F., Mullis, K. B., Horn, G. T., Erlich, H. A., and Arnheim, N. (1985). Enzymatic amplification of beta-globin genomic sequences and restriction site analysis for diagnosis of sickle cell anemia. *Science* **230**, 1350–1354.

Shaffer, L. G. and Tommerup, N. (2005). ISCN 2005: An International System for Human Cytogenetic Nomenclature.

Sherrington, R., Rogaev, E. I., Liang, Y., Rogaeva, E. A., Levesque, G., Ikeda, M. et al. (1995). Cloning of a gene bearing missense mutations in early-onset familial Alzheimer's disease. *Nature* **375**, 754–760.

Sklar, P. (2001). The genomic approach to candidate genes. *Harv Rev Psychiatry* **9**, 197–207.

Slater, H., Bruno, D., Ren, H., La, P., Burgess, T., Hills, L. et al. (2004). Improved testing for CMT1A and HNPP using multiplex ligation-dependent probe amplification (MLPA) with rapid DNA preparations: Comparison with the interphase FISH method. *Hum Mutat* **24**, 164–171.

Sleat, D. E., Donnelly, R. J., Lackland, H., Liu, C. G., Sohar, I., Pullarkat, R. K., and Lobel, P. (1997). Association of mutations in a lysosomal protein with classical late-infantile neuronal ceroid lipofuscinosis. *Science* **277**, 1802–1805.

Stenson, P. D., Ball, E. V., Mort, M., Phillips, A. D., Shiel, J. A., Thomas, N. S. et al. (2003). Human Gene Mutation Database (HGMD): 2003 update. *Hum Mutat* **21**, 577–581.

Strachan, T. and Read, A. P. (2004). *Human Molecular Genetics*. Garland Publishing, Inc., New York.

Tijo, J. H. and Levan, A. (1956). The chromosome number of man. *Hereditas* **42**, 1–6.

Turner, G., Webb, T., Wake, S., and Robinson, H. (1996). Prevalence of fragile X syndrome. *Am J Med Genet* **64**, 196–197.

Tyler-Smith, C. and Willard, H. F. (1993). Mammalian chromosome structure. *Curr Opin Genet Dev* **3**, 390–397.

van der Knaap, M. S., Pronk, J. C., and Scheper, G. C. (2006). Vanishing white matter disease. *Lancet Neurol* **5**, 413–423.

Varmus, H. (2002). Getting ready for gene-based medicine. *N Engl J Med* **347**, 1526–1527.

Vellodi, A., Bembi, B., de Villemeur, T. B., Collin-Histed, T., Erikson, A., Mengel, E. et al. (2001). Management of neuronopathic Gaucher disease: A European consensus. *J Inherit Metab Dis* **24**, 319–327.

Watson, J. D. and Crick, F. H. (1953). Molecular structure of nucleic acids; a structure for deoxyribose nucleic acid. *Nature* **171**, 737–738.

Weinreb, N. J., Charrow, J., Andersson, H. C., Kaplan, P., Kolodny, E. H., Mistry, P. et al. (2002). Effectiveness of enzyme replacement therapy in 1028 patients with type 1 Gaucher disease after 2 to 5 years of treatment: A report from the Gaucher Registry. *Am J Med* **113**, 112–119.

Wilcox, W. R., Banikazemi, M., Guffon, N., Waldek, S., Lee, P., Linthorst, G. E. et al. (2004). Long-term safety and efficacy of enzyme replacement therapy for Fabry disease. *Am J Hum Genet* **75**, 65–74.

Wolf, C. R., Smith, G., and Smith, R. L. (2000). Science, medicine, and the future: Pharmacogenetics. *BMJ* **320**, 987–990.

Wright, A. F. (2005). Neurogenetics II: Complex disorders. *J Neurol Neurosurg Psychiatry* **76**, 623–631.

Xia, H., Mao, Q., Eliason, S. L., Harper, S. Q., Martins, I. H., Orr, H. T. et al. (2004). RNAi suppresses polyglutamine-induced neurodegeneration in a model of spinocerebellar ataxia. *Nat Med* **10**, 816–820.

Yohay, K. H. (2006). The genetic and molecular pathogenesis of NF1 and NF2. *Semin Pediatr Neurol* **13**, 21–26.

Yu, S., Pritchard, M., Kremer, E., Lynch, M., Nancarrow, J., Baker, E. et al. (1991). Fragile X genotype characterized by an unstable region of DNA. *Science* **252**, 1179–1181.

Zhou, Y., Law, H. Y., Boehm, C. D., Yoon, C. S., Cutting, G. R., Ng, I. S., and Chong, S. S. (2004). Robust fragile X (CGG)n genotype classification using a methylation specific triple PCR assay. *J Med Genet* **41**, e45.

2

Neurology and Genomic Medicine

Jeffery M. Vance

I. Introduction

The practice of medicine traditionally has centered on the sick, whether acutely or chronically ill. Relatively less time, education, or payment goes toward the prevention of these illnesses. Yet for most physicians and patients, preventing or delaying disease is the ultimate goal. This is also the goal of genomic or personalized medicine.

So what is genomic medicine and how will it affect neurologists? Essentially the concept is to tailor treatment and preventative measures toward the individual, rather than a large, heterogeneous group defined by age, race, sex, or environmental activities (like smoking). It is based on the knowledge of the Human Genome Project, but doesn't just utilize information on ourselves. Knowledge of the genomes of microbes, viruses, and even environmental factors such as allergens, will all contribute to the overall approach to medicine in the future. In addition, the application of genomic techniques such as mRNA and DNA arrays are likely to be common in the future.

All of this will require a major shift in the way we practice medicine, moving at least part of medical practice away from just reacting to disease and focusing on its prevention. It will require a tremendous education shift as well, involving patients, physicians, medical administration, and all levels of medical care. For patients the benefits of genomic medicine are very clear. Patients in fact are the leading advocates for its introduction into medical care, educating themselves daily on the Internet. It is also clear that the challenges of this new burgeoning era of medicine will have to be met primarily by primary care physicians and those directly interacting with patients. Certainly neurology fits within this group. Already burdened by the immense regulatory and time constraints of clinical medicine today and the daily overload of advancements in neuroscience, to incorporate genomic medicine into daily neurological practice will clearly be a great challenge for leaders in this field.

Traditionally, neurology is a discipline rich in the study of inherited disorders and the use of genetics. Indeed, one of the first and most successful practices of genomic medicine has been the use of dietary restriction in phenylketonuria, now practiced for almost five decades. The simple routine screening of Guthrie cards (Guthrie & Susi, 1963) has identified and prevented thousands of cases from developing the severe mental retardation, microcephaly, and seizures associated with genetic mutations in the gene phenylalanine hydroxylase. Yet, because of its relative rarity, it does not reflect the majority of the diseases that the average physician experiences in his or her practice, and the overall impact on mental retardation is small. But most common disorders do have important genetic components; they just are not as obvious. Rather, each gene contributes a smaller amount of risk. It is their cumulative effect that leads to disease susceptibility. Many are thought to require environmental interactions as well to produce symptoms. It is the elucidation of these genetic elements in the common diseases of mankind that has the potential to radically change the way all physicians practice medicine in the future. In this chapter we will begin to explore some of the tools and findings that provide us the first road maps toward genomic medicine.

II. Basic Concepts

1. Single Gene Disorders (Mutations) versus Complex Disease (Polymorphisms)

There are basically two major classes of genes that contribute to disease: (1) Mendelian or single gene disorders and (2) susceptibility genes. This dichotomy is useful for discussion and conceptual understanding of the genetic contributions to disease, but it is also important to realize that like most man-made classifications, these two groups overlap. In general, Mendelian genes (named after Gregor Mendel, the monk who first realized the segregation of genetic elements) represent the gene disorders that we know from medical school, diseases such as Duchenne muscular dystrophy, spinocerebellar ataxias and Huntington's disease, to name a few. The cause of these Mendelian diseases is a "mutation" that alters the sequence in a single gene so severely that it causes the disease totally by itself, independent of environmental or other significant gene interactions. The probability of transmitting a Mendelian mutation from patient to offspring can be reliability predicted (autosomal recessive, autosomal dominant, X-linked).

However, as mentioned, the common disorders that affect most of us do not follow these well-known patterns of inheritance. When asked, we commonly say things like, "cancer runs in my family" or "I have a family history of Alzheimer's disease." We are, in essence, stating that we know we are at a higher risk for these problems than other people; that our family is somehow susceptible, though we really don't know how much risk we have. It is a complex risk, and thus the name complex genetics is applied to these genetic contributions.

The genes contributing to these familial susceptibilities are called susceptibility genes. The distinction between mutations and susceptibility genes is very important, and unfortunately is commonly confused. Again, mutations can independently cause the disease, thus their risk is predictable based on the basic laws of DNA inheritance (Mendel's laws). However, since susceptibility genes do not cause the disease by themselves, but contribute varying amounts of susceptibility to the disease, predicting the risk for a patient based on family history is not as clear. In this context, the term *variable penetrance* has been used in the past to describe those Mendelian disorders that may not always manifest in mutation carriers. The most common would be age-related penetrance such as that which occurs in Huntington's disease. The primary distinction between a variably penetrant Mendelian gene and a susceptibility gene is that the former can still cause the disease independently, but the latter does not. But the difference can be blurred in some cases.

It follows that the variations in DNA that cause susceptibility genes are not as severe as those leading to mutation. Rather they are much "milder" DNA variations that may affect the structure, efficiency, expression, forms, or how much protein is produced. These variations in DNA are numerous, from single base pair changes, di-, tri- and tetra-base repeats, to deletions and duplications (copy number variants). These normal, common variations are termed polymorphisms, and are the currency of genomic medicine and susceptibility genes. Whereas once these were defined as variables that existed in the population with a frequency greater than 1 percent, we now know this cutoff is not useful. With more sensitive techniques, we can now detect polymorphisms with a frequency much less than 1 percent. It is the multitude of interactions between the polymorphisms of different genes that make us different from each other and contribute to our susceptibility differences for every disease.

In the future, useful clinical algorithms are expected to include the use of multiple gene polymorphisms, and it will be the combination of these that provides predictive value for the physician. For the neurologist this should not be a foreign concept; we regularly make diagnoses based on the overall impression from multiple tests (an example would be multiple sclerosis).

The most common polymorphisms in the human genome are single nucleotide polymorphisms (SNPs); that is, a variation of a single base pair (usually varying between one of two DNA bases, the various possibilities are termed alleles). They are quite numerous, often every 500 to 1,000 base pairs. Which allele a person has at each SNP can be determined through various laboratory techniques, all falling within the term *genotyping*, as the two alleles at any locus in the DNA is termed a *genotype*.

2. Identifying Susceptibility Genes

There are many genetic techniques used to determine which genes contribute or cause a disease. The most commonly used is linkage analysis, which seeks to identify if a specific physical region of the genome is inherited along with a trait or disease as it travels through affected families. To locate these different areas of the human genome we use polymorphisms whose genomic location is known. Their variability between people allows them to be followed as they pass through successive family members. If the polymorphisms from a specific area of the genome always travel (or are "linked") with the disease, or are found in patients much more than we would expect by chance, then we can conclude that it is likely that the linked polymorphism lies physically near the gene causing the disease. This physical region, once identified, can be further explored to eventually identify the causal gene (see Figure 2.1).

The difference between association and linkage often is confusing to those beginning the study of complex genetics. Linkage implies causality for the physical region with the disease, but does not tell us what particular gene or variation is contributing to the disease, only that it lies within a defined physical chromosome region. This is different from association, which implicates a specific allele or variation in that region as statistically coupled with the disease. Finding an association for a polymorphism and a disease may be due to either (1) the variation is the true causal association with the disease, or (2) it lies physically near the true causal susceptibility change in the DNA and thus traveling with the true causal change in time, serving as a surrogate marker for the causal change. The latter situation is termed *linkage disequilibrium* (LD), where a marker in LD is so close to the disease locus that recombination or mutation has not yet significantly interrupted its coinheritance with the disease allele within the population under study. However, LD is a quantitative variable; that is, SNPs that are farther from the causal change may not have their alleles segregate independently, but they are not as correlated as the alleles of those SNPs lying very, very close to the causal change. The degree of LD is measured using two different variables, the traditional correlation coefficient (r^2), and the more abstract term D'. Alleles of two SNPs whose r^2 is 0.9 or greater usually are referred as being in strong LD, whereas those with an r^2 of 0.3 are only in weak LD. Thus a SNP allele traveling in a population on the same small piece of DNA as the susceptibility allele of the risk causing gene (and thus with a high r^2, in strong LD) can act as a surrogate for the presence of the actual susceptibility change. This provides a much larger area of DNA to detect for association than just trying to find the actual disease causing variation itself (see Figure 2.2). This LD phenomenon is the basis for the Hapmap project, as discussed next.

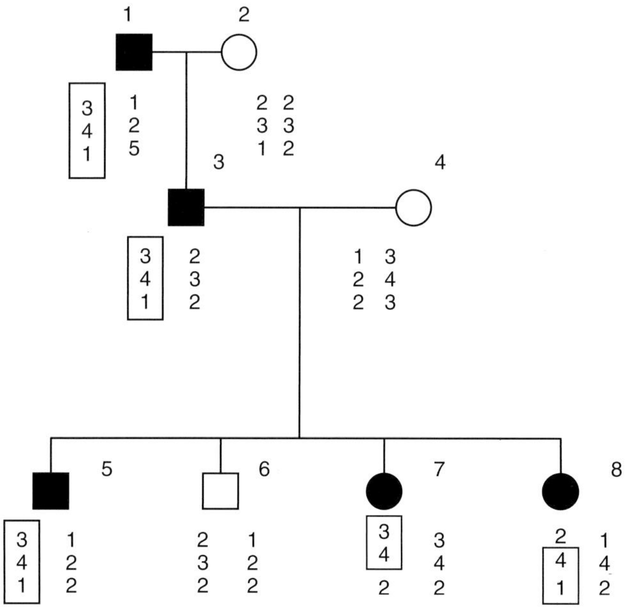

Figure 2.1 An example of genetic linkage. In this autosomal dominant pedigree affected Individuals are solid, unaffected clear. The stack of numbers under each endividual represent a three allele haplotype, one for each chromosome pair. The boxed haplotypes demonstrate the chromosome and haplotype traveling (linked) with the disease. Note recombinations occurring in individual 3 provide reduced forms of the haplotype in individuals 7 and 8. Since individuals 7 and 8 are affected, the disese gene must lie within the shared area between all three haplotype forms, i.e. between the first and last polymorphisms of the haplotype.

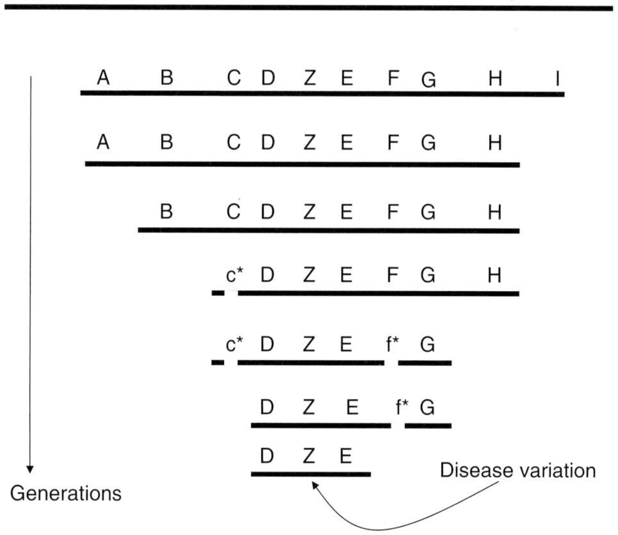

Figure 2.2 Linkage disequilibrium. The horizontal bar represents a section of a chromosome that contains polymorphisms A through I. Through multiple generations the initial haplotype (A-I) becomes reduced through recombination at the ends of the haplotype, or through mutation (C>c*) Within the conserved haplotype. However, those markers closest to the disease variation (Z) continue to travel with Z in strong linkage disequilibrium.

The term *linkage disequilibrium* is derived from the fact that physically close DNA variations (linked) do not follow the expectations of Hardy-Weinberg equilibrium ($p^2 + 2pq + q^2 = 1$, where p and q are the allele frequencies of the polymorphism), which assumes independent segregation, and thus these polymorphic variations are in Hardy-Weinberg disequilibrium. Two polymorphic alleles in strong LD form a haplotype; that is, they travel together the majority of the time. For any group of SNPs with a substantial degree of LD between them, several different haplotypes exist (see Figure 2.3). As human genetics moves toward identifying genes contributing to the susceptibility of common disorders and pharmacogenetic interactions (genetic variations that affect any biological interactions with a drug), this work has relied heavily on these genetic associations and LD relationships.

III. The Human Genome Project (HGP) and Haplotype Mapping (HapMap) Project

In 1990 a project was begun by an international consortium to determine the sequence of the human genome—the Human Genome Project (HGP) (www.ornl.gov/sci/techresources/humangenome/home.shtml). This was a considerable challenge, as it was unclear that the technology available at the time was up to the task. Although there have been many political announcements of achieved milestones, the final chromosome sequence was published (chromosome 1) in 2006. Having the human genome sequence completed provides many opportunities, not only to finally locate all the genes in humans, but also to understand the structure of the sequence that lies between genes and between exons, which represents most of the genome. It is in these regions that the regulatory mechanisms for gene expression and translation occur.

The HGP was followed by another international consortium, the HapMap project (Altshuler et al., 2005; The International HapMap Consortium, 2003). Here the goal was to take approximately 30 random individuals from several ethnic groups and determine their genotypes for hundreds of thousands of SNPs. Using this data haplotypes could then be determined for each group of SNPs. It follows that if a group of SNP alleles are traveling together in a haplotype through generations, then you would have to genotype only a few of the SNPs to know what the alleles were at the other SNP members of the haplotype. The SNPs chosen to represent all the different haplotypes in a population for a given group of SNPs are called *tagging SNPs* (see Figure 2.3). For example, instead of genotyping 20 SNPs to determine the haplotypes formed by their alleles, you may have to genotype only six to see all the haplotypes. Theoretically this should reduce significantly the cost and labor of genotyping for association studies.

IV. Family History

Without doubt, the accurate acquisition of a family history for every neurology patient is the first and most necessary step in applying genetics to common disorders. It is anticipated that computer-based algorithms will be used to screen

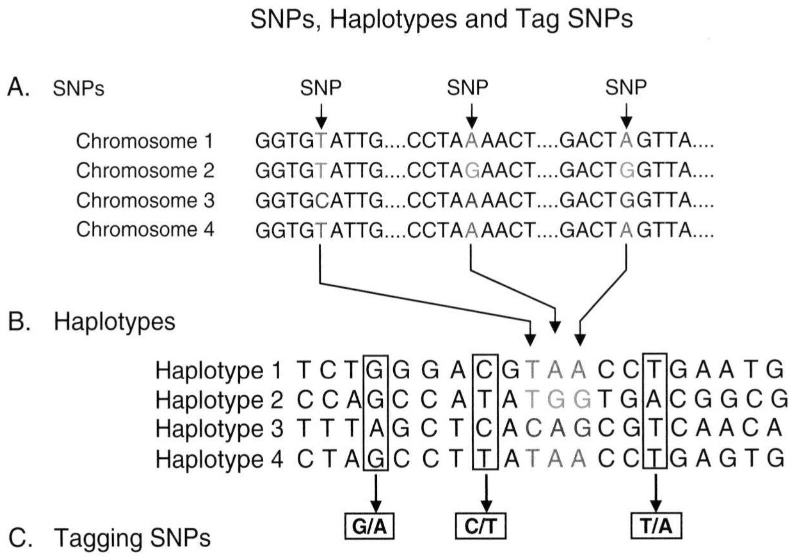

Figure 2.3 SNPs, Haplotypes and Tagging SNPs. **A.** A small area of DNA sequence from four different individuals is shown, each individuals DNA in this area identical except for three SNPs. **B.** Four haplotypes are shown comprised from twenty different SNP, including the three SNPs in 2.3A. C. Rather than genotype all twenty SNPs, haplotypes 1-4 can be differentiated by just genotyping only the three boxed SNPs. These are termed "tagging" SNPs.

patients who are at a higher risk for a disorder, and act as a triage point for the neurologist and primary care doctor to initiate additional studies. Several groups have developed initial algorithms, particularly for breast cancer, but one group, the Guilford Genomic Medicine Initiative (GGMI) (www.genomic-medicine.org), is developing a general algorithm that will then diverge into specialized areas. A recent study examined multiple computer algorithms for determining women at high risk for breast cancer (Palomaki et al., 2006). Even using only those individuals whom all the algorithms agreed were at high risk, 1 to 2 percent of the female population was identified for further screening. Similar models are only a matter of time for major neurological disorders as well.

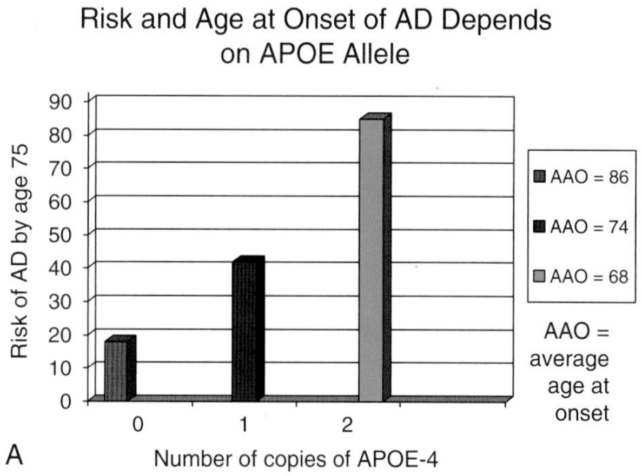

Figure 2.5 Age-at-onset (AAO) risk for Aizheimer disease determined by the number of APOE4 alleles.

V. Genetic Mechanisms

1. Susceptibility (Risk) Genes

Although by definition, susceptibility genes do not cause a disease by themselves, there are a few that will have a stronger effect than others. These are called major susceptibility genes. An excellent example of a major susceptibility gene is apolipoprotein E (Corder et al., 1993). Well accepted as a major risk factor for Alzheimer's disease, it is a polymorphism that is actually a haplotype of two different SNP polymorphisms, producing the 2, 3, and 4 alleles (see Figure 2.4). It was originally discovered using linkage analysis in only 33 families, demonstrating how strong an effect it has on susceptibility for the disease (Pericak-Vance et al., 1991). Figure 2.5 demonstrates the effects of the different alleles for Alzheimer's disease (AD), with the 4 allele being very detrimental, but the 2 allele being protective for AD. Apo E is now known to be a risk factor for additional disorders, including Parkinson disease (PD) and amyotrophic lateral sclerosis (Li et al., 2004).

VI. Pharmacogenetics

The study of how genetic variation affects the metabolism and actions of drugs is termed pharmacogenetics. It is believed that most side effects of medicines are secondary to polymorphic variations affecting drug metabolism. This is already an area of genomic medicine entering clinical use, and will likely be the first major aspect of genomic medicine that most neurologists will incorporate into their practice. The major genes involved in the metabolism of drugs are the P450 genes, and recently the first DNA chip to genotype individuals for these variations has been approved by the FDA. At this time the FDA is considering whether to recommend P450 genotyping prior to warfarin treatment, to help with dosing. The well-known lack of pain relief in 15 to 17 percent of Caucasians receiving codeine is another example of the effect of genetic variation in P450 gene cluster. Many similar variations exist, affecting drug metabolism, and it seems likely that many will eventually become part of standard clinical workups. However, this variation does not have to be confined to SNPs. The presence or absence of deletions on chromosomes 1 and 19 in the chromosome karyotype of oligodendrogliomas has been well documented to predict the efficacy of chemotherapeutic regimens (Watanabe et al., 2002). These chromosome aberrations can be measured using comparative genomic hybridization (see later), a technique that is likely eventually to replace much of today's chromosome (karyotype) analysis. There are many ways to use this technique, but essentially it measures deletions or duplications in the genome by array hybridization.

Haplotype

		bp 3937		bp 4075
	E2	T(Cys)	-	T(Cys)
APOE allele	E3	T(Cys)	-	C(Arg)
	E4	C(Arg)	-	C(Arg)

Figure 2.4 **APOE alleles.** The threee APOE alleles (E2, E3, E4) are actually products of a two SNP haptotype, creating either a cysteine (Cys) or an arginine (Arg) in the protein site. bp=basepair.

VII. Gene-Gene and Gene-Environment Interactions

The idea that gene-gene and gene-environment interactions play an important role in human biology was emphasized by Wright in 1932 (Wright, 1932). Gibson (Gibson, 1996) stresses that gene-gene and gene-environment interactions must be ubiquitous given the complexities of intermolecular interactions. The modification of genes by each other can vary from mild to complete prevention of gene expression on a phenotype. The blocking of the effect of one gene by another is termed epistasis.

First defined about 100 years ago, an excellent example is the ABO blood group. Individuals with the Bombay genotype lack the H antigen protein, which is used to form the A and B antigens. Thus, A, B, or AB individuals who also have the Bombay genotype are unable to make A, B antigens. Therefore, upon blood typing they will present as having the O blood group.

An example of gene-environmental interaction is the recent identification of an interaction with a polymorphism in the inducible Nitric Oxide gene (iNOS or NOS2A) and smoking in Parkinson disease (Hancock et al., 2006). Three laboratories now have shown that the polymorphisms in the NOS2A gene are associated with an increased risk in PD (Hague et al., 2004; Hancock et al., 2006; Levecque et al., 2003). Smoking is now accepted as a protective environmental factor for Parkinson disease, with a multitude of studies finding this phenomenon. Following up on a study by Mazzio et al. (2005) demonstrating that compounds in smoke can suppress iNOS in cultured astrocytes, Hancock et al. sought to see if there was an interaction between the associated SNPs and the protective effect of smoking. Indeed, the effect is rather profound, as shown in Table 2.1. This means that if you have the non-PD risk allele for NOS2A, then you will get the protective effect of smoking, but the carrying risk allele essentially removes the protective effect.

1. Modifier Genes

Assuming interactions are intrinsic to the expression of genetic phenotype, clearly some gene polymorphisms will be more important in effecting the progress of the disease in terms of severity, speed of progression, and extent of disease. Therapeutically these genes are extremely interesting, as they would seem to present the most likely targets for pharmaceutical therapy, as most drugs used today actually modify disease, not prevent it. Indeed, for some diseases, like AD and Parkinson disease, delaying age-of-onset could have huge therapeutic effects.

An example of such a modifier is glutathione synthtase transferase (GST) omega 1/2 gene complex (GSTO1/2), which has been shown to have an effect on the age-of-onset in both AD and PD (Li et al., 2003). Indeed, studies in fly and yeast models of alpha-synuclein overexpression, known to cause PD in humans, demonstrate that model disease phenotype can be prevented by concurrent overexpression of GSTs. Li et al. (2005) demonstrated that the GSTO1 effect in AD is equal to that of ApoE, but found in only about 25 percent of the population. This result points out the very important clinical problem of genetic heterogeneity; that is, that there may be many different genes that confer risk for the disease, but only a few may be important in any one individual. Therefore, although we discuss disorders like Alzheimer's and PD as a single disease, we know now that the clinical presentation, or phenotype, can look and present the same in affected individuals, but have different causes. One of the best examples of genetic heterogeneity in Mendelian diseases is Charcot-Marie-Tooth disease, where now over 31 genes or loci are known to cause the same basic clinical phenotype.

2. Regulatory Mechanisms

The majority of DNA in the human genome does not code for proteins. Many years ago these intronic and intergenic regions were thought to be "junk" DNA (Ohno, 1972), a useless byproduct of our evolutionary process. However, we now know that this is not the case. Indeed, what makes us the most complex of biological beings is the vast regulatory mechanisms controlling gene and protein expression, which is controlled from these nonexon-coding regions. This is an expanding field of research, but already several primary mechanisms of control are known to exist.

3. Transcription of DNA into Messenger RNA (mRNA)

Promoters are specific DNA sequences that, when activated by regulatory molecules, initiate the transcription of the gene, producing the mRNA. The up-regulation or down-regulation of these genes in specific situations or diseases can be measured by several gene expression techniques. The most common is the use of fixed chip arrays (gene expres-

Table 2.1 Interaction of NOS2A Genotype and Smoking for Risk of Developing Parkinson disease

SNP	Genotype	Cases/Controls	Odds Ratio	Lower CI	Upper CI
rs1060826	GG	94/79	0.26	0.13	0.54
	AG/AA	151/139	0.9	0.57	1.44
rs2255929	AA	46/44	0.21	0.07	0.58
	TA/TT	193/174	0.78	0.51	1.2

Two SNPs in the gene NOS2A—rs1060826 and rs2255929—are associated with Parkinson disease, with the risk allele being (A) and (T) respectively. For those individuals homozygous for the non-risk allele (GG for rs1060826 and AA for rs2255929) the protective effect of smoking is observed (odds ratio is much less than 1). However, this protective effect is lost in those individuals with the risk allele as their risk returns to the 1 (Hancock et al., 2006) sion studies), which have copies of a unique area of each gene of interest fixed on the chip (see Figure 2.6A). When mRNA from a tissue is passed over the array it binds to its specific chip sequence, and the amount of binding is directly correlated with the amount of mRNA produced by the cell. If one wishes to compare two tissues (say, normal and cancerous), two colored labels are used, red and green. Thus, the overexpression of one tissue versus the other will give that color prevalence, whereas equal expression produces a yellow color. The downside of arrays is that they can use only known genes. Serial Analysis of Gene Expression (SAGE) (Velculescu et al., 1995) is a second expression technique that does not rely on known genes to measure expression differences. Here, the mRNA is collected and then digested by specific enzymes that produce a small (10 bp) sequence known as a *tag*, which identifies that gene. These tags are ligated together in long chains and sequenced, with the sequencer acting essentially as a counter of each specific sequence. These counts can

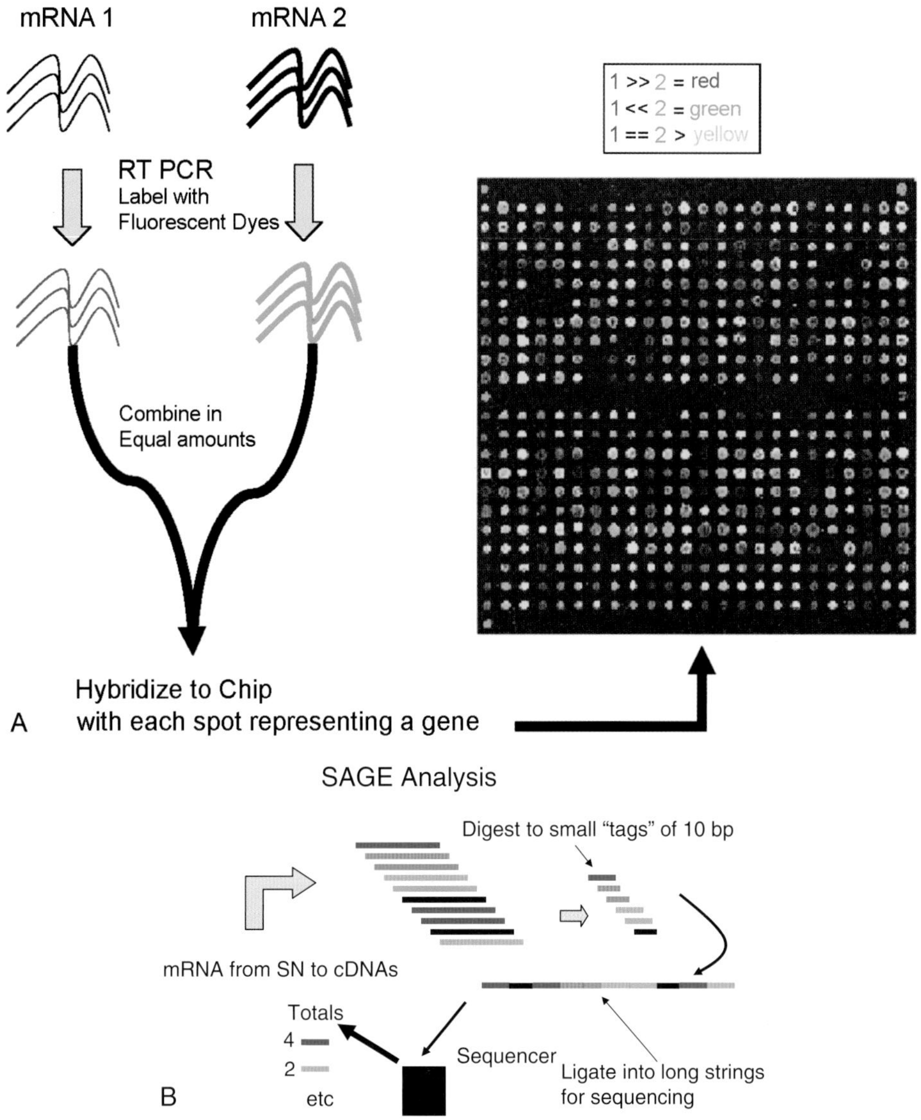

Figure 2.6 Gene Expression techniques. **A**. Microarray analysis. The messenger RNA (mRNA) from tissue 1 and tissue 2 are labeled with two different colored dyes. The relative amounts of expression of a gene in each tissue can be compared by hybridizing them to a high density array with DNA from different genes. It this example, if tissue 1 has higher expression of a specific gene than tissue 2, then the hybridization spot will be red, reflecting the greater amount of mRNA coming from tissue 1. If the amount is no different between tissues, then the green and red will be equal, producing a yellow color in the analysis. **B.** Serial Analysis of Gene Expression (SAGE). In SAGE, the comparison between two tissues is made numerically, not by hybridization. The mRNA from a single tissue is digested using restriction enzymes and a unique 10 or 13 basepair (bp) tag is created, that represents one copy of that specific mRNA. These tags are joined (ligated) into long strings of DNA, that can then be sequenced. Each time the tag's sequence is observed, that gene receives a count. These can then be compared.

then be compared (see Figure 2.6B). This has been most commonly applied to the study of cancer, but recently has been used in neurodegeneration (Hauser et al., 2003).

What could some of the clinical uses be of this technology? Well, expression data may well replace or supplement the histology reports today, providing rapid and more accurate diagnoses, with different disorders having distinct expression fingerprints. Therapy may also be directed by expression arrays, particularly in cancer. Already research suggests that chemotherapies are often most effective when certain genes have specific expression changes. It also may be useful to use expression analysis to follow the course of a patient, noting cellular changes well before clinical manifestations.

4. Splice Variants

We now know that any one human gene has multiple biological functions. One way this is accomplished is that the cell can pair various combinations of the coding exons of a gene to make a large variety of proteins. These variable mRNA products and their subsequent proteins are referred to as splice variants. An error in splicing leads to the TAU mutations that cause Frontotemporal Dementia with Parkinsonism (Hutton et al., 1998). Inefficient splicing caused by polymorphisms may also contribute to complex genetic diseases.

5. microRNA (miRNA)

This is a fascinating group of regulatory mRNAs, whose existence in biology was discovered after they were "invented" for research studies. miRNA works at the level of translation, or that process of the ribosome translating the mRNA into proteins. These miRNAs bind to the 3 untranslated regions of the mRNA and block protein formation from the mRNA, providing a whole new area of gene control. Thus, normal variation in DNA polymorphisms could easily affect how well miRNA bind to their target sites, thus producing individual variability of a cell's response to stimuli.

VIII. Comparative Genomic Hybridization (CGH)

This technique measures copy number of DNA—deletions or duplications of the genome. It is similar to gene expression arrays, but instead of placing mRNA on a chip, DNA is used, and the amount of hybridization to specific areas of DNA measured. Techniques have improved for this test, and now most of the whole genome genotyping arrays (Illumina, Affymetrix, etc.) can be used for CGH as well.

Copy number changes are known to exist in disease. CMT1A, the most common form of CMT, is primarily due to an abnormal 1.5 megabase duplication on chromosome 17. Tumors contain a large number of deletions and duplications that are coupled with chemotherapeutic response and stage of malignancy. However, it wasn't until the last few years when we realized that there is a tremendous number of deletions and duplications that exist in "normal" individuals; that is, copy number polymorphisms (Feuk et al., 2006). This is a very new and exciting area of research in medical genetics. It seems highly likely that copy number changes will be very important in complex diseases.

1. Bioinformatics

This is a term that unfortunately has become rather nonspecific, but is applied to almost any process that involves the merging of computing, programming, or databasing with genomic information. The most used sites are those that involved the storage and annotation of genomic sequencing data from various species. There are three main genomic sites: Ensembl (www.ensembl.org/index.html) located in the European Bioinformatics Institute, Cambridge, England; the "golden path" located at University of California Santa Cruz (UCSC) (http://genome.ucsc.edu/), and the National Center for Biotechnology Information (NCBI) (www.ncbi.nlm.nih.gov/). All sites are based on the same sequence data, called an *assembly*. The major difference between sites is the format presentation and the criteria used to annotate the genome (predict functional sites like genes).

Clinically useful genetic information can come from several sites. GENECLINICS (www.geneclinics.org/) is an excellent site to provide reviews of genetic disorders and a link to GENETESTS can provide laboratories performing clinical genetic testing. For a more global knowledge about all things genetic in humans, the Online Mendelian Inheritance in Man (OMIM) (www.ncbi.nlm.nih.gov/entrez/query.fcgi?db=OMIM) is the world's largest and the original database of human genes and genetic disorders. Finally, GENECARDS (www.genecards.org/index.shtml) is an automatically annotated database of genes and disorders that pulls from various outside databases to place all the data in one location and can be quite useful to the researcher and the clinician.

2. Proteomics

Proteomics is the newest "omics" in human genetics, and describes the measurement of protein variation in humans, similar to mRNA studies of gene expression research. This basically has been held back by the technical difficulty of doing large scale studies of proteins, depending on the laborious use of protein gels for many years. However, recent coupling with mass spectrometry and other techniques has begun to open up this area of research. Protein arrays, like those used in DNA and mRNA, are needed to move this field ahead to the level of activity of gene expression.

IX. Mitochondria and the Mitochondrial Genome (mtDNA)

Mitochondria are obviously extremely important in maintaining energy stores, formation of free radicals, calcium homeostasis, and apoptosis. Although most of the focus of the human genome project has focused on the nuclear genome, the mitochondrial genome, though small, is highly important in human disease. The mitochondrial genome, circular and only 16,569 base pairs in length, is important in neurodegenerative disorders like Parkinson disease and mitochondrial myopathies, and appears to have the potential to modify the clinical phenotype in a large number of diseases. It encodes only 37 genes, of which 24 are specialized transfer-RNAs and ribosomal proteins. The remaining 13 encode proteins for the electron transport chain, of which seven code specifically for complex I, the most complex protein complex in biology and the primary complex involved in PD.

Several factors separate the mitochondrial genome from the nuclear genome, besides its size and circular nature. It has a relatively poor repair system, and therefore mutations tend to remain once they occur, creating a unique signature for that mitochondrial genome. It has only one allele, thus for practical purposes it doesn't recombine. It is, of course, inherited through the cytoplasm of the egg, thus leading to maternal inheritance. It is all these qualities that have led it to be essentially a "passport" of where and who your maternal ancestors have been, making it a major tool in molecular anthropology. It is present in large quantities in the cell, constituting approximately 1 percent of the total DNA in the cell. For the most part these mitochondrial genomes are identical in the cell, which is termed homoplasty. However, there can be a mixture of mtDNA genomes, which is termed heteroplasmy.

Interestingly it also is one of the most different genetic entities between ethnic or continental groups, much more distinct than nuclear changes. These mtDNA subsets can be grouped into various haplogroups, which are similar to haplotypes of nuclear DNA. These can be used to compare whether mtDNA variations can affect disease states.

An example of an haplotype effect is the now well-documented finding that the J and K haplogroups were associated with a decreased risk for PD, particularly in females (van der Walt et al., 2003). We also know that mitochondria have copy number (deletion primarily) changes like nuclear DNA. Only these changes occur somatically, as the individual ages, and many may contribute to late-onset diseases (Bender et al., 2006; Kraytsberg et al., 2006). It seems likely that in the future we will use haplogroups information in predicting patients' response to both disease and drugs.

X. Summary

The world of medicine and neurology is clearly becoming more genetic. Neurology has the opportunity to lead this major change, but it will require knowledge and education at all levels. However, for both the patient and the neurologist, the future of genomic medicine is an exciting one.

References

Altshuler, D., Brooks, L.D., Chakravarti, A., Collins, F.S., Daly, M.J., Donnelly, P. (2005). A haplotype map of the human genome. *Nature* **437**, 1299–1320.

Bender, A., Krishnan, K.J., Morris, C.M., Taylor, G.A., Reeve, A.K., Perry, R.H. et al. (2006). High levels of mitochondrial DNA deletions in substantia nigra neurons in aging and Parkinson disease. *Nat Genet* **38**, 515–517.

Corder, E.H., Saunders, A.M., Strittmatter, W.J., Schmechel, D.E., Gaskell, P.C., Small, G.W. et al. (1993). Gene dose of apolipoprotein E type 4 allele and the risk of Alzheimer's disease in late onset families. *Science* **261**, 921–923.

Feuk, L., Marshall, C.R., Wintle, R.F., Scherer, S.W. (2006). Structural variants: Changing the landscape of chromosomes and design of disease studies. *Hum Mol Genet* **15** Spec No 1, R57–R66.

Gibson, G. (1996). Epistasis and pleiotropy as natural properties of transcriptional regulation. *Theor Popul Biol* **49**, 58–59.

Guthrie, R., Susi, A. (1963). A simple phenylalanine method for detecting phenylketonuria in large populations of newborn infants. *Pediatrics* **32**, 338–343.

Hague, S., Peuralinna, T., Eerola, J., Hellstrom, O., Tienari, P.J., Singleton, A.B. (2004). Confirmation of the protective effect of iNOS in an independent cohort of Parkinson disease. *Neurology* **62**, 635–636.

Hancock, D.B., Martin, E.R., Fujiwara, K., Stacy, M.A., Scott, B.L., Stajich, J.M. et al. (2006). NOS2A and the modulating effect of cigarette smoking in Parkinson's disease. *Ann Neurol* **60**, 366–373.

Hauser, M.A., Li, Y.J., Takeuchi, S., Walters, R., Noureddine, M., Maready, M. et al. (2003). Genomic convergence: identifying candidate genes for Parkinson's disease by combining serial analysis of gene expression and genetic linkage. *Hum Mol Genet* **12**, 671–677.

Hutton, M., Lendon, C.L., Rizzu, P., Baker, M., Froelich, S., Houlden, H. et al. (1998). Association of missense and 5'-splice-site mutations in tau with the inherited dementia FTDP-17. *Nature* **393**, 702–705.

Kraytsberg, Y., Kudryavtseva, E., McKee, A.C., Geula, C., Kowall, N.W., Khrapko, K. (2006). Mitochondrial DNA deletions are abundant and cause functional impairment in aged human substantia nigra neurons. *Nat Genet* **38**, 518–520.

Levecque, C., Elbaz, A., Clavel, J., Richard, F., Vidal, J.S., Amouyel, P. et al. (2003). Association between Parkinson's disease and polymorphisms in the nNOS and iNOS genes in a community-based case-control study. *Hum Mol Genet* **12**, 79–86.

Li, Y-J., Oliveira, S.A., Xu, P., Martin, E.R., Stenger, J.E., Scherzer, C.A. et al. (2003). Glutathione S-transferase omega-1 modifies age-at-onset of Alzheimer disease and Parkinson disease. *Hum Mol Genet* **12**, 3259–3267.

Li, Y-J., Pericak-Vance, M.A., Haines, J.L., Siddique, N., McKenna-Yasek, D., Hung, W.Y. et al. (2004). Apolipoprotein E is associated with age at onset of amyotrophic lateral sclerosis. *Neurogenetics* **5**, 209–213.

Li, Y-J., Scott, W.K., Zhang, L., Lin, P.I., Oliveira, S.A., Skelly, T. et al. (2005). Revealing the role of glutathione S-transferase omega in age-at-onset of Alzheimer and Parkinson diseases. *Neurobiol Aging* [Epub ahead of print].

Mazzio, E.A., Kolta, M.G., Reams, R., Soliman, K.F.A. (2005). Inhibitory Effects of cigarette smoke on glial inducible nitric oxide synthase and lack of protective properties against oxidative neurotoxins in vitro. *Neurotoxicology* **26**, 49–62.

Ohno, S. (1972). So much "junk" DNA in our genome. Brookhaven Symposia in Biology. Ref Type: Conference Proceeding **23**, 366–370.

Palomaki, G.E., McClain, M.R., Steinort, K., Sifri, R., LoPresti, L., Haddow, J.E. (2006). Screen-positive rates and agreement among six family history screening protocols for breast/ovarian cancer in a population-based cohort of 21- to 55-year-old women. *Genet Med* **8**, 161–168.

Pericak-Vance, M.A., Bebout, J.L., Gaskell, P.C., Yamaoka, L.H., Hung, W.Y., Alberts, M.J. et al. (1991). Linkage studies in familial Alzheimer's disease: Evidence for chromosome 19 linkage. *Am J Hum Genet* **48**, 1034–1050.

The International HapMap Consortium (2003). The International HapMap Project. *Nature* **426**, 789–796.

van der Walt, J.M., Nicodemus, K.K., Martin, E.R., Scott, W.K., Nance, M.A., Watts, R.L. et al. (2003). Mitochondrial polymorphisms significantly reduce the risk of Parkinson disease. *Am J Hum Genet* **72**, 804–811.

Velculescu, V.E., Zhang, L., Vogelstein, B., Kinzler, K.W. (1995). Serial analysis of gene expression. *Science* **270**, 484–487.

Watanabe, T., Nakamura, M., Kros, J.M., Burkhard, C., Yonekawa, Y., Kleihues, P., Ohgaki, H. (2002). Phenotype versus genotype correlation in oligodendrogliomas and low-grade diffuse astrocytomas. *Acta Neuropathol* (Berl) **103**, 267–275.

Wright, S. (1932). The roles of mutation, inbreeding, crossbreeding, and selection in evolution. Proceedings of the 6th International Congress of Genetics. Ref Type: Conference Proceeding **1**, 356–366.

3

Mitochondrial Function and Dysfunction in the Nervous System

Neggy Rismanchi and Craig Blackstone

I. Introduction

Over a billion years ago, α-protobacteria took up residence in evolving eukaryotes, eventually forming a symbiotic existence in animals as "mitochondria" and endowing cells with aerobic metabolism. Initially, the bacterial genome encoded all genes necessary for an independent organism. But as the symbiotic relationship evolved, bacterial genes were either lost or transferred to chromosomes of the host nuclear genome, such that the vestiges of the bacterial DNA, the mitochondrial DNA (mtDNA), encode merely a tiny fraction of the ~1,000 proteins of the present-day mitochondrion. In addition to their well-known role in producing energy in the form of adenosine triphosphate (ATP) through oxidative phosphorylation (OXPHOS), mitochondria are critical for a host of other functions within cells, including redox control, fatty acid oxidation, calcium homeostasis, amino acid metabolism, regulation of metabolic pathways, and physiological cell death mechanisms (McBride et al., 2006). They are also prone to dysfunction from a variety of insults, which in turn has been implicated in a number of human pathological conditions including cancer, diabetes, renal disease, ischemia-reperfusion injury, and neurodegenerative disorders.

The nervous system and skeletal muscles are particularly sensitive to manifestations of mitochondrial dysfunction, and knowledge of the basic cell biology of mitochondria is instructive in understanding how various insults can result in neurological disease. In this chapter we will outline some of the principal functions and properties of mitochondria, highlighting the unique features of highly polarized neurons that present exceptional vulnerabilities in the face of mitochondrial abnormalities. Chapter 33 will discuss the mitochondrial disorders in more detail.

II. Structure and Functions of Mitochondria

A. General Description

The term *mitochondrion* is derived from the Greek *mitos* (thread) and *khondrion* (granule), emphasizing that mitochondria exist as a mixture of different sizes and shapes within cells (see Figure 3.1). In fact, in contrast to their common depiction in schematic diagrams as sausage-shaped organelles, mitochondria within a cell comprise a dynamic reticular network of tubular structures that fuse, branch, and divide (Chan, 2006; Chen & Chan, 2006; Karbowski & Youle, 2003). They also are among the most prominent organelles within a cell, occupying around 20 to 25 percent of the total cellular volume, and their movements and distributions are tightly regulated, particularly in highly polarized cells such as neurons with their long, highly specialized processes—the axons and dendrites (Hollenbeck, 2006; Hollenbeck & Saxton, 2005; Li et al., 2004).

In all cells, mitochondria have a characteristic double-membrane structure that gives rise to four functionally important compartments: the outer mitochondrial membrane, intermembrane space, inner mitochondrial membrane, and the matrix (the center of the organelle) (see Figure 3.2). The inner mitochondrial membrane is highly invaginated, forming the cristae that dramatically increase the overall surface area of the inner membrane. Interestingly, the two membranes differ substantially in their protein and lipid composition. The outer membrane enclosing the entire organelle contains about 50 percent phospholipids and prominent proteins known as porins that form large channels permeable to small molecules <5 kDa. Larger molecules require active transport to cross, and a number of specialized proteins mediate this process at the outer membrane. Here also reside a variety of enzymes involved in such functions as elongation of fatty acids, oxidation of epinephrine, and degradation of the amino acid tryptophan (Alberts et al., 2002).

The inner mitochondrial membrane has a much higher protein-to-phospholipid ratio (approximately 3:1) and is particularly notable for its high concentration of the phospholipid cardiolipin, more characteristic of cell membranes of its ancestral bacterium. In contrast to the outer membrane, the inner membrane harbors no porins, and indeed is highly impermeable, with special transport and carrier proteins required for shuttling nearly all proteins and other molecules between the intermembrane space and the matrix (Meinecke et al., 2006). Across the inner mitochondrial membrane an electrochemical membrane potential ($\Delta\Psi_m$) is maintained that is important for driving a number of processes, most notably OXPHOS. The central core of the mitochondrion, the matrix, possesses hundreds of different enzymes along with the synthesizing machinery for proteins encoded by the mtDNA, which also resides there. Matrix enzymes are involved in a wide range of functions, including the oxidation of pyruvate and fatty acids as well as the citric acid cycle (also known as the Krebs or tricarboxylic acid cycle) (Alberts et al., 2002).

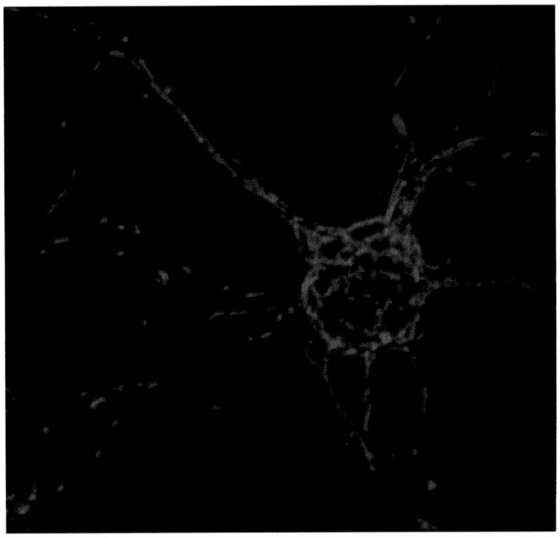

Figure 3.1 Mitochondrial distribution. Mitochondria (shown in red) are distributed as tubulovesicular structures of various sizes and shapes within the processes and soma of neurons.

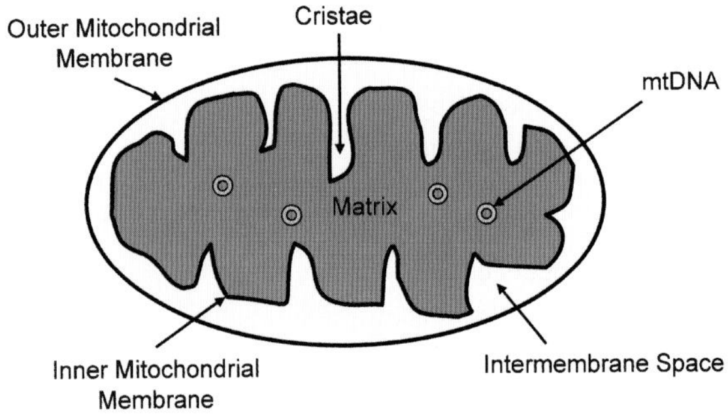

Figure 3.2 Mitochondrial structure. The mitochondrion contains an outer mitochondrial membrane as well as an inner mitochondrial membrane. The space between these two membranes is referred to as the intermembrane space. Prominent invaginations of the inner mitochondrial membrane form pockets known as cristae, often with narrow necks. The inner mitochondrial membrane encloses the mitochondrial matrix, the compartment that contains mtDNA.

Importantly, although all mitochondria share this basic structure, they can exhibit prominent cell- and tissue-specific differences. In fact, the number of mitochondria can vary substantially depending upon the cell type, a variation that correlates closely with the energy requirements of a given cell. Thus, skin fibroblasts typically contain several hundred mitochondria, neurons may contain thousands, and cardiomyocytes can harbor tens of thousands. Also, mitochondria from different tissues can show differences in size, cristae structure, and the protein/lipid composition of the membranes, with important functional sequelae. Indeed, mitochondria of cells with greater ATP demand, such as muscle cells, contain more cristae, and thus greater inner membrane surface area and ATP-generating capacity, than mitochondria from tissues with lesser demands. Furthermore, enzymes required for the metabolism of ammonia, for instance, are found only in liver mitochondria. Even the number of mtDNA molecules in the matrix can differ dramatically among mitochondria from different tissues (Alberts et al., 2002).

In spite of the variability just outlined, clearly the most prominent mitochondrial function in all cells is OXPHOS, the conversion of organic materials into cellular energy in the form of ATP to fuel a wide variety of intracellular pathways.

Even so, mitochondria participate in many other important cellular tasks with potential pathological significance, and these will be discussed by general category. We will begin by summarizing mitochondrial genetics and bioenergetics, followed by protein import mechanisms, dynamic regulation of morphology changes mediated by fission and fusion, and some unique aspects of mitochondrial distribution and function in neurons. Throughout, it is important to remember that these different functions are highly interdependent.

B. Mitochondrial Genetics

The mitochondrion is the only cellular organelle, other than the nucleus, that harbors its own DNA and machinery for synthesizing RNA and proteins; all of these processes take place within the matrix. The 16,569-bp mtDNA harbors 13 genes that encode proteins, which are all part of the respiratory chain, and 24 genes that encode two ribosomal RNAs (rRNA) and 22 transfer RNAs (tRNAs) that are necessary to produce the 13 proteins. The mtDNA is packaged into DNA-protein complexes known as mitochondrial nucleiods, which comprise several molecules of the circular mitochondrial genome along with a number of different proteins, most notably nonhistone, high mobility group proteins (Chen & Butow, 2005). Of particular relevance for neurologists is the fact that many point mutations and deletions in mtDNA have been associated with neurological diseases (see Figure 3.3).

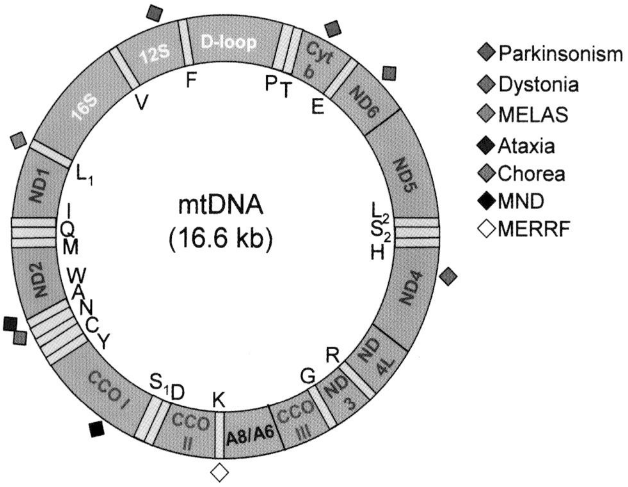

Figure 3.3 Human mitochondrial genome. The 16,569 bp circular, double-stranded mtDNA is depicted schematically. It encodes 13 polypeptides of the respiratory chain, including seven subunits of Complex I (NADH dehydrogenase-ubiquinone oxidoreductase [ND1-6 and ND4L]), one subunit of Complex III (coenzyme Q-cytochrome *c* oxidoreductase [Cyt b]), three subunits of Complex IV (cytochrome *c* oxidase [CCO I-III]), and two subunits of Complex V (ATP synthase [A8/A6]). The mtDNA also codes for the 12S and 16S ribosomal RNAs as well as 22 transfer RNAs (single letter amino acid code, shown along the inside of the ring). The D-loop is a noncoding region of the mtDNA that regulates DNA transcription and replication. Though not comprehensive, the *diamonds* depict representative mutations in mtDNA associated with the neurodegenerative conditions listed.

Important concepts relating to the genetics of mitochondria include copy number, maternal inheritance, mitotic segregation, and heteroplasmy. Unlike the nucleus, which contains two sets of chromosomes, one cell can have up to thousands of copies of mtDNA, with typically approximately five copies per mitochondrion. Mitochondria and their DNA have an almost exclusively maternal inheritance, with very rare exceptions (DiMauro & Schon, 2003). Thus, mutations in mtDNA manifest primarily as maternally inherited syndromes; a mother with a given mtDNA mutation passes it to all of her children, but only her daughters can transmit it further. Another important feature is that mitochondrial division and mtDNA replication are independent of the cell cycle in both dividing and nondividing cells. During cell division the mitochondria are split randomly between daughter cells in a mitotic segregation, and any mutant mtDNAs can thus be differentially partitioned.

The concept of heteroplasmy is particularly important for appreciating the effects of mtDNA mutations on disease. Healthy individuals have identical copies of mtDNA at birth (homoplasmy). However, mitochondria of individuals with a pathogenic mtDNA defect contain both mutated mtDNA and wild-type mtDNA within the same cell, a situation known as heteroplasmy. The amount of mutated mtDNA per mitochondrion can vary among affected patients, as well as among different cells and tissues of the individual. Because most mtDNA mutations are recessive, the affected cell can tolerate up to 70 to 90 percent mutated mtDNA before a biochemical defect in the respiratory chain, the overriding organellar phenotype of mtDNA mutations, can be detected. This contributes to a "threshold effect" that is seen with many diseases associated with mtDNA defects, with the disease threshold lower in cells highly dependent on oxidative metabolism such as those in the brain and muscle (Chinnery & Schon, 2003; DiMauro & Schon, 2003).

C. Mitochondrial Bioenergetics and Calcium Signaling

It is not surprising that all genes encoded within the mtDNA are involved in OXPHOS, since the primary function of mitochondria is to generate cellular energy in the form of ATP. This is accomplished by metabolizing NADH and pyruvate, the major products of glycolysis that occurs in the cytoplasm. Pyruvate is actively transported across the mitochondrial membranes to the matrix, where it combines with coenzyme A (CoA) to form acetyl CoA. This is fed into the citric acid cycle, which ultimately generates three molecules of NADH and one of $FADH_2$, both key cofactors in OXPHOS.

The cellular respiratory chain/OXPHOS system is made of five multimeric protein complexes (I–V) on the inner mitochondrial membrane, with basic compositions and functions as outlined in Figure 3.4. Two electron carriers, ubiquinone (also known as coenzyme Q_{10}) in the inner mitochondrial

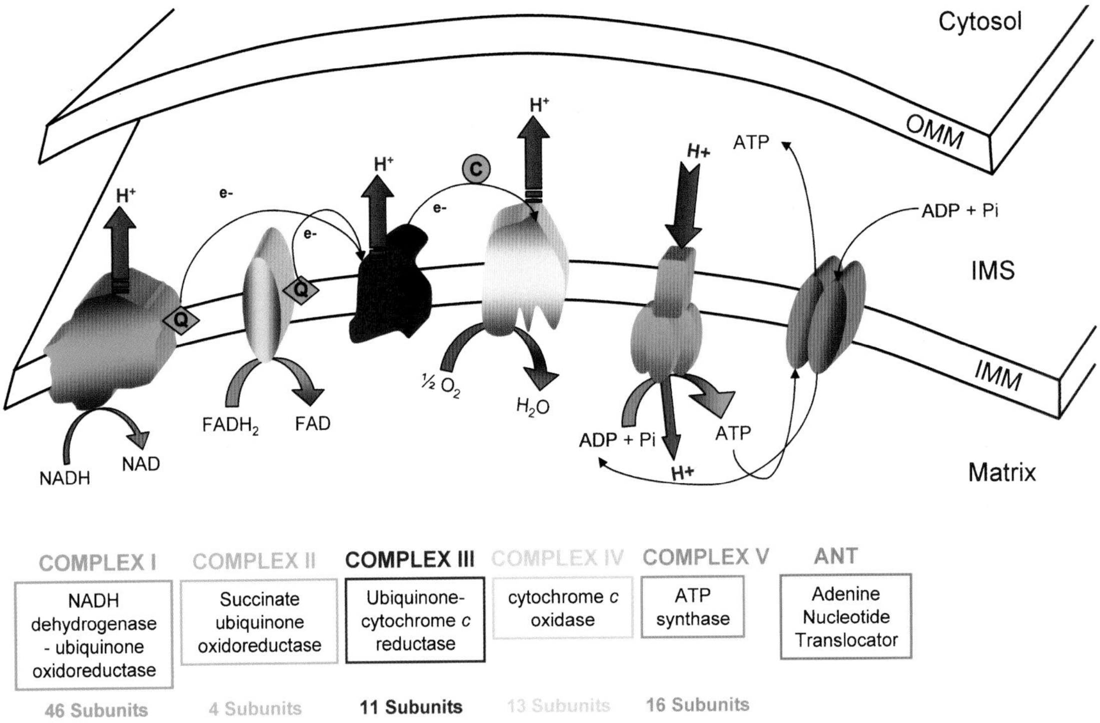

Figure 3.4 Schematic illustration of the mitochondrial respiratory chain. Electrons (e^-) are transferred to Complexes I and II by reduced cofactors (NADH and $FADH_2$) that are generated through the metabolism of carbohydrates, fats, and proteins. Electrons are subsequently transferred from these complexes to ubiquinone (Coenzyme Q [Q]) present in the inner mitochondrial membrane (IMM), then shuttled to Complex III, from which they are transferred to cytochrome *c* (C) in the intermembrane space (IMS). Cytochrome *c* transfers its electrons to Complex IV, which finally donates the electrons to oxygen (O_2) molecules to form water (H_2O). The transfer of electrons through complexes I, III, and IV generates the energy to pump protons (H^+) through these complexes from the mitochondrial matrix to the intermembrane space, creating a proton gradient. This gradient drives Complex V to synthesize ATP from ADP and inorganic phosphate (Pi). ATP is then exchanged with ADP across the inner mitochondrial membrane via the adenine nucleotide translocator (ANT).

membrane and cytochrome *c* within the mitochondrial intermembrane space, are also present between these complexes to aid in electron transfer down the chain. Briefly, the reduced cofactors produced by the citric acid cycle, NADH and $FADH_2$, donate electrons to either complexes I or II (succinate ubiquinone oxidoreductase), through two independent pathways. Electrons are then transferred among the complexes down an electrochemical gradient, carried by complex III (ubiquinone-cytochrome *c* reductase) and complex IV (cytochrome *c* oxidase) in conjunction with the two mobile electron carriers, ubiquinone and cytochrome *c*. This transfer of electrons generates energy used by complexes I, III, and IV to pump protons (H^+) out of the mitochondrial matrix into the intermembrane space, creating a proton gradient that generates the majority of the mitochondrial membrane potential ($\Delta\Psi_m$); the differential ionic distribution of Na^+, K^+, and Ca^{2+} ions across the inner mitochondrial membrane forms the chemical gradient. This electrochemical gradient is utilized by complex V (ATP synthase) to produce ATP from inorganic phosphate (P_i) and adenosine diphosphate (ADP).

In addition to this critical function in OXPHOS, the importance of the remarkable ability of mitochondria to accumulate and buffer intracellular Ca^{2+} is becoming increasingly recognized. Mitochondria can effectively buffer the cytoplasmic Ca^{2+} concentration via a set of specific transporters and pores. The $\Delta\Psi_m$ provides a large driving force for Ca^{2+}, and Ca^{2+} enters mitochondria through relatively low affinity uniporters (Ly & Verstreken, 2006). Reciprocally, Na^+-Ca^{2+} and Na^+-H^+ antiporters collaborate to extrude Ca^{2+} in a concentration-dependent manner (Ly & Verstreken, 2006; Rizzuto et al., 1994). Through these means, mitochondria are able to regulate Ca^{2+} signaling in neighboring microdomains.

So what are the functions of the intramitochondrial Ca^{2+}? For one, it has been shown to have a regulatory effect on mitochondrial enzymes, such as dehydrogenases, that are associated with citric acid cycle. As a result, the rate of carbohydrate oxidation depends on levels of Ca^{2+} within the mitochondria, which in turn can affect ATP synthesis and mitochondrial respiration (Rizzuto et al., 1994). In addition, inhibition of mitochondrial Ca^{2+} uptake attenuates mitochondrial fragmentation, prefiguring a role of intramitochondrial Ca^{2+} in the regulation of mitochondrial morphology (Breckenridge et al., 2003). Mitochondria may also act as cytosolic Ca^{2+} buffers by taking up Ca^{2+} in situations where cytosolic levels are high. Within neurons, cytosolic Ca^{2+} levels increase rapidly after depolarization, when there is an opening of voltage-gated Ca^{2+} channels. The mitochondria quickly work to take up Ca^{2+}, provoking an immediate drop in cytosolic Ca^{2+}, and then more gradually release the Ca^{2+} back into the cytosol via the Na^{+}-Ca^{2+} exchanger. Given the critical role that Ca^{2+} microdomains play in neuronal functions, for instance at the mitochondria-enriched presynaptic terminal, mitochondria may exert important influences on Ca^{2+} signaling and subsequently neuronal communication (Blackstone & Sheng, 2002; Collin et al., 2005).

D. Mitochondrial Protein Import

Though the mitochondrial matrix harbors the machinery to synthesize a small number of critical proteins for OXPHOS, approximately 1,000 other proteins encoded on nuclear DNA are required by the mitochondria to maintain a fully functional organelle. In fact, the mitochondrial rRNAs and tRNAs are encoded by mtDNA, but the proteins required for translation of even the small number of mitochondrial DNA-encoded proteins are encoded by nuclear genes (Chen & Butow, 2005; Jacobs & Turnbull, 2005). These and other nuclear-encoded mitochondrial proteins are synthesized in the cytoplasm, then targeted and delivered to the appropriate mitochondrial subcompartment largely via TIM (translocase of the inner membrane) and TOM (translocase of the outer membrane) protein complexes (Lister et al., 2005; Mokranjac & Neupert, 2005).

As can be gleaned from the number of different proteins involved (see Figure 3.5), protein import is a highly coordinated and complex process critical for the proper localization and distribution of proteins within the mitochondria. These translocase complexes have been intensively investigated in yeast, but very similar protein complexes likely exist in human mitochondria as well, and it is abundantly clear that there are multiple different pathways. Briefly, precursor proteins destined for the mitochondria are delivered through highly divergent, but specific, targeting sequences. The multisubunit TOM complex in the outer mitochondrial membrane mediates the recognition and import of proteins into the mitochondria as well as the insertion of proteins into the outer mitochondrial membrane. Tom20 and Tom70 are the primary receptors, and the other subunits (Tom40, Tom22, Tom7, Tom6, Tom5) constitute the core complex. Tom40 forms the protein conducting channel, with a proposed β-barrel topology (Rapaport, 2005).

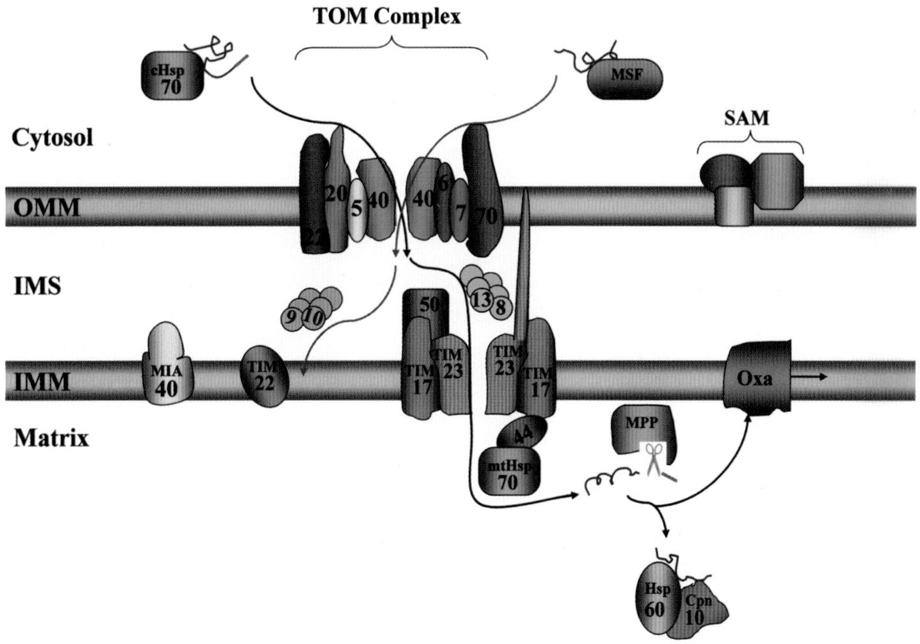

Figure 3.5 Protein translocases involved in mitochondrial protein import. Translocases of the outer mitochondrial membrane (TOM) comprise multimeric, protein complexes responsible for unfolding and shuttling proteins from the cytosol through the outer mitochondrial membrane (OMM); individual proteins are identified by number. Proteins targeted to the mitochondria with an N-terminal presequence are shuttled to the TOM complex via the heat-shock protein Hsp70 and aided through the aqueous intermembrane space (IMS) to a translocase of the inner mitochondrial membrane (TIM), the TIM23 complex, by TIM8 and TIM13. After translocation across the inner membrane into the matrix, the N-terminal presequence is cleaved off by a mitochondrial processing peptidase (MPP), and the processed protein is then refolded with the assistance of Hsp60 and Cpn10. Some proteins with an internal mitochondrial targeting sequence are carried to the TOM complex via a mitochondrial import stimulation factor (MSF). After passing through the outer membrane, these proteins traverse the intermembrane space, with the assistance of TIM9 and TIM10, to the TIM22 complex to be inserted into the inner mitochondrial membrane. Some small metal-coordinating proteins of the intermembrane space require the TOM complex and Mia40, and the TOM complex working in conjunction with the SAM complex mediates import and insertion of outer membrane β-barrel proteins. Lastly, the OXA complex assists in the import and insertion of some inner membrane proteins through the "conservative pathway." See the text for further details.

The TOM complex itself is sufficient for translocation of some outer membrane and intermembrane space proteins. For all other precursor proteins, the TOM complex cooperates with other translocases. The sorting and assembly machinery (SAM) complex, in conjunction with the TOM complex, is responsible for the assembly and import of β-barrel proteins of the outer membrane. Still other proteins, the small, metal-coordinating proteins of the intermembrane space, require cooperation between both the TOM complex and another import machinery whose first component is Mia40 (Mokranjac & Neupert, 2005).

The outer membrane TOM complexes also work in tandem with a number of inner membrane protein translocases (see Figure 3.5). A major translocase of the inner membrane is the TIM23 complex, which is used by nearly all matrix proteins and a majority of inner membrane proteins. It is dependent on ATP and $\Delta\Psi_m$, and most of its substrates have a charged presequence that is cleaved by a peptidase within the matrix. A subclass of inner mitochondrial membrane proteins, typically carrier proteins, crosses the outer membrane via the TOM complex, traverses the intermembrane space with the assistance of several small TIM proteins, and finally interacts with TIM22 to be inserted into the inner membrane. Lastly, some inner membrane proteins follow a "conservative" import pathway; they are first fully imported into the matrix via the TOM and TIM23 complexes, then exported into the inner membrane in a process that depends on the oxidase assembly (OXA) complex (Mokranjac & Neupert, 2005). Certainly, the number of proteins involved and their interdependence belies the importance of these processes in properly distributing proteins within the mitochondria, and any dysfunction in protein import could potentially interfere with multiple aspects of mitochondrial function.

E. Mitochondrial Fission and Fusion

Though in nearly all text books and review articles mitochondria are depicted as sausage-shaped organelles floating within the cytoplasmic sea, such images fail to capture the dynamic changes in mitochondrial morphology and distribution within cells that gives rise to a tubulovesicular network comprising mitochondria of various sizes and shapes (see Figure 3.1). Indeed, fusion and fission events are essential for proper mitochondrial function, and their regulation is increasingly recognized in diverse neuronal functions (Chan, 2006; Chen & Chan, 2006). These competing processes permit mitochondrial morphology to remain dynamic and flexible, and over the past several years the protein machinery for these processes has been clarified (see Figure 3.6).

Mitochondrial fission events in mammals are orchestrated by at least two proteins, Fis1 and Drp1 (dynamin-related protein-1), though a number of other proteins may play additional roles (Karbowski & Youle, 2003; Arnoult et al., 2005; Youle & Karbowski, 2005). Fis1 is an integral membrane protein of

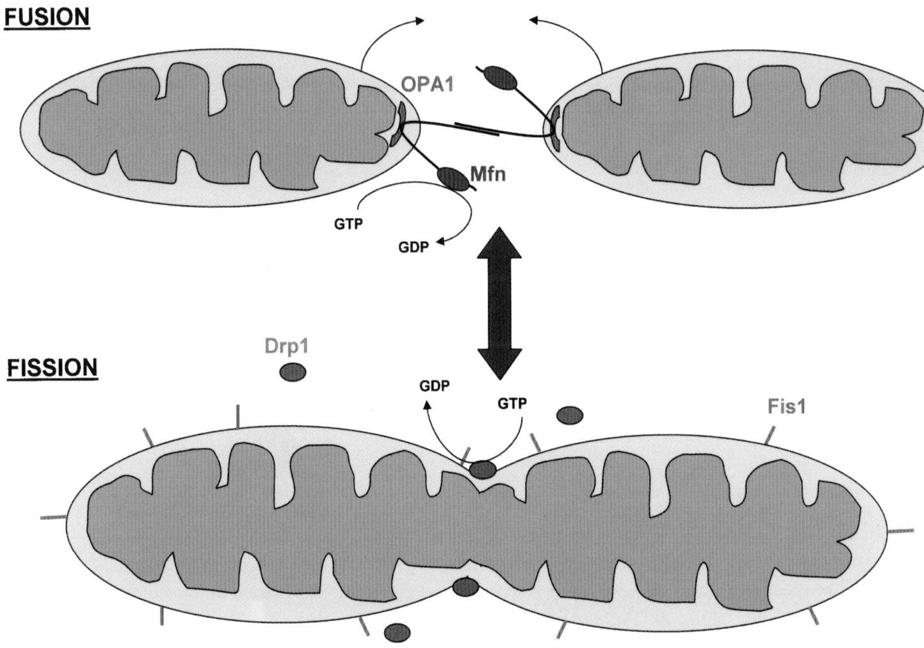

Figure 3.6 Mitochondrial fusion and fission. Proper mitochondrial fusion in humans requires several large GTPases: the intermembrane space protein OPA1 as well as integral membrane GTPases of the outer mitochondrial membrane, the mitofusins (Mfn). These mitofusins, Mfn1 and Mfn2, have the ability to form homo- and heterodimers through C-terminal coiled-coil domains that link the mitochondria together, and the hydrolysis of GTP by the mitofusin GTPases helps mitochondria fuse. In contrast, mitochondrial fission relies on Fis1, which is a small integral membrane protein fairly uniformly distributed on the mitochondrial outer membrane, and the large dynamin-related GTPase Drp1, which is mostly cytosolic with a smaller proportion localized to punctuate areas on the outer mitochondrial membrane that often represent sites of fission.

the outer mitochondrial membrane, and the Drp1 GTPase is a predominantly cytoplasmic protein that is recruited to puncta on the mitochondrial surface, where it likely functions in membrane constriction (Ingerman et al., 2005). Though the mechanisms underlying Drp1 recruitment to mitochondria are not so well understood in mammals, several critical adaptor proteins have been identified in yeast (Okamoto & Shaw, 2005), and thus additional proteins in mammals are likely to be involved (Arnoult et al., 2005).

The reciprocal process of mitochondrial fusion also requires large GTPases of the dynamin superfamily: OPA1 and the mitofusins Mfn1 and Mfn2 (Chan 2006; Chen & Chan, 2006; Praefcke & McMahon, 2004). OPA1 is present at the inner mitochondrial membrane, whereas both Mfn1 and Mfn2 are integral membrane proteins located at the outer mitochondrial membrane. The mitofusins and OPA1 have been shown to work in concert to guide the fusion process, with the ability of OPA1 to tubulate mitochondria dependent on the presence of Mfn1 (Chan, 2006; Chen & Chan, 2006).

Coordinate regulation of these competing processes facilitates the proper distribution of mitochondria within the cell and, as one might expect, these processes are critical to proper cell function and cell viability. In certain cases, the balance dramatically tips in favor of fusion; such an elongation of mitochondria may facilitate the rapid transmission of membrane potential across significant distances within a cell, a feature particularly relevant for polarized cells such as neurons (Skulachev, 2001). Mitochondrial fission events are important for normal cellular functions because they allow for the proper distribution of mtDNA to daughter cells during cell division. Also, during apoptosis, there is prominent fragmentation of the mitochondria that is due to both an increase in fission and inhibition of fusion (Karbowski & Youle, 2003; Youle & Karbowski, 2005).

F. Specializations of Mitochondrial Functions in Neurons

Within neurons the regulation of mitochondrial morphology and transport may be particularly important, and we will use distribution as a starting point for discussing the unique roles of mitochondria in neurons. It is generally believed that the localizations of mitochondria within cells are adapted to the physiological requirements in those areas, and neurons are most likely to exhibit prominent differential distributions because they are the most polarized cells in the body. Indeed, this was appreciated half a century ago, when early ultrastructural studies of synapses noted a particularly high concentration of mitochondria within the nerve terminal, where the need for both precise Ca^{2+} regulation and ATP is particularly acute because of the high physiological demands of synaptic transmission (Palay, 1956). Along myelinated axons in some neurons, mitochondria are sparse except at the nodes of Ranvier. This may reflect an increased need for ATP production in nonmyelinated areas to facilitate neurotransmission, and in fact an ATP-requiring Na^+, K^+-ATPase critical for restoring membrane potential is also concentrated at the nodes of Ranvier (Chen & Chan, 2006). Mitochondria are similarly prominent at active growth cones in developing neurons (Morris & Hollenbeck, 1993), and they play key roles in synapse development and plasticity within dendrites (Li et al., 2004). Thus, the positions and functions of mitochondria within the nerve cell are highly orchestrated and dynamic in both developing and mature neurons.

Next, we will address the issues of mitochondrial transport and localization that underlie the partitioning of mitochondria within the neuron. Mitochondria are known to be transported along microtubules in neurons, and recent studies have begun to clarify the molecular details of this process. In particular, a series of studies in the fruit fly *Drosophila* have confirmed the importance of mitochondrial distribution in neuronal function. Disruption of the mitochondrial Rho GTPase Miro causes loss of mitochondria in the axon terminals, with resulting locomotor deficits and early death. Also, loss of the Milton protein, which links Miro on the mitochondrion to the motor protein kinesin and is involved in transport along microtubules to the axon terminals, causes blindness even though synapses appear grossly normal. Lastly, mutations in the fissioning GTPase Drp1 result in elongated mitochondria mostly absent from synapses, and these flies have prolonged excitation at their neuromuscular boutons due to a defect in mobilizing reserve pool vesicles (Glater et al., 2006; Rice & Gelfand, 2006; Verstreken et al., 2005).

The other major neuronal processes are the dendrites, and mitochondrial distribution is important there as well. Mitochondria can migrate to dendritic protrusions to provide energy for local protein translation or degradation, which allows for synaptic plasticity (Steward & Schuman, 2003). Indeed, Li and colleagues (2004) found that extension or movement of mitochondria into dendritic protrusions correlated with the plasticity of the dendritic spine. Furthermore, by manipulating fission and fusion they showed that the dendritic distribution of mitochondria is essential for the support of synapses and, reciprocally, that synaptic activity modulates the fission/fusion balance and distribution of mitochondria in dendrites (Li et al., 2004). Thus, mitochondria are important regulatory players in specialized neuronal functions in a manner dependent on their distribution within the neurons.

III. Mitochondria in Mechanisms of Neuronal Cell Death and Neurological Disease

A. General Concepts

Perhaps surprisingly given their critical roles in the lives of cells, an appreciation of the role of mitochondrial dysfunction in human disease is relatively recent. The first mitochondrial disease was described in a woman with non-thyroidal hypermetabolism over 40 years ago by Luft and colleagues (1962), who observed in a skeletal muscle biopsy that the mitochondria had abnormally high ATPase activity, indicating that the woman suffered from a "futile" dissipation of energy. Over the subsequent decades, much work was done characterizing the biochemical abnormalities underlying these disorders and describing their presentations. An extensive literature was spawned on how mutations in the mtDNA can give rise to a host of widely disparate presentations; over 150 different pathogenic point mutations and an even greater number of DNA rearrangements (consisting of duplications and partial deletions) of mtDNA are associated with disease (Chinnery & Schon, 2003). More recently, with the rapid advances in human genetics, the discovery of mutations in the nuclear DNA resulting in mitochondrial dysfunction has increased dramatically, providing fresh insights into the functions of novel proteins and identifying new mitochondrial vulnerabilities relevant for neurological disease (Schapira, 2006b; Wallace, 2005).

Here we will emphasize mitochondrial dysfunction as it relates to neurological disease in particular. Although the clinical features can vary among the different neurodegenerative disorders, widely, the fact that neurons are highly dependent on energy from OXPHOS suggests a common pathogenic mechanism for neurodegeneration—dysfunctional energy metabolism by the mitochondria. Thus, the dysfunction can be due to a genetic mutation directly affecting a mitochondrial respiratory chain protein, to a defect of a mitochondrial protein important for organellar function but not directly part of the mitochondrial respiratory chain, or to a more complex malfunction that is secondary to other events in the disease state (Chinnery & Schon, 2003; Schapira, 2006b). Add to this the special challenges posed by highly polarized neurons, with their tightly coupled signaling and transport functions requiring close coordination with mitochondria, and multiple scenarios for mitochondrial dysfunction leading to disease emerge. However, rather than discuss in detail the mitochondrial diseases, which will be addressed comprehensively in Chapter 33, we will emphasize some general principles of mitochondrial dysfunction relevant for cell death mechanisms—including defects in OXPHOS, free radical formation, calcium overload, excitotoxicity, and abnormal protein import. In addition, we will discuss the key cellular roles played by mitochondria during programmed cell death.

B. Decreased ATP Production

An obvious form of mitochondrial dysfunction that can result in cell death is decreased ATP production. Neurons have a great demand for ATP but have low stores of carbohydrate and thus cannot maintain ATP synthesis long-term through glycolysis alone. In many disorders due to mutations in mtDNA, ATP production is compromised (Schapira, 2006b). This dependence on OXPHOS also becomes acutely apparent when the brain is confronted with toxins that disrupt ATP production mechanisms. For example, the basal ganglia are particularly sensitive to toxins such as cyanide, carbon monoxide, rotenone, or 3-nitroproprionic acid, which are implicated in both mitochondrial dysfunction and neurological disorders such as Parkinson's disease (Betarbet et al., 2000; Gould & Gustine, 1982; Uitti et al., 1985). Similarly, OXPHOS is impaired during other degenerative disorders such as Alzheimer's disease and also in metabolic disorders and hypoxia-ischemia. Mechanistically, this seems most likely to be due to decreased ATP production, loss of calcium buffering, increased generation of reactive oxygen species, or combinations of these (Beal, 2005; Lenaz et al., 2006; Reddy, 2006). The latter two processes could also result from loss of ATP, but may also be more directly involved in some cases.

Any discussion of mitochondria and neurological disease necessarily must address the complex and controversial issue of age-related changes in mtDNA and their effects on mitochondrial function. Such changes have been invoked to explain age-dependence of degenerative diseases, with several studies pointing to declining respiratory function, accumulation of variable degrees of mtDNA mutations and deletions, and oxidative damage to mtDNA that occurs with age (Beal, 2005). However, whether or not these mutations cause the respiratory effects is not always clear, since even when there are mutations they often appear far below the threshold effect one might expect for measurable dysfunction. On the other hand, some mtDNA disorders are extremely complex and likely not due to the mtDNA mutations alone, but rather to combinations among certain environmental factors or in conjunction with nuclear DNA mutations (Chinnery & Schon, 2003). Whether such subthreshold changes can conspire with other problems to contribute to disease is a very active area of interest and investigation, particularly for common neurodegenerative diseases such as Alzheimer's and Parkinson's disease (Howell et al., 2005).

Mitochondria play extensive roles in various metabolic processes including the citric acid cycle, fatty acid oxidation, and amino acid catabolism in addition to OXPHOS, so it is not surprising that systemic signs of metabolic dysfunction can often accompany mitochondrial disease. The most common manifestation is lactic acidosis, though a number of mitochondrial encephalomyopathies caused by mtDNA mutations have other prominent metabolic derangements as well (Chan, 2006). In many of these there are also accompanying defects in OXPHOS. However, it would not be surprising to find additional disorders due to mutations in nuclear-encoded mitochondrial proteins with more prominent metabolic components.

C. Abnormal Mitochondrial Distribution

In this and subsequent sections, we will focus more directly on several specific realms of mitochondrial dysfunction that appear strongly linked to disease, with the understanding that decreased ATP production may be a common final pathway for many of them. Disturbances in mitochondrial fusion lead to change in mitochondrial respiration and $\Delta\Psi_m$ as well as a decrease in cell survival and growth. In addition, mitochondrial fusion and fission events have an effect on apoptotic processes as well as ATP production (Perfettini et al., 2005; Youle & Karbowski, 2005). Cells lacking the mitofusins were shown to have a decrease in endogenous and uncoupled respiration rates, and this effect was even greater in cells where another large GTPase required for fusion, OPA1, was knocked down by RNA interference (Chen et al., 2005). In addition, cells lacking the mitofusins and OPA1 show a loss of $\Delta\Psi_m$ as well as a decrease in cell growth; however, all effects of loss of these proteins can be reversed by the reintroduction of the proteins (Chen et al., 2005). The importance of these mitochondrial fission and fusion proteins in neurological disease is underscored by the fact that mutations causing autosomal dominant optic atrophy type 1 and Charcot-Marie Tooth disease have been identified in the OPA1 and Mfn2 genes, respectively (Olichon et al., 2006; Züchner et al., 2006).

Despite the direct demonstration that defects in mitochondrial fusion can cause neurological disease, the dependence of fusion for normal mitochondrial function is still not well understood. One line of thought is that fusion is protective because it allows for mixing of mitochondrial membranes and contents, so that any loss of material in any single mitochondria is transient. Thus, without fusion events, mitochondrial losses would become permanent and lead to mitochondrial defects resulting in cellular dysfunction. In addition, fusion can prevent apoptosis by providing a means to repair damaged outer mitochondrial membrane (Chen & Chan, 2006).

As a result of compartment-specific needs of neurons and the fact that they have long processes, mitochondria-dependent functions cannot rely on diffusion from mitochondria present in the soma. Neurons require that the mitochondria be present throughout their processes in order to meet their immediate needs. This is particularly prominent at axon terminals, where mitochondrial are intimately associated with vesicle release during neurotransmission. At synapses, the opening of neurotransmitter and voltage-gated channels rely on ATP-driven pumps to regulate the large ionic flux that ensues. Within dendritic protrusions, mitochondria act by buffering the Ca^{2+} flux due to voltage-gated Ca^{2+} channel or NMDA receptor activation. Studies have shown that after initial uptake, Ca^{2+} is slowly released from mitochondria in dendritic spines, suggesting that a subpopulation of spines may be exposed to higher levels of Ca^{2+} based on the local presence of mitochondria, and this can activate Ca^{2+}-dependent enzymes that may not have been activated if mitochondria had not been present locally. Thus, together these studies indicate that several large GTPases, which regulate mitochondrial fission and distribution, also affect the density of synapses and spines and may be important for their development, maintenance, and functions.

D. Free Radical Formation

The previous sections emphasized pathological loss-of-function mechanisms. However, the following sections emphasize toxic gain-of-function mechanisms. A widely studied complication of mitochondrial dysfunction that can lead to disease is the production and accumulation of free radicals within a cell. During oxidative phosphorylation, the electron transfer that occurs releases single electrons as a by-product, which then results in the generation of superoxides (Balaban et al., 2005; Raha & Robinson, 2000). Toxins that inhibit the mitochondrial respiratory chain complex, such as rotenone, MPP^+, and antimycin A also are found to significantly increase the rate of superoxide formation. Thus, mitochondrial inhibition leads to oxidative stress, and an increase in oxidative stress can feedback into the respiratory chain by further inhibiting the OXPHOS complex, resulting in a detrimental cycle.

There are several features that make cells of the central nervous system particularly sensitive to damage by free radicals (Halliwell, 2006). Within neurons, there is a higher rate of OXPHOS and oxygen utilization, which results in increased levels of superoxide byproducts. Calcium movements within neurons also have been shown to be critical for normal cell function, and a change in Ca^{2+} export from the cells can result in increased oxidative stress due to increased nitric oxide synthase activity or a decrease in respiratory chain function. Also, there are many neurotransmitters, such as dopamine and norepinephrine, which can auto-oxidize to give rise to reactive quinines that can be damaging. Moreover, dopaminergic areas of the brain generate H_2O_2 as a byproduct of the breakdown of dopamine by monoamine oxidase B (MAO-B). This can be exacerbated by the fact that catalase activities are lower in brain, leading to more reliance on glutathione and glutathione peroxidase activity to remove harmful H_2O_2 present in cells. In addition, lipid peroxidase substrates (polyunsaturated fatty acids) are very high in the brain, leaving the nervous system more vulnerable to lipid peroxidation. Lastly, still another factor that renders neurons more susceptible to oxidative damage is the fact that iron levels in the brain are relatively high; however, the iron-binding capacity in cerebrospinal fluid is low. Thus, the release of iron following an injury to the central nervous system may contribute to an increase in iron-catalyzed degradation of H_2O_2 and oxidative damage.

E. Calcium Overload and Excitotoxicity

In addition to their critical roles as ATP generators and in the detoxification of oxygen free radicals, where dysfunction can have clear neuropathological consequences, impairments in the ability of mitochondria to sequester calcium similarly have important pathophysiological sequelae. Indeed, mitochondrial uptake of Ca^{2+} is important in maintaining physiological conditions, and excess cytoplasmic Ca^{2+} can prove to have deleterious effects on cellular function. This sequestration is dependent on $\Delta\Psi_m$, and the loss of $\Delta\Psi_m$ can prevent Ca^{2+} uptake into the mitochondria, which can lead to decreased OXPHOS, resulting in a further decrease in $\Delta\Psi_m$ and a drop in ATP synthesis. It is not clear if the decreased $\Delta\Psi_m$ is due to the opening of the mitochondrial permeability transition pore (which occurs during apoptosis) or because of decreased mitochondrial respiratory chain function from nitric oxide (NO) inhibition of Complexes II, III, and IV. This cell death is Ca^{2+} and NO synthase-dependent and can be prevented by depolarizing the mitochondria or via NO scavengers or NO synthase inhibitors (Dawson et al., 1996; Stout et al., 1998).

Excitotoxicity is caused by the neurotoxic effects of increased postsynaptic stimulation due to excitatory neurotransmitters such as glutamate. It is well documented that an increased exposure to glutamate *in vitro* results in neuronal death (Choi et al., 1987), and activation of *N*-methyl-D-aspartate (NMDA) ionotropic glutamate receptors plays a key role in this process. The activation of NMDA receptor in conjunction with a decrease in membrane potential causes the magnesium block of NMDA receptor to be released, allowing Ca^{2+} and Na^+ to enter the cell. This flux of ions results in the opening of voltage-gated calcium channels, allowing even more Ca^{2+} into the cell. Under normal conditions, the presence of extracellular glutamate is transient because it is quickly taken up by transporters on the presynaptic neuron or by supporting glia. However, if these mechanisms are overwhelmed, NMDA receptors may be overstimulated, with consequent increases in NO production. Subsequently, NO can inhibit OXPHOS and also react with the superoxide anion to form the oxidant species peroxynitrite (Moncada & Bolaños, 2006). Other studies suggest that a dysfunctional mitochondrial respiratory chain can lead to decreased activity of the Na^+-K^+ ATPase and a partial depolarization that is sufficient to remove the Mg^{2+} block on NMDA channels, promoting excitotoxicity.

F. Impaired Protein Import

As we have already outlined, mitochondrial protein import is a complex process, involving a number of different protein components working tightly with one another. Perhaps it is surprising, then, that import has not been described so prominently as other mitochondrial pathologic mechanisms. However, in at least one case, that of the small intermembrane space import protein DDP/TIM8 (see Figure 3.5), loss of protein import function results in the X-linked degenerative dystonia known as the Mohr-Tranebjaerg deafness-dystonia syndrome (Swerdlow et al., 2004).

G. Mitochondrial Functions during Apoptosis

Dysfunction of the mitochondrial respiratory chain can either lead to cell death by apoptosis or necrosis, depending on the level of respiratory chain inhibition. If the mitochondrial respiratory chain is significantly inhibited, resulting in a substantial drop in ATP production, severe cellular dysfunction with ensuing necrosis occurs. Slight inhibition of the respiratory chain may disrupt cellular processes and can signal cell death through a process that requires ATP, resulting in apoptosis. For example, in the case of stroke, necrosis is considered to dominate the ischemic core; however, apoptosis occurs in the cells that surround the area of insult (Charriaut-Marlangue et al., 1996). Also, apoptosis cannot only be triggered by mitochondrial dysfunction, but mitochondria play crucial roles in the process itself.

Apoptosis is a highly regulated process that is necessary for cell growth regulation required for normal embryonic development; however, it can also serve to eliminate cells that are damaged or under stress (Fadeel & Orrenius, 2005). The mechanisms underlying mitochondrial involvement in apoptosis have been the subject of much work and many recent reviews, and thus they will be addressed only briefly here. Apoptosis can proceed via a mitochondria-mediated pathway, where external stimuli cause the translocation of Bax from the cytosol to the mitochondria (Wolter et al., 1997). There, it is thought to cause permeabilization of the outer mitochondrial membrane and release of several intermembrane space proteins, including cytochrome *c*. In the cytoplasm, cytochrome *c* binds APAF-1 (apoptotic protease-activating factor 1), which then binds pro-caspase 9 forming an apoptosome. The apoptosome may then activate effector caspases, such as caspase 3, which can then activate other caspases downstream. This cascade of events results in the signature signs of apoptotic cells, which include DNA fragmentation, nuclear condensation, membrane blebbing, and autophagy of membrane-bound bodies (see Figure 3.7).

As mentioned earlier, exposure of cells to apoptotic stimuli also results in prominent mitochondrial fragmentation (Frank et al., 2001). The dynamin-like GTPase Drp1 is critical for this process, since Drp1 loss-of-function experiments result in inhibition of the mitochondrial fragmentation that normally occurs in response to an apoptotic stimulus (Frank et al., 2001). In addition, there is also a block in mitochondrial fusion, further tipping the fission/fusion balance toward fission (Youle & Karbowski, 2005). Thus, although mitochondrial fission occurs normally in healthy cells, it is the excess of fission in relation to fusion that results in mitochondrial dysfunction and can lead to cell death.

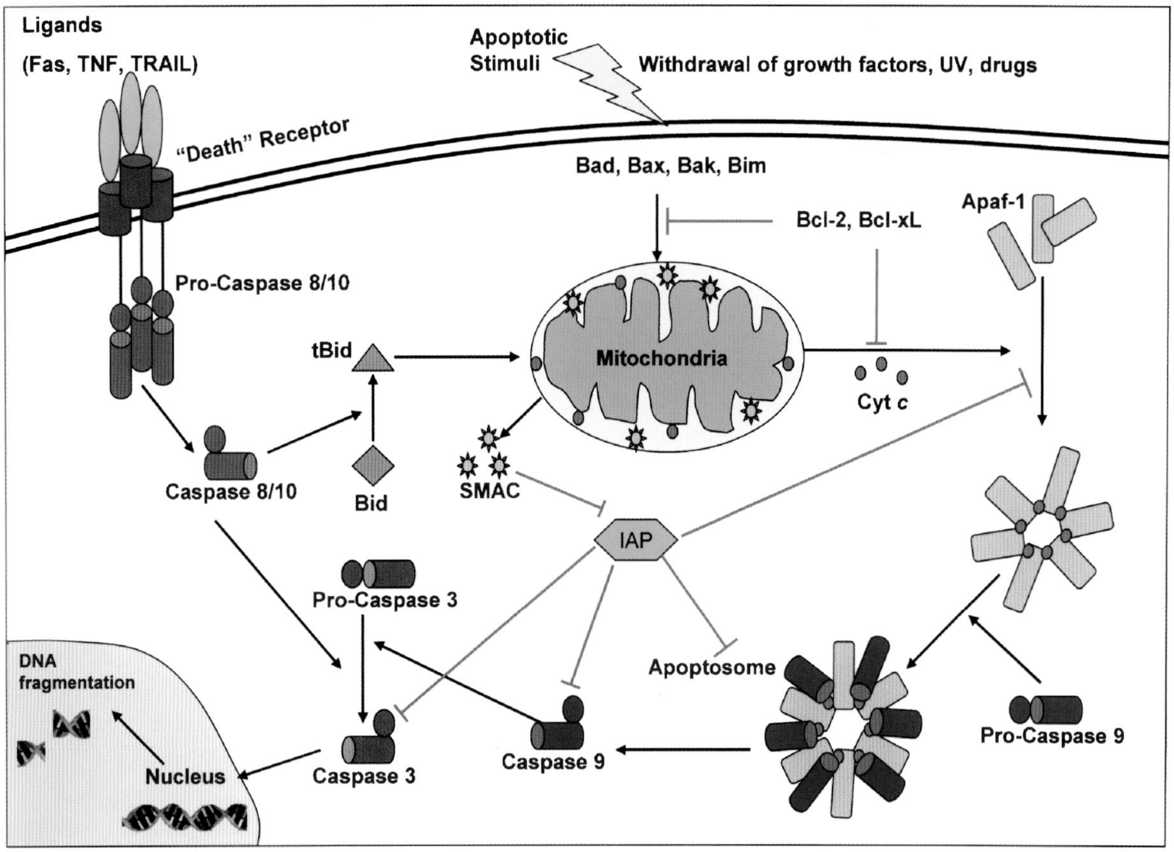

Figure 3.7 Schematic diagram of apoptosis pathways. Apoptosis can be initiated through either *intrinsic* or *extrinsic* pathways. The extrinsic pathway begins outside the cell through activation of surface receptors by ligands such as Fas, TNF, or TRAIL, which results in production of the active forms of caspases 8 or 10. Next, caspases 8 and 10 have the ability to cleave pro-caspase 3 and Bid into their active forms. Truncated Bid (tBid) can then translocate to the mitochondria to initiate release of mitochondrial intermembrane space proteins. Alternatively, intrinsic pathway activation occurs following an apoptotic stimuli, which results in activation and translocation of pro-apoptotic proteins such as Bax, Bad, and Bak, which act at the mitochondria to induce release of intermembrane space proteins. In both cases, release of proteins such as cytochrome *c* and SMAC from the mitochondria is important in downstream steps. Released cytochrome *c* forms a complex with Apaf-1 to activate caspase 9, which joins this complex to form apoptosomes. These apoptosomes further activate caspase 9, which is responsible for cleaving pro-caspase 3 to its active form, resulting in DNA fragmentation and chromatin condensation. There are cellular mechanisms for inhibition of apoptosis, for instance via the anti-apoptotic proteins Bcl-2 and Bcl-xL that are able to inhibit the actions of pro-apoptotic proteins as well as block cytochrome *c* release. There are also proteins known as inhibitors of apoptosis (IAP) that are able to block Apaf-1 complex formation as well as the function of apoptosomes and caspases. IAPs, however, can in turn be inhibited by SMAC that has been released from the mitochondria into the cytosol.

IV. Roles of Mitochondrial Dysfunction in Common Neurodegenerative Diseases

Pathogenic effects on cells of the different forms of mitochondrial dysfunction out lined in the previous section can be exacerbated by the increased demands of the highly-polarized neurons, with their long axons and dendrites. In many cases it is analysis of hereditary neurological disorders that has clarified the importance of these mitochondrial pathways in neuronal survival (Kwong et al., 2006). However, a much broader question concerns the effects of mitochondrial dysfunction on pathologic mechanisms underlying more common neurological disorders, including Parkinson's disease, amyotrophic lateral sclerosis, and Alzheimer's disease. Though this has been suggested for many years, recent evidence is lending increasing support to this concept. For Parkinson's disease in particular, models of toxicity have indicated that mitochondrial toxins reproduce many features of Parkinson's disease in rodent models, and several Parkinson's disease gene products (including PINK1, DJ-1, and parkin) have roles in mitochondrial function (Abou-Sleiman et al., 2006; Betarbet et al., 2000, 2006; Schapira, 2006a). Similarly, the case for mitochondrial dysfunction in amyotrophic lateral sclerosis and Alzheimer's disease is increasingly compelling (Hervias et al., 2006; Moreira et al., 2006; Reddy, 2006).

On the other hand, there is evidence that making mitochondria less efficient, through a mild uncoupling of OXPHOS, can keep cells under stress alive. This is through the effects of the mitochondrial uncoupling proteins, whose actions likely protect neurons from apoptotic death by reducing reactive oxygen species generation from abnormal respiration. Thus, either mild uncouplers or activation of the uncoupling proteins themselves may serve someday as therapies in neurodegenerative diseases, stroke, and other disorders characterized by oxidative mitochondrial stress.

Presently, available therapies for mitochondrial disorders are inadequate. Most treatments currently employed have been palliative and focused on an empirical use of various vitamins, cofactors, and free radical scavengers (DiMauro & Schon, 2003). However, as the previous sections have outlined, our understanding of various mitochondrial processes has increased dramatically, and has suggested several distinct pathways for possible therapeutic intervention.

V. Conclusions

In this chapter, we have outlined the complexity of such important mitochondrial processes as OXPHOS, protein import, morphological dynamics, inheritance, and apoptosis—interrelated processes where tight coordination and regulation is essential. With rapid advances in human genetics, the continuing identification of disease genes will undoubtedly further clarify both abnormal and normal functions of mitochondria in the nervous system at a brisk pace. As increasing insights into a number of mitochondrial diseases have shown, many things can go wrong, and hopefully an understanding of these processes will help guide more rational therapies. Mitochondria clearly hold great promise as targets for therapies (Bouchier-Hayes et al., 2005; Green & Kroemer, 2005), though this potential is only gradually being realized.

So what else does the future hold for mitochondrial biology? Among the most intriguing questions concerns the changes that happen to mtDNA and mitochondrial function as people age. With an aging population, any effects of these alterations on a host of diseases will become more apparent, particularly since advancing age is the most common risk factor for common neurodegenerative disease such as Alzheimer's and Parkinson's diseases, and mitochondrial dysfunction is increasingly implicated in these processes (Beal, 2005). Along these lines, particularly intriguing developments are occuring in the study of autophagy, a catalytic process that allows recycling of cytoplasmic products including organelles such as mitochondria, into basic components (Kundu & Thompson, 2005; Levine & Yuan, 2005). Is this a protective mechanism that can be stimulated to remove dysfunction or damaged mitochondria? It seems likely that it can be pro-survival under some conditions, and the future looks bright for "mitochondrial recycling." Deeper investigations into both abnormal and normal functions of mitochondria will continue to energize our appreciation of this critical organelle.

References

Abou-Sleiman, P. M., Muqit, M. M. K., and Wood, N. W. (2006). Expanding insights of mitochondrial dysfunction in Parkinson's disease. *Nat. Rev. Neurosci.* **7**, 207–219.

Alberts, B., Johnson, A., Lewis, J., Raff, M., Roberts, K., and Walter, P. (2002). *Molecular Biology of the Cell*, 4th ed. Garland Science, New York.

Arnoult, D., Rismanchi, N., Grodet, A., Roberts, R. G., Seeburg, D. P., Estaquier, J. et al. (2005). Bax/Bak-dependent release of DDP/TIMM8a promotes Drp1-mediated mitochondrial fission and mitoptosis during programmed cell death. *Curr. Biol.* **15**, 2112–2118.

Balaban, R. S., Nemoto, S., and Finkel, T. (2005). Mitochondria, oxidants, and aging. *Cell* **120**, 483–495.

Beal, M. F. (2005). Mitochondria take center stage in aging and neurodegeneration. *Ann. Neurol.* **58**, 495–505.

Betarbet, R., Sherer, T. B., MacKenzie, G., Garcia-Osuna, M., Panov, A. V., and Greenamyre, J. T. (2000). Chronic systemic pesticide exposure reproduces features of Parkinson's disease. *Nat. Neurosci.* **3**, 1301–1306.

Betarbet, R., Canet-Aviles, R. M., Sherer, T. B., Mastroberardino, P. G., McLendon, C., Kim, J.-H. et al. (2006). Intersecting pathways to neurodegeneration in Parkinson's disease: Effects of the pesticide rotenone on DJ-1, α-synuclein, and the ubiquitin-proteasome system. *Neurobiol. Dis.* **22**, 404–420.

Blackstone, C. and Sheng, M. (2002). Postsynaptic calcium signaling microdomains in neurons. *Front. Biosci.* **7**, d872–d885.

Bouchier-Hayes, L., Lartigue, L., and Newmeyer, D. D. (2005). Mitochondria: Pharmacological manipulation of cell death. *J. Clin. Invest.* **115**, 2640–2647.

Breckenridge, D. G., Stojanovic, M., Marcellus, R. C., and Shore, G. C. (2003). Caspase cleavage product of BAP31 induces mitochondrial fission through endoplasmic reticulum calcium signals, enhancing cytochrome *c* release to the cytosol. *J. Cell Biol.* **160**, 1115–1127.

Chan, D. C. (2006). Mitochondria: Dynamic organelles in disease, aging, and development. *Cell* **125**, 1241–1252.

Charriaut-Marlangue, C., Aggoun-Zouaoui, D., Represa, A., and Ben-Ari, Y. (1996). Apoptotic features of selective neuronal death in ischemia, epilepsy and gp120 toxicity. *Trends Neurosci.* **19**, 109–114.

Chen, H. and Chan, D. C. (2006). Critical dependence of neurons on mitochondrial dynamics. *Curr. Opin. Cell Biol.* **18**, 453–459.

Chen, H., Chomyn, A., and Chan, D. C. (2005). Disruption of fusion results in mitochondrial heterogeneity and dysfunction. *J. Biol. Chem.* **280**, 26185–26192.

Chen, X. J. and Butow, R. A. (2005). The organization and inheritance of the mitochondrial genome. *Nat. Rev. Genet.* **6**, 815–825.

Chinnery, P. F. and Schon, E. A. (2003). Mitochondria. *J. Neurol. Neurosurg. Psychiatry* **74**, 1188–1199.

Choi, D. W., Maulucci-Gedde, M., and Kriegstein, A. R. (1987). Glutamate toxicity in cortical cell culture. *J. Neurosci.* **7**, 357–368.

Collin, T., Marty, A., and Llano, I. (2005). Presynaptic calcium stores and synaptic transmission. *Curr. Opin. Neurobiol.* **15**, 275–281.

Dawson, V. L., Kizushi, V. M. Huang, P. L., Snyder, S. H., and Dawson, T. M. (1996). Resistance to neurotoxicity in cortical cultures from neuronal nitric oxide synthase-deficient mice. *J. Neurosci.* **16**, 2479–2487.

DiMauro, S. and Schon, E. A. (2003). Mitochondrial respiratory-chain diseases. *N. Engl. J. Med.* **348**, 2656–2668.

Fadeel, B. and Orrenius, S. (2005). Apoptosis: A basic biological phenomenon with wide-ranging implications in human disease. *J. Intern. Med.* **258**, 479–517.

Frank, S., Gaume, B., Bergmann-Leitner, E. S., Leitner, W. W., Robert, E. G., Catez, F. et al. (2001). The role of dynamin-related protein 1, a mediator of mitochondrial fission, in apoptosis. *Dev. Cell* **1**, 515–525.

Glater, E. E., Megeath, L. J., Stowers, R. S., and Schwarz, T. L. (2006). Axonal transport of mitochondria requires milton to recruit kinesin heavy chain and is light chain independent. *J. Cell Biol.* **173**, 545–557.

Gould, D. H. and Gustine, D. L. (1982). Basal ganglia degeneration, myelin alterations, and enzyme inhibition induced in mice by the plant toxin 3-nitropropanoic acid. *Neuropathol. Appl. Neurobiol.* **8**, 377–393.

Green, D. R. and Kroemer, G. (2005). Pharmacological manipulation of cell death: Clinical applications in sight? *J. Clin. Invest.* **115**, 2610–2617.

Halliwell, B. (2006). Oxidative stress and neurodegeneration: Where are we now? *J. Neurochem.* **97**, 1634–1658.

Hervias, I., Beal, M. F., and Manfredi, G. (2006). Mitochondrial dysfunction and amyotrophic lateral sclerosis. *Muscle Nerve* **33**, 598–608.

Hollenbeck, P. J. (2005). Mitochondria and neurotransmission: Evacuating the synapse. *Neuron* **47**, 331–333.

Hollenbeck, P. J. and Saxton, W. M. (2005). The axonal transport of mitochondria. *J. Cell Sci.* **118**, 5411–5419.

Howell, N., Elson, J. L., Chinnery, P. F., and Turnbull, D. M. (2005). mtDNA mutations and common neurodegenerative disorders. *Trends Genet.* **21**, 583–586.

Ingerman, E., Perkins, E. M., Marino, M., Mears, J. A., McCaffery, J. M., Hinshaw, J. E., and Nunnari, J. (2005). Dnm1 forms spirals that are structurally tailored to fit mitochondria. *J. Cell Biol.* **170**, 1021–1027.

Jacobs, H. T. and Turnbull, D. M. (2005). Nuclear genes and mitochondrial translation: A new class of genetic disease. *Trends Genet.* **21**, 312–314.

Karbowski, M. and Youle, R. J. (2003). Dynamics of mitochondrial morphology in healthy cells and during apoptosis. *Cell Death Differ.* **10**, 870–880.

Kundu, M. and Thompson, C. B. (2005). Macroautophagy *versus* mitochondrial autophagy: A question of fate? *Cell Death Differ.* **12**, 1484–1489.

Kwong, J. Q., Beal, M. F., and Manfredi, G. (2006). The role of mitochondria in inherited neurodegenerative diseases. *J. Neurochem.* **97**, 1659–1675.

Lenaz, G., Baracca, A., Fato, R., Genova, M. L., and Solaini, G. (2006). New insights into structure and function of mitochondria and their role in aging and disease. *Antioxid. Redox Signal.* **8**, 417–437.

Levine, B. and Yuan, J. (2005). Autophagy in cell death: an innocent convict? *J. Clin. Invest.* **115**, 2679–2688.

Li, Z., Okamoto, K.-I., Hayashi, Y., and Sheng, M. (2004). The importance of dendritic mitochondria in the morphogenesis and plasticity of spines and synapses. *Cell* **119**, 873–887.

Lister, R., Hulett, J. M., Lithgow, T., and Whelan, J. (2005). Protein import into mitochondria: Origins and functions today. *Mol. Membr. Biol.* **22**, 87–100.

Luft, R., Ikkos, D., Palmieri, G., Ernster, L., Afzelius, B. (1962). A case of severe hypermetabolism of nonthyroid origin with a defect in the maintenance of mitochondrial respiratory control: A correlated clinical, biochemical, and morphological study. *J. Clin. Invest.* **41**, 1776–1804.

Ly, C. V. and Verstreken, P. (2006). Mitochondria at the synapse. *Neuroscientist* **12**, 291–299.

McBride, H. M., Neuspiel, M., and Wasiak, S. (2006). Mitochondria: More than just a powerhouse. *Curr. Biol.* **16**, R551–R560.

Meinecke, M., Wagner, R., Kovermann, P., Guiard, B., Mick, D. U., Hutu, D. P. et al. (2006). Tim50 maintains the permeability barrier of the mitochondrial inner membrane. *Science* **312**, 1523–1526.

Mokranjac, D. and Neupert, W. (2005). Protein import into mitochondria. *Biochem. Soc. Trans.* **33**, 1019–1023.

Moncada, S. and Bolaños, J. P. (2006). Nitric oxide, cell bioenergetics and neurodegeneration. *J. Neurochem.* **97**, 1676–1689.

Moreira, P. I., Cardoso, S. M., Santos, M. S., and Oliveira, C. R. (2006). The key role of mitochondria in Alzheimer's disease. *J. Alzheimers Dis.* **9**, 101–110.

Morris, R. L. and Hollenbeck, P. J. (1993). The regulation of bidirectional mitochondrial transport is coordinated with axon outgrowth. *J. Cell Sci.* **104**, 917–927.

Okamoto, K. and Shaw, J. M. (2005). Mitochondrial morphology and dynamics in yeast and multicellular eukaryotes. *Annu. Rev. Genet.* **39**, 503–536.

Olichon, A., Guillou, E., Delettre, C., Landes, T., Arnauné-Pelloquin, L., Emorine, L. J. et al. (2006). Mitochondrial dynamics and disease, OPA1. *Biochim. Biophys. Acta*, **1763**, 500–509.

Palay, S. L. (1956). Synapses in the central nervous system. *J. Biophys. Biochem. Cytol.* **2**, 193–202.

Perfettini, J.-L., Roumier, T., and Kroemer, G. (2005). Mitochondrial fusion and fission in the control of apoptosis. *Trends Cell Biol.* **15**, 179–183.

Praefcke, G. J. K. and McMahon, H. T. (2004). The dynamin superfamily: Universal membrane tubulation and fission molecules? *Nat. Rev. Mol. Cell Biol.* **5**, 133–147.

Raha, S. and Robinson, B. H. (2000). Mitochondria, oxygen free radicals, disease, and ageing. *Trends Biochem. Sci.* **25**, 502–508.

Rapaport, D. (2005). How does the TOM complex mediate insertion of precursor proteins into the mitochondrial outer membrane? *J. Cell Biol.* **171**, 419–423.

Reddy, P. H. (2006). Amyloid precursor protein-mediated free radicals and oxidative damage: Implications for the development and progression of Alzheimer's disease. *J. Neurochem.* **96**, 1–13.

Rice, S. E. and Gelfand, V. I. (2006). Paradigm lost: Milton connects kinesin heavy chain to miro on mitochondria. *J. Cell Biol.* **173**, 459–461.

Rizzuto, R., Bastianutto, C., Brini, M., Murgia, M., and Pozzan, T. (1994). Mitochondrial Ca^{2+} homeostasis in intact cells. *J. Cell Biol.* **126**, 1183–1194.

Schapira, A. H. V. (2006a). Etiology of Parkinson's disease. *Neurology* **66**(Suppl. 4), S10–S23.

Schapira, A. H. V. (2006b). Mitochondrial disease. *Lancet* **368**, 70–82.

Skulachev, V. P. (2001). Mitochondrial filaments and clusters as intracellular power-transmitting cables. *Trends Biochem. Sci.* **26**, 23–29.

Steward, O. and Schuman, E. M. (2003). Compartmentalized synthesis and degradation of proteins in neurons. *Neuron* **40**, 347–359.

Stout, A. K., Raphael, H. M., Kanterewicz, B. I., Klann, E., and Reynolds, I. J. (1998). Glutamate-induced neuron death requires mitochondrial calcium uptake. *Nat. Neurosci.* **1**, 366–373.

Swerdlow, R. H., Juel, V. C., and Wooten, G. F. (2004). Dystonia with and without deafness is caused by TIMM8A mutation. *Adv. Neurol.* **94**, 147–154.

Uitti, R. J., Rajput, A. H., Ashenhurst, E. M., and Rozdilsky, B. (1985). Cyanide induced parkinsonism: A clinicopathologic report. *Neurology* **35**, 921–925.

Verstreken, P., Ly, C. V., Venken, K. J. T., Koh, T. W., Zhou, Y., and Bellen H. J. (2005). Synaptic mitochondria are critical for mobilization of reserve pool vesicles at *Drosophila* neuromuscular junctions. *Neuron* **47**, 365–378.

Wallace, D. C. (2005). A mitochondrial paradigm of metabolic and degenerative diseases, aging and cancer: A dawn for evolutionary medicine. *Annu. Rev. Genet.* **39**, 359–407.

Wolter, K. G., Hsu, Y.-T., Smith, C. L., Nechushtan, A., Xi, X.-G., and Youle, R. J. (1997). Movement of Bax from the cytosol to mitochondria during apoptosis. *J. Cell Biol.* **139**, 1281–1292.

Youle, R. J. and Karbowski, M. (2005). Mitochondrial fission in apoptosis. *Nat. Rev. Mol. Cell Biol.* **6**, 657–663.

Züchner, S., De Jonghe, P., Jordanova, A., Claeys, K. G., Guergueltcheva, V., Cherninkova, S. et al. (2006). Axonal neuropathy with optic atrophy is caused by mutations in mitofusin 2. *Ann. Neurol.* **59**, 276–281.

4

Neuronal Channels and Receptors

Alan L. Goldin

I. Introduction

The propagation of impulses in electrically excitable cells in the nervous system is determined by the coordinated function of many different ion channels. The channels are opened and closed by either voltage or ligands, and different types of channels allow the passage of different populations of cations or anions. The critical importance of ion channels to neuronal excitability is reflected in two ways with respect to neurological disease. First, mutations resulting in abnormal ion channel function can lead to a variety of neurological diseases. Because ion channels are a very diverse group, the clinical manifestations of channel dysfunction are quite variable. Second, many pharmacological agents that are used to treat neurological disorders act by modulating the function of ion channels. The purpose of this chapter is to provide an introduction to the classes of voltage-gated and ligand-gated ion channels that are most important with respect to neurological disease.

Ion channels can be divided into two major categories, those gated by voltage and those gated by ligands (Hille, 2001). The voltage-gated ion channels are responsible for the shape and propagation of action potentials, as well as modulation of resting membrane potential and excitability, whereas the ligand gated ion channels function in the transmission of impulses across the synaptic cleft. There is a large family of voltage-gated cation channels that includes those selective for specific cations such as sodium, potassium, and calcium (Yu et al., 2005). These channels share many functional and structural characteristics, including the fact that they are opened by membrane depolarization. Other members of the same channel family are not selective for a single cation and are only weakly dependent on voltage, instead being gated primarily by ligand binding. These members include cyclic nucleotide-gated (CNG) channels and hyperpolarization-activated cyclic nucleotide-gated (HCN) channels.

A completely different class of voltage-gated channels are those that are selective for anions such as chloride. Ligand-gated ion channels include receptors for acetylcholine (ACh), serotonin (5HT), glutamate (Glu), γ-amino-butyric acid (GABA), and glycine (Gly). These receptor channels are present in post-synaptic membranes and are responsible for the transmission of excitatory (ACh, 5HT, Glu) or inhibitory (GABA, Gly) impulses across the synapse. They are distinguished from ligand-gated receptors, which act through second messenger systems and have completely different structures and functional activities, even though members of both groups are activated by the same sets of ligands. Mutations in the genes encoding many voltage-gated and ligand-gated ion channels have been identified as causing human neurological diseases.

II. Nomenclature

Ion channels and receptors are named in two independent ways, referring either to the gene symbol or the protein that is expressed. The genes encoding the channels are referred to by systematic mammalian gene symbols developed by the Human Genome Organization Gene Nomenclature Committee (HGNC) (www.gene.ucl.ac.uk/nomenclature). An advantage of gene symbols is that they designate orthologous genes in all mammalian systems. A disadvantage is that they do not reflect structural or phylogenetic relationships of the proteins. To address this limitation, separate nomenclatures that refer to the actual channel proteins have been adopted. There is no universal system for naming channels, but a systematic nomenclature has been adopted for members of the voltage-gated cation gene family (www.iuphar-db.org/iuphar-ic/index.html). This nomenclature is based on phylogenetic and structural relationships among the channel proteins. A major source of confusion is that the channel nomenclature does not directly correlate with the gene symbols, which can be problematic because geneticists generally use the gene symbols, whereas physiologists generally use the channel names. Both types of nomenclature and their relationships will be presented in this chapter. Every channel and receptor has also been assigned a unique systematic receptor code by the International Union of Pharmacology (IUPHAR). Because this code rarely is used in publications describing ion channels or receptors, it has not been included in this chapter, but the receptor codes can be obtained from the IUPHAR Web site (www.iuphar-db.org/code/ReceptorCode1.pdf).

A. Voltage-Gated Ion Channels

Voltage-gated cation channels have been classified into families based on evolutionary relationships, with the names assigned in numerical order. The name consists of the chemical symbol of the principal permeating ion (Na^+, K^+, or Ca^{2+}) with the principal physiological regulator or other determinant of channel function indicated as a subscript. For example, voltage-gated sodium and potassium channels are labeled Na_v and K_v, respectively, whereas calcium-activated and inward rectifier potassium channels are labeled K_{Ca} and K_{ir}. The number following the subscript indicates the gene subfamily (e.g., K_v1, K_v2, etc.), and the number following the decimal point identifies the specific channel isoform (e.g., K_v1.1, K_v1.2, etc.). Splice variants of each family member are identified by lowercase letters following the numbers (e.g., K_v1.1a).

The primary subunit of the voltage-gated sodium channel is the α subunit, which is sufficient to form a fully functional channel. There are nine different isoforms of the α subunit, and these are classified as members of a single family called Na_v1 (see Table 4.1) (Catterall et al., 2005a). The genes are labeled *SCN1A* through *SCN11A*, with *SCN6A* and *SCN7A* missing from this list because they refer to a gene that encodes a sodium channel that is not gated by voltage (Na_x). The relationship between the gene symbols and channel names is shown in Table 4.1. There are also four sodium channel genes encoding accessory β subunits. These genes are termed *SCN1B* through *SCN4B*, and they encode subunits termed β1 through β4.

The primary subunit of the voltage-gated calcium channel is the α_1 subunit, which includes the conduction pore and gating mechanism. There are 10 isoforms of the α_1 subunit, and these are divided into three families, Ca_v1 through Ca_v3 (see Table 4.2) (Catterall et al., 2005b). There are four members of the Ca_v1 family and three members of each of the Ca_v2 and Ca_v3 families. The genes encoding the α_1 subunits are labeled *CACNA1A* through

Table 4.1 Voltage-Gated Sodium Channels

Channel Name	Gene Symbol	Subunit	Disease Syndrome
Na_v1.1	*SCN1A*	α	Generalized Epilepsy with Febrile Seizures Plus Intractable Childhood Epilepsy with Generalized Tonic-Clonic Seizures Severe Myoclonic Epilepsy of Infancy
Na_v1.2	*SCN2A*	α	Benign Familial Neonatal-Infantile Seizures
Na_v1.3	*SCN3A*	α	
Na_v1.4	*SCN4A*	α	Hyperkalemic Periodic Paralysis Hypokalemic Periodic Paralysis Paramyotonia Congenita Potassium Aggravated Myotonia
Na_v1.5	*SCN5A*	α	Long QT Syndrome Type 3 Brugada Syndrome
Na_v1.6	*SCN8A*	α	
Na_v1.7	*SCN9A*	α	Familial Erythromelalgia (Erythermalgia)
Na_v1.8	*SCN10A*	α	
Na_v1.9	*SCN11A*	α	
	SCN1B	β1	Generalized Epilepsy with Febrile Seizures Plus
	SCN2B	β2	
	SCN3B	β3	
	SCN4B	β4	

Table 4.2 Voltage-Gated Calcium Channels

Channel Name	Gene Symbol	Subunit	Current	Disease Syndrome
$Ca_v1.1$	*CACNA1S*	α_{1S}	L	Hypokalemic Periodic Paralysis Malignant Hyperthermia
$Ca_v1.2$	*CACNA1C*	α_{1C}	L	Timothy Syndrome
$Ca_v1.3$	*CACNA1D*	α_{1D}	L	
$Ca_v1.4$	*CACNA1F*	α_{1F}	L	X-Linked Congenital Stationary Night Blindness Type 2
$Ca_v2.1$	*CACNA1A*	α_{1A}	N	Familial Hemiplegic Migraine Episodic Ataxia Type 2 Spinocerebellar Ataxia Type 6 Episodic and Progressive Ataxia Absence Epilepsy with Ataxia
$Ca_v2.2$	*CACNA1B*	α_{1B}	P/Q	
$Ca_v2.3$	*CACNA1E*	α_{1E}	R	
$Ca_v3.1$	*CACNA1G*	α_{1G}	T	
$Ca_v3.2$	*CACNA1H*	α_{1H}	T	Childhood Absence Epilepsy
$Ca_v3.3$	*CACNA1I*	α_{1I}	T	
	CACNB1	β_1		
	CACNB2	β_2		
	CACNB3	β_3		
	CACNB4	β_4		Juvenile Myoclonic Epilepsy
	CACNA2D1	$\alpha_2\delta_1$		
	CACNA2D2	$\alpha_2\delta_2$		
	CACNA2D3	$\alpha_2\delta_3$		
	CACNA2D4	$\alpha_2\delta_4$		
	CACNG1	γ_1		
	CACNG2	γ_2		
	CACNG3	γ_3		
	CACNG4	γ_4		
	CACNG5	γ_5		
	CACNG6	γ_6		
	CACNG7	γ_7		
	CACNG8	γ_8		

CACNA1I and *CACNA1S*, with the final letter referring to the α_1 subunit type (a through i and s). An additional nomenclature that is used with calcium channels refers to the functional properties of the current, based on kinetics and sensitivity to different pharmacological blockers. This terminology includes high voltage-activated dihydropyridine-sensitive channels, referred to as L-type (Ca_v1); high voltage-activated dihydropyridine-insensitive channels, referred to as N-type ($Ca_v2.1$), P/Q-type ($Ca_v2.2$), and R-type ($Ca_v2.3$); and low voltage-activated channels, referred to as T-type (Ca_v3). Calcium channels contain three accessory subunits in addition to the α_1 subunit. These subunits are called β (1–4), $\alpha_2\delta$ (1–4), and γ (1–8) (see Table 4.2). The genes encoding the accessory subunits are labeled *CACNB1* through *CACNB4* (β subunit), *CACNA2D1* through *CACNA2D4* ($\alpha_2\delta$ subunit), and *CACNG1* through *CACNG8* (γ subunit).

The potassium channel gene family is the largest and most diverse group of ion channels and includes the voltage-gated potassium channels (K_v), which are structurally similar to the voltage-gated sodium and calcium channels. In addition to the voltage-activated channels, there are potassium channels that are activated by calcium (K_{Ca}), which function as inward rectifiers (K_{ir}), and which have two pores (K_{2P}). These types of potassium channels will not be discussed in this chapter. Similar to the situation with the voltage-gated sodium and calcium channels, the voltage-gated potassium channels have been divided into 12 families termed K_v1 through K_v12, each of which contains multiple subtypes (see Table 4.3) (Gutman et al., 2005). The genes encoding these channels are labeled *KCNA* through *KCND*, *KCNF* through *KCNH*, *KCNQ*, *KCNV*, and *KCNS*, with different subtypes of each gene indicated by numbers (e.g., *KCNA1* through *KCNA7*). The correspondence between channel names and gene symbols is shown in Table 4.3.

The different α subunit families associate with different types of accessory subunits. Members of the K_v1 family associate with one of two intracellular $K_v\beta$ subunits that serve to inactivate the channel. These subunits are termed $K_v\beta1$ and $K_v\beta2$, and are encoded by the genes *KCNAB1* and *KCNAB2*. K_v4 channels interact with one of four K Channel Interacting Proteins (KChIP) that enhance expression and modulate function. These subunits are termed KChIP1 through KChIP4, and are encoded by the genes *KCNIP1* through *KCNIP4*. Members of the K_v3, K_v4, K_v7, K_v10, and K_v11 families associate with one of four membrane-spanning MinK or MinK-Related Peptides (MiRP) that regulate channel function. These subunits are termed MinK and MiRP1 through MiRP3 and are encoded by the genes *KCNE1* through *KCNE4*.

The other two types of voltage-gated cation channels that will be discussed in this chapter have been assigned names that correspond closely with the HGNC gene symbols. Cyclic nucleotide-gated channels are cation selective channels that are gated by both voltage and cyclic nucleotides. They consist of two different subfamilies (see Table 4.4) (Hofmann et al., 2005). The true cyclic nucleotide-gated channels (CNG) are activated by depolarization, similar to the voltage-gated sodium, potassium, and calcium channels. There are two types of homologous subunits termed A and B, with four A subunits (CNGA1 through CNGA4) and two B subunits (CNGB1 and CNBG3). In this case, the gene symbols are the same as the channel names (*CNGA1* through *CNGA4*, *CNGB1* and *CNGB3*). The other subfamily of the CNG channels are termed Hyperpolarization-activated Cyclic

Table 4.3 Voltage-Gated Potassium Channels

Channel Name	Gene Symbol	Subunit	Disease Syndrome
K_v1.1	*KCNA1*	α	Episodic Ataxia Type 1
K_v1.2	*KCNA2*	α	
K_v1.3	*KCNA3*	α	
K_v1.4	*KCNA4*	α	
K_v1.5	*KCNA5*	α	
K_v1.6	*KCNA6*	α	
K_v1.7	*KCNA7*	α	
K_v1.8	*KCNA10*	α	
K_v2.1	*KCNB1*	α	
K_v2.2	*KCNB2*	α	
K_v3.1	*KCNC1*	α	
K_v3.2	*KCNC2*	α	
K_v3.3	*KCNC3*	α	
K_v3.4	*KCNC4*	α	
K_v4.1	*KCND1*	α	
K_v4.2	*KCND2*	α	
K_v4.3	*KCND3*	α	
K_v5.1	*KCNF1*	α	
K_v6.1	*KCNG1*	α	
K_v6.2	*KCNG2*	α	
K_v6.3	*KCNG3*	α	
K_v6.4	*KCNG4*	α	
K_v7.1	*KCNQ1*	α	Romano-Ward Syndrome Jervell and Lange-Nielsen Syndrome
K_v7.2	*KCNQ2*	α	Benign Familial Neonatal Seizures
K_v7.3	*KCNQ3*	α	Benign Familial Neonatal Seizures
K_v7.4	*KCNQ4*	α	Autosomal Dominant Nonsyndromic Deafness Type 2
K_v7.5	*KCNQ5*	α	
K_v8.1	*KCNV1*	α	
K_v8.2	*KCNV2*	α	
K_v9.1	*KCNS1*	α	
K_v9.2	*KCNS2*	α	
K_v9.3	*KCNS3*	α	
K_v10.1	*KCNH1*	α	
K_v10.2	*KCNH5*	α	
K_v11.1	*KCNH2*	α	Autosomal Dominant Long QT Syndrome
K_v11.2	*KCNH6*	α	
K_v11.3	*KCNH7*	α	
K_v12.1	*KCNH8*	α	
K_v12.2	*KCNH3*	α	
K_v12.3	*KCNH4*	α	
	KCNAB1	$K_v\beta1$	
	KCNAB2	$K_v\beta2$	
	KCNIP1	KChIP1	
	KCNIP2	KChIP2	
	KCNIP3	KChIP3	
	KCNIP4	KChIP4	
	KCNE1	MinK	
	KCNE2	MiRP1	
	KCNE3	MiRP2	Hyperkalemic Periodic Paralysis Hypokalemic Periodic Paralysis
	KCNE4	MiRP3	

Nucleotide-gated (HCN) channels. As the name implies, these channels are gated by hyperpolarization rather than depolarization. There are four types labeled HCN1 through HCN4, and the gene symbols correspond with the channel names. The final channel family consists of voltage-gated chloride channels, which are activated by depolarization but permeable to anions rather than cations. These channels consist of a single subunit, and there are seven types termed CLC-1 through CLC-7 (see Table 4.5) (Chen, 2005). The channels are encoded by genes termed *CLCN1* through *CLCN7*.

Table 4.4 Cyclic Nucleotide-Gated Channels

Channel Name	Gene Symbol	Subunit	Disease Syndrome
CNGA1	*CNGA1*	A1	Autosomal Recessive Retinitis Pigmentosa
CNGA2	*CNGA2*	A2	
CNGA3	*CNGA3*	A3	Achromatopsia and Retinal Degeneration
CNGA4	*CNGA4*	A4	
CNGB1	*CNGB1*	B1	Recessive Retinitis Pigmentosa
CNGB3	*CNGB3*	B3	Achromatopsia (Pingelapese blindness)
HCN1	*HCN1*		
HCN2	*HCN2*		
HCN3	*HCN3*		
HCN4	*HCN4*		Sick Sinus Node Disease

Table 4.5 Voltage-Gated Chloride Channels

Channel Name	Gene Symbol	Disease Syndrome
CLC-1	*CLCN1*	Myotonia Congenita (Becker's Autosomal Recessive) Myotonia Congenita (Thomsen's Autosomal Dominant)
CLC-2	*CLCN2*	Childhood Absence Epilepsy Epilepsy with Grand Mal Seizures on Awakening Idiopathic Generalized Epilepsy Juvenile Absence Epilepsy Juvenile Myoclonic Epilepsy
CLC-3	*CLCN3*	
CLC-4	*CLCN4*	
CLC-5	*CLCN5*	
CLC-6	*CLCN6*	
CLC-7	*CLCN7*	

B. Ligand-Gated Ion Channels

There are five classes of ligand-gated ion channels that will be discussed in this chapter, and there is no standardized nomenclature other than the IUPHAR receptor code. On the other hand, the names that generally are used correspond very well with the standardized HGNC gene symbols. ACh receptors include both ion channels and receptors that function through second messenger systems. The muscarinic ACh receptors comprise the second messenger receptors and will not be discussed in this chapter. The nicotinic ACh receptors comprise the ligand-gated channels and are pentamers consisting of various combinations of different subunits. There are multiple types of subunits, but the primary ones are α (1 through 10) and β (1 through 4) (see Table 4.6) (Gotti & Clementi, 2004). There are also single forms of the δ, ε, and γ subunits. The subunits are encoded by genes labeled *CHRNA1* through *CHRNA10* (α), *CHRNB1* through *CHRNB4* (β), and *CHRND* (δ), *CHRNE* (ε), and *CHRNG* (γ).

Serotonin (5HT) receptors include both ligand-gated ion channels and second messenger receptors. The only 5HT receptors that function as ion channels are the members of the 5-HT_3 subfamily. Within this group, there are three types termed 5-HT_{3A}, 5-HT_{3b}, and 5-HT_{3C} (see Table 4.7) (Hoyer et al., 2005). The subunits are encoded by genes labeled *HTR3A*, *HTR3B*, and *HTR3C*.

GABA receptors also include both ligand-gated ion channels and second messenger receptors. In this case, the $GABA_B$ receptors act through second messenger systems and will not be discussed in this chapter, whereas the $GABA_A$ receptors comprise the ligand-gated ion channels. $GABA_A$ receptors are similar to ACh receptors in that they are pentamers consisting of various combinations of different subunits, but there are even more different types of subunits than for ACh receptors (see Table 4.8) (Darlison et al., 2005). The primary subunits are α (1 through 6), β (1 through 3), γ (1 through 3), and δ. Additional subunits that are expressed in more limited regions include ε, π, θ, and ρ (1 through 3). The subunits are encoded by genes labeled *GABRA1* through *GABRA6* (α); *GABRB1* through *BAGRB3* (β);*GABRG1* through *GABRG3* (γ), *GABRD* (δ), *GABRE* (ε), *GABRP* (π), *GABRQ* (θ); and *GABRR1* through *GABRR3* (ρ).

Glycine receptors are also pentamers, but they are formed from only two different types of subunits, α (1 through 4) and β (see Table 4.9) (Lynch, 2004). These are encoded by the genes *GLRA1* through *GLRA3* (α) and *GLRB* (β).

Glutamate receptors have a more complex nomenclature than the other ligand-gated receptors because there are four different families (see Table 4.10) (Mayer, 2005). These families are named by the ligand that is most active against each type,

Table 4.6 Nicotinic Acetylcholine Receptor Channels

Subunit	Gene Symbol	Disease Syndrome
α1	*CHRNA1*	
α2	*CHRNA2*	
α3	*CHRNA3*	
α4	*CHRNA4*	Autosomal Dominant Nocturnal Frontal-Lobe Epilepsy
α5	*CHRNA5*	
α6	*CHRNA6*	
α7	*CHRNA7*	
α8	*CHRNA8*	
α9	*CHRNA9*	
α10	*CHRNA10*	
β1	*CHRNB1*	
β2	*CHRNB2*	Autosomal Dominant Nocturnal Frontal-Lobe Epilepsy
β3	*CHRNB3*	
β4	*CHRNB4*	
δ	*CHRND*	
ε	*CHRNE*	
γ	*CHRNG*	

Table 4.7 Serotonin Receptor Channels[1]

Subunit	Gene Symbol
5-HT_{3A}	*HTR3A*
5-HT_{3B}	*HTR3B*
5-HT_{3C}	*HTR3C*

[1]No disease syndromes resulting from mutations in 5-HT_3 receptor channel genes have been identified yet.

Table 4.8 $GABA_A$ Receptor Channels

Subunit	Gene Symbol	Disease Syndrome
α1	*GABRA1*	Autosomal Dominant Juvenile Myoclonic Epilepsy
α2	*GABRA2*	
α3	*GABRA3*	
α4	*GABRA4*	
α5	*GABRA5*	
α6	*GABRA6*	
β1	*GABRB1*	
β2	*GABRB2*	
β3	*GABRB3*	
γ1	*GABRG1*	
γ2	*GABRG2*	Childhood Absence Epilepsy and Febrile Seizures Generalized Epilepsy with Febrile Seizures Plus
γ3	*GABRG3*	
δ	*GABRD*	Generalized Epilepsy with Febrile Seizures Plus
ε	*GABRE*	
π	*GABRP*	
θ	*GABRQ*	
ρ1	*GABRR1*	
ρ2	*GABRR2*	
ρ3	*GABRR3*	

Table 4.9 Glycine Receptor Channels

Subunit	Gene Symbol	Disease Syndrome
α1	*GLRA1*	Dominant Hereditary Hyperekplexia
α2	*GLRA2*	
α3	*GLRA3*	
α4	*GLRA4*	
β	*GLRB*	

Table 4.10 Glutamate Receptor Channels[1]

Subunit	Receptor Family	Gene Symbol
GluR1	AMPA	*GRIA1*
GluR2	AMPA	*GRIA2*
GluR3	AMPA	*GRIA3*
GluR4	AMPA	*GRIA4*
GluR5	Kainate	*GRIK1*
GluR6	Kainate	*GRIK2*
GluR7	Kainate	*GRIK3*
KA-1	Kainate	*GRIK4*
KA-2	Kainate	*GRIK5*
NR1	NMDA	*GRIN1*
NR2A	NMDA	*GRIN2A*
NR2B	NMDA	*GRIN2B*
NR2C	NMDA	*GRIN2C*
NR2D	NMDA	*GRIN2D*
NR3A	NMDA	*GRIN3A*
NR3B	NMDA	*GRIN3B*
δ1	Orphan	*GRID1*
δ2	Orphan	*GRID2*

[1]No disease syndromes resulting from mutations in glutamate receptor channel genes have been identified yet.

and consist of AMPA (α-amino-3-hydroxy-5-methyl-4-isoxazole propionic acid), Kainate, NMDA (N-methyl-D-aspartic acid), and orphan δ receptors for which the primary ligand is unknown. The four types of AMPA receptor subunits are termed GluR1 through GluR4. There are five Kainate receptor subunits, and these are termed GluR5, GluR6, GluR7, KA-1, and KA-2. There are seven different NMDA receptor subunits divided into three subgroups, NR1, NR2 (A through D), and NR3 (A and B). Finally, the two orphan receptor subunits are termed δ1 and δ2. The subunits are encoded by genes labeled *GRIA1* through *GRIA4* (AMPA); *GRIK1* through *GRIK5* (Kainate); *GRIN1*, *GRIN2A* through *GRIN2B*, *GRIN3A* and *GRIN3B* (NMDA); and *GRID1* and *GRID2* (δ).

III. Structure and Function

A. Voltage-Gated Ion Channels

Voltage-gated sodium, calcium, and potassium channels and CNG channels share many similarities and are members of a single superfamily of voltage-gated cation channels (Catterall et al., 2002; Goldin, 2002; Hille, 2001; Yu et al., 2005). The primary, pore-forming α subunit of each channel consists of four homologous domains termed I–IV, with each domain containing six transmembrane segments called S1–S6 and a hairpin-like loop between S5 and S6 that forms part of the channel pore (see Figure 4.1). Potassium and CNG channels consist of tetramers of single-domain α subunits (see Figure 4.2), whereas sodium and calcium channels contain four homologous domains within a single α subunit (see Figure 4.1). The primary α subunit is associated with different accessory subunits for each channel type.

Voltage-gated sodium channels consist of a pore-forming α subunit that is associated in the CNS with two of four accessory subunits termed β1, β2, β3, and β4 (see Figure 4.1) (Catterall et al., 2005a; Goldin, 2001; Isom, 2001). Four of the α subunit isoforms are expressed at high levels in the CNS ($Na_v1.1$, $Na_v1.2$, $Na_v1.3$, and $Na_v1.6$) and four are expressed at high levels in the PNS ($Na_v1.6$, $Na_v1.7$, $Na_v1.8$, and $Na_v1.9$). The other two isoforms are expressed predominantly in skeletal muscle ($Na_v1.4$) and cardiac muscle ($Na_v1.5$). The β2 or β4 subunit is covalently linked to the α subunit by a disulfide bond, and the β1 or β3 subunit is noncovalently attached. The β subunits are expressed in a complementary fashion, so that α subunits are associated with either β1 or β3, and β2 or β4. Although the sequences of the sodium channels are similar enough so that there are no distinct subfamilies, some of the isoforms are more closely related to each other based on phylogeny and chromosomal localization. The genes for four isoforms ($Na_v1.1$, $Na_v1.2$, $Na_v1.3$, and $Na_v1.7$) are closely related to each other in the phylogenetic tree and are located in the same region of chromosome 2. Another three isoforms that often are referred to as tetrodotoxin-resistant channels ($Na_v1.5$, $Na_v1.8$, and $Na_v1.9$) are closely related in the phylogenetic tree, and their genes are located in one region of chromosome 3. These channels are blocked by micromolar concentrations of tetrodotoxin, in contrast to the other sodium channel isoforms that are blocked by nanomolar concentrations of tetrodotoxin. The genes for the final two isoforms ($Na_v1.4$ and $Na_v1.6$) are located on two different chromosomes, and each of these can be considered a separate group.

Voltage-gated calcium channels contain a pore-forming subunit termed α_1 that is associated with an intracellular β subunit, a disulfide-linked $\alpha_2\delta$ subunit, and a γ subunit in some tissues (see Figure 4.2) (Catterall et al., 2005b). The primary functional properties of the channel are determined by the α_1 subunit, with the accessory subunits serving to modulate the channel. Isoforms in the Ca_v2 and Ca_v3 families are expressed primarily in neurons, whereas isoforms in the Ca_v1 family are expressed in a variety of tissues including nerve ($Ca_v1.2$, $Ca_v1.3$), skeletal muscle ($Ca_v1.1$), cardiac myocytes ($Ca_v1.2$), endocrine cells ($Ca_v1.2$, $Ca_v1.3$), and the retina ($Ca_v1.4$).

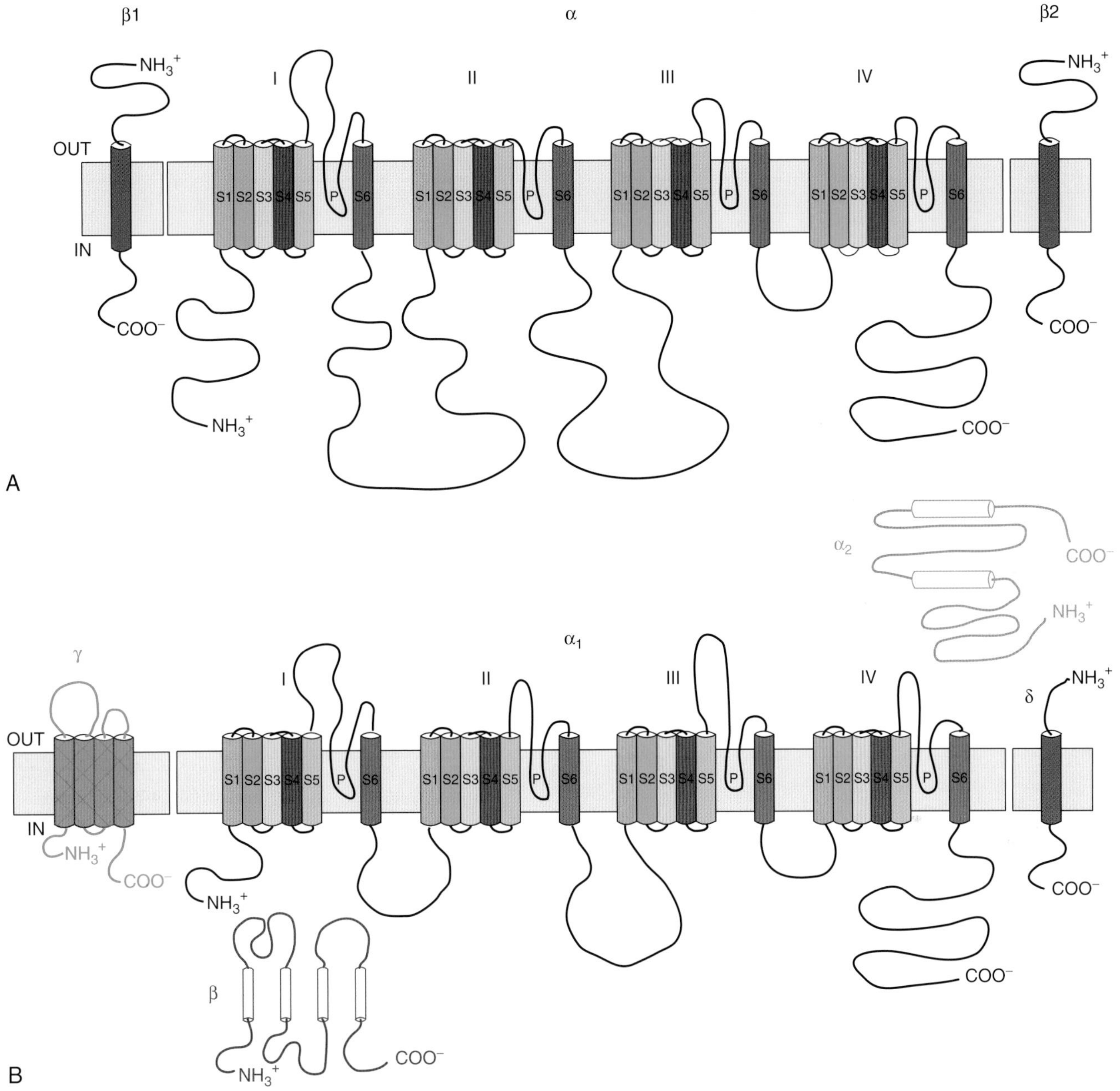

Figure 4.1 Diagram of the four domain voltage-gated ion channels. The primary pore-forming subunit of the channel consists of four homologous domains labeled I–IV, with six transmembrane spanning segments termed S1–S6 in each domain. The P region between S5 and S6 in each domain forms part of the channel pore. **A.** Voltage-gated sodium channels in the CNS contain a primary α subunit that is associated with two β subunits, shown in this figure as β1 and β2. The β1 subunit is noncovalently attached to the α subunit and the β2 subunit is covalently attached via a disulfide linkage. All the β subunits have a similar structure that consists of a small, carboxy-terminal cytoplasmic region, a transmembrane spanning region, and a larger, external amino-terminal region that contains immunoglobulin-like domains. The α subunit includes the channel pore and gating machinery, and the β subunits modulate the properties of the channel complex. The diagram was modified from Goldin (2003). **B.** The voltage-gated calcium channel α_1 subunit is comparable to the α subunit of the voltage-gated sodium channel and includes the channel pore and gating machinery. Most calcium channels also contain an intracellular β subunit (red), an extracellular α_2 subunit (blue) that is attached to a membrane-spanning δ subunit (black) by a disulfide linkage, and a membrane-spanning γ or related subunit (green). The accessory subunits modulate the properties of the channel complex. The diagram was modified from Catterall et al. (2002).

Potassium channels are the most diverse group of channels, both in terms of numbers of families and functional properties, with 12 different types of α subunits (Gutman et al., 2005). All of the α subunits have a comparable structure (see Figure 4.2). Additional diversity is created through association with different accessory subunits that modulate channel function. K_v1 channels often are associated with accessory $K_v\beta$ subunits that directly inactivate the channel. K_v4 channels associate with

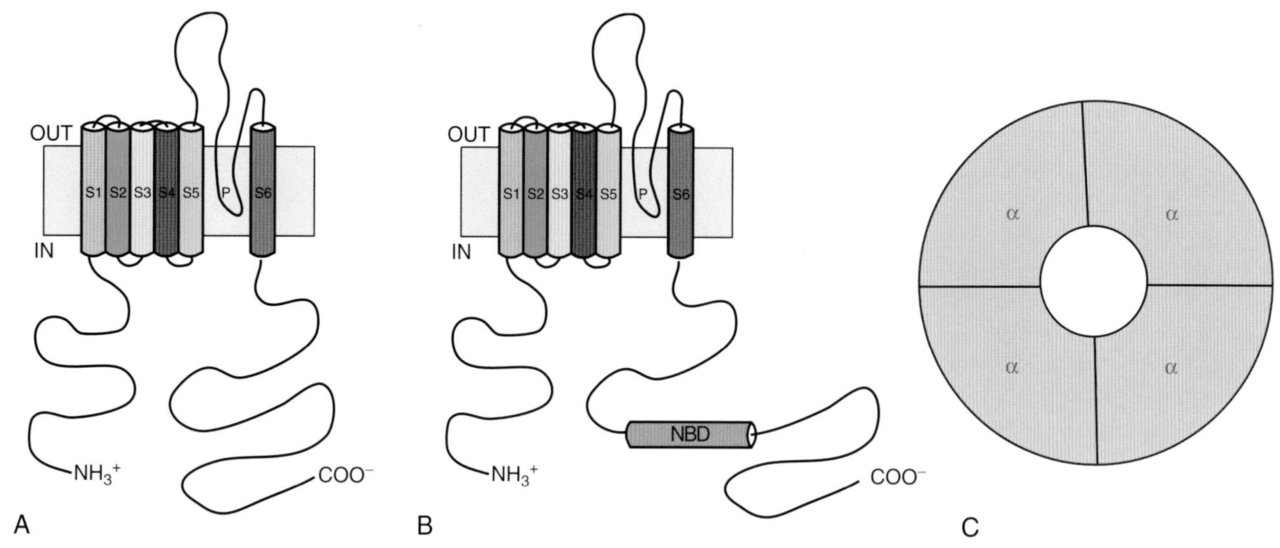

Figure 4.2 Diagram of the single domain voltage-gated cation channels. **A.** Voltage-gated potassium channels and HCN channels contain a pore-forming subunit that is comparable to one domain of the voltage-gated sodium and calcium channels. The monomer contains six transmembrane spanning segments termed S1–S6 with a region between S5 and S6 that forms part of the pore (P). Voltage-gated potassium channels associate with one of a variety of accessory cytoplasmic subunits, which are not shown in this diagram. A detailed crystal structure of the rat $K_v1.2$ voltage-gated potassium channel in association with the rat $K_v\beta2$ subunit has been determined (Long et al., 2005). **B.** CNG channel monomers have a similar structure except that there is a nucleotide binding domain (NBD) in the cytoplasmic carboxy-terminal region. **C.** Four monomers combine to form a functional channel and surround a central pore.

KChIPs, which increase the density of current, slow inactivation and accelerate recovery from inactivation (Rhodes et al., 2004). K_v7, K_v10, and K_v11 associate with MiRP subunits, which may also associate with K_v3 and K_v4 channels. The MiRP subunits modify gating, conductance, and pharmacology of the channels (McCrossan & Abbott, 2004).

CNG channels are activated by direct binding of cyclic nucleotides, including cGMP and cAMP (Hofmann et al., 2005). They are heterotetramers of A and B subunits, both of which have structures that are comparable to those of potassium channels except that the carboxy-terminal region contains a nucleotide-binding domain (see Figure 4.2). This domain is responsible for the modulation by cyclic nucleotides. CNG channels are expressed at high levels in olfactory neurons and photoreceptors, and at lower levels in many other tissues including the CNS and heart. HCN channels are members of the same gene family, but they open during hyperpolarization (Robinson & Siegelbaum, 2003). The gating is enhanced by binding of cAMP or cGMP, which shifts activation to more positive potentials. These channels serve critical roles as pacemakers in cardiac cells and some neurons. The channels form tetramers that can most likely be heteromeric.

Voltage-gated chloride channels are members of the CLC gene family (Chen, 2005; Jentsch et al., 2005). These channels consist of two identical subunits, each containing 18 α helices and an ion permeation pathway (see Figure 4.3), so that two chloride ions can pass through the channel independently. Chloride channels are present in both the cell membrane and in the membranes of intracellular organelles. They are gated by voltage and modulated by a variety of factors including anions, calcium, swelling, and phosphorylation.

B. Ligand-Gated Ion Channels

The nicotinic ACh, 5-HT, $GABA_A$, and Gly receptors are members of a large family of ligand-gated ion channels (Connolly & Wafford, 2004; Lester et al., 2004). They are composed of pentamers of different subunits, with each subunit having a comparable structure consisting of four transmembrane segments and a large extracellular amino terminus that contains the ligand binding site (see Figure 4.4). The nicotinic ACh receptor complexes are composed of two different subunits (α and β), and can consist either entirely of α subunits or of heteromers containing 2 α and 3 β subunits (Gotti & Clementi, 2004). The α1 and β1 subunits are present only in muscle nicotinic ACh receptors. The receptor forms a channel that is permeable to cations, including sodium, calcium, and potassium, so that when opened it depolarizes the membrane and initiates the firing of action potentials.

Although there is a large family of 5-HT receptors that are G-protein coupled receptors acting through a variety of second messenger systems, there is only one class that represents ligand-gated ion channels (5-HT_3 receptors) (Hoyer et al., 2005). The 5-HT_3 receptors are members of the ligand-gated ion channel receptor family, so the structure is similar to that of nicotinic ACh receptors (see Figure 4.4). The three different subunits (5-HT_{3A}, 5-HT_{3B}, and 5-HT_{3C}) co-assemble to form heteromers. 5-HT receptors are found on neurons in both the

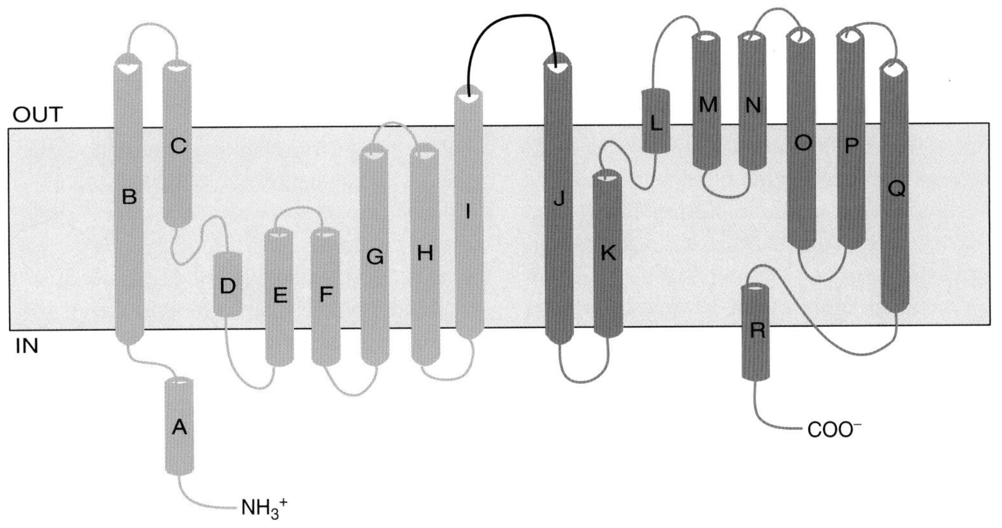

Figure 4.3 Diagram of one subunit of the voltage-gated CLC chloride channel. A subunit consists of 18 α helical segments labeled A–R with the amino-terminal half (green) being homologous to the carboxy-terminal half (blue), but in the opposite orientation in the membrane. The two halves surround a common center that forms the selectivity filter. The functional channel consists of two identical subunits and contains two independent pores through which chloride ions permeate. The diagram was modified from Dutzler et al. (2002), which also shows the detailed crystal structure of a prokaryotic CLC chloride channel.

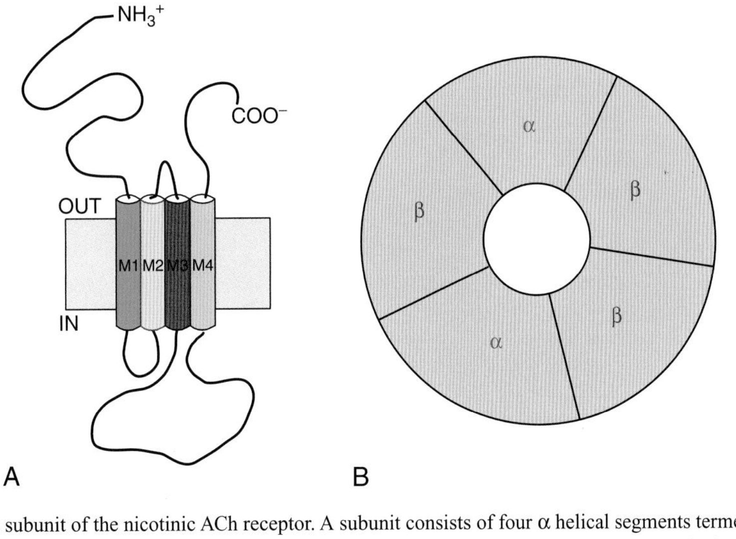

Figure 4.4 **A.** Diagram of one subunit of the nicotinic ACh receptor. A subunit consists of four α helical segments termed M1–M4, with the amino- and carboxy-terminal regions on the outside of the membrane. The ACh binding site is contained within the amino-terminal region. **B.** Five subunits combine to form a functional receptor, which can be either monomeric (all α subunits) or heteromeric (2 α and 3 β subunits). The subunits surround a central pore through which cations permeate. The predicted structures of the 5-HT_3, $GABA_A$, and Gly receptors are comparable, with each ligand binding site in the amino-terminal region of the receptor. The 5-HT_3 receptor consists of a pentameric combination that is always heteromeric and is also permeable to cations. The $GABA_A$ receptor consists of a pentamer, with the most common composition being 2 α, 2 β, and 1 γ subunit, although channels can alternatively contain a variety of other subunits. The functional Gly receptor consists of a pentamer of α and β subunits. The $GABA_A$ and Gly subunits each surround a central pore through which chloride ions permeate. The diagram was modified from Gotti and Clementi (2004).

CNS and PNS. The channel is a nonselective cation channel, permeable to sodium, potassium, and calcium, so that activation of the receptor leads to membrane depolarization.

The $GABA_A$ receptor also has a similar pentameric structure containing combinations of different subunits including α, β, γ, δ, ε, π, θ, and ρ, with most receptors consisting of α, β, and γ or α, β, and δ subunits (see Figure 4.4) (Jones-Davis & MacDonald, 2003). The receptor forms a channel that is selective for anions, mainly chloride, so that when opened it hyperpolarizes the membrane and inhibits excitability.

Gly receptors are also pentamers, formed from combinations of α and β subunits (see Figure 4.4) (Lynch, 2004). The exact stoichiometry is not known, but the receptor is assumed to consist of 3 α and 2 β subunits. The channel is permeable to chloride, which results in the unusual property that these receptors can either be excitatory or inhibitory, depending on the intracellular chloride concentration. Channel activation is usually inhibitory because the chloride equilibrium potential is generally more negative than the cell resting potential. However, the intracellular chloride concentration is significantly higher in embryonic neurons compared to adult neurons because of reduced chloride efflux resulting from lower expression of the KCC2 potassium-chloride cotransporter (Rivera et al., 2005). Therefore, the equilibrium potential for chloride is more positive during development so that activation of Gly receptors leads to depolarization and excitation of immature neurons.

The Glu receptors form a separate family of ligand-gated ion channels that are formed as tetramers rather than pentamers (see Figure 4.5). There are four families of Glu receptors termed AMPA, Kainate, NMDA, and δ (Mayer, 2005). Receptors are formed as multimers from members within each type, but not between types. NMDA receptors require both NR1 and NR2 subunits to form functional channels, whereas Kainate and AMPA receptors can be formed from homomers or heteromers. The δ subunits function as a channel only in pathological conditions (Wollmuth & Sobolevsky, 2004). All these channels are nonselective cation channels, so that activation of the receptor is excitatory.

IV. Physiological Roles

A. Voltage-Gated Ion Channels

The most important and widespread function of voltage-gated sodium channels is to mediate the depolarization phase of the action potential in electrically excitable cells. Because specific isoforms have different tissue distributions, each isoform has its own unique role (Trimmer & Rhodes, 2004). For example, $Na_v1.6$ is the isoform at nodes of Ranvier, and it is responsible for action potential transmission in myelinated axons, whereas $Na_v1.2$ is the isoform responsible for action potential transmission in unmyelinated axons. The other two CNS isoforms are present at high levels in neuronal cell bodies ($Na_v1.1$) and during embryonic and early prenatal development ($Na_v1.3$). The $Na_v1.4$ and $Na_v1.5$ isoforms are responsible primarily for action potential transmission in skeletal and cardiac muscle, respectively. Four isoforms ($Na_v1.6$, $Na_v1.7$, $Na_v1.8$, and $Na_v1.9$) mediate action potential transmission in the PNS. Sodium channels are also responsible for unique conductances in specific cell types, in which they generate repetitive firing, persistent or resurgent current.

Voltage-gated calcium channels are responsible for a wide variety of functions in electrically excitable and nonexcitable cells, with specific isoforms carrying out unique roles (Khosravani & Zamponi, 2006). One of the primary functions in neurons is to stimulate neurotransmitter release, which is mediated by $Ca_v1.4$, $Ca_v2.1$, and $Ca_v2.2$. Other Ca_v2 isoforms are located in neuronal dendrites where they are responsible

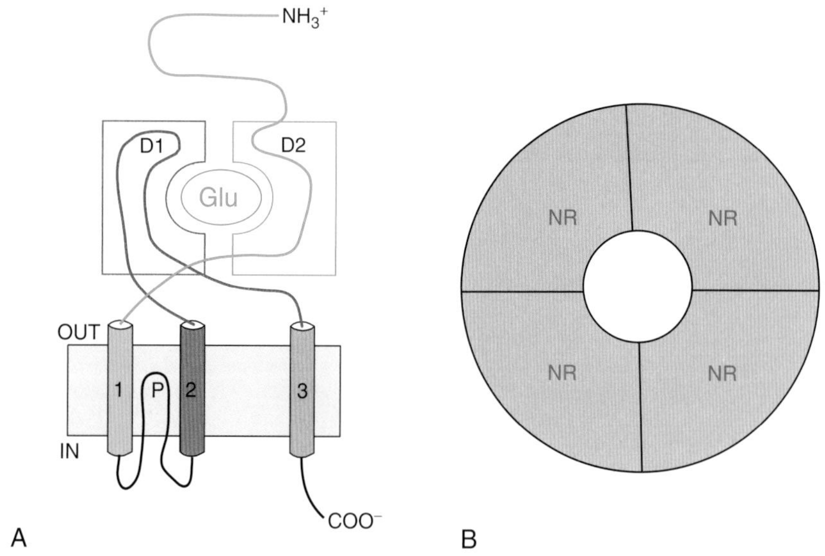

Figure 4.5 **A.** Diagram of one subunit of the Glu receptor. A subunit consists of three α helical segments termed 1–3 and a region between 1 and 2 that forms part of the pore (P). The glutamate-binding site (Glu) is comprised of two external domains (D1 and D2), with D1 located between segments 2 and 3 and D2 in the amino-terminal region. The carboxy-terminus is on the cytoplasmic side of the membrane. **B.** Four subunits combine to form a functional receptor, which can consist of either homomers or heteromers within each family (AMPA, Kainate and NMDA). There is no mixing of subunits between members of different families. The subunits surround a central pore through which cations permeate. The diagram was modified from Mayer (2005).

for transient calcium currents. Isoforms in the Ca_v3 family are important for pacemaking, repetitive firing, and shaping the action potential in cardiac muscle. The $Ca_v1.1$ and $Ca_v1.2$ isoforms play critical roles in excitation-contraction coupling in skeletal and cardiac muscle, respectively. Other important functions of calcium channel isoforms include hormone release from endocrine cells, hearing in cochlear hair cells, photoreception in retinal rod cells, and control of signaling enzymes and gene expression.

Because voltage-gated potassium channels are so diverse, it is not surprising that they play a wide variety of roles in different tissues (Gutman et al., 2005). A primary function of a number of isoforms, particularly those in the K_v1 family, is to maintain the resting membrane potential and modulate firing properties. The modulation can be quite subtle because of the wide range of potassium channel gating kinetics, so that neuronal excitability is finely tuned by the presence of different potassium channel isoforms. Other critical potassium channel functions include repolarization of the action potential in both neurons and muscle, neuronal afterhyperpolarization, and pacemaking. Some isoforms carry out specialized functions that enable high frequency firing in auditory cells and interneurons, spike-frequency adaptation, calcium signaling in lymphocytes and oligodendrocytes, and regulation of cell cycle and proliferation.

The CNG channels play a particularly important role in sensory transduction in olfactory neurons and photoreceptors, in which they are present at high concentrations (Craven & Zagotta, 2006). However, these channels are also present at lower levels in a variety of tissues including brain, heart, testis and kidney, so that it is likely that CNG channels have important physiological roles in those tissues. HCN channels are present at highest levels in two tissues, CNS neurons and cardiac muscle (Robinson & Siegelbaum, 2003). In the CNS, the channels are important for determination of the resting potential, transduction of sour taste, dendritic integration, and plasticity. In the heart, HCN channels function in determination of the resting membrane potential and in regulation of heart rate and rhythm, for which they serve as pacemaker cells.

Voltage-gated chloride channels are present in a wide variety of cells. Channels in the plasma membrane of neurons and other electrically excitable cells are important for stabilizing the resting membrane potential (Jentsch et al., 2005). Channels in the plasma membrane of epithelial and renal tissues help to regulate fluid transport and control osmotic swelling. Some chloride channel isoforms are also present in the membranes of intracellular vesicles, and those channels serve important roles in regulating volume and pH.

B. Ligand-Gated Ion Channels

The ligand-gated receptors are involved primarily in synaptic transmission. The receptors for Gly and GABA mediate inhibitory synaptic transmission and the receptors for Glu, ACh, and 5-HT mediate excitatory synaptic transmission (Connolly & Wafford, 2004; Dingledine et al., 1999). Nicotinic ACh receptors are responsible for receiving neuronal impulses across the neuromuscular junction (Hughes et al., 2006), whereas the other receptors are present on neurons. The receptors are present in varying distributions in different areas of the neuron, so that receptors on cell bodies are usually inhibitory, those on dendritic spines are primarily excitatory, and those on axon terminals function to modulate transmission. Stimulation of the receptors is integrated both temporally and spatially, so the ultimate effect is highly plastic and subject to dynamic regulation.

V. Neurological Disorders Caused by Channelopathies

Given the critical importance of ion channels to neuronal excitability, it is not surprising that abnormalities of their function lead to a variety of neurological diseases. Because ion channels are so diverse, it is also not surprising that the clinical manifestations of channel dysfunction are variable. On the other hand, similar syndromes often are caused by abnormal function of different ion channels, so that it is impossible to predict the clinical effect based on the channel abnormality. A common feature of neurological disorders caused by ion channel malfunction is that the disorders are paroxysmal, usually with long periods of normal activity punctuated by brief episodes of abnormal activity. The neurological disorders that will be introduced in this chapter result from mutations in the genes encoding either voltage-gated or ligand-gated ion channels, so that they have been called channelopathies. In all cases, the underlying abnormality is present at birth, although the clinical manifestations may not develop until later in life.

A. Epilepsy

The epilepsies are a large group of disorders characterized by abnormal electrical activity in the CNS, affecting up to 2 percent of the population (Hauser et al., 1993). Of this group, approximately 40 percent are considered idiopathic, meaning that the underlying cause is most likely a genetic abnormality (Steinlein, 2002). The specific abnormality has been identified in only a small minority of cases, most of which are dominantly inherited channelopathies characterized by defects in ion channel function (Graves, 2006; Meisler et al., 2001; Mulley et al., 2003; Scheffer & Berkovic, 2003). However, there is no simple relationship between channel type and epilepsy syndrome, so that mutations in the same gene can cause a variety of different syndromes and the same epilepsy syndrome can be caused by mutations in genes encoding different ion channels.

Many ion channel mutations that cause epilepsy have been identified in genes encoding the voltage-gated sodium channel. Mutations in two genes encoding the α subunit (*SCN1A* encoding $Na_v1.1$ and *SCN2A* encoding $Na_v1.2$) and one gene encoding the β1 subunit (*SCN1B*) cause a variety of different syndromes, including Generalized Epilepsy with Febrile Seizures Plus (GEFS+) (see Table 4.1) (Meisler & Kearney, 2005). The mechanism by which these mutations cause epilepsy is unknown, although there is a significant amount of information regarding how mutations that cause GEFS+ alter channel function. The results of those studies suggest that GEFS+ can be caused both by mutations that increase sodium channel activity and by mutations that decrease activity (Barela et al., 2006; Lossin et al., 2002, 2003; Spampanato et al., 2004). However, none of the mutations has been studied yet in neuronal cells, so the physiological relevance of the functional results is unclear.

Mutations in three $GABA_A$ receptor genes (*GABRA1* encoding α1, *GABRG2* encoding γ2, and *GABRD* encoding δ) also cause GEFS+ as well as two other epilepsy syndromes (see Table 4.8) (Macdonald et al., 2004). The functional studies concerning the effects of these mutations have been more consistent, suggesting that they all decrease channel activity (Feng et al., 2006; Harkin et al., 2002; Kang et al., 2006). Since the $GABA_A$ receptor forms a chloride channel that functions to inhibit membrane excitability, decreased activity could lead to decreased inhibition and neuronal hyperexcitability (George, Jr., 2004).

Mutations in three calcium channel subunits also cause epilepsy syndromes. Mutations in the *CACNA1H* gene encoding the $Ca_v3.2$ α_{1H} subunit cause Absence Epilepsy, mutations in the *CACNA1A* gene encoding the $Ca_v2.1$ α_{1A} subunit cause Absence Epilepsy with Ataxia, and mutations in the *CACNB4* gene encoding the accessory β_4 subunit cause Juvenile Myoclonic Epilepsy (see Table 4.2) (Khosravani & Zamponi, 2006). These mutations appear to reduce calcium currents, so that they do not directly increase action potential excitability (Imbrici et al., 2004; Khosravani et al., 2004; Vitko et al., 2005). Because calcium channels are critical mediators of neurotransmitter release, one hypothesis for the mechanism of seizure generation is that the mutations alter synaptic strength to increase synchronization. The ultimate effects vary depending on the specific synapse because different synapses use different calcium channels for neurotransmitter release (Noebels, 2003). Therefore, the effects of decreasing function of a single calcium channel isoform might specifically decrease transmission across inhibitory synapses, leading to greater synchronization and seizures.

Mutations in the *KCNQ2* and *KCNQ3* potassium genes encoding the $K_v7.2$ and $K_v7.3$ α subunits cause Benign Familial Neonatal Seizures (see Table 4.3) (Burgess, 2006; Scheffer et al., 2005). These two isoforms are delayed rectifier channels that coassemble to form the M current, which helps to determine the subthreshold excitability of neurons and to limit sustained membrane depolarization. The mutant proteins function as dominant negative subunits to reduce potassium currents, leading to prolonged membrane depolarization and seizures (George, Jr., 2004).

Mutations in the *CLCN2* gene encoding the CLC-2 chloride channel cause a variety of epilepsy syndromes (see Table 4.5) (Graves, 2006), but the mechanism by which these mutations cause epilepsy is unknown. The chloride channel is critical for maintaining the normal membrane potential, so disruption of that channel would cause membrane depolarization and hyperexcitability. Some mutations decrease activity of the channel, which is consistent with this mechanism of action. However, other mutations alter activation of the channel in a manner that should not decrease activity, so there may be multiple mechanisms by which chloride channel mutations cause epilepsy (George, Jr., 2004).

Finally, mutations in two subunits of the neuronal nicotinic ACh receptor cause a single epilepsy syndrome. Autosomal Dominant Nocturnal Frontal-Lobe Epilepsy results from mutations in the *CHRNA4* gene encoding the α4 subunit and the *CHRNB2* gene encoding the β2 subunit (see Table 4.6) (Gotti et al., 2006). All the mutations are located near the pore of the channel, although they have different effects on channel function. Some mutations inhibit receptor function either by decreasing currents or by enhancing desensitization, which closes the channel. However, other mutations increase receptor function, so the mechanism by which ACh receptor mutations cause epilepsy is unknown. One possibility is that the mutations increase the sensitivity of the receptor to ACh, enhancing neuronal excitability (Steinlein, 2004). An alternative hypothesis is that the mutations interfere with calcium modulation of the receptor, preventing a negative feedback mechanism by which glutamate receptors deplete local extracellular calcium to reduce receptor potentiation (George, Jr., 2004; Gourfinkel-An et al., 2004).

A more comprehensive discussion of the genetic basis of epilepsy is presented in Chapter 24.

B. Ataxia

Ataxia is a syndrome characterized by lack of coordination, disturbed gait, unclear speech and tremor with movement. There are numerous types of ataxia, only one of which is caused by mutations in voltage-gated ion channels. Mutations in voltage-gated potassium and calcium channels cause Episodic Ataxia Types 1 and 2 (EA-1 and EA-2) and Spinocerebellar Ataxia 6 (SCA6). These three disorders are all inherited in an autosomal dominant fashion, although the types of mutations that cause them are quite different. EA-1 results from mutations in the *KCNA1* gene encoding the $K_v1.1$ potassium channel (see Table 4.3) (Waters et al., 2006). This channel is expressed in many regions of the CNS, including the cerebellum. Most of the mutations are missense changes that affect either trafficking or function

(De Michele et al., 2004). EA-2 is caused by mutations in the *CACNA1A* gene encoding the α_{1A} subunit of the $Ca_v2.1$ calcium channel (see Table 4.2) (Wan et al., 2005). This channel is expressed at high levels in the cerebellar cortex, granule, and Purkinje cells. Most of these mutations are nonsense changes resulting in a truncated protein that is most likely nonfunctional (De Michele et al., 2004). SCA6 results from mutations in the same calcium channel gene (see Table 4.2) (Schöls et al., 2004), only in this case the mutations are expansions of the triplet sequence cytosine-adenine-guanine (CAG). Triplet expansions are the mechanism for many of the spinocerebellar ataxias, as will be discussed in Chapter 17. Relatively short expansions can cause SCA6, with as few as 20 repeats resulting in disease (normal individuals have from 4–16 triplet repeats). Longer expansions generally lead to an earlier time of onset. The triplet expansions result in a polyglutamine sequence, which causes the protein to fold incorrectly and thus become degraded (Schöls et al., 2004). Thus, both EA-2 and SCA6 result from loss of functional protein from one allele of the *CACNA1A* gene.

The ataxias caused by channelopathies result from primary cerebellar dysfunction. Since Purkinje cells are the sole output neuron from the cerebellum, the effect of these disorders is to alter the output firing of those cells. Purkinje cells have a rhythmic firing pattern and they continually fire action potentials that are inhibitory to the deep cerebellar nuclei. Decreased inhibition from the cerebellar Purkinje cells results in increased and aberrant firing from the cells in the deep cerebellar nuclei, which is the direct cause of the primary symptoms of ataxia (Orr, 2004). The ion channel mutations that cause ataxia alter this pathway either by altering channels expressed in the Purkinje cells (potassium channels in EA-1), or by altering channels expressed in cells that modulate Purkinje cell firing (calcium channels in EA-2 and SCA6). Additional symptoms in these disorders result from abnormalities in peripheral neurons, which are most likely due to aberrant ion channel function in those cells.

A more comprehensive discussion of ataxias is presented in Chapter 18.

C. Migraine

Migraine is a common headache disorder with a strong genetic component, affecting more than 10 percent of the population (Pietrobon, 2005). It can occur in two distinct forms, with and without aura, each of which is defined by the frequency of recurrent episodes that fulfill specific criteria and that cannot be attributed to another disorder. Migraine with aura involves at least two attacks and migraine without aura involves at least five attacks. Although many genes have been implicated as susceptibility loci for migraine, causative mutations have been identified for only one type, Familial Hemiplegic Migraine (FHM). FHM Type 1 is caused by mutations in the *CACNA1A* gene encoding the α_1 subunit of the $Ca_v2.1$ voltage-gated calcium channel, and FHM Type 2 is caused by mutations in the *ATP1A2* gene encoding the α_2 subunit of the Na^+,K^+-ATPase (Pietrobon, 2005; Wessmann et al., 2004).

At least 17 different mutations causing FHM1 have been identified in the *CACNA1A* gene. They are all missense mutations that are inherited in an autosomal dominant fashion, and they are located throughout the channel (Pietrobon, 2005). The functional effects of many of the mutations have been characterized after expression in heterologous systems (*Xenopus* oocytes and mammalian cell lines), in transfected neurons lacking the $Ca_v2.1$ channel, and in neurons from knock-in mice expressing one specific mutation. The effects of the mutations in heterologous cells have been variable, with alterations in numerous properties, but the one consistent finding was an increase in calcium influx that was also observed in the neurons. This observation suggests that FHM1 results from a gain-of-function of the $Ca_v2.1$ channel.

At least 17 different mutations causing FHM2 have been identified in the *ATP1A2* gene encoding the Na^+,K^+-ATPase (Pietrobon, 2005). This protein is an ion pump rather than an ion channel, and it functions to maintain the resting membrane potential of the cell. All the mutations are missense alterations that are inherited in an autosomal dominant manner, and more than half are located in a large intracellular region that is important for nucleotide binding and phosphorylation of the protein. The effects of five of the mutations have been examined, and the net effect in all cases appears to be decreased or absent function.

The pathogenesis of migraine is not understood, but the headache most likely results from activation of the trigeminovascular system (Pietrobon, 2005). Activation may result from cortical spreading depression, which is marked by a propagating wave of neuronal depolarization followed by a long-lasting neural suppression. The initiating event in this process may be an increased local concentration of potassium around the cortical neurons, leading to sustained activation. The mutations causing FHM could increase susceptibility to this effect in two ways. The *CACNA1A* mutations lead to increased calcium influx, which would lead to enhanced excitatory neurotransmitter release for a given stimulus. The *ATP1A2* mutations result in decreased transporter activity, which would reduce clearance of potassium ions from the extracellular space.

A more comprehensive discussion of ion channel mutations that cause migraine is presented in Chapter 28.

D. Pain

Voltage-gated sodium channels in sensory neurons have been implicated as playing a critical role in the initiation of pathological pain (Lai et al., 2003; Waxman & Dib-Hajj, 2005; Waxman & Hains, 2006; Wood et al., 2004). A number of sodium channel isoforms have been suggested to be involved in different pain models. The $Na_v1.8$ channel is expressed in small diameter C-type dorsal root ganglion cells, and knockout mice lacking the gene for this channel demonstrate moderate hypoalgesia in response to noxious thermal and mechanical stimuli. $Na_v1.9$ is expressed in nociceptive neurons, and it has been proposed that this channel modulates the resting potential of nociceptors because it is active at −70 mV. Although $Na_v1.3$ is expressed at highest levels in late embryonic and early postnatal stages of development, it is upregulated after nerve section or ligation in various pain models, and the activity of this channel has been suggested to cause enhanced excitability of the injured neurons.

Specific sodium channel mutations that cause pathological pain have been identified for only one syndrome, inherited erythermalgia, which is also called erythromelalgia (Waxman & Dib-Hajj, 2005). This syndrome is characterized by redness of the extremities and burning pain in response to exercise or heat. It is inherited in an autosomal dominant manner, and has been shown to be caused by missense mutations in the *SCN9A* gene encoding the $Na_v1.7$ sodium channel. The mutations alter sodium channel function, increasing the probability and duration of channel opening, resulting in hyperexcitability of the sensory neurons in which they are located.

A more comprehensive discussion of the role of ion channels in pain is presented in Chapter 27.

E. Hyperekplexia

Hyperekplexia, or startle syndrome, is characterized by an excessive reflex response to a surprising or painful stimulus (Bakker et al., 2006). The three clinical symptoms necessary for diagnosis include generalized stiffness at birth that later subsides, excessive startling that persists throughout life, and short-lasting generalized stiffness after a startle reflex. There are three groups of this disorder, but only the major category has a proven genetic basis. That syndrome is caused by mutations in the *GLRA1* gene encoding the $\alpha1$ subunit of the Gly receptor (Saul et al., 1999). The disorder usually is inherited in an autosomal dominant manner, although recessive mutations and compound heterozygosity have been reported. Both missense and nonsense mutations have been identified, with a variety of effects including loss of protein, trafficking abnormalities, and changes in ligand-binding or gating function. The origin of the abnormal startle reflex is not known and may originate from either a brainstem or cortical defect.

F. Myotonia and Periodic Paralysis

Mutations in voltage-gated ion channels expressed in skeletal muscle cause two related disorders, the periodic paralyses and the nondystrophic myotonias. Periodic paralysis is characterized by episodic attacks of flaccid weakness that often are accompanied by myotonia (delayed muscle relaxation after contraction), whereas nondystrophic myotonia is characterized by transient muscle weakness, severe myotonia, and muscle hypertrophy without paralysis (Cannon, 2002; Davies & Hanna, 2003).

Periodic paralysis is caused by mutations in genes encoding three different voltage-gated ion channels (Cannon, 2006). Mutations in the *SCN4A* gene encoding the $Na_v1.4$ sodium channel α subunit cause Hyperkalemic Periodic Paralysis (HyperPP) and Hypokalemic Periodic Paralysis (HypoPP) (see Table 4.1). The $Na_v1.4$ channel is expressed primarily in skeletal muscle and it is the only sodium channel that is highly expressed in that tissue, which explains the localization of the symptoms. All the changes are missense mutations that alter the functional properties of the channel. The same two syndromes can also be caused by mutations in the *KCNE3* gene that encodes the MiRP2 subunit of the $K_v7.1$ potassium channel (see Table 4.3). Mutations in a different potassium channel gene, *KCNJ2* encoding the $K_v2.1$ channel, cause Andersen's syndrome, in which tissues other than skeletal muscle also are affected (see Table 4.3). Mutations in the *CACNA1S* gene encoding the $Ca_v1.1$ calcium channel α_{1S} subunit cause HypoPP but not HyperPP (see Table 4.2). All these disorders are inherited in an autosomal dominant fashion.

The nondystrophic myotonias are caused by mutations in genes encoding two voltage-gated ion channels (Davies & Hanna, 2003). Mutations in the *SCN4A* sodium channel gene cause both Paramyotonia Congenita and Potassium Aggravated Myotonia. These mutations are similar to those that cause periodic paralysis, meaning that they are all missense changes that alter the functional properties of the channel. In addition, they are all inherited in an autosomal dominant fashion. There are two forms of Myotonia Congenita called Thomsen's disease, which is autosomal dominant, and Becker's generalized myotonia, which is autosomal recessive. Both forms are caused by mutations in the *CLCN1* gene encoding the CLC-1 voltage-gated chloride channel. These alterations are missense mutations that change the functional properties of the channel. There is no way to predict whether a particular mutation will cause the autosomal dominant or autosomal recessive form of myotonia.

References

Bakker, M. J., van Dijk, G., van den Maagdenberg, A. M. J. M., and Tijssen, M. A. (2006). Startle syndromes. *Lancet Neurol.* **5**, 513–524.

Barela, A. J., Waddy, S. P., Lickfett, J. G., Hunter, J., Anido, A., Helmers, S. L. et al. (2006). An epilepsy mutation in the sodium channel *SCN1A* that decreases channel excitability. *J. Neurosci.* **26**, 2714–2723.

Burgess, D. L. (2006). Neonatal epilepsy syndromes and GEFS+: Mechanistic considerations. *Epilepsia* **46**, 51–58.

Cannon, S. C. (2002). An expanding view for the molecular basis of familial periodic paralysis. *Neuromusc. Disord.* **12**, 533–543.

Cannon, S. C. (2006). Pathomechanisms in channelopathies of skeletal muscle and brain. *Annu. Rev. Neurosci.* **29**, 387–415.

Catterall, W. A., Chandy, K. G., and Gutman, G. A. (2002). *The IUPHAR compendium of voltage-gated ion channels*. IUPHAR Media, Leeds, UK.

Catterall, W. A., Goldin, A. L., and Waxman, S. G. (2005). International Union of Pharmacology. XLVII. Nomenclature and structure-function relationships of voltage-gated sodium channels. *Pharmacol. Rev.* **57**, 397–409.

Catterall, W. A., Perez-Reyes, E., Snutch, T. P., and Striessnig, J. (2005). International Union of Pharmacology. XLVIII. Nomenclature and structure-function relationships of voltage-gated calcium channels. *Pharmacol. Rev.* **57**, 411–425.

Chen, T.-Y. (2005). Structure and function of CLC channels. *Annu. Rev. Physiol.* **67**, 809–839.

Connolly, C. N. and Wafford, K. A. (2004). The cys-loop superfamily of ligand-gated ion channels: the impact of receptor structure on function. *Biochem. Soc. Trans.* **32**, 529–534.

Craven, K. B. and Zagotta, W. N. (2006). CNG and HCN channels: Two peas, one pod. *Annu. Rev. Physiol.* **68**, 375–401.

Darlison, M. G., Pahal, I., and Thode, C. (2005). Consequences of the evolution of the $GABA_A$ receptor gene family. *Cell. Mol. Neurobiol.* **25**, 607–624.

Davies, N. P. and Hanna, M. G. (2003). The skeletal muscle channelopathies: Distinct entities and overlapping syndromes. *Curr. Opin. Neurol.* **16**, 559–568.

De Michele, G., Coppola, G., Cocozza, S., and Filla, A. (2004). A pathogenetic classification of hereditary ataxias: Is the time ripe? *J. Neurol.* **251**, 913–922.

Dingledine, R., Borges, K., Bowie, D., and Traynelis, S. F. (1999). The glutamate receptor ion channels. *Pharmacol. Rev.* **51**, 7–91.

Dutzler, R., Campbell, E. B., Cadene, M., Chait, B. T., and MacKinnon, R. (2002). X-ray structure of a CLC chloride channel at 3.0 Å reveals the molecular basis of anion selectivity. *Nature* **415**, 287–294.

Feng, H.-J., Kang, J.-Q., Song, L., Dibbens, L., Mulley, J., and Macdonald, R. L. (2006). δ Subunit susceptibility variants E177A and R220H associated with complex epilepsy alter channel gating and surface expression of $\alpha4\beta2\gamma GABA_A$ receptors. *J. Neurosci.* **26**, 1499–1506.

George, A. L., Jr. (2004). Inherited channelopathies associated with epilepsy. *Epilepsy Curr.* **4**, 65–70.

Goldin, A. L. (2001). Resurgence of sodium channel research. *Annu. Rev. Physiol.* **63**, 871–894.

Goldin, A. L. (2002). Evolution of voltage-gated Na^+ channels. *J. Exp. Biol.* **205**, 575–584.

Goldin, A. L. (2003). Mechanisms of sodium channel inactivation. *Curr. Opin. Neurobiol.* **13**, 284–290.

Gotti, C. and Clementi, F. (2004). Neuronal nicotinic receptors: From structure to pathology. *Prog. Neurobiol.* **74**, 363–396.

Gotti, C., Zoli, M., and Clementi, F. (2006). Brain nicotinic acetylcholine receptors: Native subtypes and their relevance. *Trends Pharmacol. Sci.* **27**, 482–491.

Gourfinkel-An, I., Baulac, S., Nabbout, R., Ruberg, M., Baulac, M., Brice, A., and LeGuern, E. (2004). Monogenic idiopathic epilepsies. *Lancet Neurol.* **3**, 209–218.

Graves, T. D. (2006). Ion channels and epilepsy. *Quart. J. Med.* **99**, 201–217.

Gutman, G. A., Chandy, K. G., Grissmer, S., Lazdunski, M., McKinnon, D., Pardo, L. A. et al. (2005). International Union of Pharmacology. LIII. Nomenclature and molecular relationships of voltage-gated potassium channels. *Pharmacol. Rev.* **57**, 473–508.

Harkin, L. A., Bowser, D. N., Dibbens, L. M., Singh, R., Phillips, F., Wallace, R. H. et al. (2002). Truncation of the GABAA-receptor γ2 subunit in a family with generalized epilepsy with febrile seizures plus. *Am. J. Hum. Genet.* **70**, 530–536.

Hauser, W. A., Annegers, J. F., and Kurland, L. T. (1993). Incidence of epilepsy and unprovoked seizures in Rochester, Minnesota: 1935–1984. *Epilepsia* **34**, 453–468.

Hille, B. (2001). *Ion channels of excitable membranes*. Sinauer Associates, Inc., Sunderland, MA.

Hofmann, F., Biel, M., and Kaupp, U. B. (2005). International Union of Pharmacology. LI. Nomenclature and structure-function relationships of cyclic nucleotide-regulated channels. *Pharmacol. Rev.* **57**, 455–462.

Hoyer, D., Hannon, J. P., and Martin, G. R. (2005). Molecular, pharmacological and functional diversity of 5-HT receptors. *Pharmacol. Biochem. Behav.* **71**, 533–554.

Hughes, B. W., Kusner, L. L., and Kaminski, H. J. (2006). Molecular architecture of the neuromuscular junction. *Muscle & Nerve* **33**, 445–461.

Imbrici, P., Jaffe, S. L., Eunson, L. H., Davies, N. P., Herd, C., Robertson, R. et al. (2004). Dysfunction of the brain calcium channel $Ca_V2.1$ in absence epilepsy and episodic ataxia. *Brain* **127**, 2682–2692.

Isom, L. L. (2001). Sodium channel β subunits: Anything but auxiliary. *Neuroscientist* **7**, 42–54.

Jentsch, T. J., Poët, M., Furhmann, J. C., and Zdebik, A. A. (2005). Physiological functions of CLC Cl^- channels gleaned from human genetic disease and mouse models. *Annu. Rev. Physiol.* **67**, 779–807.

Jones-Davis, D. M. and MacDonald, B. T. (2003). $GABA_A$ receptor function and pharmacology in epilepsy and status epilepticus. *Curr. Opin. Pharmacol.* **3**, 12–18.

Kang, J.-Q., Shen, W., and Macdonald, R. L. (2006). Why does fever trigger febrile seizures? $GABA_A$ receptor γ2 subunit mutations associated with idiopathic generalized epilepsies have temperature-dependent trafficking deficiencies. *J. Neurosci.* **26**, 2590–2597.

Khosravani, H., Altier, C., Simms, B. A., Hamming, K., Snutch, T. P., McRory, J. E., and Zamponi, G. W. (2004). Gating effects of mutations in the $Ca_V3.2$ T-type calcium channel associated with Childhood Absence Epilepsy. *Journal of Biological Chemistry*. Ref Type: In Press.

Khosravani, H. and Zamponi, G. W. (2006). Voltage-gated calcium channels and idiopathic generalized epilepsies. *Physiological Reviews* **86**, 941–966.

Lai, J., Hunter, J. C., and Porreca, F. (2003). The role of voltage-gated sodium channels in neuropathic pain. *Curr. Opin. Neurobiol.* **13**, 291–297.

Lester, H. A., Dibas, M. I., Dahan, D. S., Leite, J. F., and Dougherty, D. A. (2004). Cys-loop receptors: New twists and turns. *Trends Neurosci.* **27**, 329–336.

Long, S. B., Campbell, E. B., and MacKinnon, R. (2005). Crystal structure of a mammalian voltage-dependent *Shaker* family K^+ channel. *Science* **309**, 897–903.

Lossin, C., Rhodes, T. H., Desai, R. R., Vanoye, C. G., Wang, D., Carniciu, S. et al. (2003). Epilepsy-associated dysfunction in the voltage-gated neuronal sodium channel SCN1A. *J. Neurosci.* **23**, 11289–11295.

Lossin, C., Wang, D. W., Rhodes, T. H., Vanoye, C.G., and George, A. L., Jr. (2002). Molecular basis of an inherited epilepsy. *Neuron* **34**, 877–884.

Lynch, J. W. (2004). Molecular structure and function of the glycine receptor chloride channel. *Physiological Reviews* **84**, 1051–1095.

Macdonald, R. L., Gallagher, M. J., Feng, H. J., and Kang, J. (2004). $GABA_A$ receptor epilepsy mutations. *Biochem. Pharmacol.* **68**, 1497–1506.

Mayer, M. L. (2005). Glutamate receptor ion channels. *Curr. Opin. Neurobiol.* **15**, 282–288.

McCrossan, Z. A. and Abbott, G. W. (2004). The MinK-related peptides. *Neuropharmacol.* **47**, 787–821.

Meisler, M. H., Kearney, J., Ottman, R., and Escayg, A. (2001). Identification of epilepsy genes in human and mouse. *Annu. Rev. Genet.* **35**, 567–588.

Meisler, M. H. and Kearney, J. A. (2005). Sodium channel mutations in epilepsy and other neurological disorders. *J. Clin. Invest.* **115**, 2010–2017.

Mulley, J. C., Scheffer, I. E., Petrou, S., and Berkovic, S. F. (2003). Channelopathies as a genetic cause of epilepsy. *Curr. Opin. Neurol.* **16**, 171–176.

Noebels, J. L. (2003). The biology of epilepsy genes. *Annu. Rev. Neurosci.* **26**, 599–625.

Orr, H. T. (2004). Into the depths of ataxia. *J. Clin. Invest.* **113**, 505–507.

Pietrobon, D. (2005). Migraine: New molecular mechanisms. *Neuroscientist* **11**, 373–386.

Rhodes, K. J., Carroll, K. I., Sung, M. A., Doliveira, L. C., Monaghan, M. M., Burke, S. L. et al. (2004). KChIPs and Kv4 α subunits as integral components of A-type potassium channels in mammalian brain. *J. Neurosci.* **24**, 7903–7915.

Rivera, C., Voipio, J., and Kaila, K. (2005). Two developmental switches in GABAergic signalling: The K^+-Cl^- cotransporter KCC2 and carbonic anhydrase CAVII. *J. Physiol. (Lond.)* **562**, 27–36.

Robinson, R. B. and Siegelbaum, S. A. (2003). Hyperpolarization-activated cation currents: From molecules to physiological function. *Annu. Rev. Physiol.* **65**, 453–480.

Saul, B., Kuner, T., Sobetzko, D., Brune, W., Hanefeld, F., Meinck, H.-M., and Becker, C.-M. (1999). Novel *GLRA1* missense mutation (P250T) in dominant hyperekplexia defines an intracellular determinant of glycine receptor channel gating. *J. Neurosci.* **19**, 869–877.

Scheffer, I. E. and Berkovic, S. F. (2003). The genetics of human epilepsy. *Trends Pharmacol. Sci.* **24**, 428–433.

Scheffer, I. E., Harkin, L. A., Dibbens, L. M., Mulley, J. C., and Berkovic, S. F. (2005). Neonatal epilepsy syndromes and generalized epilepsy with febrile seizures plus (GEFS+). *Epilepsia* **46**, 41–47.

Schöls, L., Bauer, P., Schmidt, T., Schulte, T., and Riess, O. (2004). Autosomal dominant cerebellar ataxias: Clinical features, genetics, and pathogenesis. *Lancet* **3**, 291–304.

Spampanato, J., Aradi, I., Soltesz, I., and Goldin, A. L. (2004). Increased neuronal firing in computer simulations of sodium channel mutations that cause generalized epilepsy with febrile seizures plus. *J. Neurophysiol.* **91**, 2040–2050.

Steinlein, O. K. (2002). Channelopathies can cause epilepsy in man. *Eur. J. Pain* **6** (Suppl. A), 27–34.

Steinlein, O. K. (2004). Genetics mechanisms that underlie epilepsy. *Nat. Rev. Neurosci.* **5**, 400–408.

Trimmer, J. S. and Rhodes, K. J. (2004). Localization of voltage-gated ion channels in mammalian brain. *Annu. Rev. Physiol.* **66**, 477–519.

Vitko, I., Chen, Y., Arias, J. M., Shen, Y., Wu, X.-R., and Perez-Reyes, E. (2005). Functional characterization and neuronal modeling of the effects of childhood absence epilepsy variants of *CACNA1H*, a T-type calcium channel. *J. Neurosci.* **25**, 4844–4855.

Wan, J., Khanna, R., Sandusky, M., Papazian, D. M., Jen, J. C., and Baloh, R. W. (2005). *CACNA1A* mutations causing episodic and progressive ataxia alter channel trafficking and kinetics. *Neurology* **64**, 2090–2097.

Waters, M. F., Minassian, N. A., Stevanin, G., Figueroa, K. P., Bannister, J. P. A., Nolte, D. et al. (2006). Mutations in voltage-gated potassium channel KCNC3 cause degenerative and developmental central nervous system phenotypes. *Nat. Genet.* **38**, 447–451.

Waxman, S. G. and Dib-Hajj, S. (2005). Erythermalgia: Molecular basis for an inherited pain syndrome. *Trends Mol. Med.* **11**, 555–562.

Waxman, S. G. and Hains, B. C. (2006). Fire and phantoms after spinal cord injury: Na^+ channels and central pain. *Trends Neurosci.* **29**, 207–215.

Wessmann, M., Kaunisto, M. A., Kallela, M., and Palotie, A. (2004). The molecular genetics of migraine. *Ann. Med. (Helsinki)* **36**, 462–473.

Wollmuth, L. P. and Sobolevsky, A.I. (2004). Structure and gating of the glutamate receptor ion channel. *Trends Neurosci.* **27**, 321–328.

Wood, J. N., Boorman, J. P., Okuse, K., and Baker, M. D. (2004). Voltage-gated sodium channels and pain pathways. *J. Neurobiol.* **61**, 55–71.

Yu, F. H., Yarov-Yarovoy, V., Gutman, G. A., and Catterall, W. A. (2005). Overview of molecular relationships in the voltage-gated ion channel superfamily. *Pharmacol. Rev.* **57**, 387–395.

5

Protein Misfolding, Chaperone Networks, and the Heat Shock Response in the Nervous System

Cindy Voisine, Kai Orton, and Richard I. Morimoto

I. Introduction

To function properly, newly translated proteins must fold into their native conformation. Information contained within the primary amino acid sequence dictates the three-dimensional shape of the protein (Anfinsen, 1973). Hydrophobic amino acid residues are buried within the interior of the protein, whereas hydrophilic residues are exposed on the surface, generating a thermodynamically stable structure. The pathway by which a protein achieves its unique folded state is complex. En route to its native state, a protein traverses through an ensemble of intermediate states and a myriad of potential conformations (Wolynes et al., 1995). A protein becomes misfolded when inappropriate yet energetically stable interactions occur, for example by self-association of hydrophobic residues leading to oligomerization and aggregation in the crowded cellular environment (Hartl & Hayer-Hartl, 2002) (see Figure 5.1).

Changes in the cellular environment can influence protein folding homeostasis (Morimoto et al., 1997). These include environmental stress, such as fluctuations in temperature, hydration, and nutrient balance; chemical stress caused by oxygen free radicals, and transition heavy metals; or pathophysiological states of multicellular organisms such as ischemia, viral, or bacterial infections or tissue injury. Often, this change in conformation is reversed spontaneously as the cell adapts to the stressor or upon recovery to the prestress cellular conditions. However, for some proteins, the appearance of alternate intermediate states leads to the chronic persistence of non-native misfolded off pathway intermediates (see Figure 5.1).

Mutations in the primary amino acid sequence of a protein can also alter its folding pathway. Multiple familial neurodegenerative disorders including Parkinson's disease, amyotrophic lateral sclerosis (ALS), and polyglutaminc (polyQ) diseases that include Huntington's disease and related ataxias carry mutations in disease genes that lead to protein misfolding, defining a family of diseases often referred to as Conformational Diseases (Kakizuka, 1998; Kopito & Ron, 2000; Sherman & Goldberg, 2001; Stefani & Dobson, 2003) (see Table 5.1). One hallmark of these diseases is the accumulation of protein aggregates in brain tissue of individuals

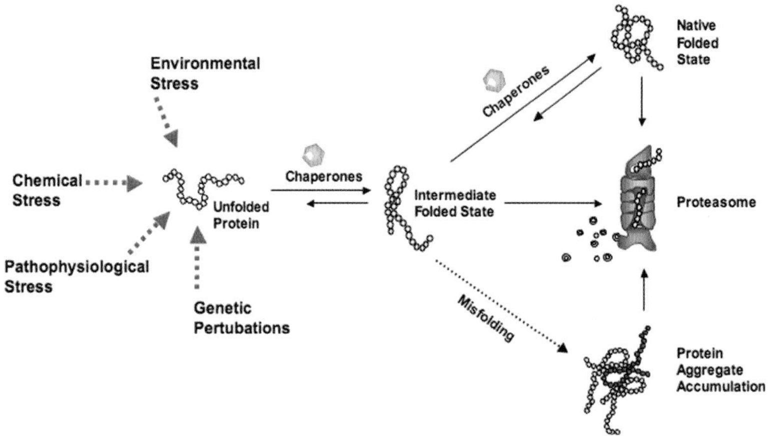

Figure 5.1 Protein Folding Homeostasis. Cellular protein folding homeostasis is a delicate balance between proper folding of proteins and efficient removal of damaged proteins. Under certain conditions, proteins misfold and adopt alternate intermediate conformations. Molecular chaperones recognize non-native states and assist in folding. The ubiquitin proteasome system degrades both normal short-lived and abnormal misfolded proteins. An impairment in protein folding homeostasis leads to accumulation of misfolded species and inclusion formation leading to cellular toxicity.

diagnosed with neurodegeneration (Muchowski & Wacker, 2005). In one well-studied family of neurodegenerative disorders, polyQ diseases, expansion of a glutamine tract within the disease protein alters its physical properties (Trottier et al., 1995), leading to formation of cytosolic and nuclear aggregates in affected tissues (DiFiglia et al., 1997). Likewise for ALS and Parkinson's disease, a genetic connection has been identified between and mutations in superoxide dismutase and α-synuclein, respectively, to formation of aggregates (Polymeropoulos et al., 1997; Rosen, 1993). As is the case with many neurodegenerative diseases, clearance of misfolded proteins caused by genetic mutations or protein damaging cellular stresses becomes a critical task for neuronal survival.

Protective mechanisms have evolved to assist in the proper folding of proteins and clearance of misfolded and damaged proteins (Goldberg, 2003). Protein homeostasis is achieved by the interplay of two complex molecular machines that maintain cellular protein integrity: the molecular chaperones that associate with nascent polypeptides and recognize misfolded proteins, and the ubiquitin proteasome system that degrades both normal short-lived and abnormal misfolded proteins (see Figure 5.1). Disruption of the protein folding quality control, therefore, leads to the appearance of non-native protein species that self-associate and form higher ordered structures such as aggregates and inclusions leading to cellular toxicity (Dobson, 2004). Despite the functional and structural diversity of proteins involved in conformational diseases, the shared characteristic of accumulation as misfolded species has led to a unifying hypothesis that an impairment of protein folding homeostasis can have severe pathogenic consequences. Cellular protein homeostasis requires a delicate balance between the proper folding of proteins by molecular chaperones and removal of damaged proteins by the proteasome. Age-related failure in quality control, therefore, may be a common element of neurodegenerative diseases (Csermely, 2001). Consequently, proper functioning of these complex molecular machines to prevent the accumulation of misfolded proteins becomes fundamental to disease prevention.

II. Role of Molecular Chaperones in Protein Folding Quality Control

Molecular chaperones are highly conserved proteins that bind transiently to stabilize non-native protein conformations (Hartl & Hayer-Hartl, 2002). Molecular chaperones are ubiquitously expressed and are present in all cell types and cellular compartments and have essential functions under normal growth conditions (Frydman, 2001). However, most molecular chaperones were first identified as stress regulated genes induced by exposure to elevated temperature (heat shock) or other environmental stress that may affect protein conformation. Due to their elevated expression, stress responsive molecular chaperones are often referred to as *heat shock proteins* (Hsps) (Gething & Sambrook, 1992). Prior to folding as an extended chain or upon unfolding or denaturation, the inappropriate association of hydrophobic amino acid residues that normally are buried within the interior of a folded protein become exposed to the aqueous environment and are prone to aggregation (Radford, 2000). Chaperones recognize and bind to these exposed hydrophobic stretches and through these interactions, have essential roles in a wide

Table 5.1

Conformational Disease	Neurons Primarily Affected in Disease	Aggregating Protein	Co-localization of chaperones w/ aggregate	Reference(s)
Alzheimer's Disease	corpus callosum	Aß peptides (Aß40, Aß42) hyper-phosphorylated tau	Hsp70, Hsp90, sHsps, ERHsp70	Hamos et al. 1991; Kakimura et al. 2002; Dou et al. 2003; Shinohara et al. 1993; Renkawek et al. 1994; Nemes et al. 2004
Parkinson's disease	dopaminergic neurons	α-synuclein	Hsp70, Hsp40	Auluck et al. 2002; McLean et al. 2002
Familial amyotrophic lateral sclerosis	motor neurons	mutant SOD1	Hsp70	Watanabe et al. 2001
Huntington's disease	striatal cortex	mutant huntingtin	Hsp70, Hsp40, ERHsp70	Waelter et al. 2001; Jana et al. 2000
Spinocerebellar ataxias SCA1-3,7	cerebellar cortex	mutant ataxin	Hsp70, Hsp40	Chai et al. 1999; Cummings et al. 1998

range of basic cellular processes (Hartl & Hayer-Hartl, 2002). Molecular chaperones are essential for protein folding, translocation of proteins across cellular membranes, assembling and disassembling multimeric complexes, preventing protein self-association, and directing misfolded proteins for degradation by the proteasome (Bukau & Horwich, 1998; Hartl, 1996) (see Figure 5.2).

During protein biogenesis, the association of molecular chaperones with client substrates provides kinetic and spatial partitioning to ensure an orderly and proper fate (see Figure 5.2). As nascent polypeptides emerge from the ribosome, chaperones bind to the nascent chain, thus preventing inappropriate interactions that may lead to protein misfolding (Deuerling et al., 1999; Teter et al., 1999; Thulasiraman et al., 1999). Once a protein is translated, chaperones can participate in their further maturation by assisting in the final folding, assembling the protein into a multimeric complex, or maintaining a protein destined for a subcellular organelle in a translocation competent state (Deuerling et al., 1999; Hartl & Hayer-Hartl, 2002; Siegers et al., 1999; Teter et al., 1999). At the site of import either into the mitochondrion or the endoplamic reticulum, the translocating protein emerges from the channel and interacts with chaperones that reside within the cellular compartment (Becker et al., 1996; Kang et al., 1990; Vogel et al., 1990). These chaperones assist in the efficient import and proper folding of the newly translocated protein into its functional conformation.

Protein damaging stresses, mutations within the primary amino acid sequences, or errors in translation can all increase the flux of misfolded proteins. Proteins beyond repair are targeted for removal by the proteasome (Goldberg, 2003) (see Figure 5.2). Due to the absence of degradative machinery within the endoplasmic reticulum, ER-damaged proteins must be recognized and retro-translocated into the cytoplasm for proteasome targeting (Romisch, 2005). The concerted efforts by molecular chaperones and the proteasome efficiently manage the normal cellular load of misfolded proteins. Overburdening either component of the protein quality control system with the chronic expression of misfolded proteins leads to off-pathway folding events that result in deposition of proteins that have deleterious consequences on cellular homeostasis.

A. Molecular Chaperone Families

Molecular chaperones are key components of the protein quality control system. The major families of chaperones are classified according to their molecular masses and sequence homology and include Hsp100, Hsp90, Hsp70, Hsp60, Hsp40, and small Hsps. The major chaperone families differ in their substrate specificity; some chaperones bind a wide range of substrates but others recognize only a few substrates. Chaperone families function with cochaperones that modulate the interaction of chaperones with substrates.

Each chaperone family varies in number, consisting of multiple members that differ in subcellular localization and expression patterns. For example, certain family members are expressed only during development in a particular cell type, whereas others are expressed only during stress in all cells. Therefore, the level and complement of chaperones expressed in various cells in multicellular organisms is specialized for that cell's proper function.

1. Hsp70 Family

The Hsp70s (heat shock proteins of the 70 kDa size) participate in diverse cellular processes of protein folding and protein translocation across membranes. Hsp70s

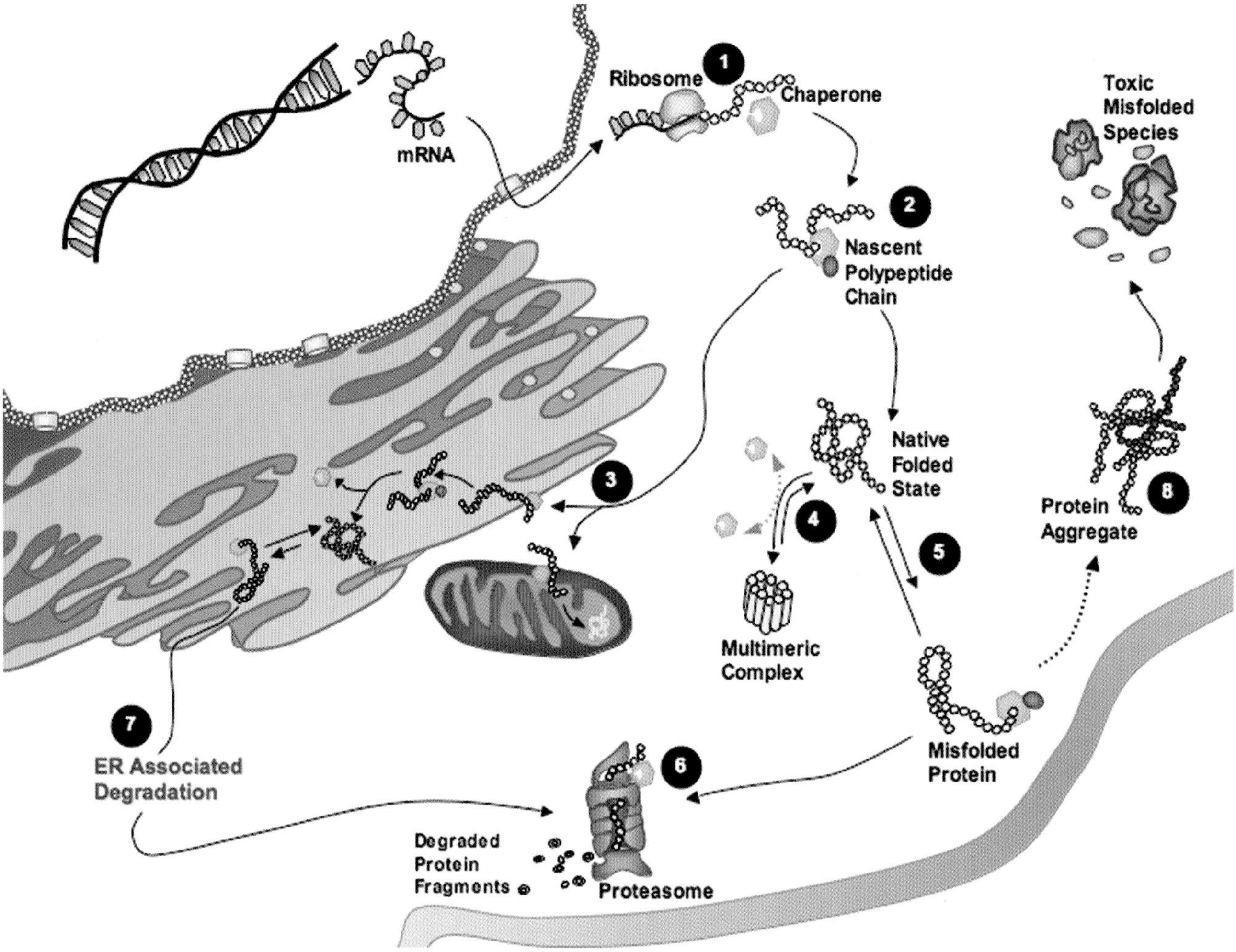

Figure 5.2 General Cellular Functions of Chaperones. Molecular chaperones participate in multiple aspects of protein biogenesis. (**1**) Newly synthesized proteins emerge from the ribosome in an extended conformation. Molecular chaperones bind exposed hydrophobic residues and prevent misfolding. (**2**) Chaperones interact transiently with nascent polypeptide chains assisting in their proper folding. (**3**) Proteins destined for subcellular compartments are maintained in a translocation competent state. At the site of import, the protein emerges from the channel and interacts with chaperones that reside within the mitochondrion or endoplasmic reticulum. These resident chaperones assist in the proper folding of the newly imported protein. (**4**) Chaperones assemble and disassemble folded proteins into multimeric complexes. (**5**) Chaperones recognize non-native states and assist in the refolding of the protein. (**6**) Proteins beyond repair are targeted for removal by the proteasome. (**7**) Due to the absence of degradative machinery within the ER, ER-damaged proteins are retro-translocated into the cytoplasm for degradation by the proteasome. (**8**) Failure of the chaperone quality control systems leads to the appearance of non-native protein species that self-associate and form higher ordered structures such as aggregates and inclusions.

correspond to a family of molecular chaperones that are highly conserved, exhibiting a high level of sequence identity (45%) across all species (Boorstein et al., 1994). With the exception of some archaea, all organisms express several different forms of Hsp70s varying in subcellular localization and time of expression (Frydman, 2001; Hartl & Hayer-Hartl, 2002). Although the expression of many Hsp70s is induced by exposure to elevated temperatures or specific stress, other Hsp70s are expressed constitutively (Lindquist & Craig, 1988).

The domain structure of Hsp70 is highly conserved and is comprised of two principal domains, the N-terminal ATPase domain (~40 kDa) and a C-terminal domain containing a substrate-binding region and a flexible lid (~25 kDa) (see Figure 5.3). The ATPase domain of Hsp70 has weak intrinsic ATPase activity and the structure shows bilobed domains with a deep cleft for binding of the adenine nucleotide (Flaherty et al., 1990). The C-terminal domain is responsible for binding to substrate proteins and includes a series of alpha helices that are thought to act as a lid, opening and closing over the peptide-binding pocket (Zhu et al., 1996). Hsp70s bind preferentially to short stretches of amino acids rich in hydrophobic and aliphatic residues in the extended conformation (Flynn et al.,

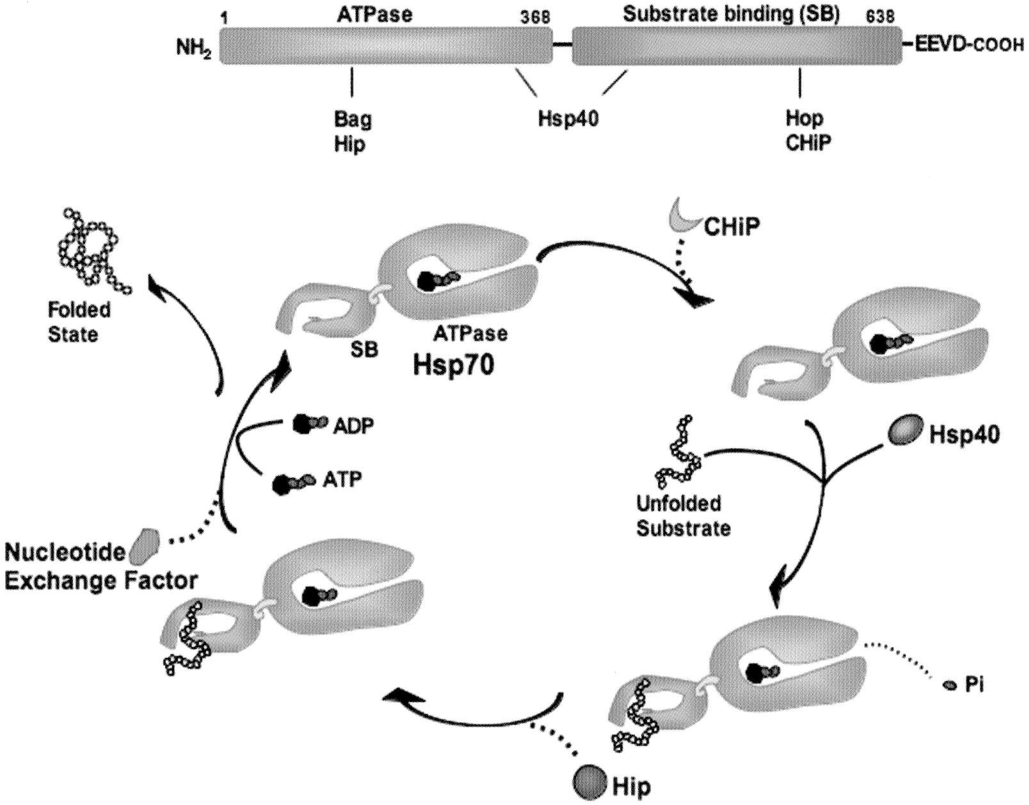

Figure 5.3 Modulators of Hsp70 Folding Cycle. Hsp70 ensures the proper folding of proteins through cycles of substrate binding and release regulated by ATP binding and hydrolysis. Hsp70s are comprised of two domains: an N-terminal ATPase domain and C-terminal substrate binding domain. Various cochaperones interact with Hsp70 in different regions to modulate its ATPase cycle. An essential component of the cycle is the cochaperone Hsp40. Hsp40 binds Hsp70 and stimulates its ATPase activity, stabilizing the interaction of Hsp70 and substrate. Additional cochaperones can further modulate the Hsp70 reaction cycle. Hip can bind the ATPase domain, stabilizing the ADP-bound state promoting stable interactions of Hsp70 with substrate. Nucleotide exchange factors, such as Bag-1, can also bind the ATPase domain of Hsp70, releasing bound ADP. The binding of ATP releases substrate, completing the cycle. In some cases, the cochaperone CHIP promotes ubiquitination and degradation of Hsp70 substrates. CHIP inhibits the Hsp40-stimulated ATPase activity of Hsp70 reducing the ability of Hsp70 to fold its substrate.

1989; Rudiger et al., 1997). Thus, Hsp70s recognize various substrates by interacting with exposed hydrophobic residues that typically are sequestered in the interior of the protein.

As molecular chaperones, Hsp70s function to prevent aggregation and ensure the proper folding of proteins through cycles of substrate binding and release regulated by ATP binding and hydrolysis (Schmid et al., 1994) (see Figure 5.3). The ATP-bound form of an Hsp70 has a rapid exchange rate for the substrate protein, thus having a low affinity for extended hydrophobic segments of unfolded proteins. When ATP is hydrolyzed, a structural rearrangement occurs in the substrate binding domain causing the alpha helical lid to close over the substrate binding site (Jiang et al., 2005; Vogel et al., 2006). The outcome is that the ADP-bound form of an Hsp70 binds and releases the substrate protein more slowly, thus having a high affinity for the substrate protein. Dissociation of ADP from the ATPase domain and rebinding of ATP releases the substrate protein to fold into its proper conformation while the ATP-bound form of the Hsp70 can participate in another round of binding. Through this cycle of nucleotide-dependent binding and release, Hsp70s prevent aberrant protein-protein interactions that may lead to the misfolding or aggregation of newly synthesized or unfolded proteins.

2. Hsp40 Family

Hsp40s (heat shock proteins of the 40 kDa size) also known as J-proteins, represent a distinct large family of structurally and functionally diverse chaperones (Craig et al., 2006; Walsh et al., 2004). Members of this family have in common the signature ~70 amino acid J domain named after the *E. coli* DnaJ protein (Kelley, 1998). In DnaJ, the J domain is followed by a glycine and phenylalanine rich region (G/F region), and a cysteine-rich region with the remaining C-terminal region being poorly conserved. The classification for J-proteins identifies Class I proteins that contain all regions described earlier; Class II proteins lack the cysteine-rich region, and Class III proteins retain only the signature J domain (Cheetham & Caplan, 1998).

J-proteins typically function in cooperation with Hsp70s as a chaperone folding machine. The J domain binds to the ATPase domain of Hsp70 and stimulates ATP hydrolysis, thus stabilizing the interaction of Hsp70 with substrate (Wittung-Stafshede et al., 2003) (see Figure 5.3). Within the J domain is a conserved motif, the histidine-proline-aspartate (HPD) tripeptide, that is essential for stimulation of Hsp70 ATPase activity (Greene et al., 1998; Tsai & Douglas, 1996; Yan et al., 1998). Some J-proteins participate in protein folding independent of Hsp70, by binding to unfolded polypeptides via the glycine/phenylalanine or cysteine-rich region and preventing aggregation (Langer et al., 1992; Lu & Cyr, 1998; Rudiger et al., 2001). A current model proposes that a J-protein first binds unfolded protein substrates and then transfers them to Hsp70 for folding.

3. Hsp90 Family

Hsp90 (heat shock proteins of the 90 kDa size) family members are critical for the regulation and maturation of diverse client proteins involved in signaling pathways and cell cycle control. Hsp90 client proteins include various transcription factors, hormone receptors, and serine/threonine kinases (Pratt & Toft, 2003). Hsp90 has three domains: an amino-terminal ATP binding domain followed by a charged region, a middle domain, and a carboxy-terminal domain. All three domains of Hsp90 have been implicated in binding of substrate polypeptides, but recent data suggests that the middle domain likely contains the client protein binding site (Pearl & Prodromou, 2000). Hsp90 recognizes folding intermediates that have near-native structure, consistent with a model in which Hsp90 acts subsequent to Hsp70 in the protein folding pathway (Freeman & Morimoto, 1996; Freeman et al., 1996). Although Hsp90 associates with a growing list of diverse client proteins (Pearl & Prodromou, 2000; Young et al., 2001), Hsp90 does not function as a general chaperone since widespread *in vivo* aggregation of newly synthesized proteins is not observed in the absence of Hsp90 function (Nathan et al., 1997).

4. Hsp60 Family

Hsp60 (heat shock proteins of the 60 kDa size) family members are also known as chaperonins. There are two classes of chaperonins based on sequence homology and the requirement of an additional subunit that functions as a regulatory lid (Kusmierczyk & Martin, 2001). One class of chaperonins is expressed in bacteria and evolutionarily derived organelles, and the second class resides in the cytoplasm of eukaryotes and in archaea. The best characterized chaperonin is *E. coli* GroEL, a 14-mer of 57.3-kDa subunits, arranged in two stacked heptameric rings (Braig et al., 1994). The GroEL monomer has an apical substrate binding domain, an intermediate domain, which mediates ATP-induced conformational shifts caused by the hydrolysis of ATP, and an equatorial domain (Braig et al., 1994; Fenton et al., 1994; Xu et al., 1997). Together, these domains assemble to form a cavity that can accommodate proteins up to 60kDa in size (Sigler et al., 1998).

GroEL interacts with folded intermediates to assist in the folding of many different proteins in an ATP-dependent manner, thus preventing their misfolding and aggregation (Buchner et al., 1991; Goloubinoff et al., 1989a, 1989b). Upon binding of ATP to all seven equatorial domains of one ring, the GroEL heptamer undergoes a major conformational rearrangement, changing the surface of the cavity from hydrophobic to hydrophilic (Xu et al., 1997). Once released from GroEL, the protein in an incompletely folded state may spontaneously refold according to its original folding pathway.

5. Hsp100 Family

Hsp100 (heat shock proteins of the 100 kDa size) family belongs to a larger family of ATPase Associated with diverse Activities (AAA+) proteins that have a conserved ATPase unit with at least one nucleotide binding domain (Beyer, 1997). Hsp100s are is composed of three domains: an N-terminal domain, which recognizes substrate protein, and two nucleotide binding domains. Hsp100 monomers are arranged in a functional hexameric ring providing chaperone activity. Although many Hsp100s associate with a protease and participate in protein degradation (Ogura & Wilkinson, 2001; Schirmer et al., 1996), specific Hsp100s, ClpB in *E. coli* or Hsp104 in *Saccharomyces cerevisiae* function with Hsp70 in reversing protein aggregation (Ben-Zvi & Goloubinoff, 2001; Goloubinoff et al., 1999). Hsp104 together with Hsp70 solubilize protein aggregates generated by *S. cerevisiae* prions (Glover & Lindquist, 1998).

6. Small HSPs

The sHsp (small heat shock protein) family is composed of many different low molecular weight proteins ranging from 15 kDa to 40 kDa (Clark & Muchowski, 2000; Sun & MacRae, 2005). Family members share sequence homology in the C-terminal α-crystallin domain and the N-terminal hydrophobic region that interacts with substrate proteins. Unlike other molecular chaperone families described thus far, sHsps do not require ATP to prevent aggregation. Under physiological conditions, sHsps form large oligomeric complexes. The N-terminal hydrophobic patches reside within the interior of the structure. When cells encounter conditions leading to protein misfolding, sHsps disassemble, form dimers, and bind aggregating proteins via their N-terminal hydrophobic region. The mechanism by which substrate proteins are released from sHsp is unclear.

B. Chaperone Machines

To increase the efficiency and specificity of the Hsp70 folding machine, a number of cochaperones act in concert to modulate various steps in the ATPase cycle (Young et al., 2003) (see Table 5.2; also see Figure 5.3). Hsp40 binds Hsp70 and stimulates its ATPase activity, stabilizing the interaction of Hsp70 and substrate (Freeman et al., 1995; Wittung-Stafshede et al., 2003). The cochaperone Hip (Hsp interacting protein) can bind the ATPase domain, thus stabilizing the ADP-bound state promoting stable interactions of Hsp70 with substrate (Hohfeld et al., 1995; Ziegelhoffer et al., 1996). Nucleotide exchange factors, such as Bag-1, can also bind the ATPase domain of Hsp70, releasing bound ADP (Demand et al., 1998; Hohfeld & Jentsch, 1997; Sondermann et al., 2001; Takayama & Reed, 2001). The binding of ATP releases substrate, completing the cycle.

Besides regulating the ATPase cycle, cochaperones play a key role coordinating interactions of chaperone machines and protein quality control systems. The cochaperone Hop (Hsp organizing protein) is a TPR (tetratricopeptide repeat) domain containing protein that couples the Hsp70 and Hsp90 folding machines (Scheufler et al., 2000). Hop functions as an adaptor protein that directs substrate transfer from Hsp70 to Hsp90. A second cochaperone CHIP (carboxyl terminus of Hsp70 interaction protein) links protein folding and degradation (Esser et al., 2004). CHIP is a chaperone-associated E3 ubiquitin ligase, promoting ubiquitination and degradation of Hsp70 substrates (Connell et al., 2001; Meacham et al., 2001). CHIP inhibits the Hsp40-stimulated ATPase activity of Hsp70 reducing the ability of Hsp70 to fold its substrate (Ballinger et al., 1999). Lowering activity allows abnormal proteins bound to Hsp70 to be degraded by the proteasome.

The low cellular levels of cochaperones as compared to Hsp70 levels provide an important regulatory feature. During each Hsp70 reaction cycle, the limiting component is the cochaperone. Specific cochaperone involvement during the ATPase cycle may be needed for various tasks to specialize the folding activity of Hsp70 for certain substrates (Mayer & Bukau, 2005).

III. Regulation of Chaperone Expression: The Heat Shock Response

It is of critical importance to have precise regulation of the individual chaperones that comprise the chaperone machines (see Figure 5.4). Members of the HSF family of transcription factors (HSFs) are key mediators of basal and stress-induced expression of chaperones. HSFs, when activated by cellular physiological stimuli and by stress stimuli, form trimers, allowing them to bind to specific consensus sequences called heat shock elements (HSEs) located in the promoters of chaperone genes. The HSE consensus sequence is defined by a 5-bp sequence repeating array of nGAAn arranged in alternating orientations (Amin et al., 1988; Xiao & Lis, 1988). The number of nGAAn repeats in a functional HSE, as well as the distance between them, varies, but typically ranges from three to six repeats. An increased number of nGAAn repeats in a gene's promoter has been shown to correspond with increased DNA binding activity of HSF trimers (Topol et al., 1985; Xiao et al., 1991). Constitutive

Table 5.2

Hsp70 co-chaperone	Sub-cellular Localization	Co-chaperone Interacting Domain	Hsp70 Interacting Domain	Influence on Hsp70 Chaperone Activity	Reference(s)
Hsp40	cytosol/nucleus ER/mitochondria	J-domain	ATPase domain and C-terminal	stimulates ATP hydrolysis domain	Walsh et al. 2004; Craig et al. 2006; Wittung-Stafshede et al. 2003; Demand et al. 1998
Hip	cytosol/nucleus	TPR & charged α-helical domain	ATPase domain	stabilizes ADP bound conformation	Hohfeld et al. 1995; Ziegelhoffer et al. 1996
Hop	cytosol/nucleus	TPR1 domain	C-terminal domain	adaptor protein links Hsp70 to Hsp90	Scheufler et al. 2000
Bag-1	cytosol/nucleus	Bag domain	ATPase domain	promotes nucleotide exchange	Takayama et al. 2001; Sondermann et al. 2001; Hohfeld et al. 1997
Chip	cytosol/nucleus	TPR1 domain	C-terminal domain	inhibits ATPase hydrolysis	Esser et al. 2004; Connell et al. 2001; Meacham et al. 2001; Ballinger et al. 1999

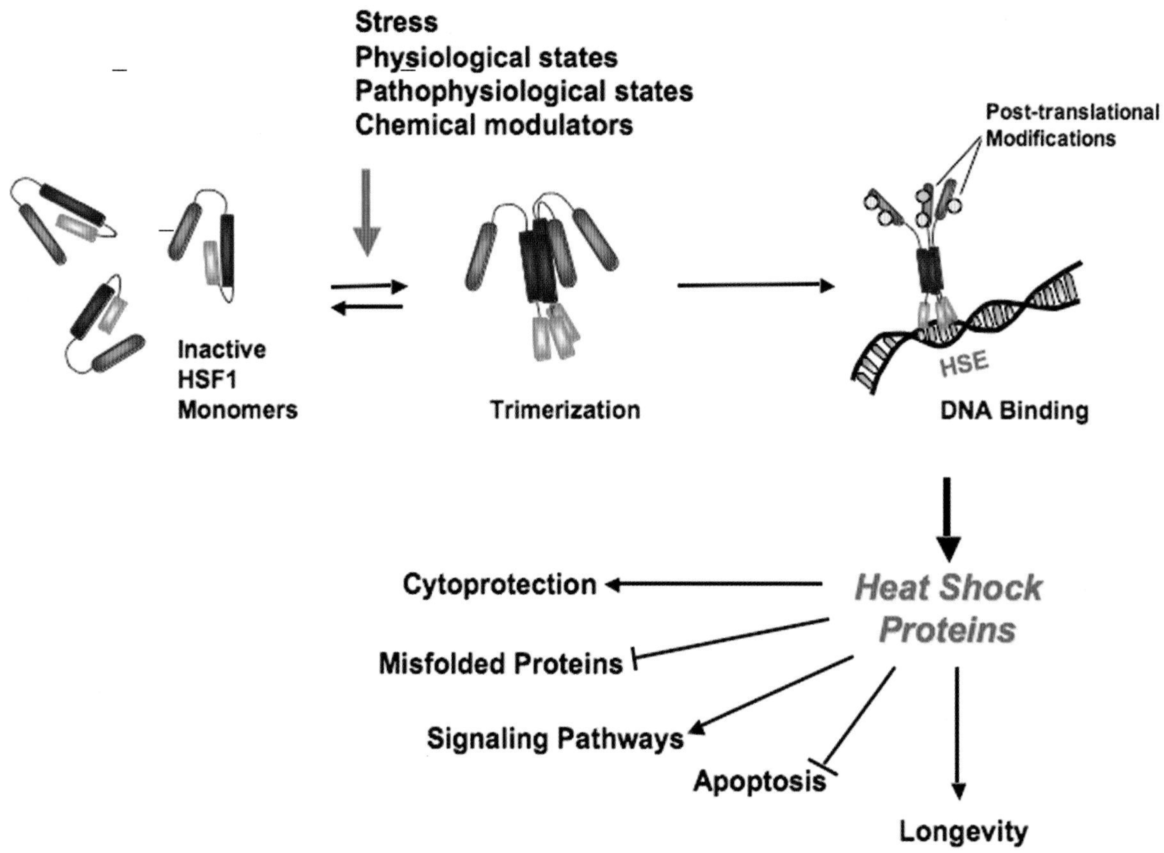

Figure 5.4 Cellular Heat Shock Response. HSF1 is the master transcriptional regulator of chaperones. Inactive HSF1 monomers are retained in the cytosol of the cell. HSF1, when activated by normal celluar stimulti or stress stimuli, or chemical modulators, becomes post-translationally modified and forms trimers, allowing binding to specific consensus sequences called heat shock elements (HSEs) located in the promoters of chaperone genes leading to the production of molecular chaperones. Molecular chaperones activate and inhibit a variety of celluar processes leading to protection from stress.

and stress-induced expression of chaperones is essential to ensure the proper folding of proteins involved in critical cellular processes.

A. HSF Family of Transcription Factors

Several members of the HSF family have been identified and are highly conserved from *S. cerevisiae* to humans. *S. cerevisiae* and *Drosophila* have only one HSF, whereas vertebrates have several HSFs, with the diversity further increased by alternative splicing of all members (Sorger & Pelham, 1988; Zimarino & Wu, 1987). In humans, there are three HSFs: HSF1, HSF2, and HSF4 (Larson et al., 1988; Nakai et al., 1997; Rabindran et al., 1991; Schuetz et al., 1991; Sistonen et al., 1992). All members of the heat shock factor family share a similar structure, comprised of an amino-terminal helix-turn-helix DNA-binding domain, an adjacent coiled-coil trimerization domain and with the exception of HSF4, a second coiled-coil domain, located in the carboxyl-terminus of the protein (Pirkkala et al., 2001). Despite their similar structure, vertebrate HSFs exhibit overlapping yet unique functions, target tissues, and activation patterns.

HSF1 is the master transcriptional regulator of chaperones and is the only family member that has been demonstrated to be active during stress (Sistonen et al., 1994). Consistent with initial studies characterizing HSFs critical function, other HSFs are unable to substitute for mammalian HSF1 or to rescue the heat shock response in HSF knockout animals (McMillan et al., 1998; Pirkkala et al., 2000; Xiao et al., 1999). Interestingly, human HSF2 has been shown to functionally complement the *S. cerevisiae* HSF in a null background and rescue the viability defect, demonstrating the functional conservation of HSFs between species and a critical role in stress gene regulation (Liu et al., 1997). It has been proposed that other HSFs may regulate additional genes that are unique from the classical heat shock protein genes, and may be involved in regulating other processes such as development or differentiation (Fujimoto et al., 2004; Sistonen et al., 1992). However, recent work indicates that there may be more overlap functionally than previously thought, as HSF2 has been found to act as a coregulator with HSF1, synergistically increasing stress-induced expression of HSPs (Alastalo et al., 2003; He et al., 2003; Mathew et al., 2001). Two HSF4 protein isoforms, HSF4a and HSF4b, have been characterized each with distinct functional properties (Tanabe et al., 1999). HSF4a exhibits a limited expression profile restricted predominantly to the eye lens where acts as an HSF4a transcriptional repressor of chaperone genes (Nakai et al., 1997; Tanabe et al., 1999; Zhang et al., 2001), where HSF4b exhibits properties of a transcriptional activator (Tanabe et al., 1999). The extent to which HSF4a influences global HSF activity or the contributions of HSF4b to stress-induced HSF activity is presently not well understood.

B. Regulation of the HSF1 Activation Cycle

HSF1 transcriptional activation involves several levels of regulation (see Figure 5.5). Under physiological conditions, human HSF1 occurs predominantly as an inactive monomer in the cytoplasm. Both Hsp70 and Hsp90 have been demonstrated as repressors of HSF1 under nonstressed conditions interacting with HSF1 to retain it in the cytosol (Abravaya et al., 1991; Shi et al., 1998). Upon stress stimuli, Hsp70 and Hsp90 interact with unfolded proteins, thereby releasing HSF1 monomers. In a rapid cascade of events, HSF1 translocates into the nucleus and trimerizes. Trimeric HSF binds to HSEs present in target chaperone gene promoters (Kroeger & Morimoto, 1994). The full activation of HSF1 trimers is further regulated by an array of post-translational modifications involving hyper-phosphorylation and sumoylation. Phosphorylation of human HSF1 on serine residues 230, 326, and 419 promotes its transcriptional activity (Pirkkala et al., 2001), whereas phosphorylation on serine residues 303, 307, 308, and 363 is involved in attenuation and repression of HSF1 activity subsequent to stress (Kline & Morimoto, 1997; Knauf et al., 1996). Sumoylation of HSF1 occurs on lysine 298 in a stress-dependent manner and may act in transcriptional repression (Anckar et al., 2006; Hietakangas et al., 2006). The transcriptional activation of HSF1 results in a robust induction of chaperone gene expression, whose protein products act to remedy protein misfolding and reestablish cellular homeostasis. Once misfolded and damaged proteins have been refolded or degraded, Hsp70 and Hsp90 interact with cytosolic HSF1, inhibiting its activation and restoring basal transcription levels of chaperone genes.

C. The HSR, HSF1, and Chaperones in the Nervous System

The HSR (Heat Shock Response) is unique in its high degree of conservation across organisms. However, within complex organisms, there may be differences in the stress response elicited by distinct tissues. One example of this has been documented in neuronal cells. Neurons exhibit unique proteomes consistent with their specialized cellular function. These unique expression profiles may confer global differences in cell-type specific responses to physiological and stressful conditions. Inherent in a cell establishing its specificity through differentiation, the composition of its proteome may reflect a specialized complement of molecular chaperones. Recent studies have identified two isoforms of the Hsp70 cochaperone, Hsj1a and Hsj1b, as being uniquely expressed in neurons (Chapple et al., 2004; Cheetham et al., 1992; Westhoff et al., 2005). Furthermore, Hsp105/Hsp110, an HSP70 family member, is enriched in neurons and other cell types of the brain (Hylander et al., 2000; Satoh et al., 1998). Thus, understanding the differences in the regulation and composition of cell-specific proteomes may suggest a molecular basis for the apparent susceptibility of neurons to the expression of aggregation prone proteins.

A number of studies reveal a disparate set of observations in neurons challenging the general assumption that the sequence of events leading to the activation of HSF1 is universal across cell types. *In vivo* and *in vitro* studies of neuronal cells indicate cell-type specific differences of the heat shock response with respect to HSF1 activation and expression, or heat shock gene expression. Batulan et al. show a failure of HSF1 to become activated specifically in motor neurons of a mixed culture, suggesting a cell-type specific negative regulation of HSF1 activation pathways (Batulan et al., 2003). Studies of intact brain sections from postnatal rats show that neuronal cells fail to induce Hsp70 protein expression following a heat shock insult despite comparable HSF1 expression when compared to glial cells (Brown & Rush, 1999). Furthermore, independent studies report significantly diminished levels of HSF1 protein expression and DNA binding activity in neuronal cell lines and primary neuronal cultures (Brown & Rush, 1999; Kaarniranta et al., 2002; Marcuccilli et al., 1996). Decreases in neuronal HSF1 levels and activation correlates with diminished HSP expression (Brown & Rush, 1999; Kaarniranta et al., 2002). Studies disrupting endogenous chaperone gene expression result in increased cellular sensitivity to a variety of stresses and increased vulnerability to proteotoxic stresses (Huang et al., 2001; Shim et al., 2002).

A general tenet of stress biology is cytoprotection: that chaperones induced during transient stress exposure protect the cell and organism against subsequent more severe stress exposure. Studies examining the effects of acute and severe stress have demonstrated that levels of HSP induction correlate with the level of cytoprotection from the stress insult (Parsell et al., 1993). Although a majority of studies on cytoprotection have been conducted in nonneuronal cells, specific studies on neuronal cell types have shown that HSPs can be induced following exposure to stress conferring protection both in culture and in intact animals. Moreover, studies of primary neuronal cultures show that HSPs are

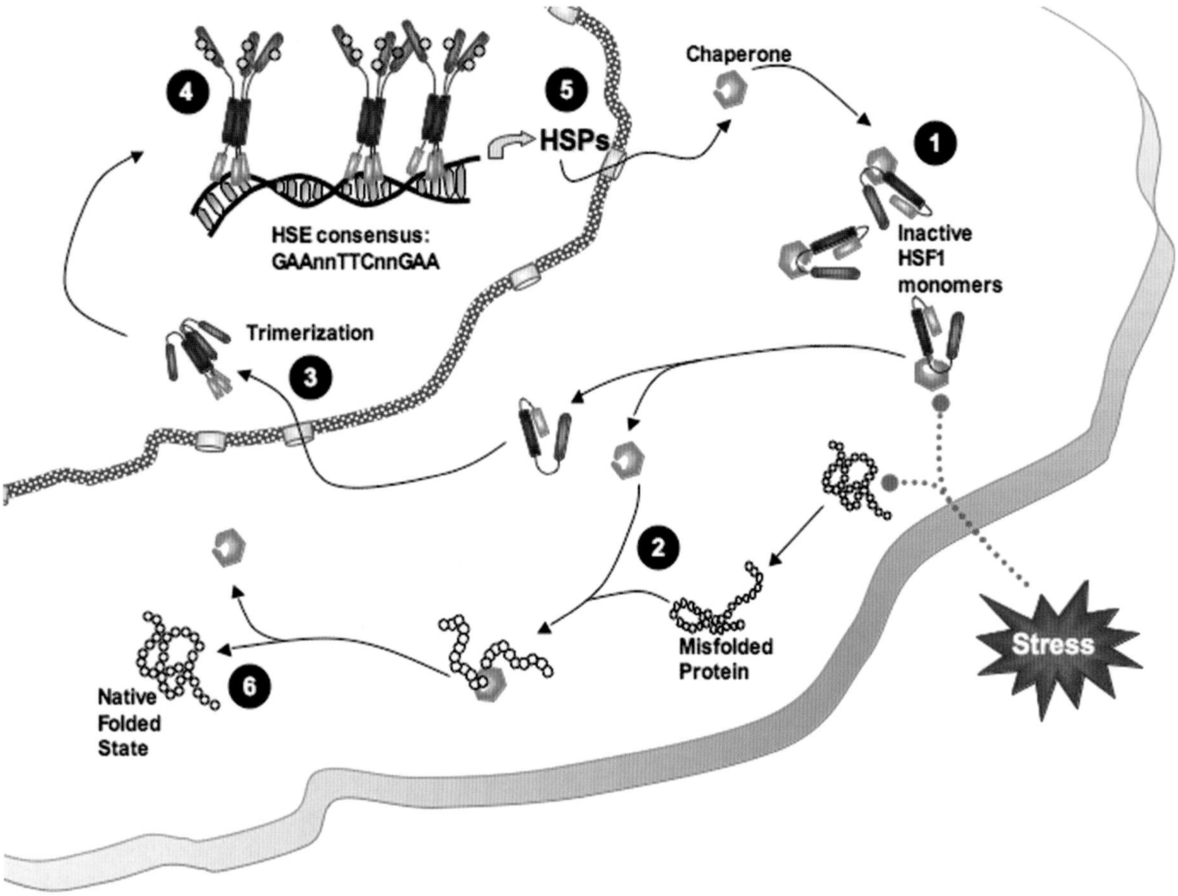

Figure 5.5 Regulation of the HSF1 Activation Cycle. (**1**) Under physiological conditions, human HSF1 occurs predominantly as an inactive monomer, retained in the cytoplasm by the molecular chaperones, Hsp70 and Hsp90. (**2**) Upon stress stimuli, chaperones bind unfolded and aggregated proteins, coinciding with the release of HSF1 monomers. (**3**) HSF1 translocates into the nucleus and trimerizes, allowing (**4**) DNA binding of post-translationally modified HSF1 to HSEs in target chaperone gene promoters. (**5**) The transcriptional activation of HSF1 results in a robust induction of chaperone gene expression. (**6**) Once misfolded and damaged proteins have been folded or cleared, chaperones act to negatively feedback on their own expression.

required for neuronal resistance to stresses such as ischemia, heat stress, and the cytotoxic effects of the neurotransmitter glutamate, as well as other stresses (Amin et al., 1995; Lowenstein et al., 1991; Rordorf et al., 1991). In general, the levels of HSF1 expression and activity as well as HSP expression are reduced in neurons in comparison to other cell types including other nonneuronal cells in the CNS, such as glia and astrocytes (Batulan et al., 2003). These observations suggest an innate susceptibility of neurons to stress and give emphasis to the significance of HSF1 and chaperones in neuroprotection.

Although we understand a great deal about stress-mediated cytoprotection as a result of exposure to various stressors, very few studies distinguish between acute and chronic disturbances on cellular homeostasis and how these correlate with stress vulnerability. In neurons, it is the chronic expression of mutant proteins resulting in the appearance of toxic aggregates that pose a persistent challenge to the protein-folding quality control machinery with deleterious consequence. Very little is known about the molecular differences in how acute and chronic stressors are sensed and remedied.

IV. Role of Molecular Chaperones in Neurodegenerative Diseases

Conformational diseases are associated with the appearance and accumulation of misfolded and aggregation-prone proteins. This has suggested a critical role for the cellular quality control system and molecular chaperones. A number of studies have demonstrated that components of protein folding and degradation machinery such as Hsp70, Hsp40, ubiquitin and proteasome complexes are associated with aggregates containing the disease-associated protein (Auluck et al., 2002; Chai et al., 1999; Cummings et al., 1998; Dou et al., 2003; Hamos et al., 1991; Jana et al., 2000; Kakimura et al., 2002; McLean et al., 2002; Nemes et al., 2004; Renkawek et al., 1994; Shinohara et al., 1993; Waelter et al., 2001; Watanabe et al., 2001). The role of Hsp70 in conformational disease has been well documented for models of polyQ expansion diseases. At least nine neurodegenerative disorders, including Huntington's disease, have been identified in which the disease locus encodes a protein containing an expanded glutamine tract (see Chapter 17). The expansion within the disease protein alters its physical properties, leading to cytoplasmic and nuclear aggregates of the polyQ containing protein (DiFiglia et al., 1997; Trottier et al., 1995). A variety of cellular and animal polyQ disease models have been generated, providing experimental systems for investigating the role of molecular chaperones in neurodegenerative diseases (see Table 5.1).

A. PolyQ Disease Models

Models for polyQ toxicity recapitulate specific aspects of the human disease. Multiple cellular models exist that express the mutant form of a polyQ containing protein in HeLa, COS7, or PC12 cells leading to the formation of large cytoplasmic or nuclear aggregates that depend on the presence of the polyQ expansion (Chai et al., 1999; Cummings et al., 1998). Immunostaining revealed that endogenous Hsp70, its cochaperone Hdj2, ubiquitin, and proteasomal subunits colocalize with aggregates, suggesting an attempt by the protein quality control system to protect the cell from the toxic effects of the disease protein (Cummings et al., 1998). Overexpression of Hsp70s and Hsp40s in various cellular models reduces polyQ aggregation formation (Chai et al., 1999; Cummings et al., 1998). In specific polyQ cell models, large aggregates in cells lead to apoptotic cell death (Kobayashi et al., 2000). Overexpression of Hsp70 or Hsp70 and Hsp40 coexpression not only suppress aggregate formation but also protect cells from polyQ-induced apoptosis (Kobayashi et al., 2000).

Model organisms have provided excellent *in vivo* systems to address genetically the mechanism of polyQ toxicity. These model systems add important criteria to evaluate the role of chaperones in the progression of the disease. Using *Drosophila melanogaster*, a polyQ neurotoxicity model was established where the disease protein was expressed in the eye leading to progressive degeneration of the eye structures and eventual retinal degeneration (Bonini, 2002; Warrick et al., 1998). Inclusions formed within the eye prior to degeneration. Furthermore, the inclusions immunostained for endogenous Hsp70 (Warrick et al., 1999). Overexpression of human constitutive Hsp70 suppressed polyQ dependent degeneration. Introduction of a dominant negative mutation in the ATPase domain (K71S) of human Hsp70 shared a similar neurodegenerative phenotype as expression of the polyQ containing protein suggesting a role for endogenous Hsp70 in neuroprotection. Additional molecular chaperones belonging to the Hsp40 family were tested in this model (Chan et al., 2000). *Drosophila* Hdj1 overexpression suppressed neurodegeneration, but *Drosophila* Hdj2 did not. Coexpression of *Drosophila* Hdj1 and Hsp70 synergistically increased the neuroprotective effects of the chaperone machine. Hsp40 family members also were identified as modifiers of polyQ neurotoxicity in forward genetic screens in multiple *Drosophila* polyQ models (Fernandez-Funez et al., 2000; Kazemi-Esfarjani & Benzer, 2000). Mutations in two J domain-containing proteins, one being a homolog of Hdj1, dramatically reduced neurodegeneration.

Overexpression of chaperones can suppress neurodegeneration in mammals also. Increasing the level of Hsp70 in a mouse polyQ model decreased neurodegeneration and improved behavioral phenotypes (Cummings et al., 2001). Taken together, *in vivo* evidence demonstrates that increasing specific chaperone levels can protect cells from polyQ toxicity in various model organisms.

Although overexpression of Hsp70 or Hsp40 delayed neurodegeneration in the *Drosophila* polyQ model, no morphological changes in aggregate structures were detected using light microscopy (Warrick et al., 1999). *In vitro* studies were needed to examine the physical and biochemical effects of chaperones on aggregate formation. Generally, the self-assembly of polyQ-containing proteins leads to the formation of SDS-resistant amyloid-like fibrils (Muchowski et al., 2000). Electron microscopy demonstrated that the addition of Hsp70 and Hsp40 inhibited self-assembly of polyQ proteins in an ATP-dependent manner. The chaperones allowed the formation of amorphous, detergent-soluble aggregates. Overexpression of Hsp70 or Hsp40 in both *S. cerevisiae* and a mammalian cellular polyQ model prevented the formation of SDS-insoluble aggregates in favor of the detergent soluble aggregates. It is likely that the chaperone-dependent alteration in aggregate structure would not change the appearance of inclusions using light microscopy. Upon closer examination in the *Drosophila* polyQ model, overexpression of Hsp70 with or without Hsp40 increased the SDS solubility of polyQ aggregates (Chan et al., 2000). *In vivo*, molecular chaperones when present at sufficient levels, altered the physical properties of the polyQ-containing aggregates, increasing detergent solubility of the disease protein resulting in neuroprotection.

Colocalization of Hsp70s with polyQ aggregates has been demonstrated in multiple model systems by immunostaining fixed cells. Furthermore, overexpression of Hsp70 suppresses the toxic effects of polyQ disease proteins. Live cell imaging techniques have been employed to investigate the behavior of chaperones at the site of aggregation (Kim et al., 2002). Fluorescence recovery after photobleaching (FRAP) is a technique that can measure the mobility of fluorescently labeled proteins. FRAP of colocalized fluorescently labeled Hsp70 and polyQ-containing aggregates showed recovery of Hsp70 with similar diffusion coefficients to an Hsp70/substrate interaction. Molecular chaperones are not trapped in aggregates but are transiently associated. Mutations in either the ATPase domain or substrate binding domain of Hsp70 reduced colocalization of the chaperone and the inclusion suggesting that a proper chaperone cycle is necessary for the dynamic interaction between Hsp70 and the polyQ aggregate.

Although the majority of studies have implicated Hsp70 and Hsp40 as prominent modifiers of neurodegeneration, the protective contributions of additional chaperone families and specialized tissue-specific chaperones has yet to be explored. Even within chaperone families, specific subclasses of molecular chaperones may be more effective in preventing toxicity. For example, *Drosophila* Hdj1 overexpression suppresses progressive polyQ-associated neurodegeneration, whereas Hdj2 overexpression had no effect (Chan et al., 2000). Although both Hsp40s reside in the cytoplasm, it is likely that the level and time of expression differ between the two Hsp40s along with their substrate specificity. With multiple members of each chaperone family residing within the same cellular compartment, the complement and level of chaperones expressed in cells becomes critical for protection from neurodegeneration. A specific set of chaperones may be required for suppression of conformational diseases. In a genome wide screen using a *Caenorhabditis elegans* polyQ model, RNAi against only two Hsp70s, one Hsp40, and six members of the Hsp60 family enhanced polyQ aggregation (Nollen et al., 2004). Understanding the complement of chaperones expressed in different cell types may reveal the susceptibility of neurons to disease proteins.

B. Parkinson Disease Models

Neurodegeneration has been associated with abnormal protein folding and toxicity for many disorders besides polyQ diseases. Aggregation of α-synuclein is associated with Parkinson disease (Polymeropoulos et al., 1997) (see Table 5.1). A small percentage of aggregates in postmortem brain tissue of Parkinson disease patients immunostain with antibodies to Hsp70 and Hsp40 (Auluck et al., 2002). To determine whether Hsp70 performs a general role in protecting cells against toxic protein aggregation, similar overexpression experiments were performed in a *Drosophila* model of Parkinson disease (Auluck & Bonini, 2002; Auluck et al., 2002). Expression of α-synuclein, driven by the *dopa decarboxylase* promoter, leads to a progressive degeneration of dopaminergic neurons in adult *Drosophila*. Overexpression of Hsp70 suppresses degeneration of the dopaminergic neurons, whereas expression of the dominant-negative Hsp70 mutant enhances neurodegeneration. Molecular chaperones suppress neurotoxicity in multiple models of Parkinson and polyQ diseases, suggesting that increasing chaperone levels is one potential therapeutic strategy for conformational diseases.

V. Chaperone Hypotheses

The molecular mechanism by which increasing chaperone levels protect against neurotoxicity of conformational diseases is unclear. Diverse disease proteins can misfold into multiple toxic conformations. Biochemical data demonstrates that molecular chaperones enhance the solubility of polyQ proteins correlating with decreased cytotoxicity (Muchowski et al., 2000). The chaperones may be maintaining the disease protein in a less toxic conformation. Identification of key toxic species of disease proteins is an active area of investigation. Once the structural nature of these detrimental conformations is determined, their downstream targets can be characterized. Increasing the solubility of polyQ proteins by chaperones may also allow more efficient turnover of the polyQ protein by the proteasome. In one study, overexpression of chaperones increased polyQ protein solubility along with significantly decreasing the half-life of the disease protein (Bailey et al., 2002). Increasing the clearance of toxic species via the cell's proteolytic machinery could also lead to neuroprotection. Thus, alterations in conformation of disease proteins by chaperones may be a fundamental step in suppressing toxicity (see Figure 5.6).

Misfolded proteins expose hydrophobic residues that can readily self-associate in the crowded cellular environment, leading to aggregation. These exposed residues on the surface of the inclusion can interact with cellular proteins, sequestering them within the aggregates leading to degeneration. In polyQ disease models, Q-rich cellular proteins, such as a key transcriptional coactivator for neuronal survival CBP (CREB-binding protein), co-localize with inclusions (McCampbell et al., 2000; Nucifora et al., 2001; Schaffar et al., 2004). Overexpression of sequestered proteins such as CBP reduces toxicity in cellular polyQ models. Hsp70 chaperones transiently associate with aggregates on the surface by interacting with exposed residues in a chaperone/substrate type of interaction (Kim et al., 2002). Overexpression of Hsp70 may protect cells from toxicity by coating the surface exposed hydrophobic residues, shielding cellular proteins from aberrant interactions with the aggregates (see Figure 5.6).

Chaperones are required for proper folding and function of many cellular proteins that participate in diverse pathways. The presence of a misfolded disease protein may interfere with chaperone function, perturbing a wide range of cellular pathways. The global consequence of expression

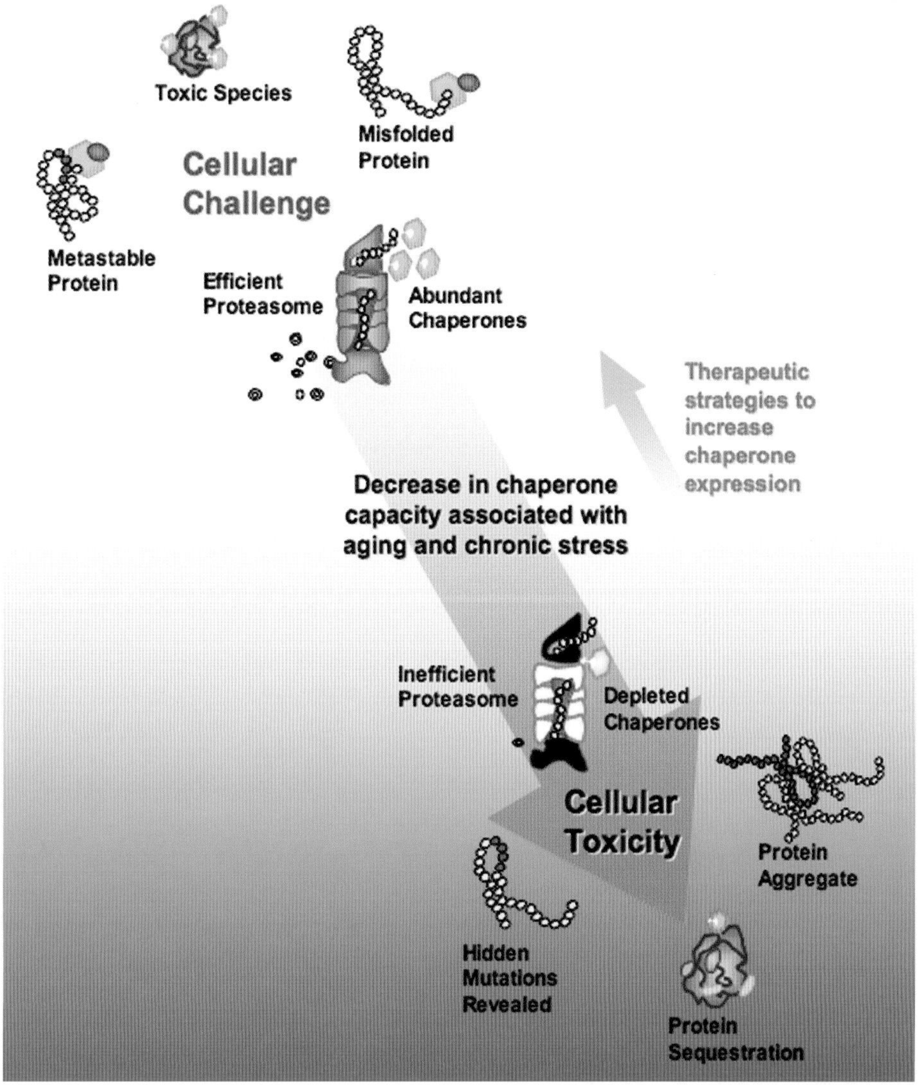

Figure 5.6 Chaperones protect neurons from disease proteins. Increasing chaperone levels may protect against the chronic expression of disease proteins in multiple ways. First, chaperones can maintain disease proteins in a soluble conformation facilitating efficient turnover. Second, chaperones can bind surface-exposed hydrophobic residues in protein deposits shielding cellular proteins from aberrant interactions and preventing sequestration. Also, metastable proteins within the genome are assisted in their folding for proper function. Changes in the capacity of the cellular protein quality control systems due to chronic expression of disease proteins or the aging process leads to the accumulation of protein aggregates. Therapeutic avenues may be directed at increasing the levels of critical chaperones needed to restore neuronal homeostasis.

of aggregation-prone proteins likely upsets the balance of the protein-folding quality control machinery. Overloading the chaperone system may reveal hidden folding mutations within the genome. Polymorphisms that lead to metastable proteins that require vigilant folding assistance to obtain their functional conformations can act as modifiers in the presence of disease proteins (Gidalevitz et al., 2006). Folding machines preoccupied with misfolded disease proteins no longer properly assist metastable protein folding. The cellular pathways in which metastable proteins function are compromised, contributing to disease progression (Bates, 2006). Overexpression of chaperones may suppress neurodegeneration by maintaining proper folding of silent mutations within the genome (see Figure 5.6). Since chaperones participate in multiple cellular pathways, we must continue to decipher which chaperone contributions are primary and which are secondary to protection against neurodegeneration.

Post-mitotic cells, such as neurons, are extremely susceptible to the problems of accumulation of toxic protein species over time. The symptoms of many conformational diseases begin late in life suggesting that, in these disorders, aged neurons are more susceptible to degeneration (Soti & Csermely, 2002). During the aging process, disease proteins continually interfere with normal chaperone function. Damaged proteins accumulating over time due to various cellular stresses further challenge the less efficient chaperone machinery. Eventually, the chaperone system is overloaded, leading to the failure of multiple cellular processes, revealing disease symptoms (see Figure 5.6).

VI. Therapeutic Avenues

A growing body of evidence suggests that the artificial manipulation of the levels of molecular chaperones may provide a promising approach for treatment of various neurodegenerative diseases. Chemical inducers of the HSR and individual HSPs recently have been demonstrated to have a significant effect in animal models of neurodegeneration (Westerheide & Morimoto, 2005). Increasing Hsp70 or Hsp40 levels in aged neurons may protect against neurodegeneration. However, cells lacking an efficient chaperone system may require more than overexpression of Hsp70 or Hsp40. Characterizing the chaperone complement present in aging neurons may identify strategies for protection from toxic protein species. Treatment for conformational diseases may be achieved through up-regulation of specific chaperone levels or activity in aged neurons (see Figure 5.6).

References

Abravaya, K., Phillips, B., and Morimoto, R. I. (1991). Attenuation of the heat shock response in HeLa cells is mediated by the release of bound heat shock transcription factor and is modulated by changes in growth and in heat shock temperatures. *Genes Dev* **5**, 2117–2127.

Alastalo, T. P., Hellesuo, M., Sandqvist, A., Hietakangas, V., Kallio, M., and Sistonen, L. (2003). Formation of nuclear stress granules involves HSF2 and coincides with the nucleolar localization of Hsp70. *J Cell Sci* **116**, 3557–3570.

Amin, J., Ananthan, J., and Voellmy, R. (1988). Key features of heat shock regulatory elements. *Mol Cell Biol* **8**, 3761–3769.

Amin, V., Cumming, D. V., Coffin, R. S., and Latchman, D. S. (1995). The degree of protection provided to neuronal cells by a pre-conditioning stress correlates with the amount of heat shock protein 70 it induces and not with the similarity of the subsequent stress. *Neurosci Lett* **200**, 85–88.

Anckar, J., Hietakangas, V., Denessiouk, K., Thiele, D. J., Johnson, M. S., and Sistonen, L. (2006). Inhibition of DNA binding by differential sumoylation of heat shock factors. *Mol Cell Biol* **26**, 955–964.

Anfinsen, C. B. (1973). Principles that govern the folding of protein chains. *Science* **181**, 223–230.

Auluck, P. K. and Bonini, N. M. (2002). Pharmacological prevention of Parkinson disease in Drosophila. *Nat Med* **8**, 1185–1186.

Auluck, P. K., Chan, H. Y., Trojanowski, J. Q., Lee, V. M., and Bonini, N. M. (2002). Chaperone suppression of alpha-synuclein toxicity in a Drosophila model for Parkinson's disease. *Science* **295**, 865–868.

Bailey, C. K., Andriola, I. F., Kampinga, H. H., and Merry, D. E. (2002). Molecular chaperones enhance the degradation of expanded polyQ repeat androgen receptor in a cellular model of spinal and bulbar muscular atrophy. *Hum Mol Genet* **11**, 515–523.

Ballinger, C. A., Connell, P., Wu, Y., Hu, Z., Thompson, L. J., Yin, L. Y., and Patterson, C. (1999). Identification of CHIP, a novel tetratricopeptide repeat-containing protein that interacts with heat shock proteins and negatively regulates chaperone functions. *Mol Cell Biol* **19**, 4535–4545.

Bates, G. P. (2006). BIOMEDICINE: One misfolded protein allows others to sneak by. *Science* **311**, 1385–1386.

Batulan, Z., Shinder, G. A., Minotti, S., He, B. P., Doroudchi, M. M., Nalbantoglu, J. et al. (2003). High threshold for induction of the stress response in motor neurons is associated with failure to activate HSF1. *J Neurosci* **23**, 5789–5798.

Becker, J., Walter, W., Yan, W., and Craig, E. A. (1996). Functional interaction of cytosolic hsp70 and a DnaJ-related protein, Ydj1p, in protein translocation in vivo. *Mol Cell Biol* **16**, 4378–4386.

Ben-Zvi, A. P. and Goloubinoff, P. (2001). Review: Mechanisms of disaggregation and refolding of stable protein aggregates by molecular chaperones. *J Struct Biol* **135**, 84–93.

Beyer, A. (1997). Sequence analysis of the AAA protein family. *Protein Sci* **6**, 2043–2058.

Bonini, N. M. (2002). Chaperoning brain degeneration. *Proc Natl Acad Sci U S A* **99 Suppl 4**, 16407–16411.

Boorstein, W. R., Ziegelhoffer, T., and Craig, E. A. (1994). Molecular evolution of the HSP70 multigene family. *J Mol Evol* **38**, 1–17.

Braig, K., Otwinowski, Z., Hegde, R., Boisvert, D. C., Joachimiak, A., Horwich, A. L., and Sigler, P. B. (1994). The crystal structure of the bacterial chaperonin GroEL at 2.8 A. *Nature* **371**, 578–586.

Brown, I. R. and Rush, S. J. (1999). Cellular localization of the heat shock transcription factors HSF1 and HSF2 in the rat brain during postnatal development and following hyperthermia. *Brain Res* **821**, 333–340.

Buchner, J., Schmidt, M., Fuchs, M., Jaenicke, R., Rudolph, R., Schmid, F. X., and Kiefhaber, T. (1991). GroE facilitates refolding of citrate synthase by suppressing aggregation. *Biochemistry* **30**, 1586–1591.

Bukau, B. and Horwich, A. L. (1998). The Hsp70 and Hsp60 chaperone machines. *Cell* **92**, 351–366.

Chai, Y., Koppenhafer, S. L., Bonini, N. M., and Paulson, H. L. (1999). Analysis of the role of heat shock protein (Hsp) molecular chaperones in polyQ disease. *J Neurosci* **19**, 10338–10347.

Chan, H. Y., Warrick, J. M., Gray-Board, G. L., Paulson, H. L., and Bonini, N. M. (2000). Mechanisms of chaperone suppression of polyQ disease: Selectivity, synergy and modulation of protein solubility in Drosophila. *Hum Mol Genet* **9**, 2811–2820.

Chapple, J. P., van der Spuy, J., Poopalasundaram, S., and Cheetham, M. E. (2004). Neuronal DnaJ proteins HSJ1a and HSJ1b: A role in linking the Hsp70 chaperone machine to the ubiquitin-proteasome system? *Biochem Soc Trans* **32**, 640–642.

Cheetham, M. E., Brion, J. P., and Anderton, B. H. (1992). Human homologues of the bacterial heat-shock protein DnaJ are preferentially expressed in neurons. *Biochem J* **284 (Pt 2)**, 469–476.

Cheetham, M. E. and Caplan, A. J. (1998). Structure, function and evolution of DnaJ: Conservation and adaptation of chaperone function. *Cell Stress Chaperones* **3**, 28–36.

Clark, J. I. and Muchowski, P. J. (2000). Small heat-shock proteins and their potential role in human disease. *Curr Opin Struct Biol* **10**, 52–59.

Connell, P., Ballinger, C. A., Jiang, J., Wu, Y., Thompson, L. J., Hohfeld, J., and Patterson, C. (2001). The co-chaperone CHIP regulates protein triage decisions mediated by heat-shock proteins. *Nat Cell Biol* **3**, 93–96.

Craig, E. A., Huang, P., Aron, R., and Andrew, A. (2006). The diverse roles of J-proteins, the obligate Hsp70 co-chaperone. *Rev Physiol Biochem Pharmacol* **156**, 1–21.

Csermely, P. (2001). Chaperone overload is a possible contributor to 'civilization diseases'. *Trends Genet* **17**, 701–704.

Cummings, C. J., Mancini, M. A., Antalffy, B., DeFranco, D. B., Orr, H. T., and Zoghbi, H. Y. (1998). Chaperone suppression of aggregation and altered subcellular proteasome localization imply protein misfolding in SCA1. *Nat Genet* **19**, 148–154.

Cummings, C. J., Sun, Y., Opal, P., Antalffy, B., Mestril, R., Orr, H. T. et al. (2001). Over-expression of inducible HSP70 chaperone suppresses neuropathology and improves motor function in SCA1 mice. *Hum Mol Genet* **10**, 1511–1518.

Demand, J., Luders, J., and Hohfeld, J. (1998). The carboxy-terminal domain of Hsc70 provides binding sites for a distinct set of chaperone cofactors. *Mol Cell Biol* **18**, 2023–2028.

Deuerling, E., Schulze-Specking, A., Tomoyasu, T., Mogk, A., and Bukau, B. (1999). Trigger factor and DnaK cooperate in folding of newly synthesized proteins. *Nature* **400**, 693–696.

DiFiglia, M., Sapp, E., Chase, K. O., Davies, S. W., Bates, G. P., Vonsattel, J. P., and Aronin, N. (1997). Aggregation of huntingtin in neuronal intranuclear inclusions and dystrophic neurites in brain. *Science* **277**, 1990–1993.

Dobson, C. M. (2004). Principles of protein folding, misfolding and aggregation. *Semin Cell Dev Biol* **15**, 3–16.

Dou, F., Netzer, W. J., Tanemura, K., Li, F., Hartl, F. U., Takashima, A. et al. (2003). Chaperones increase association of tau protein with microtubules. *Proc Natl Acad Sci U S A* **100**, 721–726.

Esser, C., Alberti, S., and Hohfeld, J. (2004). Cooperation of molecular chaperones with the ubiquitin/proteasome system. *Biochim Biophys Acta* **1695**, 171–188.

Fenton, W. A., Kashi, Y., Furtak, K., and Horwich, A. L. (1994). Residues in chaperonin GroEL required for polypeptide binding and release. *Nature* **371**, 614–619.

Fernandez-Funez, P., Nino-Rosales, M. L., de Gouyon, B., She, W. C., Luchak, J. M., Martinez, P. et al. (2000). Identification of genes that modify ataxin-1-induced neurodegeneration. *Nature* **408**, 101–106.

Flaherty, K. M., DeLuca-Flaherty, C., and McKay, D. B. (1990). Three-dimensional structure of the ATPase fragment of a 70K heat-shock cognate protein. *Nature* **346**, 623–628.

Flynn, G. C., Chappell, T. G., and Rothman, J. E. (1989). Peptide binding and release by proteins implicated as catalysts of protein assembly. *Science* **245**, 385–390.

Freeman, B. C. and Morimoto, R. I. (1996). The human cytosolic molecular chaperones hsp90, hsp70 (hsc70) and hdj-1 have distinct roles in recognition of a non-native protein and protein refolding. *Embo J* **15**, 2969–2979.

Freeman, B. C., Myers, M. P., Schumacher, R., and Morimoto, R. I. (1995). Identification of a regulatory motif in Hsp70 that affects ATPase activity, substrate binding and interaction with HDJ-1. *Embo J* **14**, 2281–2292.

Freeman, B. C., Toft, D. O., and Morimoto, R. I. (1996). Molecular chaperone machines: Chaperone activities of the cyclophilin Cyp-40 and the steroid aporeceptor-associated protein p23. *Science* **274**, 1718–1720.

Frydman, J. (2001). Folding of newly translated proteins in vivo: The role of molecular chaperones. *Annu Rev Biochem* **70**, 603–647.

Fujimoto, M., Izu, H., Seki, K., Fukuda, K., Nishida, T., Yamada, S. et al. (2004). HSF4 is required for normal cell growth and differentiation during mouse lens development. *Embo J* **23**, 4297–4306.

Gething, M. J. and Sambrook, J. (1992). Protein folding in the cell. *Nature* **355**, 33–45.

Gidalevitz, T., Ben-Zvi, A., Ho, K. H., Brignull, H. R., and Morimoto, R. I. (2006). Progressive disruption of cellular protein folding in models of polyQ diseases. *Science* **311**, 1471–1474.

Glover, J. R. and Lindquist, S. (1998). Hsp104, Hsp70, and Hsp40: A novel chaperone system that rescues previously aggregated proteins. *Cell* **94**, 73–82.

Goldberg, A. L. (2003). Protein degradation and protection against misfolded or damaged proteins. *Nature* **426**, 895–899.

Goloubinoff, P., Christeller, J. T., Gatenby, A. A., and Lorimer, G. H. (1989a). Reconstitution of active dimeric ribulose bisphosphate carboxylase from an unfolded state depends on two chaperonin proteins and Mg-ATP. *Nature* **342**, 884–889.

Goloubinoff, P., Gatenby, A. A., and Lorimer, G. H. (1989b). GroE heat-shock proteins promote assembly of foreign prokaryotic ribulose bisphosphate carboxylase oligomers in Escherichia coli. *Nature* **337**, 44–47.

Goloubinoff, P., Mogk, A., Zvi, A. P., Tomoyasu, T., and Bukau, B. (1999). Sequential mechanism of solubilization and refolding of stable protein aggregates by a bichaperone network. *Proc Natl Acad Sci U S A* **96**, 13732–13737.

Greene, M. K., Maskos, K., and Landry, S. J. (1998). Role of the J-domain in the cooperation of Hsp40 with Hsp70. *Proc Natl Acad Sci U S A* **95**, 6108–6113.

Hamos, J. E., Oblas, B., Pulaski-Salo, D., Welch, W. J., Bole, D. G., and Drachman, D. A. (1991). Expression of heat shock proteins in Alzheimer's disease. *Neurology* **41**, 345–350.

Hartl, F. U. (1996). Molecular chaperones in cellular protein folding. *Nature* **381**, 571–579.

Hartl, F. U. and Hayer-Hartl, M. (2002). Molecular chaperones in the cytosol: From nascent chain to folded protein. *Science* **295**, 1852–1858.

He, H., Soncin, F., Grammatikakis, N., Li, Y., Siganou, A., Gong, J. et al. (2003). Elevated expression of heat shock factor (HSF) 2A stimulates HSF1-induced transcription during stress. *J Biol Chem* **278**, 35465–35475.

Hietakangas, V., Anckar, J., Blomster, H. A., Fujimoto, M., Palvimo, J. J., Nakai, A., and Sistonen, L. (2006). PDSM, a motif for phosphorylation-dependent SUMO modification. *Proc Natl Acad Sci U S A* **103**, 45–50.

Hohfeld, J. and Jentsch, S. (1997). GrpE-like regulation of the hsc70 chaperone by the anti-apoptotic protein BAG-1. *Embo J* **16**, 6209–6216.

Hohfeld, J., Minami, Y., and Hartl, F. U. (1995). Hip, a novel cochaperone involved in the eukaryotic Hsc70/Hsp40 reaction cycle. *Cell* **83**, 589–598.

Huang, L., Mivechi, N. F., and Moskophidis, D. (2001). Insights into regulation and function of the major stress-induced hsp70 molecular chaperone in vivo: Analysis of mice with targeted gene disruption of the hsp70.1 or hsp70.3 gene. *Mol Cell Biol* **21**, 8575–8591.

Hylander, B. L., Chen, X., Graf, P. C., and Subjeck, J. R. (2000). The distribution and localization of hsp110 in brain. *Brain Res* **869**, 49–55.

Jana, N. R., Tanaka, M., Wang, G., and Nukina, N. (2000). PolyQ length-dependent interaction of Hsp40 and Hsp70 family chaperones with truncated N-terminal huntingtin: Their role in suppression of aggregation and cellular toxicity. *Hum Mol Genet* **9**, 2009–2018.

Jiang, J., Prasad, K., Lafer, E. M., and Sousa, R. (2005). Structural basis of interdomain communication in the Hsc70 chaperone. *Mol Cell* **20**, 513–524.

Kaarniranta, K., Oksala, N., Karjalainen, H. M., Suuronen, T., Sistonen, L., Helminen, H. J. et al. (2002). Neuronal cells show regulatory differences in the hsp70 gene response. *Brain Res Mol Brain Res* **101**, 136–140.

Kakimura, J., Kitamura, Y., Takata, K., Umeki, M., Suzuki, S., Shibagaki, K. (2002). Microglial activation and amyloid-beta clearance induced by exogenous heat-shock proteins. *Faseb J* **16**, 601–603.

Kakizuka, A. (1998). Protein precipitation: A common etiology in neurodegenerative disorders? *Trends Genet* **14**, 396–402.

Kang, P. J., Ostermann, J., Shilling, J., Neupert, W., Craig, E. A., and Pfanner, N. (1990). Requirement for hsp70 in the mitochondrial matrix for translocation and folding of precursor proteins. *Nature* **348**, 137–143.

Kazemi-Esfarjani, P. and Benzer, S. (2000). Genetic suppression of polyQ toxicity in Drosophila. *Science* **287**, 1837–1840.

Kelley, W. L. (1998). The J-domain family and the recruitment of chaperone power. *Trends Biochem Sci* **23**, 222–227.

Kim, S., Nollen, E. A., Kitagawa, K., Bindokas, V. P., and Morimoto, R. I. (2002). PolyQ protein aggregates are dynamic. *Nat Cell Biol* **4**, 826–831.

Kline, M. P. and Morimoto, R. I. (1997). Repression of the heat shock factor 1 transcriptional activation domain is modulated by constitutive phosphorylation. *Mol Cell Biol* **17**, 2107–2115.

Knauf, U., Newton, E. M., Kyriakis, J., and Kingston, R. E. (1996). Repression of human heat shock factor 1 activity at control temperature by phosphorylation. *Genes Dev* **10**, 2782–2793.

Kobayashi, Y., Kume, A., Li, M., Doyu, M., Hata, M., Ohtsuka, K., and Sobue, G. (2000). Chaperones Hsp70 and Hsp40 suppress aggregate formation and apoptosis in cultured neuronal cells expressing truncated androgen receptor protein with expanded polyQ tract. *J Biol Chem* **275**, 8772–8778.

Kopito, R. R. and Ron, D. (2000). Conformational disease. *Nat Cell Biol* **2**, E207–209.

Kroeger, P. E. and Morimoto, R. I. (1994). Selection of new HSF1 and HSF2 DNA-binding sites reveals difference in trimer cooperativity. *Mol Cell Biol* **14**, 7592–7603.

Kusmierczyk, A. R. and Martin, J. (2001). Chaperonins—Keeping a lid on folding proteins. *FEBS Lett* **505**, 343–347.

Langer, T., Lu, C., Echols, H., Flanagan, J., Hayer, M. K., and Hartl, F. U. (1992). Successive action of DnaK, DnaJ and GroEL along the pathway of chaperone-mediated protein folding. *Nature* **356**, 683–689.

Larson, J. S., Schuetz, T. J., and Kingston, R. E. (1988). Activation in vitro of sequence-specific DNA binding by a human regulatory factor. *Nature* **335**, 372–375.

Lindquist, S. and Craig, E. A. (1988). The heat-shock proteins. *Annu Rev Genet* **22**, 631–677.

Liu, X. D., Liu, P. C., Santoro, N., and Thiele, D. J. (1997). Conservation of a stress response: Human heat shock transcription factors functionally substitute for yeast HSF. *Embo J* **16**, 6466–6477.

Lowenstein, D. H., Chan, P. H., and Miles, M. F. (1991). The stress protein response in cultured neurons: characterization and evidence for a protective role in excitotoxicity. *Neuron* **7**, 1053–1060.

Lu, Z. and Cyr, D. M. (1998). The conserved carboxyl terminus and zinc finger-like domain of the co-chaperone Ydj1 assist Hsp70 in protein folding. *J Biol Chem* **273**, 5970–5978.

Marcuccilli, C. J., Mathur, S. K., Morimoto, R. I., and Miller, R. J. (1996). Regulatory differences in the stress response of hippocampal neurons and glial cells after heat shock. *J Neurosci* **16**, 478–485.

Mathew, A., Mathur, S. K., Jolly, C., Fox, S. G., Kim, S., and Morimoto, R. I. (2001). Stress-specific activation and repression of heat shock factors 1 and 2. *Mol Cell Biol* **21**, 7163–7171.

Mayer, M. P. and Bukau, B. (2005). Hsp70 chaperones: Cellular functions and molecular mechanism. *Cell Mol Life Sci* **62**, 670–684.

McCampbell, A., Taylor, J. P., Taye, A. A., Robitschek, J., Li, M., Walcott, J. et al. (2000). CREB-binding protein sequestration by expanded polyQ. *Hum Mol Genet* **9**, 2197–2202.

McLean, P. J., Kawamata, H., Shariff, S., Hewett, J., Sharma, N., Ueda, K. et al. (2002). TorsinA and heat shock proteins act as molecular chaperones: Suppression of alpha-synuclein aggregation. *J Neurochem* **83**, 846–854.

McMillan, D. R., Xiao, X., Shao, L., Graves, K., and Benjamin, I. J. (1998). Targeted disruption of heat shock transcription factor 1 abolishes thermotolerance and protection against heat-inducible apoptosis. *J Biol Chem* **273**, 7523–7528.

Meacham, G. C., Patterson, C., Zhang, W., Younger, J. M., and Cyr, D. M. (2001). The Hsc70 co-chaperone CHIP targets immature CFTR for proteasomal degradation. *Nat Cell Biol* **3**, 100–105.

Morimoto, R. I., Kline, M. P., Bimston, D. N., and Cotto, J. J. (1997). The heat-shock response: Regulation and function of heat-shock proteins and molecular chaperones. *Essays Biochem* **32**, 17–29.

Muchowski, P. J., Schaffar, G., Sittler, A., Wanker, E. E., Hayer-Hartl, M. K., and Hartl, F. U. (2000). Hsp70 and hsp40 chaperones can inhibit self-assembly of polyQ proteins into amyloid-like fibrils. *Proc Natl Acad Sci U S A* **97**, 7841–7846.

Muchowski, P. J. and Wacker, J. L. (2005). Modulation of neurodegeneration by molecular chaperones. *Nat Rev Neurosci* **6**, 11–22.

Nakai, A., Tanabe, M., Kawazoe, Y., Inazawa, J., Morimoto, R. I., and Nagata, K. (1997). HSF4, a new member of the human heat shock factor family which lacks properties of a transcriptional activator. *Mol Cell Biol* **17**, 469–481.

Nathan, D. F., Vos, M. H., and Lindquist, S. (1997). In vivo functions of the Saccharomyces cerevisiae Hsp90 chaperone. *Proc Natl Acad Sci U S A* **94**, 12949–12956.

Nemes, Z., Devreese, B., Steinert, P. M., Van Beeumen, J., and Fesus, L. (2004). Cross-linking of ubiquitin, HSP27, parkin, and alpha-synuclein by gamma-glutamyl-epsilon-lysine bonds in Alzheimer's neurofibrillary tangles. *Faseb J* **18**, 1135–1137.

Nollen, E. A., Garcia, S. M., van Haaften, G., Kim, S., Chavez, A., Morimoto, R. I., and Plasterk, R. H. (2004). Genome-wide RNA interference screen identifies previously undescribed regulators of polyQ aggregation. *Proc Natl Acad Sci U S A* **101**, 6403–6408.

Nucifora, F. C., Jr., Sasaki, M., Peters, M. F., Huang, H., Cooper, J. K., Yamada, M. et al. (2001). Interference by huntingtin and atrophin-1 with cbp-mediated transcription leading to cellular toxicity. *Science* **291**, 2423–2428.

Ogura, T. and Wilkinson, A. J. (2001). AAA+ superfamily ATPases: Common structure—Diverse function. *Genes Cells* **6**, 575–597.

Parsell, D. A., Taulien, J., and Lindquist, S. (1993). The role of heat-shock proteins in thermotolerance. *Philos Trans R Soc Lond B Biol Sci* **339**, 279–285; discussion 285–286.

Pearl, L. H. and Prodromou, C. (2000). Structure and in vivo function of Hsp90. *Curr Opin Struct Biol* **10**, 46–51.

Pirkkala, L., Alastalo, T. P., Zuo, X., Benjamin, I. J., and Sistonen, L. (2000). Disruption of heat shock factor 1 reveals an essential role in the ubiquitin proteolytic pathway. *Mol Cell Biol* **20**, 2670–2675.

Pirkkala, L., Nykanen, P., and Sistonen, L. (2001). Roles of the heat shock transcription factors in regulation of the heat shock response and beyond. *Faseb J* **15**, 1118–1131.

Polymeropoulos, M. H., Lavedan, C., Leroy, E., Ide, S. E., Dehejia, A., Dutra, A. et al. (1997). Mutation in the alpha-synuclein gene identified in families with Parkinson's disease. *Science* **276**, 2045–2047.

Pratt, W. B. and Toft, D. O. (2003). Regulation of signaling protein function and trafficking by the hsp90/hsp70-based chaperone machinery. *Exp Biol Med (Maywood)* **228**, 111–133.

Rabindran, S. K., Giorgi, G., Clos, J., and Wu, C. (1991). Molecular cloning and expression of a human heat shock factor, HSF1. *Proc Natl Acad Sci U S A* **88**, 6906–6910.

Radford, S. E. (2000). Protein folding: progress made and promises ahead. *Trends Biochem Sci* **25**, 611–618.

Renkawek, K., Bosman, G. J., and de Jong, W. W. (1994). Expression of small heat-shock protein hsp 27 in reactive gliosis in Alzheimer disease and other types of dementia. *Acta Neuropathol (Berl)* **87**, 511–519.

Romisch, K. (2005). Endoplasmic reticulum-associated degradation. *Annu Rev Cell Dev Biol* **21**, 435–456.

Rordorf, G., Koroshetz, W. J., and Bonventre, J. V. (1991). Heat shock protects cultured neurons from glutamate toxicity. *Neuron* **7**, 1043–1051.

Rosen, D. R. (1993). Mutations in Cu/Zn superoxide dismutase gene are associated with familial amyotrophic lateral sclerosis. *Nature* **364**, 362.

Rudiger, S., Germeroth, L., Schneider-Mergener, J., and Bukau, B. (1997). Substrate specificity of the DnaK chaperone determined by screening cellulose-bound peptide libraries. *Embo J* **16**, 1501–1507.

Rudiger, S., Schneider-Mergener, J., and Bukau, B. (2001). Its substrate specificity characterizes the DnaJ co-chaperone as a scanning factor for the DnaK chaperone. *Embo J* **20**, 1042–1050.

Satoh, J., Yukitake, M., and Kuroda, Y. (1998). Constitutive and heat-inducible expression of HSP105 in neurons and glial cells in culture. *Neuroreport* **9**, 2977–2983.

Schaffar, G., Breuer, P., Boteva, R., Behrends, C., Tzvetkov, N., Strippel, N. et al. (2004). Cellular toxicity of polyQ expansion proteins: Mechanism of transcription factor deactivation. *Mol Cell* **15**, 95–105.

Scheufler, C., Brinker, A., Bourenkov, G., Pegoraro, S., Moroder, L., Bartunik, H. et al. (2000). Structure of TPR domain-peptide complexes: Critical elements in the assembly of the Hsp70–Hsp90 multichaperone machine. *Cell* **101**, 199–210.

Schirmer, E. C., Glover, J. R., Singer, M. A., and Lindquist, S. (1996). HSP100/Clp proteins: A common mechanism explains diverse functions. *Trends Biochem Sci* **21**, 289–296.

Schmid, D., Baici, A., Gehring, H., and Christen, P. (1994). Kinetics of molecular chaperone action. *Science* **263**, 971–973.

Schuetz, T. J., Gallo, G. J., Sheldon, L., Tempst, P., and Kingston, R. E. (1991). Isolation of a cDNA for HSF2: Evidence for two heat shock factor genes in humans. *Proc Natl Acad Sci U S A* **88**, 6911–6915.

Sherman, M. Y. and Goldberg, A. L. (2001). Cellular defenses against unfolded proteins: A cell biologist thinks about neurodegenerative diseases. *Neuron* **29**, 15–32.

Shi, Y., Mosser, D. D., and Morimoto, R. I. (1998). Molecular chaperones as HSF1-specific transcriptional repressors. *Genes Dev* **12**, 654–666.

Shim, E. H., Kim, J. I., Bang, E. S., Heo, J. S., Lee, J. S., Kim, E. Y. et al. (2002). Targeted disruption of hsp70.1 sensitizes to osmotic stress. *EMBO Rep* **3**, 857–861.

Shinohara, H., Inaguma, Y., Goto, S., Inagaki, T., and Kato, K. (1993). Alpha B crystallin and HSP28 are enhanced in the cerebral cortex of patients with Alzheimer's disease. *J Neurol Sci* **119**, 203–208.

Siegers, K., Waldmann, T., Leroux, M. R., Grein, K., Shevchenko, A., Schiebel, E., and Hartl, F. U. (1999). Compartmentation of protein folding in vivo: Sequestration of non-native polypeptide by the chaperonin-GimC system. *Embo J* **18**, 75–84.

Sigler, P. B., Xu, Z., Rye, H. S., Burston, S. G., Fenton, W. A., and Horwich, A. L. (1998). Structure and function in GroEL-mediated protein folding. *Annu Rev Biochem* **67**, 581–608.

Sistonen, L., Sarge, K. D., and Morimoto, R. I. (1994). Human heat shock factors 1 and 2 are differentially activated and can synergistically induce hsp70 gene transcription. *Mol Cell Biol* **14**, 2087–2099.

Sistonen, L., Sarge, K. D., Phillips, B., Abravaya, K., and Morimoto, R. I. (1992). Activation of heat shock factor 2 during hemin-induced differentiation of human erythroleukemia cells. *Mol Cell Biol* **12**, 4104–4111.

Sondermann, H., Scheufler, C., Schneider, C., Hohfeld, J., Hartl, F. U., and Moarefi, I. (2001). Structure of a Bag/Hsc70 complex: convergent functional evolution of Hsp70 nucleotide exchange factors. *Science* **291**, 1553–1557.

Sorger, P. K. and Pelham, H. R. (1988). Yeast heat shock factor is an essential DNA-binding protein that exhibits temperature-dependent phosphorylation. *Cell* **54**, 855–864.

Soti, C. and Csermely, P. (2002). Chaperones and aging: Role in neurodegeneration and in other civilizational diseases. *Neurochem Int* **41**, 383–389.

Stefani, M. and Dobson, C. M. (2003). Protein aggregation and aggregate toxicity: New insights into protein folding, misfolding diseases and biological evolution. *J Mol Med* **81**, 678–699.

Sun, Y. and MacRae, T. H. (2005). Small heat shock proteins: Molecular structure and chaperone function. *Cell Mol Life Sci* **62**, 2460–2476.

Takayama, S. and Reed, J. C. (2001). Molecular chaperone targeting and regulation by BAG family proteins. *Nat Cell Biol* **3**, E237–241.

Tanabe, M., Sasai, N., Nagata, K., Liu, X. D., Liu, P. C., Thiele, D. J., and Nakai, A. (1999). The mammalian HSF4 gene generates both an activator and a repressor of heat shock genes by alternative splicing. *J Biol Chem* **274**, 27845–27856.

Teter, S. A., Houry, W. A., Ang, D., Tradler, T., Rockabrand, D., Fischer, G. et al. (1999). Polypeptide flux through bacterial Hsp70: DnaK cooperates with trigger factor in chaperoning nascent chains. *Cell* **97**, 755–765.

Thulasiraman, V., Yang, C. F., and Frydman, J. (1999). In vivo newly translated polypeptides are sequestered in a protected folding environment. *Embo J* **18**, 85–95.

Topol, J., Ruden, D. M., and Parker, C. S. (1985). Sequences required for in vitro transcriptional activation of a Drosophila hsp 70 gene. *Cell* **42**, 527–537.

Trottier, Y., Lutz, Y., Stevanin, G., Imbert, G., Devys, D., Cancel, G. et al. (1995). PolyQ expansion as a pathological epitope in Huntington's disease and four dominant cerebellar ataxias. *Nature* **378**, 403–406.

Tsai, J. and Douglas, M. G. (1996). A conserved HPD sequence of the J-domain is necessary for YDJ1 stimulation of Hsp70 ATPase activity at a site distinct from substrate binding. *J Biol Chem* **271**, 9347–9354.

Vogel, J. P., Misra, L. M., and Rose, M. D. (1990). Loss of BiP/GRP78 function blocks translocation of secretory proteins in yeast. *J Cell Biol* **110**, 1885–1895.

Vogel, M., Bukau, B., and Mayer, M. P. (2006). Allosteric regulation of Hsp70 chaperones by a proline switch. *Mol Cell* **21**, 359–367.

Waelter, S., Boeddrich, A., Lurz, R., Scherzinger, E., Lueder, G., Lehrach, H., and Wanker, E. E. (2001). Accumulation of mutant huntingtin fragments in aggresome-like inclusion bodies as a result of insufficient protein degradation. *Mol Biol Cell* **12**, 1393–1407.

Walsh, P., Bursac, D., Law, Y. C., Cyr, D., and Lithgow, T. (2004). The J-protein family: Modulating protein assembly, disassembly and translocation. *EMBO Rep* **5**, 567–571.

Warrick, J. M., Chan, H. Y., Gray-Board, G. L., Chai, Y., Paulson, H. L., and Bonini, N. M. (1999). Suppression of polyQ-mediated neurodegeneration in Drosophila by the molecular chaperone HSP70. *Nat Genet* **23**, 425–428.

Warrick, J. M., Paulson, H. L., Gray-Board, G. L., Bui, Q. T., Fischbeck, K. H., Pittman, R. N., and Bonini, N. M. (1998). Expanded polyQ protein forms nuclear inclusions and causes neural degeneration in Drosophila. *Cell* **93**, 939–949.

Watanabe, M., Dykes-Hoberg, M., Culotta, V. C., Price, D. L., Wong, P. C., and Rothstein, J. D. (2001). Histological evidence of protein aggregation in mutant SOD1 transgenic mice and in amyotrophic lateral sclerosis neural tissues. *Neurobiol Dis* **8**, 933–941.

Westerheide, S. D. and Morimoto, R. I. (2005). Heat shock response modulators as therapeutic tools for diseases of protein conformation. *J Biol Chem* **280**, 33097–33100.

Westhoff, B., Chapple, J. P., van der Spuy, J., Hohfeld, J., and Cheetham, M. E. (2005). HSJ1 is a neuronal shuttling factor for the sorting of chaperone clients to the proteasome. *Curr Biol* **15**, 1058–1064.

Wittung-Stafshede, P., Guidry, J., Horne, B. E., and Landry, S. J. (2003). The J-domain of Hsp40 couples ATP hydrolysis to substrate capture in Hsp70. *Biochemistry* **42**, 4937–4944.

Wolynes, P. G., Onuchic, J. N., and Thirumalai, D. (1995). Navigating the folding routes. *Science* **267**, 1619–1620.

Xiao, H. and Lis, J. T. (1988). Germline transformation used to define key features of heat-shock response elements. *Science* **239**, 1139–1142.

Xiao, H., Perisic, O., and Lis, J. T. (1991). Cooperative binding of Drosophila heat shock factor to arrays of a conserved 5 bp unit. *Cell* **64**, 585–593.

Xiao, X., Zuo, X., Davis, A. A., McMillan, D. R., Curry, B. B., Richardson, J. A., and Benjamin, I. J. (1999). HSF1 is required for extra-embryonic development, postnatal growth and protection during inflammatory responses in mice. *Embo J* **18**, 5943–5952.

Xu, Z., Horwich, A. L., and Sigler, P. B. (1997). The crystal structure of the asymmetric GroEL-GroES-(ADP)7 chaperonin complex. *Nature* **388**, 741–750.

Yan, W., Schilke, B., Pfund, C., Walter, W., Kim, S., and Craig, E. A. (1998). Zuotin, a ribosome-associated DnaJ molecular chaperone. *Embo J* **17**, 4809–4817.

Young, J. C., Barral, J. M., and Ulrich Hartl, F. (2003). More than folding: Localized functions of cytosolic chaperones. *Trends Biochem Sci* **28**, 541–547.

Young, J. C., Moarefi, I., and Hartl, F. U. (2001). Hsp90: A specialized but essential protein-folding tool. *J Cell Biol* **154**, 267–273.

Zhang, Y., Frejtag, W., Dai, R., and Mivechi, N. F. (2001). Heat shock factor-4 (HSF-4a) is a repressor of HSF-1 mediated transcription. *J Cell Biochem* **82**, 692–703.

Zhu, X., Zhao, X., Burkholder, W. F., Gragerov, A., Ogata, C. M., Gottesman, M. E., and Hendrickson, W. A. (1996). Structural analysis of substrate binding by the molecular chaperone DnaK. *Science* **272**, 1606–1614.

Ziegelhoffer, T., Johnson, J. L., and Craig, E. A. (1996). Chaperones get Hip. Protein folding. *Curr Biol* **6**, 272–275.

Zimarino, V. and Wu, C. (1987). Induction of sequence-specific binding of Drosophila heat shock activator protein without protein synthesis. *Nature* **327**, 727–730.

6

Metabolic Biopsy of the Brain

Ognen A. C. Petroff

In this chapter, we will review some applications of magnetic resonance spectroscopy (MRS) to the investigation of cerebral energy metabolism and neurotransmission. First we describe the use of phosphorus (31P) MRS to study high-energy phosphates of the brain including the application of the 31P-MRS in the assessment of the normal and abnormal neurophysiology of the human cerebrum *in vivo*. The phosphocreatine shuttle hypothesis, which appears to facilitate the efficient transfer of energy from mitochondria to the sodium-potassium ATPase located on the cell membranes, will be described. In subsequent parts of the chapter, dynamic measurements of lactate, glutamate, and glutamine, which are made using hydrogen (proton) spectroscopy (1H-MRS), will be discussed. We will present the astrocyte-neuron lactate shuttle hypothesis, which appears to couple glucose uptake and oxidation to energy metabolism. The applications of stable isotopes for the *in vivo*, dynamic measurement of glucose, glycogen, lactate, glutamate, glutamine, and ammonia metabolism of the human cerebrum using carbon (13C) and nitrogen (15N) spectroscopy will be described. Finally, we address the use of 13C-MRS to measure glutamatergic neurotransmission of the human neocortex and hippocampus.

I. Phosphorus Magnetic Resonance Spectroscopy

Phosphorus magnetic resonance spectroscopy (31P-MRS) provides quantitative information on ATP, phosphocreatine, and inorganic phosphate, which may be used to evaluate changes in cerebral high-energy phosphate compounds, primarily phosphocreatine and ATP, directly. Cerebral metabolites detectable by 31P-MRS include compounds related to high-energy phosphate and phospholipid metabolism, such as adenosine triphosphate (ATP), phosphomonoesters (PME), phosphodiesters (PDE), phosphocreatine (PCr), and inorganic phosphate (Pi) (see Figure 6.1). At the present time, 31P-MRS is the best method available to measure high-energy phosphates (phosphagens) because it completely avoids disrupting the energy state of the tissue being investigated. The information provided by the chemical shifts of Pi and ATP, can be used to measure brain pH and intracellular magnesium concentrations in health and disease (Barbiroli et al. 1999a; Rango et al. 2001).

A. Measurements of Intracellular pH

Inorganic phosphate has a chemical shift that is pH-dependent. The reason for this is that at neutral pH, Pi exists principally as HPO4 and H2PO4, with an acid-base dissociation constant (pK_{acid}) of 6.77 in brain tissues (Petroff et al. 1985). The chemical shift of 31P in these two molecules differs by approximately 2.4 ppm, but rapid exchange between the two forms results in only a single maximum. The resonant frequency of the peak signal amplitude is determined by the proportion of the two species present, and because the equilibrium depends on the pH of the tissue, this is reflected in the effective chemical shift of Pi. The pH measure *in vivo* reflects the weighted average of the pH of all intracellular (neurons, glia, etc.) and extracellular compartments, including blood (Kintner et al. 2000). The chemical shifts of phosphorylethanolamine, phosphorylcholine, and γ-ATP are used under special circumstances (Petroff & Prichard 1995). The buffering capacity of human brain is measured by varying the carbon dioxide content (Cadoux-Hudson et al. 1990; Jensen et al. 1988; van Rijen et al. 1989) (see Figure 6.1).

B. High-Energy Phosphate Metabolism

All cells native to the central nervous system, including neurons and glia, express creatine kinase (CK), the enzyme that catalyzes the equilibrium among phosphocreatine, ADP, creatine and ATP.

$$[\text{Phosphocreatine}] + [\text{ADP}] + [\text{H}^{+1}] + [\text{Mg}^{+2}] \xleftrightarrow{\text{creatine kinase}} [\text{ATP}] + [\text{creatine}]$$

Through the creatine kinase equilibrium, phosphocreatine preserves a high energy ratio of [ATP]/[ADP] by maintaining low ADP concentrations (Connett 1988; Neumann et al. 2003; Veech et al. 1979). The ratio of phosphocreatine/ATP therefore reflects the availability of phosphocreatine and serves as an easy-to-measure marker of the energy state of the brain using 31P-MRS, if brain pH remains constant. More detailed modeling is needed when brain pH, magnesium concentrations, or total creatine (creatine plus phosphocreatine) concentrations change. The creatine kinase equilibrium is coupled to ATP synthase and ATPase.

$$\textbf{energy} + [\text{ADP}] + [\text{Pi}] + [\text{H}^{+1}] + [\text{Mg}^{+2}] \xrightarrow{\text{ATP synthase}} [\text{ATP}]$$

$$[\text{ATP}] + \xrightarrow{\text{ATPase}} [\text{ADP}] + [\text{Pi}] + [\text{H}^{+1}] + [\text{Mg}^{+2}] + \textbf{energy}$$

Models, which assume that all three reactions are at steady state, allow calculation of the energy state. The phosphocreatine to inorganic phosphate ratio (PCr/Pi) directly mirrors alterations in energy state (thermodynamic free energy) during pathological conditions involving acidosis, low magnesium, or low total creatine content.

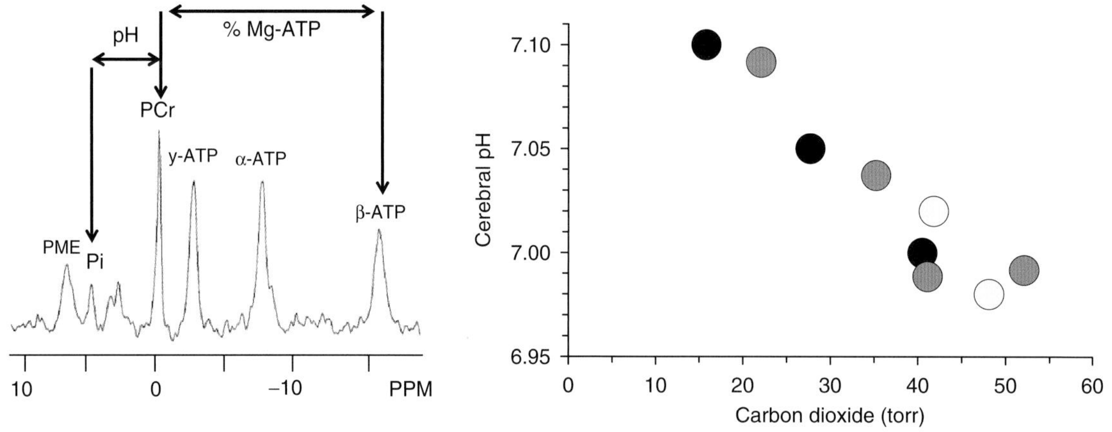

Figure 6.1 Phosphorus spectrum at 4.1-Tesla of the human cortex (adapted from Chu et al. 1996). The chemical shift of inorganic phosphate (Pi), γ-ATP and the phosphomonoesters (PME) may be used to measure cerebral pH (Petroff & Prichard 1995). For convenience, the chemical shift (frequency) of phosphocreatine (PCr) often is assigned a value of zero parts per million (PPM). The graph on the right shows the relationship between cerebral pH and the partial pressure of carbon dioxide of arterial blood during hyperventilation (data from Jensen et al. 1988, van Rijen et al. 1989, Cadoux-Hudson et al. 1990).

C. Phosphorus Spectroscopy and Intracellular Magnesium Measurements

Most intracellular ATP is complexed with magnesium (Barker et al. 1999; Halvorson et al. 1992; Iotti et al. 2005; Williams et al. 1993), which has a large effect on the chemical shift of the β-ATP resonance. This effect has been exploited to measure the available (free) intracellular magnesium concentration of the human brain. Under normal conditions, increased dietary magnesium supplementation does not increase cerebral or serum magnesium levels; the excess magnesium intake is excreted completely in the urine (Wary et al. 1999).

Low cortical magnesium is associated with migrainous headaches (Boska et al. 2002; Lodi et al. 2001; Montagna et al. 1994). Cerebral PCr levels are low; but brain pH and ATP levels are reported normal in migraine with and without aura and cluster headaches (see Figure 6.2). Studies using transcranial magnetic stimulation suggest that the phosphenes, which characterize the aura in many migraineurs, are attributable to increased cortical excitability associated with low magnesium. Cortical magnesium levels and PCr levels are lowest in patients with hemiplegic migraine or migraine with stroke. Spreading depression is thought to be the mechanism that produces the loss of vision and the stroke-like symptoms, and is attributed to a failure in astrocytic homeostasis, which controls extracellular potassium, glutamate, and magnesium concentrations (Ayata et al. 2006; Gorji 2001). Oral supplementations with magnesium, co-enzyme Q, or riboflavin are reported as effective therapies for migraine.

The findings are similar to those reported in the mitochondrial cytopathies with no clinically evident cerebral dysfunction and no abnormalities seen by clinical MRI (Barbiroli et al. 1999a, 1999b; Eleff et al. 1990; Rango et al. 2001). Cortical magnesium levels and high energy phosphate levels are decreased in the occipital lobe of patients with mitochondrial cytopathies with a normal cortical pH. Treatment with co-enzyme Q resulted in increased phosphocreatine and magnesium levels and decreased inorganic phosphate and ADP levels in all patients (see Figure 6.2). These results are consistent with the hypothesis that increased co-enzyme Q concentrations in the inner mitochondrial membrane increase the efficiency of oxidative phosphorylation, thereby improving cerebral energy state.

D. Effects of Evoked Metabolism on High-Energy Phosphates

Because 31P-MRS is completely noninvasive, serial observations are made easily, allowing dynamic measurements of the response of these highly labile phosphagens to repeated physiological perturbations (Rango et al. 1997). Cortical phosphocreatine levels drop by approximately 18 percent during the first three seconds of photic stimulation and recover completely within 17 seconds of the brief stimulation. Similarly decreases in phosphocreatine are seen in the sensory cortex of rodents during forepaw stimulation (Xu et al. 2005). With longer stimulation trains lasting more than seven minutes, phosphocreatine levels, which are measured in the human visual cortex, do not appear to change significantly (Chen et al. 1997; Rango et al. 2001, 2006). After these longer trains of 8 Hz photic stimulation, high energy phosphate (ATP and phosphocreatine) levels increase during the seven minutes after the stimulation has stopped (recovery) to about 16 percent above basal levels.

II. The Phosphocreatine Shuttle Hypothesis

In the brain, the primary role of phosphocreatine is to shuttle energy (phosphagen) from the mitochondria to areas of high

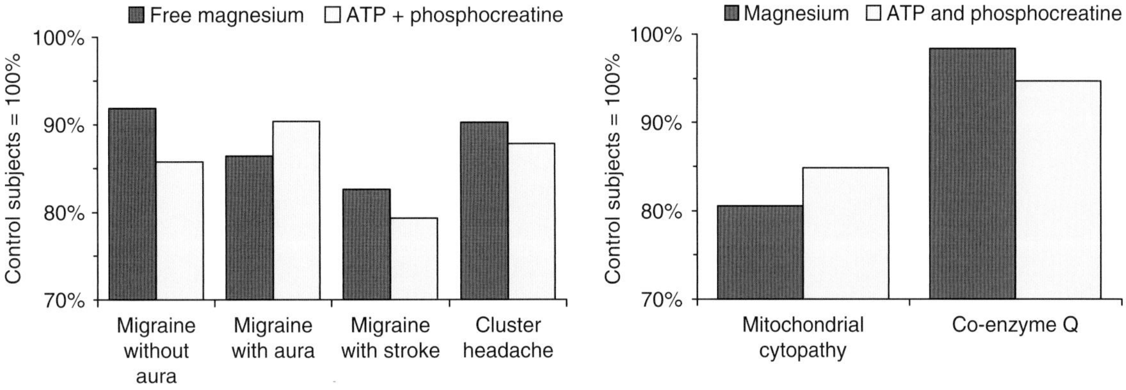

Figure 6.2 The graph on the left shows that phosphocreatine and magnesium levels are below normal in the visual cortex of patients with migraine or cluster headache (data from Montagna et al. 1994 and Lodi et al. 2001). The graph on the right shows that treatment with co-enzyme Q improves the below-normal cortical phosphocreatine and magnesium levels of patients with mitochondrial cytopathies and improves some of the neurological abnormalities (data from Barbiroli et al. 1999).

rates of energy (ATP) consumption (Neumann et al. 2003; Wyss & Kaddurah-Daouk 2000). Phosphocreatine diffuses through the cytosol considerable faster than ATP, which may be important for the neuronal dendrite/spine and the processes of astrocytes (Kekelidze et al. 2001; Li et al. 2004). Under physiological conditions, 95 percent of ATP is synthesized by oxidative metabolism by mitochondria (Attwell & Iadecola 2002; Attwell & Laughlin 2001). The final step of mitochondrial ATP synthesis is catalyzed by a complex (ATP synthasome) consisting of ATP synthase, the ADP/ATP antiporter (ANT), and the Pi carrier, which are all attached to the inner membrane of the mitochondrion (Chen et al. 2004). ATP is synthesized inside the inner membrane of the mitochondrion and transported to the intermembrane space by ANT in exchange for ADP. ATP diffuses through channels (porin) in the outer membrane of the mitochondrion, through the cytosol to the sites of energy demand, where it is hydrolyzed to ADP and Pi (see Figure 6.3). However, in brain and muscle under conditions of high energy demand, the availability of ADP in the intermembrane space limits the rate of egress of ATP through ANT (Neumann et al. 2003; Wallimann et al. 1998).

Mitochondrial creatine kinase (mi-CK), which is located on the outer surface of the inner mitochondrial membrane, forms a functional unit with ANT and porin to export phosphagen in the form of phosphocreatine out into the neuronal and glial cytosol. By controlling the intermembrane concentration of ADP, mi-CK controls mitochondria metabolism. In cells containing functional mi-CK, the rate of mitochondrial (oxidative) energy production is limited by availability of Cr and Pi, not ADP, which diffuses much more slowly than PCr, Cr, or Pi. Cytosolic ADP concentrations are maintained at very low levels by cyto-CK, which optimizes energy extraction by ATPases. By maintaining ADP concentrations at low levels, creatine kinases also buffer intracellular hydrogen ions (pH) and free magnesium concentrations, which are generated when Mg-ATP is hydrolyzed (Veech et al. 1979). Mitochondria can rapidly upregulate metabolism on demand in response to small changes in ADP, Pi, or hydrogen ion concentrations.

III. Magnetization Transfer Measurements of ATP and Phosphocreatine Synthesis

Magnetization transfer experiments are used to measure the rates of ATP and phosphocreatine synthesis and utilization using 31P-MRS (Chen et al. 1997; Lei et al. 2003a, 2003b). The 31P-MRS saturation and inversion transfer experiments involve applying magnetic label to the frequencies of the phosphate precursors of metabolic reactions and observing the appearance of the magnetic label in the product of the reaction (Lei et al. 2003a). This type of spectroscopy is used to study fast chemical reactions because the magnetic label rapidly decays after it is transferred to the new metabolite. These saturation transfer studies of the visual cortex report a unidirectional rate constant of $0.17\,s^{-1}$ for ATP synthesis and $0.24\,s^{-1}$ for the creatine kinase reaction for control subjects. The apparent unidirectional rate constant of creatine kinase in the direction of ATP synthesis from phosphocreatine increased by 34 percent in the visual cortex areas during photic stimulation without significant changes of the steady-state concentrations of high energy phosphate metabolites. Dynamic measurements of phosphocreatine, inorganic phosphate, and pH using 31P-MRSI complement EEG, evoked potentials, functional MRI, and other types of MRS studies, which can be made simultaneously or using an interleaved design.

ATP levels remained essentially the same between barbiturate anesthesia through vigorous status epilepticus (see Figure 6.4). Saturation transfer experiments in animal models show that the forward flux from phosphocreatine to ATP doubles during seizures, which indicates that mitochondrial oxidative metabolism doubled during bicuculline-induced status epilepticus (Holtzman et al. 1997; Sauter & Rudin 1993). Convulsions result in decreased phosphocreatine, increased inorganic phosphate, and produce an intracellular acidosis in adult animal models. Intracellular ADP levels, which can be calculated from the creatine kinase equilibrium, increase during seizures. Further increases in cytosolic

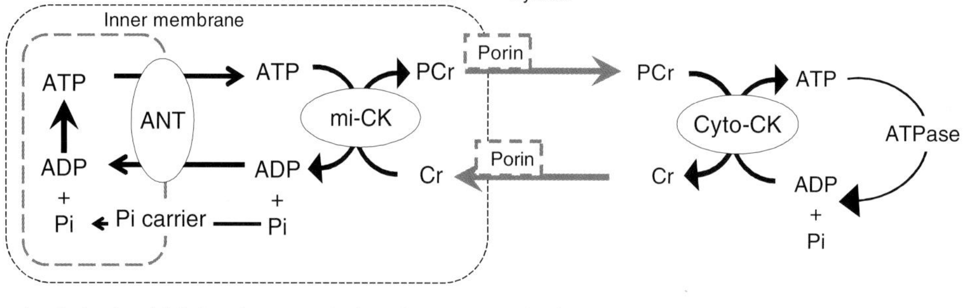

Figure 6.3 Schematic of mitochondrial phosphagen synthesis and transport mechanisms (adapted from Wallimann et al. 1998). Abbreviations: adenosine diphosphate (ADP), adenosine triphosphate (ATP), adenylate translocator (ANT), creatine (Cr), cytosolic creatine kinase (cyto-CK), mitochondrial creatine kinse (mito-CK), phosphocreatine (PCr), phosphate (Pi).

ADP levels, at the sites of intense ATP consumption (e.g., glial, neuronal, and vesicular ion transporters and enzymes requiring ATP for activity), are controlled by the cytosolic isoform of creatine kinase (CK). Elevated intracellular ADP and Pi concentrations stimulate cerebral glycolysis, glycogenolysis, and glucose oxidation (Kemp 2000).

A. Mitochondrial Cytopathies

Studies of patients with mitochondrial cytopathies and CNS symptoms reveal below-normal cerebral PCr levels and low PCr/ATP and PCr/Pi ratios measured using 31P-MRS (Antozzie et al. 1995; Argov & Arnold 2000; Matthews et al. 1991). Epilepsy can occur as the presenting sign of mitochondrial disease. Epilepsy is part of the phenotype of several inherited mitochondrial cytopathies (Kunz 2002; Simon & Johns 1999). Conversely, a number of toxins resulting in mitochondrial damage (e.g., kainic acid, pilocarpine, 3-nitropropionic acid, and cyanide) elicit seizures as phosphocreatine levels drop (Kunz 2002). Phosphocreatine levels are low, measured in the occipital lobe using 31P-MRS, and decrease further with photic stimulation in patients with MELAS (Kato et al. 1998). However, the PCr/ATP & PCr/Pi ratios and brain pH of patients with chronic progressive external ophthalmoplegia (CPEO), who did not appear to have overt CNS involvement, are normal under basal conditions (Rango et al. 2001). The deficiency in mitochondrial energy metabolism is revealed by the paradoxical response to photic stimulation with a 26 to 27 percent decrease in high energy phosphate levels and a slight alkalosis (from 7.01 to 7.09) during the recovery period after photic stimulation. The deficiency in mitochondrial energy production is unmasked by the photic stimulation that transiently places a greater metabolic stress on the visual cortex (see Figure 6.5).

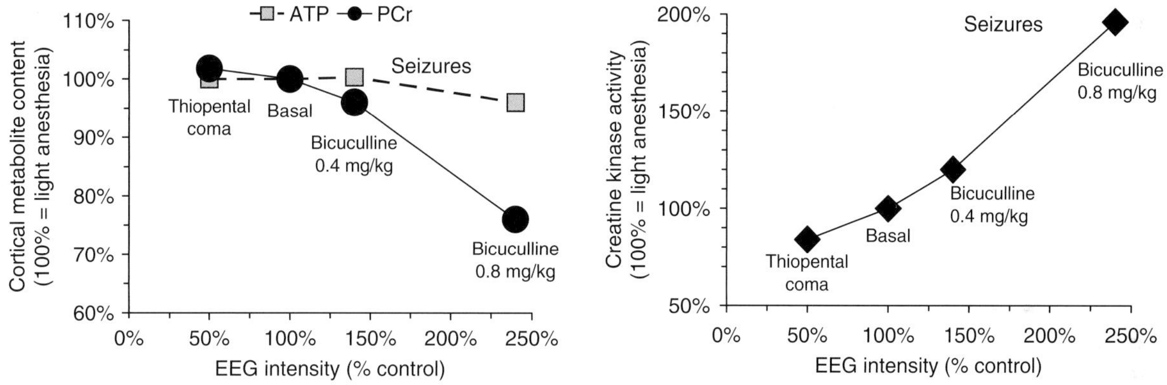

Figure 6.4 Relationship between EEG power and brain concentrations of adenosine triphosphate (ATP) and phosphocreatine (PCr) during bicuculline induced seizures (left). The PCr/ATP ratios decreased by ~20% during status epilepticus (redrawn from Sauter & Rudin 1993). The graph on the right shows the flow of high-energy phosphate through creatine kinase as a function of electroencephalogram (EEG) power. The rates of ATP synthesis and utilization doubled during status epilepticus.

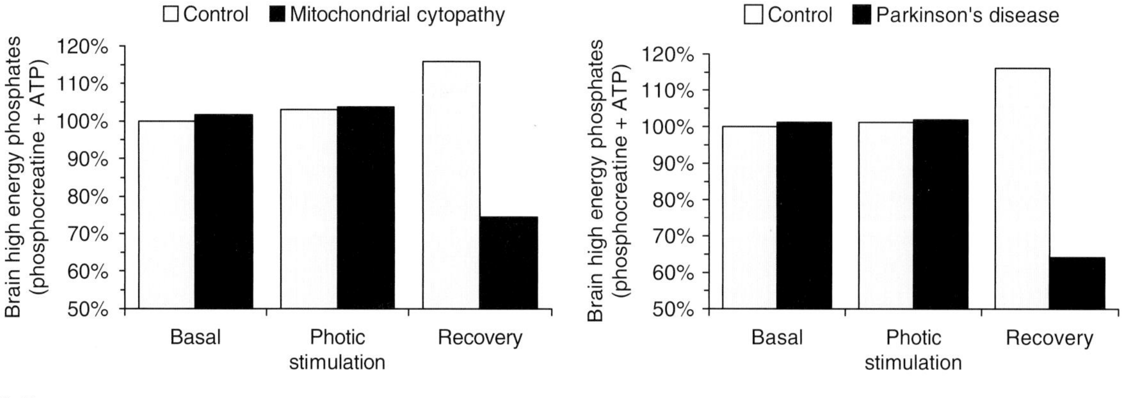

Figure 6.5 The graph of the left compares the changes in high-energy phosphate levels, primarily phosphocreatine, which were measured in the visual cortex of control subjects and patients affected by mitochondrial disease without clinical central nervous system involvement (chronic progressive external ophthalmoplegia) using 31P-MRS, in response to photic stimulation (adapted from Rango et al. 2001). The graph on the right compares the responses of control subjects and patients with Parkinson disease using the same protocol (adapted from Rango et al. 2006).

B. Parkinson Disease

MR measurements of the visual cortex comparing control subjects and patients with Parkinson disease show that after photic stimulation, phosphocreatine levels decline rather than recover similar to the response of patients with mitochondrial cytopathies (Rango et al. 2006). The responses of patients with Parkinson disease before and during photic stimulation appear to be the same as those of the control subjects. However, high energy phosphate levels decrease by 36 to 37 percent without any change in cortical pH during the recovery phase in patients and increased by 16 percent in control subjects (see Figure 6.5). These findings suggest that mitochondria in the visual cortex of patients with Parkinson disease are unable to recover quickly from a period of higher energy demand.

C. Creatine Deficiency Syndromes

Three genetic causes of creatine deficiency are known, which are associated with various forms of epilepsy (Leuzzi 2002; Schulze 2003; Wyss & Kaddurah-Daouk 2000). All are characterized by exceedingly low or absent phosphocreatine concentrations in the brain measured using 31P-MRS. Neurological symptoms associated with creatine deficiency have some variability but present overall as mental retardation, developmental regression, behavioral problems, and epilepsy, which varies from intractable seizures in guanidinoacetate methyltransferase (GAMT) deficient patients to febrile convulsions with arginine:glycine amidinotransferase (AGAT) deficiency and mild epilepsy in X-linked creatine transporter (CrT1) deficient patients. When oral creatine supplementation raises brain phosphocreatine levels, monitored using serial 31P-MRS, seizure control improves in parallel with rising brain phosphocreatine concentrations. Improvement in seizure control with creatine supplementation correlated with increased cortical PCr in an adult patient with refractory epilepsy, who was diagnosed with guanidinoacetate methyltransferase deficiency (GAMT) at age 26 years (Schulze et al. 2003).

D. Animal Models

Epilepsy can occur as the presenting sign of mitochondrial disease. Epilepsy is part of the phenotype of several inherited mitochondrial cytopathies (Kunz 2002; Simon & Johns 1999). Conversely, a number of toxins resulting in mitochondrial damage (e.g., kainic acid, pilocarpine, 3-nitropropionic acid, and cyanide) elicit seizures as phosphocreatine levels drop (Kunz 2002). The phenotype of transgenic mice lacking mi-CK, cyto-CK, and the double knockout (no CK) were characterized by a variety of methods including 1H- and 31P-MRS (in 't Zandt et al. 2004; Jost et al. 2002; Kekelidze et al. 2001). Not unexpectedly, mi-CK colocalizes with mitochondria with the highest content in neurons (grey matter) and the lowest content in white matter (glia). Cyto-CK has a more uniform distribution across brain regions and cell types. The mi-CK knockout mice had slower CK rate constants under basal conditions with normal ATP, Pi, PCr levels, and brain pH. However, under the metabolic stress induced by seizures, both PCr and ATP levels decreased significantly in animals lacking mi-CK compared with wild-type mice. Surprisingly, homozygous transgenic mice, which are deficient in both mi-CK and cyto-CK and, therefore, have low brain PCr levels, have mild behavioral problems and no convulsions, unlike patients with low PCr caused by dysfunction in creatine synthesis, who have various types of epilepsy, which can be treated successfully with creatine supplementation (Streijger et al. 2005). Mice lacking CK have normal APT, Pi, and brain pH, although they have enlarged ventricles and above-normal brain concentrations of NAA and glutamate. Apparently, brain PCr and Cr levels drop after birth; the human fetus is supplied with creatine, which is synthesized by the mother.

Recent studies showed that methyl-malonic acid, a known inhibitor of mitochondrial function, which causes coma and early death, inhibits mi-CK activity and does not affect cyto-CK activity (Schuck et al. 2004). Injection of methyl-malonic acid in a rat model lowers brain PCr levels and causes convulsions, which are successfully prevented by creatine supplementation (Royes et al. 2003). Methyl-malonic acidemia includes a group of inherited disorders (e.g., maple syrup urine disease) whose phenotype includes developmental delay, hypotonia, and seizures (Vigevano & Bartuli 2002). These observations may be reconciled by hypothesis that the post-natal loss of brain PCr maintains the seizure diathesis.

E. The Localization-related Human Epilepsies

Ictal studies in human patients are rare (Younkin et al. 1986). Seizures cause a 33 percent decrease in brain phosphocreatine and a 50 percent decrease in the phosphocreatine to inorganic phosphate (PCr/Pi) ratio. Focal seizures cause lateralized decreases in the PCr/Pi ratio; generalized seizures caused bilateral decreases. Postictal spectra show increased PCr/Pi ratios, presumably due to postictal inhibition. One patient's seizures were successfully treated with intravenously administered phenobarbital during MR data acquisition, causing an immediate increase in the PCr/Pi ratio from 0.7 to 1.2. No significant alteration in intracellular pH is seen.

Altered energy state with decreased PCr characterizes the interictal state of the involved brain regions of the localization-related epilepsies. Initial interictal patient studies using 31P-MRS reported significant alterations in inorganic phosphate (73% increase) and brain pH (mean alkalosis of 0.11 and 0.17 units) ipsilateral to the seizure focus without significant asymmetries between ipsilateral and contralateral temporal lobe concentrations of ATP, PCr, or PDE (Hugg et al. 1992; Laxer et al. 1992). In a later study of eight patients with frontal lobe epilepsy, increased pH in all eight and decreased PME in seven patients were found in the epileptogenic frontal lobes, but no alterations in Pi levels were detected (Garcia et al. 1994). These three studies indicate that energy state (PCr/Pi ratio) is reduced in the epileptogenic region. Another group reported that the PCr/Pi ratios were found to be lower in the ipsilateral than the contralateral temporal lobe of the patients, and lower in both sides than the control data (Kuzniecky et al. 1992). Subsequent studies also reported widespread alterations in bioenergetic parameters (decreased PCr/Pi and PCr/ATP ratios), but failed to show any alteration in brain pH (Chu et al. 1996, 1998; Hetherington et al. 2002). No significant decreases in ATP levels were reported.

Phosphorus spectroscopic imaging (31P-MRSI) showed that the maximal decrease in cerebral energetics appears centered in the epileptic focus. Recently, it has been shown that PCr/ATP was reduced to the greatest extent in the amygdala ipsilateral to the seizure focus followed by the ipsilateral pes, hippocampus, and thalamus with decreasing severity. A similar pattern was seen in the contralateral hemisphere, albeit to a lesser extent (Hetherington et al. 2002). These observations suggest that the epileptic state is characterized by widespread mitochondrial dysfunction, particularly in regions involved in the epileptic network as defined by EEG and alterations of cerebral glucose and oxygen metabolism.

The modest decreases (5–15%) of the PCr/ATP ratios observed in the epileptogenic human hippocampus using 31P-MRSI appears to have significant electrophysiological consequences, which can be measured using a standard brain slice preparation made from the resected tissues (Williamson et al. 2005). A significant negative correlation was seen between the ability to fire multiple spikes in response to single synaptic stimulation (number of action potentials induced by orthodromic stimulation) applied to the hippocampal slice and the PCr/ATP ratios measured before surgery. This type of increased "bursting" response is rarely seen without extensive hippocampal cell loss and synaptic reorganization, and is characteristic of granule cells obtained from patients with mesial temporal sclerosis (MTS). An increased bursting score, which is seen with low PCr/ATP ratios, reflect increased excitability of the brain slice and presumably the hippocampus *in vivo*.

The data are consistent with the hypothesis that baseline, asynchronous activity is not impaired by a modest reduction in the PCr/ATP ratio, but that evoked stimulation, which activates numerous presynaptic and postsynaptic elements and activates mitochondrial metabolism to restore ionic homeostasis, becomes uncontrolled with a 10 to 15 percent decrease in the PCr/ATP ratio. There was a strong correlation between PCr/ATP ratios and the recovery of the membrane potential following a stimulus train with low PCr/ATP being associated with prolonged recovery times. More normal PCr/ATP ratios are associated with rapid rates of recovery of membrane potential following a 10 Hz, 10 s train electrical stimulus. In short, low PCr/ATP ratios are associated with those measures of increased excitability that are associated with a high energy demand. A marker of inhibitory function of the hippocampus, inhibitory postsynaptic potential (IPSP) conductance (GIPSP), is impaired by a 5 to 10 percent reduction in the presurgical PCr/ATP ratio. The strength of inhibition is positively associated with PCr/ATP ratios. The fact that both bursting score and GIPSP correlate inversely with PCr/ATP ratios supports the hypothesis that both abnormal excitatory and inhibitory physiological responses depend on altered energetics.

Serial 31P-MRS measurements show that the ketogenic diet raises PCr levels and lowers Pi levels toward normal as seizure control improves (Pan et al. 1999). The mean PCr/Pi ratio increased by 22 percent, the PCr/ATP ratio by 14 percent, and the Pi/ATP ratio decreased by 7 percent after sustained ketosis was achieved. Whether improved mitochondrial function and energy state reflect the metabolic effects of the ketogenic diet directly or are an epiphenomenon of improved seizure control remains to be determined.

IV. Hydrogen (Proton) Spectroscopy

The principal signals, measured using proton (1H) magnetic resonance spectroscopy, include the singlet resonances of N-acetyl-aspartate (2.01 parts per million [ppm]), creatine plus phosphocreatine (3.0 ppm and 3.9 ppm), choline-containing compounds (Cho, 3.2 ppm), glutamate (GT), and myo-inositol (In). Broad signals from cytosolic macromolecules (lipids, proteins, and glycoproteins) are easily seen making the lactate resonance (centered at 1.35 ppm) difficult to measure (Behar et al. 1994). The signals from glutamate and glutamine are difficult to measure because of low signal-to-noise ratios and spectral crowding (see Figure 6.6). Investigators using 1.5 Tesla clinical spectrometers and short spin-echo time (TE) pulse sequences have used the broad signals located between 2.1 to 2.5 ppm, which are comprised of resonances from glutamate, glutamine, NAA, various macromolecules, and to a lesser extent by GABA, succinate, homocarnosine, glutathione and NAAG, as a surrogate for glutamate plus glutamine signals and referred to as GLX.

Stronger magnetic fields, reduced magnetic field inhomogeneity (shimming), better receiver design (improved filling factor and noise rejection) and advanced pulse sequences facilitate the measurement of glutamate and glutamine by improving the signal-to-noise ratio and reducing spectral crowding (De Graaf et al. 2001; Kassem & Bartha 2003; Pan et al. 2006). Pulse sequences that attenuate (filter) short T1 signals and short T2 signals (cytosolic macromolecules) are a critical first step to ensure a flat baseline (Behar et al. 1994; Kassem & Bartha 2003; Rothman et al. 1992). Because of low tissue concentrations (about 1 mM), spin-spin interactions (multiplet resonances) and resonance overlap (spectral crowding), quantitative measurements of GABA require specialized receiver coils and advance pulse techniques (Choi et al. 2005; Hanstock et al. 2002; Hyder et al. 1999; Terpstra et al. 2002). Alternatively, high-resolution analytical spectrometers may be used to measure metabolites tissue extracts or perfused brain slices (Errante et al. 2002; Lukkarinen et al. 1997; Petroff et al. 1995a).

The expansion of the white matter tracts including the corpus callosum is a feature of human and nonhuman primate cerebrum, which is not prominent in the conventional animal models (rodents, carnivores, and ungulates). The metabolite content of grey matter, which contains a high density of synapses, is higher than values for white matter, which has less water content, more myelin, and few synapses (Petroff et al. 1995a). The proportion of each voxel may be parsed into grey matter, white matter, and CSF on the basis of the MR relaxation properties of water (Pan et al. 2000, 2002a). Voxel segmentation facilitates greater precision in metabolite measurements, especially in regions, which have a convoluted anatomy (see Table 6.1).

V. Carbon Spectroscopy

In addition to hydrogen (1H) and phosphorus (31P), several other stable (nonradioactive) isotopes can be detected using MR spectroscopy including carbon (13C), nitrogen (15N), and oxygen (17O) (Henry et al. 2003; Kanamori et al. 1998; Zhu et al. 2005). *In vivo* tracer studies using substrates labeled with 13C have revolutionized our understanding of cerebral metabolism by providing new insight into the dynamics of neural-glia signaling. Most studies involving human subjects used 1-13C-glucose as a label source to measure the rate of glucose uptake (directly), glycolysis (3-13C-lactate), tricarboxylic acid (TCA cycle) (4-13C-glutamate), glutamine synthesis (4-13C-glutamine), GABA synthesis (2-13C-GABA), NAA synthesis (6-13C-NAA), and alanine synthesis (3-13C-alanine) directly (Mason GF et al. 1995; Otsuki et al. 2005; Shen et al. 1999) (see Figure 6.7).

Glucose is the preferred substrate of the CNS and is metabolized primarily in the neuronal mitochondria yielding 38 ATP per glucose molecule oxidized. The classical model

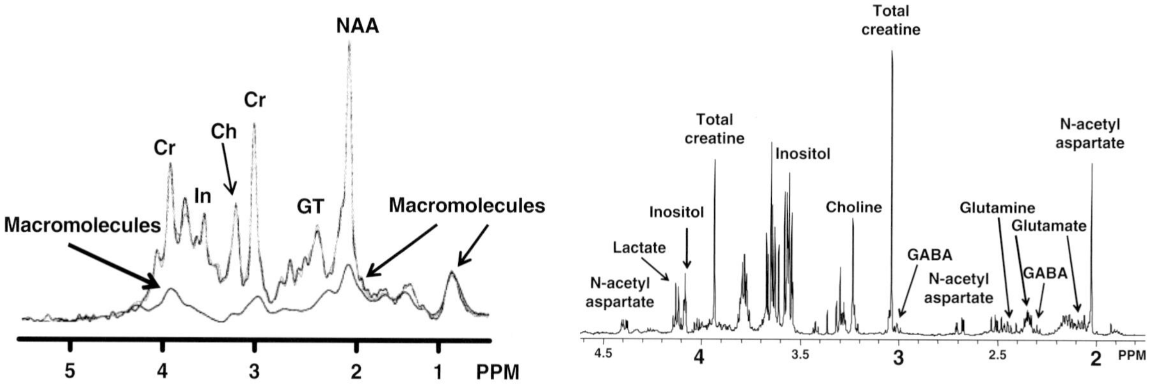

Figure 6.6 The hydrogen (proton) spectra of the human cortex on the left were made using a 4-Tesla spectrometer (see De Graaf et al. 2001). A short spin-echo sequence was used for the top spectrum, which includes all resonances; below is plotted the short T1 resonances (macromolecules), which are revealed by adding an inversion recovery pulse sequence. On the right, a proton spectrum of a perchloric acid extract of human neocortex, which was made using an 11.7-Tesla analytical spectrometer, is shown (see Petroff et al. 1995). Abbreviations: total creatine (Cr), choline (Ch), gamma-amino-butyric acid (GABA), myo-inositol (In), N-acetyl-aspartate (NAA), parts per million (PPM).

Table 6.1 Abbreviations: gamma-amino-butyric acid (GABA), N-acetyl-aspartate (NAA), not available (n.a.), not detected (n.d.) (adapted from Petroff et al. 1995; de Graaf et al. 2001; Kassen & Bartha 2003).

	Human cerebral metabolite content (millimolar)					
	Temporal lobe (biopsy)		Parietal lobe (in vivo)		Hippocampus (in vivo)	
	grey	white	grey	white	grey	white
Glutamate	10.1	5.6	9.5	6.0	8.0	5.5
Glutamine	3.3	3.2	4.1	2.3	3.9	1.8
NAA	5.8	2.8	9.9	10.5	9.2	9.4
Creatine	8.9	6.5	8.4	6.6	11.6	7.4
GABA	2.3	0.8	n.a.	n.a	1.0	n.d.
Aspartate	2.0	2.3	2.0	1.3	4.0	3.4

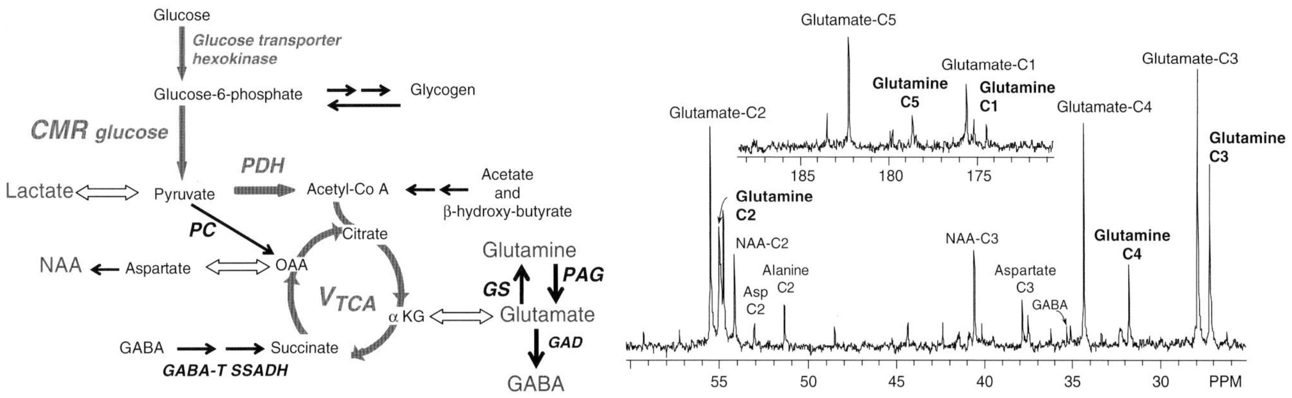

Figure 6.7 On the left is a schematic of the principal cerebral metabolic pathways, which utilize glucose, fatty acids and ketone bodies. On the right is a carbon spectrum of human brain following four hours of intravenous infusion of 2-13C-glucose. Abbreviations: aspartate (Asp), cerebral metabolic rate (CMR), gamma-amino-butyric acid (GABA), GABA-transaminase (GABA-T), glutamic acid decarboxylase (GAD), glutamine synthetase (GS), α-ketoglutarate (α-KG), N-acetyl-aspartate (NAA), oxaloacetic acid (OAA), phosphate-activated glutaminase (PAG), pyruvate carboxylase (PC), pyruvate dehydrogenase (PDH), succinic semi-aldehyde dehydrogenase (SSADH), tricarboxylic acid cycle (V_{TCA}).

suggested that neurons preferentially used glucose that was metabolized to pyruvate by the sequential enzymes of the glycolytic pathway; pyruvate served as the substrate for pyruvate dehydrogenase (PDH) complex and entered the mitochondria as acetyl-coenzyme-A for oxidation in the tricarboxylic acid (TCA) cycle. Lactate is made from pyruvate by lactate dehydrogenase (LDH) during hypoxia (anaerobic glycolysis). Cerebral glucose uptake has been studied in diabetes and other diseases, which may alter glucose uptake and metabolism (Bluml et al. 2001a; Criego et al. 2005; Mason et al. 2006; Seaquist et al. 2001). The neocortex has a lower Michaelis constant (Km 1.1 mM;) and a higher rate of uptake (Vmax 65 μmol/g/min) for glucose than white matter (Km 1.7 mM, Vmax 0.24 μmol/g/min), which is consistent with the higher cerebral metabolic rate of cortex ($CMR_{glucose}$ 0.36 μmol/g/min cortex, 0.11 white matter) (De Graaf et al. 2001; Pan et al. 2000).

VI. MR Spectroscopic Measurements of Cerebral Lactate

Serial MR spectroscopic measurements show that brain lactate rises rapidly with ischemia, reaching levels of concentrations proportionate to the available glucose (Petroff et al. 1988). This suggests that all available brain and blood glucose and glycogen stores are metabolized by glycolysis to maintain ATP levels during ischemia. However, brain lactate levels remain elevated for months following a stroke. Serial MR spectroscopic imaging shows that the regions with above-normal lactate are widespread and include areas that are not ischemic (Galanaud et al. 2003; Graham et al. 1993, 2001). Tracer studies infusing 1-13C-glucose into patients who have suffered a stroke show that the high lactate levels are maintained by continuing increased glycolytic activity, as demonstrated by the rapid synthesis and oxidation of lactate from infused 13C-labled glucose (Rothman et al. 1991). Areas containing reactive glia and macrophages and the adjacent ventricular and sulcal CSF contain above-normal concentrations of lactate (Petroff et al. 1992a). The findings suggest that increased lactate concentrations are part of the repair process, rather than a source of ongoing neuronal damage.

A. Brain Lactate Levels Are Above-Normal with Inflammation

Serial MR spectroscopic imaging shows that above-normal cerebral lactate levels are characteristic on a number of inflammatory diseases including brain abscess, multiple sclerosis, and AIDS encephalopathy (Bitsch et al. 1999; Chang et al. 1995; Confort-Gouny et al. 1993; Remy et al. 1995). Brain lactate levels increase rapidly with the onset of a new MS plaque and remain above normal for weeks, gradually decreasing as the MS flare subsides. Similarly, widespread areas of above-normal levels of lactate appear to be characteristic of HIV encephalopathy.

B. Lactate Remains Elevated Following a Seizure

Cortical lactate levels increase in the acute phase of a spontaneous seizure (Briellmann et al. 2005). The difficulty in interpretation lies in distinguishing the contribution of neuronal injury from the metabolic effects of the prolonged seizure. Lactate levels are elevated in spectra of the epileptogenic mesial temporal lobe obtained during the complex partial seizure or within four to 24 hours after the seizures (Castillo et al. 2001; Cendes et al. 1997). Lactate levels increase rapidly following brief electroshock-induced seizure discharges and remain above-normal for minutes to hours depending on the intensity of the seizure (Prichard et al. 1987). Tracer studies using serial MR spectroscopy have shown that the above-normal lactate concentration, which was raised initially by the seizure, is rapidly oxidized by the mitochondria and replaced by newly synthesized lactate. There is no metabolically inactive, "trapped" compartment containing high levels of lactate in the brain following a seizure.

Microdialysis studies of patients with temporal lobe seizures show that extracellular lactate levels slowly decline following a spontaneous seizure (Petroff et al. 1992b). Above-normal interictal extracellular lactate levels are a characteristic of the seizure focus. The data as a whole suggest that astrocytes activated by the seizure continue to release above-normal amounts of lactate for minutes to hours after a brief seizure.

C. Glycolysis

A growing body of literature suggests that lactate is an important metabolite made by astrocytes under normal conditions and exported to neurons (Hyder et al. 2006; Magistretti et al. 1999; Pellerin 2003; Pellerin & Magistretti 2004; Schurr 2006). The glycolytic enzymes aggregate in clusters associated with the external cell membrane including LDH. The first of these glycolytic enzymes, hexokinase, uses one ATP to phosphorylate glucose to glucose-6-phosphate, thereby driving glucose uptake by keeping intracellular glucose concentrations lower than the values in the extracellular fluid (ECF). Not surprisingly, the aggregate of the glycolytic enzymes including hexokinase appears to be in close physical proximity to the bidirectional glucose transporters (GLUT1 for glia and GLUT3 for neurons). Uptake of glucose by the CNS appears to be mediated primarily by GLUT1.

VII. The Astrocyte-Neuron Lactate Shuttle Hypothesis

Glucose uptake and lactate export by astrocytes to neurons are tightly coupled to neurotransmission; this model is referred to as the astrocyte to neuron lactate shuttle hypothesis (ANLSH) (Aubert et al. 2005; Bouzier-Sore et al. 2003a; Schurr 2006). Lactate is exported continually by glia to supply the energy needs of neurons. Lactate reflects increased glial metabolism; neuronal oxidative metabolism primarily reflects lactate oxidation rather than direct glucose oxidation (see Figure 6.8). The glycolytic pathway generates two NADH and four ATP molecules and consumes two NADH and two ATP for each glucose molecule, which is metabolized to two lactate molecules. As discussed in the section on high energy phosphates, ATP synthesis consumes hydrogen ions under physiological conditions. A critical review of the hypothesis was published recently by Chih & Roberts (2003).

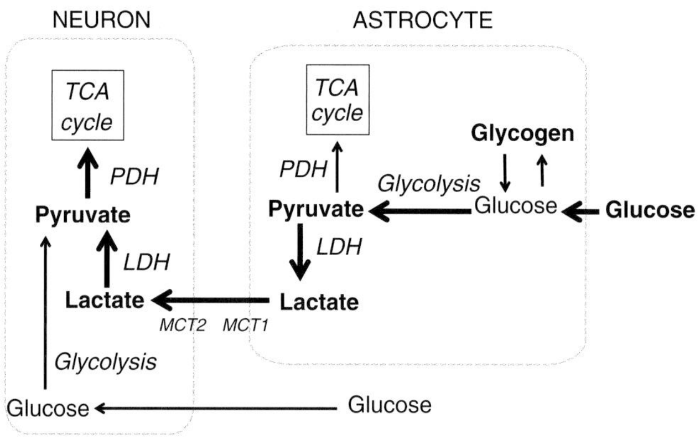

Figure 6.8 Schematic of the astrocyte-neuron lactate shuttlehypothesis. Abbreviations: lactate dehydrogenase (LDH), monocarboxylic acid transporter (MCT), pyruvate dehydrogenase (PDH), tricarboxylicacid (TCA).

A. The Critical Role of Astrocytic Glycogen in Maintaining Axonal Function

MR spectroscopy shows that glycogen has a rapid turnover in human cortex, which suggests that it is intimately associated with glycolytic complex of enzymes of glia (Oz et al. 2003). The enzymes, which synthesize and breakdown glycogen, are present in astrocytes, but not neurons (Shulman et al. 2001). MR studies indicate that glycogen turns over rapidly; it is continuously synthesized and catabolized under basal conditions (Shulman et al. 2001; Sickmann et al. 2005). Glycogen-derived lactate is exported by glia and oxidized by neurons. The ANLSH model has been extended to the energetics of the nodes of Ranvier (Brown 2004; Brown et al. 2004; Tekkok et al. 2005). Inhibition of lactate transport out of astrocytes or into axons reduces axonal function. The perinodal astrocytes supply axonal mitochondria with lactate made through glycogenolysis during periods of high-frequency firing. Decreased availability of glycogen impaired high frequency firing during normoglycaemia, which is restored by providing above-normal levels of lactate.

B. Cerebral Lactate Increases with Physiological Stimulation

Under the traditional model, brain lactate concentration usually is assumed to be stable except when pathologic conditions cause a mismatch between glycolysis and respiration. A consequence of the ANLSH model would be that physiological stimulation would transiently increase brain lactate as glial glycolytic, and glycogenolytic metabolism is activated before neuronal oxidative metabolism is fully activated. Lactate elevations of 0.3–0.9 mM occur in human visual cortex during the first few minutes of physiologic photic stimulation and decline toward basal values during continuing stimulation (Giove et al. 2003; Prichard et al. 1991). The rise in lactate was accompanied by an initial drop in cortical glucose levels (Chen et al. 1993; Frahm et al. 1996).

Patients with primary generalized epilepsy with photosensitivity showed increased lactate levels in the occipital cortex in the resting state and an increased area of visual cortical activation with photic stimulation (Hill et al. 1999). Patients with mitochondrial cytopathies (Kearns-Sayre syndrome) have elevated lactate levels during basal conditions without an additional increase with photic stimulation (Kuwabara et al. 1994). Two recent publications report increased lactate in patients with migraine with aura, but no enhanced lactate during photic stimulation in patients without aura (Sandor et al. 2005; Sarchielli et al. 2005).

Brain lactate rises under a wide variety of aerobic conditions, which stimulate glial glycolytic and glycogenolytic metabolism. Hyperventilation causes a rise in lactate acutely as the glia buffer the rising pH by excreting bicarbonate ions (Petroff et al. 1985; van Rijen et al. 1989). Hyperventilation or lactate infusion have been used to trigger panic attacks (Dager et al. 1995, 1999). Patients with panic attacks responded with high cortical lactate levels when hyperventilated for 20 minutes (see Figure 6.9). Cortical lactate reached higher levels during a 20-minute lactate infusion in a patient who experienced a panic attack, than in a patient who did not experience a panic attack, or in control subjects. Gabapentin treatment appeared to be effective in blocking a lactate-induced panic response, but did not alter the magnitude or time course of an abnormal brain lactate response to lactate infusion (Layton et al. 2001). When combined with other studies of lactate-induced panic attacks, this suggests that lactate may facilitate excitatory neurotransmission in the human brain, at least in some patients with panic attacks.

C. Blood Lactate as an Alternative Fuel for Brain Metabolism

Lactate transport across the blood-brain barrier is transporter mediated (MCT1) with a Michaelis constant (Km) of ~3.5 mM (Dalsgaard 2006). Under basal conditions, there is a slight net export of lactate from brain to blood because the blood is more alkalotic (pH ~7.4) than brain (pH ~7). Lactic acid is lipid soluble and diffuses across the endothelium, becoming trapped in the more alkalotic compartment, the blood. However, when blood lactate levels are raised to levels by infusion or during aerobic exercise to concentrations above 4 mM, there is a net uptake of lactate from the blood into the brain, which has been demonstrated using MR spectroscopy combined with parallel arterial-venous (A-V difference) concentration measurements (Young et al. 1991a). The increased availability is reflected by a decrease in the oxygen glucose index (OGI). The OGI reflects the stoichiometry; six oxygen molecules are required to oxidize completely one molecule of glucose. When glycolysis increases faster than brain oxygen utilization ($CMRO_2$), the OGI decreases. When an alternative mitochondrial fuel (e.g., lactate, β-hydroxybutyrate, acetate, glycogen, fatty acid or amino acid) is utilized and glucose consumption decreases, the OGI increases. With maximal exercise, blood lactate rises to 15–30 mM, ammonia levels increase and the blood pH will become acidotic (pH 6.8). Brain lactate uptake under those conditions will increase to nearly equal the cerebral metabolic rate of glucose (CMR glucose) and the OGI increases to ~8, which may be interpreted as showing that lactate is the preferred fuel of the brain (Bouzier-Sore et al. 2003b; Dalsgaard 2006).

D. Cerebral Glutamate and Glutamine

Glutamate is the chief excitatory neurotransmitter of the adult brain. Glutamatergic neurons possess about

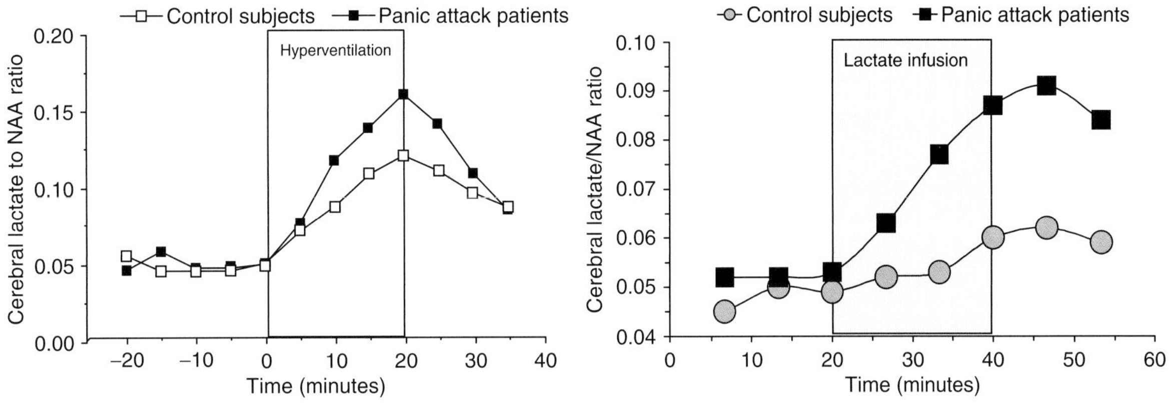

Figure 6.9 The graph on the left shows the increase of cortical lactate in response to hyperventilation in control subjects and patients with panic attacks (adapted from Dager et al. 1995). On the right, the graph shows the increase in cortical lactate during an intravenous infusion of sodium lactate in control subjects and patients with panic attacks (adapted from Dager et al. 1999).

80 to 88 percent, whereas GABAergic neurons contain about 2 to 10 percent and astrocytes about 10 percent of total tissue glutamate (Hyder et al. 2006). Decreases in tissue glutamate content, therefore, often reflect primarily the loss of glutamatergic neurons and synapses (Hertz 2006). In normal brain, glutamate is known to be in high concentration in neurons, whereas in the astrocyte, glutamate has been measured in animal models and estimated *in vivo* to be comparatively low, less than 1 mM and perhaps as low as 50 micromolar (Ottersen et al. 1992). As expected, the glutamate content of human white matter is lower than that of grey matter (de Graaf et al. 2001; Petroff et al. 1995a). This has been attributed to the increase in the fractional volume of myelin, greater density of oligodendrocytes, and the marked decrease in synapses.

E. Glutamate Synthesis

Glutamate plays several critical roles in neural functioning: it is both the primary excitatory neurotransmitter and important in oxidative metabolism. Therefore, it is important to understand how glutamate is distributed within cell populations (neurons and glia) and in the extracellular space in both normal and epileptogenic tissue (Hertz 2006; Hyder et al. 2006; Petroff et al. 1995a, 2002a). In neurons, glutamate serves as a storage form of tricarboxylic acid cycle (TCA cycle) intermediates, which are critical for mitochondrial oxidative metabolism and thus neuronal energy production. Several enzymes, aspartate transaminase (AST), GABA-transaminase (GABA-T), alanine transaminase (ALT), and branched-chain amino acid transaminase (BCAAT), which are expressed by both neurons and glia, catalyze the chemical equilibrium between alpha-ketoglutarate and glutamate (De Graaf et al. 2006; McKenna et al. 2000; Waagepetersen et al. 2000; Xu et al. 2004).

However, MR spectroscopic studies show that the primary supply for the glutamate used by neurons for neurotransmission, both glutamatergic and GABAergic, is glutamine. Phosphate-activated glutaminase (PAG), which converts glutamine to glutamate and ammonia and is expressed primarily by neurons, serves as the main enzyme employed by neurons to supply the glutamate needed for neurotransmission, as a precursor for GABA synthesis, N-acetyl-aspartyl-glutamate (NAAG), and glutathione synthesis, and maintaining the supply of alpha-ketoglutarate, which is critical for mitochondrial oxidation.

$$\text{glutamine} \xrightarrow{\text{PAG}} \text{glutamate} + \text{ammonia}$$

Glutamate lost from the neuron during neurotransmission or oxidation is replaced through phosphate activated glutaminase (PAG) acting upon glutamine synthesized in the glia and transported into neurons (Gonzalez-Gonzalez et al. 2005; Kanamori & Ross 2004; Kvamme et al. 2001; van der Gucht et al. 2003). The rate of glutamate synthesized by PAG is proportional to the rate of glutamate used by neurons for neurotransmission, because the amount of glutamate oxidized by neurons is very low. The exception occurs during prolonged hypoglycemia, when glutamate is consumed to support ATP production (Behar et al. 1985). Net glutamate oxidation is catalyzed by the enzyme glutamate dehydrogenase (GDH), which is expressed by neurons and glia (Kanamori & Ross 1995; Mastorodemos et al. 2005).

$$\text{glutamate} + \text{NAD} \xleftrightarrow{\text{GDH}} \alpha\text{-keto-glutamate} + \text{ammonia} + \text{NADH}$$

Gain of function mutations of glutamate dehydrogenase were proposed to be a cause of intractable human epilepsy (Raizen et al. 2005).

VIII. Cerebral Ammonia Metabolism

MR studies using 13C and 15N labeled substrates are useful in understanding how the brain detoxifies ammonia and synthesizes glutamate, glutamine, and GABA (Kanamori et al. 1996, 2002; Sakai et al. 2004; Shen et al. 1998; Sibson et al. 1997, 2001). Rising glial cytosolic ammonia or glutamate concentrations stimulates glutamine synthetase, which consumes one glutamate, one ATP (complexed with magnesium), and one ammonia (NH3) molecule to synthesize one glutamine, one ADP, one inorganic phosphate (Pi), one free magnesium, and releases acid, lowering cytosolic pH (Eisenberg et al. 2000).

$$\text{glutamate} + \text{ammonia} + \text{Mg-ATP} \xrightarrow{\text{glutamine synthetase}} \text{glutamate} + \text{ADP} + \text{Mg}^{+2} + \text{H}^{+1}$$

In the brain, glutamine synthetase is an enzyme of primary neurochemical importance, since it converts neurotoxic ammonia and the neurotransmitter glutamate into glutamine. Nonbrain glutamine synthetase (e.g., liver and kidney) responds to end-product (glutamine and its derivatives) feedback inhibition, whereas brain glutamine synthetase does not. Congenital dysfunction of glutamine synthetase results in a human phenotype, which includes multifocal and generalized electrographic seizures (Haberle et al. 2005). Glia express the pyruvate carboxylase (PC) and glutamine synthetase (GS), which are the enzymes needed for the *de novo* synthesis of glutamate and glutamine from glucose. Most neurons in the adult brain do not express the enzymes, pyruvate carboxylase and malic enzyme, which catalyze the *de novo* synthesis of TCA cycle intermediates from glucose (anaplerosis). During hyperammonemia, astrocytic PC, BCAAT and GS are all upregulated to detoxify the ammonia entering the brain.

A. Hyperammonemia

Above-normal blood ammonia concentrations are associated with hepatic encephalopathy caused by a variety of inherited and acquired liver diseases (Filipo & Butterworth 2002). MR studies have shown that brain glutamine content increases

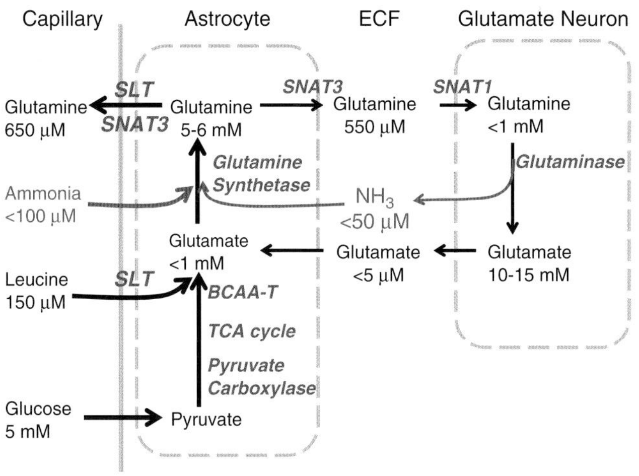

Figure 6.10 Schematic of cerebral ammonia detoxification. Abbreviations: leucine transporter (SLT), neuronal glutamine transporter (SNAT3), glial glutamine transporter (SNAT1), branched-chain amino acid transaminase (BCAA-T).

rapidly in response to rising blood ammonia (Bluml et al. 2001b; Minguez et al. 2006; Tofteng et al. 2006) (see Figure 6.10). Brain phosphocreatine levels are reduced in patients with liver disease with lower levels seen with hepatic encephalopathy (Barbiroli et al. 2002). MR studies using 1-13C-glucose as a label source indicate that above-normal ammonia levels impair brain glucose oxidation and slow glutamatergic neurotransmission (glutamate-glutamine cycle) as glia divert more and more resources to detoxify ammonia (Bluml et al. 2001b; Zwingmann & Butterworth 2005). The rapid increase in glutamine levels contribute to astrocytic swelling and brain edema and correlate with EEG slowing and cognitive decline (Balata et al. 2003; Minguez et al. 2006; Shawcross et al. 2004; Zwingmann & Butterworth 2005). Vigorous exercise regimens are known to increase blood ammonia, which may contribute to the "central fatigue" associated with maximal exercise (Nybo et al. 2005). Ammonia, which enters the brain from the blood, may serve as a powerful inhibitory neuromodulator.

B. The Effects of Antiepileptic Drugs on Cortical Glutamate and Glutamine Levels

Antiepileptic drugs could contribute to the variability in GLX, glutamate, and glutamine levels measured in the patients who use AEDs (Petroff et al. 1995b, 1999, 2000). Measurements made in the visual cortex of patients with refractory complex partial seizures, primarily temporal or frontal lobe epilepsies, suggest that glutamate levels are modestly, but significantly, above normal in patients taking carbamazepine, phenytoin, or gabapentin and significantly below normal in patients taking phenobarbital or primidone (see Figure 6.11). Whether the antiepileptic barbiturates lower tissue glutamate content remains to be determined; cortical glutamate content decreases by 16 to 28 percent with pentobarbital anesthesia (Patel et al. 2005; Sibson et al. 1998). Valproate and vigabatrin increase human brain glutamine levels by 50 to 80 percent. Valproate is known to increase blood ammonia levels through its effects on human kidney glutamine metabolism. Vigabatrin is an irreversible inhibitor of GABA-transaminase and may inhibit, albeit to a lesser degree, the other transaminases including ornithine aminotransferase, thereby raising ammonia levels (Hisama et al. 2001). Whether the use of valproate contributes to the increase of frontal lobe GLX levels in patients with idiopathic generalized epilepsy remains speculative.

C. Changes in Brain Glutamate and Glutamine Content with Physiological Stimulation

Advances in magnet technology developed very high field (11.7 Tesla) imager-spectrometers suitable for *in vivo* animal studies. The improved spectral resolution and signal-to-noise ratio allows measurements, which showed that physiological stimulation alters the glutamate, glutamine, phosphocreatine, and myo-inositol content of the somatosensory neocortex (Xu et al. 2005). The improvement in sensitivity extends the

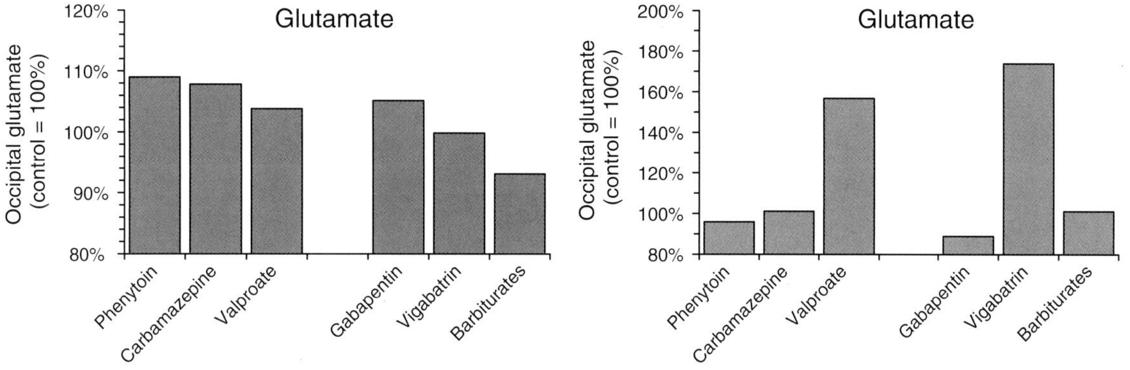

Figure 6.11 Occipital glutamate and glutamine levels of patients with epilepsy taking various antiepileptic drugs. The intracellular concentrations are normalized to the mean value of drug-free volunteers without epilepsy.

functional MRI from the water signal to the NAA signal. A significant increase in glutamine content with a reciprocal decrease in glutamate content of the somatosensory cortex were reported during 11 minutes of forepaw stimulation under α-chloralose anesthesia. The reciprocal changes during physiological stimulation (the sum of glutamate and glutamine content remained the same) were attributed to the release of neuronal glutamate, uptake by glia, and conversion into glutamine. Ammonia released by activated neurons may serve as an inhibitory neuromodulator, which may help terminate the excitatory effects of neuronal vesicular glutamate release either directly or by promoting astrocytic glutamate uptake by activating glutamine synthetase.

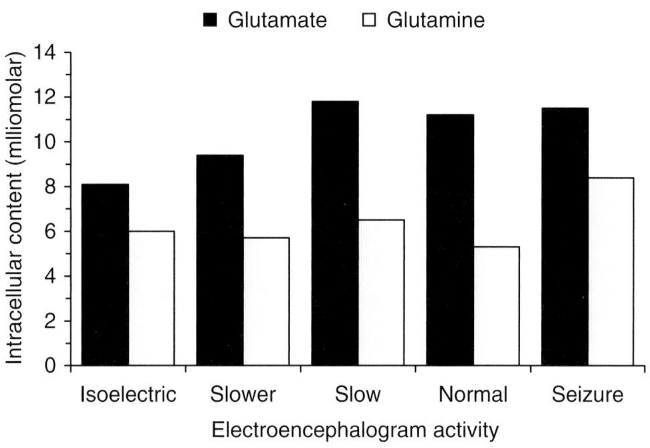

Figure 6.12 Cortical glutamate and glutamine concentrations as a function of EEG activity (adapted from Sibson et al. 1998 and Patel et al. 2004).

D. Cortical Glutamate and Glutamine and EEG Power

Cortical glutamate content appears to decrease with depth of anesthesia with a 25 to 30 percent decrease with pentobarbital coma (Sibson et al. 1998) (see Figure 6.12). Brain glutamate content increases in acute seizure models using NMDA or kainic acid and decreases with bicuculline and pentylene tetrazole (Eloqayli et al. 2003; Patel et al. 2004; Young et al. 1991b). Similarly, glutamine content of glia is greater than the neuronal content because glutamine synthetase is an exclusively glial enzyme. Changes in tissue glutamine content reflect changes in glial ammonia and glutamate concentrations content because of the critical role, that glutamine synthetase plays in detoxifying glutamate and ammonia. Glia tightly control glial intracellular ammonia and glutamate concentration by regulating the activities of glutamate synthetase and the anaplerotic enzymes (pyruvate carboxylase and branch-chain amino acid transaminase) (Kanamori et al. 1998; Kanamori & Ross 2005; Shen et al. 1998; Sibson et al. 2001). Brain glutamine content appears to increase with almost all seizure models, probably reflecting the rising ammonia released by increased glutaminase activity, which increase in proportion to sustained glutamatergic neurotransmission. Ammonia may serve as a powerful inhibitory neuromodulator when glutaminases are activated by sustained excitatory neurotransmission.

E. The Glutamate-Glutamine Cycle Measures Neurotransmission

The glutamate-glutamine cycle refers to the compartmentation of glutamate and glutamine between neurons and glia (see Figure 6.13). During glutamatergic neurotransmission neurons release glutamate into the extracellular space; the glial glutamate transporters rapidly remove the releases glutamate. To minimize the likelihood of glutamate transporter reversal during depolarization, the cell surface of glutamatergic neurons express low levels of glutamate transporters (Hertz 2006). Studies of glutamatergic synapses have shown them to be closely surrounded by glial end processes possessing high densities of glutamate transporters. Reuptake of glutamate from the extracellular space primarily by glia uses the sodium-dependent, electrogenic glutamate transporters, EAAT1 (GLAST) and EAAT2 (GLT-1) (Danbolt 2001). Under normal conditions, EAAT1 and EAAT2 are located on astrocytic membranes and terminate excitatory neurotransmission by first binding glutamate (buffering) then transporting glutamate into the astrocytic cytosol in an energy (ATP) consuming step (Cavelier et al. 2005).

Using quantitative microdialysis methods during the interictal period in patients with localization-related epilepsies, studies indicate that the basal ECF glutamate levels in the nonepileptogenic neocortex and hippocampus range from 0.5 to 5 μM (mean 2.5 μM), consistent with the measures obtained from healthy rats (Cavus et al. 2005). Low intraglial glutamate concentrations are maintained primarily by the detoxification of glutamate and ammonia into glutamine by glutamine synthetase (Eisenberg et al. 2000). Alternatively, glutamate is oxidized by glutamate dehydrogenase and enters the glial TCA cycle. Rising intracellular glutamate concentrations enhanced the transporter-mediated (SNAT3) release of glutamine, which is taken up by neurons using a sodium-gradient-dependent transporter (SNAT1) (Broer et al. 2004; Kanamori & Ross 2004). Phosphate-activated glutaminase replenishes the neuronal intracellular glutamate concentrations. This neuron-astrocyte metabolic network is called the glutamate-glutamine cycle. The glutamate-glutamine cycle is critical for (1) the rapid and efficient clearance of glutamate from the synaptic cleft and extracellular space, (2) the maintenance of neuronal mitochondrial metabolism; and (3) the detoxification of the ammonia generated by neurotransmission.

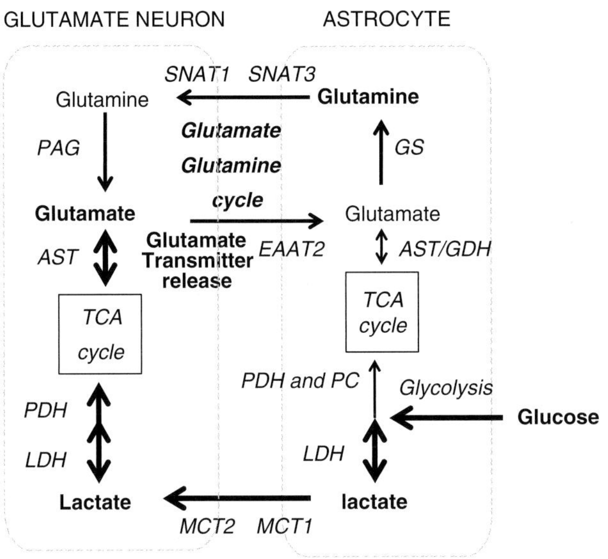

Figure 6.13 Schematic of the glutamate-glutamine cycle. Abbreviations: branched-chain amino acid transaminase (BCAA-T), aspartate transaminase (AST), glutamate dehydrogenase (GDH), glutamine synthetase (GS), lactate dehydrogenase (LDH), phosphate-activated glutaminase (PAG), pyruvate carboxylase (PC), pyruvate dehydrogenase (PDH), tricarboxylic acid (TCA), glial glutamate transporter (EAAT2), neuronal glutamine transporter (SNAT3), glial glutamine transporter (SNAT1), monocarboxylic acid transporter (MCT).

F. Glutamate Neurotransmission Is Tightly Coupled to Glucose Metabolism

Glutamatergic neurotransmission and the glutamate-glutamine cycle are tightly coupled to cerebral oxidative metabolism (de Graaf et al. 2003; Hyder et al. 2006; Magistretti et al. 1999). The large majority of cortical glutamate uptake after release is astroglial and tightly coupled to the glial sodium-potassium ATPase and therefore glial energy metabolism. One glutamate is taken up together with three sodium and one proton in exchange for one potassium consuming 1.33 ATP per glutamate (Attwell & Laughlin 2001). Rising intracellular glutamate and sodium content makes glutamate uptake energetically much more expensive. In glia, intracellular glutamate content is usually very low (values as low 50 micromolar have been proposed) to facilitate glutamate transport.

The rates of the glutamate-glutamine cycle (V_{cycle}) are linearly correlated to the TCA cycle (V_{TCA}) over the full range of cortical activity from deep pentobarbital coma with gross suppression of all cerebral rhythms by EEG through bicuculline-induced status epilepticus (Hyder et al. 2006; Shulman et al. 2004; Sibson et al. 1998; Patel et al. 2004) (see Figure 6.14). The ratio of the glutamate-glutamine cycle to the TCA cycle rates (V_{cycle}/V_{TCA}) remains about 0.44 while awake and through the various stages of anesthesia with an over three-fold variation in the rate of mitochondrial glucose oxidation. As expected, the amount of synaptic transmission and the rate of the glutamate-glutamine cycle are negligible in white matter, which has a much lower rate of glucose oxidation than grey matter (de Graaf et al. 2001; Pan et al. 2000). A ratio of the glutamate-glutamine cycle to TCA cycle rates of 0.42 was measured in the cortical and central grey matter (primarily dorsal hippocampus) in the same animals under halothane anesthesia with a ratio of 0.1 in subcortical white matter.

G. Changes in Glutamatergic Neurotransmission with Physiological Stimulation

A linear response between the change in the metabolic rate of the glutamate-glutamine cycle (V_{cycle}) and the rate of oxidative metabolism ($CMRO_2$) is maintained with physiological stimulation (evoked metabolism) (Chhina et al. 2001;

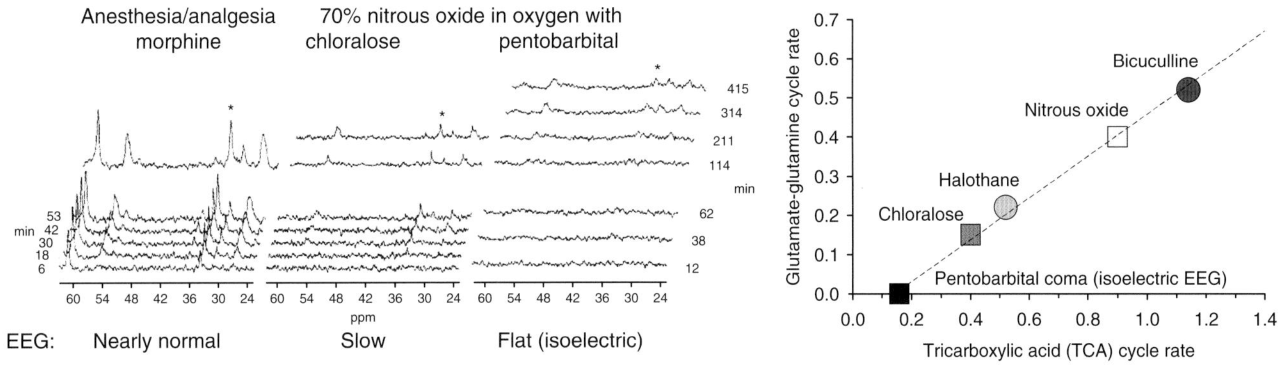

Figure 6.14 Serial carbon spectra of the rat cortex under various depths of anesthesia are shown on the left. The graph on the right shows that the metabolic rate of the glutamate-glutamine cycle and the rate of glucose oxidation are linearly related over the full range of EEG activity (adapted from Sibson et al. 1998 and Patel et al. 2004).

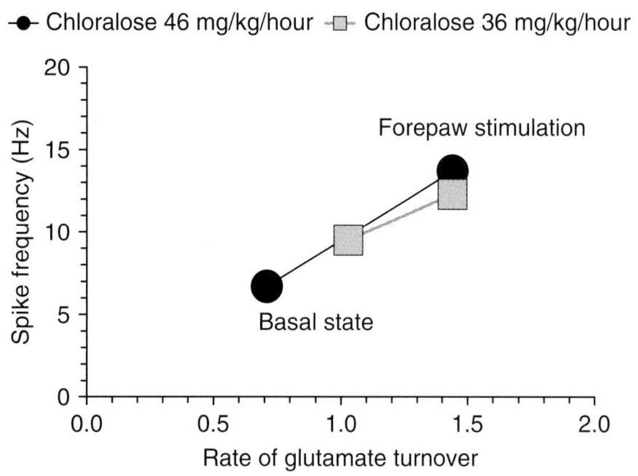

Figure 6.15 The metabolic rate of glutamate turnover and the intensity of electrical activity (spike frequency) measured in the somatosensory cortex are related linearly (data from Smith et al. 2002).

Hyder et al. 1997, 2001; Kida et al. 2006). MR spectroscopy studies show that the incremental increase in oxidative metabolism, $\Delta CMRO_2/CMRO_2$, was proportional to changes in glutamatergic neurotransmitter flux ($\Delta V_{cycle}/V_{cycle}$). Further studies show that the increase in spike frequency of the somatosensory cortex in response to forepaw stimulation was proportional to the increase in oxidative metabolism (Smith et al. 2002) (see Figure 6.15). The combined results show that $\Delta CMRO_2/CMRO_2 \approx \Delta V_{TCA}/V_{TCA} \approx \Delta$ spike frequency/spike frequency $\approx \Delta V_{cycle}/V_{cycle}$, thereby relating the energetic basis of brain activity to neuronal spiking frequency and the rate of glutamatergic neurotransmission.

H. Studies of Glutamatergic Neurotransmission of the Human Visual Cortex

Under normal conditions, 13C-glucose rapidly labels neuronal intracellular glutamate first. The human neuronal tricarboxylic acid (TCA) cycle rate (0.72 mM/minute) is five times faster than glial TCA cycle rate (0.13 mM/minute) (Shulman et al. 2004). The rate of intracellular glutamine turnover in the normal awake human brain is less than half (44–47%) of intracellular glutamate turnover and the rate of glutamine synthesis (0.23–0.36 mM/minute) is two to three times faster than the glial TCA cycle rate (De Graaf et al. 2003). Under normal conditions using 13C-glucose, the isotopic enrichment of intracellular glutamine primarily reflects the uptake and detoxification of 13C-glutamate released by neurons with only minor contributions (18–33%) from the glial TCA cycle (De Graaf et al. 2003; Lebon et al. 2002; Shen et al. 1999). The relationship between the rate of the glutamate-glutamine cycle and TCA cycle in the human cortex, which were made using 13C-glucose, 13C-acetate or 13C-β-hydroxybutyrate as the label sources, is linear (Chhina et al. 2001; Gruetter et al. 2001; Lebon et al. 2002; Pan et al. 2002b; Shen et al. 1999). The ratio of the glutamate-glutamine cycle to TCA cycle (V_{cycle}/V_{TCA}) was 0.45 in excellent agreement with homologous values measured in rodent models.

I. MR Studies of Brain Glutamate and Glutamine of Patients with Epilepsy

Initial MR studies in patients with epilepsy suggested that the sum of glutamate and glutamine (GLX) and the GLX/NAA ratios were above normal in the MRI-negative epileptogenic hippocampus or the epileptogenic areas of the nonlesional neocortical epilepsies (Savic et al. 2000; Woermann et al. 1999). Subsequent studies reported greater variability of the GLX signals and GLX/NAA ratios, which were the same as values for control subjects or above normal in various regions outside of the presumed seizure focus; that is, frontal lobes or contralateral hippocampus in TLE (Flugel et al. 2006; Simister et al. 2002; Wellard et al. 2003). The GLX signals were reported to show considerable inter- and intra-subject variability in patients with malformations of cortical development or Rasmussen encephalitis (Wellard et al. 2004; Woermann et al. 2001). Studies of patients with idiopathic generalized epilepsy suggest that the GLX signals and the GLX/NAA ratios of the frontal lobes are above normal (Simister et al. 2003).

J. Hippocampal Glutamate and Glutamine Content in Mesial TLE

Mesial temporal sclerosis (MTS) refers to a group of pathological changes in a wide area including the hippocampus, amygdala, and medial temporal cortical structures including the entorhinal cortex and parahippocampus. In a surgical series of 151 patients, there was more than a 60 percent cell loss in MTS and a 23 percent loss in nongliotic hippocampi (PTLE) (de Lanerolle et al. 2003). The large glutamatergic principle cells are primarily lost in TLE, whereas specific populations of interneurons are preserved and undergo complex synaptic reorganization (Magloczky & Freund 2005). Histopathological studies show greater cell loss and synaptic reorganization in the more anterior portions of the hippocampus.

Spectroscopic imaging at 4-Tesla of patients with MTS showed that mean hippocampal glutamate content was significantly 30 percent below normal ipsilateral to the seizure focus (Pan et al. 2006). The glutamate content of the contralateral hippocampus, which appeared normal by MRI, was 12 percent below normal. The reduction in glutamate content is surprisingly small in view of the greater than 60 percent neuron loss and greater than 80 percent increase in glial density, which are the hallmark of MTS.

Analytical MR spectroscopic studies of tissue resected in surgery showed that the glutamate content of the resected epileptogenic hippocampus with MTS is significantly less than values

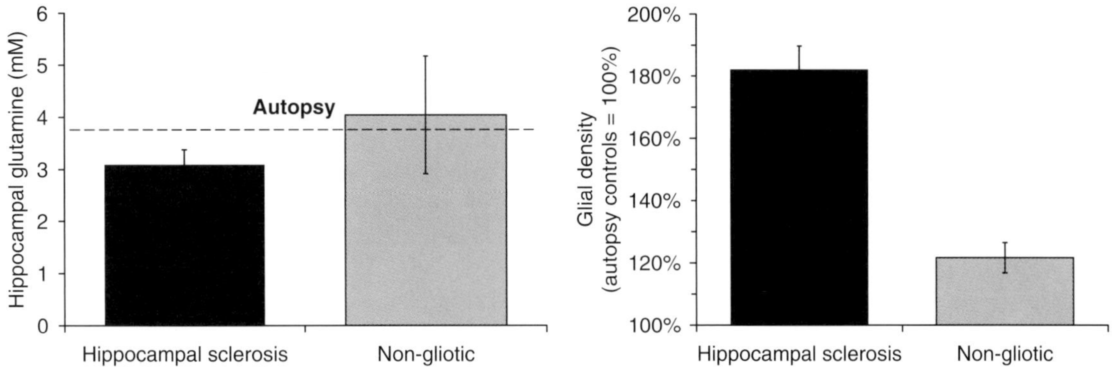

Figure 6.16 Intracellular glutamate concentrations do not reflect the loss of neurons, mainly glutamatergic neurons, in the epileptogenic human hippocampus (adapted from Petroff et al. 2002).

measured in the nongliotic ones (PTLE) (Peeling & Sutherland 1993; Petroff et al. 2002a). However, it is surprising that the glutamate content in MTS, despite the 67 percent neuronal loss, is not significantly less than the mean values measured at autopsy (8.2 mM) or *in vivo* (8.0 and 8.8 mM) of nonepileptic subjects (Kassem & Bartha 2003; Mohanakrishnan et al. 1997; Pan et al. 2006) (see Figure 6.16).

Similarly, the hippocampal glutamine content in MTS remains on the lower end of normal, despite an 82 percent increase in glial density (Petroff et al. 2003) (see Figure 6.17). The neuron to glia ratios are significantly altered in the epileptogenic hippocampus, both MTS and PTLE, yet both glutamate and glutamine concentrations remain nearly normal. Our findings suggest that the intracellular concentration of glutamate of the remaining cells must be increased markedly because of the tremendous loss of glutamatergic neurons and synaptic density, which characterizes MTS. The loss of glutamine synthetase expression and activity in the setting of increased GFAP expression in MTS suggests that astrocytes probably contain the excess glutamate (Eid et al. 2004; van der Hel et al. 2005).

K. Glutamatergic Neurotransmission in Mesial TLE

Stable-isotope studies demonstrate that glutamine synthesis is impaired in the epileptogenic human hippocampus. The V_{cycle}/V_{TCA} ratio, which was measured intra-operatively, was decreased by 56 percent in the epileptogenic hippocampus (Petroff et al. 2002b). Given the central role of glutamine synthetase in the normal functioning of the glutamate-glutamine cycle, the data suggests that glutamine synthesis rates (normalized to the rate of glucose oxidation) are decreased by more than 70 percent in the gliotic hippocampus of patients with mesial temporal sclerosis (MTS). Patient studies using microdialysis show that post-ictal glutamate reuptake is three-fold slower in the epileptogenic hippocampus than the contralateral one (During & Spencer 1993). The combined observations suggest that glial glutamate transporter and detoxification functions are slowed or deficient in the gliotic hippocampus. In the epileptic but nonepileptogenic hippocampus, glia maintained normal extracellular glutamate concentrations (~3 micromolar) interictally, and rapidly

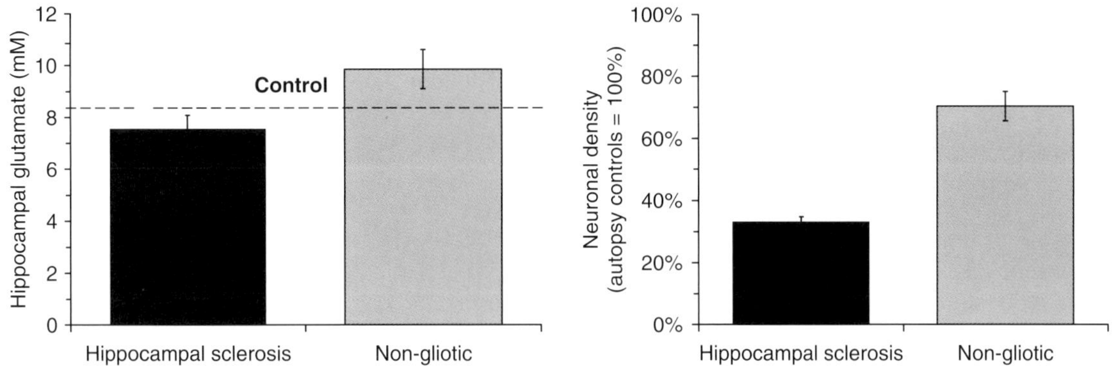

Figure 6.17 Intracellular glutamine concentrations do not reflect the increased glial density of the epileptogenic human hippocampus (adapted from Petroff et al. 2002).

restored homeostasis after seizure spread from the contralateral focus. The low rate of glutamine synthesis could account for the markedly slower rate of post-ictal glutamate clearance, which characterizes the epileptogenic state.

IX. Summary

Studies using 31P-MRS show that the interictal state in human epilepsies appears to be characterized by low phosphocreatine levels with normal ATP levels and normal pH with the lowest phosphocreatine levels in the epileptogenic regions of the brain. Low phosphocreatine levels, which affect the epileptogenic human hippocampus, are associated with specific changes in excitatory and inhibitory neuronal responses to synchronized synaptic inputs. Processes, which inhibit mitochondrial oxidative phosphorylation, are associated with epileptic phenotypes. Treatments that which raise phosphocreatine levels, improve seizure control in some patients.

Studies using 1H-MRS show that lactate is a labile metabolite, which transiently increases during cortical activity under well-oxygenated physiological conditions. A rise in glial synthesis and export of lactate appears to occur with physiological stimulation (the astrocyte neuron lactate shuttle hypothesis). Combined 1H- and 13C-MRS studies suggest that astrocytic glycogen has a rapid turnover rate; with physiological stimulation glial glycogen is converted to lactate to support neuronal high-frequency firing. Rather than contributing to neuronal damage, lactate appears to serve a neuroprotective function. Increased lactate levels appear to be a characteristic of activated glia and not merely a marker of hypoxia.

Combined MRI, 1H-, and 15N-MRS studies show that glutamine is a labile metabolite, which reflects changing cerebral ammonia concentrations. Cortical glutamine levels transiently increase with physiological stimulation. Glutamine levels also increase as blood ammonia increases; changes in brain glutamine correlate with EEG slowing and alterations of human cognitive functions. This suggests that cerebral ammonia acts as a potent inhibitory modulator of brain functions. Serial brain glutamine measurements may be used to follow fluctuations of ammonia-induced encephalopathy.

Combined 1H- and 13C-MRS studies may be used to measure glutamatergic neurotransmission (the glutamate-glutamine cycle) under physiological conditions. They show that excitatory glutamatergic neurotransmission, cerebral oxidative metabolism, EEG activity, and cortical spike frequency are all tightly coupled. The interictal state of the localization-related human epilepsies appears to be characterized by above-normal lactate levels, elevated glutamate levels, and below-normal glutamine levels. Although these changes are often widespread, they are most pronounced in the seizure-onset zone. In the epileptogenic human hippocampus, interictal glucose metabolism is reduced and associated with an even greater reduction in glutamine synthesis and the glutamate-glutamine cycle. These alterations in glutamate metabolism may maintain the dysfunctional yet hyper-excitable state within the seizure focus by maintaining the above-normal extracellular glutamate concentrations, which are characteristic of epileptogenic brain.

A. Is MRS Ready for Routine Clinical Use?

The clinical utility of MRS is limited by costs and institutional (imaging) priorities. Stronger, therefore more expensive, magnets improve spectroscopy and imaging. Signal amplitude increases in proportion to the magnetic field strength, thereby increasing spatial resolution. Upgrading from a 1.5-Tesla imager to a 3-Tesla MRI will double the resolution of the images within the 15 to 20 minutes assigned for a routine clinical MRI study. However, the typical MRS study usually requires one hour of spectrometer-imager time and dynamic studies may require several studies over periods ranging from a day to weeks. Hospitals and radiologists opt for the cost-effective solution, which is to image routinely three patients in the time required for a single spectroscopy study.

At the present time, proton MRS is used clinically in the evaluation of epilepsy and brain tumors. Spectroscopic imaging (MRSI) is useful in distinguishing recurrent tumor from phlegmon or necrosis to guide continuing therapy. Measurements of NAA levels using MRSI appears to be most useful in focal epilepsies without lesions seen using conventional MRI. The sensitivity in determining the epileptogenic hemisphere is 60 to 65 percent using single voxel methods and improves to 85 to 98 percent with spectroscopic imaging. Combined MRSI and volumetric MRI correctly predicted the surgical outcomes of 75 to 92 percent of patients with TLE who became seizure free and 63 to 72 percent of patients who did not. In these settings, MRSI is cost-effective.

Proton MRSI using a 3-Tesla or higher field system appears to be a very promising tool to monitor treatment in multiple sclerosis and other disease processes that impact axons. Abnormalities of axonal mitochondrial function are measurable in normal-appearing white matter before abnormalities become evident by MRI. Similar changes of grey and white matter may be measured in mitochondrial diseases and degenerative diseases, which impact mitochondrial functions, such as Parkinson disease, amyotrophic lateral sclerosis, Huntington's disease, frontal dementia, and Alzheimer's disease. Serial MRSI may be used to assess the effectiveness of therapies in minimally symptomatic patients or ideally presymptomatic patients. Measurements of glutamine, glutamate, and GABA using MRSI appear to be very promising for the assessment of therapy in hepatic encephalopathy, depression, anxiety disorders, and schizophrenia.

References

Antozzi, C., Franceschetti, S., Filippini, G., Barbiroli, B., Savoiardo, M., Fiacchino, F. et al. (1995). Epilepsia partialis continua associated with NADH-coenzyme Q reductase deficiency. *J Neurol Sci* **129**, 152–161.

Argov, Z., Arnold, DL. (2000). MR spectroscopy and imaging in metabolic myopathies. *Neurol Clin* **18**, 35–52.

Attwell, D., Laughlin, S. B. (2001). An energy budget for signaling in the grey matter of the brain. *J Cereb Blood Flow Metab* **21**, 1133–1145.

Attwell, D., Iadecola, C. (2002). The neural basis of functional brain imaging signals. *Trends Neurosci* **25**, 621–625.

Aubert, A., Costalat, R., Magistretti, P. J., Pellerin, L. (2002). Brain lactate kinetics: Modeling evidence for neuronal lactate uptake upon activation. *Proc Natl Acad Sci USA* **102**, 6448–6453.

Ayata, C., Jin, H., Kudo, C., Dalkara, T., Moskowitz, M. A. (2006). Suppression of cortical spreading depression in migraine prophylaxis. *Ann Neurol* **59**, 652–661.

Balata, S., Damink, S. W., Ferguson, K., Marshall, I., Hayes, P. C., Deutz, N. E. et al. (2003). Induced hyperammonemia alters neuropsychology, brain MR spectroscopy and magnetization transfer in cirrhosis. *Hepatology* **37**, 931–939.

Barbiroli, B., Iotti, S., Cortelli, P., Martinelli, P., Lodi, R., Carelli, V., Montagna, P. (1999a). Low brain intracellular free magnesium in mitochondrial cytopathies. *J Cerebr Blood Flow Metab* **19**, 528–532.

Barbiroli, B., Iotti, S., Lodi, R. (1999b). Improved brain and muscle mitochondrial respiration with CoQ. An in vivo study by 31P-MR spectroscopy in patients with mitochondrial cytopathies. *Biofactors* **9**, 253–260.

Barbiroli, B., Gaiani, S., Lodi, R., Iotti, S., Tonon, C., Clementi, V. et al. (2002). Abnormal brain energy metabolism shown by in vivo phosphorus magnetic resonance spectroscopy in patients with chronic liver disease. *Brain Res Bull* **59**, 75–82.

Barker, P. B., Butterworth, E. J., Boska, M. D., Nelson, J., Welch, K. M. A. (1999). Magnesium and pH imaging of the human brain at 3.0 tesla. *Magn Reson Med* **41**, 400–406.

Behar, K. L., den Hollander, J. A., Petroff, O. A. C., Hetherington, H., Prichard, J. W., Shulman, R. G. (1985). The effect of hypoglycemic encephalopathy upon amino acids, high-energy phosphates, and pHi in the rat brain in vivo: Detection by sequential 1H and 31P NMR spectroscopy. *J Neurochem* **44**, 1045–1055.

Behar, K. L., Rothman, D. L., Spencer, D. D., Petroff, O. A. C. (1994). Analysis of macromolecule resonances in 1H NMR spectra of human brain. *Magn Reson Med* **32**, 294–302.

Bitsch, A., Bruhn, H., Vougioukas, V., Stringaris, A., Lassmann, H., Frahm, J., Bruck, W. (1999). Inflammatory CNS demyelination: Histopathologic correlation with in vivo quantitative proton MR spectroscopy. *Am J Neuroradiol* **20**, 1619–1627.

Bluml, S., Moreno, A., Hwang, J. H., Ross, B. D. (2001a). 1-(13)C glucose magnetic resonance spectroscopy of pediatric and adult brain disorders. *NMR Biomed* **14**, 19–32.

Bluml, S., Moreno-Torres, A., Ross, B. D. (2001b). [1-13C]glucose MRS in chronic hepatic encephalopathy in man. *Magn Reson Med* **45**, 981–993.

Boska, M. D., Welch, K. M. A., Barker, P. B., Nelson, J. A., Schultz, L. (2002). Contrasts in cortical magnesium, phospholipid and energy metabolism between migraine syndromes. *Neurology* **58**, 1227–1233.

Bouzier-Sore, A. K., Serres, S., Canioni, P., Merle, M. (2003a). Lactate involvement in neuron-glia metabolic interaction: (13)C-NMR spectroscopy contribution. *Biochimie* **85**, 841–848.

Bouzier-Sore, A. K., Voisin, P., Canioni, P., Magistretti, P. J., Pellerin, L. (2003b). Lactate is a preferential oxidative energy substrate over glucose for neurons in culture. *J Cereb Blood Flow Metab* **23**, 1298–1306.

Briellmann, R. S., Wellard, R. M., Jackson, G. D. (2005). Seizure-associated abnormalities in epilepsy: Evidence from MR imaging. *Epilepsia* **46**, 760–766.

Broer, A., Deitmer, J. W., Broer, S. (2004). Astroglial Glutamine Transport by system N is upregulated by glutamate. *Glia* **48**, 298–310.

Brown, A. M., Tekkok, S. B., Ransom, B. R. (2004). Energy transfer from astrocytes to axons: The role of CNS glycogen. *Neurochem Int* **45**, 529–536.

Brown, A. M. (2004). Brain glycogen re-awakened. *J Neurochem* **89**, 537–552.

Cadoux-Hudson, T. A., Rajagopalan, B., Ledingham, J. G., Radda, G. K. (1990). Response of the human brain to a hypercapnic acid load in vivo. *Clin Sci* **79**, 1–3.

Castillo, M., Smith, J. K., Kwock, L. (2001). Proton MR spectroscopy in patients with acute temporal lobe seizures. *Am J Neuroradiol* **22**, 152–157.

Cavelier, P., Hamann, M., Rossi, D., Mobbs, P., Attwell, D. (2005). Tonic excitation and inhibition of neurons: ambient transmitter sources and computational consequences. *Prog Biophys Mol Biol* **87**, 3–16.

Cavus, I., Kasoff, W. S., Cassadey, M., Jackob, R., Gueorgieva, R., Sherwin, R. S. et al. (2005). Extracellular neurometabolites in the cortex and hippocampus of epileptic patients. *Ann Neurol* **57**, 226–235.

Cendes, F., Stanley, J. A., Dubeau, F., Andermann, F., Arnold, D. L. (1997). Proton magnetic resonance spectroscopic imaging for discrimination of absence and complex partial seizures. *Ann Neurol* **41**, 74–81.

Chang, L., Miller, B. L., McBride, D., Cornford, M., Oropilla, G., Buchthal, S. et al. (1995). Brain lesions in patients with AIDS: H-1 MR spectroscopy. *Radiology* **197**, 525–531.

Chen, C., Ko, Y., Delannoy, M., Ludtke, S. J., Chiu, W., Pedersen, P. L. (2004). Mitochondrial ATP synthasome: Three-dimensional structure by electron microscopy of the ATP synthase in complex formation with carriers for Pi and ADP/ATP. *J Biol Chem* **79**, 31761–31768.

Chen, W., Novotny, E. J., Zhu, X. H., Rothman, D. L., Shulman, R. G. (1993). Localized 1H NMR measurement of glucose consumption in the human brain during visual stimulation. *Proc Natl Acad Sci USA* **90**, 9896–9900.

Chen, W., Zhu, X-H., Adriany, G., Ugurbil, K. (1997). Increase of creatine kinase activity in the visual cortex of human brain during visual stimulation: A 31P NMR magnetization transfer study. *Magn Reson Med* **38**, 551–557.

Chhina, N., Kuestermann, E., Halliday, J., Simpson, L. J., Macdonald, I. A., Bachelard, H.S., Morris, P.G. (2001). Measurement of human tricarboxylic acid cycle rates during visual activation by 13C magnetic resonance spectroscopy. *J Neurosci Res* **66**, 737–746.

Chih, C-P., Roberts, E. L. (2003). Energy substrates for neurons during neural activity: A critical review of the astrocyte-neuron lactate shuttle hypothesis. *J Cereb Blood Flow Metab* **23**, 1263–1281.

Choi, C., Coupland, N. J., Hanstock, C. C., Ogilvie, C. J., Higgins, A. C., Gheorghiu, D., Allen, P. S. (2005). Brain gamma-aminobutyric acid measurement by proton double-quantum filtering with selective J rewinding. *Magn Reson Med* **54**, 272–279.

Chu, W. J., Hetherington, H. P., Kuzniecky, R. J., Vaughan, J. T., Twieg, D. B., Faught, R. E. et al. (1996). Is the intracellular pH different from normal in the epileptic focus of patients with temporal lobe epilepsy? A 31P NMR study. *Neurology* **47**, 756–760.

Chu, W. J., Hetherington, H. P., Kuzniecky, R. I., Simor, T., Mason, G. F., Elgavish, G. A. (1998). Lateralization of human temporal lobe epilepsy by 31P NMR spectroscopic imaging at 4.1 T. *Neurology* **51**, 472–479.

Confort-Gouny, S., Vion-Dury, J., Nicoli, F., Dano, P., Donnet, A., Grazziani, N. et al. (1993). A multiparametric data analysis showing the potential of localized proton MR spectroscopy of the brain in the metabolic characterization of neurological diseases. *J Neurol Sci* **118**, 123–133.

Connett, R. J. (1988). Analysis of metabolic control: New insights using scaled creatine kinase model. *Am J Physiol* **254**, R949–R959.

Criego, A. B., Tkac, I., Kumar, A., Thomas, W., Gruetter, R., Seaquist, E. R. (2005). Brain glucose concentrations in patients with type 1 diabetes and hypoglycemia unawareness. *J Neurosci Res* **79**, 42– 47.

Dager, S. R., Strauss, W. L., Marro, K. I., Richards, T. L., Metzger, G. D., Artru, A. A. (1995). Proton magnetic resonance spectroscopy investigation of hyperventilation in subjects with panic disorder and comparison subjects. *Am J Psychiat* **152**, 666–672.

Dager, S. R., Friedman, S. D., Heide, A., Layton, M. E., Richards, T., Artru, A. et al. (1999). Two-dimensional proton echo-planar spectroscopic imaging of brain metabolic changes during lactate-induced panic. *Arch Gen Psychiatry* **56**, 70–77.

Dalsgaard, M. K. (2006). Fuelling cerebral activity in exercising man. *J Cereb Blood Flow Metab* **26**, 731–750.

Danbolt, N. C. (2001). Glutamate uptake. *Prog Neurobiol* **65**, 1–105.

de Graaf, R. A., Pan, J. W., Telang, F., Lee, J. H., Brown, P., Novotny, E. J. et al. (2001). Differentiation of glucose transport in human brain gray and white matter. *J Cereb Blood Flow Metab* **21**, 483–492.

de Graaf, R. A., Mason, G. F., Patel, A. B., Behar, K. L., Rothman, D. L. (2003). In vivo 1H-[13C]-NMR spectroscopy of cerebral metabolism. *NMR Biomed* **16**, 339–357.

de Graaf, R. A., Patel, A. B., Rothman, D. L., Behar, K. L. (2006). Acute regulation of steady-state GABA levels following GABA-transaminase inhibition in rat cerebral cortex. *Neurochem Int* **48**, 508–514.

de Lanerolle, N. C., Kim, J. H., Williamson, A., Spencer, S. S., Zaveri, H. P., Eid, T., Spencer, D. D. (2003). A retrospective analysis of hippocampal pathology in human temporal lobe epilepsy: Evidence for distinctive patient subcategories. *Epilepsia* **44**, 677–687.

During, M. J., Spencer, D. D. (1993). Extracellular hippocampal glutamate and spontaneous seizure in the conscious human brain. *Lancet* **341**, 1607–1613.

Eid, T., Thomas, M. J., Spencer, D. D., Rundén-Pran, E., Lai, J. C. K., Malthankar, G. V. et al. (2004). Loss of glutamine synthetase in the human epileptogenic hippocampus: Possible mechanism for raised extracellular glutamate in medial temporal lobe epilepsy. *Lancet* **363**, 28–37.

Eleff, S. M., Barker, P. B., Blackband, S. J., Chatham, J. C., Lutz, N. W., Johns, D. R. et al. (1990). Phosphorus magnetic resonance spectroscopy of patients with mitochondrial cytopathies demonstrates decreased levels of brain phosphocreatine. *Ann Neurol* **27**, 626–630.

Eloqayli, H., Dahl, C. B., Gotestam, K. G., Unsgard, G., Sonnewald, U. (2004). Changes of glial-neuronal interaction and metabolism after a subconvulsive dose of pentylenetetrazole. *Neurochem Int* **45**, 739–745.

Errante, L. D., Williamson, A., Spencer, D. D., Petroff, O. A. C. (2002). Gabapentin and vigabatrin increase GABA in the human neocortical slice. *Epilepsy Res* **49**, 203–210.

Errante, L. D., Petroff, O. A. C. (2003). Acute effects of gabapentin and pregabalin on rat forebrain cellular GABA, glutamate, and glutamine concentrations. *Seizure* **12**, 300–306.

Eisenberg, D., Gill, H. S., Pfluegl, G. M. U., Rotstein, S. H. (2000). Structure-function relationships of glutamine synthetases. *Biochim Biophys Acta* **1477**, 122–145.

Felipo, V., Butterworth, R. F. (2002). Neurobiology of ammonia. *Prog Neurobiol* **67**, 259–279.

Flugel, D., McLean, M. A., Simister, R. J., Duncan, J. S. (2006). Magnetisation transfer ratio of choline is reduced following epileptic seizures. *NMR Biomed* **19**, 217–222.

Frahm, J., Kruger, G., Merboldt, K. D., Kleinschmidt, A. (1996). Dynamic uncoupling and recoupling of perfusion and oxidative metabolism during focal brain activation in man. *Magn Reson Med* **35**, 143–148.

Galanaud, D., Nicoli, F., Le Fur, Y., Guye, M., Ranjeva, J. P., Confort-Gouny, S. et al. (2003) Multimodal magnetic resonance imaging of the central nervous system. *Biochimie* **85**, 905–914.

Garcia, P. A., Laxer, K. D., van der Grond, J., Hugg, J. W., Matson, G. B., Weiner, M. W. (1994). Phosphorus magnetic resonance spectroscopic imaging in patients with frontal lobe epilepsy. *Ann Neurol* **35**, 217–221.

Giove, F., Mangia, S., Bianciardi, M., Garreffa, G., Di Salle, F., Morrone, R., Maraviglia, B. (2003). The physiology and metabolism of neuronal activation: In vivo studies by NMR and other methods. *Magn Reson Imag* **21**, 1283–1293.

Gonzalez-Gonzalez, I. M., Cubelos, B., Gimenez, C., Zafra, F. (2005). Immunohistochemical localization of the amino acid transporter SNAT2 in the rat brain. *Neuroscience* **130**, 61–73.

Gorji, A. (2001). Spreading depression: a review of the clinical relevance. *Brain Res Rev* **38**, 33–60.

Graham, G. D., Blamire, A. M., Rothman, D. L., Brass, L. M., Fayad, P. B., Petroff, O. A. C., Prichard, J. W. (1993). Early temporal variation of cerebral metabolites after human stroke: A proton magnetic resonance spectroscopy study. *Stroke* **24**, 1891–1896.

Graham, G. D., Hwang, J. H., Rothman, D. L., Prichard, J. W. (2001). Spectroscopic assessment of alterations in macromolecule and small-molecule metabolites in human brain after stroke. *Stroke* **32**, 2797–2802.

Gruetter, R., Seaquist, E. R., Ugurbil, K. (2001). A mathematical model of compartmentalized neurotransmitter metabolism in the human brain. *Am J Physiol* **281**, E100–E112.

Haberle, J., Gorg, B., Rutsch, F., Schmidt, E., Toutain, A., Benoist, J. F. et al. (2005) Congenital glutamine deficiency with glutamine synthetase mutations. *Engl J Med* **353**, 1926–1233.

Halvorson, H. R., Vande Linde, A. M. Q., Helpern, J. A., Welch, K. M. A. (1992). Assessment of magnesium concentrations by 31P NMR in vivo. *NMR Biomed* **5**, 53–58.

Hanstock, C. C., Coupland, N. J., Allen, P. S. (2002). GABA X2 multiplet measured pre- and post-administration of vigabatrin in human brain. *Magn Reson Med* **48**, 617–623.

Henry, P. G., Oz, G., Provencher, S., Gruetter, R. (2003). Toward dynamic isotopomer analysis in the rat brain in vivo: Automatic quantitation of 13C NMR spectra using LC Model. *NMR Biomed* **16**, 400–412.

Hertz, L. (2006). Glutamate, a neurotransmitter—And so much more. A synopsis of Wierzba III. *Neurochem Int* **48**, 416–425.

Hetherington, H. P., Pan, J. W., Spencer, D. D. (2002). 1H and 31P spectroscopy and bioenergetics in the lateralization of seizures in temporal lobe epilepsy. *J Magn Reson Imag* **16**, 477–483.

Hetherington, H. P., Kim, J. H., Pan, J. W., Spencer, D. D. (2004). 1H and 31P spectroscopic imaging of epilepsy: Spectroscopic and histologic correlations. *Epilepsia* **45** Suppl 4, 17–23.

Hill, R. A., Chiappa, K. H., Huang-Hellinger, F., Jenkins, B. G. (1999). Hemodynamic and metabolic aspects of photosensitive epilepsy revealed by functional magnetic resonance imaging and magnetic resonance spectroscopy. *Epilepsia* **40**, 912–920.

Hisama, F. M., Mattson, R. H., Lee, H. H., Felice, K., Petroff, O. A. C. (2001). GABA and the ornithine δ-aminotransferase gene in vigabatrin-associated visual field defects. *Seizure* **10**, 505–507.

Holtzman, D., Meyers, R., Khait, I., Jensen, F. (1997). Brain creatine kinase reaction rates and reactant concentrations during seizures in developing rats. *Epilepsy Res* **27**, 7–11.

Hugg, J. W., Laxer, K. D., Matson, G. B., Maudsley, A. A., Husted, C. A., Weiner, M. W. (1992). Lateralization of human focal epilepsy by 31P magnetic resonance imaging spectroscopy. *Neurology* **42**, 2011–2018.

Hugg, J. W., Matson, G. B., Twieg, D. B., Maudsley, A. A., Sappey-Marinier, D., Weiner, M. W. (1992) Phosphorus-31 MR spectroscopic imaging (MRSI) of normal and pathological human brains. *Magn Reson Imag* **10**, 227–243.

Hyder, F., Rothman, D. L., Mason, G. F., Rangarajan, A., Behar, K. L., Shulman, R. G. (1997). Oxidative glucose metabolism in rat brain during single forepaw stimulation: A spatially localized 1H[13C] nuclear magnetic resonance study. *J Cereb Blood Flow Metab* **17**, 1040–1047.

Hyder, F., Petroff, O. A. C., Mattson, R. H., Rothman, D. L. (1999). Localized 1H NMR measurements of 2-pyrrolidinone in human brain in vivo. *Magn Reson Med* **41**, 889–896.

Hyder, F., Kida, I., Behar, K. L., Kennan, R. P., Maciejewski, P. K., Rothman, D. L. (2001). Quantitative functional imaging of the brain: Towards mapping neuronal activity by BOLD fMRI. *NMR Biomed* **14**, 413–431.

Hyder, F., Patel, A. B., Gjedde, A., Rothman, D. L., Behar, K. L., Shulman, R. G. (2006). Neuronal–glial glucose oxidation and glutamatergic–GABAergic function. *J Cereb Blood Flow Metab* **26**, 865–877.

Iotti, S., Frassineti, C., Sabatini, A., Vacca, A., Barbiroli, B. (2005). Quantitative mathematical expressions for accurate in vivo assessment of cytosolic [ADP] and DG of ATP hydrolysis in the human brain and skeletal muscle. *Biochim Biophys Acta* **1708**, 164–177.

in 't Zandt, H. J., Renema, W. K., Streijger, F., Jost, C., Klomp, D. W., Oerlemans, F. et al. (2004). Cerebral creatine kinase deficiency influences metabolite levels and morphology in the mouse brain: A quantitative in vivo 1H and 31P magnetic resonance study. *J Neurochem* **90**, 1321–1330.

Jensen, K. E., Thomsen, C., Henriksen, O. (1988). In vivo measurement of intracellular pH in human brain during different tensions of carbon dioxide in arterial blood. A 31P-NMR study. *Acta Physiol Scand* **134**, 295–298.

Jost, C. R., Van Der Zee, C. E., in 't Zandt, H. J., Oerlemans, F., Verheij, M., Streijger, F. et al. (2002). Creatine kinase B-driven energy transfer in the brain is important for habituation and spatial learning behaviour, mossy fibre field size and determination of seizure susceptibility. *Eur J Neurosci* **15**, 1692–1706.

Kanamori, K., Ross, B. D. (1995). Steady-state in vivo glutamate dehydrogenase activity in rat brain measured by 15N NMR. *J Biol Chem* **270**, 24805–24809.

Kanamori, K., Ross, B. D., Chung, J. C., Kuo, E. L. (1996). Severity of hyperammonemic encephalopathy correlates with brain ammonia level and saturation of glutamine synthetase in vivo. *J Neurochem* **67**, 1584–1594.

Kanamori, K., Ross, B. D., Kondrat, R. W. (1998). Rate of glutamate synthesis from leucine in rat brain measured in vivo by 15N NMR. *J Neurochem* **70**, 1304–1315.

Kanamori, K., Ross, B. D., Kondrat, R. W. (2002). Glial uptake of neurotransmitter glutamate from the extracellular fluid studied in vivo by microdialysis and (13)C NMR. *J Neurochem* **83**, 682–695.

Kanamori, K., Ross, B. D. (2004). Quantitative determination of extracellular glutamine concentration in rat brain, and its elevation in vivo by system A transport inhibitor, alpha-(methylamino)isobutyrate. *J Neurochem* **90**, 203–210.

Kanamori, K., Ross, B. D. (2005). Suppression of glial glutamine release to the extracellular fluid studied in vivo by NMR and microdialysis in hyperammonemic rat brain. *J Neurochem* **94**, 74–85.

Kassem, M. N., Bartha, R. (2003). Quantitative proton short-echo-time LASER spectroscopy of normal human white matter and hippocampus at 4 Tesla incorporating macromolecule subtraction. *Magn Reson Med* **49**, 918–927.

Kato, T., Murashita, J., Shioiri, T., Terada, M., Inubushi, T., Kato, N. (1998). Photic stimulation-induced alteration of brain energy metabolism measured by 31P-MR spectroscopy in patients with MELAS. *J Neurol Sci* **155**, 182–185.

Kekelidze, T., Khait, I., Togliatti, A., Benzecry, J. M., Wieringa, B., Holtzman, D. (2001). Altered brain phosphocreatine and ATP regulation when mitochondrial creatine kinase is absent. *J Neurosci Res* **66**, 866–872 .

Kemp, G. J. (2000). Non-invasive methods for studying brain energy metabolism: What they show and what it means. *Dev Neurosci* **2**, 418–428.

Kida, I., Smith, A. J., Blumenfeld, H., Behar, K. L., Hyder, F. (2006). Lamotrigine suppresses neurophysiological responses to somatosensory stimulation in the rodent. *Neuroimage* **29**, 216–224.

Kintner, D. B., Anderson, M. K., Fitzpatrick, J. H., Sailor, K. A., Gilboe, D. D. (2000). 31P-MRS-based determination of brain intracellular and interstitial pH: Its application to in vivo H+ compartmentation and cellular regulation during hypoxic/ischemic conditions. *Neurochem Res* **25**, 1385–1396.

Ko, Y. H., Delannoy, M., Hullihen, J., Chiu, W., Pedersen, P. L. (2003). Mitochondrial ATP synthasome. Cristae-enriched membranes and a multiwell detergent screening assay yield dispersed single complexes containing the ATP synthase and carriers for Pi and ADP/ATP. *J Biol Chem* **278**, 12305–12309.

Kunz, W. S. (2002). The role of mitochondria in epileptogenesis. *Curr Opin Neurol* **15**, 179–184.

Kuwabara, T., Watanabe, H., Tanaka, K., Tsuji, S., Ohkubo, M., Ito, T. et al. (1994). Mitochondrial encephalomyopathy: elevated visual cortex lactate unresponsive to photic stimulation—a localized 1H-MRS study. *Neurology* **44**, 557–559.

Kuzniecky, R., Elgavish, G. A., Hetherington, H. P., Evanochko, W. T., Pohost, G. M. (1992). In vivo 31P nuclear magnetic resonance spectroscopy of human temporal lobe epilepsy. *Neurology* **42**, 1586–1590.

Kuzniecky, R. (1999). Magnetic resonance spectroscopy in focal epilepsy: 31P and 1H spectroscopy. *Revue Neurol* **155**, 495–498.

Kvamme, E., Torgner, I. A., Roberg, B. (2001). Kinetics and localization of brain phosphate activated glutaminase. *J Neurosci Res* **66**, 951–958.

Laxer, K. D., Hubesch, B., Sappey-Marinier, D., Weiner, M. W. (1992). Increased pH and inorganic phosphate in temporal seizure foci demonstrated by [31P]MRS. *Epilepsia* **33**, 618–623.

Layton, M. E., Friedman, S. D., Dager, S. R. (2001). Brain metabolic changes during lactate-induced panic: Effects of gabapentin treatment. *Depression & Anxiety* **14**, 251–254.

Lebon, V., Petersen, K. F., Cline, G. W., Shen, J., Mason, G. F., Dufour, S. et al. (2002). Astroglial contribution to brain energy metabolism in humans revealed by 13C nuclear magnetic resonance spectroscopy: Elucidation of the dominant pathway for neurotransmitter glutamate repletion and measurement of astrocytic oxidative metabolism. *J Neurosci* **22**, 1523–1531.

Lei, H., Ugurbil, K., Chen, W. (2003a). Measurement of unidirectional Pi to ATP flux in human visual cortex at 7T by using in vivo 31P magnetic resonance spectroscopy. *Proc Natl Acad Sci USA* **100**, 14409–14414.

Lei, H., Zhu, X-H., Zhang, X-L., Ugurbil, K., Chen, W. (2003b). In vivo 31P magnetic resonance spectroscopy of human brain at 7T: An initial experience. *Magn Reson Med* **49**, 199–205.

Leuzzi, V. (2002). Inborn errors of creatine metabolism and epilepsy: Clinical features, diagnosis, and treatment. *J Child Neurol* **17** Suppl 3, 3S89–3S97.

Li, Z., Okamoto, K-I., Hayashi, Y., Sheng, M. (2004) The importance of dendritic mitochondria in the morphogenesis and plasticity of spines and synapses. *Cell* **119**, 873–887.

Lodi, R., Iotti, S., Cortelli, P., Pierangeli, G., Cevoli, S., Clementi, V. et al. (2004). Deficient energy metabolism is associated with low free magnesium in the brains of patients with migraine and cluster headache. *Brain Res Bull* **54**, 437–441.

Lukkarinen, J., Oja, J. M., Turunen, M., Kauppinen, R. A. (1997). Quantitative determination of glutamate turnover by 1H-observed, 13C-edited nuclear magnetic resonance spectroscopy in the cerebral cortex ex vivo: interrelationships with oxygen consumption. *Neurochem Int* **31**, 95–104.

Magistretti, P. J., Pellerin, L., Rothman, D. L., Shulman, R. G. (1999). Energy on demand. *Science* **283**, 496–497.

Magloczky, Z., Freund, T. F. (2005). Impaired and repaired inhibitory circuits in the epileptic human hippocampus. *Trends Neurosci* **28**, 334–340.

Mason, G. F., Gruetter, R., Rothman, D. L., Behar, K. L., Shulman, R. G., Novotny, E. J. (1995). Simultaneous determination of the rates of the TCA cycle, glucose utilization, alpha-ketoglutarate/glutamate exchange, and glutamine synthesis in human brain by NMR. *J Cereb Blood Flow Metab* **15**, 12–25.

Mason, G. F., Petersen, K. F., Lebon, V., Rothman, D. L., Shulman, G. I. (2006). Increased brain monocarboxylic acid transport and utilization in type 1 diabetes. *Diabetes* **55**, 929–934.

Mastorodemos, V., Zaganas, I., Spanaki, C., Bessa, M., Plaitakis, A. (2005). Molecular basis of human glutamate dehydrogenase regulation under changing energy demands. *J Neurosci Res* **79**, 65–73.

Matthews, P. M., Berkovic, S. F., Shoubridge, E. A., Andermann, F., Karpati, G., Carpenter, S., Arnold, D. L. (1991). In vivo magnetic resonance spectroscopy of brain and muscle in a type of mitochondrial encephalomyopathy (MERFF). *Ann Neurol* **29**, 435–438.

McKenna, M. C., Stevenson, J. H., Huang, X., Hopkins, I. B. (2000). Differential distribution of the enzymes glutamate dehydrogenase and aspartate aminotransferase in cortical synaptic mitochondria contributes to metabolic compartmentation in cortical synaptic terminals. *Neurochem Int* **37**, 229–241.

Minguez, B., Garcia-Pagan, J. C., Bosch, J., Turnes, J., Alonso, J., Rovira, A., Cordoba, J. (2006). Noncirrhotic portal vein thrombosis exhibits neuropsychological and MR changes consistent with minimal hepatic encephalopathy. *Hepatology* **43**, 707–714.

Mohanakrishnan, P., Fowler, A. H., Vonsattel, J. P., Jolles, P. R., Husain, M. M., Liem, P. et al. (1997). Regional metabolic alterations in Alzheimer's disease: An in vitro 1H NMR study of the hippocampus and cerebellum. *J Gerontol Series A-Biol Sci Med Sci* **52**, B111–B117.

Montagna, P., Cortelli, P., Barbiroli, B. (1994). Magnetic resonance spectroscopy studies in migraine. *Cephalalgia* **14**, 184–193.

Neumann, D., Schlattner, U., Wallimann, T. (2003). A molecular approach to the concerted action of kinases involved in energy homeostasis. *Biochem Soc Trans* **31**, 169–174.

Nybo, L., Dalsgaard, M. K., Steensberg, A., Moller, K., Secher, N. H. (2005). Cerebral ammonia uptake and accumulation during prolonged exercise in humans. *J Physiology* **563**, 285–290.

Otsuki, T., Nakama, H., Kanamatsu, T., Tsukada, Y. (2005). Glutamate metabolism in epilepsy: 13C-magnetic resonance spectroscopy observation in the human brain. *NeuroReport* **16**, 2057–2060.

Ottersen, O. P., Zhang, N., Walberg, F. (1992). Metabolic compartmentation of glutamate and glutamine: Morphological evidence obtained by quantitative immunocytochemistry in rat cerebellum. *Neuroscience* **46**, 519–534.

Oz, G., Henry, P. G., Seaquist, E. R., Gruetter, R. (2003). Direct, noninvasive measurement of brain glycogen metabolism in humans. *Neurochem Int* **43**, 323–329.

Pan, J. W., Bebin, E. M., Chu, W. J., Hetherington, H. P. (1999). Ketosis and epilepsy: 31P spectroscopic imaging at 4.1 T. *Epilepsia* **40**, 703–707.

Pan, J. W., Stein, D. T., Telang, F., Lee, J. H., Shen, J., Brown, P. et al. (2000). Spectroscopic imaging of glutamate C4 turnover in human brain. *Magn Reson Med* **44**, 673–679.

Pan, J. W., Coyle, P. K., Bashir, K., Whitaker, J. N., Krupp, L. B., Hetherington, H. P. (2002a). Metabolic differences between multiple sclerosis subtypes measured by quantitative MR spectroscopy. *Multiple Sclerosis* **8**, 200–206.

Pan, J. W., de Graaf, R. A., Petersen, K. F., Shulman, G. I., Hetherington, H. P., Rothman, D. L. (2002b). [2,4-13 C2]-beta-Hydroxybutyrate metabolism in human brain. *J Cereb Blood Flow Metab* **22**, 890–898.

Pan, J. W., Venkatraman, T., Vives, K., Spencer, D. D. (2006). Quantitative glutamate spectroscopic imaging of the human hippocampus. *NMR Biomed* **19**, 209–216.

Patel, A. B., de Graaf, R. A., Mason, G. F., Kanamatsu, T., Rothman, D. L., Shulman, R. G., Behar, K. L. (2004). Glutamatergic neurotransmission and neuronal glucose oxidation are coupled during intense neuronal activation. *J Cereb Blood Flow Metab* **24**, 972–985.

Patel, A. B., de Graaf, R. A., Mason, G. F., Rothman, D. L., Shulman, R. G., Behar, K. L. (2005). The contribution of GABA to glutamate/glutamine cycling and energy metabolism in the rat cortex in vivo. *Proc Natl Acad Sci USA* **102**, 5588–5593.

Peeling, J., Sutherland, G. (1993). 1H magnetic resonance spectroscopy of extracts of human epileptic neocortex and hippocampus. *Neurology* **43**, 589–594.

Pellerin, L. (2003). Lactate as a pivotal element in neuron-glia metabolic cooperation. *Neurochem Int* **43**, 331–338.

Pellerin, L., Magistretti, P. J. (2004). Neuroscience: Let there be (NADH) light. *Science* **305**, 50–52.

Petroff, O. A., Prichard, J. W., Behar, K. L., Alger, J. R., den Hollander, J. A., Shulman, R. G. (1985). Cerebral intracellular pH by 31P nuclear magnetic resonance spectroscopy. *Neurology* **35**, 781–788.

Petroff, O. A. C., Prichard, J. W., Ogino, T., Shulman, R. G. (1988). Proton magnetic resonance spectroscopic studies of agonal carbohydrate metabolism in rabbit brain. *Neurology* **38**, 1569–1574.

Petroff, O. A. C., Graham, G. D., Blamire, A. M., Al Rayess, M., Rothman, D. L., Fayad, P. B. et al. (1992a). Spectroscopic imaging of stroke in man: Histopathology correlates of spectral changes. *Neurology* **42**, 1349–1354.

Petroff, O. A. C., Novotny, E. J., Avison, M. J., Rothman, D. L., Alger, J. R., Ogino, T. et al. (1992b). Cerebral lactate turnover after electroshock: in vivo measurements by 1H/13C magnetic resonance spectroscopy. *J Cereb Blood Flow Metab* **12**, 1022–1029.

Petroff, O. A. C., Pleban, L. A., Spencer, D. D. (1995a). Symbiosis between in vivo and in vitro NMR spectroscopy: The creatine, N-acetyl-aspartate, glutamate, and GABA content of epileptic human brain. *Magn Reson Imag* **13**, 1197–1211.

Petroff, O. A. C., Rothman, D. L., Behar, K. L., Mattson, R. H. (1995b). Initial observations on the effect of vigabatrin on the in vivo 1H spectroscopic measurements of GABA, glutamate, and glutamine in human brain. *Epilepsia* **36**, 457–464.

Petroff, O. A. C., Rothman, D. L., Behar, K. L., Hyder, F., Mattson, R. H. (1999). Effects of valproate and other antiepileptic drugs on brain glutamate, glutamine, and GABA in patients with refractory complex partial seizures. *Seizure* **8**, 120–127.

Petroff, O. A. C., Hyder, F., Rothman, D. L., Mattson, R. H. (2000). Functional imaging in the epilepsies proton MRS: GABA & glutamate. *Adv Neurol* **83**, 263–272.

Petroff, O. A., Errante, L. D., Rothman, D. L., Kim, J. H., Spencer, D. D. (2002a). Neuronal and glial metabolite content of the epileptogenic human hippocampus. *Ann Neurol* **52**, 635–642.

Petroff, O. A., Errante, L. D., Rothman, D. L., Kim, J. H., Spencer, D. D. (2002b). Glutamate-glutamine cycling in the epileptic human hippocampus. *Epilepsia* **43**, 703–710.

Petroff, O. A. C., Prichard, J. W. (1995). Measurement of cytosolic pH by nuclear magnetic resonance spectroscopy. In Kraicer, J., Dixon, S. J., Eds., *Methods in Neuroscience*, Vol 27, Chapter 11, 233–251. Academic Press, Orlando.

Prichard, J. W., Petroff, O. A. C., Ogino, T., Shulman, R. G. (1987). Cerebral lactate elevation by electroshock: A 1H magnetic resonance study. *Ann N Y Acad Sci* **508**, 54–63.

Prichard, J., Rothman, D., Novotny, E., Petroff, O., Kuwabara, T., Avison, M. et al. (1991). Lactate rise detected by 1H NMR in human visual cortex during physiologic stimulation. *Proc Natl Acad Sci USA* **88**, 5829–5831.

Raizen, D. M., Brooks-Kayal, A., Steinkrauss, L., Tennekoon, G. I., Stanley, C. A., Kelly, A. (2005). Central nervous system hyperexcitability associated with glutamate dehydrogenase gain of function mutations. *J Pediatr* **146**, 388–394.

Rango, M., Castelli, A., Scarlato, G. (1997). Energetics of 3.5 s neural activation in humans: A 31P MR spectroscopy study. *Magn Reson Med* **38**, 878–883.

Rango, M., Bozzali, M., Prelle, A., Scarlato, G., Bresolin, N. (2001). Brain activation in normal subjects and in patients affected by mitochondrial disease without clinical central nervous system involvement: A phosphorus magnetic resonance spectroscopy study. *J Cereb Blood Flow Metab* **21**, 85–91.

Rango, M., Bonifati, C., Bresolin, N. (2006). Parkinson's disease and brain mitochondrial dysfunction: A functional phosphorus magnetic resonance spectroscopy study. *J Cereb Blood Flow Metab* **26**, 283–290.

Remy, C., Grand, S., Lai, E. S., Belle, V., Hoffmann, D., Berger, F. et al. (1995). 1H MRS of human brain abscesses in vivo and in vitro. *Magn Reson Med* **34**, 508–514.

Rothman, D. L., Howseman, A. M., Graham, G. D., Petroff, O. A. C., Lantos, G., Fayad, P. B. (1991). Localized proton NMR observation of [3 13C]-lactate in stroke after [1 13C]-glucose infusion. *Magn Reson Med* **21**, 302–307.

Rothman, D. L., Hanstock, C. C., Petroff, O. A. C., Novotny, E. J., Prichard, J. W., Shulman, R. G. (1992). Localized 1H NMR measurements of glutamate in the human brain. *Magn Reson Med* **25**, 94–106.

Royes, L. F., Fighera, M. R., Furian, A. F., Oliveira, M. S., da Silva, L. G., Malfatti, C. R. et al. (2003). Creatine protects against the convulsive behavior and lactate production elicited by the intrastriatal injection of methylmalonate. *Neuroscience* **118**, 1079–1090.

Sakai, R., Cohen, D. M., Henry, J. F., Burrin, D. G., Reeds, P. J. (2004). Leucine-nitrogen metabolism in the brain of conscious rats: its role as a nitrogen carrier in glutamate synthesis in glial and neuronal metabolic compartments. *J Neurochem* **88**, 612–622.

Sandor, P. S., Dydak, U., Schoenen, J., Kollias, S. S., Hess, K., Boesiger, P., Agosti, R. M. (2005). MR-spectroscopic imaging during visual stimulation in subgroups of migraine with aura. *Cephalalgia* **25**(7), 507–518.

Sarchielli, P., Tarducci, R., Presciutti, O., Gobbi, G., Pelliccioli, G. P., Stipa, G. et al. (2005). Functional 1H-MRS findings in migraine patients with and without aura assessed interictally. *Neuroimage* **24**, 1025–1031.

Sauter, A., Rudin, M. (1993). Determination of creatine kinase kinetic parameters in rat brain by NMR magnetization transfer. Correlation with brain function. *J Biol Chem* **268**, 13166–13171.

Savic, I., Thomas, A. M., Ke, Y., Curran, J., Fried, I., Engel, J. Jr. (2000). In vivo measurements of glutamine + glutamate (Glx) and N-acetyl aspartate (NAA) levels in human partial epilepsy. *Acta Neurol Scand* **102**, 179–188.

Schuck, P. F., Leipnitz, G., Ribeiro, C. A., Dalcin, K. B., Assis, D. R., Barschak, A. G. et al. (2004). Inhibition of creatine kinase activity in vitro by methylmalonic acid in cerebral cortex of young rats. *Neurochem Int* **45**, 661–667.

Schulze, A. (2003). Creatine deficiency syndromes. *Mol Cell Biochem* **244**, 143–150.

Schulze, A., Bachert, P., Schlemmer, H., Harting, I., Polster, T., Salomons, G. S. et al. (2003). Lack of creatine in muscle and brain in an adult with GAMT deficiency. *Ann Neurol* **53**, 248–251.

Schurr, A. (2006). Lactate: The ultimate cerebral oxidative energy substrate? *J Cereb Blood Flow Metab* **26**, 142–152.

Seaquist, E. R., Damberg, G. S., Tkac, I., Gruetter, R. (2001). The effect of insulin on in vivo cerebral glucose concentrations and rates of glucose transport/metabolism in humans. *Diabetes* **50**, 2203–2209.

Shawcross, D. L., Balata, S., Olde Damink, S. W., Hayes, P. C., Wardlaw, J., Marshall, I. et al. (2004). Low myo-inositol and high glutamine levels in brain are associated with neuropsychological deterioration after induced hyperammonemia. *Am J Physiol* **287**, G503–G509.

Shen, J., Sibson, N. R., Cline, G., Behar, K. L., Rothman, D. L., Shulman, R. G. (1998). 15N-NMR spectroscopy studies of ammonia transport and glutamine synthesis in the hyperammonemic rat brain. *Dev Neurosci* **20**, 434–443.

Shen, J., Petersen, K. F., Behar, K. L., Brown, P., Nixon, T. W., Mason, G. F. et al. (1999). Determination of the rate of the glutamate-glutamine cycle in human brain by in vivo 13C NMR. *Proc Soc Natl Acad Sci USA* **96**, 8235–8240.

Shulman, R. G., Hyder, F., Rothman, D. L. (2001). Cerebral energetics and the glycogen shunt: neurochemical basis of functional imaging. *Proc Natl Acad Sci USA* **98**, 6417–6422.

Shulman, R. G., Rothman, D. L., Behar, K. L., Hyder, F. (2004). Energetic basis of brain activity: Implications for neuroimaging. *Trends Neurosci* **27**, 489–495.

Sibson, N. R., Dhankhar, A., Mason, G. F., Behar, K. L., Rothman, D. L., Shulman, R. G. (1997). In vivo 13C NMR measurements of cerebral glutamine synthesis as evidence for glutamate-glutamine cycling. *Proc Natl Acad Sci USA* **94**, 2699–2704.

Sibson, N. R., Dhankhar, A., Mason, G. F., Rothman, D. L., Behar, K. L., Shulman, R. G. (1998). Stoichiometric coupling of brain glucose metabolism and glutamatergic neuronal activity. *Proc Natl Acad Sci USA* **95**, 316–321.

Sibson, N. R., Mason, G. F., Shen, J., Cline, G. W., Herskovits, A. Z., Wall, J. E. et al. (2001). In vivo (13)C NMR measurement of neurotransmitter glutamate cycling, anaplerosis and TCA cycle flux in rat brain during. *J Neurochem* **76**, 975–989.

Sickmann, H. M., Schousboe, A., Fosgerau, K., Waagepetersen, H. S. (2005). Compartmentation of lactate originating from glycogen and glucose in cultured astrocytes. *Neurochem Res* **30**, 1295–1304.

Simister, R. J., Woermann, F. G., McLean, M. A., Bartlett, P. A., Barker, G. J., Duncan, J. S. (2002). A short-echo-time proton magnetic resonance spectroscopic imaging study of temporal lobe epilepsy. *Epilepsia* **43**, 1021–1031.

Simister, R. J., McLean, M. A., Barker, G. J., Duncan, J. S. (2003). Proton MRS reveals frontal lobe metabolite abnormalities in idiopathic generalized epilepsy. *Neurology* **61**, 897–902.

Simon, D. K., Johns, D. R. (1999). Mitochondrial disorders: Clinical and genetic features. *Annu Rev Med* **50**, 111–127.

Smith, A. J., Blumenfeld, H., Behar, K. L., Rothman, D. L., Shulman, R. G., Hyder, F. (2002). Cerebral energetics and spiking frequency: The neurophysiological basis of fMRI. *Proc Natl Acad Sci USA* **99**, 10765–10770.

Streijger, F., Oerlemans, F., Ellenbroek, B. A., Jost, C. R., Wieringa, B., Van der Zee, C. E. (2005). Structural and behavioural consequences of double deficiency for creatine kinases BCK and UbCKmit. *Behav Brain Res* **157**, 219–234.

Tekkok, S. B., Brown, A. M., Westenbroek, R., Pellerin, L., Ransom, B. R. (2005). Transfer of glycogen-derived lactate from astrocytes to axons via specific monocarboxylate transporters supports mouse optic nerve activity. *J Neurosci Res* **81**, 644–652.

Terpstra, M., Ugurbil, K., Gruetter, R. (2002). Direct in vivo measurement of human cerebral GABA concentration using MEGA-editing at 7 Tesla. *Magn Reson Med* **47**, 1009–1012.

Tofteng, F., Hauerberg, J., Hansen, B. A., Pedersen, C. B., Jørgensen, L., Larsen, F. S. (2006). Persistent arterial hyperammonemia increases the concentration of glutamine and alanine in the brain and correlates with intracranial pressure in patients with fulminant hepatic failure. *J Cereb Blood Flow Metab* **26**, 21–27.

van der Grond, J., Gerson, J. R., Laxer, K. D., Hugg, J. W., Matson, G. B., Weiner, M. W. (1998). Regional distribution of interictal 31P metabolic changes in patients with temporal lobe epilepsy. *Epilepsia* **39**, 527–536.

Van der Gucht, E., Jacobs, S., Kaneko, T., Vandesande, F., Arckens, L. (2003). Distribution and morphological characterization of phosphate-activated glutaminase-immunoreactive neurons in cat visual cortex. *Brain Res* **988**, 29–42.

van der Hel, W. S., Notenboom, R. G., Bos, I. W., van Rijen, P. C., van Veelen, C. W., de Graan, P. N. (2005). Reduced glutamine synthetase in hippocampal areas with neuron loss in temporal lobe epilepsy. *Neurology* **64**, 326–333.

van Rijen, P. C., Luyten, P. R., Berkelbach van der Sprenkel, J. W., Kraaier, V., van Huffelen, A. C., Tulleken, C. A., den Hollander, J. A. (1989). 1H and 31P NMR measurement of cerebral lactate, high-energy phosphate levels, and pH in humans during voluntary hyperventilation: associated EEG, capnographic, and Doppler findings. *Magn Reson Med* **10**, 182–193.

Veech, R. L., Lawson, J. W. R., Cornell, N. W., Krebs, H. A. (1979). Cytosolic phosphorylation potential. *J Biol Chem* **254**, 6538–6547.

Vigevano, F., Bartuli, A. (2002). Infantile epileptic syndromes and metabolic etiologies. *J Child Neurol* **17** Suppl 3, S9–S13.

Waagepetersen, H. S., Sonnewald, U., Larsson, O. M., Schousboe, A. (2000). A possible role of alanine for ammonia transfer between astrocytes and glutamatergic neurons. *J Neurochem* **75**, 471–479.

Wallimann, T., Dolder, M., Schlattner, U., Eder, M., Hornemann, T., O'Gorman, E. et al. (1998) Some new aspects of creatine kinase (CK): Compartmentation, structure, function and regulation for cellular and mitochondrial bioenergetics and physiology. *Biofactors* **8**, 229–234.

Wary, C., Brillault-Salvat, C., Bloch, G., Leroy-Willig, A., Roumenov, D., Grognet, J. M. et al. (1999). Effect of chronic magnesium supplementation on magnesium distribution in healthy volunteers evaluated by 31P-NMRS and ion selective electrodes. *Br J Clin Pharmacol* **48**, 655–662.

Wellard, R. M., Briellmann, R. S., Prichard, J. W., Syngeniotis, A., Jackson, G. D. (2003). Myoinositol abnormalities in temporal lobe epilepsy. *Epilepsia* **44**, 815–821.

Wellard, R. M., Briellmann, R. S., Wilson, J. C., Kalnins, R. M., Anderson, D. P., Federico, P. et al. (2004). Longitudinal study of MRS metabolites in Rasmussen encephalitis. *Brain* **127**, 1302–1312.

Williams, G. D., Mosher, T. J., Smith, M. B. (1993). Simultaneous determination of intracellular magnesium and pH from the three 31P NMR chemical shifts of ATP. *Anal Biochem* **214**, 458–467.

Williamson, A., Patrylo, P. R., Pan, J., Spencer, D. D., Hetherington, H. (2005). Correlations between granule cell physiology and bioenergetics in human temporal lobe epilepsy. *Brain* **128**, 1199–1208.

Woermann, F. G., McLean, M. A., Bartlett, P. A., Parker, G. J., Barker, G. J., Duncan, J. S. (1999). Short echo time single-voxel 1H magnetic resonance spectroscopy in magnetic resonance imaging-negative temporal lobe epilepsy: Different biochemical profile compared with hippocampal sclerosis. *Ann Neurol* **45**, 369–376.

Woermann, F. G., McLean, M. A., Bartlett, P. A., Barker, G. J., Duncan, J. S. (2001). Quantitative short echo time proton magnetic resonance spectroscopic imaging study of malformations of cortical development causing epilepsy. *Brain* **124**, 427–436.

Wyss, M., Kaddurah-Daouk, R. (2000). Creatine and creatinine metabolism. *Physiol Rev* **80**, 1107–1213.

Xu, S., Yang, J., Li, C. Q., Zhu, W., Shen, J. (2005). Metabolic alterations in focally activated primary somatosensory cortex of α-chloralose-anesthetized rats measured by 1H MRS at 11.7 T. *Neuroimage* **28**, 401–409.

Xu, Y., Oz, G., LaNoue, K. F., Keiger, C. J., Berkich, D. A., Gruetter, R., Hutson, S. H. (2004). Whole-brain glutamate metabolism evaluated by steady-state kinetics using a double-isotope procedure: Effects of gabapentin. *J Neurochem* **90**, 1104–1116.

Young, R. S. K., Petroff, O. A. C., Chen, B., Aquila, W. J., Gore, J. C., Yates, J. (1991a). Preferential utilization of lactate in neonatal brain: In vivo and in vitro proton NMR study. *Biol Neonate* **59**, 46–53.

Young, R. S. K., Petroff, O. A. C., Aquila, W. J., Yates, J. (1991b). Effects of glutamate, quisqualate, and N methyl D aspartate in neonatal brain. *Exptl Neurol* **111**, 362–368.

Younkin, D. P., Delivoria-Papadopoulos, M., Maris, J., Donlon, E., Clancy, R., Chance, B. (1986). Cerebral metabolic effects of neonatal seizures measured with in vivo 31P NMR spectroscopy. *Ann Neurol* **20**, 513–519.

Zhu, X. H., Zhang, N., Zhang, Y., Zhang, X., Ugurbil, K., Chen, W. (2005). In vivo 17O NMR approaches for brain study at high field. *NMR Biomed* **18**, 83–103.

Zwingmann, C., Butterworth, R. (2005). An update on the role of brain glutamine synthesis and its relation to cell-specific energy metabolism in the hyperammonemic brain: Further studies using NMR spectroscopy. *Neurochem Int* **47**, 19–30.

7

Gene Therapy Approaches in Neurology

Edward A. Burton, Joseph C. Glorioso, and David J. Fink

The recent, rapid evolutio n in understanding the molecular basis of neurological disease has fostered hope that similar progress will follow in neurological therapeutics. In this chapter, we discuss techniques for transferring exogenous genetic material into host cells (*gene transfer*) and possible clinical applications of this technology (*gene therapy*) in neurology.

I. Why Use Gene Transfer in the Development of Novel Therapies?

There are three broad reasons that a gene transfer approach to a neurological problem might be contemplated over a more conventional pharmacological approach:

▲ **Gene complementation**. Loss-of-function Mendelian single-gene disorders could be treated by delivering an intact copy of the wild-type gene. This strategy, addressing the fundamental genetic abnormality, ideally would cure the disease, in contrast to the current best practice of intermittently delivering gene products or devising some other compensatory strategy for the genetic lesion. Although the list of potential gene therapy targets in this category is very large—theoretically, any recessive genetic disease could be treated this way—there are only a few examples that have proven instructive and substantial obstacles remain.

▲ **Delivery of proteins to achieve a pharmacological action**. A variety of pathological biochemical pathways underlying both hereditary and sporadic diseases may be manipulable by effecting over-expression or ectopic expression of endogenous or engineered proteins. Delivery of the gene encoding the protein may be preferable to delivering the protein itself. Proteins that act intracellularly would need to be expressed within target cells unless an efficient means for effecting their uptake from the extracellular space was available. Gene delivery might also be advantageous for the delivery of proteins that act at the extracellular surface of cells, for example peptide growth factors, because it may be possible to achieve adequate local concentrations while avoiding the side effects that might be engendered by systemic administration of these pleiotropic macromolecules.

▲ **RNA targeting**. Mendelian dominant neurological diseases caused by over-expression of a normal gene, or expression of a mutant with a toxic gain of function or a dominant negative function, might be best tackled by targeting the abnormal mRNA using RNA interference or ribozymes, which could be delivered by gene transfer. This same approach could be applied to sporadic or infectious diseases mediated by known protein intermediaries by therapeutic down-regulation of the relevant mRNA.

II. Gene Transfer Strategies

Gene transfer can be accomplished *ex vivo* or *in vivo* (see Figure 7.1). In *ex vivo* gene transfer, cells are removed, and the gene of interest is introduced into these cells *in vitro*, followed by reimplantation of the transduced cells. This approach has a number of inherent advantages:

1. Transduced cells can be selected for expression of therapeutic or marker genes.
2. Transduced cells can be propagated prior to reimplantation, so that the number of transduced cells is increased.
3. Cells with abnormal phenotypes can be removed prior to reimplantation.

Ex vivo gene transfer to generate fibroblasts that express nerve growth factor has been tested as a treatment in Alzheimer's disease, and might be contemplated in the genetic modification of stem cells to encourage their differentiation into desired cell types in tissue engineering paradigms. However, mature neurons with complex dendritic inputs, axonal connections, glial interactions, and a delicately maintained extracellular environment cannot be removed from the brain, cultured *in vitro*, and reimplanted. Genetic modification of these cell types requires *in vivo* gene transfer, in which a gene delivery agent (*vector*) is introduced into the organism to effect genetic modification of cells in their native environment. The gene delivery aspect of this approach is more complex than that of the *ex vivo* technique, and requires that gene transfer be significantly more efficient, because there can be no selection or expansion of transduced cells after genetic modification. Whereas *ex vivo* gene transfer has been accomplished using relatively simple techniques like electroporation or liposomal transfection using naked plasmid DNA, a more sophisticated approach is necessary for *in vivo* gene transfer.

A. Types of Gene Transfer Vectors

Two broad technologies have been developed as gene transfer tools: those based on genetically engineered viruses

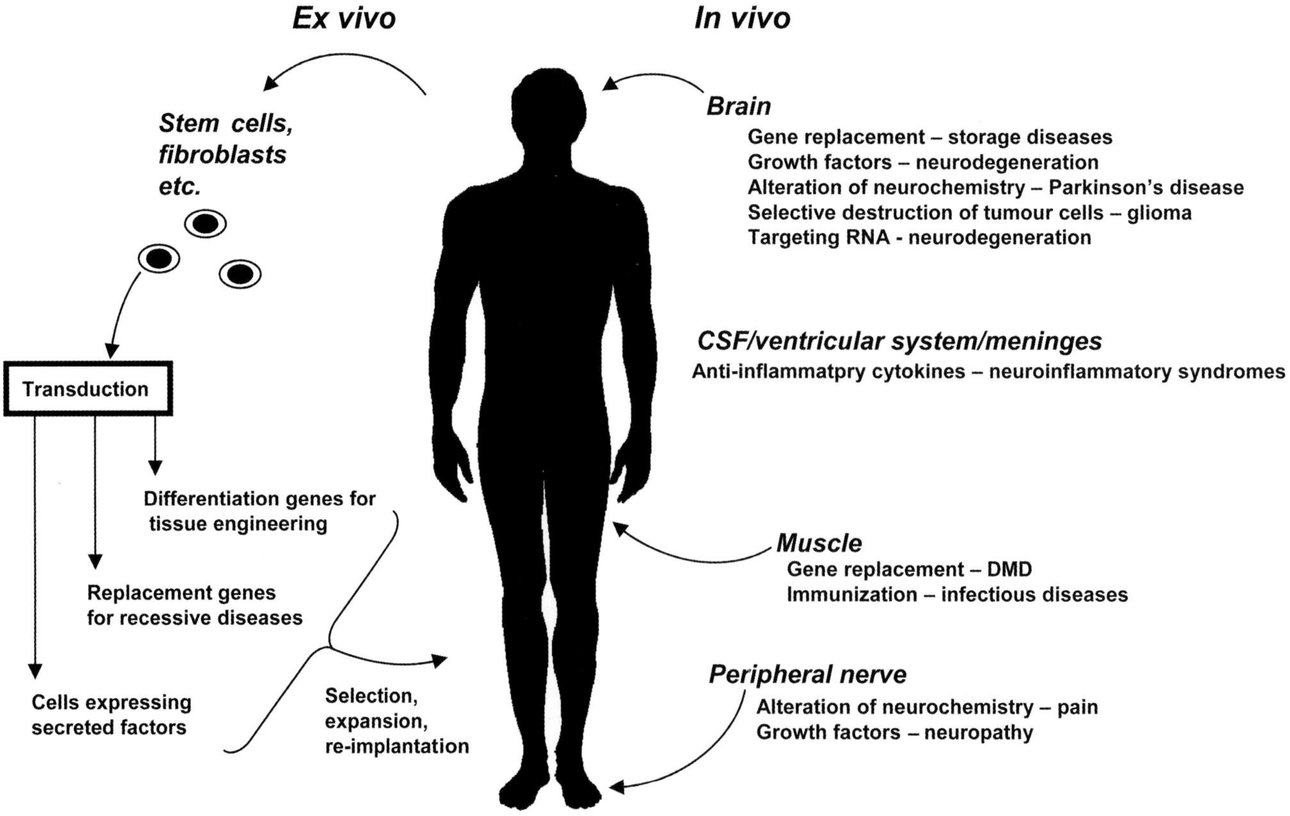

Figure 7.1 *Ex vivo* and *in vivo* gene transfer. *Ex vivo* and *in vivo* gene transfer are depicted schematically, with examples of each type of gene transfer relevant to neurological gene therapy listed next to the site of transduction. In *ex vivo* gene transfer, cells are removed before being transduced *in vitro*. Transduced cells can then be expanded or selected prior to implantation. *In vivo* gene transfer relies on inoculation of a gene transfer vector to effect genetic modification of cells *in vivo*.

(viral vectors) and those based on chemically synthesized complexes (nonviral vectors).

1. Viral Vectors

The most potent and effective gene delivery vehicles are viruses that have evolved to deliver their genetic payload to cells with great efficiency. The lifecycle of viruses, in the simplest terms, involves entry of the viral genome into cells; expression of viral genes as proteins or active RNA molecules; replication of the single- or double-stranded DNA or RNA genome; encapsidation of the genome into new virus particles; release of progeny virions by cell lysis or by budding, in which the virus capsid is enclosed in a membrane or envelope. The nascent particles are competent to invade other cells (Fields et al., 1996). These events are partially mediated by subversion of native cellular processes for viral replication, and almost invariably result in cell death. Viral vector engineering exploits the early parts of the viral lifecycle to enable efficient gene delivery and expression, while eliminating the later parts of the cascade that result in perturbation of cellular physiology, viral replication, and cell death. In general, gene transfer vectors are derived from viruses by deleting genes from the viral genome that allow viral replication and pathogenicity *in vivo*, and complementing these *in vitro* to allow vector replication for manufacture of vector particles (see Figure 7.2).

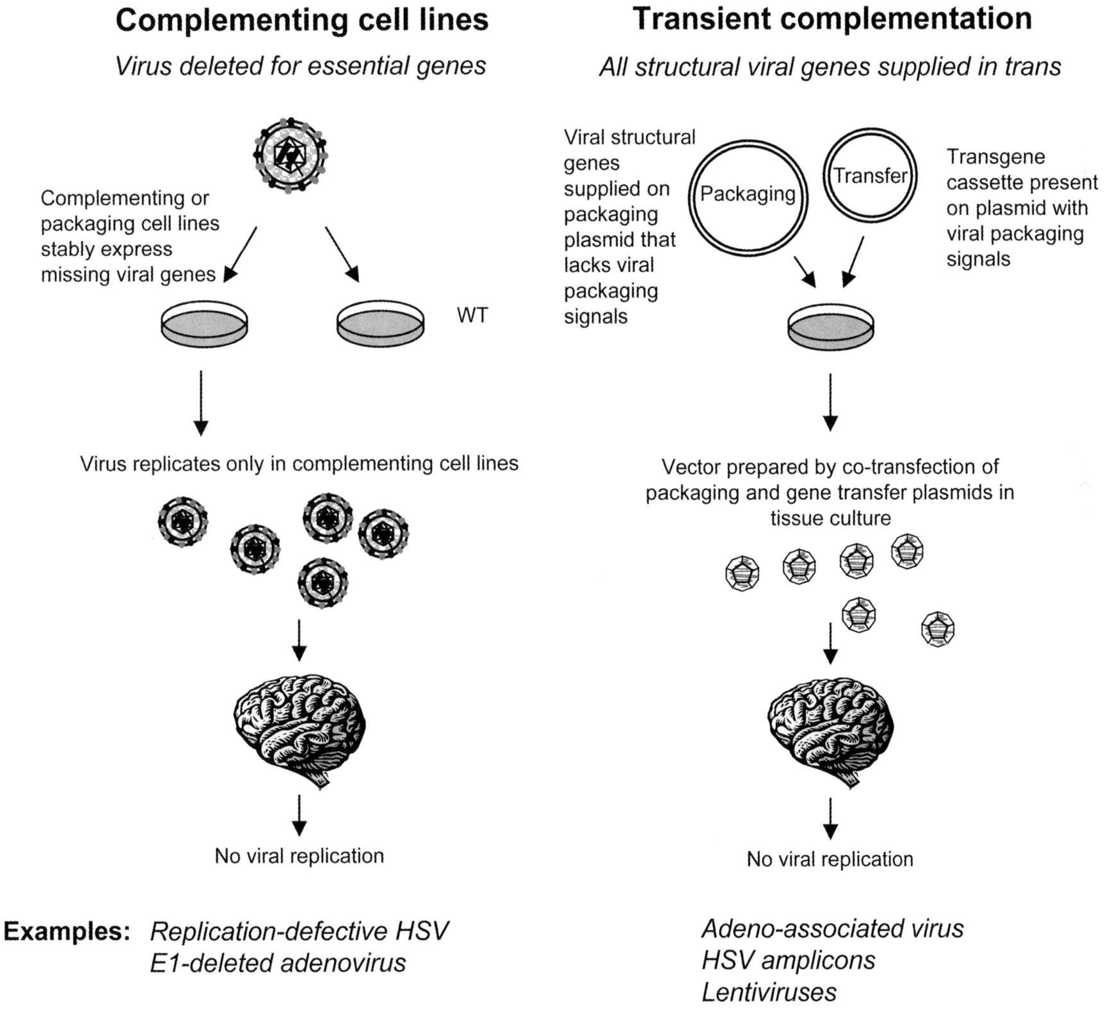

Figure 7.2 Approaches to generating replication-defective viruses as gene transfer vectors. Basic schemes are illustrated for generating viruses that can be used to transfer genetic material into cells *in vivo*, without pathogenic viral replication. Removal of key viral genes essential for replication *in vivo* allows generation of a gene transfer vector in which many viral genes are retained, but their expression is either silenced or reduced by the deletion. These vectors can be prepared to high titre in complementing cell lines that supply the missing viral functions *in trans*, allowing viral replication to take place, but do not replicate *in vivo* in the absence of essential genes. An example is provided by genomic herpes simplex virus vectors. An alternative approach is to flank a transgene expression cassette with viral packaging signals, and to provide all viral functions *in trans*, usually by cotransfection of a second plasmid that does not contain viral packaging signals. This results in generation of gene transfer particles that are devoid of viral structural genes, but is much less efficient than the use of complementing cell lines.

The choice of viral vector for a specific application depends on the biology of the wild-type virus and the properties of derivative vectors.

2. Nonviral Vectors

Nonviral vectors use a variety of lipid formulations to deliver DNA to host cells. Although this approach is technically simpler than the development of viral vectors, gene delivery by nonviral vectors is relatively inefficient compared with viral vector-mediated gene transfer, and transgene expression is highly variable and generally short-lived. There have been some exceptions to these general observations. DNA delivery to some tissues such as skeletal muscle is reasonably efficient, and nonviral DNA delivery has proven to be useful for immune priming as part of vaccine strategies. However, it has not proven useful in gene transfer to the nervous system. The majority of promising results in preclinical and clinical studies have been shown by viral gene transfer vectors, which will be discussed in the remainder of this chapter.

B. Major Types of Viral Vectors

The best characterized and most extensively studied viral gene transfer vectors for the nervous system have been generated using modified retroviruses, adenoviruses, adeno-associated viruses, and herpes simplex virus. The major properties of each of these viruses are shown in Table 7.1, and properties of the derived vectors are listed in Table 7.2. No single one of these vectors is ideal for all applications, but many successful experimental outcomes have been achieved with each of them by exploiting properties of the parental virus in the development of the derivative gene transfer vector.

1. The Retroviruses

Retroviruses (see Figures 7.3 and 7.4) are RNA viruses that integrate their genome into the host cell by using reverse transcriptase to synthesize a DNA copy of the RNA virus genome that then becomes integrated into the host DNA genome. Two principal categories of retroviruses have been modified to generate two classes of viral gene transfer vectors.

Table 7.1 Major Viruses Used to Construct Viral Vectors

	Genome	Genomic Integration?	Latent Infection?	Helper Virus?	Cellular Receptor	Disease Caused by Wild-Type Virus
Lentivirus	ssRNA; 9.5 kb	Yes	Yes	No	CD4, CXCR5 ([1]phosphatidyl serine)	AIDS
Adenovirus	dsDNA; 35 kb	No	No	No	CAR, integrin αvβ5	Respiratory infection; conjunctivitis; gastroenteritis
Adeno-associated virus	ssDNA; 5 kb	Yes	Yes	Yes	Integrin αvβ5, FGF-1R	None known
Herpes simplex virus	dsDNA; 150 kb	No	Yes	No	Glycosaminoglycan, nectin1α and HVEM	Cold sore; rarely encephalitis

[1]Lentiviral vectors are commonly pseudotyped with the VSV-G envelope glycoprotein, which binds to phosphatidylserine instead of the CD4 and CXCR5 cellular receptors bound by the native viral envelope protein.

Table 7.2 Properties of Some Viral Vectors

	Vector	Transgene Insert Capacity	Genes Removed from Vector Genome to Prevent Replication/Toxicity	Complementation of Essential Functions	Easy Preparation of High Titres and Pure Stock?	Easy Insertion of Transgene Sequences?	Toxicity
Lentivirus	HIV-based	6 kb	All except 5′ portion of gag	Transient plasmid transfection	No	Yes	Insertional mutagenesis
Adenovirus	E1, E3 deleted Ad5	8 kb	E1, E3	Complementing cell line	Yes	No	Inflammation
	'Gutless'	30 kb	All	Complementing cell line and helper virus	No	No	Inflammation
Adeno-associated virus	AAV2-based	4.5 kb	All	Helper virus or transient plasmid transfection	No	Yes	Insertional mutagenesis?
Herpes simplex virus	Genomic HSV-1 vectors	30–40 kb	ICP4, ICP27, ICP22, ICP47, ICP0, U_L41	Complementing cell line	Yes	No	Minimal
	Amplicon	120 kb	All	Transient BAC transfection	No	Yes	Minimal

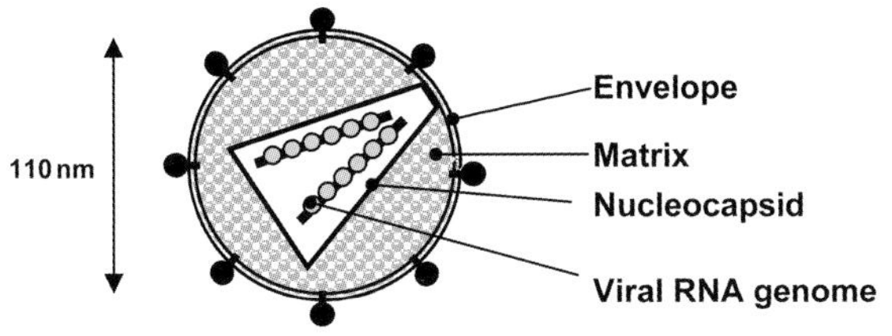

Figure 7.3 Human immunodeficiency virus. This schematic depiction of a mature HIV-1 virion illustrates key structural components of this prototypical lentivirus.

The *oncoretroviruses* were first engineered as nonreplicating gene delivery vectors in the 1980s, when it was discovered that packaging cell lines could be engineered to express the virus coat protein and envelope protein constitutively (Markowitz et al., 1988). Transfection of these packaging cells with a DNA plasmid containing only terminal repeat elements and the packaging signal from the original virus, yielded infectious particles, in which the majority of the vector genome was replaced by nonviral genes. These infectious packaged particles could be used to transduce other cells, and by virtue of their terminal repeat elements the nonviral genes were incorporated into cellular DNA (Armentano et al., 1990; Fink et al., 1990; Kantoff et al., 1986; Ledley et al., 1986; Miyanohara et al., 1988; Palmer et al., 1987; Zwiebel et al., 1989). Oncoretrovirus entry into the nucleus is largely dependent on cell division, so these viruses most efficiently infect dividing cells. This is a major limitation for neurological gene transfer, where the target cell populations are terminally differentiated and nondividing.

More recently, vectors have been constructed from modified *lentiviruses*. Lentiviral vectors have a larger capacity for foreign DNA than oncoretroviruses, and unlike oncoretroviruses, efficiently infect nondividing cells (Kim et al., 1998; Naldini et al., 1996). The most commonly used system for engineering human lentiviruses to create a vector incorporates elements of lentiviruses, an oncoretrovirus, and the envelope components of vesicular stomatitis virus (VSV-G) into three separate expression plasmids. Cotransfection of cells with these plasmids provides all the structural components to generate nonreplicating particles, which contain the transgene sequence contained within one of the plasmids (Corbeau et al., 1996; Dull et al., 1998; Miyoshi et al., 1997, 1998; Naldini et al., 1996; Reiser et al., 1996; Zufferey et al., 1997). Cotransfection of three plasmids is inherently inefficient compared with packaging cell lines, but lentiviral vectors can be concentrated by centrifugation since the VSV-G envelope component is stable (Emi et al., 1991; Naldini et al., 1996). Lentiviral vectors have proven especially useful in transduction of brain and bone marrow stem cells (Akkina et al., 1996; Desmaris et al., 2001; Kordower et al., 2000; Li et al., 1998; Mazarakis et al., 2001; Miyoshi et al., 1999; Pawliuk et al., 2001; Scherr et al., 2002; Zhang et al., 2002; Zielske & Gerson, 2002). Lentivirus is a highly efficient gene transfer vector to the brain, reflected in a number of promising preclinical studies.

An initially unrecognized shortcoming of retroviruses and lentiviruses is the propensity for the vector genome to integrate into the cell genome, preferentially near sites of active transcription (Schroder et al., 2002). This process can lead to rare insertional mutagenesis events and activation of cellular genes that have oncogenic potential. In a recent clinical trial using a retroviral vector to treat X-lined severe combined immunodeficiency (Cavazzana-Calvo et al., 2000; Hacein-Bey-Abina et al., 2002), four of the patients initially cured of the disease by gene transfer subsequently developed leukemic transformation as a result of insertional mutagenesis (Hacein-Bey-Abina et al., 2003).

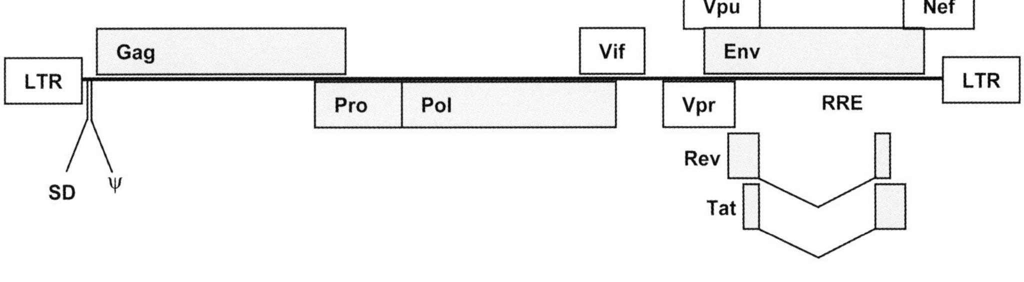

Figure 7.4 The HIV-1 genome. The HIV-1 RNA genome is illustrated diagrammatically (not to scale) to demonstrate overlapping genomic locations and splicing of viral genes. Abbreviations: LTR, long terminal repeat (containing promoter for viral transcripts); SD, major splice donor; Ψ, packaging signal; RRE, rev-response element. The names of viral genes are shown within the boxes. *gag* encodes the core antigens; *pol/pro* encodes the reverse transcriptase and protease enzymes necessary for genomic integration and post-translational processing of viral proteins, respectively; and *env* encodes the viral envelope proteins necessary for cell entry. *tat* encodes a transactivator of viral transcription and *rev* encodes a splicing/export regulator. Lentiviral gene transfer vectors are constructed by flanking a transgene expression cassette with the viral LTRs and packaging signals, and cotransfecting this gene transfer plasmid with another plasmid that contains the viral structural genes but no packaging signals.

2. The Adenoviruses

Recombinant adenoviruses (see Figures 7.5 and 7.6) lacking the tumor-causing gene, E1a, were first constructed in the early 1980s and propagated in complementing cell lines that contained the E1a gene integrated into the cell chromosome (Babiss et al., 1983; Haj-Ahmad & Graham, 1986). The deletion of E1a prevents the virus from replicating in most cell types although "leaky" expression of the other viral functions including structural proteins and the viral polymerase can be detected. Transduction of a number of target tissues including brain has been demonstrated using early adenoviral (AdV) vectors (Davidson et al., 1993; Le Gal La Salle et al., 1993; Ragot et al., 1993; Rosenfeld et al., 1991, 1992). Deletion and complementation of multiple adenoviral genes resulted in generation of more efficient and less toxic vectors (Amalfitano et al., 1998). Alterations to the vector to eliminate all the viral components from the vector genome results in a "gutless" vector with improved stability *in vivo* (Kochanek et al., 2001; Morral et al., 1999; Reddy et al., 2002; Sakhuja et al., 2003; Zou et al., 2000).

One attractive property of AdV vectors is their remarkable ability to express transgenes at very high levels. However, the robust immune response generated by even the gutless AdV vectors has proven to be a difficult problem for many proposed applications (Mercier et al., 2002; Molinier-Frenkel et al., 2000). AdV vectors cannot be administered to patients with preexisting levels of anti-AdV circulating antibody, and the vector cannot be readministered even in the absence of preexisting immunity, since vector inoculation results in antibody production. There have been attempts to protect AdV from immune recognition by coating the particle (Croyle et al., 2002; Sailaja et al., 2002), using AdVs of alternative serotypes or AdVs from other species (Mack et al., 1997; Mastrangeli et al., 1996; Seshidhar Reddy et al., 2003). The AdV particle itself, however, has the ability to induce cytokines (for example interleukin-6) that can result in elimination of the virus genome (Ben-Gary et al., 2002). A human clinical trial in which a large dose of an AdV vector was administered directly into the hepatic artery, led to the death of one patient as a result of vector-induced cytokine release and multiorgan failure (Raper et al., 2002). Strategies that take advantage of the immunogenic potential of AdV vectors, such as the use of AdVs carrying immune-activating genes in the treatment of solid tumors, and the use of recombinant AdVs for vaccination against infectious agents, are likely to be successful.

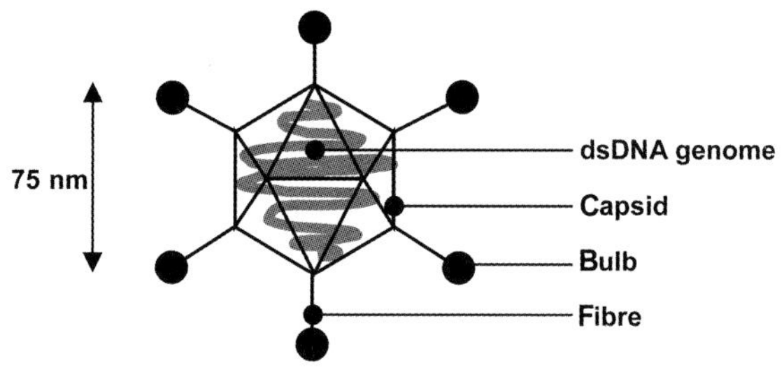

Figure 7.5 The structure of adenovirus. This schematic depiction of a mature adenovirus virion illustrates key structural components of the virus.

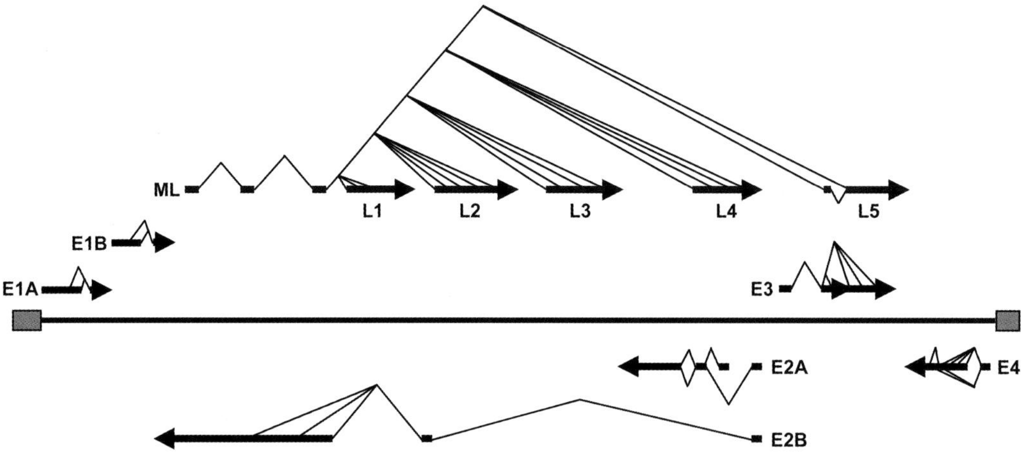

Figure 7.6 The adenovirus genome. The adenoviral genome consists of 35kb of dsDNA, in which the viral gene encoding sequence is enclosed by two inverted terminal repeats (gray boxes). Although there are only eight basic transcriptional units, approximately 40 transcripts are produced by a bewildering array of alternative splicing events. In particular, the major late promoter gives rise to transcripts that share a tripartite leader sequence, culminating in a splice boundary with one of a possible 18 coding sequences. These are grouped, according to the polyA signal at which the transcripts terminate, into L1, L2, etc. Viral transcription depends on the presence of the viral gene E1A; first-generation vectors were deleted for E1A and E1B and the genes supplied *in trans* using complementing cell lines. Later vectors are deleted for other genes, for example E3, in addition to E1. So-called "gutless" vectors contain only the packaging signals and transgene DNA, the viral genome being supplied *in trans* by use of a helper virus to express structural proteins and thus allow the generation of virion-like particles. These vectors can accommodate up to 30kb of transgene sequence, but are difficult to prepare in high titre and purity.

3. The Adeno-associated Viruses

Adeno-associated virus (AAV; see Figures 7.7 and 7.8), initially discovered as a contaminant of adenoviral preparations, has been developed into a widely used gene transfer vector. There are several advantages:

- AAV causes no known disease
- The AAV genome can be manipulated as a simple plasmid
- The small size of AAV allows the particle to infiltrate many different tissues
- Certain tissues (e.g., muscle) are transduced very efficiently

Although its small size (22 nm) (Xie et al., 2002) does not allow packaging of more than 5 kb of foreign sequence, this is sufficient to accommodate most gene coding sequences as cDNAs. *In vitro*, wild-type AAV integrates into a specific location on chromosome 19q (Giraud et al., 1994; Linden et al., 1996; Samulski et al., 1991), but this specific integration event is dependent on *replicase*, a viral gene that must be eliminated from vectors because of its toxicity to cells (Ponnazhagan et al., 1997). In the absence of *rep*, AAV particles may integrate into the genome, although infection is inefficient and cell division usually is required. In contrast to other vectors, AAV appears to require concatemerization in the nucleus for expression, so that it may take several weeks to achieve maximum transgene expression. AAV replication requires several helper functions, and these can be provided by either adenovirus or herpes simplex virus. Like adenovirus, AAV is highly susceptible to neutralization and cannot be easily repeat-dosed.

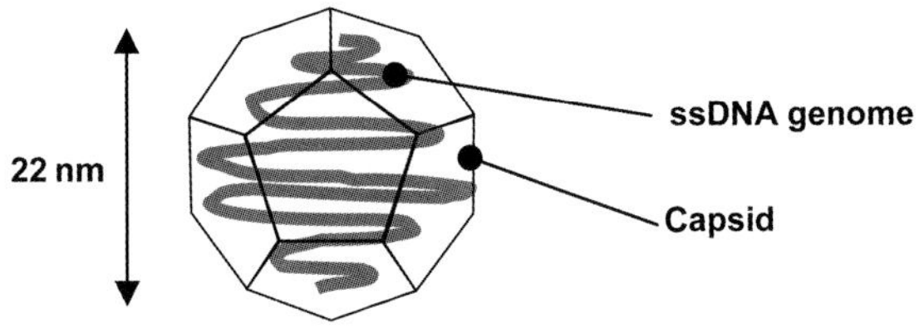

Figure 7.7 The structure of adeno-associated virus. This schematic depiction of a mature adeno-associated virus virion illustrates key structural components of the virus.

The optimal method for producing AAV is still under development. The most common approach is a triple plasmid transfection method, in which one plasmid contains packaging signals and the transgene, another expresses AAV *rep* and *cap* functions, and the final plasmid provides the adenovirus helper functions (Snyder and Flotte, 2002; Xiao et al., 1998). This method results in AAV vector particles without contamination from helper virus since the adenovirus packaging signals have been removed, but is not very efficient. The scale-up manufacture of AAV is complicated by the remarkably low transduction efficiency of AAV compared with other vectors. Despite these problems, AAV remains promising, especially for transduction and expression of genes in CNS neurons, and in muscle. Many of the current clinical trial protocols to deliver genes to the brain in the treatment of neurodegenerative diseases employ AAV-2 based vectors.

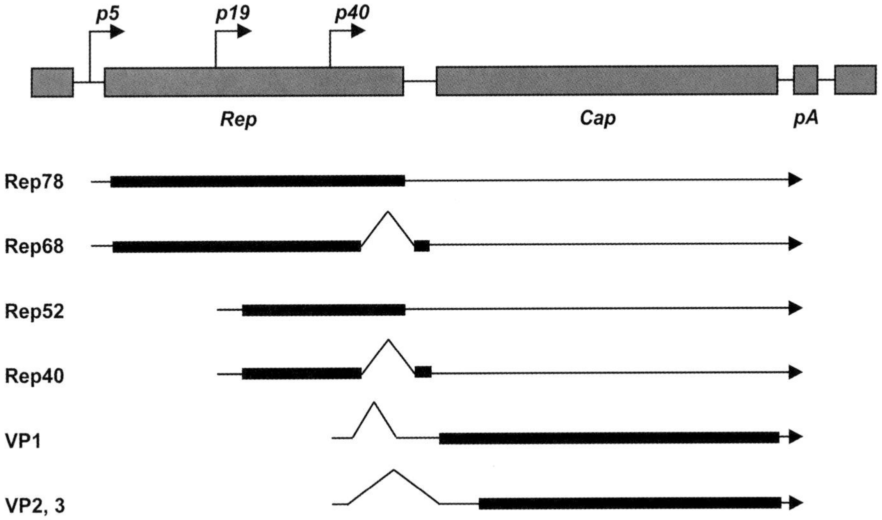

Figure 7.8 The adeno-associated virus genome. The AAV genome consists of 5 kb of ssDNA. The coding sequences are flanked by inverted terminal repeats (gray boxes). There are two major genes, *rep* (encoding proteins necessary for gene expression, genomic integration and replication) and *cap* (encoding the structural components of the virion). There are three viral promoters giving rise to six transcripts by alternative splicing. All the transcripts terminate at a single polyA signal. Seven proteins are encoded by the six transcripts, because the VP2,3 transcript can be translated using one of two different translational start codons. Gene transfer vectors are generated by deletion of *rep* and *cap*, supplying these genes *in trans* by cotransfection of a plasmid, to allow generation of particles. Unfortunately, the products of the *rep* gene are toxic to host cells, and so packaging cell lines have not been possible to generate.

4. Herpes Simplex Virus (HSV)

HSV (see Figures 7.9 and 7.10) is an enveloped double-stranded DNA virus that causes the common cold sore. Wild-type HSV is transmitted through the skin by direct contact with a lesion from an infected individual. The wild-type virus can replicate in skin and mucous membranes but does not persist in the epithelium, instead entering sensory nerve terminals from which they are transported to the nerve cell body in sensory ganglia by retrograde axonal transport (see Figure 7.11). In the sensory ganglia, the lytic viral lifecycle is curtailed, after which the wild-type virus enters a latent state, with concomitant silencing of lytic genes and expression of the viral latency locus (Gordon et al., 1988; Rock et al., 1987; Spivack & Fraser, 1987; Stevens et al., 1987). The ability to persist in neurons for life suggests that HSV may be engineered for prolonged transgene expression in nerves, and the retrograde axonal transport of virions in sensory neurons suggests that vector delivery could result from skin inoculation, or from introduction into remote sites in the CNS (Chen et al., 1995; Goins et al., 1994, 1999, 2001).

Two different strategies have been applied to the construction of HSV-based vectors. Nonreplicating genomic HSV vectors are constructed from wild-type virus by the elimination of genes that contribute to virus replication and reactivation (Burton et al., 2001a, 2001b; Krisky et al., 1997, 1998a, 1998b). HSV requires expression of two of its four immediate early genes for replication to occur, and in the absence of these genes, other viral functions fail to be expressed or are expressed at low levels. Thus, replication-defective HSV vectors can be generated by deleting the two essential immediate early genes (DeLuca et al., 1985; Sacks et al., 1985). The duration of transgene expression from HSV vectors *in vivo* depends on the choice of promoter element. Viral

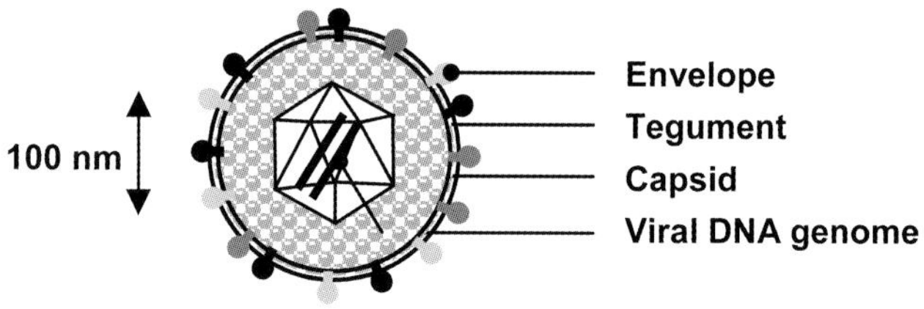

Figure 7.9 The structure of herpes simplex virus. A schematic depiction of a mature HSV-1 virion illustrates the key structural components of the viral particle.

promoters such as the cytomegalovirus immediate-early promoter or the HSV ICP4 promoter provide high levels of short-term (weeks) gene expression. The HSV latency promoter elements can be used to drive transgene expression for considerable time periods in sensory nerves (Bloom et al., 1994; Chen et al., 1995; Dobson et al., 1989, 1990, 1995; Goins et al., 1994, 2001, 2002a; Lachmann & Efstathiou, 1997; Palmer et al., 2000; Smith et al., 2000; Zwaagstra et al., 1989, 1990, 1991) and in the brain (Puskovic et al., 2004). Although multiple essential viral genes have been complemented in cell lines engineered for this purpose, a packaging cell line has not yet been produced that takes full advantage of the large genome (152 kb) for delivering foreign DNA. Current vectors can accommodate up to 50 kb.

Amplicon HSV-based vectors are constructed by packaging concatermerized copies of a plasmid *amplicon* into an HSV particle. Amplicon plasmids contain a packaging signal and an origin of replication; viral functions are supplied *in trans* by a helper virus. Since the viral genes are supplied *in trans*, the capacity of insertion of cloned DNA is large (Wade-Martins et al., 2001). Unfortunately, the *Ori* signal increases the recombination frequency, and it is therefore

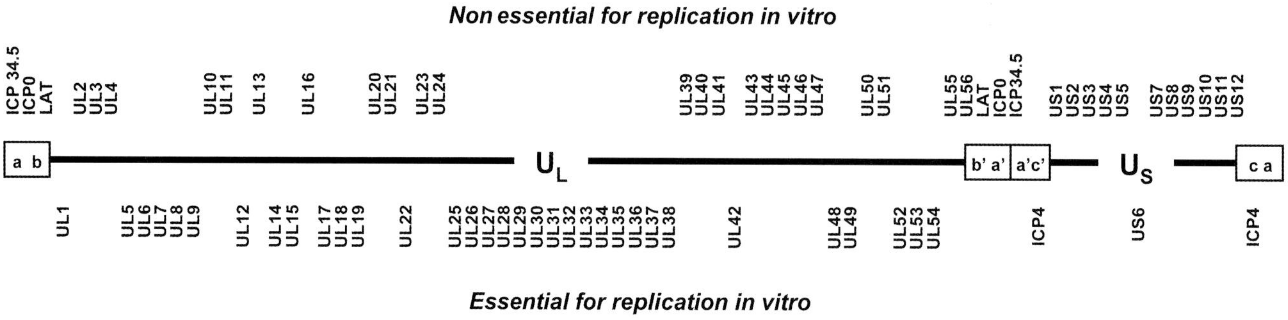

Figure 7.10 The HSV-1 genome. The HSV-1 genome is illustrated diagrammatically. The genome is 154 kb and encodes 84 genes. The genome is divided into unique long and short segments flanked by repeat sequences. Genes that are essential and nonessential for viral replication *in vitro* are indicated. Nonessential genes (which encode a variety of functions that optimize the interaction of HSV with the host organism) can be removed from the virus without compromising replication *in vitro*, allowing their replacement with transgenes of interest. The capacity for the insertion of foreign DNA sequences into the HSV genome is large; genomic vectors can accommodate around 40 kb of transgenic material and amplicons (particles that have all viral genes removed and supplied *in trans*) can accommodate 120 kb of foreign sequence. Replication-defective vectors can be prepared by removing two essential immediate-early genes from the virus, and providing these in complementing cell lines (genomic HSV vectors), or by providing all viral genes *in trans* using helper virus or a series of cosmids to package DNA containing a transgene and packaging signals (amplicon vectors).

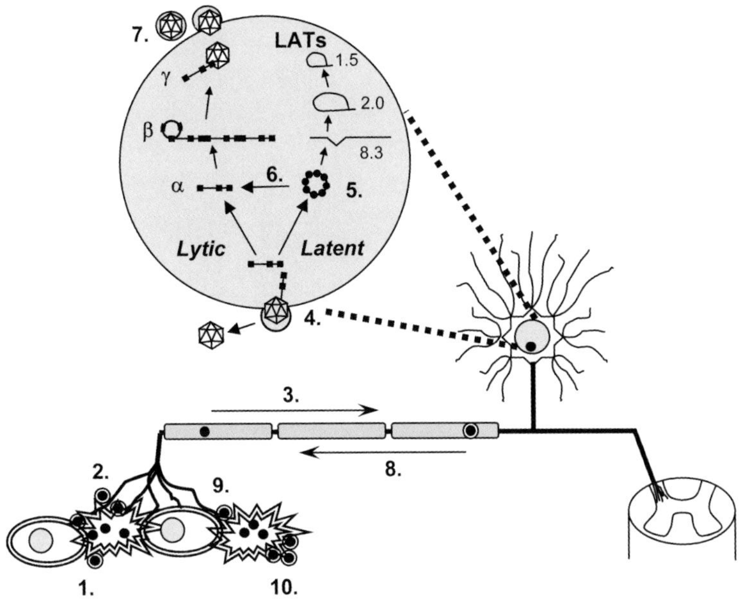

Figure 7.11 The lifecycle of HSV-1 *in vivo*. The key events occurring during infection of a human host are depicted schematically; a dorsal root ganglion sensory neuron is shown for illustration: **1**. Lytic cycle of replication at epithelial port of entry; **2**. Virions released from epithelia enter sensory nerve terminals; **3**. Nucleocapsid and tegument undergoes retrograde axonal transport to soma; **4**. Viral DNA enters neuronal nucleus and either initiates lytic cascade of gene expression or becomes latent; **5**. During latency, viral genome remains episomal and nuclear—only the LAT genes are expressed; **6**. Immunosuppression, intercurrent illness or other stimulus reactivates lytic infection; **7**. Virions formed by budding from nuclear membrane; **8**. Nucleocapsid and glycoproteins transported separately by anterograde axonal transport; **9**. Virion assembly and egress from nerve terminal; **10**. Recurrent epithelial infection at or near site of primary lesion.

The life-long latent infection in neurons, and retrograde axonal transport are features of wild-type HSV that have been exploited in the design of HSV vectors for gene transfer to neurons of the peripheral and central nervous system. Thus, elimination of replication by deletion of essential genes allows direct inoculation of vector in the vicinity of neurons (step 3) either subcutaneously, or within the CNS, with ensuing retrograde transport to the nucleus and transgene expression (steps 4 and 5) but prevents lytic infection (step 4) or reactivation (steps 6–10), resulting in a long-term latent-like infection.

difficult to avoid recombination between the amplicon and helper virus, so that completely pure vector stocks are difficult to obtain. This problem can be ameliorated to a substantial degree by using helper plasmids for amplicon packaging, but virus reconstitution may occur. Extensive experimental studies have established the utility of amplicons in gene transfer to the nervous system of rodents, but manufacture of amplicons for clinical trials would be extremely difficult, since consistent preparations are difficult to achieve.

C. Issues Common to All Viral Vectors

1. Vector Targeting

Targeting can be used to transduce cells that are not naturally infected by the wild-type virus, or to increase the uptake of vectors into a limited subset of cells. The most effective retargeting of viral vectors has involved the modification of coat protein components of adenovirus (Curiel, 1999; Dmitriev et al., 2002; Einfeld et al., 2001; Kashentseva et al., 2002; Mizuguchi et al., 2001; Wesseling et al., 2001; Wu et al., 2002). Adenovirus infects cells by binding to the coxsackie/adenovirus receptor (CAR) using the extended knob structure of the hexon capsid protein at the particle surface. The virus penton base binds to a specific integrin, $\alpha V\beta 5$ using a specific motif arginine-glutamate-aspartate (RGD). This dual receptor binding recognition is typical of many viruses and ensures specificity. The adenovirus is endocytosed and is released from the endosome as it acidifies. Strategies used to retarget adenoviruses include extension of the hexon, using an antibody ligand or a receptor ligand, and engineering the knob function directly by genetic engineering. In addition, the RGD amino acid motif has been altered, so that different integrins can be recognized. This combination has specifically altered receptor interactions that lead to adenovirus infection, in a quantitatively significant manner. Such retargeting holds promise for safe, directed *in vivo* use, especially for targeting cancer cells involved in metastasis. Enveloped viruses have been retargeted by using envelope glycoproteins from other enveloped viruses (Anderson et al., 2000; Bai et al., 2002; Burton et al., 2001a; Emi et al., 1991), or novel ligands inserted into the coding sequence of specific virus envelope genes. However, binding of enveloped virus that has been modified in these ways to target cells does not always lead to infection, since a second fusion step is required either at the cell surface or within the endosome compartment. Although efficient retargeting has not yet been achieved with AAV vectors, enhancement of entry into cells that express low levels of AAV receptor has been achieved by modification of the capsid to display a cell-specific receptor on the AAV virion (Buning et al., 2003; Girod et al., 1999; Loiler et al., 2003; Muller et al., 2003; Shi & Bartlett, 2003).

2. Regulation of Gene Expression

Many gene transfer studies have employed strong constitutively active viral promoters to express transgenes. However, there is much interest in achieving targeted, long-term, or exogenously regulated gene expression for therapeutic purposes. Many different eukaryotic promoters with tissue or cellular specificity have been isolated and some of these have been inserted into viral vectors to attempt targeted gene expression. However, promoter function in the background of a virus genome may be unpredictable; some tissue-specific promoters are complex, requiring large DNA elements that cannot be accommodated within available vectors. Considerable effort has gone into determining the critical motifs for promoter specificity and a variety of interesting sequences have been characterized including locus control regions, matrix attachment functions, and insulator elements (Chow et al., 2002; Emery et al., 2000; Miao et al., 2000). Some promoters are quite active until the virus genome becomes integrated or is subject to chromatin formation. Gene control in vectors often relies on empirical rather than rational methods. For some vectors the virus contains naturally occurring strong constitutive promoters, such as the retrovirus LTRs (Rosen et al., 1985; Starcich et al., 1985), or tissue-specific promoters such as the neuron-specific latency promoter of HSV (Chen et al., 1995; Goins et al., 1994; Zwaagstra et al., 1991) that may be exploited for transgene expression.

For many applications, it would be highly useful if a gene switch were available, with which a safe and bio-available drug could be used to turn gene expression on and off. Several systems have been developed (Hofmann et al., 1996; Hoppe et al., 2000; Iida et al., 1996; No et al., 1996; Oligino et al., 1998; Paulus et al., 1996; Rivera et al., 1996; Suzuki et al., 1996; Ye et al., 1999). The common theme shared by these systems is that a chimeric transcription factor both recognizes a specific target promoter driving the therapeutic transgene, and is activated by a drug. These systems require the constitutive expression of one or more transcriptional activator gene products. There are several problems associated with all the available systems: the promoters are usually "leaky" (low-level transcription persists even in the absence of the activating drug or presence of the repressor); the constitutively expressed chimeric transcription factors are frequently immunogenic; maintenance of activator gene transcription is as difficult as effecting long-term expression of any transgene; and the targeted promoter may become unresponsive. Despite these difficulties, evidence for gene switch function in animal models appears promising. Another strategy for achieving the same end result might be engineered gene products that are constitutively expressed, but functionally inert in the absence of an "activator" drug.

III. Development of Neurological Gene Therapy

A. The Central Nervous System

1. Viral Vector Delivery to the CNS

Gene transfer to the brain presents a number of formidable challenges. Brain tissue is inaccessible to vectors from the circulation, enclosed in dense bone, sensitive to functional disruption by mechanical manipulation, nonregenerating and physiologically vital. Various approaches have been contemplated to circumvent these barriers. It should be noted that direct intraparenchymal injection of viral vectors into the brain substance conventionally has been carried out using small volumes (1–10 μL) and low flow rates (0.2–0.4 μL/min). These parameters result in a small amount of bulk flow near the catheter tip, so that vector particles reach target cells mainly by diffusion. Consequently, transduction is at best limited to just a few millimeters around the needle tip, and sometimes along the needle track (Marconi et al., 1999). This may be adequate in few circumstances, but for most applications, exposure of a larger brain volume to the vector will be necessary.

Convection-enhanced delivery (CED) employs a catheter with a small external bore, and large volume (30–600 μL) high flow rate (0.4–4 μl/min) delivery of inoculum direct into the brain parenchyma. These infusion parameters result in bulk flow (convection) of the vector delivery solution along perivascular spaces, rather than its collection at the site of injection. Consequently, a larger volume of the brain can be exposed to the vector. CED was first demonstrated using macromolecules in the CNS (Bobo et al., 1994), but subsequently has been applied to viral delivery, including a primate model of Parkinson's disease (Bankiewicz et al., 2000). Viral particles can be delivered by this technique and it appears that their size is not a hindrance; however, the surface properties of certain viruses may impair particle transit through the brain parenchyma even when conditions for convection are satisfied (Chen et al., 2005).

2. Gene Therapy for Human CNS Diseases

Degenerative Brain Diseases There is an extensive experimental literature demonstrating the potential utility of gene transfer in the treatment of human neurodegenerative disease. In this section, we concentrate on diseases for which clinical trials have been planned or started.

Single gene disorders: Canavan disease (CD), mucopolysaccharidoses, and neuronal ceroid lipofuscinosis (NCL). Canavan disease (CD) is an autosomal recessive leukodystrophy, in which the underlying defect is loss of function of aspartoacylase, a key enzyme in the metabolism of N-acetyl-aspartate (NAA). The resulting accumulation of NAA in brain has deleterious consequences, including loss of myelin. There are two rodent models of CD, a spontaneous rat mutation with a large deletion of the aspartoacylase gene, and a mouse knockout model. In both of these models, delivery of the aspartoacylase gene using an AAV-2 vector led to biochemical and behavioral recovery. Delivery of the aspartoacylase gene to humans using a nonviral approach was not effective in preventing disease progression (Leone et al., 2000), probably as a result of inadequate transduction or expression of the transgene. Toxicity studies in primates showed that the AAV-2 aspartoacylase vector did not produce significant toxicity, and a phase I clinical trial is currently in progress, in which AAV-2 gene transfer vector coding for aspartoacylase is being used to treat patients with CD by direct intracerebral inoculation. Primary endpoints are toxicity, and biochemical, imaging, spectroscopic, and clinical outcomes.

In a preliminary report, the rAAV2 vector appeared to be well tolerated, and the low level of immune response detected in only three of 10 subjects suggested that at the dose employed and with intraparenchymal administration this approach is relatively safe (McPhee et al., 2006). Natively, NAA is thought to be produced in neurons, metabolized at the surface of oligodendrocytes, and taken up into astrocytes. In this gene therapy trial, the transgene expression cassette is targeted for neuronal aspartoacylase expression and it is hypothesized that intraneuronal NAA metabolism mediated by the transgene product will be sufficient to mitigate pathology. Whether this approach will prove effective awaits the completion of the ongoing Phase II/III trial.

There are more than 40 different lysosomal storage disorders, and approximately two-thirds are associated with neurodegeneration or demyelination in the CNS. Because lysosomal proenzymes are targeted to the lysosomes through the mannose-6-phosphate receptor, release of proenzyme from transduced cells will correct the defect in neighboring cells through uptake of the released proenzyme. Effective correction of widespread preexisting brain pathology in mucopolysaccharidosis VII has been demonstrated following intracranial inoculation of AAV (Bosch et al., 2000a) or LV (Bosch et al., 2000b) vectors coding for beta glucuronidase. More recently, Naldini and colleagues have shown that after LV vector transduction of hematopoietic stem cells *in vitro* to express arylsulfatase A, bone marrow transplantation resulted in extensive repopulation of CNS microglia and PNS endoneurial macrophages with transduced cells and correction of the histologic and behavioral phenotype of mice with metachromatic leukodystrophy (Biffi et al., 2004).

Late infantile neuronal ceroid lipofuscinosisis is a fatal, autosomal recessive–lysosomal storage disease that results from mutations in the CLN2 gene and the consequent deficiency in tripeptidyl peptidase I (TPP-I). TPP-1 deficiency leads to the accumulation of proteins in lysosomes, loss of neurons, and progressive neurologic decline. In a preliminary study in rodents and primates, Crystal and colleagues demonstrated that AAV-mediated delivery of CLN2 to brain was not associated with any obvious adverse effects (Hackett et al., 2005). Using a mouse model of neuronal ceroid lipofuscinosis, they subsequently showed that AAV-mediated gene delivery could prevent or reverse the pathologic phenotype (Passini et al., 2006). A phase I/II human trial is currently in progress (Crystal et al., 2004).

Dominant neurodegenerative diseases—RNA targeting. Huntington's disease (HD) and spinocerebellar ataxias (SCAs) are prototype autosomal dominant diseases that are caused by a toxic gain of function mechanism. The underlying problem is expansion of a polyglutamine tract within Huntington or one of the ataxins. Gene mutations also have been described that cause autosomal dominant forms of Alzheimer's disease and Parkinson's disease, due to toxic gain of function mechanisms. Over-expression of α-synuclein has been implicated in rare types of familial PD, and aberrant metabolism of mutant β-amyloid precursor protein results in over-production of amyloidogenic Aβ peptide in rare cases of familial AD. These four examples (HD, SCAs, fPD, fAD) might each be tackled by a gene delivery approach in which the abnormal allele is down-regulated through targeting its transcript. Importantly for PD and AD, since these gene products seem to play a central role in the pathogenesis of the common sporadic forms of the diseases, it is possible that a similar approach will be worthwhile in addressing the neurodegeneration underlying nonfamilial cases without genetic mutations. Two broad approaches have been tried to enable targeting of specific transcripts:

1. *RNA interference (RNAi)* is a natural mechanism found in plants and animals that is believed to provide a host defense response to intracellular pathogens and can be exploited to accomplish sequence-specific gene silencing. Long dsRNA (>200 nt) is cleaved by Dicer, a RNase III family member, into short interfering RNAs (siRNA) of approximately 22 nucleotides in length. siRNA is incorporated into a multicomponent complex, the RNA-inducing silencing complex (RISC), which mediates the endonucleolytic degradation of RNA that contains sequence complementary to the siRNA. Unlike long segments of dsRNA, short dsRNA species (<30 nt) do not induce a sequence-nonspecific interferon response in mammals, and can thus be deployed as gene targeting reagents without severe perturbation of cellular physiology. Although siRNA approaches have been highly effective in targeting gene expression in cell culture studies, efficient methods for the delivery of siRNAs *in vivo* have limited their application. Using viral vectors, expression of siRNA molecules can be

accomplished through expression of short hairpin RNAs (shRNA) that self-anneal following transcription, to form 20–30 nt stretches of dsRNA. Viral delivery of shRNA has been demonstrated in the brain *in vivo*. For example, expression of specific shRNAs using viral vectors has been shown to prevent the progression of pathology and improve the behavioral phenotype in a mouse model of SCA1 (Xia et al., 2004); and to prevent the accumulation of Aβ peptide in a mouse model of familial Alzheimer's disease (Hong et al., 2006; see Figure 7.12).

2. *Ribozymes* are a group of naturally occurring RNAs that have intrinsic enzymatic activity (Scott, 1998). At least nine ribozymes have been described. Many of these are auto-cleaving RNA virus strands; their potential application to experimental and therapeutic gene silencing arises from the ability of some of these molecules to mediate sequence-specific effects *in trans*. This means that appropriate design of therapeutic sequences allows specific RNA molecules to be targeted for cleavage by an RNA-mediated enzymatic activity. This approach has been successfully deployed *in vivo* in studies of retinitis pigmentosa (RP), a genetically heterogeneous hereditary syndrome, causing photoreceptor degeneration. One type of autosomal dominant RP is caused by a gain-of-function mutation in the rhodopsin protein, leading to deposition of insoluble protein material in the rod cells of the retina. A rat model of this disease, containing a mutant rhodopsin transgene, develops retinal photoreceptor loss and blindness. This was rescued by virally mediated expression of ribozymes targeting mutant rhodopsin in photoreceptor cells.

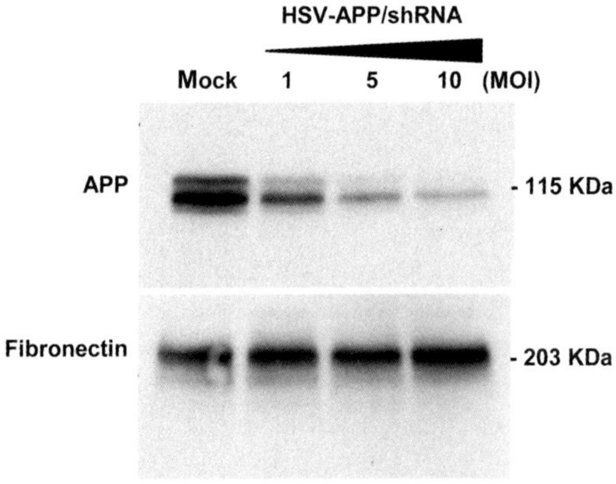

Figure 7.12 RNA targeting using a viral vector. A replication-defective HSV vector was generated that expressed a short hairpin RNA that self-annealed to form a double-stranded segment corresponding to the human amyloid precursor protein mRNA. APP is implicated in the pathogenesis of Alzheimer's disease, and targeting its gene represents a potential strategy for neuroprotection in AD. In this experiment, the virus was used to infect a cell line that overexpresses APP. Cell lysates were subjected to western blot analysis to detect APP (upper panel) or a control protein, fibronectin (lower panel). A dose-dependent reduction in APP expression was evident, indicating that the shRNA could be used to target the APP transcript, and that this can be delivered using a viral vector (Hong et al., 2006).

Sporadic neurodegenerative diseases—compensatory and regenerative strategies using therapeutic proteins. There has been much interest in the application of recombinant growth factors to arrest the progression of neurodegenerative diseases, and to promote axonal regeneration or increase neurotransmitter turnover. For example, glial cell-line derived neurotrophic factor (GDNF) and neurturin (Horger et al., 1998) both have been shown to exhibit marked trophic effects on dopaminergic neurons *in vitro*, and have been shown to protect these neurons from a diverse range of pathological insults *in vivo*. Consequently, they are being investigated as possible therapeutic agents for Parkinson disease. A phase I trial reported that chronic intraputamenal infusion of recombinant GDNF into patients with PD produced beneficial effects on functional imaging surrogates of cell integrity (Gill et al., 2003). Although a subsequent double-blind study failed to demonstrate any improvement in neurological function (Lang et al., 2006), the possibility remains that several limitations encountered in that trial might be addressed by using a vector to deliver the GDNF gene into striatum or substantia nigra. Vector-mediated delivery of GDNF using a number of different vector systems (for example, HSV; see Figure 7.13) has been demonstrated to produce a protective effect in a variety of rodent and primate models of PD (Bensadoun et al., 2003; Kordower et al., 2000; Puskovic et al., 2004; Wang et al., 2002).

An industry-sponsored human clinical trial in which an AAV vector is being used to deliver the related neuroprotective peptide neurturin is currently enrolling patients. A phase 1 trial in which autologous fibroblasts genetically modified to express human nerve growth factor (NGF) into the forebrain of eight individuals with mild Alzheimer's disease showed no long-term adverse effects of NGF and suggested improvement in the rate of cognitive decline (Tuszynski et al., 2005). A second phase I/II trial in which the NGF gene will be delivered using an AAV vector is currently enrolling patients.

An alternative approach to PD is to attempt to correct the behavioral phenotype by modifying neurotransmitter abnormal phenotyping using gene transfer. One of the characteristics of PD is overactivity of excitatory projections from the subthalamic nucleus (STN) to the substantia nigra pars reticulater and the GP. AAV-mediated transfer of the gene coding for glutamic acid decarboxylase to neurons of the STN to result in the release of the inhibitory neurotransmitter gamma amino butyric acid (GABA) in place of the excitatory neurotransmitter glutamate corrects the parkinsonian phenotype in rodents (During et al., 2001). A phase I trial of this approach in 12 patients with PD has been completed

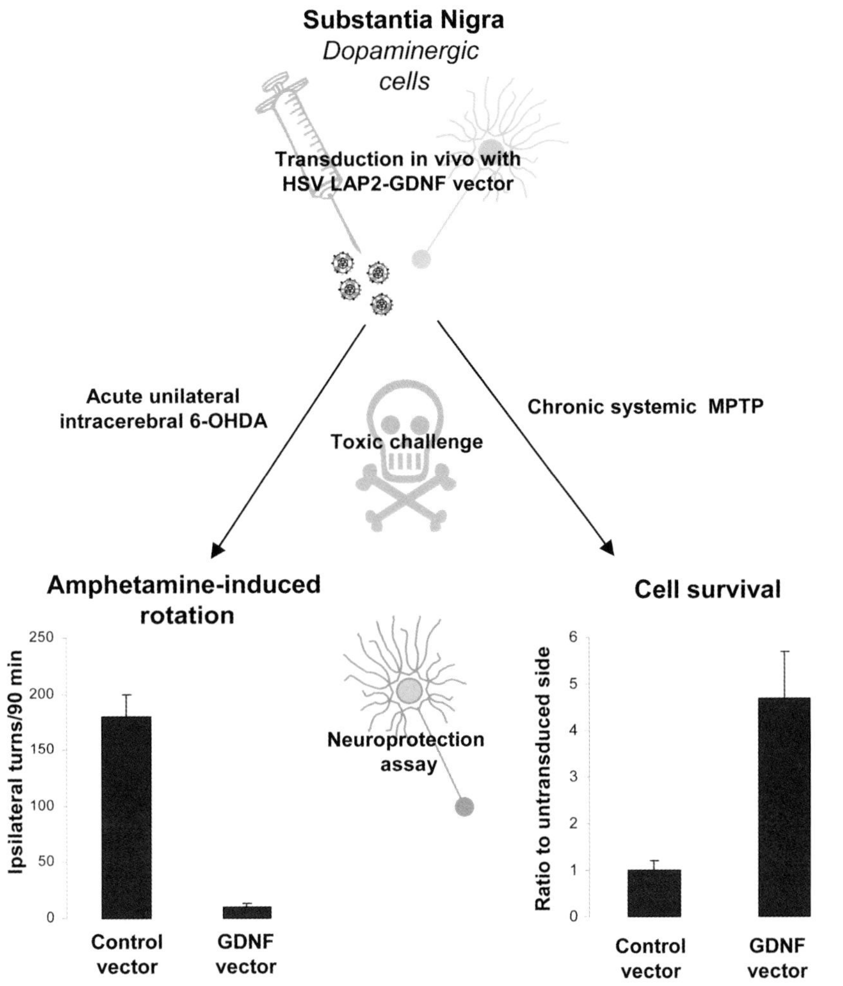

Figure 7.13 CNS neuroprotection using a replication-defective HSV vector to express a neurotrophic factor. A replication-defective HSV vector expressing the neurotrophic agent GDNF was used to demonstrate delivery of a gene encoding a biologically active neuroprotective molecule to the CNS, *in vivo*. GDNF is a powerful neurotrophic agent that is able to protect dopaminergic neurons from a variety of insults. Vector inoculation into the substantia nigra of the brain was carried out six months before a dopaminergic neuron-specific toxic challenge was presented. Two different experimental paradigms were used. First, unilateral injection of a dopamine neurotoxin, 6-OHDA, was made directly into the substantia nigra, resulting in unilateral loss of the nigrostriatal projection and circling behavior in response to amphetamine. This effect was substantially mitigated by prior inoculation with the GDNF vector. Second, chronic systemic MPTP administration resulted in bilateral loss of dopaminergic neurons. In animals injected unilaterally with the GDNF vector, cell loss was distinctly asymmetric due to protection of cells on the side of the vector inoculation. Together, these data show that GDNF-related neuroprotection occurred, in this experiment six months after the vector was introduced into the brain (Puskovic et al., 2004).

and no adverse outcomes were observed. A phase II/III trial of this approach is being planned.

Neuro-inflammatory diseases and meningeal transduction. Diseases caused by neurological inflammation are attractive targets for gene therapy approaches, because the production of anti-inflammatory cytokines from depot sites within the brain or meninges would be expected to have effects on inflammatory cells following diffusion of the cytokine and binding to specific cellular receptors at sites remote from the transduced cells. This approach has been investigated in animal models of autoimmune myelin destruction. For example, the inflammatory process of acute allergic encephalomyelitis and its pathological consequences were mitigated in monkeys by expression of human interleukin 4 (IL-4) from a transgene expressed in ependymal cells. The HSV vector encoding the transgene was introduced into the CSF and diffused throughout the CSF pathways, secreting IL4 into the CSF and adjacent brain tissue without demonstrable toxicity. CNS inflammatory perivenular infiltrates were reduced, as was demyelination, necrosis, and axonal loss (Poliani et al., 2001). Similar results were shown in a mouse model of relapsing-remitting autoimmune encephalitis by delivery of the fibroblast growth factor II (Ruffini et al., 2001) or IL4 (Furlan et al., 2001) genes. It is possible that this may be a viable way to deliver immunomodulatory molecules to the CNS.

Malignant glioma. Despite recent advances in clinical oncology and radiotherapy, malignant glioma is still associated with a poor prognosis, and many of the current best treatments are highly toxic. Gene therapy provides an attractive experimental approach to developing novel therapeutic reagents to tackle malignancy. Of human clinical gene therapy trials carried out so far, the majority have been in patients with various cancers, reflecting both the desperate need for better treatments and the perceived potential for this technology to yield favorable results.

The aims of cancer gene therapy are rather different to those of other neurological gene therapy applications; the goal is to selectively destroy target cells. This has been accomplished in a number of ways including:

1. *Suicide gene therapy.* The transgene product is toxic to expressing cells. An example is vector-mediated expression of HSV-thymidine kinase, which allows activation of a pro-drug gancicvovir in transduced cells. This interferes with DNA replication in dividing cells, such as those found in a tumor (Marconi et al., 2000).
2. *Radiosensitization.* The transgene product enhances the toxicity of gamma irradiation in the tumor. TNF-α is an example (Moriuchi et al., 1998; Niranjan et al., 2000).
3. *Immunotherapy.* The transgene product stimulates an immune response directed against tumors. Examples include CD80 (Krisky et al., 1998a), TNF α (Moriuchi et al., 1998; Niranjan et al., 2000), GM-CSF (Krisky et al., 1998a), IL-2 (Colombo et al., 1997), interferon-γ (Kanno et al., 1999), and IL-12 (Parker et al., 2000).
4. *Restoration of cellular functions altered during oncogenesis.* Many tumors acquire means of avoiding apoptosis in response to the mutations in cellular DNA that often occur during oncogenesis. A common example is the loss of p53 signaling that occurs in many types of solid tumor. Replacement of p53 function in this instance can cause the malignant cells to undergo apoptosis (Lang et al., 1999). Other examples might include delivery of other tumor suppressor genes that are mutated in the tumor, or targeting oncogene products or their related cellular pathways to suppress tumor growth.
5. *Disruption of tumor blood supply.* Most tumors excite an angiogenic response that allows continuing, and adequate, metabolic support for expanding the tumor mass. Gene therapy can be designed to interfere with this process, causing the tumor to outgrow its blood supply and undergo necrosis (Im et al., 1999; Machein et al., 1999).

The mode of action of these transgenes is to effect selective destruction of malignant cells. Various gene transfer vectors have been used in preclinical and clinical trials of cancer gene therapy, including replication-deficient lentiviruses, adenoviruses, and herpes simplex viruses. It is not yet known which vector is optimal, and it is likely that different vectors will show utility for specific applications. For example, the high-level transient expression and immunogenicity of AdV may have application for suicide gene therapy and immunotherapy applications in rapidly dividing tumors, whereas the long-term gene expression and capacity for multiple transgenes seen with HSV vectors may make them good candidates for use in slower growing tumors where multiple simultaneous pathways must be targeted to achieve tumor eradication, for example, glioblastoma (see Figure 7.14).

A different approach uses conditionally replicating HSV and adenovirus mutants. These are viruses that are engineered to replicate only in tumor cells. Consequently, they cause lysis of malignant cells, while failing to replicate in normal cells. It is likely that the resulting inflammatory response is also important to the action of these viruses. An example is provided by the HSV mutant G207. This contains a mutation in both copies of the γ34.5 neuro virulence gene, and a further mutation in the ICP6 ribonucleotide reductase gene, both of which are required for lytic neuronal infection, although dispensable for viral growth in cultured dividing cells. In animal models, this viral mutant was able to effect destruction of tumor cells without damage to neighboring neural tissue. In a phase I clinical trial of patients with malignant glioma, intracerebral doses of up to 3×10^9 plaque-forming units were tolerated well, without adverse effect, although no patient showed an objective clinical or imaging response to the treatment. Current work includes development of conditionally replicating mutants as gene transfer vectors, attempting to combine anti-tumor transgene delivery with viral oncolysis caused by lytic replication.

B. Peripheral Nervous System

1. Viral Vector Delivery to the Peripheral Nervous System

The peripheral nervous system presents different challenges to gene delivery to those posed by the brain. The target nerves are widely distributed, with long axonal processes in the periphery, but cell bodies either within the CNS or within inaccessible dorsal root or trigeminal ganglia. Systemic administration of vector to target peripheral nerves has not been demonstrated. The most successful studies have employed peripheral inoculation with retrograde axonal transport into the cell body to transduce neurons of the DRG. In this regard, vectors that are effective substrates for rapid axonal transport are ideal, and HSV in particular has shown great utility for these applications.

2. Some Examples of Gene Therapy Strategies for Treating Peripheral Nerve Diseases

Chronic Pain Therapy—Altering the Neurochemical Properties of Sensory Nerves. Chronic pain, pain that persists beyond the course of the acute insult, or pain that accompanies a chronic primary process that cannot be cured, is a significant problem.

A significant minority of patients with chronic pain are not effectively treated by currently available therapies, resulting in substantial morbidity and cost (Loeser et al., 2001). The neuroanatomic pathways involved in pain are identical to those involved in the perception of acute painful stimuli, but alterations in gene expression in neuronal and non-neuronal elements at many levels of the neuraxis produce an altered substrate that result in symptoms that are refractory to medical management (Woolf and Salter, 2000). The pharmacologic approach to chronic pain, epitomized by the search for small molecule agonists that would be selective for elements in the pain pathway, is limited because most targets are not unique to nociceptive neurotransmission. Opioid receptors for example are present at all levels of the neuraxis, including the central and peripheral terminals of primary nociceptors, second order neurons in dorsal horn of spinal cord, and nuclei in the brain and brain stem and are also found in non-neural tissues including bladder, gut, and inflammatory cells. Even though opiate drugs are highly effective, their continuous use for treatment of chronic pain is limited by side effects that occur because activation of receptors in non-nociceptive pathways results in sedation, constipation, urinary retention, and respiratory suppression. Continued use results in tolerance so that increasing doses of the drug are required to achieve maintain the therapeutic effect and abuse is a potential problem.

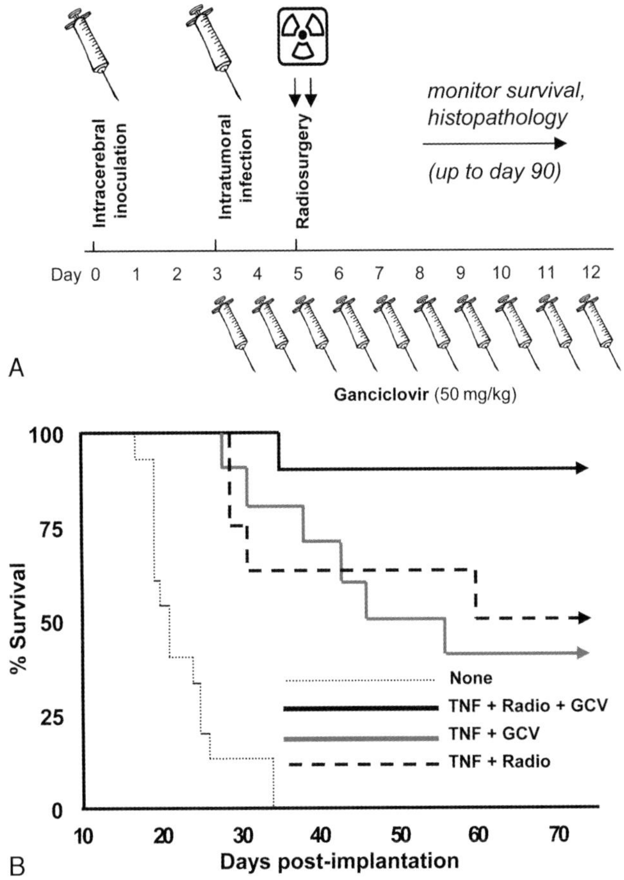

Figure 7.14 Targeting malignant glioma using a combination of gene transfer and radio surgery. A variety of approaches to targeting malignant glioma has been proposed—this figure illustrates one approach that has shown promise in preclinical studies. An orthotopic transplant model of cerebral tumor was made by implanting glioma cells into the rat striatum. Panel **A** illustrates the experimental paradigm and panel **B** shows the survival curves for cohorts of animals exposed to different interventions following tumor implantation. In this model, rats survived a median of 20 days without intervention (dotted survival curve in panel **B**). Inoculation of an HSV vector encoding thymidine kinase and tumour necrosis factor-α at day 3 was followed by administration of ganciclovir at days 3–12 (solid gray survival curve), administration of radiotherapy at day 5 (dashed survival curve) or a combination of both (solid black survival curve). Incremental improvements in survival were seen with combination therapies, illustrating that a combination of expression of multiple antitumor transgenes and other modalities of treatment may be one effective way to eradicate brain tumors using gene therapy.

Gene transfer to DRG neurons by subcutaneous inoculation of an HSV-based vector expressing proenkephalin has been shown to reduce pain-related behaviors in animal models of inflammatory pain (Goss et al., 2001; see Figure 7.15B), neuropathic pain (Hao et al., 2003a), and pain resulting from cancer in bone (Goss et al., 2002b). In a model of arthritis, HSV-mediated expression of proenkephalin in DRG not only reduced pain-related behaviors, but also reduced joint destruction (Braz et al., 2001). An HSV vector expressing glutamic acid decarboxylase effects the release of γ-amino butyric acid (GABA) from DRG neurons *in vivo* and has substantial antinociceptive effects in chronic neuropathic pain of central (Liu et al., 2004) or peripheral (Hao et al., 2005) origin. Similar results have been demonstrated with HSV vectors expressing glial cell line derived neuroptrophic factor (GDNF) (Hao et al., 2003b), interleukin-4 (Hao et al., 2006) and the soluble tumor necrosis factor receptor (Peng et al., 2006).

On the basis of the preclinical data in the several different rodent models of pain, we are moving forward with plans for a human trial. The plans for this phase I/II safety/dose-escalation trial of the proenkephalin-expressing vector was reviewed by the recombinant DNA Advisory Committee (RAC) of the NIH in June 2002, and the request for an investigational new drug (IND) waiver to the Food and Drug Administration has been submitted. The HSV vector for the human trial will be deleted for two IE HSV genes (ICP4 and ICP 27) and contains deletions in the promoters for two other IE genes (ICP22 and ICP47). The proenkephalin transgene has been placed in both copies of ICP4, so that in the unlikely event of recombination with a latent wild-type virus, the recombinants would be replication defective. In the trial we will enroll patients with cancer metastatic to a vertebral body resulting in pain refractory to maximal medical management. The primary outcome of this phase I/II trial will involve standard measures of safety assessed by common criteria. Secondary measures will include an evaluation of the focal pain using an analogue pain "thermometer,"

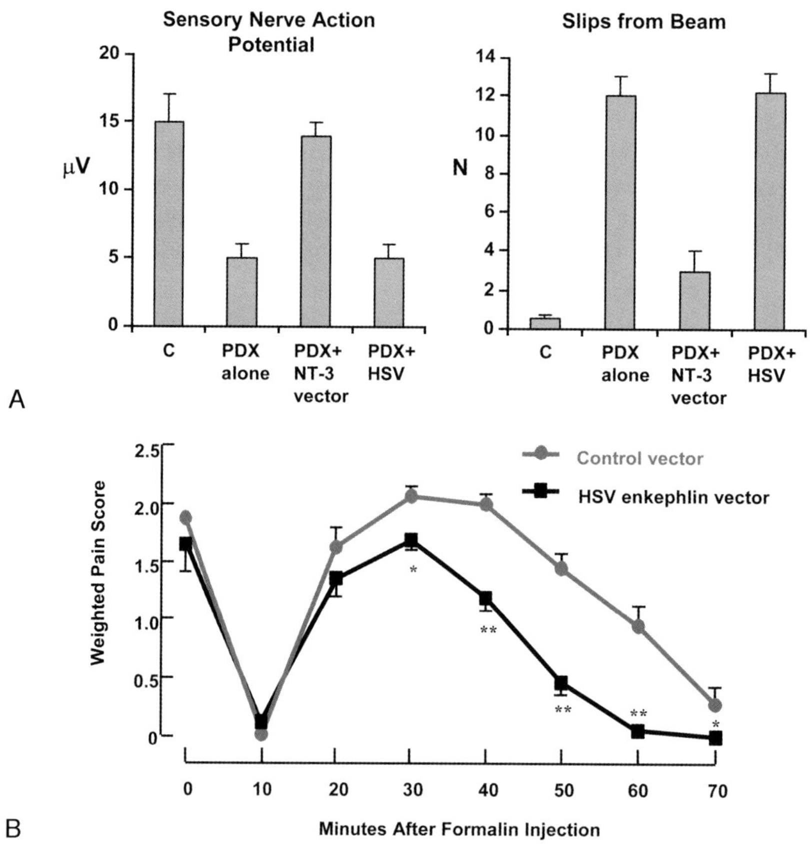

Figure 7.15 Gene transfer in models of peripheral nervous system disease. **A.** Dorsal root ganglion cells may be transduced with a replication-defective HSV vector by peripheral inoculation, allowing retrograde transport to the cell body and transgene expression. In this example, rats were inoculated with a vector that expresses neurotrophin-3, which protects DRG sensory neurons from pyridoxine toxicity. Neurophysiological (left panel) and behavioral (right panel) parameters were measured in four groups of animals (no pyridoxine exposure; pyridoxine exposure, no viral vector; pyridixone exposure, control HSV vector; pyridixone exposure, HSV-NT-3 vector). Preservation of physiological and behavioral functions following pyridoxine challenge was apparent in HSV-NT-3 treated animals compared with animals treated with the control vector or no vector. **B.** Injection of the rat footpad with formalin results in pain behavior that may be scored. Following an initial transient nociceptive response, pain behavior reappears and lasts for approximately 1 hour, reflecting a chronic pain mechanism. Pretreatment of rats with a pre-proenkephalin expressing vector altered the neurochemical properties of the transduced sensory neurons and correspondingly modulated the behavioral response to the same nociceptive challenge. In this model, the chronic pain behavior was ameliorated without loss of the initial nociceptive response.

global measures of total pain using a similar tool, as well as assessment of pain-related phenomena such as sleep and depression and the concurrent use of analgesic medication including opiate drugs. The proposed human trial will determine whether the same approach may be used in the treatment of patients, and will be the first step in the use of gene therapy for pain and polyneuropathy.

Polyneuropathy—Protection of Sensory Nerves by Trophic Factor Gene Delivery Peripheral nerve disease arising as a complication of diabetes, or during chemotherapy for malignancy is a common and debilitating problem. Although the precise mechanism underlying peripheral damage is uncertain in many neuropathic conditions, growth factors may act as trophic support signals to sick neurons. For example, nerve growth factor is able to protect neurons from a variety of insults *in vitro*. Gene delivery has provided a means to confirm these results *in vivo*, and the possibility of a therapeutic tool. In these studies, as in chronic pain, HSV has shown special utility, because of the peripheral inoculation and retrograde axonal transport and the natural tropism for sensory nerves (see Figure 7.15A). Subcutaneous inoculation of HSV vectors constructed to express nerve growth factor (NGF) can prevent the progression of neuropathy caused by diabetes (Goss et al., 2002a), or chemotherapeutic drugs (Chattopadhyay et al., 2003; Chattopadhyay et al., 2004). Similar effects have been demonstrated through HSV-mediated expression of vascular endothelial growth factor (VEGF), which has neuroprotective, as well as vascular effects (Chattopadhyay et al., 2005a). Using the HSV latency associated promoter elements, effective neuroprotective transgene expression has been detected up to six months after vector inoculation (Chattopadhyay et al., 2005b). In a model of diabetic cystopathy, the effects of chronic hyperglycemia on urinary bladder function were ameliorated by HSV-mediated NGF expression (Goins et al., 2001). Long-term use of similar vectors to treat neuropathy in patients will require the development of appropriate regulable elements to control gene expression.

C. Muscle

1. Viral Gene Delivery to Skeletal Muscle

Gene therapy of muscle disorders poses a number of very difficult challenges. The tissue compartment is vast, occupying a significant proportion of body weight, and is widely distributed. Myofibers are multinucleate cells, and many of the proteins whose delivery may be therapeutic are large and unlikely to diffuse more than a couple of nuclear domains along a myofiber, meaning that rescue of each myofiber will require transduction of multiple nuclei. Within a muscle, it will be necessary to transduce a large number of myofibers to achieve any kind of functional correction. Unfortunately, the basal lamina in muscle presents a barrier to effective diffusion of many viral vectors, meaning that inoculation at multiple sites would be necessary for transduction of a large number of myofibers. There have been two proposed solutions to this problem:

1. It appears that AAV has some highly favorable properties regarding gene transfer to muscle. In particular, AAV6 may have muscle-specific tropism. Recent studies suggest that systemic administration of AAV6 may be able to transduce a significant proportion of total muscle mass in the mouse (Gregorevic et al., 2004).
2. Delivery through the vascular compartment has been proposed as a means to expose a large volume of muscle to vector. In this scheme, the vascular bed to a limb is made more permeable than usual, either pharmacologically (Greelish et al., 1999) or by elevated venous pressure (Su et al., 2005), and the vector leaks from the circulation throughout the extracellular fluid.

These developments are relatively recent and neither has yet been tested in humans. Consequently, clinical data so far reported relate to localized transduction and do not show functional benefits, but do provide proof of concept that important protein complexes can be reassembled in muscle by gene transduction.

2. Some Examples of Gene Therapy Strategies for Treating Muscle Diseases

Muscular Dystrophies—Single Gene Disorders The muscular dystrophies are a group of inherited muscle degenerations (Burton & Davies, 2000, 2002). Many of these are recessive and caused by loss of essential protein components from muscles. These diseases might be effectively treated by genetic complementation using viral gene transfer, provided issues concerning delivery were adequately addressed. The commonest of these diseases, Duchenne muscular dystrophy (DMD) affects 1:3,500 male births. It results in progressive muscle wasting and death by the age of 15 to 20 from respiratory impairment or cardiac disease. It is caused by the loss of a muscle protein, dystrophin, which forms part of a large complex of proteins that localize to the sarcolemma of muscle fibers and function to maintain the integrity of the muscle cell membrane during contraction.

The dystrophin gene is very large (the cDNA is >10 kb) and so can be accommodated only within gutless adenovirus or herpes simplex virus vectors, both of which are inefficient vehicles for transduction of muscle. Much work has been carried out in trying to develop truncated functional versions of the dystrophin gene that may be accommodated in adeno-associated virus vectors, which appear more promising as vectors in muscle tissue, as discussed earlier. A group of conditions, the sarcoglycanopathies, are closely related to DMD. They are caused by mutations in genes encoding other components of the protein complex of which dystrophin is part. These genes are smaller and can be readily accommodated within adeno-associated virus vectors. A number of preclinical trials and studies have shown proof of concept that dystrophin and other missing proteins can be expressed transiently in dystrophic muscle using a variety of gene transfer technologies, but only the AAV-2 and AAV-6 studies discussed earlier have shown promise in addressing the difficult question of widely distributed gene delivery. A phase I gene therapy trial has been approved, in which dystrophin minigenes or sarcoglycan genes will be delivered to muscle using AAV (Stedman et al., 2000).

Myotonic Dystrophy—Targeting a Toxic RNA Myotonic dystrophy is the commonest type of hereditary myopathy. This autosomal dominant condition is caused by an expanded trinucleotide repeat sequence in the 3′ UTR of the DMPK gene, which encodes a serine/threonine kinase. The mutant mRNA is retained within myonuclei, where it aggregates in discrete foci and may interfere with export of the wild-type mRNA. The DM phenotype is reproduced in a mouse model, in which the expanded CUG repeat is placed within a heterologous transgene encoding human skeletal actin, which is unrelated to DMPK. Toxic gain of function at the RNA level may be a general mechanism resulting in this phenotype. A similar dominant disease to DM (proximal myotonic myopathy) has been described in patients with an expanded CCUG tetranucleotide repeat within the first intron of the gene encoding a zinc finger transcription factor, ZNF9. This gene is unrelated to DMPK implying that gain of RNA function is the pathogenic mechanism. A variety of RNA targeting strategies has been proposed in order to tackle this type of abnormality at a molecular level. It may be necessary to use strategies that target nuclear RNAs, as the mutant does not appear to be exported into the cytoplasm. Possible strategies include RNAi (Langlois et al., 2005), ribozymes with a nuclear retention signal (Langlois et al., 2003), and expression of antisense RNA (Furling et al., 2003). Retroviral expression of antisense RNA was shown to reduce mutant DMPK transcripts in human myotonic dystrophy myoblasts, and to restore myoblast fusion and glucose uptake (Furling et al., 2003). Similarly, a nuclear-retained hammerhead ribozyme targeting the 3′ UTR of the DMPK RNA was shown to effect a significant reduction in RNA levels and amelioration of DMPK mRNA-containing nuclear foci, *in vitro* (Langlois et al., 2003). It remains to be seen whether these gene delivery approaches will prove effective in human disease.

IV. Conclusions—Future Developments

Advances in the understanding of the molecular basis of neurological disease, coupled with advances in the technology for construction and delivery of engineered gene transfer vectors, has provided optimism that gene therapy for neurological disorders may provide treatments for presently incurable diseases. Clinical trials have commenced in several key areas.

As reviewed in this chapter, successful treatment of several single gene recessive disorders of the nervous system by gene transfer has been achieved; substantial progress with interfering RNA technologies has raised the prospect that dominantly inherited diseases caused by toxic gain of function mutations may be amenable to a similar approach; and vector-mediated targeting of neurotransmitters or neuroprotective peptides may be used to provide alternative treatments for complex multigenic or noninherited conditions.

Much work remains to be done in improving vector design to optimize the efficacy of gene delivery in humans. Because the human immune system is distinct even from that of primates, it will be important to evaluate the immune response to vector-mediated gene transfer in humans. For some applications it may be critical to develop regulatable gene expression systems, so that the amount of transgene product produced may be controlled. As these approaches move forward into the clinic, issues regarding the "scale-up" to manufacturing of adequate amounts of highly purified vector will become more important. Nonetheless, substantial progress has been made over the past decade in the study of gene transfer to the nervous system, and there is reason for cautious optimism that novel effective treatments may soon emerge.

References

Akkina, R. K., Walton, R. M., Chen, M. L., Li, Q. X., Planelles, V., and Chen, I. S. (1996). High-efficiency gene transfer into CD34+ cells with a human immunodeficiency virus type 1-based retroviral vector pseudotyped with vesicular stomatitis virus envelope glycoprotein G. *J Virol* **70**, 2581–2585.

Amalfitano, A., Hauser, M. A., Hu, H., Serra, D., Begy, C. R., and Chamberlain, J. S. (1998). Production and characterization of improved adenovirus vectors with the E1, E2b, and E3 genes deleted. *J Virol* **72**, 926–933.

Anderson, D. B., Laquerre, S., Goins, W. F., Cohen, J. B., and Glorioso, J. C. (2000). Pseudotyping of glycoprotein D-deficient herpes simplex virus type 1 with vesicular stomatitis virus glycoprotein G enables mutant virus attachment and entry. *J Virol* **74**, 2481–2487.

Armentano, D., Thompson, A. R., Darlington, G., and Woo, S. L. (1990). Expression of human factor IX in rabbit hepatocytes by retrovirus-mediated gene transfer: Potential for gene therapy of hemophilia B. *Proc Natl Acad Sci U S A* **87**, 6141–6145.

Babiss, L. E., Young, C. S., Fisher, P. B., and Ginsberg, H. S. (1983). Expression of adenovirus E1a and E1b gene products and the Escherichia coli XGPRT gene in KB cells. *J Virol* **46**, 454–465.

Bai, Q., Burton, E. A., Goins, W. F., and Glorioso, J. C. (2002). Modifying the tropism of HSV vectors for gene therapy applications. In J. Douglas and D. Curiel, Eds., *Vector targeting for therapeutic gene delivery*, Vol. in press. John Wiley and Sons, New York.

Bankiewicz, K. S., Eberling, J. L., Kohutnicka, M., Jagust, W., Pivirotto, P., Bringas, J. et al. (2000). Convection-enhanced delivery of AAV vector in parkinsonian monkeys; in vivo detection of gene expression and restoration of dopaminergic function using pro-drug approach. *Exp Neurol* **164**, 2–14.

Ben-Gary, H., McKinney, R. L., Rosengart, T., Lesser, M. L., and Crystal, R. G. (2002). Systemic interleukin-6 responses following administration of adenovirus gene transfer vectors to humans by different routes. *Mol Ther* **6**, 287–297.

Bensadoun, J. C., Pereira de Almeida, L., Fine, E. G., Tseng, J. L., Deglon, N., and Aebischer, P. (2003). Comparative study of GDNF delivery systems for the CNS: Polymer rods, encapsulated cells, and lentiviral vectors. *J Control Release* **87**, 107–115.

Biffi, A., De Palma, M., Quattrini, A., Del Carro, U., Amadio, S., Visigalli, I. et al. (2004). Correction of metachromatic leukodystrophy in the mouse model by transplantation of genetically modified hematopoietic stem cells. *J Clin Invest* **113**, 1118–1129.

Bloom, D. C., Lokensgard, J. R., Maidment, N. T., Feldman, L. T., and Stevens, J. G. (1994). Long-term expression of genes in vivo using nonreplicating HSV vectors. *Gene Ther* **1**, S36–38.

Bobo, R. H., Laske, D. W., Akbasak, A., Morrison, P. F., Dedrick, R. L., and Oldfield, E. H. (1994). Convection-enhanced delivery of macromolecules in the brain. *Proc Natl Acad Sci U S A* **91**, 2076–2080.

Bosch, A., Perret, E., Desmaris, N., and Heard, J. M. (2000a). Long-term and significant correction of brain lesions in adult mucopolysaccharidosis type VII mice using recombinant AAV vectors. *Mol Ther* **1**, 63–70.

Bosch, A., Perret, E., Desmaris, N., Trono, D., and Heard, J. M. (2000b). Reversal of pathology in the entire brain of mucopolysaccharidosis type VII mice after lentivirus-mediated gene transfer. *Hum Gene Ther* **11**, 1139–1150.

Braz, J., Beaufour, C., Coutaux, A., Epstein, A. L., Cesselin, F., Hamon, M., and Pohl, M. (2001). Therapeutic efficacy in experimental polyarthritis of viral-driven enkephalin overproduction in sensory neurons. *J Neurosci* **21**, 7881–7888.

Buning, H., Ried, M. U., Perabo, L., Gerner, F. M., Huttner, N. A., Enssle, J., and Hallek, M. (2003). Receptor targeting of adeno-associated virus vectors. *Gene Ther* **10**, 1142–1151.

Burton, E. A. and Davies, K. E. (2000). The pathogenesis of Duchenne muscular dystrophy. In M. P. Mattson, Ed., *The pathogenesis of neurodegenerative disorders*, pp. 1–46. Humana Press Inc., Totowa, NJ.

Burton, E. A. and Davies, K. E. (2002). Muscular dystrophy—Reason for optimism? *Cell* **108**, 5–8.

Burton, E. A., Bai, Q., Goins, W. F., and Glorioso, J. C. (2001a). Targeting gene expression using HSV vectors. *Adv Drug Deliv Rev* **53**, 155–170.

Burton, E. A., Wechuck, J. B., Wendell, S. K., Goins, W. F., Fink, D. J., and Glorioso, J. C. (2001b). Multiple applications for replication-defective herpes simplex virus vectors. *Stem Cells* **19**, 358–377.

Cavazzana-Calvo, M., Hacein-Bey, S., de Saint Basile, G., Gross, F., Yvon, E., Nusbaum, P. et al. (2000). Gene therapy of human severe combined immunodeficiency (SCID)-X1 disease. *Science* **288**, 669–672.

Chattopadhyay, M., Goss, J., Lacomis, D., Goins, W. C., Glorioso, J. C., Mata, M., and Fink, D. J. (2003). Protective effect of HSV-mediated gene transfer of nerve growth factor in pyridoxine neuropathy demonstrates functional activity of trkA receptors in large sensory neurons of adult animals. *Eur J Neurosci* **17**, 732–740.

Chattopadhyay, M., Goss, J., Wolfe, D., Goins, W. C., Huang, S., Glorioso, J. C. et al. (2004). Protective effect of herpes simplex virus-mediated neurotrophin gene transfer in cisplatin neuropathy. *Brain* **127**, 929–939.

Chattopadhyay, M., Krisky, D., Wolfe, D., Glorioso, J. C., Mata, M., and Fink, D. J. (2005a). HSV-mediated gene transfer of vascular endothelial growth factor to dorsal root ganglia prevents diabetic neuropathy. *Gene Ther* **12**, 1377–1384.

Chattopadhyay, M., Wolfe, D., Mata, M., Huang, S., Glorioso, J. C., and Fink, D. J. (2005b). Long-term neuroprotection achieved with latency-associated promoter-driven herpes simplex virus gene transfer to the peripheral nervous system. *Mol Ther* **12**, 307–313.

Chen, M. Y., Hoffer, A., Morrison, P. F., Hamilton, J. F., Hughes, J., Schlageter, K. S. et al. (2005). Surface properties, more than size, limiting convective distribution of virus-sized particles and viruses in the central nervous system. *J Neurosurg* **103**, 311–319.

Chen, X., Schmidt, M. C., Goins, W. F., and Glorioso, J. C. (1995). Two herpes simplex virus type 1 latency-active promoters differ in their contributions to latency-associated transcript expression during lytic and latent infections. *J Virol* **69**, 7899–7908.

Chow, C. M., Athanassiadou, A., Raguz, S., Psiouri, L., Harland, L., Malik, M. et al. (2002). LCR-mediated, long-term tissue-specific gene expression within replicating episomal plasmid and cosmid vectors. *Gene Ther* **9**, 327–336.

Colombo, F., Zanusso, M., Casentini, L., Cavaggioni, A., Franchin, E., Calvi, P., and Palu, G. (1997). Gene stereotactic neurosurgery for recurrent malignant gliomas. *Stereotact Funct Neurosurg* **68**, 245–251.

Corbeau, P., Kraus, G., and Wong-Staal, F. (1996). Efficient gene transfer by a human immunodeficiency virus type 1 (HIV-1)-derived vector utilizing a stable HIV packaging cell line. *Proc Natl Acad Sci U S A* **93**, 14070–14075.

Croyle, M. A., Chirmule, N., Zhang, Y., and Wilson, J. M. (2002). PEGylation of E1-deleted adenovirus vectors allows significant gene expression on readministration to liver. *Hum Gene Ther* **13**, 1887–1900.

Crystal, R. G., Sondhi, D., Hackett, N. R., Kaminsky, S. M., Worgall, S., Stieg, P. et al. (2004). Clinical protocol. Administration of a replication-deficient adeno-associated virus gene transfer vector expressing the human CLN2 cDNA to the brain of children with late infantile neuronal ceroid lipofuscinosis. *Hum Gene Ther* **15**, 1131–1154.

Curiel, D. T. (1999). Strategies to adapt adenoviral vectors for targeted delivery. *Ann N Y Acad Sci* **886**, 158–171.

Davidson, B. L., Allen, E. D., Kozarsky, K. F., Wilson, J. M., and Roessler, B. J. (1993). A model system for in vivo gene transfer into the central nervous system using an adenoviral vector. *Nat Genet* **3**, 219–223.

DeLuca, N. A., McCarthy, A. M., and Schaffer, P. A. (1985). Isolation and characterization of deletion mutants of herpes simplex virus type 1 in the gene encoding immediate-early regulatory protein ICP4. *J Virol* **56**, 558–570.

Desmaris, N., Bosch, A., Salaun, C., Petit, C., Prevost, M. C., Tordo, N. et al. (2001). Production and neurotropism of lentivirus vectors pseudotyped with lyssavirus envelope glycoproteins. *Mol Ther* **4**, 149–156.

Dmitriev, I. P., Kashentseva, E. A., and Curiel, D. T. (2002). Engineering of adenovirus vectors containing heterologous peptide sequences in the C terminus of capsid protein IX. *J Virol* **76**, 6893–6899.

Dobson, A. T., Margolis, T. P., Gomes, W. A., and Feldman, L. T. (1995). In vivo deletion analysis of the herpes simplex virus type 1 latency-associated transcript promoter. *J Virol* **69**, 2264–2270.

Dobson, A. T., Margolis, T. P., Sedarati, F., Stevens, J. G., and Feldman, L. T. (1990). A latent, nonpathogenic HSV-1-derived vector stably expresses beta-galactosidase in mouse neurons. *Neuron* **5**, 353–360.

Dobson, A. T., Sederati, F., Devi Rao, G., Flanagan, W. M., Farrell, M. J., Stevens, J. G. et al. (1989). Identification of the latency-associated transcript promoter by expression of rabbit beta-globin mRNA in mouse sensory nerve ganglia latently infected with a recombinant herpes simplex virus. *J Virol* **63**, 3844–3851.

Dull, T., Zufferey, R., Kelly, M., Mandel, R. J., Nguyen, M., Trono, D., and Naldini, L. (1998). A third-generation lentivirus vector with a conditional packaging system. *J Virol* **72**, 8463–8471.

During, M. J., Kaplitt, M. G., Stern, M. B., and Eidelberg, D. (2001). Subthalamic GAD gene transfer in Parkinson disease patients who are candidates for deep brain stimulation. *Hum Gene Ther* **12**, 1589–1591.

Einfeld, D. A., Schroeder, R., Roelvink, P. W., Lizonova, A., King, C. R., Kovesdi, I., and Wickham, T. J. (2001). Reducing the native tropism of adenovirus vectors requires removal of both CAR and integrin interactions. *J Virol* **75**, 11284–11291.

Emery, D. W., Yannaki, E., Tubb, J., and Stamatoyannopoulos, G. (2000). A chromatin insulator protects retrovirus vectors from chromosomal position effects. *Proc Natl Acad Sci U S A* **97**, 9150–9155.

Emi, N., Friedmann, T., and Yee, J. K. (1991). Pseudotype formation of murine leukemia virus with the G protein of vesicular stomatitis virus. *J Virol* **65**, 1202–1207.

Fields, B. N., Knipe, D. M., and Howley, P. M. (1996). *Fields virology*. Lippincott- Raven, Philadelphia.

Fink, J. K., Correll, P. H., Perry, L. K., Brady, R. O., and Karlsson, S. (1990). Correction of glucocerebrosidase deficiency after retroviral-mediated gene transfer into hematopoietic progenitor cells from patients with Gaucher disease. *Proc Natl Acad Sci U S A* **87**, 2334–2338.

Furlan, R., Poliani, P. L., Marconi, P. C., Bergami, A., Ruffini, F., Adorini, L. et al. (2001). Central nervous system gene therapy with interleukin-4 inhibits progression of ongoing relapsing-remitting autoimmune encephalomyelitis in Biozzi AB/H mice. *Gene Ther* **8**, 13–19.

Furling, D., Doucet, G., Langlois, M. A., Timchenko, L., Belanger, E., Cossette, L., and Puymirat, J. (2003). Viral vector producing antisense RNA restores myotonic dystrophy myoblast functions. *Gene Ther* **10**, 795–802.

Gill, S. S., Patel, N. K., Hotton, G. R., O'Sullivan, K., McCarter, R., Bunnage, M. et al. (2003). Direct brain infusion of glial cell line-derived neurotrophic factor in Parkinson disease. *Nat Med* **9**, 589–595.

Giraud, C., Winocour, E., and Berns, K. I. (1994). Site-specific integration by adeno-associated virus is directed by a cellular DNA sequence. *Proc Natl Acad Sci U S A* **91**, 10039–10043.

Girod, A., Ried, M., Wobus, C., Lahm, H., Leike, K., Kleinschmidt, J. et al. (1999). Genetic capsid modifications allow efficient re-targeting of adeno-associated virus type 2. *Nat Med* **5**, 1438.

Goins, W. F., Lee, K. A., Cavalcoli, J. D., O'Malley, M. E., DeKosky, S. T., Fink, D. J., and Glorioso, J. C. (1999). Herpes simplex virus type 1 vector-mediated expression of nerve growth factor protects dorsal root ganglion neurons from peroxide toxicity. *J Virol* **73**, 519–532.

Goins, W. F., Sternberg, L. R., Croen, K. D., Krause, P. R., Hendricks, R. L., Fink, D. J. et al. (1994). A novel latency-active promoter is contained within the herpes simplex virus type 1 UL flanking repeats. *J Virol* **68**, 2239–2252.

Goins, W. F., Yoshimura, N., Phelan, M. W., Yokoyama, T., Fraser, M. O., Ozawa, H. et al. (2001). Herpes simplex virus mediated nerve growth factor expression in bladder and afferent neurons: potential treatment for diabetic bladder dysfunction. *J Urol* **165**, 1748–1754.

Gordon, Y. J., Johnson, B., Romanowski, E., and Araullo Cruz, T. (1988). RNA complementary to herpes simplex virus type 1 ICP0 gene demonstrated in neurons of human trigeminal ganglia. *J Virol* **62**, 1832–1835.

Goss, J. R., Goins, W. F., Lacomis, D., Mata, M., Glorioso, J. C., and Fink, D. J. (2002a). Herpes simplex-mediated gene transfer of nerve growth factor protects against peripheral neuropathy in streptozotocin-induced diabetes in the mouse. *Diabetes* **51**, 2227–2232.

Goss, J. R., Harley, C. F., Mata, M., O'Malley, M. E., Goins, W. F., Hu, X. et al. (2002b). Herpes vector-mediated expression of proenkephalin reduces bone cancer pain. *Ann Neurol* **52**, 662–665.

Goss, J. R., Mata, M., Goins, W. F., Wu, H. H., Glorioso, J. C., and Fink, D. J. (2001). Antinociceptive effect of a genomic herpes simplex virus-based vector expressing human proenkephalin in rat dorsal root ganglion. *Gene Ther* **8**, 551–556.

Greelish, J. P., Su, L. T., Lankford, E. B., Burkman, J. M., Chen, H., Konig, S. K. et al. (1999). Stable restoration of the sarcoglycan complex in dystrophic muscle perfused with histamine and a recombinant adeno-associated viral vector. *Nat Med* **5**, 439–443.

Gregorevic, P., Blankinship, M. J., Allen, J. M., Crawford, R. W., Meuse, L., Miller, D. G. et al. (2004). Systemic delivery of genes to striated muscles using adeno-associated viral vectors. *Nat Med* **10**, 828–834.

Hacein-Bey-Abina, S., Le Deist, F., Carlier, F., Bouneaud, C., Hue, C., De Villartay, J. P. et al. (2002). Sustained correction of X-linked severe combined immunodeficiency by ex vivo gene therapy. *N Engl J Med* **346**, 1185–1193.

Hacein-Bey-Abina, S., von Kalle, C., Schmidt, M., Le Deist, F., Wulffraat, N., McIntyre, E. et al. (2003). A serious adverse event after successful gene therapy for X-linked severe combined immunodeficiency. *N Engl J Med* **348**, 255–256.

Hackett, N. R., Redmond, D. E., Sondhi, D., Giannaris, E. L., Vassallo, E., Stratton, J. et al. (2005). Safety of direct administration of AAV2(CU)hCLN2, a candidate treatment for the central nervous system manifestations of late infantile neuronal ceroid lipofuscinosis, to the brain of rats and nonhuman primates. *Hum Gene Ther* **16**, 1484–1503.

Haj-Ahmad, Y. and Graham, F. L. (1986). Development of a helper-independent human adenovirus vector and its use in the transfer of the herpes simplex virus thymidine kinase gene. *J Virol* **57**, 267–274.

Hao, S., Mata, M., Glorioso, J. C., and Fink, D. J. (2006). HSV-mediated expression of interleukin-4 in dorsal root ganglion neurons reduces neuropathic pain. *Mol Pain* **2**, 6.

Hao, S., Mata, M., Goins, W., Glorioso, J. C., and Fink, D. J. (2003a). Transgene-mediated enkephalin release enhances the effect of morphine and evades tolerance to produce a sustained antiallodynic effect in neuropathic pain. *Pain* **102**, 135–142.

Hao, S., Mata, M., Wolfe, D., Huang, S., Glorioso, J. C., and Fink, D. J. (2003b). HSV-mediated gene transfer of the glial cell-derived neurotrophic factor provides an antiallodynic effect on neuropathic pain. *Mol Ther* **8**, 367–375.

Hao, S., Mata, M., Wolfe, D., Huang, S., Glorioso, J. C., and Fink, D. J. (2005). Gene transfer of glutamic acid decarboxylase reduces neuropathic pain. *Ann Neurol* **57**, 914–918.

Hofmann, A., Nolan, G. P., and Blau, H. M. (1996). Rapid retroviral delivery of tetracycline-inducible genes in a single autoregulatory cassette. *Proc Natl Acad Sci U S A* **93**, 5185–5190.

Hong, C. S., Goins, W. F., Goss, J. R., Burton, E. A., and Glorioso, J. C. (2006). Herpes simplex virus RNAi and neprilysin gene transfer vectors reduce accumulation of Alzheimer's disease-related amyloid-beta peptide in vivo. *Gene Ther.*

Hoppe, U. C., Marban, E., and Johns, D. C. (2000). Adenovirus-mediated inducible gene expression in vivo by a hybrid ecdysone receptor. *Mol Ther* **1**, 159–164.

Horger, B. A., Nishimura, M. C., Armanini, M. P., Wang, L. C., Poulsen, K. T., Rosenblad, C. et al. (1998). Neurturin exerts potent actions on survival and function of midbrain dopaminergic neurons. *J Neurosci* **18**, 4929–4937.

Iida, A., Chen, S. T., Friedmann, T., and Yee, J. K. (1996). Inducible gene expression by retrovirus-mediated transfer of a modified tetracycline-regulated system. *J Virol* **70**, 6054–6059.

Im, S. A., Gomez Manzano, C., Fueyo, J., Liu, T. J., Ke, L. D., Kim, J. S. et al. (1999). Antiangiogenesis treatment for gliomas: Transfer of antisense-vascular endothelial growth factor inhibits tumor growth in vivo. *Cancer Res* **59**, 895–900.

Kanno, H., Hattori, S., Sato, H., Murata, H., Huang, F. H., Hayashi, A. et al. (1999). Experimental gene therapy against subcutaneously implanted glioma with a herpes simplex virus-defective vector expressing interferon-gamma. *Cancer Gene Ther* **6**, 147–154.

Kantoff, P. W., Kohn, D. B., Mitsuya, H., Armentano, D., Sieberg, M., Zwiebel, J. A. et al. (1986). Correction of adenosine deaminase deficiency in cultured human T and B cells by retrovirus-mediated gene transfer. *Proc Natl Acad Sci U S A* **83**, 6563–6567.

Kashentseva, E. A., Seki, T., Curiel, D. T., and Dmitriev, I. P. (2002). Adenovirus targeting to c-erbB-2 oncoprotein by single-chain antibody fused to trimeric form of adenovirus receptor ectodomain. *Cancer Res* **62**, 609–616.

Kim, V. N., Mitrophanous, K., Kingsman, S. M., and Kingsman, A. J. (1998). Minimal requirement for a lentivirus vector based on human immunodeficiency virus type 1. *J Virol* **72**, 811–816.

Kochanek, S., Schiedner, G., and Volpers, C. (2001). High-capacity "gutless" adenoviral vectors. *Curr Opin Mol Ther* **3**, 454–463.

Kordower, J. H., Emborg, M. E., Bloch, J., Ma, S. Y., Chu, Y., Leventhal, L. et al. (2000). Neurodegeneration prevented by lentiviral vector delivery of GDNF in primate models of Parkinson's disease. *Science* **290**, 767–773.

Krisky, D. M., Marconi, P. C., Oligino, T., Rouse, R. J., Fink, D. J., and Glorioso, J. C. (1997). Rapid method for construction of recombinant HSV gene transfer vectors. *Gene Ther* **4**, 1120–1125.

Krisky, D. M., Marconi, P. C., Oligino, T. J., Rouse, R. J., Fink, D. J., Cohen, J. B. et al. (1998a). Development of herpes simplex virus replication-defective multigene vectors for combination gene therapy applications. *Gene Ther* **5**, 1517–1530.

Krisky, D. M., Wolfe, D., Goins, W. F., Marconi, P. C., Ramakrishnan, R., Mata, M. et al. (1998b). Deletion of multiple immediate-early genes from herpes simplex virus reduces cytotoxicity and permits long-term gene expression in neurons. *Gene Ther* **5**, 1593–1603.

Lachmann, R. H. and Efstathiou, S. (1997). Utilization of the herpes simplex virus type 1 latency-associated regulatory region to drive stable reporter gene expression in the nervous system. *J Virol* **71**, 3197–3207.

Lang, A. E., Gill, S., Patel, N. K., Lozano, A., Nutt, J. G., Penn, R. et al. (2006). Randomized controlled trial of intraputamenal glial cell line-derived neurotrophic factor infusion in Parkinson disease. *Ann Neurol* **59**, 459–466.

Lang, F. F., Yung, W. K., Sawaya, R., and Tofilon, P. J. (1999). Adenovirus-mediated p53 gene therapy for human gliomas. *Neurosurgery* **45**, 1093–1104.

Langlois, M. A., Boniface, C., Wang, G., Alluin, J., Salvaterra, P. M., Puymirat, J. et al. (2005). Cytoplasmic and nuclear retained DMPK mRNAs are targets for RNA interference in myotonic dystrophy cells. *J Biol Chem* **280**, 16949–16954.

Langlois, M. A., Lee, N. S., Rossi, J. J., and Puymirat, J. (2003). Hammerhead ribozyme-mediated destruction of nuclear foci in myotonic dystrophy myoblasts. *Mol Ther* **7**, 670–680.

Le Gal La Salle, G., Robert, J. J., Berrard, S., Ridoux, V., Stratford-Perricaudet, L. D., Perricaudet, M., and Mallet, J. (1993). An adenovirus vector for gene transfer into neurons and glia in the brain. *Science* **259**, 988–990.

Ledley, F. D., Grenett, H. E., McGinnis-Shelnutt, M., and Woo, S. L. (1986). Retroviral-mediated gene transfer of human phenylalanine hydroxylase into NIH 3T3 and hepatoma cells. *Proc Natl Acad Sci U S A* **83**, 409–413.

Leone, P., Janson, C. G., Bilaniuk, L., Wang, Z., Sorgi, F., Huang, L. et al. (2000). Aspartoacylase gene transfer to the mammalian central nervous system with therapeutic implications for Canavan disease. *Ann Neurol* **48**, 27–38.

Li, X., Mukai, T., Young, D., Frankel, S., Law, P., and Wong Staal, F. (1998). Transduction of CD34+ cells by a vesicular stomach virus protein G (VSV-G) pseudotyped HIV-1 vector. Stable gene expression in progeny cells, including dendritic cells. *J Hum Virol* **1**, 346–352.

Linden, R. M., Ward, P., Giraud, C., Winocour, E., and Berns, K. I. (1996). Site-specific integration by adeno-associated virus. *Proc Natl Acad Sci U S A* **93**, 11288–11294.

Liu, J., Wolfe, D., Hao, S., Huang, S., Glorioso, J. C., Mata, M., and Fink, D. J. (2004). Peripherally delivered glutamic acid decarboxylase gene therapy for spinal cord injury pain. *Mol Ther* **10**, 57–66.

Loeser, J., Butler, S., Chapman, C., and Turk, D. (2001). *Bonica's Management of Pain*. Lippincott Williams & Wilkins, Philadelphia.

Loiler, S. A., Conlon, T. J., Song, S., Tang, Q., Warrington, K. H., Agarwal, A. et al. (2003). Targeting recombinant adeno-associated virus vectors to enhance gene transfer to pancreatic islets and liver. *Gene Ther* **10**, 1551–1558.

Machein, M. R., Kullmer, J., Fiebich, B. L., Plate, K. H., and Warnke, P. C. (1999). Vascular endothelial growth factor expression, vascular volume, and capillary permeability in human brain tumors. *Neurosurgery* **44**, 732–740.

Mack, C. A., Song, W. R., Carpenter, H., Wickham, T. J., Kovesdi, I., Harvey, B. G. et al. (1997). Circumvention of anti-adenovirus neutralizing immunity by administration of an adenoviral vector of an alternate serotype. *Hum Gene Ther* **8**, 99–109.

Marconi, P., Simonato, M., Zucchini, S., Bregola, G., Argnani, R., Krisky, D. et al. (1999). Replication-defective herpes simplex virus vectors for neurotrophic factor gene transfer in vitro and in vivo. *Gene Ther* **6**, 904–912.

Marconi, P., Tamura, M., Moriuchi, S., Krisky, D. M., Niranjan, A., Goins, W. F. et al. (2000). Connexin 43-enhanced suicide gene therapy using herpesviral vectors. *Mol Ther* **1**, 71–81.

Markowitz, D., Goff, S., and Bank, A. (1988). Construction and use of a safe and efficient amphotropic packaging cell line. *Virology* **167**, 400–406.

Mastrangeli, A., Harvey, B. G., Yao, J., Wolff, G., Kovesdi, I., Crystal, R. G., and Falck-Pedersen, E. (1996). "Sero-switch" adenovirus-mediated in vivo gene transfer: Circumvention of anti-adenovirus humoral immune defenses against repeat adenovirus vector administration by changing the adenovirus serotype. *Hum Gene Ther* **7**, 79–87.

Mazarakis, N. D., Azzouz, M., Rohll, J. B., Ellard, F. M., Wilkes, F. J., Olsen, A. L. et al. (2001). Rabies virus glycoprotein pseudotyping of lentiviral vectors enables retrograde axonal transport and access to the nervous system after peripheral delivery. *Hum Mol Genet* **10**, 2109–2121.

McPhee, S. W., Janson, C. G., Li, C., Samulski, R. J., Camp, A. S., Francis, J. et al. (2006). Immune responses to AAV in a phase I study for Canavan disease. *J Gene Med* **8**, 577–588.

Mercier, S., Gahery-Segard, H., Monteil, M., Lengagne, R., Guillet, J. G., Eloit, M., and Denesvre, C. (2002). Distinct roles of adenovirus vector-transduced dendritic cells, myoblasts, and endothelial cells in mediating an immune response against a transgene product. *J Virol* **76**, 2899–2911.

Miao, C. H., Ohashi, K., Patijn, G. A., Meuse, L., Ye, X., Thompson, A. R., and Kay, M. A. (2000). Inclusion of the hepatic locus control region, an intron, and untranslated region increases and stabilizes hepatic factor IX gene expression in vivo but not in vitro. *Mol Ther* **1**, 522–532.

Miyanohara, A., Sharkey, M. F., Witztum, J. L., Steinberg, D., and Friedmann, T. (1988). Efficient expression of retroviral vector-transduced human low density lipoprotein (LDL) receptor in LDL receptor-deficient rabbit fibroblasts in vitro. *Proc Natl Acad Sci U S A* **85**, 6538–6542.

Miyoshi, H., Blomer, U., Takahashi, M., Gage, F. H., and Verma, I. M. (1998). Development of a self-inactivating lentivirus vector. *J Virol* **72**, 8150–8157.

Miyoshi, H., Smith, K. A., Mosier, D. E., Verma, I. M., and Torbett, B. E. (1999). Transduction of human CD34+ cells that mediate long-term engraftment of NOD/SCID mice by HIV vectors. *Science* **283**, 682–686.

Miyoshi, H., Takahashi, M., Gage, F. H., and Verma, I. M. (1997). Stable and efficient gene transfer into the retina using an HIV-based lentiviral vector. *Proc Natl Acad Sci U S A* **94**, 10319–10323.

Mizuguchi, H., Koizumi, N., Hosono, T., Utoguchi, N., Watanabe, Y., Kay, M. A., and Hayakawa, T. (2001). A simplified system for constructing recombinant adenoviral vectors containing heterologous peptides in the HI loop of their fiber knob. *Gene Ther* **8**, 730–735.

Molinier-Frenkel, V., Gahery-Segard, H., Mehtali, M., Le Boulaire, C., Ribault, S., Boulanger, P. et al. (2000). Immune response to recombinant adenovirus in humans: Capsid components from viral input are targets for vector-specific cytotoxic T lymphocytes. *J Virol* **74**, 7678–7682.

Moriuchi, S., Oligino, T., Krisky, D., Marconi, P., Fink, D., Cohen, J., and Glorioso, J. C. (1998). Enhanced tumor cell killing in the presence of ganciclovir by herpes simplex virus type 1 vector-directed coexpression of human tumor necrosis factor-alpha and herpes simplex virus thymidine kinase. *Cancer Res* **58**, 5731–5737.

Morral, N., O'Neal, W., Rice, K., Leland, M., Kaplan, J., Piedra, P. A. et al. (1999). Administration of helper-dependent adenoviral vectors and sequential delivery of different vector serotype for long-term liver-directed gene transfer in baboons. *Proc Natl Acad Sci U S A* **96**, 12816–12821.

Muller, O. J., Kaul, F., Weitzman, M. D., Pasqualini, R., Arap, W., Kleinschmidt, J. A., and Trepel, M. (2003). Random peptide libraries displayed on adeno-associated virus to select for targeted gene therapy vectors. *Nat Biotechnol* **21**, 1040–1046.

Naldini, L., Blomer, U., Gallay, P., Ory, D., Mulligan, R., Gage, F. H. et al. (1996). In vivo gene delivery and stable transduction of nondividing cells by a lentiviral vector. *Science* **272**, 263–267.

Niranjan, A., Moriuchi, S., Lunsford, L. D., Kondziolka, D., Flickinger, J. C., Fellows, W. et al. (2000). Effective treatment of experimental glioblastoma by HSV vector-mediated TNF alpha and HSV-tk gene transfer in combination with radiosurgery and ganciclovir administration. *Mol Ther* **2**, 114–120.

No, D., Yao, T. P., and Evans, R. M. (1996). Ecdysone-inducible gene expression in mammalian cells and transgenic mice. *Proc Natl Acad Sci U S A* **93**, 3346–3351.

Oligino, T., Poliani, P. L., Wang, Y., Tsai, S. Y., O'Malley, B. W., Fink, D. J., and Glorioso, J. C. (1998). Drug inducible transgene expression in brain using a herpes simplex virus vector. *Gene Ther* **5**, 491–496.

Palmer, J. A., Branston, R. H., Lilley, C. E., Robinson, M. J., Groutsi, F., Smith, J. et al. (2000). Development and optimization of herpes simplex virus vectors for multiple long-term gene delivery to the peripheral nervous system. *J Virol* **74**, 5604–5618.

Palmer, T. D., Hock, R. A., Osborne, W. R., and Miller, A. D. (1987). Efficient retrovirus-mediated transfer and expression of a human adenosine deaminase gene in diploid skin fibroblasts from an adenosine deaminase-deficient human. *Proc Natl Acad Sci U S A* **84**, 1055–1059.

Parker, J. N., Gillespie, G. Y., Love, C. E., Randall, S., Whitley, R. J., and Markert, J. M. (2000). From the cover: Engineered herpes simplex virus expressing IL-12 in the treatment of experimental murine brain tumors. *Proc Natl Acad Sci U S A* **97**, 2208–2213.

Passini, M. A., Dodge, J. C., Bu, J., Yang, W., Zhao, Q., Sondhi, D. et al. (2006). Intracranial delivery of CLN2 reduces brain pathology in a mouse model of classical late infantile neuronal ceroid lipofuscinosis. *J Neurosci* **26**, 1334–1342.

Paulus, W., Baur, I., Boyce, F. M., Breakefield, X. O., and Reeves, S. A. (1996). Self-contained, tetracycline-regulated retroviral vector system for gene delivery to mammalian cells. *J Virol* **70**, 62–67.

Pawliuk, R., Westerman, K. A., Fabry, M. E., Payen, E., Tighe, R., Bouhassira, E. E. et al. (2001). Correction of sickle cell disease in transgenic mouse models by gene therapy. *Science* **294**, 2368–2371.

Peng, X. M., Zhou, Z. G., Glorioso, J. C., Fink, D. J., and Mata, M. (2006). Tumor necrosis factor-alpha contributes to below-level neuropathic pain after spinal cord injury. *Ann Neurol* **59**, 843–851.

Poliani, P. L., Brok, H., Furlan, R., Ruffini, F., Bergami, A., Desina, G. et al. (2001). Delivery to the central nervous system of a nonreplicative herpes simplex type 1 vector engineered with the interleukin 4 gene protects rhesus monkeys from hyperacute autoimmune encephalomyelitis. *Hum Gene Ther* **12**, 905–920.

Ponnazhagan, S., Erikson, D., Kearns, W. G., Zhou, S. Z., Nahreini, P., Wang, X. S., and Srivastava, A. (1997). Lack of site-specific integration of the recombinant adeno-associated virus 2 genomes in human cells. *Hum Gene Ther* **8**, 275–284.

Puskovic, V., Wolfe, D., Goss, J., Huang, S., Mata, M., Glorioso, J. C., and Fink, D. J. (2004). Prolonged biologically active transgene expression driven by HSV LAP2 in brain in vivo. *Mol Ther* **10**, 67–75.

Ragot, T., Vincent, N., Chafey, P., Vigne, E., Gilgenkrantz, H., Couton, D. et al. (1993). Efficient adenovirus-mediated transfer of a human minidystrophin gene to skeletal muscle of mdx mice. *Nature* **361**, 647–650.

Raper, S. E., Yudkoff, M., Chirmule, N., Gao, G. P., Nunes, F., Haskal, Z. J. et al. (2002). A pilot study of in vivo liver-directed gene transfer with an adenoviral vector in partial ornithine transcarbamylase deficiency. *Hum Gene Ther* **13**, 163–175.

Reddy, P. S., Sakhuja, K., Ganesh, S., Yang, L., Kayda, D., Brann, T. et al. (2002). Sustained human factor VIII expression in hemophilia A mice following systemic delivery of a gutless adenoviral vector. *Mol Ther* **5**, 63–73.

Reiser, J., Harmison, G., Kluepfel-Stahl, S., Brady, R. O., Karlsson, S., and Schubert, M. (1996). Transduction of nondividing cells using pseudotyped defective high-titer HIV type 1 particles. *Proc Natl Acad Sci U S A* **93**, 15266–15271.

Rivera, V. M., Clackson, T., Natesan, S., Pollock, R., Amara, J. F., Keenan, T. et al. (1996). A humanized system for pharmacologic control of gene expression. *Nat Med* **2**, 1028–1032.

Rock, D. L., Nesburn, A. B., Ghiasi, H., Ong, J., Lewis, T. L., Lokensgard, J. R., and Wechsler, S. L. (1987). Detection of latency-related viral RNAs in trigeminal ganglia of rabbits latently infected with herpes simplex virus type 1. *J Virol* **61**, 3820–3826.

Rosen, C. A., Sodroski, J. G., and Haseltine, W. A. (1985). The location of cis-acting regulatory sequences in the human T cell lymphotropic virus type III (HTLV-III/LAV) long terminal repeat. *Cell* **41**, 813–823.

Rosenfeld, M. A., Siegfried, W., Yoshimura, K., Yoneyama, K., Fukayama, M., Stier, L. E. et al. (1991). Adenovirus-mediated transfer of a recombinant alpha 1-antitrypsin gene to the lung epithelium in vivo. *Science* **252**, 431–434.

Rosenfeld, M. A., Yoshimura, K., Trapnell, B. C., Yoneyama, K., Rosenthal, E. R., Dalemans, W. et al. (1992). In vivo transfer of the human cystic fibrosis transmembrane conductance regulator gene to the airway epithelium. *Cell* **68**, 143–155.

Ruffini, F., Furlan, R., Poliani, P. L., Brambilla, E., Marconi, P. C., Bergami, A. et al. (2001). Fibroblast growth factor-II gene therapy reverts the clinical course and the pathological signs of chronic experimental autoimmune encephalomyelitis in C57BL/6 mice. *Gene Ther* **8**, 1207–1213.

Sacks, W. R., Greene, C. C., Aschman, D. P., and Schaffer, P. A. (1985). Herpes simplex virus type 1 ICP27 is an essential regulatory protein. *J Virol* **55**, 796–805.

Sailaja, G., HogenEsch, H., North, A., Hays, J., and Mittal, S. K. (2002). Encapsulation of recombinant adenovirus into alginate microspheres circumvents vector-specific immune response. *Gene Ther* **9**, 1722–1729.

Sakhuja, K., Reddy, P. S., Ganesh, S., Cantaniag, F., Pattison, S., Limbach, P. et al. (2003). Optimization of the generation and propagation of gutless adenoviral vectors. *Hum Gene Ther* **14**, 243–254.

Samulski, R. J., Zhu, X., Xiao, X., Brook, J. D., Housman, D. E., Epstein, N., and Hunter, L. A. (1991). Targeted integration of adeno-associated virus (AAV) into human chromosome 19. *Embo J* **10**, 3941–3950.

Scherr, M., Battmer, K., Blomer, U., Schiedlmeier, B., Ganser, A., Grez, M., and Eder, M. (2002). Lentiviral gene transfer into peripheral blood-derived CD34+ NOD/SCID-repopulating cells. *Blood* **99**, 709–712.

Schroder, A. R., Shinn, P., Chen, H., Berry, C., Ecker, J. R., and Bushman, F. (2002). HIV-1 integration in the human genome favors active genes and local hotspots. *Cell* **110**, 521–529.

Scott, W. G. (1998). RNA catalysis. *Curr Opin Struct Biol* **8**, 720–726.

Seshidhar Reddy, P., Ganesh, S., Limbach, M. P., Brann, T., Pinkstaff, A., Kaloss, M. et al. (2003). Development of adenovirus serotype 35 as a gene transfer vector. *Virology* **311**, 384–393.

Shi, W. and Bartlett, J. S. (2003). RGD inclusion in VP3 provides adeno-associated virus type 2 (AAV2)-based vectors with a heparan sulfate-independent cell entry mechanism. *Mol Ther* **7**, 515–525.

Smith, C., Lachmann, R. H., and Efstathiou, S. (2000). Expression from the herpes simplex virus type 1 latency-associated promoter in the murine central nervous system. *J Gen Virol* **3**, 649–662.

Snyder, R. O. and Flotte, T. R. (2002). Production of clinical-grade recombinant adeno-associated virus vectors. *Curr Opin Biotechnol* **13**, 418–423.

Spivack, J. G. and Fraser, N. W. (1987). Detection of herpes simplex virus type 1 transcripts during latent infection in mice. *J Virol* **61**, 3841–3847.

Starcich, B., Ratner, L., Josephs, S. F., Okamoto, T., Gallo, R. C., and Wong-Staal, F. (1985). Characterization of long terminal repeat sequences of HTLV-III. *Science* **227**, 538–540.

Stedman, H., Wilson, J. M., Finke, R., Kleckner, A. L., and Mendell, J. (2000). Phase I clinical trial utilizing gene therapy for limb girdle muscular dystrophy: Alpha-, beta-, gamma-, or delta-sarcoglycan gene delivered with intramuscular instillations of adeno-associated vectors. *Hum Gene Ther* **11**, 777–790.

Stevens, J. G., Wagner, E. K., Devi Rao, G. B., Cook, M. L., and Feldman, L. T. (1987). RNA complementary to a herpesvirus alpha gene mRNA is prominent in latently infected neurons. *Science* **235**, 1056–1059.

Su, L. T., Gopal, K., Wang, Z., Yin, X., Nelson, A., Kozyak, B. W. et al. (2005). Uniform scale-independent gene transfer to striated muscle after transvenular extravasation of vector. *Circulation* **112**, 1780–1788.

Suzuki, M., Singh, R. N., and Crystal, R. G. (1996). Regulatable promoters for use in gene therapy applications: modification of the 5-flanking region of the CFTR gene with multiple cAMP response elements to support basal, low-level gene expression that can be upregulated by exogenous agents that raise intracellular levels of cAMP. *Hum Gene Ther* **7**, 1883–1893.

Tuszynski, M. H., Thal, L., Pay, M., Salmon, D. P., U, H. S., Bakay, R. et al. (2005). A phase 1 clinical trial of nerve growth factor gene therapy for Alzheimer disease. *Nat Med* **11**, 551–555.

Wade-Martins, R., Smith, E. R., Tyminski, E., Chiocca, E. A., and Saeki, Y. (2001). An infectious transfer and expression system for genomic DNA loci in human and mouse cells. *Nat Biotechnol* **19**, 1067–1070.

Wang, L., Muramatsu, S., Lu, Y., Ikeguchi, K., Fujimoto, K., Okada, T. et al. (2002). Delayed delivery of AAV-GDNF prevents nigral neurodegeneration and promotes functional recovery in a rat model of Parkinson's disease. *Gene Ther* **9**, 381–389.

Wesseling, J. G., Bosma, P. J., Krasnykh, V., Kashentseva, E. A., Blackwell, J. L., Reynolds, P. N. et al. (2001). Improved gene transfer efficiency to primary and established human pancreatic carcinoma target cells via epidermal growth factor receptor and integrin-targeted adenoviral vectors. *Gene Ther* **8**, 969–976.

Woolf, C. J. and Salter, M. W. (2000). Neuronal plasticity: Increasing the gain in pain. *Science* **288**, 1765–1769.

Wu, H., Seki, T., Dmitriev, I., Uil, T., Kashentseva, E., Han, T., and Curiel, D. T. (2002). Double modification of adenovirus fiber with RGD and polylysine motifs improves coxsackievirus-adenovirus receptor-independent gene transfer efficiency. *Hum Gene Ther* **13**, 1647–1653.

Xia, H., Mao, Q., Eliason, S. L., Harper, S. Q., Martins, I. H., Orr, H. T. et al. (2004). RNAi suppresses polyglutamine-induced neurodegeneration in a model of spinocerebellar ataxia. *Nat Med* **10**, 816–820.

Xiao, X., Li, J., and Samulski, R. J. (1998). Production of high-titer recombinant adeno-associated virus vectors in the absence of helper adenovirus. *J Virol* **72**, 2224–2232.

Xie, Q., Bu, W., Bhatia, S., Hare, J., Somasundaram, T., Azzi, A., and Chapman, M. S. (2002). The atomic structure of adeno-associated virus (AAV-2), a vector for human gene therapy. *Proc Natl Acad Sci U S A* **99**, 10405–10410.

Ye, X., Rivera, V. M., Zoltick, P., Cerasoli, F., Jr., Schnell, M. A., Gao, G. et al. (1999). Regulated delivery of therapeutic proteins after in vivo somatic cell gene transfer. *Science* **283**, 88–91.

Zhang, X. Y., La Russa, V. F., Bao, L., Kolls, J., Schwarzenberger, P., and Reiser, J. (2002). Lentiviral vectors for sustained transgene expression in human bone marrow-derived stromal cells. *Mol Ther* **5**, 555–565.

Zielske, S. P. and Gerson, S. L. (2002). Lentiviral transduction of P140K MGMT into human CD34(+) hematopoietic progenitors at low multiplicity of infection confers significant resistance to BG/BCNU and allows selection in vitro. *Mol Ther* **5**, 381–387.

Zou, L., Zhou, H., Pastore, L., and Yang, K. (2000). Prolonged transgene expression mediated by a helper-dependent adenoviral vector (hdAd) in the central nervous system. *Mol Ther* **2**, 105–113.

Zufferey, R., Nagy, D., Mandel, R. J., Naldini, L., and Trono, D. (1997). Multiply attenuated lentiviral vector achieves efficient gene delivery in vivo. *Nat Biotechnol* **15**, 871–875.

Zwaagstra, J., Ghiasi, H., Nesburn, A. B., and Wechsler, S. L. (1989). In vitro promoter activity associated with the latency-associated transcript gene of herpes simplex virus type 1. *J Gen Virol* **70**, 2163–2169.

Zwaagstra, J. C., Ghiasi, H., Nesburn, A. B., and Wechsler, S. L. (1991). Identification of a major regulatory sequence in the latency associated transcript (LAT) promoter of herpes simplex virus type 1 (HSV-1). *Virology* **182**, 287–297.

Zwaagstra, J. C., Ghiasi, H., Slanina, S. M., Nesburn, A. B., Wheatley, S. C., Lillycrop, K. et al. (1990). Activity of herpes simplex virus type 1 latency-associated transcript (LAT) promoter in neuron-derived cells: evidence for neuron specificity and for a large LAT transcript. *J Virol* **64**, 5019–5028.

Zwiebel, J. A., Freeman, S. M., Kantoff, P. W., Cornetta, K., Ryan, U. S., and Anderson, W. F. (1989). High-level recombinant gene expression in rabbit endothelial cells transduced by retroviral vectors. *Science* **243**, 220–222.

8

Programmed Cell Death and Its Role in Neurological Disease

D. E. Bredesen, R. V. Rao, and P. Mehlen

You have conquered, and I yield. Yet, henceforward, art thou also dead—dead to the World, to Heaven, and to Hope! In me didst thou exist—and, in my death, see by this image, which is thine own, how utterly thou hast murdered thyself.

—Edgar Allan Poe

The trouble with quotes about death is that 99.999 percent of them are made by people who are still alive.

—Joshua Bruns

I. Introduction: Neurologists and Cell Death

As neurologists, many of the patients we see suffer from diseases that feature an abnormality of cell death of one sort or another; for example, developmental and neoplastic disorders of the nervous system feature dysregulation of the intrinsic cellular programs that mediate cell death. Furthermore, there is increasing evidence to suggest that such dysregulation may also occur in neurodegenerative, infectious, traumatic, ischemic, metabolic, and demyelinating disorders. These findings suggest that targeting the central biochemical controls of cell survival and death may represent a novel therapeutic approach, especially if combined with other therapeutic targets. In addition, recent results from stem cell studies suggest that the fate of neural stem cells may also play an important role in disease outcomes, and therefore cell death apparently plays a central role in many neurological diseases, and potentially in their prevention and treatment.

Classical studies of neuronal survival emphasized the status of external factors such as glucose availability, pH, and the partial pressure of oxygen; however, though these are clearly critical determinants, research over the past few decades has revealed a more active—and more plastic—role for the cell in its own decision to survive or die than was previously appreciated. In a complementary fashion, studies of the internal suicide programs of neural cells have offered new potential targets for therapeutic development.

Initial comparisons of the intrinsic suicide program in genetically tractable organisms such as the nematode *C. elegans* failed to disclose obvious relationships to genes associated with human neurodegenerative diseases (e.g., presenilin-1 does not bear an obvious relationship to the major cell death genes ced-3, ced-4, or ced-9 in *C. elegans*), but more recent studies have argued that such a relationship might indeed exist; for example, the mammalian homologues of ced-3 comprise a family of cell death proteases, the caspases, and mutation of a single caspase cleavage site in huntingtin blocks the development of the Huntington's phenotype in transgenic mice (Graham et al., 2006). A detailed understanding of the interrelationship between fundamental cell death programs and neurological disease states is still evolving, and it promises to offer novel approaches to the treatment of these diseases.

II. Cell Death: History and Classification

It has been 100 years since the first description of developmental neuronal cell death (Studnicka, 1905), and over 50 years since Levi-Montalcini showed that such physiological cell death is inhibited by soluble factors such as nerve growth factor (Levi-Montalcini, 1966). In 1964, Richard Lockshin and his colleagues introduced the term *programmed cell death* (pcd) to describe the apparently predetermined pattern by which specific cells die during insect development (Lockshin & Williams, 1964). In 1966, it was shown that this process requires protein synthesis, at least in some cases (Teta, 1966), arguing that it is the result of an active cellular suicide process. Then in 1972, John Kerr and his colleagues coined the term *apoptosis* to describe a morphologically relatively uniform set of cell deaths that occurs in many different situations, from development to insult response to cell turnover (Kerr et al., 1972).

Apoptosis has been studied extensively, with over 100,000 papers published on the subject (www.pubmed.gov). Although pcd has often been equated with apoptosis, it has become increasingly clear that nonapoptotic forms of pcd also exist (Clarke, 1990; Cunningham, 1982; Dal Canto & Gurney, 1994; Majno & Joris, 1995; Oppenheim, 1985, 1991; Pilar & Landmesser, 1976; Schwartz, 1991; Schweichel, 1972; Schweichel & Merker, 1973; Sperandio et al., 2000; Turmaine et al., 2000). For example, certain developmental cell deaths, such as "autophagic" cell death (Clarke, 1990; Lockshin & Williams, 1964; Schwartz, 1991; Schweichel, 1972; Schweichel & Merker, 1973) and "cytoplasmic" cell death (Clarke, 1990; Cunningham, 1982; Oppenheim, 1985, 1991; Pilar & Landmesser, 1976; Schweichel & Merker, 1973), do not resemble apoptosis. Furthermore, neurodegenerative diseases such as Huntington's disease and amyotrophic lateral sclerosis demonstrate neuronal cell death that does not fulfill the criteria for apoptosis (Dal Canto & Gurney, 1994; Turmaine et al., 2000). Ischemia-induced cell death may also display a nonapoptotic morphology, referred to as "oncosis" (Majno & Joris, 1995).

How many different mammalian cell death programs can be discerned, and what is their interrelationship? Morphological classifications have been proposed, but for purposes of both disease insight and therapeutic intervention, it would be preferable to construct a mechanistic taxonomy of all cell death programs, with special attention to their specific inhibitors and activators. However, the data required for such a construct are far from complete, and so the current classification will undoubtedly be revised repeatedly over time. Nonetheless, it is informative to consider, in light of currently available data, how many programs of cell death can be distinguished mechanistically.

Cell death has been divided into two general types: programmed cell death (pcd), in which the cell plays an active role; and passive (necrotic) cell death. A semantic issue has arisen with the demonstration that some forms of nonapoptotic cell death that had been labeled necrotic, and thus assumed in the past to be passive, have nonetheless turned out to be programmatic in nature; therefore, some authors have referred to these as "necrosis-like" (Vande Velde et al., 2000), whereas others have used the term "programmed necrosis" (Niquet et al., 2005; Zong & Thompson, 2006). Based on the traditional view that some term must be reserved for passive, nonprogrammatic cell death, and that necrosis has been the term historically applied to this form of cell death, the term "programmed necrosis" is an oxymoron. However, based on another characteristic attributed to necrosis—breach of the plasma membrane with consequent initiation of an inflammatory response to spilled cellular contents—"programmed necrosis" is indeed an appropriate term. Reserving the term "necrosis" for nonprogrammatic pcd, however, suggests that such programmatic cell deaths with necrotic morphology and other characteristics should be referred to as "necrosis-like," and as biochemical data accumulate for each form of pcd, it should become clear which paradigms induce necrosis-like pcd and which lead to passive, nonprogrammatic (i.e., necrotic) cell death.

III. Current Status of Programmed Cell Death Studies

A. Apoptosis

Apoptosis (from the Greek, "falling away"), which has also been referred to as nuclear or type I pcd, is far and away the most well-characterized type of pcd (see Figure 8.1). Morphologically, cells typically round up, form blebs, undergo zeiosis (an appearance of boiling), chromatin condensation, nuclear fragmentation, and the budding off of apoptotic bodies. Phosphatidylserine, normally placed asymmetrically such that it faces internally rather than externally on the plasma membrane (due to a flipase that flips

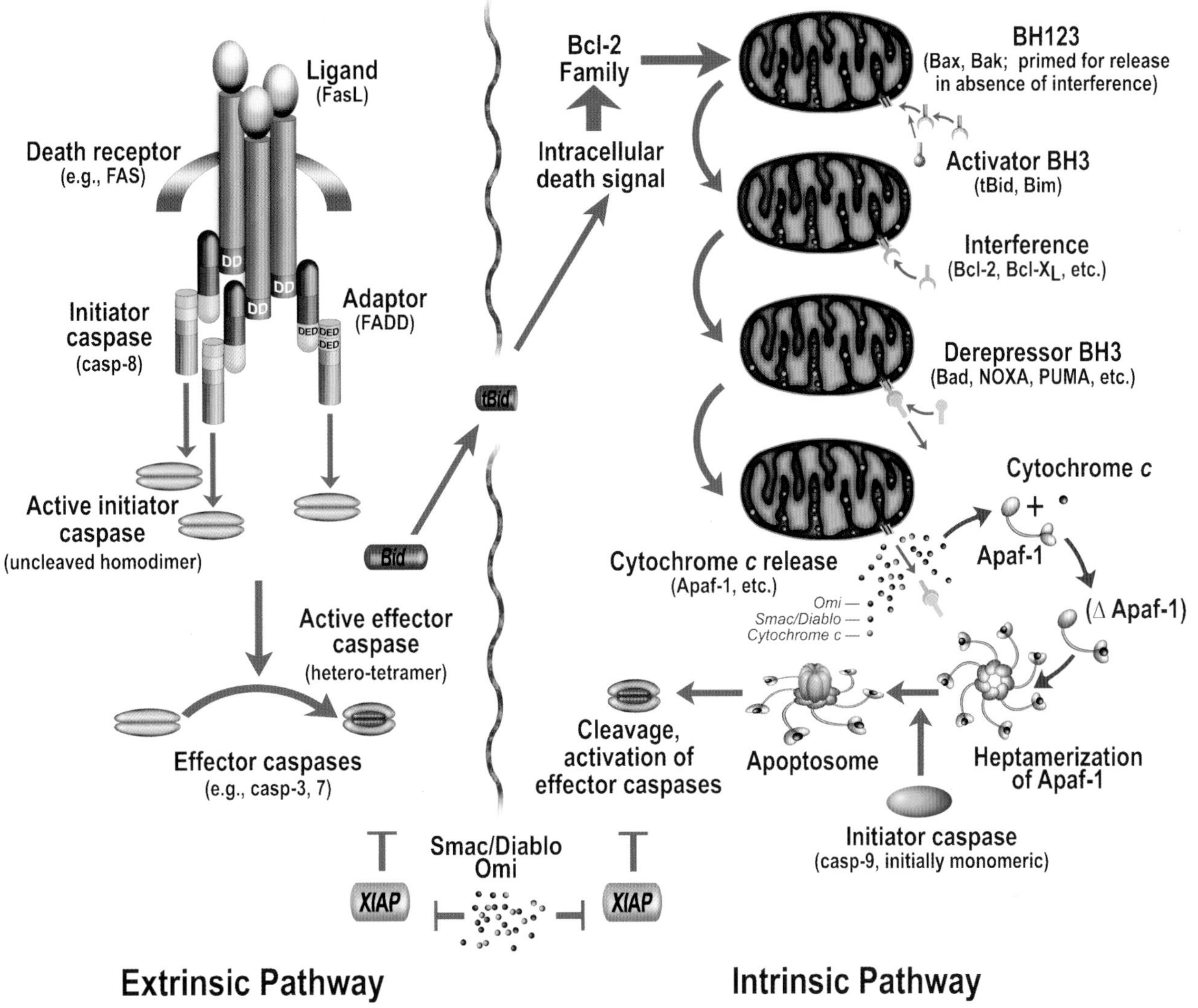

Figure 8.1 Intrinsic and extrinsic pathways of apoptosis. (Used with permission, Bredesen et al., 2006.)

the phosphatidlyserine so that it faces internally), appears externally during apoptosis (Fadok et al., 1992). These morphological and histochemical changes are largely the result of the activation of a set of cell-suicide cysteine proteases referred to as caspases (Thornberry & Lazebnik, 1998; Yuan et al., 1993). The characteristics of these proteases are described more fully next.

The biochemical activation of apoptosis occurs through two general pathways (see Figure 8.1): the intrinsic pathway, which is mediated by the mitochondrial release of cytochrome c and resultant activation of caspase-9; and the extrinsic pathway, originating from the activation of cell surface death receptors such as Fas, resulting in the activation of caspase-8 or -10 (Salvesen and Dixit, 1997). A third general pathway, which is essentially a second intrinsic pathway, originates from the endoplasmic reticulum and also results in the activation of caspase-9 (Morishima et al., 2002; Rao et al., 2001, 2002a, 2002b; Yuan & Yankner, 1999). In addition, other organelles, such as the nucleus and Golgi apparatus, also display damage sensors that link to apoptotic pathways (Green & Kroemer, 2005). Thus, damage to any of several different cellular organelles may lead to the activation of the apoptotic pathway.

Both physiological and pathological events may converge on the intrinsic pathway of apoptosis (see Figure 8.2). For example, DNA damage is sensed by a key protein involved in neurodegeneration: the protein "ataxia telangiectasia mutated" (ATM), along with ATM- and Rad-3-related (ATR), and DNA-dependent protein kinase (DNA-PK). The sensing of DNA damage by these proteins leads to the phosphorylation of the transcription factor p53, resultant stabilization of p53, and alteration in DNA binding; p53 then alters gene expression. Among the over 100 genes differentially regulated by p53, mediators of apoptosis such as Bax, Puma, Bcl-2, Scotin, Noxa, and Bid are included (Green & Kroemer, 2005; Murray-Zmijewski et al., 2006).

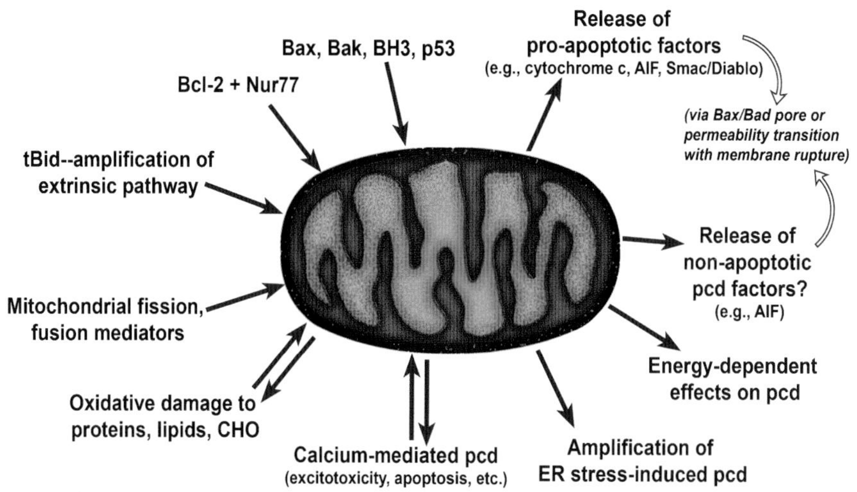

Figure 8.2 Central role of mitochondria in cell death programs. Arrows pointing toward mitochondria indicate cellular processes impinging on mitochondria, and arrows pointing away from mitochondria indicate effects of mitochondria on the extra-mitochondrial space.

The remarkably extensive nature of the p53 family response repertoire is indicated by the findings that family members p63 and p73 may also be involved in similar and related responses; that alternative splicing and alternative promoter usage generates nine different (putative) p53 isoforms, six different p63 isoforms, and 29 different p73 isoforms (Murray-Zmijewski et al., 2006); and that post-translational modifications such as phosphorylation and acetylation play key roles in protein function.

Whether triggered by DNA damage or by sensors associated with other organelles, the result is that the cell's probability of undergoing apoptosis is increased. This likelihood, which varies depending on cell type, differentiation state, cellular environment, cellular damage, and other factors, is referred to as the *apostat*. One of the critical determinants of the apostat is the balance between pro-apoptotic members of the Bcl-2 family and anti-apoptotic members, which in turn determines whether or not the mitochondria, which are targeted by the Bcl-2 family proteins, will be permeabilized and release factors such as cytochrome c that mediate caspase activation. Following damage sensing, this balance is shifted toward the pro-apoptotic members, which are of three types: (1) the multidomain members such as Bax and Bak display BH1–3 domains (Bcl–2 homology domains 1–3), and are capable of permeabilizing mitochondrial outer membranes (Kuwana et al., 2002, 2005); (2) the BH3-only activators such as tBid (truncated Bid, a product of Bid resulting from cleavage by caspases, calpains, or other proteases, and providing communication between the extrinsic and intrinsic systems (Stoka et al., 2001)) and Bim activate Bax and Bak, and may participate in their pore formation; and (3) the BH3-only derepressors such as Puma, Noxa, and Bad sequester the anti-apoptotic block created by Bcl-2, Bcl-x_L, and related anti-apoptotic proteins that display BH1–4 domains, thus freeing the BH1–3 and activators to permeabilize the mitochondrial outer membrane, allowing the release of multiple intermembrane proteins. Noteworthy among these proteins released by mitochondrial outer membrane permeabilization (MOMP) are cytochrome c, Smac/DIABLO, Omi/HtrA2, AIF (apoptosis-inducing factor), and endonuclease G. An anti-apoptotic modulator peptide, humanin, may also affect this balance, by binding and inhibiting Bax, Bid, and BimEL (Guo et al., 2003; Luciano et al., 2005; Zhai et al., 2005). Furthermore, non-Bcl-2 family members in some cases also may function to activate the multidomain protein Bax, as has been demonstrated for p53 (Chipuk et al., 2004; Schuler et al., 2000). This latter effect is mediated by the Puma-dependent release of p53 from Bcl-x_L (Chipuk et al., 2005).

Nur77/TR3, another transcription factor that mediates pcd, also has been shown to serve a distinct nontranscriptional role in apoptosis, as well (Li et al., 2000). In the case of Nur77/TR3, however, the extranuclear translocation is associated with interaction with Bcl-2, leading to the exposure of Bcl-2's pro-apoptotic BH3 domain and thus converting Bcl-2 from anti-apoptotic to pro-apoptotic (Lin et al., 2004). How many other transcriptional mediators of apoptosis may also prove to exhibit dual roles similar to those displayed by p53 and Nur77 remains an open question.

The net effect of the interplay of these Bcl-2 family and functionally related proteins is to set, along with other critical determinants, the apostat. One of the other critical determinants is the inhibition of caspases by iap (inhibitor of apoptosis) proteins. These proteins (e.g., XIAP, the X-linked iap protein) function both to inhibit the proteolytic activity of caspases directly and to mediate their degradation via an E3 ligase effect on ubiquitin-mediated degradation. However, this bimechanistic caspase inhibition may be reversed by Smac/DIABLO and possibly Omi/HtrA2, which prevent the interaction of the iap proteins with their target caspases.

In addition, one determinant that has received increasing attention recently is the fission and fusion of mitochondria (Bossy-Wetzel et al., 2003; Germain et al., 2005; Lee et al., 2004). This may be of special relevance to pcd in neurodegeneration, since mutations in the mitochondrial fusion mediator Opa1 are associated with optic atrophy (Alexander et al., 2000). Fission is mediated by Drp1 and Fis1, and inhibition of these proteins blocks apoptosis induction by staurosporine (Lee et al., 2004). Pro-apoptotic proteins may be recruited to mitochondria by Drp1, and mitochondrial remodeling may enhance the release of cytochrome c and other mitochondrial proteins. Ongoing work should clarify the roles of mitochondrial fission and fusion, and their mediators, in pcd and neurodegeneration.

Following release from the mitochondria, cytochrome c interacts with a cytosolic protein, Apaf-1, via the WD-40 repeats of Apaf-1, leading to the exposure of a (d)ATP-binding site on Apaf-1, which, when occupied, induces a conformational change resulting in heptamerization. The resultant exposure of the Apaf-1 CARD (caspase activation and recruitment domain) recruits caspase-9 into this apoptosomal complex, and the resulting induced proximity of caspase-9 molecules leads to their activation (Boatright et al., 2003). Activation of the apical caspase-9 leads to a cascade of caspase activation, including the effector caspases such as caspase-3 and caspase-7. However, the active caspases-3, 7, and 9 can be held in check by the IAP (inhibitor of apoptosis) proteins, such as XIAP (Deveraux et al., 1997), which as noted earlier may function as both direct inhibitors of caspase activity (in the case of caspase-9, by inhibiting dimerization) and as E3 ligases that mediate caspase degradation by the proteasome (Holley et al., 2002). This IAP-mediated block may itself be released by additional mitochondrially-derived proteins, Smac/DIABLO (Du et al., 2000; Verhagen et al., 2000) and Omi/HtrA2 (Martins et al., 2002; Suzuki et al., 2001). Smac (second mitochondrial activator of apoptosis), for example, binds to IAP proteins, preventing their inhibition of caspases, thus allowing caspase activation despite the presence of the otherwise inhibitory IAP proteins.

As opposed to the intrinsic pathway, which utilizes caspase-9 as its apical caspase, the extrinsic pathway utilizes caspase-8 or caspase-10. In the best characterized example, Fas (probably existing as a trimer even prior to ligand binding (Chan et al., 2000; Siegel et al., 2000)) is bound by trimeric Fas ligand. This interaction results in the interaction of an intracytoplasmic domain of Fas, dubbed the death domain, with a very similar death domain in an adaptor molecule, FADD (Fas-associated death domain protein). FADD displays, in addition to its death domain, another domain called a DED (death effector domain), and this domain interacts with a similar DED domain in caspase-8 (Muzio et al., 1996). The induced proximity of the apical caspase again leads to activation, as is the case for caspase-9. Also as for caspase-9, the initial caspase activation allows this upstream caspase to attack downstream, effector pro-caspases [somewhat analogous to what occurs in the thrombotic cascade, except that cysteine aspartyl-specific proteases (caspases) are utilized instead of serine proteases], cleaving and activating the effector caspases such as caspase-3 and caspase-7. In addition, FLIP(L) (FLICE-like inhibitory protein, long form), previously considered an inhibitor of extrinsic pathway activation, may act as a caspase-8 activator by being a more ready dimeric partner of caspase-8 than caspase-8 itself, resulting in activation by heterodimerization of what would otherwise be activated less readily by homodimerization (Boatright et al., 2004).

Both the intrinsic and extrinsic pathways of apoptosis thus converge on the activation of effector caspases by initiator caspases. Caspases are cysteine aspartyl-specific proteases that cleave with remarkable specificity at a small subset of aspartic acid residues. Their substrates, the number of which is unknown but probably somewhere between 0.5 and 5 percent of proteins, contribute to the apoptotic phenotype in several different ways; for example, following cleavage, their substrates contribute to proteolytic cascade activation, structural alterations in the cell, inactivation of repair mechanisms (e.g., DNA repair), internucleosomal DNA cleavage, phagocytic uptake signaling, mitochondrial permeabilization, and other effects. These proteases are synthesized as zymogens, but differ markedly in their activation: the initiator caspases (caspase-8, -9, and -10) exist as intracytoplasmic monomers until dimerization is effected by adaptor molecules, as noted earlier. Contrary to earlier models, cleavage of apical caspases is neither required nor sufficient for activation (Fuentes-Prior & Salvesen, 2004). The zymogenicity (ratio of activity of the active form to that of the zymogen) of these caspases is relatively low, in the range of 10 to 100 (Fuentes-Prior & Salvesen, 2004), and thus the (monomeric) zymogens themselves are actually somewhat active. These caspases display relatively large prodomains that are utilized in the protein-protein interactions that mediate activation—CARD (caspase activation and recruitment domain) in caspase-9, and DED (death effector domain) in caspase-8 and -10. The substrates of the apical caspases typically display I/L/V-E-X-D in the P4–P1 positions (the P1 position is the site of cleavage, P2 is one residue aminoterminal to P1, etc.; P1′ is one residue carboxyterminal to the cleavage site, P2′ is two residues carboxyterminal to the cleavage site, etc.), with preference for small or aromatic residues in the P1′ position (Fuentes-Prior & Salvesen, 2004).

The apical caspases activate effector caspases such as caspase-3 and -7. In contrast to the apical caspases, the effector caspases exist as dimers within the cell, display high zymogenicity (greater than 10,000 for caspase-3) and short prodomains, and are activated by cleavage rather than induced proximity. Cleavage produces a tetramer with two large subunits of 17–20 kilodaltons and two small subunits of 10–12 kilodaltons. Because of a difference in the S4 pocket (the pocket on the enzyme that interacts structurally with the P4 residue on the substrate) structure of these caspases (in comparison to the apical caspases), with similarity in the S1 and S3 pockets, their substrate preference is D-E-X-D, with a two orders of magnitude preference for Asp over Glu in the P4 position (Fuentes-Prior & Salvesen, 2004).

Caspases that do not fit clearly within these two groups include caspase-2, which displays a long prodomain like an apical caspase but has a substrate preference more similar to effector caspases (with the exception that, unlike other caspases, it has a P5 preference (for small hydrophobic residues)); caspase-6, which has a short prodomain like an effector caspase and yet substrate preference similar to apical caspases; and the inflammatory caspases (-1, -4, -5) involved in the processing of interleukin-1ß and interleukin-18. These latter are thought not to play a role in pcd; however, inhibition in some paradigms such as cerebral ischemia has indeed been associated with a reduction in infarct size (Friedlander et al., 1997b).

Caspase-12 is anomalous; in the murine system, it appears to play a role in apoptosis induced by endoplasmic reticulum (ER) stress (Morishima et al., 2002; Nakagawa et al., 2000; Rao et al., 2001). However, murine caspase-12 lacks Arg-341, which in other caspases is critical for the Asp specificity in the P1 position (Fuentes-Prior & Salvesen, 2004), and instead features a Lys in this position. Nonetheless, proteolytic activity has been reported for caspase-12 (Morishima et al., 2002), catalytically inactive caspase-12 inhibits ER stress-induced apoptosis (Rao et al., 2001), caspase-uncleavable caspase-12 also inhibits ER stress-induced apoptosis, and mice null for caspase-12 are less susceptible to amyloid-ß toxicity than wild type mice (Nakagawa et al., 2000). In the great majority of humans, however, a nonsense mutation is present in the caspase-12 gene, preventing expression of an active caspase (Fischer et al., 2002). Those without such a mutation are at increased risk for sepsis, due to the attenuation of the immune response to endotoxins such as lipopolysaccharide (Saleh et al., 2004).

B. Mechanistic Taxonomy of Cell Death: How Many Types of Programmed Cell Death Can Be Distinguished?

Classical developmental studies support the view that at least three different cell death forms are distinguishable (see Table 8.1): type I, also called nuclear or apoptotic; type II, also called autophagic; and type III, also called cytoplasmic (Clarke, 1990). These occur reproducibly within specific nuclei and with specific frequencies, at specific times of nervous system development. However, these developmental or physiological cell death pathways may also be activated by various insults, such as ischemia, DNA damage, or the accumulation of misfolded proteins. Mechanistic requirements within type I include two general groups: caspase-dependent apoptosis (extrinsic and intrinsic), and caspase-independent apoptosis (see next). Types II and III do not require caspase activation, but the possibility that they may in some cases be accompanied by caspase activation has not been excluded. The biochemical requirements for types II and III are considered further, later.

Beyond the three types of developmental cell death, other forms have been described that do not fit the criteria for any of the three types (see Table 8.1). For example, a nonapoptotic, caspase-independent form of cell death that does not resemble type II or type III developmental pcd has been described by Driscoll and her colleagues (Bianchi et al., 2004; Syntichaki et al., 2002) in *C. elegans* that express mutant channel proteins such as mec-4(d) that mediate neurodegeneration. A uniform, necrosis-like cell death ensues, characterized morphologically by membranous whorls lacking in other cell death types, triggered by calcium entry, mediated by specific calpains and cathepsins, and inhibited by calreticulin. Although it is possible that this alternative form of pcd will ultimately turn out to proceed via one of the previously described pathways (e.g., type II or type III), the morphological characteristics suggest that it is indeed a distinct form of pcd.

A fifth apparent form of pcd has been described by the Dawsons and their colleagues, who showed that a nonapoptotic form of cell death depends on the activation of poly-(ADP-ribose) polymerase (PARP) and the consequent translocation of apoptosis-inducing factor (AIF) from the mitochondria to the nucleus (Yu et al., 2002). AIF is a flavoprotein, discovered by Kroemer and his colleagues, that is involved with DNA fragmentation, along with endonuclease G and DNA fragmentation factor. This form of pcd was shown to be activated by agents that induce DNA damage, such as hydrogen peroxide, N-methyl-D-aspartate, and N-methyl-N'-nitro-N-nitrosoguanidine. Just as in the case of the calcium-activated pcd described by Driscoll and colleagues, PARP-dependent pcd displays a morphology and biochemistry that appear to be distinct from types I, II, and III pcd. It will be of interest to see how the gene expression and proteomics studies will add to the ongoing classification of pcd.

As additional data are gathered from other cell death paradigms, it is likely that novel biochemical pathways of pcd will be characterized. For example, an extensive literature on the morphological criteria for another potential form of pcd—oncosis—exists, but the biochemical underpinnings of oncosis have not yet been described. Oncosis refers to a specific morphology of cell death—cellular swelling—typically induced by ischemia, and thought to be mediated by the failure of plasma membrane ionic pumps. One potential mediator of oncosis is a calpain-family protease (possibly a mitochondrial calpain (Liu et al., 2004)). This finding suggests that oncosis may turn out to be related to, or synonymous with, the calcium-activated necrosis-like cell death described by Driscoll et al. Another potential pcd pathway has been referred to as autoschizis, a form of cell death shown to

Table 8.1 Comparison of different cell death programs. Note the difference in morphology present in each form, as well as the differences in biochemical mediators, inducers, and inhibitors. (Used with permission, Bredesen et al., 2006.)

Types →	Convicted Killers		New Suspects *(Innocent until proven guilty)*			
	Apoptosis	Autophagic	Paraptosis	Calcium-mediated	AIF/PARP-dependent	Oncosis
Characteristics ↓	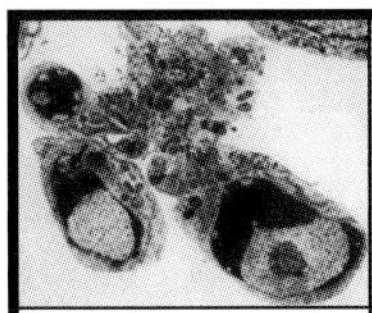	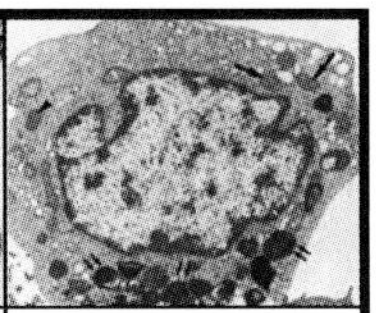	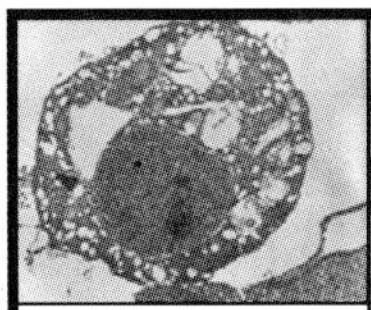	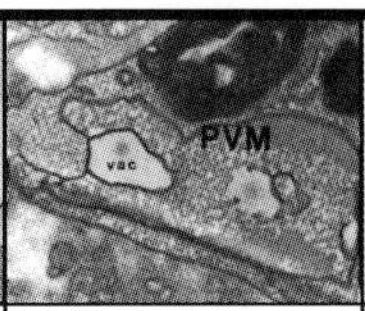	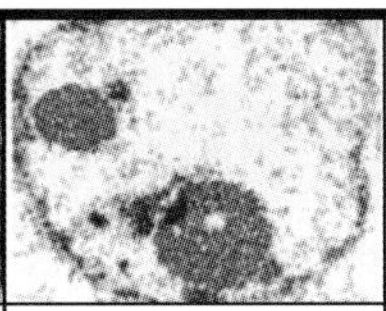	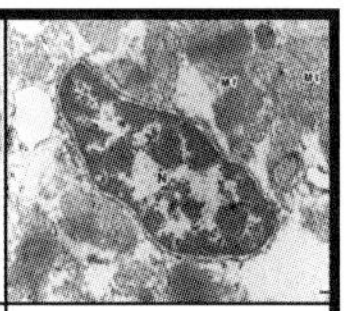
Morphology	Chromatin condensation, nuclear fragmentation, apoptotic bodies	Autophagic vacuoles	ER swelling, mitochondrial swelling	Membrane whorls	Mild chromatin condensation	Cellular swelling
Triggers	Death receptors, trophic factor withdrawal, DNA damage, viral infections, etc.	Serum, amino acid starvation, protein aggregates	Trophotoxicity	Calcium entry, deg mutants	DNA damage, glutamate, NO	Ischemia, excitotoxicity
Mediators	Caspases, BH family, etc.	JNK? MKK7? Atg orthologs	ERK2, Nur77	Calpains, cathepsins	PARP, AIF	JNK
Inhibitors	Caspase inhibitors, BH family, etc.	JNK inhibitors?	U0126 (MEK), DN Nur77	Calreticulin, some calpain inhibitors?	PARP inhibitors	JNK inhibitors
Examples	Type I pcd, nuclear pcd	Type II pcd	Type III pcd, cytoplasmic pcd	C. elegans deg mutants	Some excitotoxic pcd	Ischemic pcd

be activated in certain tumor cells following treatment with ascorbate and menadione (Gilloteaux et al., 1998).

Consideration should also be given to the biochemical pathways common to these different forms of pcd: in the intrinsic pathway of apoptosis, holocytochrome c and other pcd mediators are released from the intermembrane space of mitochondria secondary to outer membrane permeability that is induced by pro-apoptotic members of the Bcl-2 family such as Bax and Bak, in concert with BH3 proteins Bim or tBid. However, mitochondrial proteins may also be released in association with the mitochondrial membrane permeability transition (MPT), which can be subdivided into two states with different selectivities (Novgorodov & Gudz, 1996). Whether by consequent swelling and rupture of the mitochondrial outer membrane or by another mechanism, activation of the MPT by calcium, oxidants, or other activators offers a Bcl-2-independent (or at least partially independent, since Bax may interact with components of the MPT such as the adenine nucleotide translocator (Brenner et al., 2000) and the voltage-dependent anion channel (Adachi et al., 2004)), at least partially cyclophilin-D-dependent, route for the release of mitochondrially-derived pro-apoptotic factors.

Beyond these two general categories of mitochondrial pro-apoptotic factor release, more complicated scenarios are beginning to be defined: for example, recent work from Polster et al. showed that the release of AIF (apoptosis-inducing factor (Susin et al., 1999)) from mitochondria requires a combination of mitochondrial membrane permeabilization (e.g., by MOMP or by the MPT) and active calpain, and further implicated an endogenous mitochondrial calpeptin-inhibitable protease in AIF release induced by MPT (Polster et al., 2005). Such combinations raise the possibility of many multivariable-dependent paths to potentially different forms of pcd: in the simplest iteration, a 2×2 matrix could be constructed based on Bcl-2 inhibition vs. independence (implying Bcl-2 family-mediated mitochondrial outer membrane permeability, as opposed to insult-induced MPT) and caspase dependence vs. independence, then extended to 2 × 2 × 2 with the addition of calpain dependence vs. independence. Additional dimensions would then be added based on the activity of critical mediators such as PARP, AIF, and the autophagy-mediating gene products. Given such a parsing, classical apoptosis would fall predominantly into three groups: caspase dependent and Bcl-2 inhibitable (intrinsic pathway, and extrinsic pathway with amplification via the intrinsic pathway, mediated by tBid), caspase dependent and Bcl-2 resistant (some extrinsic pathway paradigms without amplification, and some MPT activators that lead to caspase-dependent pcd), and caspase independent, Bcl-2 inhibitable (e.g., some paradigms of endoplasmic reticulum stress (Egger et al., 2003) and intracellular pathogen-induced pcd (Perfettini et al., 2002)). Toxins that inactivate caspases directly or indirectly, such as diethylmaleate and buthionine sulfoximine, would fall into the group of Bcl-2 inhibitable, caspase-independent pcd (Kane et al., 1995). On the other hand, an increase in cytosolic calcium, such as occurs with the mec-4(d) mutants of *C. elegans* (Bianchi et al., 2004), could induce MPT—which would explain the Bcl-2 (ced-9) independence—and activate calpains, which would potentially inactivate caspases (Chua et al., 2000), compatible with the caspase independence of this form of pcd. The cathepsin dependence suggests lysosomal involvement, and thus a potential relationship with autophagic pcd. Adding DNA damage to the calcium entry, such as occurs with excitotoxic damage, should trigger a similar scenario with the addition of PARP activation, with the combination of calcium-activated MPT and calpain activation explaining the AIF activation (Lankiewicz et al., 2000; Polster et al., 2005).

As pointed out previously, these apparently differing cell death programs do share certain features, such as protease activation and mitochondrial release of pro-pcd factors (Green and Kroemer, 2005). Furthermore, all the alternative pathways share the common feature of caspase inhibition, be it direct (e.g., by zVAD.fmk or diethylmaleate) or indirect (e.g., via receptor tyrosine kinase or calpain activation). Nonetheless, it is likely to be of practical importance to identify the specific molecular linchpins of each of the mortal routes open to degenerating neurons, since it is likely that blocking a single path will be insufficient as a long-term therapeutic approach.

C. Autophagic Programmed Cell Death

Autophagy (from the Greek, "self eating") is a multifunctional process that occurs in diverse organisms. Autophagy includes three distinct but related processes—macroautophagy, microautophagy, and chaperone-mediated autophagy—and is a degradation pathway that complements the proteasomal pathway by degrading long-lived proteins, protein aggregates, and organelles through a regulated lysosomal pathway. Targets for degradation—for example, damaged mitochondria or aggregates of misfolded proteins—are encircled by a process that is reminiscent of phagocytosis, and the newly membrane-delimited structure—an autophagosome—then fuses with a lysosome, resulting in the degradation of the contents of the autophagosome. The molecular details of this process have been characterized best in yeast, in which a number of Atg (autophagy) genes have been identified, most with clear orthologs in higher eukaryotes.

Since the degradation of molecules and organelles by autophagy results in energy and amino acids for protein synthesis, it is a cellular protective pathway that, although active constitutively at low level, can be upregulated markedly by nutrient starvation. The process may be activated, for example, when nutrient withdrawal inactivates the TOR (target of rapamycin) protein, activating a complex that includes Atg1, 11, 13, 17, 20, 24, and Vac8 (Levine & Yuan, 2005). In a second step, vesicle nucleation is mediated by Atg6, Atg14, Vps15, and Vps34, and then, in a third step, vesicle expansion occurs (Atg3, 5, 7, 8, 12, and 16), followed finally by recycling of Atg proteins, mediated by Atg2, 9, and 18. The importance of this pathway *in vivo* has been illustrated by Atg7 conditionally null mice (Komatsu et al., 2005), which display multiple cellular abnormalities such as ubiquitin-positive aggregates and apparently damaged mitochondria. Furthermore, mice deficient in autophagy due to knockout of beclin-1 are nonviable (Yue et al., 2003).

Although the roles of the autophagic process in protein and organellar degradation, and in cellular protection during nutrient starvation, are well accepted, the role of autophagy in programmed cell death is more controversial (Levine & Yuan, 2005; Shimizu et al., 2004; Yue et al., 2003). Indeed, it has been argued that autophagic pcd is a nonentity (Levine & Yuan, 2005). This is in part because the term *autophagic cell death* has been used for two potentially distinct observations: cell death *associated with* autophagy and cell death *requiring* autophagy. The vast majority of examples of autophagic cell death represent the former rather than the latter (this is not to say that they may not prove to represent the latter as well, simply that the requirement for autophagy in the vast majority of the previous reports has not been evaluated), and the finding that a cellular protective process—be it a heat-shock response or the unfolded protein response or autophagy—is ongoing in response to an insult is not surprising, but neither is it necessarily informative about the mechanism of cell death. Somewhat surprising, however, is that, although the definition of type II programmed cell death is based on morphological criteria, and therefore is readily applicable to cell death that is simply associated with autophagy; increasing evidence suggests that the autophagic process is indeed required for at least some of what have been referred to as autophagic cell deaths. For example, haploinsufficiency (i.e., 50% reduction in gene dosage) of beclin-1 leads to a tumor predisposition phenotype, arguing that autophagy is tumor suppressive (Yue et al., 2003). Although this represents only indirect evidence that proteins associated with autophagy are mediators of a form of pcd, direct evidence has been provided in cells whose apoptotic machinery has been inhibited: mouse embryo fibroblasts (MEFs) that are null for both Bax and Bak, when treated with either of two frequently employed apoptosis inducers—staurosporine and etoposide—undergo a form of cell death that is associated with autophagosomes, dependent on autophagy genes Atg5 and beclin-1, and inhibited by the autophagy/class III PI3 kinase inhibitor, 3-methyladenine (Shimizu et al., 2004). It is important to add that, for comparison, wild-type MEFs were induced to undergo apoptosis by etoposide, and that this was not inhibited by 3-methyladenine. Although this argues that autophagy is not required for etoposide-induced death in wild-type MEFs, it does not negate the argument for the existence of an alternative, nonapoptotic, autophagy-requiring form of pcd. This autophagic pcd may, however, be a slower process than apoptosis (Lum et al., 2005), or may somehow be triggered by an apoptosis block. The latter scenario is supported by data showing that caspase inhibition by zVAD.fmk in L929 cells results in autophagy-dependent pcd (Yu et al., 2004), which has been proposed to be mediated by the selective degradation of catalase (Yu et al., 2006). On the one hand, this may alert us to the possibility that anti-apoptotic therapies carry the potential risk of inducing nonapoptotic pcd; on the other hand, it may argue that therapeutics directed at multiple cell death pathways will be required for optimal efficacy in diseases involving pcd. Since cell death associated with some neurodegenerative diseases (or disease-associated mutants, such as α-synuclein) is associated with an autophagic morphology, these implications are potentially important for the development of effective therapies to prevent or ameliorate neurodegenerative disorders (Gomez-Santos et al., 2003).

What are the mediators of autophagic pcd, and how do they interact with the autophagic machinery? Although relatively little is known in this area, JNK (c-Jun N-terminal kinase) has been implicated in a number of studies; for example, NMDA (N-methyl D-aspartate) induced excitotoxic neuronal death in hippocampal slice cultures was shown to pursue an autophagic path, with selective phosphorylation of c-Jun in regions CA1, CA3, and the dentate gyrus (Borsello et al., 2003). JNK inhibition prevented the neuronal death as well as the autophagy, suggesting that the requirement for JNK may lie upstream of the activation of autophagy.

It is clear that *in vivo* testing of the requirement of the autophagic process for any form(s) of pcd is needed, and the required genetic and pharmacological tools are increasingly available. Many questions regarding autophagic pcd remain unanswered: if autophagy is indeed a cellular protective program that—like the UPR (unfolded protein response)—at some point activates pcd, what is the signal for the switch to pcd initiation? Are there "executioners" analogous to caspases in autophagic pcd? How important is the role of autophagic pcd in neurodegeneration, ischemic pcd, and other neurological disease states? Does autophagic pcd occur *in vivo* in the absence of apoptosis inhibition, or does autophagic pcd require apoptosis inhibition? Are developing, mature, or aged neurons most susceptible to autophagic pcd? Given the common finding of protein aggregates in neurodegenerative diseases, is a defect in autophagy a common underlying problem in these diseases? If so, does this contribute to the triggering of cell death in neurodegenerative diseases?

D. Other Cell Death Programs: Fact or Fancy?

In comparison to apoptosis, relatively little is known about autophagic pcd, and even less is known about other nonapoptotic forms of pcd. Furthermore, most of what is known about these putative alternative pcd forms is based on morphological descriptions. Importantly, none of the nonapoptotic forms of pcd have gained general acceptance in the scientific community at this time, with the possible exception of autophagic pcd, as noted earlier. Nonetheless, many hints that such forms underlie diverse processes, such as development and degeneration, have emerged from morphological, genetic, and more recently, biochemical studies. Naturally occurring cell death, a label applied to physiological pcd (e.g., that occurring with development and cell turnover), typically occurs in one of three morphologies: type I (apoptosis or nuclear), II (autophagic), and III (cytoplasmic). Type III pcd, which was subdivided by Clarke (Clarke, 1990) into type A and B, is a necrosis-like form of pcd that features swelling of the endoplasmic reticulum and mitochondria, along with the absence of typical apoptotic features such as apoptotic bodies and nuclear fragmentation. Recently, it was noted that the hyperactivation of the tyrosine kinase receptor insulin-like growth factor I receptor (IGFR) induces a nonapoptotic form of cell death dubbed paraptosis (Sperandio et al., 2000). This form of cell death was shown to be programmatic in that it required transcription and translation, and was found to be indistinguishable morphologically from type III pcd, with swelling of the endoplasmic reticulum and mitochondria, and an absence of apoptotic features such as internucleosomal DNA fragmentation, apoptotic bodies, and zeiosis (however, this morphology of course is observed in many cell deaths labeled as necrosis, so the mechanistic implications of this particular morphological pattern remain an open question). Neither Bcl-2 nor caspase inhibitors block this form of pcd, nor are caspases activated, but inhibitors of ERK (extracellular signal-regulated protein kinase)—ERK2 but not ERK1—were found to inhibit paraptosis (Sperandio et al., 2004), as was AIP-1/Alix (ALG-2-interacting protein-1/ALG-2-interacting protein x). In addition, antisense oligonucleotides directed against JNK1 had a partial inhibitory effect.

The induction of pcd by hyperactivation of a trophic factor receptor—essentially "trophotoxicity"—is compatible with earlier observations that some trophic factors may increase neuronal cell death, for example induced by excitotoxicity (Koh et al., 1995). Such an effect might conceivably be protective against neoplasia, in that it may eliminate cells that would otherwise undergo autocrine loop-stimulated oncogenesis. The resulting program would necessarily be nonapoptotic, since trophic factors inactivate apoptotic signaling.

Aponecrosis is a term applied to a combination of apoptosis and necrosis (Formigli et al., 2000). Many cytotoxins induce pcd at low concentrations, but apparently necrotic cell death at higher concentrations, presumably due to the overwhelming of the cellular homeostatic processes prior to the completion of the cell death programs. In fact, this is the most common pattern seen with cellular toxins, from hydrogen peroxide and other oxidants to mitochondrial toxins such as antimycin A (Formigli et al., 2000). However, the necrotic morphology associated with aponecrosis has not been proven to be nonprogrammatic, so it is still not clear whether aponecrosis represents a combination of apoptosis and a nonapoptotic form of pcd or whether it represents a combination of apoptosis and nonprogrammatic cell death.

IV. Trophic Factors and the Concept of Cellular Dependence

Neurons, as well as other cells, for their survival depend on stimulation that is mediated by various receptors and sensors, and pcd may be induced in response to the withdrawal of trophic factors, hormonal support, electrical activity, extracellular matrix support, or other trophic stimuli (Bredesen et al., 1998). It has generally been assumed that cells dying as a result of the withdrawal of required stimuli do so because of the loss of a positive survival signal, typically mediated by receptor tyrosine kinases (Yao & Cooper, 1995). Although such positive survival signals are clearly extremely important, data obtained over the past 13 years argue for a complementary effect that is pro-apoptotic, is activated or propagated by trophic stimulus withdrawal, and is mediated by dependence receptors such as DCC (deleted in colorectal cancer) and Unc5H2 (uncoordinated gene 5 homologue 2) (Barrett & Bartlett, 1994; Barrett & Georgiou, 1996; Bordeaux et al., 2000; Bredesen & Rabizadeh, 1997; Ellerby et al., 1999b; Forcet et al., 2001; Llambi et al., 2001; Mehlen et al., 1998; Rabizadeh & Bredesen, 1994; Rabizadeh et al., 1993; Stupack et al., 2001). These receptors interact in their intracytoplasmic domains with caspases, including apical caspases such as caspase-9, and may therefore serve as sites of induced proximity and activation of these caspases. Caspase activation leads in turn to receptor cleavage, producing pro-apoptotic fragments (Ellerby et al., 1999b; Mehlen et al., 1998); conversely, mutation of the caspase cleavage sites of dependence receptors suppresses pcd mediated by the receptors (Bredesen et al., 2004; Mehlen et al., 1998). A striking example of this effect was obtained in studies of neural tube development: withdrawal of Sonic hedgehog from the developing chick spinal cord led to apoptosis mediated by its receptor, Patched, preventing spinal cord development, but apoptosis was inhibited and development partially restored by transfection of a caspase-uncleavable mutant of Patched, even in the absence of Sonic hedgehog (Thibert et al., 2003).

Thus cellular dependence on specific signals for survival is mediated, at least in part, by specific dependence receptors or addiction receptors, which induce apoptosis in the absence of the required stimulus (when unoccupied by a trophic ligand, or when bound by a competing, anti-trophic ligand), but block apoptosis following binding to their respective ligands (Bredesen et al., 1998; Rabizadeh & Bredesen, 1994; Rabizadeh et al., 1993). Expression of these dependence receptors thus creates cellular states of dependence on the associated ligands. These states of dependence are not absolute, since they can be blocked downstream in some cases by the expression of anti-apoptotic genes such as *bcl-2* or *p35* (Bredesen et al., 1998; Forcet et al., 2001; Mah et al., 1993); however, they result in a shift of the apostat (Bredesen, 1996; Salvesen & Dixit, 1997) toward an increased likelihood of triggering apoptosis. In the aggregate, these receptors may serve as a molecular integration system for trophic signals, analogous to the electrical integration system afforded by the dendritic arbors within the nervous system.

The ß-amyloid precursor protein (APP) exhibits some features reminiscent of dependence receptors: an intracytoplasmic caspase cleavage site (Asp664 (Gervais et al., 1999; Lu et al., 2000); see Figure 8.5), coimmunoprecipitation with an apical caspase (caspase-8), caspase activation, derivative pro-apoptic peptides (Aß, C99 and C31), and suppression of apoptosis induction by mutation of the caspase cleavage site (Lu et al., 2000, 2003a, 2003b). To evaluate the possibility that APP may function as a dependence receptor in the mediation of Alzheimer's disease pathogenesis, Galvan et al. compared transgenic mice modeling Alzheimer's disease with Swedish and Indiana mutations in the hAPP transgene, with vs. without a functional caspase cleavage site at Asp664; mutation at the caspase cleavage site had no effect on ß-amyloid production or plaque deposition, but it completely prevented the synapse loss, dentate gyral atrophy, astrogliosis, memory loss (assessed by Morris water maze and Y maze), neophobia (so-called, although it seems less likely that the mice "fear" novelty than that they may not recognize it, and so perhaps this might more aptly be termed "neoagnosia"), and increased neural precursor proliferation that characterize the AD model (Galvan et al., 2006). These findings suggest a model in which the Aß peptide functions as an antitrophin, binding and oligomerizing APP, recruiting and activating caspase-8, processing APP at

Asp664, and inducing neurite retraction and neuronal cell death (Galvan et al., 2006; Lu et al., 2003a; Shaked et al., 2006). An alternative possibility, however, is that the D664A mutation altered a protein-protein interaction critical to AD pathogenesis in the mouse model, and that caspase cleavage is not critical. In either case, however, the results suggest that APP signal transduction may be important in mediating Alzheimer's disease (Nishimoto, 1998), at least in the transgenic mouse model, possibly downstream from Aß oligomerization and binding of APP. The results also raise the question of what the trophic ligand for APP may be (if indeed APP proves to bind a trophic ligand), and several candidate APP interactors have been described, such as collagen (types I and IV), heparan sulfate proteoglycan, laminin, and glypican (Beher et al., 1996; Caceres & Brandan, 1997; Williamson et al., 1996).

Caspase cleavage also appears to play an important role in cytotoxicity induced by multiple polyglutamine proteins, such as huntingtin, atrophin-1, ataxin-3, and androgen receptor (Ellerby et al., 1999a, 1999b; Goldberg et al., 1996; Hermel et al., 2004; Wellington et al., 2000). In the case of huntingtin, recruitment of caspase-2 into a complex with huntingtin was found to be polyglutamine length dependent, leading to cleavage at Asp552 both *in vitro* and *in vivo*. Although huntingtin is not a surface receptor like APP, the upregulation of caspase-2 observed in Huntington's model mice correlated directly with decreased levels of brain-derived neurotrophic factor (Hermel et al., 2004), suggesting that huntingtin may indeed represent a mediator of cellular dependence on trophic support. Furthermore, as noted earlier, recent results analogous to those obtained with the caspase-uncleavable APP mutant by Galvan et al. have been obtained with the caspase-6-uncleavable huntingtin mutant by Graham et al. (2006). In that study, the YAC (yeast artificial chromosome) transgenic mouse model of Huntington's disease was utilized, and the Huntington's phenotype was prevented by mutating the caspase-6 cleavage site, but not by mutating the caspase-3 cleavage sites within the huntingtin protein.

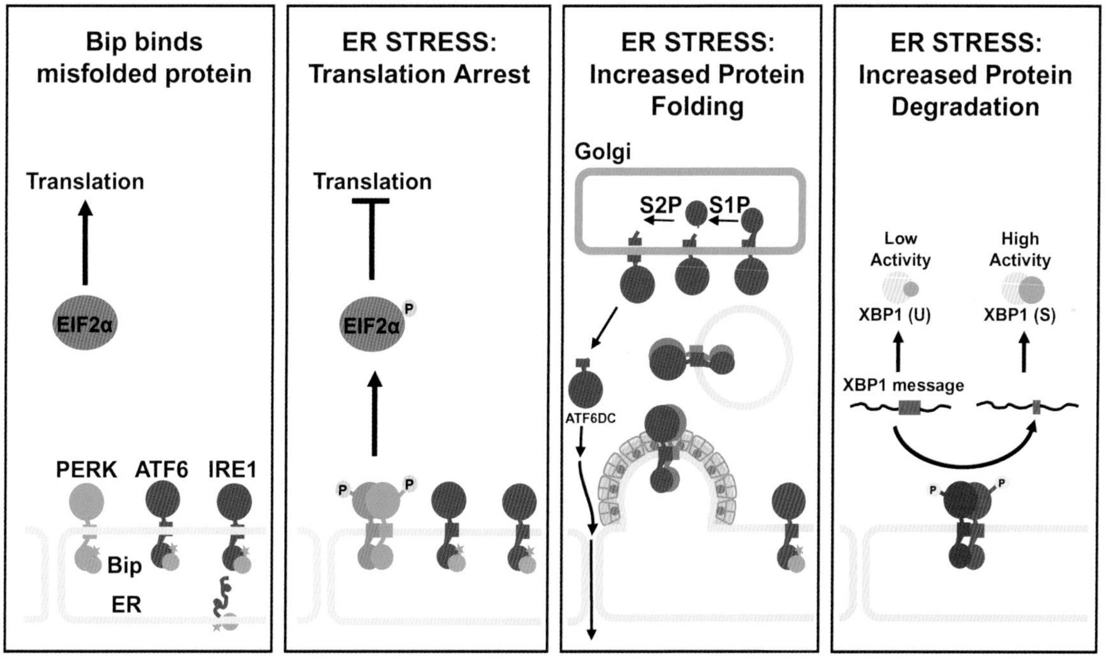

Figure 8.3 The unfolded protein response (UPR), a coordinated regulated response involving three sensor proteins: PERK (PKR-like ER kinase), ATF6 (activating transcription factor 6), and IRE1 (inositol requiring transmembrane kinase/endoribonuclease). Misfolded proteins bind Grp78/Bip, releasing it sequentially from PERK, ATF6, and IRE1. PERK undergoes oligomerization and autophosphorylation. Active PERK phosphorylates eIF2α, rendering it inactive and blocking protein translation. Inactivation of eIF2α prevents further influx of nascent proteins into the ER lumen thus limiting the incoming protein load. A selective inhibitor of eIF2alpha has been shown to block ER stress (Boyce et al., 2005). Continued accumulation leads to translocation of ATF6 to the Golgi compartment where it undergoes regulated intramembrane proteolysis (RIP) by proteases S1P and S2P, yielding a free cytoplasmic domain that triggers transcriptional upregulation of several ER resident proteins. These proteins facilitate and promote the productive folding of proteins and protein complexes, maintaining them in a folding-competent state and preventing their aggregation. UPR activation also induces homodimerization, autophosphorylation and activation of IRE1, an ER resident transmembrane serine/threonine kinase receptor protein that also possesses an intrinsic endoribonuclease activity. Activated IRE1 cleaves a preformed substrate mRNA at two sites through its endoribonuclease action, resulting in the removal of a 26-nucleotide intron from a target mRNA (Kawahara et al., 1997; Patil & Walter, 2001; Sidrauski & Walter, 1997). The two ends of the cleaved mRNA are ligated together by tRNA ligase and the newly formed mRNA encodes a transcription factor X-box binding protein (XBP-1). XBP-1 binds and activates the promoters of several ER stress-inducible target genes that facilitate retro-translocation and ER-associated degradation of misfolded proteins (Calfon et al., 2002; Lee et al., 2002; Yoshida et al., 2001). IRE1 is coupled to JNK activation through TRAF2 (Urano et al., 2000). (Used with permission, Bredesen et al., 2006.)

V. Apoptosis Induced by Unfolded, Misfolded, or Alternatively Folded Proteins

In addition to the intrinsic and extrinsic pathways of apoptosis as originally described, recent studies have suggested additional pathways to caspase activation. Misfolded proteins and other activators of endoplasmic reticulum stress trigger an alternative intrinsic pathway (see Figure 8.3) that leads to caspase-9 activation, and displays both cytochrome c/Apaf-1 independent and cytochrome c/Apaf-1 dependent activation of pcd (Rao et al., 2001, 2002a). Cell death pathways triggered by protein misfolding, unfolding, or alternative folding—and associated endoplasmic reticulum stress—are of special interest in neurodegenerative disease studies, since such pathways have been implicated in all the major neurodegenerative diseases (Rao and Bredesen, 2004). Neurodegenerative disorders such as Alzheimer's disease (AD), Parkinson disease (PD), Huntington's disease (HD), amyotrophic lateral sclerosis (ALS), and prion protein diseases all share the common feature of accumulation and aggregation of misfolded proteins (Kopito and Ron, 2000; Taylor et al., 2002). The presence of misfolded proteins elicits cellular stress responses that include an endoplasmic reticulum (ER) stress response that serves to protect cells against the toxic accumulation of misfolded proteins, and is activated by the exposure of hydrophobic protein regions that bind GRP78/BiP (glucose-regulated protein of 78 kilodaltons/binding protein), relieving its otherwise ongoing inhibition of UPR-activating proteins PERK (protein kinase R-like endoplasmic reticulum kinase), ATF6 (activating transcription factor 6), and Ire1 (inositol-requiring 1) (see Figure 8.3) (Dobson, 2003; Dobson and Ellis, 1998; Kaufman, 1999; Kopito, 2000). Accumulation of these proteins in excessive amounts, however, overwhelms the cellular "quality control" systems that induce folding, translational, degradative, and aggresomal protection, ultimately triggering cellular suicide pathways (Goldberg, 2003; Kopito and Ron, 2000; Selkoe, 2003; Sherman and Goldberg, 2001; Sitia and Braakman, 2003; Taylor et al., 2002).

Since the degradation of cellular proteins is coupled, via the ubiquitin-mediated proteasomal degradation pathway, to ER dislocation (translocating the protein targeted for degradation back out of the ER into the cytosol) (Goldberg, 2003; Kopito & Ron, 2000; Sherman and Goldberg, 2001), any conditions that block the ER retro-translocation of proteins or proteasome function may also result in the accumulation of misfolded protein substrates within the ER. Thus, misfolded proteins both within and outside the ER trigger the ER stress response. Misfolded proteins typically aggregate, initially as oligomers but ultimately as polymers that are deposited as microscopically visible inclusion bodies or plaques within cells or in extracellular spaces. These aggregates may interact with cellular targets, with several potential resulting toxic mechanisms:

- ▲ Inhibition of synaptic function
- ▲ Loss of synapses leading to disruption of neuronal functions
- ▲ Sequestration of critical cellular chaperones and vital transcription factors by misfolded proteins
- ▲ Interference with signal transduction pathways
- ▲ Alteration of calcium homeostasis
- ▲ Release of free radicals and resulting oxidative damage
- ▲ Dysfunction of the protein degradation pathway through the ubiquitin-proteasome system
- ▲ Induction of cell-death proteases

Despite these mechanisms, it is becoming increasingly clear that toxicity may be exerted prior to the appearance of aggregates, apparently with the production of small oligomers (Klein, 2002; Mucke et al., 2000).

A number of mediators of pcd induced by misfolded or unfolded proteins have recently been identified (see Figure 8.4). As for the intrinsic pathway, the Bcl-2 family proteins play a critical role in the cellular suicide decision process (Scorrano et al., 2003), with communication between the ER and the mitochondria. Strikingly, Bax/Bak double knock-out cells demonstrate no caspase activation following ER stress, arguing that these are required mediators (Ruiz-Vela et al., 2005). Bik may function to activate Bax and Bak in this pathway (Mathai et al., 2005), whereas BI-1 inhibits Bax activation and translocation to the ER (Chae et al., 2004). Other Bcl-2 family proteins are also involved; for example, the BH3 protein Puma interacts with an hsp90 (heat shock protein of 90 kilodaltons)-independent fraction of p23, which, when cleaved by caspases, releases Puma, leading to Bax interaction, oligomerization, and pcd (Rao et al., 2006). Noxa and p53 have also been implicated in this pathway (Li et al., 2006).

Not surprisingly, given the mitochondrial-ER interplay, part of the resulting apoptotic pathway is Apaf-1 dependent; however, part is also Apaf-1 independent yet caspase-9 dependent (Li et al., 2006; Rao et al., 2002a). Caspase-7 is recruited to the ER by an unknown mechanism, where it interacts with caspase-12 (at least, in the murine system; as noted earlier, most humans do not express caspase-12), caspase-12 is cleaved and released, leading to interaction with caspase-9 (Rao et al., 2001). GRP78/BiP interacts with caspase-7 (requiring the ATP-binding domain (Reddy et al., 2003)) and -12, preventing activation, but this inhibition is relieved by (d)ATP (Rao et al., 2002b). Although the upstream activation of this pathway is not certain, one candidate is the triggering of JNK activation by IRE1, via TRAF2 and ASK1 (Urano et al., 2000).

A caspase-8-dependent pathway responsive to misfolded proteins also exists: Bap31, an ER membrane protein, binds Bcl-2 (or Bcl-x_L) and a caspase-8-containing pro-apoptotic complex (Ng et al., 1997). When Bap31 is cleaved, a pro-apoptotic p20 fragment is derived, which,

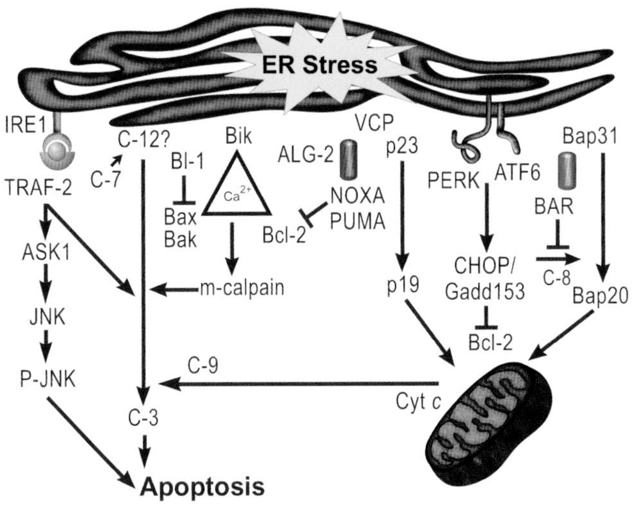

Figure 8.4 Proteins implicated in ER stress-pcd pathways. (Used with permission, Bredesen et al., 2006.)

among other effects, induces mitochondrial fission, enhancing cytochrome c release (Breckenridge et al., 2003). Conversely, BAR (bifunctional apoptosis regulator), which is expressed primarily in neurons of the CNS, also bridges Bcl-2 and caspase-8, but functions as an anti-apoptotic protein (Roth et al., 2003; Zhang et al., 2000).

Other mediators of ER stress-induced pcd have been identified, and these are depicted in Figure 8.4. One of interest in neurodegeneration is valosin-containing protein (VCP), which functions as a sensor of abnormally folded proteins and a cell death effector in polyglutamine-induced cell death (Hirabayashi et al., 2001; Kobayashi et al., 2002), as well as being a mediator of ER stress-induced pcd (Rao et al., 2004). Mutations in VCP are associated with a syndrome that includes inclusion body myositis, frontotemporal dementia, and Paget's disease of bone (Watts et al., 2004). Another is Alix/AIP-1, which is an ALG-2-interacting protein that links developmental motor neuron cell death, as well as neuronal death in a Huntington's model, to the endolysosomal system (Mahul-Mellier et al., 2006).

In addition to protein misfolding, it is worth noting recent results from Lingappa and colleagues that suggest that at least some proteins may trigger pcd via a subset of multiple physiologically relevant conformations (Hegde et al., 1998). For example, the prion protein exists in three different topologies, a secreted form, a transmembrane form with N-terminus extracellular (Ntm), and a transmembrane form with C-terminus extracellular (Ctm). The Ctm form is pro-apoptotic and associated with neurodegeneration *in vivo*, whereas the secreted form is anti-apoptotic. It is not yet clear how many proteins will display this feature, but it is possible that not only prion protein but also other neurodegeneration-associated proteins will turn out to exist in multiple physiologically relevant conformations (and perhaps even topologies), with the degenerative effect being attributed to a subset of these conformations rather than true protein misfolding.

VI. Does Programmed Cell Death Play a Role in Neurodegeneration?

Evidence for caspase activation in neurodegeneration has been generated both from the use of antibodies directed against neo-epitopes derived by caspase cleavage (Galvan et al., 2006; Tanaka et al., 2006; Yang et al., 1998) and from the inhibition of neurodegeneration by caspase inhibitors (Friedlander et al., 1997a; Ona et al., 1999). However, some neurodegenerative models and diseases clearly demonstrate nonapoptotic forms of pcd (Dal Canto & Gurney, 1994; Turmaine et al., 2000). Determining which pcd pathways are triggered in each neurodegenerative disease, which pathway accounts for each fraction of cell death, the mechanism(s) by which each pathway is triggered, and the interactions of the various pathways should shed new light on the degenerative process and its potential treatment or prevention.

One of the critical goals for dissecting the relationship between pcd and neurodegeneration is to understand the specificity of the trigger; in other words, is pcd activated in neurodegeneration as the result of a relatively nonspecific toxic effect of a peptide or protein aggregate? If so, then secondary neurodegeneration may occur due to a loss of trophic support, increased excitotoxin accumulation, or any number of other secondary effects. Alternatively, by analogy to neoplasia, are specific, physiologically relevant transduction events that underlie neurite retraction and synapse loss triggered directly by neurodegeneration-associated transcriptional and post-transcriptional events? Evidence on both sides exists. For example, numerous toxic properties have been attributed to the Aß peptide implicated in Alzheimer's disease, such as reactive oxygen species generation and metal binding, among others (Butterfield & Bush, 2004). However, signal transduction effects have also been attributed to Aß peptide, such as binding and multimerization of amyloid precursor protein, with resultant complex formation and direct caspase activation (Lu et al., 2003a).

The ability to initiate the neurodegenerative process with widely varying insults—from misfolded proteins to reactive oxygen species to caspase recruitment complexes, as well as other mechanisms—and yet produce a relatively small number of syndromes, argues for the existence of a death network, which may be entered from many different sites, but once triggered follows similar interdependent biochemical pathways, with little dependence on the point of entry. Such a notion is compatible with the findings that therapeutics aimed at different pathways (caspase activation, mitochondrial release of cytochrome c, metal binding, reactive oxygen species scavenging, etc.) all have partially salutary effects; however, it also suggests that a complete halt of the neurodegenerative process may require therapeutics that address all of the network's interacting pathways.

VII. Are Programmed Cell Death Pathways Appropriate Therapeutic Targets in Neurodegeneration?

As noted earlier, neuronal loss is a relatively late event in neurodegenerative diseases, following neuronal dysfunction, synapse loss, and often somal atrophy; nonetheless, targeting pcd has been successful in at least some model systems of neurodegeneration, and it is possible that targeting multiple pcd pathways will turn out to be even more effective. For example, caspase inhibition *in vivo* retarded degeneration in transgenic mouse models of both ALS and Huntington's disease (Friedlander et al., 1997a; Ona et al., 1999). Bcl-2 expression in a transgenic model of ALS delayed symptom onset and increased lifespan, but did not alter the disease duration (Kostic et al., 1997).

Minocycline, a second-generation tetracycline that inhibits mitochondrial cytochrome c release, effected neuroprotection in mouse models of Huntington's disease, Parkinson disease, and ALS (Chen et al., 2000; Friedlander, 2003; Wang et al., 2003). Minocycline is orally bioavailable, penetrates the blood–brain barrier, and has been proven safe for human use. It is currently being evaluated in clinical trials in patients with Huntington's disease and ALS (Friedlander, 2003).

Rasagiline, which has been approved for the treatment of Parkinson disease, has been suggested to target apoptosis (possibly via the peripheral benzodiazepine receptor (Green and Kroemer, 2005)), but since it is a potent, selective, irreversible monoamine oxidase type B inhibitor, its therapeutic effect on Parkinson's disease may have nothing to do with effects on apoptosis.

Trophic factors have multiple effects, including the inhibition of apoptosis, the stimulation of neural precursors, and the stimulation of neurite outgrowth. The literature is rife with examples of trophic factor therapy for various neurological conditions, and although many have been unsuccessful, the delivery of the correct factor(s) to the correct target in the correct concentration for the correct disease still holds great promise (although, as noted earlier, hyperactivation of at least some trophic factor receptors may induce pcd). Ellerby et al. reported interesting results with FGF2 in a mouse model of Huntington's disease: prolonged survival, improved motor performance, and reduced polyglutamine aggregates (Jin et al., 2005). A number of approaches have been taken to deliver NGF to patients with Alzheimer's disease, including the use of genetically engineered fibroblasts in a recent phase 1 trial (Tuszynski et al., 2005). In this study, improvements in cognition and positron emission tomography (fluorodeoxyglucose uptake) were documented.

The most successful—and most controversial—trophic factor therapeutic for Parkinson disease has been GDNF (glial-derived neurotrophic factor). When a Phase 2 trial was discontinued for both safety and efficacy reasons, patients and families involved with the trials requested compassionate use based on their own belief, well documented anecdotally, that GDNF was successful (Lang et al., 2006).

Huntington's disease may represent one of the best possibilities for trophic factor therapy: presymptomatic diagnosis is readily available, a number of different trophic factors have shown promise in animal models (e.g., FGF2, CNTF, NGF (Kordower et al., 1999)), and BDNF is both reduced in the disease and therapeutic in Huntington's disease models. Furthermore, BDNF concentration in the brain may be increased by the administration of cysteamine (Borrell-Pages et al., 2006).

VIII. Death and Resurrection: The Neural Stem Cell Response to Neurodegeneration

Pathological processes may stimulate neurogenesis in the brain (Gould et al., 1997; Snyder et al., 1997), and may redirect the migration of nascent neurons toward the site(s) of pathology. For example, apoptotic degeneration of corticothalamic neurons in mice is followed by the restoration of corticothalamic connections, and the connecting cells both express immature neuronal markers such as doublecortin (DCX) and Hu and can be labeled with the cell-proliferation marker bromodeoxyuridine (BrdU) (Magavi et al., 2000). In Alzheimer's disease patients and in some transgenic mouse models of AD, neurogenesis is increased (Jin et al., 2004a, 2004b). This effect is prevented by mutation of the APP caspase cleavage site (Galvan et al., 2006). Furthermore, APP_S, the soluble cleavage product of APP derived from its cleavage by α-secretase, stimulates neural stem cell proliferation (Caille et al., 2004).

In Huntington's disease and in some transgenic models of Huntington's disease, there is an increase in neurogenesis

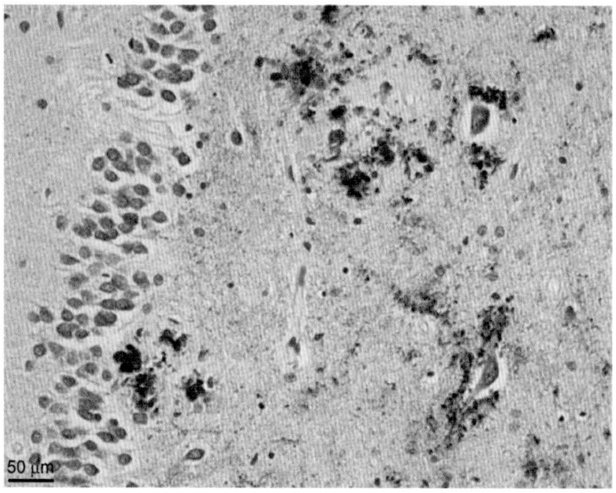

Figure 8.5 An antibody specific for the neoepitope generated by cleavage of APP at the caspase cleavage site, Asp664, demonstrates reactivity in the hippocampus of a patient with Alzheimer's disease. (Used with permission, Bredesen et al., 2006.)

(Curtis et al., 2003). As noted earlier, FGF2 administration led to increased neurogenesis as well as improved performance and survival. Interestingly, the neurogenesis included migration to the globus pallidus and differentiation into a spiny neuron phenotype—the same type of neurons preferentially lost in Huntington's disease.

These and similar results raise the question of whether "stem cell exhaustion/senescence" following prolonged stimulation may play a role in the long-term course of neurodegenerative diseases. Furthermore, although the factors linking neurodegeneration to neural stem cell proliferation and inhibition of pcd are largely undisclosed, the potential for therapy utilizing these putative factors is likely to be significant.

Acknowledgments

We thank Molly Susag, Loretta Sheridan, and Rowena Abulencia for manuscript preparation, and members of the Bredesen laboratory for discussion and critical reading of the manuscript.

References

Adachi, M., Higuchi, H., Miura, S., Azuma, T., Inokuchi, S., Saito, H. et al. (2004). Bax interacts with the voltage-dependent anion channel and mediates ethanol-induced apoptosis in rat hepatocytes. *Am J Physiol Gastrointest Liver Physiol* **287**, G695–705.

Alexander, C., Votruba, M., Pesch, U. E., Thiselton, D. L., Mayer, S., Moore, A. et al. (2000). OPA1, encoding a dynamin-related GTPase, is mutated in autosomal dominant optic atrophy linked to chromosome 3q28. *Nat Genet* **26**, 211–215.

Barrett, G. L. and Bartlett, P. F. (1994). The p75 nerve growth factor receptor mediates survival or death depending on the stage of sensory neuron development. *Proc Natl Acad Sci U S A* **91**, 6501–6505.

Barrett, G. L. and Georgiou, A. (1996). The low-affinity nerve growth factor receptor p75NGFR mediates death of PC12 cells after nerve growth factor withdrawal. *J Neurosci Res* **45**, 117–128.

Beher, D., Hesse, L., Masters, C. L., and Multhaup, G. (1996). Regulation of amyloid protein precursor (APP) binding to collagen and mapping of the binding sites on APP and collagen type I. *J Biol Chem* **271**, 1613–1620.

Bianchi, L., Gerstbrein, B., Frokjaer-Jensen, C., Royal, D. C., Mukherjee, G., Royal, M. A. et al. (2004). The neurotoxic MEC-4(d) DEG/ENaC sodium channel conducts calcium: Implications for necrosis initiation. *Nat Neurosci* **7**, 1337–1344.

Boatright, K. M., Deis, C., Denault, J. B., Sutherlin, D. P., and Salvesen, G. S. (2004). Activation of caspases-8 and -10 by FLIP(L). *Biochem J* **382**, 651–657.

Boatright, K. M., Renatus, M., Scott, F. L., Sperandio, S., Shin, H., Pedersen, I. M. et al. (2003). A unified model for apical caspase activation. *Mol Cell* **11**, 529–541.

Bordeaux, M. C., Forcet, C., Granger, L., Corset, V., Bidaud, C., Billaud, M. et al. (2000). The RET proto-oncogene induces apoptosis: A novel mechanism for Hirschsprung disease. *Embo J* **19**, 4056–4063.

Borrell-Pages, M., Canals, J. M., Cordelieres, F. P., Parker, J. A., Pineda, J. R., Grange, G. et al. (2006). Cystamine and cysteamine increase brain levels of BDNF in Huntington disease via HSJ1b and transglutaminase. *J Clin Invest.*

Borsello, T., Croquelois, K., Hornung, J. P., and Clarke, P. G. (2003). N-methyl-d-aspartate-triggered neuronal death in organotypic hippocampal cultures is endocytic, autophagic and mediated by the c-Jun N-terminal kinase pathway. *Eur J Neurosci* **18**, 473–485.

Bossy-Wetzel, E., Barsoum, M. J., Godzik, A., Schwarzenbacher, R., and Lipton, S. A. (2003). Mitochondrial fission in apoptosis, neurodegeneration and aging. *Curr Opin Cell Biol* **15**, 706–716.

Boyce, M., Bryant, K. F., Jousse, C., Long, K., Harding, H. P., Scheuner, D. et al. (2005). A selective inhibitor of eIF2-alpha dephosphorylation protects cells from ER stress. *Science* **307**, 935–939.

Breckenridge, D. G., Stojanovic, M., Marcellus, R. C., and Shore, G. C. (2003). Caspase cleavage product of BAP31 induces mitochondrial fission through endoplasmic reticulum calcium signals, enhancing cytochrome c release to the cytosol. *J Cell Biol* **160**, 1115–1127.

Bredesen, D. E. (1996). Keeping neurons alive: The molecular control of apoptosis (part I). In *The Neuroscientist*, Vol. 2, 181–190.

Bredesen, D. E., Mehlen, P., and Rabizadeh, S. (2004). Apoptosis and dependence receptors: A molecular basis for cellular addiction. *Physiol Rev* **84**, 411–430.

Bredesen, D. E., and Rabizadeh, S. (1997). p75NTR and apoptosis: Trk-dependent and Trk-independent effects. *Trends Neurosci* **20**, 287–290.

Bredesen, D. E., Rao, R. V., and Mehlen, P. (2006). Cell death in the nervous system. *Nature* **443**, 796–802.

Bredesen, D. E., Ye, X., Tasinato, A., Sperandio, S., Wang, J. J., Assa-Munt, N., and Rabizadeh, S. (1998). p75NTR and the concept of cellular dependence: Seeing how the other half die [see comments]. *Cell Death Differ* **5**, 365–371.

Brenner, C., Cadiou, H., Vieira, H. L., Zamzami, N., Marzo, I., Xie, Z. et al. (2000). Bcl-2 and Bax regulate the channel activity of the mitochondrial adenine nucleotide translocator. *Oncogene* **19**, 329–336.

Butterfield, D. A. and Bush, A. I. (2004). Alzheimer's amyloid beta-peptide (1–42): Involvement of methionine residue 35 in the oxidative stress and neurotoxicity properties of this peptide. *Neurobiol Aging* **25**, 563–568.

Caceres, J. and Brandan, E. (1997). Interaction between Alzheimer's disease beta A4 precursor protein (APP) and the extracellular matrix: Evidence for the participation of heparan sulfate proteoglycans. *J Cell Biochem* **65**, 145–158.

Caille, I., Allinquant, B., Dupont, E., Bouillot, C., Langer, A., Muller, U., and Prochiantz, A. (2004). Soluble form of amyloid precursor protein regulates proliferation of progenitors in the adult subventricular zone. *Development* **131**, 2173–2181.

Calfon, M., Zeng, H., Urano, F., Till, J. H., Hubbard, S. R., Harding, H. P. et al. (2002). IRE1 couples endoplasmic reticulum load to secretory capacity by processing the XBP-1 mRNA. *Nature* **415**, 92–96.

Chae, H. J., Kim, H. R., Xu, C., Bailly-Maitre, B., Krajewska, M., Krajewski, S. et al. (2004). BI-1 regulates an apoptosis pathway linked to endoplasmic reticulum stress. *Mol Cell* **15**, 355–366.

Chan, F. K., Chun, H. J., Zheng, L., Siegel, R. M., Bui, K. L., and Lenardo, M. J. (2000). A domain in TNF receptors that mediates ligand-independent receptor assembly and signaling. *Science* **288**, 2351–2354.

Chen, M., Ona, V. O., Li, M., Ferrante, R. J., Fink, K. B., Zhu, S. et al. (2000). Minocycline inhibits caspase-1 and caspase-3 expression and delays mortality in a transgenic mouse model of Huntington disease. *Nat Med* **6**, 797–801.

Chipuk, J. E., Bouchier-Hayes, L., Kuwana, T., Newmeyer, D. D., and Green, D. R. (2005). PUMA couples the nuclear and cytoplasmic proapoptotic function of p53. *Science* **309**, 1732–1735.

Chipuk, J. E., Kuwana, T., Bouchier-Hayes, L., Droin, N. M., Newmeyer, D. D., Schuler, M., and Green, D. R. (2004). Direct activation of Bax by p53 mediates mitochondrial membrane permeabilization and apoptosis. *Science* **303**, 1010–1014.

Chua, B. T., Guo, K., and Li, P. (2000). Direct cleavage by the calcium-activated protease calpain can lead to inactivation of caspases. *J Biol Chem* **275**, 5131–5135.

Clarke, P. G. (1990). Developmental cell death: Morphological diversity and multiple mechanisms. *Anat Embryol* **181**, 195–213.

Cunningham, T. J. (1982). Naturally occurring neuron death and its regulation by developing neural pathways. *Int Rev Cytol* **74**, 163–186.

Curtis, M. A., Penney, E. B., Pearson, A. G., van Roon-Mom, W. M., Butterworth, N. J., Dragunow, M. et al. (2003). Increased cell proliferation and neurogenesis in the adult human Huntington's disease brain. *Proc Natl Acad Sci U S A* **100**, 9023–9027.

Dal Canto, M. C. and Gurney, M. E. (1994). Development of central nervous system pathology in a murine transgenic model of human amyotrophic lateral sclerosis. *Am J Pathol* **145**, 1271–1279.

Deveraux, Q. L., Takahashi, R., Salvesen, G. S., and Reed, J. C. (1997). X-linked IAP is a direct inhibitor of cell-death proteases. *Nature* **388**, 300–304.

Dobson, C. M. (2003). Protein folding and misfolding. *Nature* **426**, 884–890.

Dobson, C. M. and Ellis, R. J. (1998). Protein folding and misfolding inside and outside the cell. *Embo J* **17**, 5251–5254.

Du, C., Fang, M., Li, Y., Li, L., and Wang, X. (2000). Smac, a mitochondrial protein that promotes cytochrome c-dependent caspase activation by eliminating IAP inhibition. *Cell* **102**, 33–42.

Egger, L., Schneider, J., Rheme, C., Tapernoux, M., Hacki, J., and Borner, C. (2003). Serine proteases mediate apoptosis-like cell death and phagocytosis under caspase-inhibiting conditions. *Cell Death Differ* **10**, 1188–1203.

Ellerby, L. M., Andrusiak, R. L., Wellington, C. L., Hackam, A. S., Propp, S. S., Wood, J. D. et al. (1999a). Cleavage of atrophin-1 at caspase site aspartic acid 109 modulates cytotoxicity. *J Biol Chem* **274**, 8730–8736.

Ellerby, L. M., Hackam, A. S., Propp, S. S., Ellerby, H. M., Rabizadeh, S., Cashman, N. R. et al. (1999b). Kennedy's disease: Caspase cleavage of the androgen receptor is a crucial event in cytotoxicity. *J Neurochem* **72**, 185–195.

Fadok, V. A., Voelker, D. R., Campbell, P. A., Cohen, J. J., Bratton, D. L., and Henson, P. M. (1992). Exposure of phosphatidylserine on the surface of apoptotic lymphocytes triggers specific recognition and removal by macrophages. *J Immunol* **148**, 2207–2216.

Fischer, H., Koenig, U., Eckhart, L., and Tschachler, E. (2002). Human caspase 12 has acquired deleterious mutations. *Biochem Biophys Res Commun* **293**, 722–726.

Forcet, C., Ye, X., Granger, L., Corset, V., Shin, H., Bredesen, D. E., and Mehlen, P. (2001). The dependence receptor DCC (deleted in colorectal cancer) defines an alternative mechanism for caspase activation. *Proc Natl Acad Sci U S A* **98**, 3416–3421.

Formigli, L., Papucci, L., Tani, A., Schiavone, N., Tempestini, A., Orlandini, G. E. et al. (2000). Aponecrosis: morphological and biochemical exploration of a syncretic process of cell death sharing apoptosis and necrosis. *J Cell Physiol* **182**, 41–49.

Friedlander, R. M. (2003). Apoptosis and caspases in neurodegenerative diseases. *N Engl J Med* **348**, 1365–1375.

Friedlander, R. M., Brown, R. H., Gagliardini, V., Wang, J., and Yuan, J. (1997a). Inhibition of ICE slows ALS in mice [letter] [published erratum appears in Nature 1998 Apr 9;392(6676):560]. *Nature* **388**, 31.

Friedlander, R. M., Gagliardini, V., Hara, H., Fink, K. B., Li, W., MacDonald, G. et al. (1997b). Expression of a dominant negative mutant of interleukin-1 beta converting enzyme in transgenic mice prevents neuronal cell death induced by trophic factor withdrawal and ischemic brain injury. *J Exp Med* **185**, 933–940.

Fuentes-Prior, P. and Salvesen, G. S. (2004). The protein structures that shape caspase activity, specificity, activation and inhibition. *Biochem J* **384**, 201–232.

Galvan, V., Gorostiza, O. F., Banwait, S., Ataie, M., Logvinova, A. V., Sitaraman, S. et al. (2006). Reversal of Alzheimer's-like pathology and behavior in human APP transgenic mice by mutation of Asp664. *Proc Natl Acad Sci U S A* **103**, 7130–7135.

Germain, M., Mathai, J. P., McBride, H. M., and Shore, G. C. (2005). Endoplasmic reticulum BIK initiates DRP1-regulated remodelling of mitochondrial cristae during apoptosis. *Embo J* **24**, 1546–1556.

Gervais, F. G., Xu, D., Robertson, G. S., Vaillancourt, J. P., Zhu, Y., Huang, J. et al. (1999). Involvement of caspases in proteolytic cleavage of Alzheimer's amyloid- beta precursor protein and amyloidogenic A beta peptide formation. *Cell* **97**, 395–406.

Gilloteaux, J., Jamison, J. M., Arnold, D., Ervin, E., Eckroat, L., Docherty, J. J. et al. (1998). Cancer cell necrosis by autoschizis: synergism of antitumor activity of vitamin C: vitamin K3 on human bladder carcinoma T24 cells. *Scanning* **20**, 564–575.

Goldberg, A. L. (2003). Protein degradation and protection against misfolded or damaged proteins. *Nature* **426**, 895–899.

Goldberg, Y. P., Nicholson, D. W., Rasper, D. M., Kalchman, M. A., Koide, H. B., Graham, R. K. et al. (1996). Cleavage of huntingtin by apopain, a proapoptotic cysteine protease, is modulated by the polyglutamine tract. *Nature Genet* **13**, 442–449.

Gomez-Santos, C., Ferrer, I., Santidrian, A. F., Barrachina, M., Gil, J., and Ambrosio, S. (2003). Dopamine induces autophagic cell death and alpha-synuclein increase in human neuroblastoma SH-SY5Y cells. *J Neurosci Res* **73**, 341–350.

Gould, E., McEwen, B. S., Tanapat, P., Galea, L. A., and Fuchs, E. (1997). Neurogenesis in the dentate gyrus of the adult tree shrew is regulated by psychosocial stress and NMDA receptor activation. *J Neurosci* **17**, 2492–2498.

Graham, R. K., Deng, Y., Slow, E. J., Haigh, B., Bissada, N., Lu, G. et al. (2006). Cleavage at the caspase-6 site is required for neuronal dysfunction and degeneration due to mutant huntingtin. *Cell* **125**, 1179–1191.

Green, D. R. and Kroemer, G. (2005). Pharmacological manipulation of cell death: Clinical applications in sight? *J Clin Invest* **115**, 2610–2617.

Guo, B., Zhai, D., Cabezas, E., Welsh, K., Nouraini, S., Satterthwait, A. C., and Reed, J. C. (2003). Humanin peptide suppresses apoptosis by interfering with Bax activation. *Nature* **423**, 456–461.

Hegde, R. S., Mastrianni, J. A., Scott, M. R., DeFea, K. A., Tremblay, P., Torchia, M. et al. (1998). A transmembrane form of the prion protein in neurodegenerative disease. *Science* **279**, 827–834.

Hermel, E., Gafni, J., Propp, S. S., Leavitt, B. R., Wellington, C. L., Young, J. E. et al. (2004). Specific caspase interactions and amplification are involved in selective neuronal vulnerability in Huntington's disease. *Cell Death Differ* **11**, 424–438.

Hirabayashi, M., Inoue, K., Tanaka, K., Nakadate, K., Ohsawa, Y., Kamei, Y. et al. (2001). VCP/p97 in abnormal protein aggregates, cytoplasmic vacuoles, and cell death, phenotypes relevant to neurodegeneration. *Cell Death Differ* **8**, 977–984.

Holley, C. L., Olson, M. R., Colon-Ramos, D. A., and Kornbluth, S. (2002). Reaper eliminates IAP proteins through stimulated IAP degradation and generalized translational inhibition. *Nat Cell Biol* **4**, 439–444.

Jin, K., Galvan, V., Xie, L., Mao, X. O., Gorostiza, O. F., Bredesen, D. E., and Greenberg, D. A. (2004a). Enhanced neurogenesis in Alzheimer's disease transgenic (PDGF-APPSw,Ind) mice. *Proc Natl Acad Sci U S A* **101**, 13363–13367.

Jin, K., LaFevre-Bernt, M., Sun, Y., Chen, S., Gafni, J., Crippen, D. et al. (2005). FGF-2 promotes neurogenesis and neuroprotection and prolongs survival in a transgenic mouse model of Huntington's disease. *Proc Natl Acad Sci U S A* **102**, 18189–18194.

Jin, K., Peel, A. L., Mao, X. O., Xie, L., Cottrell, B. A., Henshall, D. C., and Greenberg, D. A. (2004b). Increased hippocampal neurogenesis in Alzheimer's disease. *Proc Natl Acad Sci U S A* **101**, 343–347.

Kane, D. J., Ord, T., Anton, R., and Bredesen, D. E. (1995). Expression of bcl-2 inhibits necrotic neural cell death. *J Neurosci Res* **40**, 269–275.

Kaufman, R. J. (1999). Stress signaling from the lumen of the endoplasmic reticulum: Coordination of gene transcriptional and translational controls. *Genes Dev* **13**, 1211–1233.

Kawahara, T., Yanagi, H., Yura, T., and Mori, K. (1997). Endoplasmic reticulum stress-induced mRNA splicing permits synthesis of transcription factor Hac1p/Ern4p that activates the unfolded protein response. *Mol Biol Cell* **8**, 1845–1862.

Kerr, J. F., Wyllie, A. H., and Currie, A. R. (1972). Apoptosis: A basic biological phenomenon with wide-ranging implications in tissue kinetics. *Br J Cancer* **26**, 239–257.

Klein, W. L. (2002). Abeta toxicity in Alzheimer's disease: Globular oligomers (ADDLs) as new vaccine and drug targets. *Neurochem Int* **41**, 345–352.

Kobayashi, T., Tanaka, K., Inoue, K., and Kakizuka, A. (2002). Functional ATPase activity of p97/valosin-containing protein (VCP) is required for the quality control of endoplasmic reticulum in neuronally differentiated mammalian PC12 cells. *J Biol Chem* **277**, 47358–47365.

Koh, J. Y., Gwag, B. J., Lobner, D., and Choi, D. W. (1995). Potentiated necrosis of cultured cortical neurons by neurotrophins. *Science* **268**, 573–575.

Komatsu, M., Waguri, S., Ueno, T., Iwata, J., Murata, S., Tanida, I. et al. (2005). Impairment of starvation-induced and constitutive autophagy in Atg7-deficient mice. *J Cell Biol* **169**, 425–434.

Kopito, R. R. (2000). Aggresomes, inclusion bodies and protein aggregation. *Trends Cell Biol* **10**, 524–530.

Kopito, R. R. and Ron, D. (2000). Conformational disease. *Nat Cell Biol* **2**, E207–209.

Kordower, J. H., Isacson, O., and Emerich, D. F. (1999). Cellular delivery of trophic factors for the treatment of Huntington's disease: Is neuroprotection possible? *Exp Neurol* **159**, 4–20.

Kostic, V., Jackson-Lewis, V., de Bilbao, F., Dubois-Dauphin, M., and Przedborski, S. (1997). Bcl-2: Prolonging life in a transgenic mouse model of familial amyotrophic lateral sclerosis. *Science* **277**, 559–562.

Kuwana, T., Bouchier-Hayes, L., Chipuk, J. E., Bonzon, C., Sullivan, B. A., Green, D. R., and Newmeyer, D. D. (2005). BH3 domains of BH3-only proteins differentially regulate Bax-mediated mitochondrial membrane permeabilization both directly and indirectly. *Mol Cell* **17**, 525–535.

Kuwana, T., Mackey, M. R., Perkins, G., Ellisman, M. H., Latterich, M., Schneiter, R. et al. (2002). Bid, Bax, and lipids cooperate to form supramolecular openings in the outer mitochondrial membrane. *Cell* **111**, 331–342.

Lang, A. E., Gill, S., Patel, N. K., Lozano, A., Nutt, J. G., Penn, R. et al. (2006). Randomized controlled tria 1 of intraputamenal glial cell line-derived neurotrophic factor infusion in Parkinson disease. *Ann Neurol* **59**, 459–466.

Lankiewicz, S., Marc Luetjens, C., Truc Bui, N., Krohn, A. J., Poppe, M., Cole, G. M. et al. (2000). Activation of calpain I converts excitotoxic neuron death into a caspase-independent cell death. *J Biol Chem* **275**, 17064–17071.

Lee, K., Tirasophon, W., Shen, X., Michalak, M., Prywes, R., Okada, T. et al. (2002). IRE1-mediated unconventional mRNA splicing and S2P-mediated ATF6 cleavage merge to regulate XBP1 in signaling the unfolded protein response. *Genes Dev* **16**, 452–466.

Lee, Y. J., Jeong, S. Y., Karbowski, M., Smith, C. L., and Youle, R. J. (2004). Roles of the mammalian mitochondrial fission and fusion mediators Fis1, Drp1, and Opa1 in apoptosis. *Mol Biol Cell* **15**, 5001–5011.

Levi-Montalcini, R. (1966). The nerve growth factor: Its mode of action on sensory and sympathetic nerve cells. *Harvey Lect.* **60**, 217–259.

Levine, B. and Yuan, J. (2005). Autophagy in cell death: An innocent convict? *J Clin Invest* **115**, 2679–2688.

Li, H., Kolluri, S. K., Gu, J., Dawson, M. I., Cao, X., Hobbs, P. D. et al. (2000). Cytochrome c release and apoptosis induced by mitochondrial targeting of nuclear orphan receptor TR3. *Science* **289**, 1159–1164.

Li, J., Lee, B., and Lee, A. S. (2006). Endoplasmic reticulum stress-induced apoptosis: Multiple pathways and activation of p53-up-regulated modulator of apoptosis (PUMA) and NOXA by p53. *J Biol Chem* **281**, 7260–7270.

Lin, B., Kolluri, S. K., Lin, F., Liu, W., Han, Y. H., Cao, X. et al. (2004). Conversion of Bcl-2 from protector to killer by interaction with nuclear orphan receptor Nur77/TR3. *Cell* **116**, 527–540.

Liu, X., Van Vleet, T., and Schnellmann, R. G. (2004). The role of calpain in oncotic cell death. *Annu Rev Pharmacol Toxicol* **44**, 349–370.

Llambi, F., Causeret, F., Bloch-Gallego, E., and Mehlen, P. (2001). Netrin-1 acts as a survival factor via its receptors UNC5H and DCC. *Embo J* **20**, 2715–2722.

Lockshin, R. A. and Williams, C. M. (1964). Programmed cell death. II. Endocrine potentiation of the breakdown of the intersegmental muscles of silk moths. *J Insect Physiol* **10**, 643–649.

Lu, D. C., Rabizadeh, S., Chandra, S., Shayya, R. F., Ellerby, L. M., Ye, X. et al. (2000). A second cytotoxic proteolytic peptide derived from amyloid β-protein precursor [see comments]. *Nature Med* **6**, 397–404.

Lu, D. C., Shaked, G. M., Masliah, E., Bredesen, D. E., and Koo, E. H. (2003a). Amyloid beta protein toxicity mediated by the formation of amyloid-beta protein precursor complexes. *Ann Neurol* **54**, 781–789.

Lu, D. C., Soriano, S., Bredesen, D. E., and Koo, E. H. (2003b). Caspase cleavage of the amyloid precursor protein modulates amyloid beta-protein toxicity. *J Neurochem* **87**, 733–741.

Luciano, F., Zhai, D., Zhu, X., Bailly-Maitre, B., Ricci, J. E., Satterthwait, A. C., and Reed, J. C. (2005). Cytoprotective peptide humanin binds and inhibits proapoptotic Bcl-2/Bax family protein BimEL. *J Biol Chem* **280**, 15825–15835.

Lum, J. J., Bauer, D. E., Kong, M., Harris, M. H., Li, C., Lindsten, T., and Thompson, C. B. (2005). Growth factor regulation of autophagy and cell survival in the absence of apoptosis. *Cell* **120**, 237–248.

Magavi, S. S., Leavitt, B. R., and Macklis, J. D. (2000). Induction of neurogenesis in the neocortex of adult mice. *Nature* **405**, 951–955.

Mah, S. P., Zhong, L. T., Liu, Y., Roghani, A., Edwards, R. H., and Bredesen, D. E. (1993). The protooncogene bcl-2 inhibits apoptosis in PC12 cells. *J Neurochem* **60**, 1183–1186.

Mahul-Mellier, A. L., Hemming, F. J., Blot, B., Fraboulet, S., and Sadoul, R. (2006). Alix, making a link between apoptosis-linked gene-2, the endosomal sorting complexes required for transport, and neuronal death in vivo. *J Neurosci* **26**, 542–549.

Majno, G. and Joris, I. (1995). Apoptosis, oncosis, and necrosis. An overview of cell death [see comments]. *Am J Pathol* **146**, 3–15.

Martins, L. M., Iaccarino, I., Tenev, T., Gschmeissner, S., Totty, N. F., Lemoine, N. R. et al. (2002). The serine protease Omi/HtrA2 regulates apoptosis by binding XIAP through a reaper-like motif. *J Biol Chem* **277**, 439–444.

Mathai, J. P., Germain, M., and Shore, G. C. (2005). BH3-only BIK regulates BAX,BAK-dependent release of Ca2+ from endoplasmic reticulum stores and mitochondrial apoptosis during stress-induced cell death. *J Biol Chem* **280**, 23829–23836.

Mehlen, P., Rabizadeh, S., Snipas, S. J., Assa-Munt, N., Salvesen, G. S., and Bredesen, D. E. (1998). The DCC gene product induces apoptosis by a mechanism requiring receptor proteolysis. *Nature* **395**, 801–804.

Morishima, N., Nakanishi, K., Takenouchi, H., Shibata, T., and Yasuhiko, Y. (2002). An endoplasmic reticulum stress-specific caspase cascade in apoptosis. Cytochrome c-independent activation of caspase-9 by caspase-12. *J Biol Chem* **277**, 34287–34294.

Mucke, L., Masliah, E., Yu, G. Q., Mallory, M., Rockenstein, E. M., Tatsuno, G. et al. (2000). High-level neuronal expression of abeta 1–42 in wild-type human amyloid protein precursor transgenic mice: synaptotoxicity without plaque formation. *J Neurosci* **20**, 4050–4058.

Murray-Zmijewski, F., Lane, D. P., and Bourdon, J. C. (2006). p53/p63/p73 isoforms: An orchestra of isoforms to harmonise cell differentiation and response to stress. *Cell Death Differ.*

Muzio, M., Chinnaiyan, A. M., Kischkel, F. C., O'Rourke, K., Shevchenko, A., Ni, J. et al. (1996). FLICE, a novel FADD-homologous ICE/CED-3-like protease, is recruited to the CD95 (Fas/APO-1) death-inducing signaling complex. *Cell* **85**, 817–827.

Nakagawa, T., Zhu, H., Morishima, N., Li, E., Xu, J., Yankner, B. A., and Yuan, J. (2000). Caspase-12 mediates endoplasmic-reticulum-specific apoptosis and cytotoxicity by amyloid-beta. *Nature* **403**, 98–103.

Ng, F. W., Nguyen, M., Kwan, T., Branton, P. E., Nicholson, D. W., Cromlish, J. A., and Shore, G. C. (1997). p28 Bap31, a Bcl-2/Bcl-XL- and procaspase-8-associated protein in the endoplasmic reticulum. *J Cell Biol* **139**, 327–338.

Niquet, J., Liu, H., and Wasterlain, C. G. (2005). Programmed neuronal necrosis and status epilepticus. *Epilepsia* **46 Suppl 5**, 43–48.

Nishimoto, I. (1998). A new paradigm for neurotoxicity by FAD mutants of betaAPP: A signaling abnormality. *Neurobiol Aging* **19**, S33–38.

Novgorodov, S. A. and Gudz, T. I. (1996). Permeability transition pore of the inner mitochondrial membrane can operate in two open states with different selectivities. *J Bioenerg Biomembr* **28**, 139–146.

Ona, V. O., Li, M., Vonsattel, J. P. G., Andrews, L. J., Khan, S. Q., Chung, W. M. et al. (1999). Inhibition of caspase-1 slows disease progression in a mouse model of Huntington's disease. *Nature* **399**, 263–267.

Oppenheim, R. W. (1985). Naturally occurring cell death during neural development. *Trends Neurosci* **17**, 487–493.

Oppenheim, R. W. (1991). Cell death during development of the nervous system. *Annu Rev Neurosci* **14**, 453–501.

Patil, C. and Walter, P. (2001). Intracellular signaling from the endoplasmic reticulum to the nucleus: The unfolded protein response in yeast and mammals. *Curr Opin Cell Biol* **13**, 349–355.

Perfettini, J. L., Reed, J. C., Israel, N., Martinou, J. C., Dautry-Varsat, A., and Ojcius, D. M. (2002). Role of Bcl-2 family members in caspase-independent apoptosis during Chlamydia infection. *Infect Immun* **70**, 55–61.

Pilar, G. and Landmesser, L. (1976). Ultrastructural differences during embryonic cell death in normal and peripherally deprived ciliary ganglia. *J Cell Biol* **68**, 339–356.

Polster, B. M., Basanez, G., Etxebarria, A., Hardwick, J. M., and Nicholls, D. G. (2005). Calpain I induces cleavage and release of apoptosis-inducing factor from isolated mitochondria. *J Biol Chem* **280**, 6447–6454.

Rabizadeh, S. and Bredesen, D. E. (1994). Is p75NGFR involved in developmental neural cell death? *Dev Neurosci* **16**, 207–211.

Rabizadeh, S., Oh, J., Zhong, L. T., Yang, J., Bitler, C. M., Butcher, L. L., and Bredesen, D. E. (1993). Induction of apoptosis by the low-affinity NGF receptor. *Science* **261**, 345–348.

Rao, R. V. and Bredesen, D. E. (2004). Misfolded proteins, endoplasmic reticulum stress and neurodegeneration. *Curr Opin Cell Biol* **16**, 653–662.

Rao, R. V., Castro-Obregon, S., Frankowski, H., Schuler, M., Stoka, V., Del Rio, G. et al. (2002a). Coupling endoplasmic reticulum stress to the cell death program. AN Apaf-1-INDEPENDENT INTRINSIC PATHWAY. *J Biol Chem* **277**, 21836–21842.

Rao, R. V., Hermel, E., Castro-Obregon, S., del Rio, G., Ellerby, L. M., Ellerby, H. M., and Bredesen, D. E. (2001). Coupling endoplasmic reticulum stress to the cell death program: Mechanism of caspase activation. *J Biol Chem* **276**, 33869–33874.

Rao, R. V., Niazi, K., Mollahan, P., Mao, X., Crippen, D., Poksay, K. S. et al. (2006). Coupling endoplasmic reticulum stress to the cell-death program: A novel HSP90-independent role for the small chaperone protein p23. *Cell Death Differ* **13**, 415–425.

Rao, R. V., Peel, A., Logvinova, A., del Rio, G., Hermel, E., Yokota, T. et al. (2002b). Coupling endoplasmic reticulum stress to the cell death program: Role of the ER chaperone GRP78. *FEBS Lett* **514**, 122–128.

Rao, R. V., Poksay, K. S., Castro-Obregon, S., Schilling, B., Row, R. H., Del Rio, G. et al. (2004). Molecular components of a cell death pathway activated by endoplasmic reticulum stress. *J Biol Chem* **279**, 177–187.

Reddy, R. K., Mao, C., Baumeister, P., Austin, R. C., Kaufman, R. J., and Lee, A. S. (2003). Endoplasmic reticulum chaperone protein GRP78 protects cells from apoptosis induced by topoisomerase inhibitors: Role of ATP binding site in suppression of caspase-7 activation. *J Biol Chem* **278**, 20915–20924.

Roth, W., Kermer, P., Krajewska, M., Welsh, K., Davis, S., Krajewski, S., and Reed, J. C. (2003). Bifunctional apoptosis inhibitor (BAR) protects neurons from diverse cell death pathways. *Cell Death Differ* **10**, 1178–1187.

Ruiz-Vela, A., Opferman, J. T., Cheng, E. H., and Korsmeyer, S. J. (2005). Proapoptotic BAX and BAK control multiple initiator caspases. *EMBO Rep* **6**, 379–385.

Saleh, M., Vaillancourt, J. P., Graham, R. K., Huyck, M., Srinivasula, S. M., Alnemri, E. S. et al. (2004). Differential modulation of endotoxin responsiveness by human caspase-12 polymorphisms. *Nature* **429**, 75–79.

Salvesen, G. S. and Dixit, V. M. (1997). Caspases: intracellular signaling by proteolysis. *Cell* **91**, 443–446.

Schuler, M., Bossy-Wetzel, E., Goldstein, J. C., Fitzgerald, P., and Green, D. R. (2000). p53 induces apoptosis by caspase activation through mitochondrial cytochrome c release. *J Biol Chem* **275**, 7337–7342.

Schwartz, L. M. (1991). The role of cell death genes during development. *Bioessays* **13**, 389–395.

Schweichel, J. U. (1972). [Electron microscopic studies on the degradation of the apical ridge during the development of limbs in rat embryos]. *Z Anat Entwicklungsgesch* **136**, 192–203.

Schweichel, J. U. and Merker, H. J. (1973). The morphology of various types of cell death in prenatal tissues. *Teratology* **7**, 253–266.

Scorrano, L., Oakes, S. A., Opferman, J. T., Cheng, E. H., Sorcinelli, M. D., Pozzan, T., and Korsmeyer, S. J. (2003). BAX and BAK regulation of endoplasmic reticulum Ca2+: A control point for apoptosis. *Science* **300**, 135–139.

Selkoe, D. J. (2003). Folding proteins in fatal ways. *Nature* **426**, 900–904.

Shaked, G. M., Kummer, M. P., Lu, D. C., Galvan, V., Bredesen, D. E., and Koo, E. H. (2006). A{beta} induces cell death by direct interaction with its cognate extracellular domain on APP (APP 597–624). *Faseb J.*

Sherman, M. Y. and Goldberg, A. L. (2001). Cellular defenses against unfolded proteins: A cell biologist thinks about neurodegenerative diseases. *Neuron* **29**, 15–32.

Shimizu, S., Kanaseki, T., Mizushima, N., Mizuta, T., Arakawa-Kobayashi, S., Thompson, C. B., and Tsujimoto, Y. (2004). Role of Bcl-2 family proteins in a non-apoptotic programmed cell death dependent on autophagy genes. *Nat Cell Biol* **6**, 1221–1228.

Sidrauski, C. and Walter, P. (1997). The transmembrane kinase Ire1p is a site-specific endonuclease that initiates mRNA splicing in the unfolded protein response. *Cell* **90**, 1031–1039.

Siegel, R. M., Frederiksen, J. K., Zacharias, D. A., Chan, F. K., Johnson, M., Lynch, D. et al. (2000). Fas preassociation required for apoptosis signaling and dominant inhibition by pathogenic mutations. *Science* **288**, 2354–2357.

Sitia, R. and Braakman, I. (2003). Quality control in the endoplasmic reticulum protein factory. *Nature* **426**, 891–894.

Snyder, E. Y., Yoon, C., Flax, J. D., and Macklis, J. D. (1997). Multipotent neural precursors can differentiate toward replacement of neurons undergoing targeted apoptotic degeneration in adult mouse neocortex. *Proc Natl Acad Sci U S A* **94**, 11663–11668.

Sperandio, S., de Belle, I., and Bredesen, D. E. (2000). An alternative, non-apoptotic form of programmed cell death. *Proc Natl Acad Sci U S A* **97**, 14376–14381.

Sperandio, S., Poksay, K., de Belle, I., Lafuente, M. J., Liu, B., Nasir, J., and Bredesen, D. E. (2004). Paraptosis: Mediation by MAP kinases and inhibition by AIP-1/Alix. *Cell Death Differ* **11**, 1066–1075.

Stoka, V., Turk, B., Schendel, S. L., Kim, T. H., Cirman, T., Snipas, S. J. et al. (2001). Lysosomal protease pathways to apoptosis. Cleavage of bid, not pro-caspases, is the most likely route. *J Biol Chem* **276**, 3149–3157.

Studnicka, F. K. (1905). Die Parietalorgane. In A. Oppel, Ed., *Lehrbuch der vergleichende mikroskopischen Anatomie der Wirbeltiere, Vol. 5*. S.G. Fischer Verlag, Jena.

Stupack, D. G., Puente, X. S., Boutsaboualoy, S., Storgard, C. M., and Cheresh, D. A. (2001). Apoptosis of adherent cells by recruitment of caspase-8 to unligated integrins. *J Cell Biol* **155**, 459–470.

Susin, S. A., Lorenzo, H. K., Zamzami, N., Marzo, I., Snow, B. E., Brothers, G. M. et al. (1999). Molecular characterization of mitochondrial apoptosis-inducing factor. *Nature* **397**, 441–446.

Suzuki, Y., Imai, Y., Nakayama, H., Takahashi, K., Takio, K., and Takahashi, R. (2001). A serine protease, HtrA2, is released from the mitochondria and interacts with XIAP, inducing cell death. *Mol Cell* **8**, 613–621.

Syntichaki, P., Xu, K., Driscoll, M., and Tavernarakis, N. (2002). Specific aspartyl and calpain proteases are required for neurodegeneration in C. elegans. *Nature* **419**, 939–944.

Tanaka, Y., Igarashi, S., Nakamura, M., Gafni, J., Torcassi, C., Schilling, G. et al. (2006). Progressive phenotype and nuclear accumulation of an amino-terminal cleavage fragment in a transgenic mouse model with inducible expression of full-length mutant huntingtin. *Neurobiol Dis* **21**, 381–391.

Tata, J. R. (1966). Requirement for RNA and protein synthesis for induced regression of the tadpole tail in organ culture. *Dev Biol* **13**, 77–94.

Taylor, J. P., Hardy, J., and Fischbeck, K. H. (2002). Toxic proteins in neurodegenerative disease. *Science* **296**, 1991–1995.

Thibert, C., Teillet, M. A., Lapointe, F., Mazelin, L., Le Douarin, N. M., and Mehlen, P. (2003). Inhibition of neuroepithelial patched-induced apoptosis by sonic hedgehog. *Science* **301**, 843–846.

Thornberry, N. A. and Lazebnik, Y. (1998). Caspases: Enemies within. *Science* **281**, 1312–1316.

Turmaine, M., Raza, A., Mahal, A., Mangiarini, L., Bates, G. P., and Davies, S. W. (2000). Nonapoptotic neurodegeneration in a transgenic mouse model of Huntington's disease. *Proc Natl Acad Sci U S A* **97**, 8093–8097.

Tuszynski, M. H., Thal, L., Pay, M., Salmon, D. P., U, H. S., Bakay, R. et al. (2005). A phase 1 clinical trial of nerve growth factor gene therapy for Alzheimer disease. *Nat Med* **11**, 551–555.

Urano, F., Wang, X., Bertolotti, A., Zhang, Y., Chung, P., Harding, H. P., and Ron, D. (2000). Coupling of stress in the ER to activation of JNK protein kinases by transmembrane protein kinase IRE1. *Science* **287**, 664–666.

Vande Velde, C., Cizeau, J., Dubik, D., Alimonti, J., Brown, T., Israels, S. et al. (2000). BNIP3 and genetic control of necrosis-like cell death through the mitochondrial permeability transition pore [In Process Citation]. *Mol Cell Biol* **20**, 5454–5468.

Verhagen, A. M., Ekert, P. G., Pakusch, M., Silke, J., Connolly, L. M., Reid, G. E. et al. (2000). Identification of DIABLO, a mammalian protein that promotes apoptosis by binding to and antagonizing IAP proteins. *Cell* **102**, 43–53.

Wang, X., Zhu, S., Drozda, M., Zhang, W., Stavrovskaya, I. G., Cattaneo, E. et al. (2003). Minocycline inhibits caspase-independent and -dependent mitochondrial cell death pathways in models of Huntington's disease. *Proc Natl Acad Sci U S A* **100**, 10483–10487.

Watts, G. D., Wymer, J., Kovach, M. J., Mehta, S. G., Mumm, S., Darvish, D. et al. (2004). Inclusion body myopathy associated with Paget disease of bone and frontotemporal dementia is caused by mutant valosin-containing protein. *Nat Genet* **36**, 377–381.

Wellington, C. L., Singaraja, R., Ellerby, L., Savill, J., Roy, S., Leavitt, B. et al. (2000). Inhibiting caspase cleavage of huntingtin reduces toxicity and aggregate formation in neuronal and nonneuronal cells. *J Biol Chem.*

Williamson, T. L., Marszalek, J. R., Vechio, J. D., Bruijn, L. I., Lee, M. K., Xu, Z. et al. (1996). Neurofilaments, radial growth of axons, and mechanisms of motor neuron disease. *Cold Spring Harb Symp Quant Biol* **61**, 709–723.

Yang, F., Sun, X., Beech, W., Teter, B., Wu, S., Sigel, J. et al. (1998). Antibody to caspase-cleaved actin detects apoptosis in differentiated neuroblastoma and plaque-associated neurons and microglia in Alzheimer's disease. *Am J Pathol* **152**, 379–389.

Yao, R. and Cooper, G. M. (1995). Requirement for phosphatidylinositol-3 kinase in the prevention of apoptosis by nerve growth factor. *Science* **267**, 2003–2006.

Yoshida, H., Matsui, T., Yamamoto, A., Okada, T., and Mori, K. (2001). XBP1 mRNA is induced by ATF6 and spliced by IRE1 in response to ER stress to produce a highly active transcription factor. *Cell* **107**, 881–891.

Yu, L., Alva, A., Su, H., Dutt, P., Freundt, E., Welsh, S. et al. (2004). Regulation of an ATG7-beclin 1 program of autophagic cell death by caspase-8. *Science* **304**, 1500–1502.

Yu, L., Wan, F., Dutta, S., Welsh, S., Liu, Z., Freundt, E. et al. (2006). Autophagic programmed cell death by selective catalase degradation. *Proc Natl Acad Sci U S A* **103**, 4952–4957.

Yu, S. W., Wang, H., Poitras, M. F., Coombs, C., Bowers, W. J., Federoff, H. J. et al. (2002). Mediation of poly(ADP-ribose) polymerase-1- dependent cell death by apoptosis-inducing factor. *Science* **297**, 259–263.

Yuan, J., Shaham, S., Ledoux, S., Ellis, H. M., and Horvitz, H. R. (1993). The C. elegans cell death gene ced-3 encodes a protein similar to mammalian interleukin-1 beta-converting enzyme. *Cell* **75**, 641–652.

Yuan, J. and Yankner, B. A. (1999). Caspase activity sows the seeds of neuronal death. *Nat Cell Biol* **1**, E44–45.

Yue, Z., Jin, S., Yang, C., Levine, A. J., and Heintz, N. (2003). Beclin 1, an autophagy gene essential for early embryonic development, is a haplo-insufficient tumor suppressor. *Proc Natl Acad Sci U S A* **100**, 15077–15082.

Zhai, D., Luciano, F., Zhu, X., Guo, B., Satterthwait, A. C., and Reed, J. C. (2005). Humanin binds and nullifies Bid activity by blocking its activation of Bax and Bak. *J Biol Chem* **280**, 15815–15824.

Zhang, H., Xu, Q., Krajewski, S., Krajewska, M., Xie, Z., Fuess, S. et al. (2000). BAR: An apoptosis regulator at the intersection of caspases and Bcl-2 family proteins. *Proc Natl Acad Sci U S A* **97**, 2597–2602.

Zong, W. X. and Thompson, C. B. (2006). Necrotic death as a cell fate. *Genes Dev* **20**, 1–15.

9

Developmental Neurology: A Molecular Perspective

Juan M. Pascual

The clinical approach to the numerous neurological and neuromuscular diseases that manifest during development (i.e., the period of time between conception and adolescence) draws upon analytical chemistry, genetics, imaging, and pharmacology foundations. Relentless progress on molecular structure, function, and regulation steadily stimulates the prominence of pediatric (or developmental) neurology as one of the neurosciences. Underlying this scientific expansion is also an improving knowledge of genes and genetic control mechanisms. Many molecular processes and associated diseases have been related to specific genes. Even the susceptibility to infections or the recovery from stroke or trauma, which were conventionally referred to as acquired disorders to differentiate them from genetically determined diseases, can be influenced by the genome (Waters & Nicoll, 2005).

The mechanism (i.e., the integrated explanation of the relations between gene defects and variations and cellular reactions) of some developmental neurological diseases has been elucidated, as have the factors that specify their preferred age of onset and temporal evolution. A better understanding of the changes experienced by the developing nervous system and muscle has contributed to this: during infancy and childhood, several gene expression programs are played out and then become quiescent as the organism grows and matures (Thor, 1995). Thus, the phenotype associated with a pathogenic mutation may remain dormant until the mutant gene is utilized for a particular function, and this may occur after birth or well in adulthood.

An example from brain energetic metabolism illustrates this principle: in contrast with the adult, which consumes preferentially glucose, the fetal brain prefers to metabolize lipid derivatives such as ketones. At birth, the cerebral metabolic rate for glucose is small for the size of the brain, but it increases during childhood, exceeding the neonatal rate by three-fold. As adolescence approaches, glucose consumption decreases to adult levels, surpassing newborn levels by two-fold (Chugani et al., 1987). It is not surprising, then, that the neonatal brain tolerates neuroglycopenia (a state of diminished cerebral energetic substrate brought about by inadequate glucose entry) relatively well. In practice, this is underscored by the manifestations of both congenital hyperinsulinism and of glucose transporter type 1 deficiency, the former commonly due mutations in the pancreatic ATP-sensitive potassium channel (K_{ATP}) (Aguilar-Bryan et al., 2001) and the latter caused by mutation of the glucose carrier of the blood–brain barrier (Glut1) (Pascual et al., 2004). These diseases may remain unnoticed in the newborn until their symptoms become increasingly accentuated during childhood. In some occasions, the converse phenomenon is true, allowing for the restoration of a normal phenotype following a diseased state as development induces new genes to substitute for abnormal ones. This is exemplified by a transient or "reversible" form of cytochrome c oxidase deficient myopathy that manifests early in infancy with hypotonia and profound weakness, only to be followed by the return of enzyme activity in muscle and normal strength later in childhood (DiMauro et al., 1983).

Delayed onset followed by relentless progression, a pattern frequently masked by normal developmental gains during infancy and childhood, is particularly common in disorders associated with accumulation. In these storage diseases, symptoms may not emerge until a stored metabolite, degraded at excessively slow rates, interferes with other cellular reactions, or forms growing, metabolically inert aggregates inside cells. In the most common form of α-1,4-glucosidase (acid maltase) deficiency (Pompe disease, glycogen storage disease type II), infants are born normal, only to experience diminished glycogen degradation and progressive retention in the lysosomes of multiple cell types during infancy, resulting in advancing cardiomegaly, skeletal muscular enlargement, hepatomegaly, and motor neuron loss.

Yet, genotype-phenotype correlations still pose challenges, as do some genotypes, when referred to the developing nervous and neuromuscular systems. On one hand, genetic polymorphisms and variations abound in all individuals, but their function and interrelations are not well understood, as is transcriptional activity in most disease states. In addition, and in contrast with the adult, the identification of pediatric neurological phenotypes (i.e., the repertoire of neurological diseases) is necessarily restricted for as long as intelligence and volition remain immature or underdeveloped, leading to phenotypic oversimplification. These, together with difficulties in separating true phenotypic characteristics from epiphenomena (i.e., observables related only indirectly to the fundamental disease mechanism) complicate the cataloging of developmental neurological diseases. Many diseases are diagnosed by assaying metabolites generated in the course of reactions that lay distant from the original enzyme defect just because they coexist. The error and misattribution risks inherent to this diagnostic approach is not negligible, particularly when confronted with unusual or incompletely penetrant forms of otherwise frequent diseases, or especially, with rare diseases.

Genotyping is subject to additional, special considerations when applied to developmental neurological diseases; well-defined clinical entities can include phenocopies (i.e., individuals that manifest a similar phenotype due to mutation of unrelated genes), a phenomenon inevitably embedded in the pregenomic clinical literature. For example, mutation of the SCO2 gene, necessary for mitochondrial respiratory chain assembly, can reproduce the manifestations of spinal muscular atrophy (Salviati et al., 2002), a phenotype much more frequently due to SMN1 gene mutations that disrupt the formation of spliceosomes and other aspects of RNA processing (Monani, 2005). Genotyping is also influenced by uneven or skewed mutation abundance among patient tissues that contribute to the phenotype or that are sampled for diagnosis. Thus, mosaicism for mutant MeCP2 (a repressor of gene expression and regulator of RNA splicing) in the male and skewed X chromosome inactivation in the female may account for the disparate phenotypes of Rett syndrome, encompassing male neonatal death, male mental retardation, asymptomatic female carrier status or female Rett syndrome with motor and psychiatric disabilities (Zoghbi, 2005). The proportion of mutant mitochondrial DNA within the tissues of one individual is also variable, depending upon the tissue selected for genotyping (a phenomenon known as *heteroplasmy*). In practice, several DNA sources (usually blood, urinary sediment, hair follicle, and buccal smear) can be examined to detect a low-abundance mutation (Shanske et al., 2004).

Nevertheless, disease severity does not necessarily correlate with degree of cellular dysfunction or with gene conservation; in fact, the more central a reaction or process is, the less probable that its corresponding disease state will be compatible with life. Thus, human diseases are most likely to affect the least essential functions: those exhibiting redundancy, which can, at least, be partially compensated for, ultimately allowing for the survival of the organism. Mutations commonly found in highly conserved genes (noncoding polymorphisms, a frequent form of genetic alteration) may reflect evolutionary drift, but usually do not cause disease. On the other hand, higher impact mutations, such as those causing sequence variation in an important protein-coding area of a gene, are usually not tolerable and, if not compensated by the other allele, even lethal during embryogenesis.

The disorders of the developing nervous and neuromuscular systems can be classified according to several criteria such as predominant symptom, age at onset, size of the main abnormal metabolite (large polypeptides or carbohydrates, or small molecules), among others. From a molecular point of view, these disorders are best defined by cellular phenotype. It is expected that the number and features of neurological phenotypes will be expanded as novel molecular investigation methods are applied to diseased cells and organs. Cells can manifest alterations in number, structure, and fluxes arising from chemical reactions (see Table 9.1).

Abnormal cellular numbers may result from deficient or excessive cell division, or from reduced cell death during specific developmental stages. Proliferative and migrational disorders are prominent in the cerebral cortex and often cause mental retardation and epilepsy. An increasingly large group of neurological diseases associated with neurodegeneration are caused by facilitated apoptosis and selective cellular loss. Tumors, not covered in this section, are a leading cause of

Table 9.1 Phenotypes of Molecular Disorders

Cell number	Flux
Division	Electrical impulse
Death	Transport
Cell structure	Metabolism
Shape and resistance	
Movement	

childhood death and disability and represent the prototypic proliferative disorders due to excessive cell division.

Molecules primarily responsible for maintaining cell structure and for the generation of movement are vital to muscle function. Numerous myopathies are caused by deficiency or loss of function of intra- and extracellular proteins that form the muscle supporting scaffold. In these disorders, exemplified by the dystrophinopathies, the mechanical forces that the muscle normally generates disrupt muscle membranes, leading to cellular death, inflammation, and proliferation. Mutations in genes that code for contractile proteins cause myopathy by impairing molecular interactions that generate movement and force.

Flux, the transit of ions, molecules, and electrical currents constitutes the essence of cellular metabolism and communication. Fluxes are governed by reaction rates that change under the influence of cellular and external signals and metabolic fluctuations. The velocity of molecular transitions such as opening or inactivation of ion channels and of enzyme structure interconversions determines the rates of all reactions. Flux measurements across cellular compartments pose special conceptual and methodological complexities, as cellular reactions are densely interrelated while being compartmentalized, and metabolic pools and signals are frequently inaccessible to measurement in molecular time scale. Altered flux states result in abnormal nerve and muscle excitability, deficient transport of substrates across membranes, and altered metabolite formation and consumption, leading to depletion or accumulation of precursors and byproducts and to induction of alterative reactions.

The following chapters introduce the molecular principles of metabolic diseases and of neuromuscular diseases that manifest during development. Diseases associated with abnormal excitability and signaling, cell death (apoptosis), and some cellular structural disorders are explored in other chapters.

References

Aguilar-Bryan, L., Bryan, J., Nakazaki, M. (2001). Of mice and men: K(ATP) channels and insulin secretion. *Recent Prog Horm Res.* **56**, 47–68.

Chugani, H. T., Phelps, M. E., Mazziotta, J. C. (1987). Positron emission tomography study of human brain functional development. *Ann Neurol.* **22**, 487–497.

DiMauro, S., Nicholson, J. F., Hays, A. P., Eastwood, A. B., Papadimitriou, A., Koenigsberger, R., DeVivo, D. C. (1983). Benign infantile mitochondrial myopathy due to reversible cytochrome c oxidase deficiency. *Ann Neurol.* **14**, 226–234.

Monani, U. R. (2005). Spinal muscular atrophy: A deficiency in a ubiquitous protein; a motor neuron-specific disease. *Neuron* **48**, 885–896.

Pascual, J. M., Wang, D., Lecumberri, B., Yang, H., Mao, X., Yang, R., De Vivo, D. C. (2004). GLUT1 deficiency and other glucose transporter diseases. *Eur J Endocrinol.* **150**, 627–633.

Salviati, L., Sacconi, S., Rasalan, M. M., Kronn, D. F., Braun, A., Canoll, P. et al. (2002). Cytochrome c oxidase deficiency due to a novel SCO2 mutation mimics Werdnig-Hoffmann disease. *Arch Neurol.* **59**, 862–865.

Shanske, S., Pancrudo, J., Kaufmann, P., Engelstad, K., Jhung, S., Lu, J. et al. (2004). Varying loads of the mitochondrial DNA A3243G mutation in different tissues: Implications for diagnosis. *Am J Med Genet A.* **130**, 134–137.

Thor, S. (1995). The genetics of brain development: Conserved programs in flies and mice. *Neuron* **15**, 975–977.

Waters, R. J., Nicoll, J. A. (2005). Genetic influences on outcome following acute neurological insults. *Curr Opin Crit Care* **11**, 105–110.

Zoghbi, H. Y. (2005). MeCP2 dysfunction in humans and mice. *J Child Neurol.* **20**, 736–740.

10

Metabolic Diseases of the Nervous System

Juan M. Pascual

Metabolic disorders constitute an expanding group of flux diseases that includes heterogeneous conditions (see Table 10.1). Thus, a unifying definition becomes necessary. Strictly speaking, neurometabolic diseases arise from genetic deficiency of intermediary metabolism enzymes, in contrast with mutations in genes encoding cytostructural proteins or proteins involved in cell division, immunity, excitability, cell-to-cell communication, secretion or movement, which are not counted among them. Nevertheless, intermediary metabolism abnormalities can be found in virtually all these conditions, allowing for the consideration of at least some when discussing the neurometabolic diseases.

Whether involving carbohydrate, lipid, or protein metabolism, the manifestations of neurometabolic diseases are pleomorphic and can present at any time during the entire lifespan. Regardless of their age of onset and mode of presentation, the practical approach to the patient afflicted by a neurometabolic disease includes a customized but systematic series of evaluations, as well as an assessment of the ancestry and family structure aimed at identifying a possible pattern of inheritance and detecting all relatives at-risk for a potentially heritable trait. It is often possible to establish the pattern of inheritance of a familial disease on clinical grounds alone. In the case of a potentially heritable trait, apparently unaffected relatives can be found to manifest subtle abnormalities indicative of an incompletely penetrant trait. A series of analytical investigations are performed to confirm the diagnosis according to the patient's clinical syndrome. The biochemical analysis of patient tissues, such as biopsied muscle, and of cellular elements, such as cultured fibroblasts, is often necessary to confirm a specific enzyme deficiency and is followed, when available, by genotyping. In some cases, genetic test batteries or panels are available to screen genes associated with diseases that share a similar phenotype. Genotyping allows for genetic counseling, for the screening of relatives at risk and, in an increasing number of instances, for prenatal diagnosis via amniocentesis.

Table 10.1 Neurometabolic Diseases

- I. Disorders of the cell membrane
  - Transport disorders
    - Disorders of glucose transport
      - *Glucose transporter type 1 deficiency*
    - Disorders of metal transport
      - *Menkes disease*
      - *Wilson disease*
    - Other transport disorders
      - *Excitatory amino acid transporter 1 deficiency*
  - Disorders of neurotransmission
    - Neurotransmitter deficiencies
      - *Segawa disease*
      - Other disorders of amine synthesis
    - Neurotransmitter receptor diseases*
      - *Stiff baby syndrome*
- II. Disorders of intracellular organelles
  - Peroxysomal diseases
    - Disorders of peroxysome biogenesis
      - *Zellweger-spectrum diseases*
      - *Rhizomelic chondrodysplasia punctata*
    - Disorders of peroxysomal enzymes
      - *X-linked adrenoleukodystrophy*
      - *Refsum disease*
  - Mitochondrial diseases*
    - Disorders of fatty acid oxidation
      - *Carnitine transporter deficiency**
      - Carnitine palmitoyltransferase deficiencies*
      - Acyl-CoA dehydrogenase deficiencies*
    - Pyruvate metabolism disorders
      - *Pyruvate dehydrogenase deficiency*
      - *Pyruvate carboxylase deficiency*
    - Respiratory chain disorders*
    - Disorders of mitochondrial DNA maintenance and communication*
    - Krebs cycle disorders
  - Glycosylation disorders
    - Defects of oligosaccharide synthesis
    - Defects of oligosaccharide processing
  - Lysosomal diseases
    - *Gaucher disease*
    - *Niemann-Pick diseases*
    - *GM2 gangliosidoses*
    - *Sulfatide lipidoses*
    - *Krabbe disease*
    - *Mucopolysaccharidoses*
    - *Mucolipidoses*
    - *Farber disease*
    - *Fabry disease*
    - *Schindler disease*
    - Neuronal ceroid lipofuscinoses
- III. Enzyme disorders
  - Urea cycle disorders
  - *Galactosemia*
  - *Phenylketonuria*
  - *Non-ketotic hyperglycinemia*
  - Glycogen storage disorders*
  - *Maple syrup urine disease*
  - *Lesch-Nyhan disease*
  - The porphirias
  - *Glutaric acidurias I and II*
  - Other organic acidurias
  - *Sulfite oxidase deficiency*
  - *Canavan disease*
  - *Pantothenate kinase deficiency*
  - Cofactor and vitamin disorders
    - Thiamine deficiency
    - *Pyridoxine dependency*
    - Cobalamine disorders
    - Folate disorders
    - Biotine disorders
    - *Biotinidase deficiency*
  - Lipid and lipoprotein diseases
  - *Abetalipoproteinemia*
  - *Tangier disease*
  - *Smith-Lemli-Opitz syndrome*
  - *Wolman disease*
  - *Cerebrotendinous xanthomatosis*
  - *Pelizaeus-Merzbacher disease*

*Disorders covered in other chapters.

Disorders of cell membranes impact cell communication and the exchange of substances with the environment by disrupting membrane proteins or the small molecules that serve as ligands. Transport disorders, caused by primary deficiency of proteins responsible for selective permeability, cause particularly widespread cellular abnormalities that derive from secondary intracellular substrate or cofactor deficiency. Disorders that cause abnormal neurotransmission include, apart from membrane receptor diseases, neurotransmitter synthesis and recycling deficiencies that render cell membranes unexcitable or abnormally modulated. Disorders of organelles include abnormalities in organelle production or movement, or in membrane function or composition, leading to their accumulation or malformation and to the buildup of nonmetabolized compounds, or, in the special case of mitochondrial diseases, to deficient energy production. Enzymatic disorders are due to mutation of soluble enzymes or to cofactor deficiencies caused by inadequate absorption, processing, or binding, resulting in abnormal catalysis.

The cellular abnormalities brought about by soluble enzyme deficiencies tend to be morphologically modest but functionally widespread, reflecting cellular substrate diffusion, and associate with the release (or deprivation) of circulating plasmatic compounds, a feature used in the diagnosis of these diseases. A further category includes disorders of nuclear metabolism (DNA and RNA synthesis and processing, DNA methylation and repair) and of protein synthesis that are associated with broad cellular abnormalities owing to the central mission that the nucleus and the ribosome carry out in the cell. Some of these disorders are covered in other sections of this work.

In general, membrane disorders tend to be associated with milder disease phenotypes, organelle disorders with slowly accumulating abnormalities, enzyme disorders with marked biochemical abnormalities detectable in tissues and fluids, and nuclear disorders with a vast array of cellular alterations and phenotypes.

Metabolic diseases follow any imaginable temporal course even in the absence of environmental or nutritional precipitating factors, with some conditions manifesting only periodically and others exhibiting a static or apparently immutable course, in contrast with the common notion of metabolic diseases as unrelenting, continuously symptomatic processes. The basis for the apparently paradoxical static and episodic manifestations of neurometabolic diseases is provided by the compartmentalization of metabolism. Because all cellular functions are spatially limited and regulated by membranes, net catalysis occurs at rates governed not only by enzyme kinetics, but also (and often mainly) by substrate availability and product abundance. The former process, dependent on enzyme structure, is dictated by the gene; the latter, by the ability of cells and organelles to distribute and clear substrates and products, a process that is inherently dependent on membrane function (Hubert et al., 2005).

Multiple cellular compartments situated in various cell types participate in a typical metabolic pathway (Hofmeyr & Cornish-Bowden, 1991). Thus, for example, an enzymatic deficiency affecting a reaction that is constantly active may be associated with the accumulation of a substrate that interferes with other reactions, causing inhibition and resulting abnormal cell function. Such a substrate may be eliminated from the cell after it reaches a certain threshold level at a rate that is dependent on its concentration. Once the substrate first accumulates, production and elimination proceed indefinitely, maintaining a constant (elevated) concentration in the cell. Such may constitute the basis for the permanent, immutable clinical manifestations that are sometimes associated with a static metabolic abnormality. Episodic diseases can also be understood within the same mechanistic framework; a compound may accumulate silently until a threshold concentration is reached or until another slowly fluctuating cellular process renders the cell susceptible, such that additional reactions are triggered, causing a decompensation later followed by the restoration of the original (unstable) equilibrium.

Prenatal molecular diagnosis accomplished via sampling of fetal tissue obtained by chorionic biopsy is available for numerous metabolic diseases. Biochemical or molecular genetic assays of amniocytes are available for an increasing number of conditions. Yet, the most effective mode of detection is by voluntary screening of populations at risk. Several reproductive and therapeutic options are available when a pathogenic mutation is detected, including testing of an early embryo after *in vitro* fertilization before implantation, *in vitro* fertilization by a healthy donor, nuclear transfer or early initiation of therapy of an affected newborn. Newborn screening using tandem mass spectrometry applied to dried blood can detect many—if not most—metabolic disorders, including specific disorders of amino acid, organic acid, and fatty acid metabolism (Chance & Kalas, 2005). Conditions universally screened for include phenylketonuria and congenital hypothyroidism, being possible to avoid their vast impact on the developing nervous system (Therrell et al., 1992). Neurometabolic diseases can also be diagnosed postmortem using dry blood cards and skin punch biopsies from which live fibroblast cultures can be established for use in biochemical and genetic assays (Christodoulou & Wilcken, 2004).

The principles of molecular therapy are based on the use of alternative enzyme pathways, the facilitation of enzyme function by cofactor administration, enzyme replacement, and genomic modulation. Gene replacement therapy remains an elusive ideal as difficulties related to targeting, maintenance, and expression of the corrected gene construct have not been solved. Diets that diminish the utilization of a deficient metabolic pathway can be administered enterally or infused parenterally. Diets containing low protein, low carbohydrate or high glucose sometimes with extra fat supplementation meeting minimum caloric and protein and essential amino acid requirements are available for specific diseases. Cofactors and vitamins are administered at high doses when a vitamin-responsive disorder is suspected. Parenteral enzyme infusions are used with some success in lysosomal storage diseases such as Fabry disease, Gaucher disease, mucopolysaccharidosis type I, and Pompe disease (Brady & Schiffmann, 2004). Bone marrow transplantation corrects the enzymatic deficiencies of cells of hematopoietic origin in some mucopolysaccharidoses and, in some cases, the enzyme activity can be partially restored in the brain. Early transplantation may prevent progression of neurological disease but its long-term benefits are obscured by residual problems such as progression of skeletal and joint disability. Hepatic and liver–kidney transplantation have been considered in a variety of disorders with mixed overall success. Pharmacological stimulation of residual alleles or unrelated genes using histone deacetylation inhibitors, and loosening of translational fidelity by aminoglycoside antibiotic derivatives, applied to mutations that result in the generation of premature DNA termination codons.

The following sections are devoted to select neurometabolic diseases that exemplify diverse modes of transmission, mechanism, and clinical features.

I. Glucose Transporter Type 1 Deficiency

Mutations in the SLC2A1 gene, which codes for Glut1, the facilitative carrier responsible for the transport of glucose through the blood brain barrier and through astrocyte membranes, cause Glut1 deficiency syndrome. In its typical form, manifestations include infantile epilepsy, spasticity and ataxia (De Vivo et al., 1991). The disease, an example of a haploinsufficient state, can be inherited as an autosomal dominant trait from oligosymptomatic adults, who may experience only infrequent seizures or mild neuropsychological disturbances. Glut1 deficiency can lead to particularly severe neurological disability when SLC2A1 mutations (located in chromosome 1) compound in both alleles, a rare occurrence. Newly recognized phenotypes such as isolated ataxia or dystonia, both partially responsive to carbohydrate load or to a ketogenic diet, have received increased attention, as the full phenotypic spectrum of the disease continues to be expanded (Wang et al., 2005). The hallmark feature and main diagnostic parameter associated with the disease is hypoglychorrhachia: a diminished CSF glucose concentration, usually below 40 mg/dL or 2.2 mM. Supportive evidence of Glut1 deficiency is provided by a characteristic cerebral PET pattern, revealing a globally diminished uptake of fluorodeoxyglucose with marked thalamocortical depression and relative accentuation of basal ganglia tracer uptake (see Figure 10.1 and Pascual et al., 2002). The disease may be confirmed by assaying glucose uptake in patient erythrocytes, which are rich in Glut1, followed by sequence analysis of SLC2A1. The Glut1 transporter is a member of the Multiple Facilitator Superfamily (MFS), a large class of carriers that mediate the ATP hydrolysis-independent flow of diverse substrates driven by concentration gradients alone.

Three MFS members (a lactose permease, a glycerol phosphate transporter and an oxalate transporter) have been crystallized and their structures solved (Abramson et al., 2004). All three share a unique structural plan that includes 12 membrane spanning helical domains arranged in two globular, pseudosymmetrical six-helix domains. Some of the membrane helices are long and tilted and oriented such that the transporters appear to contain large water-filled extracellular and intracellular vestibules and interact directly with substrate only through a few residues contributed by the central portion of the inner core helices. The helices are thought to remain stable in place by flanking charges located at membrane boundaries. Numerous mutations have been identified and some mutational hotspots have been discovered in Glut1 deficiency patients. In particular, charged residues at helical boundaries are frequently mutated, leading to misfolding of the transporter or to the production of a carrier

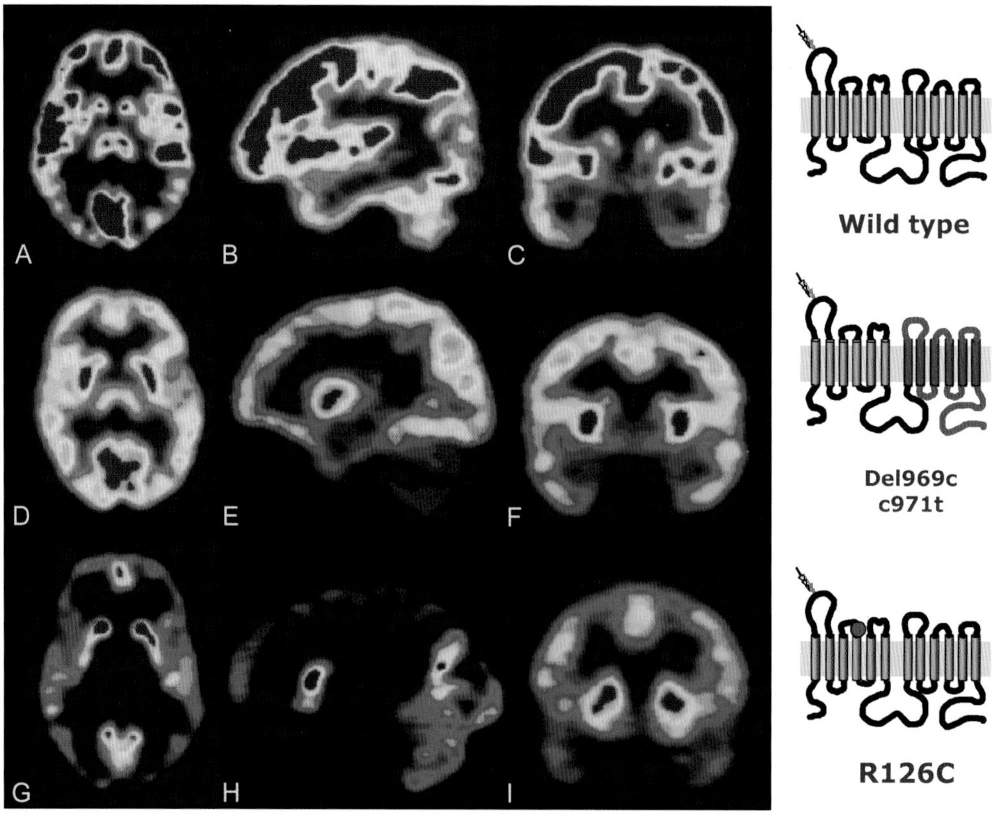

Figure 10.1 Cerebral glucose uptake in Glut1 deficiency. Left side panels **A–C**: Axial, parasagittal, and coronal PET images of a normal 20-year-old male. Physiological distribution of the radiotracer is appreciable. **D–F**: Analogous images in Glut1 deficiency in a patient 1.6 years old. **G–I**: Similar images from a 31-year-old patient. Right side panels represent the glycosylated transporter inserted in the plasma membrane of the subjects presented in the left side, colored in red and labeled according to the presence of a mutation in the transporter. Wild type: no mutation; del969c, c971t: deletion of cytosine at nucleotide 969 resulting in a premature stop codon at 971; R126C arginine 126 to cysteine substitution.
Source: author's original illustration.

incapable of undergoing normal conformational changes. More severe (homozygous) loss of Glut1 function is incompatible with embryonic survival in mice and, probably, in men. Embryonic stem cells and zebrafish neural structures devoid of Glut1 manifest increased apoptosis (Heilig et al., 2003; Jensen et al., 2006).

The treatment of Glut1 deficiency may include dietary supplementation with carbohydrate-rich compounds to increase the glucose concentration gradient across the blood brain barrier, where the residual (normal) transporters arising from the normal allele function below maximum velocity, lipoic acid, thought to enhance the expression of Glut1, or the administration of a ketogenic diet to provide alternative energetic substrates, as the brain can readily import and consume circulating ketone bodies. The ketogenic diet controls seizures, and it can be administered early in infancy, well before the cerebral metabolic rate for glucose reaches its maximum and the brain relies on glucose metabolism. Nevertheless, it appears that some of the neurological abnormalities set in very early in infancy and that, despite the responsiveness of epilepsy to the ketogenic diet, significant cognitive and motor disability persists as invariant features of the disease.

II. Menkes Disease

Mutation of the copper ATPase transporter ATP7A (see Figure 10.2), located in the trans-Golgi network and encoded by the X chromosome, causes the progressive copper deficiency disorder Menkes disease (Menkes et al., 1962). Also known as kinky hair disease, Menkes disease is associated with impaired copper flux, in contrast with Wilson's disease, which is characterized by copper excess in tissues such as brain and liver. ATP7A is ubiquitous and regulates the absorption of copper in the gastrointestinal tract (Menkes et al., 1999). Neonates affected by Menkes disease manifest seizures, hypothermia, and feeding difficulties. The infants are pale and display a characteristic kinky hair. ATP7A allows cellular copper to cross intracellular membranes and is translocated from the trans-Golgi network to the plasma membrane in the presence of extracellular copper.

Certain Menkes disease mutations specifically inhibit the copper-induced trafficking of otherwise normal functional copper transporters, a process that normally involves the formation of a phosphorylated transporter. This phosphorylated state is generated during the catalytic cycle of the pump and is recognized specifically for translocation, thus specifying

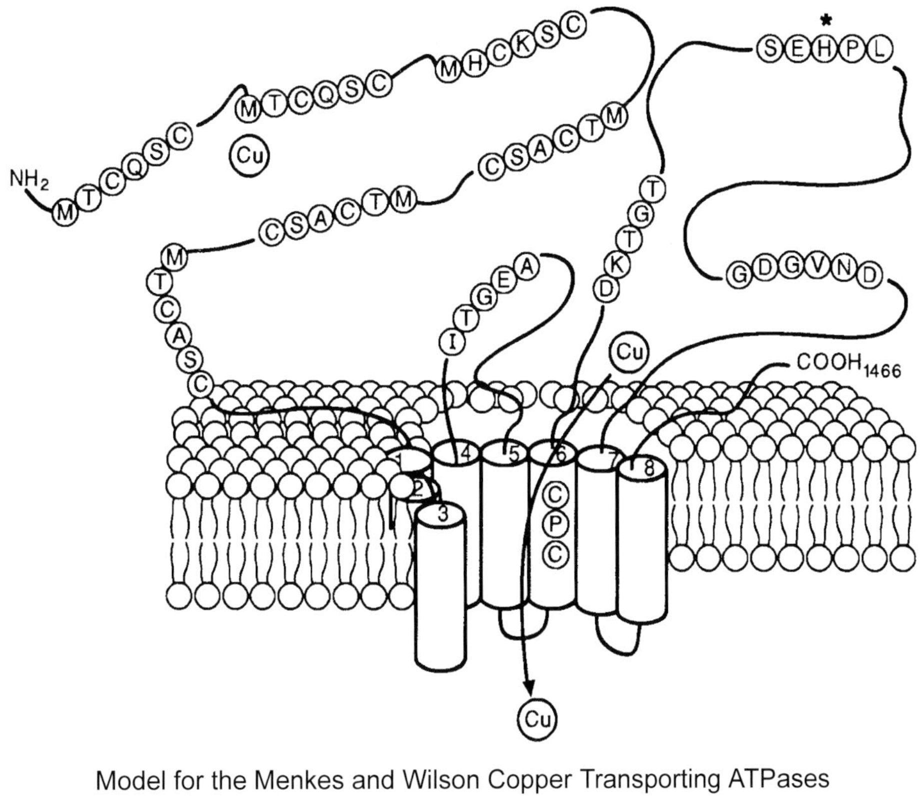

Model for the Menkes and Wilson Copper Transporting ATPases

Figure 10.2 Hypothetical structure of ATP7A based on hydropathy analysis. The protein includes eight membrane spanning domains and long cytoplasmic loops. Conserved residues are represented by circles.
From: Schaefer & Gitlin (1999).

the cellular location of the transporter. In the brain, neuronal glutamate receptor (NMDA type) activation results in trafficking of ATP7A to neuronal processes independently of intracellular copper concentration. This process is particularly important in hippocampal neurons, where a NMDA receptor-dependent, releasable pool of copper modulates neuronal activation. Copper chelation exacerbates NMDA-mediated excitotoxic cell death in hippocampal neurons, whereas the presence of copper is protective as it results in reduced cytoplasmic calcium levels after NMDA receptor activation. This phenomenon is dependent upon nitric oxide production by these neurons (Schlief et al., 2006).

In other tissues (including the brain), the fundamental abnormality is the maldistribution of copper, which is unavailable as a cofactor of several enzymes that include mitochondrial cytochrome c oxidase, lysyl oxidase, superoxide dismutase, dopamine beta-hydroxylase, and tyrosinase. Thus, the main features of the disease include mitochondrial respiratory chain dysfunction (complex IV deficiency), deficiency of collagen cross-links resulting in hair (pili torti and trichorrhexis nodosa) and vascular abnormalities (elongated cerebral vessels prone to rupture and hemorrhage), neuronal degeneration (markedly affecting Purkinje cells), and deficient melanin production. In patients, serum copper concentration is low and the ratio of urinary homovanillic acid/vanillylmandelic acid is elevated, constituting important diagnostic assays (Matsuo et al., 2005). A variety of minimally symptomatic phenotypes, including isolated ataxia or mental retardation, have been recognized. Intramuscular or subcutaneous administration of chelated copper-histidine affords protection against intellectual deterioration but is less effective in preventing other somatic complications (Christodoulou et al., 1998).

III. Segawa Disease (Dopa-Responsive Dystonia)

The neurotransmitter disorders represent an expanding group of signaling neurometabolic diseases. Several disturbances of monoamine and gamma-aminobutyric acid (GABA) metabolism presenting in infancy and childhood have been recognized and disorders involving additional neurotransmitters will probably be identified. Among the monoamine disorders, Segawa disease, aromatic L-amino acid decarboxylase deficiency, and tyrosine hydroxylase deficiency cause characteristic encephalopathies. Among the GABA disorders, pyridoxine-dependent seizures due to antiquitin gene mutations, GABA transaminase deficiency, and succinic semialdehyde dehydrogenase deficiency have been recognized (Hyland, 2003). Autosomal dominant mutations of the guanosine triphosphate cyclohydrolase I (GCH1) gene, located in chromosome 14, cause the treatable dystonic syndrome known as Segawa disease (Segawa et al., 2003). The fundamental biochemical abnormality is a decrease of tetrahydrobiopterin (BH4) associated with reduced tyrosine hydroxylase activity, leading to deficient dopaminergic transmission and extrapyramidal dysfunction. A decrease in the tyrosine hydroxylase content and activity in the ventral striatum accompanies a decrease in dopamine release that results in hypoactive D1 receptors (see Figure 10.3).

Some GCH1 mutations decrease TH activity by deleterious, dominant negative interactions between the mutated subunit and the residual normal TH from the other allele (Ichinose et al., 1999). Normally, tyrosine hydroxylase-dependent synaptic activity fluctuates throughout the day and decreases after the third decade of life; this probably accounts for the marked diurnal progression of symptoms and for the clinical stabilization observed after the fourth decade. Initial symptoms are often gait difficulties due to involuntary foot equinovarus posturing. Postural dystonia and tremor and a small stature dominate the clinical symptomatology and can be prevented by administration of levodopa. Marked intrafamilial symptom severity variability exits and nondystonic family members may suffer from major depressive disorder or obsessive-compulsive disorder responsive to enhancers of serotonergic neurotransmission and to levodopa administration. Sleep disorders, including difficulty falling asleep, excessive sleepiness, and frequent disturbing nightmares, are also features of this patient population (Van Hove et al., 2006).

Autosomal recessive Segawa syndrome is due to mutations in the tyrosine hydroxylase gene and causes early-onset parkinsonism responsive to levodopa, a more severe phenotype. Measurement of both total biopterin (most of which exists as BH4) and neopterin (the byproduct of the GTPCH1 reaction) in cerebrospinal fluid reveals that both compounds are decreased, a useful diagnostic indicator of GCH1 deficiency. Decreased activity of GCH1 in stimulated mononuclear blood cells and fibroblasts further supports the diagnosis. An oral phenylalanine load can reveal a subclinical defect in phenylalanine metabolism due to liver BH4 deficiency in patients with Segawa disease.

IV. Disorders of Pyruvate Metabolism

Defects in the pyruvate dehydrogenase (PDH) complex are an important cause of lactic acidosis. PDH is a large mitochondrial matrix enzyme complex that catalyzes the oxidative decarboxylation of pyruvate to form acetylCoA, nicotinamide adenine dinucleotide (NADH), and CO_2. Symptoms vary considerably in patients with PDH complex deficiencies, and almost equal numbers of affected males and females have been identified, despite the location of the PDH E1 alpha subunit gene (PDHA1) in the X chromosome, owing to selective female X-inactivation. Thus, the mechanisms for the clinical variation observed in E1 alpha

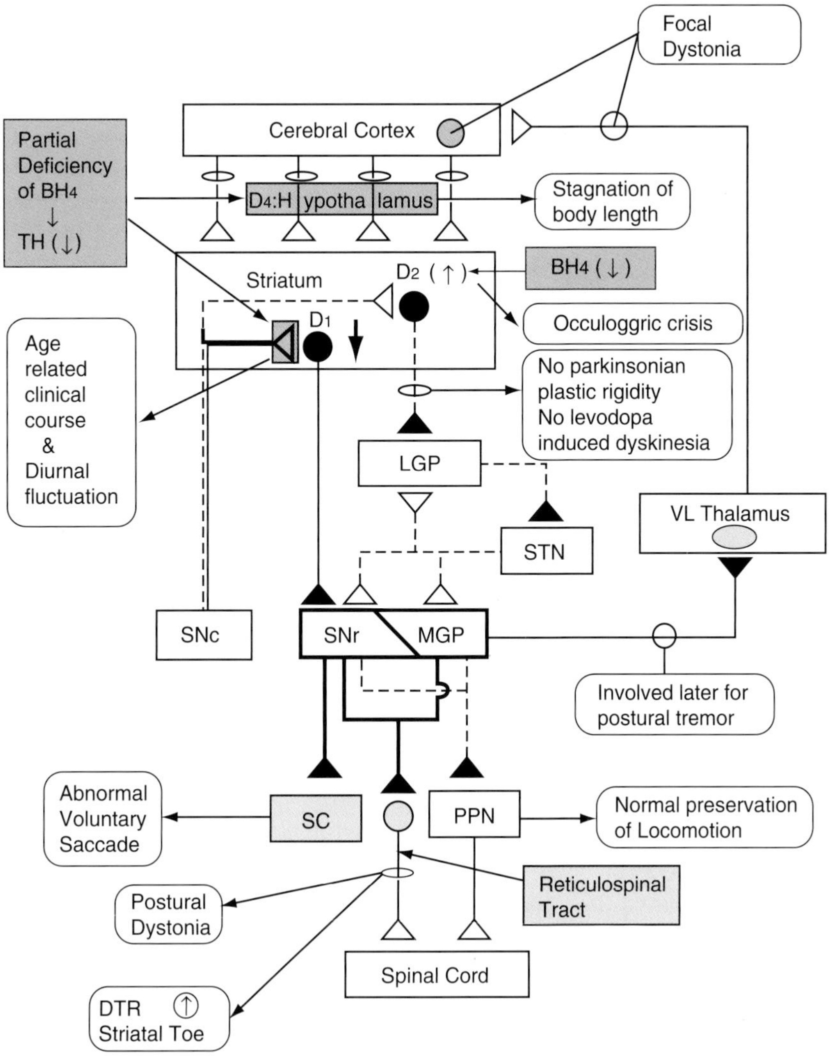

Figure 10.3 Pathophysiology of Segawa disease. Affected neural structures are enclosed in thick lines. Solid lines represent disease-involved pathways; thick lines indicate activation and thin lines inhibition. Closed triangles indicate inhibitory influences and open triangles excitation. Shaded structures and events are believed to be key for pathogenesis. TH, tyrosine hydroxylase; D1, D2 and D4, dopamine receptors subtypes; BH4, tetrahydrobiopterin; SNc, substantia nigra pars compacta; SNr, substantia nigra pars reticulata; LGP, lateral globus pallidus; MGP, medial globus pallidus; STN, subthalamic nucleus; VL thalamus, nucleus ventralis lateralis thalamus; SC, superior colliculus; PPN, pedunculopontine nucleus.
From: Segawa (2000).

deficiency patients and its resemblance to a recessive disease are mutation severity in males and the pattern of X-inactivation in females.

Several dozen PDHA1 mutations have been identified (Lissens et al., 2000). Patients harboring mutations in the E1 beta subunit, the E2 dihydrolipoyl transacetylase segment of the complex, the E3-binding protein, the lipoyl-containing protein X, and the PDH phosphatase have been reported (Maj et al., 2005). Neurodevelopmental abnormalities, microcephaly, epilepsy, and agenesis of the corpus callosum are characteristic features of E1 alpha deficiency (Nissenkorn et al., 2001). Infants may exhibit facial features of fetal alcoholic syndrome and older children can present with acute intermittent weakness and areflexia or with alternating hemiplegia. Episodic dystonia with hypotonia and ataxia and lesion of the globus pallidus can be caused by E2 deficiency. Diagnosis of these disorders requires measurements of lactate and pyruvate in plasma and cerebrospinal fluid, analysis of amino acids in plasma and organic acids in urine, as well neuroradiologic investigations, including magnetic resonance spectroscopy to detect lactate. Enzymatic analysis of fibroblast PDH activity can be performed and molecular diagnosis is available. A ketogenic diet is recommended together with thiamine supplementation, which may propitiate a substantial response (Duran & Wadman, 1985).

Pyruvate carboxylase deficiency is an autosomal recessive disease due to mutation of the PC gene, located in chromosome 11. Pyruvate carboxylase catalyzes the conversion of pyruvate to oxaloacetate in the presence of abundant acetyl-CoA, replenishing Krebs cycle intermediates in the mitochondrial matrix. The enzyme is a tetramer bound to biotin. PC is involved in gluconeogenesis, lipogenesis, and neurotransmitter synthesis. PC deficiency can presents with three degrees of phenotypic severity: an infantile form (A) with infantile moderate lactic acidosis, mental and motor deficiencies, hypotonia, pyramidal tract dysfunction, ataxia, and seizures leading to death in infancy. Episodes of vomiting, acidosis, and tachypnea can be triggered by metabolic imbalance or infection (Robinson et al., 1996). A severe neonatal form (B), manifests with severe lactic acidosis, hypoglycemia, hepatomegaly, depressed consciousness, and severely abnormal development. Abnormal limb and ocular movements are common findings. Brain MRI reveals cystic periventricular leukomalacia. Hyperammonemia and depletion of intracellular aspartate and oxaloacetate are profound. Early death is common. A rare benign form (C) causes episodic acidosis and moderate mental impairment compatible with survival and near normal neurological performance. A variety of mutations have been identified, with mosaicism probably accounting for the less severe phenotypes. Enzymatic analysis of fibroblast PC activity can be performed, but molecular diagnosis can be complicated by mosaicism. Dietary modification with triheptanoin (a triglyceride) supplementation has been attempted as a means to increase acetyl-CoA and anaplerotic propionyl-CoA (Mochel et al., 2005). Liver transplantation has also been performed, with reversal of ketoacidosis and amelioration of lactic academia (Nyhan et al., 2002).

V. Glycosylation Disorders

Glycosylation produces different glycans (or glycoconjugates) that modify the structure and function of cellular proteins and lipids. Protein glycosylation leads to the formation of N-glycans, O-glycans, and glycosaminoglycans. N-glycans are linked to asparagine residues of proteins that are part of a specific recognition motif. The degradation of proteins and glycans involves endocytosis and trafficking to lysosomes. Defects in these catabolic steps include glycosidase deficiencies that cause storage diseases like Gaucher, Niemann-Pick type C, Sandhoff, and Tay-Sachs diseases. Congenital disorders of glycosylation (CDG) are a group of autosomal recessive diseases defined by abnormal glycosylation of N-linked oligosaccharides (Freeze & Aebi, 2005). Over a dozen genes coding for enzymes involved in the N-linked oligosaccharide synthetic pathway have been found to harbor mutations, causing a variety of disease manifestations.

In some cases, the phenotypes are incompletely known, as only a small number of patients have been studied. In addition, novel enzyme deficiencies are periodically reported. Thus, genotype:phenotype correlations are preliminary. CDG-Ia, the most common type of CDG, is due to phosphomannomutase 2 deficiency leading to insufficient synthesis of the glycosylation precursor dolichol-oligosaccharide. Salient manifestations include inverted nipples, abnormal subcutaneous fat distribution, and cerebellar hypoplasia. The clinical course has been divided into an infantile multisystem stage in which all somatic organs can be affected, a late infantile and childhood ataxia-with-mental retardation stage, during which neuropathy, retinitis pigmentosa, and stroke-like episodes can manifest, and an adult stable disability stage. CDG-Ib is caused by mannose phosphate isomerase deficiency. Salient features include cyclic vomiting, hypoglycemia, hepatic fibrosis, and protein-losing enteropathy, occasionally associated with coagulation disturbances without neurologic involvement. CDG-Ic is due to deficiency of man(9)GlcNAc(2)-PP-dolichyl-alpha-1,3-glucosyltransferase and is associated with hypotonia, intellectual deficits, ataxia, strabismus, and epilepsy. The diagnosis of all types of CDG can be reached analyzing serum transferrin glycoforms by isoelectric focusing to estimate the number of sialylated N-linked oligosaccharide residues associated to the protein (Jaeken & Carchon, 2004). In select cases, molecular genetic analysis is feasible, including prenatal diagnosis. CDG-Ib is the only treatable type of CDG; mannose supplementation normalizes hypoproteinemia and coagulation defects and reverses both protein-losing enteropathy and hypoglycemia.

VI. Organic Acidurias

The organic acidemias (or organic acidurias) are disorders characterized by the urinary excretion of nonamino organic acids, which result from the abnormal amino acid catabolism of branched chain amino acids or lysine. These disorders include, but are not limited to, maple syrup urine disease (MSUD), propionic acidemia, methylmalonic acidemia, isovaleric acidemia, 3-methylcrotonyl-CoA carboxylase deficiency, 3-hydroxy-3-methylglutaryl-CoA lyase deficiency, ketothiolase deficiency, glutaric aciduria type I, and succinic semialdehyde dehydrogenase deficiency, among other less well-understood types (Ogier de Baulny & Saudubray, 2002). Specific enzymatic defects are responsible for each disorder, but several acidurias are caused by more than one enzyme deficiency. They are all inherited in an autosomal recessive fashion. The most severe and common presentation is a toxic neonatal encephalopathy. Newborns manifest vomiting, poor feeding, and progressive neurologic symptoms such as seizures, abnormal tone, and progressively depressed consciousness leading to coma.

Cerebral edema, leukoencephalopathy, perisylvian (opercular) hypotrophy, or basal ganglia necrosis are features frequently detectable in neuroimaging studies. Unrecognized children and adolescents can exhibit episodic ataxia, intellectual deficits, Reye syndrome, or psychiatric disturbances. Laboratory abnormalities include acidosis, ketosis, hyperammonemia, abnormal serum hepatic enzyme levels, hypoglycemia, and neutropenia. Secondary carnitine deficiency due to excessive excretion of acylcarnitine conjugates is common. The diagnosis is made by urine organic acid analysis, a technique that is particularly sensitive when it is performed during clinical decompensation, as the pattern of urinary excretion may be normal during symptom-free intervening periods. Analysis of plasma amino acids may also help to distinguish among specific disorders and direct enzyme activity measurements in lymphocytes or fibroblasts confirm the diagnosis. DNA sequence analysis is available for the most common disorders. Prenatal diagnosis relies on the analysis of amniotic fluid metabolites and it is simplified by DNA analysis in the context of a family in which a child has been previously diagnosed. Treatment relies on the replacement of enzyme substrates and precursors while meeting essential amino acid and caloric needs. Several special infant formulas are commercially available. Thiamine is used to treat thiamine-responsive MSUD and hydroxycobalamin to treat methylmalonic acidemia. Carnitine supplementation is used to correct secondary deficiency. In disorders of propionic acid metabolism, the periodic administration of antibiotics can reduce the production of propionate by intestinal flora (de Baulny et al., 2005). Hepatic or combined liver–renal transplantation has been attempted with moderate success in some of these disorders.

VII. Urea Cycle Disorders

The urea cycle disorders result from defects in the metabolism of nitrogen, which is predominantly produced during the breakdown of proteins and other nitrogen-containing molecules and transferred through ammonia into urea. The urea cycle is the only source of endogenous arginine and it is the main clearance mechanism for this waste nitrogen. Hyperammonemia is the defining feature of these disorders, that include deficiencies in the urea cycle enzymes carbamyl phosphate synthase I, ornithine transcarbamylase, argininosuccinic acid synthetase, argininosuccinic acid lyase and arginase, and in the cofactor producer N-acetyl glutamate synthetase (Leonard and Morris, 2002). With the exception of X-linked ornithine transcarboxylase deficiency, urea cycle disorders are inherited in an autosomal recessive fashion. These disorders manifest in the neonatal period with cerebral edema, lethargy, anorexia, hyper- or hypoventilation, hypothermia, seizures, abnormal tone, respiratory alkalosis, and coma.

In milder (or partial) urea cycle defects, ammonia accumulation may be triggered by illness, protein load, fasting, valproate administration, or by other decompensations at any later age, resulting in mild elevations of plasma ammonia accompanying cyclical vomiting, lethargy, sleep disturbances, delusions, hallucinations, and psychosis. Slowly progressive spastic paraparesis and growth retardation can be manifestations of arginase deficiency. A subset of carrier females manifest ornithine transcarboxylase deficiency owing to skewed X-inactivation, a state that may also lead to hyperammonemic crises during pregnancy or in the postpartum. In a variety of urea cycle defects, the rates of total urea synthesis and the urea cycle-specific nitrogen flux (measured by mass spectrometry using ^{15}N amide-labeled glutamine, which preferentially donates labeled nitrogen via carbamyl phosphate synthesis) correlate with phenotypic severity and predict carrier status in asymptomatic individuals (see Figure 10.4).

A specific pattern of plasma amino acid abnormalities allows a specific diagnosis. For example, glutamine, alanine, and asparagine are commonly elevated, whereas arginine may be reduced in all urea cycle disorders except in arginase deficiency, in which it is markedly elevated. Plasmatic citrulline and urinary orotic acid excretion also assist in dissecting the affected enzymatic pathway. Enzyme activity assays, usually performed in liver tissue, are reserved for confirmatory diagnosis, and DNA sequencing analysis is available for most of these disorders (Steiner & Cederbaum, 2001). The treatment during a crisis involves dialysis or other forms of plasma filtration aimed at reducing plasma ammonia concentration. Intravenous administration of arginine chloride and of the nitrogen scavengers sodium phenylacetate and sodium benzoate diminishes the accumulation of ammonia. Long-term administration of oral sodium phenylbutyrate and arginine increase the excretion of nitrogen by providing an alternative pathway (Batshaw et al., 2001). Nevertheless, dietary protein restriction constitutes the mainstay of maintenance therapy.

VIII. Galactosemia

The conversion of beta-D-galactose to glucose 1-phosphate is accomplished by the action of four enzymes that collectively constitute the Leloir pathway. In the first step of the pathway, beta-D-galactose is epimerized to alpha-D-galactose by galactose mu tarotase. The second step involves the phosphorylation of alpha-D-galactose by galactokinase to yield galactose 1-phosphate. In the fourth step, UDP-galactose is converted to UDP-glucose by UDP-galactose 4-epimerase. Classic (type I) galactosemia is caused by deficiency of the third step enzyme galactose-phosphate

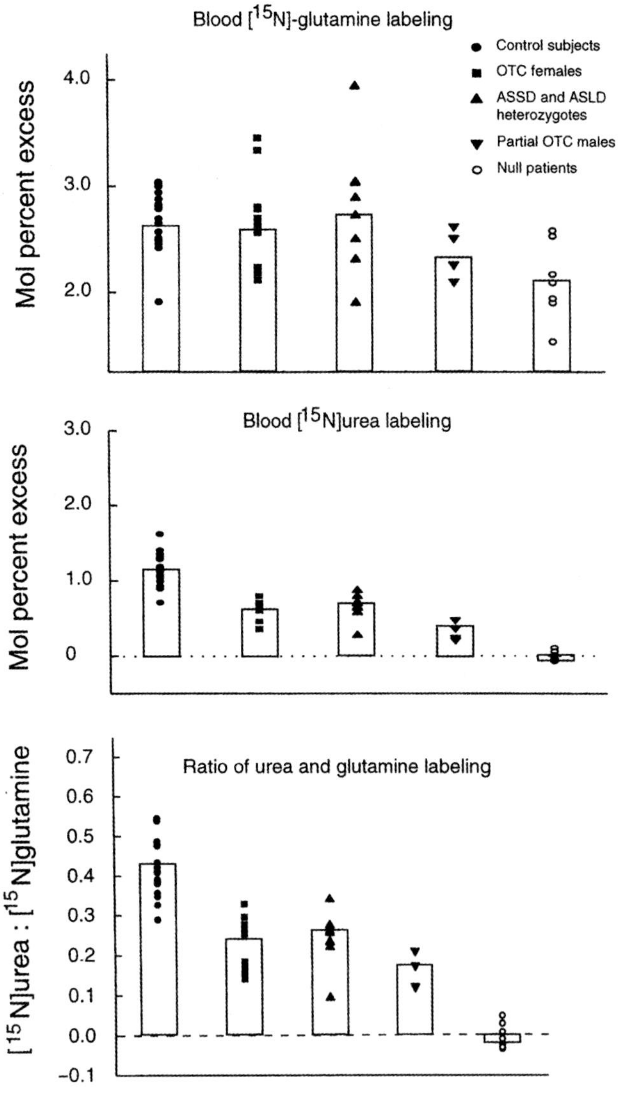

Figure 10.4 Isotopic enrichment and ^{15}N-urea:glutamine ratio in urea cycle patient and normal subjects receiving 0.4 g/kg/day protein intake. **Top**: Steady state mole percent excess enrichment of blood [^{15}N]glutamine; the mean is illustrated by bars. **Middle**: Steady state mole percent excess enrichment of blood [^{15}N]urea. **Bottom**: ^{15}N-urea:glutamine ratio representing the ratio of [^{15}N]urea and [^{15}N]glutamine labeling. OTC, ornithine transcarbamylase deficiency; ASS, argininosuccinate synthetase deficiency; ASLD, argininosuccinic aciduria.
From: Lee et al. (2000).

uridyltransferase (GALT), which catalyzes the production of glucose-1-phosphate and uridyldiphosphate (UDP)-galactose from galactose-1-phosphate and UPD-glucose (Leslie, 2003). Mutations in the genes encoding for galactokinase or epimerase can give rise to forms of galactosemia. Classic galactosemia can be inherited in an autosomal recessive fashion and is always attributable to mutations in the GALT gene in chromosome 9 (Tyfield et al., 1999). Within days of starting to feed milk or lactose-containing formulas, affected infants experience feeding difficulties, hypoglycemia, hepatic dysfunction, bleeding diathesis, jaundice and hyperammonemia. When untreated, sepsis and death may occur. Those infants who survive but continue to ingest galactose develop intellectual deficits, cortical and cerebellar tract signs. Despite early initiation of dietary therapy, the long-term outcome can include cataracts, poor growth, language dysfunction, extrapyramidal signs and ataxia, and ovarian failure.

The diagnosis is established by measuring erythrocyte GALT activity and by isoelectric focusing of the enzyme. All newborn screening programs typically include galactosemia and thus, the disease should be readily identified before becoming symptomatic. Biochemical and molecular genetic testing are widely used for heterozygote detection and prenatal diagnosis. Assay of erythrocyte galactose-1-phosphate concentration, measurement of urinary galactitol and estimation of total body oxidative flux of ^{13}C-galactose to $^{13}CO_2$ are utilized to quantify residual enzyme function and to monitor the response to dietary adjustments over time. The mainstay of therapy is lactose restriction, which rapidly reverses liver disease in newborns. Upon diagnosis, infants are immediately offered a lactose-free, soy-based formula that contains sucrose, fructose, and other nongalactose complex carbohydrates.

IX. Phenylketonuria

Classic phenylketonuria (PKU) is caused by near-complete deficiency of phenylalanine hydroxylase activity leading to hyperphenylalaninemia. The phenylalanine hydroxylase gene, PAH, is located in chromosome 12 and mutations in PAH are inherited in an autosomal recessive fashion. Over 250 missense mutations have been identified in the three domains (catalytic, regulatory, and tetramerization) of PAH (see Figure 10.5). PAH is assembled as a tetramer and as a dimer that coexist in interchangeable equilibrium. Each subunit contains an iron atom at the catalytic site. PKU mutations alter residues located at several enzyme regions: the active site, structural residues, residues involving interdomain interactions in a monomer, residues that interact with the N-terminal autoregulatory sequence that extends over the active site in the catalytic domain, and residues at the dimer or tetramer interface regions of the structure.

PKU was the first metabolic cause of mental retardation to be identified and is routinely screened for in all newborns. It is also an example of a disorder fully treatable by dietary restriction. A small proportion of infants with hyperphenylalaninemia display impaired synthesis or recycling of tetrahydrobiopterin (BH4) in the presence of a normal PAH gene, a condition that is independently treatable (Blau & Erlandsen, 2004). Classical untreated PKU leads to microcephaly, epilepsy, and severe intellectual and behavioral

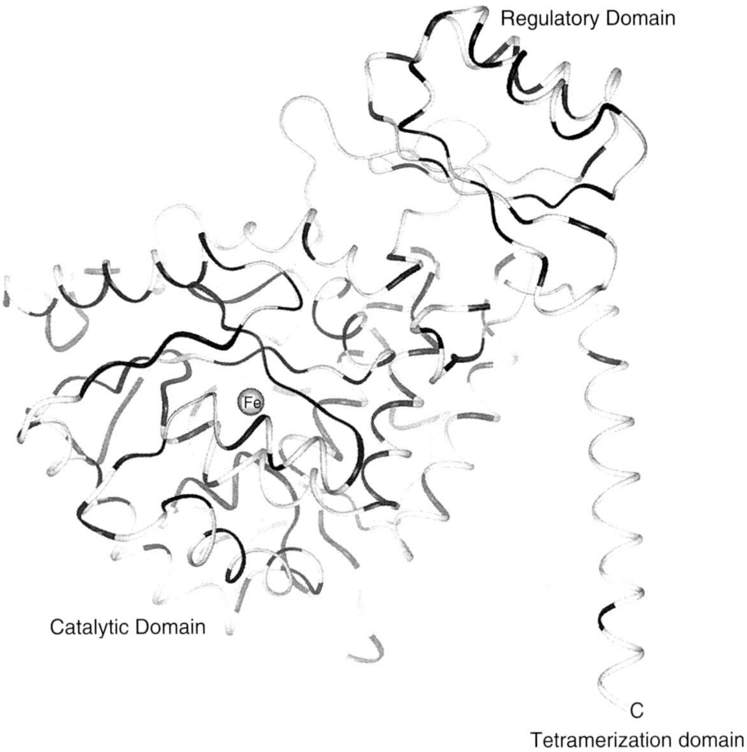

Figure 10.5 C alpha trace of a monomer of PAH. The trace is shaded for regions containing PKU mutations. The active site iron is represented as a sphere.
From: Erlandsen & Stevens (1999).

disabilities. The excretion of excessive phenylalanine and its metabolites can confer a musty odor to the skin, and the associated inhibition of tyrosinase causes decreased skin and hair pigmentation. Patients also exhibit decreased myelin formation and deficient production of dopamine, norepinephrine, and serotonin. Motor disability can be prominent later in life. Untreated maternal PKU can produce congenital heart disease, intrauterine and postnatal growth retardation, microcephaly, and mental retardation in the offspring. The diagnosis is based on plasma phenylalanine measurement and DNA sequence analysis. Prenatal diagnosis using amniocytes is available. PKU treatment consists of dietary restriction of phenylalanine. A fraction of patients with primary phenylalanine hydroxylase deficiency respond to BH4, which may act by enhancing residual enzyme function (Blau & Scriver, 2004). It is believed that the BH4 responsive form of PKU is caused by mutations that affect the enzyme Km by altering the binding affinity of BH4.

X. Lesch-Nyhan Disease

Among the inherited disorders of purine and pyrimidine metabolism, Lesch-Nyhan disease, caused by hypoxanthine-guanine phosphoribosyltransferase (HPRT) deficiency is the most common. The enzyme, encoded by the HPRT1 gene in the X chromosome, catalyzes the conversion of hypoxanthine to inosinic acid (IMP) and of guanine to guanylic acid (GMP) in the presence of phosphoribosylpyrophosphate, recycling purines derived from DNA and RNA (Nyhan, 1997). HPRT1 mutations diminish enzyme function or abundance and lead to uric acid overproduction. In addition to suffering from hyperuricemia, hyperuricuria and renal stones, male patients manifest abnormal neurological development during infancy. Hypotonia and failure to accomplish early motor milestones such as sitting, crawling, or walking can be prominent features.

Later in childhood, other symptoms emerge, including abnormal involuntary movements such as dystonia, choreoathetosis, opisthotonus, and ballismus. Pyramidal tract dysfunction includes spasticity, hyperreflexia, and Babinski signs. Profound intellectual deficits and self-injurious behavior can be prominent as are other motor compulsions. Females are carriers of HPRT1 mutations and can manifest increased uric acid excretion. They may show symptoms of the disease when nonrandom X-chromosome inactivation or skewed inactivation of the normal HPRT1 allele occur (Jinnah et al., 2000). A urinary urate-to-creatinine ratio above 2 is characteristic of the disease, as is an excessive urinary excretion of urate.

Defective HPRT enzyme activity can be measured in blood cells, fibroblasts, or lymphoblasts. DNA sequencing detects mutations in virtually all cases. Treatment aims to restrain uric acid overproduction with allopurinol, which inhibits the conversion of hypoxanthine and xanthine to uric acid mediated by xanthine oxidase. Bone marrow transplantation seems to be of only limited value both in correcting hyperuricemia and improving neurobehavioral symptoms (Deliliers & Annaloro, 2005).

XI. Pantothenate Kinase Deficiency

Also known as pantothenate kinase-associated neurodegeneration (PKAN) and formerly called Hallervorden-Spatz disease, pantothenate kinase deficiency causes neuronal degeneration associated with cerebral iron accumulation. This disorder is caused by the absence of pantothenate kinase 2, which is encoded by the PANK2 gene located in chromosome 20, and participates in coenzyme A biosynthesis, catalyzing the phosphorylation of pantothenate (vitamin B_5), N-pantothenoyl-cysteine, and pantetheine (Hayflick, 2003). Accumulation of N-pantothenoyl-cysteine and pantetheine may induce cell toxicity directly or via free radical damage by chelating iron. Deficient pantothenate kinase 2 may also be predicted to result in coenzyme A depletion and defective membrane biosynthesis in vulnerable cells such as rod photoreceptors (Johnson et al., 2004). Accumulation of iron is specific to the globus pallidus and substantia nigra. Axonal spheroids, thought to represent swollen axons secondary to defective axonal transport, appear in the pallidonigral system, in the subthalamic nucleus, and in peripheral nerves. Patients first manifest in early childhood with dystonia that interferes with ambulation, associated with dysarthria, rigidity, pigmentary retinopathy, and pyramidal tract dysfunction with spasticity and Babinski signs. Intellectual development may be variably affected. Psychiatric symptoms, including personality changes with impulsivity, depression, and emotional lability, are common (Pellecchia et al., 2005). A specific brain MRI abnormality, the eye-of-the-tiger sign, is characteristic of the disease, with rare exceptions. Hypoprebetalipoproteinemia and acanthocytosis may be additional manifestations of PANK2 mutations. The diagnosis relies on clinical and MRI features. When both are consistent with PKAN, there is a high likelihood of identifying a pathogenic mutation in PANK2 by DNA sequencing, although large chromosomal deletions affecting one allele are likely to remain undetected by this method.

XII. Smith-Lemli-Opitz Syndrome

Smith-Lemli-Opitz syndrome is a malformative autosomal recessive disorder caused by abnormal cholesterol metabolism resulting from deficiency of the enzyme 7-dehydrocholesterol reductase due to mutations of the DHCR7 gene located in chromosome 11. Decreased activity of 7-dehydrocholesterol reductase leads to a reduction in the rate of conversion of 7-dehydrocholesterol to cholesterol, causing an associated elevation in the serum concentration of 7-dehydrocholesterol or in the 7-dehydrocholesterol:cholesterol ratio (Opitz et al., 2002). A variety of proteins can be directly modulated by cholesterol, including the Sonic hedgehog and its related proteins Indian and Desert hedgehog, which require the covalent attachment of cholesterol to exert their morphogenetic function during development. Cholesterol is also an essential component of membranes and contributes to the formation of membrane lipid rafts, important for signaling. Patients manifest hypotonia and prenatal and postnatal growth retardation, microcephaly with intellectual deficiency and multiple malformations, including a characteristic facies (temporal narrowing, downslanting palpebral fissures, epicanthal folds, blepharoptosis, anteverted nares, cleft palate, and micrognathia), cardiac defects, underdeveloped external genitalia (hypospadias, bilateral cryptorchidism, and undermasculinization resulting in female external genitalia), postaxial polydactyly, and 2–3 toe syndactyly.

Holoprosencephaly can be an associated manifestation (Hennekam, 2005). Tandem mass spectrometry of dried blood card samples readily identifies patients and may be used for newborn screening. Direct analysis of the DHCR7 gene by DNA sequencing confirms the presence of a mutation in most cases. The combination of low concentrations of unconjugated estriol, HCG, and alphafetoprotein on routine maternal serum testing at 16 to 18 weeks' gestation is also suggestive of maternal carrier status and thus places the fetus at risk for Smith-Lemli-Opitz syndrome. Measurement of 7-dehydrocholesterol levels in amniotic fluid is available for prenatal diagnosis. Treatment with cholesterol supplementation and bile acids improves growth. The addition of the HMG-CoA reductase inhibitor simvastatin helps reduce serum 7-dehydrocholesterol.

References

Abramson, J., Iwata, S., Kaback, H. R. (2004). Lactose permease as a paradigm for membrane transport proteins. *Mol Membr Biol.* **21**, 227–236.

Batshaw, M. L., MacArthur, R. B., Tuchman, M. (2001). Alternative pathway therapy for urea cycle disorders: twenty years later. *J Pediatr.* **138**, S46–54.

Blau, N., Erlandsen, H. (2004). The metabolic and molecular bases of tetrahydrobiopterin-responsive phenylalanine hydroxylase deficiency. *Mol Genet Metab.* **82**, 101–111.

Blau, N., Scriver, C. R. (2004). New approaches to treat PKU: How far are we? *Mol Genet Metab.* **81**, 1–2.

Brady, R. O., Schiffmann, R. (2004). Enzyme-replacement therapy for metabolic storage disorders. *Lancet Neurol.* **3**, 752–756.

Chace, D. H., Kalas, T. A. (2005). A biochemical perspective on the use of tandem mass spectrometry for newborn screening and clinical testing. *Clin Biochem.* **38**, 296–309.

Christodoulou, J., Wilcken, B. (2004). Perimortem laboratory investigation of genetic metabolic disorders. *Semin Neonatol.* **9**, 275–280.

Christodoulou, J., Danks, D. M., Sarkar, B., Baerlocher, K. E., Casey, R., Horn, N. et al. (1998). Early treatment of Menkes disease with parenteral copper-histidine: Long-term follow-up of four treated patients. *Am J Med Genet.* **76**, 154–164.

de Baulny, H. O., Benoist, J. F., Rigal, O., Touati, G., Rabier, D., Saudubray, J. M. (2005). Methylmalonic and propionic acidaemias: management and outcome. *J Inherit Metab Dis.* **28**, 415–423.

De Vivo, D. C., Trifiletti, R. R., Jacobson, R. I., Ronen, G. M., Behmand, R. A., Harik, S. I. (1991). Defective glucose transport across the blood-brain barrier as a cause of persistent hypoglycorrhachia, seizures, and developmental delay. *N Engl J Med.* **325**, 703–709.

Deliliers, G. L., Annaloro, C. (2005). Hyperuricemia and bone marrow transplantation. *Contrib Nephrol.* **147**, 105–114.

Duran, M., Wadman, S. K. (1985). Thiamine-responsive inborn errors of metabolism. *J Inherit Metab Dis.* **8**, 70–75.

Erlandsen, H., Stevens, R. C. (1999). The structural basis of phenylketonuria. *Mol Genet Metab.* **68**, 103 –125.

Freeze, H. H., Aebi, M. (2005). Altered glycan structures: The molecular basis of congenital disorders of glycosylation. *Curr Opin Struct Biol.* **15**, 490–498.

Hayflick, S. J. (2003). Unraveling the Hallervorden-Spatz syndrome: Pantothenate kinase-associated neurodegeneration is the name. *Curr Opin Pediatr.* **15**, 572–577.

Heilig, C., Brosius, F., Siu, B., Concepcion, L., Mortensen, R., Heilig, K. et al. (2003). Implications of glucose transporter protein type 1 (GLUT1)-haplodeficiency in embryonic stem cells for their survival in response to hypoxic stress. *Am J Pathol.* **163**, 1873–1885.

Hennekam, R. C. (2005). Congenital brain anomalies in distal cholesterol biosynthesis defects. *J Inherit Metab Dis.* **28**, 385–392.

Hofmeyr, J. H., Cornish-Bowden, A. (1991). Quantitative assessment of regulation in metabolic systems. *Eur J Biochem.* **200**, 223–236.

Hulbert, A. J., Else, P. L. (2005). Membranes and the setting of energy demand. *J Exp Biol.* **208**, 1593–1599.

Hyland, K. (2003). The lumbar puncture for diagnosis of pediatric neurotransmitter diseases. *Ann Neurol.* **54**, S13–17.

Ichinose, H., Suzuki, T., Inagaki, H., Ohye, T., Nagatsu, T. (1999). Molecular genetics of dopa-responsive dystonia. *Biol Chem.* **380**, 1355–1364.

Jaeken, J., Carchon, H. (2004). Congenital disorders of glycosylation: A booming chapter of pediatrics. *Curr Opin Pediatr.* **16**, 434–439.

Jensen, P. J., Gitlin, J. D., Carayannopoulos, M. O. (2006). GLUT1 deficiency links nutrient availability and apoptosis during embryonic development. *J Biol Chem.* **281**, 13382–13387.

Jinnah, H. A., De Gregorio, L., Harris, J. C., Nyhan, W. L., O'Neill, J. P. (2000). The spectrum of inherited mutations causing HPRT deficiency: 75 new cases and a review of 196 previously reported cases. *Mutat Res.* **463**, 309–326.

Johnson, M. A., Kuo, Y. M., Westaway, S. K., Parker, S. M., Ching, K. H., Gitschier, J., Hayflick, S. J. (2004). Mitochondrial localization of human PANK2 and hypotheses of secondary iron accumulation in pantothenate kinase-associated neurodegeneration. *Ann N Y Acad Sci.* **1012**, 282–298.

Lee, B., Yu, H., Jahoor, F., O'Brien, W., Beaudet, A. L., Reeds, P. (2000). In vivo urea cycle flux distinguishes and correlates with phenotypic severity in disorders of the urea cycle. *Proc Natl Acad Sci U S A* **97**: 8021–8026.

Leonard, J. V., Morris, A. A. (2002). Urea cycle disorders. *Semin Neonatol.* **7**, 27–35.

Leslie, N. D. (2003). Insights into the pathogenesis of galactosemia. *Annu Rev Nutr.* **23**, 59–80.

Lissens, W., De Meirleir, L., Seneca, S., Liebaers, I., Brown, G. K., Brown, R. M. et al. (2000). Mutations in the X-linked pyruvate dehydrogenase (E1) alpha subunit gene (PDHA1) in patients with a pyruvate dehydrogenase complex deficiency. *Hum Mutat.* **15**, 209–219.

Maj, M. C., MacKay, N., Levandovskiy, V., Addis, J., Baumgartner, E. R., Baumgartner, M. R. et al. (2005). Pyruvate dehydrogenase phosphatase deficiency: Identification of the first mutation in two brothers and restoration of activity by protein complementation. *J Clin Endocrinol Metab.* **90**, 4101–4107.

Matsuo, M., Tasaki, R., Kodama, H., Hamasaki, Y. (2005). Screening for Menkes disease using the urine HVA/VMA ratio. *J Inherit Metab Dis.* **28**, 89–93.

Menkes, J. H., Alter, M., Steigleder, G. K., Weakley, D. R., Sung, J. H. (1962). A sex-linked recessive disorder with retardation of growth, peculiar hair, and focal cerebral and cerebellar degeneration. *Pediatrics* **29**, 764–779.

Menkes, J. H. (1999). Menkes disease and Wilson disease: Two sides of the same copper coin. Part I: Menkes disease. *Eur J Paediatr Neurol.* **3**, 147–158.

Mochel, F., DeLonlay, P., Touati, G., Brunengraber, H., Kinman, R. P., Rabier, D. et al. (2005). Pyruvate carboxylase deficiency: Clinical and biochemical response to anaplerotic diet therapy. *Mol Genet Metab.* **84**, 305–312.

Nissenkorn, A., Michelson, M., Ben-Zeev, B., Lerman-Sagie, T. (2001). Inborn errors of metabolism: A cause of abnormal brain development. *Neurology* **56**, 1265–1272.

Nyhan, W. L. (1997). The recognition of Lesch-Nyhan syndrome as an inborn error of purine metabolism. *J Inherit Metab Dis.* **20**, 171–178.

Nyhan, W. L., Khanna, A., Barshop, B. A., Naviaux, R. K., Precht, A. F., Lavine, J. E. et al. (2002). Pyruvate carboxylase deficiency—Insights from liver transplantation. *Mol Genet Metab.* **77**, 143–149.

Ogier de Baulny, H., Saudubray, J. M. (2002). Branched-chain organic acidurias. *Semin Neonatol.* **7**, 65–74.

Opitz, J. M., Gilbert-Barness, E., Ackerman, J., Lowichik, A. (2002). Cholesterol and development: The RSH ("Smith-Lemli-Opitz") syndrome and related conditions. *Pediatr Pathol Mol Med.* **21**, 153–181.

Pascual, J. M., Van Heertum, R. L., Wang, D., Engelstad, K., De Vivo, D. C. (2002). Imaging the metabolic footprint of Glut1 deficiency on the brain. *Ann Neurol.* **52**, 458–464.

Pellecchia, M. T., Valente, E. M., Cif, L., Salvi, S., Albanese, A., Scarano, V. et al. (2005). The diverse phenotype and genotype of pantothenate kinase-associated neurodegeneration. *Neurology* **64**, 1810–1812.

Robinson, B. H., MacKay, N., Chun, K., Ling, M. (1996). Disorders of pyruvate carboxylase and the pyruvate dehydrogenase complex. *J Inherit Metab Dis.* **19**, 452–462.

Schaefer, M., Gitlin, J. D. (1999). Genetic disorders of membrane transport. IV. Wilson's disease and Menkes disease. *Am J Physiol.* **276**, G311–314.

Schlief, M. L., West, T., Craig, A. M., Holtzman, D. M., Gitlin, J. D. (2006). Role of the Menkes copper-transporting ATPase in NMDA receptor-mediated neuronal toxicity. *Proc Natl Acad Sci U S A* **103**,14919–14924.

Segawa, M. (2000). Hereditary progressive dystonia with marked diurnal fluctuation. *Brain Dev.* **22** Suppl 1, S65–80.

Segawa, M., Nomura, Y., Nishiyama, N. (2003). Autosomal dominant guanosine triphosphate cyclohydrolase I deficiency (Segawa disease). *Ann Neurol.* **54**, S32–45.

Steiner, R. D., Cederbaum, S. D. (2001). Laboratory evaluation of urea cycle disorders. *J Pediatr.* **138**, S21–29.

Therrell, B. L., Panny, S. R., Davidson, A., Eckman, J., Hannon, W. H., Henson, M. A. et al. (1992). U.S. newborn screening system guidelines: Statement of the Council of Regional Networks for Genetic Services (CORN). *Screening* **1**, 135–147.

Tyfield, L., Reichardt, J., Fridovich-Keil, J., Croke, D. T., Elsas, L. J. II, Strobl, W. et al. (1999). Classical galactosemia and mutations at the galactose-1-phosphate uridyl transferase (GALT) gene. *Hum Mutat.* 13, 417–430.

Van Hove, J. L., Steyaert, J., Matthijs, G., Legius, E., Theys, P., Wevers, R. et al. (2006). Expanded motor and psychiatric phenotype in autosomal dominant Segawa syndrome due to GTP cyclohydrolase deficiency. *J Neurol Neurosurg Psychiatry* **77**, 18–23.

Wang, D., Pascual, J. M., Yang, H., Engelstad, K., Jhung, S., Sun, R. P., De Vivo, D.C. (2005). Glut-1 deficiency syndrome: clinical, genetic, and therapeutic aspects. *Ann Neurol.* **57**, 111–118.

11

Genetic Disorders of Neuromuscular Development

Juan M. Pascual

I. The Motor Unit during Development

Motor neurons are one of only a few neuronal types that project axons outside the central nervous system. During embryonic development, they are the first neurons to arise from neuroblasts, establishing a growing population of differentiating cells at the ventral aspect of the neural tube. Both the notochord (a mesodermal structure) and the floor plate (a collection of mesodermal glial cells situated at the ventral midline of the neural tube) induce motor neuron differentiation. Floor plate-related motor neuron differentiation necessitates direct cellular contact, whereas the notochord acts via the release of diffusible sonic hedgehog, a secreted precursor protein that undergoes self-cleavage (Price & Briscoe, 2004). The amino terminal product is endowed with all the inductive activities of the protein. Higher concentrations near the ventral surface of the neural tube are needed for the production of floor plate cells, while neuroblasts differentiating into motor neurons are exposed to lower concentrations as a consequence of their greater distance from the notochord. Hedgehog signaling is essential for the development of a variety of structures in addition to the nervous system, including the skeleton, the face, the gastrointestinal tract, epithelia, and bone marrow. Together with the sonic isoform, Indian and desert hedgehog comprise a family of signals transduced by patched and smoothened transmembrane receptor proteins. Abnormalities in this signaling cascade account for numerous developmental malformations such as holoprosencephaly and several neoplasms.

At least five classes of motor neurons are generated by the sonic hedgehog gradient in the neural tube. In response to sonic hedgehog, differentiating motor neurons generate several transcription factors, which are themselves the subject of cross-repressive interactions. Later, signals derived from paraxial mesoderm induce motor neuron column formation. Motor axons then reach out to their target muscles. The expression of LIM homeodomain transcription factors is associated with ventral or dorsal axonal projection choice. The last event in motor neurogenesis involves the formation of neuronal pools that control individual muscles, a process also associated with selective gene expression changes.

Both embryonic and adult muscles contain muscle stem cells, a set of undifferentiated mononuclear cells capable of division. During embryogenesis, each stem cell divides asymmetrically in the myotome portion of each somite, originating a myoblast and a replicating stem cell of progressively reduced proliferative potential (Miller et al., 2004). After birth, the muscle still contains satellite cells that contribute to the late stage of myogenesis and are probably endowed with regenerative potential for use later in life, when they will have become quiescent.

During primary myogenesis, the first stage of muscle formation, muscle stem cells differentiate into myoblasts that replicate and eventually coalesce to form multinucleated myofibers. The first myofibers formed are almost immediately contacted by approaching axons, establishing primitive but functioning neuromuscular junctions that later undergo further differentiation. A wave of secondary myogenesis occurs around the newly formed neuromuscular junctions. All newly differentiated secondary myofibers are electrically coupled to their primary myofibers. Tertiary myogenesis, which continues postnatally, originates from satellite myoblasts situated in the periphery of the developing myofiber and ultimately shapes the mature muscle.

At the conclusion of myogenesis, well-differentiated slow and fast myofibers populate all muscles at specific ratios. Slow myofibers are predominantly oxidative, contract relative more slowly and fatigue more rapidly than fast, glycolytic myofibers. Both also differ in the molecular isoforms that constitute their contractile apparatus and probably arise from different myoblast populations. Inextricably related to the set of myofibers that they innervate, motor neurons situated in the brainstem and the spinal cord maintain large myelinated axons that terminate in multiple neuromuscular junctions. A motor neuron, together with all the myofibers that it innervates is defined as a motor unit, the most elementary force-generating system controllable by the central nervous system (Buchthal & Schmalbruch, 1980).

Complex gene expression changes guide the progressive transition from myogenesis to mature motor unit activity. After myoblasts become postmitotic, muscle transcription factors activate genes that lead to patterns of stage-specific muscle protein expression. Not all muscle expression changes are irreversible, as patterns reminiscent of embryogenesis are revisited in the course of neuromuscular diseases. Several genes are susceptible to reactivation well after motor unit development. For example, utrophin, a structural membrane protein similar in function and location to dystrophin, bridges both cytoskeletal proteins and extracellular laminin only during myogenesis, disappearing upon dystrophin expression shortly before birth. Membrane dysfunction caused by dystrophinopathy is characteristically associated with the reappearance of utrophin in the myofibrillar membrane, a phenomenon of diagnostic significance. Neural cell adhesion molecule (NCAM) is produced after myoblasts differentiate into myotubes and, in mature muscle, remains confined to the neuromuscular junction. During muscle regeneration or denervation, NCAM is again expressed throughout the cell membrane.

II. Structures and Function of the Neuromuscular System

A. Motor Neurons

In vertebrates, motor neurons group in pools or clusters within motor nuclei. Neuronal clusters innervate individual muscles and are structurally and functionally separate from each other. Cells within a cluster share a set of cadherin molecules, are electrically coupled via gap junctions, and receive the same propioceptive sensory input from neurons located in the dorsal root ganglia. Three types of motor neurons populate the anterior horn. Alpha motor neurons are very large and innervate extrafusal striated muscle fibers, the predominant constituent of voluntary muscle. Gamma motor neurons are smaller and innervate specialized intrafusal muscle fibers are part of the spindle stretch receptor. Beta motor neurons innervate both types of muscle fibers. The dendrites of alpha motor neurons, the predominant motor neuron type, branch extensively and receive abundant synaptic input; yet, they channel their output through the axonal hillock, where action potentials are generated. Because of the great distance that motor axons traverse, the volume of their axoplasm well exceeds the volume of the motor neuron cell body. Myelin sheaths supplied by Schwann cells allow for great impulse conduction velocities through the anterior nerve roots of the spinal cord to the vicinity of the neuromuscular junction, where axons divide into tens to hundreds of terminal branches.

B. The Neuromuscular Junction

The flow of the information contained in the nerve impulse to the muscle takes place at the neuromuscular junction, a specialized synapse where each nerve action potential triggers a muscle action potential through the intervention of a chemical relay: the release of acetylcholine (see Figure 11.1) (Hughes et al., 2006). Each motor neuron in the brainstem and the spinal cord innervates a motor unit. The size of each motor unit depends on the precision of movement and force to which the muscle is accustomed. The entire pre- and postsynaptic apparatus, including the intercellular synaptic space, is contained in a differentiated area that spans only about 40 μm. Neural Schwann cells delimit the surface of the neuromuscular junction by deploying a protective enveloping process devoid of myelin that contains specialized membrane proteins such as neural cell adhesion molecule. Entering slightly into the muscle at the motor point, the nerve endings form boutons separated from an invaginated region of raised muscle plasma membrane by a synaptic space or cleft of 50 nm. The basal lamina of both nerve and muscle predominantly includes type IV collagen and it forms an uninterrupted continuum that also spans the synaptic cleft. The basal lamina, however, differentiates at the neuromuscular junction by

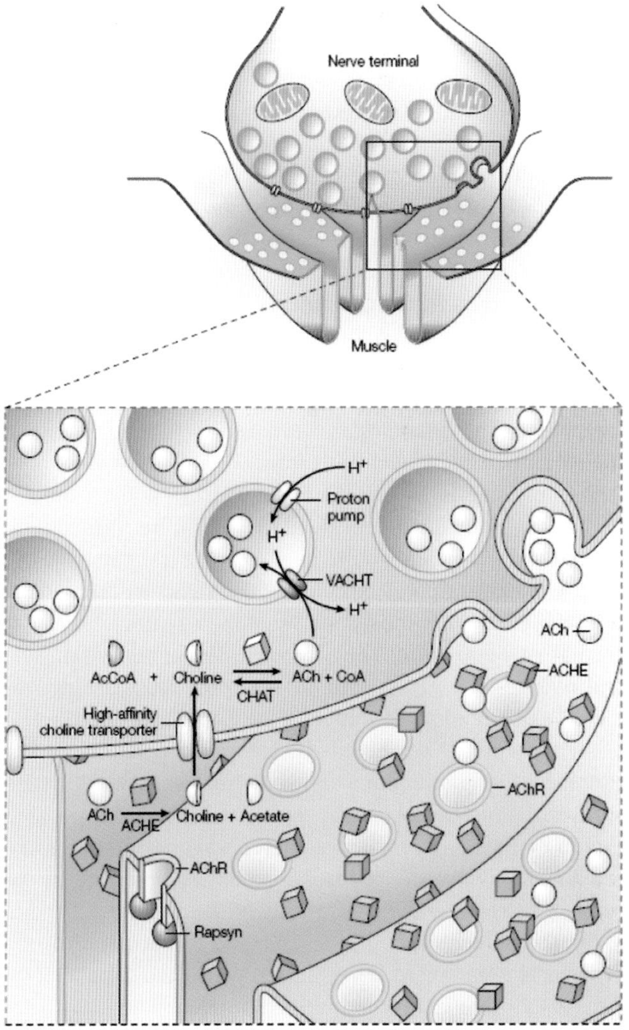

Figure 11.1 The neuromuscular endplate. **Upper panel**: Diagram of an endplate region showing the general locations of acetylcholinesterase (AChE), the acetylcholine receptor (AChR) (green), and voltage-dependent Na+ channels of the $Na_v1.4$ type (red). **Lower panel**: After exocytotic release from the nerve terminal, some acetylcholine (ACh) molecules are hydrolyzed by AChE before they bind to the AChR and the remaining ACh molecules are hydrolyzed by ACHE after dissociation from AChR. Choline is transported into the nerve terminal by a high-affinity choline transporter. ACh is resynthesized from choline and acetyl-coenzyme A (AcCoA) by choline acetyltransferase (ChAT) and is then transported into the synaptic vesicle by the vesicular ACh transporter (VACHT) in exchange for protons delivered to the synaptic vesicle by a proton pump.

harboring specialized proteins such as acetylcholinesterase, laminin, agrin, and neuregulins.

Mitochondria located in the presynaptic terminal provide ATP for the energetically demanding vesicular acetylcholine turnover process. A fraction of the synaptic vesicles in the terminal are associated with regions of increased membrane density called active or release zones. These zones contain voltage-gated Ca^{2+} channels of the P/Q type as well as calcium-activated K^+ channels in high density, and are precisely located across from the infoldings of the postsynaptic membrane. Following nerve stimulation, vesicles at these sites fuse with the plasma membrane and release acetylcholine directly over the postsynaptic receptor molecules situated on the adjacent folds. Away from the active zones, a significant fraction of synaptic vesicles is associated with actin, remaining unavailable for immediate release.

The postsynaptic muscle membrane is also highly specialized. Nicotinic acetylcholine receptors (AChR) cluster in the infoldings and directly associate with cytoplasmic rapsyn (receptor-aggregating protein at the synapse). Utrophin and α-dystrobrevin probably link the AChR complexes to actin filaments and to a much larger network of proteins, many of which are associated with specific myopathies. Agrin, a basal lamina protein associated with synaptic space laminins contributes to AChR clustering. A muscle specific kinase (MuSK), a target of autoantibodies formed in a type of immune-mediated myasthenia is part of the receptor complex for agrin.

Acetylcholine is synthesized at the nerve terminal by choline acetyltransferase (ChAT) from choline and acetyl coenzyme A—a product of glycolysis and pyruvate oxidation—during a process that does not significantly differ from that carried out at central nervous system synapses. Acetylcholine is released after the arrival of the nerve action potential to the resting terminal, whose potential is maintained by the Na^+-K^+ ATPase. Depolarization leads to activation of Ca^{2+} channels and to a surge of Ca^{2+} entry, reaching 0.1 mM for a few milliseconds. Later, Ca^{2+} is buffered in the terminal and extruded by the Na^+-Ca^{2+} exchanger, as the opening of voltage-gated K^+ channels also helps restore membrane potential. While still in the cytosol, Ca^{2+} interacts with the vesicular protein sensor synaptotagmin, priming other vesicular proteins for docking against and fusing with the plasma membrane. Acetylcholine is then freed to diffuse across the junctional cleft and interacts with AChR until it encounters AChE and is hydrolyzed into choline and acetate, later to be recovered by the terminal. AChE exists in a peculiar trimeric conformation composed of three groups of four active globular heads linked by a collagen strand named ColQ. An enzyme of great therapeutic significance, AChE is reversibly blocked by anticholinesterases of clinical use and irreversibly phosphorylated and inactivated by organophosphate insecticides, among other modulators. Following fusion and diffusion of acetylcholine, the vesicular membrane is retrieved by endocytosis of clathrin-coated vesicles and refilled with the neurotransmitter. While being refilled, vesicles are tethered to actin filaments in the nonreleasable pool via synapsin, a vesicular protein that loosens its attachment to actin after Ca^{2+}-dependent phosphorylation by calmodulin kinase.

Neuromuscular transmission relies on the generation of a junctional muscle membrane depolarization large enough to sustain the formation of a propagated action potential; the difference between the magnitude of the junctional potential and the threshold of the muscle action potential is the safety factor of neuromuscular transmission. Inexorably, impulse transmission must proceed across multiple conformational changes in the molecules that underlie neuromuscular junction excitability. An inherently stochastic (as opposed to a deterministic all-or-nothing) process, normal transmission necessitates this safety factor to operate with fidelity. Among the processes that are nondeterministic—that is, governed by probability distribution laws—are the opening of voltage gated Ca^{2+} channels, the number of acetylcholine vesicles released each time, the diffusion of acetylcholine in the cleft, the binding of two acetylcholine molecules to the AChR, the onset of desensitization and closing of the AChR, and the encounter of the neurotransmitter with AChE, to name but a few events for which their probability laws are known with some degree of certainty.

The origin of this stochastic behavior is inherent to all molecules that change conformation, but, in the case of the neuromuscular junction, it is accentuated by its limited dimensions, which allow only a relatively small number of molecules. In contrast, an enzymatic reaction carried out by an ensemble of many molecules in solution behaves, as a whole, predictably. The limitation to the size of the neuromuscular junction may be due to the large energetic expense associated with its maintenance, particularly when it is subject to frequent use and, thus, energy is spared by reducing synaptic size and by increasing the safety factor of neuromuscular transmission. In line with the importance of neuromuscular transmission for the survival of all organisms, all key molecular elements of the neuromuscular junction are encoded by highly conserved genes whose absence or severe dysfunction is incompatible with life. Instead, diseases of the neuromuscular junction tend to shift the probability distribution laws for the different conformations of its key molecules, thus decreasing the safety factor of neuromuscular transmission.

C. Muscle

The muscle cell is surrounded by a plasma membrane that, together with collagen fibrils and other connective tissue elements, forms the sarcolemma (Franzini-Armstrong, 1979). The interior of the resting cell is maintained at an electrical potential about 80 mV negative to the exterior by the combined action of pumps and channels in the plasma membrane. Unlike nerve membranes, muscle membranes exhibit a high conductance (G) to chloride ions in the resting state; G_{Cl}^- accounts for about 70 percent of the total membrane conductance. Potassium ion conductance accounts for most of the remainder, such that the membrane potential is normally close to the Nernst potential for these two ions. Asymmetrical concentration gradients for sodium and potassium ions are maintained at an energy cost by the membrane Na^+-K^+-ATPase. During the generation of a muscle action potential, a rapid and stereotyped membrane depolarization is produced by an increase in sodium ion conductance through voltage-dependent Na^+ channels. This conductance increase is self-limited and is followed by membrane repolarization induced by the delayed opening of a potassium conductance.

Action potentials originating at the neuromuscular junction spread then in nondecremental fashion over the entire surface of the muscle. Action potentials traveling on the sarcolemma penetrate the interior of the muscle cell along transverse (T) tubules that are continuous with the outer muscle membrane. These tubules are seen as openings emerging on the surface of the muscle cell from which membrane depolarization spreads to the center of the fiber. As the T-tubular network courses inward, close associations are formed with specialized elements of the sarcoplasmic reticulum (SR). At the electron microscope level, the structure formed by a single tubule interposed between two terminal SR elements is called a *triad*. The SR stores Ca^{2+} in relaxed muscle and releases it into the sarcoplasm upon depolarization of the cell membrane transmitted to the T-tubular system.

In the vicinity of the triads, electron micrographs also reveal repeating structures known as *sarcomeres*, which are separated from each other by dark lines called Z disks (see Figure 11.2). Within each sarcomere, A and I bands are seen; the A band, lying between two I bands, occupies the center of each sarcomere and is birefringent. Within the A band is a central, lighter zone, the H band, and in the center of the H band is a darker M band. The Z disk is at the center of the I band. The difference in birefringence between the A and I bands produces the characteristic striated appearance of voluntary muscle when seen with the light microscope. The repeating optical characteristics of the A and I bands in each sarcomere reflect the periodic arrangement of two sets of filaments: thin filaments, ~180 Å in diameter, that appear to attach to the Z bands and are found in the I band and part of the A band, and thick filaments, with a diameter of ~150 Å, occupying the A band and connected crosswise by material present in the M band. In cross section, the thick filaments are arranged as a hexagonal lattice with the thin filaments occupying the center of the triangles formed by the thick filaments.

The two kinds of filaments become cross-linked only upon excitation, whereas muscle contraction does not depend on the shortening of the filaments but, rather, on the relative motion of the two sets of filaments, a phenomenon termed the *sliding-filament mechanism*. Thus, the length of the muscle depends on the length of the sarcomeres, and sarcomere length variation is caused by variations in overlap between thin and thick filaments. In addition to actin and myosin, other proteins also constitute the two sets of filaments. Tropomyosin and a complex of three subunits

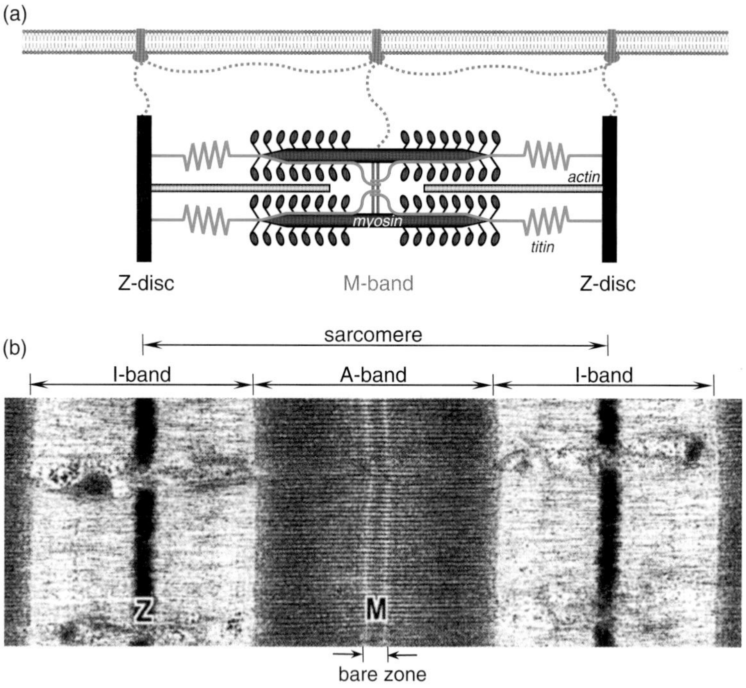

Figure 11.2 Sarcomere structure. Color code: actin filaments, dark yellow; myosin filaments, blue; titin filaments, green. (a) The transverse structures are the Z-disk (black) and the M-band (red). The extra-sarcomeric filaments (magenta) are anchored to transmembrane proteins in the sarcolemma. (b) Corresponding electron micrograph of a skeletal sarcomere. The sarcomeric borders are delineated by the Z-discs (Z) in the middle of the I-band. The M-band (M) is seen as an electron-dense transverse band in the middle of the dark A-band and encompasses part of the lighter "bare" zone, where neither myosin crossbridges nor thin filaments are present. (c) On this negatively stained cryosection (protein density appears white) of the higher-magnification view of human tibialis anterior muscle, the M-band breaks up into a series of several M-lines (arrowheads). The five strongest M-lines (or M-bridges) are designated M6', M4', M1, M4, and M6 (only numbers are shown).
From: Agarkova & Perriard (2005).

collectively called troponin are present in the thin filaments and regulate muscle contraction (Zot and Potter, 1987). The M and the Z bands (discs) contain α-actinin and desmin, as well as the enzyme creatine kinase, an important marker of muscle breakdown released into the bloodstream. Intermediate filaments containing desmin, a protein excessively accumulated in certain types of myopathy, neuropathy, and cardiopathy, maintain the myofibrillar organization within the cytoplasm. A continuous elastic network of proteins such as connectin also surrounds actin and myosin filaments and provides the muscle with a parallel elastic element. Titin, a molecular caliper, provides anchoring sites for Z disc components and for other sarcomere proteins. Actin forms the backbone of the thin filaments, arranged as linear polymers of slightly elongated, bilobar actin subunits, organized in helical fashion. Each monomer contains a single nucleotide binding site.

Actin polymerization involves the hydrolysis of ATP into ADP, a process independent of the energy consumption associated with muscle contraction. A wide variety of proteins interact with actin in both muscle and immobile cells. They may affect the polymerization and depolymerization of actin and mediate the attachment of actin to other cellular structures, including the Z disks in muscle as well as membranes in both muscle and nonmuscle cells. Myosin, the chief constituent of thick filaments, is a multisubunit protein. In contrast to actin, myosin consists of several peptide subunits. Each myosin molecule contains two heavy chains that extend throughout the entire span of the molecule. The two chains are intertwined over most of their length to form a double α-helical rod; at one end they separate, each forming an elongated globular portion. The two globular portions contain the sites responsible for ATP hydrolysis and interaction with actin. In addition to the two heavy chains, each myosin molecule contains four light chains that modulate myosin activity and can be phosphorylated. The type of heavy and light myosin chain isoforms that constitute the majority of myofibers in a muscle determines the maximum contraction rate (fast or slow) of the muscle. Myosin molecules form end-to-end aggregates involving the rod-like segments, which collectively grow into thick filaments. The polarity of the myosin molecules is reversed on either side of the central portion of the filament. The globular ends of the molecules form projections, or cross-bridges, on the aggregates that interact with actin. Conformational changes within this region, driven by ATP hydrolysis, provide the force that propels the movement of actin fibrils with respect to the myosin filament (Huxley, 1969). The ATPase activity of myosin itself is stimulated by Ca^{2+} and is low in Mg^{2+}-containing media.

The hydrolysis of ATP releases sufficient energy to induce conformational changes that signify cellular movement and mechanical work. Tropomyosin and troponin are proteins located in the thin filaments and, together with Ca^{2+}, regulate actin-myosin interactions. Tropomyosin is an α-helical protein consisting of two polypeptide chains; its structure resembles the rod portion of myosin. Troponin, in contrast, is a complex of three proteins. In the presence of the tropomyosin-troponin complex, actin cannot stimulate the ATPase activity of myosin unless the concentration of free Ca^{2+} increases substantially, whereas a system consisting solely of purified actin and myosin does not exhibit any Ca^{2+} dependence. Thus, the actin-myosin interaction is controlled by Ca^{2+} in the presence of the regulatory troponin-tropomyosin complex (Pollard and Cooper, 1986).

D. Excitation-Contraction Coupling

The transduction of muscle action potential to contraction relies on molecules that control Ca^{2+} release at the T-tubule/SR junction (Hibberd & Trentham, 1986). One protein, an integral component of the T-tubular membrane, is an L-type, dihydropyridine-sensitive, voltage-dependent calcium channel. Another is the ryanodine receptor (RyR), a large protein associated with the SR membrane in the triad that probably couples conformational changes in the Ca^{2+} channel induced by T-tubular depolarization to Ca^{2+} release from the SR (see Figure 11.3) (Campbell et al., 1987). Skeletal muscle contains a higher density of L-type Ca^{2+} channels than can be accounted for on the basis of measured voltage-dependent Ca^{2+} influx because much of the Ca^{2+} channel protein in the T-tubular membrane does not directly regulate calcium ion movement but, rather, acts as a voltage transducer that links depolarization of the T-tubular membrane to Ca^{2+} release through the ryanodine receptor in the SR membrane. The ryanodine receptor thus mediates sarcoplasmic reticulum Ca^{2+} release. This protein, which binds the plant alkaloid ryanodine, is a very large multimer comprised of four subunits (Wagenknecht et al., 1989).

Activation of the ryanodine receptor complex is coupled to events at the T-tubular membrane by direct mechanical linkage through a conformational change in the dihydropyridine receptor protein. After diffusion toward the myofibrils, Ca^{2+} reuptake in the sarcoplasmic reticulum allows the relaxation of muscle and the maintenance of a low resting intracellular Ca^{2+} concentration by means of ATP-dependent Ca^{2+}

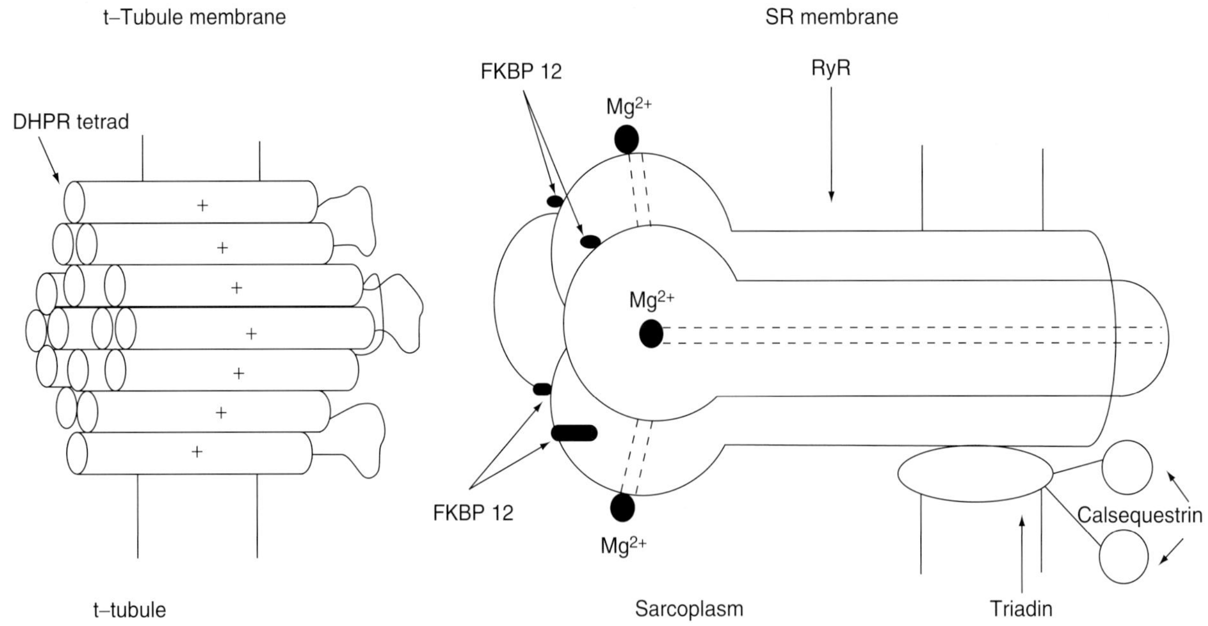

Figure 11.3 Putative arrangement of triad proteins in skeletal muscle. Each dihydropyridine receptor (DHPR)/voltage sensor consists of four homologous repeats, and four such DHPR are clustered together in a diamond-shaped arrangement (a tetrad) apposing four ryanodine receptor (RyR1) monomers that together function as a single Ca2+-release channel. The intracellular loop between repeats II and III of each DHPR is thought to activate physically in some way the adjacent RyR monomers. A 12 kDa FK 506-binding protein (FKBP 12) is associated tightly with each RyR1 monomer and may regulate interactions both within and between adjacent RyR. Calsequestrin is the main Ca2+-binding protein in the sarcoplasmic reticulum (SR) and possibly regulates the RyR. Triadin is another protein tightly associated with the RyR and has been suggested to mediate interactions with both the DHPR and calsequestrin. *From*: Lamb (2000).

pumps (SERCA) located in the SR membrane. The energy released upon ATP hydrolysis is utilized here for the concentrative uptake of Ca^{2+} into the SR vesicle through a phosphorylated enzyme intermediate. Other SR proteins assist in Ca^{2+} uptake and storage: phospholamban is expressed in cardiac muscle and slow myofibers, where phosphorylation participates in the control of Ca^{2+}-ATPase and Ca^{2+}-uptake activity. Another protein, calsequestrin, contains numerous low-affinity Ca^{2+}-binding sites and is present in the lumen of the SR, participating in Ca^{2+} storage. Fast myofibers contain a soluble Ca^{2+}-binding protein, parvalbumin, which is structurally related to troponin-C. Parvalbumin may also regulate the Ca^{2+} concentration in the initial stages of relaxation to facilitate rapid contraction.

Several genetic disorders of the developing neuromuscular system are described in the following sections according to the predominant disease locus: motor neuron, nerve, neuromuscular junction, and muscle (see Table 11.1). Several diseases are covered in other sections of this book and are noted only briefly in this chapter.

III. Diseases of Developing Nerve, the Neuromuscular Junction and Muscle

A. Spinal Muscular Atrophy as a Developmental Motor Neuron Disease

The most common form of spinal muscular atrophy (SMA), a frequent motor neuron disease, is caused by homozygous deletion of SMN1, a gene essential for the survival of the motor neuron. Other SMA phenotypes are caused by a variety of unrelated genes (see Table 11.2). The majority of the SMN protein is produced by SMN1, and one or more copies of the nearly identical SMN2 gene contribute an identical protein product at reduced efficiency owing to a single nucleotide transition (Monani, 2006). Most of the SMN2 protein product lacks a critical segment contributed by a nontranscribed exon as the consequence of the nucleotide transition and is probably rapidly degraded. SMN2, however, retains the ability to produce a small proportion of full-length SMN (see Figure 11.4). SMN2 copy number

Table 11.1 Genetic Disorders of Neuromuscular Development

I Motor neuron diseases
- *Spinal muscular atrophy*
- Other motor neuron diseases

II Disorders of nerve
- Congenital neuropathies
  - Charcot-Marie-Tooth diseases*
  - *Riley-Day syndrome*
- *Neuroaxonal dystrophy*
- Genetic neuropathies associated with encephalopathy

III Disorders of the neuromuscular junction
- Congenital myasthenic syndromes
  - *Choline acetyl transferase deficiency*
  - *Acetylcholine receptor deficiency*
  - *Rapsyn deficiency*
  - *Slow channel syndrome*
  - *Fast channel syndrome*
  - *Acetylcholinesterase deficiency*

IV Disorders of muscle
- Congenital muscular dystrophies
  - *Laminin α-2 deficiency*
  - *Fukuyama muscular dystrophy*
  - *Walker-Warburg syndrome*
  - *Muscle-Eye-Brain disease*
  - *Merosin-deficient muscular dystrophy*
- Congenital myopathies
  - *Central core disease**
  - Myotubular myopathy
  - Nemaline myopathy
  - *Desmin myopathy*

Structural myopathies
Dystrophynopathies
- *Duchenne muscular dystrophy*

Sarcoglycanopathies
Calpainopathy
Dysferlinopathy
Caveolinopathy
Emery-Dreifuss dystrophy
Bethlem myopathy
Facioscapulohumeral dystrophy
Metabolic myopathies

Disorders of fatty acid oxidation
- *Carnitine transporter deficiency*
- Carnitine palmitoyltransferase deficiencies
- Acyl-CoA dehydrogenase deficiencies

Glycogenoses
- *Acid maltase deficiency*
- *Debrancher enzyme deficiency*
- *Brancher enzyme deficiency*
- *Myophosphorylase deficiency*

Glycolytic defects
- *Phosphofructokinase deficiency*

Myoadenylate deaminase deficiency
Disorders of muscle excitability
Hyperkalemic periodic paralysis and paramyotonia congenita
Hypokalemic periodic paralysis
Andersen syndrome
Myotonic dystrophy
Congenital myotonia
Malignant hyperthermia
Brody disease

*Disorders covered in other chapters.

Table 11.2 Spinal Muscular Atrophy Phenotypes

SMA Type	Mode of Inheritance	Gene/Chromosomal Location	Phenotype	Age of Onset
Type I SMA	Autosomal recessive	SMN1;5q11.2-13.3	Proximal muscle weakness, patients nevers it unaided; death < 2 years	<6 months
Type II SMA	Autosomal recessive	SMN1;5q11.2-13.3	Proximal muscle weakness, patients sit unaided but become wheelchair bound, develop scoliosis of spine	6–18 months
Type III SMA	Autosomal recessive	SMN1;5q11.2-13.3	Proximal muscle weakness, patients walk unaided, normal lifespan	>18 months
Distal SMA	Autosomal recessive	11q13	Distal muscle weakness, diaphragmatic involvement	2 months–20 years
SMA with respiratory distress (SMARD)	Autosomal recessive	IGHMBP2;11q13.2	Distal lower limb weakness, diaphragmatic weakness, sensory, autonomic neurons also affected	1–6 months
X-linked infantile SMA	X-linked	Xp11.3-q11.2	Arthrogryposis, respiratory insufficiency, scoliosis, chest deformities, loss of anterior horn cells	at birth
Spinobulbar muscular atrophy (SBMA)	X-linked	Androgen Receptor/ Xq11.2-12	Proximal muscle weakness, lower motor neuron loss, dorsal root ganglion neuronal loss, bulbar involvement	30–50 years
Distal SMA IV	Autosomal dominant	7p15	Distal muscles affected, bilateral weakness in hands, atrophy of thenar eminence and peroneal muscle	12–36 years
Congenital SMA	Autosomal dominant	12q23-24	Arthrogryposis, nonprogressive weakness of distal muscles of lower limbs; several cases of affected pelvic girdle and truncal muscles	at birth

is variable, thus providing for significant compensation to SMN1 deletion in some individuals. Higher copy numbers of SMN2 are found in individuals afflicted by SMA type III, and low copy numbers are associated with the more severe SMA I and II phenotypes.

Patients afflicted by SMA type I may appear normal at birth but soon experience profound neck and proximal weakness, fasciculations, and areflexia due to motor neuron loss and denervation, progressing to fatal respiratory failure during infancy, if left unsupported. Apoptotic cell loss may be already apparent in the fetal spinal cord, as are reduced fetal movements toward the end of gestation. Type II and III patients suffer from less severe degrees of disability without detectable intellectual or sensory involvement. SMA is, thus, a motor neuron selective disease, despite the ubiquitous presence of the SMN protein in all human cells. SMN levels are highest during embryonic development, decreasing shortly after birth. The SMN protein is part of nuclear gem structures rich in heterogeneous nuclear ribonucleoproteins (hnRNPs). Gems are thought to host assemblysomes, complex structures that ensure pre-mRNA splicing after receiving RNAs previously bound to nuclear ribonucleoproteins in the cytoplasm. SMN is capable of polymerization, acquiring resistance to degradation.

SMN1 mutations that disrupt self-association, though rare in man, cause severe, type I SMA, presumably because of drift of free SMN monomers into the degradation pathway. SMN accumulates in human motor neurons and in axons and growth cones of other cultured neurons. Microtubule-associated granules contain both SMN and mRNA and are bidirectionally transported between cell soma and growth cones, possibly contributing to the regulation of translation inside cell processes. The SMN2 gene has proved moderately susceptible to modulation by DNA histone acetylation, a phenomenon of therapeutic significance. Inhibitors of histone deacetylation enhance transcriptional activity at multiple genomic loci, including the SMN2 promoter, and result in higher cellular SMN levels.

B. Congenital Myasthenic Syndromes as Disorders of Synaptic Transmission

These disorders result from impaired neuromuscular transmission as the consequence of mutations on either presynaptic, synaptic basal lamina-associated, or postsynaptic proteins, causing fatigable weakness in specific muscles (Engel et al., 2003a). Traditionally termed *congenital myasthenia*, they should be differentiated from acquired, immune-mediated myasthenia gravis. It is recognized that additional types of congenital myasthenic syndromes are likely to be discovered. In fact, mutations of SCN4A—the gene that encodes the $Na_V1.4$ skeletal muscle sodium channel—lead to enhanced channel inactivation causing typical myasthenic symptoms. Several disorders of cholinergic neuromuscular transmission result in overlapping phenotypes.

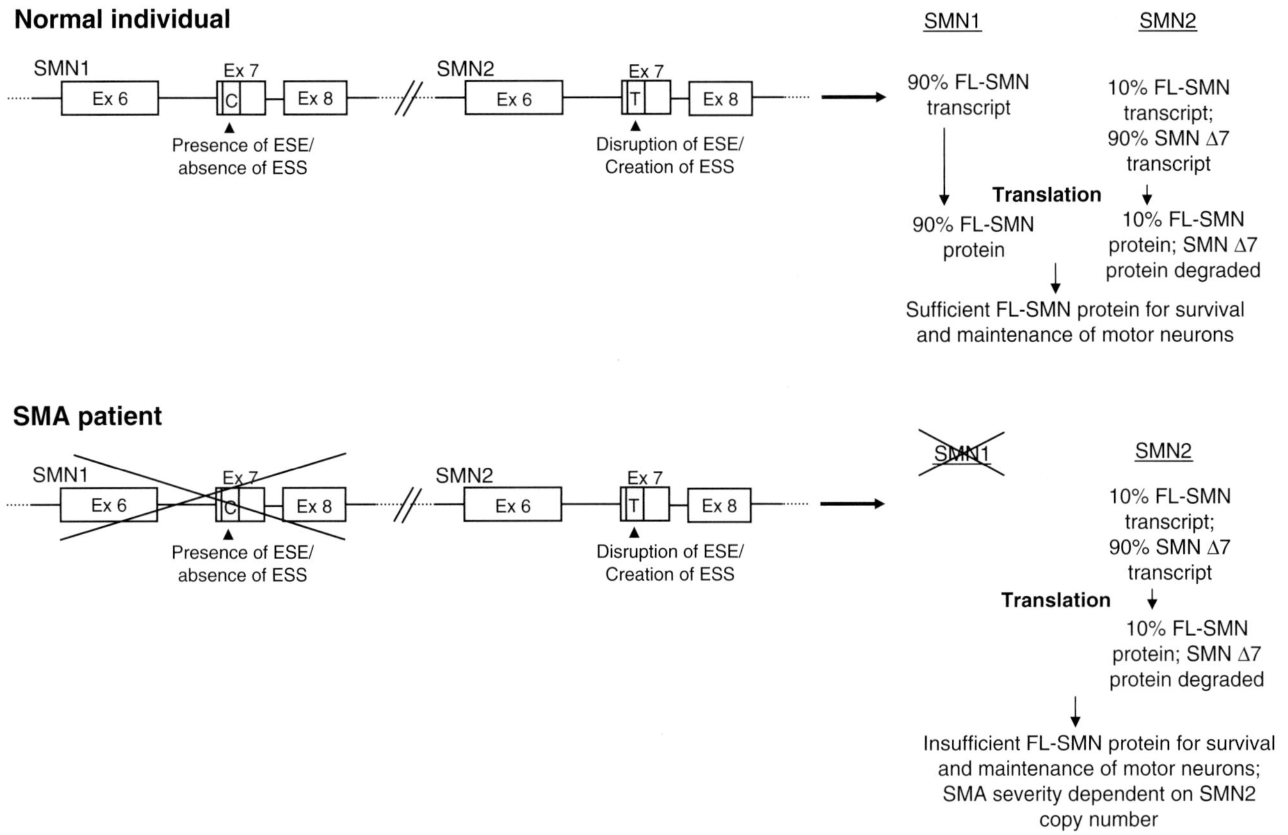

Figure 11.4 Spinal muscular atrophy genes and mechanisms.
From: Monani (2005).

1. Choline Acetyltransferase (ChAT) Deficiency

The distinguishing clinical feature is sudden episodes of severe respiratory difficulty and oropharyngeal (bulbar) weakness leading to apnea (cessation of respiration) precipitated by infections, fever, or excitement, or occurring even spontaneously. In some patients, the disease presents at birth with hypotonia and severe bulbar and respiratory weakness, requiring ventilatory support that gradually improves throughout life, but is later followed by episodes of apnea attacks and bulbar paralysis. Other patients are normal at birth and may develop myasthenic symptoms and apneic attacks after infancy or childhood. In electromyographic recordings, muscle action potentials decline abnormally when neuronal impulse flow is increased and recovers slowly, indicating an underlying defect in resynthesis or vesicular packaging of acetylcholine. Several recessive mutations have been identified, all of which decrease the abundance or the efficiency of ChAT. Although still unexplained, the selective neuromuscular—versus central nervous system—symptoms of ChAT deficiency may be due to the rate-limiting characteristics of the enzyme for the synthesis of acetylcholine at the neuromuscular junction.

2. Acetylcholine Receptor (AChR) Deficiency

The degree of disease severity may vary from mild to very severe. Patients harboring low-expression or homozygous null mutations in the AChR ε subunit may experience mild symptoms, presumably owing to the compensatory replacement of this subunit by the analogous γ subunit. In contrast, patients with low-expression mutations in the other noninterchangeable subunits (α, β, δ) are severely affected, such that no patients with null mutations in both alleles of a non-ε subunit have been observed. AChR deficiency results from mutations that cause premature termination of the translational chain by frameshift, by altering splice sites, or by generating stop codons; from point mutations in the promoter region; from chromosomal microdeletion; and even from missense mutations, some of which affect the signal peptide or residues essential for assembly of the AChR. Patients respond moderately well to cholinesterase inhibition.

3. Rapsyn Deficiency

These patients manifest AChR deficiency with decreased rapsyn as well as secondary AChR abundance and resulting impaired postsynaptic morphologic development. None of the known rapsyn mutations hinders rapsyn self-association, but all diminish clustering of AChR with this protein. In addition to weakness, some patients manifest striking facial malformations. Patients are treated with anticholinesterase agents and with 3,4-diaminopyridine, a blocker of open K^+ channels that decreases presynaptic membrane resting conductance, causing hyperpolarization and therefore resetting more Ca^{2+} channels into their closed, noninactivated state from which they can readily open, increasing Ca^{2+} influx into the terminal and facilitating a compensatory increase in acetylcholine release.

4. Slow and Fast Channel Syndromes

Abnormally long-lived openings of mutant AChR (slow) channels result in prolonged endplate currents and potentials, which in turn elicit one or more repetitive muscle action potentials of lower amplitude subject to decrement. The morphologic consequences stem from prolonged activation of the AChR channel, causing cationic overload of the postsynaptic region or endplate myopathy, with Ca^{2+} accumulation, destruction of the junctional folds, nuclear apoptosis, and vacuolar degeneration of the terminal. Some slow channel mutations in the transmembrane domain of the AChR render the channel "leaky" by stabilizing the open state even in the absence of acetylcholine. Some slow channel mutants can be opened by choline even at the concentrations normally present in serum. Quinidine, an open channel blocker of AChR is used for therapy. The clinical features of fast channel syndrome resemble those of autoimmune myasthenia gravis. Conversely to what is found in slow channel syndrome, the open state of the AChR is destabilized, manifesting as fast dissociation of acetylcholine from the receptor and/or excessively reduced open times. In most cases, the mutant allele causing the kinetic abnormality is accompanied by a null mutation in the second allele such that the kinetic mutation dominates the phenotype, but homozygous fast-channel mutations also exist. Therapy includes anticholinesterase agents and 3,4-diaminopyridine.

5. Acetylcholinesterase (AChE) Deficiency

Inhibition of the AChE results in prolonged exposure of AChR to acetylcholine, leading to prolonged endplate potentials, desensitization of AChR, and depolarization block. Endplate myopathy with loss of AChR may result. In many patients, the disease presents in the neonatal period and is highly disabling. Some patients manifest an excessively slow pupillary response to light. An array of mutations may cause reduced attachment of AChE to its collagen-derived anchor, truncation of the collagen domain rendering it incompetent for insertion, or hindrance to helical assembly by the collagen domain.

C. Muscle Sodium Channels and Hereditary Disorders of Excitability

A group of inherited muscle diseases called the periodic paralyses is characterized by intermittent episodes of skeletal muscle weakness or paralysis that occur in individuals who otherwise appear normal or are just mildly weak between attacks. The periods of paralysis often are associated with changes in serum K^+ concentration; the serum K^+ concentration may increase or diminish, but the direction of change is usually consistent for a particular family and forms one basis for classifying these diseases as either hyperkalemic or hypokalemic (Lehmann-Horn & Jurkat-Rott, 1999). A variant of periodic paralysis, in which spells of weakness are less frequent and in which a form of muscle hyperexcitability is often seen, is called *paramyotonia congenita*. Electrical recordings from muscle fibers isolated from patients during an attack of periodic paralysis have shown that the paralytic episodes are associated with acute depolarization of the sarcolemma. In all forms of periodic paralysis, this depolarization is due to an increase in membrane conductance to Na^+. In the case of hyperkalemic periodic paralysis and paramyotonia congenita, this abnormal conductance can be blocked by tetrodotoxin, a small polar molecule that is highly specific for the voltage-dependent Na^+ channel. Foods with a high K^+ content may trigger an attack, but carbohydrate-rich substances are abortive of the paralytic episode.

Single-ion channel recordings in hyperkalemic periodic paralysis have revealed that some of the muscle membrane Na^+ channels show abnormal inactivation kinetics, intermittently entering a mode in which they fail to inactivate. These channels produce a persistent, noninactivating Na^+ current that in turn produces membrane depolarization. Because normal Na^+ channels enter an inactivated state after depolarization, the net result of long-term depolarization is loss of sarcolemmal excitability and paralysis. Both hyperkalemic periodic paralysis and paramyotonia are caused by mutation of the adult skeletal muscle Na^+ channel SkM1 gene SCN4A. The hypokalemic form of periodic paralysis, however, is not linked to this Na^+ channel gene, but to a voltage gated calcium channel (Ptacek et al., 1994). Numerous mutations alter the coding region of the SCN4A gene in families with hyperkalemic periodic paralysis or paramyotonia congenita. Although these mutations are distributed through a wide span of the channel coding region, a number of them are clustered in a cytoplasmic linker region known from biophysical studies to control inactivation. Others are located near the cytoplasmic ends of transmembrane domains S5 and S6, and these residues in fact may constitute the binding site for the closing inactivation gate. Mutations in these regions may destabilize this closed conformation, leading to abnormalities in channel inactivation. Mutants associated with the paramyotonia congenita phenotype show, on the other hand, a marked slowing in the major component of fast inactivation. In some cases, the voltage dependence of the inactivation rate constant, also known as τ_h, is markedly reduced as well, and the mutations appear to uncouple the inactivation process from the voltage-dependent channel conformational changes associated with inactivation.

In some families with hyperkalemic periodic paralysis, the mutations cause a small, persistent inward Na^+ current in the myocyte that is the result of a shift in channel modal gating. Normal skeletal muscle Na^+ channels can shift between a fast and a slow inactivation gating mode and, usually, the channels are found in the fast inactivation mode. Channels with hyperkalemic periodic paralysis mutations, however, spend a greater percentage of the time in the slow inactivation gating mode, and late openings associated with this slow gating mode contribute to the persistent inward current seen in cells harboring these mutations. Under voltage-clamp conditions at the single-channel level, SkM1 channels with paramyotonia congenita mutations show multiple late openings and prolonged openings after depolarization. These late openings account for the slow inactivation of the Na^+ current observable in the cell. Hyperkalemic periodic paralysis mutations also show multiple late openings at the single-channel level, but these abnormal events are temporally clustered, consistent with an underlying shift in modal gating. Single-channel conductance is not altered by any of these mutations, but all sodium ion channel mutations in periodic paralysis produce dominant-negative effects.

Although some mutant channels only intermittently exhibit abnormal inactivation, this small population of abnormally inactivating channels can modify the behavior of the remaining mutant and normal channels present in the membrane. Unlike the CLC-1 chloride channel mutations in myotonia congenita, which produce dominant-negative effects within a single channel multimer, these Na^+ channel mutations produce dominant-negative effects that reflect the relationship of normal channel inactivation to membrane potential. In either case, the persistent inward current carried by a small population of noninactivating channels, or the prolonged inward current resulting from mutant channels with slowed inactivation rates, results in a slight but long-lasting membrane depolarization. Since the relationship between voltage and inactivation in normal channels is very steep near the resting potential, this slight depolarization can produce inactivation of normal channels. If depolarization is sufficient, too few channels will remain in the noninactivated state to satisfy the requirements for a regenerative action potential and the muscle will become paralyzed.

D. Potassium Channel Mutations in Andersen Syndrome, a Developmental Disorder

Andersen syndrome includes periodic paralysis, prolongation of the electrocardiographic QT interval causing susceptibility to cardiac ventricular arrhythmias, and characteristic physical features including low-set ears, a small jaw, and malformation of the digits, and can be inherited in an autosomal dominant fashion. The disease is unique due to the combination of both a skeletal and a cardiac muscle phenotype and may be caused by mutations in KCNJ2, a gene that encodes the inward rectifier K^+ channel Kir2.1, which is expressed in both cardiac and skeletal muscles (Plaster et al., 2001). Kir2.1 is an important contributor to the cardiac inward rectifier K^+ current, I_{K1}, which provides substantial current during the repolarization phase of the cardiac action potential. All KCNJ2 mutations cause a dominant negative effect on channel function and a reduction in I_{K1} prolongs the terminal repolarization phase, rendering the myocardium prone to repetitive ectopic action potentials. Attacks of paralysis can be associated with hypo-, hyper-, or normokalemia and, although serum potassium levels during attacks differed among kindreds, they are consistent within an individual kindred. Mutations in Kir2.1 may sufficiently reduce the muscle resting K^+ conductance such that the membrane depolarizes, leading to inactivation of Na^+ channels making them unavailable for initiation and propagation of action potentials.

E. RNA Splicing Abnormalities and Myotonic Dystrophy

Myotonic dystrophy (DM; Steinert disease), a multisystemic disorder, is one of the most common forms of muscular dystrophy in adults. In addition to hereditary muscular dystrophy and myotonia, DM causes a constellation of seemingly unrelated clinical features including cardiac conduction defects, cataracts, and endocrine and immunological abnormalities. Two clinical and genetic types of DM exist (Ranum & Day, 2004). The genetic features of DM type 1 (DM1) include variable penetrance, anticipation (a tendency for the disease to worsen in subsequent generations), and a maternal transmission bias for congenital forms despite the location of the causative gene on chromosome 19. The cause of DM1 is a $(CTG)_n$ repeat in the 3-UTR of a protein kinase (DMPK) gene. Type 2 DM, in contrast, predominantly causes pelvic girdle weakness, and often is referred to as proximal myotonic myopathy, ascribed to the genetic locus encoding zinc inger protein 9 on chromosome 3. DM2 is caused by a $(CCTG)_n$ expansion. Nevertheless, all individuals affected by DM1 and DM2 experience weakness, pain, and myotonia, and cardiac involvement may lead to conduction defects, arrhythmias, and sudden death. Endocrine abnormalities in both DM1 and DM2 result in hyperinsulinemia, hyperglycemia, and insulin insensitivity, with type 2 diabetes occurring in each disorder. Testicular failure is also common, with associated hypotestosteronism, elevated follicle-stimulating hormone (FSH) levels, and oligospermia. Other serological abnormalities in both disorders include reduced levels of immunoglobulins G and M. The brain is also affected as assessed by magnetic resonance imaging (MRI), but mental retardation is a feature of only DM1.

The multisystemic clinical parallels shared by both DM1 and DM2 suggest a similar pathogenic mechanism (see Figure 11.5). The discovery that DM2 mapped to chromosome 3 and not to the DM1 region of chromosome 19 makes it unlikely that specific gene expression defects cause the common clinical features of the disease. The discovery that a CCTG repeat expansion located on chromosome 3 that is expressed at the RNA (but not at the protein) level causes DM2, and the observation that both CUG and CCUG repeat-containing foci accumulate in affected muscle nuclei suggests that a gain-of-function RNA mechanism underlies the clinical features common to both diseases.

This is, in fact, but one of a class of disorders of post-transcriptional processing (Waxman, 2001). RNA-binding proteins, including CUG-binding protein (CUG-BP) and *muscle-blind* isoforms bind to—being sequestered—or are dysregulated by the repeat-containing RNA transcripts resulting in specific trans-alterations in pre-mRNA splicing. Specific changes in pre-mRNA splicing have been associated with several genes, including the insulin receptor, the chloride channel ClC-1, and cardiac troponin T, and probably are correlated with insulin resistance, myotonia, and cardiac abnormalities. For example, CUG-BP, which is elevated in DM1 skeletal muscle, binds to the ClC-1 pre-mRNA, causing an aberrant pattern of ClC-1 splicing. Thus, altered splicing regulation of ClC-1 decreases its abundance in the muscle plasma membrane causing hyperexcitability and leading to the DM feature of myotonia. Further, mutant RNA transcripts bind and sequester transcription factors leading to the depletion of as much as 90 percent of several of them, resulting in secondary depletion of proteins such as the chloride channel ClC-1.

F. Chloride Channel Mutations in Myotonia Congenita

In two diseases, dominant myotonia congenita (Thomsen disease) and recessive myotonia congenita (Becker myotonia), myotonia is the major presenting symptom and often

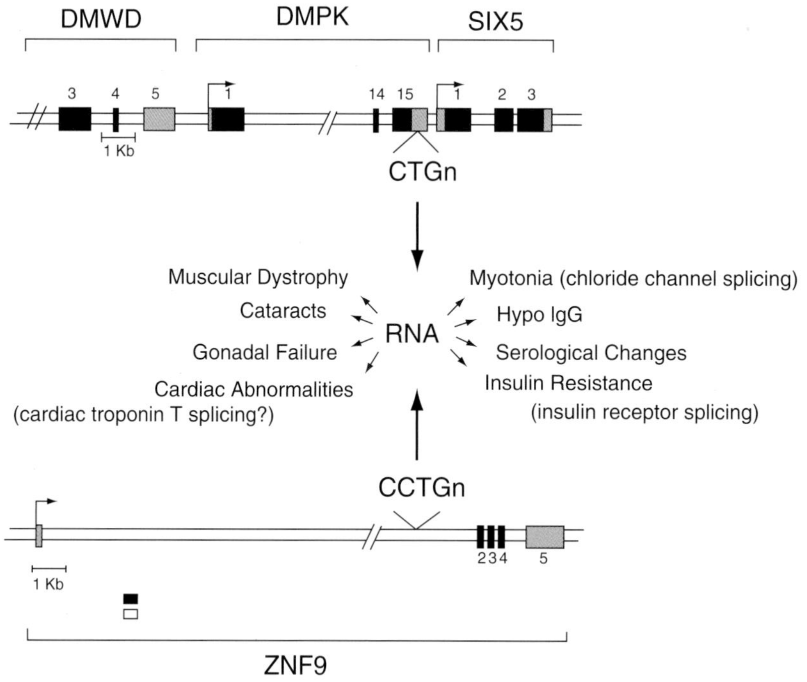

Figure 11.5 RNA in myotonic dystrophy pathogenesis. The multisystemic clinical parallels between DM1 and DM2 suggest a similar pathogenic mechanism. A CCTG repeat expansion located on chromosome 3 is expressed at the RNA but not the protein level, causing DM2, and both CUG and CCUG repeat containing foci accumulate in affected muscle nuclei, indicating a gain-of-function RNA mechanism shared by both diseases. Specific changes in pre-mRNA splicing that have been associated with several genes, including the insulin receptor, the chloride channel, and cardiac troponin T, are correlated with insulin resistance, myotonia, and cardiac abnormalities.
From: Ranum and Day (2004).

the only abnormality, although the muscles may be overdeveloped (conferring individuals a Herculean appearance) in Thomsen disease. Patients afflicted with these diseases have difficulty relaxing their muscles normally; doorknobs and handshakes are difficult to release, clumsiness is a problem and falls often occur. In both Thomsen and Becker myotonia, multiple mutations have been found in chromosome 7 and the sarcolemma exhibits a severe reduction in membrane Cl^- conductance (Lehmann-Horn & Jurkat-Rott, 1999). This locus encodes the ClC-1 skeletal muscle Cl^- channel family, whose members control anion flux in a number of tissues and are closely related in structure, forming dimers of two ClC subunits. The fact that mutations in the gene encoding ClC-1 can produce either dominant or recessive effects is surprising. Mutants that introduce frameshifts of stop codons early in the coding sequence produce a nonfunctional protein product. With one defective gene copy, wild-type channels encoded by the second allele should produce a net Cl^- conductance about 50 percent of normal. When both gene copies carry the mutation, expression of the functional channels will be very low or absent and, as a recessive disorder, the myotonia can be severe.

Point mutations can also lead to the alteration of a single amino acid in the primary structure of an otherwise full-length channel monomer, and channels formed from this protein may not function normally. The possibility also exists that channels containing even a single mutant subunit may fail to function even though the other subunits are encoded by a normal copy of the gene. Such a dominant-negative effect, which has been demonstrated for a number of myotonia congenita mutations, leads to a dominant transmission of the disease phenotype. Since mixed channels containing different numbers of mutant subunits may have different levels of residual activity, the resulting membrane Cl^- conductance may be more variable and the disease phenotype less severe than in the recessive form of the disease.

G. Muscular Dystrophy and Dystrophin Mutations

Dystrophinopathy refers to diseases caused by mutations in the locus that encodes the cytoskeletal muscle protein dystrophin, the largest gene in man located in the X chromosome (O'Brien & Kunkel, 2001). Mutations associated with

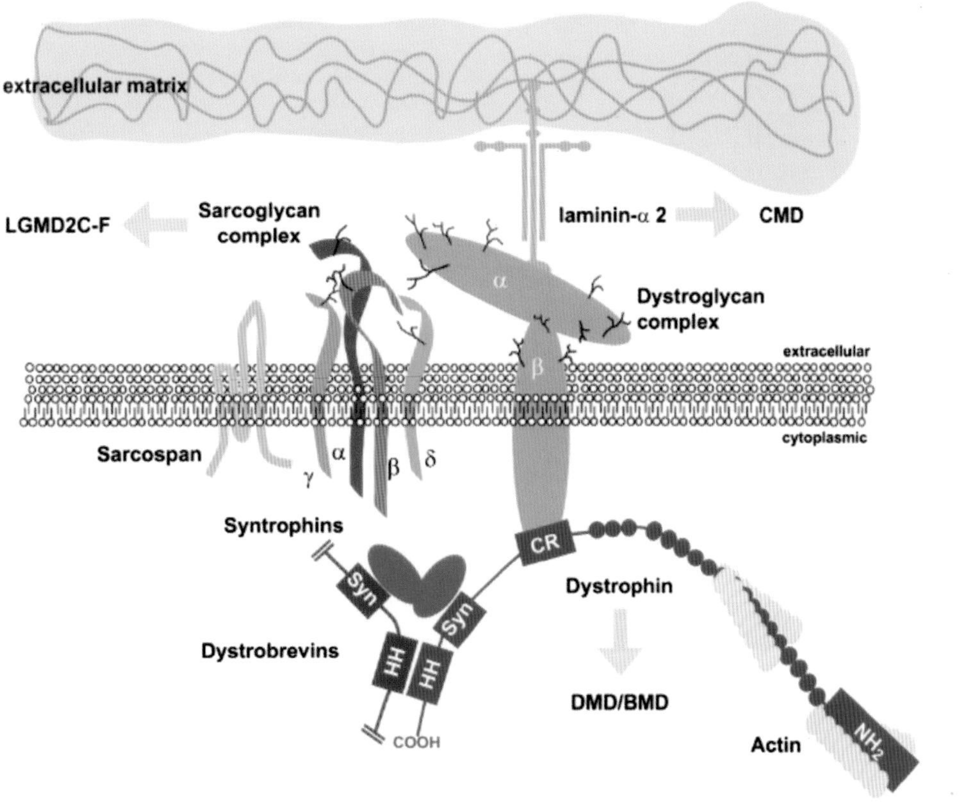

Figure 11.6 Schematic of the dystrophin-associated protein complex. Diseases caused by mutations in the genes encoding members of the complex are indicated. LGMD2C-F: types of limb girdle muscular dystrophies; CMD: congenital muscular dystrophin (merosin deficient-type); DMD/BMD: Duchenne and Becker-type dystrophinopathies.
From: O'Brien and Kunkel (2001).

severe loss of function cause Duchenne dystrophy. Children afflicted by Duchenne dystrophy experience delayed attainment of motor independence later followed by loss of ambulation, mild intellectual dysfunction, and cardiomyopathy. Muscle dystrophin content is usually severely decreased. Patients suffering from the Becker phenotype retain higher protein levels and consequently manifest milder abnormalities. Asymptomatic female carriers may manifest an elevation in plasmatic creatine kinase as the sole indicator of muscle dysfunction coexisting with a mosaic pattern of dystrophin production across myofibers. A related entity, McLeod syndrome, combines a constellation of X-linked disorders (Duchenne dystrophy, chronic granulomatosis, and retinitis pigmentosa) as the consequence of a large deletion in the X chromosome, causing a contiguous gene deletion syndrome. The dystrophin gene gives rise to several protein isoforms under the control of a variety of promoters active in selected tissues and several alternatively spliced versions of the protein also exist. The N-terminal domain of the molecule binds actin, and the central rod domain is formed by triple-helical spectrin-like repeats and by a cysteine-rich domain. Each dystrophin molecule constitutes a functional unit, associating with transmembrane proteins of the dystroglycan complex and with intracellular proteins (see Figure 11.6). At least six proteins associate with dystrophin, some of which also exhibit reduced abundance in Duchenne dystrophy muscle. Three protein complexes constitute the major molecular interactions of dystrophin:

- Extracellular α and transmembrane β dystroglycans
- Intracellular syntrophins and dystrobrevins
- Transmembrane α, β, γ, and δ sarcoglycans and sarcospan

The extracellular portion of β-dystroglycan binds to α-dystroglycan, a protein closely interactive with laminin-2, a cause of another type of muscular dystrophy, and its intracellular region is bound to dystrophin. The syntrophins and dystrobrevins are thought to function as adapters, helping bring the dystrophin complex together. Mutations in the sarcoglycans cause autosomal recessive limb girdle muscular dystrophy and, occasionally, a Duchenne-like phenotype. Several sarcoglycan protein domains are endowed with extracellular ATPase activity and may contribute to transmembrane signaling.

References

Agarkova, I., Perriard, J. C. (2005). The M-band: An elastic web that cross-links thick filaments in the center of the sarcomere. *Trends Cell Biol.* **15**, 477–485.

Buchthal, F., Schmalbruch, H. (1980). Motor unit of mammalian muscle. *Physiol Rev.* **60**, 90–142.

Campbell, K. P., Knudson, C. M., Imagawa, T. et al. (1987). Identification and characterization of the high affinity [^{3}H]ryanodine receptor of the junctional sarcoplasmic reticulum Ca^{2+} release channel. *J Biol Chem* **262**, 6460–6463.

Engel, A. G., Ohno, K., Sine, S. M. (2003a). Congenital myasthenic syndromes: A diverse array of molecular targets. *J Neurocytol.* **32**: 1017–1037.

Engel, A. G., Ohno, K., Sine, S. M. (2003b). Sleuthing molecular targets for neurological diseases at the neuromuscular junction. *Nat Rev Neurosci.* **4**, 339–352.

Franzini-Armstrong, C. (1979). Studies of the triad. I. Structure of the junction of frog twitch fibers. *J Cell Biol.* **47**, 488–499.

Hibberd, M. G., Trentham, D. R. (1986). Relationships between chemical and mechanical events during muscular contraction. *Annu Rev Biophys Biophys Chem* **15**, 119–161.

Hughes, B. W., Kusner, L. L., Kaminski, H. J. (2006). Molecular architecture of the neuromuscular junction. *Muscle Nerve* **33**, 445–461.

Huxley, H. E. (1969). The mechanism of muscle contraction. *Science* **164**, 1356–1366.

Lamb, G. (2000). Excitation–contraction coupling in skeletal muscle: Comparisons with cardiac muscle. *Clin Experimental Pharm Physiol.* **27**, 216–224.

Lehmann-Horn, F., Jurkat-Rott, K. (1999). Voltage-gated ion channels and hereditary disease. *Physiol Rev.* **79**, 1317–1372.

Miller, J. B., Schaefer, L., Dominov, J. A. (1999). Seeking muscle stem cells. *Curr Top Dev Biol.* **43**, 191–219.

Monani, U. R. (2005). Spinal muscular atrophy: A deficiency in a ubiquitous protein; a motor neuron-specific disease. *Neuron* **48**, 885–896.

O'Brien, K. F., Kunkel, L. M. (2001). Dystrophin and muscular dystrophy: Past, present, and future. *Mol Genet Metab.* **74**: 75–88.

Plaster, N. M., Tawil, R., Tristani-Firouzi, M. et al. (2001). Mutations in Kir2.1 cause the developmental and episodic electrical phenotypes of Andersen's syndrome. *Cell* **105**, 511–519.

Pollard, T. D., Cooper, J. A. (1986). Actin and actin-binding proteins. A critical evaluation of mechanisms and functions. *Annu Rev Biochem* **55**, 987–1035.

Price, S. R., Briscoe, J. (2004). The generation and diversification of spinal motor neurons: Signals and responses. *Mech Dev.* **121**, 1103–1115.

Ptacek, L., Tawil, R., Griggs, R. et al. (1994). Dihydropyridine receptor mutations cause hypokalemic periodic paralysis. *Cell* **77**, 863–898.

Ranum, L. P., Day, J. W. (2004). Myotonic dystrophy: RNA pathogenesis comes into focus. *Am J Hum Genet.* **74**, 793–804.

Wagenknecht, T., Grassucci, R., Frank, I. et al. (1989). Three-dimensional architecture of the calcium channel/foot structure of sarcoplasmic reticulum. *Nature* **338**, 167–170.

Waxman, S. G. (2001). Transcriptional channelopathies: An emerging class of disorders. *Nat Rev Neurosci.* **2**, 652–659.

Zot, A. S., Potter, J. D. (1987). Structural aspects of troponin-tropomyosin regulation of skeletal muscle contraction. *Annu Rev Biophys Biophys Chem* **16**, 535–560.

12

Molecular Mechanisms of Ischemic Brain Disease

Thomas M. Hemmen and Justin A. Zivin

I. Introduction

Ischemic stroke is a leading cause of death and disability. The molecular mechanisms leading to and caused by brain ischemia are complex. Cerebral ischemia occurs when blood flow to the brain decreases to a level where the metabolic needs of the tissue are not met. The precise flow rate below which ischemia occurs is poorly described (Siesjo et al., 1995). In recent years, new animal models, as well as imaging techniques such as functional Magnetic Resonance Imaging (fMRI) and Positron Emission Tomography (PET), have advanced the understanding of molecular changes in cerebral ischemia. In this chapter, we will review what is currently known about molecular processes in ischemic brain disease and lay out the most commonly cited theories.

II. Hypoxia/Ischemia

Brain tissue has a relatively high consumption of oxygen and glucose, and depends almost exclusively on oxidative phosphorylation for energy production. Despite extensive research for decades, our understanding of the final common pathway to cellular death in stroke remains incomplete. Current pathophysiological models differentiate global from focal ischemic injury. Global ischemia results from transient low cerebral blood flow (CBF) below 0.5 mL/100 g per minute or severe hypoxia to the entire brain. The causes are most frequently the sequela of cardiac arrest, near drowning, and hypotension, and it is sometimes seen as a consequence of surgical procedures. After a few minutes of cardiac arrest, hypoxic encephalopathy becomes irreversible. The precise duration of global ischemia necessary for irreversible neuronal damage in humans has not been fully established (Heiss, 1983; Zivin, 1997), but for the most vulnerable areas, it is about five minutes. Many animal models have confirmed cellular death within a few minutes after onset of global ischemia (Ljunggren et al., 1974). On the other hand, Hossmann et al. demonstrated that global cessation of CBF in animals of up to one hour can be followed by recovery of electrophysiological function, and in a few animals, recovery of neurological function (Hossmann et al., 1987; Hossmann and Zimmermann, 1974). It remains unclear how various classes of neurons can respond so differently after an ischemic insult. Besides the variable functional outcome after ischemia in animal models, it is well established that specific neuronal populations within an individual vary substantially in ischemic tolerance. Neurons in the CA1 region of the hippocampus and other distinct cellular populations of the caudate, thalamus, neocortex, and cerebellum are basically vulnerable to relatively brief periods of ischemia.

Brief episodes of ischemia may not lead to immediately evident neuronal cell death, but selectively vulnerable neuronal populations can develop delayed ischemic cell death. In ischemic models and human tissue studies, these populations were found in the CA1 region of the hippocampus (Squire, 1992).

In ischemic models using gerbils, CA1 cells die after five minutes, and cells in the geniculate nuclei of the thalamus and substantia nigra are irreversibly damaged after 10 to 15 minutes of ischemia (Araki et al., 1993). In other species, Purkinje cells of the cerebellum have been shown to be selectively vulnerable to ischemia.

Changes in the microcirculation, as seen in focal stroke, do not occur in global ischemia. Such changes cause variable degrees of hypoperfusion across the ischemic zone and affect tissue viability. This is especially pertinent in the reperfusion phase after transient ischemia, and explains why models of global ischemia cannot fully represent focal stroke pathophysiology (Kirino & Sano, 1984; Smith et al., 1984).

Focal ischemia results from transient or permanent reduction of CBF in a restricted vascular territory. In most cases, the reduction in blood flow is caused by occlusion of a cerebral artery by an embolus or local thrombosis. The reduction in CBF leads to hypoperfusion in the brain tissue supplied by that vessel. Complete cessation of blood flow is uncommon. Due to collateral vessels, the CBF in the ischemic core zone often remains at 5 to 15 mL per 100 g per minute and at 15 to 25 mL per 100 g per minute in the outer areas of the hypoperfused zone (Heiss, 1992).

In the landmark publication by Jones et al. (1981), unanesthetized Macaca irus monkeys subject to carotid occlusion showed mild hemiparesis starting at a CBF of approximately 22 mL/100 g per minute and complete paralysis at about 8 mL/100 g per minute or less. Clamping of the carotid artery in patients undergoing endarterectomy causes changes in EEG frequency once the blood flow is reduced below 18 mL/100 g per minute, and flattening of the EEG occurs below 12 mL/100 g per minute (Sundt et al., 1973; Trojaborg & Boysen, 1973). In decapitation models, the EEG becomes flat within 12 seconds (Swanson et al., 1989). In other studies, cerebral-evoked potentials were decreased at a CBF of 20 ml/100 g per minute and abolished below 15 mL/100 g per minute (Branston et al., 1974). The critical relationship between CBF and cerebral electrical activity has been reproduced in many experimental models and reflects neuronal dysfunction within seconds after ischemia induction (Heiss, 1992). The electrophysiological tests, however, are unable to identify the cellular causes of irreversible neuronal damage and have poor spatial correlation with long-term functional outcome (Isley et al., 1998).

Morphological changes are found in brain tissue, which is perfused at a rate of 12 mL per 100 g per minute for two to six hours (Pulsinelli et al., 1982a; Tamura et al., 1980). The degree of morphological change depends, to a considerable extent, on the CBF, the duration of reduced blood flow, the selected tolerance of the affected neurons, and the type of morphological marker (stain) that is used (Jones et al., 1981). Irreversible cellular damage varies across different neuronal populations and can be different within closely neighboring cortical areas in a single individual (Heiss & Rosner, 1983). The degree of ischemic tolerance has important implications for the potential for neurological recovery and the time window for possible treatment of acute stroke.

In addition to the selective ischemic damage in different regions of the brain, the pathophysiological changes within the ischemic tissue are heterogeneous. Usually one can differentiate a core zone of profoundly reduced CBF and an adjacent area with a lesser degree of hypoperfusion. This hypoperfused, but potentially salvageable, area is often referred to as the *penumbra*. The terminology of ischemic penumbra was first introduced by Astrup et al., and describes brain tissue that is perfused at levels between the functional and morphological threshold (Astrup et al., 1977, 1981). The pathophysiological changes in this area are complex and not fully understood. Since no marker unequivocally identifies the penumbra, it must be operationally defined in experimental stroke (Hossmann, 1994). In humans, viable tissue may be found up to 48 hours after stroke (Furlan et al., 1996), although the duration of true viability in the penumbral zone remains controversial.

III. Excitotoxicity

Although the primary effect from cerebral ischemia is the reduction in oxygen, glucose uptake and increased lactate production also contribute to neuronal cell death. The brain initially converts to anaerobic glycolysis, and when it rapidly runs out of its glucose stores, induces neuronal cell death in the most severe global ischemia or focal ischemic core. Most neurons die after transient or permanent ischemia from these secondary effects (see Figure 12.1), as demonstrated by the fact that even if blood flow is restored within minutes after transient ischemia and some hippocampal neurons survive, most CA1 neurons die. This suggests that the ischemic episode sets off a cascade of secondary events within the cell, leading to cellular demise despite blood flow restoration.

The most extensively studied process of secondary neuronal injury is excitotoxicity. Excitotoxicity plays an important role not only in cerebral ischemia, but many central nervous system (CNS) diseases, including trauma and neurodegenerative disorders (Choi, 1988). Brain tissue, with its high consumption of oxygen and glucose, depends almost exclusively on oxidative phosphorylation for energy production. Cerebral blood flow impairment reduces oxygen and glucose delivery, and impairs intracellular energy production, which is required to maintain ionic gradients (Bryan et al., 1991). The reduction of ionic gradients leads to loss of membrane potential and depolarization of neurons and glial cells (Katsura et al., 1994).

In the physiologic state, extracellular glutamate concentration is low and most is inside the cell. This gradient is maintained via cellular glutamate uptake, which is driven by transmembrane Na^+ and K^+ channels. These channels, in turn, depend on osmotic gradients driven by Na^+-K^+-ATPase. Ischemia results in the loss of ATP, thereby reducing the osmotic gradient, decreasing glutamate uptake, and increasing the extracellular glutamate concentration (Longuemare et al., 1994). This resultant increase in extracellular glutamate leads to excessive activation of postsynaptic receptors of excitatory amino acid, which is described as excitotoxicity (Olney & Sharpe, 1969).

Glutamate is the major excitatory neurotransmitter in the central nervous system and binds to a variety of receptors. The receptor subtypes have been identified using special ligands whose high affinity gave each subtype its name. Almost every glutamate receptor subtype has been implicated in excitotoxicity. The N-methyl-D-aspartate (NMDA) subtypes, however, play a major role in mediating excitotoxic cell death, mainly owing to their high calcium (Ca^{2+}) permeability. Other glutamate receptor subtypes, such as 2-amino-3-(3-hydroxy-5-methylisoxazol-4-yl) propionate (AMPA) or kainate receptors, have also been thought to play critical roles in mediating excitotoxic neuronal cell death. Although the molecular basis of glutamate toxicity is uncertain, there is general agreement that it is in large part Ca^{2+}-dependent.

The activation of NMDA receptors and other glutamate receptors leads to calcium influx via phospholipase C and Ins(1,4,5) *P*3 signaling pathways (Park et al., 1989). As a result of glutamate-mediated overactivation, Na^+ and Cl^- enter the cell and water follows passively. The influx of Na^+ and Cl^- is much larger than the efflux of K^+. The increased intracellular water leads to edema formation, which in turn, can reduce perfusion of brain regions surrounding the core of the perfusion deficit. In addition, edema can increase intracranial pressure and cause vascular compression and herniation.

Over the last decades, much clinical and preclinical research has been focused on blocking the excitotoxicity following cerebral ischemia. Despite many promising *ex vivo* and animal data, no human clinical study has shown the efficacy and safety of compounds interfering with the excitotoxic pathways. This is mainly thought to be due to toxic side effects of these compounds when used at efficacious levels and the need to apply the drug so early during the ischemic cascade that it becomes infeasible to treat patients so rapidly after stroke onset. Although NMDA receptor-blocking agents have been shown to alleviate ischemic damage in neurons, this effect usually is significant only when the drug is given before or shortly after the ischemia ensues. Most patients with cerebral ischemia, however, do not present to the medical practitioner until hours after ischemia begins.

IV. Free Radicals

The term free radical refers to reduced forms of oxygen: superoxide, peroxyl radical, nitric oxide, hydroxyl ($OH^{\bullet}$), and singlet oxygen. At low concentrations, free radicals (ROS) are produced during normal cell metabolism and are

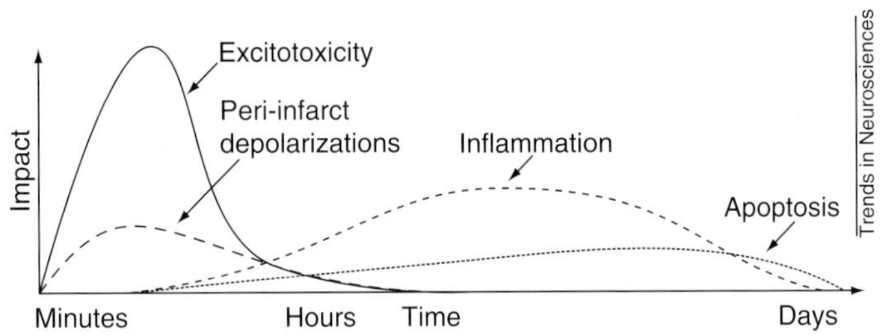

Figure 12.1 Post ischemic changes over time. Early after the onset of the focal perfusion deficit, excitotoxic mechanisms can damage neurons and glia. Later changes include peri-infarct depolarizations, inflammation, and apoptosis (Dirnagl et al., 1999).

a byproduct of the conversion of oxygen to water and ATP. When ROS interact with metals in the cytoplasm or mitochondria, hydroxyl ($OH^{\bullet}$) is formed (Halliwell, 1992). ROS attack the unsaturated bonds in fatty acids and cholesterol and produce lipid peroxides, which act as radicals, as well (Ginsberg et al., 1988). In addition, proteins and DNA are subject to ROS damage. For example, one can measure the ROS-induced injury in neuronal cells through the activity of glutamate synthetase (Floyd & Carney, 1992). Under normal conditions, most cells are protected against damage from ROS by enzymatic and nonenzymatic antioxidant pathways. Enzymatic antioxidants include cytoplasmic superoxide dismutase (CuZn-SOD) and mitochondrial Mn-SOD. Both convert O_2 to H_2O_2, which is subsequently cleared by catalase or glutathione peroxidase. Additional protection is provided by antioxidants such as vitamins C and E.

During ischemia, these clearing mechanisms are overwhelmed and oxidative damage of cellular structures occurs (Beckman, 1991). This leads to secondary cellular injury after ischemia (Palmer, 1995). Compared to other organs, the brain has a high concentration of unsaturated fatty acids and is particularly vulnerable to oxidative damage because of its high rate of oxygen consumption and low concentration of antioxidants. During ischemia, intracellular Ca^{2+} increases. Ca^{2+} functions as a universal second messenger and activates phospholipase A2; cyclooxygenase generates free-radical species that overwhelm endogenous scavenging mechanisms, resulting in lipid peroxidation and membrane damage. This damage occurs as a delayed response to ischemia and is documented by the fact that delayed treatment with free-radical scavengers may be effective in models of focal brain ischemia (Zhao et al., 1994). Overproduction of radical-scavenging enzymes protects against stroke, however animals that are deficient in these are more susceptible to cerebral ischemic damage (Kawase et al., 1999; Murakami et al., 1997). Besides their direct damage to cellular structures, ROS also serve as important signaling molecules that trigger inflammation and apoptosis.

Although most of the free radical species have clearly defined cytotoxic roles in metabolism, the role of nitric oxid (NO) is more complex. NO is synthesized by neuronal nitric-oxide synthase (nNOS), which is Ca^{2+}-dependent and leads to free radical production. NO reacts with a superoxide anion to form a highly reactive species called peroxynitrite that promotes tissue damage. In addition, NO can be formed by endothelial NOS (eNOS) and an inducible NOS (iNOS).

Endothelilial NO is vasodilatory and may play a role in protective vasodilation after ischemia; cellular NO reacts with superoxide to form peroxynitrate, which is a potent cytotoxic ROS. Animals with knockout nNOS have smaller ischemic strokes, suggesting a neuroprotective effect when intraneuronal NO is reduced (Huang et al., 1994).

Tirilazad mesylate is a 21-aminosteroid free radical scavenger and potent membrane lipid peroxidation inhibitor that showed neuroprotective promise in focal ischemia (Hal et al., 1988). It was effective mostly in primate models of transient but not permanent ischemia (Xue et al., 1992). Clinical trials in stroke have failed to demonstrate benefit from Tirilazad (RANTTAS, 1996).

The only positive clinical trial for a neuroprotective drug was the SAINT-I trial, which tested NXY-059 (Lees et al., 2006). The second Phase III trial using NXY-059, however, did not demonstrate improved clinical outcome after stroke. This drug is a nitrone-based free radical trapping agent and has been effective in multiple models testing neuroprotection in transient and permanent cerebral ischemia. In rats, NXY-059 protects the brain even when given as late as five hours after onset of transient focal ischemia (Kuroda et al., 1999). In a primate model of permanent ischemia, NXY-059 was neuroprotective, given as late as four hours after onset of ischemia (Marshall et al., 2003). The generic molecule was also effective in an embolic rabbit stroke model (Lapchak et al., 2002).

NXY-059, however, does not penetrate into brain tissue, so its effects are likely at the level of the microvasculature. Therefore, a combination of reperfusion strategies and NXY-059 may enable better brain penetration. In animal models, a combination of tenecteplase and generic NXY-059 was more effective than each compound alone when administered six hours after ischemia (Lapchak et al., 2004).

V. Inflammation

Inflammation plays an important role in the development of cerebrovascular disease (Gussekloo et al., 2000; Hallenbeck, 1996). Markers of inflammation, such as C-reactive protein (CRP), and pro-inflammatory cytokines such as Interleukin-6 (IL-6), are associated with poor outcome after stroke (Vila et al., 2000). Patients with stroke have increased levels of pro-inflammatory cytokines in peripheral blood and cerebrospinal fluid (CSF). The highest levels are found two to three days after ischemia (Tarkowski et al., 1997). Inflammatory processes play a role in subacute ischemic vessel and brain injury (see Figure 12.1) (Fagan et al., 2004). Reducing inflammatory responses has long been under investigation to lessen the cytotoxic effects of ischemia (Zhang et al., 1995).

Most inflammatory reactions are mediated by cytokines, small glycoproteins expressed by many cell types in response to acute cerebral ischemia. In a first step, endothelial adhesion molecules are upregulated by cytokine release, which leads to recruitment and activation of leukocytes, promotion of leukocyte-endothelium interaction, and conversion of the local endothelium to a prothrombotic state (Clark et al., 1995). In experimental models of transient Middle Cerebral Artery occlusion (tMCAo), pro-inflammatory cytokines (interleukin [IL]-1, tumor necrosis factor [TNF]-α, and

IL-6) are increased within the ischemic cortex (Yamasaki et al., 1995). Histopathological studies have shown that neutrophils accumulate in the infarcted zone after cerebral ischemia (Barone & Feuerstein, 1999). Blocking endothelial receptors, which mediate leukocyte adhesion such as intercellular adhesion molecule-1 (ICAM-1), reduced stroke lesion size in animals after tMCAo (Bowes et al., 1993). Intraventricular injection of pro-inflammatory cytokines such as IL-1 and TNF-α enlarges infarct volume and brain edema after MCA occlusion in rats, whereas the injection of antibodies against IL-1 and TNF-α reduces brain injury (Barone et al., 1997; Yamasaki et al., 1995).

Because of the delayed inflammatory response after ischemia, anti-inflammatory therapies were thought to block late ischemic and perfusion injury. Clinical trials of ICAM-1-blocking antibodies (Enlimomab) and others, however, failed to show improved outcome after stroke (2001). ICAM blockade leads to an increase in infectious complications, and large molecules such as antibodies may not have sufficient bioavailability at the infarct site to impact the disease process. Therapies that are applied systemically may reach infarcted tissue only after reperfusion is established. Models of combined thrombolysis and anti-inflammation demonstrated the greatest efficacy (Bowes et al., 1993). Most animal models use transient vessel occlusion, and only a minority of patients after stroke experience spontaneous reperfusion (Pessin et al., 1995). Blocking one mechanism of inflammation, as in the trial of anti-ICAM-1, may lead to upregulation of other pathways (Vuorte et al., 1999). Furthermore, the Enlimomab trial used a six-hour time-to-treatment window, and the drug was never shown to be effective in animal models more than 90 minutes after vascular occlusion (Clark et al., 1991).

Other potential treatments against inflammation in stroke are anti-CD-18 antibodies (LeukArrest), which are shorter-acting and potentially cause fewer systemic side effects than Enlimomab (Yenari et al., 1998). These newer antibodies are humanized, whereas Enlimomab is based on murine proteins, which cause less predictable immune responses and auto-antibodies.

Minocycline, a semisynthetic second-generation drug of the tetracycline group, is neuroprotective in models of focal and global ischemia models (Yrjanheikki et al., 1999). Minocycline inhibits inflammatory response after ischemia, such as microglial activation and production of other inflammation mediators (Tikka et al., 2001).

In animal models, immunosuppression with FK506 (Tacrolimus) and corticosteroids was shown to reduce stroke volume. FK506 reduces T-cell dependent immune responses, but its effect after stroke mostly stems from anti-apoptosis (Macleod et al., 2005; Noto et al., 2004).

Steroids suppress inflammation in stroke and reduce stroke volume in animal models (Limbourg et al., 2002), though earlier studies in humans failed to show clinical benefits of steroid use after ischemic stroke (Mulley et al., 1978; Norris and Hachinski, 1986). These studies, however, allowed treatment delay of 48 hours and used steroids with significant nuclear protein transcription effects. Transcriptional effects of steroids are poorly understood, but may be responsible for steroid-induced side effects. Newer synthetic steroids may avoid these effects by selectively binding to nontranscriptional receptors (Vayssiere et al., 1997).

VI. Growth Factors

Growth factors (GFs) are polypeptides that have regenerative and proliferative capacities. The main focus of GFs in cerebral ischemia research has been on the exogenous administration of GFs, whereas the function of endogenous GFs expression has been less well-studied. The most widely studied growth factors in ischemic stroke are basic fibroblast growth factor (bFGF), brain-derived neurotrophic growth factor (BDNF), insulin-like growth factor (IGF), and osteogenic protein-1 (Fisher et al., 1995; Kawamata et al., 1997). All these growth factors have neuroprotective effects when given after cerebral ischemia, attenuating the effects of excitotoxicity, improving CBF, and reducing apoptosis. GFs enhance synaptogenesis and dendritic sprouting after stroke (Kawamata et al., 1998). Receptors for GFs are found throughout the brain, and some neurons in culture were unable to survive without essential GFs. Sympathetic neurons, for example, cannot survive without NGF (Riccio et al., 1999). GFs play an important role in adaptive mechanisms and regulation of intracellular homeostasis after ischemia.

bFGF and osteogenic protein-1 have strong regenerative effects after ischemia, and in early studies, demonstrated improved behavioral outcome in animal models. Although these trials were promising, two smaller studies in humans failed. One was stopped because of safety concerns, and the other failed to show improved outcome in treated patients. These failures are at least in part due to a long treatment time window. A major obstacle to the treatment with GFs, however, is that the molecules are very large and have little or no blood–brain barrier penetration. Many early animal models, therefore, used intrathecal infusions, which are impractical in clinical applications for stroke.

Recent advances in bone marrow stromal cell (MSC) preparations may enable delivery of GFs into ischemic brain tissue. Using viral vectors, MSCs can be transfected with GFs. When intrathecally injected, MSCs survive within brain tissue and secrete GFs for a prolonged period (Azizi et al., 1998; Ikeda et al., 2005). Neurological improvement after intravenous MSC infusion also has been shown, making this technique a promising strategy for clinical application in the future (Chen et al., 2001).

In addition to neuronal, microglial, and astrocytic GFs, vascular endothelial-derived GFs (VEGF) advance vascular growth and capillary sprouting after stroke, which may aid in recovery after ischemia. Hypoxia is a major stimulant for the formation of VEGF, and animals with defective genes for VEGF have severely deficient angiogenesis and die during the neonatal phase. First attempts are being made to interfere with VEGF in cancer patients, hoping that tumor-induced vascular growth can be inhibited. Whether VEGF-mediated mechanisms may be used to advance stroke recovery is unknown.

Recent work has shown that regenerative processes through VEGF require matrix metalloproteinases 9 (MMP-9). In the early phase after stroke, MMP-9s are detrimental, playing a role in the disruption of cell matrix and homeostasis. In the later stages during brain recovery, MMP triggers elevation of VEGF, aiding in brain recovery. Medications that block MMP could be beneficial in acute stroke, but may impair recovery after stroke (Zhao et al., 2006).

VII. Gene Expression in Cerebral Ischemia

One can detect changes in gene expression within minutes of focal and global cerebral ischemia. The ischemic core and its surrounding tissue, penumbra, have different and distinct patterns of gene expression. Recent advances in the technology of genetic research, especially with the introduction of DNA micro-arrays that allow the detection of thousands of genes within one test, have advanced the research and understanding of gene expression after ischemia.

Gene expression after ischemia is time-dependent and occurs in five major waves (see Figure 12.2) (Barone and Feuerstein, 1999). Immediately after ischemia, a transient expression of transcription factors, such as c-foc and jun-B, is found. The second wave consists of heat-shock proteins (HSP). The third largely consists of increased cytokine gene expression for TNF-α and IL-1, which likely plays a role in inflammatory responses after ischemia, and increased expression of ICAM-1 and GFs (e.g., nerve growth factor, brain-derived nerve growth factor). The fourth wave includes proteolytic enzymes such as metalloproteinases (MMP), which are implicated in remodeling the extracellular matrix (Romanic et al., 1998). The fifth wave of genes is important in tissue remodeling and consists of transforming growth factor-β and osteopontin (Wang et al., 1995).

Gene expression plays a role in both cellular recovery, as well as apoptotic cell death triggered by ischemia. The first wave of gene expression commences with the early genes, which appear within minutes of cellular injury and transcribe mRNA for proteins that can be released even when protein synthesis inhibitors are administered (Morgan & Curran, 1991). The best studied early genes are c-fos and jun-B, which function as leucine zippers (Akins et al., 1996) to form dimerization products. *In situ* hybridization techniques that anatomically localize the expression of mRNA demonstrate increases in c-fos mRNA in dentate granule cells, CA1 and CA3 pyramidal neurons, neocortex, and Purkinje cells, which are known to be most susceptible to global ischemia (Wessel et al., 1991). In models of focal ischemia, gene expression is present within minutes in the immediate area of ischemia. In models of longer tMCAo or pMCAo, the upregulation of gene expression exceeds the ischemic territory (An et al., 1993). Glutamate antagonists (MK801) can reduce the early gene expression after ischemia, indicating the importance of excitatory neurotransmitter and Ca^{2+} second-messaging in gene expression.

Heat shock proteins (HSP) are induced one to two hours after ischemia and down-regulated over one to two days (Nowak et al., 1990). HSP may be neuroprotective (Sharp and Sagar, 1994). Gene expression of trophic factors, such as basic fibroblast growth factor (bFGF) and their receptors, is regulated by ischemia (Finklestein et al., 1988). The immediate early genes may be candidates for regulation of nerve growth factor expression. After focal cerebral ischemia, nerve growth factor and brain-derived neurotrophic factor mRNA can be detected within four hours of reperfusion (Lindvall et al., 1992). MMP and cellular-remodeling proteins are later genes that undergo increased levels of expression after stroke (Yong, 2005), playing a role in cellular recovery and remodeling of the infarcted tissue. Other genes, however, are involved in apoptotic cell death and are discussed later.

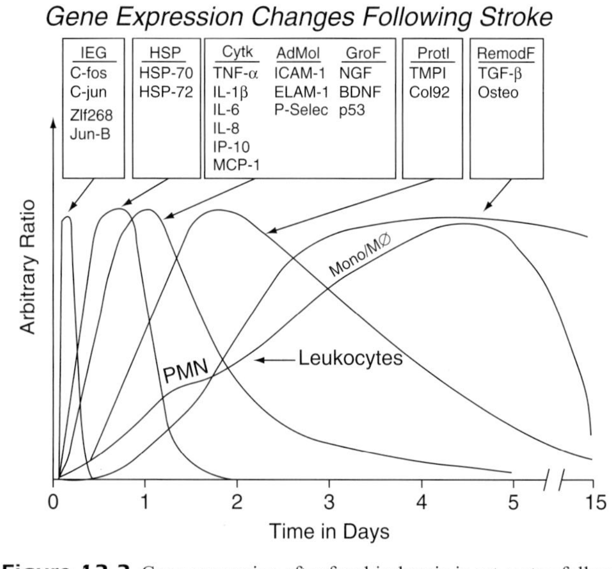

Figure 12.2 Gene expression after focal ischemia in rat cortex follows five waves: transcription factors (wave 1), heat-shock proteins (2), pro-inflammatory mediators, adhesion molecules, growth factor and oncogene expression (3), proteinase and proteinase inhibitor gene expression (4), and delayed remodeling proteins involved in resolution of the tissue injury (5). A leukocyte wave (PMN and mono/MØ) is included for comparison (Barone and Feuerstein, 1999).

The precise interaction of genes and gene products and their regulation is poorly understood.

Using DNA microarrays, one can survey thousands of genes for their differential expression patterns after ischemia. Specific genes may aid in the understanding of pathophysiological processes after ischemia. Many genes and their functions, however, are not understood, but can be detected using the microarrays. Testing for thousands of genes, one can find a certain gene expression pattern. The identification of these patterns may help not only in identifying new genes, but may also be used as diagnostic and prognostic tools. If one is able to identify a certain gene expression pattern that is highly reflective of ischemia or indicates a severity and clinical prognosis after stroke, one could use this pattern as a surrogate marker for the disease (Du et al., 2006).

VIII. Apoptosis

Cerebral ischemia leads to cell death through two major pathways: necrosis and apoptosis (Johnson et al., 1995). Necrosis is a direct result of cellular damage secondary to a loss in structural integrity of cytoplasmic structures and metabolism. Necrosis often is seen in the core region of vascular occlusion, lasting long enough to completely destroy the tissue. In contrast, apoptosis is an active pathway of programmed cell death and a physiological process in normal development. Apoptosis leads to cleavage of DNA through DNA endonucleases. The product of this cleavage is formed into DNA clumps and can be found in histological examination of tissue undergoing apoptosis (Chopp et al., 1996).

Many neurons undergo cell death even when CBF is restored, whereas surrounding, less vulnerable brain regions fail to show signs of cell death. The most prominent example is the CA1 region in the hippocampus. Cell death in pyramidal cells in the CA1 region appears over several days after ischemia and shows typical features of apoptosis (Pulsinelli et al., 1982b). Experiments have demonstrated this active process using inhibitors of protein synthesis, such as cycloheximide or anisomycin and have reduced neuronal cell death (Shigeno et al., 1990).

Transgenic mice that overexpress mitochondrial Bcl-2, which inhibits apoptotic cell death, have smaller infarcts than wild-type mice (Martinou et al., 1994). Du et al. (1996) showed that the duration of ischemia may determine whether a lesion will undergo necrosis or apoptosis. In the rat model, animals subject to 90 minutes tMCAo developed maximal infarct sizes at day one that consisted mainly of necrosis and only minimal areas of apoptosis. Animals that received 30 minutes tMCAo had no visible infarctions at day one, but developed TUNEL-positive staining at day three post-ischemia, indicating probable apoptosis. DNA prepared from the peri-infarct area of ischemic cortex showed internucleosomal fragmentation, but when the animals were treated with cycloheximide before ischemia, no delayed apoptosis was found. Apoptosis after ischemia, therefore, is process-dependent on active protein synthesis.

Apoptosis in neurons, as in many other tissues, involves the induction of death-inducing ligands (DILs) and the activation of caspases (Barinaga, 1998). DILs, such as CD95-L (APO1-L/Fas-L), TRAIL (APO2-L), and TNF-α, are transmembrane molecules that belong to the TNF-superfamily. Binding of DILs to their respective receptors activates the apoptosis pathway.

Alternative apoptosis-signaling pathways are the activation of sphingomyelinases, leading to sphingomyelin hydrolysis to generate the lipid second-messenger ceramide. Ceramides induce apoptosis that involve activation of caspases. Inhibition of ceramides through FK506 reduces infarct volume and apoptotic cell death in rats after tMCAo (Herr et al., 1999).

Caspase inhibition has been investigated and identified as a promising field of neuroprotection (Jörg B. Schulz, 1999). Currently 14 caspase families are identified. Caspase 3, in particular, plays a central role in the execution of apoptosis. During reperfusion, caspase 3 is activated both in the ischemic core and penumbra (Manabat et al., 2003). Delayed cell death is reduced in animals who received caspase 3 inhibition (Han et al., 2002). Human trials investigating the validity of this have not been performed because potent drugs of this class suitable for clinical use have not been identified.

IX. Summary

Cerebral ischemia can be separated into global and focal ischemia. The former is common after cardiac arrest, the latter results in ischemic stroke. Neuronal cells suffer ischemia after the cessation of blood flow below a critical threshold. Cell death following ischemia results from secondary changes such as excitotoxicity, inflammation, and apoptosis. The understanding of these molecular mechanisms has led to an abundance of research in the field of neuroprotection (O'Collins et al., 2006). Despite promising experimental data that has advanced our understanding of stroke pathophysiology, to date, no clinical trial has demonstrated improved clinical outcomes in the use of neuroprotective agents in stroke patients.

References

Akins, P. T., Liu, P. K., and Hsu, C. Y. (1996). Immediate early gene expression in response to cerebral ischemia: Friend or foe? *Stroke* **27**, 1682–1687.

An, G., Lin, T. N., Liu, J. S., Xue, J. J., He, Y. Y., and Hsu, C. Y. (1993). Expression of c-fos and c-jun family genes after focal cerebral ischemia. *Ann Neurol* **33**, 457–464.

Araki, T., Kanai, Y., Murakami, F., Kato, H., and Kogure, K. (1993). Postischemic changes in the binding of excitatory and inhibitory neurotransmitters in the gerbil brain. *Pharmacol Biochem Behav* **45**, 945.

Astrup, J., Siesjo, B. K., and Symon, L. (1981). Thresholds in cerebral ischemia: The ischemic penumbra. *Stroke* **12**, 723–725.

Astrup, J., Symon, L., Branston, N. M., and Lassen, N. A. (1977). Cortical evoked potential and extracellular K+ and H+ at critical levels of brain ischemia. *Stroke* **8**, 51–57.

Azizi, S. A., Stokes, D., Augelli, B. J., DiGirolamo, C., and Prockop, D. J. (1998). Engraftment and migration of human bone marrow stromal cells implanted in the brains of albino rats—Similarities to astrocyte grafts. *Proc Natl Acad Sci U S A* **95**, 3908–3913.

Barinaga, M. (1998). Stroke-damaged neurons may commit cellular suicide. *Science* **281**, 1302–1303.

Barone, F. C., Arvin, B., White, R. F., Miller, A., Webb, C. L., Willette, R. N. et al. (1997). Tumor necrosis factor-{alpha}: A mediator of focal ischemic brain injury. *Stroke* **28**, 1233–1244.

Barone, F. C. and Feuerstein, G. Z. (1999). Inflammatory mediators and stroke: New opportunities for novel therapeutics. *J Cereb Blood Flow Metab* **19**, 819–834.

Beckman, J. S. (1991). The double-edged role of nitric oxide in brain function and superoxide-mediated injury. *J Dev Physiol* **15**, 53–59.

Bowes, M. P., Zivin, J. A., and Rothlein, R. (1993). Monoclonal antibody to the ICAM-1 adhesion site reduces neurological damage in a rabbit cerebral embolism stroke model. *Exp Neurol* **119**, 215–219.

Branston, N. M., Symon, L., Crockard, H. A., and Pasztor, E. (1974). Relationship between the cortical evoked potential and local cortical blood flow following acute middle cerebral artery occlusion in the baboon. *Exp Neurol* **45**, 195–208.

Bryan, R. N., Levy, L. M., Whitlow, W. D., Killian, J. M., Preziosi, T. J., and Rosario, J. A. (1991). Diagnosis of acute cerebral infarction: Comparision of CT and MR imaging. *AJNR* **12**, 611–620.

Chen, J., Li, Y., Wang, L., Zhang, Z., Lu, D., Lu, M., and Chopp, M. (2001). Therapeutic benefit of intravenous administration of bone marrow stromal cells after cerebral ischemia in rats. *Stroke* **32**, 1005–1011.

Choi, D. W. (1988). Glutamate neurotoxicity and diseases of the nervous system. *Neuron* **1**, 623–634.

Chopp, M., Chan, P. H., Hsu, C. Y., Cheung, M. E., and Jacobs, T. P. (1996). DNA damage and repair in central nervous system injury: National Institute of Neurological Disorders and Stroke Workshop Summary. *Stroke* **27**, 363–369.

Clark, W. M., Lauten, J. D., Lessov, N., Woodward, W., and Coull, B. M. (1995). Time course of ICAM-1 expression and leukocyte subset infiltration in rat forebrain ischemia. *Mol Chem Neuropathol* **26**, 213–230.

Clark, W. M., Madden, K. P., Rothlein, R., and Zivin, J. A. (1991). Reduction of central nervous system ischemic injury by monoclonal antibody to intercellular adhesion molecule. *J Neurosurg* **75**, 623–627.

Dirnagl, U., Iadecola, C., and Moskowitz, M. A. (1999). Pathobiology of ischaemic stroke: An integrated view. *Trends in Neurosciences* **22**, 391.

Du, C., Hu, R., Csernansky, C. A., Hsu, C. Y., and Choi, D. W. (1996). Very delayed infarction after mild focal cerebral ischemia: A role for apoptosis? *J Cereb Blood Flow Metab* **16**, 195–201.

Du, X., Tang, Y., Xu, H., Lit, L., Walker, W., Ashwood, P. et al. (2006). Genomic profiles for human peripheral blood T cells, B cells, natural killer cells, monocytes, and polymorphonuclear cells: Comparisons to ischemic stroke, migraine, and Tourette syndrome. *Genomics*.

Enlimomab Acute Stroke Trial Investigators (2001). Use of anti-ICAM-1 therapy in ischemic stroke: Results of the Enlimomab Acute Stroke Trial. *Neurology* **57**, 1428–1434.

Fagan, S. C., Hess, D. C., Hohnadel, E. J., Pollock, D. M., and Ergul, A. (2004). Targets for vascular protection after acute ischemic stroke. *Stroke* **35**, 2220–2225.

Finklestein, S. P., Apostolides, P. J., Caday, C. G., Prosser, J., Philips, M. F., and Klagsbrun, M. (1988). Increased basic fibroblast growth factor (bFGF) immunoreactivity at the site of focal brain wounds. *Brain Res* **460**, 253–259.

Fisher, M., Meadows, M. E., Do, T., Weise, J., Trubetskoy, V., Charette, M., and Finklestein, S. P. (1995). Delayed treatment with intravenous basic fibroblast growth factor reduces infarct size following permanent focal cerebral ischemia in rats. *J Cereb Blood Flow Metab* **15**, 953–959.

Floyd, R. A. and Carney, J. M. (1992). Free radical damage to protein and DNA: Mechanisms involved and relevant observations on brain undergoing oxidative stress. *Ann Neurol* **32 Suppl**, S22–27.

Furlan, M., Marchal, G., Viader, F., Derlon, J. M., and Baron, J. C. (1996). Spontaneous neurological recovery after stroke and the fate of the ischemic penumbra. *Ann Neurol* **40**, 216–226.

Ginsberg, M. D., Watson, B. D., Busto, R., Yoshida, S., Prado, R., Nakayama, H. et al. (1988). Peroxidative damage to cell membranes following cerebral ischemia. A cause of ischemic brain injury? *Neurochem Pathol* **9**, 171–193.

Gussekloo, J., Schaap, M. C. L., Frolich, M., Blauw, G. J., and Westendorp, R. G. J. (2000). C-reactive protein is a strong but nonspecific risk factor of fatal stroke in elderly persons. *Arterioscler Thromb Vasc Biol* **20**, 1047–1051.

Hall, E. D., Pazara, K. E., and Braughler, J. M. (1988). 21-Aminosteroid lipid peroxidation inhibitor U74006F protects against cerebral ischemia in gerbils. *Stroke* **19**, 997–1002.

Hallenbeck, J. M. (1996). Significance of the inflammatory response in brain ischemia. *Acta Neurochir Suppl* **66**, 27.

Halliwell, B. (1992). Reactive oxygen species and the central nervous system. *J Neurochem* **59**, 1609–1623.

Han, B. H., Xu, D., Choi, J., Han, Y., Xanthoudakis, S., Roy, S. et al. (2002). Selective, reversible caspase-3 inhibitor is neuroprotective and reveals distinct pathways of cell death after neonatal hypoxic-ischemic brain injury. *J Biol Chem* **277**, 30128–30136.

Heiss, W. D. (1983). Flow thresholds of functional and morphological damage of brain tissue. *Stroke* **14**, 329.

Heiss, W. D. (1992). Experimental evidence of ischemic thresholds and functional recovery. *Stroke* **23**, 1668–1672.

Heiss, W. D. and Rosner, G. (1983). Functional recovery of cortical neurons as related to degree and duration of ischemia. *Ann Neurol* **14**, 294–301.

Herr, I., Martin-Villalba, A., Kurz, E., Roncaioli, P., Schenkel, J., Cifone, M. G., and Debatin, K. M. (1999). FK506 prevents stroke-induced generation of ceramide and apoptosis signaling. *Brain Res* **826**, 210–219.

Hossmann, K. A. (1994). Viability thresholds and the penumbra of focal ischemia. *Ann Neurol* **36**, 557–565.

Hossmann, K. A., Schmidt-Kastner, R., and Grosse Optoff, B. (1987). Recovery of integrative central nervous function after one hour global cerebro-circulatory arrest in normothermic cat. *J Neurol Sci* **77**, 305–320.

Hossmann, K. A. and Zimmermann, V. (1974). Resuscitation of the monkey brain after 1 h complete ischemia. *Brain Res* **81**, 59–74.

Huang, Z., Huang, P. L., Panahian, N., Dalkara, T., Fishman, M. C., and Moskowitz, M. A. (1994). Effects of cerebral ischemia in mice deficient in neuronal nitric oxide synthase. *Science* **265**, 1883–1885.

Ikeda, N., Nonoguchi, N., Zhao, M. Z., Watanabe, T., Kajimoto, Y., Furutama, D. et al. (2005). Bone marrow stromal cells that enhanced fibroblast growth factor-2 secretion by herpes simplex virus vector improve neurological outcome after transient focal cerebral ischemia in rats. *Stroke* **36**, 2725–2730.

Isley, M. R., Cohen, M. J., Wadsworth, J. S., Martin, S. P., and O'Callaghan, M. A. (1998). Multimodality neuromonitoring for carotid endarterectomy surgery: Determination of critical cerebral ischemic thresholds. *American Journal of Eletroneurodiagnostic Technology*, 38–122.

Johnson, E. M., Jr., Greenlund, L. J., Akins, P. T., and Hsu, C. Y. (1995). Neuronal apoptosis: Current understanding of molecular mechanisms and potential role in ischemic brain injury. *J Neurotrauma* **12**, 843–852.

Jones, T. H., Morawetz, R. B., Crowell, R. M., Marcoux, F. W., FitzGibbon, S. J., DeGirolami, U., and Ojemann, R. G. (1981). Thresholds of focal cerebral ischemia in awake monkeys. *J Neurosurg* **54**, 773–782.

Jörg B. Schulz, M. W. M. A. M. (1999). Caspases as treatment targets in stroke and neurodegenerative diseases. *Ann Neurol* **45**, 421–429.

Katsura, K., Kristian, T., and Siesjo, B. K. (1994). Energy metabolism, ion homeostasis, and cell damage in the brain. *Biochem Soc Trans* **22**, 991–996.

Kawamata, T., Dietrich, W. D., Schallert, T., Gotts, J. E., Cocke, R. R., Benowitz, L. I., and Finklestein, S. P. (1997). Intracisternal basic fibroblast growth factor enhances functional recovery and up-regulates the expression of a molecular marker of neuronal sprouting following focal cerebral infarction. *Proc Natl Acad Sci U S A* **94**, 8179–8184.

Kawamata, T., Ren, J., Chan, T. C., Charette, M., and Finklestein, S. P. (1998). Intracisternal osteogenic protein-1 enhances functional recovery following focal stroke. *Neuroreport* **9**, 1441–1445.

Kawase, M., Murakami, K., Fujimura, M., Morita-Fujimura, Y., Gasche, Y., Kondo, T. et al. (1999). Exacerbation of delayed cell injury after transient global ischemia in mutant mice with CuZn superoxide dismutase deficiency: Editorial comment. *Stroke* **30**, 1962–1968.

Kirino, T. and Sano, K. (1984). Selective vulnerability in the gerbil hippocampus following transient ischemia. *Acta Neurol pathol* **62**, 201–208.

Kuroda, S., Tsuchidate, R., Smith, M. L., Maples, K. R., and Siesjo, B. K. (1999). Neuroprotective effects of a novel nitrone, NXY-059, after transient focal cerebral ischemia in the rat. *J Cereb Blood Flow Metab* **19**, 778–787.

Lapchak, P. A., Araujo, D. M., Song, D., Wei, J., and Zivin, J. A. (2002). Neuroprotective effects of the spin trap agent disodium-[(tert-butylimino)methyl]benzene-1,3-disulfonate N-oxide (generic NXY-059) in a rabbit small clot embolic stroke model: Combination studies with the thrombolytic tissue plasminogen activator. *Stroke* **33**, 1411–1415.

Lapchak, P. A., Song, D., Wei, J., and Zivin, J. A. (2004). Coadministration of NXY-059 and tenecteplase six hours following embolic strokes in rabbits improves clinical rating scores. *Exp Neurol* **188**, 279–285.

Lees, K. R., Zivin, J. A., Ashwood, T., Davalos, A., Davis, S. M., Diener, H. C. et al. (2006). NXY-059 for acute ischemic stroke. *N Engl J Med* **354**, 588–600.

Limbourg, F. P., Huang, Z., Plumier, J. C., Simoncini, T., Fujioka, M., Tuckermann, J. et al. (2002). Rapid nontranscriptional activation of endothelial nitric oxide synthase mediates increased cerebral blood flow and stroke protection by corticosteroids. *J Clin Invest* **110**, 1729–1738.

Lindvall, O., Ernfors, P., Bengzon, J., Kokaia, Z., Smith, M., Siesjo, B. K., and Persson, H. (1992). Differential regulation of mRNAs for nerve growth factor, brain-derived neurotrophic factor, and neurotrophin 3 in the adult rat brain following cerebral ischemia and hypoglycemic coma. *PNAS* **89**, 648–652.

Ljunggren, B., Ratcheson, R. A., and Siesjo, B. K. (1974). Cerebral metabolic state following complete compression ischemia. *Brain Res* **73**, 291–307.

Longuemare, M. C., Hill, M. P., and Swanson, R. A. (1994). Glycolysis can prevent non-synaptic excitatory amino acid release during hypoxia. *Neuroreport* **5**, 1789–1792.

Macleod, M. R., O'Collins, T., Horky, L. L., Howells, D. W., and Donnan, G. A. (2005). Systematic review and metaanalysis of the efficacy of FK506 in experimental stroke. *J Cereb Blood Flow Metab* **25**, 713–721.

Manabat, C., Han, B. H., Wendland, M., Derugin, N., Fox, C. K., Choi, J. et al. (2003). Reperfusion differentially induces caspase-3 activation in ischemic core and penumbra after stroke in immature brain. *Stroke* **34**, 207–213.

Marshall, J. W., Cummings, R. M., Bowes, L. J., Ridley, R. M., and Green, A. R. (2003). Functional and histological evidence for the protective effect of NXY-059 in a primate model of stroke when given 4 hours after occlusion. *Stroke* **34**, 2228–2233.

Martinou, J. C., Dubois-Dauphin, M., Staple, J. K., Rodriguez, I., Frankowski, H., Missotten, M. et al. (1994). Overexpression of BCL-2 in transgenic mice protects neurons from naturally occurring cell death and experimental ischemia. *Neuron* **13**, 1017–1030.

Morgan, J. I., and Curran, T. (1991). Stimulus-transcription coupling in the nervous system: Involvement of the inducible proto-oncogenes fos and jun. *Annu Rev Neurosci* **14**, 421–451.

Mulley, G., Wilcox, R. G., and Mitchell, J. R. (1978). Dexamethasone in acute stroke. *Br Med J* **2**, 994–996.

Murakami, K., Kondo, T., Epstein, C. J., and Chan, P. H. (1997). Overexpression of CuZn-superoxide dismutase reduces hippocampal injury after global ischemia in transgenic mice. *Stroke* **28**, 1797–1804.

Norris, J. W. and Hachinski, V. C. (1986). High dose steroid treatment in cerebral infarction. *Br Med J (Clin Res Ed)* **292**, 21–23.

Noto, T., Ishiye, M., Furuich, Y., Keida, Y., Katsuta, K., Moriguchi, A. et al. (2004). Neuroprotective effect of tacrolimus (FK506) on ischemic brain damage following permanent focal cerebral ischemia in the rat. *Brain Res Mol Brain Res* **128**, 30–38.

Nowak, T. S., Jr., Ikeda, J., and Nakajima, T. (1990). 70-kDa heat shock protein and c-fos gene expression after transient ischemia. *Stroke* **21**, III107–111.

O'Collins, V. E., Macleod, M. R., Donnan, G. A., Horky, L. L., van der Worp, B. H., and Howells, D. W. (2006). 1,026 experimental treatments in acute stroke. *Ann Neurol* **59**, 467–477.

Olney, J. W. and Sharpe, L. G. (1969). Brain lesions in an infant rhesus monkey treated with monosodium glutamate. *Science* **166**, 386–388.

Palmer, C. (1995). Hypoxic-ischemic encephalopathy. Therapeutic approaches against microvascular injury, and role of neutrophils, PAF, and free radicals. *Clin Perinatol* **22**, 481–517.

Park, C. K., Nehls, D. G., Teasdale, G. M., and McCulloch, J. (1989). Effect of the NMDA antagonist MK-801 on local cerebral blood flow in focal cerebral ischaemia in the rat. *J Cereb Blood Flow Metab* **9**, 617–622.

Pessin, M., Zoppo, G. D., and Furlan, A. (1995). Thrombolytic treatment in acute stroke: Review and update of specific topics. In C. L. Ed., *Cerebrovascular diseases: Nineteenth Princeton Stroke Conference*, 409–418. Butterworth-Heinemann, Boston, Mass.

Pulsinelli, W. A., Brierley, J. B., and Plum, F. (1982a). Temporal profile of neuronal damage in a model of transient forebrain ischemia. *Ann Neurol* **11**, 491–498.

Pulsinelli, W. A., Brierley, J. B., and Plum, F. (1982b). Temporal profile of neuronal damage in a model of transient forebrain ischemia. *Ann Neurol* **11**, 491–498.

RANTTAS, I. (1996). A randomized trial of tirilazad mesylate in patients with acute stroke (RANTTAS). The RANTTAS Investigators. *Stroke* **27**, 1453–1458.

Riccio, A., Ahn, S., Davenport, C. M., Blendy, J. A., and Ginty, D. D. (1999). Mediation by a CREB family transcription factor of NGF-dependent survival of sympathetic neurons. *Science* **286**, 2358–2361.

Romanic, A. M., White, R. F., Arleth, A. J., Ohlstein, E. H., and Barone, F. C. (1998). Matrix metalloproteinase expression increases after cerebral focal ischemia in rats: Inhibition of matrix metalloproteinase-9 reduces infarct size. *Stroke* **29**, 1020–1030.

Sharp, F. R. and Sagar, S. M. (1994). Alterations in gene expression as an index of neuronal injury: Heat shock and the immediate early gene response. *Neurotoxicology* **15**, 51–59.

Shigeno, T., Yamasaki, Y., Kato, G., Kusaka, K., Mima, T., Takakura, K. et al. (1990). Reduction of delayed neuronal death by inhibition of protein synthesis. *Neurosci Lett* **120**, 117–119.

Siesjo, B. K., Katsura, K., Zhao, Q., Folbergrova, J., Pahlmark, K., Siesjo, P., and Smith, M. L. (1995). Mechanisms of secondary brain damage in global and focal ischemia: A speculative synthesis. *J Neurotrauma* **12**, 943–956.

Smith, M. L., Auer, R. N., and Siesjo, B. K. (1984). The density and distribution of ischemic brain injury in the rat following 2–10 min of forebrain ischemia. *Acta Neuropathol* **64**, 319–332.

Squire, L. R. (1992). Memory and the hippocampus: A synthesis from findings with rats, monkeys, and humans. *Psychol Rev* **99**, 195–231.

Sundt, T. M., Sharbrough, F. W., Anderson, R. E., and Michenfelder, J. D. (1973). Cerebral blood flow measurements and electroencephalograms during carotid endartectomy. *J Neurosurg* **41**, 310–320.

Swanson, R. A., Sagar, S. M., and Sharp, F. R. (1989). Regional brain glycogen stores and metabolism during complete global ischaemia. *Neurol Res* **11**, 24–28.

Tamura, A., Asano, T., and Sano, K. (1980). Correlation between rCBF and histological changes following temporary middle cerebral artery occlusion. *Stroke* **11**, 487–493.

Tarkowski, E., Rosengren, L., Blomstrand, C., Wikkelso, C., Jensen, C., Ekholm, S., and Tarkowski, A. (1997). Intrathecal release of pro- and anti-inflammatory cytokines during stroke. *Clin Exp Immunol* **110**, 492–499.

Tikka, T., Fiebich, B. L., Goldsteins, G., Keinanen, R., and Koistinaho, J. (2001). Minocycline, a tetracycline derivative, is neuroprotective against excitotoxicity by inhibiting activation and proliferation of microglia. *J Neurosci* **21**, 2580–2588.

Trojaborg, W. and Boysen, G. (1973). Relation between EEG, regional blood flow and internal carotid artery pressure during endarterectomy. *Electroencephalog Clin Neurophysiol* **34**, 61–69.

Vayssiere, B. M., Dupont, S., Choquart, A., Petit, F., Garcia, T., Marchandeau, C. et al. (1997). Synthetic glucocorticoids that dissociate transactivation and AP-1 transrepression exhibit antiinflammatory activity in vivo. *Mol Endocrinol* **11**, 1245–1255.

Vila, N., Castillo, J., Davalos, A., and Chamorro, A. (2000). Proinflammatory cytokines and early neurological worsening in ischemic stroke. *Stroke* **31**, 2325–2329.

Vuorte, J., Lindsberg, P. J., Kaste, M., Meri, S., Jansson, S.-E., Rothlein, R., and Repo, H. (1999). Anti-ICAM-1 monoclonal antibody R6.5 (Enlimomab) promotes activation of neutrophils in whole blood. *J Immunol* **162**, 2353–2357.

Wang, X., Yue, T. L., White, R. F., Barone, F. C., and Feuerstein, G. Z. (1995). Transforming growth factor-beta 1 exhibits delayed gene expression following focal cerebral ischemia. *Brain Res Bull* **36**, 607–609.

Wessel, T. C., Joh, T. H., and Volpe, B. T. (1991). In situ hybridization analysis of c-fos and c-jun expression in the rat brain following transient forebrain ischemia. *Brain Res* **567**, 231–240.

Xue, D., Slivka, A., and Buchan, A. M. (1992). Tirilazad reduces cortical infarction after transient but not permanent focal cerebral ischemia in rats. *Stroke* **23**, 894–899.

Yamasaki, Y., Matsuura, N., Shozuhara, H., Onodera, H., Itoyama, Y., and Kogure, K. (1995). Interleukin-1 as a pathogenetic mediator of ischemic brain damage in rats. *Stroke* **26**, 676–681.

Yenari, M. A., Kunis, D., Sun, G. H., Onley, D., Watson, L., Turner, S., et al. (1998). Hu23F2G, an antibody recognizing the leukocyte CD11/CD18 integrin, reduces injury in a rabbit model of transient focal cerebral ischemia. *Exp Neurol* **153**, 223.

Yong, V. W. (2005). Metalloproteinases: Mediators of pathology and regeneration in the CNS. *Nat Rev Neurosci* **6**, 931–944.

Yrjanheikki, J., Tikka, T., Keinanen, R., Goldsteins, G., Chan, P. H., and Koistinaho, J. (1999). A tetracycline derivative, minocycline, reduces inflammation and protects against focal cerebral ischemia with a wide therapeutic window. *Proc Natl Acad Sci U S A* **96**, 13496–13500.

Zhang, R. L., Chopp, M., Zaloga, C., Zhang, Z. G., Jiang, N., Gautam, S. C. et al. (1995). The temporal profiles of ICAM-1 protein and mRNA expression after transient MCA occlusion in the rat. *Brain Res* **682**, 182–188.

Zhao, B. Q., Wang, S., Kim, H. Y., Storrie, H., Rosen, B. R., Mooney, D. J. et al. (2006). Role of matrix metalloproteinases in delayed cortical responses after stroke. *Nat Med* **12**, 441–445.

Zhao, Q., Pahlmark, K., Smith, M. L., and Siesjo, B. K. (1994). Delayed treatment with the spin trap alpha-phenyl-N-tert-butyl nitrone (PBN) reduces infarct size following transient middle cerebral artery occlusion in rats. *Acta Physiol Scand* **152**, 349–350.

Zivin, J. A. (1997). Factors determining the therapeutic window for stroke. *Neurology* **50**, 599–603.

13

Hemorrhagic Brain Disease

Michael L. DiLuna, Kaya Bilguvar, Gamze Tanriover, and Murat Gunel

I. Introduction

Hemorrhagic brain disease or intracranial hemorrhage (ICH) accounts for up to 20 percent of all strokes, carries a 40 percent mortality rate at 30 days, and has an annual incidence of up to 20 per 100,000 (Broderick et al., 1992, 1994). Currently, with the advent of imaging techniques such as digital subtraction angiography (DSA), computed tomography (CT), and MRI, treatment options involve secondary and tertiary prevention and surgery for mass lesions. As our knowledge of the pathophysiology of this disease has evolved, so too has our capacity to treat and prevent brain hemorrhage. Diagnostic imaging has dramatically improved over the past two decades, new medications have become available, and medical devices have evolved to enable minimally invasive surgical interventions. As the paradigm has shifted from medical management, to surgical intervention, to endovascular surgery, we are now entering a third phase of treatment of hemorrhagic stroke: molecular therapy. As modern biological and genetic discovery continues to unravel the molecular and cellular underpinnings of hemorrhagic brain disease, we have not only improved our ability to treat ICH, but also have dramatically advanced our ability to shift from secondary and tertiary prevention, to primary. Though the causes of common and sporadic forms of hemorrhagic brain disease are multifactorial, incorporating genetic, environmental, and behavioral factors, Mendelian forms are monogenic and provide marked insight into the molecular pathophysiology of all hemorrhagic strokes.

As our knowledge of this devastating disease increases exponentially, our ability to alter the natural history of ICH will also dramatically improve. Hemorrhagic brain diseases can now be divided into two distinct categories: primary (lesions, vascular, or parenchymal abnormalities that are prone to hemorrhage) and secondary (metabolic syndromes, coagulation factor deficiencies, or hypertension that can lead to hemorrhage anywhere as well as systemic effects with the brain being one of many organ systems involved). The pathophysiology, diagnosis, presentation, and treatment options of both primary and secondary hemorrhagic brain diseases vary widely and are directly applicable to the more common sporadic forms. This chapter outlines the molecular biology and disease course of both primary and secondary hemorrhagic brain diseases.

II. Angiogenesis and Vasculogenesis

The cerebral vessels, arteries, and veins, develop embryologically through a complex interplay of angiogenesis around the neural tube and anastomotic connections with large vessels sprouting from the aortic arch. The fine details of this process are outside the scope of this chapter; however, understanding the basics of how the blood supply of the brain forms and how the location of "watershed" areas—areas most susceptible to damage from low or no flow—form is important to understanding how vascular anomalies and hemorrhagic stroke can occur. Furthermore, the processes of vessel formation, namely vasculogenesis and angiogenesis, are central figures to the pathology underlying hemorrhagic brain diseases.

A. Development of the Blood Supply to the Brain

The blood supply to the brain develops through direct angiogenesis of blood vessels sprouting from the pial plexus surrounding the neural tube. This process differs significantly from other organs where angioblasts invade the tissues. The nascent blood vessel sprouts penetrate the neuroepithelium and branch in all directions. Three divisions of meningeal vessels perforate the brain parenchyma: striate, cortical, and medullary. The cortical vessels supply the cortex, the medullary vessels supply the white matter, and the striate branches travel through the anterior perforated substance and supply the basal nuclei.

As the cortex develops and acquires its complicated sulcal and gyral pattern, its vascular supply concomitantly becomes more complex. The cortical blood vessels become more complex in their branching and once the fetus reaches the third trimester, the circulation dynamics switch from a circulation based in the basal nuclei through anastomoses from the striate meningeal vessels, to a predominant supply from the cortical and medullary vessels. In the fetal brain, because of the orientation and location of anastomoses, the periventricular white matter is at greatest risk. In the term infant brain, the cortical "watershed" regions are similar to those in the adult brain. In the third trimester, many of the arteries of the brain begin to form smooth muscle in their walls as well as collagen and elastin. Prior to 30 weeks' gestation, the vessels in the germinal matrix and periventricular white matter do not have smooth muscle, collagen, or elastin, and open directly into veins. These areas are highly sensitive to changes in flow and are at greatest risk postpartum to hemorrhage. The common carotid artery forms from the aortic sac and the third arch artery forms the internal carotid. From here, anastomotic connections are made that will ultimately form the circle of Willis.

B. Vasculogenesis and Angiogenesis

Vasculogenesis and angiogenesis are two mechanisms of blood vessel formation that require distinct sets of regulatory elements. Vasculogenesis refers to the process by which randomly distributed cells such as angioblasts or stem cells, with the assistance of signals from various growth factors (Vascular Endothelial Growth Factor (VEGF), its splice variants and receptors, and angiopoietin (ANG)1-4 and their receptors Tie-1 and -2), form vascular structures (see Figure 13.1). This *de novo* process, which generates the so-called primary capillary plexus, occurs primarily in the embryo. Stem cells from the mesenchyme expand, differentiate, and divide. Primitive tubes form at this stage as endothelial cells divide and mature. This process was originally thought to occur only in embryonic tissues such as the embryonic brain. It is now known that vasculogenesis is the process by which vessels form in tumors, or in scar after traumas including ischemia. This framework becomes the cornerstone for angiogenesis throughout an organism's life.

Angiogenesis is a similar process to vasculogenesis except for the fact that the nascent vessels form preexisting vessels and branch (see Figure 13.1). Angiogenesis can occur, for example, to remodel and mature vessels formed from vasculogenesis. The establishment of mature and adequate blood supply for an organism is of utmost importance to ensure the growth, development, and function of organs. The majority of angiogenesis is completed during embryologic stages; however, in the adult, it plays an important role in cell turnover, healing, and during menstruation.

Architectural elements such as smooth muscle cells and pericytes from the mesoderm and mesenchyme are recruited to give added support to the new vessels, a process that can occur at any time during either angiogenesis or vasculogenesis, thought to be triggered by the beginnings of circulation. Angiogenesis is an integral process to growth and development of an organism, but is also the focus of much attention with respect to wound healing and tumor growth and transformation to malignancy.

For the purposes of understanding the underlying molecular pathophysiology behind hemorrhagic brain disease, the basic mechanisms underlying angiogenesis need to be understood. Angiogenesis activation occurs through a series of three stages described as *initiation*, *proliferation and division*, and *maturation* (see Figure 13.1). The initiation phase, similar to the process of vasculogenesis, is defined by the process of stem cells within the extracellular matrix (EM), the circulation, and the vessel wall dividing into cells that

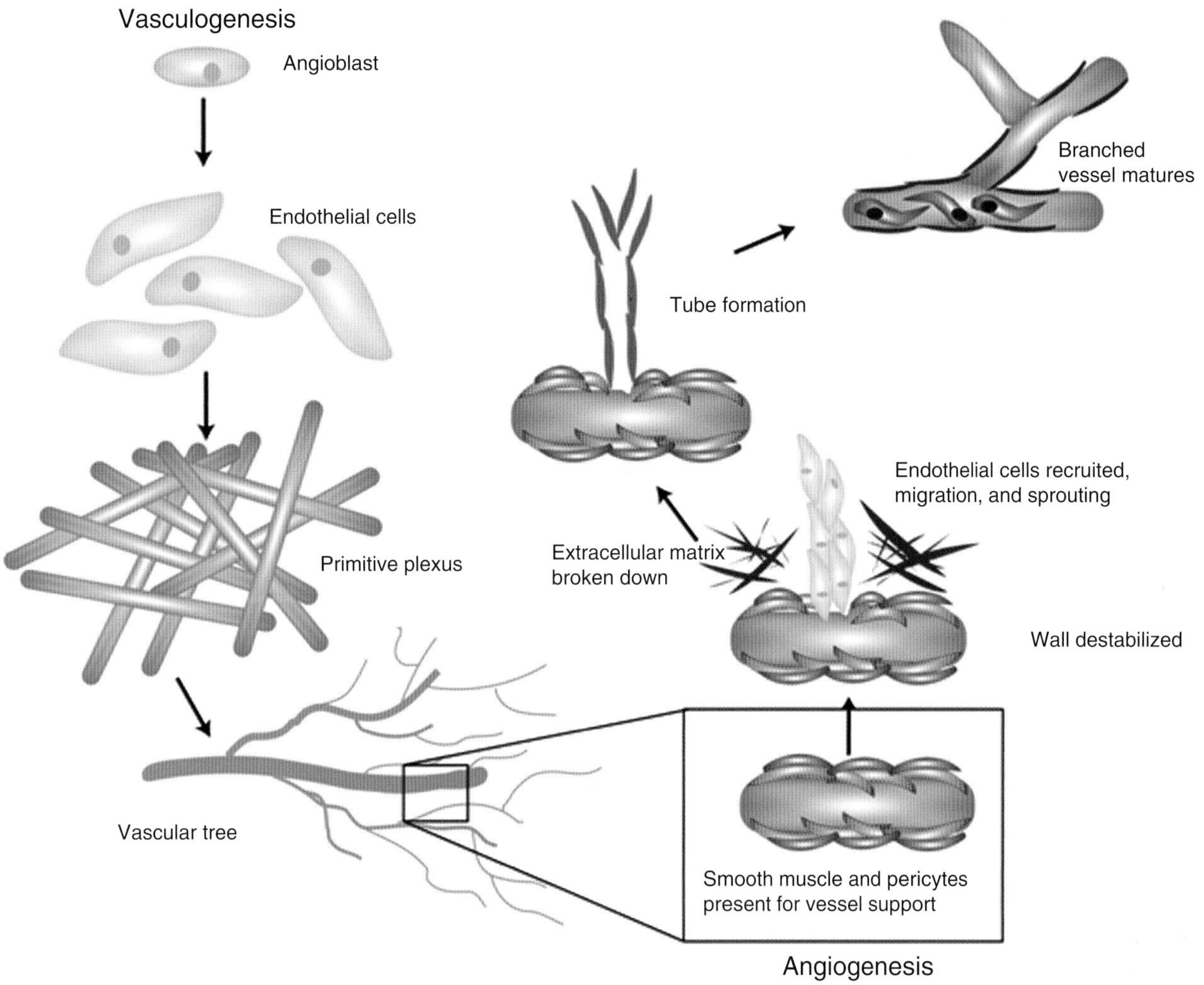

Figure 13.1 Vasculogenesis and angiogenesis.

become the new vessel walls including the building materials (endothelial cells, pericytes, and fibroblasts) important for vessel formation and support. Concomitantly, the extracellular matrix is remodeled by numerous enzymes that fall into the category of membrane-associated matrix metalloproteinases (MMPs), which break down the thick cytoarchitecture of the EM to allow for invagination of the newly sprouting vessels (Moses, 1997).

During this remodeling, cells within the wall of the vessels, specifically mural cells such as pericytes, must part to enable branching and sprouting. After invasion of the vessels, they begin to divide and sprout, as the tissue demands, to form a network. Support cells and structural components are also growing and forming their cell–cell interactions at this point and new matrix is laid down. The combination of new EM and various growth factors and cytokines will further stimulate the proliferation, migration, and organization of endothelial cells. Venous versus arterial circulation also appears to be decided at this stage as the proliferating and migrating endothelial cells express specific markers, specifically Ephrin-B2 and Eph-B4, each of which appears to direct the fate of the circulation (Wang et al., 1998). Four endogenous growth factor systems (VEGF, angiopoietins, ephrins, and Transforming Growth Factor-beta (TGF-ß)) are also integral to this stage of angiogenesis. Each plays a distinct role in inhibiting and stimulating angiogenesis, keeping the process under tight molecular control from gene transcription to translation.

Angiogenesis *initiation* is at once a simple and complex process. Numerous extracellular signals including hypoxia, injury, inflammation, and increased metabolic demand (thusly, increased consumption of metabolites and production of byproducts of metabolism) can initiate angiogenesis. Each of the aforementioned processes produces idiosyncratic signals, but they converge at paracrine signaling through the Receptor Tyrosine Kinase (RTK) pathway. Four key RTKs have been researched exhaustively as potential anti-angiogenesis drug targets, and through this research, have been shown to play an integral role in angiogenesis initiation: VEGF-R1 and VEGF-R2 (receptors for the ligand VEGF; also known as Flt1 and Flk1, respectively), Tie1and Tie2 (receptors for the proteins angiopoietin 1 and 2, respectively, though some crossover may exist) (Hanahan, 1997). An immature cell in the extracellular matrix (EM) or in a blood vessel will initiate angiogenesis (and differentiate) in response to VEGF-R1/2 activation. VEGF-R1 has a very high affinity to VEGF and is highly expressed by endothelial cells, but it does not appear to play a role in the migration or proliferation of endothelial cells (Waltenberger et al., 1994). VEGF-R2, on the contrary, is highly expressed by proliferating endothelial cells, but less so in mature vessels. Very primitive capillary tubes that immature endothelial cells form as a result of paracrine and autocrine signaling are Tie1, Tie2, and Flt1 (VEGF-R1) positive. VEGF is expressed by many different tissues at low basal levels, but when angiogenesis is being stimulated, for example in the embryo, placenta, scars, or tumors, expression is at markedly high levels. The endothelial cells form a monolayer then tertiary structures as immature tubes. These receptors then cooperate in a complex interplay of stimulation, inhibition, and regulation of new vessel growth and branching, which is paramount to the second stage of angiogenesis: *proliferation and division*. It is at this point that the specific demands of the tissue and/or microenvironment are identified and met by the newly forming blood vessels.

During the *proliferation and division* phase of angiogenesis, endothelial cells from the circulation or EM must proliferate and migrate through the EM (basal lamina) to continue the proliferation of the primitive endothelial tubes that will become the new blood vessels. Current research has shown that matrix metalloproteinases (MMPs) play a large role in the process by which the EM is broken down by cells for cellular migration. Homologous situations occur in tumor metastasis and wound healing. Urokinase (u-PA), tissue-type (t-PA) plasminogen activators, and MMPs are crucial to new vessel formation in any tissue type. The proliferation and division phase of angiogenesis has given scientists insight into not only how we pattern ourselves as organisms, but also into the processes of aging, tumor angiogenesis and metastatic behavior, organ regeneration, and apoptosis.

Though many specific cellular signaling pathways must act in concert for angiogenesis to occur properly, one specific cytokine merits mention in any description of angiogenesis. Transforming Growth Factor (TGF)-ß belongs to a superfamily of complex, highly conserved 25 kDa disulfide-linked homodimeric cytokines, which include activins, BMPs, and the Mullerian-inhibiting substance. TGF-ßs act in a similar fashion on the newly developed vessel and the proliferating endothelial cells within the mesenchyme. They exert an effect on a wide variety of tissues and appear to stimulate or inhibit cell mitosis and control cell adhesion through the EM, MMPs, MMP inhibition, and integrins. The importance of this family of cytokines in the vasculature has been established. Specifically, subtype TGF-ß1 signals within the EM for immature cells to form pericytes and smooth muscle in order that the third phase of angiogenesis, ***maturation***, can begin (Folkman, 1996; Folkman & D'Amore, 1996). TGF-ß1 dimers *in vitro* and *in vivo* inhibit endothelial cell proliferation and the monomers may play a role in deciding the venous and arterial sides of circulation. Depending upon the levels of expression, endothelial tube formation is either stimulated or inhibited by TGF-ß1 (Merwin et al., 1991a, 1991b).

Three types of cell-surface receptors are acted upon by TGF-ß: RI, RII, and RIII. Type I (RI) receptors contain an intracellular serine/threonine kinase but require the RII receptor to bind ligands. The type II (RII) receptors are constitutively active. Heteromeric binding of the RI and RII receptors is required for ligand binding (TGF-ß) and cellular signaling. The type III (RIII) receptors, which include endoglin, have no known signaling capacity and likely act as modulators of the other isoforms. TGF-ß binds the RII receptor strongly, and binding will recruit the RI receptor to form a heteromeric dimmer. This process will be explored further as we discuss one form of hemorrhagic brain disease, Hereditary Hemorrhagic Telangiectasia or HHT.

Once the vessels are appropriately organized, supernumerary vessels sclerose and die and the *maturation* phase begins. Through this phase, the final organization of the blood vessels is completed and the arterial and venous sides of the circulation are clearly defined. Cytokines recruit and organize vessel wall components like pericytes and smooth muscle to fortify the vessel walls and enable essentially a tertiary structure of the new vessels.

The Tie1 and Tie 2 receptors are highly expressed by angioblasts and endothelium (Korhonen et al., 1994; Yancopoulos et al., 1998). Based upon mouse knockout studies, Tie1 likely plays a role in the integrity and survival of the endothelium during angiogenesis, but not differentiation and proliferation as the –/– mice die from edema and hemorrhage in the presence of developed vessels (Dumont et al., 1995; Puri et al., 1995). Tie2 –/– mice also die early, however, the defect appears to be a lack of new vessels and branching of older vessels (Sato et al., 1995). These studies point to the importance of these receptors in the growth, maturation, and maintenance of newly forming blood vessels during angiogenesis. As expected, angiopoietin-1 (Ang1) deficient mice (the ligand for Tie2) show diminished sprouting and branching of vessels(Suri et al., 1996), whereas overexpression of Ang1 have stimulated vessel formation and branching (Suri et al., 1998). Interestingly, angiopoietin-2 (Ang2) also binds Tie2 and reduces branching and induces vessel breakdown (Maisonpierre et al., 1997). This is thought to be a crucial factor not only in new vessel formation, but in the process by which angiogenesis is kept under tight control.

Angiogenesis is a far more complex process than what is outlined by the aforementioned paragraphs. The key point is that angiogenesis, under normal physiological conditions, is a tightly controlled, highly ordered process encompassing cell migration, proliferation, tube formation, and vessel maturation/stabilization. There are numerous soluble proteins and membrane-bound receptors or proteins that play a large role in these processes, and any perturbation of said proteins results in vascular pathology (see Table 13.1). The understanding of the phases of angiogenesis and vasculogenesis and the genes (proteins) that contribute to these processes pertains directly to the understanding of the pathogenesis of vascular lesions that predispose patients to hemorrhagic brain diseases.

III. Mendelian Forms of Hemorrhagic Brain Disease

Twin studies (Alberts, 1990; Bak et al., 2002; Brass et al., 1992, 1996; ter Berg et al., 1992) and family history studies (Pracyk & Massey, 1989; Ronkainen et al., 1999; Ruigrok et al., 2004; Sundquist et al., 2006; Yamauchi et al., 2000) show that stroke (hemorrhagic and ischemic) has a higher incidence in families. The definition of the molecular pathophysiology of hemorrhagic stroke is still a work in progress. Some causes such as factor deficiencies and intracranial aneurysms have treatments. Others do not and the understanding of the molecular causes coupled with early diagnosis, screening, and family counseling can benefit patients greatly. The molecular and genetic causes of hemorrhagic brain diseases are complex and multifactorial. For the purposes of this chapter, we will divide this disease into two distinct subcategories: primary (lesions, vascular, or parenchymal abnormalities that are prone to hemorrhage) and secondary (metabolic syndromes, coagulation factor deficiencies, or hypertension that can lead to hemorrhage anywhere as well as systemic effects with the brain being one of many organ systems involved).

To shed light on common, sporadic diseases, science often looks toward the rarer, more severe Mendelian forms of the same disease for insight (see Table 13.2). For example, hypertension is quite common worldwide, however, severe hypertension from glucocorticoid-remediable aldosteronism (GRA) is comparably uncommon. With the sequencing of the human genome and the emergence of new technologies like DNA chips and RNAi, not only are we identifying trait loci and genes contributing to these diseases, but also the cellular pathways that may be therapeutic targets for the treatment of these diseases.

In any patient presenting with brain hemorrhage, it is important, as a part of the interview, to take a detailed family history. Whether you draw a pedigree diagram or simply write it out, this information is important to not only the physicians who care for the patient, but also will play a large role in counseling the family, if a genetic disorder is present, for implementing a screening protocol. Many hospitals now have a genetic consultation service that can assist in this role and will contact family members for interviews and screening.

Table 13.1 Angiogenesis Proteins and Their Known Functions

Factor/Receptor	Action
VEGF	Stimulates angiogenesis *in vivo*, endothelial cell proliferation, migration, and permeability.
Ang1	Stimulates endothelial stability and sprout formation.
Ang2	Destabilizes endothelium and opposes Ang1.
a/bFGF	Stimulates angiogenesis *in vivo*, endothelial cell proliferation, migration, and permeability.
PDGF	Stimulates proliferation of smooth muscle and pericytes. Stabilizes capillaries.
TGF-ß	Inhibits endothelial cell proliferation, migration, and permeability. Stabilizes vessel walls.
Tie1	Receptor. Integrity and survival of the endothelium during angiogenesis, but not differentiation and proliferation.
Tie2	Receptor for Ang1. Branching and stability of new vessels.
VEGF-RI	Receptor for VEGF. Strong affinity.
VEGF-RII	Receptor for VEGF. Weaker affinity.
VE-Cadherin	Endothelial permeability and angiogenesis.
Eph-4B/Ephrin-B2	Directs venous/arterial interface development of the embryo.
Blood flow	Promotes endothelial cell division and vessel stability.

IV. Primary Hemorrhagic Brain Diseases (Vascular Lesions)

A. Intracranial Aneurysm

Intracranial aneurysms (IA) are the second most common cause of subarachnoid hemorrhage (SAH). SAH is a true neurosurgical emergency carrying a poor prognosis. Autopsy data on patients who don't reach medical attention quote a mortality rate of approximately 12 percent (Schievink et al., 1995b, 1995c). Of those who do reach medical attention, 40 percent will die in hospital care (Huang, 1985; Juvela, 2000, 2002a, 2002b; Mayberg et al., 1994). Those who are treated successfully experience a very high mobility as many do not resume their baseline quality of life.

Intracranial aneurysms (IAs) carry a prevalence of approximately 1 to 6 percent (Inagawa & Hirano, 1990). Although quite common, the pathogenesis and molecular underpinnings of intracranial aneurysms are not completely understood. Epidemiological studies have examined the

Table 13.2 Mendelian Forms of Hemorrhagic Brain Disease

Disease	Gene	Locus	Function
CCM1	KRIT1	*7q11-q21*	Integrin binding and microtubule associated protein
CCM2	MGC4607	*7p13*	Unknown
CCM3	PDCD10	*3q26.1*	Unknown
HHT1	Endoglin	*9q34.1*	Type III TGF-β receptor
HHT2	Acvrl1/ALK1	*12q11-q14*	TGF-β receptor
HHT3	Unknown	*5q31.3-5q32*	Unknown
HSHWA-D	Beta A4 amyloid	*21q21*	Unknown
HCHWA-I	Cystatin C	*20p11.2*	Proteinase inhibitor
IA	Unknown	*11q24-25*	Unknown
		14q23-31	
		19q13.3	
		2p13	
		1p34.3-p36.13	
PKD1	Polycystin 1	*16p*	Cell–cell interactions
PKD2	Polycystin 2	*4q*	Ion channel
Moyamoya	Unknown	*3p*	Unknown
		17q25	
		8q23	
Leptomeningeal	Transthyretin	*18q11.2-12.1*	Thyroxine binding amyloidosis

various environmental and patient-specific variables and identified specific modifiable and nonmodifiable risk factors for the development and rupture of IA. For example, IA is more common among females; patients with hypertension, atherosclerosis, or diabetes; and smokers (Connolly et al., 2001; Nahed et al., 2005a). It has long been noted that family history plays a large role in the development and rupture of these lesions (see Figure 13.2).

Though the true pathogenesis is likely to be multifactorial, genetic factors play a role in 7 to 20 percent of cases reported (Schievink, 1997). Patients with a first-degree relative with aneurysmal subarachnoid hemorrhage have a four-fold increase in their relative risk of having an aneurysmal SAH (Schievink et al., 1995a; Schievink & Spetzler, 1998). Additionally, in large population-based studies and case-control studies, up to 12 percent of patients with aneurysmal SAH had a family history versus only 2 to 6 percent of controls (De Braekeleer et al., 1996; Wang et al., 1995).

Multiple additional data have emerged, further supporting the notion that genes likely play a role in the development and rupture of these lesions. For example, in families with clear transmission of the IA phenotype, lesions rupture at a younger age, the locations of the aneurysms within the circle of Willis is different than in sporadic cases, and twins or sib-pairs often have aneurysms at identical locations within the cerebrovasculature (Nahed et al., 2005b; Olson et al., 2002; Ozturk et al., 2006; Roos et al., 2004; Ruigrok et al., 2005; Schievink & Spetzler, 1998). There are many well-described, inherited disorders that predispose patients for the secondary development of IA. Among these syndromes are polycystic kidney disease (PKD), Ehlers–Danlos syndrome type IV, Marfan's syndrome, GRA, and dissection of the great arteries. For example, in PKD, the prevalence of intracranial aneurysms is as high as 25 percent, and there are specific mutations that predispose patients to develop IA (Rossetti et al., 2003; Schievink et al., 1992; Watnick et al., 1999). The identification and screening of patients for IA who have inherited disease alleles within families with these disorders is paramount.

Recently, families with isolated, inherited forms of intracranial aneurysms have been analyzed. As the molecular technologies evolved, preliminary studies focused on associations between allele inheritance and a proclivity for the development of intracranial aneurysm (Keramatipour et al., 2000). Gene chip microarray studies then focused on resected or autopsy lesion samples. Increased gene expression was noted among proteolytic enzymes, specifically MMPs, which contribute to focal degradation of the vascular extracellular matrix (Bruno et al., 1998). Specifically, up to a three-fold increase in MMP-2 is seen in the walls of aneurysm tissue (Todor et al., 1998). MMPs are a target for therapies in tumor metastases and may indeed be a therapy target for IA.

Large families and populations with histories of IA and SAH have been genotyped to look for loci that contribute to the disease (see Figure 13.2) (Farnham et al., 2004; Kuivaniemi et al., 1993; Olson et al., 2002; Onda et al., 2001, 2003; Pope et al., 1991; Takenaka et al., 1998, 1999a, 1999b, 2000; Yoneyama et al., 2003, 2004).The first studies, sibling-pair analyses from Japanese and Finnish populations, preliminarily identified candidate intervals (Olson et al., 2002; Onda et al., 2001). The Japanese sib-pair candidate intervals recently were confirmed with genome-wide linkage analysis, which found linkage to chromosomes *11q24-25*

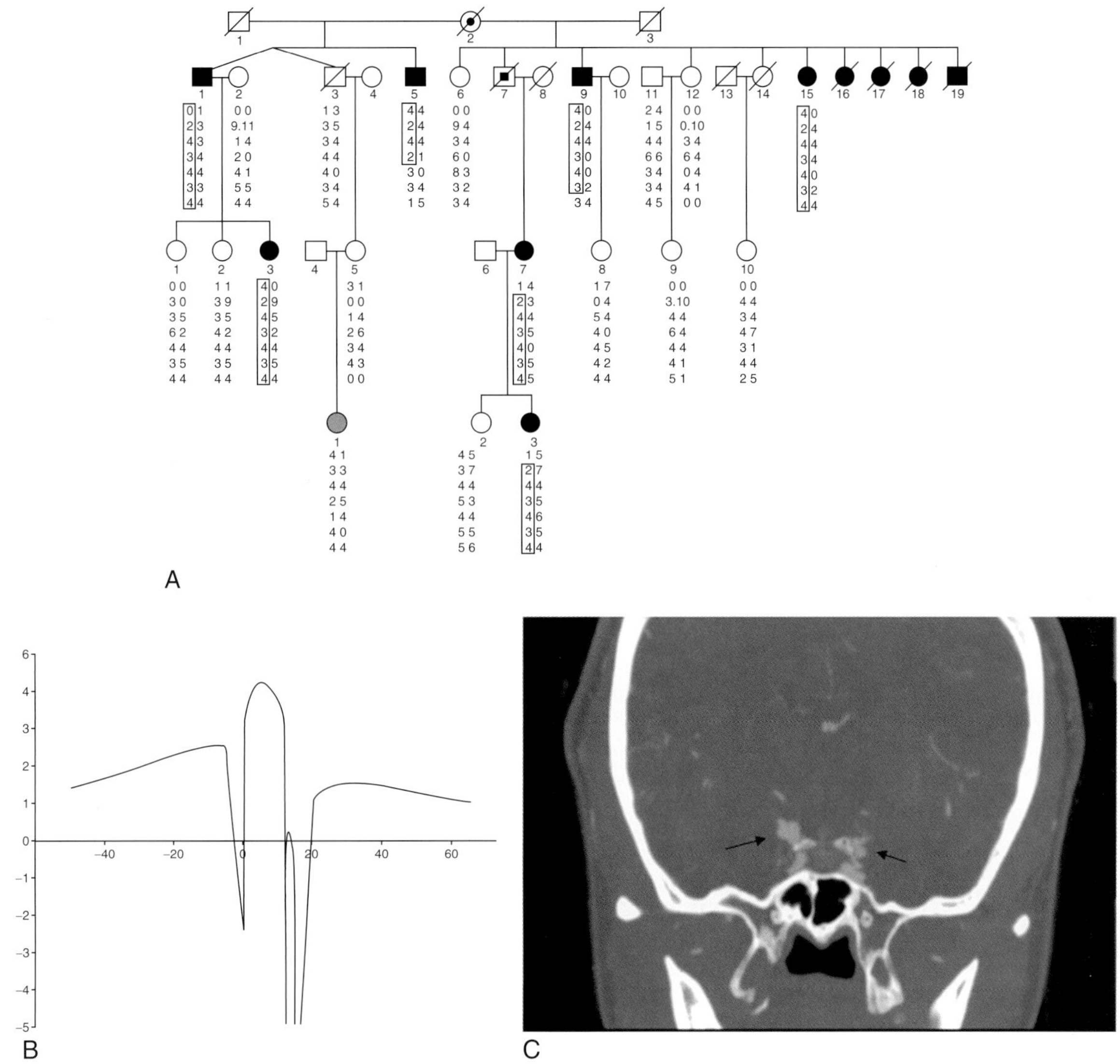

Figure 13.2 A. Ideogram depicting a family with IA. **B**. Demonstration of linkage within this family to a locus with a significant LOD score of 4.2. **C**. CT angiogram of an affected individual with familial IA. Note the presence of bilateral aneurysms of the posterior communicating arteries (arrows).

and *14q23-31* in two separate large families (Ozturk et al., 2006). The only other intervals with evidence of significant linkage is *19q13.3* in the Finnish population (van der Voet et al., 2004), *2p13* in the Dutch family (Roos et al., 2004), and *1p34.3-p36.13* in a large family from West Virginia (Nahed et al., 2005b). The hope is that one of these loci will reveal a gene mutation that will help define the molecular biology of IA. Once the underlying genes contributing to IA are identified and cellular pathways are outlined, the contribution of molecular factors to the pathophysiology of IA will be known. Combining this knowledge with the current understanding of the pathophysiology and natural history of aneurysm formation and rupture will improve diagnostic and therapeutic approaches to this disease.

B. Aneurysms in Polycystic Kidney Disease

Up to 12 percent of patients with Autosomal Dominant Polycystic Kidney Disease (ADPKD) have intracranial aneurysms, and 5.5 percent will have SAH compared to 0 percent of nonaffected family members (Belz et al., 2001). Eighty-five percent of people with ADPKD have mutations in PKD1 on chromosome 16p, and the remaining 15 percent have mutations in PKD2 on chromosome 4q. PKD1, or polycystin 1, encodes a protein involved in cell-cell interactions and PKD2, or polycystin 2, encodes an ion channel (1994; Harris et al., 1995; Mochizuki et al., 1996; Van Adelsberg & Frank, 1995). Both proteins are expressed in blood vessels, specifically in smooth muscle and endothelial cells. Mouse knockout models of *Pkd1* and *Pkd2* develop a phenotype of vascular leakage and blood-vessel rupture in homozygous fetuses from embryonic day 12·5, indicating that these proteins likely play a role in the integrity of vessel walls (Kim et al., 2000; Wu et al., 2000). The majority of patients who present with SAH or who are found to have IA have mutations in PKD1 (Rossetti et al., 2003).

Patients with ADPKD have different aneurysm and SAH characteristics than those with the sporadic form of IA. For example, patients tend to be younger at the time of onset of SAH, fewer females are affected proportionally, and the aneurysms are more commonly of the middle cerebral artery rather than the internal carotid in sporadic patients (see Figure 13.3) (Bromberg et al., 1995; Ruigrok et al., 2004, 2005). Mutation analysis in patients with ADPKD and IA have shown that the development of aneurysms is associated with mutations in the 5' end of the protein, a region also shown to correlate with a more severe renal phenotype (Rossetti et al., 2003). Current guidelines recommend screening patients over the age of 30 with ADPKD for aneurysms, especially if other affected family members have ADPKD and IA (Lieske and Toback, 1993; Loeys et al., 2006; Mariani et al., 1999; Schievink & Spetzler, 1998; ter Berg et al., 1992).

C. Moyamoya

Moyamoya (Japanese word meaning "puff of smoke") describes the angiographic image of proximal intracranial, bilateral carotid artery occlusive disease of childhood associated with telangiectatic vessels in the region of the basal ganglia. Children with this disease often present with neurological deficits, seizures, and ischemic lesions on MRI (see Figure 13.4A). With time, extensive collateral circulation develops (from the posterior and external carotid circulations). In adults, these vessels are prone to rupture and older patients will present with subarachnoid and intraparenchymal hemorrhage (see Figure 13.4B).

This disease has the highest prevalence in the Japanese population and appears to have familial phenotypes. Peak age at onset is 10 to 14 years, with a smaller peak of age at onset in the forties. The prevalence and incidence are 3.16 and 0.35 per 100,000 Japanese persons, respectively. The estimated prevalence of moyamoya disease in the entire Japanese population is less than 0.01%. The prevalence of moyamoya disease in offspring and in sibs of patients with the disease has been estimated at 2.4 and 3 percent, respectively. The incidence in both monozygotic twins is 80 percent (Goto & Yonekawa, 1992; Hung et al., 1997; Kitahara et al., 1979; Wakai et al., 1997; Yonekawa & Kahn, 2003).

Three separate loci have been mapped using parametric genetic techniques: chromosome 3p (MYMY1), 17q25 (MYMY2), and 8q23 (MYMY3). One group performed a total genome search to determine the location of a familial moyamoya disease gene, studying 16 families and assuming an unknown mode of inheritance. Linkage was found between the disease and markers located at 3p26-p24.2 with a maximum nonparametric LOD score of 3.46 obtained with marker D3S3050 (Ikeda et al., 1999). None of the genes have been identified.

D. Cerebral Cavernous Malformation (CCM)

CCMs are sporadic or inherited vascular lesions of the brain characterized by closely packed cavernous chambers that are filled with blood or thrombus, lined by a single layer of endothelial cells separated by a dysmorphic connective tissue matrix (collagen) (Clatterbuck et al., 2001; Wong et al., 2000). These lesions do not contain any intervening neural parenchyma or identifiable mature vessel wall elements such

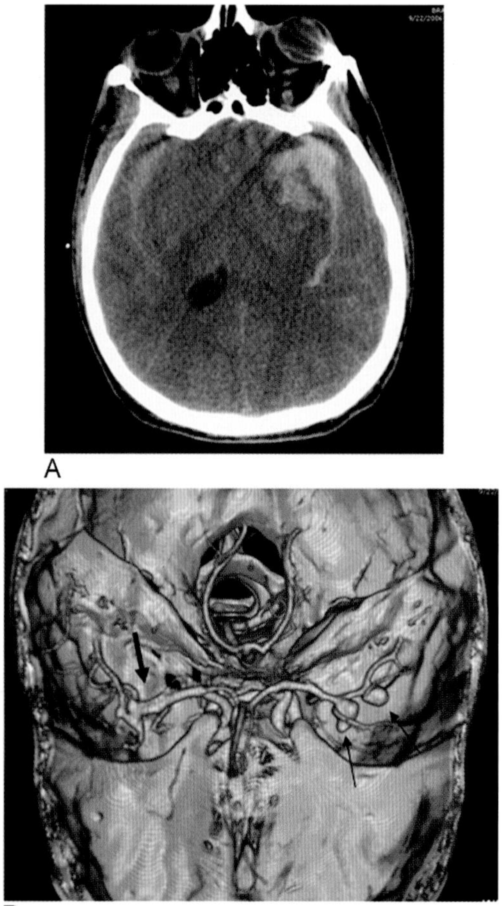

A

B

Figure 13.3 **A.** CT scan showing subarachnoid hemorrhage in a patient with ADPKD. **B.** 3D CT angiogram demonstrates multiple irregular berry aneurysms of the left MCA (thin arrows) and a fusiform aneurysm of the left internal MCA (thick arrow).

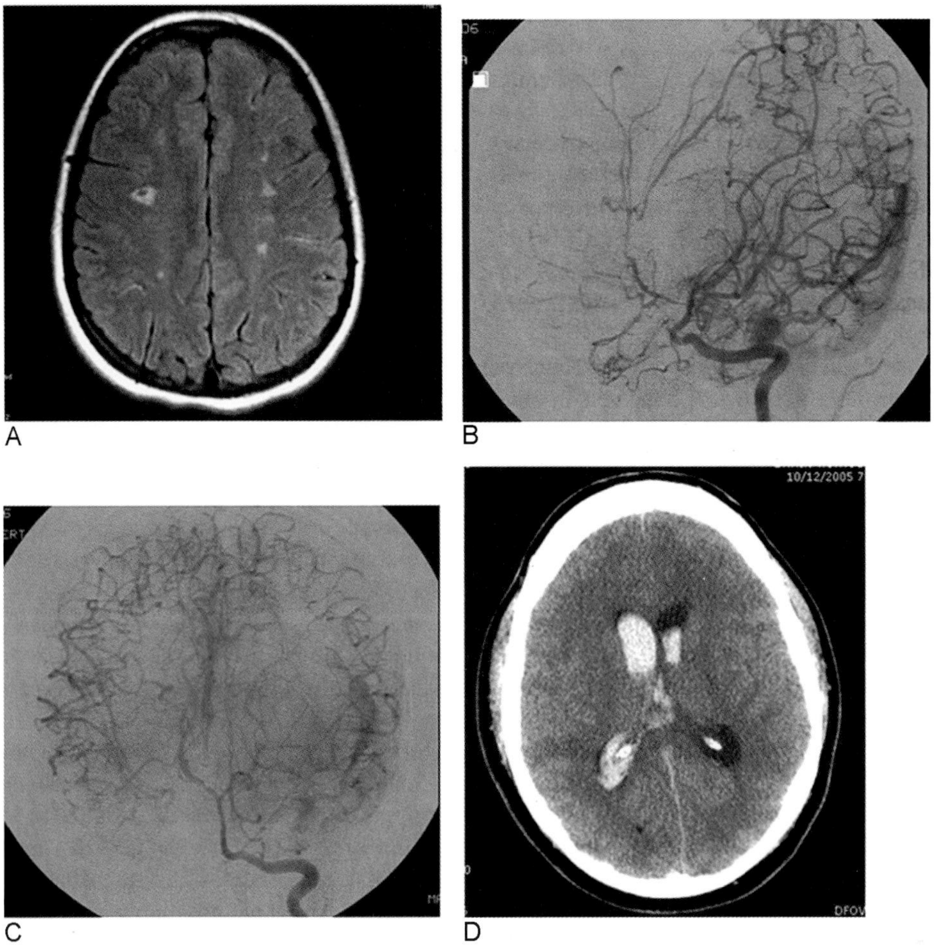

Figure 13.4 A. FLAIR sequence MRI and angiogram of a child with symptomatic moyamoya. Note the presence of multiple ischemic lesions. **B**. Angiogram of the left internal carotid artery demonstrates intracranial arterial occlusive disease and leptomeningeal collateral recruitment. Note the absence of filling of much of the ACA and distal MCA branches. **C**. Angiogram of the left vertebral artery demonstrates reconstitution of flow of the MCA and ACA circulation via the posterior circulation. **D**. CT of an adult with moyamoya who presented with intraventricular hemorrhage.

as smooth muscle or elastin (Clatterbuck et al., 2001; Wong et al., 2000). Furthermore, gaps exist between the endothelial cells, devoid of tight junctions that normally would create the blood-brain barrier, allowing for the leakage of red blood cells into the surrounding brain parenchyma, leading to heavy hemosiderin deposits (Clatterbuck et al., 2001). These lesions are found primarily in the CNS (brain and spinal cord), but may also be seen in skin, retina, vertebra, and liver. The symptoms of these lesions are caused by either mass-effect on nearby brain structures or from hemorrhage. At baseline, these lesions are "leaky," as is evidenced by the presence of hemosiderin on MRI and pathology. Occasionally, they cause hemorrhage.

Originally considered angiographically occult, this disease was found to be a common clinical entity with the advent of magnetic resonance imaging (MRI). MRIs of affected patients reveal characteristic lesions described as "popcorn-appearing" of variable signal intensity surrounded by a dark ring attributable to hemosiderin on T2-weighted images (see Figure 13.5) (Perl, 1993). Both MRI and autopsy studies estimate a prevalence of cavernous malformation of 0.5 percent, although the prevalence of symptomatic disease is much lower (Del Curling et al., 1991; Otten, 1989). Symptomatic patients typically present in the third decade of life with headache, seizures, and focal neurological deficits (Robinson et al., 1991, 1993). Treatment ranges from therapy with anti-epileptic drugs in patients with seizures, to surgical excision of accessible lesions in patients who suffer from recurrent hemorrhage or intractable seizures (Barrow, 1993a, 1993b; Robinson et al., 1991, 1993).

CCM has been widely recognized as a familial disease since its original description (Hayman et al., 1982; Kidd and Cumings, 1947; Michael and Levin, 1936). Several large kindreds affected with autosomal dominant CCM have been reported, with incomplete penetrance; the proportion of at-risk offspring of affected subjects developing clinical disease is often less than 50 percent (Hayman et al., 1982). Similar to other genetic forms of sporadic disease, the inherited form of CCM is far more severe, with numerous lesions that have a younger age of presentation. Genome-wide linkage analysis of affected families and positional cloning eventually identified three loci with significant LOD scores, CCM1 on 7q11.2-q21, CCM2 on 7p13 and CCM3 on 3q26.1, that segregate with the disease phenotype (Craig et al., 1998; Gunel et al., 1995, 1996b).

Familial cases are particularly evident among Hispanic Americans of Mexican descent, with over 50 percent of cases having another relative affected with CCM (Rigamonti et al., 1988). Analysis of these families revealed that Hispanic patients with familial CCM have inherited the same set of genetic markers, or haplotype, in the portion of 7q linked to CCM1, suggesting a "founder-effect" (Gunel et al., 1996a). Analysis of the gene mutated in CCM1, *KRIT1* (stands for Krev1 interaction trapped protein-1) revealed a specific mutation (Q455X) inherited from a common ancestor that is responsible for the majority of cases in Hispanic Americans of Mexican descent. Indeed, all affected with CCM1 have mutations in *KRIT1*. It appears as though the disease phenotype is a result of a two-hit model of disease generation, where patients with the inherited form of CCM1 are born with one mutant *KRIT1* allele and one wild-type allele, and lose the second allele through a somatic mutation, ultimately leading to complete loss of KRIT1 expression in those cells. In contrast, it is believed that patients with the sporadic form of CCM are born with two normal copies of the *KRIT1* gene and have to acquire two independent somatic mutations within the CCM molecular pathway to form lesions.

KRIT1 was identified from a yeast two-hybrid screen using Krev1/Rap1A as bait (Serebriiskii et al., 1997). Krev1 is a member of the Ras family of GTPases and inhibits GAP-mediated Ras-GTPase activity (Serebriiskii et al., 1997). Thus, Krev1 likely acts as a tumor suppressor in Ras cellular signaling. The role of KRIT1, Ras, and tumor suppressor genes in angiogenesis and the development of the CNS vasculature is unknown. Some research has suggested that GAP proteins (like Ras) are phosphorylated in response to VEGF in cultured endothelial cells to stimulate proliferation and angiogenesis (Guo et al., 1995).

Analysis of the sequence and structure of the KRIT1 protein revealed two functional domains, which pointed to a possible cellular role of the peptide. The first of these domains is a FERM (4.1 Ezrin Radixin Moesin) domain at the C-terminus of KRIT1. This domain previously has been demonstrated to play a role in between signal transduction and cytoskeletal elements such as microtubules or actin. The second domain within KRIT1 is a series of ankyrin repeats, which are key players in protein-protein interactions. The combination of these two motifs is unique to KRIT1 and points to its role in the maintenance of cell signaling and the cytoskeleton. Mutations leading to the loss of function of this protein may disrupt the regulation of cell-cell interactions, signaling, adhesion, and cyto-architecture, all of which are highly important and regulated in angiogenesis.

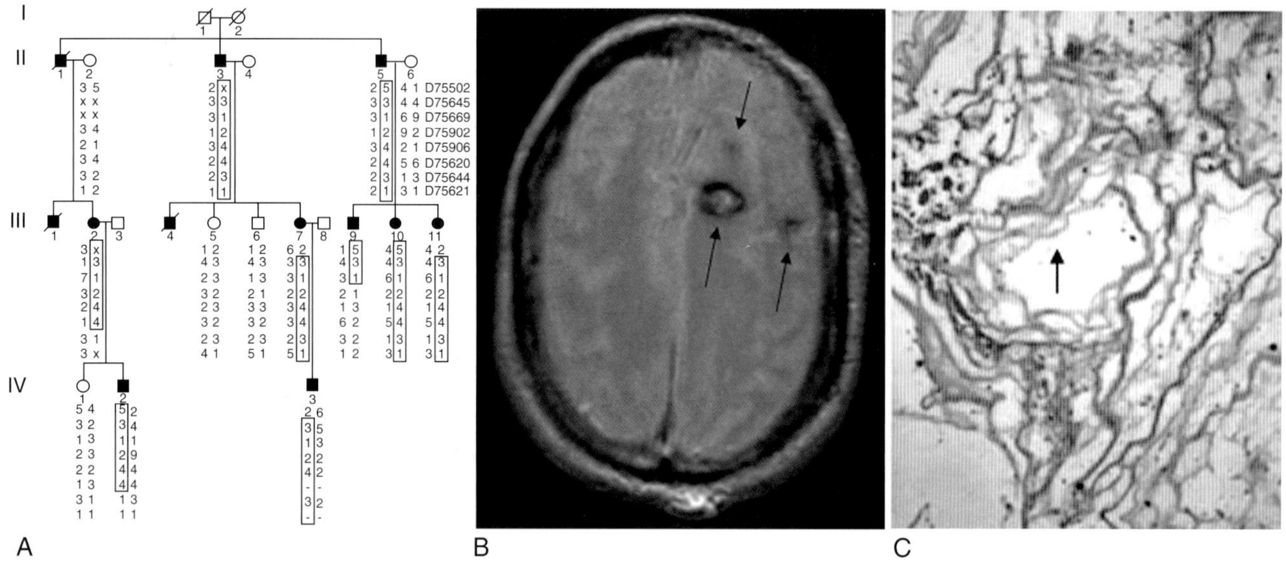

Figure 13.5 **A.** Hispanic American family with CCM1. Ideogram depicting patient's pedigree demonstrates autosomal dominant inheritance. **B.** Axial T2-weighted MRI reveals multiple CCM lesions (arrows) with characteristic hemosiderin signal surrounding the lesion. **C.** Histological specimen of a cavernous hemangioma. Note the cavern lined by a single layer of endothelial cells (arrow) separated by a dysmorphic collagen matrix.

Studies have also investigated the functional role of this protein *in vitro*. Using molecular genetic and yeast two-hybrid technologies, KRIT1 was shown to encode a microtubule-associated protein (Gunel et al., 2002), which also interacts with Integrin Cytoplasmic domain-Associated Protein-1α (ICAP-1α) (Zawistowski et al., 2002, 2005; Zhang et al., 2001). Specifically, there may be an additional functional domain at the N-terminus of the protein that interacts with integrins, an NPXY amino acid sequence, which in experiments is critical for ICAP-1 binding to β1-integrin (Zawistowski et al., 2002). Integrins are important proteins that regulate endothelial cell adhesion and migration throughout angiogenesis. They help form the new vessels' lumens and regulate the permeability of the vessel walls. Antibodies against the KRIT1 protein showed it colocalizes with beta-tubulin within the cytoplasm of endothelial cells (Gunel et al., 2002). Knockout studies in mice demonstrate that KRIT1 is ubiquitously expressed early in embryogenesis and is essential for vascular development.

Homozygous mutant embryos die in mid-gestation and the first detectable defects are exclusively vascular in nature, where the precursor vessels of the brain become dilated starting at E8.5, reminiscent of the intracranial vascular defects observed in the human disease (Whitehead et al., 2004). All the studies thus far implicate KRIT1 in integrin-mediated cell–cell signaling, endothelial cell–brain or ECM adhesion, and cell migration and shape (via microtubules). Furthermore, because CCM lesions do not contain any normal brain parenchyma and the KRIT1 protein stains both neurons, astrocytes, and endothelial cells (Guzeloglu-Kayisli et al., 2004), it is unclear if loss of this protein leads to a perturbation of the brain tissue or endothelium or both. Further research is needed to shed light on this novel cellular pathway of angiogenesis.

Using positional cloning an unknown gene, MGC4607 or *malcavernin*, was identified as the causative gene mutated in families with CCM2 (Liquori et al., 2003). This gene encodes a protein containing a putative phosphotyrosine-binding (PTB) domain, which is also found in ICAP1-alpha, KRIT1's binding partner. Little is known about the function of this gene. Histopathology analysis of CCM2 revealed similar patterns of expression in similar cell subtypes as seen with KRIT1, specifically within arterial endothelium, but not in vascular wall elements such as smooth muscle cells or the venous circulation (Seker et al., 2005). Mutations in the gene *Programmed Cell Death-10*, or *PDCD10*, were shown to be associated with the CCM3 phenotype (Guclu et al., 2005; Liquori et al., 2006; Verlaan et al., 2005). PDCD10 is an unknown protein without any functional domain found through protein sequence analysis *in silico*. The function and role of this peptide in CCM disease or KRIT1/malcavernin signaling is yet to be determined. Because the disease phenotypes caused by CCM1-3 are clinically and pathologically indistinguishable, it is highly likely that the three proteins are involved in the same pathway important for central nervous system vascular development and maintenance (see Figure 13.6).

E. Hereditary Hemorrhagic Telangiectasia (HHT)

Hereditary Hemorrhagic Telangiectasia (HHT) or Rendu-Osler-Weber syndrome is an autosomal dominant disease that causes vascular malformations prone to hemorrhage in multiple systems. Much of what we know about how errors in angiogenesis genes can lead to vascular or hemorrhagic phenotypes is a result of research in patients with this syndrome. HHT lesions demonstrate vascular dysplasia leading to epistaxis and recurrent hemorrhage of the skin, mucosa, and viscera including the lung, liver, and brain with telangiectasias and arteriovenous malformations (AVMs). Currently, clinical criteria for a diagnosis of HHT is the presence of any two of the following: a pattern of autosomal dominance inheritance, telangiectases in the nasal mucosa, recurrent epistaxis, or visceral telangiectases.

The incidence of intracranial hemorrhage in patients with HHT is believed to be low, and furthermore, patients have a good functional outcome after hemorrhage (Maher et al., 2001). HHT lesions within the CNS are mostly low-grade AVMs (Spetzler-Martin Grade I or II) and are frequently multiple in number. Seven to 12 percent of patients with HHT will have CNS lesions, and female patients are more often affected than male patients (Roman et al., 1978; Swanson et al., 1999; Willemse et al., 2000). Much controversy exists as to whether or not patients with HHT should undergo frequent screening for CNS lesions. By comparison, pulmonary arteriovenous fistulae are a much more frequent cause of neurological symptoms in this population through embolic stroke.

HHT were first noted by Osler in 1901, described as a "family form of recurring epistaxis, associated with multiple telangiectases of the skin and mucous membranes" (Osler, 1901). Histopathological and ultrastructural analysis of telangiectasias from the skin and mucosa reveals focal dilations of post capillary venules composed of prominent stress fibers and pericytes along the adluminal border (Braverman et al., 1990). There is no reported difference between mucosal and cutaneous telangiectases and those found in the viscera (Winterbauer, 1964) or the neurovasculature (Rothbart et al., 1996). As the pathologic AVM grows and develops, the abnormal vessels have multiple layers of smooth muscle often without an elastin component, and connect through one or more capillaries to dilated arterioles (Braverman et al., 1990). The natural history of the development of these lesions is known and begins with the dilation of postcapillary venules in

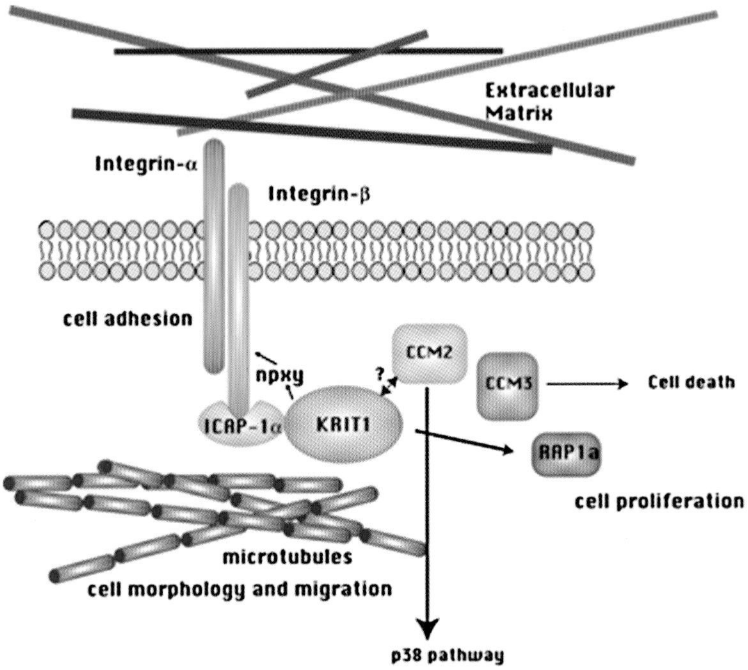

Figure 13.6 CCM pathway.

the dermis. The lumens of these vessels expand and pericytes are recruited to support the walls. The arterioles dilate next, but are still connected to the venules through a capillary bed. Once the venules become tortuous and extensive, the capillary bed disappears and direct fistulae between the arterial and venous side are seen (AVF) (Braverman et al., 1990).

Linkage studies have shown that there are two (possibly three) mutated genes that cause the disease phenotype. The first gene, HHT1, was mapped to 9q34.1 (McDonald et al., 1994), and a second group of families showed linkage to a second locus, HHT2, on 12q11-q14 (Johnson et al., 1995). A third locus was later shown within a 5.4-cM disease interval at 5q31.3-5q32 (Cole et al., 2005). Positional cloning identified that a gene coding for *endoglin*, a transforming growth factor—beta (TGF-β) binding protein—was mutated in affected individuals with HHT1 (McAllister et al., 1994). In similar fashion, loss-of-function mutations were found in the activin receptor-like kinase 1 gene, *Acvrl1 (or ALK1)*, a type I serine-threonine kinase receptor for the TGF-β superfamily of growth factors within the HHT2 locus on 12q11-q14 (Berg et al., 1997). The gene mutated in HHT3 is unknown. Both Acvrl1 and endoglin are expressed primarily by endothelial cells and are required as receptor-partners for TβRI and II to maintain the balance between the previously mentioned positive and negative biphasic effect TGF-β signaling has on angiogenesis (Shovlin et al., 1997).

F. Endoglin and ALK1

The process of angiogenesis can be described by a hierarchy of control, activation, and completion, also known as an angiogenic switch (Pepper, 1997). It is likely that ALK1, endoglin, and the TGF-ß RI and II proteins play a large role in the regulation and control of this switch. Endoglin, a Type III TGF-β receptor, is a member of a family of cell surface receptors that are among the most densely expressed receptors across a given cell's surface (Massague, 1992). Endoglin, specifically, is a membrane glycoprotein expressed in limited amounts in organ tissues, but is found in high levels predominately in vascular endothelial cells and in hematopoietic cell types (Li et al., 1999). The endoglin gene codes for a 561 amino acid protein, including the cell localization signal peptide, multiple extracellular N and O-linked glycosylation sites, and a 47 amino acid cytoplasmic domain that can vary in length due to alternative splicing events (Gougos, 1990). The native protein exists as a homodimer linked by multiple cystein–cystein bonds (Barbara et al., 1999). Endoglin nul mice (–/–) do not survive past E11.5 (11.5 days after fertilization); these knockout mice are three times smaller and exhibited a profound absence of vascular organization (Li et al., 1999). Endothelial cells in the primary capillary plexus do not mature into structured vessels and cannot recruit vessel wall elements such as smooth muscle cells. Therefore, endoglin is not necessary for vasculogenesis in the embryo; however, it is crucial to the later phases of angiogenesis.

Interestingly, in the absence of endoglin expression, Flk-1, Flt-1, Tie1, and Tie2, in addition to TGF-β expression, are all unchanged. This implies that endothelial cell differentiation from precursor cells is not solely regulated by endoglin, but given the severity of the phenotype in the nul mice, each piece of the angiogenesis puzzle is vital to vascular organization in the developing embryo. As expected, the vascular structures seen in the knockout mice are deficient in the *maturation* stage of angiogenesis as vascular smooth muscle precursors were absent from supporting structures of the nascent capillary networks. Endoglin, though without any innate signaling ability, readily binds to RI/RII or ALK1/RII dimers, likely regulating their activity.

ALK1, similar to endoglin, is an endothelial cell receptor for members of the TGF-β superfamily. ALK1, interestingly, shows little or no activity in response to all TGF-β protein subtypes. However, when bound as a chimeric heterodimer to the TGF-β Type I receptor, ALK1 signaling activity markedly increases in response to TGF-ß1 and -ß3 but not -ß2 (Lux et al., 1999). Cellular signaling dependent upon heterodimer binding specificity with TGF-β Type I receptors is a result seen with functional experiments using endoglin, lending further support to the notion that endoglin and ALK1 are in the same molecular and developmental pathway—each able to keep TGF-ß positive and negative angiogenic signaling in check (see Figure 13.7). ALK1, when activated, inhibits proliferation, migration, and adhesion of endothelial cells (Lamouille et al., 2002). The conclusion is that ALK1 is required to terminate angiogenesis. Alternatively, antisense directed against ALK1 enhances cell migration, whereas activated TGF-ß RI decreases migration, possibly arguing for a role in angiogenesis initiation (Goumans et al., 2002). It is postulated, though not yet confirmed, that mutations in endoglin or ALK1 result in unopposed or unregulated TGF-ß RI signaling and premature angiogenesis termination/resolution.

Acvrl1 (ALK1) nul (–/–) mice are also embryonic lethal (Srinivasan et al., 2003; Urness et al., 2000). At E9.5, the –/– animals failed to form distinct vitelline vessels in their yolk sacs and the vessels present were a meshwork of interconnected and homogenous endothelial tubes. Furthermore, the mice showed primary arterial-venous shunting of the central vascular tree with subsequent disruption of endothelial remodeling and vascular smooth muscle cell development (Srinivasan et al., 2003; Urness et al., 2000). Combined with the endoglin data, the process of angiogenesis and early vascular development can be broken down into a couple of distinct, but critical processes. First, proteins like VEGF, and other growth factors must differentiate the angioblasts and vascular stem cells into the endothelial tubes that form the vitelline capillary plexus of the yolk sac and the central vascular tree (Pepper, 1997). Second, the assignment of arterial-venous identity must occur to create the mature circulatory system, a process dependent on ALK1 and TGF superfamily cytokines (Urness et al., 2000). ALK1, similar to endoglin, plays a crucial role in angiogenesis, vessel turnover, and

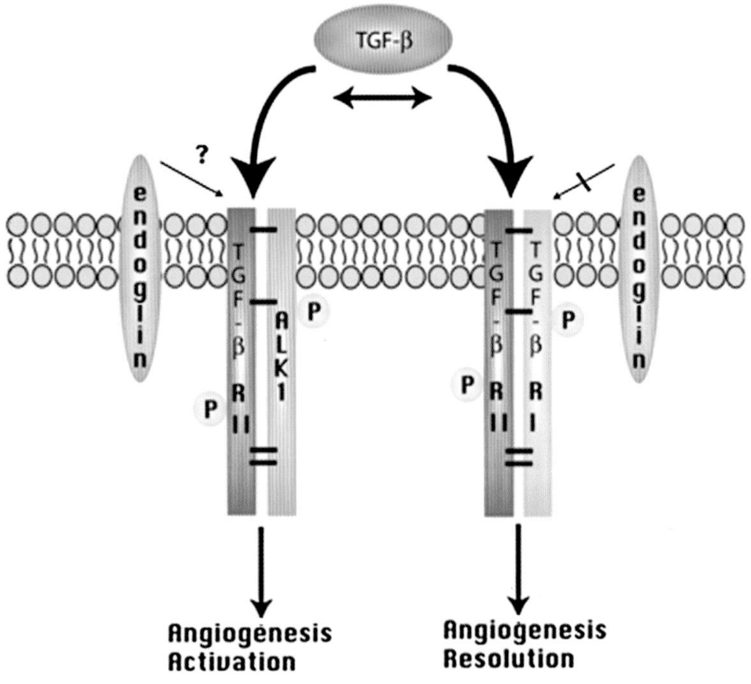

Figure 13.7 HHT/endoglin/ALK1 pathway.

venous-arterial designation through the maturation of new blood vessels, again supporting theories about the formation of AVMs.

Heterozygous mice (+/−) develop HHT lesions with age (Bourdeau et al., 1999; Srinivasan et al., 2003). These lesions are highly similar on pathology to those seen in humans, and frequently demonstrate hemorrhage. Only 40 percent of mice will develop these lesions and the development of these lesions is mouse-strain dependent. These results led investigators to speculate about the development of these lesions given an autosomal dominant inheritance pattern and tissue specific phenotypes. Although most favor a loss-of-heterozygosity model, the presence of endoglin protein in HHT1 lesions argues to the contrary (Bourdeau et al., 2000). It appears as though the lesions develop as a result of defective endothelium remodeling rather than as a proliferative or lack of angiogenesis control phenotype. More favor the argument that these lesions are "triggered" in response to inflammation, injury, stress, or hypoxia, all factors that can initiate angiogenesis and vascular remodeling. Currently, research is examining whether or not polymorphic mutations in endoglin, ALK1, and the other receptors for TGF-ß play a role in disease phenotypes such as intracranial aneurysms, pulmonary hypertension, and other disorders of the adult vascular system.

V. Secondary Hemorrhagic Brain Diseases

A. Inherited Cerebral Amyloid Angiopathies

Cerebral amyloid angiopathies are defined by the presence of amyloid deposits in the walls of cerebral blood vessels. Amyloid is the aggregation of protein in the form of ß pleated sheets that are frequently seen in aging and are insoluble, hence the deposition. In sporadic cases, adults in the sixth or seventh decades of life present with large, lobar brain hemorrhages. On MRI, prior brain hemorrhages are seen in the form of hemosiderin staining of the brain (see Figure 13.8).

Hereditary cerebral hemorrhage with cortical amyloid angiopathy (HCHWA) was one of the first Mendelian forms of stroke to be characterized on the molecular level. Hereditary cerebral hemorrhage with cortical amyloid angiopathy is subcategorized into two subsets: the Dutch type (HCHWA-D) and the Icelandic type (HCHWA-I). Serial studies of these populations showed sclerosis and amyloid-like deposits in the small arteries and arterioles in the cortex and covering arachnoid, similar to, though far more severe than the sporadic form (Jensson et al., 1987; Luyendijk et al., 1988). The pathological vessels in these two diseases often are distributed irregularly and clustered in the cortex. The Icelandic type of HSHWA is caused by the L68Q mutation (a substitution at amino acid position 68 of a glutamine for a leucine) in the cystatin C gene, a proteinase inhibitor, on chromosome *20p11.2*. Cystatin C is an abundant extracellular inhibitor of proteinases that incidentally also has reduced expression in atherosclerotic blood vessel walls and aneurysm lesions (Shi et al., 1999). In HCHWA-I, the mutant form of this protein deposits within the walls of the affected vessels and leads to the hemorrhagic stroke phenotype (Jensson et al., 1987). The glutamine substitution destabilizes alpha helical structures, exposing a tryptophan residue to a polar environment, unfolding the protein and leading to amyloid deposits. Nearly 20 percent of patients with the Icelandic type of HSHWA present with fatal hemorrhage by the age of 35, and most die by the age of 50.

The Dutch type of HSHWA is caused by the E693Q mutation in the beta A4 amyloid precursor protein gene on *21q21* (Levy et al., 1990). Two-thirds of patients with the Dutch type of HSHWA present with fatal cerebral hemorrhage, and the remaining one-third suffer from severe vascular dementia. The mechanism by which these proteins form amyloid plaques is unknown. The hope is that through further insight into these severe, inherited forms of the disease, an understanding of the common, sporadic form will be gained.

Mutations in the transthyretin gene on 18q11.2-12.1 cause familial amyloid polyneuropathy leading to amyloid deposit and destruction of small-fiber and autonomic nerves. Numerous amino acid polymorphic substitutions have been found within the transthyretin gene that cause leptomeningeal amyloidosis (Brett et al., 1999; Garzuly et al., 1996; Hirai et al., 2005; Jin et al., 2004; Uemichi et al., 1997). Similar to the amyloid deposits seen in HSHWA, amyloid deposits are seen in the leptomeningeal vessels of these patients

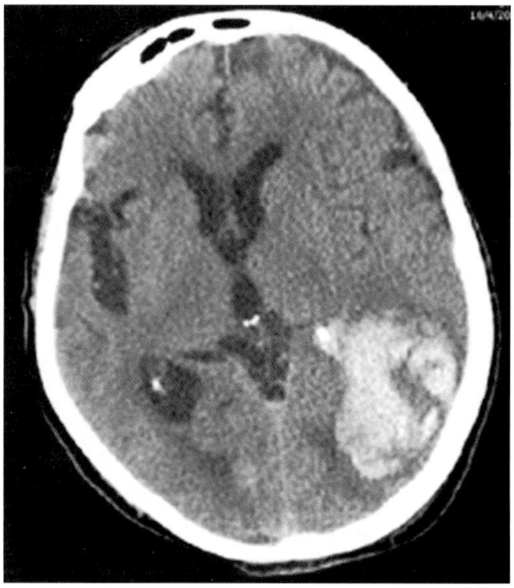

Figure 13.8 Patient with amyloid angiopathy. CT demonstrates large lobar hemorrhage in the left occipital lobe. A smaller, resolving prior hemorrhage can be seen in the right occipital lobe.

leading to small vessel disease, thrombosis, and recurrent subarachnoid hemorrhage. Again, the pathogenesis of this disease is unknown.

B. Clotting Disorders

Deficiencies in numerous clotting disorders predispose patients to various bleeding dyscrasias (see Figure 13.9A). Patients with mutations in genes that code for Factors V (parahemophilia), VII, VIII (hemophilia A), and XIII are all at risk of intracerebral hemorrhage. Often, patients with a deficiency in one of these factors will present early in childhood with bruising or uncontrollable bleeding. Occasionally, intraventricular bleeds will occur in infants at birth and spontaneous intracranial hemorrhages will occur throughout childhood. Minor traumas throughout life can result in catastrophic intracranial hemorrhage.

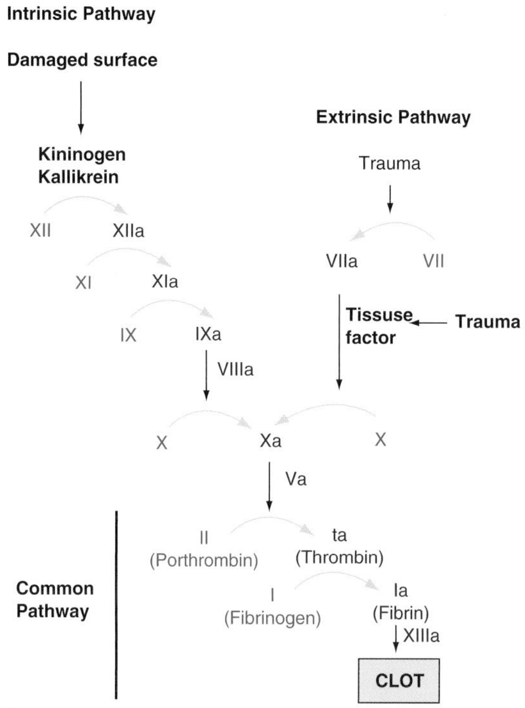

A

Genetic Defect	Test	Phenotype
Factor XII	PTT high; PT normal	Not significant
Factor XI	PTT high; PT normal	Recessive, May have excessive bleeding
Factor VIII/IX	PTT high; PT normal	VIII (hemophilia A, IX (hemophilia B). Both will experience excessive bleeding
Factor VII	PTT normal; PT high	Recessive. Associated with spontanedous CNS hemorrhage
Factor V, X	PTT high; PT high	Recessive. Associated with severe bleeding
Factor V Leiden	PTT normal; PT normal	3-15% of population is heterozygous. High risk of venous thrombosis
Fibrinogen	Low Fibrinogen	Severe bleeding
Dysfibrinogenemia	Thrombin time prolonged	Severe bleeding
Factor XIII	Clot dissolves in 5M urea	Recessive. Associated with spontaneous CNS hemorrhage

B

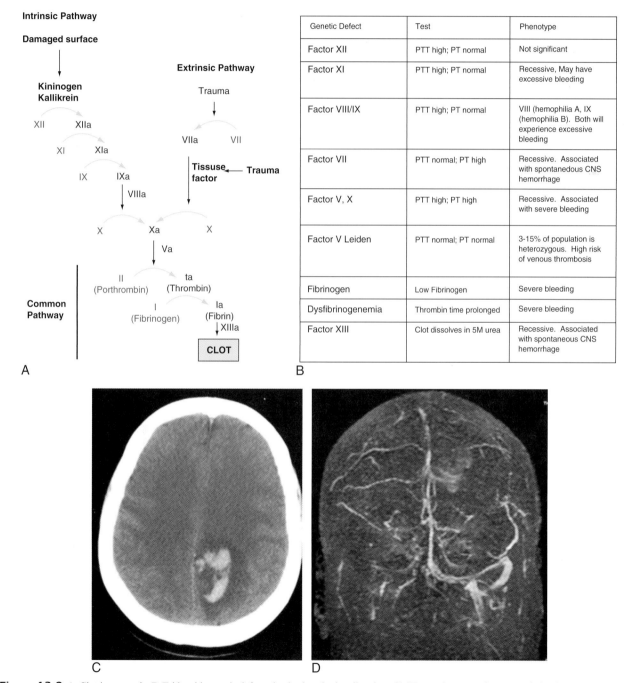

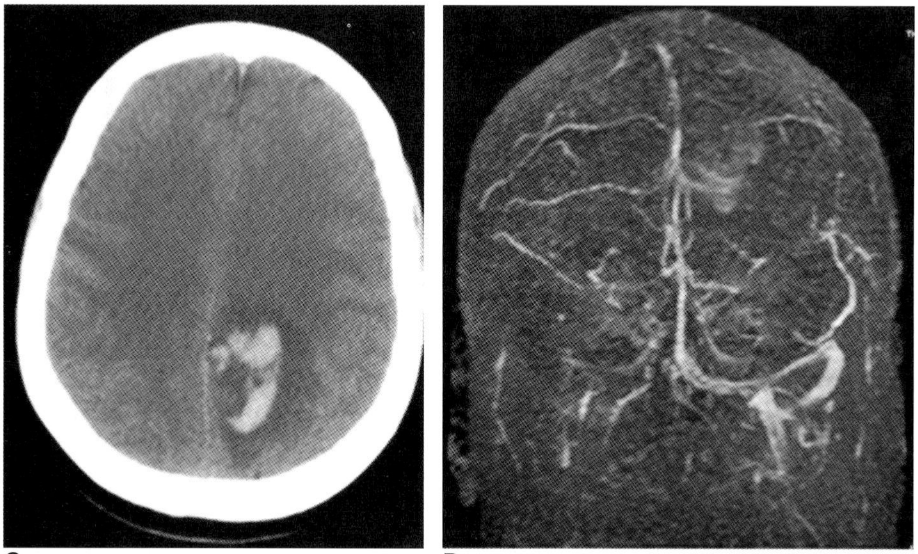

C D

Figure 13.9 **A.** Clotting cascade. **B.** Table with genetic defects that lead to clotting disorders. **C.** CT scan demonstrating venous infarction and hemorrhage. **D.** MR venogram demonstrating thrombosis of transverse, sigmoid, jugular, and portions of the sagittal sinus in a patient with Factor V Leiden mutation.

Cerebral Venous Thrombosis (CVT) is a relatively common cause of stroke and its risk factors include smoking, dehydration, use of oral contraceptive medications, and pregnancy, and family history of venous thrombosis (either of the brain or deep venous system of the extremities (DVT)) (Ciccone & Citterio, 2001). When a cortical vein thrombosis occurs within the brain, patients are at risk of venous infarction from capillary congestion and subsequent subarachnoid hemorrhage from vessel rupture (see Figure 13.9B). There are many mutations in various genes that play a role in the clotting cascade that predispose patients to clotting disorders and venous thrombosis. The most common of these mutations is the Factor V Leiden mutation. A single base-pair mutation in the Factor V gene (G1691A) destroys the proteins for an inactivating enzyme known as Protein C. Harboring this mutation puts patients at marked risk of developing not only DVT but CVT as well (Martinelli et al., 1998; Zuber et al., 1996). One study showed that the odds ratio (OR) of developing a CVT with the Factor V Leiden mutation is 7.8 (Martinelli et al., 1998).

Another mutation that predisposes patients to DVT and CVT is the G20210A mutation of prothrombin. The presence of this mutation results in an OR of 10 for CVT and with concomitant oral contraceptive use, the OR increases to 150 (Martinelli et al., 1998). Conversely, in patients who take oral contraceptives, if they harbor a mutation such as Factor V Leiden, Protein C or S deficiency, or antithrombin deficiency, their risk of CVT is 30-fold higher than women on oral contraceptives without any prothrombotic disorder (de Bruijn et al., 1998a, 1998b).

VI. Future Research

Much remains unknown in our understanding of hemorrhagic diseases of the brain. The pathways by which angiogenesis is altered or disrupted, predisposing someone to sporadic forms of vascular lesions or brain hemorrhage, is currently being uncovered. From what we know thus far, perturbations in angiogenesis, specifically endothelium structure, maintenance, and regulation can result in severe phenotypes. Given the information at hand, we can counsel our patients and their families harboring disease alleles or genes at a young age and potentially intervene early through the screening of potentially affected relatives. As we enter a new era where our ability to diagnose and treat these lesions, not only through surgical means, but also with minimally invasive techniques like stereotactic radiosurgery and endovascular technologies, the molecular data will enhance our knowledge of how these lesions grow and change through the life of a patient. The next era of medicine will usher in a biological therapy phase where we will be able to alter the clinical progression of these diseases through drugs, gene delivery, or gene modification. The ultimate goal is to initiate primary therapies soon after birth to dramatically change the natural history of hemorrhagic brain diseases and substantially reduce the morbidity and mortality associated with them.

References

Alberts, M. J. (1990). Genetics of cerebrovascular disease. *Stroke* **21**, III127–130.

Bak, S. et al. (2002). Genetic liability in stroke: A long-term follow-up study of Danish twins. *Stroke* **33**, 769–774.

Barbara, N. P. et al. (1999). Endoglin is an accessory protein that interacts with the signaling receptor complex of multiple members of the transforming growth factor-beta superfamily. *J Biol Chem* 274, 584–594.

Barrow, D., Awad, I. NA. (1993a). Conceptual overview and management strategies. In Barrow, D. Awad, I. A., Ed., *Cavernous malformations*, 205–213. AANS, Park Ridge.

Barrow, D., Krisht, A. (1993b). Cavernous malformations and hemorrhage. In Barrow, D., Awad, I. A., Ed., *Cavernous malformations*, 65–80. AANS, Park Ridge.

Belz, M. M. et al. (2001). Familial clustering of ruptured intracranial aneurysms in autosomal dominant polycystic kidney disease. *Am J Kidney Dis* **38**, 770–776.

Berg, J. N. et al. (1997). The activin receptor-like kinase 1 gene: Genomic structure and mutations in hereditary hemorrhagic telangiectasia type 2. *Am J Hum Genet* **61**, 60–67.

Bourdeau, A. et al. (1999). A murine model of hereditary hemorrhagic telangiectasia. *J Clin Invest* **104**, 1343–1351.

Bourdeau, A. et al. (2000). Endoglin-deficient mice, a unique model to study hereditary hemorrhagic telangiectasia. *Trends Cardiovasc Med* **10**, 279–285.

Brass, L. M. et al. (1992). A study of twins and stroke. *Stroke* **23**, 221–223.

Brass, L. M. et al. (1996). Importance of cerebrovascular disease in studies of myocardial infarction. *Stroke* **27**, 1173–1176.

Braverman, I. M. et al. (1990). Ultrastructure and three-dimensional organization of the telangiectases of hereditary hemorrhagic telangiectasia. *J Invest Dermatol* **95**, 422–427.

Brett, M. et al. (1999). Transthyretin Leu12Pro is associated with systemic, neuropathic and leptomeningeal amyloidosis. *Brain* **122** (Pt 2), 183–190.

Broderick, J. et al. (1994). Management of intracerebral hemorrhage in a large metropolitan population. *Neurosurgery* **34**, 882–887; discussion 887.

Broderick, J. P. et al. (1992). The risk of subarachnoid and intracerebral hemorrhages in blacks as compared with whites. *N Engl J Med* **326**, 733–736.

Bromberg, J. E. et al. (1995). Familial subarachnoid hemorrhage: Distinctive features and patterns of inheritance. *Ann Neurol* **38**, 929–934.

Bruno, G. et al. (1998). Vascular extracellular matrix remodeling in cerebral aneurysms. *J Neurosurg* **89**, 431–440.

Ciccone, A., Citterio, A. (2001). Cerebral venous thrombosis. *Lancet* **357**, 1706–1707.

Clatterbuck, R. E. et al. (2001). Ultrastructural and immunocytochemical evidence that an incompetent blood-brain barrier is related to the pathophysiology of cavernous malformations. *J Neurol Neurosurg Psychiatry* **71**, 188–192.

Cole, S. G. et al. (2005). A new locus for hereditary haemorrhagic telangiectasia (HHT3) maps to chromosome 5. *J Med Genet* **42**, 577–582.

Connolly, E. S., Jr. et al. (2001). Influence of smoking, hypertension, and sex on the phenotypic expression of familial intracranial aneurysms in siblings. *Neurosurgery* **48**, 64–68; discussion 68–69.

Consortium, T.E.P.K.D. (1994). The polycystic kidney disease 1 gene encodes a 14kb transcript and lies within a duplicated region on chromosome 16. The European Polycystic Kidney Disease Consortium. *Cell* **77**, 881–894.

Craig, H. D. et al. (1998). Multilocus linkage identifies two new loci for a mendelian form of stroke, cerebral cavernous malformation, at 7p15-13 and 3q25.2-27. *Hum Mol Genet* **7**, 1851–1858.

De Braekeleer, M. et al. (1996). A study of inbreeding and kinship in intracranial aneurysms in the Saguenay Lac-Saint-Jean region (Quebec, Canada). *Ann Hum Genet* **60**, 99–104.

de Bruijn, S. F. et al. (1998a). Case-control study of risk of cerebral sinus thrombosis in oral contraceptive users and in [correction of who are] carriers of hereditary prothrombotic conditions. The Cerebral Venous Sinus Thrombosis Study Group. *BMJ* **316**, 589–592.

de Bruijn, S. F. et al. (1998b). Increased risk of cerebral venous sinus thrombosis with third-generation oral contraceptives. Cerebral Venous Sinus Thrombosis Study Group. *Lancet* **351**, 1404.

Del Curling, O. et al. (1991). An analysis of the natural history of cavernous angiomas. *J Neurosurg* **75**, 702–708.

Dumont, D. J. et al. (1995). Vascularization of the mouse embryo: A study of flk-1, tek, tie, and vascular endothelial growth factor expression during development. *Dev Dyn* **203**, 80–92.

Farnham, J. M. et al. (2004). Confirmation of chromosome 7q11 locus for predisposition to intracranial aneurysm. *Hum Genet* **114**, 250–255.

Folkman, J. (1996). New perspectives in clinical oncology from angiogenesis research. *Eur J Cancer* **32A**, 2534–2539.

Folkman, J., D'Amore, P. A. (1996). Blood vessel formation: What is its molecular basis? *Cell* **87**, 1153–1155.

Garzuly, F. et al. (1996). Familial meningocerebrovascular amyloidosis, Hungarian type, with mutant transthyretin (TTR Asp18Gly). *Neurology* **47**, 1562–1567.

Goto, Y., Yonekawa, Y. (1992). Worldwide distribution of moyamoya disease. *Neurol Med Chir* (Tokyo) **32**, 883–886.

Gougos, A., Letarte, M. (1990). Primary structure of endoglin, an RGD-containing glycoprotein of human endothelial cells. *J Biol Chem* **265**, 8361–8364.

Goumans, M. J. et al. (2002). Balancing the activation state of the endothelium via two distinct TGF-beta type I receptors. *Embo J* **21**, 1743–1753.

Guclu, B. et al. (2005). Mutations in apoptosis-related gene, PDCD10, cause cerebral cavernous malformation 3. *Neurosurgery* **57**, 1008–1013.

Gunel, M. et al. (1995). Mapping a gene causing cerebral cavernous malformation to 7q11.2-q21. *Proc Natl Acad Sci U S A* **92**, 6620–6624.

Gunel, M. et al. (1996a). A founder mutation as a cause of cerebral cavernous malformation in Hispanic Americans. *N Engl J Med* **334**, 946–951.

Gunel, M. et al. (1996b). Genetic heterogeneity of inherited cerebral cavernous malformation. *Neurosurgery* **38**, 1265–1271.

Gunel, M. et al. (2002). KRIT1, a gene mutated in cerebral cavernous malformation, encodes a microtubule-associated protein. *Proc Natl Acad Sci U S A* **99**, 10677–10682.

Guo, D. et al. (1995). Vascular endothelial cell growth factor promotes tyrosine phosphorylation of mediators of signal transduction that contain SH2 domains. Association with endothelial cell proliferation. *J Biol Chem* **270**, 6729–6733.

Guzeloglu-Kayisli, O. et al. (2004). KRIT1/cerebral cavernous malformation 1 protein localizes to vascular endothelium, astrocytes, and pyramidal cells of the adult human cerebral cortex. *Neurosurgery* **54**, 943–949; discussion 949.

Hanahan, D. (1997). Signaling vascular morphogenesis and maintenance. *Science* **277**, 48–50.

Harris, P. C. et al. (1995). The PKD1 gene product. *Nat Med* **1**, 493.

Hayman, L. A. et al. (1982). Familial cavernous angiomas: Natural history and genetic study over a 5-year period. *Am J Med Genet* **11**, 147–160.

Hirai, T. et al. (2005). Transthyretin-related familial amyloid polyneuropathy: Evaluation of CSF enhancement on serial T1-weighted and fluid-attenuated inversion recovery images following intravenous contrast administration. *AJNR Am J Neuroradiol* **26**, 2043–2048.

Huang, T. Y. (1985). Sudden death secondary to primary dissecting aneurysm of the coronary artery. *Indiana Med* **78**, 1096–1097.

Hung, C. C. et al. (1997). Epidemiological study of moyamoya disease in Taiwan. *Clin Neurol Neurosurg* **99** Suppl 2, S23–25.

Ikeda, H. et al. (1999). Mapping of a familial moyamoya disease gene to chromosome 3p24.2-p26. *Am J Hum Genet* **64**, 533–537.

Inagawa, T., Hirano, A. (1990). Autopsy study of unruptured incidental intracranial aneurysms. *Surg Neurol* **34**, 361–365.

Jensson, O. et al. (1987). Hereditary cystatin C (gamma-trace) amyloid angiopathy of the CNS causing cerebral hemorrhage. *Acta Neurol Scand* **76**, 102–114.

Jin, K. et al. (2004). Familial leptomeningeal amyloidosis with a transthyretin variant Asp18Gly representing repeated subarachnoid haemorrhages with superficial siderosis. *J Neurol Neurosurg Psychiatry* **75**, 1463–1466.

Johnson, D. W. et al. (1995). A second locus for hereditary hemorrhagic telangiectasia maps to chromosome 12. *Genome Res* **5**, 21–28.

Juvela, S. (2000). Risk factors for multiple intracranial aneurysms. *Stroke* **31**, 392–397.

Juvela, S. (2002a). Natural history of unruptured intracranial aneurysms: Risks for aneurysm formation, growth, and rupture. *Acta Neurochir Suppl* **82**, 27–30.

Juvela, S. (2002b). Risk factors for aneurysmal subarachnoid hemorrhage. *Stroke* **33**, 2152–2153; author reply 2152–2153.

Keramatipour, M. et al. (2000). The ACE I allele is associated with increased risk for ruptured intracranial aneurysms. *J Med Genet* **37**, 498–500.

Kidd, H. A., Cumings, J. N. (1947). Cerebral angiomata in an Icelandic family. *Lancet I*, 747–748.

Kim, K. et al. (2000). Polycystin 1 is required for the structural integrity of blood vessels. *Proc Natl Acad Sci U S A* **97**, 1731–1736.

Kitahara, T. et al. (1979). Familial occurrence of moyamoya disease: Report of three Japanese families. *J Neurol Neurosurg Psychiatry* **42**, 208–214.

Korhonen, J. et al. (1994). The mouse tie receptor tyrosine kinase gene: Expression during embryonic angiogenesis. *Oncogene* **9**, 395–403.

Kuivaniemi, H. et al. (1993). Exclusion of mutations in the gene for type III collagen (COL3A1) as a common cause of intracranial aneurysms or cervical artery dissections: Results from sequence analysis of the coding sequences of type III collagen from 55 unrelated patients. *Neurology* **43**, 2652–2658.

Lamouille, S. et al. (2002). Activin receptor-like kinase 1 is implicated in the maturation phase of angiogenesis. *Blood* **100**, 4495–4501.

Levy, E. et al. (1990). Mutation of the Alzheimer's disease amyloid gene in hereditary cerebral hemorrhage, Dutch type. *Science* **248**, 1124–1126.

Li, D. Y. et al. (1999). Defective angiogenesis in mice lacking endoglin. *Science* **284**, 1534–1537.

Lieske, J. C., Toback, F. G. (1993). Autosomal dominant polycystic kidney disease. *J Am Soc Nephrol* **3**, 1442–1450.

Liquori, C. L. et al. (2003). Mutations in a gene encoding a novel protein containing a phosphotyrosine-binding domain cause type 2 cerebral cavernous malformations. *Am J Hum Genet* **73**, 1459–1464.

Liquori, C. L. et al. (2006). Low frequency of PDCD10 mutations in a panel of CCM3 probands: Potential for a fourth CCM locus. *Hum Mutat* **27**, 118.

Loeys, B. L. et al. (2006). Aneurysm syndromes caused by mutations in the TGF-beta receptor. *N Engl J Med* **355**, 788–798.

Lux, A. et al. (1999). Assignment of transforming growth factor beta1 and beta3 and a third new ligand to the type I receptor ALK-1. *J Biol Chem* **274**, 9984–9992.

Luyendijk, W. et al. (1988). Hereditary cerebral haemorrhage caused by cortical amyloid angiopathy. *J Neurol Sci* **85**, 267–280.

Maher, C. O. et al. (2001). Cerebrovascular manifestations in 321 cases of hereditary hemorrhagic telangiectasia. *Stroke* **32**, 877–882.

Maisonpierre, P. C. et al. (1997). Angiopoietin-2, a natural antagonist for Tie2 that disrupts in vivo angiogenesis. *Science* **277**, 55–60.

Mariani, L. et al. (1999). Cerebral aneurysms in patients with autosomal dominant polycystic kidney disease—to screen, to clip, to coil? *Nephrol Dial Transplant* **14**, 2319–2322.

Martinelli, I. et al. (1998). Different risks of thrombosis in four coagulation defects associated with inherited thrombophilia: a study of 150 families. *Blood* **92**, 2353–2358.

Massague, J. (1992). Receptors for the TGF-ß family. *Cell* **48**, 409–415.

Mayberg, M. R. et al. (1994). Guidelines for the management of aneurysmal subarachnoid hemorrhage. A statement for healthcare professionals from a special writing group of the Stroke Council, American Heart Association. *Stroke* **25**, 2315–2328.

McAllister, K. A. et al. (1994). Endoglin, a TGF-beta binding protein of endothelial cells, is the gene for hereditary haemorrhagic telangiectasia type 1. *Nat Genet* **8**, 345–351.

McDonald, M. T. et al. (1994). A disease locus for hereditary haemorrhagic telangiectasia maps to chromosome 9q33-34. *Nat Genet* **6**, 197–204.

Merwin, J. R. et al. (1991a). Vascular cells respond differentially to transforming growth factors beta 1 and beta 2 in vitro. *Am J Pathol* **138**, 37–51.

Merwin, J. R. et al. (1991b). Vascular cell responses to TGF-beta 3 mimic those of TGF-beta 1 in vitro. *Growth Factors* **5**, 149–158.

Michael, J. C., Levin, P. M. (1936). Multiple telangiectases of brain: A discussion of hereditary factors in their development. *Arch Neurol Psychiat* **36**, 514–536.

Mochizuki, T. et al. (1996). PKD2, a gene for polycystic kidney disease that encodes an integral membrane protein. *Science* **272**, 1339–1342.

Moses, M. A. (1997). The regulation of neovascularization of matrix metalloproteinases and their inhibitors. *Stem Cells* **15**, 180–189.

Nahed, B. V. et al. (2005a). Hypertension, age, and location predict rupture of small intracranial aneurysms. *Neurosurgery* **57**, 676–683; discussion 676–683.

Nahed, B. V. et al. (2005b). Mapping a Mendelian form of intracranial aneurysm to 1p34.3-p36.13. *Am J Hum Genet* **76**, 172–179.

Olson, J. M. et al. (2002). Search for intracranial aneurysm susceptibility gene(s) using Finnish families. *BMC Med Genet* **3**, 7.

Onda, H. et al. (2001). Genomewide-linkage and haplotype-association studies map intracranial aneurysm to chromosome 7q11. *Am J Hum Genet* **69**, 804–819.

Onda, H. et al. (2003). Endoglin is not a major susceptibility gene for intracranial aneurysm among Japanese. *Stroke* **34**, 1640–1644.

Osler, W. (1901). On a family form of recurring epistaxis, associated with multiple telangiectases of the skin and mucous membranes. *Bull Johns Hopkins Hosp* **7**, 333–337.

Otten, P., Pizzolato, G.P., Rilliet, B., Berney, J. (1989). 131 cases of cavernous angioma (cavernomas) of the CNS, discovered by retrospective analysis of 24,535 autopsies. *Neurochirurgie* **35**, 82–83.

Ozturk, A. K. et al. (2006). Molecular genetic analysis of two large kindreds with intracranial aneurysms demonstrates linkage to 11q24-25 and 14q23-31. *Stroke* **37**, 1021–1027.

Pepper, M. S. (1997). Transforming growth factor-beta: Vasculogenesis, angiogenesis, and vessel wall integrity. *Cytokine Growth Factor Rev* **8**, 21–43.

Perl, J., Ross, J. (1993). Diagnostic imaging of cavernous malformations. In I. A. Awad, Barrow, D., Eds., *Cavernous Malformations*, 37–48. AANS, Park Ridge.

Pope, F. M. et al. (1991). Type III collagen mutations cause fragile cerebral arteries. *Br J Neurosurg* **5**, 551–574.

Pracyk, J. B., Massey, J. M. (1989). Moyamoya disease associated with polycystic kidney disease and eosinophilic granuloma. *Stroke* **20**, 1092–1094.

Puri, M. C. et al. (1995). The receptor tyrosine kinase TIE is required for integrity and survival of vascular endothelial cells. *Embo J* **14**, 5884–5891.

Rigamonti, D. et al. (1988). Cerebral cavernous malformations: Incidence and familial occurrence. *N Engl J Med* **319**, 343–347.

Robinson, J. R. et al. (1991). Natural history of the cavernous angioma. *J Neurosurg* **75**, 709–714.

Robinson, J. R. et al. (1993). Factors predisposing to clinical disability in patients with cavernous malformations of the brain. *Neurosurgery* **32**, 730–735; discussion 735–736.

Roman, G. et al. (1978). Neurological manifestations of hereditary hemorrhagic telangiectasia (Rendu-Osler-Weber disease): Report of 2 cases and review of the literature. *Ann Neurol* **4**, 130–144.

Ronkainen, A. et al. (1999). Familial subarachnoid hemorrhage. Outcome study. *Stroke* **30**, 1099–1102.

Roos, Y. B. et al. (2004). Genome-wide linkage in a large Dutch consanguineous family maps a locus for intracranial aneurysms to chromosome 2p13. *Stroke* **35**, 2276–2281.

Rossetti, S. et al. (2003). Association of mutation position in polycystic kidney disease 1 (PKD1) gene and development of a vascular phenotype. *Lancet* **361**, 2196–2201.

Rothbart, D. et al. (1996). Expression of angiogenic factors and structural proteins in central nervous system vascular malformations. *Neurosurgery* **38**, 915–924; discussion 924–925.

Ruigrok, Y. M. et al. (2004). Familial intracranial aneurysms. *Stroke* **35**, e59–60; author reply e59–60.

Ruigrok, Y. M. et al. (2005). Genetics of intracranial aneurysms. *Lancet Neurol* **4**, 179–189.

Sato, T. N. et al. (1995). Distinct roles of the receptor tyrosine kinases Tie-1 and Tie-2 in blood vessel formation. *Nature* **376**, 70–74.

Schievink, W. I. (1997). Genetics of intracranial aneurysms. *Neurosurgery* **40**, 651–662; discussion 662–663.

Schievink, W. I., Spetzler, R. F. (1998). Screening for intracranial aneurysms in patients with isolated polycystic liver disease. *J Neurosurg* **89**, 719–721.

Schievink, W. I. et al. (1992). Saccular intracranial aneurysms in autosomal dominant polycystic kidney disease. *J Am Soc Nephrol* **3**, 88–95.

Schievink, W. I. et al. (1995a). Familial aneurysmal subarachnoid hemorrhage: A community-based study. *J Neurosurg* **83**, 426–429.

Schievink, W. I. et al. (1995b). Sudden death from aneurysmal subarachnoid hemorrhage. *Neurology* **45**, 871–874.

Schievink, W. I. et al. (1995c). The poor prognosis of ruptured intracranial aneurysms of the posterior circulation. *J Neurosurg* **82**, 791–795.

Seker, A. et al. (2005). CCM2 expression parallels that of CCM1. *Stroke*.

Serebriiskii, I. et al. (1997). Association of Krev-1/rap1a with Krit1, a novel ankyrin repeat-containing protein encoded by a gene mapping to 7q21-22. *Oncogene* **15**, 1043–1049.

Shi, G. P. et al. (1999). Cystatin C deficiency in human atherosclerosis and aortic aneurysms. *J Clin Invest* **104**, 1191–1197.

Shovlin, C. L. et al. (1997). Characterization of endoglin and identification of novel mutations in hereditary hemorrhagic telangiectasia. *Am J Hum Genet* **61**, 68–79.

Srinivasan, S. et al. (2003). A mouse model for hereditary hemorrhagic telangiectasia (HHT) type 2. *Hum Mol Genet* **12**, 473–482.

Sundquist, K. et al. (2006). Familial risk of ischemic and hemorrhagic stroke: A large-scale study of the Swedish population. *Stroke* **37**, 1668–1673.

Suri, C. et al. (1996). Requisite role of angiopoietin-1, a ligand for the TIE2 receptor, during embryonic angiogenesis [see comments]. *Cell* **87**, 1171–1180.

Suri, C. et al. (1998). Increased vascularization in mice overexpressing angiopoietin-1. *Science* **282**, 468–471.

Swanson, K. L. et al. (1999). Pulmonary arteriovenous fistulas: Mayo Clinic experience, 1982-(1997). *Mayo Clin Proc* **74**, 671–680.

Takenaka, K. et al. (1998). Angiotensin I-converting enzyme gene polymorphism in intracranial saccular aneurysm individuals. *Neurol Res* **20**, 607–611.

Takenaka, K. et al. (1999a). Analysis of phospholipase C gene in patients with subarachnoid hemorrhage due to ruptured intracranial saccular aneurysm. *Neurol Res* **21**, 368–372.

Takenaka, K. et al. (1999b). Polymorphism of the endoglin gene in patients with intracranial saccular aneurysms. *J Neurosurg* **90**, 935–938.

Takenaka, K. V. et al. (2000). Elevated transferrin concentration in cerebral spinal fluid after subarachnoid hemorrhage. *Neurol Res* **22**, 797–801.

ter Berg, H. W. et al. (1992). Familial intracranial aneurysms. A review. *Stroke* **23**, 1024–1030.

Todor, D. R. et al. (1998). Identification of a serum gelatinase associated with the occurrence of cerebral aneurysms as pro-matrix metalloproteinase-2. *Stroke* **29**, 1580–1583.

Uemichi, T. et al. (1997). A trinucleotide deletion in the transthyretin gene (delta V 122) in a kindred with familial amyloidotic polyneuropathy. *Neurology* **48**, 1667–1670.

Urness, L. D. et al. (2000). Arteriovenous malformations in mice lacking activin receptor-like kinase-1. *Nat Genet* **26**, 328–331.

Van Adelsberg, J. S., Frank, D. (1995). The PKD1 gene produces a developmentally regulated protein in mesenchyme and vasculature. *Nat Med* **1**, 359–364.

van der Voet, M. et al. (2004). Intracranial aneurysms in Finnish families: Confirmation of linkage and refinement of the interval to chromosome 19q13.3. *Am J Hum Genet* **74**, 564–571.

Verlaan, D. J. et al. (2005). CCM3 mutations are uncommon in cerebral cavernous malformations. *Neurology* **65**, 1982–1983.

Wakai, K. et al. (1997). Epidemiological features of moyamoya disease in Japan: Findings from a nationwide survey. *Clin Neurol Neurosurg* **99** Suppl 2, S1–5.

Waltenberger, J. et al. (1994). Different signal transduction properties of KDR and Flt1, two receptors for vascular endothelial growth factor. *J Biol Chem* **269**, 26988–26995.

Wang, H. U. et al. (1998). Molecular distinction and angiogenic interaction between embryonic arteries and veins revealed by ephrin-B2 and its receptor Eph-B4. *Cell* **93**, 741–753.

Wang, P. S. et al. (1995). Subarachnoid hemorrhage and family history. A population-based case-control study. *Arch Neurol* **52**, 202–204.

Watnick, T. et al. (1999). Mutation detection of PKD1 identifies a novel mutation common to three families with aneurysms and/or very-early-onset disease. *Am J Hum Genet* **65**, 1561–1571.

Whitehead, K. J. et al. (2004). Ccm1 is required for arterial morphogenesis: Implications for the etiology of human cavernous malformations. *Development* **131**, 1437–1448.

Willemse, R. B. et al. (2000). Bleeding risk of cerebrovascular malformations in hereditary hemorrhagic telangiectasia. *J Neurosurg* **92**, 779–784.

Winterbauer, R. H. (1964). Multiple telangiectasia, Raynaud's phenomenon, sclerodactyly and subcutaneous calcinosis: a syndrome mimicking hereditary hemorrhagic telangiectasia. *Bull Johns Hopkins Hosp* **114**, 361–383.

Wong, J. H. et al. (2000). Ultrastructural pathological features of cerebrovascular malformations: A preliminary report. *Neurosurgery* **46**, 1454–1459.

Wu, G. et al. (2000). Cardiac defects and renal failure in mice with targeted mutations in Pkd2. *Nat Genet* **24**, 75–78.

Yamauchi, T. et al. (2000). Linkage of familial moyamoya disease (spontaneous occlusion of the circle of Willis) to chromosome 17q25. *Stroke* **31**, 930–935.

Yancopoulos, G. D. et al. (1998). Vasculogenesis, angiogenesis, and growth factors: ephrins enter the fray at the border. *Cell* **93**, 661–664.

Yonekawa, Y., Kahn, N. (2003). Moyamoya disease. *Adv Neurol* **92**, 113–118.

Yoneyama, T. et al. (2003). Association of positional and functional candidate genes FGF1, FBN2, and LOX on 5q31 with intracranial aneurysm. *J Hum Genet* **48**, 309–314.

Yoneyama, T. et al. (2004). Collagen type I alpha2 (COL1A2) is the susceptible gene for intracranial aneurysms. *Stroke* **35**, 443–448.

Zawistowski, J. S. et al. (2002). KRIT1 association with the integrin-binding protein ICAP-1: A new direction in the elucidation of cerebral cavernous malformations (CCM1) pathogenesis. *Hum Mol Genet* **11**, 389–396.

Zawistowski, J. S. et al. (2005). CCM1 and CCM2 protein interactions in cell signaling: Implications for cerebral cavernous malformations pathogenesis. *Hum Mol Genet* **14**, 2521–2531.

Zhang, J. et al. (2001). Interaction between krit1 and icap1alpha infers perturbation of integrin beta1-mediated angiogenesis in the pathogenesis of cerebral cavernous malformation. *Hum Mol Genet* **10**, 2953–2960.

Zuber, M. et al. (1996). Factor V Leiden mutation in cerebral venous thrombosis. *Stroke* **27**, 1721–1723.

14

The Dawn of Molecular and Cellular Therapies for Traumatic Spinal Cord Injury

Noam Y. Harel, Yvonne S. Yang, Stephen M. Strittmatter, Jeffery D. Kocsis, and Stephen G. Waxman

There is perhaps no neurological event more sudden and devastating than traumatic spinal cord injury (SCI); one moment of misfortune becomes a lifetime of disability. Many victims of SCI exhibit great courage and hope in the face of their limitations. Over the past decade, tremendous strides have been made toward fulfilling that hope. Continued progress in understanding the mechanisms that prevent regeneration or otherwise limit functional recovery in the injured adult spinal cord has led to new therapies that may be applied not only to SCI, but to other previously untreatable neurological conditions as well. We will discuss approaches to overcoming inhibitors of axon regeneration, cellular transplantation to replace lost neurons and glia, and the molecular adaptations made by surviving neurons.

Prior to discussing these molecular advances, we will briefly review the clinical epidemiology, prognosis, pathology, and current treatment for SCI patients.

I. The Current Clinical Picture

A. Epidemiology

This year, there will be approximately 11,000 new cases of spinal cord injury in the United States alone. It is estimated that approximately 250,000 people are living with SCI in the United States today (National Spinal Cord Injury Statistical Center (NSCISC), 2006). The annual cost of spinal cord injuries in 1996 was estimated to be $9.73 billion, including $2.6 billion in lost productivity; the emotional costs are inestimable (Sekhon & Fehlings, 2001).

Motor vehicle accidents cause nearly 50 percent of spinal cord injuries, followed in frequency by falls, violence, and recreational injuries, of which diving is the most common cause (NSCISC, 2006). In some urban areas, violence is the leading cause of spinal cord injury, whereas in those 60 and older, falls account for more than 50 percent of spinal cord injuries. Since 2000, 79.6 percent of spinal cord injuries have occurred in males, although there has been a slight trend away from males over the past two decades (NSCISC, 2006). The average age at injury has risen to 37.6 years as compared to 28.7 years during the 1970s; the mode is 19 years (NSCISC, 2006).

The most common type and severity of spinal cord injury upon discharge from the hospital is incomplete tetraplegia, representing 34.5 percent of cases, followed by complete (no sparing of sacral sensation) paraplegia (23.1%), complete tetraplegia (18.4%), and incomplete paraplegia (17.5%). The most common neurological levels of injury on admission are C4–C6, whereas the most common level for paraplegia is T12 (NSCISC, 2006).

B. Prognosis

Life expectancy for the victim of SCI depends on his or her age at time of injury and the severity of the injury. For one who suffers a T12 complete injury at age 20, he or she can expect to live until age 65, 13 years less than expected for his or her noninjured peer group. If any motor function is preserved, however, he or she has a nearly normal life expectancy (NSCISC, 2006).

Several studies have shown that one can predict return to independence, mobility, and even employment with knowledge of the severity and type of SCI (Burns & Ditunno, 2001). Briefly, those with C1–C4 neurological levels of injury will need assistance with all activities of daily living (ADLs) including bed mobility and transfers. With lower neurological level comes more independence; beginning from C7 injury, those living with SCI can perform transfers independently and wheelchair propulsion becomes possible, making employment outside of the home much more likely. Even those with a complete SCI have a 70 percent likelihood of regaining one additional neurological level of function within the first year, and patients with initial grade 1 or 2 motor strength at a given neurological level have a nearly 100 percent chance of attaining strength of 3 or greater within one year (Burns & Ditunno, 2001).

Approximately 45 percent of those living with SCI become employed between one and 30 years post-injury (NSCISC, 2006). With regard to ambulation, although only 5 percent of complete paraplegics regain the ability to walk, one study reported a 76 percent return to community ambulation in incomplete paraplegics (Waters et al., 1994).

C. Pathology

The pathophysiological consequences of SCI continue to evolve after the moment of impact. These changes are generally classified into acute, early, and late phases of secondary injury.

1. Acute (seconds)

Direct trauma damages axon tracts and segmental gray matter, along with small blood vessels in the vicinity (Amar & Levy, 1999; McDonald & Sadowsky, 2002). Hemorrhage increases local hydrostatic pressure, which contributes to ischemia at the injury site (Amar & Levy, 1999). The rapid onset of ischemia contributes to the phenomenon of spinal shock, a period of flaccid areflexic paralysis that can last from hours to weeks (McDonald & Sadowsky, 2002).

2. Early (minutes to weeks)

The events triggered by the initial impact progress during the early phase. Previously sequestered electrolytes and neurotransmitters spill from dying cells, resulting in a vastly altered extracellular environment (McDonald & Sadowsky, 1999). Increased extracellular neurotransmitter concentration elicits extraneous action potential firing from surviving neurons, exacerbating already altered ionic gradients. Depolarization after injury activates a persistent sodium conductance along axons within white matter tracts which, in turn, triggers calcium-importing "reverse" sodium-calcium exchange (Imaizumi et al., 1997; Stys et al., 1992, 1993). To maintain physiologic intracellular ion concentrations and membrane potential, cells expend more energy on ion pumps and exchangers such as the Na^+/Ca^{++} ATPase (Amar &Levy, 1999). Despite these homeostatic mechanisms, membrane potential remains partially depolarized. This triggers calcium entry through NMDA receptors found not only in neurons but on oligodendrocytes as well (Amar & Levy, 1999; Wong, 2006). Increased intracellular calcium stimulates further glutamate release and reactive oxygen species (ROS) production, feeding the cycle of spreading depolarization and cellular damage (Amar & Levy, 1999). Inevitably, the combined stress of decreased energy supply and vastly increased demand leads many initially surviving neurons to suffer an excitotoxic demise (Amar & Levy, 1999).

The more that is discovered about inflammation's role in SCI, the more complex and even contradictory the interpretations become. Multiple cell types release either beneficial or harmful modulators over distinct but overlapping time courses. Though inflammation is necessary to phagocytose and otherwise dispose of debris from the injury site, significant collateral damage occurs via ROS; other cytotoxins are released by astrocytes, neutrophils, and lymphocytes; and excessive phagocytosis by macrophages and microglia (Schwab & Bartholdi, 1996). However, inflammatory responses mediate beneficial effects toward neuroregeneration as well. Preactivated macrophages foster improved recovery in animal models of SCI, and are currently in human clinical trial (Proneuron) (Knoller et al., 2005; Rapalino et al., 1998).

3. Late (months)

Further cell death, demyelination, and macroscopic structural changes can exacerbate neurological dysfunction even years after injury. The glial scar poses both mechanical and molecular barriers to regeneration, as detailed later. Oligodendrocyte cell death results in spreading demyelination, leading to conduction block in otherwise intact axons (Waxman, 1992). Furthermore, the profile of sodium and other channels normally clustered at nodes of Ranvier in uninjured axons redistributes, as described later in this chapter.

Ironically, some of the larger structural changes that occur during the late phase are less well understood than many molecular pathophysiological events. A large fluid-filled cavity, known as a syrinx, can form or enlarge months or years after injury, disfiguring portions of the cord that had previously been stable (Schurch et al., 1996).

D. Current Modes of Treatment

Essentially all modes of treatment for SCI patients to date have attempted to stabilize the injury and limit the degree of secondary damage. Pharmacologically, high-dose methylprednisolone (MP) has been the only agent widely recommended for use in acute SCI patients (Bracken et al., 1990, 1992). MP has many proposed mechanisms of action, including inhibition of lipid peroxidation, decreased ROS production, decreased production of harmful cytokines such as TNFα, and antagonizing myelin breakdown (Banik et al., 1997; Hall, 1991; Liu & McAdoo, 1993; Xu, 1998).

The landmark nature of the trials that popularized the MP protocol has steadily transformed into a land mine of controversy—debate continues over the significance of the clinical benefit derived, the risks of harmful side effects, and reproducibility of the data itself (Hurlbert, 2001). Despite the weakness of the evidence supporting the use of high-dose MP in SCI, it remains the most rigorously tested SCI treatment to date.

Aside from MP, only GM-1 ganglioside has shown promising results in multiple human clinical trials for SCI (Fehlings & Bracken, 2001; Geisler et al., 2001). GM-1 is a natural cell membrane constituent that, when provided exogenously, improves neuronal survival and outgrowth through a multitude of proposed mechanisms (Geisler et al., 2001). Several other drugs attempting to convey some of MP's more beneficial effects are being tested for SCI, including minocycline, riluzole, and polyethylene glycol (Fehlings & Baptiste, 2005).

As previously stated, none of these drugs addresses the fundamental target for reversing the damage done by SCI: facilitating the regeneration of injured spinal cord fibers and connections, and the conduction of action potentials along ascending and descending axons that have maintained continuity through the level of the lesion but fail to conduct due to demyelination. This goal was considered unattainable for millennia, but our increased understanding of the mechanisms blocking regeneration and the development of the CNS itself, and of mechanism underlying restoration of conduction along demyelinated axons, has allowed a new generation of regenerative drugs to appear on the horizon.

II. Surmounting Barriers to Axon Regeneration

A variety of barriers, both cell-autonomous and extrinsic to injured neurons, limit mammalian CNS regeneration and plasticity. The most exciting and best described recent advances are discussed here.

A. Myelin-Associated Inhibitors

Many of the most fruitful recent discoveries in SCI have dissected the differences between the central and peripheral nervous systems that underlie their contrasting abilities to regenerate. David and Aguayo resurrected a line of research originally initiated by Tello and Cajal by bridging the medulla with the thoracic spinal cord using sciatic nerve grafts (David & Aguayo, 1981). M. E. Schwab and colleagues went on to characterize the inhibitory nature of CNS myelin itself, develop a monoclonal antibody (IN-1) that blocked CNS myelin's inhibitory activity *in vitro*, and eventually identified two myelin proteins with specific nerve inhibitory activity (Caroni & Schwab, 1988). Concurrently, myelin-associated glycoprotein (MAG), found in both peripheral and central myelin, was identified as a distinct myelin-associated inhibitor (MAI) in 1994 (McKerracher et al., 1994). However, genetically MAG-deficient mice showed no improvement in SCI recovery, arguing against MAG's relevance as an MAI *in vivo* (Bartsch, 1995).

In 2000, the gene encoding the major antigen for the IN-1 antibody was identified and termed Nogo (Chen et al., 2000; GrandPre et al., 2000; Prinja et al., 2000). Nogo-A, the longest of three isoforms, blocks neurite outgrowth *in vitro* through two separate domains, Amino-Nogo and Nogo-66 (Chen et al., 2000; Fournier et al., 2001) (see Figure 14.1). The Nogo-66 Receptor-1 (NgR1) mediates Nogo-66 inhibition, whereas the receptor mediating Amino-Nogo's inhibitory potential remains unidentified (Fournier et al., 2001). NgR1 is a glycophosphatidylinositide (GPI)-linked molecule that lacks a transmembrane domain; as such, it requires a coreceptor to transduce its signal. Though many putative NgR1 coreceptors have been identified (including $p75^{NTR}$, LINGO-1, and Taj/Troy), the overall picture remains unresolved (Mi et al., 2004; Shao, et al. 2005; Wang et al., 2002).

In an unexpected development, MAG and Nogo were found in 2002 to interact with overlapping portions of NgR1 (Domeniconi et al., 2002; Liu et al., 2002). Oligodendrocyte-myelin glycoprotein (OMgp) was then identified as a third major MAI that signals via NgR1 (Wang et al., 2002). These findings were all the more surprising due to the fact that Nogo, MAG, and OMgp do not share structural similarity.

NgR1's identification as a common mediator of CNS myelin's inhibitory properties sparked tremendous optimism

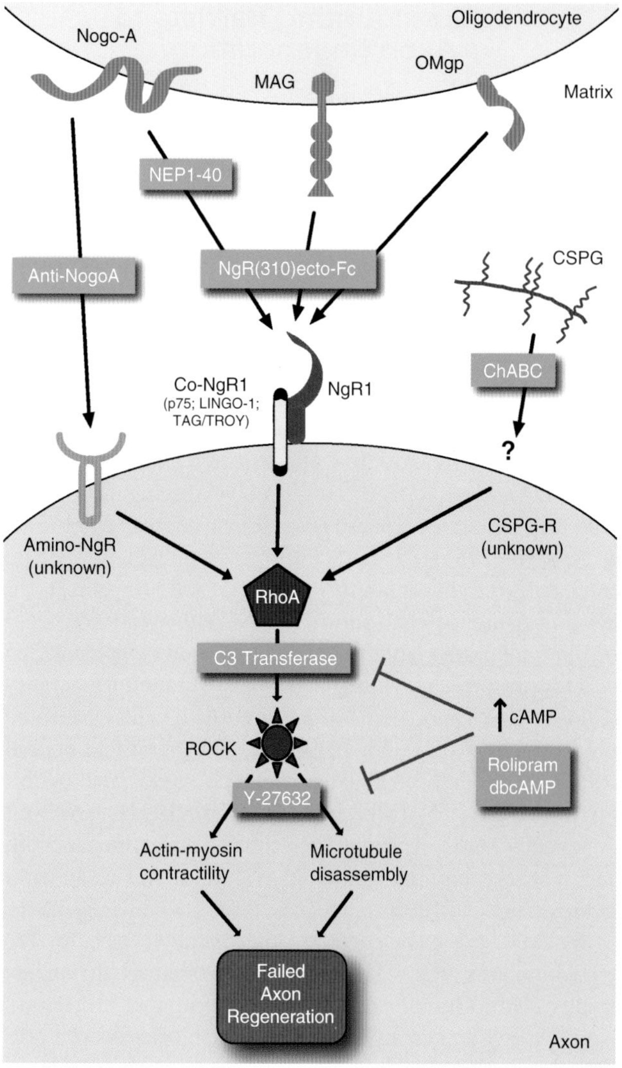

Figure 14.1 Inhibitors of Axonal Regeneration. A schematic illustrates the myelin and astroglial molecules limiting axonal regeneration after SCI. See text for detailed description. An oligodendrocyte is shown at the top expressing Nogo, MAG, and OMGP. An axonal membrane is illustrated below with receptors and signal transduction cascades leading to reduced axonal extension. The agents in the green boxes have shown efficacy to improve recovery from experimental SCI.

in targeting NgR1 with antagonists to improve regeneration following SCI. Scientists have aimed at the Nogo/NgR1 pathway with several types of weapons: genetic, immune (antibodies), and pharmacological (peptides). Most of these approaches have demonstrated substantial relief of neurite outgrowth inhibition *in vitro*; we will review *in vivo* experiments in more detail.

Genetic deletion of Nogo-A does not appear to affect overall CNS development, but resulted in improved regeneration of the corticospinal tract (CST) in mice subjected to thoracic dorsal hemisection (Simonen et al., 2003). Importantly, the Nogo-B isoform is upregulated in this line of mice (Simonen et al., 2003). Further studies comparing the effects of Nogo-A deletion in different genetic strain backgrounds confirmed significantly improved CST regeneration in mice lacking Nogo-A (Dimou et al., 2006). A separate line of mice in which both Nogo-A and Nogo-B are deleted also showed normal development and improved CST regeneration following thoracic dorsal hemisection (Kim et al., 2003). However, another Nogo-A/B knockout line showed no histological improvement following dorsal hemisection (Zheng et al., 2003). Multiple factors may underlie the discrepant results among different knockout lines created in different laboratories. Published data indicate that both strain background (Dimou et al., 2006) and age (Kim et al., 2003) influence the penetrance of the regenerative phenotype. It is also likely that specific Nogo-knockout alleles confer different phenotypes (our unpublished data). It should be noted that myelin derived from all three Nogo knockout lines shows reduced inhibitory action toward wild-type neurons *in vitro* (Kim et al., 2003; Simonen et al., 2003; Zheng et al., 2003).

Mice lacking NgR1 expression show different degrees of regeneration among different spinal cord tracts. They do not show improved CST regeneration following thoracic transection (Kim et al., 2004; Zheng et al., 2005), but improved locomotor recovery does correlate with a significant degree of rubrospinal and raphespinal tract regeneration (Kim et al., 2004). Mice lacking all three major MAIs currently are being characterized.

IN-1, the first anti-Nogo antibody, has reproducibly demonstrated beneficial effects on nerve regeneration both *in vitro* and *in vivo*. Implantation of IN-1 secreting hybridoma cells improves anatomical regeneration, locomotor performance, and electrophysiological function in multiple rat SCI models (Bregman et al., 1995; Merkler et al., 2001; Schnell & Schwab, 1990). A humanized anti-Nogo-A monoclonal antibody is effective in nonhuman primate SCI, and is entering clinical SCI trials (Freund et al., 2006).

Stimulating endogenous production of anti-Nogo antibodies via vaccination with a Nogo-A derived peptide conferred partial protection against spinal cord contusion in rats (Hauben et al., 2001). No evidence of unwanted autoimmune reactions has been observed to date using this vaccine approach (Hauben et al., 2001; Merkler et al., 2003).

In a different approach to antagonizing the Nogo-NgR1 pathway, peptides binding to or mimicking NgR1 have been engineered to act as decoys for NgR ligands. A peptide composed of the first 40 residues of the Nogo-66 domain (NEP1-40) competitively antagonizes NgR1-ligand interactions, improving axon outgrowth on myelin *in vitro* and CST regeneration following dorsal hemisection in rats (GrandPre et al., 2002). Importantly, NEP1-40 improves recovery even when administered systemically one week following rodent SCI, better replicating actual clinical scenarios (Li & Strittmatter, 2003).

Intrathecal administration of a soluble version of NgR1's ectodomain (NgR(310)ecto-Fc) has led to improved CST and raphe pinal regeneration following dorsal hemisection in rats, correlating with improved motor recovery and electrophysiological conduction (Li et al., 2004) (see Figure 14.2). NgR(310)ecto-Fc also improves recovery in a rat middle cerebral artery occlusion model of stroke (Lee et al., 2004). In the most stringent test to date, NgR(310)ecto-Fc improves recovery of motor function after a clinically relevant contusive injury even when therapy begins three days after injury (Wang et al., 2006).

Clearly, pharmacological Nogo or NgR1 blockade produces more dramatic and reproducible recovery than genetic deletion of either Nogo or NgR1. This suggests one or more nonmutually exclusive possibilities: that developmental absence of Nogo or NgR1 leads to altered expression of compensatory Nogo homologs or alternative pathways; or that pharmacological agents may mediate "off-target" inhibition of these compensating molecules (Teng & Tang, 2005). In addition, the lack of CST regeneration in NgR1 knockout mice may indicate that the Amino-Nogo receptor plays a crucial role in blocking CST regeneration; or that the major MAIs themselves may mediate "off-target" effects, either on other NgR isoforms or potentially non-NgR pathways (Kim et al., 2004; Venkatesh et al., 2005; Zheng et al., 2005).

More recently, several other proteins previously characterized as developmental axon guidance molecules have been shown to reside in adult myelin as well, contributing growth-inhibiting signals in the context of injury. EphrinB3 is a member of the two-way Ephrin-Eph signaling family that has been previously implicated in retinotectal mapping and CST development (Kullander et al., 2001). EphrinB3 is expressed by adult spinal cord myelin and inhibits neurite outgrowth *in vitro*, but has not yet been confirmed to act as an MAI *in vivo* (Benson et al., 2005). Repulsive guidance molecule (RGM), also well-characterized for its role in retinotectal pathway development, recently has been demonstrated to act as an MAI as well (Hata et al., 2006; Rajagopalan et al., 2004).

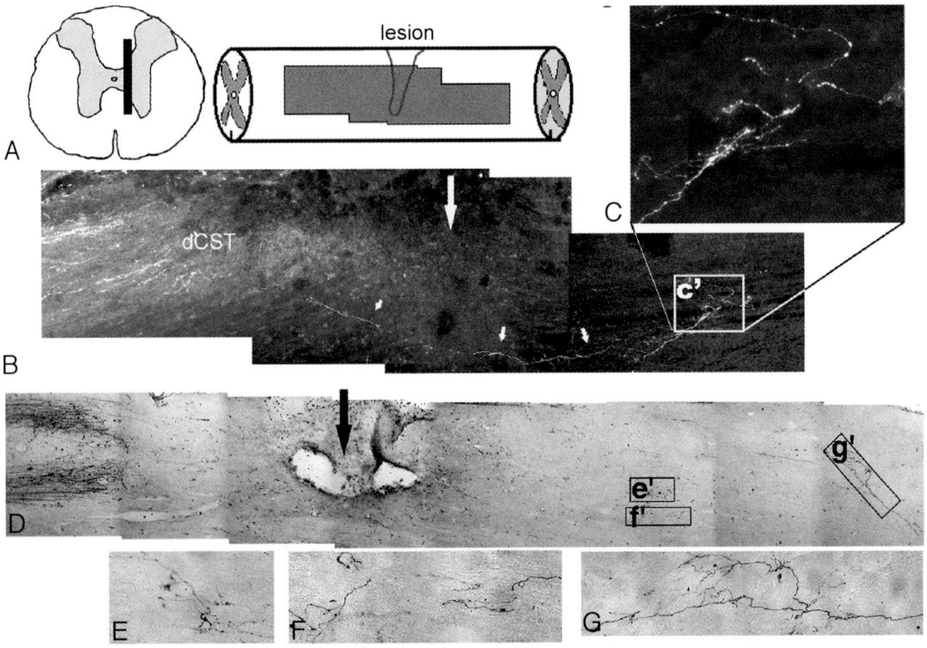

Figure 14.2 Local administration of soluble NgR(310)ecto-Fc stimulates the regeneration of CST fibers into the caudal spinal cord after dorsal hemisection. (**A**) Schematic drawing of a transverse and sagittal spinal cord demonstrates the location of spinal cord dorsal hemisection showed in panel (**B, C**). (**B, C**) Parasagittal section near the transection (large arrow) from a NgR(310)ecto-Fc-treated rat illustrates an individual fluorescently labeled regenerating dCST axon (small arrows) extending into distal spinal cord and forming branched sprouts in gray matter (higher magnification in **C**). Fibers were anterogradely labeled from cerebral cortex biotin-dextran injection sites. Dorsal is up and caudal is right. (**D-G**) Parasagittal section containing the transection site (arrow) from a different NgR(310)ecto-Fc treated rat illustrates the transection of BDA-labeled dCST fibers (dark reaction product) and some branched, sprouting fibers caudal to the hemisection site. Higher magnification of these areas in **E**, **F**, and **G** demonstrates the meandering course of the regenerating CST fibers. Dorsal is up and caudal is right. No caudal branching CST fibers similar to those in these micrographs are observed in control spinal injured rats. For details see publication by Li et al. (2003).

Anti-RGM antibodies improved CST regeneration and SCI recovery in rats (Hata et al., 2006).

B. The Glial Scar and Chondroitin Sulfate Proteoglycans

The astroglial scar is the second major extrinsic inhibitor of axon regeneration. It is a complex structure composed of astrocytes, oligodendrocytes, microglia, precursor cells, and the secreted proteins of all these cell types (Fawcett & Asher, 1999). In the wake of the CNS inflammatory response, these cells migrate into the injury site, and astrocytes weave a tight mesh of projections that mechanically bars axons from penetrating the lesion. Astrocytes, in conjunction with oligodendrocytes and O2A precursors, secrete chondroitin sulfate proteoglycans (CSPGs) into the extracellular matrix; CSPGs currently are thought to be the most important glial scar mediator of axon inhibition (Silver & Miller, 2004; Yiu & He, 2006).

CSPGs are a constituent of extracellular matrix (ECM) in all tissues. Their large, sulfated glycosaminoglycan (GAG) side chains provide structural support to the ECM. Their highly regulated spatiotemporal patterns of expression during CNS development demonstrate their crucial role in the CNS as mediators of cell migration, axon outgrowth, and limiting plasticity at the end of development (Bandtlow & Zimmermann, 2000; Pizzorusso et al., 2002).

Several groups have demonstrated the inhibitory character of CSPGs *in vitro* (Yiu & He, 2006). CSPG expression increases sharply after CNS injury (Asher et al., 2002). In assays of cultured chick dorsal root ganglion neurons on alternating stripes of CSPGs and laminin, DRG axons have a distinct preference for laminin, and turn or stop when they encounter CSPGs (Snow et al., 1990). Digesting the chondroitin sulfate (CS) side chains with the enzyme chondroitinase ABC (ChABC) releases this inhibition (Snow et al., 1990). Dou and Levine (1994) showed that NG2 inhibits neurite outgrowth from rat cerebellar cultures even when mixed with laminin or Ng-CAM/L1, a permissive substrate, and this inhibitory effect cannot be overcome until laminin is 10 times in excess of NG2. Interestingly, they found that digestion with ChABC had no effect on NG2's ability to inhibit outgrowth, and Fiedler and colleagues found that antibody inhibition of NG2 core protein increased neurite outgrowth (Fidler et al., 1999).

These findings make CSPG inactivation a target for opposing glial inhibition of axon regeneration after SCI; indeed, several recent studies have shown that ChABC administration decreases inhibition of outgrowth *in vivo*. Intrathecal ChABC administration after dorsal column crush injury in adult rats resulted in successful digestion of CS side chains, increased dorsal column and corticospinal axon outgrowth, functional recovery, and improved electrophysiological response to cortical stimulation caudal to the injury site (Bradbury et al., 2002). Caggiano and colleagues monitored locomotor function and residual bladder urine in adult rats with varying severities of contusion injury, and found improved locomotor function in the severely and moderately injured groups with ChABC administration as compared to penicillinase (Caggiano et al., 2005). ChABC administration directly to the site of injury in the nigrostriatal tract via cannula increased the number of dopaminergic axons both crossing the axotomy site and arriving at the ipsilateral striatum (Moon et al., 2001).

These successes have inspired recent experiments combining chondroitinase ABC treatment with cellular therapy. When Chau and colleagues transplanted Schwann cell-seeded Matrigel minichannels into the hemisected spinal cord of adult rat, they found increased numbers of axons growing from graft into host with ChABC delivery via osmotic pump as compared to vehicle delivery (Chau et al., 2004). Fouad and colleagues further combined Schwann cell-seeded minichannels, olfactory ensheathing glia, and ChABC administration to drastically improve locomotor function and increase the number of myelinated fibers growing through the bridge (Fouad et al., 2005). Some evidence indicates CSPGs can also have an outgrowth-promoting role *in vivo* (Jones et al., 2003), but these initial results should encourage further investigation of ChABC in conjunction with cellular transplantation as treatment for SCI.

C. cAMP

Cyclic AMP (cAMP) is an intracellular second messenger molecule involved in many signal transduction cascades. Extracellular ligands increase intracellular cAMP levels by stimulating adenylyl cyclase activity via interaction with G-protein coupled receptors; cAMP in turn activates protein kinase A.

For central nervous system cells, cAMP has been found to influence growth cone turning and neuronal response to chemoattractants (Ming et al., 1997; Nishiyama et al., 2003; Song et al., 1998). Dibutyryl cAMP, a membrane-permeant analog of cAMP, reverses inhibition of neurons grown on MAG-expressing CHO cells (Cai et al., 1999), and injection of cAMP into dorsal root ganglia of P18 rats prevents myelin-mediated inhibition of dorsal column axon regeneration in a manner similar to a conditioning peripheral nerve lesion (Neumann et al., 2002; Qui et al., 2002). Subsequently, Cui and colleagues found that intraocular cAMP injection can potentiate neurotrophin-induced regeneration of axotomized retinal ganglion cell axons in an adult rat (Cui et al., 2003).

Several groups have demonstrated the benefits of *in vivo* manipulation of cAMP levels in combination with neurotrophic and cellular approaches. cAMP elevation promotes axonal regeneration in tissue grafts after SCI. Lu and colleagues injected cAMP into lumbar DRGs, administered NT-3 at the site of dorsal column transection, and also administered NT-3 rostral to the lesion (Lu et al., 2004). They found significantly increased numbers of axons growing into the bone marrow stromal cell grafts compared to controls.

Interestingly, they found no significant functional recovery in their treated animals, perhaps due to inadequate regeneration of sensory fibers to the gracile nucleus. Pearse and colleagues (2004) administered varying combinations of Schwann cell transplant, acute rolipram administration (a phosphodiesterase inhibitor that increases cAMP levels), dibutyryl cAMP, and delayed rolipram. They found acute rolipram + Schwann cell therapy + dbcAMP increased serotonergic fiber density at the injury site, increased spared central myelinated axons, and significantly increased numbers of axons caudal to the site of injury. In addition, they found significant functional recovery with multiple assays. Nikulina and colleagues (2004) administered rolipram two weeks after spinal cord hemisection in adult rats, and found increased serotonergic fibers in the embryonic spinal tissue grafts and modest functional improvement compared to controls. Reactive gliosis was also decreased in this study. Benefit in this delayed protocol of rolipram administration is encouraging for clinical use of this compound in spinal cord injury.

D. RhoA

RhoA GTPase is one of several small GTPases involved in intracellular actin dynamics. RhoA activation (through phosphorylation by Rho kinase (ROCK)) leads to contractile actin-myosin activity (Etienne-Manneville & Hall, 2002). In the specialized neuronal cytoskeleton, this correlates with neurite retraction and growth cone collapse (Etienne-Manneville & Hall, 2002). Conforming to the theme of multiple inhibitory pathways converging on a single molecule, RhoA appears to play a key role in transducing the signal downstream not only of NgR1, but of CSPGs and other neurite outgrowth inhibitors as well (Fournier et al., 2003; Schweigreiter et al., 2004) (see Figure 14.1). Thus, RhoA antagonists (or ROCK inhibitors) hold promise as new therapeutic agents for SCI.

In vitro, the RhoA inhibitor C3 transferase and the ROCK inhibitor Y-27632 improve neurite outgrowth on a variety of myelin (Ahmed et al., 2005; Dergham et al., 2002; Fournier et al., 2003; Lehmann et al., 1999; Niederost et al., 2002) and CSPG (Borisoff et al., 2003; Dergham et al., 2002) substrates. *In vivo*, ROCK inhibition with Y-27632 leads to improved CST regeneration in rat cervical and thoracic SCI models (Chan et al., 2005; Fournier et al., 2003). A recombinant form of C3 transferase with improved CNS penetration currently is entering Phase I clinical trial for SCI (BioAxone). However, one report comparing delayed with immediate ROCK inhibitor treatment did not observe significant benefit from administration four weeks following SCI (Nishio et al., 2006). Moreover, RhoA's central role in transducing signals from multiple pathways represents both a strength and a potential weakness—systemic RhoA antagonism could interfere with many essential cytoskeletal pathways in uninjured cells and tissues.

E. Other Steps to Surmounting CNS Barriers

Scientists are enthusiastically attacking the barriers to CNS regeneration on other fronts as well. A multitude of neurotrophic factors has been tested using varying routes of administration, with varying levels of improved regeneration rates (Bregman et al., 2002). Additionally, oncomodulin and inosine appear to enhance the intrinsic growth potential of CNS axons (Benowitz et al., 1999; Yin et al., 2006). Some of the more promising agents are likely to be used in combination with other modes of treatment. Once CNS axons are successfully stimulated to regenerate, they will still need to find their way back to lost target areas. To successfully facilitate this next step in SCI recovery, the developmental pattern of guidance factor expression must be recapitulated (Harel & Strittmatter, 2006).

III. Cellular Therapies for SCI

Although regeneration of spinal cord axons remains a holy grail, the presence in many patients with nonpenetrating SCI of a population of surviving axons that do not conduct due to demyelination (Waxman, 1992) offers an alternative strategy for inducing recovery of function. One approach to this goal capitalizes on recent progress in the transplantation of cells into the injured CNS. Cellular transplantation of appropriate cells into experimental models of spinal cord injury can promote axonal regeneration, provide neuroprotective effects by secretion of neurotrophins, and remyelinate axons.

One cell of particular interest as a cell therapy candidate to both encourage axonal remyelination and regeneration is a specialized glial cell, the olfactory ensheathing cell (OEC). Adult olfactory receptor neurons continually undergo turnover from an endogenous progenitor pool, and their nascent axons grow through the olfactory nerves and cross the PNS-CNS interface, where they form new synaptic connections in the olfactory bulb (Graziadei et al., 1978). A specialized glial cell, the olfactory ensheathing cell (OEC), grossly associates with olfactory receptor neurons from their peripheral origin to their central projection in the outer nerve layer of the olfactory bulb (Doucette, 1991). This putative support role of OECs in axonal growth within the adult CNS has spawned extensive research to study the potential of OEC transplants encouraging axonal regeneration and functional recovery in SCI models (Imaizumi et al., 2000a, 2000b; Li et al., 1997, 1998; Ramon-Cueto et al., 1998, 2000).

Transplantation of OECs after SCI is associated with functional improvement even when transplantation is delayed by weeks (Keyvan-Fouladi et al., 2003; Lu et al., 2002). Although the precise mechanisms of the functional recovery after OEC transplantation are not fully understood, several mechanisms including elongative axonal regeneration, axonal sparing, sprouting, and plasticity associated with novel polysynaptic pathways, recruitment of endogenous SCs and remyelination have been proposed (Bareyre et al., 2004; Raisman, 2001; Sasaki et al., 2004). In animals with SCI and OEC transplants, myelinated axons spanning the lesion site display a characteristic peripheral pattern of myelination similar to that of Schwann cell (SC) myelination (Franklin et al., 1996, 2000a; Imaizumi et al. 1998; Li et al., 1997, 1998; Sasaki et al. 2004).

Remyelination by transplantation may be of particular importance because contusive spinal cord injury often results in loss of function from demyelination induced by spinal cord trauma. OECs can form myelin when transplanted into demyelinated spinal cord (Franklin et al., 1996; Imaizumi et al., 1998; Sasaki et al., 2004). Moreover, human OECs can form myelin in the immunosuppressed rat (Barnett et al., 2000; Kato et al., 2000), and OECs transplanted into the nonhuman primate can remyelinate spinal cord axons (Radtke et al., 2003).

We recently transplanted OECs derived from an olfactory bulb of adult GFP expressing rats into a dorsal funiculus transection lesion of rat spinal cord (Sasaki et al. 2006a). We showed that the OECs survive, integrate at the injury site, and form myelin. Figure 14.3 shows the integration of GFP-expressing OECs into a dorsal transection of the spinal cord. In plastic coronal sections through the transection site it can be seen that regenerated axons have grown through cellular bridges, and that the regenerated axons have remyelinated (see Figure 14.3B, C). Functional improvement was observed with open field locomotor testing (see Figure 14.3D). When transplanted into a chemically demyelinated lesion of the dorsal funiculus of the spinal cord, the OECs again integrate into the lesion and form myelin (see Figure 14.4). The inset in Figure 14.4A shows the close association of the transplanted GFP-expressing OECs with peripheral (P0) myelin. The axons remyelinated by the transplanted OECs form new nodes of Ranvier, which are flanked by Caspr stained paranodal regions, and the appropriate sodium channel (Nav1.6) is present at these new nodes (see Figure 14.4B–E; Sasaki et al., 2006b). Proper ion channel organization on remyelinated axons is essential for appropriate impulse conduction (Kocsis, 2001).

Neuroprotection is currently thought to play a role in functional improvement by cellular transplantation approaches.

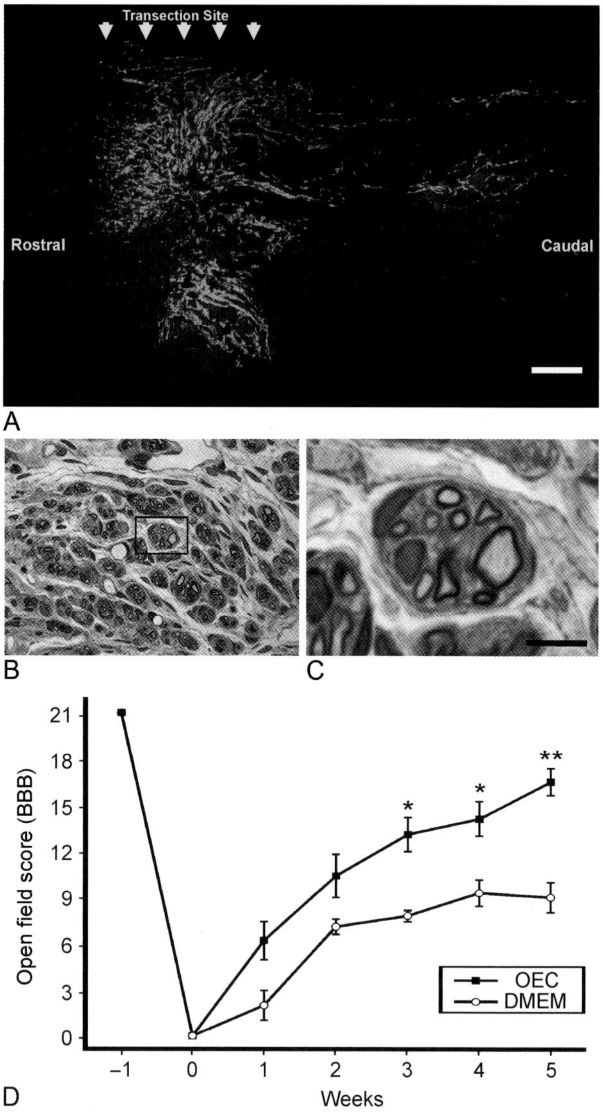

Figure 14.3 Transplanted OECs integrate into the spinal cord transection site and are associated with regenerated axons. **A**. Sagittal section through the dorsal transection site showing the transplanted OECs (green) survived and bridged the lesion zone. **B**. Plastic coronal sections through the lesion zone 4 weeks after OEC transplantation show groups of myelinated axons. **C**. Higher power micrograph from boxed region in **B**. Note the clustering of myelinated axons surrounded by a cellular element. **D**. Open field locomotor scores for OEC transplant (n = 20) and sham injection (n = 6) groups tested one week before and for 5 weeks post-transplantation. Asterisks refer to comparison between intact control animals and the experimental groups, and the pound signs refer to the significance between DMEM and OEC groups. Significance levels: $p < 0.05$ (*), $p < 0.01$(**), $p < 0.005$ (***), $p < 0.005$ (###). All values are given as means ± SEM. Scale bars: **A** = 1 mm; **B** = 30 μm; **C** = 6 μm.

Indeed, intravenous injection of bone marrow mesenchymal stem cells can be neuroprotective in spinal cord injury models (Chopp et al., 2000) and in cerebral ischemia (Lu et al., 2006). When OECs are injected into a spinal cord injury site, the apoptotic cell death of M1 cortical neurons is reduced and

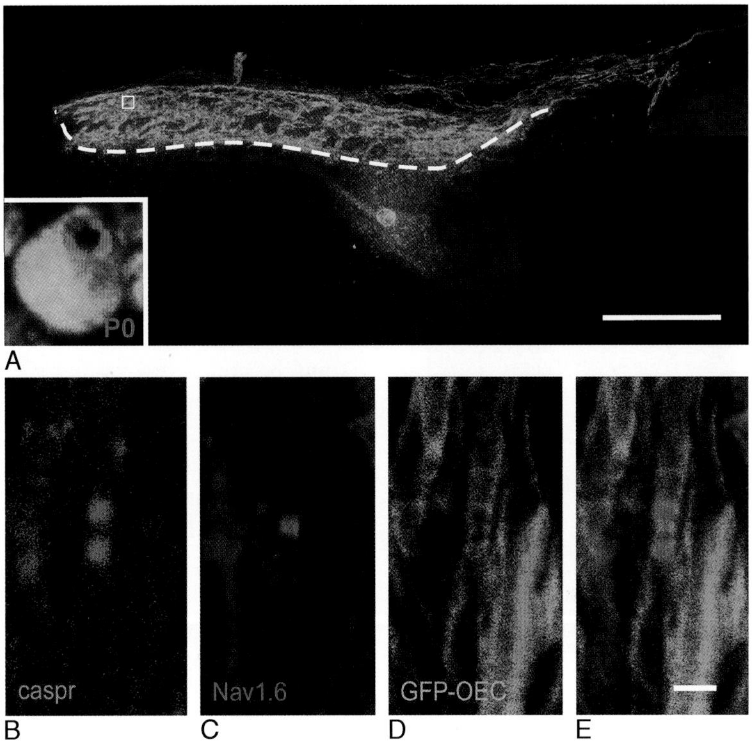

Figure 14.4 Nodes of Ranvier of axons remyelinated by transplanted OECs display appropriate sodium channel organization. **A**. Sagittal sections showing OECs distributed within a chemically demyelinated dorsal funiculus. Inset is a coronal section of an axon with peripheral-like myelin (P0) surrounded by an engrafted OEC (green). At 8 weeks post-transplantation Caspr-defined nodes flanked by GFP-OEC internodes exhibit $Na_v1.6$ immunostaining nodes. Scale bars: **A**= 1 mm; **B–E** = 10 μm. From Sasaki et al. (2006b).

cortical neuronal density is increased (Sasaki et al., 2006a; Figure 14.5). Enhanced levels of BDNF were observed in the OEC transplanted lesion site suggesting a BDNF-mediated neuroprotective effect on descending motor control systems. Taken together these results suggest that axonal regeneration, remyelination, and neuroprotection by trophic factor release of implanted cells may contribute to functional recovery. Enhanced neovascularization by release of vascular trophic factors could also contribute to improve functional outcome by transplantation of OECs or other cell types.

A. Ongoing Clinical Studies with OEC Transplantation into Spinal Cord Injury Patients

Several groups are conducting or planning clinical studies transplanting OECs into spinal cord injury patients (see discussion in Ibrahim et al., 2006; Senior 2002; Watts et al., 2005). Feron and colleagues (2005) have conducted a phase I safety study using suspensions of OECs cultured from biopsied tissue from the patient's own olfactory muscosa, thus reducing immune rejection. They report no adverse effects at 12 months post-transplantation, but no neurological improvement. Carlos Lima and colleagues (Egaz Moniz Hospital, Lisbon, Portugal) have reported a procedure where the cavity of the spinal cord injury site is filled with acutely prepared minced olfactory mucosa tissue, which includes a number of cell types in addition to OECs (Lima et al., 2006). They report that the olfactory mucosa autograft transplantation was safe and potentially beneficial, but efficacy was not clearly established.

Hongyun Huang and colleagues (Chaoyang Hospital, Beijing, China) report that several hundred patients have received transplants of cultures from human embryonic olfactory bulbs obtained from 14 to 16 fetuses (Curt & Dietz, 2005; Dobkin et al., 2006; Huang et al., 2003). Some functional improvement was reported beginning as early as a day after transplantation. Surely such an early effect is not the result of axonal regeneration or remyelination. It is important to note that the Lima and Huang studies have not carried out control studies nor have these observations been independently confirmed (Dobkin et al., 2006; Ibrahim et al., 2006).

Assessment of efficacy of putative therapeutic interventions in SCI including cell therapy approaches is difficult given that some degree of "spontaneous" functional

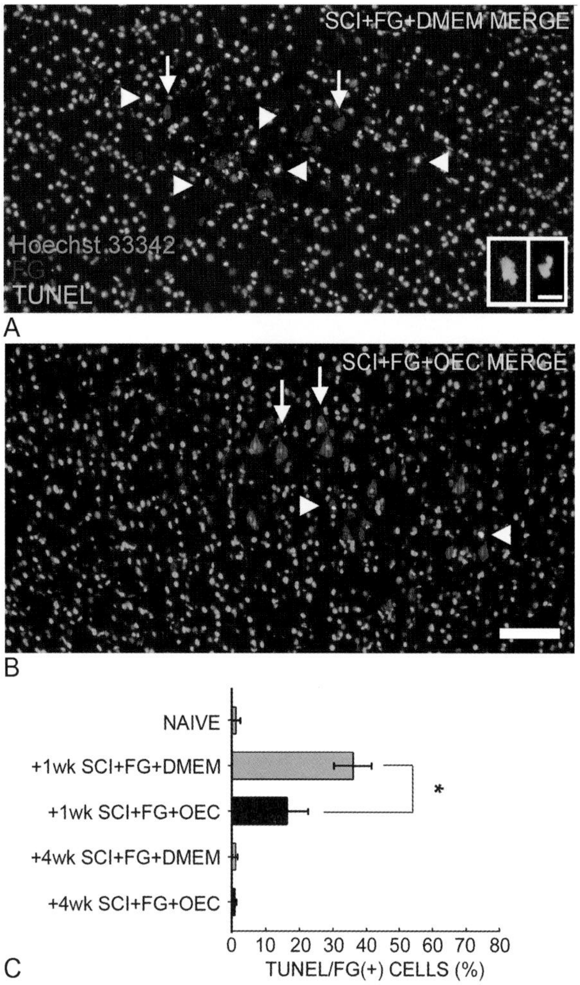

Figure 14.5 OECs transplanted into a dorsal hemisected spinal cord reduce apoptosis of corticospinal tract neurons. **A**. Hoechst 33342, Fluorogold, and TUNEL triple labeling of corticospinal neurons 1 week after injury. Hoechst staining of non-TUNEL-positive (arrows with tails) and TUNEL-positive (arrowheads) neurons with corresponding FG-backfilling are shown. **B**. In SCI+FG+OEC animals fewer TUNEL-positive FG-backfilled neurons are observed compared to SCI+FG+DMEM (A). Insets in (A) show two TUNEL-positive neurons exhibiting nuclear compartmentalization and formation of nucleosomes, hallmarks of apoptosis. Quantitation of neurons that are both TUNEL- and FG-positive (C) reveals that OEC transplantation significantly ($^*p < 0.05$) reduces apoptotic cell death at 1 week. No evidence of death was observed at any other timepoint. Scale bar **A, B** = 125 μm, **A** inset 20 μm. From Sasaki et al. (2006a).

improvement occurs in most SCI patients. Moreover, surgical intervention necessary to transplant cells in and of itself can lead to modest functional improvement. Issues related to methods of assessment of SCI patients in clinical studies currently are being discussed with an emphasis on assessing the degree of functional recovery of an individual patient (Curt et al., 2004). Clearly, the complexity of SCI and the difficulty of accurately assessing functional recovery will be a challenge for all interventional clinical studies for SCI.

Although reconstruction of appropriate spinal circuits by cell-based therapies is the ultimate long-term goal of cell transplantation research, laboratory work to date suggests that more immediate therapeutic benefits will come from neuroprotective effects and remyelination. Moreover, the most extensive functional recovery in animal models of SCI with cell transplantation is for treatment of acute and subacute SCI. Early intervention may reduce scar formation and secondary cell death by release of appropriate trophic factors by engrafted cells. Moreover, angiogenic factors released by transplanted cells could result in neovascularization, which would be critical for tissue preservation. However, if long-term SCI patients have preservation of long tracts in the spinal cord that were demyelinated, remyelination of these tracts by cell transplantation could lead to some degree of functional improvement. An important challenge that remains for cell-based therapies in SCI is to determine the optimal cell type, method of delivery, and timing of cellular intervention. Moreover the prospect of using a combinatorial approach such as cell transplantation in combination with Nogo inhibitory molecules will be important to explore.

IV. Molecular Adaptations after SCI

Another approach to functional restoration after SCI derives from our increasing understanding of mechanisms underlying recovery of action potential conduction along chronically demyelinated axons within the spinal cord. Important lessons have been learned in this regard from studies on the molecular neurobiology of multiple sclerosis (MS), where recovery of conduction along chronically demyelinated CNS axons can support clinical remissions (Waxman et al., 2004). An important first step in this direction was taken when it was recognized that in normal myelinated fibers, the voltage-gated sodium channels that support action potential electrogenesis are clustered in high density at the nodes of Ranvier (see Waxman, Chapter 22, this volume), and are present in much lower densities, too low to support secure, high-frequency conduction, in the internodal axon membrane under the myelin sheath (Ritchie & Rogart, 1997; Waxman, 1997). This observation implied that, following damage to the myelin capacitative shield, several mechanisms contribute to conduction failure: First, decreased current density due to capacitative shunting reduces safety factor, thereby reducing the security of conduction. Waxman and Brill (1978) showed that the development of closely spaced nodes of Ranvier separated by short myelin segments, as is observed at the edges of some demyelinated lesions in MS, can facilitate this impedance matching and thus promote invasion of the action potential into the demyelinated part of the axon, and Waxman and Wood (1984) subsequently showed that the development of transitional heminodes with high sodium channel densities, at the proximal end of the demyelinated region of the axon, could also facilitate action potential invasion. A second factor, however, can impede action potential conduction, even if there is successful invasion of the demyelinated zone. This is low density of sodium channels in the acutely demyelinated axon membrane.

Clinicopathological observations in MS have indicated that, notwithstanding these challenges, clinical remission (i.e., recovery of clinical function) can occur in patients with MS in whom all axons in the tract in question are demyelinated (Ulrich & Groebke-Lorenz, 1983). This clinical recovery has a basis in conduction along demyelinated axons, supported by increased expression of sodium channels within the demyelinated (previously sodium channel-poor) axon membrane. Early electrophysiological studies (Bostock & Sears, 1987) showed, in fact, that several weeks following loss of the myelin, some axons regain the capability to conduct in a continuous manner, suggesting the presence of sodium channels along demyelinated axon regions. Cytochemical (Foster et al., 1980) studies provided evidence for acquisition of relatively high densities of sodium channels along chronically demyelinated axons. Early immunocytochemical studies using pan-specific sodium channel antibodies (England et al., 1991; Novakovic et al., 1998) confirmed this finding, but could not provide information about the specific sodium channel isoforms that were deployed along demyelinated axons.

Several recent studies using immunocytochemical analysis with subtype-specific antibodies now have identified the sodium channel isoforms along demyelinated CNS axons. Craner et al. have demonstrated that Nav1.2 and Nav1.6 sodium channels are present along extended regions of demyelinated axons, including spinal cord axons, in experimental autoimmune encephalomyelitis (EAE; Craner et al., 2003, 2004a) and in MS (Craner et al., 2004b). Interestingly, Nav1.6, which is known to produce a persistent current of the type that could trigger injurious reverse sodium-calcium exchange (Rush et al., 2005), is colocalized with the sodium-calcium exchanger in axons that show signs of degeneration, whereas Nav1.2 is present in demyelinated axons that do not appear to be degenerating (Craner et al., 2004a, 2004b). The deployment of Nav1.2 channels appears to be the result of transcriptional up-regulation, since Nav1.2 mRNA levels increase in the neuronal cell bodies giving rise to axons that express it.

Studies are underway in an effort to more fully understand the up-regulation of Nav1.2 channels along chronically demyelinated axons. Although details of the control of Nav1.2 expression are not yet known, a number of neurotrophic factors have been shown to modulate sodium channel expression. By dissecting the control of Nav1.2 up-regulation, it may be possible to develop approaches that will induce it, thus supporting restoration of conduction along chronically demyelinated axons in SCI.

V. Conclusions

The increasing optimism for successfully treating SCI has been based on fundamental progress in basic neuroscience. We understand more today about the molecular nature of nervous system development, pathophysiological stages of SCI, neuronal-glial interactions, endogenous regeneration mechanisms, and recovery of conduction along demyelinated spinal cord axons. This fundamental knowledge serves as a fuel that will hopefully generate rational, safe, more effective treatments for SCI. One can envision a future in which previously disabled SCI patients no longer need to dwell in the past.

Acknowledgments

Research in the authors' laboratories has been supported by many agencies and organizations including the NIH, Dept of Veterans Affairs, PVA, Christopher Reeve Paralysis Foundation, United Spinal Organization, and NMSS.

References

Ahmed, Z. et al. (2005). Disinhibition of neurotrophin-induced dorsal root ganglion cell neurite outgrowth on CNS myelin by siRNA-mediated knockdown of NgR, p75NTR and Rho-A. *Mol Cell Neurosci* **28**, 509–523.

Amar, A. P. and Levy, M. L. (1999). Pathogenesis and pharmacological strategies for mitigating secondary damage in acute spinal cord injury. *Neurosurgery* **44**, 1027–1039; discussion 1039–1040.

Asher, R. A. et al. (2002). Versican is upregulated in CNS injury and is a product of oligodendrocyte lineage cells. *J Neurosci* **22**, 2225–2236.

Bandtlow, C. E. and Zimmermann, D. R. (2000). Proteoglycans in the developing brain: New conceptual insights for old proteins. *Physiol Rev* **80**, 1267–1290.

Banik, N. L., Matzelle, D., Terry, E., and Hogan, E. L. (1997). A new mechanism of methylprednisolone and other corticosteroids action demonstrated in vitro: Inhibition of a proteinase (calpain) prevents myelin and cytoskeletal protein degradation. *Brain Res* **748**, 205–210.

Bareyre, F. M., Kerschensteiner, M., Raineteau, O., Mettenleiter, T. C., Weinmann, O., and Schwab, M. E. (2004). The injured spinal cord spontaneously forms a new intraspinal circuit in adult rats. *Nat Neurosci* **10**, 269–277.

Bartsch, U. et al. (1995). Lack of evidence that myelin-associated glycoprotein is a major inhibitor of axonal regeneration in the CNS. *Neuron* **15**, 1375–1381.

Benowitz, L. I., Goldberg, D. E., Madsen, J. R., Soni, D., and Irwin, N. (1999). Inosine stimulates extensive axon collateral growth in the rat corticospinal tract after injury. *Proc Natl Acad Sci U S A* **96**, 13486–13490.

Benson, M. D. et al. (2005). Ephrin-B3 is a myelin-based inhibitor of neurite outgrowth. *Proc Natl Acad Sci U S A* **102**, 10694–10699.

Borisoff, J. F. et al. (2003). Suppression of Rho-kinase activity promotes axonal growth on inhibitory CNS substrates. *Mol Cell Neurosci* **22**, 405–416.

Bostock, H. and Sears, T. A. (1978). The internodal axon membrane: Electrical excitability and continuous conduction in segmental demyelination. *J Physiol* **280**, 273–301.

Bracken, M. B. et al. (1992). Methylprednisolone or naloxone treatment after acute spinal cord injury: 1-year follow-up data. Results of the second National Acute Spinal Cord Injury Study. *J Neurosurg* **76**, 23–31.

Bracken, M. B. et al. (1990). A randomized, controlled trial of methylprednisolone or naloxone in the treatment of acute spinal-cord injury. Results of the Second National Acute Spinal Cord Injury Study. *N Engl J Med* **322**, 1405–1411.

Bradbury, E. J. et al. (2002). Chondroitinase ABC promotes functional recovery after spinal cord injury. *Nature* **416**, 636–640.

Bregman, B. S. et al. (2002). Transplants and neurotrophic factors increase regeneration and recovery of function after spinal cord injury. *Prog Brain Res* **137**, 257–273.

Bregman, B. S. et al. (1995). Recovery from spinal cord injury mediated by antibodies to neurite growth inhibitors. *Nature* **378**, 498–501.

Burns, A. S. and Ditunno, J. F. (2001). Establishing prognosis and maximizing functional outcomes after spinal cord injury: A review of current and future directions in rehabilitation management. *Spine* **26**, S137–145.

Caggiano, A. O., Zimber, M. P., Ganguly, A., Blight, A. R., and Gruskin, E. A. (2005). Chondroitinase ABCI improves locomotion and bladder function following contusion injury of the rat spinal cord. *J Neurotrauma* **22**, 226–239.

Cai, D., Shen, Y., De Bellard, M., Tang, S., and Filbin, M. T. (1999). Prior exposure to neurotrophins blocks inhibition of axonal regeneration by MAG and myelin via a cAMP-dependent mechanism. *Neuron* **22**, 89–101.

Caroni, P. and Schwab, M. E. (1988). Two membrane protein fractions from rat central myelin with inhibitory properties for neurite growth and fibroblast spreading. *J Cell Biol* **106**, 1281–1288.

Caroni, P. and Schwab, M. E. (1988). Antibody against myelin-associated inhibitor of neurite growth neutralizes nonpermissive substrate properties of CNS white matter. *Neuron* **1**, 85–96.

Chan, C. C. et al. (2005). Dose-dependent beneficial and detrimental effects of ROCK inhibitor Y27632 on axonal sprouting and functional recovery after rat spinal cord injury. *Exp Neurol* **196**, 352–364.

Chau, C. H. et al. (2004). Chondroitinase ABC enhances axonal regrowth through Schwann cell-seeded guidance channels after spinal cord injury. *Faseb J* **18**, 194–196.

Chen, M. S. et al. (2000). Nogo-A is a myelin-associated neurite outgrowth inhibitor and an antigen for monoclonal antibody IN-1. *Nature* **403**, 434–439.

Chopp, M., Zhang, X. H., Li, Y., Wang, L., Chen, J., Lu, D., Lu, M., Rosenblum, M. (2000). Spinal cord injury in rat: Treatment with bone marrow stromal cell transplantation. *Neuroreport* **11**, 3001–3005.

Craner, M. J., Hains, B. C., Lo, A. C., Black, J. A., and Waxman, S. G. (2004a). Co-localization of sodium channel Nav1.6 and the sodium-calcium exchanger at sites of axonal injury in the spinal cord in EAE. *Brain* **127**, 294–303.

Craner, M. J., Lo, A. C., Black, J. A., and Waxman, S. G. (2003). Abnormal sodium channel distribution in optic nerve axons in a model of inflammatory demyelination. *Brain* **126**, 1552–1561.

Craner, M. J., Newcombe, J., Black, J. A., Hartle, C., Cuzner, M. L., and Waxman, S. G. (2004b). Molecular changes in neurons in MS: Altered axonal expression of $Na_v1.2$ and $Na_v1.6$ sodium channels and Na^+/Ca^{2+} exchanger. *Proc Natl Acad Sci* **101**, 8168–8173.

Cui, Q., Yip, H. K., Zhao, R. C., So, K. F., and Harvey, A. R. (2003). Intraocular elevation of cyclic AMP potentiates ciliary neurotrophic factor-induced regeneration of adult rat retinal ganglion cell axons. *Mol Cell Neurosci* **22**, 49–61.

Curt A. and Dietz V. (2005). Controversial treatments for spinal cord injuries. *Lancet* **365**, 841.

David, S. and Aguayo, A. J. (1981). Axonal elongation into peripheral nervous system "bridges" after central nervous system injury in adult rats. *Science* **214**, 931–933.

Dergham, P. et al. (2002). Rho signaling pathway targeted to promote spinal cord repair. *J Neurosci* **22**, 6570–6577.

Dimou, L. et al. (2006). Nogo-A-deficient mice reveal strain-dependent differences in axonal regeneration. *J Neurosci* **26**, 5591–5603.

Dobkin, B. H., Curt, A., and Guest, J. (2006). Cellular transplants in China: Observational study from the largest human experiment in chronic spinal cord injury. *Neurorehabil Neural Repair* **20**, 5–13.

Domeniconi, M. et al. (2002). Myelin-associated glycoprotein interacts with the Nogo66 receptor to inhibit neurite outgrowth. *Neuron* **35**, 283–290.

Dou, C. L. and Levine, J. M. (1994). Inhibition of neurite growth by the NG2 chondroitin sulfate proteoglycan. *J Neurosci* **14**, 7616–7628.

Doucette, R. (1991). PNS-CNS transitional zone of the first cranial nerve. *J Comp Neurol* **312**, 451–466.

England, J. D. et al. (1991). Increased numbers of sodium channels form along demyelinated axons. *Brain Res* **548**, 334–337.

Etienne-Manneville, S. and Hall, A. (2002). Rho GTPases in cell biology. *Nature* **420**, 629–635.

Fawcett, J. W. and Asher, R. A. (1999). The glial scar and central nervous system repair. *Brain Res Bull* **49**, 377–391.

Fehlings, M. G. and Baptiste, D. C. (2005). Current status of clinical trials for acute spinal cord injury. *Injury* **36 Suppl 2**, B113–122.

Fehlings, M. G. and Bracken, M. B. (2001). Summary statement: The Sygen(GM-1 ganglioside) clinical trial in acute spinal cord injury. *Spine* **26**, S99–100.

Feron, F., Perry, C., Cochrane, J., Licina, P., Nowitzke, A., Urquhart, S. et al. (2005). Autologous olfactory ensheathing cell transplantation in human spinal cord injury *Brain* **128**, 2951–2960.

Fidler, P. S. et al. (1999). Comparing astrocytic cell lines that are inhibitory or permissive for axon growth: The major axon-inhibitory proteoglycan is NG2. *J Neurosci* **19**, 8778–8788.

Foster, R. E., Whalen, C. C., and Waxman, S. G. (1980). Reorganization of the axonal membrane of demyelinated nerve fibers: Morphological evidence. *Science* **210**, 661–663.

Fouad, K. et al. (2005). Combining Schwann cell bridges and olfactory-ensheathing glia grafts with chondroitinase promotes locomotor recovery after complete transection of the spinal cord. *J Neurosci* **25**, 1169–1178.

Fournier, A. E., GrandPre, T., and Strittmatter, S. M. (2001). Identification of a receptor mediating Nogo-66 inhibition of axonal regeneration. *Nature* **409**, 341–346.

Fournier, A. E., Takizawa, B. T., and Strittmatter, S. M. (2003). Rho kinase inhibition enhances axonal regeneration in the injured CNS. *J Neurosci* **23**, 1416–1423.

Freund, P. et al. (2006). Nogo-A-specific antibody treatment enhances sprouting and functional recovery after cervical lesion in adult primates. *Nat Med* **12**, 790–792.

Geisler, F. H., Coleman, W. P., Grieco, G., and Poonian, D. (2001). The Sygen multicenter acute spinal cord injury study. *Spine* **26**, S87–98.

GrandPre, T., Li, S., and Strittmatter, S. M. (2002). Nogo-66 receptor antagonist peptide promotes axonal regeneration. *Nature* **417**, 547–551.

GrandPre, T., Nakamura, F., Vartanian, T., and Strittmatter, S. M. (2000). Identification of the Nogo inhibitor of axon regeneration as a Reticulon protein. *Nature* **403**, 439–444.

Graziadei, P. P., Levine, R. R., and Graziadei, G. A. (1978). Regeneration of olfactory axons and synapse formation in the forebrain after bulbectomy in neonatal mice. *Proc Natl Acad Sci U S A* **75**, 5230–5234.

Hains, B. C., Black, J. A., and Waxman, S. G. (2003). Primary cortical motor neurons undergo apoptosis following axotomizing spinal cord injury. *J Comp Neurol* **462**, 328–341.

Hall, E. D. (1992). The neuroprotective pharmacology of methylprednisolone. *J Neurosurg* **76**, 13–22.

Harel, N. Y. and Strittmatter, S. M. (2006). Can regenerating axons recapitulate developmental guidance during recovery from spinal cord injury? *Nat Rev Neurosci* **7**, 603–616.

Hata, K. et al. (2006). RGMa inhibition promotes axonal growth and recovery after spinal cord injury. *J Cell Biol* **173**, 47–58.

Hauben, E. et al. (2001). Vaccination with a Nogo-A-derived peptide after incomplete spinal-cord injury promotes recovery via a T-cell-mediated neuroprotective response: Comparison with other myelin antigens. *Proc Natl Acad Sci U S A* **98**, 15173–15178.

Huang, H., Chen, L., Wang, H., Xiu, B., Wang, R., Zhang, J. et al. (2003). Influence of patients' age on functional recovery after transplantation of olfactory ensheathing cells into injured spinal cord injury. *Chin Med J (Engl)* **116**, 1488–1491.

Hurlbert, R. J. (2001). The role of steroids in acute spinal cord injury: An evidence-based analysis. *Spine* **26**, S39–46.

Ibrahim, A., Li, Y., Li, D., Raisman, G., and El Masry, W. S. (2006). Olfactory ensheathing cells: Ripples of an incoming tide? *Lancet Neurol* **5(5)**, 453–457.

Imaizumi, T., Kocsis, J. D., and Waxman, S. G. (1997). Anoxic injury in the rat spinal cord: Pharmacological evidence for multiple steps in Ca^{2+}-dependent injury of the dorsal columns. *J Neurotrauma* **14**, 299–312.

Imaizumi, T., Lankford, K. L., Waxman, S. G., Greer, C. A., and Kocsis, J. D. (1998). Transplanted olfactory ensheathing cells remyelinate and enhance axonal conduction in the demyelinated dorsal columns of the rat spinal cord. *J Neurosci* **18**, 6176–6185.

Imaizumi, T., Lankford, K. L., and Kocsis, J. D. (2000a). Transplantation of olfactory ensheathing cells or Schwann cells restores rapid and secure conduction across the transected spinal cord. *Brain Res* **854**, 70–78.

Imaizumi, T., Lankford, K. L., Burton, W. V., Fodor, W. L., and Kocsis, J. D. (2000b). Xenotransplantation of transgenic pig olfactory ensheathing cells promotes axonal regeneration in rat spinal cord. *Nat Biotechnol* **18**, 949–953.

Jones, L. L., Margolis, R. U., and Tuszynski, M. H. (2003). The chondroitin sulfate proteoglycans neurocan, brevican, phosphacan, and versican are differentially regulated following spinal cord injury. *Exp Neurol* **182**, 399–411.

Keyvan-Fouladi, N., Raisman, G., Li, Y. (2003). Functional repair of the corticospinal tract by delayed transplantation of olfactory ensheathing cells in adult rats. *J Neurosci* **23**, 9428–9434.

Kim, J. E., Li, S., GrandPre, T., Qiu, D., and Strittmatter, S. M. (2003). Axon regeneration in young adult mice lacking Nogo-A/B. *Neuron* **38**, 187–199.

Kim, J. E., Liu, B. P., Park, J. H., and Strittmatter, S. M. (2004). Nogo-66 receptor prevents raphe spinal and rubrospinal axon regeneration and limits functional recovery from spinal cord injury. *Neuron* **44**, 439–451.

Knoller, N. et al. (2005). Clinical experience using incubated autologous macrophages as a treatment for complete spinal cord injury: Phase I study results. *J Neurosurg Spine* **3**, 173–181.

Kocsis, J. D. (2001). Axonal conduction and myelin. In Walz, W., Ed., *The neuronal environment: Brain homeostasis in health and disease*, 211–231. Humana Press, Totowa, New Jersey.

Kullander, K. et al. (2001). Ephrin-B3 is the midline barrier that prevents corticospinal tract axons from recrossing, allowing for unilateral motor control. *Genes Dev* **15**, 877–888.

Lee, J. K., Kim, J. E., Sivula, M., and Strittmatter, S. M. (2004). Nogo receptor antagonism promotes stroke recovery by enhancing axonal plasticity. *J Neurosci* **24**, 6209–6217.

Lehmann, M. et al. (1999). Inactivation of Rho signaling pathway promotes CNS axon regeneration. *J Neurosci* **19**, 7537–7547.

Li, S. et al. (2004). Blockade of Nogo-66, myelin-associated glycoprotein, and oligodendrocyte myelin glycoprotein by soluble Nogo-66 receptor promotes axonal sprouting and recovery after spinal injury. *J Neurosci* **24**, 10511–10520.

Li, S. and Strittmatter, S. M. (2003). Delayed systemic Nogo-66 receptor antagonist promotes recovery from spinal cord injury. *J Neurosci* **23**, 4219–4227.

Li, Y., Field, P. M., and Raisman, G. (1997). Repair of adult rat corticospinal tract by transplants of olfactory ensheathing cells. *Science* **277**, 2000–2002.

Li, Y., Field, P. M., and Raisman, G. (1998). Regeneration of adult rat corticospinal axons induced by transplanted olfactory ensheathing cells. *J Neurosci* **18**, 10514–10524.

Liu, B. P., Fournier, A., GrandPre, T., and Strittmatter, S. M. (2002). Myelin-associated glycoprotein as a functional ligand for the Nogo-66 receptor. *Science* **297**, 1190–1193.

Liu, D. and McAdoo, D. J. (1993). Methylprednisolone reduces excitatory amino acid release following experimental spinal cord injury. *Brain Res* **609**, 293–297.

Liu, H., Honmou, O., Harada, K., Nakamura, K., Houkin, K., Hamada, H., and Kocsis, J.D. (2006). Neuroprotection by PlGF gene-modified human mesenchymal stem cells after cerebral ischaemia. *Brain* 2006 Aug 10; Epub ahead of print.

Lu, J., Feron, F., Mackay-Sim, A., and Waite, P. M. (2002). Olfactory ensheathing cells promote locomotor recovery after delayed transplantation into transected spinal cord. *Brain* **125**, 14–21.

Lu, P., Yang, H., Jones, L. L., Filbin, M. T., and Tuszynski, M. H. (2004). Combinatorial therapy with neurotrophins and cAMP promotes axonal regeneration beyond sites of spinal cord injury. *J Neurosci* **24**, 6402–6409.

McDonald, J. W. and Sadowsky, C. (2002). Spinal-cord injury. *Lancet* **359**, 417–425.

McKerracher, L. et al. (1994). Identification of myelin-associated glycoprotein as a major myelin-derived inhibitor of neurite growth. *Neuron* **13**, 805–811.

Merkler, D. et al. (2001). Locomotor recovery in spinal cord-injured rats treated with an antibody neutralizing the myelin-associated neurite growth inhibitor Nogo-A. *J Neurosci* **21**, 3665–3673.

Merkler, D. et al. (2003). Rapid induction of autoantibodies against Nogo-A and MOG in the absence of an encephalitogenic T cell response: Implication for immunotherapeutic approaches in neurological diseases. *Faseb J* **17**, 2275–2277.

Mi, S. et al. (2004). LINGO-1 is a component of the Nogo-66 receptor/p75 signaling complex. *Nat Neurosci* **7**, 221–228.

Ming, G. L. et al. (1997). cAMP-dependent growth cone guidance by netrin-1. *Neuron* **19**, 1225–1235.

Moll, C. et al. (1991). Increase of sodium channels in demyelinated lesions of multiple sclerosis. *Brain Res* **556**, 311–316.

Moon, L. D., Asher, R. A., Rhodes, K. E., and Fawcett, J. W. (2001). Regeneration of CNS axons back to their target following treatment of adult rat brain with chondroitinase ABC. *Nat Neurosci* **4**, 465–466.

National Spinal Cord Injury Statistical Center. (2006). Spinal Cord Injury. Facts and figures at a glance. University of Alabama at Birmingham, PDF accessed online on 1 August 2006 (http://www.spinalcord.uab.edu/show.asp?durki=21446&site=1210&return=21816).

Neumann, S., Bradke, F., Tessier-Lavigne, M., and Basbaum, A. I. (2002). Regeneration of sensory axons within the injured spinal cord induced by intraganglionic cAMP elevation. *Neuron* **34**, 885–893.

Niederost, B., Oertle, T., Fritsche, J., McKinney, R. A., and Bandtlow, C. E. (2002). Nogo-A and myelin-associated glycoprotein mediate neurite growth inhibition by antagonistic regulation of RhoA and Rac1. *J Neurosci* **22**, 10368–10376.

Nikulina, E., Tidwell, J. L., Dai, H. N., Bregman, B. S., and Filbin, M. T. (2004). The phosphodiesterase inhibitor rolipram delivered after a spinal cord lesion promotes axonal regeneration and functional recovery. *Proc Natl Acad Sci U S A* **101**, 8786–8790.

Nishio, Y. et al. (2006). Delayed treatment with Rho-kinase inhibitor does not enhance axonal regeneration or functional recovery after spinal cord injury in rats. *Exp Neurol* **200**, 392–397.

Nishiyama, M. et al. (2003). Cyclic AMP/GMP-dependent modulation of Ca2+ channels sets the polarity of nerve growth-cone turning. *Nature* **423**, 990–995.

Novakovic, S. D., Levinson, S. R., Schachner, M., and Shrager, P. (1998). Disruption and reorganization of sodium channels in experimental allergic neuritis. *Muscle Nerve* **21**, 1019–1032.

Pearse, D. D. et al. (2004). cAMP and Schwann cells promote axonal growth and functional recovery after spinal cord injury. *Nat Med* **10**, 610–616.

Pizzorusso, T. et al. (2002). Reactivation of ocular dominance plasticity in the adult visual cortex. *Science* **298**, 1248–1251.

Prinjha, R. et al. (2000). Inhibitor of neurite outgrowth in humans. *Nature* **403**, 383–384.

Qiu, J. et al. (2002). Spinal axon regeneration induced by elevation of cyclic AMP. *Neuron* **34**, 895–903.

Raisman G. (2001). Olfactory ensheathing cells—Another miracle cure for spinal cord injury? *Nat Rev Neurosci* **2**, 369–375.

Rajagopalan, S. et al. (2004). Neogenin mediates the action of repulsive guidance molecule. *Nat Cell Biol* **6**, 756–762.

Ramon-Cueto, A., Plant, G. W., Avila, J., and Bunge, M. B. (1998). Long-distance axonal regeneration in the transected adult rat spinal cord is promoted by olfactory ensheathing glia transplants. *J Neurosci* **18**, 3803–3815.

Ramon-Cueto, A., Cordero, M. I., Santos-Benito, F. F., and Avila, J. (2000). Functional recovery of paraplegic rats and motor axon regeneration in their spinal cords by olfactory ensheathing glia. *Neuron* **25**, 425–435.

Rapalino, O. et al. (1998). Implantation of stimulated homologous macrophages results in partial recovery of paraplegic rats. *Nat Med* **4**, 814–821.

Ritchie, J. M. and Rogart, R. B. (1977). Density of sodium channels in mammalian myelinated nerve fibers and nature of the axonal membrane under the myelin sheath. *Proc Natl Acad Sci U S A* **74**, 211–215.

Rush, A. M., Dib-Hajj, S. D., and Waxman, S. G. (2005). Electrophysiological properties of two axonal sodium channels, $Na_v1.2$ and $Na_v1.6$, expressed in spinal sensory neurons. *J Physiol* **564**, **3**, 803–816.

Sasaki, M., Black, J. A., Lankford, K. L., Tokuno, H. A., Waxman, S. G., and Kocsis, J. D. (2006b). Molecular reconstruction of nodes of Ranvier after remyelination by transplanted olfactory ensheathing cells in the demyelinated spinal cord. *J Neurosci* **26**, 1803–1812.

Sasaki, M., Hains, B. C., Lankford, K. L., Waxman, S. G., and Kocsis, J. D. (2006a). Protection of corticospinal tract neurons after dorsal spinal cord transection and engraftment of olfactory ensheathing cells. *Glia* **53**, 352–359.

Sasaki, M., Lankford, K. L., Zemedkun, M., and Kocsis, J. D. (2004). Identified olfactory ensheathing cells transplanted into the transected dorsal funiculus bridge the lesion and form myelin. *J Neurosci.* **4**, 485–493.

Schnell, L. and Schwab, M. E. (1990). Axonal regeneration in the rat spinal cord produced by an antibody against myelin-associated neurite growth inhibitors. *Nature* **343**, 269–272.

Schurch, B., Wichmann, W., and Rossier, A. B. (1996). Post-traumatic syringomyelia (cystic myelopathy): A prospective study of 449 patients with spinal cord injury. *J Neurol Neurosurg Psychiatry* **60**, 61–67.

Schwab, M. E. and Bartholdi, D. (1996). Degeneration and regeneration of axons in the lesioned spinal cord. *Physiol Rev* **76**, 319–370.

Schweigreiter, R. et al. (2004). Versican V2 and the central inhibitory domain of Nogo-A inhibit neurite growth via p75NTR/NgR-independent pathways that converge at RhoA. *Mol Cell Neurosci* **27**, 163–174.

Sekhon, L. H. and Fehlings, M. G. (2001). Epidemiology, demographics, and pathophysiology of acute spinal cord injury. *Spine* **26**, S2–12.

Senior, K. (2002). Olfactory ensheathing cells to be used in spinal-cord repair trial. *Lancet Neurol* 269.

Shao, Z. et al. (2005). TAJ/TROY, an orphan TNF receptor family member, binds Nogo-66 receptor 1 and regulates axonal regeneration. *Neuron* **45**, 353–359.

Silver, J. and Miller, J. H. (2004). Regeneration beyond the glial scar. *Nat Rev Neurosci* **5**, 146–156.

Simonen, M. et al. (2003). Systemic deletion of the myelin-associated outgrowth inhibitor Nogo-A improves regenerative and plastic responses after spinal cord injury. *Neuron* **38**, 201–211.

Snow, D. M., Lemmon, V., Carrino, D. A., Caplan, A. I., and Silver, J. (1990). Sulfated proteoglycans in astroglial barriers inhibit neurite outgrowth in vitro. *Exp Neurol* **109**, 111–130.

Song, H. et al. (1998). Conversion of neuronal growth cone responses from repulsion to attraction by cyclic nucleotides. *Science* **281**, 1515–1518.

Stys, P. K., Sontheimer, H., Ransom, B. R., and Waxman, S. G. (1993). Non-inactivating, TTX-sensitive Na^+ conductance in rat optic nerve axons. *Proc Natl Acad Sci* **90**, 6976–6980.

Stys, P. K., Waxman, S. G., and Ransom, B. R. (1992). Ionic mechanisms of anoxic injury in mammalian CNS white matter: Role of Na^+ channels and Na^+-Ca^{2+} exchanger. *J Neurosci* **12**, 430–439.

Teng, F. Y. and Tang, B. L. (2005). Why do Nogo/Nogo-66 receptor gene knockouts result in inferior regeneration compared to treatment with neutralizing agents? *J Neurochem* **94**, 865–874.

Ulrich, J. and Groebke-Lorenz, W. (1983). The optic nerve in MS: A morphological study with retrospective clinicopathological correlation. *Neurol Ophthalmol* **3**, 149–159.

.Venkatesh, K. et al. (2005). The Nogo-66 receptor homolog NgR2 is a sialic acid-dependent receptor selective for myelin-associated glycoprotein. *J Neurosci* **25**, 808–822.

Wang, K. C., Kim, J. A., Sivasankaran, R., Segal, R., and He, Z. (2002). P75 interacts with the Nogo receptor as a co-receptor for Nogo, MAG and OMgp. *Nature* **420**, 74–78.

Wang, K. C. et al. (2002). Oligodendrocyte-myelin glycoprotein is a Nogo receptor ligand that inhibits neurite outgrowth. *Nature* **417**, 941–944.

.Wang, X., Baughman, K. W., Basso, D. M., and Strittmatter, S. M. (2006). Delayed Nogo receptor therapy improves recovery from spinal cord contusion. *Ann Neur*; In Press.

Waters, R. L., Adkins, R. H., Yakura, J. S., and Sie, I. (1994). Motor and sensory recovery following incomplete paraplegia. *Arch Phys Med Rehabil* **75**, 67–72.

Watts, J. (2005). Controversy in China. *Lancet* **365**, 109–110.

Waxman, S. G. (1977). Conduction in myelinated, unmyelinated, and demyelinated fibers. *Arch Neurol* **34**, 585–590.

Waxman, S. G. (1992). Demyelination in spinal cord injury and multiple sclerosis: What can we do to enhance functional recovery? *J. Neurotrauma* **9**, S105–S117.

Waxman, S. G. and Brill, M. H. (1978). Conduction through demyelinated plaques in multiple sclerosis: Computer simulations of facilitation by short internodes. *J Neurol Neurosurg Psychiatry* **41**, 408–417.

Waxman, S. G. and Wood, S. L. (1984). Impulse conduction in inhomogeneous axons: Effects of variation in voltage-sensitive ionic conductances on invasion of demyelinated axon segments and preterminal fibers. *Brain Res* **294**, 111–122.

Waxman, S. G., Craner, M., and Black, J. (2004). Sodium channel expression along axons in multiple sclerosis and its models *Trends Pharmacol Sci* **25**, 584–592.

Wong, R. (2006). NMDA receptors expressed in oligodendrocytes. *Bioessays* **28**, 460–464.

Xu, J. et al. (1998). Methylprednisolone inhibition of TNF-alpha expression and NF-kB activation after spinal cord injury in rats. *Brain Res Mol Brain Res* **59**, 135–142.

Yin, Y. et al. (2006). Oncomodulin is a macrophage-derived signal for axon regeneration in retinal ganglion cells. *Nat Neurosci* **9**, 843–852.

Yiu, G. and He, Z. (2006). Glial inhibition of CNS axon regeneration. *Nat Rev Neurosci* **7**, 617–627.

Zheng, B. et al. (2003). Lack of enhanced spinal regeneration in Nogo-deficient mice. *Neuron* **38**, 213–224.

Zheng, B. et al. (2005). Genetic deletion of the Nogo receptor does not reduce neurite inhibition in vitro or promote corticospinal tract regeneration in vivo. *Proc Natl Acad Sci U S A* **102**, 1205–1210.

15

Parkinson Disease: Molecular Insights

Joseph M. Savitt, Valina L. Dawson, and Ted M. Dawson

I. Introduction

Parkinson disease (PD) is a progressive neurodegenerative disorder that affects patients in later life. The disease occurs more frequently in men (67% vs. 33%) and has an average age of onset of 61 ± 10 years (Marras et al., 2005). PD incidence increases with age and estimates suggest that 1 percent of people over the age of 60 will be affected (Nutt & Wooten, 2005). Though symptomatic treatments for PD are available, there is no cure and thus the disease continues to progress and lead to increasing disability and suffering. In the nearly two centuries since the description of PD by James Parkinson, much has been learned about the symptoms, pathology, treatment options, and neurochemistry involved in the disease. Only recently, however, have insights into potential causes of PD been explored successfully.

II. History

James Parkinson first described this condition in detail in his 1817 monograph "An Essay on the Shaking Palsy" (Parkinson, 2002). Charcot expanded Parkinson's original description and identified the cardinal clinical features of PD that include rest tremor, rigidity, balance impairment, and slowness of movement (Goetz, 2002). Pathologic examination of brains from PD patients led to the identification of cell loss in the midbrain nucleus of the substantia nigra (SN) and the presence of eosinophilic inclusions, termed Lewy bodies, as features of the disease (Greenfield & Bosanquet, 1953; Lewy, 1912).

A pivotal discovery was made when Ehringer and Hornykiewicz determined that dopamine was deficient in the corpus striatum and SN of brains taken from those with PD (Ehringer & Hornykiewicz, 1960). A few years prior to this Carlsson had found that the drug levodopa, a precursor of dopamine, could rescue the motor symptoms of animals whose dopamine had been depleted experimentally (Carlsson et al., 1957). These studies led to initial trials of levodopa in PD patients and nothing less than a revolution in the care of those afflicted by this disease (Cotzias et al., 1969). Even today levodopa coupled with an inhibitor of its peripheral conversion to dopamine (thus reducing peripheral side effects and increasing uptake into the brain) remains the gold standard of care in treating PD. More recently, several symptomatic treatments for PD have been developed and involve methods to increase dopaminergic tone in the brain through direct stimulation of dopamine receptors using oral or injected dopamine agonists, or by preventing the breakdown of dopamine or its precursors (see later).

Another advance in symptomatic PD therapy is the reemergence of surgical interventions in the form of deep brain stimulation (DBS) (Benabid et al., 2005). DBS has become increasingly common in advanced patients whose disease is difficult to manage with medical therapy alone.

The search for the cause of PD has an interesting history, with theories often favoring either genetic or environmental factors (see Figure 15.1). Parkinson speculated as to the location of pathology, but could not identify the cause of the disease. Theories involving environmental influences including trauma, exposure to cold, emotional stress, and infection can be found in the early literature (Keppel Hesselink, 1989) but perhaps not until the occurrence of a post-encephalitic disease resembling PD did an environmental cause of the disease seem very likely. This syndrome struck patients who had previously suffered from von Economo's encephalitis lethargica in the early twentieth century, and who subsequently developed Parkinson disease-like symptoms including stooped posture, rigidity, and gait impairment (Reid et al., 2001). Further study of this post-encephalitic Parkinsonism has shown it to be clinically and pathologically distinct from idiopathic PD.

The possibility of an environmental cause for PD also was bolstered by the small, early 1980s, outbreak of a Parkinson disease-like condition that occurred secondary to accidental injection of the compound 1-methyl-4-phenyl-1,2,3,6-tetrahydropyridine (MPTP) (Langston et al., 1983). Though rarely seen clinically since the original cluster, MPTP-induced Parkinsonism continues to be a useful model of PD in animal studies. Other toxins have been associated with PD-like disease mainly in animal models and include the pesticides rotenone (Betarbet et al., 2000), paraquat, and maneb (Thiruchelvam et al., 2000), as well as the naturally occurring chemical epoxomicin (McNaught et al., 2004). The relationship between human PD and exposure to these specific chemicals is weak, though the actions of these toxins make an environmental etiology for PD seem more plausible.

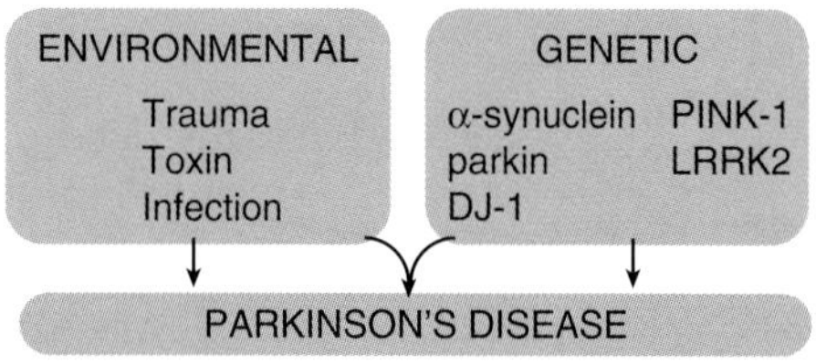

Figure 15.1 Environmental and genetic causes of PD. Parkinson's disease is largely sporadic (>90%), although there are rare familial (genetic) cases of Parkinson's disease (<10%). With the exception of the familial cases the causes of Parkinson's disease are not known. Environmental factors may be a single cause as in the case of MPTP contamination of synthetic heroin, but the common thought is that Parkinson's disease is caused by a combination of environmental and genetic factors.

On the other hand, there are studies and observations that have suggested the importance of a hereditary influence. Even as early as 1880, Leroux, a student of Charcot, believed that PD had a genetic component (Leroux, 1880). Over the past decade there has been an explosion in our understanding of the various monogenic forms of PD beginning with the identification of α-synuclein mutation in 1997 (Polymeropoulos et al., 1997) and most recently with the identification of LRRK2-associated PD and its high prevalence in certain cohorts (Lesage et al., 2006; Ozelius et al., 2006; Paisan-Ruiz et al., 2004; Zimprich et al., 2004). The variations of pathology and symptomatolgy of these forms of PD have lent insight into the mechanisms of disease but also have generated controversy regarding how we define PD. For example, is genetically determined Parkinsonian disease with atypical pathology really PD? The identification of these genes has forced us to reexamine what Parkinson's disease actually is and to determine how best to understand the various forms of the disease.

III. Diagnosis

Despite a great deal of research into the pathology of PD since the times of Parkinson and Charcot, the diagnosis of the disease continues to be made based primarily on history and physical examination. PD should be considered in the differential diagnosis of a patient who presents with at least one of the cardinal features of the disease, including rest tremor, bradykinesia, cogwheel rigidity, and in more advanced cases, postural instability. Finding a rest tremor in a single upper extremity, perhaps more than any other single feature, predicts the presence of PD, though over 20 percent of PD patients will not develop a tremor (Hughes et al., 1993). In addition to these features, a variety of other motor and nonmotor signs and symptoms can support the diagnosis. A lasting and potent response to levodopa therapy also provides strong evidence of PD. Additional features that support the diagnosis are presented in Table 15.1.

Table 15.1 Diagnosing Parkinson Disease

Very early staged PD may present with a single cardinal symptom; however, with time most of the cardinal symptoms emerge. Other symptoms are important to recognize as they may be amenable to other forms of therapy.

Cardinal features:

- Bradykinesia—slow movements and decrement of frequency and amplitude of repetitive movements
- Rest tremor—most commonly beginning in the hand
- Cogwheel rigidity
- Postural instability—seen in later stage disease

Other symptoms:

- Sustained and significant relief with levodopa
- Unilateral symptoms at onset
- Stiff gait with reduced arm swing
- Effortful standing from a low chair
- Impaired olfaction
- Soft voice
- Stooped posture
- Depression and/or anxiety
- Shuffling/festinating/freezing gait
- Difficulty turning in bed, adjusting covers
- Drooling
- Constipation
- Micrographia
- Sleep disturbance/REM sleep behavior disorder
- Seborrhea
- Later staged cognitive decline

IV. Pathology

The pathological hallmarks of PD include the loss of pigmented neurons in the SN and the presence of eosinophilic, intracellular inclusions called Lewy bodies in a subset of those midbrain neurons remaining. The loss of SN neurons leads to a striatal deficiency of dopamine that in turn generates the most recognizable motor symptoms of the disease. It long has been known, however, that neuronal pathology extends beyond the SN making the classification of PD as merely a disorder of striatal dopamine deficiency an oversimplification. Indeed in 1912 Lewy reported the presence of inclusions (later termed Lewy bodies) in the nucleus basalis and dorsal motor nucleus of the vagus of PD brains (Lewy, 1912).

More recent studies have catalogued the presence of these structures in multiple brain regions and even in peripheral neurons, including those of the enteric and sympathetic systems (Braak et al., 2006; Iwanaga et al., 1999). Pathologic examinations by Braak have attempted to stage PD pathology and determine a pattern of disease progression using α-synuclein immunostaining to identify Lewy bodies and related pathologic features called Lewy neurites as markers of disease (Braak et al., 2003). According to this study the central nervous system pathology of PD begins in the medulla, affecting the lipofuscin-rich dorsal motor nucleus of the vagus, and then progresses to the caudal raphe nucleus, the reticular formation, and the noradrenergic locus ceruleus and subceruleus. Only at the next stage (stage 3) does the pathology involve the dopamine-rich cells of the pars compacta of the SN. Stage 4 includes a dramatic cell loss in the SN and this stage, along with the subsequent stages 5 and 6, mark the increasing involvement of the cerebral cortex.

In addition to their brain localization and progression in PD, the components of the pathologic markers of the disease, the Lewy body and neurite, are of interest. Over the past decade many of the molecular characteristics of these structures have come to light, (reviewed in Schulz and Falkenburger (2004) and Shults (2006)). The Lewy body is present in the cell soma and is composed of a variety of proteins and organelles including α-synuclein, ubiquitin, UCH-L1, neurofilaments, and mitochondria, among others. Alpha-synuclein is the major component of Lewy bodies and, interestingly, mutation in the corresponding gene causes a rare form of PD (see later). Ultrastructurally, there is a granulovesicular core surrounded by a fibrillar component. Similar aggregates occurring in neuronal processes and likely forming earlier than Lewy bodies have been termed Lewy neurites. The biologic significance of both Lewy bodies and neurites is unknown and there is speculation as to whether they are mere markers of disease or are involved in damaging or, perhaps more likely, protecting the cell.

V. Etiology: Genetics and Environment

As mentioned earlier, there is support for both genetic and environmental causes of PD. Epidemiologic studies looking at environmental exposures, family history, and lifestyle choices have attempted to reveal factors that alter the risk of PD. Much attention has been focused on the role of potential risk factors such as pesticide exposure, smoking, and dietary choices. A recent review of 38 case control studies and three autopsy studies concluded that the consistent association between pesticide exposure and PD was unlikely spurious, though no clear effect of a particular agent or agent combination was apparent (Brown et al., 2006). There was no evidence to suggest that this association was necessarily causative. Similarly, behaviors such as smoking, alcohol intake, and caffeine consumption have been associated with reduced risk of the disease, though there is speculation that this may be due in part to a preexisting personality trait rather than a true protective effect (Evans et al., 2006).

Studies looking at the risk of PD with occupation are another potential way of identifying environmental risk. Several occupations including teacher, healthcare provider, scientist, farmer, clergyperson, and lawyer have been associated with increased risk of PD (Goldman et al., 2005).

Although some occupations may suggest a common etiologic feature (such as increased exposure to infectious agents in teachers and healthcare providers), a common mechanism to explain the increased risk associated with all these occupations is more difficult to discern. Certainly the effect may be explained by a premorbid personality that is drawn to these professions or an ascertainment bias based on level of education or other factors.

Another method of determining relative environmental versus genetic influence on a disease is the study of twins. Monozygotic twins should be concordant for a disease if it is caused by a fully penetrant gene and both twins survive long enough to manifest symptoms. In addition, one would expect monozygotic twins to have a higher concordance rate than dizygotic twins if genetic factors played a large role in disease risk. For the most part these types of studies have not found strong evidence of genetic influence, with one of the largest such studies finding similar rates for monzygotic and dizygotic twins overall (.155 and .111, respectively), though complete concordance was seen in the four monozygotic twin pairs where at least one of the pair was diagnosed prior to age 51 (Tanner et al., 1999). In addition, a small study using positron emission tomography (PET) to evaluate dopa uptake in twin pairs, found higher concordance rates (45% in monozygotic pairs and 29% in dizygotic pairs) than found on clinical exam alone, suggesting that despite clinical discordance, twins of affected patients do appear to have a greater risk of alterations in their dopaminergic systems (Burn et al., 1992). The overall lack of concordance suggests that environmental factors are important in determining ultimate disease expression, however the nature of PD as a late onset disease, the strong concordance seen in the small number of younger onset patients and higher concordance rates seen using functional imaging suggest that there may be a genetic influence.

The possibility of a genetic influence in PD further is suggested by the observation that PD is more common in relatives of patients than in the general population or in relatives of control groups (Maher et al., 2002b). A segregation analysis done on the families of nearly 950 PD patients supported an inheritance pattern in two models involving a gene influencing age of onset and one or the same gene influencing disease susceptibility (Maher et al., 2002a). The contribution of genetics is illustrated best, however, by the discovery of several monogenetic causes of PD or PD-like disease described next.

VI. Genes Implicated in PD

The likelihood of a genetic influence as described above and the identification of well-described kindreds with high rates of PD have spurred on the search for genes involved in the disease. To date, five genes have been convincingly linked to the generally rare, familial forms of PD. Though mutation in these genes likely plays only a small role in the incidence of "sporadic" PD, a better understanding of the pathologic process in these forms of the disease will lead to a better understanding of the PD disease process in general. Monogenetic forms of PD are listed in Table 15.2.

A. Alpha-Synuclein

In 1997 the first genetic mutation causing PD was identified and found to encode an alanine to threonine substitution (A53T) in the protein α-synuclein (Polymeropoulos et al., 1997). Subsequently two additional point mutations (A30P and E46K) in the coding region of this gene were found to cause familial PD, as were α-synuclein gene duplications and triplications (Kruger et al., 1998; Nishioka et al., 2006; Singleton et al., 2003; Zarranz et al., 2005). The severity and characteristics of the autosomal dominant PD syndrome caused by these various mutations demonstrate mutation-specific features. For example gene triplication appears to generate a more severe, earlier-onset phenotype than do gene duplications, suggesting a gene dosage effect (Eriksen et al., 2005; Nishioka et al., 2006). This also implies that gain of function α-synuclein point mutation and overexpression of wild-type protein both are pathologic, which in turn has implications for disease mechanism. Given these findings and the fact that aggregated α-synuclein is a main constituent of Lewy bodies, it is tempting to speculate that PD is a disease of α-synuclein accumulation or overexpression, and that pathology related to α-synuclein mutation is caused more by a quantitative rather than qualitative abnormality (Singleton et al., 2003). This view is further supported by data suggesting that polymorphisms in the α-synuclein promoter alter PD risk (Hadjigeorgiou et al., 2005; Pals et al., 2004; Tan et al., 2004).

Exactly how overexpression or mutation in this protein leads to PD is unknown. Since α-synuclein also is associated with sporadic disease through its presence in the Lewy body and Lewy neurites, understanding the role this protein plays in all forms of PD is important. Alpha-synuclein is a 140 amino acid, acidic and abundant phosphoprotein that is largely unstructured in solution (Weinreb et al., 1996). When present in the Lewy body, however, α-synuclein is aggregated into insoluble, hyperphosphorylated and ubiquitinated filamentous structures (Hasegawa et al., 2002; Spillantini et al., 1998). The amino terminal end of the protein contains conserved repeats thought to function in protein–protein interaction, and this region forms alpha-helices upon interaction with lipid. The carboxy-terminal end may have a chaperone-like function, whereas the hydrophobic central region likely increases the protein's tendency to aggregate.

Table 15.2 Monogenetic Parkinsonism

Genes identified whose mutation leads to a primarily Parkinsonian phenotype.

Locus/MAP SITE	Gene/Protein	Pattern	Prevalence	Pathology	Common Features	Miscellaneous
PARK 1 and PARK 4/4q21-23	α-synuclein	AD	Very rare	Lewy bodies	Early onset dementia. Presentation variable w/ mutation type.	α-synuclein is found in Lewy bodies. Mutant and abnormally high protein levels are toxic.
PARK 2/6q25-27	Parkin	AR (mostly)	18% EOPD (50% with a family history)	Rare Lewy bodies if any	Early onset, slow progression.	Parkin plays a role in the ubiquitin-proteasomal system that degrades unwanted proteins.
PARK 6/1p35-36	PINK-1	AR (carriers may be at increased risk)	2–3% EOPD	Unknown	Early onset, slow progression.	PINK1 is a protein kinase localized to mitochondria.
PARK 7/1p36-8	DJ-1	AR	<1 % EOPD	Unknown	Early onset, slow progression.	DJ-1 is involved in the cellular stress response.
PARK 8/12p 11.2-q13.1	LRRK-2	AD	Highly variable in populations (up to 40% of Parkinson cases)	Lewy bodies, variable pathology including tau pathology	Typical PD (mostly).	LRRK2 is a protein kinase. Surprisingly high prevalence is seen in some patient cohorts.

AR: autosomal recessive; AD: autosomal dominant; EOPD: early onset Parkinson disease (usually < 50 years of age).

Potential mechanisms of α-synuclein pathology focus largely on the abnormal aggregation of the protein, reviewed in Mizuno et al. (2005) and Mukaetova-Ladinska and McKeith (2006). This is thought to be a step-wise process involving the initial formation of soluble oligomers or protofibrils and subsequent assembly into insoluble, fibrillar forms (see Figure 15.2). There is evidence that the protofibril may be the most toxic form (Rochet et al., 2004). Interestingly the presence of a variety of factors has been associated with an increased propensity toward the formation of α-synuclein aggregates. These include metal ions, oxidative stress, fatty acids, septins, pesticides, and polyamines. Furthermore, dopamine has been shown to stabilize the toxic synuclein protofibril, possibly explaining the relative sensitivity of dopamine neurons to α-synuclein-mediated damage (Conway et al., 2001). Similarly the presence of a variety of post-translational modifications appears to lead to accumulation including phosphorylation (especially on serine 129), oxidation driven tyrosine cross linking, and nitration.

Alpha-synuclein is processed at its carboxy terminus by unidentified synucleinases (Li et al., 2005). This truncation of α-synuclein may correlate with its propensity to oligomerize and mice engineered to express such a truncated form of the protein in catecholaminergic areas show pathologic inclusions and reduced levels of dopamine (Tofaris et al., 2006). Finally, increased protein aggregation has been suggested as the mechanism behind the pathogenic α-synuclein mutations. Specifically, the A53T mutation tends to favor fibrillation, whereas the A30P does not. However, the decreased membrane binding seen with the A30P mutation may lead to an increased cytosolic α-synuclein concentration and more rapid protofibril formation. The E46K mutation has been shown to promote aggregation to an even greater extent than the A53T or A30P mutations (Pandey et al., 2006).

How then might α-synuclein oligomers and fibrils contribute to cellular toxicity? Protofibrils may, through an ability to form pore-like structures, permeabilize membranes and allow the leakage of toxic material such as dopamine. In addition, there is speculation that α-synuclein may interfere with the function of mitochondria by disrupting respiratory chain function, or by liberating cytochrome C or BAD proteins leading to cell death, as reviewed in Hashimoto et al. (2003). Finally, abnormal α-synuclein leads to other forms of cellular dysfunction including decreased proteasomal activity (Tanaka et al., 2001), accumulation of autophagocytic vesicles, and impaired lysosomal function (Stefanis et al., 2001).

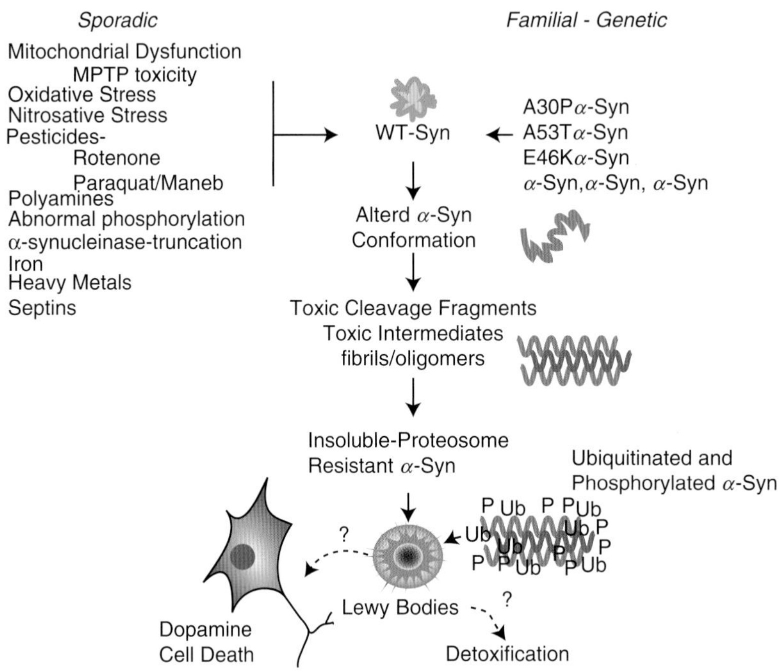

Figure 15.2 **Mechanisms of α-synuclein mediated neurodegeneration.** α-synuclein (α-syn) also is associated with sporadic and familial PD through its presence in the Lewy body and Lewy neurites, but exactly how this protein contributes to PD is not fully understood. Mechanisms of α-synuclein pathology likely involve on the abnormal aggregation of the protein that is thought to occur in a step-wise process involving the initial formation of soluble oligomers or protofibrils and subsequent assembly into insoluble, fibrillar forms. Aggregation is promoted experimentally by metal ions, oxidative stress, fatty acids, septins, pesticides, and polyamines. Additionally post-translational modifications can lead to accumulation and to oligomerization including phosphorylation, oxidation, and nitration and proteolysis by an unidentified synucleinases. Ultimately in the Lewy body, α-synuclein is aggregated into insoluble, proteosome resistant hyperphosphorylated and ubiquitinated filamentous structures.

Moreover, both ER stress contributes to α-synuclein-induced cell death (Smith et al., 2005a). As abnormal α-synuclein aggregation appears to be associated with the pathologic process in both sporadic and α-synuclein-mutation related PD, the development of ways to reduce α-synuclein expression and to impair its aggregation, modification, and truncation are being explored as possible PD therapies (Mukaetova-Ladinska & McKeith, 2006).

B. Parkin

Mutations in the protein parkin cause an autosomal recessive, early-onset form of PD (Kitada et al., 1998). The resulting clinical syndrome resembles idiopathic PD but may show several atypical features including early onset, more prominent dystonia and motor fluctuations, and a more symmetric onset of symptoms (Bonifati et al., 2001; Lohmann et al., 2003). Parkin mutations (including exon deletions, duplications, insertions, and a variety of missense and nonsense mutations) are the most common genetic cause for early-onset PD, accounting for nearly 50 percent of cases with autosomal recessive inheritance and about 20 percent of those developing early onset disease without a family history (Kubo et al., 2006). Brains from patients with parkin-related PD demonstrate neuronal cell loss and gliosis in the SN and locus ceruleus but, for the most part no Lewy bodies (Farrer et al., 2001; Mori et al., 1998; Sasaki et al., 2004). Rare cases of compound heterozygotes do appear to have Lewy body-like inclusions; and studies showing unusual tau pathology and degeneration of the spinocerebellar system also have been reported, as reviewed in Kubo et al. (2006).

The parkin protein consists of 465 amino acids with an amino terminal ubiquitin-like domain and two RING finger sequence motifs near the carboxy terminus. Despite some reports of autosomal dominant inheritance, parkin mutation most likely causes disease through a loss of function mechanism consistent with its mostly recessive mode of inheritance (Sriram et al., 2005). The protein is an E3 ligase whose function is to assist in the addition of ubiquitin molecules to target proteins, thus marking them for degradation via the ubiquitin-proteasome system (Zhang et al., 2000). A loss of normal parkin function thus could cause disease through the accumulation of potentially toxic substrates (see Figure 15.3). Several putative parkin substrates have been found to accumulate in patients with parkin-associated PD including CDCrel-1, CDCrel-2, Pael-R, cyclin E, p38/JTV-1 (also know as AIMP2), and far upstream element binding protein-1 (FBP-1), though only the latter two also have been found to accumulate in parkin knock-out mice and sporadic PD (Ko et al., 2005; Ko et al., 2006). Exactly how the accumulation of these substrates may lead to PD pathology is largely unknown.

Alternatively, parkin has been found to catalyze another form of ubiquitination that may not directly involve the proteasomal system (see Figure 15.3) (Lim et al., 2006). A link between sporadic PD and parkin function is supported by studies showing that parkin is nitrosylated in sporadic PD patients and that this modification impairs parkin's E3 ligase activity (Chung et al., 2004). Moreover, dopamine may directly inhibit parkin's activity and reduce its solubility through covalent modification (LaVoie et al., 2005).

Parkin function also may impact on the mitochondria. Mice deficient in parkin demonstrated mitochondrial dysfunction and an increase in markers of oxidative stress (Palacino et al., 2004), whereas Drosophila studies suggest that inactivation of a parkin orthologue results in mitochondrial pathology (Greene et al., 2003). Parkin thus serves as a link among the mitochondrion, the proteasome, and PD cell loss with the caveat that parkin-associated PD often lacks Lewy bodies. How to reconcile these aspects of parkin biology is an active area of research.

C. DJ-1

Mutations in DJ-1, like those in parkin, lead to early-onset, autosomal recessive PD (Bonifati et al., 2003). Patient symptoms are levodopa-responsive with asymmetric onset, slow progression, and medication-induced dyskinesias. Behavioral, psychiatric, and dystonic features can occur as well (Dekker et al., 2003). The prevalence of DJ-1 mutations is likely less than 1 percent in the early-onset patient subgroup and cause disease through a putative loss of function mechanism (Lockhart et al., 2004). The main function of DJ-1 protein is not known; however, studies suggest it may act as an oxidation/reduction sensor, anti-oxidant, molecular chaperone, and/or protease (Moore et al., 2005). It is widely expressed and appears to localize to the cytoplasm and mitochondria (Zhang et al., 2005b). DJ-1 is not found in Lewy bodies, though it has been localized to cellular inclusions present in several neurodegenerative diseases such as multiple system atrophy and various tauopathies including Alzheimer's disease, Pick's disease, and progressive supranuclear palsy (Kubo et al., 2006). When analyzed from the brains of PD patients and compared to controls, the protein appears to be more abundant, more oxidatively damaged, and less soluble in afflicted brains (Choi et al., 2006).

DJ-1 has been linked to other PD associated proteins including parkin and α-synuclein. Specifically, DJ-1 has been found to interact with parkin in an *in vitro* assay modeling oxidative stress (Moore et al., 2005) and, when oxidized to a limited degree, DJ-1 appears able to inhibit the formation of potentially toxic α-synuclein fibrils (Zhou et al., 2006). In support of its role as an anti-oxidant/neuroprotectant, studies on mice lacking DJ-1 show increased sensitivity to MPTP, and neuronal cultures derived from these mice show an increased susceptibility to oxidative stress (Kim et al., 2005). Another potential function of DJ-1 involves control of apoptosis as the protein regulates PTEN and Akt function and associates with DAXX and p54NRB to promote cell survival (reviewed in Abou-Sleiman et al. (2006)). In summary, exactly how loss of DJ-1 function leads to early onset PD is unclear, though a role for impairment of the oxidative stress response leading to reduced cell survival seems likely.

D. PINK-1

Mutations in PTEN-induced putative kinase 1 (PINK-1) lead to early-onset, autosomal recessive PD (Valente et al., 2004). Patients with this form of Parkinsonism, much like the syndromes seen with parkin and DJ-1 mutation, show early disease onset, levodopa responsiveness, and slow progression (Klein et al., 2005). Some patients possessing a single PINK-1 mutation may be at increased risk of disease with imaging studies demonstrating altered dopaminergic function in these individuals (Ibanez et al., 2006). PINK-1 mutation is found in perhaps 3 percent of patients with early-onset disease (Tan et al., 2006). The protein is localized to mitochondria and functions as a protein kinase with as yet unidentified substrates (Silvestri et al., 2005). Mutations appear to cluster around the kinase domain and/or impair total kinase activity consistent with a loss of function disease mechanism (Kubo et al., 2006). Interestingly PINK-1 is transcriptionally induced by PTEN, the same protein that down-regulates DJ-1. Mutations in PINK-1 appear to abolish the protein's ability to block apoptotic cell death in a cell culture model (Petit et al., 2005).

In addition, Drosophila lacking PINK-1 demonstrate significant mitochondrial pathology, leading to apoptotic muscle degeneration and male sterility. This phenotype can be rescued by parkin overexpression, suggesting that parkin and PINK-1 may share a common biochemical pathway likely involving the mitochondrion (Clark et al., 2006; Park et al., 2006; Tan & Dawson, 2006). Interestingly the PINK-1 and parkin Drosophila models share similar pathology; but unlike the converse, PINK-1 is unable to rescue the parkin phenotype. This suggests that PINK-1 functions upstream of parkin in a putative mitochondrial-based protection pathway.

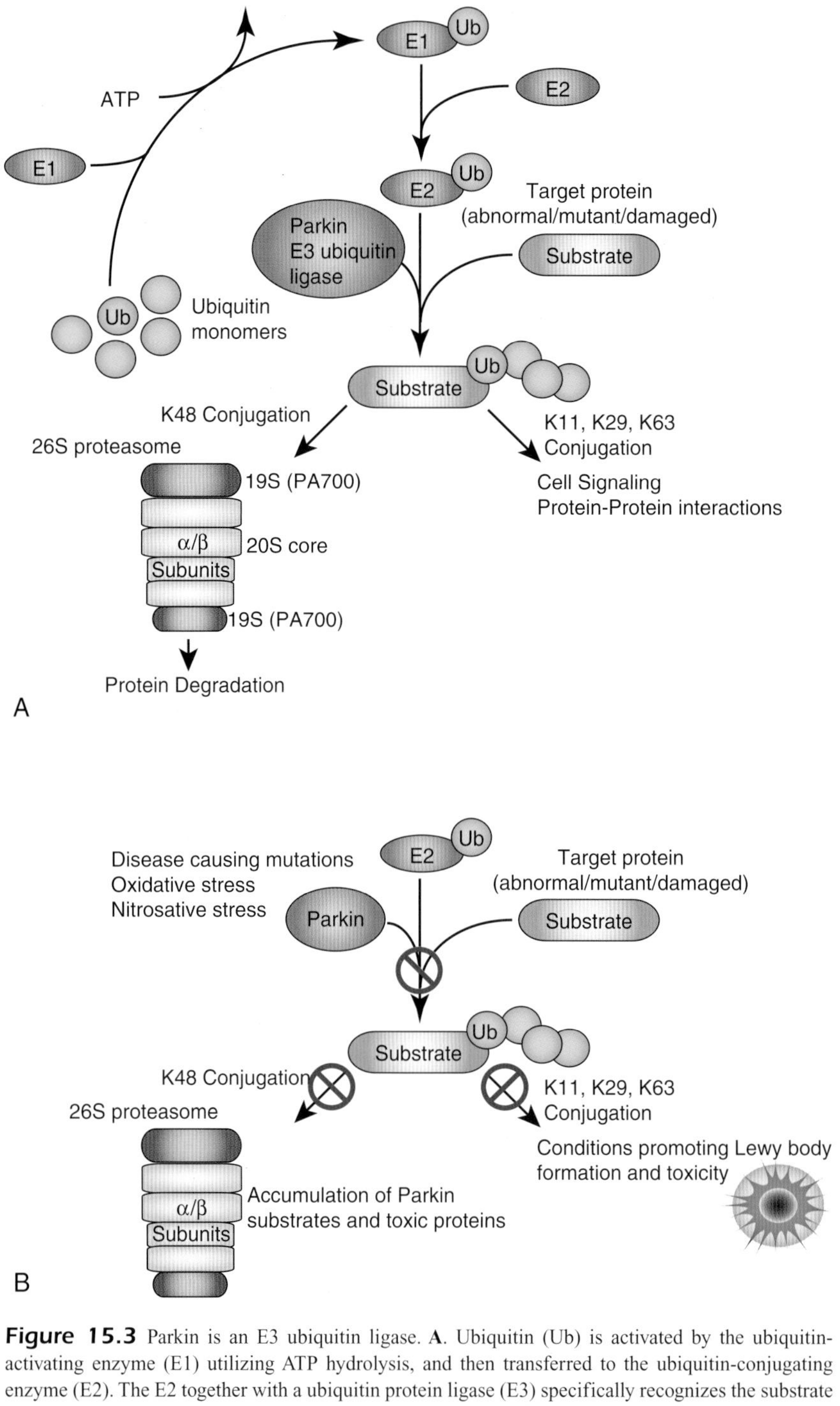

Figure 15.3 Parkin is an E3 ubiquitin ligase. **A**. Ubiquitin (Ub) is activated by the ubiquitin-activating enzyme (E1) utilizing ATP hydrolysis, and then transferred to the ubiquitin-conjugating enzyme (E2). The E2 together with a ubiquitin protein ligase (E3) specifically recognizes the substrate protein and mediates the attachment of successive ubiquitin monomers to the substrate to generate a poly-ubiquitin chain. E3 ligases, such as parkin, are therefore one of the most important factors in the regulation and selectivity of substrates to be ubiquitinated. Substrates containing lysine-48 (K48)-linked poly-ubiquitin chains are recognized by the 19S regulator subunit of the 26S proteasomal complex and degraded in an ATP-dependent manner by the 20S core to generate short peptide fragments and poly-ubiquitin chains. Alternatively, protein substrates with poly-ubiquitin chains linked through K11, K29, or K63 residues may impact cellular functions such as signaling and protein-protein interactions. **B**. Mutations in parkin, oxidative, and nitrosative stress result in abnormal parkin function. Abnormal K48 conjugation results in accumulation of substrates and toxic proteins such as p38/JTV-1 (also known as AIMP2) and far upstream element binding protein-1 (FBP-1). Abnormal K63 conjugation promotes aggregation, and Lewy-like body formation.

E. LRRK2

Mutation in leucine-rich repeat kinse-2 or LRRK2 (also called dardarin) causes an autosomal dominant, late-onset form of PD (Paisan-Ruiz et al., 2004; Zimprich et al., 2004). The majority of patients with LRRK2-related disease show the typical features of sporadic PD, though cases presenting with early onset, amyotrophy, gaze palsy, dementia, and psychiatric symptoms have been reported (Gaig et al., 2006; Paisan-Ruiz et al., 2004; Zimprich et al., 2004). The prevalence of LRRK2 mutations varies widely based on the study population and the specific mutation. Surprisingly, the G2019S mutation was found in 18 percent of all PD cases in a group of Ashkenazi Jews and in 30 percent of those patients with an affected family member (Ozelius et al., 2006). This mutation appears to have an estimated penetrance of 32 percent. Furthermore a G2019S mutation rate of 40 percent was seen in a North African PD cohort (Lesage et al., 2006). These remarkably high rates are likely due to a founder effect, with the frequency of LRRK-2 mutations in a more general European population estimated to be about 5 to 10 percent in familial cases (Brice, 2005). The pathology seen in LRRK2-related disease often mimics typical PD pathology and includes cell loss in the SN and locus ceruleus as well as the presence of Lewy bodies and Lewy neurites. There have been reports of abnormal pathology as well that include an absence of Lewy bodies, the presence of atypical inclusions, LRRK2 filled neuronal processes, and tau pathology that resembles that seen in progressive supranuclear palsy (Giasson et al., 2006; Zimprich et al., 2004).

LRRK2 is a 2,527 amino acid, multidomain structured protein that contains ROC (Ras in complex protein), COR (C-terminal of Roc), leucine-rich repeat, mixed lineage kinase, WD40 (protein-protein interaction domain), and ankryin domains (Paisan-Ruiz et al., 2004, 2005; Zimprich et al., 2004). At this time there are more than 15 possible pathogenic missense mutations present in 11 out of the 51 exons (Brice, 2005). Cellular localization studies suggest that LRRK2 is cytoplasmic, with a fraction associated with mitochondria (Gloeckner et al., 2006; West et al., 2005).

The mRNA encoding LRRK-2 has been localized to brain tissue and appears to be preferentially expressed in dopaminergic receptive fields including the striatum and olfactory tubercle rather than in the SN itself (Galter et al., 2006; Melrose et al., 2006). This suggests that LRKK2 related disease may be the result not of a cell-autonomous dysfunction in the dopaminergic neurons, but rather a primary dysfunction in the post-synaptic or target areas. It is possible this results from impaired trophic support leading to the loss of presynaptic neurons. Consistent with this notion are localization studies that find LRRK2 associated with vesicular and membranous structures implicating LRRK2 in vesicular trafficking and membrane recycling (Biskup et al., 2006). Further studies are required to verify LRRK2 localization and to better determine its overall mechanism of pathogenesis.

On a cellular level, certain pathogenic LRRK2 mutations cause an increase in its relative kinase activity (Gloeckner et al., 2006; West et al., 2005). Also, LRRK2 can bind parkin, leading to increased formation of ubiquitinated protein aggregates in a cell culture model, whereas expression of the mutated protein leads to toxicity in primary neuronal and SH-SY5Y cultures (Smith et al., 2005b). LRRK2 toxicity is directly related to it kinase activity as mutations in LRRK2 that render its kinase function inactive attenuate LRRK2 toxicity (Smith et al., 2006). Furthermore, there is speculation based on the presence of a WD40 domain that LRRK2 may participate in signal transduction, RNA processing, and transcription; cytoskeletal, and mitochondrial function; as well as vesicular formation and transport (Li and Beal, 2005; Mata et al., 2006). How perturbations in any of these functions might lead to PD is as yet unknown. Our understanding of LRRK2 biology and its relationship to PD are still in the very early stages, but given the relative high prevalence of LRRK2-related PD, further investigation should prove to be of great importance.

F. Other Candidate Genes Involved in PD

In addition to the preceding monogenetic causes of PD, several other genes have been implicated in the disease; however, the importance of these findings is less clear. Interestingly one gene that, like parkin, is involved with the ubiquitin-proteosome system has been associated with PD. An Ile93Met mutation in the protein ubiquitin carboxy-terminal hydrolase L1 (UCH-L1) has been suggested to cause PD in a small German family (Leroy et al., 1998). Casting doubt on the significance of this finding, however, is the small size of the family (only the proband and her brother's DNA were available for testing), the presence of an asymptomatic obligate carrier, and the fact that no further PD families nor individuals have been found to harbor this mutation. The protein is highly abundant in brain, is present in Lewy bodies and has ubiquitin hydrolase activity. In addition, UCH-Ll may promote α-synuclein ubiquitination (ligase activity) using lysine 63 linkages perhaps playing a role in aggregate formation or impairment of proteasomal function (Liu et al., 2002). Other studies have suggested an association between UCH-L1 genetic variants and PD risk, though a recent study has cast some doubt on this as well (Healy et al., 2006).

Variation in the gene encoding tau has been associated with PD susceptibility (Zhang et al., 2005a), and mutation in this gene is responsible for hereditary frontotemporal dementia with Parkinsonism. Interestingly α-synuclein interacts with tau and may alter its phosphorylation (Jensen et al., 1999) as well as promoting synergistic interactions promoting the aggregation of both tau and α-synuclein (Giasson et al., 2003), thereby lending additional plausibility to a tau and PD association.

Several other genetic variations have been associated with PD. These include mutation or variation in the tyrosine hydroxylase enzyme (Leu205Pro is known to cause an autosomal recessive form of infantile PD (Ludecke et al., 1996)), the transcription factor ASCL1 (Ide et al., 2005), the trophic factor fibroblast growth factor 20 (van der Walt et al., 2004), the receptor semaphorin 5A (Maraganore et al., 2005), and the transcription factor Nurrl (Grimes et al., 2006). The validity and relevance of these associations to PD are awaiting further study.

Additional genetic causes of Parkinsonism include those designated Park 3 thought to reside on chromosome 2, Park 10 on chromosome 1, Park 11 on chromosome 2, and Park 9, also a known as Kufor-Rakeb syndrome (reviewed in Morris (2005)). Additionally, several other genetic syndromes have Parkinsonism as an associated feature and include diseases such as spinocerebellar ataxia type 2, 3, 6, 12, and 17, Wilson's disease, X-linked dystonia and parkinsonism, and so on (see Farrer (2006) for a more complete review). In addition, genetic mutation in the mitochondria DNA polymerase G (POLG) can lead to progressive external ophthalmoplegia, that can also include Parkinsonian features as well as to early-onset familial Parkinsonism and the presence of multiple mitochondrial DNA deletions (Davidzon et al., 2006).

VII. Models of Pathogenesis

Clues from our understanding of the cellular pathology of PD, as well as the recent identification of monogenetic causes of PD, have led to the development of models of PD pathogenesis. In addition, the use of various toxins to mimic the disease also provides insight into possible pathogenic mechanisms (see Figure 15.4).

A. Mitochondrial Dysfunction

Multiple studies have suggested a role for mitochondrial dysfunction in the pathogenesis of PD. Indeed examination of SN and blood platelets from PD patients shows deficiencies in complex I activity of the mitochondrial respiratory chain (Shults, 2004). Complex I is a grouping of more than 40 polypeptides whose function in the inner mitochondrial membrane is to transfer electrons from NADH to ubiquinone (Abou-Sleiman et al., 2006). More than half of the protein genes expressed by mitochondrial DNA encode components of this complex. Malfunction of complex I leads to reduced activity of the respiratory chain and also to the production of reactive oxygen species that may lead to further mitochondrial and cellular dysfunction. Cultured cells in which mitochondria isolated from PD patients are used to replace endogenous mitochondria (producing cybrids) demonstrate reduced complex I activity and increased free-radical production. Interestingly, complex I of the electron transport chain also is affected by rotenone and MPTP, two toxins used to model PD (Bove et al., 2005). Specifically, rotenone produces SN cell loss, dopaminergic fiber loss, and the formation of Lewy body-like inclusions when administered to animals (Betarbet et al., 2006). MPTP exposure in patients and in animal models leads to Parkinsonian motor features and SN cell loss (see Figure 15.5) (Bove et al., 2005). Also, MPTP given chronically produces cellular inclusions containing α-synuclein (Fornai et al., 2005). Finally, 6-hydroxydopamine and paraquat induce dopamine cell loss by causing oxidative stress that may induce mitochondrial toxicity similar to that seen with rotenone and MPTP (Shults, 2004).

Genetic studies also have implicated mitochondrial dysfunction in PD. The presence of the PD-related gene products PINK-1 and DJ-1 in mitochondria supports this view as do experiments demonstrating an increased rate of mitochondrial DNA mutations in the SN of PD patients (Bender et al., 2006; Kraytsberg et al., 2006). The mutation load in these latter studies is sufficient to lead to impaired cytochrome c oxidase expression and suggests a possible chain of events leading to PD. In this scenario, oxidative stress present within the SN leads to mitochondrial DNA damage through the production of reactive molecules that, in turn, produce mitochondrial dysfunction then leading to increased reactive species. This vicious cycle continues until mitochondrial function is so impaired as to lead to cellular dysfunction and death. In addition, mutation in the mitochondrial DNA polymerase G has been found to cause SN pathology and early-onset PD in patients likely through a mechanism of increased accumulation of mitochondrial DNA deletions (Davidzon et al., 2006).

Links exist between mitochondrial pathology and other PD genes as well. For example animals lacking α-synuclein are resistant to MPTP, suggesting that this protein is involved in the pathogenic mechanism of MPTP, a known mitochondrial toxin (Dauer et al., 2002). The understanding that mitochondrial pathology and the generation of oxidative stress are important in the pathogenesis of PD, has led to trials of anti-oxidant and pro-mitochondrial substances including coenzyme Q10 (CoQ10), vitamin E, and creatine as possible neuroprotectants in PD.

B. Ubiquitin-Proteasome Dysfunction

The ubiquitin-proteasome system (UPS) plays an important role in the elimination of misfolded, damaged, or otherwise unwanted proteins. The process of protein degradation begins with the addition of ubiquitin molecules to the target protein through the action of several enzymes including an E3 ligase such as parkin (see earlier). The proteasome

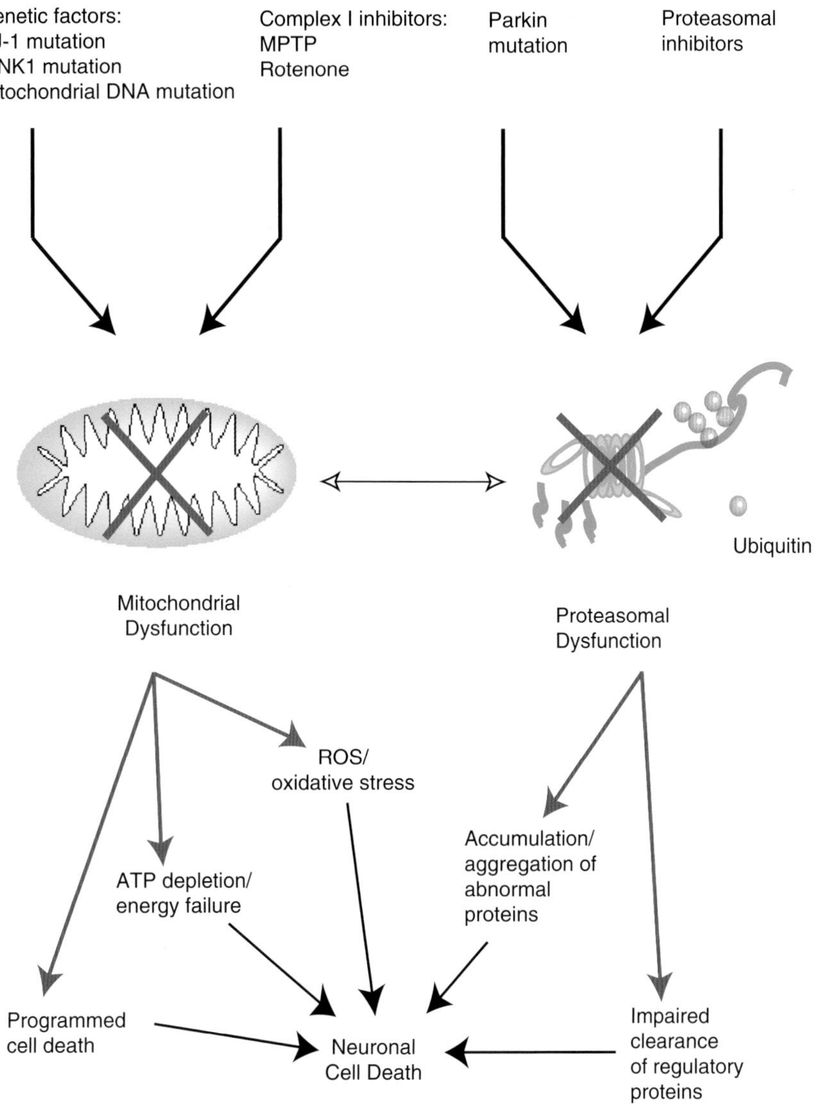

Figure 15.4 Potential mechanisms of cell death in Parkinson disease. A variety of genetic mutation and mitochondrial toxins have been implicated in causing forms of PD in patients and/or in model systems. Similarly, deficits in proteasomal function also have been associated with PD. There appears to be crosstalk between the proteasome and the mitochondria such that dysfunction in either leads to impairment in the other. These processes ultimately lead to downstream cell death mechanisms including programmed cell death, accumulation of oxidative stress, energy depletion, impaired protein clearance, and abnormal protein aggregation.

itself is composed of multiple protein subunits that form a barrel-shaped structure termed the 26S proteasome. This structure is responsible for the cleavage of unwanted proteins and is regulated by the presence of 20S and 19S complexes. The presence of the 19S component facilitates the recognition and unfolding of ubiquitinated target proteins in an ATP-dependent manner, whereas binding of the 20S component may play a larger role in the ATP and ubiquitin-independent functions of the proteasome. Impairment of proteasomal function, like all important cellular mechanisms, has broad pathologic implications for a cell. These include the abnormal accumulation of proteins, mitochondrial dysfunction, oxidative stress, possible cell-signaling abnormalities, and cell death (reviewed in McNaught et al., (2006)).

As discussed earlier, mutations in the genes encoding parkin and UCH-L1 and the PD-related modification of the parkin protein implicate ubiquitin-proteasomal system

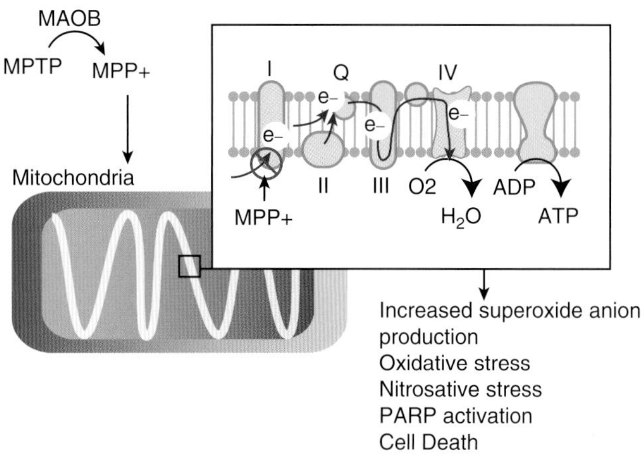

Figure 15.5 MPTP induced neurotoxicity. The toxin 1-methyl-4-phenyl-1,2,3,6-tetrahydropyridine (MPTP) induces Parkinsonism and is a useful experimental model of PD. MPTP is converted by monoamine oxidase B (MAO_B) to MPP+, which is then taken into dopamine neurons by the dopamine transporter. It acts in the mitochondria to block complex 1 (I) of the respiratory transport chain, which is responsible for maintaining the membrane potential necessary to drive the generation of ATP. Blocking complex I results in leak of electrons (e–) and the generation of superoxide anion leading to increased oxidative and nitrosative stress that activates cell death pathways.

malfunction in PD pathogenesis. In addition, direct examination of the brains of PD patients has found a reduction in both proteasomal activity and in the abundance of the alpha subunit of the 20S proteasome within the SNpc as compared to controls (McNaught et al., 2003).

Perhaps the most interesting but controversial data, however, come from studies suggesting that toxin-mediated inhibition of the proteasome can lead to dopaminergic cell death in culture and PD-like pathology in animal models that includes α-synuclein and ubiquitin accumulation (reviewed in McNaught et al. (2006)). Exactly how UPS dysfunction is involved in PD pathogenesis is unclear as is whether or not impairment of proteasomal function is a primary or secondary (or both) phenomenon. Specifically, it is not yet known whether UPS dysfunction brought about by genetic mutation or exposure to toxins leads to PD or whether it is a result of mitochondrial dysfunction, oxidative damage, protein misfolding, or other intracellular processes more upstream in a larger pathologic process. Whatever its role, UPS dysfunction does provide a target for future therapeutic interventions that stimulate or otherwise modulate the function of this vital cellular process.

C. Role of Phosphorylation

Protein phosphorylation is a vital cellular process that is particularly involved in the regulation of protein function. The discovery that mutation in two cellular kinases, namely LRRK2 and PINK-1, leads to inherited PD, suggests that abnormal phosphorylation may be involved in PD pathogenesis. Indeed, LRRK2 cellular pathogenicity has been shown to be dependent on the presence of kinase activity (Greggio et al., 2006). There is speculation that LRRK2 and PINK1 both may be involved in the Ras/Mitogen-activated protein kinase (MAPK) signaling pathway providing evidence that this particular pathway may be involved in PD pathogenesis (Shen, 2004).

The mammalian MAPK family includes p38 MAPKs, c-jun N-terminal kinases (JNKs), and extracellular signal-related kinase (ERKs) that have been implicated in cell survival and cell death mechanisms and whose expression is altered in PD and in Lewy body dementia (Ferrer et al., 2001). Upstream of the MAPK are the MAPK kinases (MAPKKs) and upstream of those are the MAPKK kinases (MAPKKKs). One important group of MAPKKKs is the mixed lineage kinases (MLKs) whose activities have been linked to cell death. In theory, cellular stress including oxidative stress may lead to a form of programmed cell death that relies on these kinase pathways. Data demonstrating the protection of dopaminergic neurons through the inhibition of the mixed lineage kinase-JNK pathway led to a clinical trial, examining the inhibitor CEP 1347 (reviewed in Silva et al. (2005)). Though this study was negative, work on the potential therapeutic value of modulators of kinase function will undoubtedly continue. More recently, the upstream kinase apoptosis signal-regulating kinase 1 (ASK1) has been implicated in dopaminergic toxin-mediated cell death (Ouyang & Shen, 2006).

Another kinase pathway implicated in PD is the phosphatidylinositol 3-kinase (PI3K)/Akt pathway. Specifically it was found that downregulation of the PD-related gene products DJ-1 and parkin in the fly leads to impaired Akt signaling and neuronal toxicity (Yang et al., 2005). In addition this pathway has been implicated in the mechanism behind the neuroprotective effects that GDNF has on hydrogen-peroxide treated cybrids (Onyango et al., 2005).

The downstream targets of LRRK2 and PINK1 are currently unknown. The downstream effects of MAPK signaling pathways include transcriptional regulation and the modification of survival factors. Interestingly, protein phosphorylation may also play a role in more direct pathogenic mechanisms. For example, α-synuclein is abnormally phosphorylated in PD and in various models of the disease. The presence of such modification can regulate the protein's toxicity and tendency to aggregate (Chen & Feany, 2005).

Our understanding of the role that changes in protein phosphorylation plays in the pathogenic mechanism of PD is rapidly growing. The implication for therapeutic intervention is great as we begin to more fully characterize these signaling pathways. Targeting seemingly distal events such as the phosphorylation of α-synuclein or more proximal events such as the MLK or ASK1 pathways may eventually lead to therapies to slow or halt progression of disease. Perhaps an even more promising area is the identification of PINK1 and

LRRK2 substrates and an understanding of how alterations in their phosphorylation status lead to clinical disease.

VIII. Therapy

A. Symptomatic Treatment

The most recognizable PD treatments including dopamine agonist therapy and L-dopa therapy address the disease primarily on the basis of the relative deficiency of dopamine. These therapies are effective in ameliorating many of the early symptoms of PD but do little or nothing to slow the progression of the disease. As the disease evolves over time many levodopa-nonresponsive symptoms develop and motor and nonmotor complications from dopaminergic therapies can significantly impair a patient's quality of life. Levodopa therapy provides dopamine precursor to the brain and the inclusion of decarboxylase, and catechol-O-methyl transferase inhibitors reduce the breakdown of levodopa thus increasing the effectiveness of a given dose of medication. Dopamine agonist therapy using agents such as pramipexole, ropinerole, and apomorphine directly stimulate dopamine receptors. In addition, the selective monoamine oxidase inhibitors selegiline and rasagiline provide symptomatic benefit likely through the inhibition of central dopamine breakdown. Common dopaminergic drugs and their sites of action are shown in Figure 15.6.

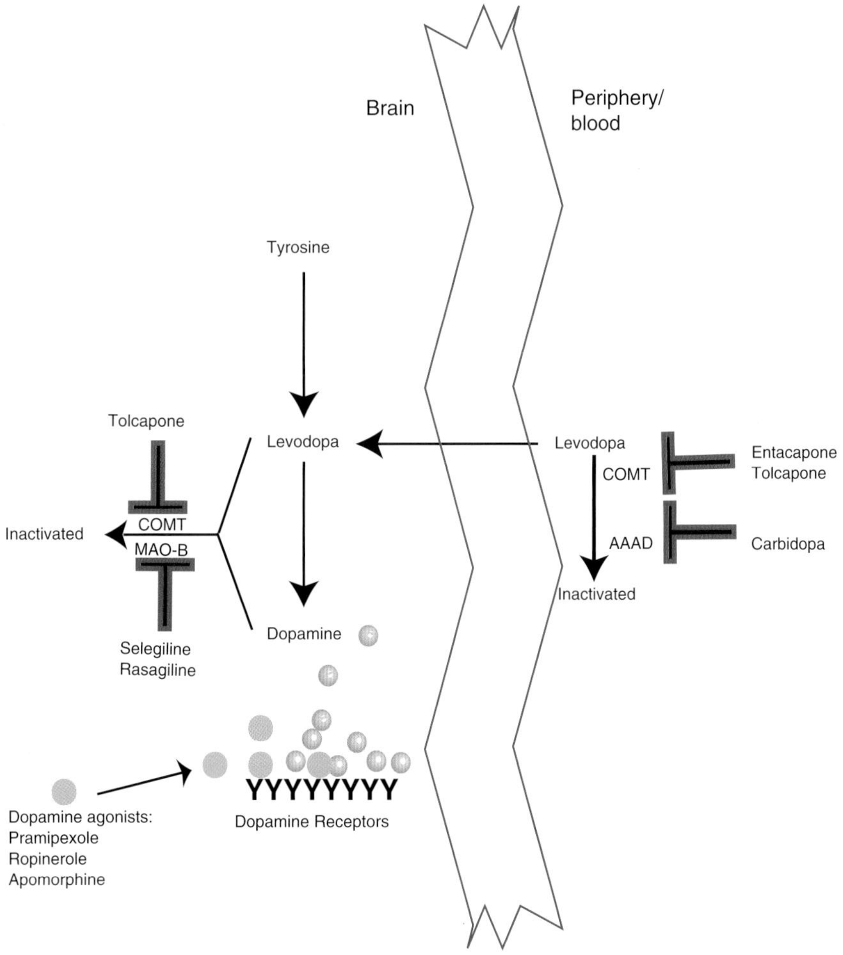

Figure 15.6 Site of action of anti-Parkinsonian symptomatic medications. Catechol-O-methyltransferase (COMT) and aromatic amino acid decarboxylase (AAAD) inhibitors prevent the peripheral metabolism of levodopa, a precursor to dopamine. Levodopa can cross the blood brain barrier where it is converted to dopamine within the brain. Dopamine inactivation in the brain can be reduced by blockade of monoamine-oxidase-B inhibitors such as selegiline and rasagiline and possibly by the COMT inhibitor tolcapone. Dopamine agonists (green ball) bind directly to dopamine receptors.

Data suggest that delaying levodopa therapy by initiating symptomatic treatment with dopamine agonists likely delays motor complications such as dyskinesias (Holloway et al., 2004; Rascol et al., 2000). These treatments are less effective than levodopa in ameliorating symptoms, and the vast majority of PD patients will eventually require levodopa therapy later in disease. The reduced dyskinesia risk using the dopamine agonists may stem from the increased half-life of these compounds relative to levodopa, providing a more continuous stimulation of dopamine receptors. This theory is supported by studies demonstrating that continuous levodopa or agonist infusion therapy has benefits relative to standard interval dosing (Colzi et al., 1998; Mouradian, 2005; Stocchi et al., 2002). When possible, patients able to tolerate dopamine agonists usually are begun on these therapies, though individual cases need to be assessed by the treating physician.

Deep brain stimulation (DBS) has emerged as a potent symptomatic therapy for advanced PD. At present this therapy is reserved for those patients whose symptoms are not adequately controlled by medication alone either due to lack of efficacy for particular symptoms (such as tremor) or an inability to tolerate the necessary doses of medication usually due to drug-related side effects such as dyskinesias. For a more detailed review see Benabid et al., 2005; Galvez-Jimenez, 2005.

B. Neuroprotection

Despite the advances in our understanding of PD, there are currently no robust agents to slow the progression of the disease. Several compounds have shown potential or have been deemed worthy of further experimentation. The monoamine oxidase inhibitors—type B including selegiline and rasagiline—have been shown in some studies to slow progression of PD, though the effect is moderate at best and the results are not universally accepted (Palhagen et al., 2006; Shults, 2005). Multiple mechanisms have been implicated in the protective effect of these compounds including inhibition of the GAPDH cell death pathway as well as other anti-apoptotic mechanisms (Hara et al., 2006).

Other agents with possible neuroprotective functions in PD include the anti-caspase, anti-inflammatory drug minocycline, and the pro-mitochondrial factors CoQ10, and creatine whose efficacy awaits further clinical studies (Shults et al., 2002).

Clinical trials building on a wealth of preclinical data have yielded mixed results using infusion therapy of glial cell line derived neurotrophic factor (GDNF). Despite positive results in previous studies, most recently a double-blind, placebo-controlled trial was discontinued early due to lack of efficacy, the development of antibodies against the drug, and concerns stemming from cerebellar toxicity seen in nonhuman primate studies (Lang et al., 2006). The mechanism of action of GDNF in preclinical models is largely unknown. Signaling through the Ret kinase pathway as well as perhaps other pathways have been implicated, but the downstream targets leading to the trophic effects of this protein on dopaminergic neurons is a field of active study.

C. Replacement Therapy

Viewing PD as a selective degeneration of dopaminergic SN neurons has led to the theory that this disease is a logical target for therapies aimed at cellular replacement. Specifically, if the correct cells could be created or isolated and transplanted into brains of patients with PD, dopamine could be specifically delivered to the deficient striatal receptors. Clinical trials using fetal-derived midbrain tissue transplanted into diseased striata have given mixed results. Two recent double-blind placebo-controlled trials did not meet their primary endpoints and were plagued by the unexpected development of dyskinesia (Freed et al., 2001; Hagell et al., 2000; Olanow et al., 2003; Piccini et al., 2005). Exactly why these studies failed to demonstrate the level of improvement seen in other open-labeled studies is unclear, though improvement in imaging characteristics and autopsy studies do provide proof of concept in that the grafted tissues appeared to survive and function on some level. Until the discrepancies between the recent controlled studies and previous uncontrolled studies are better understood and beneficial effects obtained consistently, this method of treatment cannot be recommended for routine clinical care.

Additional concerns regarding this therapy include the atopic location of the grafts. Since the tissue is not placed within the SN, it likely does not receive the normal input seen by native SN tissue. Also, transplantation of dopaminergic cells into the striata cannot be expected to treat those symptoms arising through loss of other cell types both within the central nervous system and peripherally.

One aspect of transplantation therapy that may be contributing to the lack of consistent results is variability in the preparation of source material. One potential remedy for this is the development of standardized methods to produce transplantable cells from human embryonic or adult stem cells. Various laboratories are developing protocols to produce dopaminergic cells and assaying their ability to reverse dopaminergic deficits in various animal models. For the most part these studies have demonstrated poor graft viability; however, it is likely only a matter of time until appropriate cells are produced and methods to improve survival are found (Taylor & Minger, 2005). Whether or not such cells will be able to provide meaningful therapies without risk of tumor formation, motor complications, or other complications currently is unknown.

IX. Summary

The history behind our current understanding of PD began nearly 200 years ago when the clinical syndrome was first described in detail. Since then understanding of PD pathology and the biochemical changes that led to the bulk of the motor symptoms has advanced greatly. Symptomatic treatment for the disease is a success story of modern molecular neurology and the current search for the etiology and involved pathologic processes will, in time, yield similar advances in the prevention and cure of the disease. We can hope that such advances will come soon and that methods to prevent the untimely loss of at risk neurons can be altered or reversed.

References

Abou-Sleiman, P. M., Muqit, M. M., and Wood, N. W. (2006). Expanding insights of mitochondrial dysfunction in Parkinson's disease. *Nat Rev Neurosci* **7**, 207–219.

Benabid, A. L., Chabardes, S., and Seigneuret, E. (2005). Deep-brain stimulation in Parkinson's disease: Long-term efficacy and safety—What happened this year? *Curr Opin Neurol* **18**, 623–630.

Bender, A., Krishnan, K. J., Morris, C. M., Taylor, G. A., Reeve, A. K., Perry, R. H. et al. (2006). High levels of mitochondrial DNA deletions in substantia nigra neurons in aging and Parkinson disease. *Nat Genet* **38**, 515–517.

Betarbet, R., Canet-Aviles, R. M., Sherer, T. B., Mastroberardino, P. G., McLendon, C., Kim, J. H. et al. (2006). Intersecting pathways to neurodegeneration in Parkinson's disease: Effects of the pesticide rotenone on DJ-1, alpha-synuclein, and the ubiquitin-proteasome system. *Neurobiol Dis* **22**, 404–420.

Betarbet, R., Sherer, T. B., MacKenzie, G., Garcia-Osuna, M., Panov, A. V., and Greenamyre, J. T. (2000). Chronic systemic pesticide exposure reproduces features of Parkinson's disease. *Nat Neurosci* **3**, 1301–1306.

Biskup, S., Moore, D. J., Celsi, F., Higashi, S., West, A. B., Andrabi, S. A. et al. (2006). Localization of LRRK2 indicating a role in vesicular trafficking and membrane recycling. *Ann Neurol.*

Bonifati, V., De Michele, G., Lucking, C. B., Durr, A., Fabrizio, E., Ambrosio, G. et al. (2001). The parkin gene and its phenotype. Italian PD Genetics Study Group, French PD Genetics Study Group and the European Consortium on Genetic Susceptibility in Parkinson's Disease. *Neurol Sci* **22**, 51–52.

Bonifati, V., Rizzu, P., van Baren, M. J., Schaap, O., Breedveld, G. J., Krieger, E. et al. (2003). Mutations in the DJ-1 gene associated with autosomal recessive early-onset parkinsonism. *Science* **299**, 256–259.

Bove, J., Prou, D., Perier, C., and Przedborski, S. (2005). Toxin-induced models of Parkinson's disease. *NeuroRx* **2**, 484–494.

Braak, H., de Vos, R. A., Bohl, J., and Del Tredici, K. (2006). Gastric alpha-synuclein immunoreactive inclusions in Meissner's and Auerbach's plexuses in cases staged for Parkinson's disease-related brain pathology. *Neurosci Lett* **396**, 67–72.

Braak, H., Del Tredici, K., Rub, U., de Vos, R. A., Jansen Steur, E. N., and Braak, E. (2003). Staging of brain pathology related to sporadic Parkinson's disease. *Neurobiol Aging* **24**, 197–211.

Brice, A. (2005). Genetics of Parkinson's disease: LRRK2 on the rise. *Brain* **128**, 2760–2762.

Brown, T. P., Rumsby, P. C., Capleton, A. C., Rushton, L., and Levy, L. S. (2006). Pesticides and Parkinson's disease—Is there a link? *Environ Health Perspect* **114**, 156–164.

Burn, D. J., Mark, M. H., Playford, E. D., Maraganore, D. M., Zimmerman, T. R., Jr., Duvoisin, R. C. et al. (1992). Parkinson's disease in twins studied with 18F-dopa and positron emission tomography. *Neurology* **42**, 1894–1900.

Carlsson, A., Lindqvist, M., and Magnusson, T. (1957). 3,4-Dihydroxyphenylalanine and 5-hydroxytryptophan as reserpine antagonists. *Nature* **180**, 1200.

Chen, L. and Feany, M. B. (2005). Alpha-synuclein phosphorylation controls neurotoxicity and inclusion formation in a Drosophila model of Parkinson disease. *Nat Neurosci* **8**, 657–663.

Choi, J., Sullards, M. C., Olzmann, J. A., Rees, H. D., Weintraub, S. T., Bostwick, D. E. et al. (2006). Oxidative damage of DJ-1 is linked to sporadic Parkinson's and Alzheimer's diseases. *J Biol Chem.*

Chung, K. K., Thomas, B., Li, X., Pletnikova, O., Troncoso, J. C., Marsh, L. et al. (2004). S-nitrosylation of parkin regulates ubiquitination and compromises parkin's protective function. *Science* **304**, 1328–1331.

Clark, I. E., Dodson, M. W., Jiang, C., Cao, J. H., Huh, J. R., Seol, J. H. et al. (2006). Drosophila pink1 is required for mitochondrial function and interacts genetically with parkin. *Nature.*

Colzi, A., Turner, K., and Lees, A. J. (1998). Continuous subcutaneous waking day apomorphine in the long term treatment of levodopa induced interdose dyskinesias in Parkinson's disease. *J Neurol Neurosurg Psychiatry* **64**, 573–576.

Conway, K. A., Rochet, J. C., Bieganski, R. M., and Lansbury, P. T., Jr. (2001). Kinetic stabilization of the alpha-synuclein protofibril by a dopamine-alpha-synuclein adduct. *Science* **294**, 1346–1349.

Cotzias, G. C., Papavasiliou, P. S., and Gellene, R. (1969). Modification of Parkinsonism—Chronic treatment with L-dopa. *N Engl J Med* **280**, 337–345.

Dauer, W., Kholodilov, N., Vila, M., Trillat, A. C., Goodchild, R., Larsen, K. E. et al. (2002). Resistance of alpha -synuclein null mice to the parkinsonian neurotoxin MPTP. *Proc Natl Acad Sci U S A* **99**, 14524–14529.

Davidzon, G., Greene, P., Mancuso, M., Klos, K. J., Ahlskog, J. E., Hirano, M., and DiMauro, S. (2006). Early-onset familial parkinsonism due to POLG mutations. *Ann Neurol* **59**, 859–862.

Dekker, M., Bonifati, V., van Swieten, J., Leenders, N., Galjaard, R. J., Snijders, P. et al. (2003). Clinical features and neuroimaging of PARK7-linked parkinsonism. *Mov Disord* **18**, 751–757.

Ehringer, H. and Hornykiewicz, O. (1960). Verteilung von noradrenalin und dopamin (3-hydroxytyramin) ingerhirn des menschen und ihr verhalten bei erkrankugen des extrapyramidalen systems. Klin Wochenschr **38**, 1236–1239.

Eriksen, J. L., Przedborski, S., and Petrucelli, L. (2005). Gene dosage and pathogenesis of Parkinson's disease. *Trends Mol Med* **11**, 91–96.

Evans, A. H., Lawrence, A. D., Potts, J., MacGregor, L., Katzenschlager, R., Shaw, K. et al. (2006). Relationship between impulsive sensation seeking traits, smoking, alcohol and caffeine intake, and Parkinson's disease. *J Neurol Neurosurg Psychiatry* **77**, 317–321.

Farrer, M., Chan, P., Chen, R., Tan, L., Lincoln, S., Hernandez, D. et al. (2001). Lewy bodies and parkinsonism in families with parkin mutations. *Ann Neurol* **50**, 293–300.

Farrer, M. J. (2006). Genetics of Parkinson disease: Paradigm shifts and future prospects. *Nat Rev Genet* **7**, 306–318.

Ferrer, I., Blanco, R., Carmona, M., Puig, B., Barrachina, M., Gomez, C., and Ambrosio, S. (2001). Active, phosphorylation-dependent mitogen-activated protein kinase (MAPK/ERK), stress-activated protein kinase/c-Jun N-terminal kinase (SAPK/JNK), and p38 kinase expression in Parkinson's disease and Dementia with Lewy bodies. *J Neural Transm* **108**, 1383–1396.

Fornai, F., Schluter, O. M., Lenzi, P., Gesi, M., Ruffoli, R., Ferrucci, M. et al. (2005). Parkinson-like syndrome induced by continuous MPTP infusion: convergent roles of the ubiquitin-proteasome system and alpha-synuclein. *Proc Natl Acad Sci U S A* **102**, 3413–3418.

Freed, C. R., Greene, P. E., Breeze, R. E., Tsai, W. Y., DuMouchel, W., Kao, R. et al. (2001). Transplantation of embryonic dopamine neurons for severe Parkinson's disease. *N Engl J Med* **344**, 710–719.

Gaig, C., Ezquerra, M., Marti, M. J., Munoz, E., Valldeoriola, F., and Tolosa, E. (2006). LRRK2 mutations in Spanish patients with Parkinson disease: Frequency, clinical features, and incomplete penetrance. *Arch Neurol* **63**, 377–382.

Galter, D., Westerlund, M., Carmine, A., Lindqvist, E., Sydow, O., and Olson, L. (2006). LRRK2 expression linked to dopamine-innervated areas. *Ann Neurol* **59**, 714–719.

Galvez-Jimenez, N. (2005). Advances in the surgical treatment of Parkinson's disease: emphasis on pallidotomy and deep brain stimulation. In N. Galvez-Jimenez, Ed., *Scientific basis for the treatment of Parkinson's disease, 2e*, 151–168. Taylor & Francis, London and New York.

Giasson, B. I., Covy, J. P., Bonini, N. M., Hurtig, H. I., Farrer, M. J., Trojanowski, J. Q. et al. (2006). Biochemical and pathological characterization of Lrrk2. *Ann Neurol* **59**, 315–322.

Giasson, B. I., Forman, M. S., Higuchi, M., Golbe, L. I., Graves, C. L., Kotzbauer, P. T. et al. (2003). Initiation and synergistic fibrillization of tau and alpha-synuclein. *Science* **300**, 636–640.

Gloeckner, C. J., Kinkl, N., Schumacher, A., Braun, R. J., O'Neill, E., Meitinger, T. et al. (2006). The Parkinson disease causing LRRK2 mutation I2020T is associated with increased kinase activity. *Hum Mol Genet* **15**, 223–232.

Goetz, C. G. (2002). Charcot and Parkinson's Disease. In S. Factor and W. Weiner, Eds., *Parkinson's disease diagnosis and clinical management*. Demos Medical Publishing, New York.

Goldman, S. M., Tanner, C. M., Olanow, C. W., Watts, R. L., Field, R. D., and Langston, J. W. (2005). Occupation and parkinsonism in three movement disorders clinics. *Neurology* **65**, 1430–1435.

Greene, J. C., Whitworth, A. J., Kuo, I., Andrews, L. A., Feany, M. B., and Pallanck, L. J. (2003). Mitochondrial pathology and apoptotic muscle degeneration in Drosophila parkin mutants. *Proc Natl Acad Sci U S A* **100**, 4078–4083.

Greenfield, J. G. and Bosanquet, F. D. (1953). The brain-stem lesions in Parkinsonism. *J Neurol Neurosurg Psychiatry* **16**, 213–226.

Greggio, E., Jain, S., Kingsbury, A., Bandopadhyay, R., Lewis, P., Kaganovich, A. et al. (2006). Kinase activity is required for the toxic effects of mutant LRRK2/dardarin. *Neurobiol Dis*.

Grimes, D. A., Han, F., Panisset, M., Racacho, L., Xiao, F., Zou, R. et al. (2006). Translated mutation in the Nurr1 gene as a cause for Parkinson's disease. *Mov Disord*.

Group, P. S. (2002). A controlled trial of rasagiline in early Parkinson disease: The TEMPO Study. *Arch Neurol* **59**, 1937–1943.

Hadjigeorgiou, G. M., Xiromerisiou, G., Gourbali, V., Aggelakis, K., Scarmeas, N., Papadimitriou, A., and Singleton, A. (2005). Association of alpha-synuclein Rep1 polymorphism and Parkinson's disease: Influence of Rep1 on age at onset. *Mov Disord*.

Hagell, P., Crabb, L., Pogarell, O., Schrag, A., Widner, H., Brooks, D. J. et al. (2000). Health-related quality of life following bilateral intrastriatal transplantation in Parkinson's disease. *Mov Disord* **15**, 224–229.

Hara, M. R., Thomas, B., Cascio, M. B., Bae, B. I., Hester, L. D., Dawson, V. L. et al. (2006). Neuroprotection by pharmacologic blockade of the GAPDH death cascade. *Proc Natl Acad Sci U S A* **103**, 3887–3889.

Hasegawa, M., Fujiwara, H., Nonaka, T., Wakabayashi, K., Takahashi, H., Lee, V. M. et al. (2002). Phosphorylated alpha-synuclein is ubiquitinated in alpha-synucleinopathy lesions. *J Biol Chem* **277**, 49071–49076.

Hashimoto, M., Rockenstein, E., Crews, L., and Masliah, E. (2003). Role of protein aggregation in mitochondrial dysfunction and neurodegeneration in Alzheimer's and Parkinson's diseases. *Neuromolecular Med* **4**, 21–36.

Healy, D. G., Abou-Sleiman, P. M., Casas, J. P., Ahmadi, K. R., Lynch, T., Gandhi, S. et al. (2006). UCHL-1 is not a Parkinson's disease susceptibility gene. *Ann Neurol* **59**, 627–633.

Holloway, R. G., Shoulson, I., Fahn, S., Kieburtz, K., Lang, A., Marek, K. et al. (2004). Pramipexole vs levodopa as initial treatment for Parkinson disease: A 4-year randomized controlled trial. *Arch Neurol* **61**, 1044–1053.

Hughes, A. J., Daniel, S. E., Blankson, S., and Lees, A. J. (1993). A clinicopathologic study of 100 cases of Parkinson's disease. *Arch Neurol* **50**, 140–148.

Ibanez, P., Lesage, S., Lohmann, E., Thobois, S., De Michele, G., Borg, M., Agid, Y., Durr, A., and Brice, A. (2006). Mutational analysis of the PINK1 gene in early-onset parkinsonism in Europe and North Africa. *Brain* **129**, 686–694.

Ide, M., Yamada, K., Toyota, T., Iwayama, Y., Ishitsuka, Y., Minabe, Y. et al. (2005). Genetic association analyses of PHOX2B and ASCL1 in neuropsychiatric disorders: Evidence for association of ASCL1 with Parkinson's disease. *Hum Genet* **117**, 520–527.

Investigators, N. N.-P. (2006). A randomized, double-blind, futility clinical trial of creatine and minocycline in early Parkinson disease. *Neurology* **66**, 664–671.

Iwanaga, K., Wakabayashi, K., Yoshimoto, M., Tomita, I., Satoh, H., Takashima, H. et al. (1999). Lewy body-type degeneration in cardiac plexus in Parkinson's and incidental Lewy body diseases. *Neurology* **52**, 1269–1271.

Jensen, P. H., Hager, H., Nielsen, M. S., Hojrup, P., Gliemann, J., and Jakes, R. (1999). Alpha-synuclein binds to Tau and stimulates the protein kinase A-catalyzed tau phosphorylation of serine residues 262 and 356. *J Biol Chem* **274**, 25481–25489.

Keppel Hesselink, J. M. (1989). Trauma as an etiology of parkinsonism: Opinions in the nineteenth century. *Mov Disord* **4**, 283–285.

Kim, R. H., Smith, P. D., Aleyasin, H., Hayley, S., Mount, M. P., Pownall, S. et al. (2005). Hypersensitivity of DJ-1-deficient mice to 1-methyl-4-phenyl-1,2,3,6-tetrahydropyridine (MPTP) and oxidative stress. *Proc Natl Acad Sci U S A* **102**, 5215–5220.

Kitada, T., Asakawa, S., Hattori, N., Matsumine, H., Yamamura, Y., Minoshima, S. et al. (1998). Mutations in the parkin gene cause autosomal recessive juvenile parkinsonism. *Nature* **392**, 605–608.

Klein, C., Djarmati, A., Hedrich, K., Schafer, N., Scaglione, C., Marchese, R. et al. (2005). PINK1, Parkin, and DJ-1 mutations in Italian patients with early-onset parkinsonism. *Eur J Hum Genet* **13**, 1086–1093.

Ko, H. S., Kim, S. W., Sriram, S. R., Dawson, V. L., and Dawson, T. M. (2006). Identification of far up stream element binding protein-1 as an authentic parkin substrate. *J Biol Chem*.

Ko, H. S., von Coelln, R., Sriram, S. R., Kim, S. W., Chung, K. K., Pletnikova, O. et al. (2005). Accumulation of the authentic parkin substrate aminoacyl-tRNA synthetase cofactor, p38/JTV-1, leads to catecholaminergic cell death. *J Neurosci* **25**, 7968–7978.

Kraytsberg, Y., Kudryavtseva, E., McKee, A. C., Geula, C., Kowall, N. W., and Khrapko, K. (2006). Mitochondrial DNA deletions are abundant and cause functional impairment in aged human substantia nigra neurons. *Nat Genet* **38**, 518–520.

Kruger, R., Kuhn, W., Muller, T., Woitalla, D., Graeber, M., Kosel, S. et al. (1998). Ala30Pro mutation in the gene encoding alpha-synuclein in Parkinson's disease. *Nat Genet* **18**, 106–108.

Kubo, S. I., Hattori, N., and Mizuno, Y. (2006). Recessive Parkinson's disease. *Mov Disord*.

Lang, A. E., Gill, S., Patel, N. K., Lozano, A., Nutt, J. G., Penn, R. et al. (2006). Randomized controlled trial of intraputamenal glial cell line-derived neurotrophic factor infusion in Parkinson disease. *Ann Neurol* **59**, 459–466.

Langston, J. W., Ballard, P., Tetrud, J. W., and Irwin, I. (1983). Chronic Parkinsonism in humans due to a product of meperidine-analog synthesis. *Science* **219**, 979–980.

LaVoie, M. J., Ostaszewski, B. L., Weihofen, A., Schlossmacher, M. G., and Selkoe, D. J. (2005). Dopamine covalently modifies and functionally inactivates parkin. *Nat Med* **11**, 1214–1221.

Leroux, P.-D. (1880). Contribution a l'etude des causes de la paralysie agitante. These de paris, imprimeur de la faculte de medecine.

Leroy, E., Boyer, R., Auburger, G., Leube, B., Ulm, G., Mezey, E. et al. (1998). The ubiquitin pathway in Parkinson's disease. *Nature* **395**, 451–452.

Lesage, S., Durr, A., Tazir, M., Lohmann, E., Leutenegger, A. L., Janin, S. et al. (2006). LRRK2 G2019S as a cause of Parkinson's disease in North African Arabs. *N Engl J Med* **354**, 422–423.

Lewy, F. H. (1912). Paralysis agitans. I. Pathologische Anatomic. In M. Lewandowrky, Ed., *Handbuch der neurologic,* 920–933. Springer, Berlin

Li, C. and Beal, M. F. (2005). Leucine-rich repeat kinase 2: A new player with a familiar theme for Parkinson's disease pathogenesis. *Proc Natl Acad Sci U S A* **102**, 16535–16536.

Li, W., West, N., Colla, E., Pletnikova, O., Troncoso, J. C., Marsh, L. et al. (2005). Aggregation promoting C-terminal truncation of alpha-synuclein is a normal cellular process and is enhanced by the familial Parkinson's disease-linked mutations. *Proc Natl Acad Sci U S A* **102**, 2162–2167.

Lim, K. L., Dawson, V. L., and Dawson, T. M. (2006). Parkin-mediated lysine 63-linked polyubiquitination: A link to protein inclusions formation in Parkinson's and other conformational diseases? *Neurobiol Aging* **27**, 524–529.

Liu, Y., Fallon, L., Lashuel, H. A., Liu, Z., and Lansbury, P. T., Jr. (2002). The UCH-L1 gene encodes two opposing enzymatic activities that affect alpha-synuclein degradation and Parkinson's disease susceptibility. *Cell* **111**, 209–218.

Lockhart, P. J., Lincoln, S., Hulihan, M., Kachergus, J., Wilkes, K., Bisceglio, G. et al. (2004). DJ-1 mutations are a rare cause of recessively inherited early onset parkinsonism mediated by loss of protein function. *J Med Genet* **41**, e22.

Lohmann, E., Periquet, M., Bonifati, V., Wood, N. W., De Michele, G., Bonnet, A. M. et al. (2003). How much phenotypic variation can be attributed to parkin genotype? *Ann Neurol* **54**, 176–185.

Ludecke, B., Knappskog, P. M., Clayton, P. T., Surtees, R. A., Clelland, J. D., Heales, S. J. et al. (1996). Recessively inherited L-DOPA-responsive parkinsonism in infancy caused by a point mutation (L205P) in the tyrosine hydroxylase gene. *Hum Mol Genet* **5**, 1023–1028.

Maher, N. E., Currie, L. J., Lazzarini, A. M., Wilk, J. B., Taylor, C. A., Saint-Hilaire, M. H. et al. (2002a). Segregation analysis of Parkinson disease revealing evidence for a major causative gene. *Am J Med Genet* **109**, 191–197.

Maher, N. E., Golbe, L. I., Lazzarini, A. M., Mark, M. H., Currie, L. J., Wooten, G. F. et al. (2002b). Epidemiologic study of 203 sibling pairs with Parkinson's disease: the GenePD study. *Neurology* **58**, 79–84.

Maraganore, D. M., de Andrade, M., Lesnick, T. G., Strain, K. J., Farrer, M. J., Rocca, W. A. et al. (2005). High-resolution whole-genome association study of Parkinson disease. *Am J Hum Genet* **77**, 685–693.

Marras, C., McDermott, M. P., Rochon, P. A., Tanner, C. M., Naglie, G., Rudolph, A., and Lang, A. E. (2005). Survival in Parkinson disease: Thirteen-year follow-up of the DATATOP cohort. *Neurology* **64**, 87–93.

Mata, I. F., Wedemeyer, W. J., Farrer, M. J., Taylor, J. P., and Gallo, K. A. (2006). LRRK2 in Parkinson's disease: Protein domains and functional insights. *Trends Neurosci* **29**, 286–293.

McNaught, K. S., Belizaire, R., Isacson, O., Jenner, P., and Olanow, C. W. (2003). Altered proteasomal function in sporadic Parkinson's disease. *Exp Neurol* **179**, 38–46.

McNaught, K. S., Jackson, T., JnoBaptiste, R., Kapustin, A., and Olanow, C. W. (2006). Proteasomal dysfunction in sporadic Parkinson's disease. *Neurology* **66**, S37–49.

McNaught, K. S., Perl, D. P., Brownell, A. L., and Olanow, C. W. (2004). Systemic exposure to proteasome inhibitors causes a progressive model of Parkinson's disease. *Ann Neurol* **56**, 149–162.

Melrose, H., Lincoln, S., Tyndall, G., Dickson, D., and Farrer, M. (2006). Anatomical localization of leucine-rich repeat kinase 2 in mouse brain. *Neuroscience* **139**, 791–794.

Mizuno, Y., Mochizuki, H., and Hattori, N. (2005). Alpha-synuclein, nigral degeneration and parkinsonism. In N. Galvez-Jimenez, Ed., *Scientific basis for the treatment of parkinson's disease, 2e*, 87–104. Taylor & Francis, New York and London.

Moore, D. J., Zhang, L., Troncoso, J., Lee, M. K., Hattori, N., Mizuno, Y. et al. (2005). Association of DJ-1 and parkin mediated by pathogenic DJ-1 mutations and oxidative stress. *Hum Mol Genet* **14**, 71–84.

Mori, H., Kondo, T., Yokochi, M., Matsumine, H., Nakagawa-Hattori, Y., Miyake, T. et al. (1998). Pathologic and biochemical studies of juvenile parkinsonism linked to chromosome 6q. *Neurology* **51**, 890–892.

Morris, H. R. (2005). Genetics of Parkinson's disease. *Ann Med* **37**, 86–96.

Mouradian, M. M. (2005). Should levodopa be infused into the duodenum? *Neurology* **64**, 182–183.

Mukaetova-Ladinska, E. B. and McKeith, I. G. (2006). Pathophysiology of synuclein aggregation in Lewy body disease. *Mech Ageing Dev* **127**, 188–202.

Nishioka, K., Hayashi, S., Farrer, M. J., Singleton, A. B., Yoshino, H., Imai, H. et al. (2006). Clinical heterogeneity of alpha-synuclein gene duplication in Parkinson's disease. *Ann Neurol* **59**, 298–309.

Nutt, J. G. and Wooten, G. F. (2005). Clinical practice. Diagnosis and initial management of Parkinson's disease. *N Engl J Med* **353**, 1021–1027.

Olanow, C. W., Goetz, C. G., Kordower, J. H., Stoessl, A. J., Sossi, V., Brin, M. F. et al. (2003). A double-blind controlled trial of bilateral fetal nigral transplantation in Parkinson's disease. *Ann Neurol* **54**, 403–414.

Onyango, I. G., Tuttle, J. B., and Bennett, J. P., Jr. (2005). Brain-derived growth factor and glial cell line-derived growth factor use distinct intracellular signaling pathways to protect PD cybrids from H_2O_2-induced neuronal death. *Neurobiol Dis* **20**, 141–154.

Ouyang, M. and Shen, X. (2006). Critical role of ASK1 in the 6-hydroxydopamine-induced apoptosis in human neuroblastoma SH-SY5Y cells. *J Neurochem* **97**, 234–244.

Ozelius, L. J., Senthil, G., Saunders-Pullman, R., Ohmann, E., Deligtisch, A., Tagliati, M. et al. (2006). LRRK2 G2019S as a cause of Parkinson's disease in Ashkenazi Jews. *N Engl J Med* **354**, 424–425.

Paisan-Ruiz, C., Jain, S., Evans, E. W., Gilks, W. P., Simon, J., van der Brug, M. et al. (2004). Cloning of the gene containing mutations that cause PARK8-linked Parkinson's disease. *Neuron* **44**, 595–600.

Paisan-Ruiz, C., Lang, A. E., Kawarai, T., Sato, C., Salehi-Rad, S., Fisman, G. K. et al. (2005). LRRK2 gene in Parkinson disease: Mutation analysis and case control association study. *Neurology* **65**, 696–700.

Palacino, J. J., Sagi, D., Goldberg, M. S., Krauss, S., Motz, C., Wacker, M. et al. (2004). Mitochondrial dysfunction and oxidative damage in parkin-deficient mice. *J Biol Chem* **279**, 18614–18622.

Palhagen, S., Heinonem, E., Hagglund, J., Kaugesaar, T., Maki-Ikola, O., and Palm, R. (2006). Selegiline slows the progression of the symptoms of Parkinson disease. *Neurology* **66**, 1–7.

Pals, P., Lincoln, S., Manning, J., Heckman, M., Skipper, L., Hulihan, M. et al. (2004). Alpha-synuclein promoter confers susceptibility to Parkinson's disease. *Ann Neurol* **56**, 591–595.

Pandey, N., Schmidt, R. E., and Galvin, J. E. (2006). The alpha-synuclein mutation E46K promotes aggregation in cultured cells. *Exp Neurol* **197**, 515–520.

Park, J., Lee, S. B., Lee, S., Kim, Y., Song, S., Kim, S. et al. (2006). Mitochondrial dysfunction in Drosophila PINK1 mutants is complemented by parkin. *Nature*.

Parkinson, J. (2002). An essay on the shaking palsy. *J Neuropsychiatry Clin Neurosci* **14**, 223–236; discussion 222.

Petit, A., Kawarai, T., Paitel, E., Sanjo, N., Maj, M., Scheid, M. et al. (2005). Wild-type PINK1 prevents basal and induced neuronal apoptosis, a protective effect abrogated by Parkinson disease-related mutations. *J Biol Chem* **280**, 34025–34032.

Piccini, P., Pavese, N., Hagell, P., Reimer, J., Bjorklund, A., Oertel, W. H. et al. (2005). Factors affecting the clinical outcome after neural transplantation in Parkinson's disease. *Brain* **128**, 2977–2986.

Polymeropoulos, M. H., Lavedan, C., Leroy, E., Ide, S. E., Dehejia, A., Dutra, A. et al. (1997). Mutation in the alpha-synuclein gene identified in families with Parkinson's disease. *Science* **276**, 2045–2047.

Rascol, O., Brooks, D. J., Korczyn, A. D., De Deyn, P. P., Clarke, C. E., and Lang, A. E. (2000). A five-year study of the incidence of dyskinesia in patients with early Parkinson's disease who were treated with ropinirole or levodopa. 056 Study Group. *N Engl J Med* **342**, 1484–1491.

Reid, A. H., McCall, S., Henry, J. M., and Taubenberger, J. K. (2001). Experimenting on the past: The enigma of von Economo's encephalitis lethargica. *J Neuropathol Exp Neurol* **60**, 663–670.

Rochet, J. C., Outeiro, T. F., Conway, K. A., Ding, T. T., Volles, M. J., Lashuel, H. A. et al. (2004). Interactions among alpha-synuclein, dopamine, and biomembranes: Some clues for understanding neurodegeneration in Parkinson's disease. *J Mol Neurosci* **23**, 23–34.

Sasaki, S., Shirata, A., Yamane, K., and Iwata, M. (2004). Parkin-positive autosomal recessive juvenile Parkinsonism with alpha-synuclein-positive inclusions. *Neurology* **63**, 678–682.

Schulz, J. B. and Falkenburger, B. H. (2004). Neuronal pathology in Parkinson's disease. *Cell Tissue Res* **318**, 135–147.

Shen, J. (2004). Protein kinases linked to the pathogenesis of Parkinson's disease. *Neuron* **44**, 575–577.

Shults, C. W. (2004). Mitochondrial dysfunction and possible treatments in Parkinson's disease—A review. *Mitochondrion* **4**, 641–648.

Shults, C. W. (2005). Reexamination of the TEMPO Study. *Arch Neurol* **62**, 1320; author reply 1321.

Shults, C. W. (2006). Lewy bodies. *Proc Natl Acad Sci U S A* **103**, 1661–1668.

Shults, C. W., Oakes, D., Kieburtz, K., Beal, M. F., Haas, R., Plumb, S. et al. (2002). Effects of coenzyme Q10 in early Parkinson disease: Evidence of slowing of the functional decline. *Arch Neurol* **59**, 1541–1550.

Silva, R. M., Kuan, C. Y., Rakic, P., and Burke, R. E. (2005). Mixed lineage kinase-c-jun N-terminal kinase signaling pathway: A new therapeutic target in Parkinson's disease. *Mov Disord* **20**, 653–664.

Silvestri, L., Caputo, V., Bellacchio, E., Atorino, L., Dallapiccola, B., Valente, E. M., and Casari, G. (2005). Mitochondrial import and enzymatic activity of PINK1 mutants associated to recessive parkinsonism. *Hum Mol Genet* **14**, 3477–3492.

Singleton, A. B., Farrer, M., Johnson, J., Singleton, A., Hague, S., Kachergus, J. et al. (2003). Alpha-synuclein locus triplication causes Parkinson's disease. *Science* **302**, 841.

Smith, W. W., Jiang, H., Pei, Z., Tanaka, Y., Morita, H., Sawa, A. et al. (2005a). Endoplasmic reticulum stress and mitochondrial cell death pathways mediate A53T mutant alpha-synuclein-induced toxicity. *Hum Mol Genet* **14**, 3801–3811.

Smith, W. W., Pei, Z., Jiang, H., Dawson, V. L., Dawson, T. M., and Ross, C. A. (2006). Kinase activity of mutant LRRK2 mediates neuronal toxicity. *Nat Neurosci.*

Smith, W. W., Pei, Z., Jiang, H., Moore, D. J., Liang, Y., West, A. B. et al. (2005b). Leucine-rich repeat kinase 2 (LRRK2) interacts with parkin, and mutant LRRK2 induces neuronal degeneration. *Proc Natl Acad Sci U S A* **102**, 18676–18681.

Spillantini, M. G., Crowther, R. A., Jakes, R., Hasegawa, M., and Goedert, M. (1998). Alpha-synuclein in filamentous inclusions of Lewy bodies from Parkinson's disease and dementia with lewy bodies. *Proc Natl Acad Sci U S A* **95**, 6469–6473.

Sriram, S. R., Li, X., Ko, H. S., Chung, K. K., Wong, E., Lim, K. L. et al. (2005). Familial-associated mutations differentially disrupt the solubility, localization, binding and ubiquitination properties of parkin. *Hum Mol Genet* **14**, 2571–2586.

Stefanis, L., Larsen, K. E., Rideout, H. J., Sulzer, D., and Greene, L. A. (2001). Expression of A53T mutant but not wild-type alpha-synuclein in PC12 cells induces alterations of the ubiquitin-dependent degradation system, loss of dopamine release, and autophagic cell death. *J Neurosci* **21**, 9549–9560.

Stocchi, F., Ruggieri, S., Vacca, L., and Olanow, C. W. (2002). Prospective randomized trial of lisuride infusion versus oral levodopa in patients with Parkinson's disease. *Brain* **125**, 2058–2066.

Tan, E. K., Chai, A., Teo, Y. Y., Zhao, Y., Tan, C., Shen, H. et al. (2004). Alpha-synuclein haplotypes implicated in risk of Parkinson's disease. *Neurology* **62**, 128–131.

Tan, E. K., Yew, K., Chua, E., Puvan, K., Shen, H., Lee, E. et al. (2006). PINK1 mutations in sporadic early-onset Parkinson's disease. *Mov Disord.*

Tan, J. M. and Dawson, T. M. (2006). Parkin blushed by PINK1. *Neuron* **50**, 527–529.

Tanaka, Y., Engelender, S., Igarashi, S., Rao, R. K., Wanner, T., Tanzi, R. E. et al. (2001). Inducible expression of mutant alpha-synuclein decreases proteasome activity and increases sensitivity to mitochondria-dependent apoptosis. *Hum Mol Genet* **10**, 919–926.

Tanner, C. M., Ottman, R., Goldman, S. M., Ellenberg, J., Chan, P., Mayeux, R., and Langston, J. W. (1999). Parkinson disease in twins: An etiologic study. *JAMA* **281**, 341–346.

Taylor, H. and Minger, S. L. (2005). Regenerative medicine in Parkinson's disease: Generation of mesencephalic dopaminergic cells from embryonic stem cells. *Curr Opin Biotechnol* **16**, 487–492.

Thiruchelvam, M., Richfield, E. K., Baggs, R. B., Tank, A. W., and Cory-Slechta, D. A. (2000). The nigrostriatal dopaminergic system as a preferential target of repeated exposures to combined paraquat and maneb: Implications for Parkinson's disease. *J Neurosci* **20**, 9207–9214.

Tofaris, G. K., Garcia Reitbock, P., Humby, T., Lambourne, S. L., O'Connell, M., Ghetti, B. et al. (2006). Pathological changes in dopaminergic nerve cells of the substantia nigra and olfactory bulb in mice transgenic for truncated human alpha-synuclein (1-120): Implications for Lewy body disorders. *J Neurosci* **26**, 3942–3950.

Valente, E. M., Abou-Sleiman, P. M., Caputo, V., Muqit, M. M., Harvey, K., Gispert, S. et al. (2004). Hereditary early-onset Parkinson's disease caused by mutations in PINK1. *Science* **304**, 1158–1160.

van der Walt, J. M., Noureddine, M. A., Kittappa, R., Hauser, M. A., Scott, W. K., McKay, R. et al. (2004). Fibroblast growth factor 20 polymorphisms and haplotypes strongly influence risk of Parkinson disease. *Am J Hum Genet* **74**, 1121–1127.

Weinreb, P. H., Zhen, W., Poon, A. W., Conway, K. A., and Lansbury, P. T., Jr. (1996). NACP, a protein implicated in Alzheimer's disease and learning, is natively unfolded. *Biochemistry* **35**, 13709–13715.

West, A. B., Moore, D. J., Biskup, S., Bugayenko, A., Smith, W. W., Ross, C. A. et al. (2005). Parkinson's disease-associated mutations in leucine-rich repeat kinase 2 augment kinase activity. *Proc Natl Acad Sci U S A* **102**, 16842–16847.

Yang, Y., Gehrke, S., Haque, M. E., Imai, Y., Kosek, J., Yang, L. et al. (2005). Inactivation of Drosophila DJ-1 leads to impairments of oxidative stress response and phosphatidylinositol 3-kinase/Akt signaling. *Proc Natl Acad Sci U S A* **102**, 13670–13675.

Zarranz, J. J., Fernandez-Bedoya, A., Lambarri, I., Gomez-Esteban, J. C., Lezcano, E., Zamacona, J., and Madoz, P. (2005). Abnormal sleep architecture is an early feature in the E46K familial synucleinopathy. *Mov Disord* **20**, 1310–1315.

Zhang, J., Song, Y., Chen, H., and Fan, D. (2005a). The tau gene haplotype h1 confers a susceptibility to Parkinson's disease. *Eur Neurol* **53**, 15–21.

Zhang, L., Shimoji, M., Thomas, B., Moore, D. J., Yu, S. W., Marupudi, N. I. et al. (2005b). Mitochondrial localization of the Parkinson's disease related protein DJ-1: implications for pathogenesis. *Hum Mol Genet* **14**, 2063–2073.

Zhang, Y., Gao, J., Chung, K. K., Huang, H., Dawson, V. L., and Dawson, T. M. (2000). Parkin functions as an E2-dependent ubiquitin-protein ligase and promotes the degradation of the synaptic vesicle-associated protein, CDCrel-1. *Proc Natl Acad Sci U S A* **97**, 13354–13359.

Zhou, W., Zhu, M., Wilson, M. A., Petsko, G. A., and Fink, A. L. (2006). The oxidation state of DJ-1 regulates its chaperone activity toward alpha-synuclein. *J Mol Biol* **356**, 1036–1048.

Zimprich, A., Biskup, S., Leitner, P., Lichtner, P., Farrer, M., Lincoln, S. et al. (2004). Mutations in LRRK2 cause autosomal-dominant parkinsonism with pleomorphic pathology. *Neuron* **44**, 601–607.

16

The Molecular Basis of Alzheimer's Disease

Martin Ingelsson and Bradley T. Hyman

I. Clinical Features of Alzheimer's Disease

Alzheimer's disease (AD) is the most common progressive neurodegenerative disorder, affecting at least four million people in the United States alone. Memory impairment often is seen as the first symptom and typically is followed by temporal and spatial disorientation together with various degrees of language impairment and personality change. Disease risk increases dramatically with age; 5 percent of the population above 65 years of age are affected whereas 20 to 25 percent of those over 80 years suffer from the disorder. The duration of disease typically varies between five and 20 years.

Today's drugs for treatment of AD are based on the principle of potentiating the brain's cholinergic function, one of several transmitter systems affected in the disease, and is not directed against the pathogenic mechanisms. These drugs have a limited effect on cognition, behavior, and functions related to the activities of daily living. The currently most widely used substance is donepezil (Aricept), an acetylcholine esterase inhibitor that is indicated for mild-moderate disease stages. A more recent addition is memantine (Ebixa or Namenda), an NMDA receptor antagonist indicated also for more advanced stages of the disease.

The definitive diagnosis can only be obtained after autopsy, but a preliminary diagnosis based on clinical evaluation by a skilled neurologist is accurate in up to 90 percent of all cases. Such an evaluation is supported by biochemical analyses of cerebrospinal fluid (CSF) and neuroimaging techniques, but depends mainly on a thorough medical history together with a basic neurological examination. The preliminary clinical diagnostic criteria as well as the definite neuropathological criteria are listed in Table 16.1 (Hyman, 1997; McKhann et al., 1984; Mirra et al., 1991).

The subsequent sections will describe the underlying molecular disease mechanisms and how this knowledge eventually can influence the daily lives of patients and caregivers. As of today, only symptomatic pharmacotherapies are available for the treatment of AD, but new insights into the underlying molecular mechanisms are providing strategies to prevent or even cure this devastating disorder.

II. Neuropathology of Alzheimer's Disease

Today's understanding of the molecular underpinnings in AD pathogenesis is based partly on descriptive neuropathology from the early twentieth century. On affected brains, Dr. Alois Alzheimer was able to demonstrate extracellular senile plaques and intracellular neurofibrillary tangles, lesions that still are regarded as major disease hallmarks (see Figure 16.1) (Alzheimer, 1907, 1911). Some hundred years later, many researchers in the field are still intense ly engaged in the study of these abnormal brain aggregates with the aim of understanding their exact nature and the molecular mechanisms responsible for their formation.

Apart from plaques and tangles, the AD brain displays gross atrophy as well as neuronal loss, loss of synapses, and gliosis. Dystrophic neurites and neuropil threads are most commonly in the vicinity of plaques (see Figure 16.1B). In addition, in most AD brain amyloid angiopathy is seen in the walls of large and intermediate blood vessels.

Whereas the distribution of plaques is known to vary widely, both within architectonic units and between individuals, neurofibrillary pathology evolves in a hierarchical manner. Typically, the entorhinal region is the first affected area, followed by the appearance of pathology in hippocampus before the subsequent involvement of widespread cortical association areas (Arnold et al., 1991; Braak & Braak, 1991).

Neurofibrillary tangles, rather than amyloid deposits, have been highlighted as a clinico-pathological disease correlate of dementia (Arriagada et al., 1992a; Ingelsson et al., 2004). Moreover, the extent and regional distribution of neuronal loss in AD parallels, but exceeds, tangle formation (Gomez-Isla et al., 1997). Finally, gliosis and synaptic loss, as measured by a decline of glial and synaptic protein levels, also have been shown to correlate with disease progression (Ingelsson et al., 2004).

By the mid-1980s, two research teams independently defined the biochemical composition of plaques and vessel wall deposits as the amyloid-β (Aβ) peptide (Glenner

Table 16.1 Clinical and Neuropathological Criteria for Alzheimer's Disease, According to The National Institute of Neurological and Communicative Diseases and Stroke/Alzheimer's Disease and Related Disorders Association (McKhann et al., 1984), The Consortium to Establish a Registry for Alzheimer's Disease (Mirra et al., 1991), and The National Institute on Aging-Reagan Institute Working Group (Hyman & Trojanowski, 1997).

Clinical Criteria	Neuropathological Criteria	
NINCDS-ADRDA	CERAD	NIH/Reagan
The clinical diagnosis of probable AD is supported by a combination of the following findings:	Semiquantitative assessment of neuritic plaque density from various cortical regions	All lesions (amyloid deposits, neuritic plaques, neuropil threads, and NFTs) are considered
• Dementia established by clinical examination and documented by the Mini-Mental Test, Blessed Dementia Scale, or some similar examination, and confirmed by neuropsychological tests	Plaque pathology rated as "sparse," "moderate," or "frequent." The highest value of the evaluation is compared with the age of the patient to get an "age-related plaque score."	Combination of an "age-related plaque score" according to CERAD and a topographic staging of NFT
• Deficits in two or more areas of cognition • Progressive worsening of memory and other cognitive functions	0 = no histologic evidence of AD A = histologic findings that are uncertain evidence of AD B = histologic findings suggesting AD C = histologic findings indicating the diagnosis of AD	• High likelihood of AD = CERAD plaque score is "frequent" with Braak stage V/VI • Intermediate likelihood of AD = moderate CERAD plaque score with Braak stage III/IV
• No disturbance of consciousness	The age-related plaque score is integrated with the clinical history to reach a final diagnosis of definite, neuropathologically probable, or neuropathologically possible AD as well as normal	• Low likelihood of AD = infrequent CERAD plaque score with Braak stage I/II
• Onset between ages 40 and 90, most often after the age of 65		
• Absence of systemic disorders or other brain diseases that in and of themselves could account for the progressive deficits in memory and cognition		

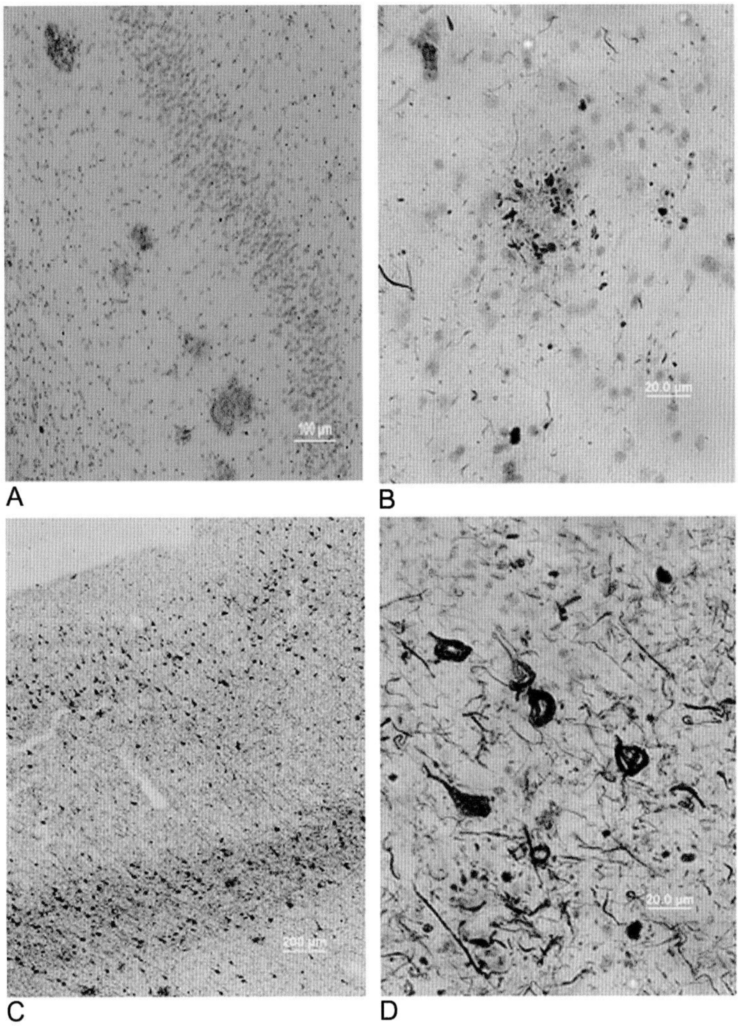

Figure 16.1 Photomicrographs of pathological changes in the Alzheimer brain by Aβ and tau immunostaining. Amyloid plaques occur frequently in the dentate gyrus molecular layer of the hippocampus **A**. A higher magnification reveals the dystrophic alterations that compose the neuritic abnormalities **B**. Laminar distribution of tau immunostaining in the cerebral cortex **C**. Which at higher power appears to be the cell bodies of pyramidal neurons and their dendritic and axonal projections **D**.

& Wong, 1984; Masters et al., 1985). A few years later, the tau protein was purified and characterized as the main constituent of neurofibrillary tangles, dystrophic neurites, and neuropil threads (Goedert et al., 1988).

III. Amyloid Biology and the Genetics of Early-Onset Alzheimer's Disease

It had been known for a long time that AD can occur as a familial disease trait with a dominant pattern of inheritance (Sjogren et al., 1952). The use of molecular genetic tools, such as linkage analysis and positional cloning, became instrumental in the hunt for the first Alzheimer gene. Researchers became especially interested in chromosome 21, as linkage analyses revealed that this part of the genome harbors a robust disease locus (Tanzi et al., 1987). In addition, individuals with Down's syndrome (trisomy 21) were known to develop Alzheimer-like pathology already in their 20s, suggesting chromosome 21 to be of particular relevance for brain function and pathology (Olson & Shaw, 1969).

In the late 1980s, the amyloid precursor protein (APP) gene, located on chromosome 21 and encoding a transmembrane protein with a largely unknown function, was successfully cloned and the first *APP* disease mutation could be identified in a Dutch family with hereditary cerebral angiopathy (Levy et al., 1990). Soon thereafter, the first AD causing

APP mutation was described in a British family (Goate et al., 1991). This so-called *London* mutation (APP^{V717I}) was hypothesized to cause disease by altering the substrate specificity for γ-secretase, one of the enzymes responsible for cleaving Aβ out of APP, resulting in increased levels of Aβ. In contrast, the *Swedish* double mutation ($APP^{K670M/N671L}$) was identified near the 5′ end of the Aβ sequence (Mullan et al., 1992) and was shown to augment Aβ levels by making APP a more suitable substrate for β-secretase, the other Aβ cleaving enzyme in the amyloidogenic pathway (Haass et al., 1995) (see Figure 16.2). In the nonamyloidogenic pathway APP instead is cleaved by α- and γ-secretases, resulting in the production of α-APP and p3 with no generation of the Aβ peptide (see Figure 16.2).

APP is one of the most abundantly expressed proteins in the central nervous system (CNS) and has been proposed to have trophic properties on neurons (Boncristiano et al., 2005). The amyloidogenic and the nonamyloidogenic pathway of APP processing are constitutively active also in nondiseased brains. However, in AD (at least for cases with the *London* and *Swedish APP* mutations) the processing has been shifted toward the amyloidogenic pathway. Interestingly, increased levels of Aβ could be demonstrated in peripheral cells from carriers with the *Swedish* mutation (Citron et al., 1994), illustrating that AD amyloidosis could be regarded as a systemic disorder. However, with cognitive malfunction as the only disease manifestation, it can be assumed that CNS neurons have a particular vulnerability for toxicity exerted by Aβ.

To date, approximately 20 pathogenic *APP* mutations have been described (see Table 16.2), a majority of which led to a clinical, biochemical, and neuropathological picture similar to that seen for sporadic AD, i.e., in cases for which no apparent heredity can be seen. The finding of disease-causing *APP* mutations, together with the observation of AD-like pathology in subjects with Down's syndrome, are two major arguments for what has become known as *the amyloid cascade hypothesis of Alzheimer's disease* (Selkoe, 1993). According to this theory, Aβ mismetabolism is driving the disease pathogenesis either directly or by eliciting putative downstream processes, such as tangle formation and inflammatory changes.

Genetic linkage to regions on chromosome 14 and 1 in families with dominantly inherited disease suggested additional Alzheimer genes. In 1995, the genes responsible were identified and named *presenilin 1 (PS1)* and *presenilin 2 (PS2)*, encoding large and previously unknown type I transmembrane proteins with a 67 percent sequence homology between each other (Levy-Lahad et al., 1995; Rogaev et al., 1995). Experimental evidence also suggested that

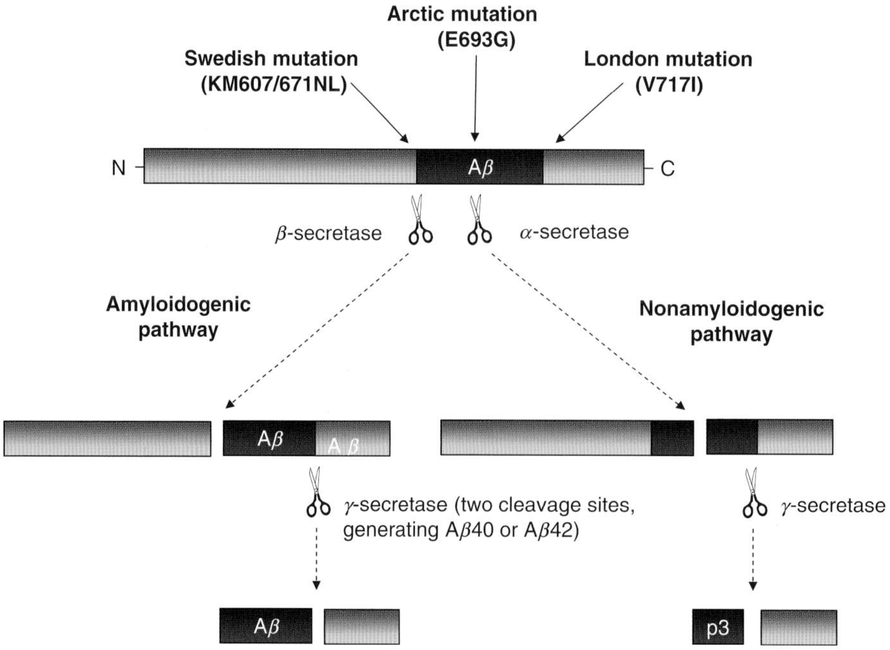

Figure 16.2 Enzymatic cleavage of the amyloid precursor protein. Cleavage by different secretases occurs via an amyloidogenic and a nonamyloidogenic pathway. In familial forms of Alzheimer's disease the amyloidogenic cleavage is augmented, leading to increased production and aggregation of the Aβ peptide.

the *presenilin* genes were involved in Aβ pathophysiology, as the mutations were shown to increase the formation of a longer form of Aβ, with 42 (Aβ42) instead of the more common form with 40 (Aβ40) amino acids (Citron et al., 1997) (see Figure 16.2). Subsequently, it has become clear that PS1 and PS2 themselves are essential components in a γ-secretase enzyme complex and that *presenilin* mutations shift the cleavage from the more common site (producing Aβ40) to an alternative site (producing Aβ42) (see Figure 16.2). In addition, carriers of various PS1 mutations were found to have increased overall Aβ plasma levels (Scheuner et al., 1996), illustrating the systemic effects of these genetic factors.

Apart from the presenilins, the γ-secretase enzyme complex consists of nicastrin, aph-1, and pen-2 (Kimberly et al., 2003). Together, these interacting molecules form a series of high-molecular mass, membrane-bound protein complexes that are necessary for cleavage of certain type I transmembrane proteins. In studies on knock-out mice lacking functional copies of any of the genes encoding γ-secretase components, it has become clear that no APP cleavage and Aβ production can occur. Hence, nicastrin, aph-1, and pen-2 are other obvious candidates in AD genetics, but to date no disease-causing mutations have been identified in any of these genes.

Altogether, the known effects of the three major AD genes highlight the significance of Aβ production and mismetabolism as the central pathological event in the AD brain. However, one has to bear in mind that the verified hereditary forms of AD seem to explain less than 1 to 2 percent of all cases, suggesting other genetic and possibly environmental factors to be of importance in disease development. In terms of nongenetic risk factors, so far only the positive correlation with previous episodes of head trauma has been consistently reproduced in epidemiological studies (Mortimer et al., 1985).

IV. Alzheimer's Disease—A Progressive Neuropathologic Syndrome

The clinical diagnosis of AD depends upon the development of memory impairments along with changes in other cognitive functions that impair day-to-day activities, leading

Table 16.2 Loci, Clinical Picture, and Biochemical/Neuropathological Features for Pathological APP Mutations Published to Date (Ancolio et al., 1999; Chartier-Harlin et al., 1991; de Jonghe et al., 2001; Eckman et al., 1997; Goate et al., 1991; Grabowski et al., 2001; Janssen et al., 2003; Hendriks et al., 1992; Kumar-Singh et al., 2000; Kwok et al., 2000; Mullan et al., 1992; Murrell et al., 1991; Nilsberth et al., 2001; Pasalar et al., 2002; Tagliavini, 1999; van Broeckhoven et al., 1990; Wakutani et al., 2004).

Mutation Name	Mutation Locus	Mutational Effect	Neuropathological Features
Swedish	Lys670Met/Asn671Leu	Increases Aβ40, Aβ42	AD pathology
	His677Arg	Unknown	AD pathology
	Asp678Asn	Unknown	N/A
Flemish	Ala692Gly	Increases Aβ40, Aβ42	Cerebral hemorrhage with amyloid angiopathy
Arctic	Glu693Gly	Increases total Aβ, increases protofibril formation	AD pathology
Dutch	Glu693Gln	Decreases Aβ42, increases protofibril formation	Cerebral hemorrhage with amyloid angiopathy
Italian	Glu692Lys	Altered fibrillization	Cerebral hemorrhage with amyloid angiopathy
Iowa	Asp694Asn	Altered fibrillization	Severe amyloid angiopathy
Iranian	Thr714Ala	Unknown	AD pathology
Austrian	Thr714Ile	Increases Aβ42	AD pathology
French	Val715Met	Decreases Aβ40, increases Aβ42	AD pathology
German	Val715Ala	Unknown	AD pathology
Florida	Ile716Val	Increases Aβ42	AD pathology
London	Val717Ile	Increases Aβ42	AD pathology
Indiana	Val717Phe	Increases Aβ42	AD pathology
	Val717Gly	Unknown	AD pathology
	Val717Leu	Increases Aβ42	AD pathology
Australian	Leu723Pro	Increases Aβ42, induces apoptosis	AD pathology

to the clinical syndrome of dementia. However, it is clear that dementia does not begin at a single moment but instead has a rather long preclinical sequence, leading to the full syndrome. This idea, of a prodromal phase of the disease, began to take shape as it was realized that cognitively normal individuals develop the same pathological hallmarks of AD in the same regional distribution, but to a quantitatively far lesser extent (Arriagada et al., 1992b; Braak et al., 2000). Imaging-based assessment showed that atrophy of the hippocampus and entorhinal cortex precede clinical onset of memory impairments by at least several years (Jack et al., 2005; Killiany et al., 2000), which was supported also by quantitative neuropathological studies (Gomez-Isla et al., 1996).

More recently, the prodromal stage of AD has been formulated as a distinct entity called *mild cognitive impairment* (MCI) (Petersen & Morris, 2005), which is viewed as a transition zone between normal cognition and dementia. In particular, one subvariant of MCI, called "amnestic minimal cognitive impairment" where the predominant symptom is a deficit in verbal memory on short-term memory tasks, appears to be commonly the clinical antecedent of AD (Markesbery et al., 2006). Thus, the definition of AD as a dementia appears to be changing and the disease is now instead beginning to be viewed as a progressive neuropathologic syndrome, which ultimately leads to end organ failure, represented clinically as dementia. In accordance with this view, more recent neuropathological criteria for the diagnosis of AD explicitly dissociate the presence of dementia from the presence of a hierarchical pattern of brain changes (Hyman & Trojanowski, 1997).

V. Tau Biology in Alzheimer's Disease

Accumulation of neurofibrillary tangles in cortical and limbic regions is the other pathological hallmark in the AD brain. Tangles are known to evolve in an anatomically stereotypical fashion with the CA1 subfield of hippocampus, subiculum, layers II/IV of the entorhinal cortex, and the perirhinal region being the earliest affected brain areas (Hirano & Zimmerman, 1962; Braak & Braak, 1991; Hyman et al., 1984). As the disease progresses, increasing numbers of NFTs occur both in these limbic areas as well as in high-order association cortices (Arriagada et al., 1992a).

As the main component of neurofibrillary tangles, dystrophic neurites, and neuropil threads, dysfunctional tau proteins self-associate into paired helical filaments. The normal function of tau is to initiate and stabilize the microtubule formation by binding to tubulin via its carboxy-terminus. Specifically, the tau-tubulin interacting region consists of four imperfect repeat sequences, corresponding to exons 9–12 of the *tau* gene. Of these, *tau* exon 10 undergoes alternative splicing and the ratio between tau isoforms with four (4R tau) and three (3R tau) tubulin-binding regions is approximately 1:1 in the healthy brain. Altogether six isoforms of tau are present in the human adult brain, as also *tau* exons 2 and 3 are regulated by splicing.

In neurofibrillary tangles and neuropil threads of the AD brain, tau is extensively phosphorylated, presumably due to an imbalance in the activity between various kinases and phosphatases. Altogether, more than 40 phosphorylatable epitopes are present throughout the tau molecule. *In vitro*, tau phosphorylation causes a weakened affinity between tau and tubulin, which impairs microtubule stability and ultimately leads to neuronal death (Grundke-Iqbal et al., 1986). Among the various kinases, cyclin-dependent kinase 5 (cdk 5), glycogen synthase kinase 3β (GSK3β), mitogen-activated protein kinase (MAPK), and MAP/microtubule affinity-regulating kinase (MARK) have been given particular attention as they all phosphorylate epitopes that seem to be important for the formation of paired helical filaments (Drewes et al., 1997). Interestingly, levels of p35, a regulatory subunit of cdk5, were found to be upregulated in the AD brain (Patrick et al., 1999), which suggests that cdk5 is constitutively activated in AD. Moreover, immunohistochemical evidence suggest that tangles are associated with GSK3β (Wang et al., 1998) and MAPK (Knowles et al., 1999). Finally, it is conceivable that also diminished activities of phosphatases such as phosphatase 2A (PP2A) and 2B (PP2B/calcineurin) could contribute to a relative dominance by kinases and, hence, to an increased degree of tau phosphorylation in the AD brain.

Surprisingly, in addition to being detrimental to tau function and stability, phosphorylation of certain epitopes seems to both cause tau dysfunction and protect paired helical formation (Schneider et al., 1999). It could therefore be hypothesized that the preservation of neuronal integrity requires a balance between tau proteins with different states of phosphorylation. Dysregulation of these processes, by elevated activation of specific kinases or downregulation of certain phosphatases, may result in loss of neuronal function and tau aggregation. Finally, recent data based on inducible expression of the human *tau* gene in a transgenic mouse model suggest that tau overexpression by itself can also be directly neurotoxic (SantaCruz et al., 2005).

Importantly, the *tau* gene does not seem to be associated with inherited AD, but *tau* mutations cause frontotemporal dementia, a different neuropathological form of neurodegeneration. To date, almost 40 *tau* mutations have been described, a majority of which cause dominantly inherited variants of frontotemporal dementia with various degrees of tau pathology. The mutational effects fall into two categories, either shifting alternative splicing of *tau* exon 10 (causing an imbalance between 4R tau and 3R tau) or altering the tubulin-binding affinity of tau. It is not fully clear whether similar mechanisms are of relevance in AD; one recent study failed to detect marked dysregulation of *tau* exon 10 in the

AD brain (Ingelsson et al., 2006). However, more subtle effects on AD development by missplicing of this or other *tau* exons cannot be ruled out.

VI. Genetics of Late-Onset Alzheimer's Disease

As dominantly inherited AD is relatively rare, researchers have been exploring the possibility of genetic factors contributing also to the vast majority of late-onset, seemingly sporadic disease cases.

Intragenic variation is a common feature of all human genes and such polymorphic changes may sometimes have a functional impact, either by altering protein conformation or by affecting alternative mRNA splicing. Polymorphic changes may occur as deletions, inserts, or inversions, but more frequently as single nucleotide polymorphisms (SNPs). In SNPs, one nucleotide has been exchanged by another, which may be either silent or result in translation of an alternative amino acid. For several disorders, SNPs have been shown to modulate disease risk, and genes known to harbor such SNPs are defined as vulnerability genes.

The apolipoprotein E gene (*APOE*) has three common alleles, *ε2, ε3, ε4*, and it was observed that carriers of the *APOE ε4* allele has a substantially increased risk to develop AD (Strittmatter et al., 1993). Heterozygotes (i.e., carriers of one *ε4* allele) have a three-fold increased risk whereas homozygotes, carriers of two *ε4* alleles, have a ten-fold increase in risk (Saunders et al., 1993). On the contrary, the *APOE ε2* allele conveys a protective effect against AD development (Corder et al., 1 994).

Numerous other genes have been implicated as disease modulators, but none of these have been consistently shown to confer a robust risk increase. An example of such a gene is *tau*, for which several SNPs are inherited together as two common haplotypes, *H1* and *H2*. The *H1* haplotype has in numerous studies been shown to confer an increased risk for progressive supranuclear palsy, another neurodegenerative disorder with tau pathology, and recently it has been proposed that a *H1* subhaplotype may increase the risk for AD (Myers et al., 2005).

However, sporadic disease may result as a consequence of a large number of genetic alterations that by themselves only exert small effects, but in combination may lead to a clinically relevant modulation of disease risk. Such genetic factors may not be detectable in association studies based on hundreds or thousands of cases and controls, but may require sample sizes of at least 10,000 individuals in each group. Alternatively, meta-analyses can be performed on the available data to evaluate the overall risk effects of certain genetic factors across different samples. In a recent initiative, researchers initiated a project that aims to catalog, summarize, and meta-analyze all genetic association studies performed on AD phenotypes that are published in peer-reviewed journals in English, and make these data publicly available in an online database (www.alzgene.org). This continuously updated resource currently includes details of nearly 1,000 studies investigating more than 400 genes (Bertram et al., 2007) (see Table 16.3).

Table 16.3 Examples of Genes Currently (March 2007) Showing Significant Signals in Systematic Meta-Analyses Performed as Part of the "AlzGene" Database

Gene	Chromosome (Polymorphism [Alias])	Protein Name	Proposed Functional Relevance of Protein / Polymorphism to AD Pathogenesis
APOE	19q13 (ε2, ε3, ε4)	Apolipoprotein E	May influence Aβ metabolism. Risk allele (ε4) causes 3x increased disease risk
APOE (promoter)	19q13 (rs 449647 and others)	Apolipoprotein E	May influence Aβ metabolism
APOC1	19q13 (HpaI ins/del)	Apolipoprotein C1	Unknown
ACE	17q23 (Intron 16 ins/del and others)	Angiotensin converting enzyme	Unknown
CST3	20p11 (rs1064039 [A25T] and others)	Cystatin C	Unknown
ESR1	6q25 (PvuII (rs2234693))	Estrogen receptor 1	May confer a disease protective effect (as suggested by epidemiological studies)
IDE	10q23 (rs2251101 [IDE_7])	Insulin degrading enzyme	Degrades Aβ; risk allele may render the enzyme inefficient
PRNP	20p13 (rs1799990 [M129V])	Prion protein	Unknown
PS1	14q24 (rs 165932 [intron 8])	Presenilin 1	A component of the γ-secretase complex in the amyloidogenic pathway; risk allele may make the enzyme more efficient
TF	3q22 (rs1049296 [P570S])	Transferrin	Unknown

For up-to-date summaries and meta-analyses of these and over 400 additional putative AD genes, please visit the AlzGene web site at www.alzgene.org (Bertram et al., 2007).

VII. Animal Models of Alzheimer's Disease

A. APP-based Models

The use of transgenic mice has become instrumental in the pursuit to develop model systems to mimic the disease. However, in spite of plaque formation and behavioral disturbances in transgenic *APP* mice, macroscopic atrophy and cell loss have not been typical brain features of these models (Irizarry et al., 1997). Nevertheless, the same genetic alterations causing plaque deposition in humans led to Aβ aggregation also in transgenic mice.

Among the various *APP* models, transgenic mice using the *London* and the *Swedish* mutations have been the most frequently studied. PDAPP mice, based on the *London* mutation, were found to have numerous Aβ brain deposits, as well as synaptic loss, astrocytosis, and microgliosis (Games et al., 1995). Tg2576 mice, expressing human APP with the *Swedish* mutation, also display Aβ plaque formation and several related neuropathological features as well as cognitive disturbances from the age of nine months (Hsiao et al., 1996). Both of these animal models have been used for the purpose of studying related biochemical abnormalities, assessing associated alterations in gene expression as well as monitoring effects of drug intervention (Lesne et al., 2006).

As expected, a more pronounced phenotype in parallel with a more severe neuropathological picture is observed in mice expressing mutated forms of both *APP* and *presenilin 1*. The observations on such double transgenic mice indicate that mutant PS1 accelerates Aβ deposition in brain (Borchelt et al., 1997). Moreover, when mice were engineered to express mutant forms of AD-causing *APP* and *PS1* in combination with a *tau* mutation (causing frontotemporal dementia), neurofibrillary tangles were also generated (Oddo et al., 2003). Another interesting feature of this triple transgenic model is the appearance of intracellular Aβ concomitant with an impairment of long-term synaptic plasticity, suggesting the (still controversial) view that Aβ's toxicity may stem from its intracellular influence rather than (or in addition to) the extracellular plaques. The importance of intracellular Aβ in the disease pathogenesis has been further corroborated in a recent double transgenic mouse model, in which the *Arctic* and the *Swedish APP* mutations were expressed together (Lord et al., 2005). The *Arctic/Swedish APP* mice deposit intraneuronal Aβ in neocortex and hippocampus at the age of two months, long before the appearance of extracellular Aβ deposits. With increasing age intraneuronal Aβ disappears, whereas the plaque burden gets more pronounced.

Moreover, animal models have also been crucial to gain a better understanding of the mechanisms regulating levels of Aβ and tau in the brain. Aβ is produced continuously and experimental evidence suggest that its concentration is determined in part by the activities of several degrading enzymes, such as neprilysin (NEP), insulin-degrading enzyme (IDE), endothelin-converting enzyme-1 (ECE-1), ECE-2, and transthyretin (reviewed in Eckman & Eckman (2005)). Hence, decreased activity of any of these enzymes may cause an abnormal accumulation and aggregation of Aβ. Conversely, increased enzyme expression may confer a protective effect. Finally, transport mechanisms removing Aβ from the brain, including apolipoprotein E (Fryer et al., 2005) and low-density lipoprotein receptor-related protein (LRP) mediated mechanisms, may also impact Aβ deposition (Shibata et al., 2000).

The relation between NEP and Aβ has been particularly well investigated. Originally, it was observed in mice that degradation of intracerebrally injected Aβ was inhibited by the selective NEP inhibitor thiorphan, but not by most other classes of protease inhibitors (Iwata et al., 2000). Next, mice lacking the *NEP* gene displayed the same feature, that is, a reduction in Aβ degradation with a concomitant accumulation of both Aβ40 and Aβ42 in brain (Iwata et al., 2001). These findings led to the hypothesis that levels and/or activity of NEP are decreased in demented brains and several studies have indeed found age-related reductions in *NEP* expression in amyloid-vulnerable regions (Carpentier et al., 2002; Wang et al., 2005).

IDE is another protease that seems to play an important role in regulating brain Aβ. Similar to *NEP*, *IDE* knock-out mice also demonstrated increased endogenous brain levels of both Aβ40 and Aβ42 (Farris et al., 2003). Moreover, in cultured neurons from a rat model of type II diabetes mellitus, carrying two *IDE* missense mutations, both insulin and Aβ degradation were found to be impaired (Farris et al., 2004). Interestingly, Aβ brain levels in these rats were not altered, indicating that other mechanisms of Aβ clearance may compensate for decreased IDE function. However, in the human brain one could envision that a partial loss of function of IDE (e.g., due to polymorphic influence) may affect long-term Aβ accumulation and result in disease over a longer time span. In accordance with this view, albeit controversial, are epidemiological observations that individuals with diabetes may be at higher risk of developing AD (Arvanitakis et al., 2004).

B. Tau-based Models

Although not leading to AD, the *tau* mutations have been essential to generate animal models in which the dynamics of tangle formation can be studied. Several different models, based on both wild-type and mutant forms of tau, have been engineered. Mice carrying *P301L*, the most common *tau* mutation causing familial frontotemporal dementia, develop motor- and behavior-related symptoms and tangle formation in brain. However, in the initial models, utilizing the mouse prion promoter, mainly midbrain and brainstem pathology could be seen (Lewis et al., 2000). Also overexpression of normal, wild-type human tau can cause neuropathology in transgenic mice.

One such model, expressing human tau on a mouse tau null background displayed substantial neuronal loss as well as abundant tangle-like inclusions with hyperphosphorylated tau protein (Andorfer et al., 2003). In a recent study, the dynamics of *tau* expression were analyzed in a mouse model expressing *tau P301L* with the forebrain-specific Ca(2+) calmodulin kinase II promoter system under the control of a tetracycline-operon responsive element. By feeding the mice doxycyclin, the expression of *tau* could be turned off. Neuronal death was prevented and memory dysfunction reversed in these mice if the expression of mutant *tau* was switched off after the onset of pathology. Intriguingly, tangles seemed to continue regardless of whether *tau* expression continued or not (SantaCruz et al., 2005).

VIII. What Makes the Neurons Die?

Although Aβ clearly is playing a central role in AD pathogenesis, it is still unclear how the peptide causes neurodegeneration and dementia. The lessons learned from clinicopathological studies rather point toward tau and tangles as being the mediators of toxicity in affected brains. Possibly, Aβ might not be neurotoxic by itself, but instead initiate secondary processes of tau fibrillization and tangle formation, which may disrupt vital functions such as axonal transport, causing neuronal dysfunction and degeneration. However, even if certain species of Aβ were toxic and if tau pathology would not have a primary influence on the disease process, a mechanistic relation between the two molecules could be envisioned. Tangles should not simply be regarded as the inevitable end-stage of a neurodegenerative process, as several brain disorders genetically linked to tau (including numerous FTD phenotypes) display neuronal loss without tangle formation. The molecular link between Aβ and tau could thus be viewed as a very central and still unresolved question in AD research.

As tau is not known to exist outside of cells in the brain parenchyma, the identification of intracellular Aβ is an important step toward elucidating the Aβ-tau relationship. The first model-based evidence for a direct relationship came from the demonstration of aggravated tangle pathology in *tau* transgenic mice by intracerebral injections of Aβ42 (Götz et al., 2001). More recently, tau and Aβ were found to form intermolecular complexes both in human brain tissue and *in vitro* as synthetic peptide fragments (Guo et al., 2006), implying that such an association may be of functional relevance in the disease process.

As overall amyloid plaque load does not correlate well with the severity of AD (Ingelsson et al., 2004), it has been assumed that instead soluble prefibrillar species of Aβ may confer toxicity in the AD brain. Conflicting results exist as to whether levels of soluble Aβ are continuously elevated during disease progression (Ingelsson et al., 2004; Näslund et al., 2000). However, in cell-based models Aβ has been shown to depress synaptic function along with a reduction in dendritic density and decreased numbers of synaptic NMDA and AMPA receptors (Kamenetz et al., 2003), and loss of spines near plaques is evident in transgenic mice as well (Spires et al., 2005).

Increasing attention is now being paid to particular intermediates in the amyloid forming process. Initially, it was found that the *Arctic* APP mutation (APP^{E693G}) causes increased levels of oligomeric Aβ forms (Nilsberth et al., 2001). Subsequently, it has been shown that such Aβ protofibrils, even consisting of wild type Aβ, can impair electrophysiological properties in cell-based models (Walsh et al., 2002) as well as lead to behavioral changes when injected intracerebrally in mice (Kayed et al., 2003).

An oligomeric Aβ species, with particularly toxic properties and consisting of twelve Aβ moieties, was recently characterized and purified from the tg2576 transgenic mouse model (Lesne et al., 2006). This peptide variant, named Aβ*56, caused memory impairment when injected into the lateral ventricle of healthy rats. Although this Aβ dodecamer is not likely to be solely responsible for all the adverse effects of Aβ in brain, the findings reflect interesting differences in the degree of toxicity between various prefibrillar Aβ species and suggest an intriguing target for therapeutic intervention.

The growing awareness of the role of prefibrillar Aβ species in AD pathophysiology has led to an alternative view of plaques as pathological "tombstones" (or even as reservoirs for deleterious Aβ peptides/protofibrils). However, amyloid plaques appear in different variants and it is possible that a subset of them, determined by size or composition, exert toxicity whereas a majority may be less harmful. From this perspective, plaques might contribute to widespread lesions in the neuropil that, although inconsequential individually, might act in concert to disrupt distributed neural systems in critical cortical regions. Finally, it is conceivable that whereas oligomers and/or protofibrils induce synaptic dysfunction and membrane disruption, the mature fibrils may be responsible for associated inflammatory reactions.

It is still unclear whether Aβ is degraded mainly by mechanisms intrinsic to the brain or whether the peptide first is transported out of the CNS and subsequently undergoes degradation in peripheral tissues. Lessons learned from recent Aβ vaccination trials (see Section XI) indicate that surrounding glial cells may be able to clear Aβ deposits. In favor of the latter belief is the *sink hypothesis*, claiming that an excess of Aβ can be transported out of CNS compartments, such as cerebrospinal fluid (CSF) via the choroid plexus, into the peripheral circulation (DeMattos et al., 2002). Such a mechanism would offer an explanation for the observation that levels of Aβ42 are diminished in CSF among AD patients (see Section X).

IX. Abnormal Protein Conformation—A Unifying Factor in Neurodegeneration?

Conformational changes and aggregation of certain proteins are central features in AD as well as in several other neurodegenerative diseases. Inherited forms of dementias (AD, frontotemporal dementia, and dementia with Lewy bodies) as well as movement disorders (Parkinson's disease, amyotrophic lateral sclerosis, Huntington's disease) and prion diseases (Creutzfeldt's Jakob's disease) have all been shown to be caused by mutations in specific genes. Even though the respective disease processes have been characterized for these rare disease forms, there is substantial support for the molecular mechanisms being similar also for the vast majority of sporadic disease cases. In AD, cases with identified APP- or PS-mutations have similar neuropathology and clinical manifestations as patients for which no genetic background is known (Gomez-Isla et al., 1999).

A unifying feature for several of the mutations causing neurodegenerative disease is that they lead to expression of protein variants with an increased aggregation propensity. By a gradual oligomerization of conformationally changed proteins fully developed fibrils are eventually formed. The extreme view is that the toxic, disease-causing effect may be the soluble oligomeric forms, whereas the extra- and intracellular aggregates seen in light microscopy in fact may be relatively harmless or even protective remnants of the disease process. More likely, both forms are deleterious by various mechanisms.

X. Emerging Diagnostic Tools

With the introduction of computerized tomography, it became possible to visualize the neurodegenerative process with respect to different brain regions. An even more detailed high-resolution picture was obtained with magnetic resonance imaging (MRI), but as these examinations reflect only loss of tissue, techniques that more closely mirror the ongoing disease process have been greatly needed.

Functional MRI (fMRI) is one of several promising techniques that may identify persons at risk for AD before the onset of symptoms (Sperling et al., 2003). When challenged with a cognitive task, individuals at risk (i.e., those carrying the *APOE ε4* allele and/or having a positive family history) were found to have a more pronounced change in blood oxygenation in regions commonly associated with AD pathology (Bookheimer et al., 2000; Fleisher et al., 2005). In addition, follow-up analyses have revealed that these fMRI changes indeed correlated with subsequent deterioration in memory functions. These findings support a theory of altered neuronal memory systems in people at risk for AD many years before the typical age at disease onset.

Also positron emission tomography (PET) has been demonstrated as a useful tool to visualize disease-related processes. With classical PET, using ligands such as 18-fluorodeoxyglucose (FDG), decreased glucose metabolism can be measured in regions affected by the disease. By developing novel ligands with affinity to more disease-specific features, other causes for impaired metabolism can hopefully be ruled out. A promising candidate is the so-called Pittsburgh compound (PIB), a derivative of Congo red and belonging to a class of compounds that bind amyloid plaques and other biological structures with a beta-sheet conformation. The use of PIB on transgenic animal models has confirmed that the substance readily crosses the blood–brain barrier and binds plaques *in vivo* (Bacskai et al., 2003).

The first trial with PIB-PET on humans was successful insofar that AD patients displayed a higher signal retention as compared to control subjects (Klunk et al., 2004). However, the signal intensity did not correlate very well to the severity of dementia and a follow-up study did not show any increase in signals after two years in spite of clear clinical progression (Engler et al., 2006). These results are in accord with clinical-neuropathological studies, suggesting good correlation of amyloid deposition with diagnosis of AD, but poor correlation of amount of Aβ deposits with clinical symptoms (Ingelsson et al., 2004). Overall, these observations support the view of amyloid plaques as steady state structures.

Biochemical disease markers are greatly needed for early diagnosis and will become even more important once efficient preventive pharmaceuticals are available. Today, measurements on CSF of Aβ and tau levels are used routinely to support the clinical diagnosis. In approximately 20 published studies, a 50 percent average decrease of Aβ42 has been detected in AD patients, whereas in more than 50 studies, there is an average 320 percent increase of CSF tau (Blennow and Hampel, 2003). Each of these measures has between 80 and 90 percent mean sensitivity and close to 90 percent mean specificity. Combining the two measures was reported to improve sensitivity to 95 percent and had a high predictive value in a recent study on subjects with AD and MCI (Hansson et al., 2006).

Even though the combined CSF Aβ and tau measure is reasonably sensitive, improved biomarkers are greatly needed because of a certain overlap with other dementias, such as dementia with Lewy bodies and vascular dementia. Different enzyme-linked immunosorbent assays (ELISAs), utilizing antibodies against various tau phosphoepitopes (Thr181/Thr231, Thr181, Thr231, Thr199, Ser396/Ser404), have therefore been evaluated and assays based on Thr181 and Thr231 were found to slightly improve on the specificity (Hampel et al., 2004; Hansson et al., 2006).

Yet other approaches to improve on the quality of preclinical biomarkers for AD are currently being evaluated. In particular, systems with antibodies specific against certain conformational epitopes of Aβ and tau may be promising as new diagnostic tools.

Finally, the ambition to develop a diagnostic test based on blood samples was strongly encouraged by the finding of increased Aβ42 plasma levels among carriers of *PS1* and *APP* mutations (Scheuner et al., 1996). However, most AD cases do not display elevated Aβ plasma levels (Fukumoto et al., 2003) and ongoing studies are aimed at modifying the test design with the goal that also sporadic AD in the future can be predicted and monitored by a simple blood test.

XI. Disease-Modifying Strategies of Tomorrow

A detailed understanding of the molecular pathogenesis is a prerequisite for the development of efficient treatment strategies in AD and several promising pharmacotherapeutic approaches, based on such insight, are now being pursued.

With the amyloid cascade hypothesis in mind, the most attractive approach would be to pharmacologically inhibit either β- or γ-secretase or, alternatively, to stimulate α-secretase (see Figure 16.2). However, the putative physiological function of these enzymes (and of Aβ itself) has to be considered and such a drug should probably not inhibit the formation of Aβ completely, as the peptide may also have important physiological functions. In addition, it is important to keep in mind that an altered enzymatical activity could have an adverse influence on other proteins, such as the signaling molecule Notch.

The perhaps most promising attempt to find an efficacious therapy for AD is based on the idea of inducing an immunological response against the Aβ aggregates. Transgenic APP mice that were actively immunized with Aβ, before the expected onset of plaque formation, displayed virtually no plaques in brain (Schenk et al., 1999). Moreover, the plaque load was considerably reduced when treating mice that already had developed plague pathology (Bacskai et al., 2001). Finally, behavioral symptoms in the immunized mice were significantly ameliorated (Schenk et al., 1999).

In a clinical phase II multicenter trial, AD patients were actively immunized, but unfortunately the study had to be halted after 6 percent of the participating patients developed meningoencephalitis (Gilman et al., 2005; Orgogozo et al., 2003). However, brains from several immunized cases have now come to autopsy, all of which display a far less advanced amyloid pathology than expected (Ferrer et al., 2004; Masliah et al., 2005; Nicoll et al., 2003). In addition, immunized subjects were found to have lower levels of tau in CSF, indicating that the treatment also may have reduced the extent of neuronal damage. As the study was discontinued, the clinical effects are hard to assess, but Gilman and colleagues reported a modest positive effect on a subset of cognitive tests (Gilman et al., 2005). Intense efforts are now underway to design an AD vaccine that does not elicit a damaging T-cell response. As an alternative, passive immunization with monoclonal Aβ antibodies could be a more successful strategy and such studies have recently been initiated.

Other amyloid-related therapeutic strategies, currently being evaluated on cell- and animal models, are transcriptional or pharmacological activation of Aβ-degrading enzymes, such as ECE, NEP, and IDE. Finally, a novel way of applying stem cells for therapeutic purposes has been described in a study, in which bone marrow cells were transplanted into the brains of transgenic mice before onset of pathology. In such mice, a local inflammatory response was seen in parallel with a decrease in Aβ-burden (Malm et al., 2005).

Both genetic and epidemiological data support a role of cholesterol metabolism in AD pathogenesis (reviewed in Hartmann (2001)), and some clinical studies have suggested that statins lower the risk of developing AD (Wolozin, 2004). It is known that both statins and inhibitors of cholesterol acyltransferase (ACAT), an acyl-coenzyme A regulating intracellular cholesterol homeostasis, inhibit cholesterol generation and lower intracellular cholesterol levels. Interestingly, it could be demonstrated that an ACAT inhibitor reduced accumulation of amyloid plaques as well as levels of soluble Aβ by 83 to 99 percent in brains from transgenic mice carrying the *Swedish* or the *London* APP mutations (Hutter-Paier et al., 2004).

Yet other common drugs, used for different disorders, are currently being evaluated for the prevention and/or treatment of Alzheimer's disease. Initially, indomethacin was found to protect from cognitive decline in a pilot six-month clinical trial (Rogers et al., 1993). In the large epidemiological Rotterdam study, a substantial number of AD patients and nondemented control subjects were retrospectively assessed and long-term use of nonsteroidal inflammatory drugs (NSAIDs) appeared to reduce the risk of AD (In't Veld et al., 1998). Experimental evidence for the putative effect of NSAIDs include the observation that certain NSAIDs (ibuprofen, indomethacin, sulindac sulphide) affect γ-secretase cleavage leading to a decreased Aβ42/Aβ40 ratio (Weggen et al., 2001). Moreover, evidence suggests that this effect on the γ-secretase complex is due to a change in PS 1 conformation (Lleo et al., 2004).

Epidemiological evidence indicates also that dietary habits influence the risk for AD. In a prospective community-based study over seven years, 815 residents were followed and it was found that those who consumed fish once per week or more had 60 percent less risk of developing AD (Morris et al., 2003). In accordance with these findings is the observation that a diet enriched with the omega-3 fatty acid docosahexaenoic acid (DHA) reduces levels of Aβ and amyloid plaque burden by more than 70 percent and 40 percent, respectively, in brains of transgenic mice with the *Swedish APP* mutation (Lim et al., 2005).

Also, tau is being considered as a pharmacological target and compounds supposed to stop tau aggregation by inhibiting GSK3β are currently being evaluated in clinical trials (reviewed in Churcher (2006)). Such drugs, if efficacious, could be useful not only for AD but also for frontotemporal dementia, progressive supranuclear palsy, and other disorders with tau brain pathology.

XII. Summary and Conclusions

In the last 15 years the genetic basis of rare cases of autosomal dominant AD has focused attention on the generation and deposition of Aβ as the major pathophysiological event in the brain. However, many uncertainties remain; how Aβ is related to neurofibrillary lesions and to neuronal death is unknown and the etiology of Aβ accumulation in sporadic AD is speculative. The emergence of animal models of both plaque and tangle pathology have allowed further dissection of genetic modifiers and of therapeutic approaches. Attempts to prevent or treat AD currently are focused on anti-Aβ synthesis therapies (directed toward either gamma- or beta secretase), improved clearance of Aβ by active or passive vaccination as well as general neuronal protection strategies.

Similarly, advances in the genetics of frontotemporal dementia have suggested that the microtubule associated molecule tau is a critical therapeutic target in neurodegenerative illnesses marked by massive neuronal loss. Exactly how tau malfunction in these disorders relates to the alterations in tau that lead to neurofibrillary tangles in AD remains a central question in neurodegeneration. However, neuroprotective strategies, focusing especially on kinases that modify tau phosphorylation, are beginning to show promise in experimental systems.

As further advances are made in understanding the genetic risk factors—and protective factors—that define an individual's risk for AD, and as better biomarkers become available, we shall hopefully be entering an era of preventive therapy where individuals at risk can be given therapeutics targeting their specific risk factors.

References

Alzheimer, A. (1911). Über eigenartige Krankheitsfälle des späteren Alters. *Zschr Ges Neurol Psychiatr* **4**, 356–385.

Ancolio, K., Dumanchin, C., Barelli, H., Warter, J., Brice, A., Campion, D. et al. (1999). Unusual phenotypic alteration of β-amyloid precursor protein (β-APP) maturation by a new Val 715–Met ßAPP 770 mutation. *Proc Natl Acad Sci U S A* **96**, 4119–4124.

Andorfer, C., Kress, Y., Espinoza, M., de Silva, R., Tucker, K. L., Barde, Y. A. et al. (2003). Hyperphosphorylation and aggregation of tau in mice expressing normal human tau isoforms. *J Neurochem* **86**, 582–590.

Arnold, S., Hyman, B. T., Flory, J., Damasio, A., and van Hoesen, G. (1991). The topographical and neuroanatomical distribution of neurofibrillary tangles and neuritic plaques in the cerebral cortex of patients with Alzheimer's disease. *Cerebral Cortex* **1**, 103–116.

Arriagada, P. V., Growdon, J. H., Hedley-Whyte, E. T., and Hyman, B. T. (1992a). Neurofibrillary tangles but not senile plaques parallel duration and severity of Alzheimer's disease. *Neurology* **42**, 631–639.

Arriagada, P. V., Marzloff, K., and Hyman, B. T. (1992b). Distribution of Alzheimer-type pathologic changes in nondemented elderly individuals matches the pattern in Alzheimer's disease. *Neurology* **42**, 1681–1688.

Arvanitakis, Z., Wilson, R. S., Bienias, J. L., Evans, D. A., and Bennett, D. A. (2004). Diabetes mellitus and risk of Alzheimer disease and decline in cognitive function. *Arch Neurol* **61**, 661–666.

Bacskai, B. J., Hickey, G. A., Skoch, J., Kajdasz, S. T., Wang, Y., Huang, G. F. et al. (2003). Four-dimensional multiphoton imaging of brain entry, amyloid binding, and clearance of an amyloid-beta ligand in transgenic mice. *Proc Natl Acad Sci U S A* **100**, 12462–12467.

Bacskai, B. J., Kajdasz, S. T., Christie, R. H., Carter, C., Games, D., Seubert, P. et al. (2001). Imaging of amyloid-β deposits in brains of living mice permits direct observation of clearance of plaques with immunotherapy. *Nat Med* **7**, 369–372.

Bertram, L., McQueen, M. B., Mullin, K., Blacker, D., and Tanzi, R. E. (2007). Systematic meta-analyses of Alzheimer disease genetic association studies: The AlzGene database. *Nat Genet* **39**, 17–23.

Blennow, K. and Hampel, H. (2003). CSF markers for incipient Alzheimer's disease. *Lancet Neurol* **2**, 605–613.

Boncristiano, S., Calhoun, M. E., Howard, V., Bondolfi, L., Kaeser, S. A., Wiederhold, K. H. et al. (2005). Neocortical synaptic bouton number is maintained despite robust amyloid deposition in APP23 transgenic mice. *Neurobiol Aging* **26**, 607–613.

Bookheimer, S. Y., Strojwas, M. H., Cohen, M. S., Saunders, A. M., Pericak-Vance, M. A., Mazziotta, J. C., and Small, G. W. (2000). Patterns of brain activation in people at risk for Alzheimer's disease. *N Engl J Med* **343**, 450–456.

Borchelt, D., Ratovitski, T., van Lare, J., Lee, M., Gonzales, V., Jenkins, N. et al. (1997). Accelerated amyloid deposition in the brains of transgenic mice coexpressing mutant presenilin 1 and amyloid precursor proteins. *Neuron* **19**, 939–945.

Braak, H. and Braak, E. (1991). Neuropathological stageing of Alzheimer-related changes. *Acta Neuropathol* **82**, 239–259.

Braak, H., Del Tredici, K., Bohl, J., Bratzke, H., and Braak, E. (2000). Pathological changes in the parahippocampal region in select non-Alzheimer's dementias. *Ann N Y Acad Sci* **911**, 221–239.

Carpentier, M., Robitaille, Y., DesGroseillers, L., Boileau, G., and Marcinkiewicz, M. (2002). Declining expression of neprilysin in Alzheimer disease vasculature: possible involvement in cerebral amyloid angiopathy. *J Neuropathol Exp Neurol* **61**, 849–856.

Chartier-Harlin, M.-C., Crawfort, F., Houlden, H., Warren, A., Hughes, D., Fidani, L. et al. (1991). Early-onset Alzheimer's disease caused by mutations at codon 717 of the β-amyloid precursor protein gene. *Nature* **353**, 844–846.

Churcher, I. (2006). Tau therapeutic strategies for the treatment of Alzheimer's disease. *Curr Top Med Chem* **6**, 579–595.

Citron, M., Vigo-Pelfrey, C., Teplow, D. B., Miller, C., Schenk, D., Johnston, J. et al. (1994). Excessive production of amyloid β-protein by peripheral cells of symptomatic and presymptomatic patients carrying the Swedish familial Alzheimer disease mutation. *Proc Natl Acad Sci USA* **91**, 11993–11997.

Citron, M., Westaway, D., Xia, W., Carlson, G., Diehl, T., Levesque, G. et al. (1997). Mutant presenilins of Alzheimer's disease increase production of 42-residue amyloid beta-protein in both transfected cells and transgenic mice. *Nat Med* **3**, 67–72.

Corder, E. H., Saunders, A. M., Risch, N. J., Strittmatter, W. J., Schmechel, D. E., Gaskell, P. C. et al. (1994). Protective effect of apolipoprotein E type 2 allele for late onset Alzheimer disease. *Nat Genet* **7**, 180–184.

de Jonghe, C., Esselens, C., Kumar-Singh, S., Craessaerts, K., Serneels, S., Checler, F. et al. (2001). Pathogenic APP mutations near the gamma-secretase cleavage site differentially affect Abeta secretion and APP C-terminal fragment stability. *Hum Mol Genet* **10**, 1665–1671.

DeMattos, R. B., Bales, K. R., Parsadanian, M., O'Dell, M. A., Foss, E. M., Paul, S. M., and Holtzman, D. M. (2002). Plaque-associated disruption of CSF and plasma amyloid-beta (Abeta) equilibrium in a mouse model of Alzheimer's disease. *J Neurochem* **81**, 229–236.

Drewes, G., Ebneth, A., Preuss, U., Mandelkow, E.-M., and Mandelkow, E. (1997). MARK, a novel family of protein kinases that phosphorylate microtubule-associated proteins and trigger microtubule disruption. *Cell* **89**, 297–308.

Eckman, C., Mehta, N., Crook, R., Perez-Tur, J., Prihar, G., Pfeiffer, E. et al. (1997). A new pathogenic mutation in the APP gene (I716V) increases the relative proportion of Abeta 42(43). *Hum Mol Genet* **6**, 2087–2089.

Eckman, E. A. and Eckman, C. B. (2005). Abeta-degrading enzymes: Modulators of Alzheimer's disease pathogenesis and targets for therapeutic intervention. *Biochem Soc Trans* **33**, 1101–1105.

Engler, H., Forsberg, A., Almkvist, O., Blomquist, G., Larsson, E., Savitcheva, I. et al. (2006). Two-year follow-up of amyloid deposition in patients with Alzheimer's disease. *Brain* **129**, 2856–2866.

Farris, W., Mansourian, S., Chang, Y., Lindsley, L., Eckman, E. A., Frosch, M. P. et al. (2003). Insulin-degrading enzyme regulates the levels of insulin, amyloid beta-protein, and the beta-amyloid precursor protein intracellular domain in vivo. *Proc Natl Acad Sci U S A* **100**, 4162–4167.

Farris, W., Mansourian, S., Leissring, M. A., Eckman, E. A., Bertram, L., Eckman, C. B. et al. (2004). Partial loss-of-function mutations in insulin-degrading enzyme that induce diabetes also impair degradation of amyloid beta-protein. *Am J Pathol* **164**, 1425–1434.

Ferrer, I., Boada Rovira, M., Sanchez Guerra, M. L., Rey, M. J., and Costa-Jussa, F. (2004). Neuropathology and pathogenesis of encephalitis following amyloid-beta immunization in Alzheimer's disease. *Brain Pathol* **14**, 11–20.

Fleisher, A. S., Houston, W. S., Eyler, L. T., Frye, S., Jenkins, C., Thal, L. J., and Bondi, M. W. (2005). Identification of Alzheimer disease risk by functional magnetic resonance imaging. *Arch Neurol* **62**, 1881–1888.

Fryer, J. D., Simmons, K., Parsadanian, M., Bales, K. R., Paul, S. M., Sullivan, P. M., and Holtzman, D. M. (2005). Human apolipoprotein E4 alters the amyloid-beta 40:42 ratio and promotes the formation of cerebral amyloid angiopathy in an amyloid precursor protein transgenic model. *J Neurosci* **25**, 2803–2810.

Fukumoto, H., Tennis, M., Locascio, J. J., Hyman, B. T., Growdon, J. H., and Irizarry, M. C. (2003). Age but not diagnosis is the main predictor of plasma amyloid beta-protein levels. *Arch Neurol* **60**, 958–964.

Games, D., Adams, D., Alessandrini, R., Barbour, R., Berthelette, P., Blackwell, C. et al. (1995). Alzheimer-type neuropathology in transgenic mice overexpressing V717F β-amyloid precursor protein. *Nature* **373**, 523–527.

Gilman, S., Koller, M., Black, R., Jenkins, L., Griffith, S., Fox, N. et al. (2005). Clinical effects of Abeta immunization (AN1792) in patients with AD in an interrupted trial. *Neurology* **64**, 1553–1562.

Glenner, G. G. and Wong, C. W. (1984). Alzheimer's disease: Initial report of the purification and characterization of a novel cerebrovascular amyloid protein. *Biochem Biophys Res Comm* **120**, 885–890.

Goate, A., Chartier-Harlin, M. C., Mullan, M., Brown, J., Crawford, F., Fidani, L. et al. (1991). Segregation of a missense mutation in the amyloid precursor protein gene with familial Alzheimer's disease. *Nature* **349**, 704–706.

Goedert, M., Wischik, C. M., Crowther, R. A., Walker, J. E., and Klug, A. (1988). Cloning and sequencing of the cDNA encoding a core protein of the paired helical filament of Alzheimer disease: Identification as the microtubule-associated protein tau. *Proc Natl Acad Sci USA* **85**, 4051–4055.

Gomez-Isla, T., Growdon, W. B., McNamara, M. J., Nochlin, D., Bird, T. D., Arango, J. C. et al. (1999). The impact of different presenilin 1 and presenilin 2 mutations on amyloid deposition, neurofibrillary changes and neuronal loss in the familial Alzheimer's disease brain: Evidence for other phenotype-modifying factors. *Brain* **122 (Pt 9)**, 1709–1719.

Gomez-Isla, T., Hollister, R., West, H., Mui, S., Growdon, J., Petersen, R. et al. (1997). Neuronal loss correlates with but exceeds neurofibrillary tangles in Alzheimer's disease. *Ann Neurol* **41**, 17–24.

Gomez-Isla, T., Price, J. L., McKeel, D. W., Jr., Morris, J. C., Growdon, J. H., and Hyman, B. T. (1996). Profound loss of layer II entorhinal cortex neurons occurs in very mild Alzheimer's disease. *J Neurosci* **16**, 4491–4500.

Götz, J., Chen, F., van Dorpe, J., and Nitsch, R. (2001). Formation of neurofibrillary tangles in P301l tau transgenic mice induced by Abeta 42 fibrils. *Science* **293**, 1491–1495.

Grabowski, T., Cho, H., Vonsattel, J., Rebeck, G., and Greenberg, S. (2001). Novel amyloid precursor protein mutation in an Iowa family with dementia and severe cerebral amyloid angiopathy. *Ann Neurol* **49**, 697–705.

Grundke-Iqbal, I., Iqbal, K., Tung, Y. C., Quinlan, M., Wisniewski, H. M., and Binder, L. I. (1986). Abnormal phosphorylation of the microtubule-associated protein tau (tau) in Alzheimer cytoskeletal pathology. *Proc Natl Acad Sci U S A* **83**, 4913–4917.

Guo, J. P., Arai, T., Miklossy, J., and McGeer, P. L. (2006). Abeta and tau form soluble complexes that may promote self aggregation of both into the insoluble forms observed in Alzheimer's disease. *Proc Natl Acad Sci U S A* **103**, 1953–1958.

Haass, C., Lemere, C., Capell, A., Citron, M., Seubert, P., Schenk, D., Lannfelt, L., and Selkoe, D. (1995). The Swedish mutation causes early-onset Alzheimer's disease by beta-secretase cleavage within the secretory pathway. *Nat Med* **1**, 1291–1296.

Hampel, H., Buerger, K., Zinkowski, R., Teipel, S. J., Goernitz, A., Andreasen, N. et al. (2004). Measurement of phosphorylated tau epitopes in the differential diagnosis of Alzheimer disease: A comparative cerebrospinal fluid study. *Arch Gen Psychiatry* **61**, 95–102.

Hansson, O., Zetterberg, H., Buchhave, P., Londos, E., Blennow, K., and Minthon, L. (2006). Association between CSF biomarkers and incipient Alzheimer's disease in patients with mild cognitive impairment: a follow-up study. *Lancet Neurol* **5**, 228–234.

Hartmann, T. (2001). Cholesterol, Abeta and Alzheimer's disease. *Trends Neurosci* **24**, S45–S48.

Hendriks, L., van Duijn, C. M., Cras, P., Cruts, M., Van Hul, W., Van Harskamp, F. et al. (1992). Presenile dementia and cerebral haemorrhage linked to a mutation at codon 692 of the β-amyloid precursor protein gene. *Nat Genet* **1**, 218–221.

Hirano, A. and Zimmerman, H. (1962). Alzheimer's neurofibrillary changes—a topographical study. *Arch Neurol* **7**, 227–242.

Hsiao, K., Chapman, P., Nilsen, S., Ekman, C., Harigaya, Y., Younkin, S. et al. (1996). Correlative memory deficits, Aβ elevation, and amyloid plaques in transgenic mice. *Science* **274**, 99–102.

Hutter-Paier, B., Huttunen, H., Puglielli, L., Eckman, C., Kim, D., Hofmeister, A. et al. (2004). The ACAT inhibitor CP-113,818 markedly reduces amyloid pathology in a mouse model of Alzheimer's disease. *Neuron* **44**, 227–238.

Hyman, B. and Trojanowski, J. (1997). Consensus recommendations for the postmortem diagnosis of Alzheimer disease from the National Institute on Aging and the Reagan Institute Working Group on diagnostic criteria for the neuropathological assessment of Alzheimer disease. *J Neuropathol Exp Neurol* **56**, 1095–1097.

Hyman, B., Van Horsen, G., Damasio, A., and Barnes, C. (1984). Alzheimer's disease: Cell-specific pathology isolates the hippocampal formation. *Science* **225**, 1168–1170.

Ingelsson, M., Fukumoto, H., Newell, K., Growdon, J. H., Hedley-Whyte, T. E., Albert, M. S. et al. (2004). Early Abeta accumulation and progressive synaptic loss, gliosis and tangle formation in AD brain. *Neurology* **62**, 925–931.

Ingelsson, M., Ramasamy, K., Cantuti-Castelvetri, I., Skoglund, L., Matsui, T., Orne, J. et al. (2006). No alteration in tau exon 10 alternative splicing in tangle-bearing neurons of the Alzheimer's disease brain. *Acta Neuropathol*. In Press.

In't Veld, B., Launer, L., Hoes, A., Ott, A., Hofman, A., Breteler, M., and Stricker, B. (1998). NSAIDs and incident Alzheimer's disease. The Rotterdam study. *Neurobiol Aging* **19**, 607–611.

Irizarry, M., McNamara, M., Fedorchak, K., Hsiao, K., and Hyman, B. (1997). APPSw transgenic mice develop age-related A beta deposits and neuropil abnormalities, but no neuronal loss in CA1. *J Neuropathol Exp Neurol* **56**, 965–973.

Iwata, N., Tsubuki, S., Takaki, Y., Shirotani, K., Lu, B., Gerard, N. P. et al. (2001). Metabolic regulation of brain Abeta by neprilysin. *Science* **292**, 1550–1552.

Iwata, N., Tsubuki, S., Takaki, Y., Watanabe, K., Sekiguchi, M., Hosoki, E. et al. (2000). Identification of the major Abeta1-42-degrading catabolic pathway in brain parenchyma: Suppression leads to biochemical and pathological deposition. *Nat Med* **6**, 143–150.

Jack, C. R., Jr., Shiung, M. M., Weigand, S. D., O'Brien, P. C., Gunter, J. L., Boeve, B. F. et al. (2005). Brain atrophy rates predict subsequent clinical conversion in normal elderly and amnestic MCI. *Neurology* **65**, 1227–1231.

Janssen, J. C., Beck, J. A., Campbell, T. A., Dickinson, A., Fox, N. C., Harvey, R. J. et al. (2003). Early onset familial Alzheimer's disease: Mutation frequency in 31 families. *Neurology* **60**, 235–239.

Kamenetz, F., Tomita, T., Hsieh, H., Seabrook, G., Borchelt, D., Iwatsubo, T. et al. (2003). APP processing and synaptic function. *Neuron* **37**, 925–937.

Kayed, R., Head, E., Thompson, J., McIntire, T., Milton, S., Cotman, C., and Glabe, C. (2003). Common structure of soluble amyloid oligomers implies common mechanism of pathogenesis. *Science* **18**, 486–489.

Killiany, R. J., Gomez-Isla, T., Moss, M., Kikinis, R., Sandor, T., Jolesz, F. et al. (2000). Use of structural magnetic resonance imaging to predict who will get Alzheimer's disease. *Ann Neurol* **47**, 430–439.

Kimberly, W. T., LaVoie, M. J., Ostaszewski, B. L., Ye, W., Wolfe, M. S., and Selkoe, D. J. (2003). Gamma-secretase is a membrane protein complex comprised of presenilin, nicastrin, Aph-1, and Pen-2. *Proc Natl Acad Sci U S A* **100**, 6382–6387.

Klunk, W. E., Engler, H., Nordberg, A., Wang, Y., Blomqvist, G., Holt, D. P. et al. (2004). Imaging brain amyloid in Alzheimer's disease with Pittsburgh Compound-B. *Ann Neurol* **55**, 306–319.

Knowles, R., Chin, J., Ruff, C., and Hyman, B. T. (1999). Demonstration by fluorescence resonance energy transfer of a close association between activated MAP kinase and neurofibrillary tangles: Implications for MAP kinase activation in Alzheimer disease. *J Neuropathol Exp Neurol* **58**, 1090–1098.

Kumar-Singh, S., De Jonghe, C., Cruts, M., Kleinert, R., Wang, R., Mercken, M. et al. (2000). Nonfibrillar diffuse amyloid deposition due to a (42)-secretase site mutation points to an essential role for N-truncated Aβ(42). *Hum Mol Genet* **9**, 2589–2598.

Kwok, J., Li, Q., Hallupp, M., Whyte, S., Ames, D., Beyreuther, K. et al. (2000). Novel Leu723Pro amyloid precursor protein mutation increases amyloid β42(43) peptide levels and induces apoptosis. *Ann Neurol* **47**, 249–253.

Lesne, S., Koh, M. T., Kotilinek, L., Kayed, R., Glabe, C. G., Yang, A. et al. (2006). A specific amyloid-beta protein assembly in the brain impairs memory. *Nature* **440**, 352–357.

Levy, E., Carman, M., Fernandez-Madrid, I., Power, M., Lieberburg, I., van Duinen, S. et al. (1990). Mutation of the Alzheimer's disease amyloid gene in hereditary cerebral hemorrhage, Dutch type. *Science* **248**, 1124–1126.

Levy-Lahad, E., Wasco, W., Poorkaj, P., Romano, D. M., Oshima, J., Pettingell, W. H. et al. (1995). Candidate gene for the chromosome 1 familial Alzheimer's disease locus. *Science* **269**, 973–976.

Lewis, J., McGowan, E., Rockwood, J., Melrose, H., Nacharaju, P., Van Slegtenhorst, M. et al. (2000). Neurofibrillary tangles, amyotrophy and progressive motor disturbance in mice expressing mutant (P301L) tau protein. *Nat Genet* **25**, 402–405.

Lim, G. P., Calon, F., Morihara, T., Yang, F., Teter, B., Ubeda, O. et al. (2005). A diet enriched with the omega-3 fatty acid docosahexaenoic acid reduces amyloid burden in an aged Alzheimer mouse model. *J Neurosci* **25**, 3032–3040.

Lleo, A., Berezovska, O., Herl, L., Raju, S., Deng, A., Bacskai, B. et al. (2004). Nonsteroidal anti-inflammatory drugs lower Abeta42 and change presenilin 1 conformation. *Nat Med* **10**, 1065–1066.

Lord, A., Kalimo, H., Eckman, C., Zhang, X.-Q., Lannfelt, L., and Nilsson, L. (2005). The Arctic Alzheimer mutation facilitates early intraneuronal Aß aggregation and senile plaque formation in transgenic mice. *Neurobiol Aging* **27**, 67–77.

Malm, T. M., Koistinaho, M., Parepalo, M., Vatanen, T., Ooka, A., Karlsson, S., and Koistinaho, J. (2005). Bone-marrow-derived cells contribute to the recruitment of microglial cells in response to beta-amyloid deposition in APP/PS1 double transgenic Alzheimer mice. *Neurobiol Dis* **18**, 134–142.

Markesbery, W. R., Schmitt, F. A., Kryscio, R. J., Davis, D. G., Smith, C. D., and Wekstein, D. R. (2006). Neuropathologic substrate of mild cognitive impairment. *Arch Neurol* **63**, 38–46.

Masliah, E., Hansen, L., Adame, A., Crews, L., Bard, F., Lee, C. et al. (2005). Abeta vaccination effects on plaque pathology in the absence of encephalitis in Alzheimer disease. *Neurology* **64**, 129–131.

Masters, C. L., Multhaup, G., Simms, G., Pottgiesser, J., Martins, R. N., and Beyreuther, K. (1985). Neuronal origin of a cerebral amyloid: Neurofibrillary tangles of Alzheimer's disease contain the same protein as the amyloid of plaque cores and blood vessels. *The EMBO Journal* **4**, 2757–2763.

McKhann, G., Drachman, D., and Folstein, M. (1984). Clinical diagnosis of Alzheimer's disease: Report of NINCDS-ADRDA Work Group under the auspices of department of health and human services task forces on Alzheimer's disease. *Neurology* **34**, 939–944.

Mirra, S. S., Heyman, A., and McKeel, D. (1991). The Consortium to establish a registry for Alzheimer's Disease (CERAD). Part II. Standardization of the neuropathologic assessment of Alzheimer's disease. *Neurology* **41**, 479–486.

Morris, M., Evans, D., Bienias, J., Tangney, C., Bennett, D., Wilson, R., Aggarwal, N., and Schneider, J. (2003). Consumption of fish and n-3 fatty acids and risk of incident Alzheimer disease. *Arch Neurol* **60**, 940–946.

Mortimer, J. A., French, L. R., Hutton, J. T., and Schuman, L. M. (1985). Head injury as a risk factor for Alzheimer's disease. *Neurology* **35**, 264–267.

Mullan, M., Crawford, F., Axelman, K., Houlden, H., Lilius, L., Winblad, B., and Lannfelt, L. (1992). A pathogenic mutation for probable Alzheimer's disease in the APP gene at the N-terminus of beta-amyloid. *Nat Genet* **1**, 345–347.

Murrell, J., Farlow, M., Ghetti, B., and Benson, M. D. (1991). A mutation in the amyloid precursor protein gene associated with hereditary Alzheimer's disease. *Science* **254**, 97–99.

Myers, A. J., Kaleem, M., Marlowe, L., Pittman, A. M., Lees, A. J., Fung, H. C. et al. (2005). The H1c haplotype at the MAPT locus is associated with Alzheimer's disease. *Hum Mol Genet* **14**, 2399–2404.

Näslund, J., Haroutunian, V., Mohs, R., Davis, K. L., Davies, P., Greengard, P., and Buxbaum, J. (2000). Correlation between elevated levels of amyloid ß-peptide in the brain and cognitive decline. *JAMA* **283**, 1571–1577.

Nicoll, J. A., Wilkinson, D., Holmes, C., Steart, P., Markham, H., and Weller, R. O. (2003). Neuropathology of human Alzheimer disease after immunization with amyloid-beta peptide: A case report. *Nat Med* **9**, 448–452.

Nilsberth, C., Westlind-Danielsson, A., Eckman, C. B., Condron, M. M., Axelman, K., Forsell, C. et al. (2001). The "Arctic" (E693G) mutation in the Aβ region of APP causes Alzheimer's disease by increasing Aβ protofibril formation. *Nat Neurosci* **4**, 887–893.

Oddo, S., Caccamo, A., Shepherd, J., Murphy, M., Golde, T., Kayed, R. et al. (2003). Triple-transgenic model of Alzheimer's disease with plaques and tangles: Intracellular Abeta and synaptic dysfunction. *Neuron* **31**, 409–421.

Olson, M. I. and Shaw, C. M. (1969). Presenile dementia and Alzheimer's disease in mongolism. *Brain* **92**, 147–156.

Orgogozo, J. M., Gilman, S., Dartigues, J. F., Laurent, B., Puel, M., Kirby, L. C. et al. (2003). Subacute meningoencephalitis in a subset of patients with AD after Abeta42 immunization. *Neurology* **61**, 46–54.

Pasalar, P., Najmabadi, H., Noorian, A. R., Moghimi, B., Jannati, A., Soltanzadeh, A. et al. (2002). An Iranian family with Alzheimer's disease caused by a novel APP mutation (Thr714Ala). *Neurology* **58**, 1574–1575.

Patrick, G. N., Zukerberg, L., Nikolic, M., de la Monte, S., Dikkes, P., and Tsai, L.-H. (1999). Conversion of p35 to p25 deregulates Cdk5 activity and promotes neurodegeneration. *Nature* **402**, 615–622.

Petersen, R. C. and Morris, J. C. (2005). Mild cognitive impairment as a clinical entity and treatment target. *Arch Neurol* **62**, 1160–1163; discussion 1167.

Rogaev, E. I., Sherrington, R., Rogaeva, E. A., Levesque, G., Ikeda, M., Liang, Y. et al. (1995). Familial Alzheimer's disease in kindreds with missense mutations in a gene on chromosome 1 related to the Alzheimer's disease type 3 gene. *Nature* **376**, 775–778.

Rogers, J., Kirby, L. C., Hempelman, S. R., Berry, D. L., McGeer, P. L., Kaszniak, A. W. et al. (1993). Clinical trial of indomethacin in Alzheimer's disease. *Neurology* **43**, 1609–1611.

SantaCruz, K., Lewis, J., Spires, T., Paulson, J., Kotilinek, L., Ingelsson, M. et al. (2005). Tau suppression in a neurodegenerative mouse model improves memory function. *Science* **309**, 476–481.

Saunders, A. M., Strittmatter, W. J., Schmechel, D., George-Hyslop, P. H., Pericak-Vance, M. A., Joo, S. H. et al. (1993). Association of apolipoprotein E allele epsilon 4 with late-onset familial and sporadic Alzheimer's disease. *Neurology* **43**, 1467–1472.

Schenk, D., Barbour, R., Dunn, W., Gordon, G., Grajeda, H., Guido, T. et al. (1999). Immunization with amyloid-beta attenuates Alzheimer-disease-like pathology in the PDAPP mouse. *Nature* **400**, 173–177.

Scheuner, D., Eckman, C., Jensen, M., Song, X., Citron, M., Suzuki, N. et al. (1996). Secreted amyloid β-protein similar to that in the senile plaques of Alzheimer's disease is increased in vivo by the presenilin 1 and 2 and APP mutations linked to familial Alzheimer's disease. *Nature Med* **2**, 864–870.

Schneider, A., Biernat, J., von Bergen, M., Mandelkow, E., and Mandelkow, E. (1999). Phosphorylation that detaches tau protein from microtubules (Ser262, Ser214) also protects it against aggregation into Alzheimer paired helical filaments. *Biochem* **38**, 3549–3558.

Selkoe, D. (1993). Physiological production of the ß-amyloid protein and the mechanism of Alzheimer's Disease. *Trends Neurosci* **16**, 403–409.

Shibata, M., Yamada, S., Kumar, S. R., Calero, M., Bading, J., Frangione, B. et al. (2000). Clearance of Alzheimer's amyloid-ss(1-40) peptide from brain by LDL receptor-related protein-1 at the blood-brain barrier. *J Clin Invest* **106**, 1489–1499.

Sjogren, T., Sjogren, H., and Lindgren, A. G. (1952). Morbus Alzheimer and morbus Pick; a genetic, clinical and patho-anatomical study. *Acta Psychiatr Neurol Scand Suppl* **82**, 1–152.

Sperling, R. A., Bates, J. F., Chua, E. F., Cocchiarella, A. J., Rentz, D. M., Rosen, B. R. et al. (2003). fMRI studies of associative encoding in young and elderly controls and mild Alzheimer's disease. *J Neurol Neurosurg Psychiatry* **74**, 44–50.

Spires, T. L., Meyer-Luehmann, M., Stern, E. A., McLean, P. J., Skoch, J., Nguyen, P. T. et al. (2005). Dendritic spine abnormalities in amyloid precursor protein transgenic mice demonstrated by gene transfer and intravital multiphoton microscopy. *J Neurosci* **25**, 7278–7287.

Strittmatter, W. J., Saunders, A. M., Schmechel, D., Pericak-Vance, M., Enghild, J., Salvesen, G. S., and Roses, A. D. (1993). Apolipoprotein E: High-avidity binding to β-amyloid and increased frequency of type 4 allele in late-onset familial Alzheimer disease. *Proc Natl Acad Sci U S A* **90**, 1977–1981.

Tagliavini, F. et al. (1999). A new APP mutation related to hereditary cerebral haemorrhage. *Alzh Report*, 28.

Tanzi, R., Gusella, J., Watkins, P., Bruns, G., St George-Hyslop, P., Van Keuren, M. et al. (1987). Amyloid beta protein gene: cDNA, mRNA distribution, and genetic linkage near the Alzheimer locus. *Science* **20**, 880–884.

van Broeckhoven, C., Haan, J., Bakker, E., Hardy, J., Van Hul, W., Wehnert, A. et al. (1990). Amyloid beta protein precursor gene and hereditary cerebral hemorrhage with amyloidosis (Dutch). *Science* **248**, 1120–1122.

Wakutani, Y., Watanabe, K., Adachi, Y., Wada-Isoe, K., Urakami, K., Ninomiya, H. et al. (2004). Novel amyloid precursor protein gene missense mutation (D678N) in probable familial Alzheimer's disease. *J Neurol Neurosurg Psychiatry* **75**, 1039–1042.

Walsh, D., Klyubin, I., Fadeeva, J., Cullen, W., Anwyl, R., Wolfe, M. et al. (2002). Naturally secreted oligomers of amyloid-beta protein potently inhibit hippocampal long-term potentiation in vivo. *Nature* **416**, 535–539.

Wang, D. S., Lipton, R. B., Katz, M. J., Davies, P., Buschke, H., Kuslansky, G. et al. (2005). Decreased neprilysin immunoreactivity in Alzheimer disease, but not in pathological aging. *J Neuropathol Exp Neurol* **64**, 378–385.

Wang, J., Wu, Q., Smith, A., Grundke-Iqbal, I., and Iqbal, K. (1998). Tau is phosphorylated by GSK-3 at several sites found in Alzheimer disease and its biological activity markedly inhibited only after it is prephosphorylated by A-kinase. *FEBS Lett* **436**, 28–34.

Weggen, S., Eriksen, J., Das, P., Sagi, S., Wang, R., Pietrzik, C., et al. (2001). A subset of NSAIDs lower amyloidogenic A42 independently of cyclooxygenase activity. *Nature* **414**, 212–216.

Wolozin, B. (2004). Cholesterol and the biology of Alzheimer's disease. *Neuron* **41**, 7–10.

17

Polyglutamine Disorders Including Huntington's Disease

Sokol V. Todi, Aislinn J. Williams, and Henry L. Paulson

I. Introduction

At least nine inherited neurodegenerative disorders are caused by expansion of a CAG repeat in the protein-coding region of the respective disease genes (see Table 17.1). Because the repeated CAG codon is in frame to encode the amino acid glutamine (Q), the disease protein in each case carries an expanded stretch of polyglutamine, hence the name polyglutamine (polyQ) diseases. To date, the list includes Huntington's disease (HD), dentatorubral-pallidoluysian atrophy (DRPLA), spinobulbar muscular atrophy (SBMA, also known as Kennedy's disease), and six spinocerebellar ataxias (SCAs 1, 2, 3, 6, 7, and 17). SCA8 is another dominant ataxia that may be due to expression of an expanded polyQ tract, though this is less clearly established than for the other polyQ diseases (Moseley et al., 2006).

Evidence strongly suggests that polyQ diseases share a common pathogenic mechanism involving a toxic gain-of-function acquired by the disease protein. Nevertheless the disorders are clinically distinct, indicating that the specific features of each disease are not determined exclusively by the polyQ tract but also by the protein context in which the expansion occurs (Michalik & Van Broeckhoven, 2003).

The polyQ diseases share common features that reflect the special nature of polyQ repeats (Everett & Wood, 2004). First, all are slowly progressive neurodegenerative diseases that tend to manifest in the adult years. Second, the length of a pathogenic expanded repeat varies among affected individuals in each polyQ disease, with longer expansions causing more severe disease with earlier onset, sometimes even in childhood if the repeat is very large. Third, the phenotype can vary greatly, even within a family; this variability largely reflects differences in repeat size. And fourth, eight of the nine known polyQ diseases are autosomal dominant; the outlier is the motor neuron disorder, SBMA, which is an X-linked condition that essentially affects only males.

PolyQ disease proteins are widely expressed throughout the body and brain, from early development to later stages of life. Despite this ubiquitous expression pattern, only certain brain regions degenerate in each disease. Thus, though expanded polyQ is intrinsically neurotoxic, its protein context helps

Table 17.1 The Proteins Involved in the Known PolyQ Diseases

Disorder	Protein Name	Relative Protein Size & polyQ Position	Protein Function
Spinal and Bulbar Muscular Atrophy (SBMA)	Androgen Receptor	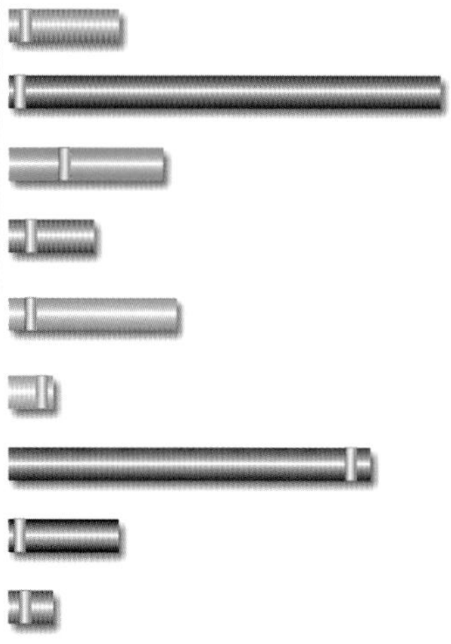	Testosterone-activated steroid receptor
Huntington's Disease	Huntingtin		Essential for brain development; proposed involvement in gene expression regulation, vesicular transport, mitochondrial function
Dentatorubral-Pallidoluysian atrophy (DRPLA)	Atrophin-1		Proposed involvement in gene transcription regulation
Spinocerebellar Ataxia 1 (SCA1)	Ataxin-1		Involved in transcription regulation, cell specification, synaptic activity
Spinocerebellar Ataxia 2 (SCA2)	Ataxin-2		Binds mRNA and may function as a translational regulator; linked to cell specification, apoptosis, secretion, receptor-mediated signaling
Spinocerebellar Ataxia 3 (SCA3)	Ataxin-3		Deubiquitinating enzyme involved in protein quality control
Spinocerebellar Ataxia 6 (SCA6)	P/Q type calcium channel Subunit α1A		Voltage-sensitive calcium channel subunit
Spinocerebellar Ataxia 7 (SCA7)	Ataxin-7		Involved in transcriptional regulation, component of histone acetyl-transferase complex TFTC/STAGA
Spinocerebellar Ataxia 17 (SCA17)	TATA-box binding protein		Basal transcription factor

Each confirmed polyQ disease is listed with the protein in which the mutation occurs. A schematic shows the relative size of the disease proteins and the location of the polyQ domain within them. For each protein, a brief list of confirmed or suggested functions is provided.

determine where in the brain its toxic property is most manifest (La Spada & Taylor, 2003). Although the diseases share a common disease mechanism, the individual host proteins clearly have divergent functions. Here, we present a brief clinical overview of the nine (possibly 10) known polyQ diseases, and discuss possible, shared pathogenic mechanisms supported by current research.

II. Clinical and Genetic Features

A. Spinobulbar Muscular Atrophy (SBMA or Kennedy's Disease)

This rare motor neuron disorder was the first identified polyQ disease (La Spada et al., 1991). It afflicts males nearly exclusively and often is accompanied by signs of feminization, including gynecomastia and some degree of infertility. In SBMA the disease-causing CAG expansion occurs in the androgen receptor (AR) gene, which encodes the male steroid receptor, a hormone-activated transcription factor. Normal alleles range from approximately 9 to 36 repeats, whereas disease repeats range from about 40 to 62. Unlike many other dynamic repeat diseases, SBMA does not show much anticipation from one generation to the next. In genetics, the term "anticipation" means the tendency of repeats to increase in size when transmitted from generation to generation, which causes an increase in disease severity.

AR is expressed in motor neurons of the spinal cord and brainstem, where it mediates neurotrophic responses to androgen. Female carriers do not develop the disease though some experience muscle cramps and subclinical electrophysiological abnormalities; they may be protected from neurodegeneration by low circulating androgens. Findings in transgenic mice are consistent with this: castrated male mice do not develop disease (Chevalier-Larsen et al., 2004).

B. Huntington's Disease (HD)

The most common polyQ disease in the western hemisphere, HD is characterized by cognitive impairment, psychiatric disturbance, and chorea. Although symptoms of HD typically appear in adulthood, people with the longest polyQ expansions can develop symptoms much earlier, sometimes even in the first decade of life. Such juvenile-onset cases have more stiffness and slowness with less chorea. More than 95 percent of persons who develop HD come from families with a family history of HD (i.e., previous generations affected). Occasionally, sporadic cases occur in the absence of family history. In these cases, nonpathogenic intermediate-size CAG repeats in an unaffected parent expand into a pathogenic repeat length in the affected offspring.

In 1993, the HD mutation was discovered in a novel gene encoding a large protein, huntingtin (Group, 1993). The length of the CAG repeat is 6 to 35 in normal individuals, and 36 to 121 CAGs in affected individuals. Intermediate repeats of 36 to 39 are not fully penetrant, but repeat lengths greater than 40 are invariably associated with disease.

The neuronal functions of the huntingtin protein are still unclear. Huntingtin is essential for brain development and has been implicated in many cellular processes including vesicle trafficking, synaptic activity, mitochondrial function, and regulation of neuronal gene expression (Landles & Bates, 2004). Research in animal models of HD has identified compounds and genes that can ameliorate the toxicity of the mutant protein. These modifiers include caspase inhibitors, minocycline, creatine, cystamine, histone deacetylase inhibitors, and genes encoding various molecular chaperones.

C. Dentatorubral-Pallidoluysian Atrophy (DRPLA)

Uncommon in the United States but relatively common in Japan, DRPLA is characterized by progressive ataxia, cognitive changes, chorea, and seizure disorder. Compared clinically to the other polyQ diseases, DRPLA is something of a hybrid between HD and the spinocerebellar ataxias discussed next. In DRPLA, the CAG repeat expansion occurs in the gene encoding atrophin-1, which is thought to be a transcriptional corepressor that regulates gene expression (Zhang et al., 2002).

Members of a large North Carolina family who initially were diagnosed with a unique disorder, Haw River syndrome, were later discovered to have the DRPLA mutation (Burke et al., 1994). Haw River syndrome typically presents with brain calcifications and generalized seizures, signs quite distinct from DRPLA. The marked clinical difference between families with the same mutation indicates the importance of other genetic and/or environmental factors in the manifestation of the DRPLA mutation.

D. Spinocerebellar Ataxia 1 (SCA1)

SCA1 was the first dominantly inherited ataxia for which the gene defect was identified. Neuropathological findings include neuronal loss in the cerebellum (principally Purkinje cells) and brainstem, and degeneration of spinocerebellar tracts. Like most SCAs, SCA1 begins as a gait ataxia, evolving to severe four-limb ataxia with dysarthria and leaving most patients wheelchair-bound within 15 years. There are no cognitive defects associated with SCA1.

SCA1 is caused by an expanded CAG repeat in the *SCA1* gene that encodes the protein ataxin-1 (Banfi et al., 1994). The *SCA1* repeat is normally 6 to 39 repeats long and usually interrupted by one or a few non-CAG triplets; in diseased alleles, the repeat is expanded from 41 up to 82 uninterrupted glutamines. As with other polyQ disorders, repeat length is positively correlated with disease severity. Although the exact function of ataxin-1 remains unknown, it has been shown to regulate the activity of specific transcription factors (Mizutani et al., 2005).

E. Spinocerebellar Ataxia 2 (SCA2)

Initially described in a large Cuban family, SCA2 is one of the most common dominant ataxias. It is typically characterized by ataxia, dysarthria, slow saccades, and neuropathy. In most respects the clinical features are similar to those of SCA1 and SCA3 with extensive cerebellar and brainstem involvement.

The expanded polyQ repeat in SCA2 occurs in ataxin-2, a cytoplasmic protein that has been shown to bind mRNA and is thought to be a translational regulator (Satterfield & Pallanck, 2006). Ataxin-2 function has also been linked to secretion, cell specification, actin filament formation, and apoptotic and receptor-mediated signaling (Satterfield & Pallanck, 2006). Normal alleles are between 15 to 32 repeats in length, and expanded alleles are 35 to 77 repeats in length. A "zone of reduced penetrance" exists for repeats of 32 to 34; not all persons with this repeat size will develop signs of disease. Rarely, expansions in ataxin-2 have presented as autosomal dominant Parkinsonism (Simon-Sanchez et al., 2005).

F. Spinocerebellar Ataxia 3/Machado Joseph Disease (SCA3/MJD)

Perhaps the most common dominant ataxia, SCA3/MJD was identified independently in two clinically and ethnically distinct populations (Kawaguchi et al., 1994; Stevanin et al., 1995a, 1995b). Often, SCA3/MJD begins as a progressive ataxia accompanied by difficulties with eye movements, including bulging eyes and ophthalmoparesis, as well as dysarthria and dysphagia. Neuropathological findings include widespread degeneration of cerebellar pathways, pontine and dentate nuclei, substantia nigra, globus pallidus interna, cranial motor nerve nuclei, and anterior horn cells.

SCA3/MJD is caused by a CAG expansion in the *MJD1* gene that is normally 12 to 42 repeats in length and is expanded to approximately 52 to 84 repeats in disease. The common adult-onset ataxic presentation of disease is caused by mid-sized expansions, whereas longer repeats cause earlier onset disease that is characterized by more rigidity and dystonia, and the shortest expanded repeats can manifest at older age as a motor neuronopathy with ataxia. Some patients develop parkinsonism responsive to dopamine. A few individuals have been described with intermediate alleles (~45-55 repeats), which can cause restless legs syndrome.

The polyQ expansion occurs in the disease protein, ataxin-3. Ataxin-3 is a ubiquitin-binding protein and de-ubiquitinating enzyme that may function in the ubiquitin-proteasome protein degradation pathway (Burnett et al., 2003). In *Drosophila*, wild-type ataxin-3 suppresses polyQ neurodegeneration caused by other polyQ disease proteins (Warrick et al., 2005). This suppression requires its enzymatic activity, suggesting that ataxin-3 normally participates in protein quality control in neurons. Interestingly, ataxin-3 is unique among polyQ proteins in that the *non*pathogenic form of the protein colocalizes to inclusions formed in other polyQ diseases. Likely explanations for this unique feature are (1) ataxin-3 binds ubiquitin chains, which are often found in polyQ inclusions, and (2) ataxin-3 regulates the production of protein accumulations known as aggresomes (Burnett & Pittman, 2005).

G. Spinocerebellar Ataxia 6 (SCA6)

In contrast to most SCAs, SCA6 usually manifests late in life as a "pure" cerebellar ataxia accompanied only by dysarthria and gaze-evoked nystagmus. Disease progresses more slowly than other SCAs, and is usually compatible with a normal lifespan. Uncommon noncerebellar symptoms may include impaired upward gaze, spasticity, and hyperreflexia. In some populations it is fairly common, accounting for 31 percent and 22 percent of the ataxic families in Japan and Germany, respectively.

SCA6 is caused by a much smaller polyQ repeat expansion than are the other polyQ diseases. Normal SCA6 alleles range from three to 17 CAG repeats, and disease alleles contain 21 to 30 repeats (Zhuchenko et al., 1997). Thus, pathogenic SCA6 alleles lie in a range that, in other polyQ diseases, would be nonpathogenic. The expansion occurs in the gene for a P/Q-type voltage-dependent calcium channel subunit (*CACNA1A*); other, nonrepeat mutations in this gene have been shown to cause two distinct neurological conditions, episodic ataxia type 2 and familial hemiplegic migraine (Ophoff et al., 1996). Because this polyQ disease protein is the only membrane protein among the polyQ disease proteins, and causes disease with much smaller repeat sizes, the pathogenic mechanism in SCA6 may differ from that in other polyQ diseases.

H. Spinocerebellar Ataxia 7 (SCA7)

SCA7 is unique among the SCAs. It is the only SCA characterized by retinal degeneration as well as ataxia. In other respects, SCA7 resembles the other common SCAs (SCA1, SCA2, SCA3) in being characterized by ataxia and brainstem findings. The defect is an expanded CAG repeat in the *SCA7* gene encoding the protein ataxin-7 (David et al., 1997). Ataxin-7 is a nuclear protein that is thought to regulate gene transcription (Helmlinger et al., 2004). The SCA7 expansion is highly variable, ranging from 34 to greater than 200 repeats, whereas the normal allele repeat ranges from seven to 17. Although gonadal repeat instability is common in polyQ diseases, it is particularly impressive in SCA7 (David et al., 1997). Instability is especially striking with paternal transmission, leading in some cases to massive expansions that cause disease in infancy or *in utero*.

I. Spinocerebellar Ataxia 8 (SCA8)

Koob and colleagues (1999) identified the genetic defect in SCA8 as an unstable repeat expansion in the *SCA8* gene. In one direction in the genome, this repeat is transcribed into RNA but not translated into protein. This led to the original hypothesis that SCA8 is due to a dominant toxic RNA effect as occurs in the myotonic dystrophies. Recently, however, it was discovered that this repeat can also be transcribed in the opposite direction (Moseley et al., 2006). The resultant CAG repeat expansion can be translated into an expanded polyQ protein without any adjacent protein sequence. Examination of a single SCA8 brain to date suggests that inclusions containing polyQ are indeed a neuropathological feature of this disease.

The assignment of SCA8 to the polyQ camp, however, is still controversial. But this finding does raise the possibility that repeat expansions may, in some cases, be transcribed in both directions; thus, a single repeat expansion in the DNA could perhaps act through two mechanisms rather than just one. It is important to recognize that some individuals with very large *SCA8* expansions do *not* develop disease, which suggests that *SCA8* repeat expansions are incompletely penetrant or—less likely—the actual mutation lies elsewhere but is closely linked to the expansion. Because of this uncertainty, SCA8 gene test results should be interpreted with caution.

SCA8 clinically resembles most other SCAs: adult onset ataxia with variable brainstem signs. In a study of a large family, the main clinical symptoms were prominent gait and limb ataxia accompanied by abnormalities of swallowing, speech, and eye movements. In contrast to many other SCAs in which paternal transmission is more prone to result in further repeat expansions, there is marked maternal bias in SCA8 disease transmission. This may reflect contractions of the expanded repeat during spermatogenesis.

J. Spinocerebellar Ataxia 17 (SCA17)

Originally described in Japan, SCA17 is quite rare in the United States. The ataxia of SCA17 is accompanied by more intellectual decline and extrapyramidal features (for example, bradykinesia and dystonia) than in the other SCAs. It is also more likely to show widespread cerebral degeneration. In some cases, SCA17 can even resemble HD in certain respects. SCA17 is caused by a CAG repeat expansion in the TATA binding protein gene (TBP or TFIID), which encodes a critical component of the basal transcription machinery (Koide et al., 1999). TBP has a rather large, normal-sized repeat of up to 44 repeats in length, which becomes expanded in disease to 48 or greater.

III. Protein Misfolding and Failures in Protein Quality Control

Protein quality control (QC) is tightly regulated in all cells. Its importance becomes especially clear in the polyQ diseases, where protein conformational defects result in the detrimental accumulation of aggregated protein. Long-lived, post-mitotic neurons are particularly vulnerable to the accumulation of abnormal proteins because they are unable to disperse misfolded, aggregated protein through cell division.

The amino acid sequence of a protein contains the necessary information for nascent proteins to assume their proper tertiary structure. However, the crowded intracellular environment presents significant obstacles to error-free protein folding (Muchowski & Wacker, 2005). Accordingly, cells have evolved highly conserved chaperone proteins that facilitate protein folding and promote degradation of misfolded or incorrectly assembled proteins (see also Chapter 5).

One family of chaperone proteins, the heat shock proteins (HSPs), has been studied extensively in the context of polyQ disease. Most HSPs function as molecular chaperones. They are divided into six classes, denoted by the approximate size of the protein in kilodaltons: HSP100, 90, 70, 60, 40, and smaller HSPs of less than 40 kD (Macario, 1995).

If a protein does not fold properly despite the assistance of chaperones, it often is then degraded by the ubiquitin proteasome system (UPS), the key protein degradation pathway in the cell. Certain proteins link UPS degradation to the protein folding chaperone system. Another major route for abnormal protein elimination is lysosome-mediated autophagy, which engulfs and destroys damaged cellular components ranging from individual proteins to entire organelles. Although we will focus here on the UPS as a major clearance mechanism for polyQ disease proteins, we remind you that increasing evidence suggests that autophagy also plays a role in the destruction of polyQ proteins in at least some diseases (Ravikumar & Rubinsztein, 2004).

The following sections focus on abnormally expanded polyQ proteins, their propensity to aggregate, the efforts of the cellular QC system to handle these misfolded proteins, and on studies that have aimed to identify precisely what is toxic about expanded polyQ proteins (see Figure 17.1).

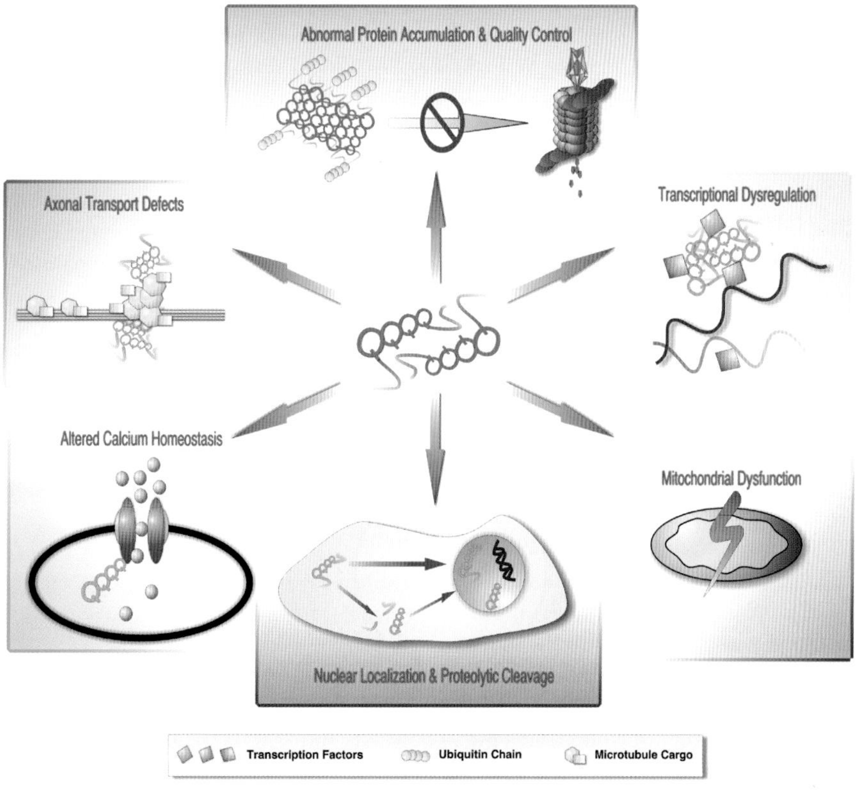

Figure 17.1 Potential mechanisms to explain polyglutamine toxicity. As shown in the center, a toxic conformation of expanded polyQ, either monomeric or oligomeric, is thought to underlie pathogenesis. As shown from the top, all polyQ diseases demonstrate accumulation of toxic protein and some degree of impairment in protein quality control, either in proteasomal degradation (as pictured) or in chaperone-mediated refolding (not shown). Moving clockwise, evidence suggests transcriptional dysregulation in many polyQ diseases, pictured here as polyQ disease proteins abnormally interacting with transcription factors. PolyQ-mediated mitochondrial dysfunction may occur through a variety of pathways, though these are currently less well defined than those for other potential mechanisms listed. Nuclear localization of full-length, mutant protein or cleavage fragments (as in HD) —a recurrent theme in polyQ diseases—may contribute to transcriptional dysregulation and/or interference with other normal nuclear functions. Altered calcium homeostasis may be an intrinsic feature of SCA6, as the mutation is in a calcium channel, although there is also evidence for its importance in HD. Lastly, axonal transport defects are an important cause of neurodegeneration beyond polyQ disorders; the extent to which they represent a pathogenic feature in polyQ diseases is under investigation.

A. Aggregation and PolyQ Diseases

A pathological hallmark of all polyQ disorders is the presence of intraneuronal inclusions of the mutant protein. Expanded polyQ tracts tend to perturb the native folding of the protein in which they reside (Ignatova & Gierasch, 2006) increasing the probability that the disease protein will misfold, oligomerize, and aggregate. The tendency of polyQ proteins to misfold correlates with the length of the repeat: longer repeats are more likely to misfold and form aggregates. Post-translational modifications, such as phosphorylation, ubiquitination, and SUMOylation, also influence the propensity of the protein to aggregate (Humbert et al., 2002; Steffan et al., 2004). Protein aggregates are consolidated into inclusions by microtubule-based cellular processes. Such inclusion bodies formed by aggregated polyQ proteins can reside in the nucleus (SCA1, SCA7, SCA17), cytoplasm (SCA2 and SCA6), or both (HD, SCA3, DRPLA, SBMA).

The relationship of proteinaceous inclusions to disease pathogenesis is unclear. One possibility is that nuclear inclusions are responsible for cellular demise. However, the causal link between inclusions and neurodegeneration has been challenged by several reports. Some studies indicate that inclusions may actually play a protective role; cells that form inclusions tend to survive longer than ones that form inclusions slowly or not at all (Arrasate et al., 2004; Bowman et al., 2005; Saudou et al., 1998; Slow et al., 2005; Watase et al., 2002; Yoo et al., 2003). The current view is that oligomeric or protofibrillar intermediates are most likely the cytotoxic polyQ species rather than large mature inclusions (Muchowski & Wacker, 2005; Poirier et al., 2002; Sanchez et al., 2003). Indeed, evidence is accumulating that soluble polyQ oligomers may be most toxic to the cell (Arrasate et al., 2004; Bucciantini et al., 2002).

The apparent contradictions about whether inclusions are protective or harmful may be partially resolved by understanding that the biochemical process of "aggregation" is not identical to "inclusion formation." In this chapter, we define *aggregates* as biochemically detectable protein complexes, which are detergent- and protease-resistant. *Inclusions* are visible protein structures generated by regulated cellular processes that sequester these proteins away from other cellular components.

PolyQ protein aggregates are detergent-insoluble and protease-resistant. Mature aggregates have a fibrillar, β-sheet rich amyloid-like structure that can bind lipophilic dyes such as Congo Red and thioflavin. The aggregation pathway for expanded polyQ proteins is likely to be complex, with oligomeric, globular, and fibrillar intermediates appearing before mature amyloid fibers are present (Ross & Poirier, 2004). Several studies support a seeding model for aggregate formation, whereby a misfolded monomer acts as a nidus for subsequent misfolded proteins to bind and form oligomers (Bhattacharyya et al., 2005; Chen et al., 2002).

Aggregated polyQ proteins in neuronal inclusions are ubiquitinated in most polyQ disorders, the sole exception being SCA6 (Ishikawa et al., 1999). Immunocytochemical analyses of inclusions in disease tissues and animal models indicate that mutant polyQ proteins are sequestered into inclusions along with components of the UPS, certain HSPs, and other proteins. The presence of UPS components and chaperones in inclusions suggests that the polyQ proteins are indeed misfolded, and that the neuronal QC machinery has been recruited in an effort to refold or degrade the expanded polyQ proteins. Although some studies suggest that polyQ proteins can impair UPS function (Bence et al., 2001; Diaz-Hernandez et al., 2006; Khan et al., 2006), there are conflicting data about whether the UPS in fact is impaired by polyQ proteins *in vivo* (Gatchel & Zoghbi, 2005).

In summary, neuronal inclusions are a hallmark of polyQ disease but their relation to pathology is still under debate. The involvement of the UPS in polyQ protein handling, however, has been clearly established and molecular chaperones may be a viable target for therapeutic considerations for persons afflicted by polyQ disorders.

B. Toxic Species in PolyQ Neurodegeneration

During the past decade, as the genes responsible for polyQ disorders were being discovered, a central question arose: Does polyQ repeat expansion cause a dominant loss of protein function or a toxic gain of function? Studies in various models have established that polyQ disorders are due primarily to a toxic gain-of-function (Zoghbi & Orr, 2000). For example, knock-in mouse models recapitulate disease features (Watase et al., 2002; Yoo et al., 2003), whereas knock-out mouse models for the various disease genes do not (Matilla et al., 1998). Nevertheless, for some polyQ diseases a partial loss of function may contribute to the disease phenotype. In SBMA, for example, partial loss-of-function of the androgen receptor (AR) explains some clinical features of the disease including gynecomastia and testicular atrophy. Moreover, polyQ expansion in AR suppresses its normal transcriptional activity (Katsuno et al., 2006). Similarly, in SCA6 the expansion in a subunit of the P/Q-type calcium channel alters channel function, leading to a decrease in calcium influx into neurons (Matsuyama et al., 1999; Toru et al., 2000). Therefore, although polyQ expansion generally can be considered a toxic gain-of-function mutation, there are important exceptions to this generalization.

1. Nuclear Localization of Disease Protein

Attempts to identify toxic species in polyQ diseases have determined that nuclear localization of mutant polyQ protein is necessary for neurodegeneration in several polyQ diseases (Klement et al., 1998; Kordasiewicz et al., 2006; Saudou et al., 1998). For example, when the nuclear localization signal in ataxin-1 is mutated, expanded ataxin-1 is no longer transported to the nucleus and no longer causes Purkinje cell degeneration and ataxia in transgenic mice (Cummings et al., 1998; Klement et al., 1998). Furthermore, a cellular model of HD also indicated that nuclear localization is necessary for expanded huntingtin-dependent apoptotic cell death (Saudou et al., 1998). Lastly, in cell models of SCA6, toxicity of the expanded protein is dependent upon its nuclear localization (Kordasiewicz et al., 2006). Thus, as indicated by these examples, nuclear localization of mutant polyQ species may be necessary for pathogenesis in many polyQ diseases.

2. Toxic Protein Fragments

Another point of interest in polyQ disorders is whether full-length proteins or proteolytic polyQ-containing fragments are the primary toxic species. The answer may vary depending upon the disease protein in question. In HD, for example, the generation of proteolytic fragments of huntingtin may be important for pathogenesis (Graham et al., 2006; Mangiarini et al., 1996). Several caspases and other proteases can cleave expanded huntingtin, atrophin-1, ataxin-1, and ataxin-3 (Goldberg et al., 1996; Wellington & Hayden, 2000; Wellington et al., 2000), and some evidence suggests that proteolytic cleavage is required for ataxin-3 aggregation (Haacke et al., 2006). Inhibition of caspase 1 activity delays pathogenesis in a mouse model of HD (Ona et al., 1999), and caspase activation has been observed in human HD brains (Ona et al., 1999; Sanchez et al., 1999). Evidence in SBMA suggests that cleavage products of expanded AR are toxic to cells (Cowan et al., 2003; Li et al., 1998; Merry et al., 1998). Likewise, cleavage of the P/Q-type calcium channel subunit may be important in SCA6 pathogenesis (Kordasiewicz et al., 2006). In contrast, there is no evidence that proteolysis of ataxin-1 plays a role in SCA1 pathogenesis. Therefore, cleavage products may contribute to pathogenesis in only a subset of polyQ disorders.

3. Post-Translational Modifications and Importance of Protein Context

Many lines of evidence have shown that although pure, expanded polyQ domains without flanking amino acids are toxic, protein context also greatly influences pathogenesis. The fact that each disease manifests its own specific clinical and neuropathological features despite having the "same" mutation means that disease protein context must help to specify the range and degree of neurodegeneration. In keeping with the central role of protein context in pathogenesis, the toxicity of several polyQ disease proteins has been shown to be regulated by phosphorylation at distant points in the protein, far removed from the polyQ domain. In the case of expanded huntingtin, phosphorylation of its serine 421 residue by Akt (a Ser/Thr kinase) abrogates its pro-apoptotic activity *in vivo* and *in vitro* (Humbert et al., 2002). Akt-dependent phosphorylation of ataxin-1 also modulates neurodegeneration in SCA1 models. In a fruit fly model of SCA1, Akt-dependent ataxin-1 phosphorylation is required for the stabilization of polyQ-expanded ataxin-1, which enhances neurodegeneration (Chen et al., 2003). Consistent with this finding, phosphorylation of ataxin-1 at serine 776 is critically important for pathogenesis in SCA1 transgenic mice (Emamian et al., 2003). Akt also acts on AR, though the effects of this modification on SBMA neurodegeneration are unclear (Lin et al., 2001).

Postranslational modifications with ubiquitin or the small ubiquitin-like modifier, SUMO, also can modulate toxicity in a protein-context dependent manner. For example, in cell and fruit fly models of HD, a toxic cleavage product of huntingtin is readily SUMOylated (Steffan et al., 2004). This stabilizes the fragment, increases its solubility, and enhances its ability to repress transcription and cause degeneration (Steffan et al., 2004). In contrast, ubiquitination of this huntingtin fragment at the same lysine residues suppresses degeneration (Steffan et al., 2004).

There is also evidence that nonpolyQ regions of polyQ disease proteins are important in pathogenesis independent of post-translational modifications. SCA1 models indicate that overexpression of the AXH domain of ataxin-1 contributes to neurodegeneration. When overexpressed, the AXH domain reduces the expression of proteins important for cellular survival. The expansion of polyQ in ataxin-1 enhances the effect of the AXH domain by rendering ataxin-1 less degradable (Tsuda et al., 2005). And in SBMA, the AR ligand testosterone is necessary for pathogenesis since full-blown disease does not occur in women who are SBMA gene carriers or even homozygous for the mutation (Schmidt et al., 2002), or in male transgenic mice that have been castrated surgically or pharmacologically (Katsuno et al., 2002, 2006). These findings support the view that protein regions outside of the polyQ domain contribute to pathogenesis.

In summary, it has become apparent that (1) nuclear localization of some polyQ proteins is intimately linked to neurodegeneration; (2) polyQ protein cleavage likely plays an important role in some disorders; and (3) post-translational modifications and disease protein context can modulate toxicity.

C. Evidence That Protein Quality Control Pathways Are Involved in Disease

Considerable evidence indicates that protein QC systems, particularly the UPS system and chaperones, are important in the cellular response to expanded polyQ protein. Although these systems function cooperatively, we will discuss the evidence for each separately.

Several lines of evidence suggest that the UPS is involved in polyQ disease. PolyQ inclusions often are ubiquitinated, suggesting that they have been targeted for degradation, and proteasomal subunits often localize to polyQ protein inclusions. Ubiquitination of expanded huntingtin fragments abrogates neurodegeneration presumably by facilitating proteasomal degradation (Steffan et al., 2004). Conversely, inhibition of proteasomal function exacerbates toxicity in polyQ disease models (Bailey et al., 2002; Chan et al., 2002; Cummings et al., 1999). Deleting components of the UPS produces a similar exacerbation of polyQ toxicity. For example, a SCA1 mouse model engineered to lack the ubiquitin ligase E6-AP shows reduced inclusion formation but increased Purkinje cell degeneration (Cummings et al., 1999). This suggests that ubiquitin dependent pathways—presumably including the UPS—are critical for handling expanded polyQ proteins.

As mentioned earlier, the HSP family of chaperones has been studied extensively in polyQ disease models. HSP levels in mouse models of HD and SBMA are decreased in affected brain areas (Hay et al., 2004; Katsuno et al., 2005), suggesting that QC is impaired in these areas. Additionally, HSP70 and HSP40 colocalize with ataxin-1, 2, 3, 7, and AR inclusions (Bailey et al., 2002; Chai et al., 1999a, 1999b; Chen et al., 2003; Cummings et al., 1998; Katsuno et al., 2006; Muchowski & Wacker, 2005), implying that the cellular protein refolding machinery has been recruited to inclusions to handle mutant polyQ protein.

Many studies show that overexpression of HSPs can ameliorate the cellular dysfunction caused by expanded polyQ proteins (Hsu et al., 2003; Meriin et al., 2002; Muchowski et al., 2000; Warrick et al., 1998, 1999). For example, expression of an expanded ataxin-3 fragment in the fruit fly eye leads to severe photoreceptor degeneration that can be rescued by overexpression of wild-type HSP70 and enhanced by overexpression of a dominant-negative form of HSP70 (Warrick et al., 1998, 1999). Overexpression of HSP70 also improves behavioral and neuropathological phenotypes in a model of SCA1 (Cummings et al., 2001); similar results have been obtained in a study of SBMA mice overexpressing HSP70, wherein HSP70 ameliorates pathology, possibly by enhancing the degradation of the mutant AR (Adachi et al., 2003; Bailey et al., 2002). On the other hand, overexpression of HSP70 in a mouse model of HD has no discernible effect (Hansson et al., 2003). Pharmacological induction of HSP70, HSP90, and HSP105 in the central nervous system by the small compound geranylgeranylacetone inhibits nuclear accumulation of AR and ameliorates pathology in an SBMA mouse model (Katsuno et al., 2005).

Although chaperones can clearly affect polyQ toxicity, their role in modulating aggregation is less clear. Overexpression of HSP40 and HSP70 consistently reduces polyQ-induced toxicity, and HSP27 has similar effects (Wyttenbach et al., 2002), but the extent to which these proteins affect polyQ protein inclusions varies. HSP40 has been shown to reduce inclusion formation, enhance proteasomal degradation, and reduce toxicity of expanded polyQ proteins in SCA1, 2, 3, 7, and SBMA models (Adachi et al., 2003; Bailey et al., 2002; Chai et al., 1999a, 1999b; Chan et al., 2002; Fu et al., 2005; Jana et al., 2000; Kobayashi et al., 2000; Kobayashi & Sobue, 2001). The chaperones Hdj2 and Hsc70 colocalize with huntingtin inclusions and, when overexpressed, reduce their formation (Everett & Wood, 2004). In contrast, HSP70 reduces toxicity in various model systems but does not always reduce inclusion formation (Zhou et al., 2001). And studies in *S. cerevisiae* and *C. elegans* indicate that HSPs directly regulate steps in protein inclusion (Hsu et al., 2003; Meriin et al., 2002), supporting an evolutionarily conserved role for HSPs in protein quality control and clearance.

In yeast models of polyQ disease, certain chaperones are required for inclusion formation. Hsp104 is required for aggregation of polyQ protein in a yeast model of HD (Krobitsch & Lindquist, 2000), and is also critically important for the handling of aggregated prion proteins (Chernoff et al., 1995). Overexpression of yeast Hsp104 in a mouse model of HD results in amelioration of the phenotype and a decrease in inclusions of expanded huntingtin (Vacher et al., 2005). Since a definite mammalian homolog of Hsp104 has not been found, however, the full implications of the yeast data are unclear.

Researchers have long wondered whether polyQ protein misfolding affects only proteins that interact with polyQ or more globally affects cellular protein QC. Recent experiments in temperature-sensitive mutant strains of *C. elegans* revealed that expression of expanded polyQ proteins promotes misfolding of unrelated temperature-sensitive proteins even at the normally permissive temperature (Gidalevitz et al., 2006). Thus, not only do polyQ proteins misfold but their presence in the cell also leads to the misfolding of other proteins, presumably due to a global perturbation in QC. It is possible that localization of proteasomal subunits and chaperones to polyQ aggregates has similar detrimental effects on QC in neurons.

In conclusion, evidence from a wide range of studies indicates that components of QC are intimately involved in processing expanded polyQ proteins. HSPs, in particular, can help ameliorate pathology stemming from polyQ proteins in many, but not necessarily all, disease models. QC itself may also be impeded by the presence of polyQ proteins, amplifying the detrimental effects of misfolded polyQ proteins on the cell.

IV. Transcriptional Dysregulation

Many recent reports support the view that expanded polyQ proteins perturb gene expression in the brain. At least four lines of data implicate transcriptional dysregulation in polyQ neurodegeneration:

▲ Several polyQ disease proteins normally function in transcription pathways

- ▲ Nuclear localization is tightly linked to pathogenesis in many polyQ disorders
- ▲ Transcription factors can be sequestered by polyQ inclusions or by soluble complexes of polyQ disease proteins
- ▲ Transcriptional changes seem to be an early event in disease pathogenesis, often preceding the onset of neurodegeneration

Several models have been proposed to explain how expanded polyQ proteins affect transcription. It is important, first, to recognize that several polyQ disease proteins are themselves transcription factors or cofactors including AR, ataxin-7, atrophin-1, and TBP. To the extent that polyQ expansion alters the conformation of these transcriptional components, it could directly affect expression of target genes. Other polyQ proteins also are known to interact with gene regulatory complexes, including ataxin-1, ataxin-3, and huntingtin. A second important point is that a polyQ domain can directly regulate transcription: polyQ repeats activate transcription when fused to a simple DNA binding domain, and the repeat length determines the degree of transcriptional activation (Gerber et al., 1994). Another possibility is that when polyQ is expanded, polyQ proteins engage in novel interactions with components of the transcriptional machinery (Schaffar et al., 2004). Such an interaction need not be at the level of insoluble inclusion formation. In a SCA7 mouse model, for example, transcriptional changes occur before inclusions are observed (Yoo et al., 2003). Finally, transcriptional dysregulation could result from recruitment of other transcription factors and cofactors that themselves contain polyQ stretches (e.g., CBP, TAF4, TP53, and Sp1) through a homotypic polyQ-polyQ domain interaction. One or more of these mechanisms may contribute to transcriptional dysregulation in various polyQ diseases.

What is the evidence for altered transcription in cells expressing expanded polyQ proteins? Expanded polyQ proteins have been shown to increase or decrease transcription levels of various genes (Luthi-Carter et al., 2002a, 2002b; McCampbell et al., 2000, 2001; Nucifora et al., 2001; Steffan et al., 2000; Sugars et al., 2004). For example, genes for certain neuropeptides such as enkephalin and trophic factors are decreased, but some stress-related proteins are increased in HD and DRPLA mouse models (Luthi-Carter et al., 2002a, 2002b). Mutant huntingtin can affect transcription by interacting with nuclear response elements that regulate the expression of genes encoding BDNF, NMDA receptor subunits, PSD-95, and α-actinin (Luthi-Carter et al., 2002a, 2002b, 2003; Zuccato et al., 2001, 2003). Moreover, CBP and CREB-dependent transcription is decreased in cells expressing polyQ expanded huntingtin, and overexpression of CBP or CREB restores levels of transcription (Nucifora et al., 2001; Sugars et al., 2004).

Importantly, recent investigations reveal significant gene expression changes in brain areas with extensive neurodegeneration such as the caudate nucleus in HD (Hodges et al., 2006). One study showed that expression levels of 81 percent of genes normally enriched in the striatum were decreased in HD mice (Desplats et al., 2006). These mRNA changes occurred before signs of degeneration or gliosis, and in many instances before mice showed any symptoms. For some genes, transcription levels were also reduced in caudate tissue from postmortem HD brain (Desplats et al., 2006). These data suggest that normal neuronal function in HD is affected by polyQ-dependent changes in gene expression and that transcriptional dysregulation contributes to pathogenesis.

Transcriptional dysregulation also has been observed in other polyQ diseases. Tsuda and colleagues (2005) recently provided evidence that the AXH domain of ataxin-1 interacts with the *Drosophila* protein *Senseless* and its mammalian growth factor ortholog, Gfi-1. *Senseless* is required for the peripheral nervous system development in *Drosophila* and Gfi-1 is necessary for survival of Purkinje cells in mice (Tsuda et al., 2005). Overexpression of wild-type and expanded ataxin-1 reduces the levels of *Senseless* in flies and Gfi-1 in mouse Purkinje cells in a proteasome-dependent manner (Tsuda et al., 2005). Importantly, loss of Gfi-1 reproduces aspects of the SCA1 phenotype (Tsuda et al., 2005), arguing that loss of Gfi-1 plays an important role in SCA1 pathogenesis. Expansion of the polyQ domain renders ataxin-1 more resistant to degradation, increasing its cellular levels and possibly its effects on Gfi-1, leading to neuropathological changes.

Ataxin-1 may also affect transcription by binding the cerebellum-enriched transcription regulator PQBP-1 in a polyQ-dependent fashion (Okazawa et al., 2002; Waragai et al., 1999). A mechanism has been proposed by which the ataxin-1/PQBP-1 interaction impedes RNA polymerase II function, especially in the cerebellum (severely affected in SCA1). Since PQBP-1 can interact with other polyQ proteins, a similar repression of RNA polymerase II could contribute to neurodegeneration in other polyQ diseases.

Evidence for transcriptional dysregulation also has come from SCA7 models. Recently, Helminger and collaborators reported that ataxin-7 is a subunit of the GCN5 histone acetyl-transferase complex TFTC/STAGA. However, though ataxin-7 recruitment into the TFTC/STAGA complex appears unaffected by polyQ expansion (Helmlinger et al., 2004), investigations in a SCA7 model provided evidence of chromatin decondensation, H3 histone hyperacetylation, and surprisingly, rather global transcriptional downregulation in rod photoreceptors (Helmlinger et al., 2006). On a similar note, in a SCA7 mouse model, retinal degeneration could be linked to altered function of a specific transcription factor, CRX, in the presence of expanded ataxin-7 (La Spada et al., 2001).

The transcriptional changes observed in SCA17 come as no real surprise given that the polyQ expansion resides in the TATA binding protein (TBP), a basal transcription factor that binds the TATA box, a core promoter element present in most genes. In SCA17 models, expanded TBP was shown to interfere with CREB-dependent transcription (Reid et al.,

2003). TBP is also recruited to NI in other polyQ disorders including SCA1, 2, 3, HD, and DRPLA (Huang et al., 1998; Perez et al., 1998; Uchihara et al., 2001).

Lastly, in SBMA polyQ expansion resulted in weaker transactivation of AR target genes both through changes in AR activity and through decreased ability of AR to interact with its coactivator, ARA (Hsiao et al., 1999; Kazemi-Esfarjani et al., 1995; Mhatre et al., 1993). This may ultimately lead to reduced androgen sensitivity and neurodegeneration.

In summary, selective yet significant alterations in gene expression are observed in several polyQ diseases and in animal models. Such changes may directly contribute to early events in the pathogenic cascade.

V. Mitochondrial Dysfunction

As metabolically active cells with high energy demands, neurons are particularly sensitive to mitochondrial dysfunction. Evidence that mitochondrial dysfunction contributes to polyQ neurodegeneration comes primarily from studies of HD. One model of HD pathogenesis posits that mitochondrial dysfunction renders striatal neurons especially sensitive to excitotoxicity from glutamatergic cortical inputs. In addition, because protein QC pathways are ATP-dependent, impaired mitochondrial function could exacerbate problems in protein folding or decrease the degradation of expanded polyQ proteins.

In HD patients, metabolic anomalies have been detected before obvious striatal atrophy (Grafton et al., 1990; Hayden et al., 1986; Mazziotta et al., 1987). More recent studies have revealed decreased glucose metabolism in HD brain, altered skeletal muscle metabolism, increased lactate/pyruvate ratio in CSF, and increased lactate levels in some brain areas (Antonini et al., 1996; Brouillet et al., 2005; Cooper & Schapira, 1997; Jenkins et al., 1993; Koroshetz et al., 1997; Leegwater-Kim & Cha, 2004). Not all studies, however, are in agreement (Hoang et al., 1998).

Additional support for mitochondrial dysfunction in HD comes from studies showing reduced levels of enzymes important in oxidative phosphorylation, including mitochondrial complexes II, III, and IV, but not complex I (Browne et al., 1997; Butterworth et al., 1985; Gu et al., 1996; Tabrizi et al., 2000). HD mitochondria are less capable of handling calcium (Panov et al., 2002), supporting the hypothesis that excitotoxicity and mitochondrial dysfunction are linked. In fact, administration of mitochondrial inhibitors (e.g., 3-NP or malonate) in mammals leads to HD-like neurodegeneration (Brouillet et al., 1995; Henshaw et al., 1994) that can be prevented by NMDA receptor antagonists (Henshaw et al., 1994). There is evidence for mitochondrial depolarization associated with decreased ATP production in HD specimens, and a knock-in mouse model of HD shows increased levels of AMP (Gines et al., 2003), implying deficits in energy production. Finally, some wild-type huntingtin in cells localizes to the outer mitochondrial membrane and expanded huntingtin alters membrane permeability (Choo et al., 2004). Based on this collective evidence, mitochondrial dysfunction is likely an important element in HD pathogenesis.

Mitochondrial dysfunction also has been observed in SBMA models, where mitochondria are recruited into AR inclusions, leading in some cases to sequestration of the entire mitochondrial pool (Simeoni et al., 2000; Stenoien et al., 1999). In addition, wild-type and expanded AR interact differentially with the mitochondrial electron transport chain complex, cytochrome c oxidase (Beauchemin et al., 2001). AR inclusions sequester subunits of cytochrome c oxidase, which may disrupt mitochondrial function and activate apoptotic pathways (Beauchemin et al., 2001).

Mitochondrial dysfunction may also contribute to SCA1 pathogenesis: mitochondrial dysfunction occurs in cells expressing mutant ataxin-1 (Kim et al., 2003) and in patients suffering from SCA1 (Kish et al., 1999). In one SCA1 mouse study, supplemental creatine extended the life of Purkinje cells (Kaemmerer et al., 2001), consistent with the view that low ATP levels may be involved in SCA1 neurodegeneration. Similarly, defective mitochondrial oxidative phosphorylation in muscles was reported in human cases of DRPLA (Lodi et al., 2000).

There are relatively few reports regarding mitochondrial dysfunction in the other polyQ diseases. In two cell models of SCA3, expanded ataxin-3 expression led to apoptosis by activating pro-apoptotic mitochondrial pathways, including cytochrome c release and changes in levels of pro-apoptotic and anti-apoptotic proteins of the Bcl-2 family (Chou et al., 2006; Tsai et al., 2004; Wang et al., 2006). A study assessing mitochondrial dysfunction in SCA3/MJD patients, however, did not yield conclusive evidence for mitochondrial abnormalities (Matsuishi et al., 1996).

In summary, mitochondrial defects likely contribute to the disease process in at least some polyQ disorders, most notably in HD. However, mitochondrial defects may not occur in all polyQ disorders, and a direct causal relationship between mitochondrial defects and polyQ toxicity has not been firmly established.

VI. Excitotoxicity and Calcium Homeostasis

Excitotoxicity and abnormal calcium homeostasis also have been proposed as a mechanism of polyQ neurodegeneration, largely based on studies of HD models (DiFiglia, 1990). There is increased sensitivity to NMDA receptor activation in HD brains of transgenic mice (Levine et al., 1999; Zeron et al., 2002), leading to increased calcium currents and mitochondrial membrane depolarization (Zeron et al., 2004). This could reflect, in part, reduced mitochondrial function in neurons. Glutamate recycling may also be affected since levels of glial glutamate transporters are reduced in HD mice (Behrens et al., 2002) and expanded huntingtin inhibits glutamate reuptake (Li et al., 2000). In support, Desplats and colleagues (2006) recently described striatal-specific reduction in the expression of genes involved in calcium homeostasis in HD models. Together with a report showing striatal apoptosis due to huntingtin-induced calcium disturbances (Tang et al., 2005), these data indicate that dysregulation of calcium levels may be an important component in HD pathogenesis.

Data from SCA1 also implicate perturbations in calcium homeostasis in polyQ pathogenesis. Several genes central to calcium homeostasis are reduced in SCA1 mice (Lin et al., 2000) and, in SCA1 brains, IP3-mediated calcium release and metabolism are significantly lower than in control brains (Desaiah et al., 1991; Vig et al., 1992, 2001). In SCA6, decreased calcium influx through the P/Q-type calcium channel (Matsuyama et al., 1999; Toru et al., 2000) may underlie neurodegeneration. Thus, calcium imbalance may also contribute to neurotoxicity in SCA1, SCA6, and possibly other polyQ disorders.

VII. Axonal Transport Defects

Neuronal function is heavily dependent on the transport of organelles and signaling molecules between cell bodies

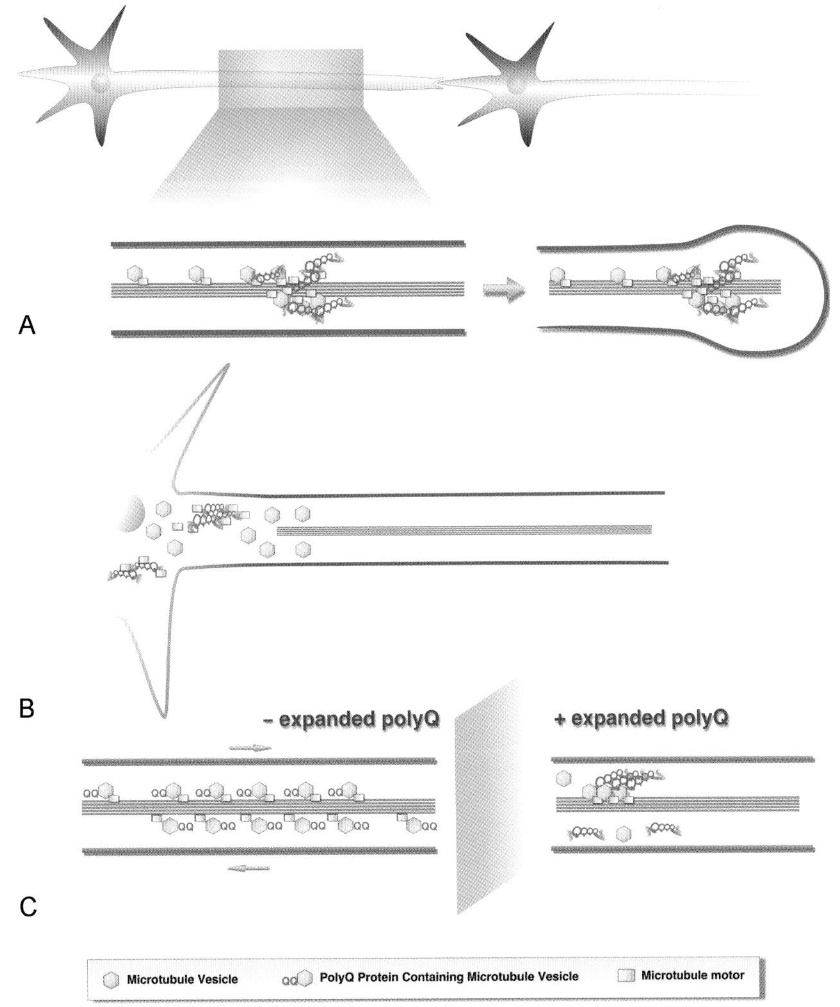

Figure 17.2 Potential mechanisms to explain polyglutamine-induced axonal transport defects. Three potential mechanisms are illustrated here that are not necessarily mutually exclusive. **A**. PolyQ proteins attached to vesicles or motor proteins are "sticky," binding to various proteins in the axon to create protein clumps that physically block the axon. If the blockage is severe, this leads to degeneration of the axon. **B**. PolyQ proteins bind to motor proteins, sequestering them away from microtubules and thereby impairing cargo movement along the axon. **C**. Certain wild-type polyQ proteins, including huntingtin, may normally participate in axonal transport. When polyQ is expanded, the protein can no longer function properly, and axonal transport of specific cargo or interacting proteins is impaired.

and distal processes. Expanded polyQ proteins can perturb this efficient, orderly process. Disruption of axonal transport causes neuronal degeneration in various mouse models and in human disease (Gauthier et al., 2004). Figure 17.2 illustrates three ways that polyQ disease proteins could promote axonal transport deficits. PolyQ proteins might:

1. Self-assemble or assemble with other axonal proteins, resulting in physical blockage of narrow axons.
2. Directly bind motor proteins and effectively titrate them out of the soluble protein pool, leaving the axon functionally depleted of motor proteins.
3. For some polyQ proteins that might normally participate in axonal transport, polyQ expansion could directly interfere with this normal axonal function (Gunawardena & Goldstein, 2005).

Findings exist to support all three hypotheses in various models. It is unclear, however, which of these best explains polyQ-mediated deficits in axonal transport, and whether these deficits are broadly relevant to the pathogenesis of polyQ diseases.

Supporting the first hypothesis, mutant AR forms neuritic inclusions in motor neuron cell lines, leading to abnormal accumulations of mitochondria and the motor protein kinesin in axonal swellings (Piccioni et al., 2002). Similarly, in *Drosophila* an expanded huntingtin fragment forms neuritic inclusions that sequester other polyQ proteins and cause abnormal accumulations of the synaptic protein synaptotagmin and axonal widening near the inclusions (Lee et al., 2004). Furthermore, HD brains develop neuropil inclusions (Jackson et al., 1995) and SCA6 brains have axonal accumulations that contain neurofilaments and other proteins (Gunawardena & Goldstein, 2005) that may block axonal transport of other cargoes.

Supporting the second hypothesis, aggregated huntingtin can titrate motor proteins away from their normal cellular localization (Gunawardena & Goldstein, 2005). In addition, several expanded polyQ proteins cause decreased movement of vesicles in isolated axoplasm preparations, suggesting a deficit in fast axonal transport independent of physical blockage of a narrow axonal lumen (Szebenyi et al., 2003).

Evidence supporting the third hypothesis comes from studies of BDNF trafficking in neurons. Wild-type huntingtin increases the efficiency of BDNF vesicular transport along microtubules, whereas mutant huntingtin reduces BDNF transport efficiency in neurons; this effect is huntingtin-specific, as neither ataxin-3 nor ataxin-7 has similar effects (Gauthier et al., 2004). These data also indirectly support the second hypothesis, as expanded huntingtin was found to bind certain motor proteins more efficiently than wild-type huntingtin, preventing those motor proteins from binding to microtubules. Interestingly, an N-terminal fragment of huntingtin generated in HD patient brains cannot stimulate BDNF transport, suggesting that cleavage of expanded huntingtin is another route by which the expansion may cause axonal transport problems (Gauthier et al., 2004).

In summary, the question of whether axonal transport defects play a more general role in polyQ disease remains unanswered. However, current data indicate that axonal transport may be perturbed in various polyQ diseases by several mechanisms that are not mutually exclusive (see Figure 17.2).

VIII. Neuronal Dysfunctions versus Neuronal Cell Death

The striking degree of brain atrophy in polyQ diseases indicates that neuronal cell death is a major feature of the pathogenic process. Consistent with this, many cell-based studies have confirmed that expression of expanded polyQ disease proteins leads to caspase activation and apoptosis (Chou et al., 2006; Gafni et al., 2004; Gervais et al., 2002; Kouroku et al., 2002; Sanchez et al., 1999; Wellington & Hayden, 2000). It is unlikely, however, that neuronal cell death is exclusively responsible for symptoms early in the course of disease. Increasing evidence suggests that neuronal cell death is preceded by a long period of neuronal dysfunction.

As with much of our knowledge about polyQ diseases, considerable data on this issue have been collected from HD patients and animal models. Rather early in disease, neuronal function at the synaptic level appears to be perturbed in HD patients. Decreased dopamine D1 and D2 receptor mRNA levels have been reported in human HD brains and similar observations have been made in animal models (Leegwater-Kim & Cha, 2004).

Evidence from several HD mouse models suggests that neuronal dysfunction can lead to symptoms of HD in the absence of neuronal cell death. There are indications of transcriptional dysregulation, activation of cellular stress responses, reduction of mitochondrial function, reduction in levels of synaptic proteins, and alterations of synaptic proteins that affect endocytosis and exocytosis in neurons in these mice (Bogdanov et al., 2001; Desplats et al., 2006; Luthi-Carter et al., 2000; Mangiarini et al., 1996; Smith et al., 2005; Tabrizi et al., 2000). Neuronal circuitry is compromised in HD mice (Li et al., 2003; Smith et al., 2005), as is neuronal activity (Rebec et al., 2006), again supporting the theory that major pathophysiological changes occur well before cell death. At least in HD, neuronal dysfunction appears to be intimately linked to pathology. Neuronal cell death—whether apoptotic or nonapoptotic—occurs when neurons can no longer tolerate irreversible and widespread dysfunction of multiple cellular pathways.

Neuronal dysfunction likely underlies early features of disease in other polyQ disorders as well. In SCA1, for example, there is evidence that neuronal transmission may be affected; there is downregulation of mRNAs enriched in

Purkinje cell dendritic arbors, including genes involved with glutamate signaling and calcium release from the smooth endoplasmic reticulum (Serra et al., 2004). Expression of some of these genes also is decreased before symptom onset in a mouse model of SCA1 (Lin et al., 2000).

Studies from SBMA models also indicate that neuronal dysfunction can precede neuronal degeneration. An SBMA mouse model with manifest symptoms and nuclear inclusions does not show signs of neuronal loss, suggesting that symptoms displayed by the mice reflect motor neuron dysfunction rather than cell death (Adachi et al., 2001). Finally, in SCA6 alterations in calcium channel function may compromise calcium influx into neurons, resulting in dysfunction of Purkinje cells followed by cell death (Matsuyama et al., 1999; Toru et al., 2000).

IX. Cell Autonomous versus Non-cell Autonomous Effects

An unsettled question in polyglutamine disorders is whether neurotoxicity occurs in a cell autonomous manner. All but one of the nine polyQ disease proteins are expressed in the cytoplasm or the cell nucleus, the exception being the P/Q-type calcium channel in SCA6, an integral membrane protein. This fact leads one to assume that polyQ toxicity is primarily cell autonomous—that is, a neuron gets sick and dies due to the toxic polyQ protein expressed within it, independent of the health of surrounding neurons. This assumption, however, may not be correct.

A recent report provides clear evidence of mutant huntingtin causing progressive motor deficits and cortical neuropathology in a noncell autonomous manner (Gu et al., 2005). Gu and colleagues generated mouse models of HD expressing mutant huntingtin in specific brain cell populations. When mutant huntingtin was expressed only in cortical pyramidal neurons and glia, neuropathology was very mild. However, when mutant huntingtin was expressed in all neurons of the cortex and striatum, including cortical interneurons, neuropathology and reactive gliosis were observed (Gu et al., 2005). Whereas aggregation proved to be cell autonomous as one would expect, the development of disease required interneuronal effects. Reduced inhibitory input to pyramidal neurons from interneurons may explain this observed noncell autonomous toxicity.

Whatever the mechanism proves to be, noncell autonomous toxicity in polyQ disorders likely reflects perturbations in normal neuronal connectivity. In light of this, Huntingtin's putative roles in transcriptional regulation, synaptic activity, and vesicular transport in neurons seem particularly relevant. Studies focusing on the effects of mutant huntingtin on transcription indicate that levels of brain derived neurotrophic factor (BDNF) are decreased in HD models of disease (Luthi-Carter et al., 2002a, 2002b; Zuccato et al., 2001). BDNF is required for the survival of striatal medium spiny neurons, which preferentially die in HD. HD mouse models show signs of global intracellular transport alterations, and dysregulated transport of BDNF vesicles; in contrast, mutant ataxin-3 and ataxin-7 do not affect BDNF transport (Gauthier et al., 2004; Gunawardena et al., 2003; Lee et al., 2004). BDNF supplementation ameliorates pathology in HD models (Bemelmans et al., 1999; Zala et al., 2005), supporting the view that huntingtin regulates BDNF transport and/or release. Huntingtin also is enriched at synaptic terminals where it may function in synaptic membrane regulation, and there is evidence that synaptic transmission is functionally impaired in HD models.

In a very different way, mutant ataxin-1 may indirectly modulate synaptic plasticity (Zoghbi & Orr, 2000). The function of ataxin-1 remains unknown though it is thought to participate in transcriptional regulation of specific genes (Lin et al., 2000; Okazawa et al., 2002; Serra et al., 2004; Waragai et al., 1999). Genes that are highly expressed in Purkinje cell dendritic arbors appear to be downregulated in SCA1; therefore, abnormal synaptic communication between cells could contribute to SCA1 pathogenesis.

Compelling findings consistent with noncell autonomous actions of polyQ proteins also have been observed in a mouse model of SCA7. In this model, expanded ataxin-7 is expressed in many brain regions but not in Purkinje cells, which are known to be a major target in SCA7. Despite this lack of ataxin-7 expression in Purkinje cells of SCA7 mice, Purkinje cells still degenerate in these mice over a time course that closely mirrors the onset of the behavioral phenotype (Garden et al., 2002). Thus, in SCA7, the effects of expanded ataxin-7 are not limited to cells in which it is expressed.

In some polyQ disorders, therefore, neurons that are not directly compromised by expanded polyQ proteins can become involved due to improper communication from, or interaction with, surrounding cells. These surrounding cells may be other neurons in a network or neighboring glial support cells.

X. Conclusion

This chapter has addressed several possible pathogenic mechanisms for polyQ diseases, as illustrated in Figure 17.1. Evidence discussed here strongly suggests that polyQ-related pathogenesis involves a "toxic gain-of-function" acquired by the disease protein. But since the disorders are distinct from each other, specific features of each disease must be partly determined by the protein context in which the polyQ expansion occurs.

Several mechanisms have been proposed for polyQ-dependent neurodegeneration. Each mechanism is supported by findings in various model systems. These potential mechanisms are not mutually exclusive nor do they necessarily apply to all polyQ diseases. All of them, however, can plausibly be linked to a fundamental problem of polyQ-induced protein misfolding and to the direct or indirect effects that this has on essential biological pathways in neurons. It remains to be seen whether any one, potential pathogenic mechanism will prove to be more crucial than the others to the development of therapy for these incurable and fatal disorders.

References

Adachi, H., Katsuno, M., Minamiyama, M., Sang, C., Pagoulatos, G., Angelidis, C. et al. (2003). Heat shock protein 70 chaperone overexpression ameliorates phenotypes of the spinal and bulbar muscular atrophy transgenic mouse model by reducing nuclear-localized mutant androgen receptor protein. *J Neurosci* **23**, 2203–2211.

Adachi, H., Kume, A., Li, M., Nakagomi, Y., Niwa, H., Do, J. et al. (2001). Transgenic mice with an expanded CAG repeat controlled by the human AR promoter show polyglutamine nuclear inclusions and neuronal dysfunction without neuronal cell death. *Hum Mol Genet* **10**, 1039–1048.

Antonini, A., Leenders, K. L., Spiegel, R., Meier, D., Vontobel, P., Weigell-Weber, M. et al. (1996). Striatal glucose metabolism and dopamine D2 receptor binding in asymptomatic gene carriers and patients with Huntington's disease. *Brain* **119 (Pt 6)**, 2085–2095.

Arrasate, M., Mitra, S., Schweitzer, E. S., Segal, M. R., and Finkbeiner, S. (2004). Inclusion body formation reduces levels of mutant huntingtin and the risk of neuronal death. *Nature* **431**, 805–810.

Bailey, C. K., Andriola, I. F., Kampinga, H. H., and Merry, D. E. (2002). Molecular chaperones enhance the degradation of expanded polyglutamine repeat androgen receptor in a cellular model of spinal and bulbar muscular atrophy. *Hum Mol Genet* **11**, 515–523.

Banfi, S., Servadio, A., Chung, M. Y., Kwiatkowski, T. J., Jr., McCall, A. E., Duvick, L. A. et al. (1994). Identification and characterization of the gene causing type 1 spinocerebellar ataxia. *Nat Genet* **7**, 513–520.

Beauchemin, A. M., Gottlieb, B., Beitel, L. K., Elhaji, Y. A., Pinsky, L., and Trifiro, M. A. (2001). Cytochrome c oxidase subunit Vb interacts with human androgen receptor: A potential mechanism for neurotoxicity in spinobulbar muscular atrophy. *Brain Res Bull* **56**, 285–297.

Behrens, P. F., Franz, P., Woodman, B., Lindenberg, K. S., and Landwehrmeyer, G. B. (2002). Impaired glutamate transport and glutamate-glutamine cycling: Downstream effects of the Huntington mutation. *Brain* **125**, 1908–1922.

Bemelmans, A. P., Horellou, P., Pradier, L., Brunet, I., Colin, P., and Mallet, J. (1999). Brain-derived neurotrophic factor-mediated protection of striatal neurons in an excitotoxic rat model of Huntington's disease, as demonstrated by adenoviral gene transfer. *Hum Gene Ther* **10**, 2987–2997.

Bence, N. F., Sampat, R. M., and Kopito, R. R. (2001). Impairment of the ubiquitin-proteasome system by protein aggregation. *Science* **292**, 1552–1555.

Bhattacharyya, A. M., Thakur, A. K., and Wetzel, R. (2005). Polyglutamine aggregation nucleation: Thermodynamics of a highly unfavorable protein folding reaction. *Proc Natl Acad Sci U S A* **102**, 15400–15405.

Bogdanov, M. B., Andreassen, O. A., Dedeoglu, A., Ferrante, R. J., and Beal, M. F. (2001). Increased oxidative damage to DNA in a transgenic mouse model of Huntington's disease. *J Neurochem* **79**, 1246–1249.

Bowman, A. B., Yoo, S. Y., Dantuma, N. P., and Zoghbi, H. Y. (2005). Neuronal dysfunction in a polyglutamine disease model occurs in the absence of ubiquitin-proteasome system impairment and inversely correlates with the degree of nuclear inclusion formation. *Hum Mol Genet* **14**, 679–691.

Brouillet, E., Hantraye, P., Ferrante, R. J., Dolan, R., Leroy-Willig, A., Kowall, N. W., and Beal, M. F. (1995). Chronic mitochondrial energy impairment produces selective striatal degeneration and abnormal choreiform movements in primates. *Proc Natl Acad Sci U S A* **92**, 7105–7109.

Brouillet, E., Jacquard, C., Bizat, N., and Blum, D. (2005). 3-Nitropropionic acid: A mitochondrial toxin to uncover physiopathological mechanisms underlying striatal degeneration in Huntington's disease. *J Neurochem* **95**, 1521–1540.

Browne, S. E., Bowling, A. C., MacGarvey, U., Baik, M. J., Berger, S. C., Muqit, M. M. et al. (1997). Oxidative damage and metabolic dysfunction in Huntington's disease: Selective vulnerability of the basal ganglia. *Ann Neurol* **41**, 646–653.

Bucciantini, M., Giannoni, E., Chiti, F., Baroni, F., Formigli, L., Zurdo, J. et al. (2002). Inherent toxicity of aggregates implies a common mechanism for protein misfolding diseases. *Nature* **416**, 507–511.

Burke, J. R., Wingfield, M. S., Lewis, K. E., Roses, A. D., Lee, J. E., Hulette, C. et al. (1994). The Haw River syndrome: Dentatorubropalli doluysian atrophy (DRPLA) in an African-American family. *Nat Genet* **7**, 521–524.

Burnett, B., Li, F., and Pittman, R. N. (2003). The polyglutamine neurodegenerative protein ataxin-3 binds polyubiquitylated proteins and has ubiquitin protease activity. *Hum Mol Genet* **12**, 3195–3205.

Burnett, B. G. and Pittman, R. N. (2005). The polyglutamine neurodegenerative protein ataxin 3 regulates aggresome formation. *Proc Natl Acad Sci U S A* **102**, 4330–4335.

Butterworth, J., Yates, C. M., and Reynolds, G. P. (1985). Distribution of phosphate-activated glutaminase, succinic dehydrogenase, pyruvate dehydrogenase and gamma-glutamyl transpeptidase in post-mortem brain from Huntington's disease and agonal cases. *J Neurol Sci* **67**, 161–171.

Chai, Y., Koppenhafer, S. L., Bonini, N. M., and Paulson, H. L. (1999a). Analysis of the role of heat shock protein (Hsp) molecular chaperones in polyglutamine disease. *J Neurosci* **19**, 10338–10347.

Chai, Y., Koppenhafer, S. L., Shoesmith, S. J., Perez, M. K., and Paulson, H. L. (1999b). Evidence for proteasome involvement in polyglutamine disease: Localization to nuclear inclusions in SCA3/MJD and suppression of polyglutamine aggregation in vitro. *Hum Mol Genet* **8**, 673–682.

Chan, H. Y., Warrick, J. M., Andriola, I., Merry, D., and Bonini, N. M. (2002). Genetic modulation of polyglutamine toxicity by protein conjugation pathways in Drosophila. *Hum Mol Genet* **11**, 2895–2904.

Chen, H. K., Fernandez-Funez, P., Acevedo, S. F., Lam, Y. C., Kaytor, M. D., Fernandez, M. H. et al. (2003). Interaction of Akt-phosphorylated ataxin-1 with 14-3-3 mediates neurodegeneration in spinocerebellar ataxia type 1. *Cell* **113**, 457–468.

Chen, S., Ferrone, F. A., and Wetzel, R. (2002). Huntington's disease age-of-onset linked to polyglutamine aggregation nucleation. *Proc Natl Acad Sci U S A* **99**, 11884–11889.

Chernoff, Y. O., Lindquist, S. L., Ono, B., Inge-Vechtomov, S. G., and Liebman, S. W. (1995). Role of the chaperone protein Hsp104 in propagation of the yeast prion-like factor [psi+]. *Science* **268**, 880–884.

Chevalier-Larsen, E. S., O'Brien, C. J., Wang, H., Jenkins, S. C., Holder, L., Lieberman, A. P., and Merry, D. E. (2004). Castration restores function and neurofilament alterations of aged symptomatic males in a transgenic mouse model of spinal and bulbar muscular atrophy. *J Neurosci* **24**, 4778–4786.

Choo, Y. S., Johnson, G. V., MacDonald, M., Detloff, P. J., and Lesort, M. (2004). Mutant huntingtin directly increases susceptibility of mitochondria to the calcium-induced permeability transition and cytochrome c release. *Hum Mol Genet* **13**, 1407–1420.

Chou, A. H., Yeh, T. H., Kuo, Y. L., Kao, Y. C., Jou, M. J., Hsu, C. Y. et al. (2006). Polyglutamine-expanded ataxin-3 activates mitochondrial apoptotic pathway by upregulating Bax and downregulating Bcl-xL. *Neurobiol Dis* **21**, 333–345.

Cooper, J. M. and Schapira, A. H. (1997). Mitochondrial dysfunction in neurodegeneration. *J Bioenerg Biomembr* **29**, 175–183.

Cowan, K. J., Diamond, M. I., and Welch, W. J. (2003). Polyglutamine protein aggregation and toxicity are linked to the cellular stress response. *Hum Mol Genet* **12**, 1377–1391.

Cummings, C. J., Mancini, M. A., Antalffy, B., DeFranco, D. B., Orr, H. T., and Zoghbi, H. Y. (1998). Chaperone suppression of aggregation and altered subcellular proteasome localization imply protein misfolding in SCA1. *Nat Genet* **19**, 148–154.

Cummings, C. J., Reinstein, E., Sun, Y., Antalffy, B., Jiang, Y., Ciechanover, A. et al. (1999). Mutation of the E6-AP ubiquitin ligase reduces nuclear inclusion frequency while accelerating polyglutamine-induced pathology in SCA1 mice. *Neuron* **24**, 879–892.

Cummings, C. J., Sun, Y., Opal, P., Antalffy, B., Mestril, R., Orr, H. T., Dillmann, W. H., and Zoghbi, H. Y. (2001). Over-expression of inducible HSP70 chaperone suppresses neuropathology and improves motor function in SCA1 mice. *Hum Mol Genet* **10**, 1511–1518.

David, G., Abbas, N., Stevanin, G., Durr, A., Yvert, G., Cancel, G. et al. (1997). Cloning of the SCA7 gene reveals a highly unstable CAG repeat expansion. *Nat Genet* **17**, 65–70.

Desaiah, D., Vig, P. J., Subramony, S. H., and Currier, R. D. (1991). Inositol 1,4,5-trisphosphate receptors and protein kinase C in olivopontocerebellar atrophy. *Brain Res* **552**, 36–40.

Desplats, P. A., Kass, K. E., Gilmartin, T., Stanwood, G. D., Woodward, E. L., Head, S. R. et al. (2006). Selective deficits in the expression of striatal-enriched mRNAs in Huntington's disease. *J Neurochem* **96**, 743–757.

Diaz-Hernandez, M., Valera, A. G., Moran, M. A., Gomez-Ramos, P., Alvarez-Castelao, B., Castano, J. G. et al. (2006). Inhibition of 26S proteasome activity by huntingtin filaments but not inclusion bodies isolated from mouse and human brain. *J Neurochem*.

DiFiglia, M. (1990). Excitotoxic injury of the neostriatum: A model for Huntington's disease. *Trends Neurosci* **13**, 286–289.

Emamian, E. S., Kaytor, M. D., Duvick, L. A., Zu, T., Tousey, S. K., Zoghbi, H. Y. et al. (2003). Serine 776 of ataxin-1 is critical for polyglutamine-induced disease in SCA1 transgenic mice. *Neuron* **38**, 375–387.

Everett, C. M. and Wood, N. W. (2004). Trinucleotide repeats and neurodegenerative disease. *Brain* **127**, 2385–2405.

Fu, L., Gao, Y. S., and Sztul, E. (2005). Transcriptional repression and cell death induced by nuclear aggregates of non-polyglutamine protein. *Neurobiol Dis* **20**, 656–665.

Gafni, J., Hermel, E., Young, J. E., Wellington, C. L., Hayden, M. R., and Ellerby, L. M. (2004). Inhibition of calpain cleavage of huntingtin reduces toxicity: Accumulation of calpain/caspase fragments in the nucleus. *J Biol Chem* **279**, 20211–20220.

Garden, G. A., Libby, R. T., Fu, Y. H., Kinoshita, Y., Huang, J., Possin, D. E. et al. (2002). Polyglutamine-expanded ataxin-7 promotes non-cell-autonomous purkinje cell degeneration and displays proteolytic cleavage in ataxic transgenic mice. *J Neurosci* **22**, 4897–4905.

Gatchel, J. R. and Zoghbi, H. Y. (2005). Diseases of unstable repeat expansion: Mechanisms and common principles. *Nat Rev Genet* **6**, 743–755.

Gauthier, L. R., Charrin, B. C., Borrell-Pages, M., Dompierre, J. P., Rangone, H., Cordelieres, F. P. et al. (2004). Huntingtin controls neurotrophic support and survival of neurons by enhancing BDNF vesicular transport along microtubules. *Cell* **118**, 127–138.

Gerber, H. P., Seipel, K., Georgiev, O., Hofferer, M., Hug, M., Rusconi, S., and Schaffner, W. (1994). Transcriptional activation modulated by homopolymeric glutamine and proline stretches. *Science* **263**, 808–811.

Gervais, F. G., Singaraja, R., Xanthoudakis, S., Gutekunst, C. A., Leavitt, B. R., Metzler, M. et al. (2002). Recruitment and activation of caspase-8 by the Huntingtin-interacting protein Hip-1 and a novel partner Hippi. *Nat Cell Biol* **4**, 95–105.

Gidalevitz, T., Ben-Zvi, A., Ho, K. H., Brignull, H. R., and Morimoto, R. I. (2006). Progressive disruption of cellular protein folding in models of polyglutamine diseases. *Science* **311**, 1471–1474.

Gines, S., Seong, I. S., Fossale, E., Ivanova, E., Trettel, F., Gusella, J. F. et al. (2003). Specific progressive cAMP reduction implicates energy deficit in presymptomatic Huntington's disease knock-in mice. *Hum Mol Genet* **12**, 497–508.

Goldberg, Y. P., Nicholson, D. W., Rasper, D. M., Kalchman, M. A., Koide, H. B., Graham, R. K. et al. (1996). Cleavage of huntingtin by apopain, a proapoptotic cysteine protease, is modulated by the polyglutamine tract. *Nat Genet* **13**, 442–449.

Grafton, S. T., Mazziotta, J. C., Pahl, J. J., St George-Hyslop, P., Haines, J. L., Gusella, J. et al. (1990). A comparison of neurological, metabolic, structural, and genetic evaluations in persons at risk for Huntington's disease. *Ann Neurol* **28**, 614–621.

Graham, R. K., Deng, Y., Slow, E. J., Haigh, B., Bissada, N., Lu, G. et al. (2006). Cleavage at the caspase-6 site is required for neuronal dysfunction and degeneration due to mutant huntingtin. *Cell* **125**, 1179–1191.

Group, T. H. s. D. C. R. (1993). A novel gene containing a trinucleotide repeat that is expanded and unstable on Huntington's disease chromosomes. The Huntington's Disease Collaborative Research Group. *Cell* **72**, 971–983.

Gu, M., Gash, M. T., Mann, V. M., Javoy-Agid, F., Cooper, J. M., and Schapira, A. H. (1996). Mitochondrial defect in Huntington's disease caudate nucleus. *Ann Neurol* **39**, 385–389.

Gu, X., Li, C., Wei, W., Lo, V., Gong, S., Li, S. H. et al. (2005). Pathological cell-cell interactions elicited by a neuropathogenic form of mutant Huntingtin contribute to cortical pathogenesis in HD mice. *Neuron* **46**, 433–444.

Gunawardena, S. and Goldstein, L. S. (2005). Polyglutamine diseases and transport problems: Deadly traffic jams on neuronal highways. *Arch Neurol* **62**, 46–51.

Gunawardena, S., Her, L. S., Brusch, R. G., Laymon, R. A., Niesman, I. R., Gordesky-Gold, B. et al. (2003). Disruption of axonal transport by loss of huntingtin or expression of pathogenic polyQ proteins in Drosophila. *Neuron* **40**, 25–40.

Haacke, A., Broadley, S. A., Boteva, R., Tzvetkov, N., Hartl, F. U., and Breuer, P. (2006). Proteolytic cleavage of polyglutamine-expanded ataxin-3 is critical for aggregation and sequestration of non-expanded ataxin-3. *Hum Mol Genet* **15**, 555–568.

Hansson, O., Nylandsted, J., Castilho, R. F., Leist, M., Jaattela, M., and Brundin, P. (2003). Overexpression of heat shock protein 70 in R6/2 Huntington's disease mice has only modest effects on disease progression. *Brain Res* **970**, 47–57.

Hay, D. G., Sathasivam, K., Tobaben, S., Stahl, B., Marber, M., Mestril, R. et al. (2004). Progressive decrease in chaperone protein levels in a mouse model of Huntington's disease and induction of stress proteins as a therapeutic approach. *Hum Mol Genet* **13**, 1389–1405.

Hayden, M. R., Martin, W. R., Stoessl, A. J., Clark, C., Hollenberg, S., Adam, M. J. et al. (1986). Positron emission tomography in the early diagnosis of Huntington's disease. *Neurology* **36**, 888–894.

Helmlinger, D., Hardy, S., Abou-Sleymane, G., Eberlin, A., Bowman, A. B., Gansmuller, A. et al. (2006). Glutamine-expanded ataxin-7 alters TFTC/STAGA recruitment and chromatin structure leading to photoreceptor dysfunction. *PLoS Biol* **4**, e67.

Helmlinger, D., Hardy, S., Sasorith, S., Klein, F., Robert, F., Weber, C. et al. (2004). Ataxin-7 is a subunit of GCN5 histone acetyltransferase-containing complexes. *Hum Mol Genet* **13**, 1257–1265.

Henshaw, R., Jenkins, B. G., Schulz, J. B., Ferrante, R. J., Kowall, N. W., Rosen, B. R., and Beal, M. F. (1994). Malonate produces striatal lesions by indirect NMDA receptor activation. *Brain Res* **647**, 161–166.

Hoang, T. Q., Bluml, S., Dubowitz, D. J., Moats, R., Kopyov, O., Jacques, D., and Ross, B. D. (1998). Quantitative proton-decoupled 31P MRS and 1H MRS in the evaluation of Huntington's and Parkinson's diseases. *Neurology* **50**, 1033–1040.

Hodges, A., Strand, A. D., Aragaki, A. K., Kuhn, A., Sengstag, T., Hughes, G. et al. (2006). Regional and cellular gene expression changes in human Huntington's disease brain. *Hum Mol Genet* **15**, 965–977.

Hsiao, P. W., Lin, D. L., Nakao, R., and Chang, C. (1999). The linkage of Kennedy's neuron disease to ARA24, the first identified androgen receptor polyglutamine region-associated coactivator. *J Biol Chem* **274**, 20229–20234.

Hsu, A. L., Murphy, C. T., and Kenyon, C. (2003). Regulation of aging and age-related disease by DAF-16 and heat-shock factor. *Science* **300**, 1142–1145.

Huang, C. C., Faber, P. W., Persichetti, F., Mittal, V., Vonsattel, J. P., MacDonald, M. E., and Gusella, J. F. (1998). Amyloid formation by mutant huntingtin: Threshold, progressivity and recruitment of normal polyglutamine proteins. *Somat Cell Mol Genet* **24**, 217–233.

Humbert, S., Bryson, E. A., Cordelieres, F. P., Connors, N. C., Datta, S. R., Finkbeiner, S. et al. (2002). The IGF-1/Akt pathway is neuroprotective in Huntington's disease and involves Huntingtin phosphorylation by Akt. *Dev Cell* **2**, 831–837.

Ignatova, Z. and Gierasch, L. M. (2006). Extended polyglutamine tracts cause aggregation and structural perturbation of an adjacent beta barrel protein. *J Biol Chem* **281**, 12959–12967.

Ishikawa, K., Watanabe, M., Yoshizawa, K., Fujita, T., Iwamoto, H., Yoshizawa, T. et al. (1999). Clinical, neuropathological, and molecular study in two families with spinocerebellar ataxia type 6 (SCA6). *J Neurol Neurosurg Psychiatry* **67**, 86–89.

Jackson, M., Gentleman, S., Lennox, G., Ward, L., Gray, T., Randall, K. et al. (1995). The cortical neuritic pathology of Huntington's disease. *Neuropathol Appl Neurobiol* **21**, 18–26.

Jana, N. R., Tanaka, M., Wang, G., and Nukina, N. (2000). Polyglutamine length-dependent interaction of Hsp40 and Hsp70 family chaperones with truncated N-terminal huntingtin: Their role in suppression of aggregation and cellular toxicity. *Hum Mol Genet* **9**, 2009–2018.

Jenkins, B. G., Koroshetz, W. J., Beal, M. F., and Rosen, B. R. (1993). Evidence for impairment of energy metabolism in vivo in Huntington's disease using localized 1H NMR spectroscopy. *Neurology* **43**, 2689–2695.

Kaemmerer, W. F., Rodrigues, C. M., Steer, C. J., and Low, W. C. (2001). Creatine-supplemented diet extends Purkinje cell survival in spinocerebellar ataxia type 1 transgenic mice but does not prevent the ataxic phenotype. *Neuroscience* **103**, 713–724.

Katsuno, M., Adachi, H., Kume, A., Li, M., Nakagomi, Y., Niwa, H. et al. (2002). Testosterone reduction prevents phenotypic expression in a transgenic mouse model of spinal and bulbar muscular atrophy. *Neuron* **35**, 843–854.

Katsuno, M., Adachi, H., Waza, M., Banno, H., Suzuki, K., Tanaka, F. et al. (2006). Pathogenesis, animal models and therapeutics in spinal and bulbar muscular atrophy (SBMA). *Exp Neurol.*

Katsuno, M., Sang, C., Adachi, H., Minamiyama, M., Waza, M., Tanaka, F. et al. (2005). Pharmacological induction of heat-shock proteins alleviates polyglutamine-mediated motor neuron disease. *Proc Natl Acad Sci U S A* **102**, 16801–16806.

Kawaguchi, Y., Okamoto, T., Taniwaki, M., Aizawa, M., Inoue, M., Katayama, S. et al. (1994). CAG expansions in a novel gene for Machado-Joseph disease at chromosome 14q32.1. *Nat Genet* **8**, 221–228.

Kazemi-Esfarjani, P., Trifiro, M. A., and Pinsky, L. (1995). Evidence for a repressive function of the long polyglutamine tract in the human androgen receptor: Possible pathogenetic relevance for the (CAG)n-expanded neuronopathies. *Hum Mol Genet* **4**, 523–527.

Khan, L. A., Bauer, P. O., Miyazaki, H., Lindenberg, K. S., Landwehrmeyer, B. G., and Nukina, N. (2006). Expanded polyglutamines impair synaptic transmission and ubiquitin-proteasome system in Caenorhabditis elegans. *J Neurochem* **98**, 576–587.

Kim, S. J., Kim, T. S., Kim, I. Y., Hong, S., Rhim, H., and Kang, S. (2003). Polyglutamine-expanded ataxin-1 recruits Cu/Zn-superoxide dismutase into the nucleus of HeLa cells. *Biochem Biophys Res Commun* **307**, 660–665.

Kish, S. J., Mastrogiacomo, F., Guttman, M., Furukawa, Y., Taanman, J. W., Dozic, S. et al. (1999). Decreased brain protein levels of cytochrome oxidase subunits in Alzheimer's disease and in hereditary spinocerebellar ataxia disorders: A nonspecific change? *J Neurochem* **72**, 700–707.

Klement, I. A., Skinner, P. J., Kaytor, M. D., Yi, H., Hersch, S. M., Clark, H. B. et al. (1998). Ataxin-1 nuclear localization and aggregation: Role in polyglutamine-induced disease in SCA1 transgenic mice. *Cell* **95**, 41–53.

Kobayashi, Y., Kume, A., Li, M., Doyu, M., Hata, M., Ohtsuka, K., and Sobue, G. (2000). Chaperones Hsp70 and Hsp40 suppress aggregate formation and apoptosis in cultured neuronal cells expressing truncated androgen receptor protein with expanded polyglutamine tract. *J Biol Chem* **275**, 8772–8778.

Kobayashi, Y. and Sobue, G. (2001). Protective effect of chaperones on polyglutamine diseases. *Brain Res Bull* **56**, 165–168.

Koide, R., Kobayashi, S., Shimohata, T., Ikeuchi, T., Maruyama, M., Saito, M. et al. (1999). A neurological disease caused by an expanded CAG trinucleotide repeat in the TATA-binding protein gene: A new polyglutamine disease? *Hum Mol Genet* **8**, 2047–2053.

Koob, M. D., Moseley, M. L., Schut, L. J., Benzow, K. A., Bird, T. D., Day, J. W., and Ranum, L. P. (1999). An untranslated CTG expansion causes a novel form of spinocerebellar ataxia (SCA8). *Nat Genet* **21**, 379–384.

Kordasiewicz, H. B., Thompson, R. M., Clark, H. B., and Gomez, C. M. (2006). Carboxyl termini of P/Q-type Ca2+ channel {alpha}1A subunits translocate to nuclei and promote polyglatumine-mediated toxicity. *Hum Mol Genet.*

Koroshetz, W. J., Jenkins, B. G., Rosen, B. R., and Beal, M. F. (1997). Energy metabolism defects in Huntington's disease and effects of coenzyme Q10. *Ann Neurol* **41**, 160–165.

Kouroku, Y., Fujita, E., Jimbo, A., Kikuchi, T., Yamagata, T., Momoi, M. Y. et al. (2002). Polyglutamine aggregates stimulate ER stress signals and caspase-12 activation. *Hum Mol Genet* **11**, 1505–1515.

Krobitsch, S. and Lindquist, S. (2000). Aggregation of huntingtin in yeast varies with the length of the polygltuamine expansion and the expression of chaperone proteins. *Proc Natl Acad Sci U S A* **97**, 1589–1594.

La Spada, A. R., Fu, Y. H., Sopher, B. L., Libby, R. T., Wang, X., Li, L. Y. et al. (2001). Polyglutamine-expanded ataxin-7 antagonizes CRX function and induces cone-rod dystrophy in a mouse model of SCA7. *Neuron* **31**, 913–927.

La Spada, A. R. and Taylor, J. P. (2003). Polyglutamines placed into context. *Neuron* **38**, 681–684.

La Spada, A. R., Wilson, E. M., Lubahn, D. B., Harding, A. E., and Fischbeck, K. H. (1991). Androgen receptor gene mutations in X-linked spinal and bulbar muscular atrophy. *Nature* **352**, 77–79.

Landles, C. and Bates, G. P. (2004). Huntingtin and the molecular pathogenesis of Huntington's disease. Fourth in molecular medicine review series. *EMBO Rep* **5**, 958–963.

Lee, W. C., Yoshihara, M., and Littleton, J. T. (2004). Cytoplasmic aggregates trap polyglutamine-containing proteins and block axonal transport in a Drosophila model of Huntington's disease. *Proc Natl Acad Sci U S A* **101**, 3224–3229.

Leegwater-Kim, J. and Cha, J. H. (2004). The paradigm of Huntington's disease: Therapeutic opportunities in neurodegeneration. *NeuroRx* **1**, 128–138.

Levine, M. S., Klapstein, G. J., Koppel, A., Gruen, E., Cepeda, C., Vargas, M. E. et al. (1999). Enhanced sensitivity to N-methyl-D-aspartate receptor activation in transgenic and knock-in mouse models of Huntington's disease. *J Neurosci Res* **58**, 515–532.

Li, H., Li, S. H., Johnston, H., Shelbourne, P. F., and Li, X. J. (2000). Amino-terminal fragments of mutant huntingtin show selective accumulation in striatal neurons and synaptic toxicity. *Nat Genet* **25**, 385–389.

Li, J. Y., Plomann, M., and Brundin, P. (2003). Huntington's disease: A synaptopathy? *Trends Mol Med* **9**, 414–420.

Li, M., Miwa, S., Kobayashi, Y., Merry, D. E., Yamamoto, M., Tanaka, F. et al. (1998). Nuclear inclusions of the androgen receptor protein in spinal and bulbar muscular atrophy. *Ann Neurol* **44**, 249–254.

Lin, H. K., Yeh, S., Kang, H. Y., and Chang, C. (2001). Akt suppresses androgen-induced apoptosis by phosphorylating and inhibiting androgen receptor. *Proc Natl Acad Sci U S A* **98**, 7200–7205.

Lin, X., Antalffy, B., Kang, D., Orr, H. T., and Zoghbi, H. Y. (2000). Polyglutamine expansion down-regulates specific neuronal genes before pathologic changes in SCA1. *Nat Neurosci* **3**, 157–163.

Lodi, R., Schapira, A. H., Manners, D., Styles, P., Wood, N. W., Taylor, D. J., and Warner, T. T. (2000). Abnormal in vivo skeletal muscle energy metabolism in Huntington's disease and dentatorubropallidoluysian atrophy. *Ann Neurol* **48**, 72–76.

Luthi-Carter, R., Apostol, B. L., Dunah, A. W., DeJohn, M. M., Farrell, L. A., Bates, G. P. et al. (2003). Complex alteration of NMDA receptors in transgenic Huntington's disease mouse brain: Analysis of mRNA and protein expression, plasma membrane association, interacting proteins, and phosphorylation. *Neurobiol Dis* **14**, 624–636.

Luthi-Carter, R., Hanson, S. A., Strand, A. D., Bergstrom, D. A., Chun, W., Peters, N. L. et al. (2002a). Dysregulation of gene expression in the R6/2 model of polyglutamine disease: Parallel changes in muscle and brain. *Hum Mol Genet* **11**, 1911–1926.

Luthi-Carter, R., Strand, A., Peters, N. L., Solano, S. M., Hollingsworth, Z. R., Menon, A. S. et al. (2000). Decreased expression of striatal signaling genes in a mouse model of Huntington's disease. *Hum Mol Genet* **9**, 1259–1271.

Luthi-Carter, R., Strand, A. D., Hanson, S. A., Kooperberg, C., Schilling, G., La Spada, A. R. et al. (2002b). Polyglutamine and transcription: Gene expression changes shared by DRPLA and Huntington's disease mouse models reveal context-independent effects. *Hum Mol Genet* **11**, 1927–1937.

Macario, A. J. (1995). Heat-shock proteins and molecular chaperones: Implications for pathogenesis, diagnostics, and therapeutics. *Int J Clin Lab Res* **25**, 59–70.

Mangiarini, L., Sathasivam, K., Seller, M., Cozens, B., Harper, A., Hetherington, C. et al. (1996). Exon 1 of the HD gene with an expanded CAG repeat is sufficient to cause a progressive neurological phenotype in transgenic mice. *Cell* **87**, 493–506.

Matilla, A., Roberson, E. D., Banfi, S., Morales, J., Armstrong, D. L., Burright, E. N. et al. (1998). Mice lacking ataxin-1 display learning deficits and decreased hippocampal paired-pulse facilitation. *J Neurosci* **18**, 5508–5516.

Matsuishi, T., Sakai, T., Naito, E., Nagamitsu, S., Kuroda, Y., Iwashita, H., and Kato, H. (1996). Elevated cerebrospinal fluid lactate/pyruvate ratio in Machado-Joseph disease. *Acta Neurol Scand* **93**, 72–75.

Matsuyama, Z., Wakamori, M., Mori, Y., Kawakami, H., Nakamura, S., and Imoto, K. (1999). Direct alteration of the P/Q-type Ca2+ channel property by polyglutamine expansion in spinocerebellar ataxia 6. *J Neurosci* **19**, RC14.

Mazziotta, J. C., Phelps, M. E., Pahl, J. J., Huang, S. C., Baxter, L. R., Riege, W. H. et al. (1987). Reduced cerebral glucose metabolism in asymptomatic subjects at risk for Huntington's disease. *N Engl J Med* **316**, 357–362.

McCampbell, A., Taye, A. A., Whitty, L., Penney, E., Steffan, J. S., and Fischbeck, K. H. (2001). Histone deacetylase inhibitors reduce polyglutamine toxicity. *Proc Natl Acad Sci U S A* **98**, 15179–15184.

McCampbell, A., Taylor, J. P., Taye, A. A., Robitschek, J., Li, M., Walcott, J. et al. (2000). CREB-binding protein sequestration by expanded polyglutamine. *Hum Mol Genet* **9**, 2197–2202.

Meriin, A. B., Zhang, X., He, X., Newnam, G. P., Chernoff, Y. O., and Sherman, M. Y. (2002). Huntington toxicity in yeast model depends on polyglutamine aggregation mediated by a prion-like protein Rnq1. *J Cell Biol* **157**, 997–1004.

Merry, D. E., Kobayashi, Y., Bailey, C. K., Taye, A. A., and Fischbeck, K. H. (1998). Cleavage, aggregation and toxicity of the expanded androgen receptor in spinal and bulbar muscular atrophy. *Hum Mol Genet* **7**, 693–701.

Mhatre, A. N., Trifiro, M. A., Kaufman, M., Kazemi-Esfarjani, P., Figlewicz, D., Rouleau, G., and Pinsky, L. (1993). Reduced transcriptional regulatory competence of the androgen receptor in X-linked spinal and bulbar muscular atrophy. *Nat Genet* **5**, 184–188.

Michalik, A. and Van Broeckhoven, C. (2003). Pathogenesis of polyglutamine disorders: aggregation revisited. *Hum Mol Genet* **12 Spec No. 2**, R173–186.

Mizutani, A., Wang, L., Rajan, H., Vig, P. J., Alaynick, W. A., Thaler, J. P., and Tsai, C. C. (2005). Boat, an AXH domain protein, suppresses the cytotoxicity of mutant ataxin-1. *Embo J* **24**, 3339–3351.

Moseley, M. L., Zu, T., Ikeda, Y., Gao, W., Mosemiller, A. K., Daughters, R. S. et al. (2006). Bidirectional expression of CUG and CAG expansion transcripts and intranuclear polyglutamine inclusions in spinocerebellar ataxia type 8. *Nat Genet* **38**, 758–769.

Muchowski, P. J., Schaffar, G., Sittler, A., Wanker, E. E., Hayer-Hartl, M. K., and Hartl, F. U. (2000). Hsp70 and hsp40 chaperones can inhibit self-assembly of polyglutamine proteins into amyloid-like fibrils. *Proc Natl Acad Sci U S A* **97**, 7841–7846.

Muchowski, P. J. and Wacker, J. L. (2005). Modulation of neurodegeneration by molecular chaperones. *Nat Rev Neurosci* **6**, 11–22.

Nucifora, F. C., Jr., Sasaki, M., Peters, M. F., Huang, H., Cooper, J. K., Yamada, M. et al. (2001). Interference by huntingtin and atrophin-1 with cbp-mediated transcription leading to cellular toxicity. *Science* **291**, 2423–2428.

Okazawa, H., Rich, T., Chang, A., Lin, X., Waragai, M., Kajikawa, M. et al. (2002). Interaction between mutant ataxin-1 and PQBP-1 affects transcription and cell death. *Neuron* **34**, 701–713.

Ona, V. O., Li, M., Vonsattel, J. P., Andrews, L. J., Khan, S. Q., Chung, W. M. et al. (1999). Inhibition of caspase-1 slows disease progression in a mouse model of Huntington's disease. *Nature* **399**, 263–267.

Ophoff, R. A., Terwindt, G. M., Vergouwe, M. N., van Eijk, R., Oefner, P. J., Hoffman, S. M. et al. (1996). Familial hemiplegic migraine and episodic ataxia type-2 are caused by mutations in the Ca2+ channel gene CACNL1A4. *Cell* **87**, 543–552.

Panov, A. V., Gutekunst, C. A., Leavitt, B. R., Hayden, M. R., Burke, J. R., Strittmatter, W. J., and Greenamyre, J. T. (2002). Early mitochondrial calcium defects in Huntington's disease are a direct effect of polyglutamines. *Nat Neurosci* **5**, 731–736.

Perez, M. K., Paulson, H. L., Pendse, S. J., Saionz, S. J., Bonini, N. M., and Pittman, R. N. (1998). Recruitment and the role of nuclear localization in polyglutamine-mediated aggregation. *J Cell Biol* **143**, 1457–1470.

Piccioni, F., Pinton, P., Simeoni, S., Pozzi, P., Fascio, U., Vismara, G. et al. (2002). Androgen receptor with elongated polyglutamine tract forms aggregates that alter axonal trafficking and mitochondrial distribution in motor neuronal processes. *Faseb J* **16**, 1418–1420.

Poirier, M. A., Li, H., Macosko, J., Cai, S., Amzel, M., and Ross, C. A. (2002). Huntingtin spheroids and protofibrils as precursors in polyglutamine fibrilization. *J Biol Chem* **277**, 41032–41037.

Ravikumar, B. and Rubinsztein, D. C. (2004). Can autophagy protect against neurodegeneration caused by aggregate-prone proteins? *Neuroreport* **15**, 2443–2445.

Rebec, G. V., Conroy, S. K., and Barton, S. J. (2006). Hyperactive striatal neurons in symptomatic Huntington R6/2 mice: Variations with behavioral state and repeated ascorbate treatment. *Neuroscience* **137**, 327–336.

Reid, S. J., Rees, M. I., van Roon-Mom, W. M., Jones, A. L., MacDonald, M. E., Sutherland, G. et al. (2003). Molecular investigation of TBP allele length: A SCA17 cellular model and population study. *Neurobiol Dis* **13**, 37–45.

Ross, C. A. and Poirier, M. A. (2004). Protein aggregation and neurodegenerative disease. *Nat Med* **10 Suppl**, S10–17.

Sanchez, I., Mahlke, C., and Yuan, J. (2003). Pivotal role of oligomerization in expanded polyglutamine neurodegenerative disorders. *Nature* **421**, 373–379.

Sanchez, I., Xu, C. J., Juo, P., Kakizaka, A., Blenis, J., and Yuan, J. (1999). Caspase-8 is required for cell death induced by expanded polyglutamine repeats. *Neuron* **22**, 623–633.

Satterfield, T. F. and Pallanck, L. J. (2006). Ataxin-2 and its Drosophila homolog, ATX2, physically assemble with polyribosomes. *Hum Mol Genet* **15**, 2523–2532.

Saudou, F., Finkbeiner, S., Devys, D., and Greenberg, M. E. (1998). Huntingtin acts in the nucleus to induce apoptosis but death does not correlate with the formation of intranuclear inclusions. *Cell* **95**, 55–66.

Schaffar, G., Breuer, P., Boteva, R., Behrends, C., Tzvetkov, N., Strippel, N. et al. (2004). Cellular toxicity of polyglutamine expansion proteins: Mechanism of transcription factor deactivation. *Mol Cell* **15**, 95–105.

Schapira, A. H. (1997). Mitochondrial function in Huntington's disease: Clues for pathogenesis and prospects for treatment. *Ann Neurol* **41**, 141–142.

Schmidt, B. J., Greenberg, C. R., Allingham-Hawkins, D. J., and Spriggs, E. L. (2002). Expression of X-linked bulbospinal muscular atrophy (Kennedy disease) in two homozygous women. *Neurology* **59**, 770–772.

Serra, H. G., Byam, C. E., Lande, J. D., Tousey, S. K., Zoghbi, H. Y., and Orr, H. T. (2004). Gene profiling links SCA1 pathophysiology to glutamate signaling in Purkinje cells of transgenic mice. *Hum Mol Genet* **13**, 2535–2543.

Simeoni, S., Mancini, M. A., Stenoien, D. L., Marcelli, M., Weigel, N. L., Zanisi, M. et al. (2000). Motoneuronal cell death is not correlated with aggregate formation of androgen receptors containing an elongated polyglutamine tract. *Hum Mol Genet* **9**, 133–144.

Simon-Sanchez, J., Hanson, M., Singleton, A., Hernandez, D., McInerney, A., Nussbaum, R. et al. (2005). Analysis of SCA-2 and SCA-3 repeats in Parkinsonism: Evidence of SCA-2 expansion in a family with autosomal dominant Parkinson's disease. *Neurosci Lett* **382**, 191–194.

Slow, E. J., Graham, R. K., Osmand, A. P., Devon, R. S., Lu, G., Deng, Y. et al. (2005). Absence of behavioral abnormalities and neurodegeneration in vivo despite widespread neuronal huntingtin inclusions. *Proc Natl Acad Sci U S A* **102**, 11402–11407.

Smith, R., Brundin, P., and Li, J. Y. (2005). Synaptic dysfunction in Huntington's disease: A new perspective. *Cell Mol Life Sci* **62**, 1901–1912.

Steffan, J. S., Agrawal, N., Pallos, J., Rockabrand, E., Trotman, L. C., Slepko, N., Illes, K. et al. (2004). SUMO modification of Huntingtin and Huntington's disease pathology. *Science* **304**, 100–104.

Steffan, J. S., Kazantsev, A., Spasic-Boskovic, O., Greenwald, M., Zhu, Y. Z., Gohler, H. et al. (2000). The Huntington's disease protein interacts with p53 and CREB-binding protein and represses transcription. *Proc Natl Acad Sci U S A* **97**, 6763–6768.

Stenoien, D. L., Cummings, C. J., Adams, H. P., Mancini, M. G., Patel, K., DeMartino, G. N. et al. (1999). Polyglutamine-expanded androgen receptors form aggregates that sequester heat shock proteins, proteasome components and SRC-1, and are suppressed by the HDJ-2 chaperone. *Hum Mol Genet* **8**, 731–741.

Stevanin, G., Cancel, G., Didierjean, O., Durr, A., Abbas, N., Cassa, E. et al. (1995a). Linkage disequilibrium at the Machado-Joseph disease/spinal cerebellar ataxia 3 locus: Evidence for a common founder effect in French and Portuguese-Brazilian families as well as a second ancestral Portuguese-Azorean mutation. *Am J Hum Genet* **57**, 1247–1250.

Stevanin, G., Cassa, E., Cancel, G., Abbas, N., Durr, A., Jardim, E. et al. (1995b). Characterisation of the unstable expanded CAG repeat in the MJD1 gene in four Brazilian families of Portuguese descent with Machado-Joseph disease. *J Med Genet* **32**, 827–830.

Sugars, K. L., Brown, R., Cook, L. J., Swartz, J., and Rubinsztein, D. C. (2004). Decreased cAMP response element-mediated transcription: An early event in exon 1 and full-length cell models of Huntington's disease that contributes to polyglutamine pathogenesis. *J Biol Chem* **279**, 4988–4999.

Szebenyi, G., Morfini, G. A., Babcock, A., Gould, M., Selkoe, K., Stenoien, D. L. et al. (2003). Neuropathogenic forms of huntingtin and androgen receptor inhibit fast axonal transport. *Neuron* **40**, 41–52.

Tabrizi, S. J., Workman, J., Hart, P. E., Mangiarini, L., Mahal, A., Bates, G. et al. (2000). Mitochondrial dysfunction and free radical damage in the Huntington R6/2 transgenic mouse. *Ann Neurol* **47**, 80–86.

Tang, T. S., Slow, E., Lupu, V., Stavrovskaya, I. G., Sugimori, M., Llinas, R. et al. (2005). Disturbed Ca2+ signaling and apoptosis of medium spiny neurons in Huntington's disease. *Proc Natl Acad Sci U S A* **102**, 2602–2607.

Toru, S., Murakoshi, T., Ishikawa, K., Saegusa, H., Fujigasaki, H., Uchihara, T. et al. (2000). Spinocerebellar ataxia type 6 mutation alters P-type calcium channel function. *J Biol Chem* **275**, 10893–10898.

Tsai, H. F., Tsai, H. J., and Hsieh, M. (2004). Full-length expanded ataxin-3 enhances mitochondrial-mediated cell death and decreases Bcl-2 expression in human neuroblastoma cells. *Biochem Biophys Res Commun* **324**, 1274–1282.

Tsuda, H., Jafar-Nejad, H., Patel, A. J., Sun, Y., Chen, H. K., Rose, M. F. et al. (2005). The AXH domain of Ataxin-1 mediates neurodegeneration through its interaction with Gfi-1/Senseless proteins. *Cell* **122**, 633–644.

Uchihara, T., Fujigasaki, H., Koyano, S., Nakamura, A., Yagishita, S., and Iwabuchi, K. (2001). Non-expanded polyglutamine proteins in intranuclear inclusions of hereditary ataxias—Triple-labeling immunofluorescence study. *Acta Neuropathol (Berl)* **102**, 149–152.

Vacher, C., Garcia-Oroz, L., and Rubinsztein, D. C. (2005). Overexpression of yeast hsp104 reduces polyglutamine aggregation and prolongs survival of a transgenic mouse model of Huntington's disease. *Hum Mol Genet* **14**, 3425–3433.

Vig, P. J., Subramony, S. H., Currier, R. D., and Desaiah, D. (1992). Inositol 1,4,5-trisphosphate metabolism in the cerebella of Lurcher mutant mice and patients with olivopontocerebellar atrophy. *J Neurol Sci* **110**, 139–143.

Vig, P. J., Subramony, S. H., and McDaniel, D. O. (2001). Calcium homeostasis and spinocerebellar ataxia-1 (SCA-1). *Brain Res Bull* **56**, 221–225.

Wang, H. L., Yeh, T. H., Chou, A. H., Kuo, Y. L., Luo, L. J., He, C. Y. et al. (2006). Polyglutamine-expanded ataxin-7 activates mitochondrial apoptotic pathway of cerebellar neurons by upregulating Bax and downregulating Bcl-x(L). *Cell Signal* **18**, 541–552.

Waragai, M., Lammers, C. H., Takeuchi, S., Imafuku, I., Udagawa, Y., Kanazawa, I. et al. (1999). PQBP-1, a novel polyglutamine tract-binding protein, inhibits transcription activation by Brn-2 and affects cell survival. *Hum Mol Genet* **8**, 977–987.

Warrick, J. M., Chan, H. Y., Gray-Board, G. L., Chai, Y., Paulson, H. L., and Bonini, N. M. (1999). Suppression of polyglutamine-mediated neurodegeneration in Drosophila by the molecular chaperone HSP70. *Nat Genet* **23**, 425–428.

Warrick, J. M., Morabito, L. M., Bilen, J., Gordesky-Gold, B., Faust, L. Z., Paulson, H. L., and Bonini, N. M. (2005). Ataxin-3 suppresses polyglutamine neurodegeneration in Drosophila by a ubiquitin-associated mechanism. *Mol Cell* **18**, 37–48.

Warrick, J. M., Paulson, H. L., Gray-Board, G. L., Bui, Q. T., Fischbeck, K. H., Pittman, R. N., and Bonini, N. M. (1998). Expanded polyglutamine protein forms nuclear inclusions and causes neural degeneration in Drosophila. *Cell* **93**, 939–949.

Watase, K., Weeber, E. J., Xu, B., Antalffy, B., Yuva-Paylor, L., Hashimoto, K. et al. (2002). A long CAG repeat in the mouse Sca1 locus replicates SCA1 features and reveals the impact of protein solubility on selective neurodegeneration. *Neuron* **34**, 905–919.

Wellington, C. L. and Hayden, M. R. (2000). Caspases and neurodegeneration: On the cutting edge of new therapeutic approaches. *Clin Genet* **57**, 1–10.

Wellington, C. L., Leavitt, B. R., and Hayden, M. R. (2000). Huntington disease: New insights on the role of huntingtin cleavage. *J Neural Transm Suppl*, 1–17.

Wyttenbach, A., Sauvageot, O., Carmichael, J., Diaz-Latoud, C., Arrigo, A. P., and Rubinsztein, D. C. (2002). Heat shock protein 27 prevents cellular polyglutamine toxicity and suppresses the increase of reactive oxygen species caused by huntingtin. *Hum Mol Genet* **11**, 1137–1151.

Yoo, S. Y., Pennesi, M. E., Weeber, E. J., Xu, B., Atkinson, R., Chen, S. et al. (2003). SCA7 knock-in mice model human SCA7 and reveal gradual accumulation of mutant ataxin-7 in neurons and abnormalities in short-term plasticity. *Neuron* **37**, 383–401.

Zala, D., Benchoua, A., Brouillet, E., Perrin, V., Gaillard, M. C., Zurn, A. D. et al. (2005). Progressive and selective striatal degeneration in primary neuronal cultures using lentiviral vector coding for a mutant huntingtin fragment. *Neurobiol Dis* **20**, 785–798.

Zeron, M. M., Fernandes, H. B., Krebs, C., Shehadeh, J., Wellington, C. L., Leavitt, B. R. et al. (2004). Potentiation of NMDA receptor-mediated excitotoxicity linked with intrinsic apoptotic pathway in YAC transgenic mouse model of Huntington's disease. *Mol Cell Neurosci* **25**, 469–479.

Zeron, M. M., Hansson, O., Chen, N., Wellington, C. L., Leavitt, B. R., Brundin, P. et al. (2002). Increased sensitivity to N-methyl-D-aspartate receptor-mediated excitotoxicity in a mouse model of Huntington's disease. *Neuron* **33**, 849–860.

Zhang, S., Xu, L., Lee, J., and Xu, T. (2002). Drosophila atrophin homolog functions as a transcriptional corepressor in multiple developmental processes. *Cell* **108**, 45–56.

Zhou, H., Li, S. H., and Li, X. J. (2001). Chaperone suppression of cellular toxicity of huntingtin is independent of polyglutamine aggregation. *J Biol Chem* **276**, 48417–48424.

Zhuchenko, O., Bailey, J., Bonnen, P., Ashizawa, T., Stockton, D. W., Amos, C. et al. (1997). Autosomal dominant cerebellar ataxia (SCA6) associated with small polyglutamine expansions in the alpha 1A-voltage-dependent calcium channel. *Nat Genet* **15**, 62–69.

Zoghbi, H. Y. and Orr, H. T. (2000). Glutamine repeats and neurodegeneration. *Annu Rev Neurosci* **23**, 217–247.

Zuccato, C., Ciammola, A., Rigamonti, D., Leavitt, B. R., Goffredo, D., Conti, L. (2001). Loss of huntingtin-mediated BDNF gene transcription in Huntington's disease. *Science* **293**, 493–498.

Zuccato, C., Tartari, M., Crotti, A., Goffredo, D., Valenza, M., Conti, L. et al. (2003). Huntingtin interacts with REST/NRSF to modulate the transcription of NRSE-controlled neuronal genes. *Nat Genet* **35**, 76–83.

18

Friedreich's Ataxia and Related DNA Loss-of-Function Disorders

Massimo Pandolfo and Chantal Depondt

I. Summary

First described in 1863 by Nicholaus Friedreich, Professor of Medicine in Heidelberg, Friedreich ataxia (FA) is an autosomal recessive disease characterized, in its typical form, by onset around puberty, progressive limb and trunk ataxia, pyramidal weakness, loss of deep tendon reflexes, hypertrophic cardiomyopathy, and increased risk of diabetes mellitus. The identification of the FA gene (FRDA), encoding the mitochondrial protein frataxin (FXN), and of its most common mutation, the unstable hyperexpansion of a GAA triplet repeat sequence (TRS) allowed researchers to better define the clinical and pathological spectrum of the disease, to investigate pathogenesis, and eventually to propose novel treatments for this so far incurable disease.

Other recessive, loss-of-function disorders may cause early onset ataxia, which in some cases may clinically resemble FA. Ataxia with oculomotor apraxia (AOA) types 1 and 2, ataxia with vitamin E deficiency (AVED), and the autosomal recessive ataxia of Charlevoix and Saguenay (ARSACS) are the best characterized disorders in this group.

II. Epidemiology

FA is the most common recessive ataxia in Caucasians and in people from the Middle East and the Indian subcontinent, with an estimated birth incidence of up to 1:25,000 and a carrier frequency of up to 1:90 in Western Europe (Romeo et al., 1983). The disease is absent in other populations, suggesting that a single mutation event originated the pathogenic GAA repeat expansion, a hypothesis also supported by haplotype analysis (Labuda et al., 2000). Other recessive ataxias are very rare disorders, except in areas where a founder effect exists. AVED is probably more common around the Mediterranean, and in Tunisia it may represent a significant proportion of cases with an FA-like phenotype. AOA1 and AOA2 first were described in Japan, where AOA1 may be the most common recessive ataxia, as well as in Portugal other cases have been found elsewhere. ARSACS is common in the region of Quebec (after which the disease is named), where carrier frequency is around 1:20, but cases have been identified around the world, including the rest of North America, Europe, North Africa, and Japan.

III. Pathology of Friedreich Ataxia

A. Nervous System

The degeneration of the posterior columns of the spinal cord is the hallmark of the disease (Koeppen, 1998). The longest fibers are most affected. Atrophy also involves Clarke's column and the spinocerebellar tracts. In the brainstem, affected structures include the gracilis and cuneate nuclei; the medial lemnisci; the sensory nuclei and entering roots of the V, IX, and X nerves; the descending trigeminal tracts; the solitary tracts; and the accessory cuneate nuclei, corresponding to Clarke's column in the spinal cord. Overall, Friedreich ataxia severely affects the sensory systems' ability to provide information to the brain and cerebellum about the position and speed of body segments. The motor system is directly affected as well: the long crossed and uncrossed corticospinal motor tracts are atrophic, more so distally, suggesting a "dying back" process (Said et al., 1986). There is a variable loss of pyramidal neurons in the motor cortex, whereas motor neurons in the brainstem and in the ventral horns of the spinal cord are less affected.

In the cerebellum, cortical atrophy occurs only late and is mild, whereas atrophy of the dentate nuclei and of superior cerebellar pedunculi is prominent. Quantitative analysis of synaptic terminals indicates a loss of contacts over Purkinje cell bodies and proximal dendrites (Koeppen, 1998).

The auditory system and the optic nerves and tracts are variably affected. The external pallidus and subthalamic nuclei may show a moderate cell loss. Finally, since many patients with Friedreich ataxia die as a consequence of heart disease, widespread hypoxic changes and focal infarcts are often found in the CNS.

In the peripheral nervous system, the major abnormailities occur in the dorsal root ganglia (DRG), where a loss of large primary sensory neurons is observed, accompanied by proliferation of capsule cells that form clumps called *Residualknötchen of Nageotte* (Hughes et al., 1968). Loss of large myelinated sensory fibers is prominent in peripheral nerves, whereas the fine, unmyelinated fibers are less involved (Pasternac et al., 1980). Interstitial connective tissue is increased.

B. Heart

A hypertrophic cardiomyopathy is typical of FA. Ventricular walls and interventricular septum are thickened. Hypertrophic cardiomyocytes are found early in the disease and are intermingled with normal appearing ones. Then, atrophic, degenerating, even necrotic fibers progressively appear, there is diffuse and focal inflammatory cell infiltration, and connective tissue increases. In the late stage of the disease, with extensive fibrosis, the cardiomyopathy becomes dilatative (Casazza & Morpurgo, 1996). A variable number of cardiomyocytes, from less than 1 percent to more than 10 percent, show intracellular iron deposits (Lamarche et al., 1993). This is a specific finding in Friedreich ataxia and a direct consequence of the basic biochemical defect.

C. Other Organs

More than three-quarters of the patients have kyphoscoliosis, mostly as a double thoracolumbar curve resembling the idiopathic form (Aronsson et al., 1994). *Pes cavus*, *pes equinovarus*, and clawing of the toes are found in about half the cases. Patients with diabetes (about 10%) often show a loss of islet cells in the pancreas, which is not accompanied by the autoimmune inflammatory reaction found in type I diabetes (Schoenle et al., 1989).

IV. Clinical Phenotype of Friedreich Ataxia

A. Neurological Symptoms

The phenotype has been redefined and extended after the discovery in 1996 of the disease gene (Campuzano et al., 1996). The typical age of onset is around puberty, but it may vary substantially even within a sibship. Late-onset cases exist (late-onset FA or LOFA), even in the sixth decade and later, a phenomenon now in part explained by the dynamic nature of the underlying mutation. Gait instability or generalized clumsiness are typical presenting symptoms (Filla et al., 1990; Harding, 1981). Scoliosis may already be present when neurological symptoms appear. In rare cases, hypertrophic cardiomyopathy is diagnosed before the onset of ataxia. Ataxia is of mixed cerebellar and sensory type, affecting the trunk, with swaying, imbalance and falls, and the limbs, with increasing difficulty in activities such as writing, dressing, and handling utensils. Ataxia is progressive and unremitting, though periods of stability are frequent at the beginning of the illness. With progression, gait becomes broad-based, with frequent losses of balance requiring intermittent, then constant support.

Fine motor skills deteriorate, dysmetria and intention tremor become evident. Mild to moderate limb weakness of central origin is a constant feature of advanced FA. Some amyotrophy, particularly in the hands, is also very common. On average 10 to 15 years after onset, patients lose the ability to walk, stand, and sit without support. Evolution is variable, however, with mild cases who are still ambulatory decades after onset and severe cases that are wheelchair-bound in a few years. Dysarthria consisting of slow, jerky speech with sudden utterances is a relatively early feature and progresses until speech becomes almost unintelligible. Dysphagia, particularly for liquids, appears with advancing disease, requiring modified foods and eventually a nasogastric tube or gastrostomy feedings. Cognitive functions are generally well preserved. However, because of substantial physical disability, FA may have a substantial impact on academic, professional, and personal development.

B. Neurological Examination

The neurological examination of a typical case of Friedreich ataxia is very characteristic and reflects the underlying pathology. A sensory neuropathy in dorsal root ganglia, accompanied by loss of sensory fibers in peripheral nerves and degeneration of the posterior columns in the spinal cord, is a hallmark of the disease. The larger neurons carrying proprioceptive information are mostly affected, resulting in loss of position and vibration sense and abolished reflexes. Perception of light touch, pain, and temperature only decreases with advancing disease. A progressive degeneration of pyramidal tracts usually determines extensor plantar responses and progressive muscular weakness. Some patients with more limited sensory neuronopathy have retained reflexes (FA with retained reflexes or FARR). In these cases pyramidal involvement may lead to hyperreflexia and spasticity, to the point of mimicking a spastic paraplegia (Badhwar et al., 2004).

The relative impact of sensory neuropathy and of pyramidal tract degeneration clearly varies from patient to patient, resulting most often in the typical picture of areflexia associated with extensor plantar responses, but sometimes one component prevails. Such partial pictures usually are observed in milder cases of the disease. Fixation instability with square-wave jerks is the typical oculomotor abnormality in FA, whereas gaze-evoked nystagmus is uncommon and ophthalmoparesis does not occur. About 30 percent of the patients develop optic atrophy, with or without visual impairment, and about 20 percent develop sensorineural hearing loss.

C. Heart Disease

Heart disease is most commonly asymptomatic, but it may contribute to disability and cause premature death in a significant minority of patients, particularly those with earlier age of onset. Shortness of breath and palpitations are the usual initial symptoms. The electrocardiogram shows inverted T-waves in essentially all patients, ventricular hypertrophy in most, conduction disturbances in about 10 percent, occasionally supraventricular ectopic beats, and atrial fibrillation. Echocardiography demonstrates concentric hypertrophy of the ventricles or asymmetric septal hypertrophy (Morvan et al., 1992).

D. Diabetes Mellitus and Other Signs and Symptoms

About 10 percent of Friedreich ataxia patients develop diabetes mellitus, and 20 percent have carbohydrate intolerance. The mechanisms are complex, with beta-cell dysfunction (Schoenle et al., 1989) as well as peripheral insulin resistance (Fantus et al., 1993), resembling other mitochondrial disorders.

Kyphoscoliosis may cause pain and cardiorespiratory problems. *Pes cavus* and *pes equinovarus* may further affect ambulation. Autonomic disturbances, most commonly cold and cyanotic legs and feet, become increasingly frequent as the disease advances. Sphincteric problems are rare.

E. Variant Phenotypes

A positive molecular test for Friedreich ataxia is found in up to 10 percent of patients with recessive or sporadic ataxia who do not fulfill the classical clinical diagnostic criteria for the disease. A reanalysis of these cases revealed that no clinical finding or combination of findings characterizes exclusively or is necessarily present in positive cases. Even neurophysiological evidence of axonal sensory neuropathy is very rarely absent in patients with a proven molecular diagnosis. Overall, however, the classic features of Friedreich ataxia, including cardiomyopathy, are highly predictive of a positive test. Absence of cardiomyopathy and moderate to severe cerebellar cortical atrophy, demonstrated by MRI, appear to be the best predictors of a negative test. In the author's opinion, a molecular test for Friedreich ataxia, along with MRI and neurophysiological investigations, is indicated in the initial workup of all cases of sporadic or recessive degenerative ataxia, regardless of whether or not they fulfill the diagnostic criteria.

F. Quantifying Friedreich Ataxia for Clinical Studies

A measure of the severity of neurological dysfunction is necessary to study the natural history of the FA and to perform clinical treatment trials. However, because of the complex clinical presentation of the disease, its variable age of onset and rate of progression, the variety of possibly affected neural systems, the development of reliable rating scales has been difficult. Some items of the general ataxia scale ICARS (Trouillas et al., 1997) have been useful for the long-term follow up of patients. However, ICARS was not tailored on the clinical spectrum of FA and has not been validated in this disease. FARS, a neurologic rating scale specifically developed for FA, was validated in 2005 (Subramony et al., 2005). FARS includes assessments of stance, gait, upper and lower limb coordination, speech, proprioception, and strength. A unctional composite scale that includes several performance measures such as the nine-hole peg test, a timed 25-foot walk, and low-contrast letter acuity has also been proposed (Lynch et al., 2005).

G. Prognosis

Friedreich ataxia patients become wheelchair-bound on average 15 years after onset, but variability is very large. Early onset and left ventricular hypertrophy predict a faster rate of progression. The burden of neurological impairment, cardiomyopathy, occasionally diabetes, shortens life expectancy. Older studies found that most patients died in their thirties, but survival may be significantly prolonged with treatment of cardiac symptoms, particularly arrhythmias, by antidiabetic treatment, and by preventing and controlling complications resulting from prolonged neurological disability. Carefully assisted patients may live several more decades.

V. Ancillary Tests

A. Neuroimaging

Structural neuroimaging, by computed tomography (CT) or magnetic resonance imaging (MRI), typically shows an atrophic cervical spinal cord; cerebellar atrophy is mild and commonly seen only in advanced cases (Wullner et al., 1993). Newer MRI approaches, including tractography, may have the potential to reveal degeneration of specific fiber systems, but they have not yet been systematically evaluated in this disease. Preliminary studies with magnetic resonance spectroscopy (MRS) are compatible with mitochondrial dysfunction, as were older metabolic imaging studies with fluorodeoxyglucose (FDG) positron emission tomography (PET) that showed a diffusely increased FDG uptake in the brain in early FA, suggesting inadequate utilization of energy sources (Gilman et al., 1990).

B. Clinical Neurophysiology

Neurophysiological studies of the peripheral nervous system (PNS) reveal a severe reduction or complete loss of sensory nerve action potentials (SNAPs), with motor and sensory (when measurable) nerve conduction velocities (NCVs) within or just below the normal range (Peyronnard et al., 1976). These neurophysiological abnormalities do not appear to progress with time (Santoro et al., 1999). In the central nervous system (CNS), degeneration of both peripheral and central sensory fibers also results in dispersion and delay of somatosensory evoked potentials (SEPs) and brainstem auditory evoked potentials (BAEPs) (Vanasse et al., 1988). Visual evoked potentials (VEPs) commonly are reduced in amplitude, but not delayed. Central motor conduction velocity, reflecting pyramidal tract function and determined by cortical magnetic stimulation, is slower than normal and, contrary to the sensory involvement, this slowing becomes more severe with increasing disease duration (Santoro et al., 2000).

C. Biochemical Investigations

A number of biochemical tests usually are performed to exclude the diagnoses of abeta- or hypobetalipoproteinemia, isolated vitamin E deficiency, mitochondrial disorders, adrenoleukodystrophy or adrenomyeloneuropathy, Refsum disease, and a number of lysosomal storage diseases, all of which can be differential diagnoses, particularly at the early stage of the disease. These tests include lipoproteins, vitamin E, lactate, pyruvate, urinary organic acids, serum very long chain fatty acids (VLCFA), serum phytanic acid, and leukocyte and/or fibroblast lysosomal enzymes, which are all normal in Friedreich ataxia.

Attention currently is focused on iron metabolism and oxidative stress. Blood iron, iron binding capacity, and ferritin are normal, but there is an increase in circulating transferrin receptors. In addition, several markers of oxidative stress are increased, including urinary 8-hydroxy-2′-deoxyguanosine (8OH2′dG), a marker of oxidative DNA damage (Schulz et al., 2000), and plasma malondialdheyde, a marker of lipid peroxidation (Emond et al., 2000).

VI. Frataxin Gene Structure and Expression

The Friedreich ataxia gene (FRDA) is localized in the proximal long arm of chromosome 9 (Chamberlain et al., 1988). The main mRNA has a size of 1.3 Kb and corresponds to five exons, numbered 1 to 5a (Campuzano et al., 1996). The encoded protein, predicted to contain 210 amino acids, was called frataxin (FXN) and was shown to localize in mitochondria (Babcock et al., 1997).

The gene is expressed in all cells, but at variable levels in different tissues and during development that can be accounted for only in part by differences in mitochondrial content. In adult humans, FXN mRNA is most abundant in the heart and spinal cord, followed by liver, skeletal muscle, and pancreas (Campuzano et al., 1996). In mouse embryos, expression starts in the neuroepithelium at embryonic day 10.5 (E10.5); it reaches its highest level at E14.5 and into the postnatal period. In developing mice, the highest levels of FXN mRNA are found in the spinal cord, particularly at the thoracolumbar level, and in the dorsal root ganglia. The developing brain is also very rich in FXN mRNA, which is abundant in the proliferating neural cells in the periventricular zone, in the cortical plates, and in the ganglionic eminence. Expression is also high in the heart, liver, and brown fat (Jiralerspong et al., 1997; Koutnikova et al., 1997).

VII. DNA Mutations

A. The Intronic GAA Triplet Repeat Expansion in Friedreich Ataxia

FA is the consequence of FXN deficiency. Most patient are homozygous for the expansion of a GAA triplet repeat sequence (TRS) in the first intron (see Figure 18.1), and 5 percent are heterozygous for a GAA expansion and a point mutation in the FXN gene. Repeats in normal chromosomes contain up to approximately 40 triplets, 90 to >1,000 triplets in Friedreich ataxia chromosomes. As in other TRS expansion diseases, pathological alleles show meiotic and mitotic instability.

Two major hypotheses have been proposed to explain the reduced FXN expression in the presence of lengths of the GAA TRS that correspond to pathological FRDA alleles. The first hypothesis is based on the observation that such repeats can adopt a triple helical structure in physiological conditions. Two triplexes may associate to form a novel DNA structure called "sticky DNA" (Sakamoto et al., 1999), a phenomenon that may occur intramolecularly when the TRS is sufficiently long (Vetcher et al., 2002). This structure has been shown to inhibit transcription *in vitro* (Koutnikova et al., 1997) and in *in vivo* models (Ohshima et al., 1998), providing a mechanism to explain FXN deficiency in FA.

The second hypothesis proposes that the GAA repeat, possibly by recruiting specific binding proteins, triggers modifications in DNA packaging proteins that eventually result in chromatin condensation and gene silencing (Saveliev et al., 2003). Chromatin condensation is a well-known mechanism to regulate gene expression. It is largely determined by post-translational modification of histones, including acetylation, methylation, and ubiquitination. The GAA TRS resembles known heterochromatin-associated DNA sequences like satellite III in Drosophila, whose repeating sequence is GAGAA, and it may trigger such modifications

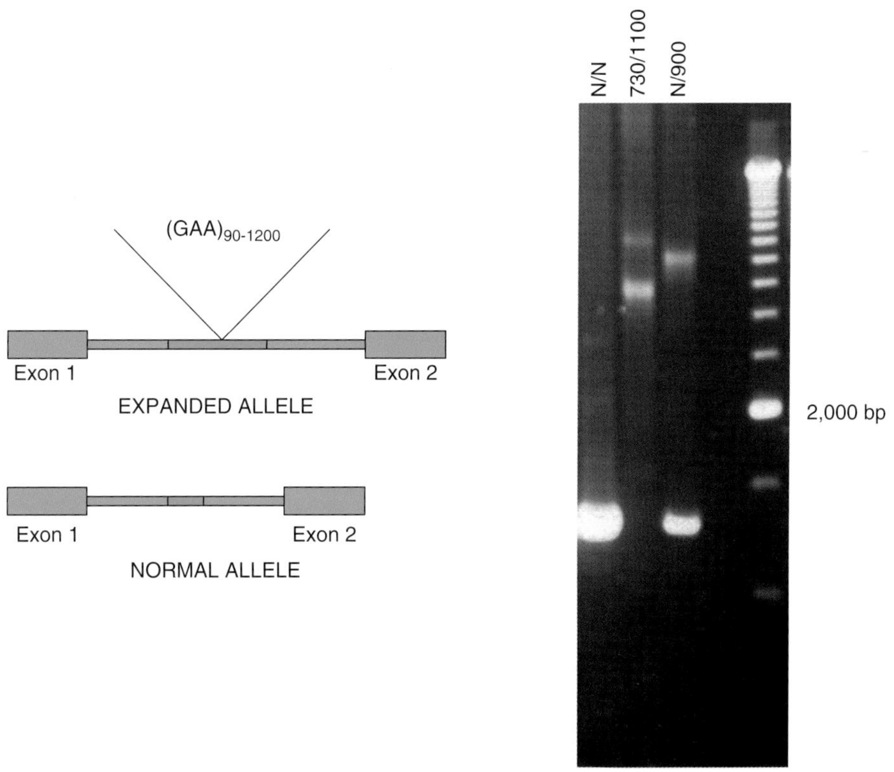

Figure 18.1 Left panel: Schematic representation of the GAA repeat expansion in the first intron of the frataxin gene causing Friedreich ataxia. Right panel: Agarose gel electrophoresis of PCR products containing the Friedreich ataxia GAA repeat in a control, noncarrier individual (first lane), a Friedreich ataxia patient (second lane), a carrier individual (third lane).

by recruiting specific DNA binding proteins. Experimental evidence supporting this hypothesis mostly comes from studies with transgenic mice. When a reporter gene is associated with GAA TRS in a construct used to create transgenic mice it will be silenced in a proportion of cells, a phenomenon called position effect variegation that is observed in genes localized close to heterochomatic regions (Saveliev et al., 2003). Recent data show that changes in histone acetylation and methylation indeed occur in the vicinity of the expanded GAA TRS in lymphocyte DNA from FA patients (Herman et al., 2006).

B. Frataxin Point Mutations

About 2 percent of Friedreich ataxia chromosomes carry GAA repeat of normal length, but have missense, nonsense, or splice site mutations ultimately affecting the frataxin coding sequence. Friedreich ataxia patients carrying a frataxin point mutation are in all cases compound heterozygotes for a GAA repeat expansion. Therefore, as GAA expansion homozygotes, they express a small amount of normal frataxin, which is thought to be essential for survival (see later, mouse models). Mutations have been found to frequently affect the initiation ATG codon in exon 1, where changes involving each of the nucleotides were identified. A stretch of four C's near the end of exon 1 is another hot spot for mutations, with insertions or deletions detected in several unrelated families. Missense mutations so far have been identified only in the C- terminal portion of the protein corresponding to the mature intramitochondrial form of frataxin (Cossée et al., 1999).

C. DNA Tests

DNA testing may be requested to confirm a clinical diagnosis, for carrier detection and for prenatal diagnosis. Polymerase chain reaction (PCR) (see Figure 18.1) or Southern blot can be used to directly detect the intronic GAA expansion present on 98 percent of Friedreich ataxia chromosomes.

The smallest pathological expansion reported so far contained 66 triplets. Interrupted or variant repeats occasionally are found and may be nonpathological up to the equivalent of 130 triplets.

A search for a point mutation in the FXN gene is performed when a clinically affected subject is heterozygous for the GAA repeat expansion. Sometimes, however, no mutation is found after sequencing the exons and flanking intronic sequences of the gene. Heterozygous deletions involving one or more exons have been found in some such cases, others may be associated with deep intronic or promoter mutations, a few may be affected with a different disease and be FA carriers by chance.

Carrier detection can be requested by relatives of affected individuals and their spouses. A test for an expanded GAA repeat can readily be performed in such cases; FXN point mutations usually are screened only in family members of a known carrier.

Presymptomatic diagnosis in a sibling of an affected individual is not currently performed, and is clearly excluded for minors. However, the situation may soon change with the identification of effective treatments that may delay or prevent certain disease manifestations.

D. Genotype-Phenotype Correlation and Modifying Alleles

As expected by the experimental finding that smaller expansions allow a higher residual gene expression, expansion sizes have an influence on the severity of the phenotype (see Figure 18.2). A direct correlation has been firmly established between the size of GAA repeats and earlier age of onset, earlier age when confined in wheelchair, more rapid rate of disease progression, and presence of nonobligatory disease manifestations indicative of more widespread degeneration (Dürr et al., 1996; Filla et al., 1996; Montermini et al., 1997;). However, differences in GAA expansions account for only about 50 percent of the variability in age of onset, indicating that other factors influence the phenotype. These may include somatic mosaicism for expansion sizes,

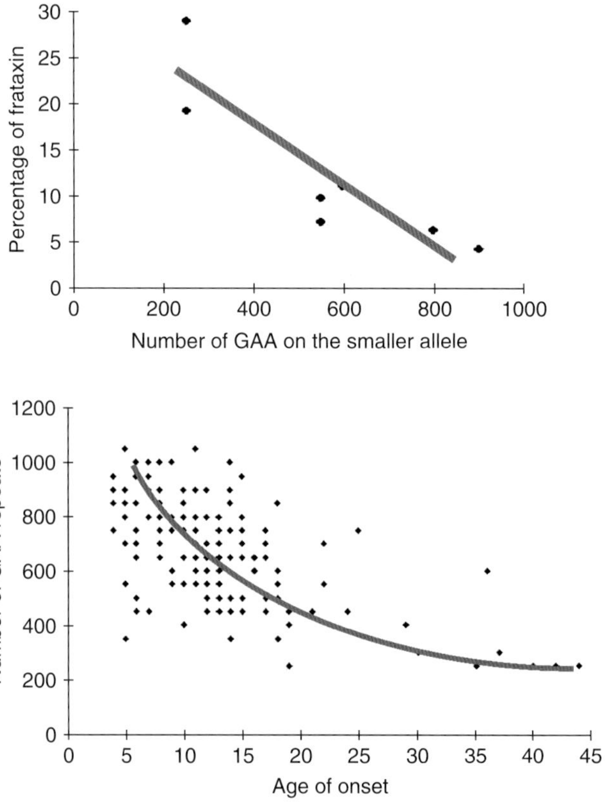

Figure 18.2 The size of the GAA repeat expansion in Friedreich ataxia correlates with the residual frataxin level in fibroblasts from Friedreich ataxia patients (upper panel), and, more loosely, with age of onset (lower panel).

possibly variation in the TRS sequence, modifier genes, and environmental factors. Variations in mitochondrial DNA, revealed by haplogroup analysis, have been shown to affect the FA phenotype. In particular, haplogroup U carrying patients seem to have a delay in the disease onset and a lower rate of cardiomyopathy (Giacchetti et al., 2004).

Nonsense and most missense FXN mutations result in a typical Friedreich's ataxia phenotype, and a few missense mutations are associated with milder atypical phenotypes with slow progression, suggesting that the mutated proteins preserve some residual function. The G130V mutation in particular is associated with early onset but slow progression, no dysarthria, mild limb ataxia, and retained reflexes. For unclear reasons, optic atrophy is more frequent in patients with point mutations of any kind (50%) (Cossée et al., 1999).

VIII. Frataxin Function

FXN does not resemble any protein of known function. Its amino acid sequence predicts a small soluble protein with no transmembrane domain. It is highly conserved during evolution, with homologs in mammals, invertebrates, yeast, and plants. FXN is imported into the mitochondria, where an N-terminal peptide of about 40 amino acids is removed, generating the mature, functional protein.

Much information on FXN function comes from yeast studies. The yeast frataxin homolog (yfh1) is thought to share function with human FXN, not only because of high sequence similarity and corresponding mitochondrial localization, but also because the latter can fully complement a yfh1 deletion. Most yfh1 knock-out yeast strains, called Δ*yfh1*, lose the ability to carry out oxidative phosphorylation and form *petite* colonies with defects or loss of mitochondrial DNA that cannot grow on nonfermentable substrates. Most remarkably, iron accumulates in Δyfh1 mitochondria, averaging a more than 10-fold excess compared to the corresponding wild-type yeast (Babcock et al., 1997). Loss of respiratory competence requires the presence of iron in the culture medium, and occurs more rapidly as iron concentration in the medium is increased, suggesting that permanent mitochondrial damage is the consequence of iron toxicity (Radisky et al., 1999). Ferrous iron reacts with reactive oxygen species (ROS) and can form the highly toxic hydroxyl radical ($OH^{\cdot}$) through the Fenton reaction with H_2O_2, a process likely to occur in Δ*yfh1* yeast, as suggested by its enhanced sensitivity of to H_2O_2. In mitochondria, H_2O_2 is produced by the mitochondrial form of the enzyme superoxide dismutase (SOD) from superoxide (O_2^-), in turn generated by the reaction of oxygen with electrons prematurely leaking from the respiratory chain. FXN deficiency likely establishes a vicious cycle involving damage to the respiratory chain, increased ROS production, and iron toxicity. Accordingly, a role of oxidative stress in the pathogenesis of FA has been proposed.

Excess iron in Δ*yfh1* yeast has been related to a marked induction (10- to 50-fold) of the high-affinity iron transport system on the cell membrane, normally not expressed in yeast cells that are iron replete. This induction has been recently identified as a response to a deficient mitochondrial synthesis of iron-sulfur clusters (ISCs) (Chen et al., 2004). ISCs are complexes of iron and sulfur atoms that act as prosthetic groups in several enzymes of different function and cellular localization. They are synthesized in mitochondria; cytosolic synthesis has been proposed but is somewhat controversial (Rouault & Tong, 2005). ISC-containing enzymes, as respiratory chain complexes I, II, and III and aconitase, are impaired in Δ*yfh1* yeast.

FXN is thought to be involved in an early step of ISC synthesis, through its interaction with the scaffold protein IscU, where the first ISC assembly takes place, probably facilitating iron incorporation (Muhlenhoff et al., 2003; Yoon & Cowan, 2003). Iron enters mitochondria through specific transporters, called Mrs3 and 4 in yeast. In this process, sulfur is provided by the cysteine desulfurase Nsf1, which also appears to interact with FXN. FXN may be a mitochondrial iron chaperone with a more general function to protect this metal from ROS and making it bioavailable. *In vitro*, FXN binds Fe^{2+} with low affinity, as it would be expected for a chaperone, and has the capacity to oxidize it to the much less reactive Fe^{3+} form when its concentration increases (Park et al., 2003). A mitochondrial form of ferritin (m. ferr) may also be involved in this detoxification process. Some data suggest that FXN also acts as an iron chaperone in heme synthesis, providing iron to ferrochelatase (FCH), the enzyme that inserts iron into protoporphyrin IX to form heme (Yoon & Cowan, 2004) (see Figure 18.3). A higher affinity of FXN for the heme synthesis enzyme ferrochelatase than for IscU may explain why heme synthesis is resistant to low FXN level and essentially unaffected in FA patients. FXN has also been implicated in the modulation of aconitase (AC) activity through assembly and disassembly of its ISC (Bulteau et al., 2004).

IX. Animal Models

A mouse model of FA has been difficult to generate because complete loss of FXN, as in FXN knock-out (KO) mice, causes early embryonic lethality (Cossée et al., 2000). Viable mouse models were first obtained through a conditional gene targeting approach, by crossing animals with a LoxP-flanked frataxin exon 2 with transgenic animals expressing Cre recombinase under the control of different promoters (Puccio et al., 2001). Striated muscle-, neuron-, pancreas- and liver-restricted exon 2 deletions have been obtained this way. Mutants with neuron-specific enolase

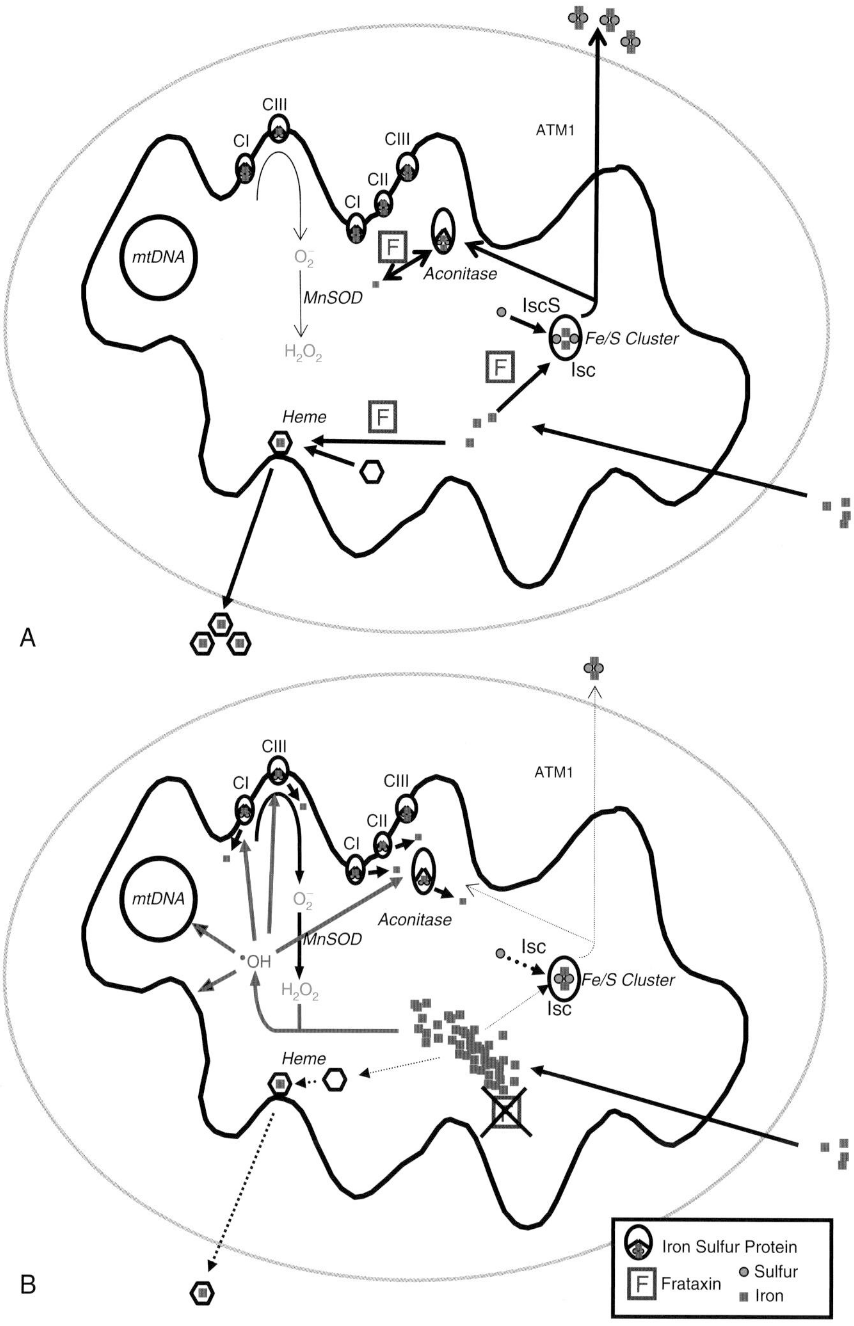

Figure 18.3 Schematic representation of the postulated function of frataxin and of the consequences of its deficiency on mitochondrial iron metabolism and free radical production. **Panel A** shows the normal situation, with frataxin acting as an iron chaperon that donates this metal to IscU for iron-sulfur cluster synthesis, to ferrochelatase for heme synthesis, and to aconitase for the maintenance of its iron-sulfur cluster. Lack of frataxin causes insufficient iron-sulfur cluster synthesis, defective maintenance of the aconitase iron-sulfur cluster, defective heme synthesis (when deficiency is complete, as in the yeast model), increased free radical production and further damage to iron-sulfur clusters, proteins and DNA (**panel B**).

(NSE) promoter-directed Cre expression have a low birth weight and develop a progressive neurological phenotype with the average onset of ataxia at 12 days, hunched stance, and loss of proprioception. Mutants with muscle creatin kinase (MCK) promoter-directed Cre expression show cardiac hypertrophy with thickening of the walls of the left ventricle, and show myocardial degeneration with cytoplasmic vacuolization in the myocytes, and evidence of necrosis and post-necrotic fibrosis.

Loss of activity of ISC-containing enzymes is an early finding in these models. The MCK mutants accumulate iron in heart mitochondria at later stages. Using a similar conditional knock-out approach, but with a tamoxifen-inducible Cre recombinase under the control of a neuron-specific prion protein promoter, two different lines that exhibit a progressive neurological phenotype with slow evolution that recreates the neurological features of the human disease were developed (Simon et al., 2004). An autophagic process was detected in the DRG, leading to removal of mitochondrial debris and apparition of lipofuscin deposits.

To have a more faithful model of the human disease, knock-in mice that carry a $(GAA)_{230}$ repeat in the endogenous frataxin gene (*frda*) were generated. This repeat, when in homozygosity, leads to a decrease in frataxin expression to only 75 percent of wild-type levels. To further decrease frataxin expression, frataxin knock-in–knock-out mice (*frda*$^{-/230GAA}$) were generated, expressing approximately 25 to 30 percent of wild-type frataxin levels, a reduction associated with mild, but clinically evident FA in humans (Miranda et al., 2002). These mice have no frank pathology, only minor motor abnormalities, but they show clear changes in gene expression in the CNS, heart, and muscle (Coppola et al., 2006). Such modifications may represent a basic response to frataxin deficiency not yet blurred by secondary changes due to degeneration and cell loss. Genes identified are involved in nucleic acid and protein metabolism, signal transduction, stress response, and nucleic acid binding, with an over-representation of mitochondria-related transcripts.

Recently, two lines of human FXN YAC transgenic mice were obtained and crossbred with heterozygous frataxin knock-out mice. The resultant mice express only low levels of human frataxin mRNA and protein, show decreased aconitase activity and oxidative stress, and develop progressive neurodegenerative and cardiac pathological phenotypes (Al-Mahdawi et al., 2006). Coordination deficits are present, as measured by accelerating rotarod analysis, together with a progressive decrease in locomotor activity and increase in weight. Large vacuoles are detected within neurons of the dorsal root ganglia (DRG), predominantly within the lumbar regions in six-month-old mice, but spreading to the cervical regions after 1 year of age. Secondary demyelination of large axons is also detected within the lumbar roots of older mice. Lipofuscin deposition is increased in both DRG neurons and cardiomyocytes, and iron deposition is detected in cardiomyocytes after one year of age.

X. Pathogenesis

Altered iron metabolism, free radical damage, and mitochondrial dysfunction all occur in FA patients, suggesting that the information derived from the investigations on FXN function and from the yeast and animal models is relevant for the pathogenesis of the human disease. Oxidative stress is revealed by increased plasma levels of malondialdehyde (Emond et al., 2000), a lipid peroxidation product, increased urinary 8-hydroxy-2′-deoxyguanosine, a marker of oxidative DNA damage (Schulz et al., 2000), decreased plasma-free glutathione, and increased plasma glutathione-S-transferase activity (Tozzi et al., 2002). Increased free radical production could be demonstrated directly in cultured cells engineered to produce reduced levels of FXN (Santos et al., 2001). In addition, patients' fibroblasts are sensitive to low doses of H_2O_2, which induce apoptosis at lower doses than in control fibroblasts, suggesting that even nonaffected cells are at an "at risk" status for oxidative stress as a consequence of the primary genetic defect (Wong et al., 1999).

FA fibroblasts also show abnormal antioxidant responses—in particular, a blunted increase in mitochondrial SOD triggered by iron and by oxidants in control cells (Chantrel-Groussard et al., 2001; Jiralerspong et al., 2001). Intriguingly, no evidence of oxidative stress was obtained in studies on conditional knock-out mouse models (Seznec et al., 2005). However, interfering factors may have affected these results, including the admixture of cells with normal FXN with progressively disappearing cells with no FXN, and the almost complete shut down of the respiratory chain in the latter.

Mitochondrial dysfunction has been proven to occur *in vivo* in FA patients. Phosphorus-MRS analysis of skeletal muscle and heart shows a reduced rate of ATP synthesis (Lodi et al., 1999). Finally and most importantly, the same multiple ISC-containing enzyme dysfunctions found in Δ*yfh1* yeast and in mouse models also are found in affected tissues from FA patients (Rotig et al., 1997).

Activation of stress pathways, triggered by mitochondrial dysfunction causing energy deficit and oxidative stress, occurs in FA and likely plays an important role in cell atrophy and death. Studies on cultured PC12 cells, rat pheochromocytoma cells that can be differentiated into neurons by adding nerve growth factor (NGF), showed in particular an increased expression and activity of the MKK4-JNK kinase pathway, which may be at first a protective response, but also eventually may trigger apoptosis (Pianese et al., 2002). Different vulnerable cell types may activate different pathways, as suggested by the observation of the specific occurrence of autophagic vacuoles only in primary sensory neurons in the inducible conditional knock-out mouse model.

XI. Treatment: Directions and Perspectives

All FA patients carry at least one expanded GAA repeat and therefore make an insufficient amount of otherwise normal FXN. If it were possible to increase their FXN production even to levels that are similar to those of healthy carriers, one could possibly stop the course of the disease and maybe even induce some improvement. An increased FXN production could be obtained:

- Through gene replacement therapy; that is, by introducing a FXN gene without the GAA expansion into the patient cells
- By giving FXN directly; the protein should, however, be modified in such a way as to be able to reach the nerve cells affected by the disease and the mitochondria inside these cells
- By intervening on the GAA expansion with molecules that can destabilize the triple helical structure and shift the equilibrium toward the physiological double helix that allows FXN expression

Though still in their infancy, all these approaches are under study. Recently, encouraging results have been obtained for gene replacement therapy, with partial correction of the oxidative stress hypersensitivity of FA fibroblasts by FXN-encoding adeno-associated virus and lentivirus vectors (Fleming et al., 2005), and for some approaches aimed to reactivate FXN transcription by using histode deacetylase inhibitors (Herman et al., 2006).

Further possibilities to treat FA may come from the studies on the function of FXN. Based on these findings, therapeutic approaches aimed to free radical control and respiratory chain activation may be proposed. Concerning anti-oxidant molecules and respiratory chain stimulants, some coenzyme Q derivatives (idebenone, CoQ-10) already have produced promising results, not only in experimental models (Seznec et al., 2004), but also in clinical trials, at least on the FA cardiomyopathy (Buyse et al., 2003; Mariotti et al, 2003). An intriguing possibility would be the identification of small molecules capable of effectively replacing FXN by binding mitochondrial iron and increasing its bio-availability.

Cellular therapies, in particular the use of stem cells, may be explored to promote regeneration and function recovery. The widespread nature of neurodegeneration in FA would require to diffusely deliver cells in the central nervous system of the patients, a major difficulty at present, but the object of intense research.

In conclusion, thanks to the remarkable progress in understanding the pathogenesis of FA since the responsible gene was discovered in 1996, the perspective of a treatment for this so-far incurable neurodegenerative disease now appears a realistic goal.

XII. Related Loss-of-Function Disorders

Recessive ataxias comprise a very heterogeneous group of disorders. Only those diseases with symptoms that show a resemblance to FA are discussed here (Table 18.1)

A. Ataxia with Vitamin E Deficiency (AVED)

1. Clinical and Pathological Features

Ataxia with vitamin E deficiency (AVED) is a rare disorder with a symptomatology that closely resembles that of Friedreich's ataxia. The condition is more common in patients of Mediterranean and North African origin.

Age at onset ranges from two to 52 years, but usually is less than 20 years old (Cavalier et al., 1998). Typical symptoms include progressive gait and limb ataxia, dysarthria, loss of deep tendon reflexes, extensor plantar reflexes, muscle weakness, and decreased proprioception. Symptoms may also include scoliosis and *pes cavus* (Mariotti et al., 2004). Cardiomyopathy is encountered but is less frequent than in FRDA. On the other hand, head titubation, dystonia, and retinopathy are more common. The disease is not known to be associated with diabetes. The biochemical hallmark of the disease is a very low vitamin E plasma level (usually below 3 mg/l) with normal lipid and lipoprotein profiles (Mariotti et al., 2004). Brain imaging shows mild cerebellar atrophy in about half of cases (Mariotti et al., 2004). Peripheral nerve conduction studies show slight to moderate axonal sensory neuropathy and somatosensory evoked potentials show increased central conduction times and increased latencies of cortical responses (Zouari et al., 1998).

The importance of this condition lies in the fact that administration of vitamin E supplements results in arrest of progression of neurological symptoms and signs, and in amelioration of established neurological abnormalities in a number of patients (Kohlschutter et al., 1988; Mariotti et al., 2004; Yokota et al., 1997). Response is best when therapy is started early in the course of the disease (Gabsi et al., 2001). The usual dosage needed in adults to restore normal vitamin E plasma levels is between 1,000 and 2,400 mg of the RRR stereo-isomer of α-tocopherol in two daily doses, with fat-containing meals (Cavalier et al., 1998; Mariotti et al., 2004). In children, the usual dosage is 40 mg/kg body weight. Plasma vitamin E concentration should be measured regularly during therapy, especially in children. Ideally the plasma concentration should be in the high-normal range. If therapy is initiated in presymptomatic individuals, neurological signs do not develop (Mariotti et al., 2004). Hence all relatives, especially younger siblings of a proband, should be evaluated for vitamin E deficiency.

2. Gene Structure and Expression

The *TTPA* gene, encoding α-tocopherol transfer protein, localized on chromosome 8q13, is the only gene associated

Table 18.1 Summary of the Main Clinical and Genetic Features of Friedreich Ataxia and Related Loss-of-Function Disorders

Disease	Usual Age of Onset	Cerebellar Atrophy	Pyramidal Signs	Peripheral Neuropathy	Other Signs and Symptoms	Cardiomyopathy	Gene (Protein) and Mutations	Postulated Pathogenic Mechanism
Friedreich ataxia	<20 (2–>50)	+/–	+/–	+ (sensory axonal)	kyphoscoliosis; pes cavus; optic atrophy; deafness; diabetes	+	FRDA (frataxin). GAA repeat expansion, rare point mutations (always in heterozygosity with GAA repeat expansion)	Mitochondrial dysfunction
Ataxia with vitamin E deficiency (AVED)	<20 (2–52)	+/–	–	+ (sensory axonal)	head titubation; dystonia; retinopathy	–	TTPA (alpha-tocopherol transfer protein). Missense, nonsense, frameshift, and splice-site mutations, insertions and deletions.	Vitamin E deficiency (oxidative stress?)
Ataxia with oculomotor apraxia Type 1 (AOA1)	<7 (2–young adult)	+	–	+ (motor axonal)	oculomotor apraxia; chorea; dystonia; low albumin, high cholesterol	–	APTX (aprataxin). Missense, nonsense, frameshift, and splice-site mutations.	Defective single-strand DNA break repair
Ataxia with oculomotor apraxia Type 2 (AOA2)	10–22	+	+	+ (sensorimotor axonal)	oculomotor apraxia; dystonia; chorea; tremor; cognitive impairment; high alpha-fetoprotein, CK, IgG, IgA, cholesterol	–	SETX (senataxin), loss-of-function missense, nonsense, truncating mutations, and large-scale rearrangements	Defective double-strand DNA break repair
Autosomal recessive spastic ataxia of charlevoix-saguenay (ARSACS)	12–18 months (may be later outside Québec)	+	+	+	myelinated optic nerve fibers in the retina; scoliosis; pes cavus	– (but mitral valve prolapse common)	SACS (sacsin), missense, deletions or insertion	Impaired response to misfolded proteins

with AVED (Ouahchi et al., 1995). The 25.1 kb gene is composed of five uniformly spliced exons with an open reading frame of 834 bp. The 278 amino acid, 31.7 kD protein product is expressed mainly in hepatocytes (Sato et al., 1993), but also in kidney, cerebellar pyramidal cells (Copp et al., 1999; Hosomi et al., 1998) and in placenta (Kaempf-Rotzoll et al., 2003; Muller-Schmehl et al., 2004).

3. Mutations, DNA Tests, Genotype–Phenotype Correlations

At least 20 different mutations in the *TTPA* gene have been reported. Missense and splice-site mutations, as well as deletions, insertions, and indeletions have been reported (Cavalier et al., 1998). Most mutations are private, except for the 744delA and the 513insTT mutations, which are more common in individuals of North African and Mediterranean ancestry, respectively.

Sequencing of the five exons and flanking intron sequences detects mutations in more than 90 percent of individuals with AVED (Cavalier et al., 1998; Ouahchi et al., 1995). Most patients are homozygous or compound heterozygotes for one of the known mutations. Penetrance is almost complete.

There is a clinico-molecular correlation, with patients with truncating or nonconservative missense mutations showing earlier age at onset and more severe phenotype as compared to patients with semiconservative missense mutations. Similarly, it has been shown that *TTPA* missense mutations that result in a clear impairment in binding and transfer activity of TTPA are associated with severe, early onset disease; mutations with biochemical properties similar to wild-type protein are associated with milder, late-onset forms (Morley et al., 2004). The H101Q mutation, which appears to be restricted to Japan, is associated with a late-onset, mild phenotype and is more commonly associated with retinopathy.

4. Animal Model

Terasawa et al. (2000) generated a *Ttpa* knock-out mouse model. *Ttpa–/–* mice were generally healthy and showed no obvious signs of neurological disease at 18 months. They showed that in atherosclerosis-susceptible apolipoprotein E (*ApoE)* knock-out mice, vitamin E deficiency caused by disruption of Ttpa increased the severity of atherosclerotic lesions in the proximal aorta. Thus, vitamin E deficiency appears to modulate rather than cause atherosclerosis, which may explain why AVED patients do not exhibit an increased incidence of atherosclerosis.

5. Pathogenesis

AVED is caused by defective incorporation of vitamin E into very low density lipoproteins (VLDL) by the α-tocopherol transfer protein, leading to vitamin E deficiency (Traber et al., 1990). There is no malabsorption.

The α-tocopherol transfer protein is a cytosolic liver protein, which maintains normal vitamin E plasma concentrations through preferential binding of the RRR-α-tocopherol isomer to VLDL proteins.

B. Ataxia with Oculomotor Apraxia Type 1 (AOA1)

1. Clinical and Pathological Features

Ataxia with oculomotor apraxia type 1 (AOA1) is an early-onset, autosomal recessive ataxia resembling ataxia telangiectasia but lacking the extraneurological features. It is encountered worldwide, although the prevalence varies according to ethnic origin. AOA1 seems to be the most frequent cause of autosomal recessive ataxia in Japan. Onset of disease symptoms is between ages two and 16 years, with 50 percent of individuals presenting before the age of seven years, although disease onset in young adults has been reported (Criscuolo et al., 2004).

AOA1 is characterized by progressive cerebellar ataxia followed by oculomotor apraxia and severe motor neuropathy, and hypoalbuminemia. The first manifestation is usually slowly progressive gait ataxia, followed by dysarthria, then upper-limb dysmetria with mild intention tremor. Oculomotor apraxia is the most striking feature in the disorder. It is characterized by an inability to visually track objects without turning the head. It is present in 80 percent of individuals with AOA1 and is followed by progressive external ophthalmoplegia. All affected individuals develop peripheral axonal motor neuropathy with severe weakness and wasting. Neuropathy represents the main cause of motor disability in advanced stages of the disease, with most individuals becoming wheelchair bound by adolescence. *Pes cavus* is present in 30 percent of patient and scoliosis in a few. The Babinski sign is absent. Other symptoms include chorea, present in the majority of patients at onset, but disappearing in almost 50 percent over the course of the disease (Le Ber et al., 2003), upper limb dystonia, present in about 50 percent of patients, and variable degrees of cognitive impairment. Although the disease leads to severe motor disability, it is compatible with long survival (Barbot et al., 2001; Le Ber et al., 2003).

Typical laboratory findings include hypoalbuminemia (<3.8 g/l) and hypercholesterolemia (>5.6 mmol) in individuals with disease duration of more than 10 to 15 years. Brain MRI shows cerebellar atrophy in all affected individuals and rarely brainstem atrophy. EMG shows signs of axonal motor neuropathy in 100 percent of affected individuals with advanced disease. Treatment of the disease is purely symptomatic. A high-protein, low-cholesterol diet is advised.

2. Gene Structure and Expression

The *APTX* gene, encoding aprataxin, localized on chromosome 9p13.3, is the only gene associated with AOA1

(Date et al., 2001; Moreira et al., 2001). The 17.47 kb gene consists of seven exons. The gene encodes 2 major mRNA transcripts resulting from alternative splicing of exon 3: a short form of 168 amino acids and the first ATG codon in exon 3, and a long form of 342 amino acids and the first ATG codon in exon 1. The long transcript is the major form and is ubiquitously expressed in human cell lines, whereas the shorter form is present in lower amounts. Aprataxin is expressed mainly in the nucleus (Gueven et al., 2004; Sano et al., 2004).

3. Mutations, DNA Tests, Genotype–Phenotype Correlations

At least 16 different mutations have been identified. Missense, nonsense, frameshift, and splice-site mutations, as well as complete gene deletion, have been reported. Mutation detection rates have not yet been reported for sequence analysis; however, mutation scanning has revealed mutations in about 41 percent of Portuguese individuals and in 100 percent of Japanese individuals with AOA1 (Moreira et al., 2001). Patients can be homozygotes or compound heterozygotes. There is no clear correlation between specific genotypes and age of onset of the disease, nor with survival. Similarly, cognitive impairment has been reported with a range of different mutations, but there seems to be no clear correlation between specific genotypes and degree of cognitive impairment. Two compound heterozygotes for the R199H missense mutation and an unidentified second mutation had an atypical presentation with marked dystonia and mask-like faces in addition to the AOA1 clinical picture (Moreira et al., 2001). The mutation A198V is associated with predominant, more severe, and persistent chorea (Le Ber et al., 2003).

4. Animal Model

No AOA1 animal model has been reported to date.

5. Pathogenesis

The aprataxin protein is a member of the histidine-triad (HIT) (His-X-His-X-His-X-X, where X is a hydrophobic amino acid) superfamily of nucleotide hydrolases and transferases. The long form of aprataxin is composed of three domains:

1. The N-terminal PANT (PNKP-AOA1 N-terminal) domain, which shares distant homology with the N-terminal domain of polynucleotide kinase 3′-phosphatase (PNKP) and interacts with DNA polymerase b, DNA ligase III and XRCC1 (x-ray repair cross-complementing group 1) protein, forming the single-strand break repair (SSBR) complex, following exposure to ionizing radiation and reactive oxygen species (Sano et al., 2004; Whitehouse et al., 2001).
2. The HIT domain (middle domain), defined by the HIT motif, for nucleotide binding and hydrolysis. Fragile-HIT proteins (FHIT), which have the highest homology to aprataxin, cleave diadenosine tetraphosphate (Ap4A), which potentially is produced during activation of the DNA SSBR complex (Moreira et al., 2001).
3. The C-terminal domain, containing a divergent zinc-finger motif (Moreira et al., 2001), which could allow binding to DNA and or RNA (Kijas et al., 2006). These findings suggest that aprataxin is part of a multiprotein complex involved in single-strand DNA break repair, reminiscent of other DNA repair diseases presenting with cerebellar ataxia. Inactivating mutations are thought to result in impairment of DNA and RNA integrity, leading to accumulation of damaged DNA, resulting in selective neuronal loss.

C. Ataxia with Oculomotor Apraxia Type 2 (AOA2)

1. Clinical and Pathological Features

Ataxia with oculomotor apraxia Type 2 (AOA2) is an adolescent-onset, autosomal recessive ataxia syndrome characterized by progressive cerebellar ataxia followed by oculomotor apraxia and sensorimotor neuropathy, and elevated serum levels of alpha-fetoprotein. It is encountered worldwide. One study suggested that AOA2 accounts for 8 percent of all non-Friedreich autosomal recessive ataxias, making it the most frequent cause of autosomal recessive ataxia after Friedreich's ataxia in adults (Le Ber et al., 2004). Onset is usually between 10 and 22 years of age.

The first symptom consists of slowly progressive gait ataxia (Moreira et al., 2004). About 90 percent of individuals develop an axonal, sensorimotor neuropathy. Oculomotor apraxia develops in up to 50 percent of patients. Up to 44 percent develop movement disorders, including dystonic posturing of the hands, chorea, and head or postural tremor, which may persist or not with disease progression (Criscuolo et al., 2006; Le Ber et al., 2004; Nemeth et al., 2000). Other symptoms include mild cognitive impairment, extensor plantar reflexes, sphincter disturbances, and skeletal and foot deformities (Criscuolo et al., 2006; Le Ber et al., 2004). There are no extraneurological features. AOA2 is a slowly progressive disorder with reported disease duration of up to 51 years (Moreira et al., 2004).

Serum levels of alpha-fetoprotein typically are increased (>20 ng/ml) (Moreira et al., 2004). Additional laboratory findings may include hypercholesterolemia (>5.6 mmol/l), increased levels of serum creatine kinase, and elevated IgG and IgA immunoglobulins (Le Ber et al., 2004; Watanabe et al., 1998). Brain MRI shows cerebellar atrophy with normal brain stem in all affected individuals (Moreira et al., 2004).

Treatment of the disease is purely symptomatic.

2. Gene Structure and Expression

The *SETX* gene, encoding senataxin, localized on chromosome 9q34, is the only gene associated with AOA1

(Moreira et al., 2004). The 91 kb gene consists of 24 exons with an open reading frame of 8,031 bp. The gene gives rise to two transcripts of 11.5 and 9 kb length, respectively. The 303 kD, 2677 amino acid protein is ubiquitously expressed.

3. Mutations, DNA Tests, Genotype–Phenotype Correlations

At least 20 different mutations have been identified. Missense, nonsense, and frameshift mutations, as well as large-scale rearrangements have been reported (Criscuolo et al., 2006; Duquette et al., 2005; Moreira et al., 2004). Mutation detection is through gene sequencing. Mutation detection rates have not yet been reported.

No obvious genotype–phenotype correlations have been identified. Truncations and missense mutations may result in a similar phenotype (Criscuolo et al., 2006). Interestingly, specific missense mutations in the *SETX* gene also have been identified in ALS4, an autosomal dominant form of juvenile amyotrophic sclerosis (Chen et al., 2004).

4. Animal Model

No AOA2 animal model has been reported to date.

5. Pathogenesis

The senataxin protein contains at its C-terminus a classical seven-motif domain found in the superfamily 1 of helicases (Moreira et al., 2004). Senataxin was named for its homology to the fungal Sen1p protein, which has RNA helicase activity encoded by its C-terminal domain. Senataxin shares significant similarity with two other members of the superfamily of helicases: RENT1/Upf1, involved in nonsense-mediated RNA decay, and IGHMBP2, which is mutant in spinal muscular atrophy with respiratory distress-1, a motor neuron disease, and in mouse neuromuscular degeneration. Upf1 proteins have RNA helicase activity, but IGHMBP2 initially was identified as a DNA binding protein with transcriptional transactivating properties. It is therefore thought that senataxin has both RNA and DNA helicase activity and that it acts in a DNA repair pathway. Alternatively, senataxin might be a nuclear RNA helicase with a role in the splicing machinery.

D. ARSACS

1. Clinical and Pathological Features

Autosomal recessive spastic ataxia of Charlevoix-Saguenay (ARSACS) is an ataxia syndrome described originally in the Saguenay-Lac-St-Jean area of North eastern Quebec in French Canada (Bouchard et al., 1978). Subsequently, the disease was also reported in individuals of Italian, Japanese, Tunisian, and Turkish ancestry (El Euch-Fayache et al., 2003; Grieco et al., 2004; Mrissa et al., 2000; Ogawa et al., 2004; Shimazaki et al., 2005; Richter et al., 2004). The estimated carrier frequency of ARSACS in the Saguenay-Lac-St-Jean region of Quebec is 1/21 and the birth incidence is 1/1,932 (De Braekeleer et al., 1993). A founder effect, which could date back to 1650, is largely suspected as the root cause of the high regional prevalence of ARSACS in northeastern Quebec. The true worldwide incidence of ARSACS remains unknown as underdiagnosis is still likely. Onset is between 12 and 18 months in individuals born in Quebec province, but often is delayed until the juvenile or even adult age in individuals born outside Quebec.

Typical symptoms include progressive gait ataxia, dysarthria, spasticity, extensor plantar reflexes, distal muscle wasting, distal sensorimotor neuropathy, and horizontal gaze nystagmus. Ankle reflexes are gradually lost, but other reflexes remain brisk. Individuals originating from Quebec also show yellow streaks of hypermyelinated fibers radiating from the optic fundi in the retina. These changes may appear late in the course of the disease. They are usually absent in individuals not born in Quebec. Other features include *pes cavus*, hammertoes, scoliosis, and mitral valve prolapse, but cardiomyopathy does not occur. Although IQ levels tend to be in the lower range of normal, cognitive skills tend to be preserved into late adult life. ARSACS is a slowly progressive disorder with death usually occurring in the sixth decade. Neuroimaging typically reveals atrophy of the superior vermis with sparing of other cerebellar structures. Nerve conduction studies typically show absent sensory nerve conduction potentials and moderately decreased motor nerve velocities. Treatment of the disease is purely symptomatic.

2. Gene Structure and Expression

The *SACS* gene, encoding sacsin, localized on chromosome 13q12, is the only gene associated with ARSACS (Engert et al., 2000). The 39.7 kb gene contains one gigantic exon spanning 12.8 kb, encoding an open reading frame of 11,487 bp, and eight recently identified exons located upstream of the gigantic one (Ouyang et al., 2006). The open reading frame is conserved in both the human and mouse. The gene gives rise to at least three different transcripts. The 437 kD, 3829 amino acid protein is expressed in the central nervous system and various other tissues.

3. Mutations, DNA Tests, Genotype–Phenotype Correlations

Two founder mutations account for about 96 percent of mutations in individuals with ARSACS from northeastern Quebec: 92.6 percent are homozygous for the g.6594delT mutation and 3.7 percent are compound heterozygotes for the g.6594delT deletion and the g.5254C→T nonsense mutation (Richter et al., 1999). The g.6594delT mutation results in a frameshift and the introduction of a stop codon. The g.5254C→T results in substitution of a stop codon for an arginine. Both mutations predict truncation of the sacsin protein. Targeted mutation analysis for these mutations is available on a clinical basis.

At least 20 other mutations have been reported in individuals with ARSACS from Italy, Japan, Tunisia, Turkey, and Spain (El Euch-Fayache et al., 2003; Grieco et al., 2004; Mrissa et al., 2000; Ogawa et al., 2004; Shimazaki et al., 2005; Richter et al., 2004; Okawa et al., 2006; Yamamoto et al., 2005; Yamamoto et al., 2006; Criscuolo et al., 2005; Hera et al., 2005). These include missense mutations, deletions, and insertions. Except for two (Ouyang et al., 2006), all identified mutations are located in the gigantic downstream exon of *SACS*. Molecular genetic testing for mutations observed in other populations is through direct DNA analysis and is available on a research basis only. Little intra- and extrafamilial phenotypic variability has been observed among individuals born in Quebec. In patients from other populations the clinical features are more heterogeneous and there is no clear genotype-phenotype correlation.

4. Animal Model

The recessive mutation *tumbler* (tb), causing ataxia in mice, previously was mapped to chromosome 1, which contains the mouse *sacs* gene (Dickie, 1965). The tb mouse line died out, but Engert et al. (2000) speculated that it harbored a mutation in *sacs* (Engert et al., 2000). No knock-out transgenic models of ARSACS are yet available.

5. Pathogenesis

Structure prediction programs suggest the presence of two leucine zippers, three coiled-coils, and one hydrophilic domain within the carboxy-terminal half of the protein (Engert et al., 2000). Sacsin was found to have no extensive similarity to any known protein by sequence comparison analyses. The C-terminal portion of the predicted protein, contains a DnaJ motif, which has the potential to interact with members of the HSP70 family of heat shock proteins and stimulate its ATPase activity. The N-terminus has extensive homology for HSP90, a subtype of heat shock protein that can act as a chaperone molecule important in the regulation of protein folding. These data suggest that sacsin may function in chaperone-mediated protein folding.

Individuals homozygous for the g.6594dlT deletion have complete loss of sacsin immunocytochemical and western blot expression in skin fibroblasts. It is likely that major deletions result in complete suppression of sacsin expression, including the CNS. It is postulated that *SACS* mutations may interfere with protein folding and lead to significant loss of function in key signaling pathways even at an embryonic stage. Compound heterozygotes for less extensive deletions or point mutations will result in the synthesis of a truncated sacsin molecule that may not be able to interact normally with other proteins.

E. Other Recessive Ataxias

One report described two sib pairs from two families with typical symptoms of Friedreich's ataxia but no mutations in the FRDA gene (Kostrzewa et al., 1997). The FRDA2 locus was subsequently mapped to chromosome 9p23-p11 in a large consanguineous family, but the responsible gene has not been identified yet (Christodoulou et al., 2001).

Spinocerebellar ataxia with axonal neuropathy (SCAN1) is characterized by mild ataxia, axonal sensorimotor neuropathy, distal muscle atrophy, *pes cavus*, and steppage gait as seen in Charcot-Marie-Tooth neuropathy (Takashima et al., 2002). Other features include mild hypercholesterolemia and borderline hypoalbuminemia. A homozygous 1478A→G transition mutation in the tyrosyl-DNA phosphodiesterase 1 (TDP1) gene on chromosome 14q31-q32 was identified.

TDP1 is required for repair of chromosomal single-strand breaks arising independently of DNA replication from abortive topoisomerase-1 (TOP1) activity or oxidative stress (El-Khamisy et al., 2005).

Infantile-onset spinocerebellar ataxia (IOSCA) is a disorder described exclusively in a small founder group of Finnish patients (Koskinen et al., 1994). Onset is between one and two years of age. Features include progressive ataxia, athetosis, hypotonia, areflexia, ophthalmoplegia, optic atrophy, hearing loss, and sensory neuropathy. Female hypogonadism and epilepsy are late manifestations. The IOSCA locus was mapped to 10q24, but the responsible gene has not been identified so far (Nikali et al., 1995).

An infantile-onset, nonprogressive ataxia with hypotonia, slow speech development, hyperreflexia, occasional mild spasticity, *pes planus*, and short stature was linked to chromosome 20q11-q13 (Tranebjaerg et al., 2003).

An early-onset cerebellar ataxia with pyramidal signs, deep sensory loss, and postural tremor was described in a nonconsanguineous family from the Netherlands (Breedveld et al., 2004). Severity of symptoms and progression were very variable. The disorder was linked to chromosome 11p15.

A spastic ataxia with variable age of onset recently was described in several families from the same region in Quebec (Thiffault et al., 2006). Other features may include scoliosis, dystonia, and cognitive impairment. One in two individuals showed white matter changes on brain MRI. Linkage analysis showed that all families were linked to the same region on chromosome 2q33-34.

Behr syndrome is a childhood spastic ataxia with optic atrophy, mental retardation, and deep sensory loss (Thomas et al., 1984). No genetic linkage has been reported so far.

References

Al-Mahdawi, S., Pinto, R. M., Varshney, D., et al. (2006). GAA repeat expansion mutation mouse models of Friedreich ataxia exhibit oxidative stress leading to progressive neuronal and cardiac pathology. *Genomics* **88**, 580–590.

Aronsson, D. D., Stokes, I. A., Ronchetti, P. J. et al. (1994). Comparison of curve shape between children with cerebral palsy, Friedreich's ataxia, and adolescent idiopathic scoliosis. *Dev Med Child Neurol* **36**, 412–418.

Babcock, M., de Silva, D., Oaks, R. et al. (1997). Regulation of mitochondrial iron accumulation by Yfh1, a putative homolog of frataxin. *Science* **276**, 1709–1712.

Badhwar, A., Jansen, A., Andermann, F. et al. (2004). Striking intrafamilial phenotypic variability and spastic paraplegia in the presence of similar homozygous expansions of the FRDA1 gene. *Mov Disord* **19**, 1424–1431.

Barbot, C., Coutinho, P., Chorao, R., Ferreira, C., Barros, J., Fineza, I. et al. (2001). Recessive ataxia with ocular apraxia: Review of 22 Portuguese patients. *Arch Neurol* **58**, 201–205.

Bouchard, J. P., Barbeau, A., Bouchard, R., and Bouchard, R. W. (1978). Autosomal recessive spastic ataxia of Charlevoix-Saguenay. *Can J Neurol Sci* **5**, 61–69.

Breedveld, G. J., van Wetten, B., te Raa, G. D., Brusse, E., van Swieten, J. C., Oostra, B. A., and Maat-Kievit, J. A. (2004). A new locus for a childhood onset, slowly progressive autosomal recessive spinocerebellar ataxia maps to chromosome 11p15. *J Med Genet* **41**, 858–866.

Bulteau, A. L., O'Neill, H. A., Kennedy, M. C. et al. (2004). Frataxin acts as an iron chaperone protein to modulate mitochondrial aconitase activity. *Science* **305**, 242–245.

Buyse, G., Mertens, L., Di Salvo, G. et al. (2003). Idebenone treatment in Friedreich's ataxia: Neurological, cardiac, and biochemical monitoring. *Neurology* **60**, 1679–1681.

Campuzano, V., Montermini, L., Moltó, M. D. et al. (1996). Friedreich ataxia: Autosomal recessive disease caused by an intronic GAA triplet repeat expansion. *Science* **271**, 1423–1427.

Casazza, F., Morpurgo, M. (1996). The varying evolution of Friedreich's ataxia cardiomyopathy. *Am J Cardiol* **77**, 895–898.

Cavalier, L., Ouahchi, K., Kayden, H. J., Di Donato, S., Reutenauer, L., Mandel, J. L., and Koenig, M. (1998). Ataxia with isolated vitamin E deficiency: Heterogeneity of mutations and phenotypic variability in a large number of families. *Am J Hum Genet* **62**, 301–310.

Chamberlain, S., Shaw, J., Rowland, A., Wallis, J., South, S., Nakamura, Y. et al. (1988). Mapping of mutation causing Friedreich's ataxia to human chromosome 9. *Nature* **334**, 248–250.

Chantrel-Groussard, K., Geromel, V., Puccio, H. et al. (2001). Disabled early recruitment of antioxidant defenses in Friedreich's ataxia. *Hum Mol Genet* **10**, 2061–2067.

Charlevoix-Saguenay (2006). *J Neurol Neurosurg Psychiatry* **77**, 280–282.

Chen, O. S., Crisp, R. J., Valachovic, M. et al. (2004). Transcription of the yeast iron regulon does not respond directly to iron but rather to iron-sulfur cluster biosynthesis. *J Biol Chem* **279**, 29513–29518.

Chen, Y. Z., Bennett, C. L., Huynh, H. M., Blair, I. P., Puls, I., Irobi, J. et al. (2004). DNA/RNA helicase gene mutations in a form of juvenile amyotrophic lateral sclerosis (ALS4). *Am J Hum Genet* **74**, 1128–1135.

Christodoulou, K., Deymeer, F., Serdaroglu, P., Ozdemir, C., Poda, M., Georgiou, D. M. et al. (2001). Mapping of the second Friedreich's ataxia (FRDA2) locus to chromosome 9p23–p11: Evidence for further locus heterogeneity. *Neurogenetics* **3**, 127–132.

Copp, R. P., Wisniewski, T., Hentati, F., Larnaout, A., Ben Hamida, M., and Kayden, H. J. (1999). Localization of alpha-tocopherol transfer protein in the brains of patients with ataxia with vitamin E deficiency and other oxidative stress related neurodegenerative disorders. *Brain Res* **822**, 80–87.

Coppola, G., Choi, S. H., Santos, M. M. et al. (2006). Gene expression profiling in frataxin deficient mice: Microarray evidence for significant expression changes without detectable neurodegeneration. *Neurobiol Dis* **22**, 302–311.

Cossée, M., Dürr, A., Schmitt, M. et al. (1999). Frataxin point mutations and clinical presentation of compound heterozygous Friedreich ataxia patients. *Ann Neurol* **45**, 200–206.

Cossée, M., Puccio, H., Gansmuller, A. et al. (2000). Inactivation of the Friedreich ataxia mouse gene leads to early embryonic lethality without iron accumulation. *Hum Mol Genet* **9**, 1219–1226.

Criscuolo, C., Chessa, L., Di Giandomenico, S., Mancini, P., Sacca, F., Grieco, G. S. et al. (2006). Ataxia with oculomotor apraxia type 2: A clinical, pathologic, and genetic study. *Neurology* **66**, 1207–1210.

Criscuolo, C., Mancini, P., Sacca, F., De Michele, G., Monticelli, A., Santoro, L. et al. (2004). Ataxia with oculomotor apraxia type 1 in Southern Italy: Late onset and variable phenotype. *Neurology* **63**, 2173–2175.

Criscuolo, C., Sacca, F., De Michele, G. et al. (2005). Novel mutation of SACS gene in a Spanish family with autosomal recessive spastic ataxia. *Mov Disord* **20**,1358–1361.

Date, H., Onodera, O., Tanaka, H., Iwabuchi, K., Uekawa, K., Igarashi, S. et al. (2001). Early-onset ataxia with ocular motor apraxia and hypoalbuminemia is caused by mutations in a new HIT superfamily gene. *Nat Genet* **29**, 184–188.

De Braekeleer, M., Giasson, F., Mathieu, J., Roy, M., Bouchard, J. P., and Morgan, K. (1993). Genetic epidemiology of autosomal recessive spastic ataxia of Charlevoix-Saguenay in northeastern Quebec. *Genet Epidemiol* **10**, 17–25.

Dickie, M. M. (1965). Tumbler, tb. *Mouse News Lett.* **32**, 45.

Duquette, A., Roddier, K., McNabb-Baltar, J., Gosselin, I., St-Denis, A., Dicaire, M. J., et al. (2005). Mutations in senataxin responsible for Quebec cluster of ataxia with neuropathy. *Ann Neurol* **57**, 408–414.

Dürr, A., Cossée, M., Agid, Y. et al. (1996). Clinical and genetic abnormalities in patients with Friedreich's ataxia. *N Engl J Med* **335**, 1169–1175.

El Euch-Fayache, G., Lalani, I., Amouri, R. et al. (2003). Phenotypic features and genetic findings in sacsin-related autosomal recessive ataxia in Tunisia. *Arch Neurol* **60**, 982–988.

El-Khamisy, S. F., Saifi, G. M., Weinfeld, M., Johansson, F., Helleday, T., Lupski, J. R., and Caldecott, K. W. (2005). Defective DNA single-strand break repair in spinocerebellar ataxia with axonal neuropathy-1. *Nature* **434**, 108–113.

Emond, M., Lepage, G., and Vanasse, M. (2000). Increased levels of plasma malondialdehyde in Friedreich ataxia. *Neurology* **55**, 1752–1753.

Engert, J. C., Berube, P., Mercier, J., Dore, C., Lepage, P., Ge, B. et al. (2000). ARSACS, a spastic ataxia common in northeastern Quebec, is caused by mutations in a new gene encoding an 11.5-kb ORF. *Nat Genet* **24**, 120–125.

Fantus, I. G., Seni, M. H., and Andermann, E. (1993). Evidence for abnormal regulation of insulin receptors in Friedreich's ataxia. *J Clin Endocrinol Metab* **76**, 60–63.

Filla, A., De Michele, G., Caruso, G. et al. (1990). Genetic data and natural history of Friedreich's disease: A study of 80 Italian patients. *J Neurol* **237**, 345–351.

Filla, A., De Michele, G., Cavalcanti, F. et al. (1996). The relationship between trinucleotide (GAA) repeat length and clinical features in Friedreich ataxia. *Am J Hum Genet* **59**, 554–560.

Fleming, J., Spinoulas, A., Zheng, M. et al. (2005). Partial correction of sensitivity to oxidant stress in Friedreich ataxia patient fibroblasts by frataxin-encoding adeno-associated virus and lentivirus vectors. *Hum Gene Ther* **16**, 947–956.

Gabsi, S., Gouider-Khouja, N., Belal, S., Fki, M., Kefi, M., Turki, I. et al. (2001). Effect of vitamin E supplementation in patients with ataxia with vitamin E deficiency. *Eur J Neurol* **8**, 477–481.

Giacchetti, M., Monticelli, A., De Biase, I. et al. (2004). Mitochondrial DNA haplogroups influence the Friedreich's ataxia phenotype. *J Med Genet* **41**, 293–295.

Gilman, S., Junck, L., Markel, D.S. et al. (1990). Cerebral glucose hypermetabolism in Friedreich's ataxia detected with positron emission tomography. *Ann Neurol* **28**, 750–757.

Grieco, G. S., Malandrini, A., Comanducci, G. et al. (2004). Novel SACS mutations in autosomalrecessive spastic ataxia of Charlevoix-Saguenay type. *Neurology* **62**, 103–106.

Gueven, N., Becherel, O. J., Kijas, A. W., Chen, P., Howe, O., Rudolph, J. H. et al. (2004). Aprataxin, a novel protein that protects against genotoxic stress. *Hum Mol Genet* **13**, 1081–1093.

Hara, K., Onodera, O., Endo, M. et al. (2005). Sacsin-related autosomal recessive ataxia without prominent retinal myelinated fibers in Japan. *Mov Disord* **20**, 380–382.

Harding, A. E. (1981). Friedreich's ataxia: A clinical and genetic study of 90 families with an analysis of early diagnosis criteria and intrafamilial clustering of clinical features. *Brain* **104**, 589–620.

Herman, D., Jenssen, K., Burnett, R., Soragni, E., Perlman, S. L., Gottesfeld, J. M. (2006). Histone deacetylase inhibitors reverse gene silencing in Friedreich's ataxia. *Nat Chem Biol* **2**, 551–558.

Hosomi, A., Goto, K., Kondo, H., Iwatsubo, T., Yokota, T., Ogawa, M. et al. (1998). Localization of alpha-tocopherol transfer protein in rat brain. *Neurosci Lett* **256**, 159–162.

Hughes, J. T., Brownell, B., Hewer, R. L. (1968). The peripheral sensory pathway in Friedreich's ataxia: An examination by light and electron microscopy of the posterior nerve roots, posterior root ganglia, and peripheral sensory nerves in cases of Friedreich's ataxia. *Brain* **91**, 803–818.

Jiralerspong, S., Liu, Y., Montermini, L. et al. (1997). Frataxin shows developmentally regulated tissue-specific expression in the mouse embryo. *Neurobiol Dis* **4**, 103–113.

Jiralerspong, S., Ge, B., Hudson, T. J. et al. (2001). Manganese superoxide dismutase induction by iron is impaired in Friedreich ataxia cells. *FEBS Letters* **509**, 101–105.

Kaempf-Rotzoll, D. E., Horiguchi, M., Hashiguchi, K., Aoki, J., Tamai, H., Linderkamp, O., and Arai, H. (2003). Human placental trophoblast cells express alpha-tocopherol transfer protein. *Placenta* **24**, 439–444.

Kijas, A. W., Harris, J. L., Harris, J. M., and Lavin, M. F. (2006). Aprataxin forms a discrete branch in the HIT (histidine triad) superfamily of proteins with both DNA/RNA binding and nucleotide hydrolase activities. *J Biol Chem* **281**, 13939–13948.

Koeppen, A. H. (1998). The hereditary ataxias. *J Neuropathol Exp Neurol* **57**, 531–543.

Kohlschutter, A., Hubner, C., Jansen, W., and Lindner, S. G. (1988). A treatable familial neuromyopathy with vitamin E deficiency, normal absorption, and evidence of increased consumption of vitamin E. *J Inherit Metab Dis* **11 Suppl 2**, 149–152.

Koskinen, T., Santavuori, P., Sainio, K., Lappi, M., Kallio, A. K., and Pihko, H. (1994). Infantile onset spinocerebellar ataxia with sensory neuropathy: A new inherited disease. *J Neurol Sci* **121**, 50–56.

Kostrzewa, M., Klockgether, T., Damian, M. S., and Muller, U. (1997). Locus heterogeneity in Friedreich ataxia. *Neurogenetics* **1**, 43–47.

Koutnikova, H., Campuzano, V., Foury, F. et al. (1997). Studies of human, mouse and yeast homologues indicate a mitochondrial function for frataxin. *Nat Genet* **16**, 345–351.

Labuda, M., Labuda, D., Miranda, C. et al. (2000). Unique origin and specific ethnic distribution of the Friedreich ataxia GAA expansion. *Neurology* **54**, 2322–2324.

Lamarche, J. B., Shapcott, D., Côté, M. et al. (1993). Cardiac iron deposits in Friedreich's ataxia. In Lechtenberg, R., Ed., *Handbook of Cerebellar Diseases*, 453–458. New York: Marcel Dekker.

Le Ber, I., Bouslam, N., Rivaud-Pechoux, S., Guimaraes, J., Benomar, A., Chamayou, C. et al. (2004). Frequency and phenotypic spectrum of ataxia with oculomotor apraxia 2: A clinical and genetic study in 18 patients. *Brain* **127**, 759–767.

Le Ber, I., Moreira, M. C., Rivaud-Pechoux, S., Chamayou, C., Ochsner, F., Kuntzer, T. et al. (2003). Cerebellar ataxia with oculomotor apraxia type 1: Clinical and genetic studies. *Brain* **126**, 2761–2772.

Lodi, R., Cooper, J. M., Bradley, J. L. et al. (1999). Deficit of in vivo mitochondrial ATP production in patients with Friedreich ataxia. *Proc Natl Acad Sci USA* 96, 11492–11495.

Lynch, D. R., Farmer, J. M., Wilson, R. L., Balcer, L. J. (2005). Performance measures in Friedreich ataxia: Potential utility as clinical outcome tools. *Mov Disord* **20**, 777–782.

Mariotti, C., Gellera, C., Rimoldi, M., Mineri, R., Uziel, G., Zorzi, G. et al. (2004). Ataxia with isolated vitamin E deficiency: neurological phenotype, clinical follow-up and novel mutations in TTPA gene in Italian families. *Neurol Sci* **25**, 130–137.

Mariotti, C., Solari, A., Torta, D. et al. (2003). Idebenone treatment in Friedreich patients: One-year-long randomized placebo-controlled trial. *Neurology* **60**, 1676–1679.

Miranda, C. J., Santos, M. M., Ohshima, K., et al. (2002). Frataxin knockin mouse. *FEBS Lett* **512**, 291–297.

Montermini, L., Richter, A., Morgan, K. et al. (1997). Phenotypic variability in Friedreich ataxia: Role of the associated GAA triplet repeat expansion. *Ann Neurol* **41**, 675–682.

Moreira, M. C., Barbot, C., Tachi, N., Kozuka, N., Uchida, E., Gibson, T. et al. (2001). The gene mutated in ataxia-ocular apraxia 1 encodes the new HIT/Zn-finger protein aprataxin. *Nat Genet* **29**, 189–193.

Moreira, M. C., Klur, S., Watanabe, M., Nemeth, A. H., Le Ber, I., Moniz, J. C. et al. (2004). Senataxin, the ortholog of a yeast RNA helicase, is mutant in ataxia-ocular apraxia 2. *Nat Genet* **36**, 225–227.

Morley, S., Panagabko, C., Shineman, D., Mani, B., Stocker, A., Atkinson, J., and Manor, D. (2004). Molecular determinants of heritable vitamin E deficiency. *Biochemistry* **43**, 4143–4149.

Morvan, D., Komajda, M., Doan, L. D. et al. (1992). Cardiomyopathy in Friedreich's ataxia: A Doppler-echocardiographic study. *Eur Heart J* **3**, 1393–1398.

Mrissa, N., Belal, S., Hamida, C. B. et al. (2000). Linkage to chromosome 13q11–12 of an autosomal recessive cerebellar ataxia in a Tunisian family. *Neurology* **54**, 1408–1414.

Muhlenhoff, U., Gerber, J., Richhardt, N. et al. (2003). Components involved in assembly and dislocation of iron-sulfur clusters on the scaffold protein Isu1p. *EMBO J* **22**, 4815–4825.

Muller-Schmehl, K., Beninde, J., Finckh, B., Florian, S., Dudenhausen, J. W., Brigelius-Flohe, R., and Schuelke, M. (2004). Localization of alpha-tocopherol transfer protein in trophoblast, fetal capillaries' endothelium and amnion epithelium of human term placenta. *Free Radic Res* **38**, 413–420.

Nemeth, A. H., Bochukova, E., Dunne, E., Huson, S. M., Elston, J., Hannan, M. A. et al. (2000). Autosomal recessive cerebellar ataxia with oculomotor apraxia (ataxia-telangiectasia-like syndrome) is linked to chromosome 9q34. *Am J Hum Genet* **67**, 1320–1326.

Nikali, K., Suomalainen, A., Terwilliger, J., Koskinen, T., Weissenbach, J., and Peltonen, L. (1995). Random search for shared chromosomal regions in four affected individuals: The assignment of a new hereditary ataxia locus. *Am J Hum Genet* **56**, 1088–1095.

Ogawa, T., Takiyama, Y., Sakoe, K., et al. (2004). Identification of a SACS gene missense mutation in ARSACS. *Neurology* **62**, 107–109.

Okawa, S., Sugawara, M., Watanabe, S., Imota, T., Toyoshima, I. (2006). A novel sacsin mutation in a Japanese woman showing clinical uniformity of autosomal recessive spastic ataxia of Charlevoix-Saguenay. *J Neurol Neurosurg Psychiatry* **77**, 280–282.

Ohshima, K., Montermini, L., Wells, R. D. et al. (1998). Inhibitory effects of expanded GAA•TTC triplet repeats from intron 1 of Friedreich's ataxia gene on transcription and replication in vivo. *J Biol Chem* **273**, 14588–14595.

Ouahchi, K., Arita, M., Kayden, H., Hentati, F., Ben Hamida, M., Sokol, R. et al. (1995). Ataxia with isolated vitamin E deficiency is caused by mutations in the alpha-tocopherol transfer protein. *Nat Genet* **9**, 141–145.

Ouyang, Y., Takiyama, Y., Sakoe, K., Shimazaki, H., Ogawa, T., Nagano, S. et al. (2006). Sacsin-related ataxia (ARSACS): Expanding the genotype upstream from the gigantic exon. *Neurology* **66**, 1103–1104.

Park, S., Gakh, O., O'Neill, H. A. et al. (2003). Yeast frataxin sequentially chaperones and stores iron by coupling protein assembly with iron oxidation. *J Biol Chem* **278**, 31340–31351.

Pasternac, A., Krol, R., Petitclerc, R. et al. (1980). Hypertrophic cardiomyopathy in Friedreich's ataxia: Symmetric or asymmetric? *Can J Neurol Sci* **7**, 379–382.

Peyronnard, J. M., Bouchard, J. P., Lapointe, M. (1976). Nerve conduction studies and electromyography in Friedreich's ataxia. *Can J Neurol Sci* **3**, 313–317.

Pianese, L., Busino, L., De Biase, I. et al. (2002). Up-regulation of c-Jun N-terminal kinase pathway in Friedreich's ataxia cells. *Hum Mol Genet* **11**, 2989–2996.

Puccio, H., Simon, D., Cossee, M. et al. (2001). Mouse models for Friedreich ataxia exhibit cardiomyopathy, sensory nerve defect and Fe-S enzyme deficiency followed by intramitochondrial iron deposits. *Nat Genet* **27**, 181–618.

Radisky, D. C., Babcock, M. C., Kaplan, J. (1999). The yeast frataxin homologue mediates mitochondrial iron efflux. Evidence for a mitochondrial iron cycle. *J Biol Chem* **274**, 4497–4499.

Richter, A. M., Ozgul, R. K., Poisson, V. C., Topaloglu, H. (2004). Private SACS mutations in autosomal recessive spastic ataxia of Charlevoix-Saguenay (ARSACS) families from Turkey. *Neurogenetics* **5**, 165–170.

Richter, A., Rioux, J. D., Bouchard, J. P., Mercier, J., Mathieu, J., Ge, B., et al. (1999). Location score and haplotype analyses of the locus for autosomal recessive spastic ataxia of Charlevoix-Saguenay, in chromosome region 13q11. *Am J Hum Genet* **64**, 768–775.

Romeo, G., Menozzi, P., Ferlini, A. et al. (1983). Incidence of Friedreich ataxia in Italy estimated from consanguinous marriages. *Am J Hum Genet* **35**, 523–529.

Rötig, A., deLonlay, P., Chretien, D. et al. (1997). Frataxin gene expansion causes aconitase and mitochondrial iron-sulfur protein deficiency in Friedreich ataxia. *Nature Genet* **17**, 215–217.

Rouault, T. A., Tong, W. H. (2005). Iron-sulphur cluster biogenesis and mitochondrial iron homeostasis. *Nat Rev Mol Cell Biol* **6**, 345–351.

Said, G., Marion, M. H., Selva, J. et al. (1986). Hypotrophic and dying-back nerve fibers in Friedreich's ataxia. *Neurology* **36**, 1292–1299.

Sakamoto, N., Chastain, P. D., Parniewski, P. et al. (1999). Sticky DNA: Self-association properties of long GAA•TTC repeats in R•R•Y triplex structures from Friedreich ataxia. *Mol Cell* **3**, 465–475.

Sano, Y., Date, H., Igarashi, S., Onodera, O., Oyake, M., Takahashi, T. et al. (2004). Aprataxin, the causative protein for EAOH is a nuclear protein with a potential role as a DNA repair protein. *Ann Neurol* **55**, 241–249.

Santoro, L., De Michele, G., Perretti, A. et al. (1999). Relation between trinucleotide GAA repeat length and sensory neuropathy in Friedreich's ataxia. *J Neurol Neurosurg Psychiatry* **66**, 99–96.

Santoro, L., Perretti, A., Lanzillo, B. et al. (2000). Influence of GAA expansion size and disease duration on central nervous system impairment in Friedreich's ataxia: Contribution to the understanding of the pathophysiology of the disease. *Clin Neurophysiol* **111**, 1023–1030.

Santos, M., Ohshima, K., Pandolfo, M. (2001). Frataxin deficiency enhances apoptosis in cells differentiating into neuroectoderm. *Hum Mol Genet* **10**, 1935–1944.

Sato, Y., Arai, H., Miyata, A., Tokita, S., Yamamoto, K., Tanabe, T., and Inoue, K. (1993). Primary structure of alpha-tocopherol transfer protein from rat liver. Homology with cellular retinaldehyde-binding protein. *J Biol Chem* **268**, 17705–17710.

Saveliev, A., Everett, C., Sharpe, T., Webster, Z., Festenstein, R. (2003). DNA triplet repeats mediate heterochromatin-protein-1-sensitive variegated gene silencing. *Nature* **422**, 909–913.

Schoenle, E. J., Boltshauser, E. J., Baekkeskov, S. et al. (1989). Preclinical and manifest diabetes mellitus in young patients with Friedreich's ataxia: No evidence of immune process behind the islet cell destruction. *Diabetologia* **32**, 378–381.

Schulz, J. B., Dehmer, T., Schöls, L. et al. (2000). Oxidative stress in patients with Friedreich ataxia. *Neurology* **55**, 1719–1721.

Seznec, H., Simon, D., Bouton, C. et al. (2005). Friedreich ataxia: The oxidative stress paradox. *Hum Mol Genet* **14**, 463–474.

Seznec, H., Simon, D., Monassier, L., et al. (2004). Idebenone delays the onset of cardiac functional alteration without correction of Fe-S enzymes deficit in a mouse model for Friedreich ataxia. *Hum Mol Genet* **13**, 1017–1024. Epub 2004 Mar 17.

Shimazaki, H., Takiyama, Y., Sakoe, K., Ando, Y., Nakano, I. (2005). A phenotype without spasticity in sacsin-related ataxia. *Neurology* **64**, 2129–2131.

Simon, D., Seznec, H., Gansmuller, A. et al. (2004). Friedreich ataxia mouse models with progressive cerebellar and sensory ataxia reveal autophagic neurodegeneration in dorsal root ganglia. *J Neurosci* **24**, 1987–1995.

Subramony, S. H., May, W., Lynch, D. et al. (2005). Measuring Friedreich ataxia: Interrater reliability of a neurologic rating scale. *Neurology* **64**, 1261–1262.

Takashima, H., Boerkoel, C. F., John, J., Saifi, G. M., Salih, M. A., Armstrong, D. et al. (2002). Mutation of TDP1, encoding a topoisomerase I-dependent DNA damage repair enzyme, in spinocerebellar ataxia with axonal neuropathy. *Nat Genet* **32**, 267–272.

Terasawa, Y., Ladha, Z., Leonard, S. W., Morrow, J. D., Newland, D., Sanan, D. et al. (2000). Increased atherosclerosis in hyperlipidemic mice deficient in alpha-tocopherol transfer protein and vitamin E. *Proc Natl Acad Sci U S A* **97**, 13830–13834.

Thiffault, I., Rioux, M. F., Tetreault, M., Jarry, J., Loiselle, L., Poirier, J. et al. (2006). A new autosomal recessive spastic ataxia associated with frequent white matter changes maps to 2q33-34. *Brain* **129**, 2332–2340.

Thomas, P. K., Workman, J. M., and Thage, O. (1984). Behr's syndrome. A family exhibiting pseudodominant inheritance. *J Neurol Sci* **64**, 137–148.

Tozzi, G., Nuccetelli, M., Lo Bello, M. et al. (2002). Antioxidant enzymes in blood of patients with Friedreich's ataxia. *Arch Dis Child* **86**, 376–379.

Traber, M. G., Sokol, R. J., Burton, G. W., Ingold, K. U., Papas, A. M., Huffaker, J. E., and Kayden, H. J. (1990). Impaired ability of patients with familial isolated vitamin E deficiency to incorporate alpha-tocopherol into lipoproteins secreted by the liver. *J Clin Invest* **85**, 397–407.

Tranebjaerg, L., Teslovich, T. M., Jones, M., Barmada, M. M., Fagerheim, T., Dahl, A. et al. (2003). Genome-wide homozygosity mapping localizes a gene for autosomal recessive non-progressive infantile ataxia to 20q11-q13. *Hum Genet* **113**, 293–295.

Trouillas, P., Takayanagi, T., Hallett, M., et al. (1997). International Cooperative Ataxia Rating Scale for pharmacological assessment of the cerebellar syndrome. The Ataxia Neuropharmacology Committee of the World Federation of Neurology. *J Neurol Sci* **145**, 205–211.

Vanasse, M., Garcia-Larrea, L., Neuschwander, P. (1988). Evoked potential studies in Friedreich's ataxia and progressive early onset cerebellar ataxia. *Can J Neurol Sci* **15**, 292–298.

Vetcher, A. A., Napierala, M., Iyer, R. R., et al. (2002). Sticky DNA, a long GAA.GAA.TTC triplex that is formed intramolecularly, in the sequence of intron 1 of the frataxin gene. *J Biol Chem* **277**, 39217–39227.

Watanabe, M., Sugai, Y., Concannon, P., Koenig, M., Schmitt, M., Sato, M. et al. (1998). Familial spinocerebellar ataxia with cerebellar atrophy, peripheral neuropathy, and elevated level of serum creatine kinase, gamma-globulin, and alpha-fetoprotein. *Ann Neurol* **44**, 265–269.

Whitehouse, C. J., Taylor, R. M., Thistlethwaite, A., Zhang, H., Karimi-Busheri, F., Lasko, D. D. et al. (2001). XRCC1 stimulates human polynucleotide kinase activity at damaged DNA termini and accelerates DNA single-strand break repair. *Cell* **104**, 107–117.

Wong, A., Yang, J., Cavadini, P. et al. (1999). The Friedreich ataxia mutation confers cellular sensitivity to oxidant stress which is rescued by chelators of iron and calcium and inhibitors of apoptosis. *Hum Mol Genet* **8**, 425–430.

Wullner, U., Klockgether, T., Petersen, D. et al. (1993). Magnetic resonance imaging in hereditary and idiopathic ataxia [see comments]. *Neurology* **43**, 318–325.

Yamamoto, Y., Hiraoka, K., Araki, M. et al. (2005). Novel compound heterozygous mutations in sacsin-related ataxia. *J Neurol Sci* **239**, 101–104.

Yamamoto, Y., Nakamori, M., Konaka, K. et al. (2006). Sacsin-related ataxia caused by the novel nonsense mutation Arg4325X. *J Neurol.*

Yokota, T., Shiojiri, T., Gotoda, T., Arita, M., Arai, H., Ohga, T., Kanda, T., Suzuki, J. et al. (1997). Friedreich-like ataxia with retinitis pigmentosa caused by the His101Gln mutation of the alpha-tocopherol transfer protein gene. *Ann Neurol* **41**, 826–832.

Yoon, T., Cowan, J. A. (2003). Iron-sulfur cluster biosynthesis. Characterization of frataxin as an iron donor for assembly of [2Fe-2S] clusters in ISU-type proteins. *J Am Chem Soc* **125**, 6078–6084.

Yoon, T., Cowan, J. A. (2004). Frataxin-mediated iron delivery to ferrochelatase in the final step of heme biosynthesis. *J Biol Chem* **279**, 25943–25946.

Zouari, M., Feki, M., Ben Hamida, C., Larnaout, A., Turki, I., Belal, S. et al. (1998). Electrophysiology and nerve biopsy: Comparative study in Friedreich's ataxia and Friedreich's ataxia phenotype with vitamin E deficiency. *Neuromuscul Disord* **8**, 416–425.

19

DYT1, An Inherited Dystonia

Susan B. Bressman and Laurie Ozelius

Dystonia or torsion dystonia refers to repetitive, often sustained muscle contractions that cause directional or twisting movements and postures (Fahn, 1988). There are many causes for dystonia and various classification schemes have been employed to help organize the diverse etiologies (see Table 19.1a). One classification (see Table 19.1b) proposes two main etiologic categories (Bressman, 2003): primary dystonia and secondary dystonia. Primary dystonia is defined as a syndrome in which dystonia is the only or predominant clinical manifestation, and there is no evidence of neuronal degeneration or an acquired cause. Secondary dystonia consists of inherited disorders that typically produce clinical signs in addition to dystonia, degenerative disorders with complex or unknown etiologies, and acquired causes.

The clinical spectrum of primary dystonia is remarkably broad. Symptoms may begin at any age from early-childhood to senescence and severity ranges from involvement of a single muscle to generalized contractions of the limb, axial, and cranial muscles. Age at onset distribution for primary dystonia is bimodal, with modes at age 9 (early-onset) and 45 (late-onset), divided by a nadir at age 27 (Bressman, 1989). Further, there is a relationship between the age at onset of symptoms, body region first affected, and clinical progression of signs (Fahn, 1987; Marsden, 1974; O'Dwyer, 2005). When primary dystonia begins in childhood or adolescence, it often starts in a leg or arm, and then progresses over five to 10 years to involve multiple body regions. When it begins in adult years, symptoms first involve the neck (cervical dystonia or torticollis), arm (writer's cramp), or cranial muscles (e.g., blepharospasm). Unlike early-onset, adult or late-onset dystonia tends to remain localized with a focal or segmental anatomic distribution.

Primary dystonia is considered the third most common movement disorder after Parkinson disease and essential tremor. In Rochester, Minnesota, prevalence was 330 per million, with late-onset focal disease being nine times more common than early-onset generalized (Nutt, 1988). A more recent analysis of European cases found a lower frequency of about 152 per million, and again focal cases constituted the majority (117 per million) (ESDE Collaborative Group, 2000). Both of these clinically based studies are likely underestimates of the true frequency, because a significant proportion of disease is not diagnosed (Muller, 2002; Risch,

Table 19.1a. Classification of Dystonia

By Age Onset
- Early (< 26)
- Late

By Distribution
- Focal (single body region)
- Segmental (contiguous regions)
- Multifocal (non-contiguous regions)
  - Hemi (a type of multifocal=ipsilateral arm and leg)
- Generalized (leg+trunk+one other region or both legs +/− trunk+one other region)

By Cause
- Primary - dystonia is only sign, except tremor, and no acquired/exogenous cause or degenerative disorder
- Secondary
  - due to inherited and/or degenerative disorders – signs other than dystonia and/or brain degeneration distinguish from primary dystonia
  - due to acquired or exogenous causes

Table 19.1b. Causes of Dystonia

- Primary
  - Early limb onset (DYT1, other genes to be determined)
  - Mixed phenotype (DYT6, DYT13, other genes to be determined)
  - Late Focal (DYT7, other genes to be determined)
- Secondary Dystonia
  - Dystonia Plus
    - Dopa-Responsive Dystonia (DYT5, DYT14, other biopterin deficiencies)
    - Myoclonus – Dystonia (DYT11, DYT15)
    - Rapid-Onset Dystonia Parkinsonism (DYT12)
  - Heredodegenerative
    - Autosomal Dominant (eg, Huntington's disease, SCAs especially SCA3)
    - Autosomal Recessive (eg, Wilson's, NBIA1, GM1 and GM2, Parkin)
    - X-linked (eg, X-linked Dystonia-Parkinsonism/Lubag, DDP)
    - Mitochondrial
    - Complex/unknown
    - Parkinson's disease, multisystem atrophy, progressive supranuclear palsy
  - Acquired
    - drug-induced, perinatal injury, head trauma, cervical trauma, peripheral trauma, infectious and post infectious, tumor, stroke, multiple sclerosis

1995). Ethnic differences in disease frequency for the various primary phenotypic subtypes have been reported and are known or suspected to be due to founder effects. Several large studies, including the earliest epidemiological studies, reported that early-onset primary dystonia is five to 10 times more common in Ashkenazi Jews than in non-Jews or non-Ashkenazi Jews (Zeman, 1967; Zilber, 1984). Prevalence estimates of focal dystonia also describe ethnic differences, ranging from 6.1/100,000 in Yonago, Japan (Nakashima, 1995) to 225 per 100,000 in Bruneck, Italy (Muller, 2002).

Although pathology involving the basal ganglia and defective dopaminergic transmission is present in many secondary dystonias, consistent diagnostic morphological or biochemical brain changes have not been found in primary dystonia (Zeman, 1967, 1970; Zweig, 1988). Several recent studies utilizing a variety of techniques, including imaging and novel antibodies, have demonstrated abnormalities in primary dystonia subtypes. These include MRI voxel-based morphometric increases in gray matter of primary somatosensory and motor cortices in the patients with focal hand dystonia (Garraux, 2004), an increase in striatal 3-,4-dihydroxyphenylacetic acid/dopamine ratio of early-onset brains with mutations in DYT1 (see later) (Augood, 2002), increased copper and deficient Menkes protein in the lentiform nucleus of three patients with adult onset focal dystonia (Becker, 1999; Berg, 2001), and perinuclear inclusion bodies in the midbrain reticular formation and periaqueductal in four DYT1 brains (McNaught, 2004). The latter studies did not find evidence for neuronal loss, but the number of brains studied is small. Thus, primary dystonia usually is classified as a *non*degenerative disorder, although current studies suggest anatomic changes that are associated with neuronal dysfunction. As this field evolves, changes that are specific to primary dystonia clinical or genetic subtypes may emerge.

There is more known about the neurophysiological correlates of dystonia. A growing body of evidence suggests loss of normal inhibitory mechanisms resulting in excessive movement. Beginning with early studies showing deficient reciprocal inhibition at the spinal level (Rothwell, 1983), a host of spinal and brainstem reflexes have been studied and a common finding is that inhibitory processes are reduced (Hallett, 2004). Other studies show hyperexcitability of the motor cortex and loss of inhibition in cortical processing. One hypothesis to help explain these findings is impairment of the normal surround inhibition that is mediated by the basal ganglia (Mink, 1996). This could occur secondary to many different perturbations, including dopaminergic dysfunction, especially involving the D2 receptor and indirect pathway (Perlmutter, 2004). Supporting this explanation are positron emission tomography (PET) studies showing decreased D2 binding (Asanuma, 2005; Perlmutter, 1997). Sensory abnormalities also are found in primary dystoniam including disorganized cortical representation in patients with focal hand dystonia (Bara-Jimenez, 1998), abnormal kinesthesia (Grunewald, 1997), and abnormal spatial or temporospatial sensory discrimination (Sanger, 2001). The import of these sensory abnormalities in pathogenesis is unclear but may also be related to basal ganglia dysfunction in integrating sensory input and motor output for learned motor acts (Kaji, 2001, 2004).

There are few effective treatments for primary dystonia and therapeutic response is seldom complete (Bressman, 1990, 2000b; Greene, 1988). Current treatments for dystonia include anticholinergic medications, botulinum toxin injections, and surgical interventions such as pallidal (deep brain) stimulation. Botulinum toxin is particularly effective for focal dystonias but injections need to be regularly repeated. Deep brain stimulation of the globus pallidus appears to control dystonia effectively in many disabled patients with primary generalized dystonia. This form of treatment is relatively new and efficacy over time is still being studied, but brings great hope to this group of patients. Future treatments under study include approaches based on evolving insight into genetic subtypes and mechanisms (see later).

I. Early-Onset Primary Dystonia and Identifying DYT1

Childhood and adolescent-onset PTD (also known as dystonia musculorum deformans or Oppenheim's disease) is transmitted in an autosomal dominant fashion with reduced penetrance of 30 to 40 percent and as stated earlier, is more common in Ashkenazi Jews. Because of this reduced penetrance, large multiplex families with this phenotype are uncommon. One such large North American non-Jewish family with 13 affected members was ascertained, and this allowed for mapping of the first primary dystonia locus (DYT1) to chromosome 9q32-34 (Ozelius, 1989). Subsequently linkage to the same 9q region in clinically similar Ashkenazi and non-Jewish families was found (Kramer, 1990, 1994). Linkage disequilibrium was then noted among Ashkenazim, with sharing of a common haplotype of 9q alleles at marker loci spanning about 2 cM (Bressman, 1994a; Ozelius, 1992; Risch, 1995).

The finding of linkage disequilibrium supports the idea that a single mutational event is responsible for most cases of early onset PTD in the Ashkenazi population. The presence of very strong linkage disequilibrium at a relatively large genetic distance of about 2 cM also suggests that the mutation is recent. From this haplotype data, Risch and colleagues calculated that the mutation was introduced into the Ashkenazi population about 350 years ago and probably originated in Lithuania or Byelorussia (Risch, 1995). They also argued that the current high prevalence of the disease in Ashkenazim (estimated to be about 1:3,000–1:9,000 with a gene frequency of about 1:2,000–1:6,000) is due to the tremendous growth of that population in the eighteenth century from a small reproducing founder population (Risch, 1995). A founder mutation and genetic drift (changes in gene frequency due to chance events such as migrations, population expansions), rather than a heterozygote advantage (i.e., nonpenetrant DYT1 carriers have some advantage that leads to carriers being more prevalent), is probably responsible for the high frequency of DYT1 dystonia in Ashkenazim.

Using linkage disequilibrium and recombinations among Ashkenazi families to limit the candidate region, Ozelius identified the DYT1 gene in 1997 (Ozelius, 1997). An in-frame GAG deletion was identified in the coding sequence of one of four genes within the region (see Figure 19.1). This deletion originally was found in both Ashkenazi and North American non-Jewish PTD families (Ozelius, 1997) and subsequently identified in families of diverse ethnic background (Ikeuchi, 1999; Lebre, 1999; Major, 2001; Slominsky, 1999; Valente, 1998). Analyses of haplotypes indicate that current deletions in the non-Ashkenazi population originated from multiple independent mutation events, including *de novo* mutations (Klein, 1998). In contrast, among the great majority of Ashkenazim, the GAG deletion derives from the same founder mutation. The reason for the GAG deletion's singular disease-causing status is not known but it is hypothesized that genetic instability due to an imperfect tandem 24 bp repeat in the region of the deletion may lead to an increased frequency of the mutation (Klein, 1998).

Despite extensive screening (Leung, 2001; Ozelius, 1999; Tuffery-Giraud, 2001), the GAG deletion is the only definitive DYT1 disease mutation identified to date. Three other variations in torsinA have been found that change the amino acid sequence and none have been unequivocally associated with disease. An 18 bp deletion causes loss of residues 323–328 (Leung, 2001). This deletion was found in a family that included affected individuals with both myoclonus and dystonia who subsequently were found to have a mutation in the epsilon-sarcoglycan gene (Klein, 2002), thus casting doubt on whether it contributed to disease. A 4 bp deletion causes a frameshift and truncation starting at residue 312 and was found in a single control blood donor who was not examined neurologically (Kabakci, 2004). Finally, a polymorphism in the coding sequence for residue 216 encodes aspartic acid

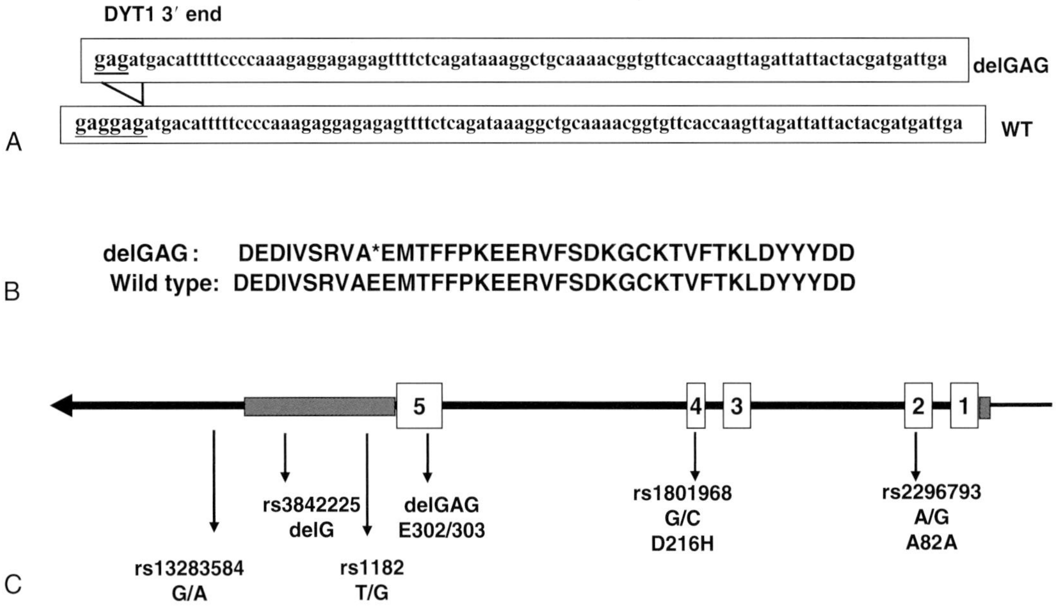

Figure 19.1 Deletion mutation and other single nucleotide polymorphisms (SNPs) in DYT1/torsinA. **A**. DNA sequence from exon 5 of the DYT1 gene showing the GAG deleted and the normal sequence. **B**. The carboxy terminal 40 amino acid sequence of the GAG deleted torsinA and the normal torsinA. The * represents the deleted glutamic acid (E) amino acid. **C**. A schematic (not drawn to scale) of the exon:intron structure of DYT1 showing the position of the GAG deletion in exon 5 as well as several SNPs used in association and functional studies described in the text. Figure is adapted from Leung (2001).

in 88 percent and histidine in 12 percent of alleles in control populations (Ozelius, 1997). Its potential disease modifying effects are discussed later.

II. Gene and Protein Properties

The *DYT1* (also known as *TOR1A*) cDNA is 998 bp long with two poly-A addition sites in the 3′ untranslated region. As a consequence of the two poly-A sites, two ubiquitously expressed messages of size 1.8 kb and 2.2 kb are seen by Northern analysis (Ozelius, 1997). Sequence analysis of the human genome reveals three other genes that are highly homologous to *DYT1*: *TOR1B*, *TOR2A*, and *TOR3A*. At both the DNA and protein level, *TOR1B* is 70 percent identical to *DYT1*. They each have five exons and their splice sites are conserved. The genes are located adjacent to each other in a tail-to-tail orientation on chromosome 9q34 and presumably arose from a tandem duplication of an evolutionary precursor gene (Ozelius, 1999). *TOR2A* and *TOR3A* share about 50 percent homology with *DYT1* at the amino acid level (Dron, 2002; Ozelius, 1999; unpublished results). *TOR2A* is located on chromosome 9q34 about 10 cM centromeric to *DYT1* and *TOR1B*, has five exons, and encodes a protein of 321 amino acids. *TOR3A* was independently cloned by virtue of transcriptional regulation in response to alpha-interferon and given the alternative name *ADIR1* (ATP-dependent interferon responsive gene) (Dron, 2002). It is located on chromosome 1q24 and has alternative splicing of exon 6 resulting in two protein products of 397 amino acids or 336 amino acids (*ADIR2*) (Dron, 2002). By Northern analysis *TOR1B*, *TOR2A*, and *TOR3A* are all ubiquitously expressed (Dron, 2002; Ozelius, 1997; unpublished data). Genomic database searches have revealed torsin-like genes, in mouse, rat, nematode, fruit fly, pig, cow, zebrafish, chicken, hamster, and Xenopus (Ozelius, 1999).

The protein encoded by the *DYT1* gene is called torsinA. It is 332 amino acids long (~37kD), with a signal sequence and membrane-spanning region in the N-terminus as well as a glycosylation site and putative phosphorylation sites (Ozelius, 1997). The GAG deletion in the *DYT1* gene results in the loss of one of a pair of glutamic acid residues in the C-terminal region of the protein (Ozelius, 1997). TorsinA is a member of a superfamily of ATPases associated with a variety of cellular activities (AAA^+) (Lupas, 1997; Neuwald, 1999; Ozelius, 1997). These proteins typically form six-membered homomeric ring structures, possess Mg++-dependent ATPase activity, and share a secondary structure (Neuwald, 1999). This superfamily of chaperone proteins mediate conformational changes in target proteins and perform a variety of functions, including correct folding of nascent proteins, degradation of denatured proteins, cytoskeletal dynamics, membrane trafficking, vesicle fusion, and organelle movement (Hanson, 2005; Vale, 2000). Studies carried out both *in vivo* and *in vitro* and documented later suggest several of these are plausible functions for torsinA.

TorsinA is widely expressed in most cells in the body. Examination of torsinA in the normal adult brain reveals that it is widely distributed throughout the brain with intense expression in substantia nigra dopamine neurons, cerebellar Purkinje cells, the thalamus, globus pallidus, hippocampal formation, and cerebral cortex (Augood, 1998, 1999, 2003; Konakova, 2001a; Shashidharan, 2000a). Both the mRNA and protein are localized to neurons and not to glia, whereas the protein studies also showed torsinA in neuronal processes. Labeling was predominantly present in cytoplasm with some perinuclear staining (Konakova, 2001a; Shashidharan, 2000a). A similar widespread pattern of expression was seen in both rat (Shashidharan, 2000a; Walker, 2001) and mouse (Konakova, 2001b) brains. Ultrastructural studies in human adult and macaque striatum revealed torsinA immunostaining of small vesicles in the presynaptic terminals consistent with a role in modulating striatal signaling (Augood, 2003).

Developmental expression of torsinA also has been explored in human, mouse, and rat brains. In humans, torsinA immunoreactivity is first seen between four and eight weeks postnatal in the four regions tested—cerebellum, substantia nigra, hippocampus, and basal ganglia (Siegert, 2005). In both mice and rats, torsinA is most highly expressed during prenatal and early postnatal development (Vasudevan, 2006; Xiao, 2004). Significant regional differences were noted with the highest level of expression in the cerebral cortex from embryonic day 15 (E-15)-E17, in the striatum from E17-P7, from P0-P7 in the thalamus and from P7-P14 in the cerebellum (Vasudevan, 2006; Xiao, 2004).

Finally, two studies have further implicated torsinA in dopamine transmission with the finding of torsinA and alpha-synuclein immunoreactivity colocalized in Lewy bodies (Sharma, 2001; Shashidharan, 2000b).

III. Neuropathology

Early studies on brains from patients with dystonia report no consistent neuropathological changes (Hedreen, 1988; Zeman, 1970). However, with the identification of the *DYT1* gene, researchers have now been able to test dystonia brains for the 3 bp deletion and identify *DYT1* specific brains. Studying a single *DYT1* brain, nigral cellularity was normal as was the striatal dopamine and homovanillic acid levels except in the rostal portions of the putamen and caudate nucleus where they were slightly decreased compared to controls (Furukawa, 2000). Although this suggests that the *DYT1* mutation is not associated with significant damage to the nigrostriatal dopaminergic system, a second study involving four *DYT1* brains noted an increase in the ratio of dopamine metabolites to dopamine when compared to controls (Augood, 2002), and studies in various *DYT1* mouse models implicate the dopaminergic system (see later) (Dang, 2005, 2006; Shashidharan, 2005). At the protein level, several studies have found no differences in the immunostaining pattern of torsinA between DYT1- positive, DYT1-negative, and control brains (Rostasy, 2003; Walker, 2002). However, comparing DYT1-positive and DYT1-negative brains to controls, enlarged and closely spaced nigral dopaminergic neurons were identified in DYT1-positive dystonia brains as compared to controls. No evidence was found for neuronal loss, suggesting a functional rather than degenerative etiology (Rostasy, 2003). A further study examining four DYT1-positive brains found ubiquitin positive perinuclear inclusions in the midbrain reticular formation and the periaqueductal gray but not in the substantia nigra, striatum, hippocampus, or select regions of the cerebral cortex (McNaught, 2004). Similar inclusions have been reported in several *DYT1* mouse models (Dang, 2005; Shashidharan, 2005).

IV. Cellular and Animal Models of Disease

In Vitro

Cellular studies on torsinA indicate that the majority of this protein is localized in the lumen of the endoplasmic reticulum (ER) consistent with the deduced signal sequence and the observed high mannose content (Hewett, 2000; Kustedjo, 2000). These and other studies suggest torsinA is associated with the ER membrane through its hydrophobic N-terminal region (Hewett, 2000; Kustedjo, 2000, 2003; Liu, 2003). However, a recent report determined that torsinA is associated peripherally with the ER membrane possibly through an interaction with an integral membrane protein (Callan, 2006). Both the 3 bp deletion in torsinA found in *DYT1* patients and the introduction of an E171Q mutation in the ATP binding domain, which does not allow ATP hydrolysis, lead to a striking redistribution of torsinA to the nuclear envelope (NE). This enrichment presumably is due to prolonged interaction of torsinA with substrate(s) at the NE (Goodchild, 2004; Naismith, 2004). Possible substrates include LAP1, a NE localized protein that interacts to a greater extent with mutant torsinA than wild-type protein; and LULL1, a novel lumenal ER membrane protein that is related to LAP1 (Goodchild, 2005). Expression of mutant torsinA is associated with apparent thickening and abnormal morphology of the NE including altered connections between the inner and outer membranes, as well as generation of whorled membrane inclusions that appear to "spin off" the ER/NE (Bragg, 2004; Gonzalez-Alegre, 2004; Goodchild, 2004; Naismith, 2004). Immunoreactive for VMAT2, a protein important for bioactive monoamines in neurons, is associated with the membrane inclusions again relating torsinA to dopamine (Misbahuddin, 2005).

Expression studies using an ADIR-EGFP (*TOR3A*) fusion protein expressed in HeLa cells also showed an association with the ER suggesting that other torsin-family members may share similar functions (Dron, 2002).

TorsinA also is found associated with neurite varicosities and vesicles, and extends to the ends of processes (Ferrari-Toninelli, 2004; Hewett, 2000). This is supported by the localization of torsin in neuronal processes and at synaptic endings in association with vesicles in human and nonhuman primate brain (Augood, 2003). Consistent with a role for torsinA in intracellular trafficking and association with the cytoskeleton, torsinA has been shown to interact with the kinesin light chain 1 (*KLC1*) (Kamm, 2004) as well as vimentin (*VIM*) (Hewett, 2006) and to regulate the cellular trafficking of the dopamine transporter and other polytopic membrane bound proteins (Torres, 2004).

Several different groups have looked at the role of torsinA in cellular stress. These include studies showing that overexpression of wild-type but not mutant torsinA suppresses alpha-synuclein aggregation in cells (McLean, 2002); in PC12 cells levels of endogenous torsinA increase and the protein redistributes in response to oxidative stress (Hewett, 2003); overexpression of torsinA in both COS-1 and PC12 cells protects against cell death when cells are exposed to a variety of toxic insults (Kuner, 2003; Shashidharan, 2004). Taken together with the *in vivo* experiments described later (Caldwell, 2003; Cao, 2005), these studies point to a chaperone function for torsinA.

V. Invertebrates

In addition to the mammalian torsin family members, there are three torsin-related genes in nematodes and a single torsin-like gene in Drosophila and zebra fish (Ozelius, 1999). One of the nematode genes, *OOC-5*, is critical for rotation of the nuclear-centrosome complex during embryogenesis and when defective leads to misorientation of the mitotic spindle and disruption of asymmetric cell division and cell fate determination (Basham & Rose, 2001). Taken together with the information that torsinA interacts with *KLC1* (Kamm, 2004) and *VIM* (Hewett, 2006), these suggest an interaction of torsinA with the cytoskeleton and a role in membrane movement. The OOC-5 protein also is found in the ER, suggesting that some essential ER-related function has been conserved throughout evolution in the torsin proteins.

Studies involving a second *Caenorhabditis elegans* torsin-like gene, *TOR-2*, show that wild-type torsin has the ability to suppress polyglutamine-induced protein aggregation (Caldwell, 2003) as well as protect dopaminergic neurons from cellular stress after treatment with the neurotoxin 6-hydroxydopamine (6-OHDA) (Cao, 2005). In both cases, expression of mutant torsin does not show these effects. A similar study in mice found that torsinA expression was increased significantly in the brains of mice several hours after treatment with another dopaminergic toxin, MPTP (Kuner, 2004). These results are similar to what has been described in cell culture studies earlier (Kuner, 2003; McLean, 2002; Shashidharan, 2004), and provide evidence that torsinA has a role in protein folding and degradation.

Two Drosophila models of torsin have been described. Koh et al. (2004) found that overexpression of mutant human torsinA but not wild-type elicited locomotor defects in the flies. In neurons they identified enlarged synaptic boutons of irregular shape with reduced vesicle content and also found dense torsinA-immunoreactive bodies associated with synaptic densities and the nuclear envelope consistent with the increased perinuclear staining seen in cultured cells overexpressing this protein (see earlier) (Goodchild, 2004; Naismith, 2004). Overexpression of human or fly Smad2, a downstream effector of the TGF-beta signaling pathway, suppressed both the locomotor and cellular defects, suggesting that TGF-beta signaling might be involved in early-onset dystonia (Koh, 2004). The second Drosophila model used RNA interference (RNAi) and overexpression to analyze the function of the endogenous fly torsin, torp4a. Using the eye as a model, down regulation of torp4a caused degeneration of the retina, whereas overexpression protected the retina from age-related neural degeneration (Muraro, 2006). Torp4a was expressed largely in the ER but also found at the NE consistent with cellular and mouse studies. A genetic screen to identify enhancers of torp4a demonstrated an association with components of the AP-3 adaptor complex, a protein related to myosin II function, and the superoxide dismutase 1 (SOD1) gene (Muraro, 2006).

VI. Mouse

A number of genetic models are available for *DYT1* dystonia in mice including both overexpressing transgenic models, where the human gene has been randomly inserted into the mouse genome, and several engineered lines where the endogenous mouse locus (*Tor1a*) has been modified (for review see Jinnah, 2005).

In a transgenic model, where the human mutant torsinA is overexpressed using the neuron-specific enolase (NSE) promoter, about 40 percent of the mice show hyperactivity, circling, and abnormal movement (Shashidharan, 2005). These mice also demonstrate abnormal levels of dopamine metabolites as well as aggregates in the brainstem similar to what was reported in *DYT1* human brains (Shashidharan, 2005). A second transgenic model expressing human mutant torsinA under the control of the CMV promoter does not show an overt movement disorder but exhibits impaired motor sequence learning on the rotorod (Sharma, 2005) reminiscent of the motor learning difficulties reported in human *DYT1* nonmanifesting mutation carriers (Ghilardi, 2003). Recently, recording the activity of striatal cholinergic interneurons in slice preparations from these animals in the presence of quinpirole, an increase in firing rate was observed in the mutant animals that was mediated by a greater inhibition of N-type calcium currents (Pisani, 2006). An imbalance between striatal dopaminergic and cholinergic signaling in DYT1 dystonia is suggested by this study.

Three different types of engineered mice have been published to date. Knock-in (KI) mice bearing the 3 bp deletion in the heterozygous state, analogous to the human DYT1 dystonia, manifest hyperactivity in the open field, difficulty in beam walking, possess abnormal levels of dopamine metabolites, but no overt dystonic posturing (Dang, 2005). These mice also have neuronal aggregates in neurons in the brainstem consistent with human pathologic data (McNaught, 2004). In contrast, mice that are either homozygous KI or knock-out (KO) for the deletion die at birth with apparently normal morphology, but with postmigratory neurons showing abnormalities of the nuclear membranes (Goodchild, 2005) comparable to what is seen in cell culture experiments (Goodchild, 2004; Naismith, 2004). The fact that both the homozygous KO and KI animals display the same lethal phenotype suggests that DYT1 dystonia results from a loss of function of the torsinA protein. The knock-down (KD) mouse model in which a reduced level of torsinA protein is expressed, displays a phenotype very similar to the heterozygous KI mice showing both deficits in motor control as well as dopamine metabolite levels (Dang, 2006). This mouse also supports a loss of function model because no deleted torsinA is necessary to produce the phenotype; however it is also consistent with a dominant negative model, whereby the mutant protein interferes with the wild-type protein to cause the loss of function.

Although the exact function of torsinA remains elusive, evidence presented earlier suggests a role in protein folding and degradation (Caldwell, 2003; Cao, 2005; Hewett, 2003; Kuner, 2003; McLean, 2002; Shashidharan, 2004; Torres, 2004) and/or membrane movement within cells (Basham & Rose, 2001; Hewett, 2006; Kamm, 2004). From both the animal and cellular models, it seems clear that DYT1 dystonia results from a loss of function (Dang, 2006; Goodchild, 2005; Torres, 2004). The fact that AAA+ proteins usually form oligomeric complexes may explain how a loss of function can be associated with the dominant inheritance of DYT1 dystonia through a dominant negative mechanism. The carboxy terminus of AAA+ proteins are important for the oligomerization (Whiteheart, 1994) as well as for the binding of interacting proteins (Akiyama, 1994; Missiakas, 1996). If mutant torsinA interact with wild-type torsinA forming inactive multimers (Breakefield, 2001), then suboptimal levels of functional torsinA might result. Alternately, mutant torsinA could block binding to interacting partners or bind to and sequester partner proteins, either way, interfering with their functions.

VII. DYT1 Role in Focal Dystonia

Recent studies implicate involvement of other variations in the *DYT1/TORB* genomic region in late onset, mainly focal dystonias (see Figure 19.1). In dystonia patients from Iceland, a significant association was observed with a haplotype spanning the DYT1 gene (Clarimon, 2005). Two studies from Germany failed to replicate this association (Hague, 2006; Sibbing, 2003). However, a study involving Italian and North American cohorts revealed an association in the Italian group with the same risk allele as was seen in Iceland, but no association in the American group (Clarimon, 2006). Finally, a group of Austrian and German patients with predominantly focal dystonia showed a strong association with two single nucleotide polymorphisms (SNPs) in the 3′, untranslated region of the gene; however, rather than being a risk haplotype as shown in the previous populations, the SNPs showed a strong protective effect (Kamm, 2006). Whether these opposing results reflect population difference or instead indicate that the tested SNPs are in strong linkage disequilibrium with a real causal variant(s) is unknown. Nevertheless, the combined results strongly support a role for genetic variability in the *DYT1* genomic region as a contributing factor in the risk of developing late onset, focal dystonia.

As discussed, when mutant torsinA is overexpressed in cells, it forms membrane inclusions that are thought to derive from the ER/NE (Bragg, 2004; Gonzalez-Alegre, 2004; Goodchild, 2004; Naismith, 2004). The only non-synonymous coding variant in the *DYT1* gene is located in exon 4 and replaces an aspartic acid (D) at position 216 with a histidine (H) in about 12 percent of normal alleles (Ozelius, 1997). It has been shown recently that when the H allele is overexpressed in cells, similar membrane inclusions result (Kock, 2006a). However, when the H allele is co-overexpressed with a construct carrying the GAG deleted torsinA, fewer inclusions are formed suggesting that the two alleles have a compensatory effect (Kock, 2006a). A possible role for this variant in the reduced penetrance associated with *DYT1* dystonia or in causing other forms of dystonia should be examined. In two of the studies examining the role of *DYT1* in focal dystonia, this D216H SNP was examined but in both cases no associations were identified (Kamm, 2006; Sibbing, 2003).

VIII. DYT1 Phenotype and Endophenotype

With the identification of DYT1, it has become possible to return to the clinical domain to determine the phenotypic spectrum and role of DYT1 in the dystonia population. Clinical expression is extraordinarily broad, even within families; 70 percent of gene carriers have no definite signs of dystonia, and among the remaining 30 percent, dystonia ranges from focal to severe generalized (Gasser, 1998; Opal, 2002). There are, however, common DYT1 clinical characteristics that have been described across ethnic groups (Bressman, 2000; Gambarin, 2006; Im, 2004; Lin, 2006; Valente, 1998; Yeung, 2005). The great majority of people with dystonia due to DYT1 have early onset (before 26 years) that first affects an arm or leg. About 65 percent progress to a generalized or multifocal distribution, the rest having segmental (10%) or only focal (25%) involvement. When viewed in terms of body regions ultimately involved, one or more limbs almost always are affected (over 95% have an affected arm). The trunk and neck may also be affected (about 25–35%) and they may be the regions producing the greatest disability (Chinnery 2002); the cranial muscles are less likely to be involved (<15–20%). Rarely, affected family members have late-onset (up to age 64 years) (Opal, 2002). Also, although the arm is the body region most commonly affected in those with focal disease, the neck or cranial muscles have been reported as isolated affected sites (Bressman, 2000; Leube, 1999; Tuffery-Giraud, 2001).

Because of the founder effect, the DYT1 GAG deletion is more important in the Ashkenazi population, where it accounts for about 80 percent of early (less than 26 years) onset cases (Bressman, 1994, 2000); this compares with 16 to 53 percent in early-onset non-Jewish populations (Brassat, 2000; Bressman, 2000; Lebre, 1999; Slominsky, 1999; Valente, 1998; Zorzi, 2002). Thus, a significant proportion of early-onset cases, especially among non-Ashkenazim, is not due to *DYT1*; other causes, including proposed autosomal dominant and recessive genes, are implicated (Gambarin, 2006; Moretti, 2005).

Another avenue opened by *DYT1* identification is a further exploration of the range of clinical expression in addition to dystonia and also the exploration of DYT1 endophenotypes that use imaging, electrophysiological, and other techniques to measure subclinical traits. Nonmanifesting family members (i.e., those without overt dystonia), a group constituting 70 percent of mutation carriers, can be studied; they can be compared to their noncarrier family members as well as those manifesting dystonia. Using this strategy, psychiatric expression of *DYT1* was investigated. The same increased risk for early-onset recurrent major depression was found in *both* manifesting and nonmanifesting gene carriers compared to their noncarrier-related family members (Heiman, 2004); differences in OCD frequency, a psychiatric feature associated with other movement disorders including tics and myoclonus dystonia were not observed (Heiman, 2006). Other subtle clinical abnormalities noted in nonmanifesting carriers are deficiencies in sequence learning (Ghilardi, 2003) and "probable" dystonia. The latter, although increased in carriers compared to noncarriers, is not 100 percent specific, raising concerns about using family members with only probable dystonia in genetic linkage studies (Bressman, 2002).

DYT1 endophenotypes have been investigated using various imaging and neurophysiological approaches. Eidelberg and colleagues demonstrated a characteristic pattern of glucose utilization with ^{18}F-fluorodeoxyglucose positron emission tomography (PET) and network analysis. There are covarying metabolic increases in the basal ganglia, cerebellum, and supplementary motor cortex (SMA) in both manifesting and nonmanifesting gene carriers (Carbon, 2004; Eidelberg, 1998). Other imaging studies of DYT1 gene carriers, including nonmanifesting carriers, have found decreased striatal D2 receptor binding (Asanuma, 2005), and microstructural changes involving the subgyral white matter of the sensorimotor cortex (Carbon, 2004). Electrophysiological analyses also have identified genotype associated abnormalities, namely reduced intracortical inhibition and a shortened cortical silent period (Edwards, 2003) as well as higher tactile and visuotactile temporal discrimination thresholds and temporal order judgments (Fiorio, 2006). These studies strongly support the presence of wider clinical gene expression, abnormal brain processing, and associated structural brain changes in gene carriers regardless of overt motor signs of dystonia, expanding the notion of penetrance and phenotype.

IX. Future Directions

There has been a veritable explosion in understanding of the genetic underpinnings of primary dystonia over the last 20 years; yet much remains unknown. Only one gene for primary dystonia has been identified, *DYT1*. Further, although a mutation in *DYT1* is responsible for a significant proportion of early-onset generalized dystonia, *DYT1* is a very rare cause of adult onset dystonia, which constitutes the great majority of primary cases. So the hunt for dystonia genes continues.

Our understanding of DYT1 normal and mutated protein, torsinA, is widening and no doubt will continue to progress along current paths, as cellular and animals models are further explored. Especially important will be investigations not only focusing on the striatum, but also assessing anatomic and functional changes elsewhere in the brain. Various lines of study suggest that the thalamus, brainstem, and cerebellum (Jinnah, 2006; McNaught, 2004) need closer scrutiny. Also, there are only a handful of human DYT1 neuropathological studies and confirmation and elaboration of the crucial findings of McNaught et al. (2004) are needed. Other lines of investigation that hold great promise include the search for DYT1 modifiers, genetic and environmental. Only 30 percent of GAG deletion carriers ever manifest dystonia and clinical expression ranges from severe generalized dystonia to barely discernable action dystonias. Understanding the natural occurring modulators of disease expression will shed light on the pathogenic steps that take human motor control across a threshold into clinical dysfunction.

Finally new avenues of research that hold the promise for targeted treatments of DYT1 dystonia are just being initiated. These derive from several different approaches including the search for DYT1 modifiers, better understanding of the neuro-physiological correlates of DYT1, and cellular and animal models that not only shed light on disease mechanism, but also allow for drug or other interventional screening. One such novel approach uses RNA interference (RNAi) in cell culture systems overexpressing the mutant torsin protein to block aggregate formation and restore normal distribution of wild-type torsinA (Gonzalez, 2005; Kock, 2006b). These results support the dominant negative model for torsinA function but also suggest RNAi could be used therapeutically.

References

Akiyama, Y., Shirai, Y., Ito, K. (1994). Involvement of FtsH in protein assembly into and through the membrane. *J Biol Chem* **269**, 5225–5229.

Asanuma, K., Ma, Y., Huang, C., Carbon-Correll, M., Edwards, C., Raymond, D. et al. (2005). Decreased striatal D2 receptor binding in non-manifesting carriers of the DYT1 dystonia mutation. *Neurology* **64(2)**, 347–349.

Augood, S. J., Penney, J. B., Friberg, I., Breakefield, X. O., Young, A., Ozelius, L. J., Standaert, D. G. (1998). Expression of the early-onset torsion dystonia gene (DYT1) in human brain. *Ann Neurol* **43**, 669–673.

Augood, S. J., Martin, D. M., Ozelius, L. J., Breakefield, X. O., Penney, J. B. J., Standaert, D. G. (1999). Distribution of the mRNAs encoding torsinA and torsinB in the adult human brain. *Ann Neurol* **46**, 761–769.

Augood, S. J., Hollginsworth, Z., Albers, D., Yang, L., Leung, J., Muller, B. et al. (2002). Dopamine transmission in DYT1 dystonia: A biochemical and autoradiographical study. *Neurol* **59**, 445–448.

Augood, S. J., Keller-McGandy, C. E., Siriani, A., Hewett, J., Ramesh, V., Sapp, E. et al. (2003). Distribution and ultrastructural localization of torsinA immunoreactivity in the human brain. *Brain Res* **986**, 12–21.

Bara-Jimenez, W., Catalan, M. J., Hallett, M., Gerloff, C. (1998). Abnormal somatosensory homunculus in dystonia of the hand. *Ann Neurol* **44(5)**, 828–831.

Basham, S. E., Rose, L. S. (2001). The Caenorhabditis elegans polarity gene ooc-5 encodes a Torsin-related protein of the AAA ATPase superfamily. *Development* **128**, 4645–4656.

Becker, G., Berg, D., Rausch, W. D., Lange, H. K., Riederer, P., Reiners, K. (1999). Increased tissue copper and manganese content in the lentiform nucleus in primary adult-onset dystonia. *Ann Neurol* **46(2)**, 260–263.

Berg, D., Herrmann, M. J., Muller, T. J., Strik, W. K., Aranda, D., Koenig, T. et al. (2001). Cognitive response control in writer's cramp. *Eur J Neurol* **8(6)**, 587–594.

Bragg, D. C., Kaufman, C. A., Kock, N., Breakefield, X. O. (2004a). Inhibition of N-linked glycosylation prevents inclusion formation by the dystonia-related mutant form of torsinA. *Mol Cell Neurosci* **27**, 417–426.

Brassat, D., Camuzat, A., Vidailhet, M., Feki, I., Jedynak, P., Klap, P. et al. (2000). Frequency of the DYT1 mutation in primary torsion dystonia without family history. *Arch Neurol* **57(3)**, 333–335.

Breakefield, X. O., Kamm, C., Hanson, P. I. (2001). TorsinA: Movement at many levels. *Neuron* **31**, 9–12.

Bressman, S. B. (2003). Dystonia: phenotypes and genotypes. *Rev Neurol* (Paris) **159(10 Pt 1)**, 849–856.

Bressman, S. B., de Leon, D., Brin, M. F., Risch, N., Burke, R. E., Greene, P. E. et al. (1989). Idiopathic torsion dystonia among Ashkenazi Jews: Evidence for autosomal dominant inheritance. *Ann Neurol* **26**, 612–620.

Bressman, S. B., Greene, P. E. (1990). Treatment of hyperkinetic movement disorders. *Neurologic Clinics* **8**, 51–75.

Bressman, S. B., de Leon, D., Kramer, P. L., Ozelius, L. J., Brin, M. F., Greene, P. E. et al. (1994). Dystonia in Ashkenazi Jews: Clinical characterization of a founder mutation. *Ann Neurol* **36(5)**, 771–777.

Bressman, S. B., Sabatti, C., Raymond, D. et al. (2000). The DYT1 phenotype and guidelines for diagnostic testing. *Neurology* **54**, 1746–1752.

Bressman, S. B., Greene, P. E. (2000). Dystonia. *Curr Treat Options Neurol* **2(3)**, 275–285.

Bressman, S. B., Raymond, D., Wendt, K., Saunders-Pullman, R., de Leon, D., Fahn, S., Ozelius, L., Risch, N. (2002). Diagnostic criteria for dystonia in DYT1 families. *Neurology* **59**, 1780–1782.

Caldwell, G. A., Cao, S., Sexton, E. G., Gelwix, C. C., Bevel, J. P., Caldwell, K. A. (2003). Suppression of polyglutamine-induced protein aggregation in Caenorhabditis elegans by torsin proteins. *Hum Mol Genet* **12**, 307–319.

Callan, A. C., Bunning, S., Jones, O. T., High, S., Swanton, E. (2006). Biosynthesis of the dystonia-associated AAA + ATPase torsinA at the endoplasmic reticulum. *Biochem J* (Epub ahead of print).

Cao, S., Gelwix, C. C., Caldwell, K. A., Caldwell, G. A. (2005). Torsin-mediated protection from cellular stress in the dopaminergic neurons of *Caenorhabditis elegans*. *J Neurosci* **25**, 380–13812.

Carbon, M., Kingsley, P. B., Su, S., Smith, G. S., Spetsieris, P., Bressman, S., Eidelberg, D. (2004). Microstructural white matter changes in carriers of the DYT1 gene mutation. *Ann Neurol* **56(2)**, 283–286.

Chinnery, P. F., Reading, P. J., McCarthy, E. L., Curtis, A., Burn, D. J. (2002). Late-onset axial jerky dystonia due to the DYT1 deletion. *Mov Disord* **17(1)**, 196–198.

Clarimon, J., Asgeirsson, H., Singleton, A., Jakobsson, F., Hjaltason, H., Hardy, J., Sveinbjornsdottir, S. (2005). TorsinA haplotype predisposes to idiopathic dystonia. *An Neurol* **57**, 765–767.

Clarimon, J., Brancati, F., Peckham, E., Valente, E. M., Dallapiccola, B., Abruzzese, G. et al. (2006). Assessing the role of DRD5 and DYT1 in two different case-control series with primary blepharospasm. *Mov Disord* (Epub ahead of print).

Dang, M. T., Yokoi, F., Pence, M. A., Li, Y. (2006). Motor deficits and hyperactivity in Dyt1 knockdown mice. *Neurosci Res* **56**, 470–474.

Dang, M. T., Yokoi, F., McNaught, K. S., Jengelley, T. A., Jackson, T., Li, J., Li, Y. (2005). Generation and characterization of Dyt1 DeltaGAG knock-in mouse as a model for early-onset dystonia. *Exp Neurol* **196**, 452–463.

Dron, M., Meritet, J. F., Dandoy-Dron, F., Meyniel, J. P., Maury, C., Tovey, M. G. (2002). Molecular cloning of ADIR, a novel interferon responsive gene encoding a protein related to the torsins. *Genomics* **79**, 315–325.

Edwards, M. J., Huang, Y. Z., Wood, N. W., Rothwell, J. C., Bhatia, K. P. (2003). Different patterns of electrophysiological deficits in manifesting and non-manifesting carriers of the DYT1 gene mutation. *Brain* **126**, 2074–2080.

Eidelberg, D., Moeller, J. R., Antonini, A., et al. (1998). Functional brain networks in DYT1 dystonia. *Ann Neurol* **44**, 303–312.

The Epidemiological Study of Dystonia in Europe (ESDE) Collaborative Group. (2000). A prevalence study of primary dystonia in eight European countries. *J Neurol* **247**, 787–792.

Fahn, S. (1987). Systemic therapy of dystonia. *Can J Neurol Sci* **14** (3 Suppl), 528–532.

Fahn, S. (1988). Concept and classification of dystonia. *Adv Neurol* **50**, 1–8.

Ferrari-Toninelli, G., Paccioretti, S., Francisconi, S., Uberti, D., Memo, M. (2004). TorsinA negatively controls neurite outgrowth of SH-SY5Y human neuronal cell line. *Brain Res* **1012(1–2)**, 75–81.

Fiorio, M., Gambarin, M., Valente, E. M., Liberini, P., Loi, M., Cossu, G. et al. (2006). Defective temporal processing of sensory stimuli in DYT1 mutation carriers: A new endophenotype of dystonia? *Brain* (Epub ahead of print).

Furukawa, Y., Hornykiewicz, O., Fahn, S., Kish, S. J. (2000). Striatal dopamine in early-onset primary torsion dystonia with the DYT1 mutation. *Neurology* **54**, 1193–1195.

Gambarin, M., Valente, E. M., Liberini, P., Barrano, G., Bonizzato, A., Padovani, A. et al. (2006). Atypical phenotypes and clinical variability in a large Italian family with DYT1-primary torsion dystonia. *Mov Disord* **21(10)**, 1782–1784.

Garraux, G., Bauer, A., Hanakawa, T., Wu, T., Kansaku, K., Hallett, M. (2004). Changes in brain anatomy in focal hand dystonia. *Ann Neurol.* **55(5)**, 736–739.

Gasser, T., Windgassen, K., Bereznai, B., Kabus, C., Ludolph, A. C. (1998). Phenotypic expression of the DYT1 mutation: A family with writer's cramp of juvenile onset. *Ann Neurol* **44(1)**, 126–128.

Ghilardi, M. R., Carbon, M., Silvestri, G., Dhawan, V., Tagliati, M., Bressman, S. et al. (2003). Impaired sequence learning in carriers of the DYT1 dystonia mutation. *Ann Neurol* **54**, 102–109.

Gonzalez-Alegre, P., Paulson, H. L. (2004). Aberrant cellular behavior of mutant torsinA implicates nuclear envelope dysfunction in DYT1 dystonia. *J Neurosci* **24**, 2593–2601.

Gonzalez-Alegre, P., Bode, N., Davidson, B. L., Paulson, H. L. (2005). Silencing primary dystonia: Lentiviral-mediated RNA interference therapy for DYT1 dystonia. *J Neurosci* **25**, 10502–10509.

Goodchild, R. E., Dauer, W. T. (2004). Mislocalization of the nuclear envelope: An effect of the dystonia-causing torsinA mutation. *Proc Natl Acad Sci USA* **1001**, 847–852.

Goodchild, R. E., Dauer, W. T. (2005). The AAA+ protein torsinA interacts with a conserved domain present in LAP1 and a novel ER protein. *Cell Biol* **168**, 855–862.

Goodchild, R. E., Kim, C. E., Dauer, W. T. (2005b). Loss of the dystonia-associated protein torsinA selectively disrupts the neuronal nuclear envelope. *Neuron* **48**, 923–932.

Greene, P., Shale, H., Fahn, S. (1988). Experience with high dosages of anticholinergic and other drugs in the treatment of torsion dystonia. *Adv Neurol* **50**, 547–556.

Grunewald, R. A., Yoneda, Y., Shipman, J. M., Sagar, H. J. (1997). Idiopathic focal dystonia: A disorder of muscle spindle afferent processing? *Brain* **120** (Pt 12), 2179–2185.

Hague, S., Klaffke, S., Clarimon, J., Hemmer, B., Singleton, A., Kupsch, A., Bandmann, O. (2006). Lack of association with TorsinA haplotype in German patients with sporadic dystonia. *Neurology* **66**, 951–952.

Hallett, M. (2004). Dystonia: Abnormal movements result from loss of inhibition. *Adv Neurol* **94**, 1–9.

Hanson, P. I., Whiteheart, S. W. (2005). AAA+ proteins: Have engine, will work. *Nat Rev Mol Cell Biol* **6**, 519–529.

Hedreen, J. C., Zweig, R. M., DeLong, M. R., Whitehouse, P. J., Price, D. L. (1988). Primary dystonias: A review of the pathology and suggestions for new directions of study. *Adv Neurol* **50**, 123–132.

Heiman, G. A., Ottman, R., Saunders-Pullman, R. J., Ozelius, L. J., Risch, N. J., Bressman, S. B. B. (2004). Increased risk for recurrent major depression in DYT1 dystonia mutation carriers. *Neurology* **63(4)**, 631–637.

Heiman, G. A., Ottman, R., Saunders-Pullman, R. J., Ozelius, L. J., Risch, N. J., Bressman, S. B. (2006). Obsessive-compulsive disorder is not a clinical manifestation of the DYT1 dystonia gene. *Am J Med Genet B Neuropsychiatr Genet* (Epub ahead of print).

Hewett, J., Gonzalez-Agosti, C., Slater, D., Li, S., Ziefer, P., Bergeron, D. et al. (2000). Mutant torsinA, responsible for early onset torsion dystonia, forms membrane inclusions in cultured neural cells. *Hum Mol Genet* **22**, 1403–1413.

Hewett, J., Ziefer, P., Bergeron, D., Naismith, T., Boston, H., Slater, D. et al. (2003). TorsinA in PC12 cells: Localization in the endoplasmic reticulum and response to stress. *J Neurosci Res* **72**, 158–168.

Hewett, J. W., Zeng, J., Niland, B. P., Bragg, D. C., Breakefield, X. O. (2006). Dystonia-causing mutant torsinA inhibits cell adhesion and neurite extension through interference with cytoskeletal dynamics. *Neurobiol Dis* **22**, 98–111.

Ikeuchi, T., Shimohata, T., Nakano, R., Koide, R., Takano, H., Tsuji, S. (1999). A case of primary torsion dystonia in Japan with the 3-bp (GAG) deletion in the DYT1 gene with a unique clinical presentation. *Neurogenetics* **2(3)**, 189–190.

Im, J. H., Ahn, T. B., Kim, K. B., Ko, S. B., Jeon, B. S. (2004). DYT1 mutation in Korean primary dystonia patients. *Parkinsonism Relat Disord* **10(7)**, 421–423.

Jinnah, H. A., Hess, E. J., Ledoux, M. S., Sharma, N., Baxter, M. G., Delong, M. R. (2005). Rodent models for dystonia research: Characteristics, evaluation, and utility. *Mov Disord* **20**, 283–292.

Jinnah, H. A., Hess, E. J. (2006). A new twist on the anatomy of dystonia: The basal ganglia and the cerebellum? *Neurology* **67(10)**, 1740–1741.

Kabakci, K., Hedrich, K., Leung, J. C., Mitterer, M., Vieregge, P., Lencer, R. et al. (2004). Mutations in DYT1: Extension of the phenotypic and mutational spectrum. *Neurology* **62(3)**, 395–400.

Kaji, R., Murase, N. (2001). Sensory function of basal ganglia. *Mov Disord* **16(4)**, 593–594.

Kaji, R., Murase, N., Urushihara, R., Asanuma, K. (2004). Sensory deficits in dystonia and their significance. *Adv Neurol* **94**, 11–17.

Kamm, C., Boston, H., Hewett, J., Wilbur, J., Corey, D. P., Hanson, P. I., Ramesh, V., Breakefield, X. O. (2004). The early onset dystonia protein torsinA interacts with kinesin light chain 1. *J Biol Chem* **279**, 19882–19892.

Kamm, C., Asmus, F., Mueller, J., Mayer, P., Sharma, M., Muller, U. J. et al. (2006). Strong genetic evidence for association of TOR1A/TOR1B with idiopathic dystonia. *Neurology* **67**, 1857–1859.

Klein, C., Pramstaller, P. P. et al. (1998). Clinical and genetic evaluation of a family with a mixed dystonia phenotype from South Tyrol. *Ann Neurol* **44(3)**, 394–398.

Klein, C., Hedrich, K. et al. (2002). Exon deletions in the GCHI gene in two of four Turkish families with dopa-responsive dystonia. *Neurology* **59(11)**, 1783–1786.

Kock, N., Naismith, T. V., Boston, H. E., Ozelius, L. J., Corey, D. P., Breakefield, X. O., Hanson, P. I. (2006a). Effects of genetic variations in the dystonia protein torsinA: Identification of polymorphism at residue 216 as protein modifier. *Hum Mol Genet* **15**, 1355–1364.

Kock, N., Allchorne, A. J., Sena-Esteves, M., Woolf, C. J., Breakefield, X. O. (2006b). RNAi blocks DYT1 mutant torsinA inclusions in neurons. *Neurosci Lett* **395**, 201–205.

Koh, Y. H., Rehfeld, K., Ganetzky, B. (2004). A Drosophila model of early onset torsion dystonia suggests impairment in TGF-beta signaling. *Hum Mol Genet* **13**, 2019–2030.

Konakova, M., Huynh, D. P., Yong, W., Pulst, S. M. (2001a). Cellular distribution of torsin A and torsin B in normal human brain. *Arch Neurol* **58**, 921–927.

Konakova, M., Pulst, S. M. (2001b). Immunocytochemical characterization of torsin proteins in mouse brain. *Brain Res* **922**, 1–8.

Kramer, P. L., deLeon, D., Ozelius, L. O., Risch, N., Bressman, S. B., Brin, M. F. et al. (1990). Dystonia gene in Ashkenazi Jewish population located on chromosome 9q32–34. *Ann Neurol* **27**, 114–120.

Kramer, P. L., Heiman, G., Gasser, T., Ozelius, L., deLeon, D., Brin, M. F. et al. (1994). The DYT1 gene on 9q34 is responsible for most cases of early-onset idiopathic torsion dystonia in non-Jews. *Am J Hum Genet* **55**, 468–475.

Kuner, R., Teismann, P., Trutzel, A., Naim, J., Richter, A., Schmidt, N. et al. (2003). TorsinA protects against oxidative stress in COS-1 and PC12 cells. *Neurosci Lett* **350**, 153–156.

Kuner, R., Teismann, P., Trutzel, A., Naim, J., Richter, A., Schmidt, N. et al. (2004). TorsinA, the gene linked to early-onset dystonia, is upregulated by the dopaminergic toxin MPTP in mice. *Neurosci Lett* **355**, 126–130.

Kustedjo, K., Bracey, M. H., Cravatt, B. F. (2000). Torsin A and its torsion dystonia-associated mutant forms are lumenal glycoproteins that exhibit distinct subcellular localizations. *J Biol Chem* **275**, 27933–27939.

Kustedjo, K., Deechongkit, S., Kelly, J. W., Cravatt, B. F. (2003). Recombinant expression, purification, and comparative characterization of torsinA and its torsin dystonia-associated variant Delta E-torsinA. *Biochemistry* **42**, 15333–15341.

Lebre, A. S., Durr, A. et al. (1999). DYT1 mutation in French families with idiopathic torsion dystonia. *Brain* **122** (Pt 1), 41–45.

Leube, B., Kessler, K. R., Ferbert, A. et al. (1999). Phenotypic variability of the DYT1 mutation in German dystonia patients. *Acta Neurol Scand* **99**, 248–251.

Leung, J. C., Klein, C., Friedman, J., Vieregge, P., Jacobs, H., Doheny, D. et al. (2001). Novel mutation in the TOR1A (DYT1) gene in atypical early onset dystonia and polymorphisms in dystonia and early onset parkinsonism. *Neurogenetics* **3(3)**, 133–143.

Lin, Y. W., Chang, H. C., Chou, Y. H., Chen, R. S., Hsu, W. C., Wu, W. S. et al. (2005). DYT1 mutation in a cohort of Taiwanese primary dystonias. *Parkinsonism Relat Disord* **12(1)**, 15–19. (Epub 2005 Sep 29.)

Liu, Z., Zolkiewska, A., Zolkiewska, M. (2003). Characterization of human torsinA and its dystonia-associated mutant form. *Biochem J* **374**, 117–122.

Lupas, A., Flanagan, J. M., Tamura, T., Baumeister, W. (1997). Self-compartmentalization proteases. *Trends Biochem Sci* **22**, 399–404.

Major, T., Svetel, M., Romac, S., Kostic, V. S. (2001). DYT1 mutation in primary torsion dystonia in a Serbian population. *J Neurol* **248(11)**, 940–943.

Marsden, C. D., Harrison, M. J. (1974). Idiopathic torsion dystonia (dystonia musculorum deformans). A review of forty-two patients. *Brain* **97(4)**, 793–810.

McLean, P. J., Kawamata, H., Shariff, S., Hewett, J., Sharma, N., Ueda, K. et al. (2002). TorsinA and heat shock proteins act as molecular chaperones: Suppression of alpha-synuclein aggregation. *J Neurochem* **83**, 846–854.

McNaught, K. S., Kapustin, A., Jackson, T., Jengelley, T. A., Jnobaptiste, R., Shashidharan, P. et al. (2004). Brainstem pathology in DYT1 primary torsion dystonia. *Ann Neurol* **56**, 540–547.

Mink, J. W. (1996). The basal ganglia: Focused selection and inhibition of competing motor programs. *Prog Neurobiol* **50(4)**, 381–425.

Misbahuddin, A., Placzek, M. R., Taanman, J. W., Gschmeissner, S., Schiavo, G., Cooper, J. M., Warner, T. T. (2005). Mutant torsinA, which causes early-onset primary torsion dystonia, is redistributed to membranous structures enriched in vesicular monoamine transporter in cultured human SH-SY5Y cells. *Mov Disord* **20**, 432–440.

Missiakas, D., Schwager, F., Betton, J-M., Georgopoulos, C., Raina, J. (1996). Identification and characterization of HS1V HS1U (ClpQ ClpY) proteins involved in overall proteolysis of misfolded proteins in Eschericia coli. *EMBO J* **15**, 6899–6909.

Muller, J., Kiechl, S., Wenning, G. K., et al. (2002). The prevalence of primary dystonia in the general community. *Neurology* **59**, 941–943.

Moretti, P., Hedera, P., Wald, J., Fink, J. (2005). Autosomal recessive primary generalized dystonia in two siblings from a consanguineous family. *Mov Disord* **20(2)**, 245–247.

Muraro, N. I., Moffat, K. G. (2006). Down-regulation of torp4a, encoding the Drosophila homologue of torsinA, results in increased neuronal degeneration. *J Neurobiol* **66**, 1338–1353.

Naismith, T. V., Heuser, J. E., Breakefield, X. O., Hanson, P. I. (2004). TorsinA in the nuclear envelope. *Proc Natl Acad Sci U S A* **101**, 7612–7617.

Nakashima, K., Kusumi, M., Inoue, Y., Takahashi, K. (1995). Prevalence of focal dystonias in the western area of Tottori Prefecture in Japan. *Mov Disord* **10(4)**, 440–443.

Neuwald, A. F., Aravind, L., Spouge, J. L., Koonin, E. V. (1999). AAA+: A class of chaperone-like ATPases associated with the assembly, operation, and disassembly of protein complexes. *Genome Res* **9**, 27–43.

Nutt, J. G., Muenter, M. D., Aronson, A., Kurland, L. T., Melton, L. J., 3rd. (1988). Epidemiology of focal and generalized dystonia in Rochester, Minnesota. *Mov Disord* **3(3)**, 188–194.

O'Dwyer, J. P., O'Riordan, S., Saunders-Pullman, R., Bressman, S. B., Molloy, F., Lynch, T., Hutchinson, M. (2005). Sensory abnormalities in unaffected relatives in familial adult-onset dystonia. *Neurology* **65(6)**, 938–940.

Opal, P., Tintner, R., Jankovic, J. et al. (2002). Intrafamilial phenotypic variability of the DYT1 dystonia: From asymptomatic TOR1A gene carrier status to dystonic storm. *Mov Disord* **17(2)**, 339–345.

Ozelius, L. O., Kramer, P. L., Moskowitz, C. B., Kwiatkowski, D. J., Brin, M., Bressman, S. B. et al. (1989). Human gene for torsion dystonia located on chromosome 9q32-q34. *Neuron* **2**, 1427–1434.

Ozelius, L. J., Kramer, P. L., de Leon, D., Risch, N., Bressman, S. B., Schuback, D. E. et al. (1992). Strong allelic association between the torsion dystonia gene (DYT1) and loci on chromosome 9q34 in Ashkenazi Jews. *Am J Hum Genet* **50(3)**, 619–628.

Ozelius, L. J., Hewett, J., Page, C., Bressman, S., Kramer, P., Shalish, C. et al. (1997). The early-onset torsion dystonia gene (*DYT1*) encodes an ATP-binding protein. *Nat Genet* **17**, 40–48.

Ozelius, L. J., Page, C. E., Klein, C., Hewett, J. W., Mineta, M., Leung, J. et al. (1999). The TOR1A (DYT1) gene family and its role in early onset torsion dystonia. *Genomics* **62**, 377–384.

Perlmutter, J. S., Tempel, L. W., Black, K. J., Parkinson, D., Todd, R. D. (1997). MPTP induces dystonia and parkinsonism. Clues to the pathophysiology of dystonia. *Neurology* **49(5)**, 1432–1438.

Perlmutter, J. S., Mink, J. W. (2004). Dysfunction of dopaminergic pathways in dystonia. *Adv Neurol* **94**, 163–170.

Pisani, A., Martella, G., Tscherter, A., Bonsi, P., Sharma, N., Bernardi, G., Standaert, D. G. (2006). ltered responses to dopaminergic D2 receptor activation and N-type calcium currents in striatal cholinergic interneurons in a mouse model of DYT1 dystonia. *Neurobiol Dis* **24**, 318–325.

Risch, N., de Leon, D., Ozelius, L., et al. (1995). 3. Genetic analysis of idiopathic torsion dystonia in Ashkenazi Jews and their recent descent from a small founder population. *Nature Genet* **9**, 152–159.

Rostasy, K., Augood, S. J., Hewett, J. W., Leung, J. C., Sasaki, H., Ozelius, L. J et al. (2003). TorsinA protein and neuropathology in early onset generalized dystonia with GAG deletion. *Neurobiol Dis* **12**, 11–24.

Rothwell, J. C., Obeso, J. A., Day, B. L., Marsden, C. D. (1983). Pathophysiology of dystonias. *Adv Neurol* **39**, 851–863.

Sanger, T. D., Tarsy, D., Pascual-Leone, A. (2001). Abnormalities of spatial and temporal sensory discrimination in writer's cramp. *Mov Disord* **16(1)**, 94–99.

Sharma, N., Hewett, J., Ozelius, L. J., Ramesh, V., McLean, P. J., Breakefield, X. O., Hyman, B. T. (2001). A close association of torsinA and alpha-synuclein in Lewy bodies: A fluorescence resonance energy transfer study. *Am J Pathol* **159**, 339–344.

Sharma, N., Baxter, M. G., Petravicz, J., Bragg, D. C., Schienda, A., Standaert, D. G., Breakefield, X. O. (2005). Impaired motor learning in mice expressing torsinA with the DYT1 dystonia mutation. *J Neurosci* **25**, 5351–5355.

Shashidharan, P., Kramer, C., Walker, R., Olanor, C. W., Brin, M. F. (2000a). Immunohistochemical localization and distribution of torsinA in normal human and rat brain. *Brain Res* **853**, 197–206.

Shashidharan, P., Good, P. F., Hsu, A., Perl, D. P., Brin, M. F., Olanow, C. W. (2000b). TorsinA accumulation in Lewy bodies in sporadic Parkinson's disease. *Brain Res* **877**, 379–381.

Shashidharan, P., Paris, N., Sandu, D., Karthikeyan, L., McNaught, K. S., Walker, R. H., and Olanow, C. W. (2004). Overexpression of torsinA in PC12 cells protects against toxicity. *J Neurochem* **88**, 1019–1025.

Shashidharan, P., Sandu, D., Potla, U., Armata, I. A., Walker, R. H., McNaught, K. S. et al. (2005). Transgenic mouse model of early-onset DYT1 dystonia. *Hum Mol Genet* **14**, 125–133.

Sibbing, D., Asmus, F., Konig, I. R., Tezenas du Montcel, S., Vidailhet, M., Sangla, S. et al. (2003). Candidate gene studies in focal dystonia. *Neurology* **61**, 1097–1101.

Siegert, S., Bahn, E., Kramer, M., Schulz-Schaeffe, W., Hewett, J., Breakefield, X. O., Hedreen, J., Rostasy, K. (2005). TorsinA expression is detectable in human infants as old as four weeks old. *Brain Res Dev Brain Res* **157**, 19–26.

Slominsky, P. A., Markova, E. D., Shadrina, M. I., Illarioshkin, S. N., Miklina, N. I., Limborska, S. A., Ivanova-Smolenskaya, I. A. (1999). A common 3-bp deletion in the DYT1 gene in Russian families with early-onset torsion dystonia. *Hum Mutat* **14(3)**, 269.

Torres, G. E., Sweeney, A. L., Beaulieu, J. M., Shashidharan, P., Caron, M. G. (2004). Effect of torsinA on membrane proteins reveals a loss of function and a dominant negative phenotype of the dystonia-associated {Delta}E-torsinA mutant. *Proc Natl Acad Sci U S A* **101**, 15650–15655.

Tuffery-Giraud, S., Cavalier, L., Roubertie, A., Guittard, C., Carles, S., Calvas, P. et al. (2001). No evidence of allelic heterogeneity in DYT1 gene of European patients with early onset torsion dystonia. *J Med Genet* **38**, e35.

Vale, R. D. (2000). AAA proteins: Lords of the ring. *J Cell Biol* **150**, F13–F19.

Valente, E. M., Warner, T. T., Jarman, P. R. et al. (1998). The role of primary torsion dystonia in Europe. *Brain* **121**, 2335–2339.

Vasudevan, A., Breakefield, X. O., Bhide, P. G. (2006). TorsinA and torsinB expression in the developing mouse brain. *Mol Brain Res* 139–145, 1073–1074.

Walker, R. H., Brin, M. F., Sandu, D., Gujjari, P., Hof, P. R., Warren Olanow, C., Shashidharan, P. (2001). Distribution and immunohistochemical characterization of torsinA immunoreactivity in rat brain. *Brain Res* **900**, 348–354.

Walker, R. H., Brin, M. F., Sandu, D., Good, P. F., Shashidharan, P. (2002). TorsinA immunoreactivity in brains of patients with DYT1 and non-DYT1 dystonia. *Neurology* **58**, 120–124.

Whiteheart, S. W., Rossnagel, K., Buhrow, S. A., Brunner, M., Jaenicke, R., Rothman, J. E. (1994). N-ethylmaleimids-sensitive fusion protein: A trimeric ATPase whose hydrolysis of ATP is required for membrane fusion. *J Cell Biol* **125**, 945–954.

Xiao, J., Gong, S., Zhao, Y., LeDoux, M. S. (2004). Developmental expression of rat torsinA transcript and protein. *Brain Res Dev Brain Res* **152**, 47–60.

Yeung, W. L., Lam, C. W., Cheng, W. T., Sin, N. C., Wong, W. K., Wong, C. N. et al. (2005). Early-onset primary torsional dystonia in a 4-generation Chinese family with a mutation in the DYT1 gene. *Chin Med J* (Engl). **118(10)**, 873–876.

Zeman, W., Dyken, P. (1967). Dystonia musculorum deformans. Clinical, genetic and pathoanatomical studies. *Psychiatr Neurol Neurochir* **70(2)**, 77–121.

Zeman, W. (1970). Pathology of the torsion dystonias (dystonia musculorum deformans). *Neurology* **20** (No. 11 Part 2), 79–88.

Zilber, N., Korczyn, A. D., et al. (1984). Inheritance of idiopathic torsion dystonia among Jews. *J Med Genet* **21(1)**, 13–20.

Zorzi, G., Garavaglia, B., et al. (2002). Frequency of DYT1 mutation in early onset primary dystonia in Italian patients. *Mov Disord* **17(2)**, 407–408.

Zweig, R. M., Hedreen, J. C., Jankel, W. R., Casanova, M. F., Whitehouse, P. J., Price, D. L. (1988). Pathology in brainstem regions of individuals with primary dystonia. *Neurology* **38(5)**, 702–706.

20

Motor Neuron Disease: Amyotrophic Lateral Sclerosis

Nicholas J. Maragakis and Jeffrey D. Rothstein

I. ALS Background

Amyotrophic lateral sclerosis (ALS) is the most common form of adult motor neuron disease. ALS differs from other motor neuron disorders such as spinal muscular atrophy and poliomyelitis in that motor neurons of the motor cortex also are affected. This results in unique clinical and biological implications distinct from spinal muscular atrophy and polio. ALS is the focus of this chapter given its higher incidence and prevalence than these other two disorders.

ALS is an uncommon, but not rare disease with an incidence of 1–3/100,000 (Yoshida et al., 1986) individuals and has a greater incidence in males than females with a male:female ratio of 1.4–2.5 (Mitsumoto et al., 1998). The mean duration of disease from onset to death or ventilator dependence is two to five years, although a significant percentage (19–39%) survives five years and a smaller percentage (8–22%) survives 10 years without ventilator use. Factors suggested as predictors of survival include age at onset, gender, clinical presentation (bulbar vs. spinal), and rate of disease progression. Age at onset appears to be a powerful predictor of disease duration with younger patients surviving longer (Eisen et al., 1993a; Haverkamp et al., 1995).

The majority (~95%) of patients have sporadic disease. The remainder of the ALS population has inherited the disease, and a number of genetic forms of ALS have been identified. The first gene abnormalities associated with ALS were mutations in superoxide dismutase (SOD1) (Rosen et al., 1993). This autosomally dominant form of ALS has led to an increased understanding of the disease particularly through the development of mouse models of ALS harboring this same mutation. More recently, other causal mutations have been found in subunits of dynactin, also successfully recreated in mouse models.

II. Clinical Manifestations of ALS

ALS is characterized by upper and lower motor neuron dysfunction accompanied by progression (worsening) of disability. The hallmark of the disease is weakness and a relative sparing of sensation and autonomic function. ALS may present with prominent bulbar dysfunction (bulbar ALS) including slurring of speech, shortness of breath, or swallowing abnormalities with subsequent progression of weakness to the limbs. Spinal ALS is characterized by initial weakness and muscle atrophy in the limbs, often asymmetric, which progresses from limb to limb prior to involvement of bulbar muscles. Most patients die from respiratory dysfunction as a result of diaphragmatic weakness. Despite these two categorizations, the clinical presentation of ALS is often heterogeneous with an initial predominance of lower motor neuron signs and symptoms (muscle weakness, atrophy, fasciculations) or, conversely, upper motor neuron signs or symptoms (spasticity and hyperreflexia) with a paucity of lower motor neuron features. This heterogeneity often results in a delay in diagnosis. Although most associate ALS with neuromuscular weakness, it has become apparent that ALS is a neurodegenerative disease. This designation is more appropriate and places ALS in the category of other neurodegenerative diseases including Parkinson's disease, Alzheimer's disease, and Huntington's disease, among others. There are numerous molecular and pathophysiological features shared by each of these diseases, and although each has a prominent clinical feature associated with it, there is significant overlap between these features as diseases progress. For example, cognitive function traditionally has been "spared" in ALS but investigators now believe that there is a significant association of mild cognitive abnormalities with ALS (Lomen-Hoerth et al., 2002, 2003).

The diagnosis of ALS is made on the basis of upper and lower motor neuron signs and symptoms accompanied by progression over a 12-month period. Additional studies that may be helpful in excluding other diseases include electrophysiological studies (electromyography and nerve conduction studies), imaging studies (MRI) of the brain and spinal cord, cerebrospinal fluid, and blood analyses for disorders such as Lyme disease, West Nile encephalitis, B_{12} deficiency, and neurosyphilis. Other potential exclusionary diagnoses are dependent upon the clinical presentation and are beyond the scope of this chapter (Mitsumoto et al., 1998).

III. Animal Models of Motor Neuron Diseases

Mouse models of motor neuron diseases have provided insights into the pathophysiology of motor neuron diseases and highlighted the observations that a variety of mutations in different genes produce a clinical phenotype of motor neuron disease (see Table 20.1). Among the lessons from these models is the heterogeneity in the pathology, site and speed of disease onset, and other behavioral phenotypes. Furthermore, these models have been used as preclinical tools for studying potential therapeutic interventions in motor neuron diseases.

A. Transgenic Mice with Mutations in Superoxide Dismutase (SOD1)

SOD1 is a free radical scavenging enzyme that forms a major component in guarding against oxygen radical species produced during cellular metabolism. SOD1 is an ubiquitously expressed 153 amino acid protein that functions as a homodimer that binds copper and zinc. SOD1 catalyzes the conversion of superoxide to hydrogen peroxide and oxygen in two asymmetrical steps utilizing an essential copper atom in the active site of the enzyme.

What is the mechanism behind mutant SOD1 (mSOD1) toxicity? More than 100 mutations in the SOD1 gene have been reported and this figure grows as more patients are analyzed. These mutations, at last count, involve 54 of the 153 amino acid residues that comprise the protein.

Shortly following the identification of SOD1 mutations in familial ALS in humans, a number of different transgenic mouse models were described. The most widely used are the $SOD1^{G93A}$, $SOD1^{G85R}$, $SOD1^{G37R}$ mutations (Bruijn et al., 1997b; Gurney et al., 1994b; Wong et al., 1995). Each of these three transgenic mouse models is characterized by progressive hindlimb weakness and ultimately, progression of weakness to the forelimbs and death, presumably from respiratory compromise. Phenotypically, these animals differ in the ages at which hindlimb weakness occurs and their overall time of survival. Pathologically all three of these models show dramatic loss of ventral horn motor neurons with SOD1 and neurofilament inclusions. This is accompanied by extensive astrogliosis as the disease progresses. The cortices do not appear to be affected despite ubiquitous expression of the mSOD1 protein. Some, but not all, of these models also develop varying degrees of vacuole formation thought to represent swollen mitochondria. A loss of enzymatic antioxidant function from mSOD1 was initially thought to lead to an increase in free radical-mediated injury by superoxide. However, studies in these mice show elevated or unchanged wild-type SOD1 (Bruijn et al., 1998; Gurney et al., 1994a; Wong et al., 1995;). Furthermore, SOD1-deficient mice live to adulthood without spontaneously developing symptoms of motor neuron disease (Reaume et al., 1996).

B. PMN Mouse

The progressive motor neuronopathy (pmn) mouse is an autosomal recessive model of motor neuron disease in which homozygous mice develop paralysis of the hindlimbs during

Table 20.1

Mouse Model	Pathology	Site of Disease Onset	Disease Onset	Survival
SOD1G93A	Motor neuron degeneration, astrogliosis, vacuole formation, Microglial activation, intracellular protein aggregates	Hindlimb weakness	>100 days of age	129 days
SOD1G85R	Motor neuron degeneration, astrogliosis, Microglial activation, intracellular protein aggregates	Hindlimb weakness	~2 weeks prior to endstage	~345 days
SOD1G37R	Motor neuron degeneration, astrogliosis, vacuole formation, Microglial activation, intracellular protein aggregates	Hindlimb weakness, axial tremors	4–6 months	~365 days
Als 2 knockout	Purkinje cell degeneration, Axon loss in corticospinal tracts	N/A	N/A	Normal
Wobbler	Motor neuron loss, astrogliosis, microglial activation, neurofilament accumulation, vacuole formation	Forelimb weakness and facial muscle atrophy	3–4 weeks	4 months–1year
pmn	Early-onset motor neuron degeneration, Axonal swellings	Hindlimb weakness	3 weeks	7 weeks
Dynein/Dynactin Mutations				
Transgenic overexpression of dynamitin	Motor neuron loss, Neurofilament accumulation	Hindlimb weakness, spastic tremors	5–9 months of age	Normal
Loa mice (dynein missense mutations)	Motor neuron loss, Abnormalities of facial motor neuron migration, Protein aggregates	Motor abnormalities notable in hindlimbs	1 month of age	Normal

the third week of life. This is followed by forelimb weakness and death within six to seven weeks after birth. Axonal degeneration apparently starts at the endplates and is prominent in the sciatic and phrenic nerves. Interestingly the brain does not show any histological abnormalities (Schmalbruch et al., 1991). The pmn mutation appears to result from a Trp-524Gly substitution at the last residue of the tubulin-specific chaperone (Tbce), a protein that leads to decreased protein (microtubule) stability. Pathological analyses also show reductions in microtubules (Martin et al., 2002).

C. Wobbler Mouse

One of the oldest and most well-characterized models of motor neuron disease is the wobbler mouse (Boillee et al., 2003). This autosomal recessive mouse model is characterized by neurodegeneration and male sterility. The first motor symptoms occur at three to four weeks of age and, unlike the mSOD1 mice, first develop in the forelimbs rather than the hindlimbs. This forelimb weakness is accompanied by weakness in neck muscles and atrophy of facial muscles. The hindlimbs are not significantly affected until 12 weeks of age. There is a significant variation in lifespan from four months to one year in most models. Pathologically, motor neuron loss in the cervical spinal cord and brainstem is most dramatic. This unique anatomical and phenotypic pattern of motor neuron degeneration recently was linked to a mutation in Vps54 (Vacuolar-vesicular protein sorting) factor involved in vesicular trafficking in eukaryotic cells (Schmitt-John et al., 2005). In exon 23 of Vps54, wr/wr genomic DNA contains an A-T transversion in the second position of codon 967 that results in the amino acid substitution L967Q.

D. Dynein/Dynactin Mutant Mice

Mutations in the dynein protein were found to cause a progressive motor neuron disorder in mice (Hafezparast et al., 2003) and overexpression of the protein dynamitin (part of the dynein-dynactin complex of microtubule transport) resulted in the development of late-onset motor neuron disease in a transgenic mouse model (LaMonte et al., 2002). Subsequent studies of the dynein-dynactin complex with mutant SOD1 demonstrated that an inhibition of microtubule transport from muscle to cell body occurred (Ligon et al., 2005). Taken together, these data potentially are redefining definitions of motor neuron disease and suggest an expansion of the biology related to motor neuron degeneration outside the cell body.

E. ALS2 Knockout

The Als2 gene is ubiquitously expressed and encodes the protein alsin, a guanine nucleotide exchange factor (GEF) known to activate small guanosine triphosphatase (GTPase) belonging to the Ras superfamily. Deletion mutations in this gene have been associated with ALS2—a juvenile onset form of ALS primarily associated with upper motor neuron findings. Als2 knockout mice do not produce an obvious clinical phenotype comparable to the human disease (Cai et al., 2005; Hadano et al., 2006). Pathologically, Purkinje cell loss and axon degeneration in the corticospinal tracts have been observed (Hadano et al., 2006). This model is of interest because unlike other mouse models of motor neuron disease where alpha motor neurons degenerate, this model provides a molecular link between a primarily upper motor neuron form of ALS in humans, and a mouse model showing subtle but potentially correlative pathology and phenotypes.

IV. Molecular Hypotheses in ALS

A. Oxidative Stress

The toxic gain of function observed in mSOD1 mice has raised the possibility that the mutant SOD1 enzyme may produce aberrant substrates including peroxynitrite and hydrogen peroxide. The spontaneous reaction of superoxide with nitric oxide yields peroxynitrite, which SOD1 can utilize for tyrosine nitration of proteins. Studies have reported increased levels of free nitrotyrosine in the spinal cords of both familial and sporadic ALS patients, but as of yet no specific nitration targets have yet been identified (Beal et al., 1997). However, oxidative damage within a closed cellular compartment, such as mitochondria, could result in downstream damage to the cell by interfering with organelle function, and might escape traditional antioxidant therapies.

In the case of hydrogen peroxide there is the potential to produce the highly reactive hydroxyl radical. A normal reaction cycle releases hydrogen peroxide and an oxidized form of the enzyme. The use of peroxide as a substrate by the enzyme in a reduced form generates the hydroxyl radical, which can initiate a cascade of peroxidation. *In vivo* it is unclear whether higher peroxidation will occur by such a mechanism as elevated products were only found in $SOD1^{G93A}$ (Andrus et al., 1998; Hall et al., 1998) mice and in no other transgenic mouse models at any stage of disease (Bruijn et al., 1997a).

B. Glutamate Excitotoxicity

Neuronal degeneration has been linked to glutamate excitotoxicity in a variety of models both *in vitro* and *in vivo*. Elevations in synaptic glutamate or abnormalities in glutamate receptors are two mechanisms by which glutamate can cause motor neuron cell death. In large studies of human ALS patients, elevated levels of glutamate have been observed in the cerebrospinal fluid of up to 40 percent of all patients. This observation has suggested that increases in synaptic glutamate may be the result of glutamate transporter dysfunction—a central function of astrocytes. The astrocyte glutamate transporter EAAT2 (GLT1 in rodents) is responsible for more than 90 percent of glutamate transport in the brain. Studies of the EAAT2 transporter in sporadic ALS tissue showed that in some ALS patients, a marked loss of up to 95 percent of astroglial EAAT2 protein and activity in affected areas was observed (Bristol & Rothstein, 1996). One mechanism for glutamate transporter (EAAT2) reduction or dysfunction was the finding of aberrant EAAT2 RNA species. The production of truncated EAAT2 protein by aberrant RNA splicing shows that truncated mutants have less ability to transport glutamate and may lead to the retention of normal EAAT2 protein within the cytoplasm. This may be due to disrupted trafficking of normal EAAT2 to the cell membrane and the formation of protein aggregates comprised of a mixture of truncated and normal EAAT2 proteins (Lin et al., 1998).

A consistent observation observed in all mutant SOD1 mice is the reduction in GLT1 (EAAT2). The mechanism of this reduction is not yet known. One mechanism behind glutamate transporter dysfunction may be related to inactivation of the EAAT2 protein by mutant SOD1 (Trotti et al., 1999). Such an inactivation could lead to a rise in synaptic glutamate and contribute to glutamate neurotoxicity. In support for a contributing role of EAAT2 in mutant SOD1 biology was the finding that overexpression of EAAT2 by either transgenic means or by pharmaceuticals in these mice resulted in a delay in grip strength decline and motor neuron loss and, in some cases, survival (Guo et al., 2003; Rothstein et al., 2005).

Glutamate neurotoxicity also is mediated through glutamate receptors. AMPA receptors are the primary glutamate receptors on motor neurons. Investigators overexpressed the AMPA subunit GluB-(N) (a subunit with a particularly high permeability to calcium) in a transgenic mouse model. When crossed with the $SOD1^{G93A}$ mouse, the resultant offspring had a more rapid decline in motor performance and a shortened lifespan (Kuner et al., 2005). Interestingly, the AMPA antagonist NBQX was neuroprotective in this model (Van Damme et al., 2003). The calcium impermeable GluR2 subunit was found to be reduced on motor neurons in mutant SOD1 mice with a concomitant increase in the more calcium-permeable GluR3 subunit (Tortarolo et al., 2006). Similar findings with reduced GluR2 expression also have been observed in human ALS spinal cord (Kawahara et al., 2003). These data suggest that abnormalities in glutamate neurotransmission may not play the central role in the initiation of the disease, but may play a contributory role in disease propagation.

C. Mitochondrial Dysfunction

Mitochondrial dysfunction could contribute to ALS pathogenesis through a variety of pathways including calcium homeostasis, a factor in apoptotic cascades, and through the generation of ROS.

Ultrastructural changes in mitochondria including vacuolization have been observed in the axons, dendrites, and soma of motor neurons in SOD1 mouse models. The significance of these mitochondrial structural abnormalities is not clear since mitochondrial vacuolization is not a prominent part of ALS pathology. Abnormalities in oxidative phosphorylation have been observed in mitochondrial preparations from both ALS mouse models as well as human ALS patients although the data in the latter have not always been reproducible (Bacman et al., 2006).

In the SOD1^{G93A} mutant, cytochrome c is released from mitochondria followed by the activation of caspase 9, which is believed to be an effector for the subsequent activation of caspases 3 and 7 (Guegan et al., 2001). Interestingly, some data suggest that mitochondrial-specific changes may account for the specific spinal cord pathology in mutant SOD1 mice. A selective recruitment of mutant SOD1 to spinal cord mitochondria, but not to mitochondria in unaffected tissue has been observed (Liu et al., 2004a). Whether these changes in fact contribute to disease pathogenesis, or reflect alterations that occur after the disease damages cells is not known.

D. Apoptotic Cascades

Apoptosis involves a variety of regulated pathways through the actions of various factors (genetic regulation, death receptors and pro/anti-apoptotic proteins), which eventually lead to programmed cell death.

Bcl-2 family members have been examined in the transgenic SOD1^{G93A} mouse model of ALS. Expression of the anti-apoptotic proteins, Bcl-2 and Bcl-xL, and pro-apoptotic proteins, Bad and Bax, were found to be similar in both asymptomatic SOD1^{G93A} and normal mice. However with the onset of the disease in the SOD1^{G93A} mice, a decrease in the expression of Bcl-2 and Bcl-xL was noted with an increase in the expression levels of Bad and Bax (Vukosavic et al., 1999). In conjunction with these findings, the overexpression of Bcl-2 in SOD1^{G93A} mice resulted in a slowing of disease onset, and increased survival somewhat. However, to date, no manipulation (drug or genetic) of apoptotic cascades have halted disease (Kostic et al., 1997).

Caspase activity is also a group of proteases activated in apoptotic cascades. In both motor neurons and astrocytes, activation of caspase 3 plays a central role in the cell death mediated by mutant SOD1 at the time of earliest onset in three mouse models, the SOD1^{G93A}, SOD1^{G37R}, and SOD1^{G85R} mice (Li et al., 2000; Pasinelli et al., 2000 Vukosavic et al., 2000). Cytochrome c release from mitochondria followed by the activation of caspase 9, which is believed to be an effector for the subsequent activation of caspases 3 and 7, has also been implicated (Guegan et al., 2001).

V. Axonal Pathology

ALS has been labeled both a "neurodegenerative disease" and a "motor neuron disease," suggesting that the primary pathophysiological features are related to motor neuron cell death. These features distinguish ALS from other disorders where degeneration is a more distal phenomenon; that is, common polyneuropathies or "dying back" neuropathies where clinical symptoms such as numbness and weakness occur in the most distal aspects of the limbs. However, it is becoming increasingly recognized that distal changes either at the synapse or axon are part of the spectrum of cellular abnormalities in the disease.

Abnormal accumulations of neurofilaments (NF) are a pathological hallmark in both mutant SOD1 mouse models and in ALS. Neurofilaments provide structural support for neurons. Three distinct neurofilament protein subunits exist differing in molecular weight: NF-heavy, NF-medium, and NF-light. Neurofilament accumulations are noted pathologically in ALS cases with some speculation that abnormal phosphorylation of these proteins potentially playing a role in disease pathogenesis (Manetto et al., 1988; Munoz et al., 1988; Sobue et al., 1990). However, it does not appear that mutations in neurofilament genes themselves are initiators of sporadic ALS (Garcia et al., 2006). Whether these accumulations affect neurons through loss of structural integrity or result in other physiological abnormalities such as abnormal axon transport is not yet known. However, in mouse models of ALS, similar accumulations of neurofilaments are also observed suggesting a possible molecular link between mSOD1 biology and neurofilament abnormalities (Gurney et al., 1994b; Morrison et al., 1996 Tu et al., 1996).

More direct molecular evidence for the importance of neurofilaments in modifying ALS pathobiology has come from important manipulations of the different neurofilament subunits in mutant SOD1 mouse models. Disruption of the NF-L gene in SOD1 mice removed all axonal neurofilaments leading to the accumulation of NF-M and NF-H subunits within neurons. Although a reduction in motor neurons was observed postnatally, these mice had a delay in the onset of disease as well as a prolonged survival (Williamson et al., 1998). Manipulation of the NF-H subunit through overexpression also resulted in reduced axonal neurofilament organization and extended survival in mutant SOD1 mice as well (CouillardDespres et al., 1998; Nguyen et al., 2001). More recently, targeting the removal of the phosphorylated tail domains of NF-M and NF-H was shown to delay disease onset and prolong survival. This results in a reduction in axon caliber and may increase the movement of molecules involved in slow axonal transport and reduce the amount of cross-linking between NF and microtubules thus altering axonal structure (Lobsiger et al., 2005).

The concept of ALS as a "distal axonopathy" as an early event is difficult to assess in humans since those who die of ALS have significant motor neuron loss in the motor cortex as well as the spinal cord. In a single ALS case, pathologically examined early in disease, suggested that distal axonal changes may proceed frank lower motor neuron degeneration (Fischer et al., 2004). However, mutant SOD1 mice may offer a clue as to the abundance of axonal pathology in early stages of the disease. Both immunohistochemical measures showing the loss of fast-firing neuromuscular synapses as early as 50 days (before disease onset in mutant SOD1 mice) of age (Frey et al., 2000) as well as electrophysiological measures of progressive loss of motor unit numbers preceding motor neuron cell death are intriguing (Kennel et al., 1996). This is mirrored pathologically that the number of ventral roots axon loss clearly predates motor neuron loss in these same mouse models (Fischer et al., 2004).

In another model, the *Wld^S* mouse is a spontaneous mutant with the remarkable phenotype of prolonged survival of injured axons (Lunn et al., 1989). The gene for *Wld^S* is created by the splicing of fragments of two genes, *Ube4b* and *Nmnat1*, within an 85 kb triplication on chromosome 4 that creates a new open reading frame coding for a novel 42-kDa protein (Coleman et al., 1998; Conforti et al., 2000). However, the mechanism for axonal protection by *Wld^S* remains unknown. This model has been shown to be neuroprotective in a number of other axonopathy models. In the PMN model of motor neuron disease, the *Wld^S* gene product attenuates symptoms, extends lifespan, prevents axon degeneration, rescues motor neuron number and size, and delays retrograde transport deficits in these mice (Ferri et al., 2003). This same gene product in mSOD1 mice either produced a minimal effect on disease progression (Fischer et al., 2005) or none at all (Vande et al., 2004).

Impaired axonal transport may play a role in motor neuron disease. Support for these hypotheses was in the transgenic mutant SOD1 mouse where deficits in slow axonal transport occurred early in the disease course (Williamson & Cleveland, 1999). A link to human motor neuron disease was then made following the identification of a point mutation in the p150 subunit of dynactin. This protein complex is required for dynein-mediated retrograde transport of vesicles and organelles along microtubules. Clinically this was manifest in adult patients with vocal fold paralysis, progressive facial weakness, and weakness in the hands. Distal lower limb weakness occurred later in the course of disease (Puls et al., 2003). A potential relationship between dynactin mutations was later made in ALS patients suggesting that allelic variants in the dynactin gene may confer a genomic risk factor for the development of ALS (Munch et al., 2004).

Mutations in the dynein protein were found to cause a progressive motor neuron disorder in mice (Hafezparast et al., 2003) and overexpression of the protein dynamitin (part of the dynein-dynactin complex of microtubule transport) resulted in the development of late-onset motor neuron disease in a transgenic mouse model (LaMonte et al., 2002). Subsequent studies of the dynein-dynactin complex with mutant SOD1 demonstrated that an inhibition of microtubule transport from muscle to cell body occurred (Ligon et al., 2005). Similar defects in retrograde transport have also been described in the PMN mouse model of motor neuron disease with a more severe phenotype of clinical weakness when compared with other mutations affecting retrograde transport (Jablonka et al., 2004). Taken together, these data are potentially redefining definitions of "motor neuron disease" and suggest that mutations in neurofilaments and microtubules may influence the onset and severity of weakness previously attributed only to motor neuron loss.

A. Growth Factor Dysregulation

Vascular endothelial cell growth factor (VEGF) is a critical factor that controls the growth and permeability of blood vessels. Under conditions of hypoxia, VEGF can maintain and restore vascular perfusion of normal tissues as well as stimulate the growth of new blood vessels. The induction of VEGF in such situations is governed through transcription factors that react to low oxygen tension. The discovery of a possible role in ALS for VEGF stems from the genetic manipulation of the control mechanism responsible for the expression of inducible VEGF gene in mice. Studies in a transgenic mouse model in which the VEGF gene had the specific hypoxia-response element deleted reported that although mice maintained normal baseline levels of VEGF expression, there was a severe reduction in the ability to induce VEGF during bouts of hypoxia.

In a subset of these altered mice surviving through early development, profound and gradually increasing motor deficits were observed. They progressed to display all the hallmark features of ALS (i.e., accumulation of neurofilaments in motor neurons, degeneration of motor axons and muscle denervation). When a VEGF mutant mouse was crossed with the SOD1^{G93A} mouse, the course of the disease was accelerated, thus suggesting a potential neuroprotective role for

VEGF. The effects of VEGF on motor neurons could be in part through a direct action on these cells, serving as a neurotrophic or neuroprotective factor. *In vitro* studies have demonstrated that VEGF can support the survival of primary motor neurons and protect against cell death induced by hypoxia or serum deprivation (Oosthuyse et al., 2001). Alternatively it may act indirectly by regulating the blood supply to motor neurons, which consume high levels of energy to sustain a high rate of electrical firing. In ALS, VEGF was found to potentially be a modifier of the disease with patients who were homozygous for some VEGF haplotypes showing an increased risk of developing the disease (Lambrechts et al., 2003) although these findings do not appear to be consistent among all groups (Van Vught et al., 2005). VEGF levels were also found to be decreased in the CSF of ALS patients (Devos et al., 2004) and lack of VEGF upregulation in hypoxemic ALS patients suggesting VEGF dysregulation (Moreau et al., 2006).

VI. Neuroinflammation

Traditionally, a strong inflammatory response that is seen with infectious encephalitides has not been ascribed to ALS. Autoimmune mechanisms and modulators have found more reliable success in disorders such as multiple sclerosis and myasthenia gravis among others. However, it is now more evident that neuroinflammation, primarily in the form of microglial and astroglial activation, likely plays at least some role in ALS pathobiology. Although unlikely to be an initiating factor in the development of this disease, neuroinflammation may result in the propagation of disease following an initial insult. Evidence for the role of microglia comes both from human ALS tissues as well as mouse models of ALS. Activated microglia produce numerous inflammatory, proliferative, oxidative, and excitatory compounds with potential neuroprotective as well as neurotoxic effects. These compounds can initiate or become part of a cascade with interactions not only on neurons but other cell types as well. Comparison of postmortem spinal cord tissue from ALS patients shows that activated microglia are more abundant in ALS tissues (Henkel et al., 2004; Kawamata et al., 1992). The presence of activated microglia in the motor cortex, dorsolateral prefrontal cortex, thalamus, and pons of living ALS patients, and their absence in healthy controls, was shown by positron emission tomography (PET), using PK11195; ligand for the "peripheral benzodiazepine binding site" expressed by activated microglia (Turner et al., 2004). It is not only the presence of activated microglia present in ALS tissue but also the finding that a number of genes or gene-products associated with microglia are also increased in a variety of ALS tissues that implicates these cells and their cellular pathways in motor neuron degeneration (Sargsyan et al., 2005). How important are microglia to motor neuron pathology? A number of studies have documented elevated levels of microglial factors including interleukins, TNF, TGF, COX2, and interferons in mouse models of ALS. These factors seem to be increased in abundance and variety as disease progresses (Sargsyan et al., 2005); consistent with the hypothesis that these cells and their factors are part of a cascade following initial injury.

VII. Cell Autonomy in ALS—Contributions from Nonneuronal Cells

Although ALS is classified as a motor neuron disease it is more accurately a neurodegenerative disease. Is the progressive nature of this disorder purely a result of inherent motor neuron abnormalities or do other cell types play a role in the initiation or propagation of the disease? Studies of astrocytes proteins (e.g., EAAT2) and microglial protein have suggested that pathogenic cascades are not restricted to motor neurons. Furthermore, pathological studies in postmortem tissue have provided evidence for substantial loss of small interneurons in the cortex and spinal cord.

To determine whether neuron-specific expression of mutant SOD1 is sufficient to produce such an ALS phenotype, transgenic animals carrying the $SOD1^{G37R}$ mutation under the neurofilament light chain promoter were created. Although the transgenic animals expressed high levels of the human SOD1 protein in neuronal tissues, including the large motor neurons of the spinal cord, no apparent motor deficit was observed, suggesting that neuron-specific expression of ALS-associated mutant human SOD1 may not be sufficient for the development of the disease in mice (Pramatarova et al., 2001). Using a different neuron-specific promoter (Thy1), the hSOD1 (G93A) and hSOD1 (G85R) mutations were expressed in neurons. Neither of these mutations expressed solely in neurons resulted in signs of spinal cord pathology or disease in these transgenic mice (Lino et al., 2002). Conversely, the expression of mSOD1 in astrocytes under the GFAP promoter did not lead to motor neuron death but astrogliosis was noted around regions of intact motor neurons suggesting some abnormal pathology (Gong et al., 2000).

Chimeric mice (WT/mSOD1) reveal that simple expression of the mutant SOD1 in neurons/motor neurons was not sufficient to lead to neuronal death—a concomitant expression in glia was necessary. Furthermore, the chimeric animals WT/mSOD1 lived longer than nonchimeric mSOD1 mice to a degree proportional to their chimerism (Clement et al., 2003). Another powerful observation from this study was that wild-type motor neurons appeared to undergo degeneration, and the development of ubiquitinated inclusions (not typically seen in wild-type neurons) when surrounded by mSOD1 astroglia.

More specifically, the selective reduction in mSOD1 in motor neurons and microglia was carried out using Cre-Lox technology. To examine the role of mSOD1 expressed only in motor neurons to disease, LoxSOD1^{G37R} mice were mated to mice carrying a Cre-encoding sequence under control of the promoter from the Islet-1 transcription factor. This recombination was sufficient to substantially reduce mSOD1 accumulation in most motor axons of L5 motor roots and lumbar motor neurons of presymptomatic Isl1-Cre^{+}/LoxSOD1^{G37R} animals. This resulted in a modest delay in onset and progression of disease in these animals. To test the role of mSOD1 in microglia, mice expressing Cre selectively in these cells was carried out using the CD11b promoter. The resulting mice had little delay in the onset or early stages of disease progression but a more dramatic slowing of the later course of the disease—suggesting that microglial pathways are clearly contributors after disease onset, at least in the mouse model of the familial disease (Boillee et al., 2006).

VIII. Regional Differences in ALS and SOD1 Pathophysiology

Why does ALS present either with prominent upper motor neuron findings or predominantly lower motor neuron findings prior to progression to other regions?

An intriguing observation in mSOD1 biology is the dichotomy between the robust pathology consistently observed in mSOD1 G93A lumbar spinal cord when compared to very little pathology noted in the brain. This is particularly notable since mSOD1 is a ubiquitously expressed protein in the CNS. Examples of mSOD1 astrocyte-specific properties are notable for the lack of significant changes in the glutamate transporter subtype GLT1 levels previously described in SOD1^{G93A} cortex (Alexander et al., 2000), whereas a reduction in GLT1 is noted in the spinal cord of mSOD1 mice. Other investigators have ascribed selective damage in this model from the action of spinal cord-specific factors that recruit mutant SOD1 to spinal mitochondria. Thus, it appears that not only is there selectivity for certain cell subtypes but also highlights *regional* influences/differences between cells (astrocytes, motor neurons, and potentially other cell types) (Liu et al., 2004b).

Interestingly, the delivery of siRNA (targeting human SOD1) to motor neurons by injecting into muscles of mSOD1 mice and allowing for retrograde transport resulted in an improvement (but not complete sparing) in disease onset and prolongation of survival (Miller et al., 2005; Ralph et al., 2005). These findings, though noted to have the potential for therapeutic interventions, also highlight that mSOD1 mice continue to develop disease and again emphasize a role for other cell types besides motor neurons.

IX. Targeting Therapies to Molecular Pathways

This chapter has highlighted the heterogeneity of ALS in its clinical presentation and pathophysiology (see Table 20.2). A number of animal models of ALS have been developed, but it is becoming increasingly evident that many cell types are involved in either the initiation and/or propagation of the disease course. Nevertheless, our appreciation of the multiple molecular pathways in ALS using both animal models and human tissues is significantly greater than even a decade ago. It has also become clear that a single cure for the disease is unlikely to present itself, much like there is no single cure for most cancers. Rather, molecular targets for a variety of pathways have been developed and are being studied in both animal models and ALS patients. Some of these potential targets are reviewed next.

A. Modulating Glutamatergic Pathways in Neurotoxicity

To date, only one drug modulating a single pathway has been shown to be efficacious in ALS treatment—riluzole. Riluzole is currently the only FDA approved drug for treating ALS and appears to have several mechanisms of action including the inhibition of glutamic acid release, blockade of amino acid receptors, and inhibition of voltage-dependent sodium channels on dendrites and cell bodies (Doble, 1996).

Riluzole's efficacy was established in two important ALS clinical trials. In the first trial, a more robust effect in survival was seen in patients with bulbar-onset disease. The significance of this finding is not well-understood but the effect was clear. In the riluzole-treated group, 74 percent of patients were alive at 12 months compared with 58 percent in the placebo treated group (Bensimon et al., 1994). The second human clinical trial was much larger with 959 ALS patients treated for 18 months. The most efficacious dose of riluzole also was determined in this study. This study showed that at 18 months, survival rates were 50.4 percent for placebo and 56.8 percent for 100 mg/day riluzole. Adjustment for baseline prognostic factors showed a 35 percent decreased risk of death with the 100 mg dose compared with placebo (Lacomblez et al., 1996). The consistent results in these human clinical trials have led to the use of riluzole as a standard-of-care in the pharmacological treatment of ALS. The relative efficacy of riluzole spawned the study of other drugs with some degree of antiglutamate actions including topiramate, lamotrigine (Eisen et al., 1993b), dextromethorphan (Gredal et al., 1997), and gabapentin (Miller et al., 2001), none of which has proven beneficial in human clinical trials. A major problem with some of the failed trials was the use of inappropriate drug doses or poor trial design.

Table 20.2

Pathways	Molecular Mechanisms	Targeted Molecular Therapeutics
Free radical formation (oxidative stress)	Peroxynitrite and hydrogen peroxide with hydroxyl radicals	
Glutamate excitotoxicity	Reduced Glutamate transporter expression/dysfunction	Riluzole, Ceftriaxone
	Glutamate receptor abnormalities	
Apoptosis	Increases in caspase activation	Sodium phenylbutyrate
Axonal pathology	Neurofilament accumulation	
Growth factor dysregulation	Abnormalities in VEGF regulation	VEGF, IGF-1,
Mitochondrial dysfunction	Mitochondrial swelling, apoptotic cascades, energy failure	Creatine
Neuroinflammation	Elevated cytokines, interleukins, TNF-alpha	Minocycline, Thalidomide

B. Growth Factors

Neurotrophic factors have been studied extensively in mouse models of ALS and led to some of the first trials in ALS therapeutics. Trials of the neurotrophic factors brain-derived neurotrophic factor (BDNF), glial-derived neurotrophic factor (GDNF), and ciliary neurotrophic factor (CNTF) did not show an improvement in survival and had a significant number of side effects most notably debilitating anorexia, nausea, and vomiting (1999; Miller et al., 1996).

Insulin-like growth factor (IGF-1) is notable because its investigation in ALS is still ongoing. An initial study of (IGF-1) in ALS patients showed a beneficial effect (Lai et al., 1997) but a subsequent follow-up study failed to support a significant impact on survival (Borasio et al., 1998). One of the major concerns with delivery of any compound, particularly growth factors, is the delivery of these large molecular weight agents to the target cells. For ALS, this means crossing the blood–brain barrier or circumventing it. Some have speculated that the lack of effect in some studies highlight this point. In light of this, innovative investigations for the delivery of IGF-1 using adeno-associated viral vector injection into hindlimb and respiratory muscles with the subsequent retrograde transport into motor neurons showed both a delay in onset and prolonged survival in mutant SOD1 mice (Kaspar et al., 2003).

In vitro studies have demonstrated that VEGF can support the survival of primary motor neurons and protect against cell death induced by hypoxia or serum deprivation (Oosthuyse et al., 2001). Investigators also have shown that the delivery of VEGF to muscles with a lentiviral vector with subsequent retrograde transport to the spinal cord resulted in a significant prolongation in survival of the $SOD1^{G93A}$ mouse (Azzouz et al., 2004). Intracerebroventricular delivery of VEGF into an $SOD1^{G93A}$ rat model also resulted in a delay in hindlimb paralysis and prolonged survival—offering yet another method for more directed delivery of therapeutics of interest (Storkebaum et al., 2005).

C. Antioxidants

The antioxidant vitamin E delayed disease onset and slowed progression in mutant SOD1 mice but did not produce changes in survival (Gurney et al., 1996). In a human clinical trial, vitamin E delayed the progression of ALS from a mild to a more severe state but failed to increase survival (Desnuelle et al., 2001). N-acetyl-L-cysteine (NAC), an over-the-counter agent that reduces free radical damage, significantly prolonged survival and delayed onset of motor impairment in G93A mice treated with NAC compared to control mice (Andreassen et al., 2000). Its use in a human clinical trial, however, did not demonstrate an effect on survival or progression of the disease (Louwerse et al., 1995).

D. Neuroinflammation Modulators

Evidence for modulation of these neuroinflammatory properties was supported by three trials of minocycline in mouse models of motor neuron disease. Minocycline is known to block microglial activation. Administration of minocycline to transgenic mutant SOD1 mice resulted in a reduction in microglial activation and prolonged survival (Kriz et al., 2002; Van Den et al., 2002; Zhu et al., 2002). Clinical trials to study minocycline are underway.

Celecoxib, a COX2 inhibitor, was also effective in prolonging survival in the $SOD1^{G93A}$ mouse (Drachman et al., 2002). Studies showing increased levels of prostaglandin E2 (PGE2) in a small number of ALS cerebrospinal fluid (Almer et al., 2002; Ilzecka, 2003) specimens resulted in a large trial of celecoxib in ALS patients, which did not show any benefit in slowing ALS disease progression or survival although CSF prostaglandin levels were not found to be elevated in this study.

The neuroinflammatory mediators TNF-alpha and FasL have been shown to be elevated in mutant SOD1 mice presymptomatically. Thalidomide and lenalidomide inhibit the production of both TNF-alpha and FasL. Treatment of mSOD1 mice was beneficial in not only reducing the levels of these molecules but also resulted in an improvement of motor performance and survival. Thalidomide currently has approval for the treatment of severe erythema nodosum leprosum but has a history of causing severe birth defects, which had limited its use.

X. Targeting ALS Subgroups Using RNAi and Antisense Technologies

The ability to silence specific genes has been used in animal models to study the role of gene products in the development of disease. However, if the disease-causing gene is known (as with mutant SOD1 in some forms of familial ALS), then the selective reduction of these proteins may have therapeutic benefit as well. Because the use of RNAi and antisense is a relatively new technology, investigators have turned to animal models first to demonstrate the potential therapeutic benefits of such a strategy. Investigators using a lentiviral vector to deliver interfering RNA (RNAi) to spinal motor neurons of the $SOD1^{G93A}$ mouse by injection into the muscle showed that the RNAi was retrogradely transported into the spinal cord, resulting in a downregulation of SOD1 and an impressive prolongation of motor strength and survival (Ralph et al., 2005). Similar results were obtained using adeno-associated virus delivery of siRNA to muscle (Miller et al., 2005).

The direct intraspinal injection of RNAi using a lentivirus also demonstrated an improvement in survival in mutant SOD1 mice (Raoul et al., 2005). Experiments to date, however, were conducted well before animals developed disease symptoms; it is not yet known if disease can be substantially circumvented with delivery of drugs after disease onset—comparable to human treatment. Although this selective method does not have a broad therapeutic potential for sporadic ALS, it may serve those patients with SOD1 mutations and as a result, provide an important understanding about the potential for reversing clinical symptoms resulting from motor neurons that carry mutations, are dysfunctional, but not dead.

Antisense nucleotides are also capable of targeting specific RNA sequences of interest. Investigators targeted the GluR3 subunit of the AMPA receptor in the SOD1 mouse (Rembach et al., 2004) using antisense technology. The calcium-permeable GluR3 subunit appears to be upregulated in SOD1 tissues when compared with the less permeable GluR2 subunit. Following targeted antisense delivery against GluR3, a modest 10 person increase in survival for these mice was observed although the reduction in GluR3 protein could only be demonstrated *in vitro*.

XI. Predictive Value of Preclinical Models

The mouse models just described present great progress in translating genetic discoveries from ALS patients into research tools. Although the exact mechanism for each gene mutation has yet to be identified, the animal represents a powerful tool to discover the disease-causing events that result from the mutant proteins. Similarly, the animals allow us to test new therapeutic avenues, in hopes of more quickly discovering effective therapeutics. To date, all preclinical rodent therapeutic studies have been carried out in mutant SOD1 mice (or rats). The very first therapeutic study, of riluzole, demonstrated that the modest effect of the drug in humans could also be observed in rodents. Unfortunately, many subsequent studies, revealing far more potent therapies in mice, have not yielded clinically effective drugs in humans. There are many variables that may account for this disappointing discrepancy (e.g., lack of appropriate drug dosing in humans) and the possibility that drug potency in familial ALS models does not translate to the more common sporadic disease. The use of other mouse models (e.g., dynactin mutations) along with novel *in vitro* drug screening strategies to mimic molecular and biochemical pathways implicated in ALS are underway.

References

(1999). A controlled trial of recombinant methionyl human BDNF in ALS: The BDNF Study Group (Phase III). *Neurology* **52**, 1427–1433.

Alexander, G. M., Deitch, J. S., Seeburger, J. L., Del Valle, L., Heiman-Patterson, T. D. (2000). Elevated cortical extracellular fluid glutamate in transgenic mice expressing human mutant (G93A) Cu/Zn superoxide dismutase. *J Neurochem* **74**, 1666–1673.

Almer, G., Teismann, P., Stevic, Z., Halaschek-Wiener, J., Deecke, L., Kostic, V., Przedborski, S. (2002). Increased levels of the pro-inflammatory prostaglandin PGE2 in CSF from ALS patients. *Neurology* **58**, 1277–1279.

Andreassen, O. A., Dedeoglu, A., Klivenyi, P., Beal, M. F., Bush, A. I. (2000). N-acetyl-L-cysteine improves survival and preserves motor performance in an animal model of familial amyotrophic lateral sclerosis. *Neuroreport* **11**, 2491–2493.

Andrus, P. K., Fleck, T. J., Gurney, M. E., Hall, E. D. (1998). Protein oxidative damage in a transgenic mouse model of familial amyotrophic lateral sclerosis. *J Neurochem* **71**, 2041–2048.

Azzouz, M., Ralph, G. S., Storkebaum, E., Walmsley, L. E., Mitrophanous, K. A., Kingsman, S. M. et al. (2004) VEGF delivery with retrogradely transported lentivector prolongs survival in a mouse ALS model. *Nature* **429**, 413–417.

Bacman, S. R., Bradley, W. G., Moraes, C. T. (2006). Mitochondrial involvement in amyotrophic lateral sclerosis: trigger or target? *Mol Neurobiol* **33**, 113–131.

Beal, M. F., Ferrante, R. J., Browne, S. E., Matthews, R. T., Kowall, N. W., Brown, R. H. (1997). Increased 3-nitrotyrosine in both sporadic and familial amyotrophic lateral sclerosis. *Ann Neurol* **42**, 644–654.

Bensimon, G., Lacomblez, L., Meininger, V., The ALS/Riluzole Study Group (1994). A controlled trial of riluzole in amyotrophic lateral sclerosis. *N Engl J Med* **330**, 585–591.

Boillee, S., Peschanski, M., Junier, M. P. (2003). The wobbler mouse: A neurodegeneration jigsaw puzzle. *Mol Neurobiol* **28**, 65–106.

Boillee, S., Yamanaka, K., Lobsiger, C. S., Copeland, N. G., Jenkins, N. A., Kassiotis, G. et al. (2006). Onset and progression in inherited ALS determined by motor neurons and microglia. *Science* **312**, 1389–1392.

Borasio, G. D., Robberecht, W., Leigh, P. N., Emile, J., Guiloff, R. J., Jerusalem, F. et al. (1998). A placebo-controlled trial of insulin-like growth factor-I in amyotrophic lateral sclerosis. European ALS/IGF-I Study Group. *Neurology* **51**, 583–586.

Bristol, L. A., Rothstein, J. D. (1996). Glutamate transporter gene expression in amyotrophic lateral sclerosis motor cortex. *Ann Neurol* **39**, 676–679.

Bruijn, L. I., Beal, M. F., Becher, M. W., Schultz, J. B., Wong, P. C., Price, D. L., Cleveland, D. W. (1997a). Elevated free levels, but not protein-bound nitrotyrosine or hydroxyl radicals, throughout amyotrophic lateral sclerosis (ALS)-like disease implicate tyrosine nitration as an aberrant property of one familial ALS-linked superoxide dismutase 1 mutant. *Proc Natl Acad Sci* **94**, 7606–7611.

Bruijn, L. I., Becher, M. W., Lee, M. K., Anderson, K. L., Jenkins, N. A., Copeland, N. G. et al. (1997b). ALS-linked SOD1 mutant G85R mediates damage to astrocytes and promotes rapidly progressive disease with SOD1-containing inclusions. *Neuron* **18**, 327–338.

Bruijn, L. I., Houseweart, M. K., Kato, S., Anderson, K. L., Anderson, S. D., Ohama, E. et al. (1998). Aggregation and motor neuron toxicity of an ALS-linked SOD1 mutant independent from wild-type SOD1. *Science* **281**, 1851–1854.

Cai, H., Lin, X., Xie, C., Laird, F. M., Lai, C., Wen, H. et al. (2005). Loss of ALS2 function is insufficient to trigger motor neuron degeneration in knock-out mice but predisposes neurons to oxidative stress. *J Neurosci* **25**, 7567–7574.

Clement, A. M., Nguyen, M. D., Roberts, E. A., Garcia, M. L., Boillee, S., Rule, M. et al. (2003). Wild-type nonneuronal cells extend survival of SOD1 mutant motor neurons in ALS mice. *Science* **302**, 113–117.

Coleman, M. P., Conforti, L., Buckmaster, E. A., Tarlton, A., Ewing, R. M., Brown, M.C. et al. (1998). An 85-kb tandem triplication in the slow Wallerian degeneration (Wlds) mouse. *Proc Natl Acad Sci U S A* **95**, 9985–9990.

Conforti, L., Tarlton, A., Mack, T. G., Mi, W., Buckmaster, E. A., Wagner, D. et al. (2000). A Ufd2/D4Cole1e chimeric protein and overexpression of Rbp7 in the slow Wallerian degeneration (WldS) mouse. *Proc Natl Acad Sci U S A* **97**, 11377–11382.

CouillardDespres, S., Zhu, Q. Z., Wong, P. C., Price, D. L., Cleveland, D. W., Julien, J. P. (1998). Protective effect of neurofilament heavy gene overexpression in motor neuron disease induced by mutant superoxide dismutase. *Proc Natl Acad Sci U S A* **95**, 9626–9630.

Desnuelle, C., Dib, M., Garrel, C., Favier, A. (2001). A double-blind, placebo-controlled randomized clinical trial of alpha-tocopherol (vitamin E) in the treatment of amyotrophic lateral sclerosis. ALS riluzole-tocopherol Study Group. *Amyotroph Lateral Scler Other Motor Neuron Disord* **2**, 9–18.

Devos, D., Moreau, C., Lassalle, P., Perez, T., De Seze, J., Brunaud-Danel, V. et al. (2004). Low levels of the vascular endothelial growth factor in CSF from early ALS patients. *Neurology* **62**, 2127–2129.

Doble, A. (1996). The pharmacology and mechanism of action of riluzole. *Neurology* **47**, S233–S241.

Drachman, D. B., Frank, K., Dykes-Hoberg, M., Teismann, P., Almer, G., Przedborski, S., Rothstein, J. D. (2002). Cyclooxygenase 2 inhibition protects motor neurons and prolongs survival in a transgenic mouse model of ALS. *Ann Neurol* **52**, 771–778.

Eisen, A., Schulzer, M., MacNeil, M., Pant, B., Mak, E. (1993a). Duration of amyotrophic lateral sclerosis is age dependent. *Muscle Nerve* **16**, 27–32.

Eisen, A., Stewart, H., Schulzer, M., Cameron, D. (1993b). Anti-glutamate therapy in amyotrophic lateral sclerosis: A trial using lamotrigine. *Can J Neurol Sci* **20**, 297–301.

Ferri, A., Sanes, J. R., Coleman, M. P., Cunningham, J. M., Kato, A. C. (2003). Inhibiting axon degeneration and synapse loss attenuates apoptosis and disease progression in a mouse model of motoneuron disease. *Curr Biol* **13**, 669–673.

Fischer, L. R., Culver, D. G., Davis, A. A., Tennant, P., Wang, M., Coleman, M. et al. (2005). The WldS gene modestly prolongs survival in the SOD1G93A fALS mouse. *Neurobiol Dis* **19**, 293–300.

Fischer, L. R., Culver, D. G., Tennant, P., Davis, A. A., Wang, M., Castellano-Sanchez, A. et al. (2004). Amyotrophic lateral sclerosis is a distal axonopathy: Evidence in mice and man. *Exp Neurol* **185**, 232–240.

Frey, D., Schneider, C., Xu, L., Borg, J., Spooren, W., Caroni, P. (2000). Early and selective loss of neuromuscular synapse subtypes with low sprouting competence in motoneuron diseases. *J Neurosci* **20**, 2534–2542.

Garcia, M. L., Singleton, A. B., Hernandez, D., Ward, C. M., Evey, C., Sapp, P. A. et al. (2006). Mutations in neurofilament genes are not a significant primary cause of non-SOD1-mediated amyotrophic lateral sclerosis. *Neurobiol Dis* **21**, 102–109.

Gong, Y. H., Parsadanian, A. S., Andreeva, A., Snider, W. D., Elliott, J. L. (2000). Restricted expression of G86R Cu/Zn superoxide dismutase in astrocytes results in astrocytosis but does not cause motoneuron degeneration. *J Neurosci* **20**, 660–665.

Gredal, O., Werdelin, L., Bak, S., Christensen, P.B., Boysen, G., Kristensen, M. O. et al. (1997). A clinical trial of dextromethorphan in amyotrophic lateral sclerosis. *Acta Neurol Scand* **96**, 8–13.

Guegan, C., Vila, M., Rosoklija, G., Hays, A. P., Przedborski, S. (2001). Recruitment of the mitochondrial–dependent apoptotic pathway in amyotrophic lateral sclerosis. *J Neurosci* **21**, 6569–6576.

Guo, H., Lai, L., Butchbach, M. E., Stockinger, M. P., Shan, X., Bishop, G. A., Lin, C. L. (2003). Increased expression of the glial glutamate transporter EAAT2 modulates excitotoxicity and delays the onset but not the outcome of ALS in mice. *Hum Mol Genet* **12**, 2519–2532.

Gurney, M. E., Cuttings, F. B., Zhai, P., Doble, A., Taylor, C. P., Andrus, P. K., Hall, E. D. (1996). Benefit of vitamin E, riluzole, and gabapentin in a transgenic model of familial amyotrophic lateral sclerosis. *Ann Neurol* **39**, 147–157.

Gurney, M. E., Pu, H., Chiu, A. Y., Dal Canto, M. C., Polchow, C. Y., Alexander, D.D. et al. (1994a). Motor neuron degeneration in mice that express a human Cu,Zn superoxide dismutase mutation [see comments]. *Science* **264**, 1772–1775.

Gurney, M. E., Pu, H., Chiu, A. Y., Dal Canto, M. C., Polchow, C. Y., Alexander, D. D. et al. (1994b). Motor neuron degeneration in mice that express a human Cu,Zn superoxide dismutase mutation. *Science* **264**, 1772–1775.

Hadano, S., Benn, S. C., Kakuta, S., Otomo, A., Sudo, K., Kunita, R. et al. (2006). Mice deficient in the Rab5 guanine nucleotide exchange factor ALS2/alsin exhibit age-dependent neurological deficits and altered endosome trafficking. *Hum Mol Genet* **15**, 233–250.

Hafezparast, M. et al. (2003). Mutations in dynein link motor neuron degeneration to defects in retrograde transport. *Science* **300**, 808–812.

Hall, E. D., Andrus, P. K., Oostveen, J. A., Fleck, T. J., Gurney, M. E. (1998). Relationship of oxygen radical-induced lipid peroxidative damage to disease onset and progression in a transgenic model of familial ALS. *J Neurosci Res* **53**, 66–77.

Haverkamp, L. J., Appel, V., Appel, S. H. (1995). Natural history of amyotrophic lateral sclerosis in a database population. Validation of a scoring system and a model for survival prediction. *Brain* **118**, 707–719.

Henkel, J. S., Engelhardt, J. I., Siklos, L., Simpson, E. P., Kim, S. H., Pan, T. et al. (2004). Presence of dendritic cells, MCP-1, and activated microglia/macrophages in amyotrophic lateral sclerosis spinal cord tissue. *Ann Neurol* **55**, 221–235.

Ilzecka, J. (2003). Prostaglandin E2 is increased in amyotrophic lateral sclerosis patients. *Acta Neurol Scand* **108**, 125–129.

Jablonka, S., Wiese, S., Sendtner, M. (2004). Axonal defects in mouse models of motoneuron disease. *J Neurobiol* **58**, 272–286.

Kaspar, B. K., Llado, J., Sherkat, N., Rothstein, J. D., Gage, F. H. (2003). Retrograde viral delivery of IGF-1 prolongs survival in a mouse ALS model. *Science* **301**, 839–842.

Kawahara, Y., Kwak, S., Sun, H., Ito, K., Hashida, H., Aizawa, H. et al. (2003). Human spinal motoneurons express low relative abundance of GluR2 mRNA: An implication for excitotoxicity in ALS. *J Neurochem* **85**, 680–689.

Kawamata, T., Akiyama, H., Yamada, T., McGeer, P. L. (1992). Immunologic reactions in amyotrophic lateral sclerosis brain and spinal cord tissue. *Am J Pathol* **140**, 691–707.

Kennel, P. F., Finiels, F., Revah, F., Mallet, J. (1996). Neuromuscular function impairment is not caused by motor neurone loss in FALS mice: An electromyographic study. *Neuroreport* **7**, 1427–1431.

Kostic, V., Jackson-Lewis, V., de Bilbao, F., Dubois-Dauphin, M., Przedborski, S. (1997). Bcl-2: Prolonging life in a transgenic mouse model of familial amyotrophic lateral sclerosis. *Science* **277**, 559–562.

Kriz, J., Nguyen, M. D., Julien, J. P. (2002). Minocycline slows disease progression in a mouse model of amyotrophic lateral sclerosis. *Neurobiol Dis* **10**, 268–278.

Kuner, R., Groom, A. J., Bresink, I., Kornau, H. C., Stefovska, V., Muller, G. et al. (2005). Late-onset motoneuron disease caused by a functionally modified AMPA receptor subunit. *Proc Natl Acad Sci U S A*.

Lacomblez, L., Bensimon, G., Leigh, P. N., Guillet, P., Powe, L., Durrleman, S. et al. (1996). A confirmatory dose-ranging study of riluzole in ALS. ALS/Riluzole Study Group-II. *Neurology* **47**, S242–S250.

Lai, E. C., Felice, K. J., Festoff, B. W., Gawel, M. J., Gelinas, D. F., Kratz, R. et al. (1997). Effect of recombinant human insulin-like growth factor-I on progression of ALS. A placebo-controlled study. The North America ALS/IGF-I Study Group. *Neurology* **49**, 1621–1630.

Lambrechts, D. et al. (2003). VEGF is a modifier of amyotrophic lateral sclerosis in mice and humans and protects motoneurons against ischemic death. *Nat Genet* **34**, 383–394.

LaMonte, B. H., Wallace, K. E., Holloway, B. A., Shelly, S. S., Ascano, J., Tokito, M. et al. (2002). Disruption of dynein/dynactin inhibits axonal transport in motor neurons causing late-onset progressive degeneration. *Neuron* **34**, 715–727.

Li, M., Ona, V. O., Guegan, C., Chen, M., Jackson-Lewis, V., Andrews, L. J. et al. (2000). Functional role of caspase-1 and caspase-3 in an ALS transgenic mouse model. *Science* **288**, 335–339.

Ligon, L. A., LaMonte, B. H., Wallace, K. E., Weber, N., Kalb, R. G., Holzbaur, E. L. (2005). Mutant superoxide dismutase disrupts cytoplasmic dynein in motor neurons. *Neuroreport* **16**, 533–536.

Lin, C. G., Bristol, L. A., Jin, L., DykesHoberg, M., Crawford, T., Clawson, L., Rothstein, J. D. (1998). Aberrant RNA processing in a neurodegenerative disease: The cause for absent EAAT2 a glutamate transporter, in amyotrophic lateral sclerosis. *Neuron* **20**, 589–602.

Lino, M. M., Schneider, C., Caroni, P. (2002). Accumulation of SOD1 mutants in postnatal motoneurons does not cause motoneuron pathology or motoneuron disease. *J Neurosci* **22**, 4825–4832.

Liu, J., Lillo, C., Jonsson, P. A., Vande, V. C., Ward, C. M., Miller, T. M. et al. (2004a). Toxicity of familial ALS-linked SOD1 mutants from selective recruitment to spinal mitochondria. *Neuron* **43**, 5–17.

Liu, J., Lillo, C., Jonsson, P. A., Vande, V. C., Ward, C. M., Miller, T. M. et al. (2004b). Toxicity of familial ALS-linked SOD1 mutants from selective recruitment to spinal mitochondria. *Neuron* **43**, 5–17.

Lobsiger, C. S., Garcia, M. L., Ward, C. M., Cleveland, D. W. (2005). Altered axonal architecture by removal of the heavily phosphorylated neurofilament tail domains strongly slows superoxide dismutase 1 mutant-mediated ALS. *Proc Natl Acad Sci U S A* **102**, 10351–10356.

Lomen-Hoerth, C., Anderson, T., Miller, B. (2002). The overlap of amyotrophic lateral sclerosis and frontotemporal dementia. *Neurology* **59**, 1077–1079.

Lomen-Hoerth, C., Murphy, J., Langmore, S., Kramer, J. H., Olney, R. K., Miller, B. (2003). Are amyotrophic lateral sclerosis patients cognitively normal? *Neurology* **60**, 1094–1097.

Louwerse, E. S., Weverling, G. J., Bossuyt, P. M., Meyjes, F. E., de Jong, J. M. (1995). Randomized, double-blind, controlled trial of acetylcysteine in amyotrophic lateral sclerosis. *Arch Neurol* **52**, 559–564.

Lunn, E. R., Perry, V. H., Brown, M. C., Rosen, H., Gordon, S. (1989). Absence of wallerian degeneration does not hinder regeneration in peripheral nerve. *Eur J Neurosci* **1**, 27–33.

Manetto, V., Sternberger, N. H., Perry, G., Sternberger, L. A., Gambetti, P. (1988). Phosphorylation of neurofilaments is altered in amyotrophic lateral sclerosis. *J Neuropathol Exp Neurol* **47**, 642–653.

Martin, N., Jaubert, J., Gounon, P., Salido, E., Haase, G., Szatanik, M., Guenet, J. L. (2002). A missense mutation in Tbce causes progressive motor neuronopathy in mice. *Nat Genet* **32**, 443–447.

Miller, R. G. et al. (2001). Phase III randomized trial of gabapentin in patients with amyotrophic lateral sclerosis. *Neurology* **56**, 843–848.

Miller, R. G., Petajan, J. H., Bryan, W. W., Armon, C., Barohn, R. J., Goodpasture, J. C. et al. (1996). A placebo-controlled trial of recombinant human ciliary neurotrophic (rhCNTF) factor in amyotrophic lateral sclerosis. rhCNTF ALS Study Group. *Ann Neurol* **39**, 256–260.

Miller, T. M., Kaspar, B. K., Kops, G. J., Yamanaka, K., Christian, L. J., Gage, F. H., Cleveland, D. W. (2005). Virus-delivered small RNA silencing sustains strength in amyotrophic lateral sclerosis. *Ann Neurol* **57**, 773–776.

Mitsumoto, H., Chad, D., Pioro, E. P. (1998). Amyotrophic lateral sclerosis. Philadelphia: F. A. Davis Company.

Moreau, C., Devos, D., Brunaud-Danel, V., Lefebvre, L., Perez, T., Destee, A. et al. (2006). Paradoxical response of VEGF expression to hypoxia in CSF of patients with ALS. *J Neurol Neurosurg Psychiatry* **77**, 255–257.

Morrison, B. M., Gordon, J. W., Ripps, M. E., Morrison, J. H. (1996). Quantitative immunocytochemical analysis of the spinal cord in G86R superoxide dismutase transgenic mice: Neurochemical correlates of selective vulnerability. *J Comp Neurol* **373**, 619–631.

Munch, C., Sedlmeier, R., Meyer, T., Homberg, V., Sperfeld, A. D., Kurt, A. et al. (2004). Point mutations of the p150 subunit of dynactin (DCTN1) gene in ALS. *Neurology* **63**, 724–726.

Munoz, D. G., Greene, C., Perl, D. P., Selkoe, D. J. (1988). Accumulation of phosphorylated neurofilaments in anterior horn motoneurons of amyotrophic lateral sclerosis patients. *J Neuropathol Exp Neurol* **47**, 9–18.

Nguyen, M. D., Lariviere, R. C., Julien, J. P. (2001). Deregulation of Cdk5 in a mouse model of ALS: Toxicity alleviated by perikaryal neurofilament inclusions. *Neuron* **30**, 135–147.

Oosthuyse, B. et al. (2001). Deletion of the hypoxia-response element in the vascular endothelial growth factor promoter causes motor neuron degeneration. *Nat Genet* **28**, 131–138.

Pasinelli, P., Houseweart, M. K., Brown, R. H., Jr., Cleveland, D. W. (2000). Caspase-1 and -3 are sequentially activated in motor neuron death in Cu,Zn superoxide dismutase-mediated familial amyotrophic lateral sclerosis. *Proc Natl Acad Sci U S A* **97**, 13901–13906.

Pramatarova, A., Laganiere, J., Roussel, J., Brisebois, K., Rouleau, G. A. (2001). Neuron-specific expression of mutant superoxide dismutase 1 in transgenic mice does not lead to motor impairment. *J Neurosci* **21**, 3369–3374.

Puls, I., Jonnakuty, C., LaMonte, B. H., Holzbaur, E. L., Tokito, M., Mann, E. et al. (2003). Mutant dynactin in motor neuron disease. *Nat Genet* **33**, 455–456.

Ralph, G. S., Radcliffe, P. A., Day, D. M., Carthy, J. M., Leroux, M. A., Lee, D. C. et al. (2005). Silencing mutant SOD1 using RNAi protects against neurodegeneration and extends survival in an ALS model. *Nat Med* **11**, 429–433.

Raoul, C., Abbas-Terki, T., Bensadoun, J. C., Guillot, S., Haase, G., Szulc, J. et al. (2005). Lentiviral-mediated silencing of SOD1 through RNA interference retards disease onset and progression in a mouse model of ALS. *Nat Med* **11**, 423–428.

Reaume, A. G., Elliott, J. L., Hoffman, E. K., Kowall, N. W., Ferrante, R. J., Siwek, D. F. et al. (1996). Motor neurons in Cu/Zn superoxide dismutase-deficient mice develop normally but exhibit enhanced cell death after axonal injury. *Nat Genet*.

Rembach, A., Turner, B. J., Bruce, S., Cheah, I. K., Scott, R. L., Lopes, E. C. et al. (2004). Antisense peptide nucleic acid targeting GluR3 delays disease onset and progression in the SOD1 G93A mouse model of familial ALS. *J Neurosci Res* **77**, 573–582.

Rosen, D. R. et al. (1993). Mutations in Cu/Zn superoxide dismutase gene are associated with familial amyotrophic lateral sclerosis. *Nature* **362**, 59–62.

Rothstein, J. D., Patel, S., Regan, M. R., Haenggeli, C., Huang, Y. H., Bergles, D. E. et al. (2005). Beta-lactam antibiotics offer neuroprotection by increasing glutamate transporter expression. *Nature* **433**, 73–77.

Sargsyan, S. A., Monk, P. N., Shaw, P. J. (2005). Microglia as potential contributors to motor neuron injury in amyotrophic lateral sclerosis. *Glia* **51**, 241–253.

Schmalbruch, H., Jensen, H. J., Bjaerg, M., Kamieniecka, Z., Kurland, L. (1991). A new mouse mutant with progressive motor neuronopathy. *J Neuropathol Exp Neurol* **50**, 192–204.

Schmitt-John, T., Drepper, C., Mussmann, A., Hahn, P., Kuhlmann, M., Thiel, C. et al. (2005). Mutation of Vps54 causes motor neuron disease and defective spermiogenesis in the wobbler mouse. *Nat Genet* **37**, 1213–1215.

Sobue, G., Hashizume, Y., Yasuda, T., Mukai, E., Kumagai, T., Mitsuma, T., Trojanowski, J. Q. (1990). Phosphorylated high molecular weight neurofilament protein in lower motor neurons in amyotrophic lateral sclerosis and other neurodegenerative diseases involving ventral horn cells. *Acta Neuropathol (Berl)* **79**, 402–408.

Storkebaum, E. et al. (2005). Treatment of motoneuron degeneration by intracerebroventricular delivery of VEGF in a rat model of ALS. *Nat Neurosci* **8**, 85–92.

Tortarolo, M., Grignaschi, G., Calvaresi, N., Zennaro, E., Spaltro, G., Colovic, M. et al. (2006). Glutamate AMPA receptors change in motor neurons of SOD1G93A transgenic mice and their inhibition by a noncompetitive antagonist ameliorates the progression of amytrophic lateral sclerosis-like disease. *J Neurosci Res* **83**, 134–146.

Trotti, D., Rolfs, A., Danbolt, N. C., Brown, R. H., Jr., Hediger, M. A. (1999). SOD1 mutants linked to amyotrophic lateral sclerosis selectively inactivate a glial glutamate transporter [In Process Citation]. *Nat Neurosci* **2**, 427–433.

Tu, P. H., Raju, P., Robinson, K. A., Gurney, M. E., Trojanowski, J. Q., Lee, V. M. Y. (1996). Transgenic mice carrying a human mutant superoxide dismutase transgene develop neuronal cytoskeletal pathology resembling human amyotrophic lateral sclerosis lesions. *Proc Natl Acad Sci U S A* J1 - PNAS **93**, 3155–3160.

Turner, M. R., Cagnin, A., Turkheimer, F. E., Miller, C. C., Shaw, C. E., Brooks, D. J. et al. (2004). Evidence of widespread cerebral microglial activation in amyotrophic lateral sclerosis: An [11C](R)-PK11195 positron emission tomography study. *Neurobiol Dis* **15**, 601–609.

Van Damme, P., Leyssen, M., Callewaert, G., Robberecht, W., Van Den, B. L. (2003). The AMPA receptor antagonist NBQX prolongs survival in a transgenic mouse model of amyotrophic lateral sclerosis. *Neurosci Lett* **343**, 81–84.

Van Den, B. L., Tilkin, P., Lemmens, G., Robberecht, W. (2002). Minocycline delays disease onset and mortality in a transgenic model of ALS. *Neuroreport* **13**, 1067–1070.

Van Vught, P. W., Sutedja, N. A., Veldink, J. H., Koeleman, B. P., Groeneveld, G. J., Wijmenga, C. et al. (2005). Lack of association between VEGF polymorphisms and ALS in a Dutch population. *Neurology* **65**, 1643–1645.

Vande, V. C., Garcia, M. L., Yin, X., Trapp, B. D., Cleveland, D. W. (2004). The neuroprotective factor Wlds does not attenuate mutant SOD1-mediated motor neuron disease. *Neuromolecular Med* **5**, 193–203.

Vukosavic, S., Dubois-Dauphin, M., Romero, N., Przedborski, S. (1999). Bax and Bcl-2 interaction in a transgenic mouse model of familial amyotrophic lateral sclerosis. *J Neurochem* **73**, 2460–2468.

Vukosavic, S., Stefanis, L., Jackson-Lewis, V., Guegan, C., Romero, N., Chen, C. se retardation in a transgenic mouse model of amyotrophic lateral sclerosis. *J Neurosci* **20**, 9119–9125.

Williamson, T. L., Bruijn, L. I., Zhu, Q., Anderson, K. L., Anderson, S. D., Julien, J. P., Cleveland, D. W. (1998). Absence of neurofilaments reduces the selective vulnerability of motor neurons and slows disease caused by a familial amyotrophic lateral sclerosis-linked superoxide dismutase 1 mutant. *Proc Natl Acad Sci U S A* **95**, 9631–9636.

Williamson, T. L., Cleveland, D. W. (1999). Slowing of axonal transport is a very early event in the toxicity of ALS-linked SOD1 mutants to motor neurons. *Nat Neurosci* **2**, 50–56.

Wong, P. C., Pardo, C. A., Borchelt, D. R., Lee, M. K., Copeland, N. G., Jenkins, N. A. et al. (1995). An adverse property of a familial ALS-linked SOD1 mutation causes motor neuron disease characterized by vacuolar degeneration of mitochondria. *Neuron* **14**, 1105–1116.

Yoshida, S., Mulder, D. W., Kurland, L. T., Chu, C. P., Okazaki, H. (1986). Follow-up study on amyotrophic lateral sclerosis in Rochester, Minn., 1925 through 1984. *Neuroepidemiology* **5**, 61–70.

Zhu, S., Stavrovskaya, I. G., Drozda, M., Kim, B. Y., Ona, V., Li, M. et al. (2002). Minocycline inhibits cytochrome c release and delays progression of amyotrophic lateral sclerosis in mice. *Nature* **417**, 74–78.

21

Genetic Disorders of the Autonomic Nervous System

Stephen J. Peroutka

I. Introduction

The autonomic nervous system (ANS) regulates physiological actions and functions that are not under conscious control. It consists of two major subdivisions, the efferent and afferent systems. The efferent portion of the ANS is further subdivided to the sympathetic, parasympathetic, and enteric nervous systems. Each of these four major components of the ANS plays a specific role in autonomic function.

Recent genetic findings have advanced the molecular understanding of the ANS. Two major types of advances have occurred within the last decade. In the first case, rare genetic variants have been identified that led to developmental abnormalities of the ANS. These disorders affect very few people but define the role of the specific genes in development of the ANS. In the second case, more common genetic variants have been identified that explain some of the variation in autonomic function within the general population.

This chapter therefore is divided into two major sections. The first and largest section reviews the rare genetic mutations that give rise to developmental abnormalities of the ANS that are usually identified at, or shortly after, birth. The second section reviews human genetic variants, some of which are fairly common, that alter the function of the ANS in both children as well as adults. In general, common polymorphisms (i.e., those occurring in >1% of the general population) cause variations in autonomic function that are relatively mild. Since molecular genetic advances in elucidating ANS development and function continue to occur rapidly, this chapter should be considered a brief overview of a dynamic subject. A much more thorough and continually updated listing of molecular genetic data can be obtained via a search of Online Mendelian Inheritance in Man at the following Internet address: www.ncbi.nlm.nih.gov/entrez/query.fcgi?db=OMIM&cmd=Limits.

II. Developmental Abnormalities of the Autonomic Nervous System

A. Developmental Disorders that Include the ANS

Multiple specific genetic mutations have been identified that impair the normal development of the ANS (see Table 21.1) in addition to a wide variety of other organ defects. For example, Hirschsprung Disease (also known as aganglionic megacolon) is a congenital disorder characterized by

Table 21.1 Human Molecular Genetic Variations Affecting Autonomic Nervous System Function

Gene Symbol	Gene Name	CHR	Polymorphism	Estimated Frequency	Autonomic Clinical Features
ECE1	Endothelin converting enzyme	1p36.1	ARG742CYS	< 0.0002%	Neural Crest Developmental Abnormalities (e.g., Hirschsprung disease, cardiac defects, essential hypertension)
EDN3	Endothelin 3	20q13.2–13.3	multiple	< 0.0002%	Neural Crest Developmental Abnormalities (e.g., Hirschsprung disease, Waardenburg-Shah syndrome, central hypoventilation syndrome)
EDNRB	Endothelin receptor B	13q22	multiple	< 0.0002%	Neural Crest Developmental Abnormalities (e.g., Hirschsprung disease, Waardenburg-Shah syndrome, ABCD syndrome)
SOX10	SRY-BOX 10	22q13	multiple	< 0.0002%	Neural Crest Developmental Abnormalities (e.g., Hirschsprung disease, Waardenburg-Shah syndrome, peripheral demyleinating neuropathy)
PHOX2B	Paired-like Homeobox 2B	4p12	multiple	< 0.0002%	Neural Crest Developmental Abnormalities (e.g., Hirschsprung disease, neuroblastoma, central hypoventilation syndrome)
RET	Rearranged transfection proto-oncogene (tyrosine kinase receptor for GDNF)	10q11.2	multiple	< 0.0002%	Neural Crest Developmental Abnormalities (e.g., Hirschsprung disease, central hypoventilation syndrome, multiple endocrine neoplasia syndrome, thyroid carcinoma)
GDNF	Glial derived neurotrophic factor	5p13.1–p12	multiple	< 0.0002%	Neural Crest Developmental Abnormalities (e.g., Hirschsprung disease)
SPTLC1	Serine palmitoyltransferase, long chain base subunit 1	9q22.1–q22.3	multiple	< 0.0002%	Hereditary Sensory Neuropathy, Type I, HSN1; HSAN1 (sensorimotor axonal neuropathy)
HSN2	HSN2	12p13.33	multiple	< 0.0002%	HSN2 (large and small axon loss)
IKBKAP	Inhibitor of kappa light polypeptide gene enhancer in B cells, kinase complex-associated protein	9q31	multiple	< 0.0002%	Hereditary Sensory and Autonomic Neuropathy, Type III, HSAN3; Familial dysautonomia; Riley-Day syndrome (large and small axon loss)
NTRK1	Neurotrophic tyrosine kinase receptor, type 1	1q21–q22	multiple	< 0.0002%	Congenital Insensitivity to Pain with Anhidrosis (CIPA); also called Hereditary Sensory and Autonomic Neuropathy, Type IV, HSAN4, familial dysautonomia, type II, congenital sensory neuropathy with anhidrosis (C-fiber loss)
NGFB	Nerve Growth Factor, Beta subunit	1p13.1	ARG211TRP	< 0.0002%	Hereditary Sensory and Autonomic Neuropathy, Type V, HSAN5 (C-fiber and A-delta fiber loss)
SCN9A	Sodium Channel Voltage-gated Type IX, Alpha Subunit	2q24	I848Tand L858H	< 0.0002%	Hereditary Erythermalgia (erythromelagia)
ADRA2B	Alpha-2B-adrenergic receptor	2	3 glutamic acid deletion	13% homozygotes	Increased SNS activity
ADRB2	Beta-2-adrenergic receptor	5q32–q34	THR164ILE	4%	Reduced response to beta-2-adrenoreceptor agonist; increased mortality from congestive heart failure
DBH	Dopamine beta-hydroxylase	9q34	multiple	< 0.0002%	Norepinephrine deficiency secondary to inability to convert endogenous dopamine to norpepinephrine
DRD4	Dopamine receptor D4	11p15.5	13-bp deletion	2%	Autonomic hyperactivity
TTR	Transthyretin	18q11.2–q12.1	VAL30MET and multiple others	1.5%	Amyloid polyneuropathy
SLC6A2	Solute carrier family 6 (neurotransmitter transporter, noradrenaline), member 2	16q12.2	ALA457PRO	< 0.0002%	Orthostatic intolerance

the absence of the enteric ganglia along a variable length of the intestine, leading to functional bowel obstruction and distension shortly after birth. Hirschsprung disease (which occurs in ~1 in 5,000 births) is the most common hereditary cause of intestinal obstruction. However, it shows considerable variation and a complex inheritance pattern.

The inheritance pattern of Hirschsprung disease is multifactorial since the condition can result from a mutation in any one or more of several genes that play key roles in the development of the ANS from the embryonic neural crest. Specifically, coding sequence mutations in a variety of genes (e.g., ECE1, EDN3, EDNRB, PHOX2B, and SOX10) are involved in the pathogenesis of Hirschsprung Disease as well as a variety of other developmental abnormalities. Mutations in these genes can result in dominant, recessive, or polygenic patterns of inheritance (Carrasquillo et al., 2002; Hofstra et al., 1997; McCallion et al., 2003). Aganglionic megacolon also is observed in some cases of trisomy 21 (Down syndrome).

1. Hirschsprung Disease, Cardiac Defects, Essential Hypertension, and Endothelin Converting Enzyme 1 (ECE1)

The endothelins are a family of potent vasoactive peptides whose effects are mediated via G protein-coupled receptors. Endothelin-converting enzyme-1 is involved in the physiological processing of endothelin-1 (Valdenaire et al., 1995). A sequence variant was observed in a patient with Hirschsprung Disease, cardiac defects (i.e., ductus arteriosus, small subaortic ventricular septal defect, and small atrial septal defect), craniofacial abnormalities (i.e., cupped ears that were immature and posteriorly rotated and small nose with a high bridge and bulbous tip), other dysmorphic features (tapered fingers with hyperconvex nails, a single left palmar crease, contractures at the interphalangeal joint of the thumbs, proximal interphalangeal joints of the fingers bilaterally, and micropenis), and autonomic dysfunction (episodes of severe agitation in association with significant tachycardia, hypertension, and core temperatures as high as 40.5°C, and status epilepticus) (Hofstra et al., 1999). It has been suggested that the ARG742CYS mutation is responsible for, or at least contributed to, the phenotype of this patient because of an overlap in phenotypic features of between mouse models of this variant and those of the patient (Hofstra et al., 1999). Moreover, the mutation was thought to lead to the phenotype by resulting in reduced levels of EDN1 and EDN3 (Hofstra et al., 1999). Other mutations of ECE1 have been associated with essential hypertension.

2. Hirschsprung Disease, Waardenburg-Shah Syndrome, Central Hypoventilation Syndrome, and Endothelin 3 (EDN3; EDNRB)

The endothelins are a family of potent vasoactive peptides consisting of three isopeptides: EDN1, EDN2, and EDN3. EDN3 exerts a dose-dependent stimulation of proliferation and melanogenesis in neural crest cells (Nagy & Goldstein, 2006). Mice lacking the EDN3 gene display a phenotype similar to that seen in humans with Waardenburg-Hirschsprung syndrome. In humans, at least seven different EDN3 mutations have been found in patients with Hirschsprung disease (Chen et al., 2006; McCallion & Chakravarti, 2001; Puri & Shinkai, 2004). Genetic mutations in the EDN3 gene also have been associated with the Waardenburg-Shah syndrome (i.e., hearing loss, dystopia canthorum, and pigmentary abnormalities of the hair, skin, and eyes with Hirschsprung Disease) and the central hypoventilation syndrome (i.e., Ondine's curse).

3. Hirschsprung Disease, Type 2, Waardenburg-Shah Syndrome, ABCD Syndrome, and Endothelin Receptor, Type B (EDNRB)

The endothelins mediate their effects via G protein-coupled receptors. Mutations within the EDNRB gene are the molecular basis of Hirschsprung Disease, Type 2 (Amiel et al., 1996; Auricchio et al., 1999; Kusafuka et al., 1996; Puffenberger et al., 1994; Svensson et al., 1998). It has been estimated that EDNRB mutations account for approximately 5 percent of cases of Hirschsprung Disease (Chakravarti, 1996). However, penetrance is not 100 percent for some of the mutations, even in homozygotes with the mutation (Duan et al., 2003; Puffenberger et al., 1994a). These data indicate that the EDNRB gene is an important modifier gene for the development of Hirschsprung Disease, Type 2. Genetic mutations in the EDN3 gene also have been associated with the Waardenburg-Shah syndrome and the ABCD syndrome.

4. Hirschsprung Disease, Waardenburg-Shaw Syndrome, and SRY-Box 10 (SOX10)

The SRY-Box 10 gene (SOX10) plays a key role in neural crest development. In Waardenburg-Shah syndrome, patients have deafness, pigmentary abnormalities, and Hirschsprung disease, all caused by the failure of embryonic neural crest development. In patients with Waardenburg-Shah syndrome, mutations in the SOX10 gene have been identified (Pingault et al., 1998; Southard-Smith et al., 1999; Verheij et al., 2006).

5. Hirschsprung Disease (short segment), Neuroblastoma, Central Hypoventilation Syndrome, and Paired-like Homeobox 2B (PHOX2B)

The paired-like homeobox 2B gene (PHOX2B) is involved in the early development of the ANS from the neural crest (Pattyn et al., 1999). PHOX2B is one of multiple genes needed for the differentiation and survival of subsets of autonomic neurons. In mice with a targeted deletion of the PHOX2B gene, all autonomic ganglia and the three cranial sensory ganglia (that are part of the autonomic reflex circuits) fail to develop properly and eventually degenerate (Pattyn et al., 1999). In humans, mutations in the PHOX2B gene have been associated with Hirschsprung Disease (short segment) (Benailly et al., 2003), both with and without associated and neuroblastoma (Mosse et al., 2004; Trochet et al., 2004), as well as with central hypoventilation syndrome (Trochet et al., 2005).

B. Developmental Disorders of the Efferent Portion of the ANS

1. The Role of the GRFA-RET Receptor System in Autonomic Nervous System Development

A multicomponent receptor system consisting of the RET tyrosine kinase and a group of associated glycosyl-phosphatidylinositol-anchored coreceptors (designated GFRA1–4) constitute the GFRA/RET signaling pathway (see Figure 21.1) (Baloh et al., 1998). The GFRA-RET receptor complexes are activated selectively by a number of growth factors that are related structurally to transforming growth factor beta (TGF-beta). Thus, glial cell line-derived neurotrophic factor (GDNF) neurturin (NTN), artemin (ARTN), and persephin (PSPN) activate GFRA1–4, respectively (Baloh et al., 1998). This system plays a crucial role in the development of the peripheral ANS, central motor and dopamine neurons, the renal system, and the regulation of spermatogonia differentiation (Takahashi, 2001).

2. Hirschsprung Disease, Multiple Endocrine Neoplasia, Medullary Thyroid Carcinoma, and Rearranged During Transfection Protooncogene (RET)

The RET gene was identified first in 1985 as an oncogene activated by DNA rearrangement (Takahashi et al., 1985). Mutations in the RET gene are associated with a number of neural crest derived organ dysfunctions including Hirschsprung Disease, multiple endocrine neoplasia, type IIA, multiple endocrine neoplasia, type IIB, and medullary thyroid carcinoma. High levels of RET expression also are detected in human tumors of neural crest origin such as neuroblastoma, pheochromocytoma, and medullary thyroid carcinoma (Kapur, 2005; Lantieri et al., 2006).

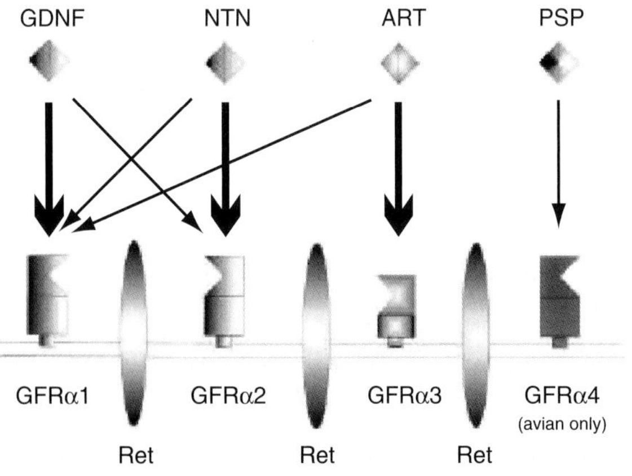

Figure 21.1 Schematic diagram of ligand–receptor interactions in the GDNF ligand family (reprinted with permission from Baloh et al., 1998).

In both embryonic and adult tissues, RET is highly expressed in peripheral enteric, sympathetic, and sensory neurons as well as central motor, dopamine, and adrenergic neurons. In mice, *Ret* gene knockouts result in animals that lack enteric neurons and superior cervical ganglia and also have renal agenesis or dysgenesis. Thus, RET activation is involved in the migration and differentiation of enteric ganglion cells, sympathetic neurons, and melanocytes from the neural crest during embryogenesis. RET is therefore critical for normal ANS formation and development.

Frameshift and missense mutations that disrupt or change the structure of the RET protein have been identified in Hirschsprung Disease (Chakravarti, 1996). Initially, it was believed that approximately 50 percent of familial HSCR and 30 percent of isolated Hirschsprung Disease resulted from mutations in the RET gene (Attie et al., 1994; Edery et al., 1994; Gabriel et al., 2002; Pasini et al., 1996). However, more recent data indicate that only about 3 percent of isolated HSCR cases have RET mutations (Borrego, Fernandez et al., 2003). In addition, certain genotypes comprising specific combinations of RET polymorphisms are highly associated with isolated Hirschsprung Disease (Borrego, Wright et al., 2003). Penetrance is greater in males than in females, in keeping with the higher frequency of the disorder in males.

RET mutations that cause a gain in function also cause several human diseases. Examples of such genetic variants include papillary thyroid carcinoma and multiple endocrine neoplasia types 2A and 2B (Takahashi, 2001).

3. Hirschsprung Disease, Renal Agenesis, and the Glial Cell Line-derived Neurotrophic Factor (GDNF)/GFRA1 System

Glial cell line-derived neurotrophic factor (GDNF) was first discovered as a potent survival factor for midbrain dopaminergic neurons in the CNS (Saarma & Sariola, 1999). During the development of the enteric nervous system, GDNF plays a key role in inducing vagal neural crest cells to enter the gut, in retaining neural crest cells within the gut, and in promoting the migration of neural crest cells along the gut. GDNF also plays a major role in kidney development (Tomac et al., 2000). A link between Hirschsprung Disease and the GDNF gene was first established when *Gdnf–/–* mice were found to have congenital intestinal aganglionosis and renal agenesis (Moore et al., 1996). Heterozygous *Gdnf+/–* mice also develop Hirschsprung-type intestinal obstruction (Shen et al., 2002). Of note is the fact that *Gdnf–/–* or *Gfra1–/–* mice have normal superior cervical ganglia (unlike *Ret–/–* mice and humans with certain RET mutations) (Tomac et al., 2000).

In humans, a direct screen of 106 unrelated Hirschsprung Disease patients for mutations in GDNF identified one familial GDNF missense mutation (ARG93TRP) in a Hirschsprung Disease patient with a known mutation in the RET gene (Angrist et al., 1998). At least four different

GDNF mutations have been found in Hirschsprung Disease although these mutations do not reduce the activation of RET (Borghini et al., 2002; Eketjall & Ibanez, 2002). These data suggest that GDNF is either a minor contributor to Hirschsprung Disease susceptibility or that human inactivating mutations may be lethal. GFRA1 mutations have not yet been found in Hirschsprung Disease patients (Borrego, Fernandez et al., 2003) but might be expected to exist.

4. The Parasympathetic Nervous System and the Neurturin (NRTN)/GFRA2 System

Neurturin and its receptor GFRA2 appears to play a primary role in the development of the parasympathetic nervous system (Wanigasekara & Keast, 2005). As might therefore be expected based on the expression patterns, *Gfra2*- and *Nrtn*-deficient mice show defects in the development of the parasympathetic ganglia (Heuckeroth et al., 1999; Rossi et al., 1999), unlike *Gfra1–/–* and *Gfra3–/–* mice. *Gfra2*-deficient mice have ptosis similar to *Gfra3–/–* mice (see later) although the ptosis in the *Gfra2–/–* mice most likely results from a reduction in lacrimal secretion caused by the defects in parasympathetic ganglia (Rossi et al., 1999) as opposed to the sympathetic defect observed in *Gfra3–/–* mice.

Minimal human mutation data are available for the NRTN/GFRA2 system. A heterozygous NRTN mutation has been observed in a family in which 4 of 8 siblings were affected with aganglionosis extending up to the small intestine (Doray et al., 1998). The ala96-to-ser (A96S) mutation was absent in normal controls and led to substitution of a neutral hydrophobic amino acid (alanine) by a hydrophilic amino acid (serine) in a basic region of the protein. However, a RET mutation had previously been identified in the same family (Attie et al., 1994) so the clinical significance of the coincident NRTN mutation cannot be determined.

5. The Sympathetic Nervous System and the Artemin (ARTN)/GFRA3 System

The ARTN/GFRA3 system regulates key aspects of SNS development as ganglion cells of the sympathetic system develop from the cells of the neural crests. Transgenic mice overexpressing *Artn* throughout development exhibit systemic autonomic neural lesions including fusion of adrenal medullae with adjacent paraganglia, adrenal medullary dysplasia and marked enlargement of sympathetic (i.e., SCG and sympathetic chain ganglia). *Artn* supplementation in wild-type adult mice results in hyperplasia or neuronal metaplasia at the adrenal corticomedullary junction (Bolon et al., 2004).

The rostral migration of cells to form the superior cervical ganglion (SCG) and the extension of axons along blood vessels involves ARTN activation of the GFRA3/RET receptor complex. ARTN is expressed by vascular smooth muscle cells that are innervated by normal sympathetic terminals. Both *Artn*- and *Gfra3*-deficient mice share abnormalities in the migration and axonal projection pattern of the entire sympathetic nervous system, resulting in abnormal innervation of target tissues due to lack of neurotrophic support (Honma et al., 2002). There were no sensory system deficits observed in either the *Artn*- and *Gfra3*-deficient mice (Honma et al., 2002).

Gfra3 is expressed during development in the adrenal medulla, sensory, and autonomic ganglia and their projections. Thus, *Gfra3–/–* mice exhibit severe defects in the superior cervical ganglia (SCG), whereas nonsympathetic ganglia appear normal (Nishino et al., 1999). Specifically, parasympathetic ganglia are normal in *Gfra3–/–* mice, an observation that is consistent with the fact that *Gfra3* is not highly expressed in developing parasympathetic ganglia. In addition, the enteric nervous system of the *Gfra3–/–* animals is normal (Nishino et al., 1999). The total number of neurons in the DRG also appears normal in *Gfra3–/–* mice at all developmental stages examined. The numbers of DRG neurons positive for calcitonin gene–related protein (CGRP), neurofilaments, or *Ret* also did not differ between wild-type and mutant animals, suggesting the absence of abnormalities in the various subpopulations of DRG neurons in *Gfra3–/–* mice. Furthermore, the trigeminal ganglion in *Gfra3–/–* -deficient mice appeared normal with regard to cell number and location. In the developing trigeminal ganglion of both wild-type and *Gfra3–/–* mice, *Ret* was widely expressed, whereas *Gfra1* and *Gfra2* were expressed in restricted populations of cells. Phenotypically, *Gfra–/–* mice appear to be normal with the exception that they exhibit ptosis.

SCG precursor cells in the *Gfra3–/–* embryos fail to migrate to the correct position, and they subsequently fail to innervate the target organs. In wild-type embryos, *Gfra3* was expressed in migrating SCG precursors, and artemin was expressed in and near the SCG. After birth, SCG neurons in the *Gfra3–/–* mice undergo progressive cell death. These observations suggest that ARTN/GFRA3-mediated signaling is required both for the development and survival of SCG and perhaps other sympathetic neurons. Human mutations in this system have not yet been reported but might be expected to lead to sympathetic nervous system dysfunction.

C. Developmental Disorders of the Sensory Portion of the ANS

1. Hereditary Sensory Neuropathy, Type 1 (HSN1; also called Hereditary Sensory and Autonomic Neuropathy, Type 1, HSAN1) and Serine Palmitoyltransferase, Long-Chain Base Subunit 1 (SPTLC1)

Serine palmitoyltransferase is the key enzyme in sphingolipid biosynthesis. Mutations in the gene can lead to increased activation of the enzyme, causing an increased production of glucosyl ceramide. Increased levels of ceramide, resulting from catabolism of sphingomylelin, can mediate cell death. Multiple mutations within the SPTLC1 gene have been associated with Hereditary Sensory Neuropathy, Type 1 (i.e., a dominantly inherited sensorimotor axonal neuropathy) (Bejaoui et al., 2001; Bejaoui et al., 2002; Dawkins et al., 2001).

2. Hereditary Sensory and Autonomic Neuropathy, Type II (HSAN2; also called Hereditary Sensory Neuropathy, Type II, HSN2), and the HSN2 Gene (HSN2)

Hereditary Sensory Neuropathy, Type II (HSN2), is a rare autosomal recessive disorder that usually begins in childhood and is characterized by an impairment of pain, temperature, and touch sensation as a result of a reduction or absence of peripheral sensory neurons. Symptoms often include inflammation of the digits, especially around the nails, usually accompanied by infection of the fingers and on the soles of the feet. A novel gene, designated HSN2, has been identified on CHR 12p13.33 that cosegregates with the disease (Coen et al., 2006; Lafreniere et al., 2004; Riviere et al., 2004). It has been suggested that the HSN2 protein may play a role in the development and/or maintenance of peripheral sensory neurons or their supporting Schwann cells (Lafreniere et al., 2004). The molecular mechanism by which this gene induces sensory neuronal death remains unknown.

3. Hereditary Sensory Neuropathy, Type III (HSN3; also known as Familial Dysautonomia, Riley–Day syndrome) and Inhibitor of Kappa Light Polypeptide Gene Enhancer in B Cells, Kinase Complex-associated Protein (IKBKAP)

An autosomal recessive clinical syndrome consisting of a congenital lack of tearing, emotional lability, paroxysmal hypertension, increased sweating, cold hands and feet, corneal anesthesia, erythematous skin blotching, and drooling has been termed the Riley–Day syndrome or hereditary sensory and autonomic neuropathy, type III (Brunt & McKusick, 1970). Autonomic dysfunction is the principal symptom due to decreased populations of sensory, sympathetic, and parasympathetic neurons. However, the clinical manifestations are variable and may also include absence of fungiform papillae of the tongue, severe scoliosis, and neuropathic joints. The individuals, thus far all Ashkenazi Jewish patients, also have an enhanced response to pressor agents due to a denervation supersensitivity to catecholamines (Bickel et al., 2002). Excretion of both dopamine and noradrenaline metabolites is reduced and plasma DBH levels are low.

Linkage analysis (Blumenfeld et al., 1993) led to the identification of the inhibitor of kappa light polypeptide gene enhancer in B cells, kinase complex-associated protein (IKBKAP) as the causative gene located on CHR 9q31–q33 (Slaugenhaupt et al., 2001). IKBKAP is a scaffold protein and a regulator of at least three different kinases involved in proinflammatory signaling. The major haplotype mutation was located at the donor splice site of intron 20. In patients with HSAN3, wild-type IKBKAP transcripts are present, although to varying extents, in all cell lines, blood, and postmortem tissues (Cuajungco et al., 2003). However, the relative wild-type-to-mutant IKBKAP RNA levels are highest in cultured patient lymphoblasts and lowest in postmortem central and peripheral nervous tissues. The authors suggested that the relative inefficiency of wild-type IKBKAP mRNA production from the mutant alleles in the nervous system underlies the selective degeneration of sensory and autonomic neurons in HSAN3 (Cuajungco et al., 2003).

4. Hereditary Sensory and Autonomic Neuropathy, Type IV (HSAN4) and Neurotrophic Tyrosine Kinase Receptor, Type 1 (NTRK1)

Neurotrophins and their receptors regulate the development and maintenance of both the central and the peripheral nervous systems. The neurotrophin, type 1, receptor gene (NTRK1) is located at CHR 1q32–q41 and is a primary receptor for nerve growth factor (NGF). Defects in the NGF signal transduction pathway lead to a failure to support the survival of primary sensory neurons. In addition, NGF acts as an immunoregulatory cytokine on monocytes.

Congenital insensitivity to pain with anhidrosis (CIPA or HSAN4) is an autosomal recessive disorder characterized by the absence of pain sensation often leading to self-mutilation, anhidrosis, recurrent episodic high fever, and mental retardation (Indo, 2001, 2002). Sweating cannot be elicited by thermal, painful, emotional or chemical stimuli despite the presence of normal sweat glands. The clinical syndrome has also been called Hereditary Sensory and Autonomic Neuropathy IV (HSAN IV), familial dysautonomia, type II and congenital sensory neuropathy with anhidrosis. HSAN4 patients have a hereditary developmental defect of nerve outgrowth (Verze et al., 2000).

In three unrelated patients with HSAN4, different mutations within the NTRK1 gene were identified (Indo et al., 1996). Specifically, a single base C deletion at nucleotide 1726 in exon C causes a frameshift and premature termination of the receptor protein in two individuals with congenital insensitivity to pain. The individuals are homozygous for the mutation. The authors suggested that the clinical symptomatology results from the fact that the NGF–NTRK1 system plays a critical role in the development and function of the peripheral pain and temperature systems (Indo et al., 1996). Over the past decade, approximately 40 specific mutations have been identified in the NTRK1 gene that are involved in the extracellular domain of the receptor that interacts with the binding of NGF as well as in the intracellular signal transduction domain of the receptor (Indo, 2001). Many of these mutations have been identified in HSAN4 patients.

5. Hereditary Sensory and Autonomic Neuropathy, Type V (HSAN5) and Nerve Growth Factor, Beta S subunit (NGFB)

HSAN5 is characterized by a loss of deep pain and temperature sensation with a severe reduction of unmyelinated nerve fibers and moderate loss of thin myelinated nerve fibers

(Einarsdottir et al., 2004). In contrast to HSAN4, mental abilities and most other neurologic functions remain intact. A candidate gene analysis of a large family from Sweden revealed a mutation in the coding region of the NGFB gene (ARG211TRP) that cosegregated with the disease phenotype (Einarsdottir et al., 2004). Neurotrophins such as NGF regulate the development and maintenance of both the central and the peripheral nervous systems and also play an immunoregulatory role. The authors noted that this NGFB mutation seems to separate the effects of NGF involved in development of central nervous system functions from those involved in peripheral pain pathways (Einarsdottir et al., 2004). Indeed, this conclusion is consistent with the observation that NGF plays a key role in the maintenance and regeneration of peripheral sensory neurons (Ramer et al., 2000) but may be less important in the development and maintenance of the CNS.

6. Hereditary Erythermalgia (erythromelagia) and Sodium Channel Voltage-gated Type IX, Alpha Subunit (SCN9A; Nav1.7)

Primary erythermalgia (also termed erythromelalgia) is an autosomal dominant neuropathy characterized by the childhood onset of episodic symmetrical redness of the skin, vasodilatation, and burning pain of the feet and lower legs provoked by mild warmth, exercise, and/or prolonged standing. Conversely, cooling of the extremities can reverse the symptoms. The first erythermalgia missense mutations to be identified were in the SCN9A gene (I848Tand L858H) that encodes the voltage-gated sodium channel $Na_v1.7$ were identified (Michiels et al., 2005 ;Yang et al., 2004). Subsequently, multiple additional mutations in the SCN9A gene have been identified in individuals with erythermalgia (Waxman & Dib-Hajj, 2005).

The SCN9A sodium channel is expressed predominantly in primary sensory and sympathetic neurons. SCN9A sodium channel mutations represent the first genetic variants that cause, rather than ameliorate, pain. The identified SCN9A mutations produce changes in channel physiology that enhance the response of SCN9A channels to small stimuli (Waxman & Dib-Hajj, 2005). Within dorsal root ganglion (DRG) neurons, expression of the mutant channels results in both a lower current threshold for generation of single action potentials, and a higher frequency of firing at graded stimulus intensities (i.e., hallmarks of neuronal hyperexcitability) (Dib-Hajj et al., 2005). Interestingly, although this mutation depolarizes resting membrane potential in both sensory neurons and sympathetic neurons, it renders sensory neurons hyperexcitable and sympathetic neurons hypoexcitable (Rush et al., 2006). It has been suggested that the selective presence of the Nav1.8 channel in sensory, but not sympathetic neurons, is the major determinant of these opposing physiological effects (Rush et al., 2006). Thus, the SCN9A mutations identified in erythermalgia provide a mechanistic explanation for the role of this channel in pain signaling by DRG and sympathetic neurons (Waxman & Dib-Hajj, 2005).

III. Functional Abnormalities of the Autonomic Nervous System

A. Increased Sympathetic Nervous System Activity and the Alpha-2B-Adrenergic Receptor (ADRA2B)

A polymorphism in the ADRA2B gene leading to the deletion of three glutamic acids from a glutamic acid repeat element (glu12, amino acids 297 to 309) in the third intracellular loop of the receptor protein has been evaluated (Heinonen et al., 1999). This repeat element had been shown to be important for agonist-dependent alpha-2B-adrenergic receptor desensitization. In subjects homozygous for the short allele compared to subjects with two long alleles, the basal metabolic rate was 6 percent lower ($p = 0.009$) than in controls. Since a lower basal metabolic rate is a risk factor for obesity, the authors concluded that this polymorphism of the ADRA2B subtype could partly explain the variation in basal metabolic rate in an obese population and may therefore contribute to the pathogenesis of obesity (Heinonen et al., 1999).

In addition, the same ADRA2B 3-glutamic acid deletion polymorphism was evaluated in 381 healthy Japanese males using electrocardiogram R-R interval power spectral analysis (Suzuki et al., 2003). In R-R spectral analysis of heart rate variability, homozygous carriers of the short allele (13% of the study group) had significantly greater low frequency and very low frequency than did homozygous carriers of the long allele, as well as a higher sympathetic nervous system index. In general, low frequencies of heart rate variability are associated with both SNS and PNS activities and the very low frequencies of heart rate variability reflect thermoregulatory control of SNS activity. Thus, these findings suggested that the ADRA2B deletion polymorphism is associated with increased SNS activity and lower PNS activity in healthy, young males (Suzuki et al., 2003). The authors hypothesized that the increased SNS activity might be secondary to the decreased basal metabolic rate observed by others (Heinonen et al., 1999). Alternatively, the ADRA2B deletion may result in increased SNS activity with a secondary decrease in basal metabolic rate.

B. Reduced Response to Beta-2-Adrenoreceptor Agonists and the Beta-2-Adrenergic Receptor (ADRB2)

The Thr164Ile polymorphism of the beta2-adrenoceptor is present in approximately 4 percent of the population and is associated with a significant decrease in response to beta-2-adrenergic agonists (Brodde et al., 2001; Dishy et al., 2001). Altered adrenergic vascular sensitivity may contribute to the decreased survival observed in patients with congestive heart failure carrying the Ile164 allele. In a study of patients with congestive heart failure, those with the Ile164 polymorphism displayed a striking difference in survival with a relative risk of death or cardiac transplant of 4.81 ($p < 0.001$) compared with those with the wild-type Thr at this position (Liggett et al., 1998). The one-year survival for Ile164 patients was 42 percent compared with 76 percent for patients with the wild-type beta-2-adrenoreceptor.

The dose of isoproterenol required to achieve 50 percent venodilation is significantly higher in individuals with the Ile164 allele than those without although the maximal response to isoproterenol does not differ (Dishy et al., 2001). Conversely, the dose of phenylephrine needed to induce 50 percent venoconstriction is significantly lower in individuals with the Ile164 allele than those without. Thus, the Thr164Ile polymorphism of the beta2-adrenergic receptor is associated with a five-fold reduction in sensitivity to beta2 receptor agonist-mediated vasodilation while vasoconstrictor sensitivity is increased. The overall effect of the Thr164Ile polymorphism is to shift the balance of adrenergic vascular tone toward vasoconstriction. This suggests a mechanistic explanation for the clinical observation of decreased survival in patients with congestive heart failure heterozygous for the Thr164Ile polymorphism (Dishy et al., 2001).

C. Norepinephrine Deficiency and Dopamine Beta-Hydroxylase (DBH)

DßH converts dopamine to norepinephrine in postganglionic sympathetic neurons. The DßH gene is located at CHR 9q34. Several patients have been analyzed with congenital DßH deficiency (Man in 't Veld et al., 1987a; Mathias et al., 1990; Robertson et al., 1986; Robertson et al., 1991). The individuals have noradrenergic denervation and adrenomedullary failure but baroreflex afferents, cholinergic innervation, and adrenocortical function are normal. Norepinephrine, epinephrine, and their breakdown products are not detectable in plasma, urine, and cerebrospinal fluid but dopamine levels are increased significantly. Physiological and pharmacological stimuli of sympathetic nervous system activity cause increases in dopamine but not norepinephrine.

DßH deficient infants have a delay in the opening of the eyes (2 weeks in one case) and ptosis has been observed in almost all DßH deficient infants. Some of the infants have been reported to be sickly in the neonatal stage and survival is often believed to be unlikely. Hypotension, hypoglycemia, and hypothermia are present early in life. Postural hypotension is exhibited during exercise in childhood and is marked by an increase in heart rate as the blood pressure falls. Syncopal episodes may be misinterpreted as epilepsy in the children, leading to the treatment with anticonvulsants. In general, symptoms worsen during adolescence and severely limit the function of the individual.

Clinical features during adolescence and adulthood include reduced exercise tolerance, skeletal muscle hypotonia, recurrent hypoglycemia, ptosis of the eyelids, nasal stuffiness, and prolonged or retrograde ejaculation. The severe postural hypotension is attributed to the impairment of sympathetic vasoconstrictor function. Symptoms in DßH deficiency have also been reported to worsen in the morning, after exercise, and in warm weather (Mathias et al., 1990). No other neurological or psychiatric abnormalities have been reported.

The symptoms of DßH deficiency respond well from treatment with dihydroxyphenylserine (DOPS; also known as Droxidopa) (Biaggioni & Robertson, 1987; Man in 't Veld et al., 1987b; Mathias et al., 1990; Thompson et al., 1995). This molecule is converted to norepinephrine by decarboxylation of the terminal carboxyl group. Treatment with DOPS (150–600 mg/day) leads to a reduction in orthostatic hypotension, increased plasma levels of norepinephrine, and, in the males, the ability to ejaculate.

Specific mutations within the DBH gene have been identified that appear to be the cause of the enzymatic deficiency (Kim et al., 2002). Specifically, seven novel variants, including four potentially pathogenic mutations in the human DBH gene, were identified from an analysis of two unrelated patients with DBH deficiency and their families. Both patients were found to be compound heterozygotes for variants affecting expression of DBH. Each patient carried one copy of a T→C transversion in the splice donor site of DBH intron 1, creating a premature stop codon. One patient also had a missense mutation in DBH exon 2 while the other had missense mutations in exons 1 and 6. The authors propose that NE deficiency is an autosomal recessive disorder resulting from heterogeneous molecular lesions with the DBH gene (Kim et al., 2002).

D. Autonomic Hyperactivity and Dopamine 4 Receptor (DRD4)

The dopamine D4 receptor gene (DRD4) is one of the five known G protein-coupled receptors for which dopamine is the primary neurotransmitter. The gene is located at CHR 11p15.5 and codes for a receptor protein of 387 amino acids (Van Tol et al., 1991). At least three common polymorphic variants of the gene exist in the human population based as a result of known variations in a 48-base-pair sequence in the third cytoplasmic loop of the receptor (Van Tol et al., 1992).

A 13-base-pair deletion of bases 235–247 in the DRD4 gene has been identified in approximately 2 percent of the general population (Nothen et al., 1994). The deletion alters the reading frame from amino acid 79 in the receptor and generates a stop codon 20 amino acids downstream, thereby truncating the receptor to an abnormally short 98 amino acids.

No major neuropsychiatric disturbances have been observed in heterozygotes with this mutation (Nothen et al., 1994). However, in a single homozygous individual with this mutation, autonomic hyperactivity was observed. Specifically, the 50-year-old (at the time of the study) male reported severe dermatographism and excessive sweating. These symptoms were exacerbated in social gatherings and moderately warm temperatures. The individual denied feeling anxious in these situations but characterized himself as nervous and explosive since early adulthood. He has had severe migrainous headaches since adolescence, successfully treated with a tricyclic antidepressant. He has been obese since adolescence. He had an acoustic neuroma removed at age 38, with a negative family history, and a recurrence removed at age 44. Pulse-rate fluctuations leading to intermittent sinus tachycardia had been treated with beta-blockers since approximately age 40. A consistently reduced body temperature (35.4°C) was documented (Nothen et al., 1994). The authors speculated that at least some of these autonomic disturbances could be attributed directly to the absence of a functional DRD4 receptor (Nothen et al., 1994).

E. Amyloid Polyneuropathy and Transthyretin (TTR)

Transthyretin is a prealbumin protein of 127 amino acids and is a primary transport protein for thyroxine and retinol (vitamin A). The protein is a common constituent of neuritic plaques and micro-angiopathic lesions related to amyloid deposition. More than 70 different mutations associated with amyloid deposition have been identified within the TTR gene located at CHR 18q11.2–q12.1 (Saraiva, 1995, 2001). Amyloidogenic mutations in this gene leads to decreased stability of the TTR protein, with the VAL30MET variant (also called the Andrade or Portuguese type) being the most common allelic variant. The majority of the mutations result in an amyloid polyneuropathy, which involves small, unmyelinated fibers. The neuropathy disproportionately affects pain and temperature sensation although significant clinical variation exists between the various mutations (Ikeda, 2002). Indeed, there is a much larger prevalence of TTR mutations than recognized disease due to the wide variation in phenotypes.

F. Orthostatic Intolerance and Solute Carrier Family 6 (Neurotransmitter Transporter, Noradrenaline), Member 2 (SLC6A2)

Orthostatic intolerance (OI) is a clinical syndrome consisting of group of symptoms that can occur after assuming an upright posture. Although a variety of OI definitions have been used in the medical literature, OI can best be defined as the development of lightheadedness or dizziness, as well as visual changes and other symptoms, upon arising from the supine position to an upright position. In general, OI can be defined as a standing heart rate increase of at least 30 beats per minute, without orthostatic hypotension. Most patients with a diagnosis of orthostatic intolerance are women between the ages of 20 and 50 years (Low et al., 1995). This syndrome often has been described by a number of other names such as the postural orthostic tachycardia syndrome (POTS), soldiers heart, neurocirculatory asthenia, and mitral valve prolapse syndrome. It may also play a role in Chronic Fatigue Syndrome (Schondorf & Freeman, 1999).

The reuptake of norepinephrine into sympathetic terminals occurs via a specific Na(+)- and Cl(−)-dependent transport system mediated by the solute carrier family 6 (neurotransmitter transporter, norepinephrine), member 2 gene (SLC6A2). In a patient with orthostatic intolerance, an elevated mean plasma norepinephrine concentration was observed while standing (Shannon et al., 2000). Analysis of the norepinephrine-transporter gene revealed that the proband was heterozygous for a ALA457PRO mutation that resulted in more than a 98 percent loss of function as compared with that of the wild-type gene. The authors concluded that the impairment of synaptic norepinephrine clearance may result in a syndrome characterized by excessive sympathetic activation in response to physiologic stimuli.

IV. Future Directions

Molecular genetics allows for an unprecedented mechanistic analysis of the ANS. Moreover, molecular genetic data offer the longer term potential to provide immediate diagnostic, prognostic, and therapeutic guidance to clinicians. The coupling of molecular genetic diagnoses and rational therapeutic approaches based on these data should have a progressively significant impact on the diagnosis and management of patients with autonomic dysfunction.

References

Amiel, J., Attie, T., Jan, D., Pelet, A., Edery, P., Bidaud, C. et al. (1996). Heterozygous endothelin receptor B (EDNRB) mutations in isolated Hirschsprung disease. *Hum Mol Genet* **5**, 355–357.

Angrist, M., Jing, S., Bolk, S., Bentley, K., Nallasamy, S., Halushka, M. et al. (1998). Human GFRA1: Cloning, mapping, genomic structure, and evaluation as a candidate gene for Hirschsprung disease susceptibility. *Genomics* **48**, 354–362.

Attie, T., Edery, P., Lyonnet, S., Nihoul-Fekete, C., Munnich, A. (1994). [Identification of mutation of RET proto-oncogene in Hirschsprung disease]. *C R Seances Soc Biol Fil* **188**, 499–504.

Auricchio, A., Griseri, P., Carpentieri, M. L., Betsos, N., Staiano, A., Tozzi, A. et al. (1999). Double heterozygosity for a RET substitution interfering with splicing and an EDNRB missense mutation in Hirschsprung disease. *Am J Hum Genet* **64**, 1216–1221.

Baloh, R. H., Gorodinsky, A., Golden, J. P., Tansey, M. G., Keck, C. L., Popescu, N. C. et al. (1998). GFRalpha3 is an orphan member of the GDNF/neurturin/persephin receptor family. *Proc Natl Acad Sci U S A* **95**, 5801–5806.

Baloh, R. H., Tansey, M. G., Lampe, P. A., Fahrner, T. J., Enomoto, H., Simburger, K. S. et al. (1998). Artemin, a novel member of the GDNF ligand family, supports peripheral and central neurons and signals through the GFRalpha3-RET receptor complex. *Neuron* **21**, 1291–1302.

Bejaoui, K., Uchida, Y., Yasuda, S., Ho, M., Nishijima, M., Brown, R. H., Jr. et al. (2002). Hereditary sensory neuropathy type 1 mutations confer dominant negative effects on serine palmitoyltransferase, critical for sphingolipid synthesis. *J Clin Invest* **110**, 1301–1308.

Bejaoui, K., Wu, C., Scheffler, M. D., Haan, G., Ashby, P., Wu, L. et al. (2001). SPTLC1 is mutated in hereditary sensory neuropathy, type 1. *Nat Genet* **27**, 261–262.

Benailly, H. K., Lapierre, J. M., Laudier, B., Amiel, J., Attie, T., De Blois, M. C. et al. (2003). PMX2B, a new candidate gene for Hirschsprung's disease. *Clin Genet* **64**, 204–209.

Biaggioni, I., Robertson, D. (1987). Endogenous restoration of noradrenaline by precursor therapy in dopamine-beta-hydroxylase deficiency. *Lancet* **2**, 1170–1172.

Bickel, A., Axelrod, F. B., Schmelz, M., Marthol, H., Hilz, M. J. (2002). Dermal microdialysis provides evidence for hypersensitivity to noradrenaline in patients with familial dysautonomia. *J Neurol Neurosurg Psychiatry* **73**, 299–302.

Blumenfeld, A., Slaugenhaupt, S. A., Axelrod, F. B., Lucente, D. E., Maayan, C., Liebert, C. B. et al. (1993). Localization of the gene for familial dysautonomia on chromosome 9 and definition of DNA markers for genetic diagnosis. *Nat Genet* **4**, 160–164.

Bolon, B., Jing, S., Asuncion, F., Scully, S., Pisegna, M., Van, G. Y. et al. (2004). The candidate neuroprotective agent artemin induces autonomic neural dysplasia without preventing peripheral nerve dysfunction. *Toxicol Pathol* **32**, 275–294.

Borghini, S., Bocciardi, R., Bonardi, G., Matera, I., Santamaria, G., Ravazzolo, R., Ceccherini, I. (2002). Hirschsprung associated GDNF mutations do not prevent RET activation. *Eur J Hum Genet* **10**, 183–187.

Borrego, S., Fernandez, R. M., Dziema, H., Niess, A., Lopez-Alonso, M., Antinolo, G., Eng, C. (2003). Investigation of germline GFRA4 mutations and evaluation of the involvement of GFRA1, GFRA2, GFRA3, and GFRA4 sequence variants in Hirschsprung disease. *J Med Genet* **40**, e18.

Borrego, S., Wright, F. A., Fernandez, R. M., Williams, N., Lopez-Alonso, M., Davuluri, R. et al. (2003). A founding locus within the RET proto-oncogene may account for a large proportion of apparently sporadic Hirschsprung disease and a subset of cases of sporadic medullary thyroid carcinoma. *Am J Hum Genet* 72, **88**–100.

Brodde, O. E., Buscher, R., Tellkamp, R., Radke, J., Dhein, S., Insel, P. A. (2001). Blunted cardiac responses to receptor activation in subjects with Thr164Ile beta(2)-adrenoceptors. *Circulation* **103**, 1048–1050.

Brunt, P. W., McKusick, V. A. (1970). Familial dysautonomia. A report of genetic and clinical studies, with a review of the literature. *Medicine* (Baltimore) **49**, 343–374.

Carrasquillo, M. M., McCallion, A. S., Puffenberger, E. G., Kashuk, C. S., Nouri, N., Chakravarti, A. (2002). Genome-wide association study and mouse model identify interaction between RET and EDNRB pathways in Hirschsprung disease. *Nat Genet* **32**, 237–244.

Chakravarti, A. (1996). Endothelin receptor-mediated signaling in Hirschsprung disease. *Hum Mol Genet* **5**, 303–307.

Chen, W. C., Chang, S. S., Sy, E. D., Tsai, M.C. (2006). A De Novo novel mutation of the EDNRB gene in a Taiwanese boy with Hirschsprung disease. *J Formos Med Assoc* **105**, 349–354.

Coen, K., Pareyson, D., uer-Grumbach, M., Buyse, G., Goemans, N., Claeys, K. G. et al. (2006). Novel mutations in the HSN2 gene causing hereditary sensory and autonomic neuropathy type II. *Neurology* **66**, 748–751.

Cuajungco, M. P., Leyne, M., Mull, J., Gill, S. P., Lu, W., Zagzag, D. et al. (2003). Tissue-specific reduction in splicing efficiency of IKBKAP due to the major mutation associated with familial dysautonomia. *Am J Hum Genet* **72**, 749–758.

Dawkins, J. L., Hulme, D. J., Brahmbhatt, S. B., uer-Grumbach, M., Nicholson, G. A. (2001). Mutations in SPTLC1, encoding serine palmitoyltransferase, long chain base subunit-1, cause hereditary sensory neuropathy type I. *Nat Genet* **27**, 309–312.

Dib-Hajj, S. D., Rush, A. M., Cummins, T. R., Hisama, F. M., Novella, S., Tyrrell, L. et al. (2005). Gain-of-function mutation in Nav1.7 in familial erythromelalgia induces bursting of sensory neurons. *Brain* **128**, 1847–1854.

Dishy, V., Sofowora, G. G., Xie, H. G., Kim, R. B., Byrne, D. W., Stein, C. M., Wood, A. J. (2001). The effect of common polymorphisms of the beta2-adrenergic receptor on agonist-mediated vascular desensitization. *N Engl J Med* **345**, 1030–1035.

Doray, B., Salomon, R., Amiel, J., Pelet, A., Touraine, R., Billaud, M. et al. (1998). Mutation of the RET ligand, neurturin, supports multigenic inheritance in Hirschsprung disease. *Hum Mol Genet* **7**, 1449–1452.

Duan, X. L., Zhang, X. S., Li, G. W. (2003). Clinical relationship between EDN-3 gene, EDNRB gene and Hirschsprung's disease. *World J Gastroenterol* **9**, 2839–2842.

Edery, P., Lyonnet, S., Mulligan, L. M., Pelet, A., Dow, E., Abel, L. et al. (1994). Mutations of the RET proto-oncogene in Hirschsprung's disease. *Nature* **367**, 378–380.

Einarsdottir, E., Carlsson, A., Minde, J., Toolanen, G., Svensson, O., Solders, G. et al. (2004). A mutation in the nerve growth factor beta gene (NGFB) causes loss of pain perception. *Hum Mol Genet* **13**, 799–805.

Eketjall, S., Ibanez, C. F. (2002). Functional characterization of mutations in the GDNF gene of patients with Hirschsprung disease. *Hum Mol Genet* **11**, 325–329.

Gabriel, S. B., Salomon, R., Pelet, A., Angrist, M., Amiel, J., Fornage, M. et al. (2002). Segregation at three loci explains familial and population risk in Hirschsprung disease. *Nat Genet* **31**, 89–93.

Heinonen, P., Koulu, M., Pesonen, U., Karvonen, M. K., Rissanen, A., Laakso, M. et al. (1999). Identification of a three-amino acid deletion in the alpha2B-adrenergic receptor that is associated with reduced basal metabolic rate in obese subjects. *J Clin Endocrinol Metab* **84**, 2429–2433.

Heuckeroth, R. O., Enomoto, H., Grider, J. R., Golden, J. P., Hanke, J. A., Jackman, A. et al. (1999). Gene targeting reveals a critical role for neurturin in the development and maintenance of enteric, sensory, and parasympathetic neurons. *Neuron* **22**, 253–263.

Hofstra, R. M., Osinga, J., Buys, C. H. (1997). Mutations in Hirschsprung disease: When does a mutation contribute to the phenotype. *Eur J Hum Genet* **5**, 180–185.

Hofstra, R. M., Valdenaire, O., Arch, E., Osinga, J., Kroes, H., Loffler, B. M. et al. (1999). A loss-of-function mutation in the endothelin-converting enzyme 1 (ECE-1) associated with Hirschsprung disease, cardiac defects, and autonomic dysfunction. *Am J Hum Genet* **64**, 304–308.

Honma, Y., Araki, T., Gianino, S., Bruce, A., Heuckeroth, R., Johnson, E., Milbrandt, J. (2002). Artemin is a vascular-derived neurotropic factor for developing sympathetic neurons. *Neuron* **35**, 267–282.

Ikeda, S. (2002). Clinical picture and outcome of transthyretin-related familial amyloid polyneuropathy (FAP) in Japanese patients. *Clin Chem Lab Med* **40**, 1257–1261.

Indo, Y. (2001). Molecular basis of congenital insensitivity to pain with anhidrosis (CIPA): Mutations and polymorphisms in TRKA (NTRK1) gene encoding the receptor tyrosine kinase for nerve growth factor. *Hum Mutat* **18**, 462–471.

Indo, Y. (2002). Genetics of congenital insensitivity to pain with anhidrosis (CIPA) or hereditary sensory and autonomic neuropathy type IV. Clinical, biological and molecular aspects of mutations in TRKA(NTRK1) gene encoding the receptor tyrosine kinase for nerve growth factor. *Clin Auton Res* **12** Suppl 1, I20–I32.

Indo, Y., Tsuruta, M., Hayashida, Y., Karim, M. A., Ohta, K., Kawano, T. et al. (1996). Mutations in the TRKA/NGF receptor gene in patients with congenital insensitivity to pain with anhidrosis. *Nat Genet* **13**, 485–488.

Kapur, R. P. (2005). Multiple endocrine neoplasia type 2B and Hirschsprung's disease. *Clin Gastroenterol Hepatol* **3**, 423–431.

Kim, C. H., Zabetian, C. P., Cubells, J. F., Cho, S., Biaggioni, I., Cohen, B. M. et al. (2002). Mutations in the dopamine beta-hydroxylase gene are associated with human norepinephrine deficiency. *Am J Med Genet* **108**, 140–147.

Kusafuka, T., Wang, Y., Puri, P. (1996). Novel mutations of the endothelin-B receptor gene in isolated patients with Hirschsprung's disease. *Hum Mol Genet* **5**, 347–349.

Lafreniere, R. G., MacDonald, M. L., Dube, M. P., MacFarlane, J., O'Driscoll, M., Brais, B. et al. (2004). Identification of a novel gene (HSN2) causing hereditary sensory and autonomic neuropathy type II through the Study of Canadian Genetic Isolates. *Am J Hum Genet* **74**, 1064–1073.

Lantieri, F., Griseri, P., Ceccherini, I. (2006). Molecular mechanisms of RET-induced Hirschsprung pathogenesis. *Ann Med* **38**, 11–19.

Liggett, S. B., Wagoner, L. E., Craft, L. L., Hornung, R. W., Hoit, B. D., McIntosh, T. C., Walsh, R. A. (1998). The Ile164 beta2-adrenergic receptor polymorphism adversely affects the outcome of congestive heart failure. *J Clin Invest* **102**, 1534–1539.

Low, P. A., Opfer-Gehrking, T. L., Textor, S. C., Benarroch, E. E., Shen, W. K., Schondorf, R. et al. (1995). Postural tachycardia syndrome (POTS). *Neurology* **45**, S19–S25.

Man in 't Veld, A. J., Boomsma, F., Moleman, P., Schalekamp, M. A. (1987a). Congenital dopamine-beta-hydroxylase deficiency. A novel orthostatic syndrome. *Lancet* **1**, 183–188.

Man in 't Veld, A. J., Boomsma, F., van den Meiracker, A. H., Schalekamp, M. A. (1987b). Effect of unnatural noradrenaline precursor on sympathetic control and orthostatic hypotension in dopamine-beta-hydroxylase deficiency. *Lancet* **2**, 1172–1175.

Mathias, C. J., Bannister, R. B., Cortelli, P., Heslop, K., Polak, J. M., Raimbach, S. et al. (1990). Clinical, autonomic and therapeutic observations in two siblings with postural hypotension and sympathetic failure due to an inability to synthesize noradrenaline from dopamine because of a deficiency of dopamine beta hydroxylase. *Q J Med* **75**, 617–633.

McCallion, A. S., Chakravarti, A. (2001). EDNRB/EDN3 and Hirschsprung disease type II. *Pigment Cell Res* **14**, 161–169.

McCallion, A. S., Emison, E. S., Kashuk, C. S., Bush, R. T., Kenton, M., Carrasquillo, M. M. et al. (2003). Genomic variation in multigenic traits: Hirschsprung disease. *Cold Spring Harb Symp Quant Biol* **68**, 373–381.

Michiels, J. J., te Morsche, R. H., Jansen, J. B., Drenth, J. P. (2005). Autosomal dominant erythermalgia associated with a novel mutation in the voltage-gated sodium channel alpha subunit Nav1.7. *Arch Neurol* **62**, 1587–1590.

Moore, M. W., Klein, R. D., Farinas, I., Sauer, H., Armanini, M., Phillips, H. et al. (1996). Renal and neuronal abnormalities in mice lacking GDNF. *Nature* **382**, 76–79.

Mosse, Y. P., Laudenslager, M., Khazi, D., Carlisle, A. J., Winter, C. L., Rappaport, E., Maris, J. M. (2004). Germline PHOX2B mutation in hereditary neuroblastoma. *Am J Hum Genet* **75**, 727–730.

Nagy, N., Goldstein, A. M. (2006). Endothelin-3 regulates neural crest cell proliferation and differentiation in the hindgut enteric nervous system. *Dev Biol* **293**, 203–217.

Nishino, J., Mochida, K., Ohfuji, Y., Shimazaki, T., Meno, C., Ohishi, S. et al. (1999). GFR alpha3, a component of the artemin receptor, is required for migration and survival of the superior cervical ganglion. *Neuron* **23**, 725–736.

Nothen, M. M., Cichon, S., Hemmer, S., Hebebrand, J., Remschmidt, H., Lehmkuhl, G. et al. (1994). Human dopamine D4 receptor gene: Frequent occurrence of a null allele and observation of homozygosity. *Hum Mol Genet* **3**, 2207–2212.

Pasini, B., Ceccherini, I., Romeo, G. (1996). RET mutations in human disease. *Trends Genet* **12**, 138–144.

Pattyn, A., Morin, X., Cremer, H., Goridis, C., Brunet, J. F. (1999). The homeobox gene Phox2b is essential for the development of autonomic neural crest derivatives. *Nature* **399**, 366–370.

Pingault, V., Bondurand, N., Kuhlbrodt, K., Goerich, D. E., Prehu, M. O., Puliti, A. et al. (1998). SOX10 mutations in patients with Waardenburg-Hirschsprung disease. *Nat Genet* **18**, 171–173.

Puffenberger, E. G., Hosoda, K., Washington, S. S., Nakao, K., deWit, D., Yanagisawa, M., Chakravart, A. (1994). A missense mutation of the endothelin-B receptor gene in multigenic Hirschsprung's disease. *Cell* **79**, 1257–1266.

Puri, P., Shinkai, T. (2004). Pathogenesis of Hirschsprung's disease and its variants: Recent progress. *Semin Pediatr Surg* **13**, 18–24.

Ramer, M. S., Priestley, J. V., McMahon, S.B. (2000). Functional regeneration of sensory axons into the adult spinal cord. *Nature* **403**, 312–316.

Riviere, J. B., Verlaan, D. J., Shekarabi, M., Lafreniere, R. G., Benard, M., Der, K. V. et al. (2004). A mutation in the HSN2 gene causes sensory neuropathy type II in a Lebanese family. *Ann Neurol* **56**, 572–575.

Robertson, D., Goldberg, M. R., Onrot, J., Hollister, A. S., Wiley, R., Thompson, J. G. Jr., Robertson, R. M. (1986). Isolated failure of autonomic noradrenergic neurotransmission. Evidence for impaired beta-hydroxylation of dopamine. *N Engl J Med* **314**, 1494–1497.

Robertson, D., Haile, V., Perry, S. E., Robertson, R. M., Phillips, J. A. III, Biaggioni, I. (1991). Dopamine beta-hydroxylase deficiency. A genetic disorder of cardiovascular regulation. *Hypertension* **18**, 1–8.

Rossi, J., Luukko, K., Poteryaev, D., Laurikainen, A., Sun, Y. F., Laakso, T. et al. (1999). Retarded growth and deficits in the enteric and parasympathetic nervous system in mice lacking GFR alpha2, a functional neurturin receptor. *Neuron* **22**, 243–252.

Rush, A. M., Dib-Hajj, S. D., Liu, S., Cummins, T. R., Black, J. A., Waxman, S. G. (2006). A single sodium channel mutation produces hyper- or hypoexcitability in different types of neurons. *Proc Natl Acad Sci U S A* **103**, 8245–8250.

Saarma, M., Sariola, H. (1999). Other neurotrophic factors: Glial cell line-derived neurotrophic factor (GDNF). *Microsc Res Tech* **45**, 292–302.

Saraiva, M. J. (1995). Transthyretin mutations in health and disease. *Hum Mutat* **5**, 191–196.

Saraiva, M. J. (2001). Transthyretin amyloidosis: A tale of weak interactions. *FEBS Lett* **498**, 201–203.

Schondorf, R., Freeman, R. (1999). The importance of orthostatic intolerance in the chronic fatigue syndrome. *Am J Med Sci* **317**, 117–123.

Shannon, J. R., Flattem, N. L., Jordan, J., Jacob, G., Black, B. K., Biaggioni, I. et al. (2000). Orthostatic intolerance and tachycardia associated with norepinephrine-transporter deficiency. *N Engl J Med* **342**, 541–549.

Shen, L., Pichel, J. G., Mayeli, T., Sariola, H., Lu, B., Westphal, H. (2002). Gdnf haploinsufficiency causes Hirschsprung-like intestinal obstruction and early-onset lethality in mice. *Am J Hum Genet* **70**, 435–447.

Slaugenhaupt, S. A., Blumenfeld, A., Gill, S. P., Leyne, M., Mull, J., Cuajungco, M. P. et al. (2001). Tissue-specific expression of a splicing mutation in the IKBKAP gene causes familial dysautonomia. *Am J Hum Genet* **68**, 598–605.

Southard-Smith, E. M., Angrist, M., Ellison, J. S., Agarwala, R., Baxevanis, A. D., Chakravarti, A., Pavan, W. J. (1999). The Sox10(Dom) mouse: Modeling the genetic variation of Waardenburg-Shah (WS4) syndrome. *Genome Res* **9**, 215–225.

Suzuki, N., Matsunaga, T., Nagasumi, K., Yamamura, T., Shihara, N., Moritani, T. et al. (2003). Alpha(2B)-adrenergic receptor deletion polymorphism associates with autonomic nervous system activity in young healthy Japanese. *J Clin Endocrinol Metab* **88**, 1184–1187.

Svensson, P. J., Anvret, M., Molander, M. L., Nordenskjold, A. (1998). Phenotypic variation in a family with mutations in two Hirschsprung-related genes (RET and endothelin receptor B). *Hum Genet* **103**, 145–148.

Takahashi, M. (2001). The GDNF/RET signaling pathway and human diseases. *Cytokine Growth Factor Rev* **12**, 361–373.

Takahashi, M., Ritz, J., Cooper, G. M. (1985). Activation of a novel human transforming gene, ret, by DNA rearrangement. *Cell* **42**, 581–588.

Thompson, J. M., O'Callaghan, C. J., Kingwell, B. A., Lambert, G. W., Jennings, G. L., Esler, M. D. (1995). Total norepinephrine spillover, muscle sympathetic nerve activity and heart-rate spectral analysis in a patient with dopamine beta-hydroxylase deficiency. *J Auton Nerv Syst* **55**, 198–206.

Tomac, A. C., Grinberg, A., Huang, S. P., Nosrat, C., Wang, Y., Borlongan, C. et al. (2000). Glial cell line-derived neurotrophic factor receptor alpha1 availability regulates glial cell line-derived neurotrophic factor signaling: evidence from mice carrying one or two mutated alleles. *Neuroscience* **95**, 1011–1023.

Trochet, D., Bourdeaut, F., Janoueix-Lerosey, I., Deville, A., de, P. L., Schleiermacher, G. et al. (2004). Germline mutations of the paired-like homeobox 2B (PHOX2B) gene in neuroblastoma. *Am J Hum Genet* **74**, 761–764.

Trochet, D., Hong, S. J., Lim, J. K., Brunet, J. F., Munnich, A., Kim, K. S. et al. (2005). Molecular consequences of PHOX2B missense, frameshift and alanine expansion mutations leading to autonomic dysfunction. *Hum Mol Genet* **14**, 3697–3708.

Valdenaire, O., Rohrbacher, E., Mattei, M. G. (1995). Organization of the gene encoding the human endothelin-converting enzyme (ECE-1). *J Biol Chem* **270**, 29794–29798.

Van Tol, H. H., Bunzow, J. R., Guan, H. C., Sunahara, R. K., Seeman, P., Niznik, H. B., Civelli, O. (1991). Cloning of the gene for a human dopamine D4 receptor with high affinity for the antipsychotic clozapine. *Nature* **350**, 610–614.

Van Tol, H. H., Wu, C. M., Guan, H. C., Ohara, K., Bunzow, J. R., Civelli, O. et al. (1992). Multiple dopamine D4 receptor variants in the human population. *Nature* **358**, 149–152.

Verheij, J. B., Sival, D. A., van der Hoeven, J. H., Vos, Y. J., Meiners, L. C., Brouwer, O. F., van Essen, A. J. (2006). Shah-Waardenburg syndrome and PCWH associated with SOX10 mutations: A case report and review of the literature. *Eur J Paediatr Neurol* **10**, 11–17.

Verze, L., Viglietti-Panzica, C., Plumari, L., Calcagni, M., Stella, M., Schrama, L. H., Panzica, G. C. (2000). Cutaneous innervation in hereditary sensory and autonomic neuropathy type IV. *Neurology* **55**, 126–128.

Wanigasekara, Y., Keast, J. R. (2005). Neurturin has multiple neurotrophic effects on adult rat sacral parasympathetic ganglion neurons. *Eur J Neurosci* **22**, 595–604.

Waxman, S. G., Dib-Hajj, S. (2005). Erythermalgia: Molecular basis for an inherited pain syndrome. *Trends Mol Med* **11**, 555–562.

Yang, Y., Wang, Y., Li, S., Xu, Z., Li, H., Ma, L. et al. (2004). Mutations in SCN9A, encoding a sodium channel alpha subunit, in patients with primary erythermalgia. *J Med Genet* **41**, 171–174.

22

Multiple Sclerosis as a Neurodegenerative Disease

Stephen G. Waxman

Multiple sclerosis (MS), the most common neurological cause of neurological disability in young adults in industrialized societies, traditionally has been classified as a demyelinating disease. Indeed, the remarkable relapsing–remitting pattern of clinical deficits that often is seen in MS is due, in significant part, to demyelination of white matter followed by restoration of conduction in a subpopulation of demyelinated axons. Importantly, however, patients with progressive forms of MS acquire an increasing burden of persistent neurological deficits that develop as a result of incomplete remission following relapses, or in a pattern of progression without remission, which begins at the onset. It is now clear that clinical progression of MS reflects pathologic changes within axons and possibly within the neuronal cells that give rise to them.

Recognition of axonal drop-out as a cause of disability in MS presents an important paradigm shift, since it presents new therapeutic targets. A number of lines of research, including epidemiological investigations, provide evidence suggesting that the progression of disability over the long term in MS may occur independently of, or depend only partially on, acute inflammatory attack (see, for example, Confavreux, 2006). Thus, whereas demyelination and the inflammatory processes leading to it may contribute substantially to its pathophysiology, neuronal injury including degeneration of axons and the processes leading up to it also appear to be crucial. Recognition that MS is, at least in part, a neurodegenerative disorder implies that by understanding *neurodegenerative mechanisms* that contribute to processes such as neuronal injury and axonal degeneration, it may be possible to devise protective interventions that will preserve function in MS. Because neurodegenerative and inflammatory processes may be driven by different mechanisms within neurons and immune cells, respectively, it has been suggested that multimodal therapy, aimed on the one hand at neurons and neuronal molecules, and on the other at inflammatory cells and the molecules associated with them, may prove more effective than either approach alone. Some molecules, such as sodium channels, may play important roles in both the neurodegenerative and immune cascades, suggesting the possibility that it may be possible to devise multimodal therapies that use single drugs.

This chapter will review recent progress in understanding neurodegenerative aspects of MS. As an example of a line of research that has begun to identify molecular targets that participate in neurodegeneration in MS, this chapter will focus on voltage-gated sodium channels, which play pivotal roles not only in normal axonal function which is compromised in MS, but also in the intraneuronal cascade leading to axonal degeneration, and possibly in the activity of immune cells such as macrophages and microglia which can injure axons in MS.

I. Focal Distribution of Sodium Channels in Myelinated Axons

Voltage-gated sodium channels are an important prerequisite for generation and conduction of action potentials along mammalian axons. Sodium channels activate, producing an inward flow of sodium ions and a consequent depolarization, in response to membrane depolarization, and provide the current underlying the upstroke of the action potential. In contrast to nonmyelinated axons (e.g., the giant axon of the squid and nonmyelinated axons of the mammalian PNS and CNS), which display a moderate density (~100/μm^2) of sodium channels that are sprinkled in a relatively uniform distribution along the entire length of the axon, normal myelinated axons display a focal distribution of sodium channels, which are aggregated at a high density (approximately 1,000/μm^2) in the axon membrane at the nodes of Ranvier, but are sparse (less than 25 channels/μm^2) in the internodal and paranodal axon under the myelin (Ritchie & Rogart, 1976; Waxman, 1977). The clustering of sodium channels at nodes matches the pattern of distribution of the insulating myelin (see Figure 22.1A) and thus supports saltatory conduction, and illustrates the elegance of the architecture of the normal myelinated axon. The nonuniform distribution of sodium channels is, however, less well matched to functional needs of demyelinated axons, where the myelin capacitative shield is lost and dissipation of current through Na^+ channel-poor (formerly myelinated) regions of the axon membrane contributes to conduction failure (see Figure 22.1B), and thereby to clinical deficits.

II. Demyelination in Multiple Sclerosis

Demyelination traditionally has been considered to be a pathological hallmark of MS and indeed, it occurs commonly in MS, in lesions that are disseminated both in space and in time. There is little if any endogenous remyelination within the cores of most of these lesions following loss of the myelin in MS. However, clinico-pathological studies have demonstrated that remissions can occur in the absence of remyelination—that is, in the context of persistent demyelination. As just one example, patients with demyelination that affects a substantial length (a centimeter or more, thus encompassing multiple myelin segments along each axon) of all the axons within the optic nerve have been reported to recover functionally useful vision after episodes of optic neuritis (Ulrich & Groebke-Lorenz, 1983). Functional recovery in cases such as this depends, at least in part, on restoration of secure action potential conduction along at least a subpopulation of demyelinated axons. Even before the "molecular era" (i.e., before the cloning of voltage-gated sodium channels and subsequent identification of multiple channel isoforms), longitudinal current analysis (Bostock & Sears, 1978) indicated that chronically demyelinated axons can reorganize so as to recover the capability to conduct action potentials; this conduction proceeds along the axon in a continuous, rather than saltatory manner, suggesting that demyelinated (formerly internodal) parts of the axon have acquired the capability for electrogenesis (see Figure 22.1C). Early electron microscopic studies suggested a molecular basis for this functional restoration by showing that, after demyelination, the denuded axon membrane can express higher-than-normal densities of sodium channels (Foster et al., 1980).

III. Axonal Degeneration in Multiple Sclerosis

Although demyelination is emphasized as the hallmark of MS in most textbook descriptions, it was appreciated even by early observers such as Charcot that axonal degeneration commonly occurs in MS. Axonal degeneration appears to be a frequent occurrence in acute MS lesions; for example, Trapp et al. (1998) estimated that there were approximately 11,000 degenerating axons per mm^3 in the MS lesions they described, and Craner et al. (2004a) reported 7,500 per mm^3. Recent studies have underscored the functional importance of axonal loss in MS (Kuhlmann et al., 2002; Trapp et al., 1998). The available evidence suggests that axonal degeneration begins early in the course of MS (Filippi et al., 2003). From a clinical perspective, axonal loss assumes significant importance because it has been shown, both in animal models and in human MS, to be accompanied by the acquisition of persistent, nonremitting neurological deficits (Davie et al., 1995; Wujek et al., 2002).

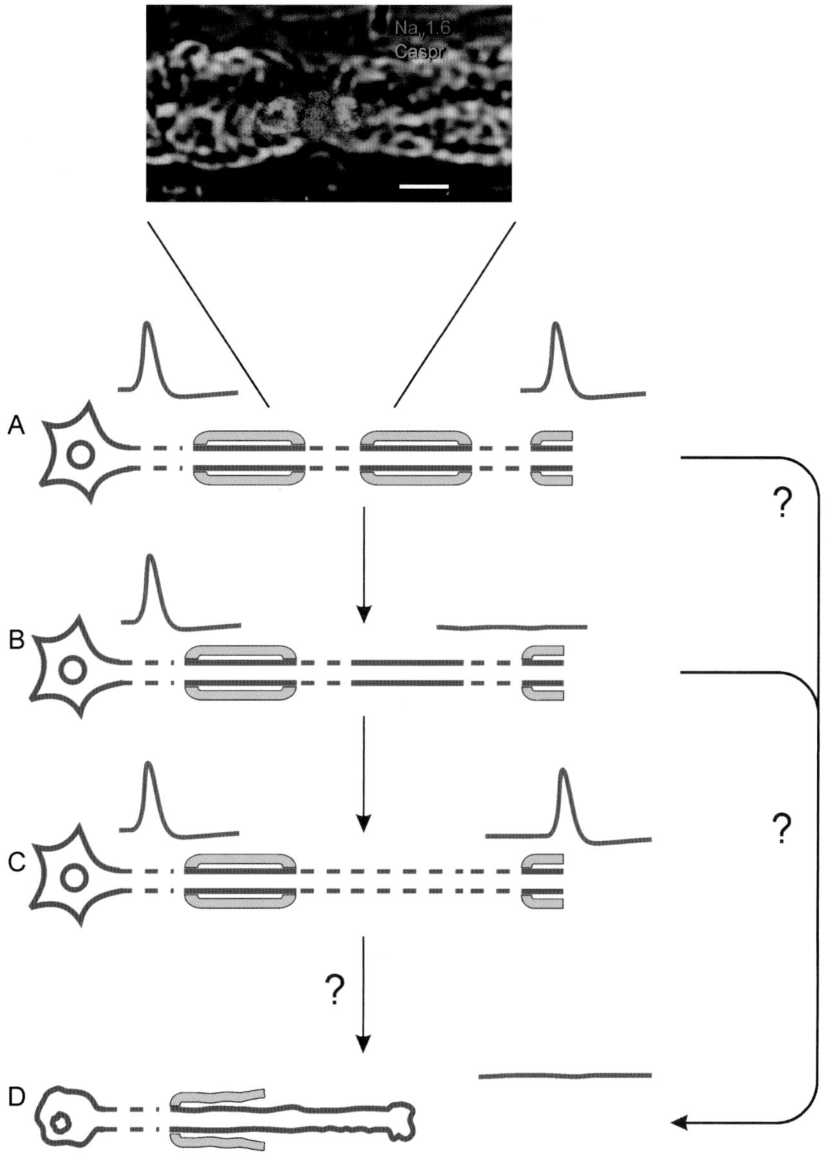

Figure 22.1 (**A**) Voltage-gated Na^+ channels are clustered at a high density within the normal axon membrane at nodes of Ranvier, but are sparse in the paranodal and internodal axon membrane under the myelin. Upper inset: $Na_v1.6$ channels (red) bounded by Caspr (a constituent of the paranodal apparatus) in paranodal regions (green), at a node of Ranvier in a normal myelinated axon. The fluorescence image of the node was merged with a differential contrast image to show the myelin sheath. Scale bar 5 μm. (**B**) Demyelination in multiple sclerosis (MS) initially exposes axon membrane with low Na^+ channel density producing conduction block. (**C**) Some demyelinated axons acquire higher than normal densities of Na^+ channels in demyelinated (formerly paranodal and/or internodal) regions, supporting the recovery of action potential conduction that contributes to clinical remissions. (**D**) Degeneration of axons also occurs in MS, and produces nonremitting, permanent loss of function. Color inset (top) from Black et al. (2002).

IV. Sodium Channels and Axonal Injury

Because of its role in producing nonremitting deficits, there is now substantial interest in the development of therapeutic approaches aimed at slowing or halting axonal degeneration within the brain and spinal cord in MS. As with demyelination, a growing body of evidence indicates that sodium channels within the axon membrane play an important role in axonal degeneration, but their role in axonal degeneration is different. Early studies used anoxia as a reproducible, model insult and showed that anoxic injury to axons in CNS white matter is dependent on Na^+ influx, which drives reverse Na^+–Ca^{2+} exchange that in turn imports damaging levels of Ca^{2+} into axons (Stys et al., 1992a) (see Figure 22.2). The timing of Na^+ influx and the protracted protective effect of the sodium channel blocker tetrodotoxin (TTX) throughout periods of anoxia lasting for an hour or more both suggested the involvement of a persistent (noninactivating) Na^+ current (Stys et al., 1993). Electron microprobe studies on anoxic myelinated axons, moreover, have demonstrated a continuous rise in intra-axonal Na^+, paralleled by a rise in Ca^{2+} levels within the axon throughout the period of anoxia (Lopachin & Stys, 1995). Supporting the idea that noninactivating sodium channels carry the injurious Na^+ influx, a TTX-sensitive persistent sodium conductance is present along the trunks of CNS myelinated axons (Stys et al., 1993).

Also supporting a role for sodium channels in axonal injury, pharmacological block of sodium channels with blockers that include tetrodotoxin, saxitoxin, lidocaine, procaine, phenytoin, and carbamazepine have a protective effect within anoxic CNS white matter (Fern et al., 1993; Stys et al., 1992a, 1992b). As discussed later, recent studies have shown that the Na^+ channel blockers phenytoin (Lo et al., 2003) and flecainide (Bechtold et al., 2004) prevent degeneration of CNS axons, maintain axonal conduction, and improve clinical outcome in experimental autoimmune encephalomyelitis (EAE), an animal model of MS.

V. Energetics, Ionic Homeostasis, and Axonal Injury

Although the early studies on axonal injury did not directly demonstrate energy failure, they suggested that inadequacy of the supply of ATP might be important in the pathophysiology of axonal injury since insufficient levels of activity of Na/K-ATPase (which normally extrudes Na^+ ions while importing K^+ in an energy-dependent manner) would be expected to exacerbate any increase in intra-axonal Na^+. An experimental link to energy supply was provided by Kapoor et al. (2003), who showed that sodium channel blockers can protect axons from NO-induced injury and suggested that NO, which is present at increased concentrations within acute MS lesions (Smith & Lassmann, 2002) can injure axons by damaging mitochondria within them, thereby producing axonal energy failure that impairs Na/K-ATPase activity and limits the ability of axons to extrude Na^+. Further supporting the notion that NO-mediated mitochondrial dysfunction dampens ATP-dependent extrusion of Na^+ and thereby contributes to axonal injury, Garthwaite et al. (2002) showed that TTX preserves ATP levels concurrent with protecting white matter axons from NO-induced injury.

More recently Dutta et al. (2006) have provided evidence for reductions in mitochondrial gene expression in human MS lesions. Using global transcript profiling, they demonstrated

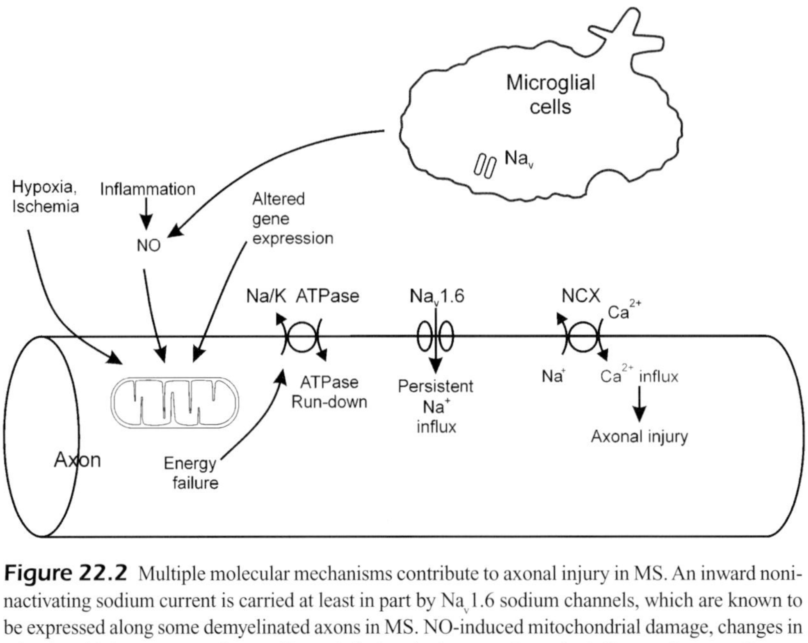

Figure 22.2 Multiple molecular mechanisms contribute to axonal injury in MS. An inward noninactivating sodium current is carried at least in part by $Na_v1.6$ sodium channels, which are known to be expressed along some demyelinated axons in MS. NO-induced mitochondrial damage, changes in mitochondrial gene expression, and hypoxia/ischemia due to perivascular inflammation contribute to energy failure, which in turn leads to loss of function of Na/K ATPase, compromising the ability of the axon to extrude Na^+ and maintain membrane potential. Increased intra-axonal and Na^+ depolarization drive the Na^+/Ca^{2+} exchanger to operate in a "reverse" mode where it imports damaging levels of Ca^{2+}. Sodium channels (Nav) also are involved in phagocytosis by microglia and macrophages, and inactivation of these cells, which contribute to production of NO.

decreased mRNA levels for nuclear-encoded mitochondrial genes within these lesions; they observed decreased activities of respiratory gene complex I and III, which suggested that the changes are functionally significant. They also observed pathological changes including fragmented neurofilaments, depolymerized microtubules, and reduced organelle content within residual demyelinated axons within MS lesions, suggestive of calcium-mediated injury. On the basis of these findings they proposed that mismatch between ATP supply and energy demand contributes to axonal degeneration in MS by impairing Na/K-ATPase activity, thereby limiting or preventing extrusion of increased axoplasmic Na^+.

VI. Molecular Identity of Axonal Sodium Channels

It is now clear that at least nine genes encode distinct voltage-gated sodium channels (termed, according to consensus nomenclature, $Na_v1.1$–$Na_v1.9$), with a shared overall motif but with different amino acid sequences, voltage-dependencies, and kinetics (for review see Catterall et al. 2005). $Na_v1.4$ and $Na_v1.5$ are expressed predominantly in muscle and cardiac tissue, respectively (although low levels of $Na_v1.5$ have been detected in some parts of the CNS). $Na_v1.1$–$Na_v1.3$ and $Na_v1.6$–$Na_v1.9$ are expressed predominantly within neurons. $Na_v1.1$, $Na_v1.2$, and $Na_v1.6$ channels are expressed widely within the adult CNS; as described later, $Na_v1.2$ and $Na_v1.6$ are the predominant sodium channel isoforms along CNS axons ($Na_v1.2$ along nonmyelinated CNS axons and $Na_v1.6$ as the predominant subtype at nodes of Ranvier). In contrast, the peripheral nerve channels $Na_v1.7$, $Na_v1.8$, and $Na_v1.9$ are expressed preferentially within dorsal root ganglion and trigeminal ganglion neurons (and in the case of $Na_v1.7$, within sympathetic ganglion neurons) and, although present along the axons of these peripheral neurons, are not usually detected along CNS axons. As noted later, $Na_v1.2$ and $Na_v1.6$ are both expressed along demyelinated axons in experimental models of MS and in MS.

VII. Sodium Channels and Recovery of Conduction in Demyelinated Axons

Early electrophysiological studies on demyelinated peripheral nerves (Bostock & Sears, 1978) showed that, weeks after loss of the myelin, some denuded axons recover the capability to conduct action potentials. Action potentials propagate along these chronically demyelinated axons in a continuous manner, suggesting the presence of sodium channels along extensive, formerly myelinated (and therefore previously sodium channel-poor) domains of the demyelinated axons. Early cytochemical (Foster et al., 1980) and immunocytochemical (England et al., 1991; Novakovic et al., 1998) studies using pan-specific Na^+ channel antibodies demonstrated the acquisition of higher than normal numbers of Na^+ channels in chronically demyelinated axons in experimental model systems, providing a molecular basis for this recovery of conduction (see Figure 22.1C). Also suggesting the expression of new sodium channels along demyelinated axons, a four-fold increase in saxitoxin binding sites was observed within demyelinated white matter from MS patients (Moll et al., 1991). These early studies, which were carried out prior to discovery of the sodium channel isoforms that are expressed along axons, could not reveal, of course, the molecular identity of the new axonal channels that restore conduction.

VIII. $Na_v1.2$ and $Na_v1.6$ in Normal and Dysmyelinated Axons

More recent studies have utilized isoform-specific antibodies, generated against specific subtypes of sodium channels, to determine the molecular identities of the sodium channels along normal, demyelinated, and dysmyelinated axons. These studies show that $Na_v1.2$ and $Na_v1.6$ sodium channels are present in normal myelinated axons and their premyelinated precursors within the CNS, but tend to be expressed at different stages of development. At early developmental stages prior to glial ensheathment, there is low density of Na^+ channels that are distributed relatively uniformly along the entire trajectory of premyelinated CNS axons (Waxman et al., 1989); these channels appear to support action potential conduction, which is known to occur along premyelinated axons, prior to myelination (Foster et al., 1982; Rasband et al., 1990). Initially $Na_v1.2$ channels are present. With progression of myelination, there is a loss of $Na_v1.2$ channels and an aggregation of $Na_v1.6$ channels as nodes of Ranvier mature (Boiko et al., 2001; Kaplan et al., 2001) so that, at fully formed nodes of Ranvier, $Na_v1.6$ is the predominant sodium channel isoform (see Figure 22.1, top inset) (Caldwell et al., 2000). The $Na_v1.6$ channels are tightly clustered at the nodes; in contrast to the nodal membrane where $Na_v1.6$ channels are aggregated, $Na_v1.6$ channels are not detectable in the paranodal and internodal axon membrane under the myelin of mature myelinated fibers.

Interestingly, $Na_v1.2$ channels continue to be expressed along axons that do not acquire myelin sheaths in dysmyelinated mutants where myelin fails to form as a result of abnormalities within glial cells. For example, $Na_v1.2$ channels continue to be expressed (Boiko et al., 2001; Westenbroek et al., 1989), whereas $Na_v1.6$ channels are not detectable along dysmyelinated axons within the *Shiverer* mutant rat, which lack compact myelin. Together with the observations on myelinated axons in different stages of development, these observations on dysmyelinated axons suggest a relationship between myelin formation and the sequential expression of $Na_v1.2$ and then $Na_v1.6$; these studies, however, do not provide information about the types of Na^+ channels that are expressed in demyelinated axons.

IX. Axonal Sodium Channels in Demyelinated Axons: Lessons from EAE

Experimental autoimmune encephalitis (EAE), in which experimental animals develop an inflammatory disorder that often includes demyelination as well as axonal degeneration after they are inoculated with components of white matter, is commonly studied as a model of MS. Recent studies using immunocytochemical methods and *in situ* hybridization have identified the Na^+ channel isoforms expressed along demyelinated axons in EAE. In contrast to normal white matter where most nodes of Ranvier express Na$_v$1.6, both Na$_v$1.2 and Na$_v$1.6 channels are present along myelinated (or remyelinated) axons in EAE where there is an increased frequency of Na$_v$1.2 channel-positive nodes and a reduction in the frequency of Na$_v$1.6 channel-positive nodes within white matter (Craner et al., 2003). An increase in the overall number of nodes in EAE suggests the formation of some new myelin sheaths and of some new nodes at formerly internodal sites, but whether the Na$_v$1.2 channel-expressing nodes have been formed in association with remyelinated axons or by replacement of Na$_v$1.6 channels with Na$_v$1.2 channels at preexisting nodes is not known. Moreover, it is not known whether Na$_v$1.2 is expressed in a stable manner, or transiently, in EAE. In other models of CNS demyelination, the ethidium bromide model, clustering of Na$_v$1.6 is reestablished at most nodes found during remyelination by transplanted olfactory ensheathing cells (Sasaki et al., 2006) and by Schwann cells (Black et al., 2006a).

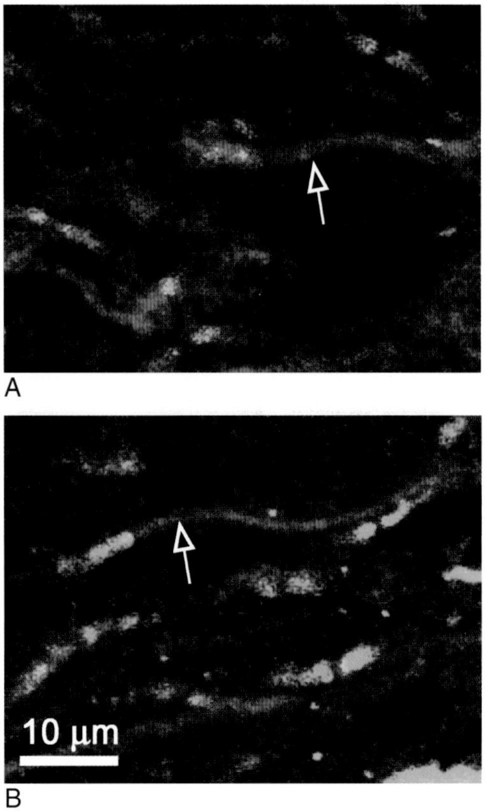

Figure 22.3 Extensive regions of in Na$_v$1.6 (A) and Na$_v$1.2 channel (B) expression along optic nerve axons in EAE. These diffuse zones of sodium channel expression can extend for tens of microns. Scale bar 10 μm. Reproduced, with permission, from Craner et al. (2003).

In regions where they are demyelinated, some axons in EAE display extensive regions of immunostaining for Na$_v$1.2 and Na$_v$1.6, running tens of microns along the fiber axis (much longer than the length of a node of Ranvier, and thus encompassing axon regions that have lost their myelin) (see Figure 22.3). The presence of these extensive zones of Na$_v$1.2 and Na$_v$1.6 expression along axons in the optic nerve, where all the axons are normally myelinated, shows unequivocally that these channels have been expressed along demyelinated axons, and not merely along normal nonmyelinated axons (Craner et al., 2003).

Increased neuronal transcription of the gene encoding Na$_v$1.2 channels appear to contribute to increased expression of Na$_v$1.2 protein along demyelinated optic nerve axons in EAE, since it is paralleled by upregulated Na$_v$1.2 channel mRNA levels within retinal ganglion cells, the cells of origin of these axons (Craner et al., 2003). The signal that triggers this gene activation and the intracellular processes that determine whether Na$_v$1.2 or Na$_v$1.6 are expressed, are not yet understood.

X. Axonal Sodium Channels in Injured Axons: EAE

As described earlier, a sustained Na^+ influx through persistently activated sodium channels can produce Ca^{2+}-mediated injury of white matter axons, by driving the Na^+–Ca^{2+} exchanger to operate in a reverse mode in which it imports Ca^{2+} (Stys et al., 1992a). Recent work indicates that Na$_v$1.6 channels are a source of this persistent sodium influx; these channels have been shown to produce a persistent current in a spectrum of cell types, and it is larger than the persistent current produced by Na$_v$1.2 channels (Rush et al., 2005). Thus expression of Na$_v$1.6 channels together with the Na^+–Ca^{2+} exchanger in demyelinated axons would be expected to poise them to import injurious levels of Ca^{2+}.

To determine whether Na$_v$1.6 channels and the Na^+–Ca^{2+} exchanger are, in fact, expressed in close proximity within degenerating axons in EAE, Craner et al. (2004b) used double-label immunocytochemistry to localize these molecules and β-amyloid precursor protein (β-APP), a marker of axonal injury. This study showed that more than 90 percent of β-APP-positive axons in EAE express Na$_v$1.6 channels (either alone (56%) or together with Na$_v$1.2 channels

(36%)); in contrast, less than 2 percent of β-APP-positive axons were observed to express only $Na_v1.2$ channels. Triple-labeling immunohistochemistry showed coexpression of $Na_v1.6$ channels and the Na^+–Ca^{2+} exchanger within 74 percent of β-APP-positive axons, in contrast to only 4 percent of β-APP-negative axons that exhibited such coexpression (see Figure 22.4). These results suggest that colocalization of $Na_v1.6$ channels and the Na^+–Ca^{2+} exchanger may to be associated with axonal injury in EAE.

XI. Axonal Sodium Channels in MS

Brain tissue is now rarely biopsied in patients with MS or with suspected MS (due, in part, to availability of imaging methods such as MRI, which have made diagnosis more straightforward), and the time lag between death and removal of tissue for postmortem study, which is usually at least hours, results in degradation of channel proteins and mRNA and thus limits the utility of molecular analysis. Lesion-to-lesion differences, both between patients and within patients, further complicate analysis (Lassmann, 2005). Nevertheless, some clues have been provided by a postmortem analysis (Craner et al., 2004a) of spinal cord and optic nerve from patients who died with a diagnosis of disabling secondary progressive MS.

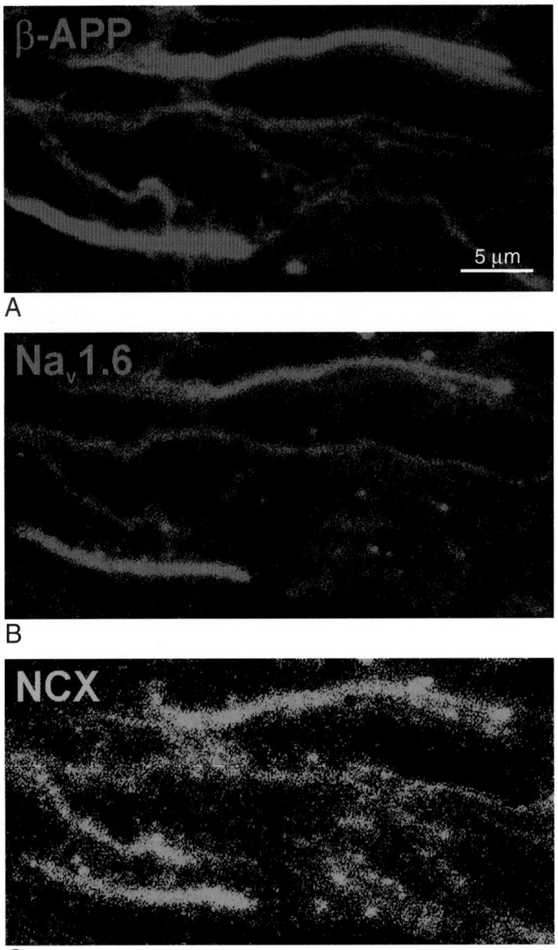

Figure 22.4 Coexpression of $Na_v1.6$ channels (B) and the Na^+–Ca^{2+} exchanger (C) in degenerating spinal cord axons in EAE. β-APP is a marker of axonal injury (**A**). Modified, with permission, from Craner et al. (2004b).

Analysis of this tissue suggests a pattern of Na^+ channel expression within acute MS lesions that is similar to the pattern seen in EAE (Craner et al., 2004a). Control white matter, from patients with no neurological disease, displayed abundant myelin basic protein (MBP; a marker for myelin) and the expected focal pattern of expression of $Na_v1.6$ at nodes of Ranvier. In contrast, acute MS plaques (which could be identified on the basis of attenuated MBP immunostaining, evidence of inflammation and recent phagocytosis of myelin) displayed $Na_v1.6$ and $Na_v1.2$ along extensive regions, often running tens of microns, along demyelinated axons (see Figure 22.5). Demyelinated axonal regions expressing $Na_v1.6$ or $Na_v1.2$ in some cases were bounded by damaged myelin (see Figure 22.5E, F) or Caspr (see Figure 22.5G, H), a constituent of the paranodal apparatus (Bhat et al., 2001; Einheber et al., 1997), confirming the identity of these profiles as demyelinated axons.

Within the lesions examined, almost all β-APP-immunopositive axons showed extensive regions of $Na_v1.6$ expression, whereas few β-APP-immunopositive axons expressed $Na_v1.2$ (Craner et al., 2004a). As shown in Figure 22.6, $Na_v1.6$ channels and the Na^+–Ca^{2+} exchanger tended to be colocalized within β-APP-positive axons within these MS lesions. $Na_v1.2$ channels and the Na^+–Ca^{2+} exchanger tended to be expressed, in contrast, in β-APP-negative axons. Thus, in these acute MS lesions, there was an association between coexpressions of $Na_v1.6$ and the Na^+–Ca^{2+} exchanger and axonal injury.

XII. $Na_v1.2$ Channels in Demyelinated Axons: Functional Role

The diffuse distribution of $Na_v1.2$ channels for tens of microns along demyelinated but apparently uninjured axons in EAE and MS is similar to the continuous pattern of distribution of $Na_v1.2$ channels along premyelinated (Boiko et al., 2001) and nonmyelinated CNS axons (Boiko et al., 2003 Gong et al., 1999; Westenbroek et al., 1989; Whitaker et al., 2000). Action potential conduction is known to occur along premyelinated axons well in advance of myelination (Foster et al., 1982), and is presumably supported by these channels.

$Na_v1.2$ channels differ from $Na_v1.6$ channels in terms of a number of physiological parameters, including activation and availability (steady-state inactivation) that are more depolarized (Rush et al., 2005). However, $Na_v1.2$ channels show greater accumulation of inactivation at high frequencies (20–100 Hz)

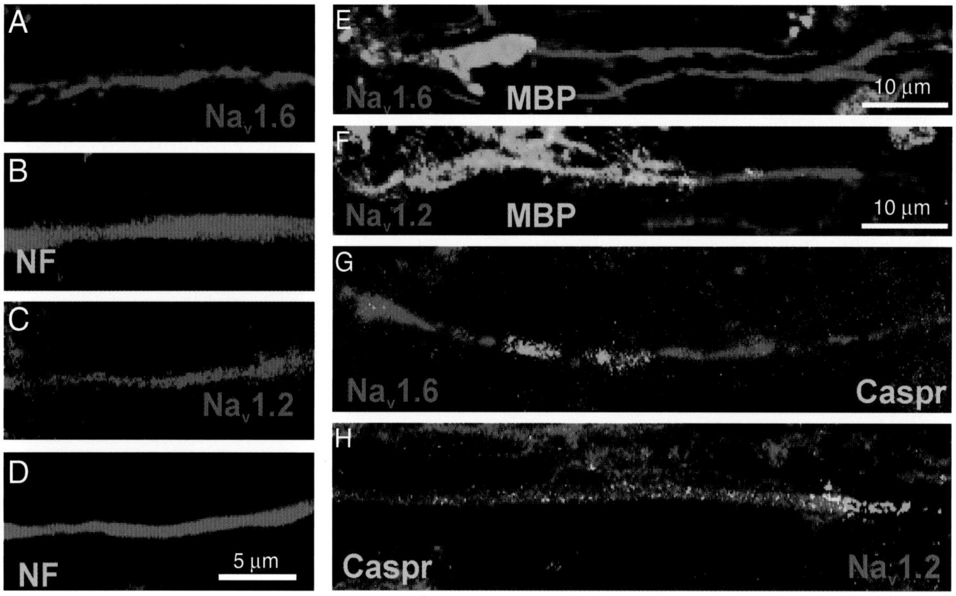

Figure 22.5 Changes in $Na_v1.6$ and $Na_v1.2$ channel expression along demyelinated axons within active lesions from patients with disabling secondary progressive MS. $Na_v1.6$ (**A**) and $Na_v1.2$ (**C**) are present along extensive regions of denerated axons, which also express neurofilament protein (**B,D**). Panels (**E**) and (**F**) show edges of active MS lesions within spinal cord, with residual damaged myelin (green) next to extensive regions of diffuse expression of $Na_v1.6$ channels (**E**) (red) and $Na_v1.2$ channels (F) (red). In some cases extensive regions of $Na_v1.6$ (red, **panel G**) or $Na_v1.2$ channels (red, **panel H**) are bounded by Caspr (green), without overlap, consistent with the expression of $Na_v1.6$ and $Na_v1.2$ channels within the demyelinated axon membrane. Abbreviation: MBP, myelin basic protein. From Craner et al. (2004a).

compared to $Na_v1.6$. The persistent current produced by $Na_v1.2$ channels is smaller than for $Na_v1.6$ channels, and the activation and availability characteristics of $Na_v1.2$ predict a smaller "window" current than produced by $Na_v1.6$ (Rush et al., 2005). These physiological characteristics suggest that $Na_v1.2$ channels might play an adaptive role along demyelinated axons, supporting the conduction of action potentials after demyelination, at least at lower frequencies, while limiting the degree of sustained Na^+ influx. Nonetheless, it is possible that the expression of $Na_v1.2$ along demyelinated axons may contribute to some clinical deficits. Inactivation, including closed-state inactivation, has a slow onset in $Na_v1.2$ channels (Rush et al., 2005), and this would be expected to increase the sensitivity of these channels to slow depolarizations (Cummins et al., 1998), a factor that might contribute to ectopic firing or unstable patterns of firing after demyelination that could produce paraesthesia or similar phenomena (see Baker, 2005; Kapoor et al., 1997).

XIII. $Na_v1.6$ Channels in Demyelinated Axons: Functional Role

Direct comparison of $Na_v1.6$ and $Na_v1.2$ (Rush et al., 2005) shows that $Na_v1.6$ channels produce a larger persistent current than $Na_v1.2$ channels throughout the potential domain between 0 and −80 mV. The biophysical properties of $Na_v1.6$ channels predict a persistent "window" current that can carry sodium inward between −65 and −40 mV suggesting that, in axons that are depolarized after injury, $Na_v1.6$ channels can drive reverse Na/Ca exchange even in the absence of action potential activity. Moreover, there is recent evidence suggesting that persistent current through $Na_v1.6$ channels might be further increased by secondary proteolytic injury to the channel itself, particularly to its inactivation mechanism, triggered by the rise in intra-axonal Ca^{2+} levels early in the course of injury, introducing a feed-forward process that would further increase $Na_v1.6$ channel current amplitude (Iwata et al., 2004). Whether this occurs in MS is not yet clear.

The physiological and immunolocalization results support the proposal that $Na_v1.6$ channels, when coexpressed with the Na^+–Ca^{2+} exchanger along demyelinated axons, can contribute to axonal injury as shown in Figure 22.2. Consistent with this proposal, it is known that dysmyelinated CNS axons express $Na_v1.2$ channels rather than $Na_v1.6$ channels (Boiko et al., 2001; Westenbroek et al., 1989), and it also has been shown that dysmyelinated axons are significantly less sensitive than myelinated axons (which express $Na_v1.6$ at their nodes) to injury triggered by activity of Na^+ channels (Waxman et al., 1990).

It is possible that some axons may degenerate in MS without becoming demyelinated. DeLuca et al. (2006) found only a poor correlation between plaque load and axonal

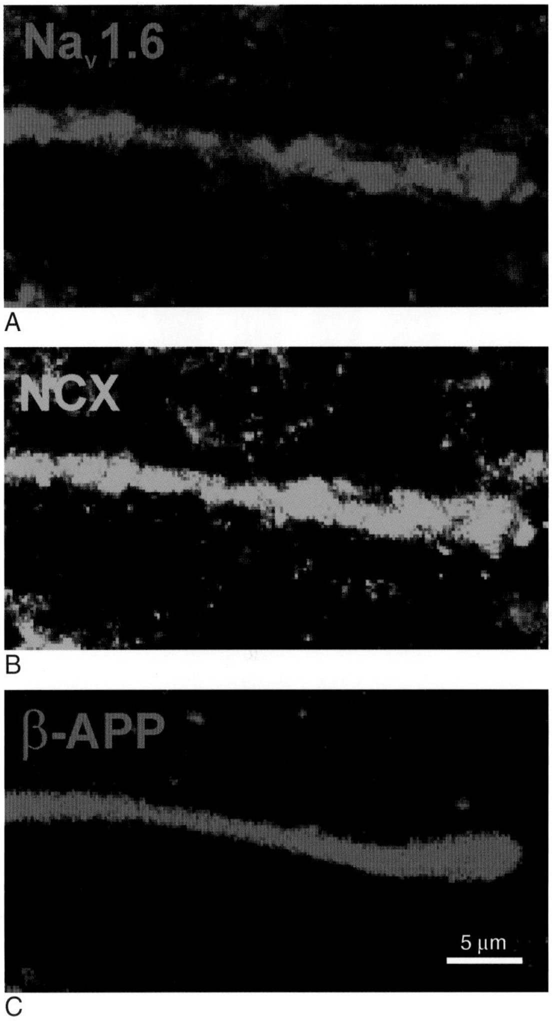

Figure 22.6 The Na^+–Ca^{2+} exchanger and Nav1.6 channels are coexpressed within injured (β-APP-positive) axons in MS. Axons within MS spinal cord white matter immunostained for (**A**) Nav1.6 (red), (**B**) the Na^+/Ca^{2+} exchanger (green), and (**C**) β-APP (blue). From Craner et al. (2004a).

loss in postmortem MS tissue, and interpret their results as suggesting that demyelination may not be a primary determinant of axonal degeneration. Steffensen et al. (1997) showed that the Na^+–Ca^{2+} exchanger is present at nodes of Ranvier, where Na$_v$1.6 channels are present. If the compromise of ATP supply described by Dutta et al. (2006) occurs in axons (or regions of axons) that are not demyelinated in MS, it could predispose axons with intact myelin to degeneration.

XIV. Sodium Channels in Microglia and Macrophages

Although most research to date has focused on the roles of sodium channels within the intraneuronal cascade that leads to axonal degeneration in MS, recent evidence suggests that members of this family of channels may also contribute to axonal degeneration via a second mechanism, by regulating the activity of microglia and macrophages, which in turn can injure axons (Craner et al., 2005). The association between microglia and macrophages and degenerating axons in MS has been well established (Furguson et al., 1997; Kornek et al., 2000; Trapp et al., 1998). These inflammatory cells appear to injure axons by multiple mechanisms that include phagocytosis (Li et al., 1996), induction of proliferation of CD4+ T-cells (Cash & Rott, 1994), production of pro-inflammatory cytokines (Renno et al., 1995) and NO (deGroot et al., 1997; Hooper et al., 1997), and antigen presentation (Matsumoto et al., 1992). The presence of voltage-gated sodium channels within microglia was demonstrated in early patch clamp studies (Korotzer & Cotman, 1992; Norenberg et al., 1994).

Craner et al. (2005) recently used immunocytochemical methods to show that, indeed, Na$_v$1.6 sodium channels are present in microglia (see Figure 22.7). They observed up-regulation of Na$_v$1.6 that was associated with microglial activation in EAE (see Figure 22.7A). Supporting the hypothesis that sodium channels might play a role in microglial activation, Craner et al. (2005) observed that block of sodium channels in cultured microglia with tetrodotoxin (TTX) results in a 40 percent reduction in phagocytic activity of these cells (see Figure 22.8A, B), and noted that administration of phenytoin to rats with EAE results in a 75 percent decrease in the number of inflammatory cells (see Figure 22.8C, D). Craner et al. (2005) also observed that activation of microglia from *med* mice (which lack functional Na$_v$1.6 channels) is decreased compared to wild-type mice in which Na$_v$1.6 is present, and showed that the suppressing effect of TTX on microglial activation is not present in *med* mice.

Extending these observations to human MS tissue, Craner et al. also demonstrated an up-regulation of Na$_v$1.6 expression within macrophages and microglia in acute MS lesions, with a 1.6-fold increase in Na$_v$1.6 immunosignal in resting, and a four-fold increase within activated microglia, compared to microglia (almost all of which have resting characteristics) in control patients without neurological disease (see Figure 22.7B, C).

Further studies are examining the mechanism by which Na$_v$1.6 channels (and possibly other sodium channel isoforms) participate in the function of microglia and macrophages. Irrespective of the full repertoire of sodium channel isoforms involved in the function of these cells and the intracellular mechanisms that mediate the activity of these channels, the available observations suggest that, in addition to a neuroprotective effect that acts directly on axons, sodium channel blockade may have a direct effect on some immune cells, and might possibly attenuate axonal injury in MS and its animal models via a second, parallel mechanism that limits inflammatory activity.

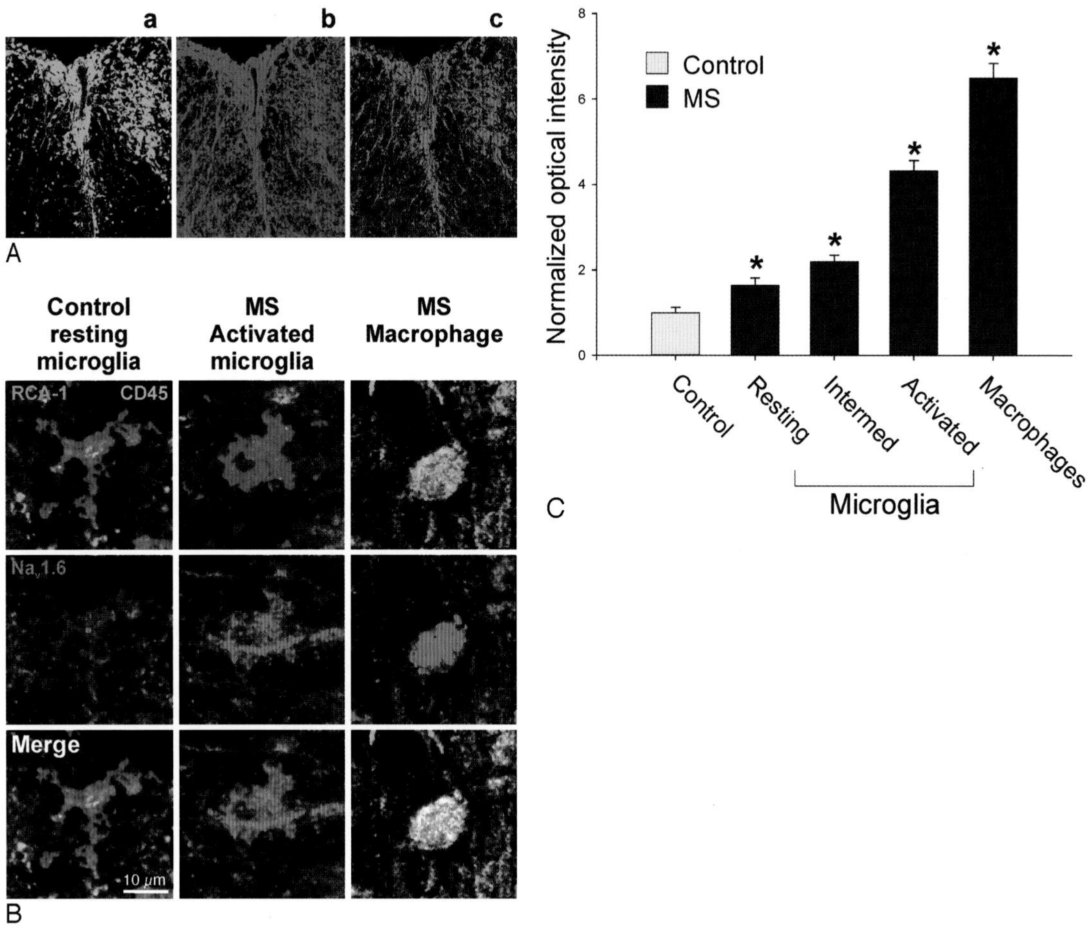

Figure 22.7 Sodium channels are present within, and contribute to microglia/macrophage activation and function, in EAE and MS. **A.** Sections stained for CD45 (**a**), OX-42 (**b**), and $Na_v1.6$ (**c**) from spinal cord from a mouse with EAE. $Na_v1.6$ is present within OX-42-, and CD45-positive inflammatory cells. **B.** Up-regulation of $Na_v1.6$ protein in activated microglia and macrophages within acute MS lesions. Microglia and macrophages were identified by staining with biotinylated RCA-1 (blue) and anti-CD45 (green) immunostaining. Within control tissue, microglia demonstrate a resting morphology (left) with small cell bodies and thin, branched processes, and only very low levels of expression of $Na_v1.6$ protein (red). Activated microglia within white matter, at the edges of acute MS lesions, display changes in morphology into a rounded, ameboid shape associated with activated, phagocytic phenotype. Transformation to an activated phenotype is associated with an up-regulation of $Na_v1.6$ protein (red). Right column: Macrophage in an active MS plaque showing rounded morphology and robust immunostaining for RCA-1, CD45, and $Na_v1.6$ protein. **C.** Histogram demonstrating progressive up-regulation of $Na_v1.6$ protein with activation of microglia and macrophages in acute MS lesions, compared with resting microglia in controls with no neurological disease. $^*p < 0.001$.

XV. From Neurodegeneration to Neuroprotection?

On the basis of these results, nonspecific sodium channel blockers or isoform-specific blockers of $Na_v1.6$ channels (or, preferably, of the persistent component of the current produced by $Na_v1.6$ channels) would be predicted to be protective in MS, acting directly on axons to prevent axonal degeneration, or attenuating activation of, and phagocytosis by, microglia and macrophages. Subtype-specific blockers are not yet available, but neuroprotective effects of the nonspecific Na^+ channel blockers phenytoin (Lo et al., 2003) and flecainide (Bechtold et al., 2004) have been demonstrated in EAE, where these agents reduce the degree of axonal degeneration (see Figure 22.9). Lo et al., (2003) showed that phenytoin reduces the loss of dorsal corticospinal axon from 63 to 25 percent, and of cuneate fasciculus axons from 43 to 17 percent after 28 days of EAE (Lo et al., 2003). Electrophysiological recordings demonstrate that axonal conduction is maintained in a significant number of the surviving axons after treatment with these drugs (see Figure 22.9B, C). Importantly, treatment with these sodium channel blockers improves clinical outcome (see Figure 22.9D). Similar results were reported by Bechtold et al. (2004) in a study employing flecainide. Although these initial studies demonstrated protection of axons and improved clinical status for animals fol-

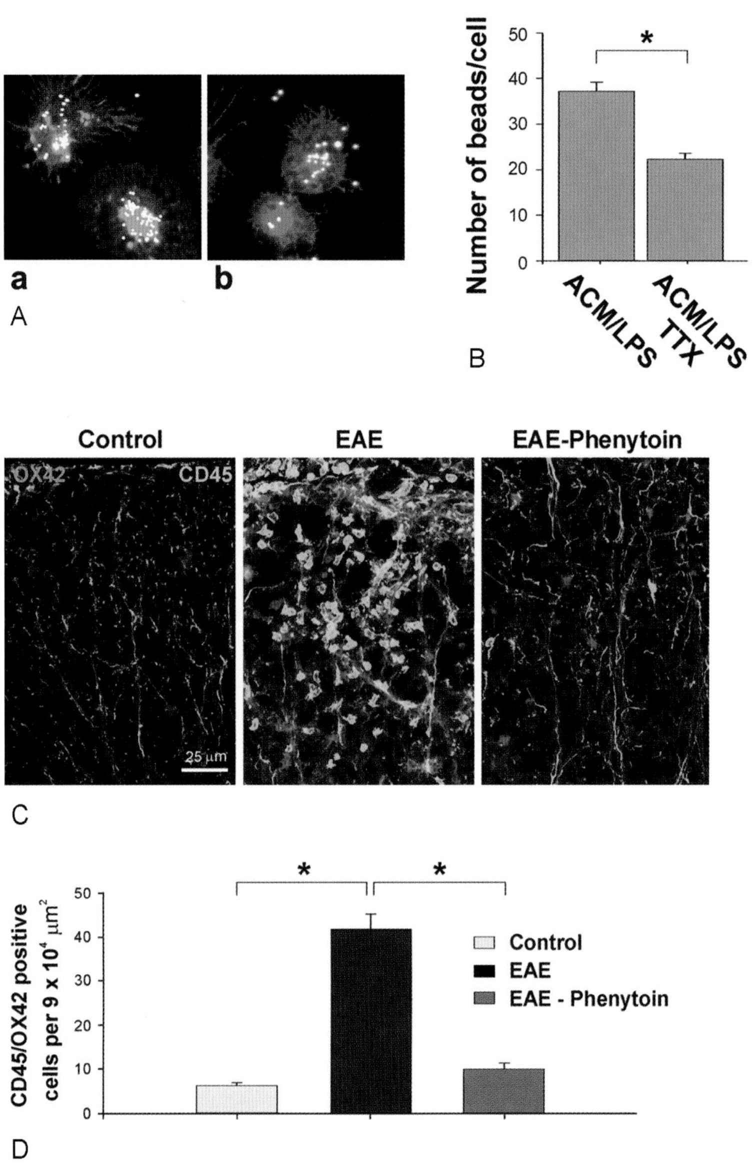

Figure 22.8 **A**. Activation and phagocytic function of microglia/macrophages are attenuated by sodium channel blockade. (**a**) Administration of TTX (**b**) to lipopolysaccharide (LPS) – stimulated microglia *in vitro* reduces phagocytotic function, indicated by decreased number of latex particles phagocytized per cell compared to cells not treated with TTX (**a**). **B**. Particle counts demonstrating significant reduction in the degree of phagocytosis after treatment with TTX. **C**. Phenytoin reduces inflammatory infiltrate in EAE. **a**, **b**, and **c** show control, untreated EAE, and phenytoin-treated EAE spinal cord immunostained for anti-CD45 (green) and anti-OX42 (blue). Phenytoin results in a marked reduction in inflammatory infiltrate. **D**. Histogram showing number of CD45 and/or OX42 immunopositive cells per 9×10^4 μm^2. There is a significant increase in the number of immune cells in EAE. The number of immune cells in EAE is reduced by treatment with phenytoin. *,$P < 0.005$. From Craner et al. (2005).

lowed for 28 to 30 days with MS, a more recent study (Black et al., 2006b) has shown that the protective effect and clinical improvement persist in mice with monophasic EAE treated continuously with phenytoin for as long as 180 days, and in mice with chronic-relapsing EAE treated with phenytoin followed for 120 days; that is, for a length of time that is a substantial portion of the mouse lifespan.

Whether these sodium channel blockers exert their protective effect via a direct action on axons, or via an action on immune cells such as microglia or macrophages, or via

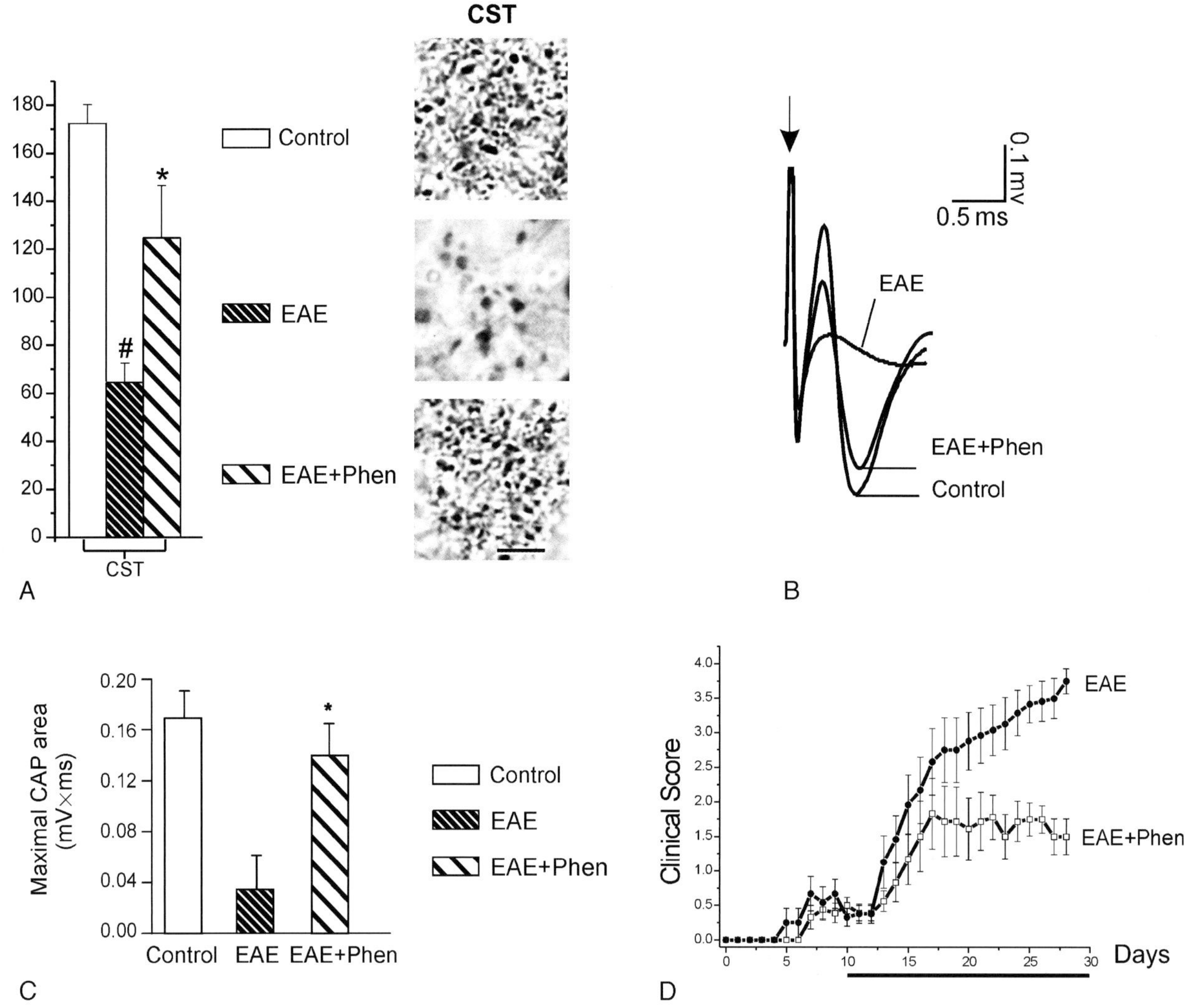

Figure 22.9 Phenytoin, orally administered at doses that achieve levels in the human therapeutic range, protects axons, preserves axonal conduction, and results in improved neurological function in mice with EAE. **A, left**: Cross-sections through the mid-cervical dorsal corticospinal tract stained for neurofilaments, showing increase in the number of surviving axons in phenytoin-treated EAE. **A, right**: Quantification of dorsal corticospinal tract axons in control mice, untreated EAE, and phenytoin-treated EAE (axon numbers/500 μm^2) showing protective effect of phenytoin. **B**. Superimposed supramaximal compound action potentials (CAP) from representative phenytoin-treated control, untreated EAE, and phenytoin-treated EAE. *Arrow* = stimulus artifact. The CAP is highly attenuated in untreated EAE, and is partially restored in phenytoin-treated EAE. **C**. Average supramaximal CAP area in phenytoin-treated control, untreated EAE, and phenytoin-treated EAE ($^{*}p < 0.05$, phenytoin-treated EAE compared with untreated EAE. **D**. Phenytoin treatment improves neurological status in EAE. Clinical scores (0–6 scale, with 0 = normal; 6 = death) showing improved neurological function in EAE as a result of treatment with phenytoin. Oral administration of phenytoin was started on day 10, as indicated by the horizontal bar. From Lo et al. (2003).

both mechanisms is not yet definitively known. On the basis of the protective effect of flecainide on neurological symptoms early in the course of EAE (10–13 days post-disease induction), Bechtold et al. (2004) suggested an immunomodulatory action. Craner et al. (2005), as discussed earlier, observed that treatment with phenytoin ameliorates the inflammatory cell infiltrate in EAE by 75 percent, and observed that TTX reduces the phagocytic function of activated microglia by 40 percent. Yet sodium channel blockers are also protective *in vitro*, in assays in which white matter axons are subjected to injury under conditions where inflammatory or immune processes are minimized (Stys et al., 1992a,b; Fern et al., 1993). In these models the sodium channel blockers must be acting on neural cells. Thus it appears likely that sodium channels blockers prevent axonal degeneration via a dual mechanism, involving both a direct action on axons and an immunomodulatory action on microglia and/or macrophages.

Whether $Na_v1.6$ channel-specific blockade, or nonspecific Na^+ channel blockade, will be useful clinically as a neuroprotective strategy in MS remains to be determined. Clinical trials of sodium channel blocking agents in MS are now being planned by some investigators, although others are carrying out additional animal studies prior to making a go–no-go decision about clinical trials.

Although this chapter has focused primarily on sodium channels, it is possible that elements within the intra-axonal energy chain may also represent tractable therapeutic targets in MS. Strategies for energy repletion within injured neurons might, for example, prove to be efficacious. Alternatively, mitochondrial molecules, or the injurious processes (such as production of NO that damages them), might be targeted. Development of therapeutic approaches of this type might benefit from prior attempts within the neurological community to target energy depletion in disorders such as Parkinson disease.

Even if sodium channel blockers do not prove to be effective in halting or slowing clinical progression in MS, or if they are shown to have a mechanism of action that operates on immune cells as well as neurons, the preclinical data provide proof-of-principle that suggests that, by targeting *neuronal* molecules, it may be possible to prevent or slow neuronal degeneration in ways that may be relevant to MS. In this regard, sodium channels can be viewed, at a minimum, as providing a prototype family of target molecules, since other neuronal molecules—some of which are involved in degenerative processes—are also amenable to pharmacologic manipulation. Hopefully we will learn in the relatively near future whether neurodegeneration in MS can be abated via drugs that act directly on neurons.

Acknowledgments

Research described in the author's laboratory has been supported, in part, by grants from the National Multiple Sclerosis Society and the Medical Research Service and Rehabilitation Research Service, Department of Veteran Affairs, and by gifts from Destination Cure and the Nancy Davis Foundation. The Neuroscience and Regeneration Research Center is a Collaboration of the Paralyzed Veterans of America and the United Spinal Association with Yale University.

References

Baker, M. D. (2005). Ion currents and axonal oscillators: A possible basis for positive signs in MS. In S. G. Waxman, Ed., *Multiple sclerosis as a neuronal disease*, 131–143. Elsevier, San Diego.

Bechtold, D. A., Kapoor, R., and Smith, K. J. (2004). Axonal protection using flecainide in experimental autoimmune encephalomyelitis. *Ann Neurol* **55**, 607– 616.

Bhat, M. A. et al. (2001). Axon-glia interactions and the domain organization of myelinated axons requires neurexin IV/Caspr/ Paranodin. *Neuron* **30**, 369–383.

Black, J. A., Waxman, S. G., and Smith, K. J. (2006a). Remyelination of dorsal column axons by endogenous Schwann cells restores the normal pattern of $Na_v1.6$ and Kvl.2 at nodes of Ranvier. *Brain* **129**, 1319–1329.

Black, J. A., Liu, S., Hains, B. C., Saab, C. Y., and Waxman, S. G. (2006b). Long-term protection of central axons with phenytoin in monophasic and chronic-relapsing EAE. *Brain*. In press.

Black, J. A., Renganathan, M., and Waxman, S. G. (2002). Sodium channel $Na_v1.6$ is expressed along nonmyelinated axons and it contributes to conduction. *Molec Brain Research* **105**, 19–28.

Boiko, T. et al. (2001). Compact myelin dictates the differential targeting of two sodium channel isoforms in the same axon. *Neuron* **30**, 91–104.

Boiko, T., Van Wart, A., Caldwell, J. H, Levinson, S. R, Trimmer, J. S, and Matthews, G. (2003). Functional specialization of the axon initial segment by isoform-specific sodium channel targeting. *J Neurosci* **23**, 2306–2313.

Bostock, H. and Sears, T. A. (1978). The internodal axon membrane: Electrical excitability and continuous conduction in segmental demyelination. *J Physiol* **280**, 273–301.

Caldwell, J. H., Schaller, K. L., Lasher, R. S., Peles, E., and Levinson, S. D. (2000). Sodium channel Na(v)1.6 is localized at nodes of Ranvier, dendrites, and synapses. *Proc Natl Acad Sci U S A* **97**, 5616–5620.

Cash, E. and Rott, O. (1994). Microglial cells qualify as the stimulators of unprimed $CD4^+$ and $CD8^+$ T lymphocytes in the cenetral nervous system. *Clin Exp Immunol* **98**, 313–318.

Catterall W. A., Goldin A. L, and Waxman, S. G. (2005). International Union of Pharmacology. XLVII. Nomenclature and structure-function relationships of voltage-gated sodium channels. *Pharmacol Rev* **57**, 397–409.

Confavreux, C. and Vukusic, S. (2006). Natural history of multiple sclerosis: A unifying concept. *Brain* **129**, 606–616.

Craner, M. J. et al. (2005). Sodium channels contribute to microglia/macrophage activation and function in EAE and MS. *Glia* **49**, 220–229.

Craner, M. J., Hains, B. C., Lo, A. C., Black, J. A., and Waxman S. G. (2004b). Colocalization of sodium channel $Na_v1.6$ and the sodium-calcium exchanger at sites of axonal injury in the spinal cord in EAE. *Brain* **127**, 294–303.

Craner, M. J., Lo, A. C., Black, J. A., and Waxman, S. G. (2003). Abnormal sodium channel distribution in optic nerve axons in a model of inflammatory demyelination. *Brain* **126**, 1552–1562.

Craner, M. J., Newcombe, J., Black, J. A., Hartle, C., Cuzner, M. L., and Waxman, S. G. (2004a). Molecular changes in neurons in MS: Altered axonal expression of $Na_v1.2$ and $Na_v1.6$ sodium channels and Na^+/Ca^{2+} exchanger. *Proc Natl Acad Sci U S A* **101**, 8168–8173.

Cummins, T. R., Howe, J. R., Waxman, S. G. (1998). Slow closed-state inactivation: A novel mechanism underlying ramp currents in cells expressing the hNE/PN1 sodium channel. *J Neurosci* **18**, 9606–9619.

Davie, C. et al. (1995). Functional deficit in multiple sclerosis and autosomal dominant cerebellar ataxia is associated with axon loss. *Brain* **118**, 1583–1592.

DeGroot, C. J., et al. (1997). Immunocytochemical characterization of the expression of inducible and constitutive isoforms of nitric oxide synthase in demyelinating multiple sclerosis lesions. *Neuropathol Exp Neurol* **56**, 10–20.

DeLuca, G. C., Williams, K., Evangelou, N., Ebers, G. C., and Esiri, M. M. (2006). The contribution of demyelination to axonal loss in multiple sclerosis. *Brain* **129**, 1507–1516.

Dutta, R. et al. (2006). Mitochondrial dysfunction as a cause of axonal degeneration in multiple sclerosis patients. *Ann Neurol* **59**, 478–489.

Einheber, S. et al. (1997). The axonal membrane protein Caspr, a homologue of neurexin IV, is a component of the septate-like paranodal junctions that assemble during myelination. *J Cell Biol* **139**, 1495–1506.

England, J. D., Gamboni, F., and Levinson, S. R. (1991). Increased numbers of sodium channels form along demyelinated axons. *Brain Res* **548**, 334–337.

Ferguson B., Matyszak, M. K., Esiri, M. M., and Perry, V. H. (1997). Axonal damage in acute multiple sclerosis lesions. *Brain* **120**, 393–399.

Fern, R., Ransom, B. R., Stys, P. K., and Waxman, S. G. (1993). Pharmacological protection of CNS white matter during anoxia: Actions of phenytoin, carbamazepine and diazepam. *J Pharmacol Exper Ther*. **266**, 1549–1555.

Filippi, M., Bozzali, M., Rovaris, M., Gonen, O., Kesavadas, C., Ghezzi, A., et al. (2003). Evidence for widespread axonal damage at the earliest clinical stage of multiple sclerosis. *Brain* **126**, 433–437.

Foster, R. E., Connors, B. R., and Waxman, S. G. (1982). Rat optic nerve: Electrophysiological, pharmacological and anatomical studies during development. *Dev Brain Res* **3**, 361–376.

Foster, R. E., Whalen, C. C., and Waxman, S. G. (1980). Reorganization of the axonal membrane of demyelinated nerve fibers: Morphological evidence. *Science* **210**, 661–663.

Garthwaite, G., Goodwin, D. A., Batchelor, A. M., Leeming, K., and Garthwaite, J. (2002). Nitric oxide toxicity in CNS white matter: An in vitro study using rat optic nerve. *Neuroscience* **109**, 145–155.

Gong, B., Rhodes, J., Bekele-Arcuri, Z., and Trimmer, J. S. (1999). Type I and type II Na^+ channel alpha-subunit polypeptides exhibit distinct spatial and temporal patterning, and association with auxiliary subunits in rat brain. *J Comp Neurol* **412**, 342–352.

Hooper, D. C., et al. (1997). Prevention of experimental allergic encephalomyelitis by targeting nitric oxide and peroxynitrite: Implications for the treatment of multiple sclerosis. *Proc Natl Acad Sci U S A* **94**, 2528–2533.

Iwata, A. et al. (2004). Traumatic axonal injury induces proteolytic cleavage of the voltage-gated sodium channels modulated by tetrodotoxin and protease inhibitors. *J Neurosci* **24**, 4605–4613.

Kaplan, M. R. et al. (2001). Differential control of clustering of the sodium channels Na(v)1.2 and Na(v)1.6 at developing CNS nodes of Ranvier. *Neuron* **30**, 105–119.

Kapoor, R., Davies, M., Blaker, P. A., Hall, S. M., and Smith, K. J. (2003). Blockers of sodium and calcium entry protect axons from nitric oxide-mediated degeneration. *Ann Neurol* **53**, 174–180.

Kapoor, R., Li, S. G., and Smith, K. J. (1997). Slow sodium-dependent potential oscillations contribute to ectopic firing in mammalian demyelinated acons. *Brain* **120**, 647–652.

Kornek, B., et al. (2000). Multiple sclerosis and chronic autoimmunoencephalomyelitis: A comparative quantitative study of axonal injury in active, inactive and remyelinated lesions. *Am J Pathol* **157**, 267–276.

Korotzer, A. R. and Cotman, C. W. (1992). Voltage-gated currents expressed by rat microglia in culture. *Glia* **6**, 81–88.

Kuhlmann, T., Lingfeld, G., Bitsch, A., Schuchardt, J., and Brück, W. (2002). Acute axonal damage in multiple sclerosis is most extensive in early disease stages and decreases over time. *Brain* **125**, 2202–2212.

Lassmann, H. (2005). Pathology of neurons in multiple sclerosis. In S. G. Waxman, Ed., *Multiple sclerosis as a neuronal disease*, 153–165. Elsevier, Amsterdam.

Li, H., Cuzner, M. L., and Newcombe, J. (1996). Microglia-derived macrophages in early multiple sclerosis plaques. *Neuropathol Appl Neurobiol* **22**, 207–215.

Lo, A. C., Black, J. A., and Waxman, S. G. (2003). Phenytoin protects spinal cord axons and preserves axonal conduction and neurological function in a model of neuroinflammation in vivo. *J Neurophysiol* **90**, 3566–3572.

LoPachin, R. M., Jr. and Stys, P. K. (1995). Elemental composition and water content of rat optic nerve myelinated axons and glial cells: Effects of in vitro anoxia and reoxygenation. *J Neurosci* **15**, 6735–6746.

Matsumoto, Y., Ohmori, Y., and Fujiwara, K. (1992). Immune regulation by brain cells in the central nervous system: Microglia but not astrocytes present myelin basic protein to encephalitogenic T cells under in vivo-mimicking conditions. *Immunology* **76**, 209–216.

Moll, C., Mouvre, C., Lazdunski, M., and Ulrich, J. (1991). Increase of sodium channels in demyelinated lesions of multiple sclerosis. *Brain Res* **556**, 311–316.

Nörenberg W., Illes, P., and Gebicke-Haeter, P. J. (1994). Sodium channels in isolated human brain macrophages (microglia). *Glia* **10**, 65–172.

Novakovic, S. D., Levinson, S. R., Schachner, M., and Shrager, P. (1998). Disruption and reorganization of sodium channels in experimental allergic neuritis. *Muscle Nerve* **21**, 1019–1032.

Rasband, M. N., Peles, E., Trimmer, J. S., Levinson, S. R., Lux, S. E., and Shrager, P. (1999). Dependence of nodal sodium channel clustering on paranodal axo-glial contact in the developing CNS. *J Neurosci* **19**, 7516–7528.

Renno, T., Krakowski, M., Piccirillo, C., Lin, J. Y., and Owens, T. (1995). TNF-alpha expression by resident microglia and infiltrating leukocytes in the central nervous system of mice with experimental allergic encephalomyelitis. Regulation by Th1 cytokines. *J Immunol* **154**, 944–953.

Ritchie, J. M. and Rogart, R. B. (1977). The density of sodium channels in mammalian myelinated nerve fibers and the nature of the axonal membrane under the myelin sheath. *Proc Natl Acad Sci U S A* **74**, 211–215.

Rush, A. M., Dib-Hajj, S. D., and Waxman, S. G. (2005). Electrophysiological properties of two axonal sodium channels, $Na_v1.2$ and $Na_v1.6$, expressed in spinal sensory neurons. *J Physiol* **564**, 803–816.

Sasaki, M., Black, J. A., Lankford, K. L., Tokuno, H. A., Waxman, S. G., and Kocsis, J. D. (2006). Molecular reconstruction of nodes of Ranvier after remyelination by transplanted olfactory ensheaathing cells in the demyelinated spinal cord. *J Neurosci* **26**, 1803–1812.

Smith, K. J. and Lassmann, H. (2002). The role of NO in multiple sclerosis. *Lancet Neurol* **1**, 232–241.

Steffensen, I., Waxman, S. G., Mills, L., and Stys, P. K. (1997). Immunolocalization of the Na^+-Ca^{2+} exchanger in mammalian myelinated axons. *Brain Research* **776**, 1–9.

Stys, P. K., Waxman, S. G., and Ransom, B. R. (1992a). Ionic mechanisms of anoxic injury in mammalian CNS white matter: Role of Na^+ channels and Na^+-Ca^{2+} exchanger. *J Neurosci* **12**, 430–439.

Stys, P. K., Ransom, B. R., and Waxman, S. G. (1992b). Tertiary and quaternary local anesthetics protect CNS white matter from anoxic injury at concentrations that do not block excitability. *J Neurophysiol* **67**, 236–240.

Stys, P. K., Sontheimer, H., Ransom, B. R., and Waxman, S. G. (1993). Noninactivating, tetrodotoxin-sensitive Na^+ conductance in rat optic nerve axons. *Proc Natl Acad Sci U S A* **90**, 6976–6980.

Trapp, B. D., Peterson, J., Ransohoff, R. M., Rudick, R., Mörk, J., and Böö, L. (1998). Axonal transection in the lesions of multiple sclerosis. *N Engl J Med* **338**, 278–285.

Ulrich, J. and Groebke-Lorenz, W. (1983). The optic nerve in MS: A morphological study with retrospective clinicopathological correlation. *Neurol Ophthalmol* **3**, 149–159.

Waxman, S. G. (1977). Conduction in myelinated, unmyelinated, and demyelinated fibers. *Arch Neurol* **34**, 585–590.

Waxman, S. G., Black, J. A., Kocsis, J. D., and Ritchie, J. M. (1989). Low density of sodium channels supports action potential conduction in axons of neonatal rat optic nerve. *Proc Natl Acad Sci U S A* **86**, 1406–1410.

Waxman, S. G., Davis, P. K., Black, J. A., and Ransom, B. R. (1990). Anoxic injury of mammalian central white matter: Decreased susceptibility in myelin-deficient optic nerve. *Ann Neurol* **28**, 335–340.

Westenbroek, R. E., Merrick, D. K., and Catteral, W. A. (1989). Differential subcellular localization of the RI and RII Na^+ channel subtypes in central neurons. *Neuron* **3**, 695–704.

Whitaker, W. R., Clares, J. J., Powell, A. J., Chen, Y. H., Faull, R. L., and Emson, P. C. (2000). Distribution of voltage-gated sodium channel alpha-subunit and beta-subunit mRNAs in human hippocampal formation, cortex, and cerebellum. *J Comp Neurol* **422**, 123–139.

Wujek, J. R. et al. (2002). Axon loss in the spinal cord determines permanent neurological disability in an animal model of multiple sclerosis. *J Neuropathol Exp Neurol* **61**, 23–32.

23

Acquired Epilepsy: Cellular and Molecular Mechanisms

Christopher B. Ransom and Hal Blumenfeld

I. Introduction

Seizures may be defined in several ways, but the definition put forth by Hughlings Jackson in 1870 of "an occasional, an excessive, and a disorderly discharge of cerebral nervous tissue on muscles," remains an accurate description (Jackson, 1931). Modern neuroscience has demonstrated that seizures are the result of abnormal, synchronized paroxysms of electrical activity in a population of neurons (an epileptogenic focus) that are able to rapidly recruit other parts of the brain to share in its rhythmic, self-sustaining electrical discharge. The clinical manifestations of seizures are varied. Seizures often spread to involve much of the brain, producing loss of consciousness and convulsions, the generalized tonic-clonic seizure most familiar to the laity. Seizures may also remain highly localized, or *focal*, producing only abnormal sensations or motor activity in a single limb depending on the neuroanatomical substrate generating the epileptiform electrical activity.

Epilepsy, the condition of recurrent unprovoked seizures, is among the most common neurologic disorders with a prevalence of 0.5 to 1 percent (Theodore et al., 2006). Epilepsy occurs in idiopathic forms, as a result of congenital abnormalities in brain development (fetal stroke, cortical malformations, neuronal migration disorders), genetic alterations in brain metabolism, or excitability proteins (see Chapter 24), and from insults that derange the anatomy and physiology of a previously normal brain. Acquired, *localization-related epilepsies*, are the most common epilepsy syndromes with temporal lobe epilepsy comprising the majority (Zarelli et al., 1997). Not only is temporal lobe epilepsy (TLE) the most common seizure disorder in adults, it is often refractory to medical therapy.

The cellular and molecular underpinnings of acquired epilepsy, the study of which lends itself to experimental models, are becoming increasingly well-understood. A large literature exists describing changes in growth factor expression, cellular connectivity, ligand- and voltage-gated ion channel properties and expression, neurotransmitter transporters, and ionic homeostasis that occur in human epilepsy, and following experimental manipulations known to produce seizures. In this chapter we will review the diverse molecular and cellular changes that accompany acquired epilepsies, with particular focus on temporal lobe epilepsy as a prototype. Table 23.1 lists the causes of acquired epilepsy.

A. Experimental Models

Much of what is known regarding the cellular and molecular basis of epileptogenesis has been learned from animal models (see Table 23.2). As enumerated by other authors (White, 2002), an appropriate animal model of epileptogenesis should share characteristics with human epilepsy, including

- Similar pathology
- A latent period following initial insult
- Chronic hyperexcitability
- Spontaneous seizures

Manipulations that produce prolonged seizures (status epilepticus) in experimental animals have proven to satisfy all these criteria. Intraperitoneal injection of pilocarpine or kainic acid frequently are employed to induce status epilepticus in experimental animals. Direct electrical stimulation has been used to trigger status epilepticus, and animals treated thus likewise develop spontaneous seizures. *Kindling* is another model of epileptogenesis involving subthreshold electrical stimulation; repeated stimuli on consecutive days leads to incremental neuronal responses until a previously subthreshold stimulus will reliably produce seizures reflecting plastic changes leading to chronic hyperexcitability.

Traumatic head injury is an important cause of acquired epilepsy that has been studied in several manners: chronic isolated cortex, focal iron-induced epilepsy, and fluid percussion injury. Chronic isolated cortex, produced by lesions of underlying white matter, recapitulates the diffuse axonal injury and cavitation seen in human injury and leads to hyperexcitable cortex (Li & Prince, 2002). The risk of epilepsy is increased significantly with injuries accompanied by intracerebral hemorrhage, and the effects of blood products has been studied using direct injection of iron chloride (Wilmore et al., 1978). Fluid percussion to the dura of experimental animals produces typical pathologic changes and chronic hyperexcitability but usually does not cause spontaneous seizures (Santhakumar et al., 2001).

Additional models of acquired epilepsy are based on the introduction of a focal cortical lesion and include the freeze-lesion, alumina gel lesion models, and cerebral hypoxia/ischemia. *In vitro* models of epileptiform activity involve exogenous treatments that increase neuronal excitability. Although these *in vitro* treatments do not strictly relate to the acquired epilepsies, they highlight mechanisms contributing to hyperexcitability and include $GABA_A$ receptor blockade (bicuculline, penicillin), potassium channel inhibition with 4-aminopyridine (4-AP), elevated extracellular potassium, and zero magnesium-containing solutions.

Table 23.1 Etiologies of Acquired Epilepsy

Status epilepticus	Autoimmune/Inflammatory disease
Ischemic stroke	Neoplasm
Hemorrhagic stroke	Infection
Trauma/Intracranial hemorrhage	Febrile seizure

Table 23.2 Experimental Models of Acquired Epilepsy

Status epilepticus	Kainic acid
	Pilocarpine
	Electrical stimulation
Traumatic injury	Fluid percussion injury
	Cortical undercut
	Alumina gel lesion
	Freeze lesion
	Ferric chloride injection
Kindling model	Repeated, subthreshold electrical stimulation
Hypoxic-Ischemic brain injury	Carotid artery occlusion
	Hypoxia
Febrile seizures	Hyperthermia

II. Temporal Lobe Epilepsy

Temporal lobe epilepsy (TLE) usually is manifested clinically by complex-partial seizures, and pathologically by *mesial temporal sclerosis*. Mesial temporal sclerosis is characterized by hippocampal atrophy with neuronal loss and reactive gliosis, and often is detectable clinically with modern MRI neuroimaging. Reactive gliosis, as will be discussed further in this chapter, is a feature common to many acquired epilepsies. The neuronal loss, hippocampal atrophy, and gliosis seen in TLE likely *result from* and *contribute to* the seizures in this condition. The development of TLE is believed to result from an initial precipitating insult to the brain that, either due to its severity or occurrence during a susceptible developmental state, leads to progressive molecular and cellular changes favoring hyperexcitability and seizures (Mathern et al., 1995). Initial precipitating injury of the central nervous system is more likely to later produce TLE when the insult occurs between ages 1 and 5 years (Marks et al., 1992, 1995; Sagar & Oxbury, 1987). Types of injury associated with later development of TLE include status epilepticus, febrile seizures, head trauma, and CNS infections. Status epilepticus, the condition of prolonged or repetitive seizures without interval recovery, is an ominous presentation of a first seizure and often predicts development of epilepsy. Febrile seizures deserve special comment; 53 percent of patients with epilepsy and mesial temporal sclerosis have a history of febrile seizures (French et al., 1993) and patients with complex febrile seizures (focal onset, multiple seizures, prolonged seizures) have increased risk of developing epilepsy (Annegers et al., 1987).

A prominent feature of temporal lobe epilepsy, and all acquired epilepsies, is a latency of months to years between the initial insult and the development of spontaneous seizures, the so-called "silent period." Although some recent work suggests that epileptiform activity is not truly silent during this interval, it is during this period of relative quiescence that deleterious changes in cellular organization/connectivity and molecular expression develop to the point of overt spontaneous seizure generation. Experimental and clinical evidence suggests that "seizures beget seizures," implying a continued pattern of injury and reinforcement of the maladaptive processes producing the epileptic condition (Elwes et al., 1988; Shorvon & Reynolds, 1982; *but see* Berg & Shinnar, 1997). The pioneering British neurologist Gowers appreciated this in 1881, "The tendency of the disease is toward self-perpetuation; each attack facilitates the occurrence of another by increasing the instability of the nerve elements" (Gowers, 1881). Thus, TLE may be viewed as an acquired disease resulting from a myriad of insults, the development of which is time-dependent, dependent on genetic susceptibility, the developmental state at the time of insult, and the character of the injury (see Figure 23.1).

A. Changes in Gene Expression

It is self-evident that the development of durable changes in brain leading to an epileptic state will be governed by changes in gene expression. Of particular interest are those genes activated by seizure activity that are predicted to produce ongoing regulation of other genes directly related to excitability, including ion channels, neurotransmitter receptors, synaptic structural proteins, and neurotransmitter transporters. These are now well-described. The cellular immediate-early genes c-fos and c-jun, transcription factors that regulate expression of delayed-response genes, are activated in the hippocampus of rats following seizure induction with kainic acid, pilocarpine, electrical stimulation, or hilus lesion (Elliott & Gall, 2000; Labiner et al., 1993; Murray et al., 1998; Scharfman et al., 2002; Smeyne et al., 1993). Following seizures the most prominent c-fos activation is seen in the dentate gyrus of hippocampus, the locus of many epileptogenic cellular and molecular events (White & Gall, 1987). The activation of immediate early gene transcription factors following seizures have different temporal sequences in different parts of the brain and even within different cell types in a given brain region. For example, c-fos expression occurs within minutes of spontaneous seizures in dentate granule cells whereas c-fos activation in interneurons of dentate gyrus is delayed several hours, peaking at a time when c-fos immunolabelling of granule cells has returned to baseline levels (Peng & Houser, 2005) (see Figure 23.2).

The stimulus-transcription coupling mediated by c-fos following seizures, and other immediate early genes, is followed by increased expression of growth factors. Among the first growth factors identified to be induced by seizure activity was nerve growth factor (NGF); enhanced expression of NGF mRNA was seen in limbic structures including hippocampus within one hour of seizure activity (Gall & Isackson, 1989). Further work has identified increases in neurotrophins (NT-3), vascular endothelial growth factor

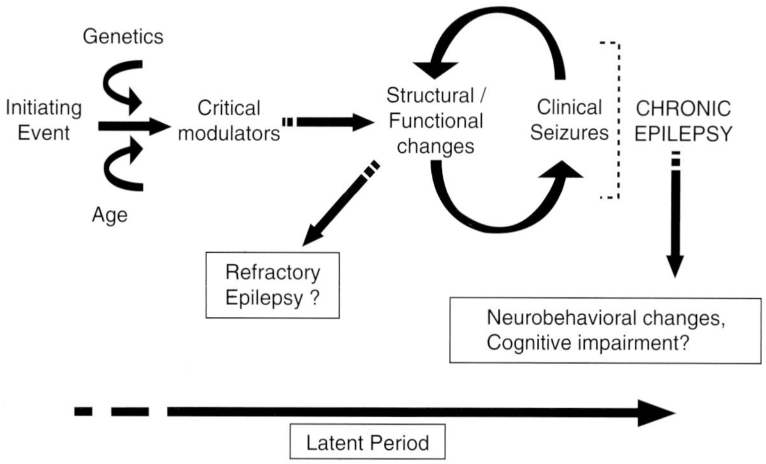

Figure 23.1 Schematic of the events and factors leading to acquired epilepsy. A CNS insult (status epilepticus, febrile seizures, infection, stroke, trauma, autoimmune encephalitis), modulated by genetics and age, produces changes in the nervous system that lead to hyperexcitability and seizures. Seizures reinforce the maladaptive processes leading to chronic epilepsy (White, 2002).

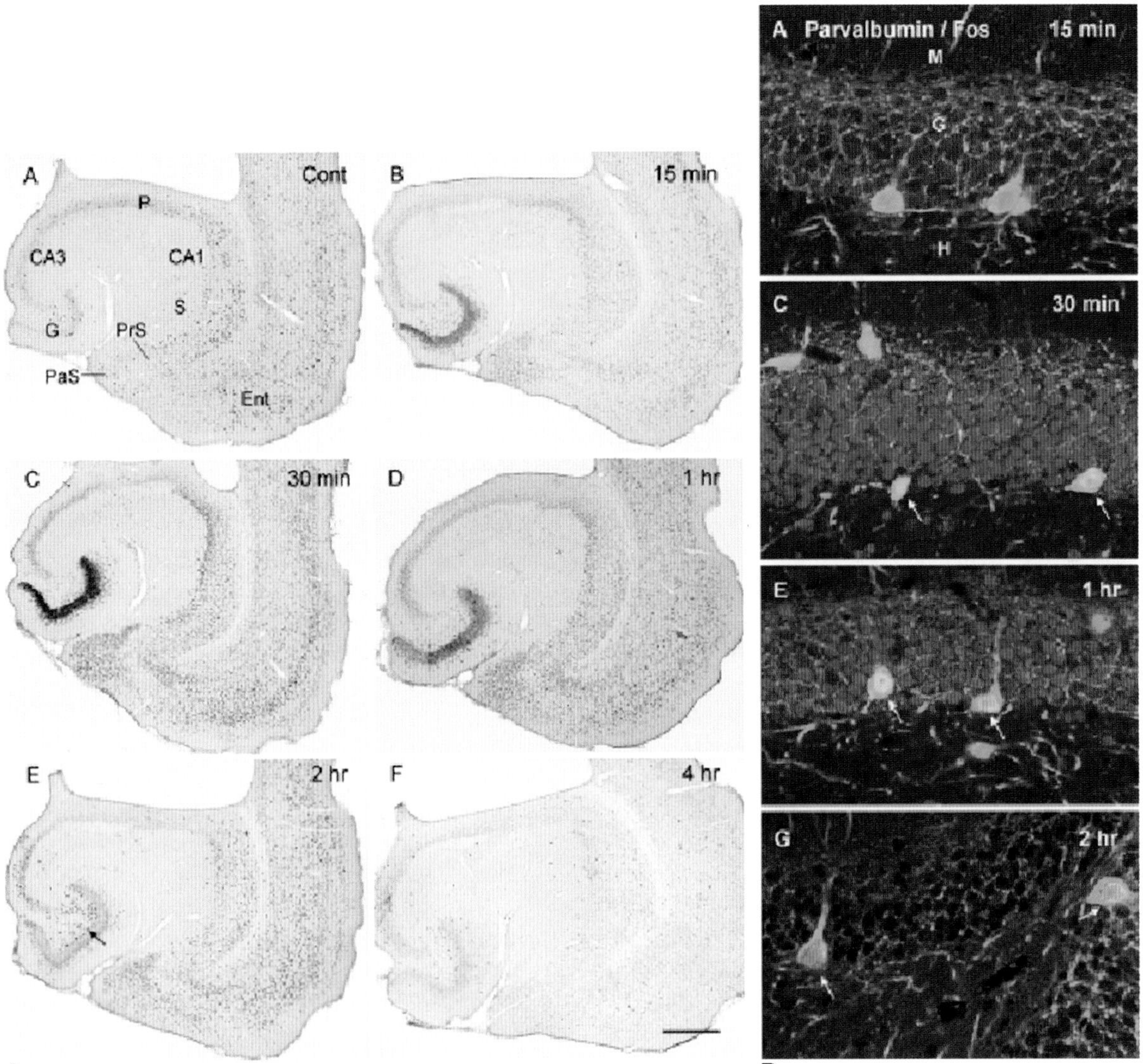

Figure 23.2 Seizure-induced changes in gene expression. **A**. Expression of the cellular immediate early gene c-Fos is transiently increased following spontaneous seizures in pilocarpine-treated rat hippocampus, most prominently in dentate gyrus granule cell layer. **B**. Time course of c-fos expression differs between granule cells and interneurons in dentate gyrus. Parvalbumin-stained interneurons (green) do not label for c-fos (red) 30 minutes after spontaneous seizures, a time when robust labeling is seen in granule cells. In contrast, at 2h after spontaneous seizures interneurons are double-labeled for both parvalbumin and c-fos, and granule cells have lost their c-fos immunoreactivity (Peng & Houser, 2005).

(VEGF), fibroblast growth factor (FGF), and brain-derived neurotrophic factor (BDNF) in several models of epileptogenesis (Gall, 1993b) (see Figure 23.3A). BDNF is of particular interest. BDNF protein is increased within 10h of seizures and exogenous BDNF has been shown to increase synapse number of CA1 neurons, increase neurotransmitter release probability, facilitate the induction of synaptic long term potentiation, and can directly activate sodium currents (Binder et al., 2001; Gall, 1993a; Kafitz et al., 1999; Kramar et al., 2004; Tyler & Pozzo-Miller, 2001, 2003). Most importantly, antagonism of BDNF's action on its cognate receptor trkB delays the development of hyperexcitability in the kindling model, demonstrating a role for BDNF in epileptogenesis (Binder et al., 1999) (see Figure 23.3B).

The increase in immediate-early gene and growth factor expression following seizures likely influences multiple signal transduction pathways within neurons, leading to maladaptive modulation of excitability proteins in addition to altering expression of single genes. Microarray gene chips have been applied to study changes in gene expression following experimental status epilepticus. One group identified increased expression of 40 genes, many of which are related to signal transduction (Hunsberger et al., 2005). The microarray profiles of rat dentate gyrus

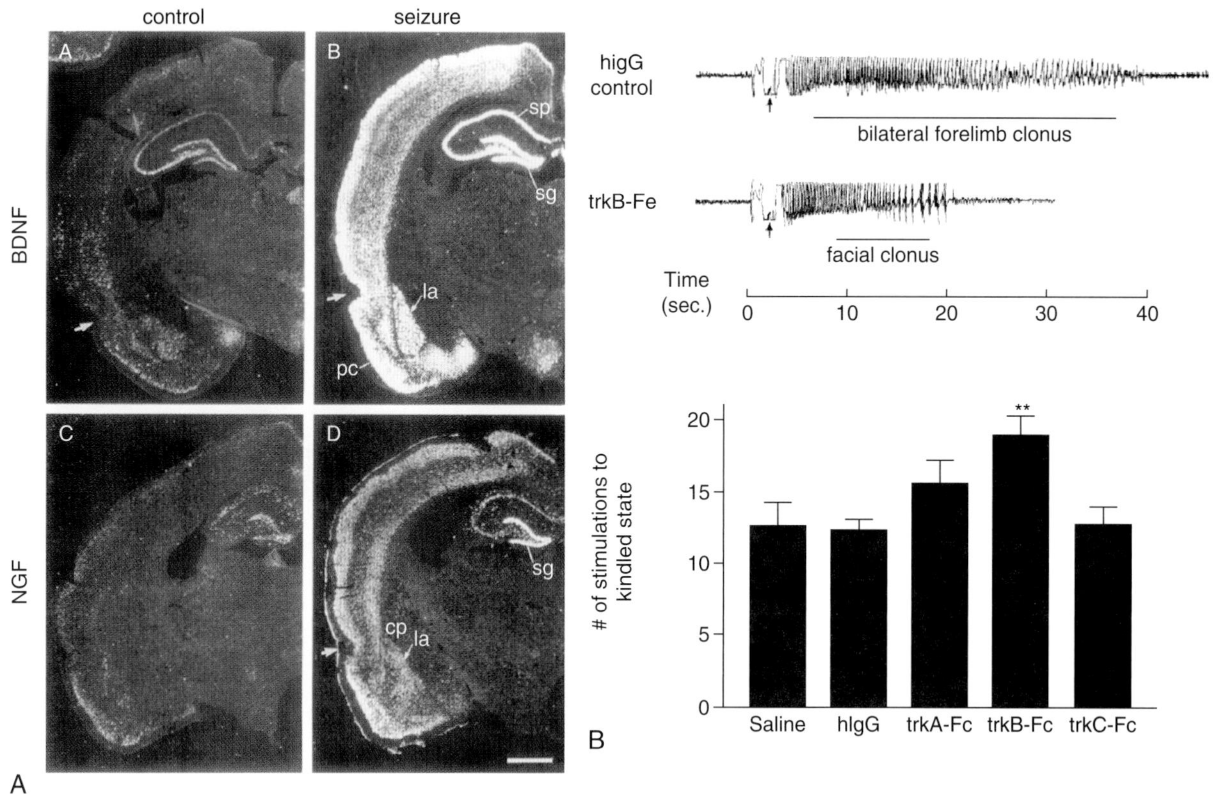

Figure 23.3 Neurotrophin expression following seizures. **A**. *In situ* hybridization reveals increased expression of brain-derived neurotrophic factor (BDNF, top panel) and nerve growth factor (NGF, bottom panel) mRNA 6h and 24h, respectively, following seizures. BDNF and NGF expression is increased in hippocampus and cortical regions including the entorhinal cortex following seizures (Gall, 1993a). **B**. Intraventricular injection of antibodies against trkB growth factor receptors antagonizes the development of the fully kindled state. The top panel shows electroencephalograms of animals during their twelfth simulation. Whereas the hIgG-treated animal had a 36 sec seizure discharge with a clonic motor component of 30 sec, the trkB–Fc-treated animal had only a 19 sec seizure discharge with facial clonus. Bottom panel shows effect on number of stimulations to reach fully kindled state. Human IgG and antibodies against trkA and trkC receptors did not block kindling (Binder et al., 1999).

during development and epileptogenesis identify 600 regulated genes, 37 of which are common to both processes and notable for downregulation of hippocalcin (a neuronal calcium-sensor protein) and $GABA_A$ receptor delta subunits (Elliott et al., 2003). Both proconvulsant/neurotoxic (e.g., substance P) and anticonvulsant/neuroprotective (e.g., neuropeptide Y) peptides are increased in hippocampi of kainate-treated rats (Wilson et al., 2005). Additionally, the changes in expression of some genes after seizures is age-dependent (Wilson et al., 2005).

The precise role, relative contributions, and interactions of this host of seizure-activated genes to the development of epilepsy remains to be shown but it is expected that they influence the expression and modulation of genes and proteins contributing to hyperexcitability and the epileptic state. The consequences of changes in the expression of ion channels, neurotransmitter receptors, and neurotransmitter transporters in human and experimental epilepsy will be discussed later.

B. Cellular Morphology, Synaptogenesis, and Neurogenesis

Requisite for the generation of epileptiform discharges and seizures is synchronous activation of populations of neurons. This implies structural and functional reorganization of neuronal connectivity in the development of epilepsy. In TLE dramatic structural and cellular changes occur in the hippocampus. These include neuronal loss, seen most prominently in the dentate gyrus and in area CA1, profound mossy fiber sprouting, and neurogenesis. Mossy fibers are the glutamatergic, efferent axons of dentate granule cells projecting to pyramidal neurons in area CA3. Mossy fibers collateralize to inhibitory interneurons within the dentate hilus and in CA3 as well as "mossy cells," the glutamatergic interneurons in dentate gyrus (see Figure 23.4A) (Scharfman, 2002). This arrangement allows both feedback- and feed-forward-inhibition of dentate granule cell output (Lawrence & McBain 2003). During epileptogenesis there is robust

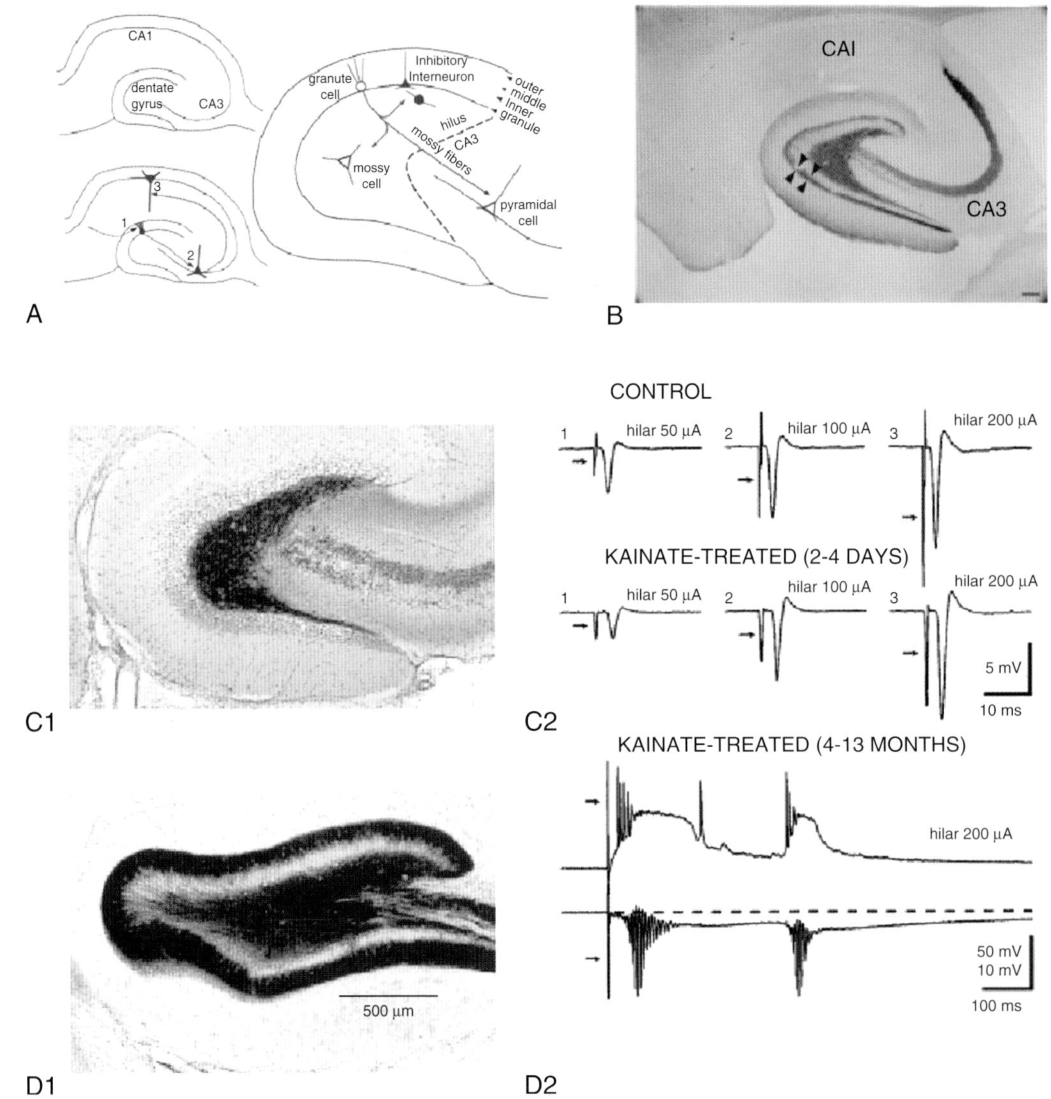

Figure 23.4 Mossy fiber sprouting in temporal lobe epilepsy leads to recurrent excitatory inputs. **A**. Schematic of hippocampal dentate gyrus connectivity. Granule cells located in the dentate gyrus receive input from entorhinal cortex via the perforant path (1) and project mossy fibers to area CA3 (2). Granule cell mossy fibers innervate GABAergic interneurons in the dentate hilus that provide feed-back inhibition. Pyramidal neurons in area CA3 project to neurons in area CA1 via the Schaffer collateral pathway (3). The apical dendrites of granule cells are located in the molecular layer (outer, middle, and inner layers). **B**. Mossy fiber sprouting in hippocampus of pilocarpine-treated rat. Ligh t micrograph of hippocampus stained with antibody against neuropeptide Y, a peptide primarily expressed in mossy fibers. The dark staining seen in the hilus, area CA3, and the inner molecular layer (arrows) reflects mossy fiber sprouting. **C, D**. Left-hand panels show Timm-stained rat hippocampus of control animals (**C1**) and 4–13 months after treatment with kainate (**D1**). The dark Timm-stained areas in (**D**) are due to mossy fiber sprouting, mossy fibers are preferentially labeled with this technique due to their high Zn^{2+} content. Dark staining is seen throughout hilus and molecular layer. **C2**. Synaptic field potentials recorded during antidromic hilar stimulation. No epileptiform discharges are seen in control animals or 2–4 days after kainate-treatment. **D2**. Intracellular recording from a granule cell (top) and synaptic field potentials (bottom) recorded during antidromic hilar stimulation 4–13 months after kainate-treatment. At time points when mossy fiber sprouting has occurred, antidromic stimulation produces intracellularly-recorded paroxysmal depolarizing shift with overriding bursts of action potentials that occur spontaneously after initial stimulus. These intracellular events correlate with epileptiform activity recorded extracellularly (bottom panel). (Panels **A, B** from Scharfman, 2002; panels **C, D** from Wuarin and Dudek, 1998).

sprouting of mossy fibers, these sprouted mossy fibers result in increased connectivity to CA3 pyramidal cells as well as recurrent excitatory inputs to granule cells themselves (Shao & Dudek, 2005; Wuarin & Dudek, 1996). The high Zn^{2+} content and neuropeptide Y expression of mossy fibers has allowed investigators to visualize changes in mossy fiber distribution with the Timm stain or immunohistochemistry (see Figure 23.4B).

Figure 23.4C, D illustrates the development of prominent mossy fiber sprouting within the molecular layer and hilus of dentate gyrus after pilocarpine-induced status epilepticus detected with Timm stain. At earlier times following status epilepticus, when no mossy fiber sprouting is seen, antidromic stimulation of dentate granule cells in the presence of elevated potassium evokes single population spikes recorded extracellularly (see Figure 23.4C1–C2). In contrast, intracellular recording from dentate granule cells four to 13 months after status epilepticus demonstrates epileptiform paroxysmal depolarizing shifts with bursts of action potentials and spontaneous after-discharges in response to antidromic stimulation (see Figure 23.4D) (Wuarin & Dudek, 1996). Mossy fiber sprouting is seen in human TLE and has been described in pilocarpine- and kainate-treated animals (status epilepticus model), kindling model, hypoxic-ischemic injury, and after experimental febrile seizures (Bender et al., 2003a; Cronin et al., 1992; Mathern et al., 1996; Watanabe et al., 1996; Williams et al., 2004). These changes, particularly the formation of recurrent collaterals, are well-suited to contribute to synaptically driven, synchronous, and rhythmic neuronal activation. The observations that epileptic dentate granule cells form monosynaptic excitatory connections, have increased dendritic spine densities, and form "basal" dendrites that receive mossy fiber input highlight the excitatory influence provided by mossy fiber sprouting (Isokawa, 2000; Ribak et al., 2000; Scharfman et al., 2003; Shao & Dudek, 2005).

Although fewer interneurons than granule cells receive input from sprouted mossy fibers, mossy fiber collateralization to interneurons is predicted to increase feedback inhibition of granule cells, a situation that could limit recurrent excitation of granule cells (Buckmaster et al., 2002). Indeed, hyperinhibition of granule cells, assessed by paired-pulse suppression of granule cell responses during perforant path stimulation, is seen to develop in behaving, chronically epileptic rats with a time course paralleling the development of mossy fiber sprouting (Harvey & Sloviter, 2005; Sloviter et al., 2006). Granule cell hyperinhibition can be overcome during seizures that originate elsewhere. The issue of granule cell inhibition is made more complex by the preferential death of some interneurons, including "mossy cells." Mossy cells provide excitatory afferents to basket cells, inhibitory neurons that innervate the soma of principal cells. The "dormant basket cell hypothesis" put forth by Sloviter indicts mossy cell death and impaired activation of surviving basket cells as a cause of disinhibition in some epilepsy models (Sloviter, 1987, 1991). The pattern of epileptiform activity during status epilepticus, and of the resultant cellular injury, likely contributes to the degree of inhibition of granule cells seen; interneuron death and granule cell disinhibition was observed only in kainate-treated animals that experienced continuous granule cell discharges during status (Sloviter et al., 2003).

In the past decade the work of many investigators has demonstrated *de novo* neurogenesis in adult brain, most notably in the dentate gyrus. Neurogenesis is increased in experimental and human epilepsy (Parent et al., 2006). The primary afferents to granule cells normally arise from entorhinal cortex and enter the hippocampus in the *perforant path*. Newly born granule cells project mossy fibers to area CA3 and interact with other granule cells by receiving and producing recurrent collaterals. Seizures significantly influence the connectivity and behavior of newly born granule cells; seizures increase neurogenesis, enhance dendritic arborization within the molecular layer of dentate gyrus, and lead to functional integration of newly born granule cells into hippocampal circuits (Overstreet-Wadiche et al., 2006) (see Figure 23.5). Transgenic animals expressing green fluorescent protein under the control of a cell cycle regulated gene (pro-opiomelanocortin) has allowed *a priori* identification of newly born granule cells during electrophysiologic recording from brain slices.

These experiments demonstrate that only after seizures does perforant path stimulation activate AMPA- and NMDA-mediated synaptic currents on newly born granule cells (see Figure 23.5C). Increased neurogenesis and functional integration of newly born granule cells is evident within 14 days of experimental status epilepticus. The perforant path innervation of newly born granule cells, coupled with the recurrent collaterals produced by mossy fiber sprouting, is a compelling mechanism to reinforce, amplify, and synchronize granule cell activity. Many of the newly formed dentate granule cells, although identical with respect to biophysical and biochemical properties, are located outside of the granule cell layer. These ectopic granule cells, found in the inner molecular layer and in the dentate hilus, receive mossy fiber and perforant path excitatory synaptic input and project to CA3 pyramidal cells as well as forming recurrent collaterals (Scharfman, 2002).

C. Neurotransmitter Systems—Glutamate

In normal brain there exists a balance between excitatory and inhibitory neurotransmitter systems, represented primarily by glutamate and GABA, respectively, that keeps electrical impulses in check to allow effective signaling and avoid excessive, self-perpetuating discharges (i.e., epileptiform discharges and seizures). Changes in hippocampal circuitry, as discussed earlier, is one manner by which the excitation-inhibition balance is shifted toward excitation in TLE. Alterations in neurotransmitter release, neurotransmitter removal, and neurotransmitter receptors that occur in TLE likewise are predicted to shift the balance toward excitation. Interestingly, an emerging theme to the changes in neurotransmitter receptor expression seen during epileptogenesis is a reversion to patterns seen in early development—a period of increased seizure susceptibility. Likewise, the development and reinforcement of hyperexcitability during epileptogenesis recapitulates and exploits many of the same mechanisms in place for normal plasticity, especially activation of glutamate receptors (Scharfman, 2002). NMDA receptor antagonists abrogate the development of hyperexcitability in

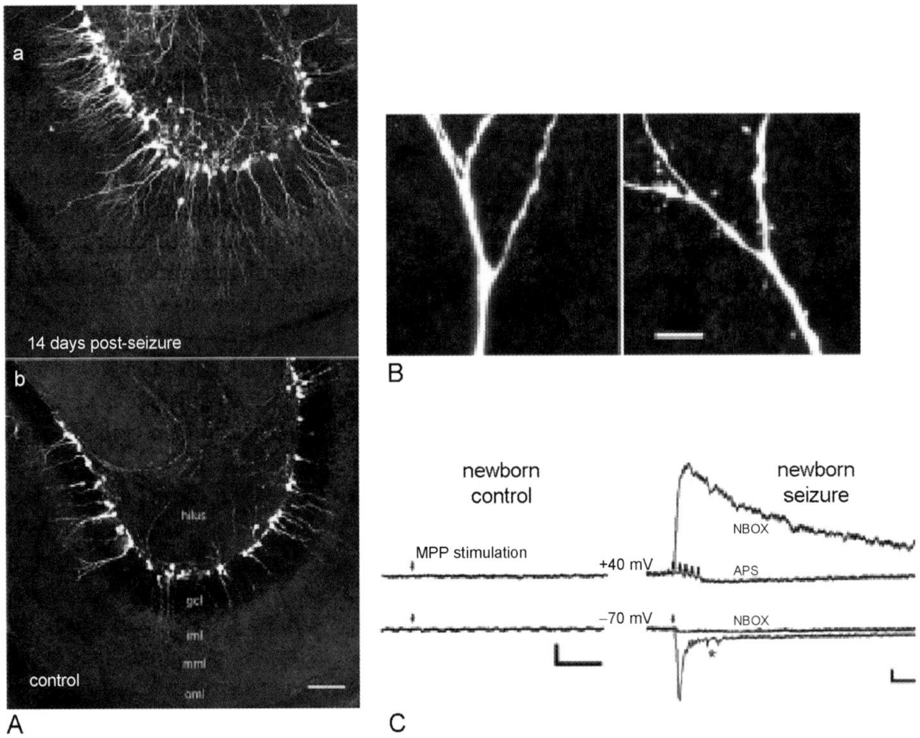

Figure 23.5 Seizures increase dentate gyrus granule cell neurogenesis, dendritic arborization, and formation of functional synaptic connections. **A**. Confocal fluorescent microscopy of dentate gyrus of transgenic animals expressing green-fluorescent protein (GFP) under the promoter of pro-opiomelanocortin (POMC), a cell-cycle regulated gene. Only newly born cells express GFP. Top panel shows increased numbers of newly born cells with robust dendritic arbors in the molecular layer and GFP+ processes (mossy fibers and basal dendrites) in the hilus 14 days after seizures induced with kainate. In contrast, control animals (bottom panel) have relatively few newly born granule cells and sparse processes. **B**. Higher power view shows dendritic spines on newly born granule cells of kainate-treated animals but not in controls (right and left panels, respectively). **C**. Whole-cell patch clamp recordings from newly born granule cells during medial perforant path (MPP) stimulation. Medial perforant path stimulation demonstrates functional glutamatergic synapses with AMPA-R and NMDA-R mediated responses in newly born granule cells only from kainate-treated animals (Overstreet-Wadiche et al., 2006).

kindled animals (Durmuller et al., 1994; McNamara et al., 1988; Sutula et al., 1996).

Glutamate affects fast, excitatory neurotransmission via NMDA, AMPA, and kainate receptors at multiple hippocampal synapses; perforant path-dentate gyrus granule cell (DGGC) synapses, DGGC–CA3 pyramidal cell synapses, principal cell (DGGC and pyramidal cells)-interneuron synapses, and CA3-CA1 synapses via the Schaffer collateral pathway. The excitation mediated by these synapses can be enhanced by:

- ▲ Increasing their number
- ▲ Increasing the neurotransmitter release probability
- ▲ Increasing the receptor number per synapse
- ▲ Changing/modulating receptor properties
- ▲ Prolonging neurotransmitter duration of action/activation of extrasynaptic receptors

We have already discussed the increase in number of synapses and temporal summation of synaptic currents produced by mossy fiber sprouting and recurrent collateralization. Transmitter release probability is increased in the entorhinal cortex and at perforant path-DGCC synapses of pilocarpine-treated animals (Scimemi et al., 2006; Yang et al., 2006). In the entorhinal cortex this increase in transmitter release probability is mediated by presynaptic NR2B subunit-containing NMDA receptors (NMDA-R), a situation seen during development but normally lost in adult animals. In experimental and human epilepsy, several investigators have identified an increase of slow, long-duration synaptic responses mediated by NMDA receptors and molecular experiments have demonstrated elevated levels of the NR2A and NR2B subunits (Behr et al., 2000; Isokawa & Levesque, 1991; Isokawa & Mello, 1991; Kraus & McNamara, 1998; Kraus et al., 1994; Mathern et al., 1997;). Increased NMDA-R expression has the potential to induce plastic changes in synaptic strength and extend the period for temporal summation due to their relatively slow deactivation and desensitization kinetics. Additionally, NMDA-R activation during high-frequency stimulation leads to transient reductions of $GABA_A$-mediated inhibition, perhaps via the Ca-dependent kinase calcineurin (Isokawa, 1998; Sanchez et al., 2005).

Changes in AMPA receptor expression are seen following seizures in hippocampus and other brain regions, again with a pattern reminiscent of development. Reduction of the GluR2-subunit is of special interest, AMPA receptors lacking this subunit have increased single-channel conductance and calcium-permeability. The reduction of GluR2 mRNA and protein is demonstrable at 48 h and 96 h following status epilepticus, respectively, and is persistent into adulthood accompanied by appreciable increases in calcium permeability (Sanchez et al., 2001; Zhang et al., 2004a). Thus, the absence of GluR2-subunits has the capacity to reinforce signal transduction cascades initiated with the initial insult as well as confer chronic hyperexcitability.

In other animal studies GluR1- and GluR2-subunits were decreased and GluR3-subunits were increased. However, in epileptic human tissue an overall reduction in GluR2 immunoreactivity was observed but this was *increased* in the molecular layer of dentate gyrus when normalized for degree of cell loss (Blumcke et al., 1996). The precise roles and relative contributions of AMPA-R subunit alterations to the production and maintenance of hyperexcitability remain an area of active investigation.

Metabotropic glutamate receptors (mGluRs) are classified into three groups based on their associated signal transduction cascades: group I receptors (mGluR1 and mGluR5) activate phospholipase C and groups II and III (comprising all other mGluRs) are negatively linked to adenylate cyclase. The changes in expression reported for mGluRs in experimental and human TLE are varied. Increased mGluR5 immunoreactivity, but not mGluR1, has been reported for one series of human TLE, whereas another study found increased mGluR1, but not mGluR5 (Blumcke et al., 2000; Notenboom et al., 2006). Both of these group I receptors have the potential to alter gene expression and signal transduction by triggering release of Ca^{2+} from intracellular stores. Elevated levels of mGluR4 immunoreactivity have also been reported in human TLE (Lie et al., 2000). In rats, kainate-induced status epilepticus resulted in a 2.6-fold increase of mGluR5 expression assayed by gene chip microarray and functional reduction of presynaptic group III mGluR activity was seen in kindled animals (Hunsberger et al., 2005; Klapstein et al., 1999).

The functional consequences of these findings are not well-established, indeed mGluRs can have many anti- and pro-convulsant effects. Group I mGluR agonists can depress both excitatory and inhibitory synaptic transmission, or reduce voltage-gated calcium currents (VGCC) in human epileptic hippocampus, but also are capable of generating intrinsic epileptiform bursting of CA1 pyramidal cells recorded in the presence of the AMPA-R and NMDA-R antagonists (Burke & Hablitz, 1994, 1995; Chuang et al., 2001; Nagerl & Mody, 1998; Schumacher et al., 2000; Traub et al., 2005). Doherty and Dingledine (2001) described a reduction of excitatory inputs to GABAergic interneurons in dentate gyrus of pilocarpine-treated rats dependent on group II mGluR activation, a net disinhibitory effect normally present only in juvenile animals.

Glutamate transporters of both neurons (EAAC1/EAAT3) and glia (GLT-1, GLAST) play pivotal roles in maintaining appropriate baseline levels of glutamate and terminating synaptic events, thereby preventing excitotoxic injury. Under electrochemically unfavorable conditions these transporters can reverse direction and become a glutamate source. Irrespective of changes in the biophysical or biochemical properties of glutamate receptors in epilepsy, failure of extracellular glutamate homeostasis due to transporter dysfunction will have deleterious effects, including seizures (Campbell & Hablitz, 2004, 2005). Reductions in the glial glutamate transporters EAAT2, GLT-1, and GLAST have been found in experimental and human TLE (Mathern et al., 1999; Ueda et al., 2001; van der Hel et al., 2005).

In kainate-treated rats, persistent reductions of GLT-1 and GLAST protein have been identified. In these same animals microdialysis measurements of extracelluar glutamate were markedly different from control animals; basal glutamate levels were increased; and potassium-induced glutamate increases were doubled in amplitude and prolonged in duration (Ueda et al., 2001) (see Figure 23.6). It must be noted that not all studies have identified reductions in glial glutamate transporters. Reductions of glutamine synthetase in glia, the enzyme converting cytosolic glutamate to glutamine, have been reported in human TLE. This represents a mechanism by which the electrochemical gradients favoring glutamate uptake could be compromised (Eid et al., 2004; van der Hel et al., 2005). In contrast, expression of neuronal glutamate transporters is increased in experimental and human TLE; it is postulated that this is a protective compensation by neurons to limit local glutamate accumulation and to obtain substrate for GABA synthesis (Crino et al., 2002; Eid et al., 2004; Ghijsen et al., 1999; van der Hel et al., 2005).

D. Neurotransmitter Systems—GABA

Inhibitory signaling in the hippocampus is mediated by GABAergic interneurons and basket cells located throughout the dentate hilus and CA1/CA2/CA3 subfields. These cells receive input from mossy fibers and pyramidal cells and project to dentate granule cells and pyramidal cells, thereby providing feed-back and feed-forward inhibition. Derangements of this system are predicted to result in unchecked excitation. Although GABAergic interneurons receive excitatory inputs due to mossy fiber sprouting, these may not be equal in number to recurrent inputs to granule cells, favoring net

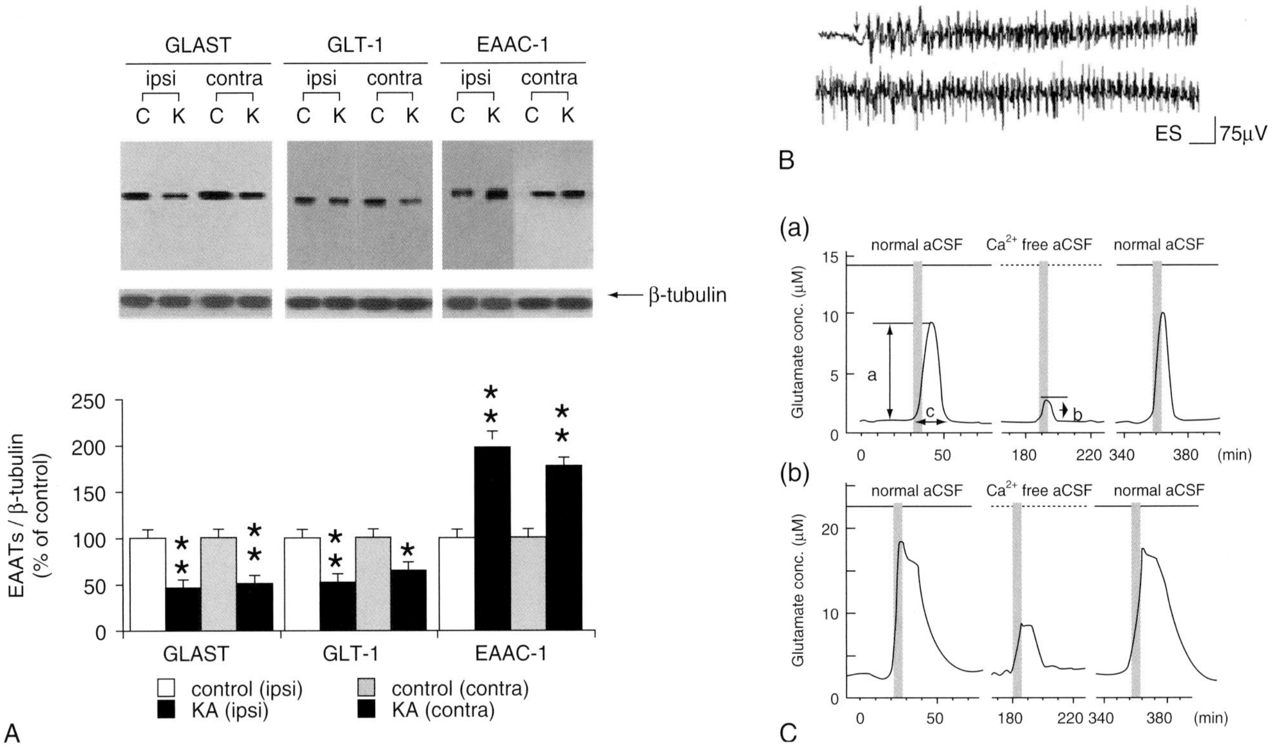

Figure 23.6 Altered glutamate transporter expression and extracellular glutamate regulation during seizures in experimental TLE. **A**. Western blots for excitatory amino acid transporters in kainate-treated rats (K) vs. controls (C) (top panel), and quantification of these data normalized to beta-tubulin (bottom panel). "Ipsi" and "contra" refer to location of a microdialysis catheter. The glial glutamate transporters GLT-1 and GLAST were downregulated after experimental status epilepticus but expression of the neuronal glutamate transporter EAAC1 was increased. **B**. EEG recorded with implanted electrodes in ventral hippocampus during injection of artificial cerebrospinal fluid with 40 mM KCl. Kainate-treated animals, but not controls, developed spike wave discharges in response to high potassium. **C**. Glutamate measurements from microdialysis catheter in response to high potassium in control (top panel) and kainate-treated animals. In kainate-treated animals extracellular glutamate concentrations reached higher peak levels and recovered more slowly than controls following high potassium stimulation (Ueda et al., 2001).

excitation (see earlier). Recent data suggests mossy fibers, in addition to their usual excitatory release of glutamate, acquire the ability to directly release GABA onto CA3 pyramidal cells after seizures; this novel pathway for monosynaptic feed-forward inhibition likely would be overwhelmed during periods of high-frequency granule cell firing (e.g., seizures) because of potent inhibition mediated by group III mGluRs (Gutierrez, 2005; Trevino & Gutierrez, 2005). In human and experimental TLE feed-back inhibition is compromised by interneuron death, enhanced short-term depression at granule cell-interneuron synapses, and depression of inhibitory inputs to granule cells by activation of pre- and post-synaptic glutamate receptors (Behr et al., 2001; Doherty & Dingledine, 2001; Isokawa, 1996, 1998; Sanchez et al., 2005; Sloviter, 1987). GABAergic neurons within dentate gyrus and area CA1 of hippocampus are lost in pilocarpine-treated animals but surviving cells have increased expression of glutamic acid decarboxylase, the synthetic enzyme for GABA. This may represent a compensatory mechanism to quell unchecked excitation (Esclapez & Houser, 1999; Houser & Esclapez, 1996; Obenaus et al., 1993).

Independent of synaptic modulation, $GABA_A$ receptors are affected by changes in subunit expression. Rapid and persistent reductions of sensitivity to type 1 and type 2 benzodiazepines (clonazepam and zolpidem, respectively) after seizures reflect changes in receptor properties, and could compromise clinical treatment of status epilepticus with benzodiazepines and barbiturates (Brooks-Kayal et al., 1998; Kapur & Macdonald, 1997) (see Figure 23.7A). Indeed, animals with prolonged (45 minutes) status epilepticus required larger doses of diazepam to control their seizures than those in status epilepticus for only 10 minutes (see Fig 23.7B). The molecular basis of altered $GABA_A$-R pharmacology in epilepsy is at least in part due to a downregulation of alpha-1 and upregulation of alpha-4 subunits, developmentally regulated subunits minimally expressed in early postnatal animals (Bouilleret et al., 2000; Brooks-Kayal et al., 1998, 2000, 2001, 2005; Loup et al., 2000; Peng et al., 2004). However, seizure-induced changes in $GABA_A$-R subunit expression and distribution show age-dependence, time-dependence, and regional variation within the hippocampus (Nishimura et al., 2005; Schwarzer et al., 1997; Sperk et al., 1998; Tsunashima et al., 1997; Zhang et al., 2004b).

Using patch-clamp techniques and single-cell PCR, two populations of epileptic human dentate granule cells with respect to $GABA_A$-R expression were identified; those cells

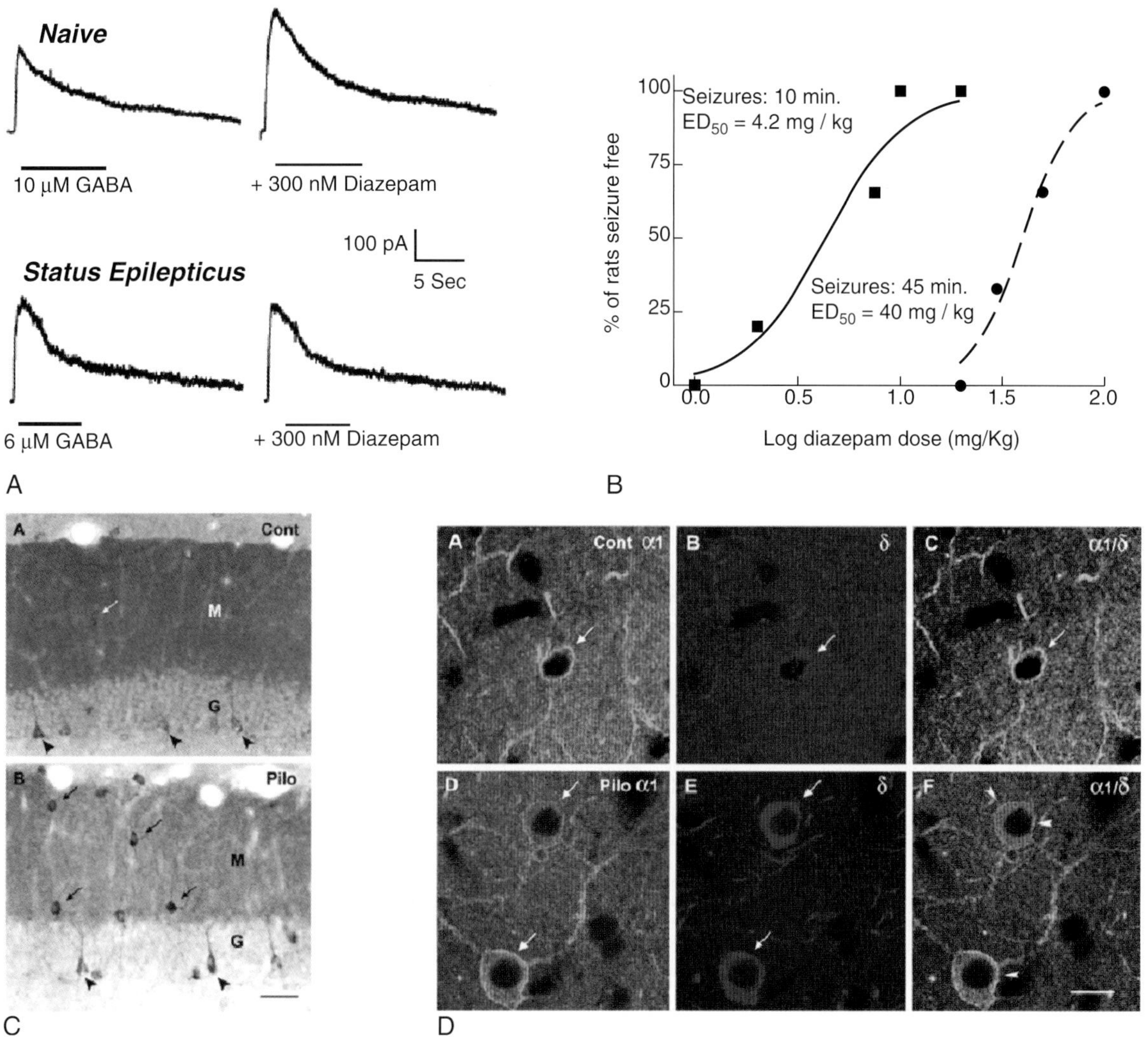

Figure 23.7 Changes in $GABA_A$ receptors seen in experimental TLE. **A**. Whole-cell patch-clamp recordings of GABA-evoked currents in dentate granule cells from control animals and animals subjected to status epilepticus by Li-pilocarpine. $GABA_A$ receptors from animals with status epilepticus have reduced benzodiazepine sensitivity. **B**. Dose-response curve of diazepam for seizure termination in animals with brief (10 minutes) or prolonged (45 minutes) status epilepticus. Prolonged status epilepticus was associated with reduced benzodiazepine sensitivity (Kapur & Macdonald, 1997). **C**. Changes in $GABA_A$ receptor delta-subunit immunoreactivity in dentate gyrus in control (top) and pilocarpine-treated (bottom) animals. Diffuse staining in granule cell layer (G) and molecular layer (M) is reduced after seizures and staining of interneurons (indicated with arrows) is increased. **D**. Double-immunolabeling of dentate interneurons with antibodies against alpha1- and delta-subunits of $GABA_A$ receptors (alpha1, green; delta, red). Top panel is under control conditions, showing staining for alpha (left), delta (middle), and overlay (right). Clear membrane expression of alpha1 is seen, there is diffuse delta signal within the neuropil but little within the cytosol of interneurons, and the overlay does not show any membrane expression of delta subunits. Bottom panel is from pilocarpine-treated animals. After status epilepticus, there is increased alpha1 signal in the cytosol, robust delta expression within interneurons, and the overlay shows colocalized surface expression of delta and alpha1 subunits (arrows) (Peng et al., 2004).

expressing alpha1- and gamma2-subunits demonstrated $GABA_A$ currents augmented by benzodiazepines, and alpha4-, beta2-, and delta-subunit expression correlated with minimal benzodiazepine sensitivity (Brooks-Kayal et al., 1999). Zn^{2+} sensitivity of $GABA_A$ receptors is enhanced in experimental TLE, in conjunction with benzodiazepine resistance (Brooks-Kayal et al., 1998; Kapur & Macdonald, 1997). Recurrent mossy fibers have high concentrations of Zn^{2+}, which is released synaptically and, owing to the molecular reorganization of $GABA_A$ receptors and resultant Zn^{2+}-sensitivity, may abrogate feedback inhibition in epilepsy (Buhl et al., 1996).

Ambient extracellular GABA concentrations, controlled by synaptic spillover and GABA transporter thermodynamics, can activate $GABA_A$-R resulting in "tonic inhibition." Tonic activation of $GABA_A$ receptors is mediated primarily by extrasynaptic receptors containing delta-subunits with high affinity and slow desensitization (Mtchedlishvili & Kapur, 2006; Semyanov et al., 2004; Wei et al., 2003).

Hyperpolarization and reductions of input resistance due to tonic activation of $GABA_A$ receptors represent powerful inhibitory mechanisms. In pilocarpine-treated mice, delta-subunit expression is altered in a manner predicted to lead to disinhibition; delta-subunit gene expression and immunoreactivity is decreased in granule cells and increased in inhibitory interneurons (Elliott et al., 2003; Peng et al., 2004) (see Figure 23.7C,D). Alpha5-subunits of $GABA_A$ receptors also contribute to extrasynaptic GABA receptors, and are downregulated in epileptic animals (Houser & Esclapez, 2003; Houser et al., 1986; Scimemi et al., 2005). GABAergic inhibition may also be compromised by internalization of receptors following seizures (Goodkin et al., 2005).

GABA transporters are additional targets of anticonvulsant drugs and their function may change in epilepsy. In kainate-treated animals, no changes in either mRNA or protein of the neuronal-type presynaptic GABA transporter (GAT-1) was observed, but the astrocytic GABA transporter (GAT-3) was decreased (Ueda et al., 2001). In human TLE, increases of both GAT-1 and GAT-3 are reported (Mathern et al., 1999). Others have suggested that both forward- and reverse-GABA transport is functionally compromised in experimental and human TLE (Patrylo et al., 2001). Tonic activation of extrasynaptic $GABA_A$ receptors (tonic inhibition) is largely dependent on ambient extracellular GABA levels determined by the GABA transporters and the prevailing thermodynamic conditions; increases or decreases of GABA transporter expression have the potential to affect tonic inhibition (Richerson & Wu, 2004).

An important, emerging concept is that of $GABA_A$ receptor-mediated *excitation*. The voltage-changes produced by $GABA_A$ receptor activation are dependent on the equilibrium potential for Cl^- ions (E_{Cl}) established by the chloride cotransporters KCC2 (potassium-coupled chloride cotransporter) and NKCC1 (sodium/potassium-coupled chloride cotransporter). KCC2 normally extrudes Cl^- ions, making E_{Cl} more negative and $GABA_A$-R responses hyperpolarizing. NKCC1 leads to Cl^- import, positive shifts in E_{Cl}, and depolarizing $GABA_A$-R responses. The expression of NKCC1 and KCC2 are developmentally regulated in a manner that renders GABAergic signaling excitatory in the early postnatal rat brain (DeFazio et al., 2000; Dzhala et al., 2005). Although not strictly related to TLE, NKCC1 expression by immature hippocampal neurons may underlie the enhanced seizure susceptibility of young animals and subsequent development of epilepsy (Dzhala et al., 2005).

Increases in NKCC1 expression in hippocampus following seizures have been reported (Kang et al., 2002; Okabe et al., 2002), a potentially maladaptive response that could render GABA excitatory. GABAergic signaling may also be converted from inhibitory to excitatory in a dynamic manner by exogenous or activity-dependent increases in extracellular potassium that can limit Cl^- extrusion mediated by KCC2 (DeFazio et al., 2000; Dzhala & Staley, 2003; Bihi et al., 2005). This is notable because activity-dependent increases of extracellular potassium likely are accentuated during seizures and epilepsy (see more, later, under Glia, brain microenvironment, and epilepsy).

E. Voltage-gated Ion Channels

Seizure activity is a manifestation of a hyperexcitable, interconnected network of cells and as we have seen there are descriptions of numerous mechanisms mediating this hyperexcitability. Voltage-gated ion channels can potentiate this hyperexcitability and are the targets of many anticonvulsant drugs. Increases in voltage-gated Na^+ currents and voltage-gated calcium currents or decreases in voltage-gated potassium currents are predicted to increase excitability and the identification of genetic epilepsy syndromes due to single gene mutations of ion channels suggests that such changes are sufficient to produce an epileptic state. In human and experimental TLE upregulation of persistent, voltage-gated Na^+ currents has been identified (Agrawal et al., 2003; Lampert et al., 2005; Vreugdenhil et al., 2004) (see Figure 23.8A). These persistent Na^+ currents have lower thresholds for activation and due to reduced inactivation have a larger window of "steady-state" activation, thereby providing a powerful depolarizing influence to neurons, promoting burst firing (Yue et al., 2005).

Voltage-gated calcium currents (VGCC) directly contribute to excitability and by increasing intracellular calcium can influence signal transduction and excitotoxic injury. Bursting behavior of CA1 neurons in kainate-treated animals is sensitive to Ni^{2+} but not agatoxin or omega-conotoxin, suggesting a specific role for t-type VGCC in this form of hyperexcitability (Su et al., 2002) (see Figure 23.8B). Other investigators have found enhanced Ca^{2+}-dependent inactivation of VGCC; although this effect may counteract hyperexcitability, it illustrates the diminished Ca^{2+}-buffering capacity of epileptic neurons (Beck et al., 1999; Nagerl & Mody, 1998).

Transient, A-type K^+ channel function of dentate granule cells following status epilepticus is reduced. Specifically, suppression of Kv4.2 activity, due to decreased expression and increased phosphorylation, results in efficient conduction of action potentials from dendrites to the soma (Bernard et al., 2004) (see Figure 23.8C). The facilitation of "active" dendritic responses by Kv4.2 downregulation is a mechanism that could potentiate the synaptic reorganization and neurotransmitter receptor changes see in TLE.

Alterations in hyperpolarization gated cation currents (I_h) may play an important role in both idiopathic generalized epilepsy, and acquired epilepsies. Changes in I_h have been studied in several models of acquired epilepsy (Bender et al., 2003b; Brewster et al., 2002, 2005; Poolos et al., 2006; Shah et al., 2004), but our discussion here will focus on febrile seizures.

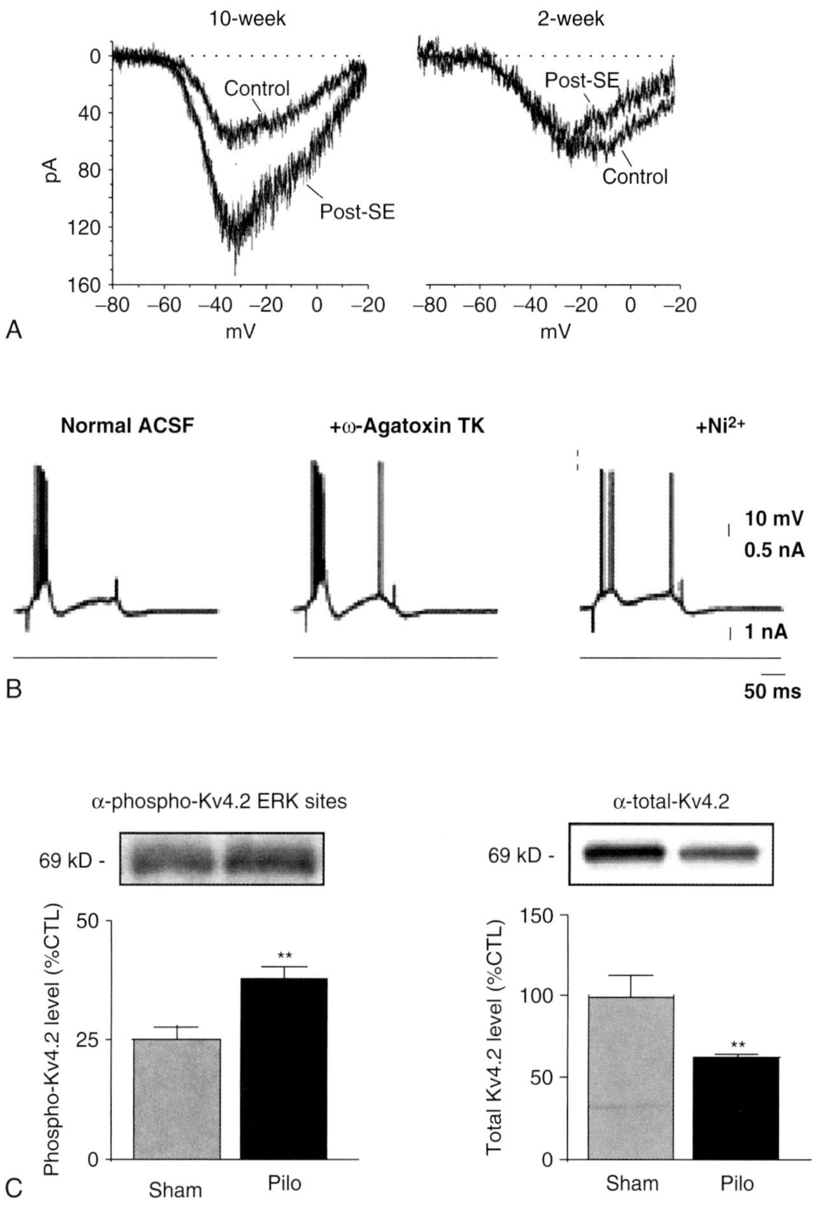

Figure 23.8 Changes in voltage-gated ion channels in experimental TLE. **A**. Low-threshold, persistent voltage-gated sodium currents evoked with slow voltage ramps from neurons in rat entorhinal cortex. At 10 weeks after status epilepticus (SE) induced with pilocarpine, an increase in the amplitude of low-threshold, persistent sodium current is seen. These changes are not apparent at 2 weeks post-SE (Agrawal et al., 2003). **B**. Voltage responses of CA1 pyramidal cells in response to current injection. Bursting behavior is observed in neurons from kainate-treated animals that is insensitive to ω-Agatoxin TK but blocked by Ni^{2+}, indicating a role of T-type voltage-gated calcium channels in this form of hyperexcitability (Su et al., 2002). **C**. Western blots of phosphorylated Kv4.2 and total Kv4.2 protein. The transient voltage-gated potassium channel subunit Kv4.2 has increased phosphorylation and decreased expression in pilocarpine treated animals. Both of these changes diminish Kv4.2 function and enhance conduction of action potentials from dendrites to the soma (Bernard et al., 2004).

F. Febrile Seizures

Febrile seizures are common and represent an important risk factor for the development of TLE, and as such they warrant special attention. Febrile seizures occur in 2 to 4 percent of children in the United States, whereas they are seen in up to 14 percent of Japanese children, speaking to a genetic susceptibility (Hauser, 1994). Febrile seizures occur between the ages of 6 months and 5 years, a period of profound neural plasticity associated with immature patterns of receptor expression and intracellular chloride regulation. Why do febrile seizures produce epilepsy? It has become clear that seizures, but not hyperthermia alone, are the trigger for molecular changes that lead to epilepsy following hyperthermia-induced seizures (Dube et al., 2006). Importantly, hyperthermia-induced seizures, but not hyperthermia itself, leads to breakdown of blood-brain barrier and exposure of brain to serum proteins, a potent epileptogenic stimulus (Ilbay et al., 2003; Seiffert et al., 2004). Exogenous interleukin-1beta (IL-1beta), a pyrogenic inflammatory cytokine whose expression in brain is increased during fever from any cause, lowers seizure threshold during hyperthermic challenge, and IL-1 receptor knockout mice have *higher* seizure thresholds (Dube et al., 2005). IL-1beta may influence seizure thresholds through activation of Src family kinases with subsequent phosphorylation and modulation of NMDA receptors (Viviani et al., 2003).

Baram and colleagues have identified persistent changes in excitability of hippocampal neurons following experimental hyperthermia-induced seizures. These changes include increases in the frequency of spontaneous inhibitory post-synaptic currents (IPSCs) and amplitudes of evoked IPSCs (Chen et al., 1999). On first evaluation this would be predicted to have an anticonvulsant effect. However, subsequent work has identified modification of hyperpolarization and cyclic-nucleotide gated cation currents (I_h) following hyperthermia-induced seizures. The modified I_h has slowed deactivation kinetics, such that I_h activation during a train of IPSCs produces rebound excitation (Chen et al., 2001) (see Figure 23.9A). This rebound excitation is blocked by the specific I_h inhibitor ZD7288 (see Figure 23.9B). Altered expression of the I_h subunits HCN1 and HCN2 has been identified in hippocampus of experimental animals and patients with hippocampal sclerosis, greater than 50 percent of whom had history of prolonged febrile seizures (Brewster et al., 2002, Bender et al., 2003b) (see Figure 23.9C). Additionally, the I_h subunits HCN1 and HCN2 form heteromeric channels following experimental febrile seizures which may account for the persistently modified kinetic properties of I_h and post-inhibitory excitability (Brewster et al., 2005).

III. Post-Traumatic and Post-Stroke Epilepsy

Closed head trauma with loss of consciousness is an important cause of temporal lobe epilepsy, as well as seizures

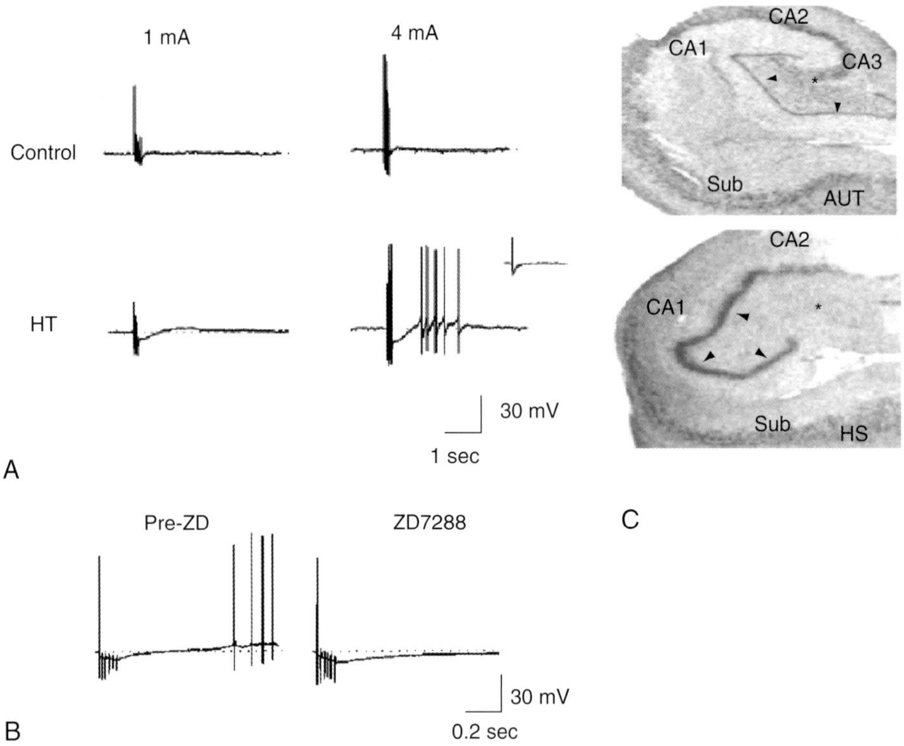

Figure 23.9 Hyperthermia-induced seizures. **A**. Current-clamp recordings from dentate granule cells in response to evoked IPSPs. In cells from animals subjected to experimental febrile seizures (HT), a volley of IPSPs generates a volley of action potentials as a result of rebound excitation. **B**. Rebound excitation is inhibited by the specific hyperpolarization-activated cation channel inhibitor ZD7288 (Chen et al., 2001). **C**. *In situ* hybridization for the hyperpolarization-activated cation channel subunit HCN1. Staining is increased in dentate granule cells (arrows) of patients with intractable epilepsy, many of whom had history of febrile seizures, compared to controls (Bender et al., 2003b).

originating in other areas of the brain, accounting for up to 5 to 13 percent of adult-onset epilepsy (French et al., 1993; Zarrelli et al., 1999). Penetrating injuries or those associated with contra-coup contusions and hemorrhage lead to cortical injury that increases the risk of later epilepsy. Post-traumatic epilepsy is a useful clinical example of the latent or "silent" in the development of epilepsy; late seizures experienced less than two months following trauma, but no seizures experienced at the time of impact, are predictive of the development of epilepsy.

The Vietnam Head Injury study has shown that patients continue to manifest new-onset epilepsy up to 20 years after their injury (Salazar et al., 1985). Post-traumatic epilepsy has been studied with the fluid-percussion injury model, which produces hyperexcitability with the pathologic hallmarks of TLE. Other direct injuries to neocortex, including ferric chloride injection, alumina gel injection, and freeze lesions, have served as models for focal epilepsy such as that seen after traumatic injury. Wilmore and colleagues (1978a, 1978b) have used ferric chloride injection as a model of intracranial hemorrhage due to trauma or hemorrhagic stroke. Animals treated in this manner develop spontaneous seizures, and a number of molecular changes are seen including reductions in glutamate transporters, increased GABA transporters, and NMDA receptor upregulation (Doi et al., 2001; Ueda & Willmore, 2000).

Diffuse axonal injury can be seen as a consequence of closed head trauma, which may lead to deafferentation of cortex. Cortical epilepsy can be produced experi mentally by introduction of lesions that undercut cortex producing isolated, deafferented cortical islands. Mechanisms of hyperexcitability in undercut neocortex include enhanced Ach sensitivity, loss of inhibitory interneurons, and most importantly, axonal sprouting and formation of recurrent excitatory synaptic connections (Jin et al., 2006; Li & Prince, 2002; Salin et al., 1995). Passive and active membrane properties also are affected in undercut cortex to favor hyperexcitability; increases are seen in the input resistance, and in the input-output relationship of spike frequency to injected depolarizing current (Jacobs et al., 2000). Neocortical pyramidal cells in undercut cortex display impaired extrusion of Cl^- ions, suggesting a potential role for $GABA_A$ receptor-mediated excitation in post-traumatic epilepsy (Hablitz & DeFazio, 1998; Jin et al., 2005). This last mechanism could reconcile the increased GABAergic inhibitory synaptic currents observed in undercut cortex with the occurrence of spontaneous seizures (Prince & Jacobs, 1998). However, a loss of GABAergic terminals is reported in other models (alumina gel lesioned cortex) of post-traumatic epilepsy (Houser et al., 1986).

In the freeze lesion model, alterations in NMDA-R and $GABA_A$-R subunit expression similar to those occurring in TLE are observed. Specifically, an upregulation of NR2B subunits and a downregulation of alpha1-subunits of GABA-R lead to decreased modulation by ifendropil and benzodiazepines, respectively (DeFazio & Hablitz, 2000; Hablitz & DeFazio, 2000).

Stroke is a common affliction affecting 700,000 Americans every year. Post-stroke epilepsy resembles post-traumatic epilepsy in many ways. In both situations cortical injury and remodeling lead to hyperexcitability, late seizures are predictive of the development of epilepsy, hemorrhage increases the risk of epilepsy, and a long latent period is common (Bladin et al., 2000). The cellular and molecular changes contributing to post-stroke epilepsy and perinatal hypoxia-induced seizures have been studied using models of hypoxic-ischemic injury. Downregulation of GluR2-subunit expression is seen in response to hypoxia, and this is predicted to increase single-channel conductance and the calcium permeability of glutamate receptors, changes that would facilitate excitation at glutamatergic synapses (Sanchez et al., 2001). Glutamatergic neurotransmission is capable of calcineurin-dependent reduction of GABA-mediated inhibition following hypoxic-ischemic injury (Sanchez et al., 2005). The occurrence of early seizures during stroke may relate to cortical depolarization, but the development of epilepsy is likely to involve chronic changes similar to those seen in TLE, including gliosis and mossy fiber sprouting (Williams et al., 2004). Stroke occurring *in utero* can lead to abnormalities in cortical development, and cortical malformations including polymicrogyria, with subsequent epilepsy.

VI. Rasmussen's Syndrome

Rasmussen's syndrome is characterized pathologically by a chronic encephalitis and clinically by an average age of onset of 6 years, intractable epilepsy, and progressive neurologic deficits. Pathology shows changes typical of encephalitis, including neuronal loss, reactive gliosis, inflammatory cell infiltrate (primarily CD3+ T lymphocytes), microglial nodules, and neuronophagia. The truly fascinating and unexplained feature of Rasmussen's syndrome is that it affects a single hemisphere producing a hemiparesis, hemisensory loss, and a progressive aphasia if the dominant hemisphere is involved. The frequency of seizures can be astoundingly high, and by themselves produce an epileptic encephalopathy. Seizures can be of multiple types including complex-partial, secondarily generalized, or simple partial seizures, and a common presentation is *epilepsia partialis continua* (focal status epilepticus). An inflammatory, autoimmune pathogenesis with breakdown of blood–brain barrier is apparent but the inciting events and predisposing factors have been elusive. Viral causes are suggested but attempts at isolation of viral particles or viral genome have been inconsistent (Freeman, 2005).

A novel pathogenic mechanism for epilepsy was identified in patients with Rasmussen's encephalitis following the observation that mice inoculated with the GluR3b subunit during attempts to raise monoclonal antibodies developed epilepsy with pathologic features similar to Rasmussen's encephalitis. This led to subsequent isolation of antibodies directed against

GluR3b-subunits of glutamate receptors that are capable of receptor activation from the sera of patients (Rogers et al., 1994). These antibodies appeared to be clinically important as patients treated with plasmapheresis experienced transient remissions. Additional experimental work has demonstrated that immunization of rats with GluR3b-subunits leads to production of receptor-activating, excitotoxic antibodies (Levite et al., 1999) (see Figure 23.10).

A mechanism linking the cell-mediated immune mechanisms (evidenced by CD3+/CD8+ T-lymphocyte infiltrate in pathology specimens) to the production of humoral autoimmunity against glutamate receptors has been identified. Granzyme b, a serine-threonine protease, released onto neurons by CD3+/CD8+ T cells leads to neuronal apoptosis and liberation of the autoimmunogenic fragment of GluR3b-subunit (Bien et al., 2002; Gahring et al., 2001). Thus, CD3+ T cells directly kill neurons and facilitate production of excitotoxic and potentially epileptogenic anti-GluR3b antibodies. Subsequent human studies have demonstrated that GluR3b autoantibodies are neither sensitive nor specific for Rasmussen's syndrome, and are seen in other focal epilepsies (Ganor et al., 2004; Wiendl et al., 2001). The relationship of GluR3b autoantibodies to the pathogenesis of other focal epilepsies is not established. Antibodies directed against other ion channels, including voltage-gated K^+ channels, are associated with noninfectious limbic encephalitis and seizures (Buckley et al., 2001; Harrower et al., 2006). As is the case with post-traumatic epilepsy, post-stroke epilepsy, and temporal lobe epilepsy, neuronal injury and death in Rasmussen's is likely to lead to developmental patterns of receptor expression, synaptic reorganization, and gliosis that predispose to hyperexcitability and seizures.

V. Post-Infectious Epilepsy

Infections of the central nervous system can acutely cause seizures, including status epilepticus, and represent an important risk factor for the subsequent development of epilepsy. In the acute setting, bacterial or viral meningitis may trigger seizures through blood–brain barrier breakdown, associated fever, or proconvulsant inflammatory mediators (Vezzani & Granata, 2005). IL-1 modulation of NMDA receptors and arachidonic acid downregulation of A-type potassium channel function are but two examples of how inflammatory mediators produce hyperexcitability (Keros & McBain, 1997; Viviani et al., 2003). Many of the chronic cellular and molecular mechanisms described for TLE are seen in post-infectious epilepsies, and TLE is a common form of epilepsy seen following encephalitis or meningitis as the initial precipitating injury.

Neurocysticercosis, due to larvae of the tapeworm *Taenia solium* encysted in brain parenchyma, with an associated inflammatory reaction, is perhaps the single greatest cause of

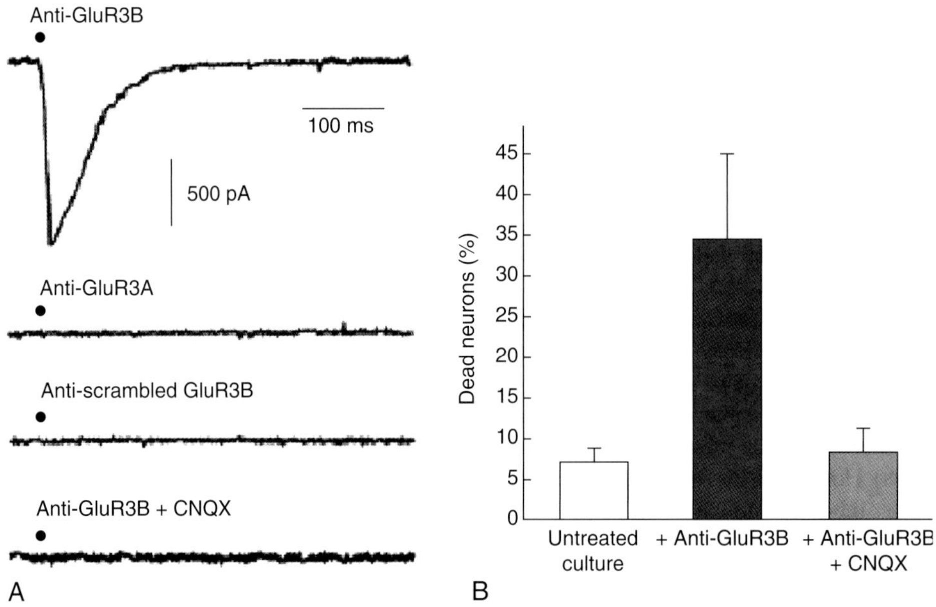

Figure 23.10 Rasmussen's encephalitis. **A**. Voltage-clamp recordings from neurons in response to application of anti-GluR3b antibodies. Antibodies directed against the GluR3 subunit of AMPA receptors, such as those found in the sera in patient's with Rasmussen's encephalitis with intractable epilepsy, activate GluR3-containing receptors. Scrambled antibodies have no effect and the response is completely blocked by CNQX. **B**. Histogram of plot of neuronal death. Anti-GluR3b antibodies results in excitotoxic neuronal death that is antagonized by CNQX (Levite et al., 1999).

acquired epilepsy worldwide (Garcia et al., 2005). Cerebral toxoplasmosis and cerebral abscesses are additional infectious causes of seizures and epilepsy that result from blood–brain barrier breakdown, inflammation, and derangement of cortical architecture.

Herpes simplex virus (HSV1) infection causes a hemorrhagic meningoencephalitis, consistently involving the limbic cortex. Fever, mental status changes, aphasia, and seizures are seen acutely with sequelae of cognitive impairment and epilepsy. The underlying substrate for this type of epilepsy may include neuronal death, mossy fiber sprouting, reduced GABAergic inhibition, and altered biophysical properties of CA3 neurons (depolarized resting membrane potentials and increased input resistance) (Chen et al., 2004). In organotypic hippocampal slice cultures, the antiviral drug acyclovir was effective in limiting these pathologic changes and abrogated hyperexcitability (Chen et al., 2004).

Other viral causes of encephalitis such as EBV and CMV may also result in epilepsy. CMV and HSV viral particles have been identified in cases of chronic (Rasmussen's) encephalitis with intractable epilepsy (Jay et al., 1995), although a causative role has not been confirmed. HIV-infected individuals have a high incidence of seizures, most frequently associated with an identifiable cause of meningitis or structural brain lesion, but up to 24 percent of these will be unexplained (Modi et al., 2000). Experiments with the feline immunodeficiency virus have provided a clue to these seizures without an identifiable cause by demonstrating neuronal loss and mossy fiber sprouting in the dentate gyrus (Mitchell et al., 1998). Subacute sclerosing pan encephalitis, a devastating form of post-infectious epilepsy and dementia, has fortunately become rare since the introduction of the measles vaccine.

VI. Epilepsy Caused by Neoplasms and Other Mass Lesions

Brain tumors and other structural abnormalities such as vascular malformations, or subdural hematomas, are important causes of acquired epilepsy. Common primary brain tumors include glial-derived tumors (oligodendroglioma, low- and high-grade astrocytoma), meningioma, and primary CNS lymphoma. Metastatic brain tumors as well as malignant and benign primary brain tumors can cause seizures and epilepsy. Twenty to 45 percent of patients with primary brain tumors will develop seizures (Schaller & Ruegg, 2003). Conceptually, structural mass lesions abutting cortical tissue create an "irritative focus" for seizure generation. The reality is much more complex with a variety of growth factors, immune cells, and alterations of blood–brain barrier related to neo-angiogenesis contributing to neuronal alterations and excitability.

Metastatic brain tumors, primary brain tumors, and space-occupying infections (i.e., toxoplasmosis, cerebral abscesses) "enhance" with intravenous contrast material during clinical neuroimaging studies indicating blood–brain barrier breakdown at the lesion-parenchymal interface or within the lesion itself. As noted earlier, blood–brain barrier breakdown is a potent epileptogenic event (Seiffert et al., 2004). Inflammatory changes at sites of blood–brain barrier breakdown are predicted to contribute to hyperexcitability.

Derangements of brain microenvironment near tumors, most notably increased K^+ and H^+ concentrations, is suspected to participate in tumor-related seizures (Schaller & Ruegg, 2003). Changes in NMDA receptor subunits, $GABA_A$ receptor subunits, and glutamic acid decarboxylase are altered in peritumoral neurons in patients with epilepsy due to neoplasm (Wolf et al., 1996). These changes are heterogeneous but suggest alterations in the excitation-inhibition balance in epilepsy associated with neoplasms. Physiological data have identified unique membrane properties in neurons adjacent to vascular malformations; these neurons had higher firing rates in response to synaptic input than those neurons in control tissue or adjacent to glial tumors (Williamson et al., 2003).

VII. Glia, Brain Microenvironment, and Epilepsy

Of all the cellular and molecular changes occurring in the acquired epilepsies, glial proliferation, or *gliosis*, associated with neuronal injury is arguably the most consistent finding. Gliosis is seen in humans and experimental animals following insults to brain tissue including status epilepticus, trauma, stroke, neoplasm, infection, and inflammatory conditions. Glial cells influence neuronal excitability by regulation of the ionic composition of the extracellular space (H^+, K^+ ions), uptake of excitatory neurotransmitters, modulation of synaptic transmission, and provision of energy substrate (i.e., lactate) to neurons (Newman, 2005; Ransom et al., 2003).

In the setting of gliosis the ability of glia to perform these functions is thought to be altered through modifications of glial gene expression, and structural/anatomical changes (Mandell & VandenBerg, 1999; Westenbroek et al., 1998). We have already discussed how changes in glial glutamate transporter expression can lead to dysregulation of extracellular glutamate, thereby influencing the stimulation of neurons in a potentially injurious manner (Samuelsson et al., 2000; Tessler et al., 1999; Ueda & Willmore, 2000; Ueda et al., 2001). Another established role for glia relevant to hyperexcitability and epileptogenesis is regulation of K^+ ions released by neurons into the extracellular space during action potentials. This is accomplished in part via barium-sensitive inwardly rectifying K^+ channels (Kir 4.1). In epilepsy there is a downregulation of these channels and derangement of normal extracellular potassium homeostasis, particularly in sclerotic hippocampus with prominent astrogliosis (Bordey & Spencer, 2004; Bordey et al., 2001; Gabriel et al., 1998; Heinemann et al., 2000; Hinterkeuser et al., 2000; Samuelsson et al., 2000). Potassium channels do not work in isolate to regulate extracellular potassium; gap junctional coupling, membrane transporters/pumps, and aquaporin water channels contribute to potassium redistribution and uptake.

In normal tissue, aquaporin-4 (AQP4) is exquisitely localized to perivascular astrocytic endfeet and this subcellular localization is lost in TLE with hippocampal sclerosis (Eid et al., 2005) (see Figure 23.11A). Additional studies have demonstrated that the absence or mislocalization of AQP4 leads to abnormal extracellular potassium regulation (higher peak levels of extracellular K^+ accumulation and delayed recovery kinetics) and increased seizure duration (Amiry-Moghaddam et al., 2003; Binder et al., 2006) (see Figure 23.11B, C). It is now clear that glia do not function purely as housekeepers, but are capable of initiating paroxysmal neurologic events such as spreading depression and depolarizing shifts in neurons by release of glutamate through anion channels (Basarsky et al., 1998, 1999; Tian et al., 2005). Glial calcium waves, mediated by gap junctional coupling, are the trigger for this glial glutamate release and are facilitated by proconvulsant manipulations and inhibited by anticonvulsant drugs (Tian et al., 2005). The contribution of these mechanisms to the acquired epilepsies is not established, but they are expected to be altered in human epilepsy where glial gap junctional coupling is enhanced (Lee et al., 1995).

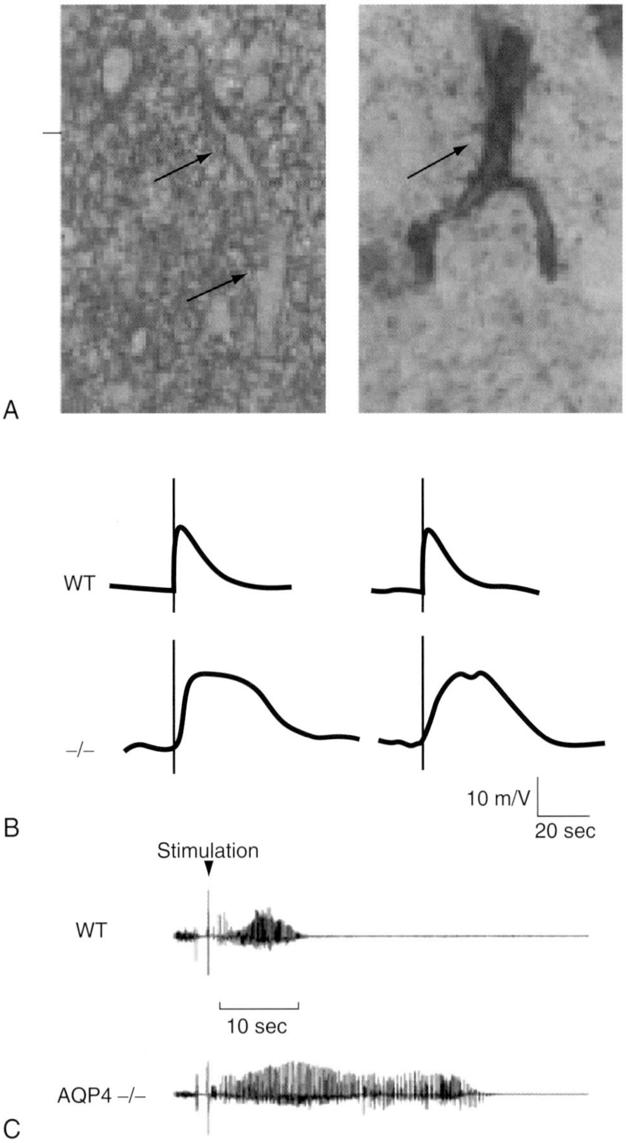

Figure 23.11 Loss of AQP4 localization in astrocytes results in abnormal potassium regulation and prolonged seizure duration. **A.** Immunhistochemistry for AQP4 in human tissue. AQP4 is normally exquisitely localized to perivascular astrocytic endfeet (right panel, arrows) and this localization is lost in tissue from patients with intractable temporal lobe epilepsy and hippocampal sclerosis (left panel) (Eid et al., 2005). **B.** Activity-dependent changes of extracellular potassium concentration recorded with ion-sensitive microelectrodes from hippocampus of wild-type (WT) animals and AQP4 –/– animals. Activity-dependent changes in extracellular potassium is increased with slowed recovery kinetics in AQP4 knockouts. **C.** Electrographic seizures elicited by electrical stimulation. AQP4 –/– animals had longer seizure durations than WT animals (Binder et al., 2006).

VIII. Summary and Conclusions

Localization-related epilepsy is the most common type of epilepsy in adults and may result from febrile seizures, infection, stroke, trauma, neoplasm, or inflammatory conditions. The development of epilepsy following an initial precipitating injury is dependent on changes in gene expression and durable structural and molecular changes that shift the excitation-inhibition balance to support spontaneous seizures. Our understanding of these proconvulsant neural modifications is considerable and is opening the door to new therapeutic targets. Despite this, approximately 30 percent of patients with epilepsy will be refractory to best medical therapy, and at least 62 percent of patients undergoing surgery for intractable epilepsy will have seizure recurrence (Wiebe et al., 2001). Identification of the relative contribution of the multitude of cellular and molecular mechanisms seen in epilepsy (alterations in neurotrophin expression, axonal sprouting, molecular reorganization of neurotransmitter receptors, changes in ion channel properties, alterations in extracellular neurotransmitter and ion regulation, inflammation) to both the progressive changes occurring during the latent period and the occurrence of seizures in individual cases, currently beyond the reach of clinical neurology, has the potential to allow improved selection of antiepileptic drug regimens.

Table 23.3 Cellular and Molecular Mechanisms in Acquired Epilepsy

Cell loss	• Neuronal death Dentate gyrus granule cell loss Pyramidal cell loss • Dentate gyrus interneuron death • Mossy cell death, basket cell survival
Axonal sprouting/Synaptogenesis	• Mossy fiber sprouting Recurrent excitation of granule cells Hyper-inhibition of granule cells • Axonal sprouting in cortex
Neurogenesis	• Dentate granule cell proliferation • Ectopic granule cells
Alterations of glutamate signaling	• Increased synapse number/axonal sprouting • Changes in transmitter release probability • Molecular reorganization of glutamate receptors Loss of GluR2 subunit NR2A and NR2B expression • Changes in mGluR expression • Extracellular glutamate dysregulation Decreased in glial glutamate transport Increased neuronal glutamate transport
Alterations of GABA signaling	• Interneuron death • Molecular reorganization of $GABA_A$ receptors Alpha1 subunit downregulation Alpha4 subunit upregulation Loss of delta subunits, change in tonic inhibition Reduced benzodiazepine sensitivity Increased Zn^{2+} sensitivity • Alterations in chloride transport/GABA-mediated excitation • Modulation of GABAergic transmission by NMDA-R, mGluR • Receptor internalization
Voltage-gated ion channels	• Persistent sodium current upregulation • A-type potassium channel downregulation • Ca^{2+} channel upregulation • Hyperpolarization-activated cation current reorganization
Autoimmune/inflammatory mechanisms	• Modulation of ion channels • Receptor-activating autoantibodies • Blood-brain barrier breakdown
Brain microenvironment	• Glutamate dysregulation • Potassium dysregulation

In the future, advances in neuroimaging techniques and expanded invasive testing of patients with refractory epilepsy at epilepsy surgery centers could provide greater understanding of seizures in specific patients that will impact their therapy. The use of gene chip microarray is an exciting and promising approach to the study of epilepsy with aptitude to identify heretofore underappreciated mechanisms and novel approaches to treatment. As we learn more about the basic mechanisms of acquired epilepsy and predisposing genetic substrates, patients suffering from this debilitating disease are likely to benefit from improved medical and surgical evaluations and treatments.

References

Agrawal, N., Alonso, A., Ragsdale, D. S. (2003). Increased persistent sodium currents in rat entorhinal cortex layer V neurons in a post-status epilepticus model of temporal lobe epilepsy. *Epilepsia* **44**, 1601–1604.

Amiry-Moghaddam, M., Williamson, A., Palomba, M., Eid, T., de Lanerolle, N. C., Nagelhus, E. A. et al. (2003). Delayed K+ clearance associated with aquaporin-4 mislocalization: Phenotypic defects in brains of alpha-syntrophin-null mice. *Proceedings of the National Academy of Sciences of the U S A* **100**, 13615–13620.

Basarsky, T. A., Feighan, D., MacVicar, B. A. (1999). Glutamate release through volume-activated channels during spreading depression. *Journal of Neuroscience* **19**, 6439–6445.

Basarsky, T. A., Duffy, S. N., Andrew, R. D., MacVicar, B. A. (1998). Imaging spreading depression and associated intracellular calcium waves in brain slices. *Journal of Neuroscience* **18**, 7189–7199.

Beck, H., Steffens, R., Heinemann, U., Elger, C. E., Beck, H., Steffens, R. et al. (1999). Ca(2+)-dependent inactivation of high-threshold Ca(2+) currents in hippocampal granule cells of patients with chronic temporal lobe epilepsy. *Journal of Neurophysiology* **82**, 946–954.

Behr, J., Heinemann, U., Mody, I. (2001). Kindling induces transient NMDA receptor-mediated facilitation of high-frequency input in the rat dentate gyrus. *Journal of Neurophysiology* **85**, 2195–2202.

Behr, J., Heinemann, U., Mody, I., Behr, J., Heinemann, U., Mody, I. (2000). Glutamate receptor activation in the kindled dentate gyrus. *Epilepsia* **41 Suppl 6**, S100–103.

Bender, R. A., Dube, C., Gonzalez-Vega, R., Mina, E. W., Baram, T. Z. (2003a). Mossy fiber plasticity and enhanced hippocampal excitability, without hippocampal cell loss or altered neurogenesis, in an animal model of prolonged febrile seizures. *Hippocampus* **13**, 399–412.

Bender, R. A., Soleymani, S. V., Brewster, A. L., Nguyen, S. T., Beck, H., Mathern, G. W., Baram, T. Z. (2003b). Enhanced expression of a specific hyperpolarization-activated cyclic nucleotide-gated cation channel (HCN) in surviving dentate gyrus granule cells of human and experimental epileptic hippocampus. *Journal of Neuroscience* **23**, 6826–6836.

Berg, A. T., Shinnar, S. (1997). Do seizures beget seizures? An assessment of the clinical evidence in humans. *Journal of Clinical Neurophysiology* **14**, 102–110.

Bernard, C., Anderson, A., Becker, A., Poolos, N. P., Beck, H., Johnston, D. (2004). Acquired dendritic channelopathy in temporal lobe epilepsy [see comment]. *Science* **305**, 532–535.

Bien, C. G., Bauer, J., Deckwerth, T. L., Wiendl, H., Deckert, M., Wiestler, O. D. et al. (2002). Destruction of neurons by cytotoxic T cells: A new pathogenic mechanism in Rasmussen's encephalitis. *Annals of Neurology* **51**, 311–318.

Bihi, R. I., Jefferys, J. G., Vreugdenhil, M. (2005). The role of extracellular potassium in the epileptogenic transformation of recurrent GABAergic inhibition. *Epilepsia* **5**, 64–71.

Binder, D. K., Croll, S. D., Gall, C. M., Scharfman, H. E. (2001). BDNF and epilepsy: Too much of a good thing? [see comment]. *Trends in Neurosciences* **24**, 47–53.

Binder, D. K., Routbort, M. J., Ryan, T. E., Yancopoulos, G. D., McNamara, J. O. (1999). Selective inhibition of kindling development by intraventricular administration of TrkB receptor body. *Journal of Neuroscience* **19**, 1424–1436.

Binder, D. K., Yao, X., Zador, Z., Sick, T. J., Verkman, A. S., Manley, G. T. (2006). Increased seizure duration and slowed potassium kinetics in mice lacking aquaporin-4 water channels. *Glia* **53**, 631–636.

Bladin, C. F., Alexandrov, A.V., Bellavance, A., Bornstein, N., Chambers, B., Cote, R. et al. (2000). Seizures after stroke: A prospective multicenter study. *Archives of Neurology* **57**, 1617–1622.

Blumcke, I., Beck, H., Scheffler, B., Hof, P. R., Morrison, J. H., Wolf, H. K. et al. (1996). Altered distribution of the alpha-amino-3-hydroxy-5-methyl-4-isoxazole propionate receptor subunit GluR2(4) and the N-methyl-D-aspartate receptor subunit NMDAR1 in the hippocampus of patients with temporal lobe epilepsy. *Acta Neuropathologica* **92**, 576–587.

Blumcke, I., Becker, A. J., Klein, C., Scheiwe, C., Lie, A. A., Beck, H. et al. (2000). Temporal lobe epilepsy associated up-regulation of metabotropic glutamate receptors: Correlated changes in mGluR1 mRNA and protein expression in experimental animals and human patients. *Journal of Neuropathology & Experimental Neurology* **59**, 1–10.

Bordey, A., Spencer, D. D. (2004). Distinct electrophysiological alterations in dentate gyrus versus CA1 glial cells from epileptic humans with temporal lobe sclerosis. *Epilepsy Research* **59**, 107–122.

Bordey, A., Lyons, S. A., Hablitz, J. J., Sontheimer, H. (2001). Electrophysiological characteristics of reactive astrocytes in experimental cortical dysplasia. *Journal of Neurophysiology* **85**, 1719–1731.

Bouilleret, V., Loup, F., Kiener, T., Marescaux, C., Fritschy, J. M. (2000). Early loss of interneurons and delayed subunit-specific changes in GABA(A)-receptor expression in a mouse model of mesial temporal lobe epilepsy. *Hippocampus* **10**, 305–324.

Brewster, A., Bender, R. A., Chen, Y., Dube, C., Eghbal-Ahmadi, M., Baram, T. Z. (2002). Developmental febrile seizures modulate hippocampal gene expression of hyperpolarization-activated channels in an isoform- and cell-specific manner. *Journal of Neuroscience* **22**, 4591–4599.

Brewster, A. L., Bernard, J. A., Gall, C. M., Baram, T. Z. (2005). Formation of heteromeric hyperpolarization-activated cyclic nucleotide-gated (HCN) channels in the hippocampus is regulated by developmental seizures. *Neurobiology of Disease* **19**, 200–207.

Brooks-Kayal, A. R. (2005). Rearranging receptors. *Epilepsia* **7**, 29–38.

Brooks-Kayal, A. R., Shumate, M. D., Jin, H., Rikhter, T. Y., Coulter, D. A. (1998). Selective changes in single cell GABA(A) receptor subunit expression and function in temporal lobe epilepsy [see comment] [erratum appears in Nat Med 1999 May;5(5):590]. *Nature Medicine* **4**, 1166–1172.

Brooks-Kayal, A. R., Shumate, M. D., Jin, H., Rikhter, T. Y., Kelly, M. E., Coulter, D. A. (2000). Gamma-aminobutyric acid(A) receptor subunit expression predicts functional changes in hippocampal dentate granule cells during postnatal development. *Journal of Neurochemistry* **77**, 1266–1278.

Brooks-Kayal, A. R., Shumate, M. D., Jin, H., Rikhter, T. Y., Kelly, M. E., Coulter, D. A. (2001). Gamma-aminobutyric acid(A) receptor subunit expression predicts functional changes in hippocampal dentate granule cells during postnatal development. *Journal of Neurochemistry* **77**, 1266–1278.

Brooks-Kayal, A. R., Shumate, M. D., Jin, H., Lin, D. D., Rikhter, T. Y., Holloway, K. L., Coulter, D. A. (1999). Human neuronal gamma-aminobutyric acid(A) receptors: Coordinated subunit mRNA expression and functional correlates in individual dentate granule cells. *Journal of Neuroscience* **19**, 8312–8318.

Buckley, C., Oger, J., Clover, L., Tuzun, E., Carpenter, K., Jackson, M., Vincent, A. (2001). Potassium channel antibodies in two patients with reversible limbic encephalitis. *Annals of Neurology* **50**, 73–78.

Buckmaster, P. S., Zhang, G. F., Yamawaki, R. (2002). Axon sprouting in a model of temporal lobe epilepsy creates a predominantly excitatory feedback circuit. *Journal of Neuroscience* **22**, 6650–6658.

Buhl, E. H., Otis, T. S., Mody, I. (1996). Zinc-induced collapse of augmented inhibition by GABA in a temporal lobe epilepsy model. *Science* **271**, 369–373.

Burke, J. P., Hablitz, J. J. (1994). Metabotropic glutamate receptor activation decreases epileptiform activity in rat neocortex. *Neuroscience Letters* **174**, 29–33.

Burke, J. P., Hablitz, J. J. (1995). Modulation of epileptiform activity by metabotropic glutamate receptors in immature rat neocortex. *Journal of Neurophysiology* **73**, 205–217.

Campbell, S., Hablitz, J. J. (2005). Modification of epileptiform discharges in neocortical neurons following glutamate uptake inhibition. *Epilepsia* **46 Suppl 5**, 129–133.

Campbell, S. L., Hablitz, J. J. (2004). Glutamate transporters regulate excitability in local networks in rat neocortex. *Neuroscience* **127**, 625–635.

Chen, K., Baram, T. Z., Soltesz, I. (1999). Febrile seizures in the developing brain result in persistent modification of neuronal excitability in limbic circuits [see comment]. *Nature Medicine* **5**, 888–894.

Chen, K., Aradi, I., Thon, N., Eghbal-Ahmadi, M., Baram, T. Z., Soltesz, I. (2001). Persistently modified h-channels after complex febrile seizures convert the seizure-induced enhancement of inhibition to hyperexcitability. *Nature Medicine* **7**, 331–337.

Chen, S. F., Huang, C. C., Wu, H. M., Chen, S. H., Liang, Y. C., Hsu, K. S. (2004). Seizure, neuron loss, and mossy fiber sprouting in herpes simplex virus type 1-infected organotypic hippocampal cultures. *Epilepsia* **45**, 322–332.

Chuang, S-C., Bianchi, R., Kim, D., Shin, H-S., Wong, R. K. S. (2001). Group I metabotropic glutamate receptors elicit epileptiform discharges in the hippocampus through PLC{beta}1 signaling. *Journal of Neuroscience* **21**, 6387–6394.

Crino, P. B., Jin, H., Shumate, M. D., Robinson, M. B., Coulter, D. A., Brooks-Kayal, A. R. (2002). Increased expression of the neuronal glutamate transporter (EAAT3/EAAC1) in hippocampal and neocortical epilepsy. *Epilepsia* **43**, 211–218.

Cronin, J., Obenaus, A., Houser, C. R., Dudek, F. E. (1992). Electrophysiology of dentate granule cells after kainate-induced synaptic reorganization of the mossy fibers. *Brain Research* **573**, 305–310.

DeFazio, R. A., Hablitz, J. J. (2000). Alterations in NMDA receptors in a rat model of cortical dysplasia. *Journal of Neurophysiology* **83**, 315–321.

DeFazio, R. A., Keros, S., Quick, M. W., Hablitz, J. J. (2000). Potassium-coupled chloride cotransport controls intracellular chloride in rat neocortical pyramidal neurons. *Journal of Neuroscience* **20**, 8069–8076.

Doherty, J., Dingledine, R. (2001). Reduced excitatory drive onto interneurons in the dentate gyrus after status epilepticus. *Journal of Neuroscience* **21**, 2048–2057.

Doi, T., Ueda, Y., Tokumaru, J., Mitsuyama, Y., Willmore, L. J. (2001). Sequential changes in AMPA and NMDA protein levels during Fe(3+)-induced epileptogenesis. *Brain Research Molecular Brain Research* **92**, 107–114.

Dube, C., Vezzani, A., Behrens, M., Bartfai, T., Baram, T. Z. (2005). Interleukin-1beta contributes to the generation of experimental febrile seizures [see comment] [erratum appears in Ann Neurol. 2005 Apr;57(4):609]. *Annals of Neurology* **57**, 152–155.

Dube, C., Richichi, C., Bender, R. A., Chung, G., Litt, B., Baram, T. Z. (2006). Temporal lobe epilepsy after experimental prolonged febrile seizures: Prospective analysis. *Brain* **129**, 911–922.

Durmuller, N., Craggs, M., Meldrum, B. S. (1994). The effect of the non-NMDA receptor antagonist GYKI 52466 and NBQX and the competitive NMDA receptor antagonist D-CPPene on the development of amygdala kindling and on amygdala-kindled seizures. *Epilepsy Research* **17**, 167–174.

Dzhala, V. I., Staley, K. J. (2003). Excitatory actions of endogenously released GABA contribute to initiation of ictal epileptiform activity in the developing hippocampus. *Journal of Neuroscience* **23**, 1840–1846.

Dzhala, V. I., Talos, D. M., Sdrulla, D. A., Brumback, A. C., Mathews, G. C., Benke, T. A. et al. (2005). NKCC1 transporter facilitates seizures in the developing brain [see comment]. *Nature Medicine* **11**, 1205–1213.

Eid, T., Thomas, M. J., Spencer, D. D., Runden-Pran, E., Lai, J. C., Malthankar, G. V. et al. (2004). Loss of glutamine synthetase in the human epileptogenic hippocampus: Possible mechanism for raised extracellular glutamate in mesial temporal lobe epilepsy [see comment]. *Lancet* **363**, 28–37.

Eid, T., Lee, T. S., Thomas, M. J., Amiry-Moghaddam, M., Bjornsen, L. P., Spencer, D. D. et al. (2005). Loss of perivascular aquaporin 4 may underlie deficient water and K+ homeostasis in the human epileptogenic hippocampus. *Proceedings of the National Academy of Sciences of the United States of America* **102**, 1193–1198.

Elliott, R. C., Gall, C. M. (2000). Changes in activating protein 1 (AP-1) composition correspond with the biphasic profile of nerve growth factor mRNA expression in rat hippocampus after hilus lesion-induced seizures. *Journal of Neuroscience* **20**, 2142–2149.

Elliott, R. C., Miles, M. F., Lowenstein, D. H. (2003). Overlapping microarray profiles of dentate gyrus gene expression during development- and epilepsy-associated neurogenesis and axon outgrowth. *Journal of Neuroscience* **23**, 2218–2227.

Elwes, R. D., Johnson, A. L., Reynolds, E. H. (1988). The course of untreated epilepsy. *BMJ* **297**, 948–950.

Esclapez, M., Houser, C. R. (1999). Up-regulation of GAD65 and GAD67 in remaining hippocampal GABA neurons in a model of temporal lobe epilepsy. *Journal of Comparative Neurology* **412**, 488–505.

Freeman, J. M. (2005). Rasmussen's syndrome: Progressive autoimmune multi-focal encephalopathy. *Pediatric Neurology* **32**, 295–299.

French, J. A., Williamson, P. D., Thadani, V. M., Darcey, T. M., Mattson, R. H., Spencer, S. S., Spencer, D. D. (1993). Characteristics of medial temporal lobe epilepsy: I. Results of history and physical examination. *Annals of Neurology* **34**, 774–780.

Gabriel, S., Kivi, A., Kovacs, R., Lehmann, T. N., Lanksch, W. R., Meencke, H. J. (1998). Effects of barium on stimulus-induced changes in [K+]o and field potentials in dentate gyrus and area CA1 of human epileptic hippocampus. *Neuroscience Letters* **249**, 91–94.

Gahring, L., Carlson, N. G., Meyer, E. L., Rogers, S. W. (2001). Granzyme B proteolysis of a neuronal glutamate receptor generates an autoantigen and is modulated by glycosylation. *Journal of Immunology* **166**, 1433–1438.

Gall, C. M. (1993a). Seizure-induced changes in neurotrophin expression: Implications for epilepsy. *Experimental Neurology* **124**, 150–166.

Gall, C. M. (1993b). Seizure-induced changes in neurotrophins: Implications for epilepsy. *Experimental Neuology* **124**, 150–166.

Gall, C. M., Isackson, P. J. (1989). Limbic seizures increase neuronal production of messenger RNA for nerve growth factor. *Science* **245**, 758–761.

Ganor, Y., Goldberg-Stern, H., Amromd, D., Lerman-Sagie, T., Teichberg, V. I., Pelled, D. et al. (2004). Autoimmune epilepsy: some epilepsy patients harbor autoantibodies to glutamate receptors and dsDNA on both sides of the blood-brain barrier, which may kill neurons and decrease brain fluids after hemispherotomy. *Clinical & Developmental Immunology* **11**, 241–252.

Garcia, H. H., Del Brutto, O. H., Cysticercosis Working Group in P. (2005). Neurocysticercosis: Updated concepts about an old disease. *Lancet Neurology* **4**, 653–661.

Ghijsen, W. E., da Silva Aresta Belo, A. I., Zuiderwijk, M., Lopez da Silva, F. H. (1999). Compensatory change in EAAC1 glutamate transporter in rat hippocampus CA1 region during kindling epileptogenesis. *Neuroscience Letters* **276**, 157–160.

Goodkin, H. P., Yeh, J. L., Kapur, J. (2005). Status epilepticus increases the intracellular accumulation of GABAA receptors. *Journal of Neuroscience* **25**, 5511–5520.

Gowers, W. R. (1881). Epilepsy and other chronic convulsive diseases: Their causes, symptoms and treatment. London: J & A Churchill.

Gutierrez, R. (2005). The dual glutamatergi-GABAergic phenotype of hippocampal granule cells. *Trends in Neurosciences* **28**, 297–303.

Hablitz, J. J., DeFazio, T. (1998). Excitability changes in freeze-induced neocortical microgyria. *Epilepsy Research* **32**, 75–82.

Hablitz, J. J., DeFazio, R. A. (2000). Altered receptor subunit expression in rat neocortical malformations. *Epilepsia* **41 Suppl 6:**S82–85.

Harrower, T., Foltynie, T., Kartsounis, L., De Silva, R. N., Hodges, J. R. (2006). A case of voltage-gated potassium channel antibody-related limbic encephalitis. *Nature Clinical Practice Neurology* **2**, 339–343.

Harvey, B. D., Sloviter, R. S. (2005). Hippocampal granule cell activity and c-Fos expression during spontaneous seizures in awake, chronically epileptic, pilocarpine-treated rats: Implications for hippocampal epileptogenesis. *Journal of Comparative Neurology* **488**, 442–463.

Hauser, W. A. (1994). The prevalence and incidence of convulsive disorders in children. *Epilepsia* **35**.

Heinemann, U., Gabriel, S., Jauch, R., Schulze, K., Kivi, A., Eilers, A. et al. (2000). Alterations of glial cell function in temporal lobe epilepsy. *Epilepsia* **41 Suppl 6**, S185–189.

Hinterkeuser, S., Schroder, W., Hager, G., Seifert, G., Blumcke, I., Elger, C. E. et al. (2000). Astrocytes in the hippocampus of patients with temporal lobe epilepsy display changes in potassium conductances. *European Journal of Neuroscience* **12**, 2087–2096.

Houser, C. R., Esclapez, M. (1996). Vulnerability and plasticity of the GABA system in the pilocarpine model of spontaneous recurrent seizures. *Epilepsy Research* **26**, 207–218.

Houser, C. R., Esclapez, M. (2003). Downregulation of the alpha5 subunit of the GABA(A) receptor in the pilocarpine model of temporal lobe epilepsy. *Hippocampus* **13**, 633–645.

Houser, C. R., Harris, A. B., Vaughn, J. E. (1986). Time course of the reduction of GABA terminals in a model of focal epilepsy: A glutamic acid decarboxylase immunocytochemical study. *Brain Research* **383**, 129–145.

Hunsberger, J. G., Bennett, A. H., Selvanayagam, E., Duman, R. S., Newton, S. S. (2005). Gene profiling the response to kainic acid induced seizures. *Brain Research Molecular Brain Research* **141**, 95–112.

Ilbay, G., Sahin, D., Ates, N. (2003). Changes in blood-brain barrier permeability during hot water-induced seizures in rats [see comment]. *Neurological Sciences* **24**, 232–235.

Isokawa, M. (1996). Decrement of GABAA receptor-mediated inhibitory postsynaptic currents in dentate granule cells in epileptic hippocampus. *Journal of Neurophysiology* **75**, 1901–1908.

Isokawa, M. (1998). Modulation of GABAA receptor-mediated inhibition by postsynaptic calcium in epileptic hippocampal neurons. *Brain Research* **810**, 241–250.

Isokawa, M. (2000). Remodeling dendritic spines of dentate granule cells in temporal lobe epilepsy patients and the rat pilocarpine model. *Epilepsia* **41**.

Isokawa, M., Levesque, M. F. (1991). Increased NMDA responses and dendritic degeneration in human epileptic hippocampal neurons in slices. *Neuroscience Letters* **132**, 212–216.

Isokawa, M., Mello, L. E. (1991). NMDA receptor-mediated excitability in dendritically deformed dentate granule cells in pilocarpine-treated rats. *Neuroscience Letters* **129**, 69–73.

Jackson, J. H. (1931). Selected writings. London: Staples Press

Jacobs, K. M., Graber, K. D., Kharazia, V. N., Parada, I., Prince, D. A. (2000). Postlesional epilepsy: The ultimate brain plasticity. *Epilepsia* **41**, S153–S161.

Jay, V., Becker, L. E., Otsubo, H., Cortez, M., Hwang, P., Hoffman, H. J., Zielenska, M. (1995). Chronic encephalitis and epilepsy (Rasmussen's encephalitis): Detection of cytomegalovirus and herpes simplex virus 1 by the polymerase chain reaction and in situ hybridization. *Neurology* **45**, 108–117.

Jin, X., Huguenard, J. R., Prince, D. A. (2005). Impaired Cl^- extrusion in layer V pyramidal neurons of chronically injured epileptogenic neocortex. *Journal of Neurophysiology* **93**, 2117–2126.

Jin, X., Prince, D. A., Huguenard, J. R. (2006). Enhanced excitatory synaptic connectivity in layer v pyramidal neurons of chronically injured epileptogenic neocortex in rats. *Journal of Neuroscience* **26**, 4891–4900.

Kafitz, K. W., Rose, C. R., Thoenen, H., Konnerth, A. (1999). Neurotrophin-evoked rapid excitation through TrkB receptors [see comment]. *Nature* **401**, 918–921.

Kang, T. C., An, S. J., Park, S. K., Hwang, I. K., Bae, J. C., Suh, J. G. et al. (2002). Changes in Na(+)-K(+)-Cl(−) cotransporter immunoreactivity in the gerbil hippocampus following spontaneous seizure. *Neuroscience Research* **44**, 285–295.

Kapur, J., Macdonald, R. L. (1997). Rapid seizure-induced reduction of benzodiazepine and Zn2+ sensitivity of hippocampal dentate granule cell GABAA receptors. *Journal of Neuroscience* **17**, 7532–7540.

Keros, S., McBain, C. J. (1997). Arachidonic acid inhibits transient potassium currents and broadens action potentials during electrographic seizures in hippocampal pyramidal and inhibitory interneurons. *Journal of Neuroscience* **17**, 3476–3487.

Klapstein, G. J., Meldrum, B. S., Mody, I. (1999). Decreased sensitivity to Group III mGluR agonists in the lateral perforant path following kindling. *Neuropharmacology* **38**, 927–933.

Kramar, E. A., Lin, B., Lin, C. Y., Arai, A. C., Gall, C. M., Lynch, G. (2004). A novel mechanism for the facilitation of theta-induced long-term potentiation by brain-derived neurotrophic factor. *Journal of Neuroscience* **24**, 5151–5161.

Kraus, J. E., McNamara, J. O. (1998). Measurement of NMDA receptor protein subunits in discrete hippocampal regions of kindled animals. *Brain Research Molecular Brain Research* **61**, 114–120.

Kraus, J. E., Yeh, G. C., Bonhaus, D. W., Nadler, J. V., McNamara, J. O. (1994). Kindling induces the long-lasting expression of a novel population of NMDA receptors in hippocampal region CA3. *Journal of Neuroscience* **14**, 4196–4205.

Labiner, D. M., Butler, L. S., Cao, Z., Hosford, D. A., Shin, C., McNamara, J. O. (1993). Induction of c-fos mRNA by kindled seizures: Complex relationship w ith neuronal burst firing. *Journal of Neuroscience* **13**, 744–751.

Lampert, A., Klein, J. P., Mission, J. F., Rivera, M., Chen, M., Hains, B. C. et al. (2005). The role of altered sodium channel expression in kindling epileptogenesis. *Epilepsia* **46**, 292.

Lee, S. H., Magge, S., Spencer, D. D., Sontheimer, H., Cornell-Bell, A. H. (1995). Human epileptic astrocytes exhibit increased gap junction coupling. *GLIA* **15**, 195–202.

Levite, M., Fleidervish, I. A., Schwarz, A., Pelled, D., Futerman, A. H. (1999). Autoantibodies to the glutamate receptor kill neurons via activation of the receptor ion channel. *Journal of Autoimmunity* **13**, 61–72.

Li, H., Prince, D. A. (2002). Synaptic activity in chronically injured, epileptogenic sensory-motor neocortex. *Journal of Neurophysiology* **88**, 2–12.

Lie, A. A., Becker, A., Behle, K., Beck, H., Malitschek, B., Conn, P. J. et al. (2000). Up-regulation of the metabotropic glutamate receptor mGluR4 in hippocampal neurons with reduced seizure vulnerability. *Annals of Neurology* **47**, 26–35.

Loup, F., Wieser, H. G., Yonekawa, Y., Aguzzi, A., Fritschy, J. M. (2000). Selective alterations in GABAA receptor subtypes in human temporal lobe epilepsy. *Journal of Neuroscience* **20**, 5401–5419.

Mandell, J. W., VandenBerg, S. R. (1999). ERK/MAP kinase is chronically activated in human reactive astrocytes. *Neuroreport* **10**, 3567–3572.

Marks, D. A., Kim, J., Spencer, D. D., Spencer, S. S. (1992). Characteristics of intractable seizures following meningitis and encephalitis [see comment]. *Neurology* **42**, 1513–1518.

Marks, D. A., Kim, J., Spencer, D. D., Spencer, S. S. (1995). Seizure localization and pathology following head injury in patients with uncontrolled epilepsy. *Neurology* **45**, 2051–2057.

Mathern, G. W., Babb, T. L., Vickrey, B. G., Melendez, M., Pretorius, J. K. (1995). The clinical-pathogenic mechanisms of hippocampal neuron loss and surgical outcomes in temporal lobe epilepsy. *Brain* **118**, 105–118.

Mathern, G. W., Babb, T. L., Leite, J. P., Pretorius, K., Yeoman, K. M., Kuhlman, P. A. (1996). The pathogenic and progressive features of chronic human hippocampal epilepsy. *Epilepsy Research* **26**, 151–161.

Mathern, G. W., Pretorius, J. K., Kornblum, H. I., Mendoza, D., Lozada, A., Leite, J. P. et al. (1997). Human hippocampal AMPA and NMDA mRNA levels in temporal lobe epilepsy patients. *Brain* **120**, 1937–1959.

Mathern, G. W., Mendoza, D., Lozada, A., Pretorius, J. K., Dehnes, Y., Danbolt, N. C. et al. (1999). Hippocampal GABA and glutamate transporter immunoreactivity in patients with temporal lobe epilepsy [see comment]. *Neurology* **52**, 453–472.

McNamara, J. O., Russell, R. D., Rigsbee, L., Bonhaus, D. W. (1988). Anticonvulsant and antiepileptogenic actions of MK-801 in the kindling and electroshock models. *Neuropharmacology* **27**, 563–568.

Mitchell, T. W., Buckmaster, P. S., Hoover, E. A., Whalen, L. R., Dudek, F. E. (1998). Axonal sprouting in hippocampus of cats infected with feline immunodeficiency virus (FIV). *Journal of Acquired Immune Deficiency Syndromes & Human Retrovirology* **17**, 1–8.

Modi, G., Modi, M., Martinus, I., Saffer, D. (2000). New-onset seizures associated with HIV infection. *Neurology* **55**, 1558–1561.

Mtchedlishvili, Z., Kapur, J. (2006). High-affinity, slowly desensitizing GABAA receptors mediate tonic inhibition in hippocampal dentate granule cells. *Molecular Pharmacology* **69**, 564–575.

Murray, K. D., Hayes, V. Y., Gall, C. M., Isackson, P. J. (1998). Attenuation of the seizure-induced expression of BDNF mRNA in adult rat brain by an inhibitor of calcium/calmodulin-dependent protein kinases. *European Journal of Neuroscience* **10**, 377–387.

Nagerl, U. V., Mody, I. (1998). Calcium-dependent inactivation of high-threshold calcium currents in human dentate gyrus granule cells. *Journal of Physiology* **509**, 39–45.

Newman, E. A. (2005). Glia and synaptic transmission. In Kettenmann, H. R., Ed., *Neuroglia, 2e*, 355–366. New York: Oxford University Press.

Nishimura, T., Schwarzer, C., Gasser, E., Kato, N., Vezzani, A., Sperk, G. (2005). Altered expression of GABA(A) and GABA(B) receptor subunit mRNAs in the hippocampus after kindling and electrically induced status epilepticus. *Neuroscience* **134**, 691–704.

Notenboom, R. G., Hampson, D. R., Jansen, G. H., van Rijen, P. C., van Veelen, C. W., van Nieuwenhuizen, O., de Graan, P. N. (2006). Up-regulation of hippocampal metabotropic glutamate receptor 5 in temporal lobe epilepsy patients. *Brain* **129**, 96–107.

Obenaus, A., Esclapez, M., Houser, C. R. (1993). Loss of glutamate decarboxylase mRNA-containing neurons in the rat dentate gyrus following pilocarpine-induced seizures. *Journal of Neuroscience* **13**, 4470–4485.

Okabe, A., Ohno, K., Toyoda, H., Yokokura, M., Sato, K., Fukuda, A. (2002). Amygdala kindling induces upregulation of mRNA for NKCC1, a Na(+), K(+)-2Cl(−) cotransporter, in the rat piriform cortex. *Neuroscience Research* **44**, 225–229.

Overstreet-Wadiche, L. S., Bromberg, D. A., Bensen, A. L., Westbrook, G. L. (2006). Seizures accelerate functional integration of adult-generated granule cells. *J Neurosci* **26**, 4095–4103.

Parent, J. M., Elliott, R. C., Pleasure, S. J., Barbaro, N. M., Lowenstein, D. H. (2006). Aberrant seizure-induced neurogenesis in experimental temporal lobe epilepsy. *Annals of Neurology* **59**, 81–91.

Patrylo, P. R., Spencer, D. D., Williamson, A., Patrylo, P. R., Spencer, D. D., Williamson, A. (2001). GABA uptake and heterotransport are impaired in the dentate gyrus of epileptic rats and humans with temporal lobe sclerosis. *Journal of Neurophysiology* **85**, 1533–1542.

Peng, Z., Houser, C. R. (2005). Temporal patterns of fos expression in the dentate gyrus after spontaneous seizures in a mouse model of temporal lobe epilepsy. *Journal of Neuroscience* **25**, 7210–7220.

Peng, Z., Huang, C. S., Stell, B. M., Mody, I., Houser, C. R. (2004). Altered expression of the delta subunit of the GABAA receptor in a mouse model of temporal lobe epilepsy. *Journal of Neuroscience* **24**, 8629–8639.

Poolos, N. P., Bullis, J. B., Roth, M. K. (2006). Modulation of h-channels in hippocampal pyramidal neurons by p38 mitogen-activated protein kinase. *Journal of Neuroscience* **26**, 7995–8003.

Prince, D. A., Jacobs, K. (1998). Inhibitory function in two models of chronic epileptogenesis. *Epilepsy Research* **32**, 83–92.

Ransom, B., Behar, T., Nedergaard, M. (2003). New roles for astrocytes (stars at last). *Trends in Neurosciences* **26**, 520–522.

Ribak, C. E., Tran, P. H., Spigelman, I., Okazaki, M. M., Nadler, J. V. (2000). Status epilepticus-induced hilar basal dendrites on rodent granule cells contribute to recurrent excitatory circuitry. *Journal of Comparative Neurology* **428**, 240–253.

Richerson, G. B., Wu, Y. (2004). Role of the GABA transporter in epilepsy. *Advances in Experimental Medicine & Biology* **548**, 76–91.

Rogers, S. W., Andrews, P. I., Gahring, L. C., Whisenand, T., Cauley, K., Crain, B. et al. (1994). Autoantibodies to glutamate receptor GluR3 in Rasmussen's encephalitis. *Science* **265**, 648–651.

Sagar, H. J., Oxbury, J. M. (1987). Hippocampal neuron loss in temporal lobe epilepsy: Correlation with early childhood convulsions. *Annals of Neurology* **22**, 334–340.

Salazar, A. M., Jabbari, B., Vance, S. C., Grafman, J., Amin, D., Dillon, J. D. (1985). Epilepsy after penetrating head injury. I. Clinical correlates: A report of the Vietnam Head Injury Study. *Neurology* **35**, 1406–1414.

Salin, P., Tseng, G. F., Hoffman, S., Parada, I., Prince, D. A. (1995). Axonal sprouting in layer V pyramidal neurons of chronically injured cerebral cortex. *Journal of Neuroscience* **15**, 8234–8245.

Samuelsson, C., Kumlien, E., Flink, R., Lindholm, D., Ronne-Engstrom, E. (2000). Decreased cortical levels of astrocytic glutamate transport protein GLT-1 in a rat model of posttraumatic epilepsy. *Neuroscience Letters* **289**, 185–188.

Sanchez, R. M., Dai, W., Levada, R. E., Lippman, J. J., Jensen, F. E. (2005). AMPA/kainate receptor-mediated downregulation of GABAergic synaptic transmission by calcineurin after seizures in the developing rat brain. *Journal of Neuroscience* **25**, 3442–3451.

Sanchez, R. M., Koh, S., Rio, C., Wang, C., Lamperti, E. D., Sharma, D. et al. (2001). Decreased glutamate receptor 2 expression and enhanced epileptogenesis in immature rat hippocampus after perinatal hypoxia-induced seizures. *Journal of Neuroscience* **21**, 8154–8163.

Santhakumar, V., Ratzliff, A. D., Jeng, J., Toth, Z., Soltesz, I. (2001). Long-term hyperexcitability in the hippocampus after experimental head trauma [see comment]. *Annals of Neurology* **50**, 708–717.

Schaller, B., Ruegg, S. J. (2003). Brain tumor and seizures: Pathophysiology and its implications for treatment revisited [see comment] [retraction in Fisher, R. S. *Epilepsia*. 2003 Nov;44(11):1463; PMID: 14636358]. *Epilepsia* **44**, 1223–1232.

Scharfman, H. E. (2002). Epilepsy as an example of neural plasticity. *Neuroscientist* **8**, 154–173.

Scharfman, H. E., Sollas, A. L., Goodman, J. H. (2002). Spontaneous recurrent seizures after pilocarpine-induced status epilepticus activate calbindin-immunoreactive hilar cells of the rat dentate gyrus. *Neuroscience* **111**, 71–81.

Scharfman, H. E., Sollas, A. L., Berger, R. E., Goodman, J. H. (2003). Electrophysiological evidence of monosynaptic excitatory transmission between granule cells after seizure-induced mossy fiber sprouting. *Journal of Neurophysiology* **90**, 2536–2547.

Schumacher, T. B., Beck, H., Steffens, R., Blumcke, I., Schramm, J., Elger, C. E. et al. (2000). Modulation of calcium channels by group I and group II metabotropic glutamate receptors in dentate gyrus neurons from patients with temporal lobe epilepsy. *Epilepsia* **41**, 1249–1258.

Schwarzer, C., Tsunashima, K., Wanzenbock, C., Fuchs, K., Sieghart, W., Sperk, G. (1997). GABA(A) receptor subunits in the rat hippocampus II: Altered distribution in kainic acid-induced temporal lobe epilepsy. *Neuroscience* **80**, 1001–1017.

Scimemi, A., Schorge, S., Kullmann, D. M., Walker, M. C. (2006). Epileptogenesis is associated with enhanced glutamatergic transmission in the perforant path. *Journal of Neurophysiology* **95**, 1213–1220.

Scimemi, A., Semyanov, A., Sperk, G., Kullmann, D. M., Walker, M. C. (2005). Multiple and plastic receptors mediate tonic GABAA receptor currents in the hippocampus. *Journal of Neuroscience* **25**, 10016–10024.

Seiffert, E., Dreier, J. P., Ivens, S., Bechmann, I., Tomkins, O., Heinemann, U. et al. (2004). Lasting blood-brain barrier disruption induces epileptic focus in the rat somatosensory cortex. *Journal of Neuroscience* **24**, 7829–7836.

Semyanov, A., Walker, M. C., Kullmann, D. M., Silver, R. A. (2004). Tonically active GABA A receptors: Modulating gain and maintaining the tone. *Trends in Neurosciences* **27**, 262–269.

Shah, M. M., Anderson, A. E., Leung, V., Lin, X., Johnston, D. (2004). Seizure-induced plasticity of h channels in entorhinal cortical layer III pyramidal neurons. *Neuron* **44**, 495–508.

Shao, L. R., Dudek, F. E. (2005). Detection of increased local excitatory circuits in the hippocampus during epileptogenesis using focal flash photolysis of caged glutamate. *Epilepsia* **5**, 100–106.

Shorvon, S. D., Reynolds, E. H. (1982). Early prognosis of epilepsy. *British Medical Journal Clinical Research Ed* **285**, 1699–1701.

Sloviter, R. S. (1987). Decreased hippocampal inhibition and a selective loss of interneurons in experimental epilepsy. *Science* **235**, 73–76.

Sloviter, R. S. (1991). Permanently altered hippocampal structure, excitability, and inhibition after experimental status epilepticus in the rat: The "dormant basket cell" hypothesis and its possible relevance to temporal lobe epilepsy. *Hippocampus* **1**, 41–66.

Sloviter, R. S., Zappone, C. A., Harvey, B. D., Frotscher, M. (2006). Kainic acid-induced recurrent mossy fiber innervation of dentate gyrus inhibitory interneurons: Possible anatomical substrate of granule cell hyper-inhibition in chronically epileptic rats. *Journal of Comparative Neurology* **494**, 944–960.

Sloviter, R. S., Zappone, C. A., Harvey, B. D., Bumanglag, A. V., Bender, R. A., Frotscher, M. (2003). "Dormant basket cell" hypothesis revisited: Relative vulnerabilities of dentate gyrus mossy cells and inhibitory interneurons after hippocampal status epilepticus in the rat. *Journal of Comparative Neurology* **459**, 44–76.

Smeyne, R. J., Vendrell, M., Hayward, M., Baker, S. J., Miao, G. G., Schilling, K. et al. (1993). Continuous c-fos expression precedes programmed cell death in vivo [see comment] [erratum appears in *Nature* 1993 Sep 16;365(6443):279]. *Nature* **363**, 166–169.

Sperk, G., Schwarzer, C., Tsunashima, K., Kandlhofer, S. (1998). Expression of GABA(A) receptor subunits in the hippocampus of the rat after kainic acid-induced seizures. *Epilepsy Research* **32**, 129–139.

Su, H., Sochivko, D., Becker, A., Chen, J., Jiang, Y., Yaari, Y., Beck, H. (2002). Upregulation of a T-type Ca2+ channel causes a long-lasting modification of neuronal firing mode after status epilepticus. *Journal of Neuroscience* **22**, 3645–3655.

Sutula, T., Koch, J., Golarai, G., Watanabe, Y., McNamara, J. O. (1996). NMDA receptor dependence of kindling and mossy fiber sprouting: Evidence that the NMDA receptor regulates patterning of hippocampal circuits in the adult brain. *Journal of Neuroscience* **16**, 7398–7406.

Tessler, S., Danbolt, N. C., Faull, R. L., Storm-Mathisen, J., Emson, P. C. (1999). Expression of the glutamate transporters in human temporal lobe epilepsy. *Neuroscience* **88**, 1083–1091.

Theodore, W. H., Spencer, S. S., Wiebe, S., Langfitt, J. T., Ali, A., Shafer, P. O. et al. (2006). Epilepsy in North America: A report prepared under the auspices of the Global Campaign against Epilepsy, the International Bureau for Epilepsy, the International League Against Epilepsy, and the World Health Organization. *Epilepsia* **47**, 1700–1722.

Tian, G. F., Azmi, H., Takano, T., Xu, Q., Peng, W., Lin, J. et al. (2005). An astrocytic basis of epilepsy [see comment]. *Nature Medicine* **11**, 973–981.

Traub, R. D., Pais, I., Bibbig, A., Lebeau, F. E., Buhl, E. H., Garner, H. et al. (2005). Transient depression of excitatory synapses on interneurons contributes to epileptiform bursts during gamma oscillations in the mouse hippocampal slice. *Journal of Neurophysiology* **94**, 1225–1235.

Trevino, M., Gutierrez, R. (2005). The GABAergic projection of the dentate gyrus to hippocampal area CA3 of the rat: Pre- and postsynaptic actions after seizures. *Journal of Physiology* **567**, 939–949.

Tsunashima, K., Schwarzer, C., Kirchmair, E., Sieghart, W., Sperk, G. (1997). GABA(A) receptor subunits in the rat hippocampus III: Altered messenger RNA expression in kainic acid-induced epilepsy. *Neuroscience* **80**, 1019–1032.

Tyler, W. J., Pozzo-Miller, L. D. (2001). BDNF enhances quantal neurotransmitter release and increases the number of docked vesicles at the active zones of hippocampal excitatory synapses. *Journal of Neuroscience* **21**, 4249–4258.

Tyler, W. J., Pozzo-Miller, L. D. (2003). Miniature synaptic transmission and BDNF modulate dendritic spine growth and form in rat CA1 neurones. *Journal of Physiology* **553**, 497–509.

Ueda, Y., Willmore, L. J. (2000). Sequential changes in glutamate transporter protein levels during Fe(3+)-induced epileptogenesis. *Epilepsy Research* **39**, 201–209.

Ueda, Y., Doi, T., Tokumaru, J., Yokoyama, H., Nakajima, A., Mitsuyama, Y. et al. (2001). Collapse of extracellular glutamate regulation during epileptogenesis: Down-regulation and functional failure of glutamate transporter function in rats with chronic seizures induced by kainic acid. *Journal of Neurochemistry* **76**, 892–900.

van der Hel, W. S., Notenboom, R. G., Bos, I. W., van Rijen, P. C., van Veelen, C. W., de Graan, P. N. (2005). Reduced glutamine synthetase in hippocampal areas with neuron loss in temporal lobe epilepsy. *Neurology* **64**, 326–333.

Vezzani, A., Granata, T. (2005). Brain inflammation in epilepsy: Experimental and clinical evidence. *Epilepsia* **46**, 1724–1743.

Viviani, B., Bartesaghi, S., Gardoni, F., Vezzani, A., Behrens, M. M., Bartfai, T. et al. (2003). Interleukin-1 beta enhances NMDA receptor-mediated intracellular calcium increase through activation of the Src family of kinases. *Journal of Neuroscience* **23**, 8692–8700.

Vreugdenhil, M., Hoogland, G., van Veelen, C. W., Wadman, W. J. (2004). Persistent sodium current in subicular neurons isolated from patients with temporal lobe epilepsy. *European Journal of Neuroscience* **19**, 2769–2778.

Watanabe, Y., Johnson, R. S., Butler, L. S., Binder, D. K., Spiegelman, B. M., Papaioannou, V. E., McNamara, J. O. (1996). Null mutation of c-fos impairs structural and functional plasticities in the kindling model of epilepsy. *Journal of Neuroscience* **16**, 3827–3836.

Wei, W., Zhang, N., Peng, Z., Houser, C. R., Mody, I. (2003). Perisynaptic localization of delta subunit-containing GABA(A) receptors and their activation by GABA spillover in the mouse dentate gyrus. *Journal of Neuroscience* **23**, 10650–10661.

Westenbroek, R. E., Bausch, S. B., Lin, R. C., Franck, J. E., Noebels, J. L., Catterall, W. A. (1998). Upregulation of L-type Ca2+ channels in reactive astrocytes after brain injury, hypomyelination, and ischemia. *Journal of Neuroscience* **18**, 2321–2334.

White, H. S. (2002). Animal models of epileptogenesis. *Neurology* **59**, S7–S14.

White, J. D., Gall, C. M. (1987). Differential regulation of neuropeptide and proto-oncogene mRNA content in the hippocampus following recurrent seizures. *Brain Research* **427**, 21–29.

Wiebe, S., Blume, W. T., Girvin, J. P., Eliasziw, M.: Effectiveness and Efficiency of Surgery for Temporal Lobe Epilepsy Study G (2001). A randomized, controlled trial of surgery for temporal-lobe epilepsy [see comment]. *New England Journal of Medicine* **345**, 311–318.

Wiendl, H., Bien, C. G., Bernasconi, P., Fleckenstein, B., Elger, C. E., Dichgans, J. et al. (2001). GluR3 antibodies: Prevalence in focal epilepsy but no specificity for Rasmussen's encephalitis. *Neurology* **57**, 1511–1514.

Williams, P. A., Dou, P., Dudek, F. E. (2004). Epilepsy and synaptic reorganization in a perinatal rat model of hypoxia-ischemia. *Epilepsia* **45**, 1210–1218.

Williamson, A., Patrylo, P. R., Lee, S., Spencer, D. D., Williamson, A., Patrylo, P. R. et al. (2003). Physiology of human cortical neurons adjacent to cavernous malformations and tumors. *Epilepsia* **44**, 1413–1419.

Willmore, L. J., Sypert, G. W., Munson, J. B. (1978a). Recurrent seizures induced by cortical iron injection: A model of posttraumatic epilepsy. *Annals of Neurology* **4**, 329–336.

Willmore, L. J., Sypert, G. W., Munson, J. V., Hurd, R. W. (1978b). Chronic focal epileptiform discharges induced by injection of iron into rat and cat cortex. *Science* **200**, 1501–1503.

Wilson, D. N., Chung, H., Elliott, R. C., Bremer, E., George, D., Koh, S. (2005). Microarray analysis of postictal transcriptional regulation of neuropeptides. *Journal of Molecular Neuroscience* **25**, 285–298.

Wolf, H. K., Roos, D., Blumcke, I., Pietsch, T., Wiestler, O. D. (1996). Perilesional neurochemical changes in focal epilepsies. *Acta Neuropathologica* **91**, 376–384.

Wuarin, J. P., Dudek, F. E. (1996). Electrographic seizures and new recurrent excitatory circuits in the dentate gyrus of hippocampal slices from kainate-treated epileptic rats. *Journal of Neuroscience* **16**, 4438–4448.

Yang, J., Woodhall, G. L., Jones, R. S., Yang, J., Woodhall, G. L., Jones, R. S. G. (2006). Tonic facilitation of glutamate release by presynaptic NR2B-containing NMDA receptors is increased in the entorhinal cortex of chronically epileptic rats. *Journal of Neuroscience* **26**, 406–410.

Yue, C., Remy, S., Su, H., Beck, H., Yaari, Y. (2005). Proximal persistent Na+ channels drive spike after depolarizations and associated bursting in adult CA1 pyramidal cells. *Journal of Neuroscience* **25**, 9704–9720.

Zarrelli, M. M., Beghi, E., Rocca, W. A., Hauser, W. A. (1999). Incidence of epileptic syndromes in Rochester, Minnesota: 1980–1984. *Epilepsia* **40**, 1708–1714.

Zhang, G., Raol, Y. S., Hsu, F. C., Brooks-Kayal, A. R. (2004a). Long-term alterations in glutamate receptor and transporter expression following early-life seizures are associated with increased seizure susceptibility. *Journal of Neurochemistry* **88**, 91–101.

Zhang, G., Raol, Y. H., Hsu, F. C., Coulter, D. A., Brooks-Kayal, A. R. (2004b). Effects of status epilepticus on hippocampal GABAA receptors are age-dependent. *Neuroscience* **125**, 299–303.

24

Genetic Epilepsies

Ingo Helbig, Ingrid E. Scheffer, and Samuel F. Berkovic

I. Introduction: A Genetic Approach to Seizure Disorders

> I am about to discuss the disease called sacred. It is not, in my opinion, any more divine or more sacred than any other diseases, but has a natural cause … Its origin, like that of other diseases, lies in heredity.
>
> —Hippocrates of Kos 470–410 BC (Riggs & Riggs, 2005)

Seizures are the result of a pathological hypersynchrony within the central nervous system. Epilepsy is diagnosed once a patient has suffered at least two unprovoked seizures. With a cumulative lifetime incidence of 2 percent or more, epilepsies are among the most common neurological disorders.

Genetic research has an important role in understanding seizure disorders. In general, the etiology of the epilepsies is multifactorial, with interacting genetic and environmental factors. Nevertheless, from a population perspective, genetic factors explain a significant proportion of the variation in the tendency to develop seizures. This is true even in seizure syndromes that superficially appear environmental such as febrile seizures or posttraumatic epilepsy. Genetic variations are probably the most important risk factor for seizure disorders and understanding how genes translate into phenotypes will give important insight into the biology of the epilepsies. This research is the prerequisite for the development of novel diagnostic and therapeutic strategies, particularly because current treatment options are often insufficient.

A. Many Genes Cause Seizures but Few Are "Epilepsy Genes"

More than 300 Mendelian disorders are known that have seizures as associated features (Online Mendelian Inheritance in Man). In addition, seizures are common in genetic brain malformations and many chromosomal aberrations. Even though remarkable progress has been made in understanding the biological consequences of the gene mutations involved, these findings have contributed only little to painting a coherent picture of the pathophysiology of human epilepsy per se. The explanation for this discrepancy is at hand. Seizures can be an outcome of many different pathogenic processes within the brain.

The focus of this chapter will be the idiopathic epilepsy syndromes. Idiopathic seizure disorders arise spontaneously without any extrinsic cause. Idiopathic seizure disorders are epilepsies in their purest form.

B. Not All Epilepsies Are the Same: Classifying Epilepsies

Epilepsy syndromes are traditionally conceptualized according to the double dichotomy of partial versus generalized and symptomatic versus idiopathic epilepsies in the classification of the International League Against Epilepsy (Wolf, 2005). In partial epilepsies, seizures arise from a relatively restricted cortical focus; in generalized epilepsies, both hemispheres are involved in seizure generation.

Symptomatic seizure disorders are due to an identifiable external (i.e., brain injury) or internal cause (metabolic or chromosomal disorder). Idiopathic epilepsies, in contrast, occur in the absence of any identifiable causative factor and usually are not associated with neurological deficits. These disorders are considered as essentially genetic.

Gene findings in monogenic idiopathic epilepsy syndromes have challenged the distinction into these subgroups. Dravet syndrome (formerly Severe Myoclonic Epilepsy of Infancy) shares features of symptomatic epilepsies because it is associated with intellectual impairment and neurological features such as ataxia or spasticity. It is, however, part of the broader spectrum of channelopathies, which are the molecular hallmark of the idiopathic epilepsies. Therefore, Dravet syndrome will also be discussed in this chapter.

C. Not All Epilepsies Are the Same: The Importance of Phenotyping

In addition to the broad subdivisions just discussed, there are over 50 epilepsy syndromes described. In attempting to dissect out the molecular determinants of these, careful clinical analysis is required. There can be cooccurrence of different syndromes in one family, and clinical genetic study is required to consider whether the syndromes are genetically related, or are occurring together by chance as epilepsy is reasonably common. Focusing on the clinical details of epilepsy syndromes may appear disproportionate when trying to understand basic molecular principles of human seizure disorders, particularly in the age of rapidly evolving technologies for genetic research. However, it was detailed clinical analysis that has enabled the molecular discoveries to date.

For example, if heterogeneous phenotypes that are included are genetically unlikely to be related to the familial mutation when analyzing a large apparently monogenic pedigree, linkage analysis may fail to identify the correct critical region. As the mutation status is not known *a priori*, this creates a dilemma for the investigator. Excluding cases reduces power, yet including them may mislead the investigation. This dilemma is resolved by careful clinical research on similar pedigrees whereby the most likely scenario regarding the clinical genetic relationship between different syndromes can be identified. Consequently, sufficient resources need to be allocated to clinical phenotyping in genetic studies of seizure disorders.

D. Complicating Matters: Phenotypic and Genetic Heterogeneity

Many familial idiopathic epilepsy syndromes demonstrate considerable phenotypic and genetic heterogeneity (see Figure 24.1). Genetic heterogeneity is present when similar phenotypes are caused by mutations in different genes. Phenotypic heterogeneity, in contrast, refers to the variability in the phenotype that is caused by mutations in the same gene. Phenotypic variability is present in some monogenic epilepsy syndromes such as the Generalized Epilepsy with Febrile Seizures Plus (GEFS+) spectrum (Scheffer & Berkovic, 1997). In other monogenic epilepsy syndromes such as Benign Familial Neonatal Seizures (BFNS) or Autosomal Dominant Partial Epilepsy with Auditory Features (ADPEAF), the phenotype is relatively uniform and clinically distinct (Biervert et al., 1998b; Kalachikov et al., 2002). The factors responsible for phenotypic heterogeneity within monogenic syndromes are not understood.

II. Familial Idiopathic Epilepsy Syndromes

A. Disorders of Infancy

Case Study 1: "Suddenly, his eyes rolled upward and he seized …"

D.S. was four months when he had his first seizure. D.S.'s eyes suddenly rolled upward, his face became pale and his jaw started trembling. D.S. was floppy and unresponsive for about two minutes. His parents rushed him to the nearby Emergency Department where he had the next event 20 minutes later. The presentation of this event was similar and D.S. had a third event later the same day.

A cluster of seizures in an infant is alarming. It can be the presentation of an acute life-threatening illness or a sign of an underlying structural brain malformation. In D.S.'s case, the events occurred in the absence of fever and no provocative factor could be elucidated. Numerous investigations including structural imaging and electroencephalography revealed no abnormalities. D.S. was treated with antiepileptic medication and had no further seizures.

D.S. was seen by a pediatric neurologist at the age of 11 months and presented as a developmentally normal child. Evaluation of the family history revealed that D.S.'s paternal

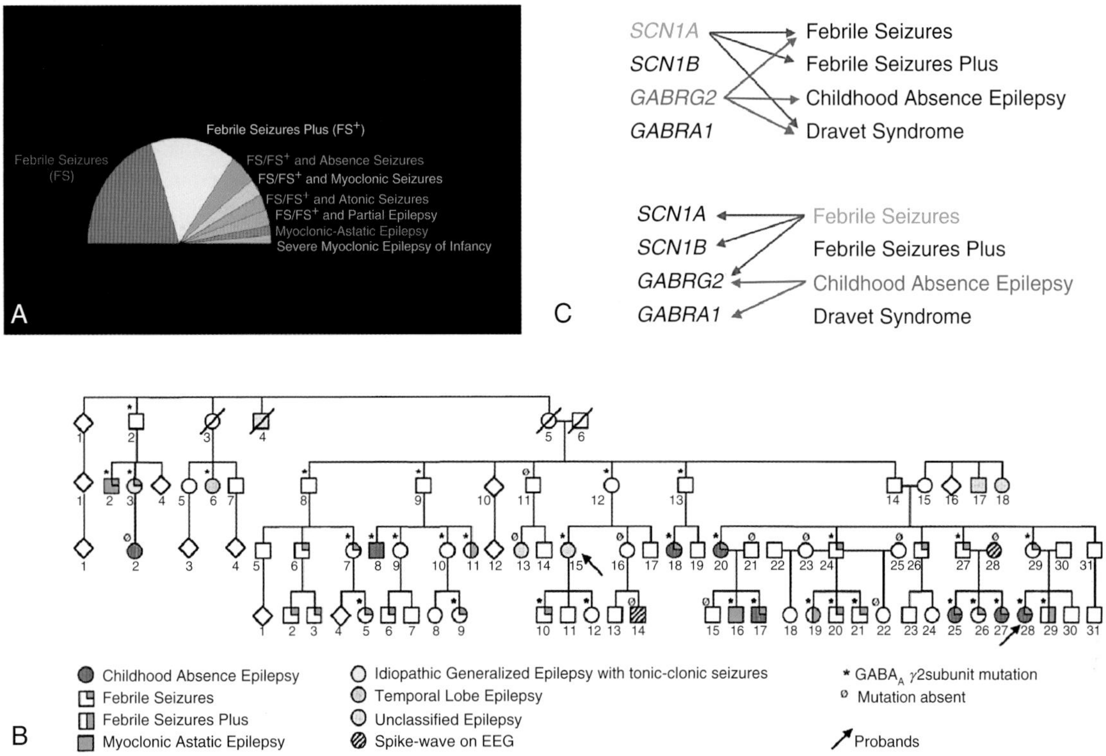

Figure 24.1 Phenotypic and Genetic Heterogeneity in Familial Idiopathic Epilepsy Syndromes. **A**. The spectrum of phenotypes caused by mutations in *SCN1A* (Mulley et al., 2005a). Familial mutations in *SCN1A* are responsible for a wide variety of phenotypes from self-limiting Febrile Seizures to the severe epilepsy syndromes of Myoclonic Astatic Epilepsy (MAE) and Dravet Syndrome. **B**. Pedigree of a large family with a mutation in *GABRG2* (Marini et al., 2003). The mutations in *GABRG2* result in Febrile Seizures and Childhood Absence Epilepsy (CAE). **C**. Schematic representation of phenotypic and genetic heterogeneity. Mutations in *SCN1A* or *GABRG2* can cause a multitude of different phenotypes (*phenotypic* heterogeneity). Likewise, a specific epilepsy syndrome such as Febrile Seizures or Childhood Absence Epilepsy can be caused by defects in different genes (*genetic* heterogeneity).

grandmother had suffered from similar events at the age of two to three months. Molecular genetic analysis revealed a mutation in the SCN2A gene and confirmed the diagnosis of Benign Familial Neonatal-Infantile Seizures (BFNIS).

Idiopathic seizure syndromes in infancy challenge classification into either partial or generalized seizure syndromes as the immature central nervous system of the neonate and infant is not capable of generating truly generalized seizures. Seizures in these syndromes usually have focal features and occur in a previously well and developmentally normal infant. These disorders are characterized by clusters of seizures that usually last only for a few days or rarely weeks. Prognosis in these disorders is good, but a minority of patients continue to have febrile seizures or epilepsy. The familial idiopathic seizure disorders of infancy consist of three autosomal-dominant disorders, two of which are known to be caused by ion channel mutations. Clinically, these disorders have been separated by their age of onset.

1. Benign Familial Neonatal Seizures (BFNS)

In this syndrome a previously healthy neonate suddenly develops a flurry of seizures over a few days, which then spontaneously settle. Focal features in these convulsions are usually present, which sometimes prompts structural imaging to exclude lesions or developmental abnormalities.

Benign Familial Neonatal Seizures literally have shaped the history of epilepsy genetics as they have laid the foundation for the channelopathy concept of idiopathic epilepsies. Rett and Teubel (1964) first described a familial form of Benign Neonatal Seizures in 1964, which followed an autosomal dominant inheritance pattern. Analysis by Leppert and coworkers (1989) then resulted in the first successful linkage in idiopathic epilepsies. The causative genes eventually were identified as *KCNQ2* and *KCNQ3*, coding for central nervous system M-type potassium channels (Biervert et al., 1998a; Charlier et al., 1998; Singh et al., 1998). Most families with BFNS have mutations in *KCNQ2*, whereas *KCNQ3* mutations have been detected only in rare families. In addition to the described familial cases, *de novo* mutations in *KCNQ2* have been identified in sporadic cases (Claes et al., 2004).

BFNS usually remits in the first year of life, but additional clinical features have been described in several families with *KCNQ2/KCNQ3* potassium channel mutations. Five percent of infants go on to develop Febrile Seizures and 11 percent later develop frank epilepsy (Plouin & Anderson, 2002). Although these small numbers seem reassuring in the light of the sudden and frightening presentation of this condition, it emphasizes the role of M-type potassium channels in the regulation of central nervous system excitability. Identification of the genetic and nongenetic factors that confer risk for a later seizure disorder in these benign conditions will give valuable insight into gene–gene or gene–environment interaction in monogenic disorders.

Myokymia has been reported in a single family with BFNS (Dedek et al., 2001). M-type potassium channels are key regulators of membrane excitability in the node of Ranvier in addition to the axon initial segment in central neurons (Devaux et al., 2004). This highlights that ion channel disorders in humans are a common pathophysiological principle that result in different clinical conditions based upon spatial and temporal expression of the mutated channel.

2. Benign Familial Neonatal-Infantile Seizures (BFNIS)

This disorder is challenging on a conceptual level. BFNIS is a family diagnosis in families where both neonatal and infantile seizure onsets are observed and was first described by Kaplan and Lacey in 1983 (Kaplan & Lacey, 1983). Whereas the subdivision between BFNS, BFNIS, and Benign Familial Infantile Seizures (BFIS) appears artificial at first glance, it has proven crucial in identifying the underlying causative gene. In 2002, mutations in *SCN2A* coding for the alpha 2 subunit of the neuronal sodium channel were reported in families with BFNIS (Heron et al., 2002). To date, several families with mutations in *SCN2A* have been described, including the original North American Family by Kaplan and Lacey (Berkovic et al., 2004). In contrast to the predictable age of onset in families with BFNS, BFNIS families show remarkable variability. Onset may vary from 2 days to 6 months with a mean onset of 11 weeks. In contrast to BFNS, Febrile Seizures and Epilepsy are rarely sequelae of BFNIS.

Mutations in *SCN2A* have been sought in a variety of epilepsy syndromes and BFNIS is the only condition where mutations have been regularly identified. A single patient has been reported with a severe epilepsy syndrome and a mutation in *SCN2A* (Kamiya et al., 2004).

3. Benign Familial Infantile Seizures (BFIS)

This syndrome usually starts between four and eight months and typically presents with focal features such as head and eye deviation or unilateral clonic features that may alternate sides between events. As in BFNS, the paroxysmal events occur in clusters and settle spontaneously.

Even though families with autosomal inheritance were described in 1992 (Vigevano et al., 1992), the causative gene in this disorder is still elusive. Linkage to a large pericentromeric region on chromosome 16 has been described, but identification of the underlying mutation has been unsuccessful, possibly due to a number of gene duplications in this region.

Additionally, BFIS can occur in association with other paroxysmal disorders of the central nervous system. Infantile Convulsions and Choreoathetosis is a well-described entity that is consistent with autosomal dominant inheritance (Szepetowski et al., 1997). Furthermore, mutations in *ATP1A2* have been described in a family with hemiplegic migraine and benign infantile seizures (Vanmolkot et al., 2003). Both findings demonstrate a relationship between apparently distinct idiopathic paroxysmal disorders of the central nervous system.

4. Molecular Mechanism in Idiopathic Infantile Seizures Syndromes

BFNS, BFNIS, and BFIS share two striking features: They have clinical onsets in specific time windows and are self-limiting. The mechanisms responsible for this phenomenon are largely unknown.

In vitro functional studies suggest that the mutations impair channel function through various mechanisms. For example, in BFNS, impaired potassium-dependent repolarization can be caused by a reduction in the maximal current carried by the KCNQ2/KCNQ3 M-channels (Lerche et al., 1999) or through slower opening and faster closing kinetics in combination with a decreased voltage sensitivity (Castaldo et al., 2002). Mutations in *SCN2A* in BFNIS have been found to increase subthreshold and action sodium currents through various mechanisms (Scalmani et al., 2006). These changes led to hyperexcitability in cultured pyramidal and bipolar neocortical neurons.

In addition, KCNQ2/KCNQ3 heteromers as well as SCN2A possess Ankyrin-G binding domains that might be involved in targeting to the axon initial segment (Pan et al., 2006). BFNS mutations in *KCNQ2/KCNQ3* inhibit selective targeting to the axon, suggesting that impairment of intracellular transport may contribute to the aetiology of BFNS (Chung et al., 2006). Disturbance of ion channels at the axon initial segment might be the common pathophysiological mechanism for both BFNS and BFNIS that sets these disorders apart from other channelopathies.

B. Febrile Seizures and Related Disorders

Case Study 2: "It was the vaccination that caused it …"

O.G. had his first seizure nine hours after his third triple vaccination (pertussis, diphtheria, and tetanus) at six months of age. He was prone to prolonged seizures thereafter. At eight months, he presented with status epilepticus with fever, which lasted for two hours. At nine months, a right hemiclonic seizure lasted for 20 minutes. In his first two years of life, O.G. had 15 episodes of status epilepticus. His development plateaued and he showed regression after prolonged episodes of status. At the age of 15 months, O.G. started having myoclonic jerks and absence seizures. Partial seizures were also noted.

The family and their family doctor firmly believed that the triple vaccination at six months had caused O.G.'s debilitating seizure disorder. It had not. O.G. and his parents saw a pediatric neurologist at the age of 2 years and he was diagnosed with Dravet Syndrome (Severe Myoclonic Epilepsy of Infancy SMEI). Mutational analysis revealed a de novo truncation mutation in the SCN1A gene.

Dravet Syndrome sometimes masquerades as so-called "vaccine encephalopathy" because the clinical onset coincides with the vaccination (Berkovic et al., 2006a). Proper clinical analysis rather than false attribution is necessary to determine the correct diagnosis and provide the child with optimal treatment options.

1. Generalized Epilepsy with Febrile Seizures Plus (GEFS+)

Febrile Seizures occur in 3 percent of all children between six months and six years of age and are the most common seizure disorders in man. A considerable genetic contribution to Febrile Seizures has been known for decades and families with monogenic inheritance are well described. Despite this favorable setting for gene discoveries, genes for Febrile Seizures have been elusive.

It took a conceptual advance to identify the first genes for Febrile Seizures, the recognition of the familial syndrome of Generalized Epilepsy with Febrile Seizure Plus (GEFS+) (Scheffer and Berkovic, 1997). In families with GEFS+ there is a spectrum of seizure syndromes in individuals including Febrile Seizures, Febrile Seizures Plus (FS+), as well as FS+ with other seizure types. FS+ refers to subjects who have seizures with fever that occur outside the age boundaries of six months to six years or are accompanied by afebrile convulsions. Under the assumption that these somewhat variable phenotypes within large pedigrees are caused by the same genetic defect, several genes for GEFS+ have been identified.

To date, mutations in *SCN1A*, *SCN1B*, and *GABRG2* have been identified in multiple families with GEFS+ (Baulac et al., 2001; Escayg et al., 2000b; Wallace et al., 1998, 2001). Missense mutations in *SCN1A* are the most frequent genetic alterations found to date, but account for only a minority of GEFS+ families.

2. Febrile Seizures

Gene findings in GEFS+ have allowed for inferences about the genetic alterations in families with simple Febrile Seizures without associated GEFS+ features. A family with pure Febrile Seizures and a mutation in *SCN1A* has been described (Mantegazza et al., 2005), and a mutation in *GABRG2* has been found that segregates with Febrile Seizures (Audenaert et al., 2006). Whereas these findings suggest that familial Febrile Seizures in these families may be regarded as a milder version of GEFS+ on a molecular level, it must be cautioned that despite these initial findings, most genes causing Febrile Seizures in families as well as in sporadic patients are unknown.

3. Dravet Syndrome: Severe Myoclonic Epilepsy of Infancy (SMEI)

The patient described in the case vignette suffers from Dravet Syndrome, formerly known as Severe Myoclonic Epilepsy of Infancy. Dravet Syndrome presents in the first year of life in a previously healthy infant. Seizures usually are triggered by fever and febrile status is common. The child develops afebrile seizures with different seizure types in the next few years, including partial, absence, and myoclonic seizures. Intellectual development is initially normal, but regression then occurs with resulting considerable intellectual impairment (Dravet et al., 2002).

Approximately 70 percent of children with Dravet Syndrome have mutations of the *SCN1A* gene and this syndrome demarcates the severe end of the GEFS+ spectrum. Mutations occur *de novo* in 95 percent of cases (Mulley et al., 2005b). It is considered a rare clinical syndrome, but is probably underdiagnosed.

4. Phenotypic Heterogeneity in Sodium Channelopathies

Epileptic sodium channelopathies are characterized by considerable phenotypic heterogeneity. Mutations in *SCN1A* can cause several Febrile Seizure-related syndromes, spanning from simple Febrile Seizures to the devastating clinical picture of Dravet Syndrome. What determines whether a child with a mutation in *SCN1A* presents with self-limiting Febrile Seizures or progresses to a severe, disabling epilepsy?

About half the children with Dravet Syndrome have truncation mutations. Missense mutations occur with simple Febrile Seizures, GEFS+, and some children with Dravet Syndrome. Presumably the site and type of missense mutation is important, but genotype–phenotype correlations have not yet been disentangled (see Figure 24.2). Furthermore, in some but not all studies, there is a higher than expected frequency of Febrile Seizures and epilepsy in relatives of patients with *de novo* mutations in *SCN1A*, suggesting that Dravet Syndrome is the result of an *SCN1A* mutation on an "epileptogenic" genetic background.

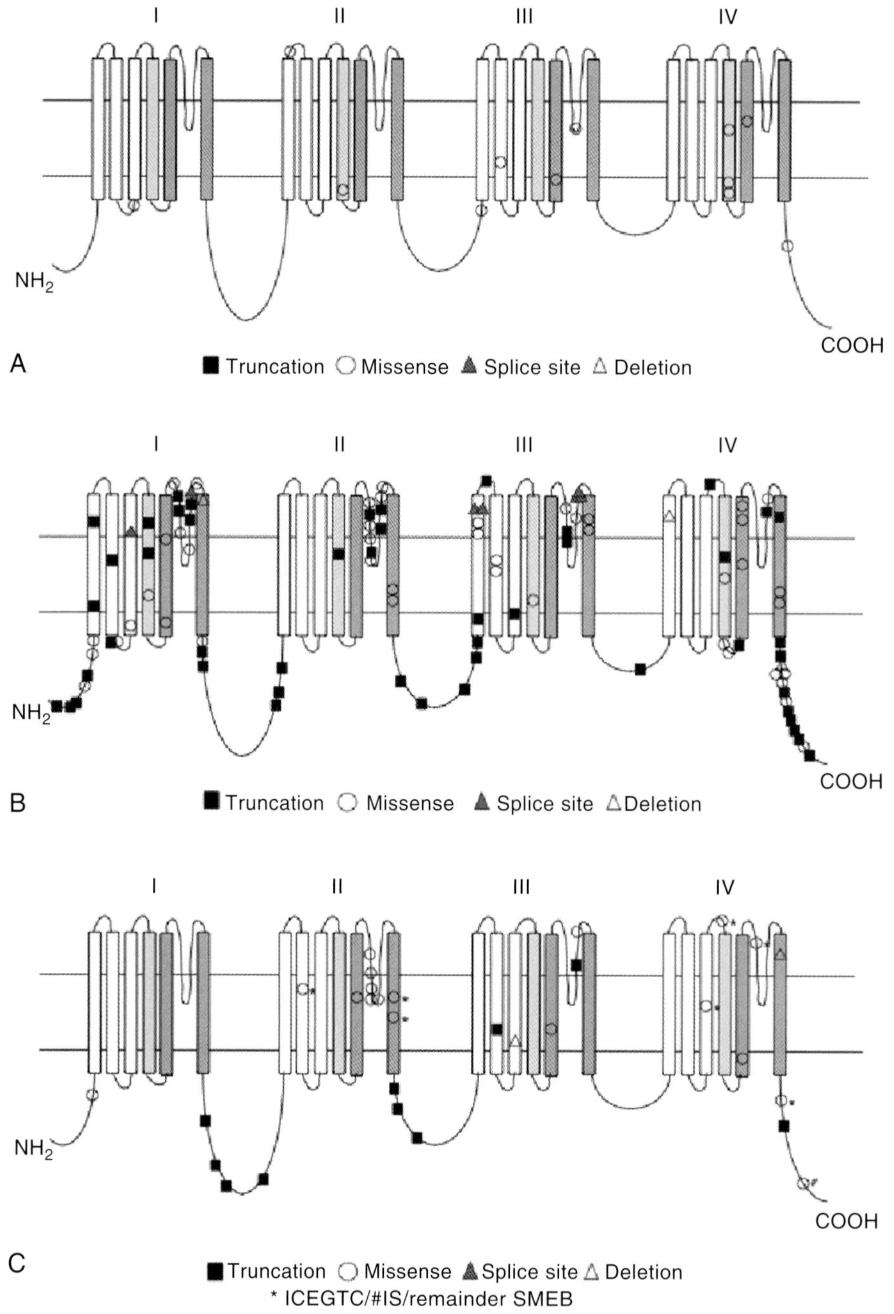

Figure 24.2 Genotype-phenotype correlation in genetic epilepsies caused by mutations in *SCN1A* (Mulley et al., 2005b). **A**. SCN1A mutations found in pedigrees with GEFS+. In these families, only missense mutations can be detected. **B**. Mutations found in patients with Dravet Syndrome (Severe Myoclonic Epilepsy of Infancy SMEI). The mutations found in Dravet Syndrome are mainly truncation mutations. However, missense mutations are also detected, which does not allow a clear distinction from the mutations found in GEFS+. The presence of deletions and splice site mutations further complicate the picture. **C**. Mutations found in other severe epilepsy syndromes caused by mutations in *SCN1A*. These disorders emcompass ICEGTC (Intractable Childhood Epilepsy with Generalised Tonic Clonic Seizures), Infantile Spasms, and SMEB (Severe Myoclonic Epilepsy Borderland). In these conditions, truncation mutations, deletions as well as missense mutations can be found.

5. Why Does Fever Trigger Seizures?

Despite the fact that Febrile Seizures are the most common seizures in humans, little is known about the mechanisms that link the rise of body temperature with the tendency to convulse. Research has concentrated on two starting points to investigate the pathophysiology of this phenomenon; the cytokines associated with the inflammatory response, and the ion channels that have been identified in GEFS+ families.

Proinflammatory cytokines are important pyrogens and mediate the elevation of the body temperature in fever. In Febrile Seizures a role for Interleukin-1 beta (IL-1β) has been suggested. Mice lacking the IL-1β receptor are resistant to Febrile Seizures despite an unaltered cytokine profile. In addition, high doses of IL-1β can generate seizures in the absence of fever (Dube et al., 2005). This suggests that IL-1β signaling might—at least partly—be the link between hyperthermia and neuronal hyperexcitability.

Despite this evidence in an animal model of Febrile Seizures, translation to the situation in humans has proven difficult. It is debated whether children with Febrile Seizures have an increased IL-1β production compared to controls (Haspolat et al., 2002; Lahat et al., 1997; Matsuo et al., 2006; Tutuncuoglu et al., 2001; Virta et al., 2002b). Furthermore, despite initial reports of a positive association of Febrile Seizures with polymorphisms in the Interleukin-1 beta gene (*IL-1β*) and the Interleukin-1 receptor antagonist gene (*IL-1RN*) (Tsai et al., 2002; Virta et al., 2002a), these findings could not be confirmed (Haspolat et al., 2005; Matsuo et al., 2006).

A different line of thinking about the pathogenesis of Febrile Seizures concentrates on a possible dysfunction of ion channels in the setting of elevated temperature. Preliminary evidence suggests that GEFS+ associated mutations in *GABRG2* result in a temperature-dependent block of intracellular transport, which ultimately leads to a reduced cell–surface expression of the GABA receptor subunit (Kang et al., 2006). Although this finding requires further confirmation, it raises an interesting principle for ion channel dysfunction in the setting of fever.

C. The Idiopathic Generalized Epilepsies

Case Study 3: Constance and Kathryn

Constance and Kathryn are identical twins with Childhood Absence Epilepsy (CAE). The EEG trace shows the characteristic 3 Hz generalized spike-wave activity. Initially investigated by Dr. William Lennox in the 1950s, Constance and Kathryn have become the faces of epilepsy genetics. They developed absence seizures at the age of six years and had a very similar clinical course. Twin studies elegantly demonstrate the importance of genetic factors in seizure disorders. Among the other epilepsy syndromes, the Idiopathic Generalized Epilepsies including Childhood Absence Epilepsy stand out because of their strong if not exclusive genetic component (see Figure 24.3).

1. Childhood Absence Epilepsy (CAE)

Childhood Absence Epilepsy typically presents with typical absence seizures in primary school children. Absence seizures (previously known as *petit mal*) occur many times a day and present as staring spells. The diagnosis is made in conjunction with the EEG recording, which shows the characteristic 3 Hz generalized spike-wave activity. Data from twin and family studies strongly suggest that CAE is a genetic disorder and no evidence for environmental factors has been found.

Despite the strong clinical evidence for a genetic causation, finding genes has been difficult because most cases of CAE are not caused by a single major gene. However, several genes have been identified that contribute to the complex genetic architecture of CAE. Mutations in *GABRG2* coding for the gamma 2 subunit of the GABA(A) receptor have been found in pedigrees with Febrile Seizures and CAE (Wallace et al., 2001). Mutations in the *CLCN2* gene, coding for the neuronal chloride channel CIC-2, are associated with different syndromes within the IGE spectrum including CAE (Haug et al., 2003). Mutations in *CACNA1A* cause several neurological disorders including spinocerebellar ataxia type 6 (SCA-6), episodic ataxia type 2, and familial hemiplegic migraine. Mutations in *CACNA1A* were also found in a family with pure CAE (Popa et al., 2005). In addition, a mutation in *JRK/JH8* has been detected in a single patient with CAE evolving to Juvenile Myoclonic Epilepsy. *JRK/JH8* is the homologue of the mouse *jerky* gene (Moore et al., 2001). A *de novo* mutation in *GABRA1* coding for the alpha 1 subunit of the GABA(A) receptor has been identified in a single patient with CAE (Maljevic et al., 2006). It seems unlikely that these findings are relevant to the majority of children with this common epilepsy.

2. Juvenile Myoclonic Epilepsy (JME)

Juvenile Myoclonic Epilepsy is the most important adolescent syndrome among the Idiopathic Generalized Epilepsies. This disorder is characterized by myoclonic jerks that characteristically occur on awakening as well as generalized tonic clonic seizures. Whereas JME is responsive to antiepileptic medication in the vast majority of cases, it is considered a life-long condition that requires long-term pharmacotherapy. The genetic relationship between CAE and JME within the Idiopathic Generalized Epilepsies is complex. Evidence points toward shared as well as distinct genetic influences in these subsyndromes (Winawer et al., 2003; Winawer et al., 2005). CAE can evolve to JME and there is considerable overlap in multiplex families. Besides the genes implicated in CAE as well as JME, several large pedigrees with pure JME have been described. Mutations in *GABRA1* and *EFHC1* have been identified

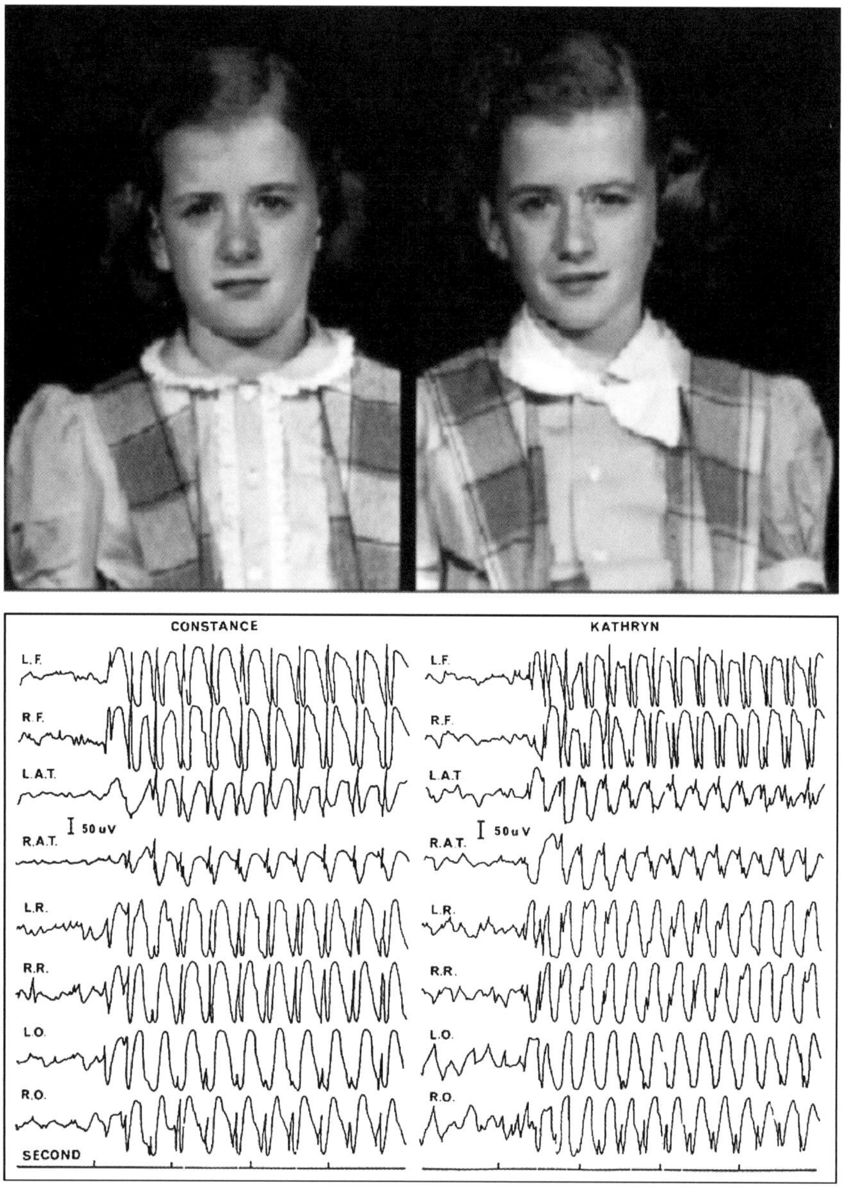

Figure 24.3 Constance and Kathryn (Lennox, 1960). Constance and Kathryn are identical twins with Childhood Absence Epilepsy (CAE). The EEG trace shows 3 Hz generalized spike-wave activity and is almost identical in the twins; "as alike as fingerprints." Identical twins with Idiopathic Generalized Epilepsy are mostly concordant (both twins affected) in contrast to nonidentical twins. This emphasizes the importance of genetic factors in IGE.

in large pedigrees with JME (Cossette et al., 2002; Suzuki et al., 2004). Whereas *GABRA1* codes for the alpha 1 subunit of the GABA(A) receptor, *EFHC1* codes for an EF-hand domain protein of unknown function. A truncation mutation in *CACNB4* has been identified in a patient with JME (Escayg et al., 2000a), but the significance of this is uncertain. *CACNB4* is the human orthologue of the *lethargic* gene, which leads to epilepsy and ataxia in mice.

3. The Genetics of Generalized Spike Wave—An Outlook

The thalamocortical system is a major circuit of the human central nervous system. Generalized spike-wave activity, the electrographic hallmark of the Idiopathic Generalized Epilepsies, is the consequence of a functional disturbance of this network on different levels (Blumenfeld, 2005).

So how do mutations in *GABRG2*, *GABRA1*, and *EFHC1* contribute to the pathogenesis of Idiopathic Generalized Epilepsy? The effect of mutations involved in Idiopathic Generalized Epilepsy has been investigated mostly at a molecular level, suggesting alteration of channel function through several mechanisms involving ion channel conductances, kinetics, or intracellular transport. Little work has been done to put these findings in context at a system level to understand the effect of a pathogenic mutation within the neuronal network. The complex network activity within the thalamocortical system can be modeled in simulations, which integrate electrophysiological findings into a coherent picture (Traub et al., 2005). Enriching these models with data from human mutations might shed light onto the underlying disturbances. Whereas the involvement of GABA receptors indicate a dysfunction of GABAergic transmission, the nature and pathophysiology of this disturbance is not well understood. Involvement of non-ion channel genes such as *EFHC1* raise the possibility that other—still unknown—pathways are involved in the pathological thalamocortical hypersynchrony seen in Idiopathic Generalized Epilepsies.

D. The Idiopathic Focal Epilepsies

Case Study 4: "Can't you hear the radio?"

S.W. was spending an afternoon at his girlfriend's house when he suddenly heard a "radio sound" in his right ear. He asked his girlfriend whether she was aware of this sound as well, but he could not understand her answer. S.W. then lost consciousness and convulsed. He presented with a second seizure several months later. S.W. was started on carbamazepine and remained seizure free.

Interestingly, his mother had seizures starting at the same age. Furthermore, his sister had her first seizure at the age of eight years and frequently reported auras where she would see colorful pictures and hear noises "like a familiar song." These episodes sometimes occurred seconds before she had a generalized tonic-clonic seizure.

S.W. and his family have the familial syndrome of Autosomal Dominant Partial Epilepsy with Auditory Features (ADPEAF), caused by mutations in the LGI1 gene. ADPEAF is a fascinating condition that presents with partial seizures in the lateral temporal lobe that are perceived as auditory and sometimes visual features. D'Orsi and Tinuper recently hypothesized that the voices and visions of Jean d'Arc, the maid of Orleans, share a certain similarity with this syndrome (d'Orsi & Tinuper, 2006).

1. Autosomal Dominant Nocturnal Frontal Lobe Epilepsy (ADNFLE)

Vigorous nocturnal motor seizures, often with retained awareness, characterize the rare familial syndrome of ADNFLE, which can be misdiagnosed as a sleep disturbance. ADNFLE was the first familial epilepsy syndrome to be understood at a molecular genetic level and three genes for ADNFLE have been established. *CHRNA4*, *CHRNB2*, and *CHRNA2* code for subunits of the neuronal nicotinic acetylcholine receptor (Aridon et al., 2006; Fusco et al., 2000; Phillips et al., 2001; Steinlein et al., 1995). It also has been suggested that mutations of *CRH*, the gene for the corticotropin-releasing hormone, may play a role (Combi et al., 2005).

The thalamus and cortex are innervated by cholinergic neurons arising from the brainstem and basal forebrain. An imbalance between excitation and inhibition that is pronounced in sleep is thought to underlie the pathogenesis of ADNFLE.

2. Autosomal Dominant Partial Epilepsy with Auditory Features (ADPEAF)

Temporal Lobe Epilepsy is the most common epilepsy of adulthood, and despite a considerable genetic impact, few genetic alterations are known. ADPEAF is a familial temporal lobe epilepsy syndrome that is characterized by auditory features as described in the case vignette. Mutations in *LGI1* coding for the leucine-rich, glioma inactivated 1 protein have been identified in families with ADPEAF (Kalachikov et al., 2002). The role of the putative tumor suppressor gene *LGI1* in the pathogenesis of human epilepsy remains poorly understood. However, increasing evidence suggests that the LGI1 protein is involved in presynaptic function and possibly constitutes a potassium channel subunit (Fukata et al., 2006; Schulte et al., 2006). Even though it remains unsolved why dysfunction of the LGI1 protein can lead to a specific and localized syndrome such as ADPEAF, it provides the first molecular insight into the biology of Temporal Lobe Epilepsy. A truncation mutation in *KCND2* coding for the voltage gated potassium channel Kv4.2 has been identified in a patient with Temporal Lobe Epilepsy (Singh et al., 2006) and Temporal Lobe Epilepsy is an occasional phenotype of familial GEFS+ (Scheffer et al., 2007). These findings further suggest that genetic alterations in ion channels or ion channel related genes might predispose to Temporal Lobe Epilepsy.

Familial Temporal Lobe Epilepsy also has been described as a feature of choreoancanthocytosis in families with mutations in the chorein gene *VPS13A* (Al-Asmi et al., 2005).

III. The Channelopathy Concept of Idiopathic Epilepsies

A. The Eye of the Needle for Signal Transduction: The Neuronal Cell Membrane

Even though it had been known for a long time that membrane biology was critical to signal transduction in the brain, it was not clear how individual genetic defects might result in seizure disorders. The discovery of ion channel mutations in human epilepsy provided the explanation. The term *channelopathy*, first used by Ptacek for hyperkalemic periodic paralysis (Ptacek & Fu, 2004; Ptacek et al., 1991) was adopted for this pathogenic mechanism. The beauty of the channelopathy concept lies in the fact that it unites the fields of genetics and electrophysiology. With the working concept of channelopathies in mind, it became possible to reduce the causative factor of certain epilepsies to a single gene, transfer its pathogenic effect to a model system, and investigate the functional consequences.

Related genetic disorders are likely to be caused by genes with related biological function (Oti et al., 2006; van Driel et al., 2006). Whereas the idiopathic epilepsies diverge on a phenotypic and genetic level, they may converge at the level of gene function and gene ontology. Ion channels are the molecular basis of monogenic idiopathic epilepsies and will be the standard that future gene findings will be measured against (see Table 24.1).

Recent evidence suggests that the channelopathy concept might extend to acquired epilepsies where the role of ion channel dysregulation increasingly is recognized. In a model of chronic epilepsy, the increased dendritic excitability in CA1 pyramidal cells was found to be caused by a decreased expression of A-type potassium channels on the cell surface (Bernard et al., 2004). Transcriptional as well as post-translational mechanisms were implicated. Furthermore, ample evidence exists for dysregulation of other ion channels and neurotransmitter receptors in epilepsy models. It is therefore tempting to hypothesize that ion channels represent the central substrate for seizure disorders along the neurobiological spectrum (Berkovic et al., 2006b) (see Figure 24.4). Future research will tell whether the channel-opathy concept of epilepsy will lead to the development of novel diagnostic and therapeutic strategies.

IV. The Next Step: Monogenic Disorders as Model Systems for Common Epilepsies

A. Understanding the Genetic Architecture of Seizure Disorders

In contrast to the monogenic idiopathic epilepsy syndromes, most epilepsies are not caused by a single major gene. Many common epilepsies such as the Idiopathic Generalized Epilepsies are under a strong genetic influence, which is likely due to the additive effect of several susceptibility genes. Identification of susceptibility genes for common epilepsies is the major goal of epilepsy research in the next decade.

A number of genes have been suggested as susceptibility genes for Idiopathic Generalized Epilepsy including *BRD2 (RING3)* coding for the bromodomain containing 2 protein and *ME2* coding for Malic Enzyme 2 (Greenberg et al., 2005; Pal et al., 2003). *GABRD*, the gene for the delta subunit of the GABA(A) receptor, may be a susceptibility gene for GEFS+ (Dibbens et al., 2004). Whereas confirmatory evidence for a role of these genes is still lacking, other susceptibility genes for Idiopathic Generalized Epilepsy have been found in at least two laboratories. *CACNA1H* coding for the alpha 1H subunit of the voltage-dependent calcium channel and *KCNJ10* coding for the inwardly rectifying potassium channel Kir1.2 may predispose to CAE (Buono

Table 24.1 Some Ion Channels Implicated in Human Idiopathic Epilepsy Syndromes

	Monogenic Disorders	Susceptibility Genes
Monogenic Epilepsies of Infancy		
BFNS	*KCNQ2*	
	KCNQ3	
BFNIS	*SCN2A*	
Febrile Seizures and GEFS+		
Generalized Epilepsy with Febrile Seizures Plus (GEFS+)	*SCN1A*	
	SCN1B	
	GABRG2	
		GABRD
Dravet Syndrome	*SCN1A*	
Idiopathic Generalized Epilepsies		
Childhood Absence Epilepsy (CAE)	*GABRG2*	
	CACNA1A	
		CACNA1H
Juvenile Myoclonic Epilepsy (JME)	*GABRA1*	
Idiopathic Partial Epilepsies		
Autosomal Dominant Nocturnal Frontal Lobe Epilepsy (ADNFLE)	*CHRNA4*	
	CHRNA2	
	CHRNB2	
Autosomal Dominant Partial Epilepsy with Auditory Features (ADPEAF)	*LGI1*	

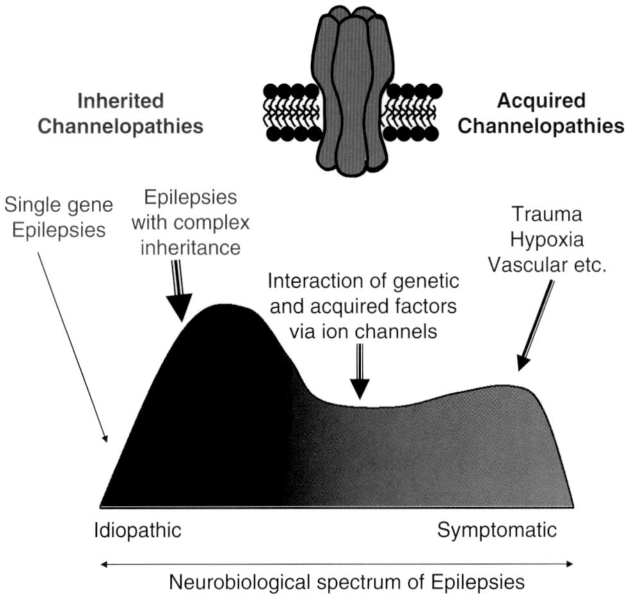

Figure 24.4 The ion channel concept of genetic and acquired epilepsies (Berkovic et al., 2006). The neurobiological spectrum of seizure disorders comprises idiopathic and symptomatic epilepsies. The gray shading represents the genetic component, which is strong in idiopathic epilepsies, but decreases in symptomatic epilepsies. Ion channel dysfunction is inherited mainly in idiopathic epilepsies and acquired mainly in symptomatic seizure disorders. In complex epilepsies, both inherited and acquired ion channel dysfunctions interact.

et al., 2004; Chen et al., 2003; Lenzen et al., 2005; Liang et al., 2006). *CX36* coding for the neuronal gap junction protein Connexin 36 has been reported as a susceptibility gene for JME in two studies (Hempelmann et al., 2006; Mas et al., 2004). The presence of false positive association due to limited sample size and ascertainment bias has long been a major setback in association studies of epilepsy susceptibility genes (Tan et al., 2004). With the realization that sufficient sample size, independent replication, and stringent statistical methods are a prerequisite, more susceptibility genes for IGE will be discovered in the near future. Other common epilepsy syndromes such as Temporal Lobe Epilepsy may follow.

B. Controversies in Epilepsy Genetics: Benign Rolandic Epilepsy

Benign Rolandic Epilepsy is the most common epilepsy of childhood. The characteristic rolandic seizures are partial motor seizures, which affect one side of the face and the upper limb. The electrographic hallmark of Benign Rolandic Epilepsy is the EEG with the characteristic centro-temporal spikes.

The genetics of Benign Rolandic Epilepsy has been an enigma. Whereas families with monogenic inheritance have been reported, additional neurological symptoms such as speech dyspraxia or dystonic features are usually present in these families. In sporadic cases of BRE, the genetic component appears to be small. Identical twins with BRE are consistently discordant, that is, only one twin is affected (Vadlamudi, 2006; Vadlamudi et al., 2004), suggesting that the heritability of BRE is very low.

Nevertheless, the underlying EEG trait of centro-temporal spikes probably has a strong genetic component. The example of BRE illustrates the intricacy between genes, endophenotypes (EEG), environment, and phenotypes in complex seizure disorders with interacting environmental and genetic factors.

The genetic architecture of most seizure disorders is complex and far from being understood. The analysis of monogenic idiopathic seizure disorders has revealed valuable insight into the biology of the epilepsies. The ion channel concept has provided a valuable theoretical foundation to understand the epilepsies on the level of molecular and cellular dysfunction. It is the hope of epilepsy research to translate these findings into new therapies to fight one of the most common disorders of the central nervous system.

References

Al-Asmi, A., Jansen, A. C., Badhwar, A., Dubeau, F., Tampieri, D., Shustik, C. et al. (2005). Familial temporal lobe epilepsy as a presenting feature of choreoacanthocytosis. *Epilepsia* **46**, 1256–1263.

Aridon, P., Marini, C., Di Resta, C., Brilli, E., De Fusco, M., Politi, F. et al. (2006). Increased sensitivity of the neuronal nicotinic receptor alpha 2 subunit causes familial epilepsy with nocturnal wandering and ictal fear. *Am J Hum Genet* **79**, 342–350.

Audenaert, D., Schwartz, E., Claeys, K. G., Claes, L., Deprez, L., Suls, A. et al. (2006). A novel GABRG2 mutation associated with febrile seizures. *Neurology* **67**, 687–690.

Baulac, S., Huberfeld, G., Gourfinkel-An, I., Mitropoulou, G., Beranger, A., Prud'homme, J. F. et al. (2001). First genetic evidence of GABA(A) receptor dysfunction in epilepsy: A mutation in the gamma2-subunit gene. *Nat Genet* **28**, 46–48.

Berkovic, S. F., Harkin, L., McMahon, J. M., Pelekanos, J. T., Zuberi, S. M., Wirrell, E. C. et al. (2006a). De-novo mutations of the sodium channel gene SCN1A in alleged vaccine encephalopathy: a retrospective study. *Lancet Neurol* **5**, 488–492.

Berkovic, S. F., Heron, S. E., Giordano, L., Marini, C., Guerrini, R., Kaplan, R. E. et al. (2004). Benign familial neonatal-infantile seizures: Characterization of a new sodium channelopathy. *Ann Neurol* **55**, 550–557.

Berkovic, S. F., Mulley, J. C., Scheffer, I. E., and Petrou, S. (2006b). Human epilepsies: Interaction of genetic and acquired factors. *Trends Neurosci* **29**, 391–397.

Bernard, C., Anderson, A., Becker, A., Poolos, N. P., Beck, H., and Johnston, D. (2004). Acquired dendritic channelopathy in temporal lobe epilepsy. *Science* **305**, 532–535.

Biervert, C., Schroeder, B. C., Kubisch, C., Berkovic, S. F., Propping, P., Jentsch, T. J., and Steinlein, O. K. (1998). A potassium channel mutation in neonatal human epilepsy. *Science* **279**, 403–406.

Blumenfeld, H. (2005). Cellular and network mechanisms of spike-wave seizures. *Epilepsia* **46 Suppl 9**, 21–33.

Buono, R. J., Lohoff, F. W., Sander, T., Sperling, M. R., O'Connor, M. J., Dlugos, D. J. et al. (2004). Association between variation in the human KCNJ10 potassium ion channel gene and seizure susceptibility. *Epilepsy Res* **58**, 175–183.

Castaldo, P., del Giudice, E. M., Coppola, G., Pascotto, A., Annunziato, L., and Taglialatela, M. (2002). Benign familial neonatal convulsions caused by altered gating of KCNQ2/KCNQ3 potassium channels. *J Neurosci* **22**, RC199.

Charlier, C., Singh, N. A., Ryan, S. G., Lewis, T. B., Reus, B. E., Leach, R. J., and Leppert, M. (1998). A pore mutation in a novel KQT-like potassium channel gene in an idiopathic epilepsy family. *Nat Genet* **18**, 53–55.

Chen, Y., Lu, J., Pan, H., Zhang, Y., Wu, H., Xu, K. et al. (2003). Association between genetic variation of CACNA1H and childhood absence epilepsy. *Ann Neurol* **54**, 239–243.

Chung, H. J., Jan, Y. N., and Jan, L. Y. (2006). Polarized axonal surface expression of neuronal KCNQ channels is mediated by multiple signals in the KCNQ2 and KCNQ3 C-terminal domains. *Proc Natl Acad Sci U S A* **103**, 8870–8875.

Claes, L. R., Ceulemans, B., Audenaert, D., Deprez, L., Jansen, A., Hasaerts, D. et al. (2004). De novo KCNQ2 mutations in patients with benign neonatal seizures. *Neurology* **63**, 2155–2158.

Combi, R., Dalpra, L., Ferini-Strambi, L., and Tenchini, M. L. (2005). Frontal lobe epilepsy and mutations of the corticotropin-releasing hormone gene. *Ann Neurol* **58**, 899–904.

Cossette, P., Liu, L., Brisebois, K., Dong, H., Lortie, A., Vanasse, M. et al. (2002). Mutation of GABRA1 in an autosomal dominant form of juvenile myoclonic epilepsy. *Nat Genet* **31**, 184–189.

d'Orsi, G. and Tinuper, P. (2006). "I heard voices …": from semiology, a historical review, and a new hypothesis on the presumed epilepsy of Joan of Arc. *Epilepsy Behav* **9**, 152–157.

Dedek, K., Kunath, B., Kananura, C., Reuner, U., Jentsch, T. J., and Steinlein, O. K. (2001). Myokymia and neonatal epilepsy caused by a mutation in the voltage sensor of the KCNQ2 K+ channel. *Proc Natl Acad Sci U S A* **98**, 12272–12277.

Devaux, J. J., Kleopa, K. A., Cooper, E. C., and Scherer, S. S. (2004). KCNQ2 is a nodal K+ channel. *J Neurosci* **24**, 1236–1244.

Dibbens, L. M., Feng, H. J., Richards, M. C., Harkin, L. A., Hodgson, B. L., Scott, D. et al. (2004). GABRD encoding a protein for extra- or peri-synaptic GABAA receptors is a susceptibility locus for generalized epilepsies. *Hum Mol Genet* **13**, 1315–1319.

Dravet, C., Bureau, M., Oguni, H., Fukuyama, Y., and Cokar, O. (2002). Severe myoclonic epilepsy in infancy (Dravet syndrome). In J. Roger, M. Bureau, C. Dravet, P. Genton, C. A. Tassinari, P. Wolf, Eds., *Epileptic Syndromes in Infancy, Childhood and Adolescence, 3e* (, 81–103. John Libbey & Co. Ltd. Eastleigh.

Dube, C., Vezzani, A., Behrens, M., Bartfai, T., and Baram, T. Z. (2005). Interleukin-1 beta contributes to the generation of experimental febrile seizures. *Ann Neurol* **57**, 152–155.

Escayg, A., De Waard, M., Lee, D. D., Bichet, D., Wolf, P., Mayer, T. et al. (2000a). Coding and noncoding variation of the human calcium-channel beta 4-subunit gene CACNB4 in patients with idiopathic generalized epilepsy and episodic ataxia. *Am J Hum Genet* **66**, 1531–1539.

Escayg, A., MacDonald, B. T., Meisler, M. H., Baulac, S., Huberfeld, G., An-Gourfinkel, I. et al. (2000b). Mutations of SCN1A, encoding a neuronal sodium channel, in two families with GEFS+2. *Nat Genet* **24**, 343–345.

Fukata, Y., Adesnik, H., Iwanaga, T., Bredt, D. S., Nicoll, R. A., and Fukata, M. (2006). Epilepsy-related ligand/receptor complex LGI1 and ADAM22 regulate synaptic transmission. *Science* **313**, 1792–1795.

Fusco, M. D., Becchetti, A., Patrignani, A., Annesi, G., Gambardella, A., Quattrone, A. et al. (2000). The nicotinic receptor β2 subunit is mutant in nocturnal frontal lobe epilepsy. *Nat Genet* **26**, 275–276.

Greenberg, D. A., Cayanis, E., Strug, L., Marathe, S., Durner, M., Pal, D. K. et al. (2005). Malic enzyme 2 may underlie susceptibility to adolescent-onset idiopathic generalized epilepsy. *Am J Hum Genet* **76**, 139–146.

Haspolat, S., Baysal, Y., Duman, O., Coskun, M., Tosun, O., and Yegin, O. (2005). Interleukin-1 alpha, interleukin-1 beta, and interleukin-1 Ra polymorphisms in febrile seizures. *J Child Neurol* **20**, 565–568.

Haspolat, S., Mihci, E., Coskun, M., Gumuslu, S., Ozben, T., and Yegin, O. (2002). Interleukin-1 beta, tumor necrosis factor-alpha, and nitrite levels in febrile seizures. *J Child Neurol* **17**, 749–751.

Haug, K., Warnstedt, M., Alekov, A. K., Sander, T., Ramirez, A., Poser, B. et al. (2003). Mutations in CLCN2 encoding a voltage-gated chloride channel are associated with idiopathic generalized epilepsies. *Nat Genet* **33**, 527–532.

Hempelmann, A., Heils, A., and Sander, T. (2006). Confirmatory evidence for an association of the connexin-36 gene with juvenile myoclonic epilepsy. *Epilepsy Res* **71**, 223–228.

Heron, S. E., Crossland, K. M., Andermann, E., Phillips, H. A., Hall, A. J., Bleasel, A. et al. (2002). Sodium-channel defects in benign familial neonatal-infantile seizures. *Lancet* **360**, 851–852.

Kalachikov, S., Evgrafov, O., Ross, B., Winawer, M., Barker-Cummings, C., Martinelli Boneschi, F. et al. (2002). Mutations in LGI1 cause autosomal-dominant partial epilepsy with auditory features. *Nat Genet* **30**, 335–341.

Kamiya, K., Kaneda, M., Sugawara, T., Mazaki, E., Okamura, N., Montal, M. et al. (2004). A nonsense mutation of the sodium channel gene SCN2A in a patient with intractable epilepsy and mental decline. *J Neurosci* **24**, 2690–2698.

Kang, J. Q., Shen, W., and Macdonald, R. L. (2006). Why does fever trigger febrile seizures? GABAA receptor gamma 2 subunit mutations associated with idiopathic generalized epilepsies have temperature-dependent trafficking deficiencies. *J Neurosci* **26**, 2590–2597.

Kaplan, R. E. and Lacey, D. J. (1983). Benign Familial Neonatal-Infantile Seizures. *Am J Med Genet* **16**, 595–599.

Lahat, E., Livne, M., Barr, J., and Katz, Y. (1997). Interleukin-1 beta levels in serum and cerebrospinal fluid of children with febrile seizures. *Pediatr Neurol* **17**, 34–36.

Lennox, W.G.L., M.A. (1960). *Epilepsy and related disorders*. Little, Brown & Co., Boston.

Lenzen, K. P., Heils, A., Lorenz, S., Hempelmann, A., Hofels, S., Lohoff, F. W. et al. (2005). Supportive evidence for an allelic association of the human KCNJ10 potassium channel gene with idiopathic generalized epilepsy. *Epilepsy Res* **63**, 113–118.

Leppert, M., Anderson, V. E., Quattlebaum, T., Stauffer, D., O'Connell, P., Nakamura, Y. (1989). Benign familial neonatal convulsions linked to genetic markers on chromosome 20. *Nature* **337**, 647–648.

Lerche, H., Biervert, C., Alekov, A. K., Schleithoff, L., Lindner, M., Klinger, W. et al. (1999). A reduced K+ current due to a novel mutation in KCNQ2 causes neonatal convulsions. *Ann Neurol* **46**, 305–312.

Liang, J., Zhang, Y., Wang, J., Pan, H., Wu, H., Xu, K. et al. (2006). New variants in the CACNA1H gene identified in childhood absence epilepsy. *Neurosci Lett* **406**, 27–32.

Maljevic, S., Krampfl, K., Cobilanschi, J., Tilgen, N., Beyer, S., Weber, Y. G. et al. (2006). A mutation in the GABA(A) receptor alpha(1)-subunit is associated with absence epilepsy. *Ann Neurol* **59**, 983–987.

Mantegazza, M., Gambardella, A., Rusconi, R., Schiavon, E., Annesi, F., Cassulini, R. R. et al. (2005). Identification of an Nav1.1 sodium channel (SCN1A) loss-of-function mutation associated with familial simple febrile seizures. *Proc Natl Acad Sci U S A* **102**, 18177–18182.

Marini, C., Harkin, L. A., Wallace, R. H., Mulley, J. C., Scheffer, I. E., and Berkovic, S. F. (2003). Childhood absence epilepsy and febrile seizures: A family with a GABA(A) receptor mutation. *Brain* **126**, 230–240.

Mas, C., Taske, N., Deutsch, S., Guipponi, M., Thomas, P., Covanis, A. et al. (2004). Association of the connexin 36 gene with juvenile myoclonic epilepsy. *J Med Genet* **41**, e93.

Matsuo, M., Sasaki, K., Ichimaru, T., Nakazato, S., and Hamasaki, Y. (2006). Increased IL-1 beta production from dsRNA-stimulated leukocytes in febrile seizures. *Pediatr Neurol* **35**, 102–106.

Moore, T., Hecquet, S., McLellann, A., Ville, D., Grid, D., Picard, F. et al. (2001). Polymorphism analysis of JRK/JH8, the human homologue of mouse jerky, and description of a rare mutation in a case of CAE evolving to JME. *Epilepsy Res* **46**, 157–167.

Mulley, J. C., Scheffer, I. E., Petrou, S., Dibbens, L. M., Berkovier, S. F. and Harkin, L. A. (2005a). Suceptibility genes for complex epilepsy. *Hum Mol Genet 14 Spec No. 2*, R 243–249.

Mulley, J. C., Scheffer, I. E., Petrou, S., Dibbens, L. M., Berkovic, S. F., and Harkin, L. A. (2005b). SCN1A mutations and epilepsy. *Hum Mutat* **25**, 535–542.

Online Mendelian Inheritance in Man, O. Online Mendelian Inheritance in Man, OMIM. McKusick-Nathans Institute for Genetic Medicine, Johns Hopkins University (Baltimore, MD) and National Center for Biotechnology Information, National Library of Medicine (Bethesda, MD) World Wide Web URL: http://www.ncbi.nlm.nih.gov/omim/.

Oti, M., Snel, B., Huynen, M. A., and Brunner, H. G. (2006). Predicting disease genes using protein-protein interactions. *J Med Genet* **43**, 691–698.

Pal, D. K., Evgrafov, O. V., Tabares, P., Zhang, F., Durner, M., and Greenberg, D. A. (2003). BRD2 (RING3) is a probable major susceptibility gene for common juvenile myoclonic epilepsy. *Am J Hum Genet* **73**, 261–270.

Pan, Z., Kao, T., Horvath, Z., Lemos, J., Sul, J. Y., Cranstoun, S. D. et al. (2006). A common ankyrin-G-based mechanism retains KCNQ and NaV channels at electrically active domains of the axon. *J Neurosci* **26**, 2599–2613.

Phillips, H. A., Favre, I., Kirkpatrick, M., Zuberi, S. M., Goudie, D., Heron, S. E. et al. (2001). CHRNB2 is the second acetylcholine receptor subunit associated with autosomal dominant nocturnal frontal lobe epilepsy. *Am J Hum Genet* **68**, 225–231.

Plouin, P. and Anderson, V. E. (2002). Benign familial and non-familial neonatal seizures. In J. Roger, M. Bureau, C. Dravet, P. Genton, C. A. Tassinari, P. Wolf, Eds., *Epileptic Syndromes in Infancy, Childhood and Adolescence, 3e*, 3–13. John Libbey & Co Ltd, Eastleigh.

Popa, M. O., Cobilanschi, J., Maljevic, S., Heils, A., and Lerche, H. (2005). Pure childhood absence epilepsy associated with a gain-of-function mutation in CACNA1A affecting G-protein modulation of P/Q-type Ca2+ channels. *Epilepsia* **46** (Suppl 6), 80.

Ptacek, L. J. and Fu, Y. H. (2004). Channels and disease: Past, present, and future. *Arch Neurol* **61**, 1665–1668.

Ptacek, L. J., George, A. L., Jr., Griggs, R. C., Tawil, R., Kallen, R. G., Barchi, R. L. et al. (1991). Identification of a mutation in the gene causing hyperkalemic periodic paralysis. *Cell* **67**, 1021–1027.

Rett, A. and Teubel, R. (1964). *Neugeborenenkrämpfe im Rahmen einer epileptisch belasteten familie. Wiener Klinische Wochenschrift* **76**, 609–613.

Riggs, A. J. and Riggs, J. E. (2005). Epilepsy's role in the historical differentiation of religion, magic, and science. *Epilepsia* **46**, 452–453.

Scalmani, P., Rusconi, R., Armatura, E., Zara, F., Avanzini, G., Franceschetti, S., and Mantegazza, M. (2006). Effects in neocortical neurons of mutations of the Na(v)1.2 Na+ channel causing benign familial neonatal-infantile seizures. *J Neurosci* **26**, 10100–10109.

Scheffer, I. E. and Berkovic, S. F. (1997). Generalized epilepsy with febrile seizures plus. A genetic disorder with heterogeneous clinical phenotypes. *Brain* **120**, 479–490.

Scheffer, I. E., Harkin, L. A., Grinton, B. E., Dibbens, L. M., Turner, S. J., Zielinski, M. A. et al. (2007). Temporal lobe epilepsy and GEFS+ phenotypes associated with SCN1B mutations. *Brain* **130**, 100–109.

Schulte, U., Thumfart, J. O., Klocker, N., Sailer, C. A., Bildl, W., Biniossek, M. et al. (2006). The epilepsy-linked Lgi1 protein assembles into presynaptic Kv1 channels and inhibits inactivation by Kv beta 1. *Neuron* **49**, 697–706.

Singh, B., Ogiwara, I., Kaneda, M., Tokonami, N., Mazaki, E., Baba, K. et al. (2006). A K(v)4.2 truncation mutation in a patient with temporal lobe epilepsy. *Neurobiol Dis* **24**, 245–253.

Singh, N. A., Charlier, C., Stauffer, D., DuPont, B. R., Leach, R. J., Melis, R. et al. (1998). A novel potassium channel gene, *KCNQ2*, is mutated in an inherited epilepsy of newborns. *Nature Genetics* **18**, 25–29.

Steinlein, O. K., Mulley, J. C., Propping, P., Wallace, R. H., Phillips, H. A., Sutherland, G. R. et al. (1995). A missense mutation in the neuronal nicotinic acetylcholine receptor alpha 4 subunit is associated with autosomal dominant nocturnal frontal lobe epilepsy. *Nat Genet* **11**, 201–203.

Suzuki, T., Delgado-Escueta, A. V., Aguan, K., Alonso, M. E., Shi, J., Hara, Y. et al. (2004). Mutations in EFHC1 cause juvenile myoclonic epilepsy. *Nat Genet* **36**, 842–849.

Szepetowski, P., Rochette, J., Berquin, P., Piussan, C., Lathrop, G. M., and Monaco, A. P. (1997). Familial infantile convulsions and paroxysmal choreoathetosis: A new neurological syndrome linked to the pericentromeric region of human chromosome 16. *American Journal of Human Genetics* **61**, 889–898.

Tan, N. C., Mulley, J. C., and Berkovic, S. F. (2004). Genetic association studies in epilepsy: "the truth is out there." *-Epilepsia* **45**, 1429–1442.

Traub, R. D., Contreras, D., Cunningham, M. O., Murray, H., LeBeau, F. E., Roopun, A. et al. (2005). Single-column thalamocortical network model exhibiting gamma oscillations, sleep spindles, and epileptogenic bursts. *J Neurophysiol* **93**, 2194–2232.

Tsai, F. J., Hsieh, Y. Y., Chang, C. C., Lin, C. C., and Tsai, C. H. (2002). Polymorphisms for interleukin 1 beta exon 5 and interleukin 1 receptor antagonist in Taiwanese children with febrile convulsions. *Arch Pediatr Adolesc Med* **156**, 545–548.

Tutuncuoglu, S., Kutukculer, N., Kepe, L., Coker, C., Berdeli, A., and Tekgul, H. (2001). Proinflammatory cytokines, prostaglandins and zinc in febrile convulsions. *Pediatr Int* **43**, 235–239.

Vadlamudi, L. (2006). Analysing the etiology of benign rolandic epilepsy: A multicentre twin collaboration. *Epilepsia* **47**, 550–555.

Vadlamudi, L., Harvey, A. S., Connellan, M. M., Milne, R. L., Hopper, J. L., Scheffer, I. E., and Berkovic, S. F. (2004). Is benign rolandic epilepsy genetically determined? *Ann Neurol* **56**, 129–132.

van Driel, M. A., Bruggeman, J., Vriend, G., Brunner, H. G., and Leunissen, J. A. (2006). A text-mining analysis of the human phenome. *Eur J Hum Genet* **14**, 535–542.

Vanmolkot, K. R., Kors, E. E., Hottenga, J. J., Terwindt, G. M., Haan, J., Hoefnagels, W. A. et al. (2003). Novel mutations in the Na+, K+-ATPase pump gene ATP1A2 associated with familial hemiplegic migraine and benign familial infantile convulsions. *Ann Neurol* **54**, 360–366.

Vigevano, F., Fusco, L., Di Capua, M., Ricci, S., Sebastianelli, R., and Lucchini, P. (1992). Benign infantile familial convulsions. *Eur J Pediatr* **151**, 608–612.

Virta, M., Hurme, M., and Helminen, M. (2002a). Increased frequency of interleukin-1 beta ($^{-51}$ 1) allele 2 in febrile seizures. *Pediatr Neurol* **26**, 192–195.

Virta, M., Hurme, M., and Helminen, M. (2002b). Increased plasma levels of pro- and anti-inflammatory cytokines in patients with febrile seizures. *Epilepsia* **43**, 920–923.

Wallace, R. H., Marini, C., Petrou, S., Harkin, L. A., Bowser, D. N., Panchal, R. G. et al. (2001). Mutant GABA(A) receptor gamma 2-subunit in childhood absence epilepsy and febrile seizures. *Nat Genet* **28**, 49–52.

Wallace, R. H., Wang, D. W., Singh, R., Scheffer, I. E., George, A. L., Jr., Phillips, H. A. et al. (1998). Febrile seizures and generalized epilepsy associated with a mutation in the Na+-channel beta 1 subunit gene SCN1B. *Nat Genet* **19**, 366–370.

Winawer, M. R., Marini, C., Grinton, B. E., Rabinowitz, D., Berkovic, S. F., Scheffer, I. E., and Ottman, R. (2005). Familial clustering of seizure types within the idiopathic generalized epilepsies. *Neurology* **65**, 523–528.

Winawer, M. R., Rabinowitz, D., Pedley, T. A., Hauser, W. A., and Ottman, R. (2003). Genetic influences on myoclonic and absence seizures. *Neurology* **61**, 1576–1581.

Wolf, P. (2005). Historical aspects of idiopathic generalized epilepsies. *Epilepsia* **46 Suppl 9**, 7–9.

25

Tourette's Syndrome

James F. Leckman and Michael H. Bloch

Tic disorders are transient or chronic conditions associated with difficulties in self-esteem, family life, social acceptance, or school or job performance that are directly related to the presence of motor and/or phonic tics. Although tic symptoms have been reported since antiquity, systematic study of individuals with tic disorders dates only from the nineteenth century with the reports of Itard (1825) and Gilles de la Tourette (1885). Gilles de la Tourette, in his classic study of 1885, described nine cases characterized by motor "incoordinations" or tics, "inarticulate shouts accompanied by articulated words with echolalia and coprolalia" (Gilles de la Tourette, 1885). In addition to identifying the cardinal features of severe tic disorders, his report noted an association between tic disorders and obsessive-compulsive symptoms as well as the hereditary nature of the syndrome in some families.

Despite the overt nature of tics and decades of scientific scrutiny, our ignorance remains great. Notions of cause have ranged from "hereditary degeneration" to the "irritation of the motor neural systems by toxic substances, of a self-poisoning bacteriological origin" to "a constitutional inferiority of the subcortical structures ... [that] renders the individual defenseless against overwhelming emotional and dynamic forces" (Kushner, 1999). Predictably, each of these etiological explanations has prompted new treatments and new ways of relating to families.

In addition to tics, individuals with tic disorders may present with a broad array of behavioral difficulties including disinhibited speech or conduct, impulsivity, distractibility, motoric hyperactivity, and obsessive–compulsive symptoms (Leckman & Cohen, 1998). Alternatively, a sizable portion of children and adolescents with tics will be free of coexisting developmental or emotional difficulties. Scientific opinion has been divided on how broadly to conceive the spectrum of maladaptive behaviors associated with Tourette's syndrome (TS) (Comings, 1988; Shapiro et al., 1988). This controversy is fueled in part by the genuine frustration that parents and educators encounter when they attempt to divide an individual child's repertoire of problem behaviors into those that are "Tourette-related" and those that are not. Population-based epidemiological studies and family-genetic studies have begun to clarify these issues, but much work remains to be done.

In this chapter, a presentation of the natural history precedes a review of the etiology and the neurobiological substrates of these conditions before closing with a set of integrative hypotheses. The general perspective that will be presented is that TS and related disorders are model neurobiological disorders in which to study multiple interactive genetic and environmental (epigenetic) mechanisms that interact over the course of development to produce a distinctive range of complex syndromes of varying severity.

I. Clinical Description and Natural History

A. Phenomenology of Tics

A *tic* is a sudden, repetitive movement, gesture, or utterance that typically mimics some aspect or fragment of normal behavior. Usually of brief duration, individual tics rarely last more than a second. Individual tics can occur singly or together in an orchestrated pattern. They vary in their intensity or forcefulness. Typically the disorder begins in early childhood, on average between the ages of three and seven years, with transient bouts of simple motor tics such as eye blinking or head jerks. These tics may initially come and go, but eventually they become persistent and begin to have adverse effects on the child and his family. The repertoire of motor tics can be vast, incorporating virtually any voluntary movement by any portion of the body. Although some patients have a rostral-caudal progression of motor tics (head, neck, shoulders, arms, torso), this course is not predictable. As the syndrome develops, complex motor tics may appear. Typically, they accompany simple motor tics. Often they have a camouflaged or purposive appearance (e.g., brushing hair away from the face with an arm), and can only be distinguished as tics by their repetitive character. They can involve dystonic movements. In a small fraction of cases (<5%), complex motor tics have the potential to be self-injurious and to further complicate management. These self-injurious symptoms may be relatively mild (e.g., slapping or tapping), or quite dangerous (e.g., punching one side of the face, biting a wrist, or gouging eyes to the point of blindness).

On average, phonic tics begin one to two years after the onset of motor symptoms and are usually simple in character, for example, throat clearing, grunting, and squeaks. More complex vocal symptoms such as echolalia (repeating another's speech), palilalia (repeating one's own speech), and coprolalia (obscene or socially unacceptable speech) occur in a minority of cases. Other complex phonic symptoms include dramatic and abrupt changes in rhythm, rate, and volume of speech.

Clinicians characterize tics by their anatomical location, number, frequency, and duration. The intensity or "forcefulness" of the tic can also be an important characteristic as some tics call attention to themselves simply by virtue of the exaggerated fashion in which they are performed or uttered. Finally, tics vary in terms of their complexity. Complexity usually refers to how simple or involved a movement or sound is, ranging from brief, meaningless, abrupt fragments (simple tics) to ones that are longer, more involved, and seemingly more goal-directed in character (complex tics). Each of these elements has been incorporated into clinician rating scales that have proven to be useful in monitoring tic severity (Leckman et al., 1989).

Motor and phonic tics tend to occur in bouts with brief inter-tic intervals (Peterson & Leckman, 1998). Their frequency ranges from nonstop bursts that are virtually uncountable (>100 tics per minute) to rare events that occur only a few times a week (see Figure 25.1).

By the age of 10 years, most individuals with tics are aware of premonitory urges that may either be experienced as a focal perception in a particular body region where the tic is about to occur (like an itch or a tickling sensation) or as a mental awareness (Lang, 1991; Leckman et al., 1993). A majority of patients also report a fleeting sense of relief after a bout of tics has occurred. These premonitory and consummatory phenomena contribute to an individual's sense that tics are a habitual, yet partially intentional, response to unpleasant stimuli. Indeed, most adolescent and adult subjects describe their tics as either voluntary or as having both voluntary and involuntary aspects. In contrast, many young children are oblivious to their tics and experience them as wholly involuntary movements or sounds. Most tics can also be suppressed for brief periods of time. The warning given by premonitory urges may contribute to this phenomenon.

Tic disorders tend to improve in late adolescence and early adulthood. In many instances, the phonic symptoms become increasingly rare or may disappear altogether, and the motor tics may be reduced in number and frequency. Complete remission of both motor and phonic symptoms has also been reported (Goetz et al., 1992; Leckman et al., 1998; Shapiro et al., 1988). In contrast, adulthood is also the period when the most severe and debilitating forms of tic disorder can be seen. The factors that influence the continuity of tic disorders from childhood to adolescence to adulthood are not well understood but likely involve the interaction of normal maturational processes occurring in the central nervous system (CNS) with the neurobiological mechanisms responsible for TS, the exposure to cocaine, other CNS stimulants, androgenic steroids, and the amount of intramorbid emotional trauma and distress experienced by affected individuals during childhood and adolescence. In addition, it should be emphasized that tic disorders may be etiologically separable so that some of these factors, such as activation of the immune system or exposure to heat stress, may influence the pathogenesis and intramorbid course for some tic disorders but not others. Other factors, such as psychological stress, may have a more uniform impact.

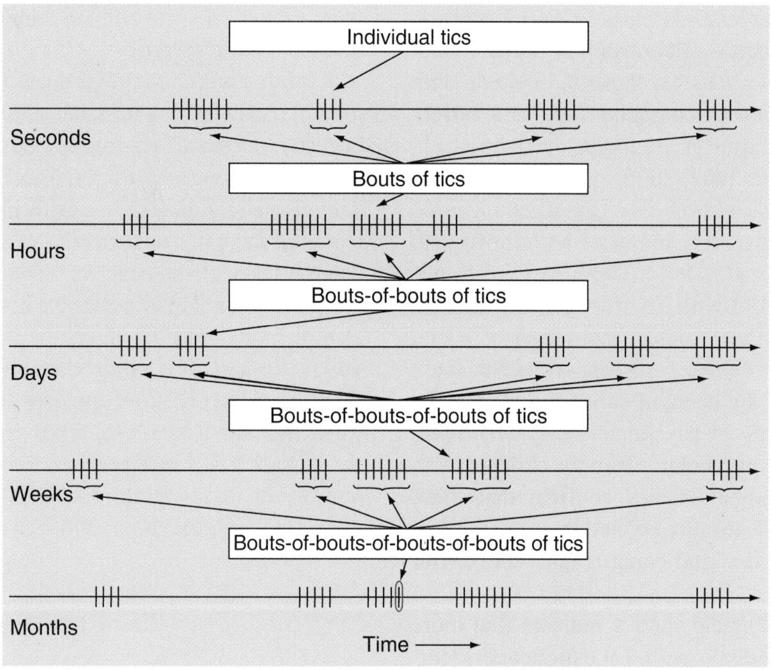

Figure 25.1 Fractal character of temporal occurrence of tics. Progressively longer time scales (seconds to months) are depicted. Adapted from Leckman (2002).

The factors that determine the degree of disability and handicap versus resiliency are largely unknown. They are likely to include the presence of additional developmental, mental, and behavioral disorders; the level of support and understanding from parents, peers, and educators; and the presence of special abilities (as in sports) or personal attributes (intelligence, social abilities, and personality traits). Behavioral and emotional problems frequently complicate TS and range from impulsive, disinhibited, and immature behavior to compulsive touching or sniffing. At present, there are no clear dividing lines between these disruptive behaviors and complex tics on the one hand and comorbid conditions of attention deficit hyperactivity disorder and obsessive-compulsive disorder on the other. As described later, some of these conditions may be alternate expressions of the same underlying vulnerability such as obsessive-compulsive disorder, others may be intimately related by virtue of shared pathophysiological mechanisms such as attention deficit hyperactivity disorder, and still others may be the consequence of having a chronic disorder that is socially disfiguring such as affective and anxiety syndromes. Defining the limits of tic disorders vis-à-vis other forms of psychopathology remains one of the most controversial and difficult areas for families, clinicians, and researchers. Some investigators believe that the spectrum of TS includes attentional deficits, impulsivity, hyperactivity, disruptive behavior, learning disabilities, pervasive developmental disorders, affective and anxiety disorders, as well as tics and obsessive-compulsive disorder (Comings, 1988).

Although most children with TS are loving and affectionate, maintaining age-appropriate social skills appears to be a particularly difficult area for many of them (Bawden et al., 1998; Dykens et al., 1990; Stokes et al., 1991). Whether this is due to the stigmatizing effects of the tics, the patients' own uneasiness, or some more fundamental difficulty linked to the neurobiology of this disorder is unknown.

II. Coexisting Conditions

The past decade has seen a renewed emphasis on the range of neurological and psychiatric symptoms seen in TS patients (Leckman & Cohen, 1998). In both clinical and epidemiological samples TS alone is the exception rather than the rule (Bloch et al., 2006b; Khalifa & von Knorring, 2005). Symptoms associated with attention deficit hyperactivity disorder (ADHD) and obsessive–compulsive disorder (OCD) have received the most attention. It is also becoming clear that the more severe the tic severity the greater the likelihood of detecting coexisting conditions, even in representative, population-based samples (Khalifa & von Knorring, 2006).

Both clinical and epidemiological studies vary according to setting and established referral patterns, but it is not uncommon to see reports of 30 to 50 percent of children with TS diagnosed with comorbid ADHD (Khalifa & von Knorring, 2005; Scahill et al., 2006; Walkup et al., 1998). Of interest, TS has the highest rate of ADHD relative to the other lesser variants of tic disorders ((Khalifa & von Knorring, 2005). Although the etiological relationship between TS and ADHD is in dispute, it is clear that those individuals with both TS and ADHD are at a much greater risk for a variety of untoward outcomes (Carter et al., 2000; Peterson et al., 2001a; Sukhodolsky et al., 2003, 2005). Uninformed peers frequently tease individuals with TS. They are often regarded as less likeable, more aggressive, and more withdrawn than their classmates (Stokes et al., 1991). These social difficulties are amplified in a child with TS who also has ADHD (Bawden et al., 1998). In such cases, their level of social skill is often several years behind their peers (Dykens et al., 1990). Negative appraisal by peers in childhood is a strong predictor of global indices of psychopathology (Hinshaw, 1994). This appears to be particularly true for children with TS and ADHD. Longitudinal studies confirm that these individuals are at high risk for anxiety and mood disorders, oppositional defiant disorder, and conduct disorder (Carter et al., 1994, 2000).

Clinical and epidemiological studies indicate that more than 40 percent of individuals with TS experience recurrent obsessive-compulsive (OC) symptoms (Hounie et al., 2006; Khalifa & von Knorring, 2006; Leckman et al., 1994, 1997b). Genetic, neurobiological, and treatment response studies suggest that there may be qualitative differences between tic-related forms of OCD and cases of OCD in which there is no personal or family history of tics. Specifically, compared with non-tic-related OCD, tic-related OCD has a male preponderance, an earlier age of onset, a poorer level of response to standard anti-obsessional medications, and a greater likelihood of first-degree family members with a tic disorder (Hounie et al., 2006). Symptomatically the most common obsessive-compulsive symptoms encountered in TS patients are (1) obsessions about aggression, sex, religion, and the body, as well as related checking compulsions; and (2) obsessions concerning a need for symmetry or exactness, repeating rituals, counting compulsions, and ordering/arranging compulsions (Leckman et al., 1997a). Also, obsessive-compulsive symptoms, when present in children with TS, appear more likely to persist into adulthood than the tics themselves (Bloch et al., 2006b). An increased childhood level of intelligence may also herald a higher risk of developing more severe OCD symptoms in adulthood (Bloch et al., 2006b).

The cooccurrence of depression and anxiety symptoms with TS are commonplace and may reflect the cumulative psychosocial burden of having tics or shared biological diatheses or both (Coffey et al., 2000; Khalifa & von Knorring, 2006; Robertson & Orth, 2006). In a recent longitudinal study, investigators found that antecedent depressive symptoms do predict modest increases in tic severity (Lin et al., 2006) consistent with earlier predictions (Robertson et al., 2006). However, this study also documented that future depression severity is associated more closely with antecedent worsening of psychosocial stress and OC symptoms than it is with future measures of tic severity.

Children with a range of developmental disorders appear to be at increased risk for tic disorders. Kurlan and colleagues (1994) reported a four-fold increase in the prevalence of tic disorders among children in special educational settings in a single school district in upstate New York. These children were not mentally retarded but did have significant learning disabilities or other speech or physical impairments. Children with autism and other pervasive developmental disorders are also at higher risk for developing TS (Baron-Cohen et al., 1999). Remarkably, in a recent epidemiological study from Sweden, three (12%) of the 25 children diagnosed with TS were also found to a have a pervasive developmental disorder (Khalifa & von Knorring, 2005). It is also clear that a number of neurological disorders can cause tics or mimic them (Jankovic & Mejia, 2006).

III. Prevalence

Once thought to be rare, current estimates of the prevalence of TS vary 100-fold, from 2.9 per 10,000 (Caine et al., 1988) to 299 per 10,000 (Mason et al., 1998). Recently, Khalifa and von Knorring (2003, 2005) studied a total population of 4,479 children aged 7 to 15 years and their parents in a town in central Sweden. A three-stage procedure was used. Twenty-five subjects were identified as having TS, yielding a prevalence estimate of 5.6 per 1,000 pupils in this age group. Scahill et al. (2006), using a two-stage design in a slightly younger cohort of children (aged 6 and 12 years), estimated the prevalence to be 3.3 per 1,000 pupils in this age group.

IV. Neuropsychological Findings

Although motor and phonic tics constitute the core elements of the diagnostic criteria for TS, perceptual and cognitive difficulties are also common. These neuropsychological symptoms are potentially informative about the pathobiology of the disorder. Moreover, these associated difficulties can be more problematic for school and social adjustment than the primary motor symptoms. Neuropsychological studies of TS have focused on a broad array of functions. Review of the literature suggests that the most consistently observed deficits occur on tasks requiring the accurate copying of geometric designs, that is, "visual-motor integration" or "visual-graphic" ability (see Schultz et al. (1998) for a review). There was no evidence to suggest that comorbid ADHD or depressive symptomatology could account for the observed group differences. Even after controlling statistically for visual-perceptual skill, intelligence, and fine motor control, children with TS continued to perform worse than controls on the visual-motor tasks, suggesting that the integration of visual inputs and organized motor output is a specific area of weakness in individuals with TS. Perhaps the most striking observation is the recent finding that poorer performance with the dominant hand on the Purdue Pegboard test during childhood is associated with worse adulthood tic severity (Bloch et al., 2006a).

More recently, a specific deficit in probabilistic learning has been documented in individuals with TS (Keri et al., 2002; Marsh et al., 2004, 2005). In each of these studies of procedural memory (habit learning), subjects with severe tic symptoms had a higher level of impairment than individuals with less severe symptoms. In contrast, no deficits were seen in declarative memory functioning (Marsh et al., 2004, 2005).

V. Etiology and Pathogenesis

During the course of the past decade, TS and related conditions have emerged as model disorders for researchers interested in the interaction of genetic, neurobiological, and environmental (epigenetic) factors that shape clinical outcomes from health to chronic disability over the lifespan.

A. Genetic Factors

Twin and family studies provide evidence that genetic factors are involved in the vertical transmission within families of a vulnerability to TS and related disorders (Pauls & Leckman, 1986). The concordance rate for TS among monozygotic twin pairs is greater than 50 percent, and the concordance of dizygotic twin pairs is about 10 percent (Hyde et al., 1992; Price et al., 1985). If cotwins with chronic motor tic disorder are included, these concordance figures increase to 77 percent for monozygotic and 30 percent for dizygotic twin pairs. Differences in the concordance of monozygotic and dizygotic twin pairs indicate that genetic factors play an important role in the etiology of TS and related conditions. These figures also suggest that nongenetic factors are critical in determining the nature and severity of the clinical syndrome.

Other studies indicate that first degree family members of TS probands are at substantially higher risk for developing TS, chronic motor tic disorder, and obsessive-compulsive disorder than unrelated individuals (Hebebrand et al., 1997; McMahon et al., 2003; Pauls et al., 1991; Walkup et al., 1996). Overall, the risk to male first-degree family members approximates 50 percent (18% TS, 31% chronic motor tics, and 7% obsessive-compulsive disorder), and the overall risk to females is less (5% TS, 9% chronic motor tics, and 17% obsessive-compulsive disorder). These rates are substantially higher than might be expected by chance in the general population, and greatly exceed the rates for these disorders among the relatives of individuals with other psychiatric disorders except obsessive-compulsive disorder.

The pattern of vertical transmission among family members has led several groups of investigators to test whether or not mathematical models of specific genetic hypotheses could be rejected. Although not definitive, most segregation analyses could not rule out models of autosomal transmission (Pauls & Leckman, 1986; Seuchter et al., 2000; Walkup et al., 1996). These studies prompted the identification of large multigenerational families for genetic linkage studies. However, subsequent efforts to identify susceptibility genes within these high-density families using traditional linkage strategies have met with little success (Barr & Sandor, 1998).

More recently, nonparametric approaches using families in which two or more siblings are affected with TS have been undertaken (The Tourette Syndrome International Consortium for Genetics, 1999). This sib-pair approach is suited for diseases with an unclear mode of inheritance and has been used successfully in studies of other complex disorders such as diabetes mellitus and essential hypertension. In this study, two areas are suggestive of linkage to TS, one on chromosome 4q and another on chromosome 8p. This international consortium of researchers is actively completing high-density maps of several genomic regions in an effort to refine and extend their preliminary results. However, the data at present do not replicate the initial linkage results.

Identity-by-descent approaches, a technique that assumes that a few founder individuals contributed the vulnerability genes that are now distributed within a much larger population, have been used to study TS populations in South Africa and Costa Rica, and French Canada (Mathews et al., 2004). They implicate regions near the centromere of chromosome 2, as well as 6p, 8q, 11q, 14q, 20q, 21q (Simonic et al., 1998), and X (Diaz-Anzaldua et al., 2004). Another large pedigree study in the UK, involving linkage analysis of TS patients, is suggestive of linkage at loci on chromosomes 5, 10, and 13 (Curtis et al., 2004).

In addition, a number of cytogenetic abnormalities have been reported in TS families (3 [3p21.3], 7 [7q35-36], 8 [8q21.4] 9 [9pter] and 18 [18q22.3]) (Cuker et al., 2004; State et al., 2003). Among the more recent findings, Verkerk and colleagues (2003) reported the disruption of the contactin-associated protein 2 gene on chromosome 7. This gene encodes a membrane protein located at nodes of Ranvier of axons that may be important for the distribution of the K(+) channels, which would affect signal conduction along myelinated neurons. In addition, using a candidate gene approach identified by chromosomal anomalies, Abelson et al. (2005) identified and mapped a *de novo* chromosome 13 inversion in a patient with TS. The gene SLITRK1 was identified as a brain expressed candidate gene mapping approximately 350 kilobases from the 13q31 breakpoint. Mutation screening of 174 patients with TS was undertaken with the resulting identification of a truncating frame-shift mutation in a second family affected with TS (see Figure 25.2). In addition, two examples of a rare variant were identified in a highly conserved region of the 3′ untranslated region of the gene corresponding to a brain expressed micro-RNA binding domain. None of these anomalies were demonstrated in 3600 controls. *In vitro* studies showed that both the frame-shift and the miRNA binding site variant had functional potential and were consistent with a loss-of-function mechanism. Studies of both SLITRK1 and the micro-RNA predicted to bind in the variant-containing 3′ region showed expression in multiple neuroanatomical areas implicated in TS neuropathology including the cortical plate, striatum, globus pallidus, thalamus, and subthalamic nucleus. Future progress is anticipated and clarity about the nature and normal expression of even a few of the TS susceptibility genes is likely to provide a major step forward in understanding the pathogenesis of this condition.

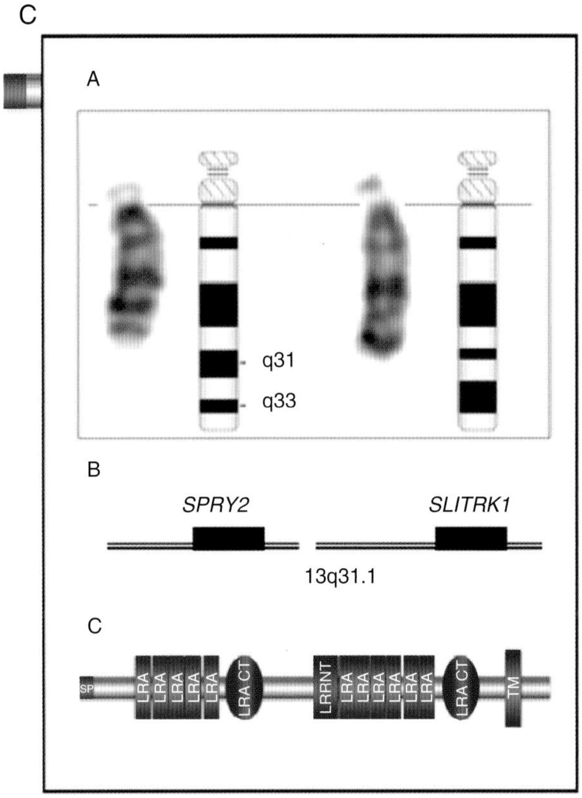

Figure 25.2 Chromosomal location of the *SLITRK1* gene and protein characterization. **A.** G-banded metaphase chromosomes 13. The ideogram for the normal (left) and inverted (right) chromosomes are presented. **B.** Diagram of the interval surrounding the spanning BAC RP11-375K12 at 13q31.1. *SLITRK1* (National Center for Biotechnology Information accession code NM_052910 [GenBank]) maps approximately 350 kb telomeric, and *SPRY2* (NM_005842 [GenBank]) maps more than 3 million base pairs centromeric, to the breakpoint. **C.** Diagram of the normal SLITRK1 protein (http://smart.embl-heidelberg.de/). SP, signal peptide; LRRNT, LRR N-terminal domain; LRRCT, LRR C-terminal domain; TM, transmembrane domain. Adapted from Abelson et al. (2005).

B. Neural Circuits

Investigators interested in procedural learning, habit formation, and internally and externally guided motor control have focused their attention on multisynaptic neural circuits or loops that link the cerebral cortex with several subcortical regions (Albin & Mink, 2006; Graybiel & Canales, 2001; Haber, 2003; Middleton & Strick, 2000, 2003). These circuits direct information from the cerebral cortex to the basal ganglia, cerebellum, and thalamus, and then back to specific regions of the cortex, thereby forming multiple cortical-subcortical loops.

1. Neuroanatomy

Cortical neurons projecting to the striatum outnumber striatal (MSs) spiny neurons by about a factor of 10 (Zheng & Wilson, 2002). These convergent cortical efferent neurons project to the dendrites of MSs within two structurally similar, but neurochemically distinct compartments in the striatum: striosomes and matrix. These two compartments differ by their cortical inputs, with the striosomal MS projection mainly receiving convergent limbic and prelimbic inputs, and neurons in the matrix mainly receiving convergent input from ipsilateral primary motor and sensory motor cortices and contralateral primary motor cortices.

Several other less abundant striatal cell types probably have a key role in habit learning, including cholinergic tonically active neurons (TANs) and fast-spiking GABAergic interneurons (FSNs) (Gonzalez-Burgos et al., 2005; Jog et al., 1999). TANs are very sensitive to salient perceptual cues because they signal the networks within the cortico-basal ganglia learning circuits when these cues arise (Raz et al., 1996). Specifically, they are responsive to dopaminergic inputs from the substantia nigra, and these signals likely participate in calculation of the perceived salience (reward value) of perceptual cues along with excitatory inputs from midline thalamic nuclei.

The FSNs of the striatum receive direct cortical inputs predominantly from lateral cortical regions, including the primary motor and somatosensory cortex, and they are highly sensitive to cortical activity in these regions (McGeorge & Faull, 1989). They are also known to be electrically coupled via gap junctions that connect adjacent dendrites. Once activated, these FSNs can inhibit many nearby striatal projection neurons synchronously via synapses on cell bodies and proximal dendrites (Koos & Tepper, 1999). These FSNs are also very sensitive to cholinergic drugs, suggesting that they are functionally related to the TANs active neurons (Koos & Tepper, 2002).

2. Neuropathological Data

Although neuropathological studies of postmortem TS brains are few in number, a recent stereological study indicates that there is a marked alteration in the number and density of GABAergic parvalbumin-positive cells in basal ganglia structures (see Figure 25.3; Kalanithi et al., 2005). In the caudate there was a greater than 50 percent reduction in the GABAergic FSNs and a 30 to 40 percent reduction of these same cells in the putamen. This same study found a reduction of the GABAergic parvalbumin-positive projection neurons in the external segment globus pallidus (GPe) as well as a dramatic increase (>120%) in the number and proportion of GABAergic projection neurons of the internal segment of the globus pallidus (GPi). These alterations are consistent with a developmental defect in tangential migration of some GABAergic neurons. Further studies are needed to confirm and extend these findings toward a more complete understanding of how the different striatal interneurons are affected, and to determine how alterations in GABAergic interneurons and GPi projection neurons could lead to a form of thalamocortical dysrhythmia (Leckman et al., 2006b; Llinas et al., 2005).

3. Structural Brain Imaging

Volumetric magnetic resonance imaging (MRI) studies of basal ganglia in individuals with TS are largely consistent with these postmortem results—with the finding of a slight reduction in caudate volume (Hyde et al., 1995; Peterson et al., 2003; Singer et al., 1993). For example, in the largest and most recent study of basal ganglia volume involving a total of 154 subjects with TS and 130 healthy controls, Peterson et al. (2003) found a significant decrease in the volume of the caudate nucleus in both the child and adult age groups. Although there was no correlation between symptom severity and caudate volumes in this cross-sectional study, Bloch and colleagues (2005) found an inverse correlation between caudate volume in childhood and tic-severity in early adulthood.

In the same group of subjects the cerebrums, corpus callosa, and ventricles were isolated and then parcellated into subregions using standard anatomical landmarks. Individuals with TS were found to have larger volumes in dorsal prefrontal regions, larger volumes in parieto-occipital regions, and smaller inferior occipital volumes (Peterson et al., 2001b). Regional cerebral volumes were significantly associated with the severity of tic symptoms in orbitofrontal, midtemporal, and parieto-occipital regions. There also appears to be an age-dependent decrease in the cross-sectional area of the corpus callosum in children aged 6 to 16 (Baumgardner et al., 1996; Peterson et al., 1994). This may reflect the natural history of the disorder, or brain compensations, or may yet be explained by other subtle confounding variables, such as the manner in which brain volumes are determined in different studies. In a recent report, Plessen and colleagues (2004) show a decrease in corpus callosum size in children as well as an increase in size in adults with TS indicating that changes in white matter tracks in this disorder. In addition, Lee and colleagues (2006) used volumetric MRI methods to compare thalamic volume in 18 treatment-naïve boys versus

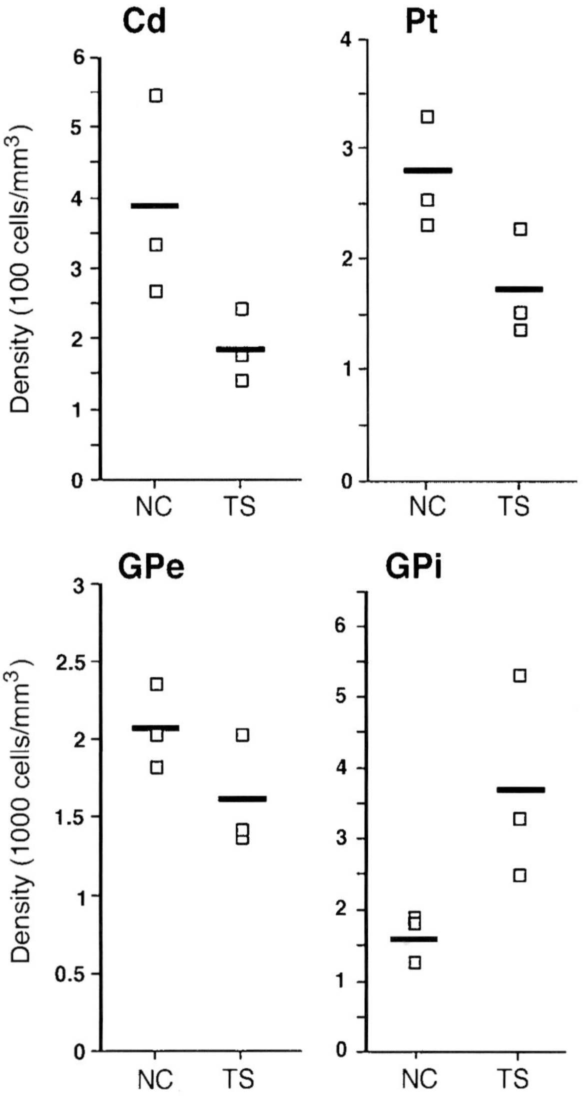

Figure 25.3 Altered number of fast-spiking GABAergic interneurons in the basal ganglia of individuals with Tourette's syndrome. Unbiased estimates of parvalbumin-positive cell density in three TS and three NC brains using stereological analyses. Each small square represents a single subject. Differences in PV neuronal density in the caudate (Cd), Putamen (Pt), and Globus Pallidus pars internus (GPi) of TS patients were statistically significant ($P < 0.025$, Mann-Whitney U test). Adapted from Kalanithi et al. (2005).

16 healthy control subjects finding larger left thalamic volume in the TS cases. Clearly, more volumetric studies using comparable methods across all implicated brain regions are needed to clarify the brain morphology of TS.

4. Functional Brain Imaging

Thus far, there have been only a few published studies of TS using functional magnetic resonance imaging (fMRI), which takes advantage of state-dependant blood oxygenation as a measure of brain activity. In adults with TS, Peterson and colleagues (1998b) compared brain activity during blocks of time, during which tics were voluntarily suppressed or not suppressed. During tic suppression, prefrontal cortical, thalamic, and basal ganglia areas were activated. These activations were inversely correlated with tic severity (i.e., lower activation was associated with higher tic severity). This finding suggests that a greater ability of basal ganglia to suppress cortical activity might be linked with decreased tic severity, and is in agreement with PET and SPECT studies that suggest involvement of the basal ganglia in TS (Braun et al., 1995; Butler et al., 2006). Some investigators have sought to alter the activity of the prefrontal areas with magnetic fields in an effort to enhance the voluntary control of tics, with mixed results (George et al., 2001). In another fMRI study, Serrien et al. (2002) mapped brain activity during motor tasks compared to baseline in three control and three TS patients. TS subjects had considerably reduced activations in premotor and parietal cortices, as well as basal ganglia and thalamus. Most recently, Bohlhalter et al. (2006) studied the neural correlates of tics and associated urges using an event-related fMRI protocol. On the basis of synchronized video/audio recordings, fMRI activities were analyzed two seconds before and at tic onset. A brain network of paralimbic areas including the anterior cingulate and insular cortex, supplementary motor area, and parietal operculum was found to be activated before tic onset. In contrast, at the beginning of tic action, significant fMRI activities were found in sensorimotor areas including superior parietal lobule bilaterally and cerebellum. The results of this study indicate that paralimbic and sensory association areas are critically implicated in tic generation.

Investigators also have examined the correlation of metabolic activity across various brain regions and found that changes in the coupling of the putamen and ventral striatum with a number of other brain regions differentiated TS patients from controls. For example in PET studies, Jeffries and colleagues (2002) noted a reversal in the pattern of cortico-striato-thalamo-cortical circuit interactions in motor and lateral orbitofrontal cortices. Similarly, Stern and colleagues (2000) found that increased activity in a set of neocortical, paralimbic, and subcortical regions (including supplementary motor, premotor, anterior cingulate, dorsolateral-rostral prefrontal, and primary motor cortices; Broca's area, insula, claustrum, putamen, and caudate) were highly correlated with tic behavior. Perhaps not surprisingly in the one patient with prominent coprolalia, the vocal tics were associated with increased activity in prerolandic and postrolandic language regions, insula, caudate, thalamus, and cerebellum.

C. Neurophysiology

Noninvasive *in vivo* neurophysiological research in TS has led to several areas of significant progress. The first concerns the use of a startle paradigm to measure inhibitory deficits by monitoring the reduction in startle reflex magnitude. Swerdlow and colleagues (2001) recently have confirmed and extended earlier findings indicating that TS patients have deficits in sensory gating across a number of sensory modalities. Although prepulse inhibition (PPI) abnormalities have been observed across a variety of neuropsychiatric populations, including those with schizophrenia, OCD, Huntington's disease, nocturnal enuresis, attention deficit disorder, Asperger's syndrome, and TS, perhaps some final common pathways mediate abnormal PPI in all these diseases. With respect to TS, these deficits in inhibitory gating are consistent with the idea that there is some diminished ability to appropriately manage or "gate" sensory inputs to motor programs, which are released as tics (Swerdlow & Sutherland, 2006).

A second advance has been the investigation of motor system excitability by means of single and paired pulse transcranial magnetic stimulation (TMS). Studies to date in groups of patients with TS have indicated that the cortical silent period (a period of decreased excitability following stimulation) is shortened in TS. This intracortical excitability is seen frequently in children with ADHD comorbid with a tic disorder (Moll et al., 1999; Ziemann et al., 1997). This heightened level of cortical excitability may be related to the possible reduction in the number of GABAergic interneurons in the cortex.

Third, Serrien and coworkers (2005) recently identified similar sensorimotor-frontal connections involved in the acute suppression of involuntary tics as evidenced by increased EEG coherence in the alpha frequency band (8–12 Hz) range during suppression of voluntary movements in individuals with TS compared with healthy subjects during a Go-NoGo task. This finding taken with the findings from the Peterson et al. (1998a) report suggest fairly clearly that the frontal lobes may play an important compensatory role in tic suppression and that coherence in the alpha band may be part of this process.

Finally, the preliminary findings that ablation (or high frequency stimulation using deep brain electrodes) in regions of the GPi and/or the midline thalamic nuclei can ameliorate tics in severe, persistent cases of TS (Hassler & Dieckmann, 1973; Vandewalle et al., 1999) powerfully support the view that electrophysiological studies and interventions hold promise just as they do for disorders like Parkinson's disease. The original electrode placement for DBS surgery was placed in the medial part of the thalamus, based on the results of previous lesioning studies (Hassler & Dieckmann, 1973). Four patients received bilateral medial thalamic DBS surgery, and all four experienced a substantial improvement in tic severity following the procedure (Ackermans et al., 2006; Visser-Vandewalle et al., 2003). Subsequent case reports demonstrated comparable efficacy in bilateral palladial stimulation compared with bilateral medial thalamic stimulation (Diederich et al., 2005; Houeto et al., 2005).

However this was not the case with electrode placements within the anterior internal capsule (Flaherty et al., 2005).

Prospective longitudinal studies with higher resolution will be needed to examine fully the developmental processes, sexual dimorphisms, and possible effects of medication on critical cell compartments. It will also be important to confirm if any of these volumetric and functional findings in childhood are predictive of later clinical outcomes. The combination of imaging techniques with real-time neurophysiological techniques, such as electroencephalography or magnetoencephalography, may help to determine whether any brain imaging findings in TS contribute to the production of tics or whether they constitute a compensatory response (Llinas et al., 2005; Segawa, 2003).

In addition, reciprocal connections between midbrain sites (periaqueductal gray, substantia nigra, and the ventral tegmental area), portions of the hypothalamus, and structures in the basal ganglia and the amygdala are likely to play a critical role in the genesis and maintenance of the symptoms of TS. These connections may also contribute to stress sensitivity, including sensitivity to thermal stress observed in a limited number of subjects, and to the more frequent expression of TS in males than females, as many of these structures contain receptors for gonadal steroids and are responsive to alterations in their hormonal environment (see later).

D. Neurochemical and Neuropharmacological Data

Extensive immunohistochemical studies of the basal ganglia have demonstrated the presence of a wide spectrum of classic neurotransmitters, neuromodulators, and neuropeptides (Graybiel, 1990; Parent, 1986). The functional status of a number of these systems has been evaluated in TS (see Albin, 2006; Singer & Wendlandt, 2001 for reviews). In the following section, the current status of a select group of these compounds is reviewed with special attention to those that are related to medications currently used to treat tics and related conditions.

As with habits and stereotypes, ascending dopaminergic pathways are likely to play a role in the consolidation and performance of tics (Aosaki et al., 1994). Evidence for abnormal dopamine neurotransmission in TS is inferred from two clinical observations. First, blockade of dopamine receptors by neuroleptic drugs suppresses tics in a majority of patients. In addition, dopamine-releasing drugs precipitate or exacerbate tics (Scahill et al., 2001). Indeed, it has been shown that TS patients release more dopamine in response to amphetamine compared to normal controls at dopaminergic synapses (Singer et al., 2002). Second, the importance of dopamine in TS is supported by brain imaging using dopamine ligands in single photon emission computed tomography (SPECT) and positron emission tomography (PET). For example, in one twin study involving five pairs, tic severity was related to D_2 dopamine receptor binding in the head of the caudate (Wolf et al., 1996).

Inputs from ascending dopamine pathways originating in the substantia nigra, pars compacta, play a crucial role in coordinating the output from the striatum (Aosaki et al., 1994). Explicit "dopamine" hypotheses for TS posit either an excess of dopamine or an increased sensitivity of D_2 dopamine receptors. These hypotheses are consistent with multiple lines of empirical evidence as well as emerging data from animal models of habit formation. First, data implicating central dopaminergic mechanisms include the results of double-blind clinical trials in which haloperidol, pimozide, tiapride, and other neuroleptics that preferentially block dopaminergic D_2 receptors have been found to be effective in the temporary suppression of tics for a majority of patients (see Scahill et al., 2003 for a review). Second, tic suppression has also been reported following administration of agents such as tetrabenazine that reduce dopamine synthesis (see Kenney & Jankovic 2006 for a review). Third, increased tics have been reported following withdrawal of neuroleptics or following exposure to agents that increase central dopaminergic activity such as L-dopa and CNS stimulants, including cocaine (see Anderson et al., 1998, for a review). Fourth, using imaging techniques a number of investigators also have found increased levels of dopaminergic innervation of the striatum in TS subjects compared with controls (Albin et al., 2003; Cheon et al., 2004; Malison et al., 1995; Muller-Vahl et al., 2000; Serra-Mestres et al., 2004). Although not all studies agree, the strongest evidence suggests density of dopaminergic terminals in the ventral striatum (Albin et al., 2003). Finally, postmortem brain studies have reported alterations in the number or affinity of presynaptic dopamine carrier sites in the striatum of TS subjects (Singer et al., 1991). In sum, although the evidence that dopaminergic pathways are intimately involved in the pathobiology of TS is compelling, the exact nature of the abnormality remains to be elucidated but likely involves the potentiation of limbic inputs to the striatum.

1. Excitatory and Inhibitory Amino Acid Systems

The excitatory neurotransmitter, glutamate, is released upon depolarization by the corticostriatal, corticosubthalamic, subthalamic, and thalamocortical projection neurons. As such, these excitatory neurons are key players in the functional anatomy of the basal ganglia and the CSTC loops. Although the activity of these neurons is likely to be important in TS, only limited data are available to evaluate their role in this disease. For example, Anderson and colleagues (1992) have hypothesized that the disinhibition of thalamocortical projection neurons may be due in part to a failure of the subthalamic nucleus to activate the inhibitory output neurons of the basal ganglia. However, this speculation is based on the results of an examination of just four postmortem brain specimens.

Neurons containing inhibitory amino acid neurotransmitters, particularly gamma aminobutyric acid (GABA), also form major portions of CSTC loops and are also present in key sets of interneurons in the cortex, striatum, and thalamus. These include GABAergic MSs of the striatum that project to the internal segment of the globus pallidus (GP) and the pars reticulata of the substantia nigra within the "direct pathway." GABAergic neurons are also present in the "indirect pathway" that relays information from the striatum to the external segment of the GP and from there to the internal segment of the GP. An imbalance between the direct and indirect pathways has been hypothesized in TS (Leckman et al., 1991b). However, there is very little direct evidence available to support this hypothesis. More promising are the results of a recent postmortem study suggesting that in cases of persistent refractory TS there are major changes in the number of FSNs in the caudate and an increase in the number of GABAergic projection neurons in the GPi (Kalanithi et al., 2005). This finding requires replication and it is unclear if this reflects part of the pathobiology of TS or if it is an adaptive response. There is also increasing evidence that agents that influence GABA transmission may have a role to play in the management of TS (Awaad et al., 2005; Singer et al., 2001).

2. Other Neurotransmitters and Neuromodulators

The large aspiny cholinergic interneurons found throughout the striatum are likely to be critically involved in the coordination of striatal response through interactions with central dopaminergic and GABAergic neurons (Aosaki et al., 1994; Wang & McGinty, 1997). Cholinergic projections from the basal forebrain are found throughout the cortex and within key structures of the basal ganglia and mesencephalon, including the internal segment of the GP, the pars reticulata of the substantia nigra, and the locus coeruleus. Evidence of cholinergic involvement in the pathobiology of TS concerns the reported potentiation of D_2 dopamine receptor blocking agents through the use of transdermal nicotine or nicotine gum (Sanberg et al., 1997). Finally, measurements in cortical postmortem brain tissue have failed to find a difference between cases and controls (Singer et al., 1990).

Endogenous opioid peptides (EOP), localized in structures of the extrapyramidal system, are known to interact with central dopaminergic and GABAergic neurons, and are likely to be importantly involved in the gating of motor functions. Two of the three families of EOPs, dynorphin and met-enkephalin, are highly concentrated and similarly distributed in the basal ganglia and substantia nigra. In addition, significant levels of opiate receptor binding have been detected in both primate and human neostriatum and substantia nigra.

EOPs have been directly implicated in the pathophysiology of TS. Haber and coworkers reported decreased levels of dynorphin A(1-17) immunoreactivity in striatal fibers projecting to the GP in postmortem material from a small number of TS patients (Haber et al., 1986). This observation, coupled with the neuroanatomic distribution of dynorphin, its broad range of motor and behavioral effects, and its modulatory interactions with striatal dopaminergic systems, suggested that dynorphin might have a key role in the pathobiology of TS. However, subsequent studies have failed to confirm these initial observations (Anderson et al., 1998; van Wattum et al., 1999).

Noradrenergic projections from the locus coeruleus project widely to the prefrontal and other cortical regions. Noradrenergic pathways are also likely to indirectly influence central dopaminergic pathways via projections to areas near the ventral tegmental area (Grenhoff & Svensson, 1989). Speculation that noradrenergic mechanisms might be relevant to the pathobiology of TS was based initially on the beneficial effects of alpha-2 adrenergic agonists including clonidine (Cohen et al., 1979). Although clonidine is one of the most widely prescribed agents for the treatment of tics, its effectiveness remains controversial (Goetz et al., 1987; Leckman et al., 1991a). In open trials another related alpha-2 adrenergic agonist, guanfacine, also has been reported to reduce tics and improve ADHD symptoms (Chappell et al., 1995). At the level of receptor function, clonidine traditionally has been viewed as a selective α_2 adrenoceptor agonist active at presynaptic sites, and its primary mode of action may be its ability to reduce the firing rate and the release of norepinephrine from central noradrenergic neurons. However, evidence of heterogeneity among the alpha-2 class of adrenoceptors and their distinctive distribution within relevant brain regions adds further complexity. Specifically, differential effects in cortical regions mediated by specific receptor subtypes may account for the differential responsiveness of particular behavioral features of this syndrome to treatment with clonidine versus guanfacine (Arnsten, 2000).

The involvement of the noradrenergic pathways may be one of the mechanisms by which stressors may influence tic severity. For example, a series of adult TS patients were found to have elevated levels of CSF norepinephrine (Leckman et al., 1995) and to have excreted high levels of urinary norepinephrine in response to the stress of the lumbar puncture (Chappell et al., 1994a). These elevated levels of CSF norepinephrine may also contribute to the elevation in CSF corticotopin releasing factor levels seen in some TS patients (Chappell et al., 1996).

Ascending serotonergic projections from the dorsal raphe have been invoked repeatedly as playing a role in the pathophysiology of both TS and OCD. The most compelling evidence relates to OCD and is based largely on the well-established efficacy of potent serotonin reuptake inhibitors (RUIs) such as clomipramine and fluvoxamine in the treatment of OCD (Greist & Jefferson, 1998). However, some investigators have reported that the serotonin RUIs are less effective in treating tic-related OCD compared to other forms of OCD (McDougle et al., 1993). It is also doubtful that treatment with serotonin RUIs diminishes tic symptoms (Scahill et al., 1997). Additional evidence has come from pharmacological challenge studies in which serotonergic agonists such as m-chlorophenylpiperazine (mCPP) were found to exacerbate OC symptoms in some patients (Zohar et al., 1987). However, not all OCD patients show this response (Goodman et al., 1995) and this agent does not appear to exacerbate tics in TS patients (Cath et al., 1999). Finally, preliminary postmortem brain studies in TS have suggested that serotonin and the related compounds tryptophan (TRP) and 5-hydroxy-indoleacetic acid (5-HIAA) may be globally decreased in the basal ganglia and other areas receiving projections from the dorsal raphe (Anderson et al., 1992).

E. Gender-Specific Endocrine Factors

Males are more frequently affected with TS than females (Shapiro et al., 1988). Although this could be due to genetic mechanisms, frequent male-to-male transmissions within families appear to rule out the presence of an X-linked vulnerability gene. This observation has led us to hypothesize that androgenic steroids act at key developmental periods to influence the natural history of TS and related disorders (Peterson et al., 1992). These developmental periods include the prenatal period when the brain is being formed, adrenarche when adrenal androgens first appear at age 5 to 7 years, and puberty. Androgenic steroids may be responsible for these effects or they may act indirectly through estrogens formed in key brain regions by the aromatization of testosterone.

Surges in testosterone and other androgenic steroids during critical periods in fetal development are known to be involved in the production of long-term functional augmentation of subsequent hormonal challenges (as in adrenarche and during puberty) and in the formation of structural CNS dimorphisms (Sikich & Todd, 1988). In recent years several sexually dimorphic brain regions have been described, including portions of the amygdala (and related limbic areas) and the hypothalamus (including the medial preoptic area that mediates the body's response to thermal stress) (Boulant, 1981). These regions contain high levels of androgen and estrogen receptors and are known to influence activity in the basal ganglia both directly and indirectly (Fehrbach et al., 1985). Indeed, a proportion of TS patients appears to be uniquely sensitive to thermal stress so that when their core body temperature increases, and they begin to sweat, their tics increase (Lombroso et al., 1991). It is also of note that some of the neurochemical and neuropeptidergic systems implicated in TS and related disorders, such as dopamine, serotonin, and the opioids, are involved with these regions and appear to be regulated by sex-specific factors.

Further support for a role for androgens comes from anecdotal reports of tic exacerbation following androgen use (Leckman & Scahill, 1990) and from trials of antiandrogens in patients with severe TS and/or OCD (Altemus et al., 1999; Casas et al., 1986; Peterson et al., 1998b). In the most rigorous study to date, Peterson and colleagues (1998) found that the therapeutic effects of the antiandrogen, flutamide, were modest in magnitude and that these effects were short-lived, possibly because of physiologic compensation for androgen receptor blockade. This view, that sex steroids of gonadal origin organize the neural circuits of the developing brain, recently has been challenged by the finding that independent of the masculinizing effects of gonadal secretions XY and XX brain cells have different patterns of gene expression that influence their differentiation and function. Remarkably, one of the male specific genes, Sry, is specifically expressed in the dopamine containing cells of the substantia nigra (Dewing et al., 2006). Based on this study it appears that Sry directly affects the function of these dopaminergic neurons and the specific motor behaviors they control. Since these results demonstrate a direct male-specific effect on the brain by a gene encoded only in the male genome, without any mediation by gonadal hormones, it is possible that allelic variation or epigenetic effects at this locus may be important in the pathophysiology of TS.

F. Perinatal Risk Factors

The search for nongenetic factors that mediate the expression of a genetic vulnerability to TS and related disorders also has focused on the role of adverse perinatal events. This interest dates from the report of Pasamanick and Kawi (1956), who found that mothers of children with tics were 1.5 times more likely to have experienced a complication during pregnancy than the mothers of children without tics. Other investigations have reported that among monozygotic twins discordant for TS, the index twins with TS had lower birth weights than their unaffected cotwins (Hyde et al., 1992; Leckman et al., 1987). Severity of maternal life stress during pregnancy, and severe nausea and/or vomiting during the first trimester have also emerged as potential risk factors in the development of tic disorders (Leckman et al., 1990). In 1997, Whitaker and coworkers reported that premature and low birth weight children are at increased risk of developing tic disorders and ADHD. This appears to be especially true of children who had ischemic parenchymal brain lesions.

More recently, Burd and colleagues (1999) presented the results of a case control study in which low Apgar scores at 5 minutes and *more* prenatal visits were associated with a higher risk of TS. Finally, there is limited evidence that smoking and alcohol use, as well as forceps delivery, can predispose individuals with a vulnerability to TS to develop comorbid OCD (Santangelo et al., 1994). However, investigations into the effects of perinatal complications into the later development of TS are hampered by the possibility that recall bias likely influences results in the retrospective case-control studies addressing this question, and by multiple hypothesis testing without appropriate statistical correction. The only nested case-control study to date that examined TS cases arising in a Swedish community sample found that the mothers of children with TS were two times more likely to have had complications during pregnancy and were younger than control mothers when they gave birth to the index child (Khalifa & von Knorring, 2005).

G. Post-Infectious Autoimmune Mechanisms

It is well established that group A beta hemolytic streptococci (GABHS) can trigger immune-mediated disease in genetically predisposed individuals (Bisno, 1991). Speculation concerning a post-infectious (or at least a post-rheumatic fever) etiology for tic disorder symptoms dates from the late 1800s (Kushner, 1999). Acute rheumatic fever (RF) is a delayed sequela of GABHS, occurring approximately three weeks following an inadequately treated upper respiratory tract infection. RF is characterized by inflammatory lesions involving the heart (rheumatic carditis), joints (polymigratory arthritis), and/or central nervous system (Sydenham's chorea (SC)). The immune response in the CNS of SC patients appears to involve molecular mimicry between streptococcal antigens and self-antigens (Kirvan et al., 2003). SC and TS, OCD, and ADHD share common anatomic targets—the basal ganglia of the brain and the related cortical and thalamic sites (Husby et al., 1976). Furthermore, SC patients frequently display motor and vocal tics, obsessive-compulsive, and ADHD symptoms suggesting the possibility that at least in some instances these disorders share a common etiology (Mercadante et al., 1997; Swedo et al., 1989).

It has been proposed that Pediatric Autoimmune Neuropsychiatric Disorder Associated with Streptococcal infection (PANDAS) represents a distinct clinical entity, and includes SC and some cases of TS and OCD (Swedo et al., 1998). The most compelling evidence of an etiological link between these disorders and GABHS infection comes from a recently published case-control study that found an increased proportion of GABHS infections (odds ratio = 3.05) within the preceding three months in children newly diagnosed with TS compared to well-matched controls (Mell et al., 2005). A larger odds ratio (OR = 12.1 for TS) was demonstrated when subjects were required to have multiple GABHS infections in the previous year, suggesting a dose-dependent effect of exposure. The PANDAS hypothesis is indirectly supported by the presence of high levels of anti-streptococcal antibodies in some patients with TS (Cardona & Orefici, 2001; Church et al., 2003; Muller-Vahl et al., 2000).

Prospective longitudinal studies provide a mixed picture concerning the ability of GABHS infections to induce future tic exacerbations. For example, a recent longitudinal study found no greater temporal relationship between GABHS infection and symptom exacerbations than would be expected by chance alone among 40 children with TS and/or OCD (Luo et al., 2004). Among this study sample there was no consistent change in D8/17-reactive cells with symptom exacerbation (Luo et al., 2004). Another prospective cohort study failed to demonstrate any increase in the occurrence of *de novo* PANDAS symptoms in the three months following a pediatric visit among 411 children ages 4 to 11 that experienced a GABHS infection and 403 uninfected controls (Perrin et al., 2004). However, since these patients were all treated with antibiotics it is possible that a PANDAS presentation occurs only in untreated or predisposed hosts.

A number of studies also have been completed in an effort to detect and characterize the putative cross reactive epitopes. Although it appears that TS patients may have higher levels of circulating antineural antibodies (Church et al., 2003; Kiessling et al., 1993; Morshed et al., 2001; Singer et al., 1998), no specific epitopes have been unequivocally identified (Dale et al., 2004; Kirvan et al., 2006; Singer et al., 2005). Animal models that involve the injection of patient sera high in levels of antineural antibodies into tic-related basal ganglia areas have also led to equivocal results (Singer et al., 2005; Taylor et al., 2002). However, there is evidence that tic and OC symptoms may improve following plasma exchange in selected patients with the PANDAS phenotype (Allen et al., 1995; Perlmutter et al., 1999).

Additional evidence that acute exacerbations of TS and OCD could also be triggered by GABHS comes from four independent reports demonstrating that the majority of patients with childhood-onset TS or OCD have elevated expression of a stable B-cell marker (Hoekstra et al., 2001; Luo et al., 2004; Murphy et al., 1997; Swedo et al., 1997). The D8/17 marker identifies close to 100 percent of RF patients (with or without SC) but is present at low levels of expression in healthy control populations. The identity of the D8/17 epitope is not yet known, but it can be expressed by several non-B-cell types (Kemeny et al., 1994). Although this may emerge as a useful biomarker, more recent evidence suggests that these reports may reflect an increased expression of receptors for the constant parts of IgM molecules on B cells rather than the D8/17 epitope (Hoekstra et al., 2004).

The bias for studies addressing the role of B cell immunity in TS pathogenesis has been driven by the common view that GABHS is an extracellular bacteria and that its clearance is mediated primarily by antibodies. However, GABHS can be internalized in human cells with the same frequency as the classical intracellular pathogens *Listeria* or *Salmonella* species (LaPenta et al., 1994) and activates human T lymphocytes (Degnan et al., 1997). A pathogenic role of T-cell immunity has been clearly demonstrated in post-streptococcal rheumatic heart disease (Cunningham et al., 1997; Guilherme & Kalil, 2004). However, the role of T cells in TS is just beginning to be addressed with the finding that there may be a reduced number of regulatory T cells in selected patients with TS (Kawikova et al., 2007). Likewise there is a single report that suggests that individuals with TS may have elevated levels of pro-inflammatory cytokines (Leckman et al., 2005).

In sum, a substantial body of circumstantial evidence exists that links post-infectious autoimmune phenomena with TS, OCD, and ADHD. However, these data are not compelling with regard to specific immunological mechanisms (Giovannoni, 2006; Singer & Williams, 2006; Swedo & Grant, 2005), nor do they establish where in the sequence of causal events these immune changes occur. These potentially important findings require replication in independent samples and warrant more intensive investigation using prospective longitudinal designs.

H. Psychological Factors

Tic disorders have long been identified as "stress-sensitive" conditions (Jagger et al., 1982; Shapiro et al., 1988; Silva et al., 1995). Typically, symptom exacerbations follow in the wake of stressful life events. As noted by Shapiro and colleagues (1988), these events need not be adverse in character. Clinical experience suggests that in some unfortunate instances a vicious cycle can be initiated in which tic symptoms are misunderstood by the family and teachers, leading to active attempts to suppress the symptoms by punishment and humiliation. These efforts can lead to a further exacerbation of symptoms and further increase the stress in the child's interpersonal environment. Unchecked, this vicious cycle can lead to the most severe manifestations of TS and dysthymia as well as maladaptive characterological traits. Although psychological factors are insufficient to cause TS, the intimate association of the content and timing of tic behaviors and dynamically important events in the lives of children make it difficult to overlook their contribution to the intramorbid course of these disorders (Carter et al., 1994, 2000).

Prospective longitudinal studies have begun to examine more carefully the effect of intramorbid stress. These studies indicate that patients with TS experience more stress than matched healthy controls (Findley et al., 2003; Lin et al., 2007) and that antecedent stress may play a role in subsequent tic exacerbation (Lin et al., 2007). Increases in depressive symptoms also have emerged as a significant predictor of future tic severity (Lin et al., 2007).

In addition to the intramorbid effects of stress, anxiety, and depression that are beginning to be well characterized, premorbid stress may also play an important role as a sensitizing agent in the pathogenesis of TS among vulnerable individuals (Leckman et al., 1984). It is likely that the immediate family environment (e.g., parental discord) and the coping abilities of family members play some role (Leckman et al., 1990), and this may lead to a sensitization of stress responsive biological systems such as the hypothalamic-pituitary-adrenal axis (see earlier; Chappell et al., 1994b, 1996; Leckman et al., 1995).

VI. Integrative Hypotheses—Neural Oscillations

Our current hypotheses concerning the etiology of TS focus on the putative role of aberrant neural oscillations (Leckman et al., 2006). Although there are several mechanisms capable of generating oscillatory rhythms in central neurons (Llinas et al., 2005), the manner in which they are transmitted and transformed is likely to be fundamentally important with regard to the emergence of goal-directed behavior and the formation of habits. As with respiration and sleep-wake-cycle, the expression of habits and tics is likely to depend upon the quasi-random oscillatory activity of groups of neurons. Their synchronization with phase-locking and frequency stabilization can be seen as a resonance phenomenon of the brain. Selectively distributed oscillatory systems of the brain exist as resonant communication networks through large populations of neurons, which work in parallel and are interwoven with sensory, motor, cognitive, and emotional functions (Buzsaki & Draguhn, 2004). Oscillatory activity has long been recognized in the EEG, in which synchrony between thalamus and cortex can be observed in different frequencies. These oscillations are usually subdivided into types on the basis of their characteristic frequencies such as *delta* (<2Hz), *theta* (~2–7 Hz), *alpha* (~8–12 Hz), *beta* (~15–30 Hz), and *gamma* (~30–80 Hz).

In the case of TS, the prevailing theory is that there is an increased activity of the inhibitory direct pathway connecting the basal ganglia striatal input to its output in the GPi and decreased activity in the indirect pathway via the GPe, and the STN. The net effect of this imbalance between the pathways is to reduce the firing rate of inhibitory neurons in GPi that project to the thalamus. The resulting disinhibition of a discrete set of thalamo-cortical projection neurons provides the anatomical basis for the emergence of tics (Mink, 2001). A prediction of this model of TS and other hyperkinetic movement disorders (such as dystonia and chorea) is that these disorders are associated with low firing rates in GPi output neurons. However, recent advances in basal ganglia research (Graybiel, 2005) and the observation that the electrical activity GPi seen in dystonia and tic disorders is similar to that in Parkinson's disease (Hutchison et al., 2003; Marsden & Obeso, 1994; Zhuang et al., 2004a, 2004b) have challenged the prevailing wisdom. Similarly, the therapeutic inactivation of GPi, either by high-frequency stimulation or ablative surgery, improves tics and dyskinesias as well as Parkinson's symptoms (Marsden & Obeso, 1994). These contradictions, taken together with the evident functional importance of *oscillatory processes*, led us to hypothesize that one or more of these oscillatory processes are aberrant in TS and related disorders. Moreover, the possibility must be examined that as in the case of Parkinson disease, the abnormal activity in fact may be due to excess inhibition giving rise to thalamic oscillatory activity due to the activation of cationic current I(h) (Bal & McCormick, 1997) leading to calcium-dependent currents producing rebound low threshold calcium spikes (Llinas & Jahnsen, 1982).

A. Hypothesis 1: Loss of Basal Ganglia Control

In TS, we hypothesize first, that "at rest" clusters of MSs associated with tics in the dorsal lateral striatum become disengaged from their usual episodic strong entrainment to synchronized oscillatory activity. This disengagement may mediate in part the relentless drumbeat of sensory information that besieges the conscious mind of the individual with TS. We propose that this vulnerability to autonomous disengagement may be due to a number of factors including a relative loss of FSNs in the striatum leading to an increased effect of cortical sensory inputs to the MSs.

Preliminary anatomical, neurosurgical, and psychopharmacological evidence from TS patients as well as data from animal models provide support for our first hypothesis. The quantitative anatomical examination of the brains of three individuals severely affected with TS revealed altered numbers of FSNs in the striatum, among other abnormalities (Kalanithi et al., 2005). The number of FSNs (identified by their immunoreactivity for the calcium-binding protein parvalbumin (PV)) was reduced by 54 percent in the striatum and by 39 percent in the caudate and putamen (see Figure 25.3). In accordance with our first hypothesis, the loss of these FSNs cells would allow clusters of MSs within the somatotopic areas associated with tics to become disengaged from the high-voltage spindle oscillations and to become relatively autonomous, giving rise to tics.

Physiological studies demonstrate that the cerebral cortex uses FSNs to exert powerful feed-forward inhibition upon MSs (Mallet et al., 2005), suggesting that this inhibitory system with its ensuing oscillatory activity may be used to suppress and filter out of consciousness unattended patterns of activity. Hence, we also predict that with the disruption of the basal ganglia oscillation many TS patients would be easily distracted and would have difficulty performing the saccade tasks similar to those described by Courtemanche et al. (2003). This, indeed, appears to be the case as documented by Nomura et al. (2003) in their study of saccades in more than 100 children and adolescents with TS.

A decreased inhibitory influence of FSNs over MSs could allow the cortical sensorimotor inputs to more easily activate MSs, eliciting the premonitory urges experienced by individuals with TS (Banaschewski et al., 2003; Leckman et al., 1993). This speculation is consistent with the hypothesis that sensory inputs guide the basal ganglia in action selection and that such sensory inputs can switch selected regions of the cortex-basal ganglia circuitry from an "idling state" characterized by the high voltage spindles to an active state characterized by gamma band synchrony (Brown & Marsden, 1998). It would also seem likely that because the FSNs play a key role in the flexible entrainment of a variety of oscillatory patterns in the basal ganglia, inappropriate tic-related oscillations (in the gamma band) involving the sensorimotor cortex and the basal ganglia, would be more readily established when fewer inhibitory FSNs are present. Over time, tics may lead to fixed patterns of synchronous oscillations within the gamma band (possibly because of aberrant patterns of synaptic plasticity?). These tic-specific patterns would be likely to interfere with normal functioning of these circuits (e.g., procedural memory tasks), and it has been documented that individuals with TS have deficits in this arena of cognitive function (Keri et al., 2002; Marsh et al., 2004).

Premonitory sensory phenomena are quite often subjectively located in peripheral muscle groups, indicating that there exists some subliminal sensory registration, which may serve as feedback to the brain. Such thalamocortical circuit activity may increase slowly in amplitude and finally give rise to a tic-specific fixed action pattern of thalamocortical oscillatory activity that becomes increasingly difficult to inhibit once a certain intensity has been reached. It seems likely that the premonitory urges arise at the thalamic level (perhaps as a result of faulty GPi activity), and it is probable that other circuits active during central-peripheral feedback processes may play a role in tic occurrence (Rothenberger, 1991).

In addition to these neuropathological findings, neurosurgical procedures used to treat refractory cases of TS also support our first hypothesis. It has been recently reported that deep brain stimulation (or the lesioning) of the intralaminar nuclei of the thalamus may work by reestablishing more normal, tic-suppressing oscillatory patterns via its effect on TANs and the indirect stimulation of the FSNs in the striatum (Temel & Visser-Vandewalle, 2004).

Although ideal anti-tic treatments are not presently available, there is a large body of data that indicates that agents that potently block postsynaptic D_2 receptors can be helpful in achieving a reduction in the severity of tics in some cases. Pimozide, haloperidol, sulpiride, and tiapride are the most frequently used typical neuroleptics, whereas risperidone and ziprasidone are two atypical neuroleptics with proven tic-suppressant efficacy (Scahill et al., 2003). Remarkably, intrastriatal injections of dopamine D_2 receptor antagonists (but not D_1 receptor antagonists) *greatly increase* the incidence of the cortical high-voltage spindles (Buzsaki et al., 1990; Semba & Komisaruk, 1984). The beneficial effects of alpha adrenergic agonists on tics are more controversial (Scahill et al., 2003), but these agents have also been reported to *increase* the incidence of the anti-kinetic cortical high-voltage spindles (Riekkinen et al., 1992).

Finally, the functional significance of the loss of these neurons is manifest in an animal model of idiopathic paroxysmal dystonia in which there is a 30 to 50 percent loss of the FSN population (Gernert et al., 2002). Remarkably, the phenotype of these animals includes facial contortions, hyperextension of limbs, and other dystonic postures associated with cocontractions in opposing muscle groups (Loscher et al., 1989), all features seen in severe cases of TS. In addition, these motor symptoms show an age-dependent reduction in severity that is similar to the natural history of TS (Gernert et al., 2002). Hence, maturational effects of basal ganglia and frontal lobes leading to better entrainment of the FSN inhibitory system in cortex and striatum might explain to a certain extent the reduction of tics in late adolescence.

B. Hypothesis 2: Thalamocortical Dysrhythmia

Next, we hypothesize that the normal patterns of discharge from the basal ganglia output nuclei are finely modulated oscillations, and that these oscillations are disrupted in TS similarly to what is seen in patients with dystonia. The irregular bursting of the GPi projection neurons would transiently hyperpolarize selected thalamocortical neurons, causing them to transiently increase the amplitude of their high-frequency membrane potential oscillations (20–80 Hz) which in turn would lead to the ectopic activation of selected cortical pyramidal neurons, leading to the overt and/or subliminal perception of premonitory urges and the performance of tics.

Circumstantial evidence derived from intra-operative recordings from patients with severe dystonia and patients with refractory TS, postmortem brain studies, and animal models of hyperkinetic movement disorders provide limited support for our second hypothesis. Single unit intra-operative recordings have been performed in both dystonic and severe tic disorder patients with similar results—the clear presence of altered neuronal slow activity including grouped discharges in the GPi that were in turn correlated with the EMG signals at the frequency of the abnormal movements in the range of 0.12 to 0.84 Hz (Vitek et al., 1999; Zhuang et al., 2004a, 2004b). Remarkably, abolishing this activity through electrolytic lesions in the GPi resulted in an immediate improvement of tics (Zhuang et al., 2004a) as well as of severe dystonia (Lozano et al., 1997; Vitek et al., 1998). Similar benefits have been reported following bilateral deep brain stimulation of the GPi in a single TS patient (Houeto et al., 2005).

The synchronous ultra-slow activity in the multisecond range (2–60 sec and longer) that is found in the GP of experimental animals (Ruskin et al., 2003) may also be dysregulated in TS (see Figure 25.4; Moran et al., 2005).

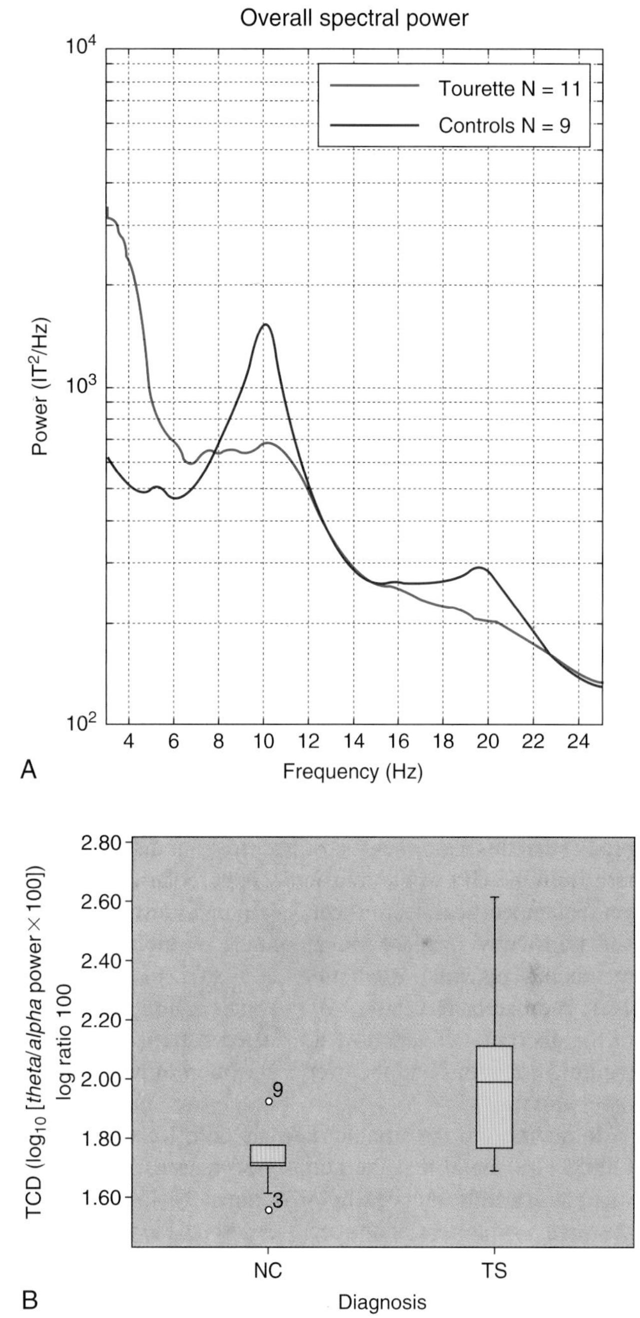

Figure 25.4 Altered overall power spectra in Tourette's syndrome. **A.** Eleven TS patients show a marked loss of power in the *alpha* range and increased power at *theta* frequencies (red line) relative to nine normal control subjects (blue line). **B.** Box plot comparing $\log_{10}$ transformation of the *theta/alpha* power ratio in nine healthy adult subjects (NC) vs. 11 patients with TS. Adapted from Moran et al. (2005).

These oscillations are very sensitive to the presence of dopamine agonists and substantial evidence exists for a heightened dopaminergic innervation of the basal ganglia in some individuals with TS (Albin et al., 2003; Cheon et al., 2004;

Heinz et al., 1998; Malison et al., 1995; Singer et al., 1991, 2002; Stamenkovic et al., 2001; Wolf et al., 1996). Remarkably, chronic administration of apomorphine in rats for one year is associated with multiweek oscillations in the frequency of stereotypes (Csernansky et al., 1986). These temporal patterns bear a resemblance to the occurrence of tics in bouts over seconds to minutes as well as to the waxing and waning of tic symptoms over weeks to months (Leckman et al., 1998; Peterson & Leckman, 1998). These multisecond oscillations also recently have been implicated in the variability of neuropsychological performance of children with ADHD. Future research is needed to determine whether these apparent similarities are based on a common set of processes (Castellanos et al., 2005).

In addition, recent preliminary postmortem studies of the brains from individuals with severe TS provide further indirect evidence that the electrical activity in the pallidum may be aberrant. The most extraordinary finding from the same postmortem study discussed earlier (Kalanithi et al., 2005) was the *120 percent increase* in the PV-positive GABAergic projection neurons in the GPi (coexisting with the decrease of striatal interneurons; see earlier) in the three patients with extremely severe TS. The functional consequences of this increase are as yet unclear, but it would not be surprising if this increase led to the "irregular neuronal activity with grouped discharges" described by Zhuang et al. (2004b). We predict that this abnormally slow and irregular discharge pattern from the GPi would transiently hyperpolarize their target thalamocortical neurons causing them to reset the phase and transiently increase the amplitude of high-frequency membrane potential oscillations (20–80 Hz) (see Figure 25.1; Pedroarena & Llinas, 1997), which in turn would lead to the aberrant activation of a selected pattern of cortical pyramidal neurons and the overt perception of premonitory urges and tics.

In addition to the striatum and the GPi, Kalanithi et al. (2005) also found that the number of neurons in the GPe was significantly reduced by approximately 45 percent in the three postmortem brain specimens. Since the GPe inhibitory neurons directly project to the GPi, this reduction likely could contribute to the abnormal electrical activity in GPi noted previously.

In dt^{sz} hamsters, whose phenotype closely resembles severe TS and is characterized by a loss of striatal FSNs, there is an age-dependent shift toward grouped-irregular and burst-like firing in the EPN but not in the SNr (Gernert et al., 2002). Unfortunately, whether these animals show an increase in PV-positive neurons in the EPN has not been determined. We also note that these same animals show a marked increase in their symptoms following systemic or striatal injections of amphetamine or dopamine D_2 agonists and a moderate decrease in symptoms (~33%) following the systemic administration of the D_2 antagonist raclopride (Rehders et al., 2000). These findings are not too dissimilar to those seen in TS patients treated with D_2 antagonists (Scahill et al., 2003). These animals also show an increase in D_2 receptor binding in the shell of the nucleus accumbens (Nobrega et al., 1996), a finding that may be consistent with the observations of Wolf and colleagues (1996).

C. Hypothesis 3: Frontal Lobe Compensation

We hypothesize that compensatory systems originating in the prefrontal cortex (PFC) adaptively modulate the misguided striatal and thalamocortical oscillations that are characteristic of TS. Repeated activation of these frontal systems can lead not only to tic suppression, but also to a willful alteration of the character of the movements involved.

Surface EEG recordings, functional and structural neuroimaging studies, as well as the recent success of Habit Reversal Training in the treatment of mild to moderate tics all provide circumstantial support for our third hypothesis. Serrien and coworkers (2005) recently identified sensorimotor-frontal neuronal activity apparently involved in the acute suppression of involuntary tics, as evidenced by increased EEG coherence in the alpha frequency band (8–12 Hz) range during suppression of voluntary movements in individuals with TS compared with healthy subjects during a Go-NoGo task. Unfortunately, the specificity of this oscillatory system is unclear, since it was the only one investigated. Earlier, Rothenberger (1990, 1995) found a frontal shift of *Bereitschaftspotential*, a movement-related cognitive potential indicative of motor preparation, when tic patients performed a voluntary movement, which was normalized after intake of D_2 receptor blockers. Both of these clinical studies point to a dynamic circuit involved in normal inhibition of movement that is increased in activity in TS and that can be further activated acutely in the voluntary suppression of tics.

Tic suppression also has been studied using functional MRI (Peterson et al., 1998a). This study revealed a consistent pattern of activations (right frontal cortex, right caudate nucleus, and regions of the cingulate and temporal cortices) and deactivations bilaterally in subcortical areas (putamen, GP, and thalamus). Remarkably, increased activity in the right frontal cortex was associated with increased activity in the right caudate nucleus, and increased activity in the right caudate nucleus in turn was associated with greater decreases in activity of the GP, the putamen, and the thalamus during tic suppression (see Figure 25.1). The decreased activities of the GP and the putamen were correlated, and both positively predicted thalamic activity. Equally impressive is the fact that the levels of activation (right caudate) or deactivation (putamen, GP, and thalamus) during periods of tic suppression were correlated with the individual's current level of tic severity outside the magnet.

The brain regions that are activated during tic suppression are virtually identical to those that have been described as belonging to a distributed neural circuit that participates in the inhibition of unwanted impulses (Goldman-Rakic, 1987). This circuit consists of the prefrontal, parietal, temporal, and cingulate cortices, which are thought to modulate the activity in the basal ganglia and thalamus. Precisely how this circuit inhibits inappropriate responses allowing the desired behavioral responses in the motor cortex to occur is not entirely clear. We suggest that this regulation depends upon the integration and cross talk of the prefrontal circuits with the sensorimotor circuits in the basal ganglia and thalamus leading to an enhanced parallel synchronization of different oscillatory systems in motor cortex and basal ganglia depending on the task. This is more difficult in children with tics, since they show a general motor circuit inhibitory deficit, which in adults is confined to the very circuit related to the observable tics (Moll et al., 2005).

Persistence of tic symptoms into adulthood may reflect a failure of this circuitry to operate properly because of anatomical limitations. For example, two groups of investigators have independently documented that smaller caudate volumes are a trait marker for TS (Hyde et al., 1995; Peterson et al., 2003). We note in passing that the animal model that most closely resembles severe TS also has a 13 percent reduction in striatal volume (Gernert et al., 2000). It appears that the smaller an individual's caudate is in childhood, the more likely he or she is to have persistent tics into adulthood (Bloch et al., 2005). The nature of this volume reduction is being studied in postmortem tissue using unbiased stereological methods. Whether there are other findings besides the 50 percent reduction in the number of PV-positive FSNs awaits future investigations.

One would also speculate that frontal abnormalities such as those seen in ADHD might limit an individual's ability to mobilize this inhibitory system (Booth et al., 2005; Filipek et al., 1997; Fredericksen et al., 2002). In the individuals that are able to compensate, one might also expect to see prefrontal hypertrophy in an effort to regulate tic symptoms (Peterson et al., 2001a, 2001b) as well as other adaptive changes in other structures including the corpus callosum (von Plessen et al., 2004).

We also posit that it is via this distributed neural circuit that treatments such as Habit Reversal Training work in some cases of TS (Deckersbach et al., 2005; Wilhelm et al., 2003). The presence of premonitory sensations distinguishes TS from other movement disorders such as Parkinson disease, Huntington's chorea, and hemiballismus. Acutely aware of their premonitory urges, many adult patients report (or can be taught) that they can inhibit tics or change the tic through a variety of behavioral strategies like *competing motor responses* (alternative actions that are by their nature incompatible with the performance of a specific tic) made in response to the premonitory urges.

In agreement with Woods et al. (2003), we regard the competing response as a voluntary act that requires attention. We further contend that the prefrontal cortex (PFC), applying a competing response contingent on actual tic occurrence over time, uncouples the discriminative stimulus (i.e., the premonitory urge) from the performance of the tic. How this willful act affects this outcome is unclear and raises metaphysical questions beyond the scope of this annotation (Nagel, 1974; Schwartz & Begley, 2002; Schwartz et al., 2005). By increasing the input from prefrontal areas it may be possible to alter the functional association of premonitory urge and tic expression and diminish the power of the urge to initiate the vicious cycle of tics and urges via the various nigral, pallidal, and cerebellar channels. Lastly, relaxation training is conducted to reduce the amplitude of peripheral-central sensorimotor feedback, which probably diminishes limbic afferent inputs to the striatum and decreases dopaminergic tone from the SNc. We predict that if surface EEG recordings (analyzed with recent three-dimensional localization algorithms) or functional MRI studies of tic suppression were completed before and after successful Habit Reversal Treatment we would see both a further increase in frontomesial EEG coherence in the alpha frequency band (8–12 Hz) after training, as well as a larger area of neural activation both in the right prefrontal area and in the right caudate nucleus, similar to what is seen after neurofeedback training in ADHD (Heinrich et al., 2004). Further, reduction of tics in patients with TS plus ADHD while taking either methylphenidate (a dopamine agonist) or atomoxetine (a norepinephrine agonist) might be explained by an increase of dopamine in the frontal cortex by both substances (Banaschewski et al., 2003).

D. Are These Hypotheses Testable?

We posit that tics are governed by a multiplicity of factors that influence oscillatory patterns within basal ganglia networks, particularly those that interface with frontal, limbic, and cerebellar circuits. This model has generated three hypotheses that are consistent with the available scientific literature and that contain many testable elements.

First, studies using magneto-encephalography (MEG) and dense array EEG offer considerable promise particularly when applying these techniques before, during, and after various cognitive-behavioral, pharmacological, and neurosurgical treatments. One key prediction from our first hypothesis is that, at baseline, the number of high voltage spindles (or the beta band oscillations) will be reduced in the basal ganglia of individuals with TS, and that when present, smaller regions of the frontal and somatosensory cortices will participate in these oscillations. We predict that the number of high voltage spindles will increase following successful treatment with both dopamine D_2 antagonists and alpha-2 adrenergic agonists.

Our second hypothesis can also be tested directly using MEG and dense array EEG recordings (Jeanmonod et al.,

2003). Compared to age-matched controls, we would expect to see evidence of thalamocortical dysrhythmia as evidenced by an increase of MEG power at the theta-delta interface and an associated increase of coherence both within this domain and between it and the beta-range (Llinas et al., 2005). We also predict that individuals whose tics remit in adulthood will show less thalamocortical dysrhythmia than adults with active TS. Similarly, recordings of neural oscillations before, during, and after neurosurgical interventions for adults with severe, persistent, refractory, self injurious tics are likely to lead to a better understanding of these phenomena.

Our third hypothesis is best tested in the context of Habit Reversal Training. We predict that successful Habit Reversal Training will lead to enhanced alpha coherence between prefrontal, mesiofrontal, sensorimotor, and motor cortical regions as measured using MEG and dense array EEG. We also anticipate that charting the developmental time course of measures of alpha coherence, and the volume and myelinization status of various cortical regions, particularly the dorsal lateral and mesial prefrontal regions, may advance our understanding of the neurobiological determinants of the natural history of TS.

VII. Future Directions

Along with a deepening appreciation of the clinical phenomenology of TS and related disorders, recent progress in genetics, neuroanatomy, systems neuroscience, and functional *in vivo* neuroimaging has set the stage for a major advance in our understanding of TS and related disorders. Success in this area will lead to the targeting of specific brain circuits for more intensive study. Diagnostic and prognostic advances can also be anticipated; for example, which circuits are involved and to what degree? How does that degree of involvement affect the patient's symptomatic course and outcome?

Given this potential, TS can be considered a model disorder for the study of the dynamic interplay of neurobiological systems during development. It is likely that the research paradigms utilized in these studies and many of the empirical findings resulting from them, will be relevant to other disorders of childhood onset and will enhance our understanding of normal development.

Acknowledgments

We are indebted to Rodolfo Llinas, MD, PhD, and Kerry Walton, PhD, and to the Yale Tourette's Syndrome and Obsessive-Compulsive Disorder Research Group. Current and former members of this research group include Donald J. Cohen, MD; Lawrence Scahill, MSN, PhD; Robert A. King, MD; Matthew T. State, MD, PhD; Flora M. Vaccarino, MD; Heping Zhang, PhD; Haiqun Lin, PhD; Paul J. Lombroso, MD; Robert T. Schultz, PhD; Denis Sukhodolsky, PhD; Heidi Grantz, MSW; Liliya Katsovich, MBA; George M. Anderson, PhD; Diane B. Findley, PhD; James E. Swain, MD, PhD; Linda C. Mayes, MD; Michael Crowley, PhD; Kenneth K. Kidd, PhD; David L. Pauls, PhD; Bradley S. Peterson, MD; John T. Walkup, MD; Alice S. Carter, PhD; Debra E. Bessen, PhD; Syed A. Morshed, MD, PhD; Salina Parveen, MD, PhD; Yukiko Kano, MD, PhD; Marcos Mecadante, MD, PhD; and Maria C. do Rosario Campos, MD, PhD.

Portions of the research described in this review were supported by grants from the National Institutes of Health: MH18268, MH49351, MH30929, HD03008, NS16648, and RR00125, as well as by the Tourette Syndrome Association. Portions of this chapter have appeared elsewhere. The three hypotheses proposed in the final section of the chapter, were first articulated in Leckman et al. (2006).

References

Abelson, J. F., Kwan, K. Y., O'Roak, B. J., Baek, D. Y., Stillman, A. A., Morgan, T. M. et al. (2005). Sequence variants in SLITRK1 are associated with Tourette's syndrome. *Science* **310**, 317–320.

Ackermans, L., Temel, Y., Cath, D. C., van der Linden, C., Bruggeman, R., Kleijer, M. et al. (2006). Deep brain stimulations in Tourette's syndrome: Two targets? *Movement Disorders* **21**, 709–713.

Albin, R. L. (2006). Neurobiology of basal ganglia and Tourette syndrome: Striatal and dopamine function. *Adv Neurol* **99**, 99–106.

Albin, R. L., Koeppe, R. A., Bohne, A., Nichols, T. E., Meyer, P., Wernette, K. et al. (2003). Increased ventral striatal monoaminergic innervation in Tourette sydnrome. *Neurology* **61**, 310–315.

Albin, R. L. and Mink, J. W. (2006). Recent advances in Tourette syndrome research. *Trends Neurosci* **29**, 175–182.

Allen, A. J., Leonard, H. L., and Swedo, S. E. (1995). Case study: A new infection-triggered, autoimmune subtype of pediatric OCD and Tourette's syndrome. *J Am Acad Child Adolesc Psychiatry* **34**, 307–311.

Altemus, M., Greenberg, B. D., Keuler, D., Jacobson, K. R., and Murphy, D. L. (1999). Open trial of flutamide for treatment of obsessive-compulsive disorder. *J Clin Psychiatry* **60**, 442–445.

Anderson, G. M., Leckman, J. F., and Cohen, D. J. (1998). Neurochemical and neuropeptide systems. In J. F. Leckman and D. J. Cohen, Eds., *Tourette's syndrome tics, obsessions, compulsions—Developmental psychopathology and clinical care*, 261–281. John Wiley and Sons, New York.

Anderson, G. M., Polack, E. S., Chatterjee, D., Leckman, J. F., Riddle, M. A., and Cohen, D. J. (1992). Postmortem analysis of subcortical monoamines and aminoacids in Tourette syndrome. In T. N. Chase, A. J. Friedhoff, and D. J. Cohen, Eds., *Tourette syndrome: Genetics, neurobiology, and treatment*, 253–262. Raven Press, Ltd., New York.

Aosaki, T., Graybiel, A. M., and Kimura, M. (1994). Effect of the nigrostriatal dopamine system on acquired neural responses in the striatum of behaving monkeys. *Science* **265**, 412–415.

Arnsten, A. F. (2000). Through the looking glass: Differential noradrenergic modulation of prefrontal cortical function. *Neural Plast* **7**, 133–146.

Awaad, Y., Michon, A. M., and Minarik, S. (2005). Use of levetiracetam to treat tics in children and adolescents with Tourette syndrome. *Mov Disord* **20**, 714–718.

Bal, T. and McCormick, D. A. (1997). Synchronized oscillations in the inferior olive are controlled by the hyperpolarization-activated cation current I(h). *J Neurophysiol* **77**, 3145–3156.

Banaschewski, T., Woerner, W., and Rothenberger, A. (2003). Premonitory sensory phenomena and suppressibility of tics in Tourette syndrome: Developmental aspects in children and adolescents. *Dev Med Child Neurol* **45**, 700–703.

Baron-Cohen, S., Mortimore, C., Moriarty, J., Izaguirre, J., and Robertson, M. (1999). The prevalence of Gilles de la Tourette's syndrome in children and adolescents with autism. *J Child Psychol Psychiatry* **40**, 213–218.

Barr, C. L. and Sandor, P. (1998). Current status of genetic studies of Gilles de la Tourette syndrome. *Can J Psychiatry* **43**, 351–735.

Baumgardner, T. L., Singer, H. S., Denckla, M. B., Rubin, M. A., Abrams, M. T., Colli, M. J., and Reiss, A. L. (1996). Corpus callosum morphology in children with Tourette syndrome and attention deficit hyperactivity disorder. *Neurology* **47**, 477–482.

Bawden, H. N., Stokes, A., Camfield, C. S., Camfield, P. R., and Salisbury, S. (1998). Peer relationship problems in children with Tourette's disorder or diabetes mellitus. *J Child Psychol Psychiatry* **39**, 663–668.

Bisno, A. L. (1991). Group A streptococcal infections and acute rheumatic fever. *N Engl J Med* **325**, 783–793.

Bloch, M. H., Landeros-Weisenberger, A., Kelmendi, B., Coric, V., Bracken, M. B., and Leckman, J. F. (2006a). A systematic review: antipsychotic augmentation with treatment refractory obsessive-compulsive disorder. *Mol Psychiatry*. **11**, 622–632.

Bloch, M. H., Leckman, J. F., Zhu, H., and Peterson, B. S. (2005). Caudate volumes in childhood predict symptom severity in adults with Tourette syndrome. *Neurology* **65**, 1253–1258.

Bloch, M. H., Peterson, B. S., Scahill, L., Otka, J., Katsovich, L., Zhang, H., and Leckman, J. F. (2006b). Adulthood outcome of tic and obsessive-compulsive symptom severity in children with Tourette syndrome. *Arch Pediatr Adolesc Med* **160**, 65–69.

Bohlhalter, S., Goldfine, A., Matteson, S., Garraux, G., Hanakawa, T., Kansaku, K. et al. (2006). Neural correlates of tic generation in Tourette syndrome: An event-related functional MRI study. *Brain* **129**, 2029–2037.

Booth, J. R., Burman, D. D., Meyer, J. R., Lei, Z., Thrommer, B. L., Davenport, N. D. et al. (2005). Larger deficits in brain networks for response inhibition than for visual selective attention in attention deficit hyperactivity disorder (ADHD). *Child Psychol Psychiatry*, 94–111.

Boulant, J. A. (1981). Hypothalamic mechanisms in thermoregulation. *Fed Proc* **40**, 2843–2850.

Braun, A. R., Randolph, C., Stoetter, B., Mohr, E., Cox, C., Vladar, K. et al. (1995). The functional neuroanatomy of Tourette's syndrome: An FDG-PET Study. II: Relationships between regional cerebral metabolism and associated behavioral and cognitive features of the illness. *Neuropsychopharmacology* **13**, 151–168.

Brown, P. and Marsden, C. D. (1998). What do the basal ganglia do? *Lancet*, 1801–1804.

Burd, L., Severud, R., Klug, M. G., and Kerbeshian, J. (1999). Prenatal and perinatal risk factors for Tourette disorder. *J Perinat Med* **27**, 295–302.

Butler, T., Stern, E., and Silbersweig, D. (2006). Functional neuroimaging of Tourette syndrome: Advances and future directions. *Adv Neurol* **99**, 115–129.

Buzsaki, G. and Draguhn, A. (2004). Neuronal oscillations in cortical networks. *Science* **304**, 1926–1929.

Buzsaki, G., Smith, A., Berger, S., Fisher, L. J., and Gage, F. H. (1990). Petit mal epilepsy and parkinsonian tremor: Hypothesis of a common pacemaker. *Neuroscience* **36**, 1–14.

Caine, E. D., McBride, M. C., Chiverton, P., Bamford, K. A., Rediess, S., and Shiao, J. (1988). Tourette's syndrome in Monroe County school children. *Neurology* **38,** 472–475.

Cardona, F. and Orefici, G. (2001). Group A streptococcal infections and tic disorders in an Italian pediatric population. *J Pediatr* **138**, 71–75.

Carter, A. S., O'Donnell, D. A., Schultz, R. T., Scahill, L., Leckman, J. F., and Pauls, D. L. (2000). Social and emotional adjustment in children affected with Gilles de la Tourette's syndrome: Associations with ADHD and family functioning. Attention Deficit Hyperactivity Disorder. *J Child Psychol Psychiatry* **41**, 215–223.

Carter, A. S., Pauls, D. L., Leckman, J. F., and Cohen, D. J. (1994). A prospective longitudinal study of Gilles de la Tourette's syndrome. *J Am Acad Child Adolesc Psychiatry* **33**, 377–385.

Casas, M., Alvarez, E., Duro, P., Garcia-Ribera, C., Udina, C., Velat, A. et al. (1986). Antiandrogenic treatment of obsessive-compulsive neurosis. *Acta Psychiatr Scand* **73**, 221–222.

Castellanos, F. X., Sonuga-Barke, E. J., Scheres, A., Di Martino, A., Hyde, C., and Walters, J. R. (2005). Varieties of attention-deficit/hyperactivity disorder-related intra-individual variability. *Biol Psychiatry* **57**, 1416–1423.

Cath, D. C., Gijsman, H. J., Schoemaker, R. C., van Griensven, J. M., Troost, N., van Kempen, G. M., and Cohen, A. F. (1999). The effect of m-CPP on tics and obsessive-compulsive phenomena in Gilles de la Tourette syndrome. *Psychopharmacology (Berl)* **144**, 137–143.

Chappell, P., Leckman, J., Goodman, W., Bissette, G., Pauls, D., Anderson, G. et al. (1996). Elevated cerebrospinal fluid corticotropin-releasing factor in Tourette's syndrome: comparison to obsessive compulsive disorder and normal controls. *Biol Psychiatry* **39**, 776–783.

Chappell, P. B., McSwiggan-Hardin, M. T., Scahill, L., Rubenstein, M., Walker, D. E., Cohen, D. J., and Leckman, J. F. (1994a). Videotape tic counts in the assessment of Tourette's syndrome: Stability, reliability, and validity. *J Am Acad Child Adolesc Psychiatry* **33**, 386–393.

Chappell, P. B., Riddle, M., Anderson, G., Scahill, L., Hardin, M., Walker, D. et al. (1994b). Enhanced stress responsivity of Tourette syndrome patients undergoing lumbar puncture. *Biol Psychiatry* **36**, 35–43.

Chappell, P. B., Riddle, M. A., Scahill, L., Lynch, K. A., Schultz, R., Arnsten, A. et al. (1995). Guanfacine treatment of comorbid attention-deficit hyperactivity disorder and Tourette's syndrome: Preliminary clinical experience. *J Am Acad Child Adolesc Psychiatry* **34**, 1140–1146.

Cheon, K. A., Ryu, Y. H., Namkoong, K., Kim, C. H., Kim, J. J., and Lee, J. D. (2004). Dopamine transporter density of the basal ganglia assessed with [123] IPT SPECT in drug-naive children with Tourette's disorder. *Psychiatry Reseach* **130**, 85–95.

Church, A. J., Dale, R. C., Lees, A. J., Giovannoni, G., and Robertson, M. M. (2003). Tourette's syndrome: A cross sectional study to examine the PANDAS hypothesis. *J Neurol Neurosurg Psychiatry* **74**, 602–607.

Coffey, B. J., Biederman, J., Smoller, J. W., Geller, D. A., Sarin, P., Schwartz, S., and Kim, G. S. (2000). Anxiety disorders and tic severity in juveniles with Tourette's disorder. *J Am Acad Child Adolesc Psychiatry* **39**, 562–568.

Cohen, D. J., Young, J. G., Nathanson, J. A., and Shaywitz, B. A. (1979). Clonidine in Tourette's syndrome. *Lancet* **2**, 551–553.

Comings, D. E. (1988). *Tourette syndrome and human behavior*. Hope Press, Daurte, CA.

Courtemanche, R., Fujii, N., and Graybiel, A. M. (2003). Synchronous, focally modulated beta-band oscillations characterize local field potential activity in the striatum of awake behaving monkeys. *Journal of Neurosciences*, 11741–11752.

Csernansky, J. G., Csernansky, C. A., King, R., and Hollister, L. E. (1986). Oscillations in apomorphine-induced stereotypes during 1 year of apomorphine administration. *Biol Psychiatry* **21**, 402–405.

Cuker, A., State, M. W., King, R. A., Davis, N., and Ward, D. C. (2004). Candidate locus for Gilles de la Tourette syndrome/obsessive compulsive disorder/chronic tic disorder at 18q22. *Am J Med Genet A* **130**, 37–39.

Cunningham, M. W., Antone, S. M., Smart, M., Liu, R., and Kosanke, S. (1997). Molecular analysis of human cardiac myosin-cross-reactive B- and T-cell epitopes of the group A streptococcal M5 protein. *Infect Immun* **65**, 3913–3923.

Curtis, D., Brett, P., Dearlove, A. M., McQuillin, A., Kalsi, G., Robertson, M. M., and Gurling, H. M. (2004). Genome scan of Tourette syndrome in a single large pedigree shows some support for linkage to regions of chromosomes 5, 10 and 13. *Psychiatr Genet* **14**, 83–87.

Dale, R. C., Candler, P., Church, A. R., and Pocock, J. G. (2004). Glycolytic enzymes on neuronal membranes are candidate auto-antigens in post-streptococcal neuropsychiatric disorders. *Movement Disorders* **9**, 19.

Deckersbach, T., Rauch, S., Buhlmann, U., and Wilhelm, S. (2005). Habit reversal versus supportive psychotherapy in Tourette's disorder: A randomized controlled trial and predictors of treatment response. *Behav Res Ther*.

Degnan, B. A., Kehoe, M. A., and Goodacre, J. A. (1997). Analysis of human T cell responses to group A streptococci using fractionated Streptococcus pyogenes proteins. *FEMS Immunol Med Microbiol* **17**, 161–170.

Dewing, P., Chiang, C. W., Sinchak, K., Sim, H., Fernagut, P. O., Kelly, S. et al. (2006). Direct regulation of adult brain function by the male-specific factor SRY. *Curr Biol* **16**, 415–420.

Diaz-Anzaldua, A., Joober, R., Riviere, J. B., Dion, Y., Lesperance, P., Richer, F. et al. (2004). Tourette syndrome and dopaminergic genes: A family-based association study in the French Canadian founder population. *Mol Psychiatry* **9**, 272–277.

Diederich, N. J., Kalteis, K., Stamenkovic, M., Pieri, V., and Alesch, F. (2005). Efficient internal pallidal stimulation in Gilles de la Tourette syndrome: A case report. *Mov Disord* **20**, 1496–1499.

Dykens, E., Leckman, J., Riddle, M., Hardin, M., Schwartz, S., and Cohen, D. (1990). Intellectual, academic, and adaptive functioning of Tourette syndrome children with and without attention deficit disorder. *J Abnorm Child Psychol* **18**, 607–615.

Fahrbach, S. E., Morell, J. I., and Pfaff, D. W. (1985). Identification of medial preoptic neurons that concentrate estradiol and project to the midbrain in the rat. *Comp Neuro* **247**, 364–382.

Filipek, P. A., Semrud-Clikeman, M., Steingard, R. J., Renshaw, P. F., Kennedy, D. N., and Biederman, J. (1997). Volumetric MRI analysis comparing attention deficit hyperactivity disorder and normal controls. *Neurology* **48**, 589–601.

Findley, D. B., Leckman, J. F., Katsovich, L., Lin, H., Zhang, H., Grantz, H. et al. (2003). Development of the Yale Children's Global Stress Index (YCGSI) and its application in children and adolescents with Tourette's syndrome and obsessive-compulsive disorder. *J Am Acad Child Adolesc Psychiatry* **42**, 450–457.

Flaherty, A. W., Williams, Z. M., Amirnovin, R., Kasper, E., Rauch, S. L., Cosgrove, G. R., and Eskandar, E. N. (2005). Deep brain stimulation of the anterior internal capsule for the treatment of Tourette syndrome: Technical case report. *Neurosurgery* **57**, E403; discussion E403.

Fredericksen, K. A., Cutting, L. E., Kates, W. R., Mostofsky, S. H., Singer, H. S., Cooper, K. L. et al. (2002). Disproportionate increases of white matter in right frontal lube in Tourette syndrome. *Neurology*, 85–89.

George, M. S., Sallee, F. R., Nahas, Z., Oliver, N. C., and Wassermann, E. M. (2001). Transcranial magnetic stimulation (TMS) as a research tool in Tourette syndrome and related disorders. *Adv Neurol* **85**, 225–235.

Gernert, M., Bennay, M., Fedrowitz, M., Rehders, J. H., and Richter, A. (2002). Altered discharge pattern of basal ganglia output neurons in an animal model of idiopathic dystonia. *J Neurosci* **22**, 7244–7253.

Gernert, M., Hamann, M., Bennay, M., Loscher, W., and Richter, A. (2000). Deficit of striatal parvalbumin-reactive GABAergic interneurons and decreased basal ganglia output in a genetic rodent model of idiopathic paroxysmal dystonia. *Journal of Neurosciences*, 7052–7258.

Gilles de la Tourette, G. (1885). Étude sur une affection nerveuse caractérisé par de l' incoordination motrice accompagne'e d'echolalie et de copralalie. *Archive Neurologie* **9**, 19–42, 158–200.

Giovannoni, G. (2006). PANDAS: Overview of the hypothesis. *Adv Neurol* **99**, 159–165.

Goetz, C. G., Tanner, C. M., Stebbins, G. T., Leipzig, G., and Carr, W. C. (1992). Adult tics in Gilles de la Tourette's syndrome: Description and risk factors. *Neurology* **42**, 784–788.

Goetz, C. G., Tanner, C. M., Wilson, R. S., Carroll, V. S., Como, P. G., and Shannon, K. M. (1987). Clonidine and Gilles de la Tourette's syndrome: Double-blind study using objective rating methods. *Ann Neurol* **21**, 307–310.

Goldman-Rakic, P. S. (1987). Motor control function of the prefrontal cortex. *Ciba Found Symp* **132**, 187–200.

Gonzalez-Burgos, G., Krimer, L. S., Povysheva, N. V., Barrionuevo, G., and Lewis, D. A. (2005). Functional properties of fast spiking interneurons and their synaptic connections with pyramidal cells in primate dorsolateral prefrontal cortex. *J Neurophysiol* **93**, 942–953.

Goodman, W. K., McDougle, C. J., Price, L. H., Barr, L. C., Hills, O. F., Caplik, J. F. et al. (1995). m-Chlorophenylpiperazine in patients with obsessive-compulsive disorder: Absence of symptom exacerbation. *Biol Psychiatry* **38**, 138–149.

Graybiel, A. M. (1990). Neurotransmitters and neuromodulators in the basal ganglia. *Trends Neurosci* **13**, 244–254.

Graybiel, A. M. (2005). The basal ganglia: Learning new tricks and loving it. *Curr Opin Neurobiol* **15**, 638–644.

Graybiel, A. M. and Canales, J. J. (2001). The neurobiology of repetitive behaviors: Clues to the neurobiology of Tourette syndrome. *Adv Neurol* **85**, 123–131.

Greist, J. H. and Jefferson, J. W. (1998). Pharmacotherapy for obsessive-compulsive disorder. *Br J Psychiatry Suppl*, 64–70.

Grenhoff, J. and Svensson, T. H. (1989). Clonidine modulates dopamine cell firing in rat ventral tegmental area. *Eur J Pharmacol* **165**, 11–18.

Guilherme, L. and Kalil, J. (2004). Rheumatic fever: From sore throat to autoimmune heart lesions. *Int Arch Allergy Immunol* **134**, 56–64.

Haber, S. N. (2003). The primate basal ganglia: parallel and integrative networks. *J Chem Neuroanat* **26**, 317–330.

Haber, S. N., Kowall, N. W., Vonsattel, J. P., Bird, E. D., and Richardson, E. P., Jr. (1986). Gilles de la Tourette's syndrome. A postmortem neuropathological and immuohistochemical study. *J Neurol Sci.* **75**, 225–241.

Hassler, R. and Dieckmann, G. (1973). Relief of obsessive-compulsive disorders, phobias and tics by stereotactic coagulations of the rostral intralaminar and medial-thalamic nuclei. In L. Laitinen and K. Livingston, Eds., *Surgical approaches in psychiatry: Proceedings of the Third International Congress of Psychosurgery*, 206–212. Garden City Press, Cambridge.

Hebebrand, J., Klug, B., Fimmers, R., Seuchter, S. A., Wettke-Schafer, R., Deget, F. et al. (1997). Rates for tic disorders and obsessive compulsive symptomatology in families of children and adolescents with Gilles de la Tourette syndrome. *J Psychiatr Res* **31**, 519–530.

Heinrich, H., Gevensleen, G., Freisleder, F. J., Moll, G. H., and Rothenberger, A. (2004). Training for slow cortical potentials in ADHD children: Evidence from positive behavioral and neurophysiological effects. *Biological Psychiatry* 772–775.

Heinz, A., Knable, M. B., Wolf, S. S., Jones, D. W., Gorey, J. G., Hyde, T. M., and Weinberger, D. R. (1998). Tourette syndrome: [^{123}I]B-CIT SPECT correlates of vocal tic severity. *Neurology* 1069–1074.

Hinshaw, S. P. (1994). Conduct disorder in childhood: Conceptualization, diagnosis, comorbidity, and risk status for antisocial functioning in adulthood. *Prog Exp Pers Psychopathol Res* 3–44.

Hoekstra, P. J., Bijzet, J., Limburg, P. C., Steenhuis, M. P., Troost, P. W., Oosterhoff, M. D. et al. (2001). Elevated D8/17 expression on B lymphocytes, a marker of rheumatic fever, measured with flow cytometry in tic disorder patients. *Am J Psychiatry* **158**, 605–610.

Hoekstra, P. J., Steenhuis, M. P., Kallenberg, C. G., and Minderaa, R. B. (2004). Association of small life events with self reports of tic severity in pediatric and adult tic disorder patients: A prospective longitudinal study. *J Clin Psychiatry* **65**, 426–431.

Houeto, J. L., Karachi, C., Mallet, L., Pillon, B., Yelnik, J., Mesnage, V. et al. (2005). Tourette's syndrome and deep brain stimulation. *J Neurol Neurosurg Psychiatry* **76**, 992–995.

Hounie, A. G., do Rosario-Campos, M. C., Diniz, J. B., Shavitt, R. G., Ferrao, Y. A., Lopes, A. C. et al. (2006). Obsessive-compulsive disorder in Tourette syndrome. *Adv Neurol* **99**, 22–38.

Husby, G., van de Rijn, I., Zabriskie, J. B., Abdin, Z. H., and Williams, R. C., Jr. (1976). Antibodies reacting with cytoplasm of subthalamic and caudate nuclei neurons in chorea and acute rheumatic fever. *J Exp Med* **144**, 1094–1110.

Hutchison, W. D., Lang, A. E., Dostrovsky, J. O., and Lozano, A. M. (2003). Pallidal neuronal activity: Implications for models of dystonia. *Annals Neurol* **53**, 480–488.

Hyde, T. M., Aaronson, B. A., Randolph, C., Rickler, K. C., and Weinberger, D. R. (1992). Relationship of birth weight to the phenotypic expression of Gilles de la Tourette's syndrome in monozygotic twins. *Neurology* **42**, 652–658.

Hyde, T. M., Aaronson, B. A., Randolph, C., and Weinberger, D. R. (1995). Cerebral morphometric abnormalities in Tourette syndrome: A quantitative MRI study of monozygotic twins. *Neurology*, 1176–1182.

Itard, J. M. G. (1825). Memoire sur quelques fonctions involontaries ses appareils de la locomotion de la prehension et de la voix. *Archives Generales de Medecine* **8**, 385–407.

Jagger, J., Prusoff, B. A., Cohen, D. J., Kidd, K. K., Carbonari, C. M., and John, K. (1982). The epidemiology of Tourette's syndrome: A pilot study. *Schizophr Bull* **8**, 267–278.

Jankovic, J. and Mejia, N. I. (2006). Tics associated with other disorders. *Adv Neurol* **99**, 61–68.

Jeanmonod, D., Schulman, J., Ramirez, R. R., Cancro, R., Lanz, M., Morel, A. et al. (2003). Neuropsychiatric thalamocortical dysrhythmia: Surgical implications. *Neurosurgical Clinics of North America*, 251–165.

Jeffries, K. J., Schooler, C., Schoenbach, C., Herscovitch, P., Chase, T. N., and Braun, A. R. (2002). The functional neuroanatomy of Tourette's syndrome: An FDG PET study III: Functional coupling of regional cerebral metabolic rates. *Neuropsychopharmacology* **27**, 92–104.

Jog, M. S., Kubota, Y., Connolly, C. I., Hillegaart, V., and Graybiel, A. M. (1999). Building neural representations of habits. *Science* **286**, 1745–1749.

Kalanithi, P. S., Zheng, W., Kataoka, Y., DiFiglia, M., Grantz, H., Saper, C. B. et al. (2005). Altered parvalbumin-positive neuron distribution in basal ganglia of individuals with Tourette syndrome. *Proc Natl Acad Sci U S A* **102**, 13307–13312.

Kawikova, I., Leckman, J. F., Kronig, H., Katsovich, L., Bessen, D. E., Ghebremichael, M., and Bothwell, A. (2007). Decreased number of regulatory T cells suggests impaired immune tolerance in children with Tourette's syndrome. *Biological Psychiatry* **61**, 273–278.

Kemeny, E., Husby, G., Williams, R. C., Jr., and Zabriskie, J. B. (1994). Tissue distribution of antigen(s) defined by monoclonal antibody D8/17 reacting with B lymphocytes of patients with rheumatic heart disease. *Clin Immunol Immunopathol* **72**, 35–43.

Kenney, C. and Jankovic, J. (2006). Tetrabenazne in the treatment of hyperkinetic movement disorders. *Expert Rev Neurother* **6**, 7–17.

Keri, S., Szlobodnyik, C., Benedek, G., Janka, Z., and Gadoros, J. (2002). Probabilistic classification learning in Tourette syndrome. *Neuropsychologia* **40**, 1356–1362.

Khalifa, N. and von Knorring, A. L. (2003). Prevalence of tic disorders and Tourette syndrome in a Swedish school population. *Dev Med Child Neurol* **45**, 315–319.

Khalifa, N. and von Knorring, A. L. (2005). Tourette syndrome and other tic disorders in a total population of children: Clinical assessment and background. *Acta Paediatr* **94**, 1608–1614.

Khalifa, N. and Von Knorring, A. L. (2006). Psychopathology in a Swedish population of school children with tic disorders. *J Am Acad Child Adolesc Psychiatry* **45 (11)**: 1346–53.

Kiessling, L. S., Marcotte, A. C., and Culpepper, L. (1993). Antineuronal antibodies in movement disorders. *Pediatrics* **92**, 39–43.

Kirvan, C. A., Swedo, S. E., Heuser, J. S., and Cunningham, M. W. (2003). Mimicry and autoantibody-mediated neuronal cell signaling in Sydenham chorea. *Nat Med* **9**, 914–920.

Kirvan, C. A., Swedo, S. E., Kurahara, D., and Cunningham, M. W. (2006). Streptococcal mimicry and antibody-mediated cell signaling in the pathogenesis of Sydenham's chorea. *Autoimmunity* **39**, 21–29.

Koos, T. and Tepper, J. M. (1999). Inhibitory control of neostriatal projection neurons by GABAergic interneurons. *Nat Neurosci* **2**, 467–472.

Koos, T. and Tepper, J. M. (2002). Dual cholinergic control of fast-spiking interneurons in the neostriatum. *J Neurosci* **22**, 529–535.

Kurlan, R., Whitmore, D., Irvine, C., McDermott, M. P., and Como, P. G. (1994). Tourette's syndrome in a special education population: A pilot study involving a single school district. *Neurology* **44**, 699–702.

Kushner, H. I. (1999). *A crusing brain? The histories of Tourette Syndrome*. Harvard University Press, Cambridge, MA.

Lang, A. (1991). Patient perception of tics and other movement disorders. *Neurology* **41**, 223–228.

LaPenta, D., Rubens, C., Chi, E., and Cleary, P. P. (1994). Group A streptococci efficiently invade human respiratory epithelial cells. *Proc Natl Acad Sci U S A* **91**, 12115–12119.

Leckman, J. F. (2002). Tourette's syndrome. *Lancet* **360**, 1577–1586.

Leckman, J. F. and Cohen, D. J. (1998). *Tourette's Syndrome—Tics, obsessions, compulsions. developmental psychopathology and clinical care*. John Wiley and Sons., New York.

Leckman, J. F., Cohen, D. J., Goetz, C. G., and Jankovic, J. (2001). Tourette Syndrome—Pieces of the puzzle. In D. J. Cohen, C. Goetz, and J. Jankovic, Eds., *Tourette Syndrome and associated disorders*. Lippincott, Williams & Wilkins, New York.

Leckman, J. F., Cohen, D. J., Price, R. A., Minderaa, R. B., Anderson, G. M., and Pauls, D. L. (1984). The pathogenesis of Gilles de la Tourette's syndrome. A review of data and hypothesis. In A. B. Shah, N. S. Shah, and A. G. Donald, Eds., *Movement disorders*. Plenum, New York.

Leckman, J. F., Dolnansky, E. S., Hardin, M. T., Clubb, M., Walkup, J. T., Stevenson, J., and Pauls, D. L. (1990). Perinatal factors in the expression of Tourette's syndrome: An exploratory study. *J Am Acad Child Adolesc Psychiatry* **29**, 220–226.

Leckman, J. F., Goodman, W. K., Anderson, G. M., Riddle, M. A., Chappell, P. B., McSwiggan-Hardin, M. T. et al. (1995). Cerebrospinal fluid biogenic amines in obsessive compulsive disorder, Tourette's syndrome, and healthy controls. *Neuropsychopharmacology* **12**, 73–86.

Leckman, J. F., Grice, D. E., Boardman, J., Zhang, H., Vitale, A., Bondi, C. et al. (1997a). Symptoms of obsessive-compulsive disorder. *Am J Psychiatry* **154**, 911–917.

Leckman, J. F., Hardin, M. T., Riddle, M. A., Stevenson, J., Ort, S. I., and Cohen, D. J. (1991a). Clonidine treatment of Gilles de la Tourette's syndrome. *Arch Gen Psychiatry* **48**, 324–328.

Leckman, J. F., Katsovich, L., Kawikova, I., Lin, H., Zhang, H., and Kronig, H. (2005). Increased serum levels of inerleukin-12 and tumor necrosis factor-alpha in Tourette's syndrome. *Biol Psychiatry* **57**, 667–673.

Leckman, J. F., King, R. A., Scahill, L., Findley, D., Ort, S., and Cohen, D. J. (1998). Yale approach to assessment and treatment. In J. F. Leckman and D. J. Cohen, Eds., *Tourette's Syndrome tics, obsessions, compulsions—Developmental psychopathology and clinical care*, 285–309. John Wiley and Sons, New York.

Leckman, J. F., Knorr, A. M., Rasmusson, A. M., and Cohen, D. J. (1991b). Basal ganglia research and Tourette's syndromes. *Trends Neurosci* **14**, 94.

Leckman, J. F., Peterson, B. S., Anderson, G. M., Arnsten, A. F., Pauls, D. L., and Cohen, D. J. (1997b). Pathogenesis of Tourette's syndrome. *J Child Psychol Psychiatry* **38**, 119–142.

Leckman, J. F., Price, R. A., Walkup, J. T., Ort, S., Pauls, D. L., and Cohen, D. J. (1987). Nongenetic factors in Gilles de la Tourette's syndrome. *Arch Gen Psychiatry* **44**, 100.

Leckman, J. F., Riddle, M. A., Hardin, M. T., Ort, S. I., Swartz, K. L., Stevenson, J., and Cohen, D. J. (1989). The Yale Global Tic Severity Scale: Initial testing of a clinician-rated scale of tic severity. *J Am Acad Child Adolesc Psychiatry* **28**, 566–573.

Leckman, J. F. and Scahill, L. (1990). Possible exacerbation of tics by androgenic steroids. *N Engl J Med* **322**, 1674.

Leckman, J. F., Vaccarino, F. M., Kalanithi, P. S., and Rothenberger, A. (2006). Tourette syndrome: A relentless drumbeat. *Child Psychol Psychiatry* **47**, 537–550.

Leckman, J. F., Walker, D. E., and Cohen, D. J. (1993). Premonitory urges in Tourette's syndrome. *Am J Psychiatry* **150**, 98–102.

Leckman, J. F., Walker, D. E., Goodman, W. K., Pauls, D. L., and Cohen, D. J. (1994). "Just right" perceptions associated with compulsive behavior in Tourette's syndrome. *Am J Psychiatry* **151**, 675–680.

Lee, J. S., Yoo, S. S., Cho, S. Y., Ock, S. M., Lim, M. K., and Panych, L. P. (2006). Abnormal thalamic volume in treatment-naive boys with Tourette syndrome. *Acta Psychiatr Scand* **113,** 64–67.

Lin, H., Katsovich, L., Ghebremichael, M., Findley, D., Grantz, H., Lombroso, P. J. et al. (In press). Psychosocial stress predicts future symptom severities in children and adolescents with Tourette Syndrome and/or Obsessive-Compulsive Disorder. *Child Psychol Psychiatry* **48**, 157–166.

Llinas, R. and Jahnsen, H. (1982). Electrophysiology of mammalian thalamic neurones in vitro. *Nature* **297**, 406–408.

Llinas, R., Urbano, F. J., Leznik, E., Ramirez, R. R., and van Marle, H. J. (2005). Rhythmic and dysrhythmic thalamocortical dynamics: GABA systems and the edge effect. *Trends Neurosci* **28**, 325–333.

Lombroso, P. J., Mack, G., Scahill, L., King, R. A., and Leckman, J. F. (1991). Exacerbation of Gilles de la Tourette's syndrome associated with thermal stress: A family study. *Neurology* **41**, 1984–1987.

Loscher, W., Fisher, J. E., Jr., Schmidt, D., Fredow, G., Honack, D., and Siturrian, W. B. (1989). The sz mutant hamster: A genetic model of spilepsy or of paroxysmal dystonia? *Mov Disord* 219–232.

Lozano, A. M., Kumar, R., Gross, R. E., Giladi, N., Hutchison, W. D., Dostrovsky, J. O., and Lang, A. E. (1997). Globus pallidus internus pallidotomy for generalized dystonia. *Mov Disord* **12**, 865–870.

Luo, F., Leckman, J. F., Katsovich, L., Findley, D., Grantz, H., Tucker, D. M. et al. (2004). Prospective longitudinal study of children with tic disorders and/or obsessive-compulsive disorder: Relationship of symptom exacerbations to newly acquired streptococcal infections. *Pediatrics* **113**, e578–585.

Malison, R. T., McDougle, C. J., van Dyck, C. H., Scahill, L., Baldwin, R. M., Seibyl, J. P. et al. (1995). [123I]beta-CIT SPECT imaging of striatal dopamine transporter binding in Tourette's disorder. *Am J Psychiatry* **152**, 1359–1361.

Mallet, N., LeMoine, C., Charpier, S., and Gonon, F. (2005). Feed forward inhibition of projection neurons by fast-spiking GABA interneurons in the rat striatum in vivo. *J Neurosci* **25**, 3857–3869.

Marsden, C. D. and Obeso, J. A. (1994). The functions of the basal ganglia and the paradox of stereotaxic surgery in Parkinson's disease. *Brain* **117** (Pt 4), 877–897.

Marsh, R., Alexander, G. M., Packard, M. G., Zhu, H., and Peterson, B. S. (2005). Perceptual-motor skill learning in Gilles de la Tourette syndrome. Evidence for multiple procedural learning and memory systems. *Neuropsychologia* **43**, 1456–1465.

Marsh, R., Alexander, G. M., Packard, M. G., Zhu, H., Wingard, J. C., Quackenbush, G., and Peterson, B. S. (2004). Habit learning in Tourette syndrome: A translational neuroscience approach to a developmental psychopathology. *Arch Gen Psychiatry* **61**, 1259–1268.

Mason, A., Banerjee, S., Eapen, V., Zeitlin, H., and Robertson, M. M. (1998). The prevalence of Tourette syndrome in a mainstream school population. *Dev Med Child Neurol* **40**, 292–296.

Mathews, C. A., Reus, V. I., Bejarano, J., Escamilla, M. A., Fournier, E., Herrera, L. D. et al. (2004). Genetic studies of neuropsychiatric disorders in Costa Rica: A model for the use of isolated populations. *Psychiatr Genet* **14**, 13–23.

McDougle, C. J., Goodman, W. K., Leckman, J. F., Barr, L. C., Heninger, G. R., and Price, L. H. (1993). The efficacy of fluvoxamine in obsessive-compulsive disorder: Effects of comorbid chronic tic disorder. *J Clin Psychopharmacol* **13**, 354–358.

McGeorge, A. J. and Faull, R. L. (1989). The organization of the projection from the cerebral cortex to the striatum in the rat. *Neuroscience* **29**, 503–537.

McMahon, W., Carter, A. S., Fredine, N., and Pauls, D. L. (2003). Children at familial risk for Tourette's disorder: Child and parent diagnoses. *Neuropsychiatr Genet* **121**, 105–111.

Mell, L. K., Davis, R. L., and Owens, D. (2005). Association between streptococcal infection and obsessive-compulsive disorder, Tourette's syndrome, and tic disorder. *Pediatrics* **116**, 56–60.

Mercadante, M. T., Campos, M. C., Marques-Dias, M. J., Miguel, E. C., and Leckman, J. (1997). Vocal tics in Sydenham's chorea. *J Am Acad Child Adolesc Psychiatry* **36**, 305–306.

Middleton, F. A. and Strick, P. L. (2000). Basal ganglia and cerebellar loops: Motor and cognitive circuits. *Brain Res Brain Res Rev* **31**, 236–250.

Middleton, F. A. and Strick, P. L. (2003). Basal-ganglia 'projections' to the prefrontal cortex of the primate. *Cereb Cortex*, 929–935.

Mink, J. (2001). Basal Ganglia dysfunction in Tourette's syndrome: A new hypothesis. *Pediatr Neurol*, 190–198.

Moll, G. H., Heinrich, H., Gevensleben, H., and Rothenberger, A. (2005). *Entwicklungsverlauf inhibitor-ischer Prozesse im sensomotorischen Regelkreis bbei Tic-Storungen*. Vandenhoeck & Ruprecht, Gottingen.

Moll, G. H., Wischer, S., Heinrich, H., Tergau, F., Paulus, W., and Rothenberger, A. (1999). Deficient motor control in children with tic disorder: Evidence from transcranial magnetic stimulation. *Neurosci Lett* **272**, 37–40.

Moran, K. A., Leckman, J. F., Vaccarino, F. M., Walton, K., and Llinas, R. R. (2005). Neuromagnetic correlates of Gilles de la Tourette syndrome. In *Abstract, 35th Annual Meeting*, Washington, DC.

Morshed, S. A., Parveen, S., Leckman, J. F., Mercadante, M. T., Bittencourt Kiss, M. H., Miguel, E. C. et al. (2001). Antibodies against neural, nuclear, cytoskeletal, and streptococcal epitopes in children and adults with Tourette's syndrome, Sydenham's chorea, and autoimmune disorders. *Biol Psychiatry* **50**, 566–577.

Muller-Vahl, K. R., Berding, G., Brucke, T., H., K., Meyer, G. J., Hundeshagen, H., Dengler, R. et al. (2000). Dopamine transporter binding in Gilles de la Tourette syndrome. *J Neurol* **247**, 514–520.

Murphy, T. K., Goodman, W. K., Fudge, M. W., Williams, R. C., Jr., Ayoub, E. M., Dalal, M. et al. (1997). B lymphocyte antigen D8/17: A peripheral marker for childhood-onset obsessive-compulsive disorder and Tourette's syndrome? *Am J Psychiatry* **154**, 402–407.

Nagel, T. (1974). What is it like to be a bat? *Philosophical Review* LXXXIII, 435–450.

Nobrega, J. N., Richter, A., Tozman, N., Jiwa, D., and Loscher, W. (1996). Quantitative autoradiography reveals regionally selective changes in dopamine D1 and D2 receptor binding in the genetically dystonic hamster. *Neuroscience* **71**, 927–937.

Nomura, Y., Fukuda, H., Terao, Y., Hikosaka, O., and Segawa, M. (2003). Abnormalities of voluntary saccades in Gilles de la Tourette's syndrome: Pathophysiological consideration. *Brain Dev* **25 Suppl 1**, S48–54.

Parent, A. (1986). *Comparative neurobiology of the basal ganglia*. John Wiley and Sons, New York.

Pasamanick, B. and Kawi, A. (1956). A study of the association of prenatal and paranatal factors with the development of tics in children; a preliminary investigation. *J Pediatr* **48**, 596–601.

Pauls, D. L. and Leckman, J. F. (1986). The inheritance of Gilles de la Tourette's syndrome and associated behaviors. Evidence for autosomal dominant transmission. *N Engl J Med* **315**, 993–997.

Pauls, D. L., Raymond, C. L., Stevenson, J. M., and Leckman, J. F. (1991). A family study of Gilles de la Tourette syndrome. *Am J Hum Genet* **48**, 154–163.

Pedroarena, C. and Llinas, R. (1997). Dendritic calcium conductances generate high-frequency oscillation in thalamocortical neurons. *Proc Natl Acad Sci U S A* **94**, 724–728.

Perlmutter, S. J., Leitman, S. F., Garvey, M. A., Hamburger, S., Feldman, E., Leonard, H. L., and Swedo, S. E. (1999). Therapeutic plasma exchange and intravenous immunoglobulin for obsessive-compulsive disorder and tic disorders in childhood. *Lancet* **354**, 1153–1158.

Perrin, E. M., Murphy, M. L., Casey, J. R., Pichichero, M. E., Runyan, D. K., Miller, W. C. et al. (2004). Does group A beta-hemolytic streptococcal infection increase risk for behavioral and neuropsychiatric symptoms in children? *Arch Pediatr Adolesc Med* **158**, 848–856.

Peterson, B. S. and Leckman, J. F. (1998). The temporal dynamics of tics in Gilles de la Tourette syndrome. *Biol Psychiatry* **44**, 1337–1348.

Peterson, B. S., Leckman, J. F., Duncan, J. S., Wetzles, R., Riddle, M. A., Hardin, M. T., and Cohen, D. J. (1994). Corpus callosum morphology from magnetic resonance images in Tourette's Syndrome. *Psychiatry Res* **55**, 85–99.

Peterson, B. S., Leckman, J. F., Scahill, L., Naftolin, F., Keefe, D., Charest, N. J., and Cohen, D. J. (1992). Steroid hormones and CNS sexual dimorphisms modulate symptom expression in Tourette's syndrome. *Psychoneuroendocrinology* **17**, 553–563.

Peterson, B. S., Pine, D. S., Cohen, P., and Brook, J. S. (2001a). Prospective, longitudinal study of tic, obsessive-compulsive, and attention-deficit/hyperactivity disorders in an epidemiological sample. *J Am Acad Child Adolesc Psychiatry* **40**, 685–695.

Peterson, B. S., Skudlarski, P., Anderson, A. W., Zhang, H., Gatenby, J. C., Lacadie, C. M. et al. (1998a). A functional magnetic resonance imaging study of tic suppression in Tourette syndrome. *Arch Gen Psychiatry* **55**, 326–333.

Peterson, B. S., Staib, L., Scahill, L., Zhang, H., Anderson, C., Leckman, J. F. et al. (2001b). Regional brain and ventricular volumes in Tourette syndrome. *Arch Gen Psychiatry* **58**, 427–440.

Peterson, B. S., Thomas, P., Kane, M. J., Scahill, L., Zhang, H., Bronen, R. et al. (2003). Basal ganglia volumes in patients with Gilles de la Tourette syndrome. *Arch Gen Psychiatry* **60**, 415–424.

Peterson, B. S., Zhang, H., Anderson, G. M., and Leckman, J. F. (1998b). A double-blind, placebo-controlled, crossover trial of an antiandrogen in the treatment of Tourette's syndrome. *J Clin Psychopharmacol* **18**, 324–331.

Plessen, K. J., Wentzel-Larsen, T., Hugdahl, K., Feineigle, P., Klein, J., Staib, L. H. et al. (2004). Altered interhemispheric connectivity in individuals with Tourette's disorder. *Am J Psychiatry* **161**, 2028–2037.

Price, R. A., Kidd, K. K., Cohen, D. J., Pauls, D. L., and Leckman, J. F. (1985). A twin study of Tourette syndrome. *Arch Gen Psychiatry* **42**, 815–820.

Raz, A., Feingold, A., Zelanskaya, V., Vaadia, E., and Bergman, H. (1996). Neuronal synchronization of tonically active neurons in the striatum of normal and parkinsonian primates. *J Neurophysiol* **76**, 2083–2088.

Rehders, J. H., Loscher, W., and Richter, A. (2000). Evidence for striatal dopaminergic overactivity in paroxysmal dystonia indicated by microinjections in a genetic rodent model. *Neuroscience* **97**, 267–277.

Riekkinen, P., Jr, Riekkinen, M., Sirwio, J., Riekkinen, P. (1992). Neurophysiological consequences of combined cholinergic and noradrenergic lesions. *Exp Neurol* **116(1)**: 64–8.

Robertson, M. M. and Orth, M. (2006). Behavioral and affective disorders in Tourette syndrome. *Adv Neurol* **99**, 39–60.

Robertson, M. M., Williamson, F., and Eapen, V. (2006). Depressive symptomatology in young people with Gilles de la Tourette Syndrome—A comparison of self-report scales. *J Affect Disord* **91**, 265–268.

Rothenberger, A. (1990). The role of the frontal lobes in child psychiatric disorders. In A. Rothenberger, Ed., *Brain and behavior in child psychiatry*, 34–58. Springer, Berlin, Heidelberg.

Rothenberger, A. (1991). *Wenn Kinder Tics entwicckeln-Storung*. Fisher, New York.

Rothenberger, A. (1995). Electrical brain activity in children with hyperkinetic syndrome: Evidence of frontal cortical dysfunction. In J. S., Ed., *Eunethydis—European approaches to hyperkinetic disorder*, 255–270. Trumpi, Zurich.

Ruskin, D. N., Bergstrom, D. A., Tierney, P. L., and Walters, J. R. (2003). Correlated multisecond oscillations in firing rate in the basal ganglia: Modulation by dopamine and the subthalamic nucleus. *Neuroscience* 427–438.

Sanberg, P. R., Silver, A. A., Shytle, R. D., Philipp, M. K., Cahill, D. W., Fogelson, H. M., and McConville, B. J. (1997). Nicotine for the treatment of Tourette's syndrome. *Pharmacol Ther* **74**, 21–25.

Santangelo, S. L., Pauls, D. L., Goldstein, J. M., Faraone, S. V., Tsuang, M. T., and Leckman, J. F. (1994). Tourette's syndrome: What are the influences of gender and comorbid obsessive-compulsive disorder? *J Am Acad Child Adolesc Psychiatry* **33**, 795–804.

Scahill, L., Chappell, P. B., Kim, Y. S., Schultz, R. T., Katsovich, L., Shepherd, E. et al. (2001). A placebo-controlled study of guanfacine in the treatment of children with tic disorders and attention deficit hyperactivity disorder. *Am J Psychiatry* **158**, 1067–1074.

Scahill, L., Leckman, J. F., Schultz, R. T., Katsovich, L., and Peterson, B. S. (2003). A placebo-controlled trial of risperidone in Tourette syndrome. *Neurology* **60**, 1130–1135.

Scahill, L., Riddle, M. A., King, R. A., Hardin, M. T., Rasmusson, A., Makuch, R. W., and Leckman, J. F. (1997). Fluoxetine has no marked effect on tic symptoms in patients with Tourette's syndrome: A double-blind placebo-controlled study. *J Child Adolesc Psychopharmacol* **7**, 75–85.

Scahill, L., Williams, S., Schwab-Stone, M., Applegate, J., and Leckman, J. F. (2006). Disruptive behavior problems in a community sample of children with tic disorders. *Adv Neurol* **99**, 184–190.

Schultz, R. T., Carter, A. S., Gladstone, M., Scahill, L., Leckman, J. F., Peterson, B. S. et al. (1998). Visual-motor integration functioning in children with Tourette syndrome. *Neuropsychology* **12**, 134–145.

Schwartz, J. M. and Begley, S. (2002). *The mind and the brain: Neuroplasticity and the power of mental force*. Harper Collins Publishers, Inc., New York.

Schwartz, J. M., Stapp, H. P., and Beauregard, M. (2005). Quantum physics in neuroscience and psychology: A neurophysical model of mind/brain interaction. *Philosophical Transactions of the Royal Society of London, Series B, Biological Sciences*, 1309–1327.

Segawa, M. (2003). Neurophysiology of Tourette's syndrome: Pathophysiological considerations. *Brain Development* **Suppl. 1**, S62–S69.

Semba, K. and Komisaruk, B. R. (1984). Neural substrates of two diff erent rhythmical vibrissal movements in the rat. *Neuroscience* **12**, 761–774.

Serra-Mestres, J., Ring, H. A., Costa, D. C., Gacinovic, S., Walker, Z., Lees, A. J. et al. (2004). Dopamine transporter binding in Gilles de la Tourette syndrome: A [123I]FP-CIT/SPECT study. *Acta Psychiatr Scand* **109**, 140–146.

Serrien, D. J., Nirkko, A. C., Loher, T. J., Lovblad, K. O., Burgunder, J. M., and Wiesendanger, M. (2002). Movement control of manipulative tasks in patients with Gilles de la Tourette syndrome. *Brain* **125**, 290–300.

Serrien, D. J., Orth, M., Evans, A. H., Lees, A. J., and Brown, P. (2005). Motor inhibition in patients with Gilles de la Tourette syndrome: Functional activation patterns as revealed by EEG coherence. *Brain* **128**, 116–125.

Seuchter, S. A., Hebebrand, J., Klug, B., Knapp, M., Lehmkuhl, G., Poustka, F. et al. (2000). Complex segregation analysis of families ascertained through Gilles de la Tourette syndrome. *Genet Epidemiol* **18**, 33–47.

Shapiro, A. K., Shapiro, E. S., Young, J. G., and Freinberg, T. E. (1988). *Gilles de la Tourette Syndrome, 2e*. Raven Press, New York.

Sikich, L. and Todd, R. D. (1988). Are the neurodevelopmental effects of gonadal hormones related to sex differences in psychiatric illnesses? *Psychiatr Dev* **6**, 277–309.

Silva, R. R., Munoz, D. M., Barickman, J., and Friedhoff, A. J. (1995). Environmental factors and related fluctuation of symptoms in children and adolescents with Tourette's disorder. *J Child Psychol Psychiatry* **36**, 305–312.

Simonic, I., Gericke, G. S., Ott, J., and Weber, J. L. (1998). Identification of genetic markers associated with Gilles de la Tourette syndrome in an Afrikaner population. *Am J Hum Genet* **63**, 839–846.

Singer, H. S., Giuliano, J. D., Hansen, B. H., Hallett, J. J., Laurino, J. P., Benson, M., and Kiessling, L. S. (1998). Antibodies against human putamen in children with Tourette syndrome. *Neurology* **50**, 1618–1624.

Singer, H. S., Hahn, I. H., Krowiak, E., Nelson, E., and Moran, T. (1990). Tourette's syndrome: A neurochemical analysis of postmortem cortical brain tissue. *Ann Neurol* **27**, 443–446.

Singer, H. S., Hahn, I. H., and Moran, T. H. (1991). Abnormal dopamine uptake sites in postmortem striatum from patients with Tourette's syndrome. *Ann Neurol* **30**, 558–562.

Singer, H. S., Mink, J. W., Loiselle, C. R., Burke, K. A., Ruchkina, I., Morshed, S. et al. (2005). Microinfusion of antineuronal antibodies into rodent striatum: Failure to differentiate between elevated and low titers. *J Neuroimmunol* **163**, 8–14.

Singer, H. S., Reiss, A. L., Brown, J. E., Aylward, E. H., Shih, B., Chee, E. et al. (1993). Volumetric MRI changes in basal ganglia of children with Tourette's syndrome. *Neurology* **43**, 950–956.

Singer, H. S., Szymanski, S., Giuliano, J., Yokoi, F., Dogan, A. S., Brasic, J. R. et al. (2002). Elevated intrasynaptic dopamine release in Tourette's syndrome measured by PET. *Am Psychiatry* **159**, 1329–1336.

Singer, H. S., Wendlandt, J., Krieger, M., and Giuliano, J. (2001). Baclofen treatment in Tourette syndrome: A double-blind, placebo-controlled, crossover trial. *Neurology* **56**, 599–604.

Singer, H. S. and Wendlandt, J. T. (2001). Neurochemistry and synaptic neurotransmission in Tourette syndrome. *Adv Neurol* **85**, 163–178.

Singer, H. S. and Williams, P. N. (2006). Autoimmunity and pediatric movement disorders. *Adv Neurol* **99**, 166–178.

Stamenkovic, M., Schindler, S. D., Asenbaum, S., Neumeister, A., Willeit, M., Willinger, U. et al. (2001). No change in striatal dopamine re-uptake site density in psychotropic drug naive and in currently treated Tourette disorder patients: A [^{123}I] beta-CIT SPECT-study. *European Journal of Neuropsychopharmacology*, 69–74.

State, M. W., Greally, J. M., Cuker, A., Bowers, P. N., Henegariu, O., Morgan, T. M. et al. (2003). Epigenetic abnormalities associated with a chromosome 18(q21-q22) inversion and a Gilles de la Tourette syndrome phenotype. *Proc Natl Acad Sci U S A* **100**, 4684–4689.

Stern, E., Silbersweig, D. A., Chee, K. Y., Holmes, A., Robertson, M. M., Trimble, M. et al. (2000). A functional neuroanatomy of tics in Tourette syndrome. *Arch Gen Psychiatry* **57**, 741–748.

Stokes, A., Bawden, H. N., Camfield, P. R., Backman, J. E., and Dooley, J. M. (1991). Peer problems in Tourette's disorder. *Pediatrics* **87**, 936–942.

Sukhodolsky, D. G., do Rosario-Campos, M. C., Scahill, L., Katsovich, L., Pauls, D. L., Peterson, B. S. et al. (2005). Adaptive, emotional, and family functioning of children with obsessive-compulsive disorder and comorbid attention deficit hyperactivity disorder. *Am J Psychiatry* **162**, 1125–1132.

Sukhodolsky, D. G., Scahill, L., Zhang, H., Peterson, B. S., King, R. A., Lombroso, P. J. et al. (2003). Disruptive behavior in children with Tourette's syndrome: Association with ADHD comorbidity, tic severity, and functional impairment. *J Am Acad Child Adolesc Psychiatry* **42**, 98–105.

Swedo, S. E. and Grant, P. J. (2005). Annotation: PANDAS: A model for human autoimmune disease. *J Child Psychol Psychiatry* **46**, 227–234.

Swedo, S. E., Leonard, H. L., Garvey, M., Mittleman, B., Allen, A. J., Perlmutter, S. et al. (1998). Pediatric autoimmune neuropsychiatric disorders associated with streptococcal infections: Clinical description of the first 50 cases. *Am J Psychiatry* **155**, 264–271.

Swedo, S. E., Leonard, H. L., Mittleman, B. B., Allen, A. J., Rapoport, J. L., Dow, S. P. et al. (1997). Identification of children with pediatric autoimmune neuropsychiatric disorders associated with streptococcal infections by a marker associated with rheumatic fever. *Am J Psychiatry* **154**, 110–112.

Swedo, S. E., Rapoport, J. L., Cheslow, D. L., Leonard, H. L., Ayoub, E. M., Hosier, D. M., and Wald, E. R. (1989). High prevalence of obsessive-compulsive symptoms in patients with Sydenham's chorea. *Am J Psychiatry* **146**, 246–249.

Swerdlow, N. R., Karban, B., Ploum, Y., Sharp, R., Geyer, M. A., and Eastvold, A. (2001). Tactile prepuff inhibition of startle in children with Tourette's syndrome: In search of an "fMRI-friendly" startle paradigm. *Biol Psychiatry* **50**, 578–585.

Swerdlow, N. R. and Sutherland, A. N. (2006). Preclinical models relevant to Tourette syndrome. *Adv Neurol* **99**, 69–88.

Taylor, J. R., Morshed, S. A., Parveen, S., Mercadante, M. T., Scahill, L., Peterson, B. S. et al. (2002). An animal model of Tourette's syndrome. *Am J Psychiatry* **159**, 657–660.

Temel, Y. and Visser-Vandewalle, V. (2004). Surgery in Tourette syndrome. *Mov Disord* **19**, 3–14.

The Tourette Syndrome International Consortium for Genetics. (1999). A complete genome screen in sib pairs affected by Gilles de la Tourette syndrome. The Tourette Syndrome Association International Consortium for Genetics. *Am J Hum Genet* **65**, 1428–1436.

van Wattum, P. J., Anderson, G. M., Chappell, P. B., Goodman, W. K., Riddle, M. A., and Leckman, J. F. (1999). Cerebrospinal fluid dynorphin A[1-8] and beta-endorphin levels in Tourette's syndrome are unaltered. *Biol Psychiatry* **45**, 1527–1528.

Vandewalle, V., Van der Linden, C., Groenewegen, H. J., and Caemaert, J. (1999). Stereotactic treatment of Gilles de la Tourette syndrome by high frequency stimulation of thalamus. *Lancet* **27**, 724.

Verkerk, A. J., Mathews, C. A., Joosse, M., Eussen, B. H., Heutink, P., and Oostra, B. A. (2003). CNTNAP2 is disrupted in a family with Gilles de la Tourette syndrome and obsessive compulsive disorder. *Genomics* **82**, 1–9.

Visser-Vandewalle, V., Temel, Y., Boon, P., Vreeling, F., Colle, H., Hoogland, G. et al. (2003). Chronic bilateral thalamic stimulation: a new therapeutic approach in intractable Tourette syndrome. Report of three cases. *J Neurosurg* **99**, 1094–1100.

Vitek, J. L., Chockkan, V., Zhang, J. Y., Kaneoke, Y., Evatt, M., DeLong, M. R. et al. (1999). Neuronal activity in the basal ganglia in patients with generalized dystonia and hemiballismus. *An Neurol* 22–35.

Vitek, J. L., Zhang, J. Y., Evatt, M., Mewes, K., DeLong, M. R., Hashimoto, T. et al. (1998). GPi pallidotomy for dystonia: Clinical outcome and neuronal activity. In S. F., Ed., *Advances in Neurology: Dystonia 3*, 211–219. Lippincott-Raven, Philadelphia.

von Plessen, K., Wentzel-Larsen, T., Hugdahi, K., Feineigle, P., Klein, J., Staib, L. et al. (2004). Altered interhemispheric connecticity in individuals with Tourette syndrome. *Am Psychiatry* 2028–2037.

Walkup, J. T., Khan, S., Schuerholz, L., Paik, Y.-S., Leckman, J. F., and Schultz, R. T. (1998). Phenomenology and natural history of Tic-related ADHD and learning disabilities. In J. F. Leckman and D. J. Cohen, Eds., *Tourette's Syndrome tics, obsessions, compulsions—Developmental psychopathology and clinical care*, 63–79. John Wiley and Sons, New York.

Walkup, J. T., LaBuda, M. C., Singer, H. S., Brown, J., Riddle, M. A., and Hurko, O. (1996). Family study and segregation analysis of Tourette syndrome: Evidence for a mixed model of inheritance. *Am J Hum Genet* **59**, 684–693.

Wang, J. Q. and McGinty, J. F. (1997). Intrastriatal injection of a muscarinic receptor agonist and antagonist regulates striatal neuropeptide mRNA expression in normal and amphetamine-treated rats. *Brain Res* **748**, 62–70.

Wilhelm, S., Deckersbach, T., Coffey, B. J., Bohne, A., Peterson, A. L., and Baer, L. (2003). Habit reversal versus supportive psychotherapy for Tourette's disorder: A randomized controlled trial. *Am J Psychiatry* **160**, 1175–1177.

Wolf, S. S., Jones, D. W., Knable, M. B., Gorey, J. G., Lee, K. S., Hyde, T. M. et al. (1996). Tourette syndrome: Prediction of phenotypic variation in monozygotic twins by caudate nucleus D2 receptor binding. *Science* **273**, 1225–1227.

Woods, D. W., Twohig, M. P., Roloff, T., and Flessner, C. (2003). Treatment of vocal tics in children with Tourette syndrome: investigating the efficacy of habit reversal. *App Behav Anal* **36**, 109–112.

Zheng, T. and Wilson, C. J. (2002). Corticostriatal combinatorics: the implications of corticostriatal axonal arborizations. *J Neurophysiol* **87**, 1007–1017.

Zhuang, P., Hallett, M., Zhang, X. H., and Li, Y. (2004a). Neuronal activity in the globus pallidus internus in patients with tics. Program no. 183.7. In *Society for Neuroscience*. Washington, DC.

Zhuang, P., Li, Y., and Hallett, M. (2004b). Neuronal activity in the basal ganglia and thalamus in patients with dystonia. *Clin Neurophysiology* **115**, 2542–2557.

Ziemann, U., Paulus, W., and Rothenberger, A. (1997). Decreased motor inhibition in Tourette's disorder: Evidence from transcranial magnetic stimulation. *Am J Psychiatry* **154**, 1277–1284.

Zohar, J., Mueller, E. A., Insel, T. R., Zohar-Kadouch, R. C., and Murphy, D. L. (1987). Serotonergic responsivity in obsessive-compulsive disorder. Comparison of patients and healthy controls. *Arch Gen Psychiatry* **44**, 946–951.

26

Disorders of Sleep and Circadian Rhythms

Sebastiaan Overeem, Gert Jan Lammers,
and Mehdi Tafti

I. Introduction

Virtually all life forms on planet Earth cyclically adapt their behavior to a 24-hour period, in close conjunction with the 24-hour changes in the environment imposed by the rotation of our planet around its axis (Turek & van Reeth, 1996). The most obvious example of such a 24-hour rhythm is the sleep/wake cycle: both sleep and wake episodes normally occur only at specific times of the day. However, there are many more such circadian rhythms, for example in body temperature and the levels of several hormones.

Although the presence of circadian rhythms has long been noted, only relatively recently the concept of an *internal* timekeeping system in the brain has been established (Panda, Hogenesch & Kay, 2002; Schwartz, 1997). Experiments in both animals and humans showed that in an environment devoid of all external time cues, day/night rhythms persist with a period close to 24 hours (hence the name *circadian*, meaning *approximately 1 day*). Besides being governed by the circadian timekeeper, the sleep/wake cycle is influenced by other factors such as homeostatic systems, which becomes readily apparent when we skip a night of sleep (Dijk & Lockley, 2002).

Perhaps because sleep takes up such a large part of our lives, we tend to take normal sleep/wake regulation for granted. This adds to the huge burden imposed on the life of patients when sleep is *not* following its normal pattern. Sleep and circadian disorders affect the personal and social lives of patients, but also their professional opportunities and their safety.

Our understanding of sleep and its disorders has gained a lot of momentum with the advance of molecular biology. Not only has this resulted in the elucidation of the molecular basis of sleep and circadian mechanisms in various animal models, but also opened up the way to understand human (patho)physiology. It is exciting to note that very recent discoveries have already been implemented in clinical practice, for example in new diagnostic techniques but also in the development of specific, rational therapeutic strategies.

In this chapter we give a brief overview of sleep and circadian neurology. Furthermore, we will describe our current understanding of the biological clock, and of the brain's control of sleep and wakefulness. Finally, to illustrate the leaps in knowledge we have made, not only in animal models but also toward human pathophysiology, we will describe the advances in two prototypical disorders: narcolepsy and the advanced sleep phase syndrome.

II. General Aspects of Sleep and Circadian Rhythms

A. What Is Sleep, and Why Do We Sleep?

Although we spent about one third of our lifespan asleep, it is actually not easy to clearly define what sleep is. An often used definition of sleep is the following: "Sleep is a state of immobility with greatly reduced responsiveness, which can be distinguished from coma or anesthesia by its rapid reversibility" (Siegel, 2005). An additional characteristic is the typical alternation between sleep and its counterpart, wakefulness. However, this timing mechanism can easily be overcome by homeostatic factors: When sleep is prevented for a while, the organism tries to recover the lost amount. Finally, sleep is not a uniform state varying only in depth. Halfway through the twentieth century a curious fact was discovered: Parts of the sleep period are characterized by high levels of brain activity, accompanied by vigorous eye movements (Aserinsky & Kleitman, 1953). Rapid eye movement (REM) sleep, as it was called, is so distinct that the other sleep stages are collectively called non-REM (NREM) sleep.

We are confronted with the need for sleep on a daily basis. Like hunger and thirst, sleep is a basic biological drive we cannot ignore. Although being a compulsive need, the fact is that we do not exactly know *why* we sleep. The lack of an answer does not mean that there is a lack of research. However, most of this research is based on the study of physiological and behavioral changes resulting when an organism is deprived of sleep. This poses an inherent methodological problem: It is impossible to remove sleep without influencing numerous other physiological processes.

It is clear that sleep, or sleep-like states in lower animals, is essential for life: Total sleep deprivation in rodents and flies can cause death more quickly than food deprivation (Everson, Bergmann & Rechtschaffen, 1989). This is also true when selective sleep deprivation is applied, for example only removing REM sleep. Currently, there are a number of theories pertaining to the function of sleep (Siegel, 2005). The fact that sleep is characterized by a lowered body metabolism (evidenced in a drop in oxygen consumption, heart and respiratory rates, and body temperature) suggest a role in energy conservation as well as in body recuperation. The latter may be even more specific in the form of nervous system recuperation, or neocortical maintenance. Sleep is also implied in more specific functions, such as brain development and emotional regulation. The role of sleep in the consolidation of memory is perhaps most widely known, although there still is a lively debate on this topic (Stickgold & Walker, 2005; Vertes & Siegel, 2005). Whether there are functions of sleep that are specific for REM or for NREM sleep is another area of uncertainty.

B. Sleep in the Animal Kingdom

Sleep is not a unique human behavior, and is widely present throughout the animal kingdom. In lower animals however, it becomes difficult to establish if real sleep is present, and the behavior is better described as a rest/activity cycle. However, in most vertebrates there is a behavioral state that has obvious characteristics of sleep. Within the animal kingdom, there is a fascinating variety of sleep duration and sleep patterns, of which the relevance is not completely understood (Siegel, 2005). For example, carnivores sleep more than omnivores, which sleep more than herbivores. Herbivores show an inverse correlation between body mass and sleep duration that is not found in carnivores and omnivores.

Within comparable species, there may be large differences in the amount and depth of sleep, depending for example on living environment. Macaque monkeys live in densely forested areas and are not easily visible in their surroundings. Macaques display a long and consolidated sleep period. Baboons, however, sleep in trees, but in large open spaces. For their safety, baboons choose to sleep in trees as high as possible, and have a very light sleep, easily disturbed by small signs of imminent danger. Both bats and shrews are small animals with a very high metabolic rate during wakefulness. Bats sleep over 20 hours per day with a clear drop in metabolism. Shrews, on the contrary, sleep much less, and are almost continuously active. Interestingly, shrews have an average lifespan of only two years, compared to over 15 years in bats, perhaps supporting the energy conservation hypothesis.

C. Circadian Rhythms

The rotation of the earth results in a predictable and constant change in our environment, through cyclic changes in light and temperature. The various daily rhythms displayed in animal and human behaviors are not just the *result* of these environmental changes however. Instead, natural selection has favored the evolution of endogenous, autonomic circadian clocks regulating various body functions. By keeping track of time, these internal clock mechanisms enable organisms to anticipate daily environmental changes and thus tailor behavior and physiology to the requirements of that particular time of the day.

The sleep/wake cycle is the most obvious example of a circadian rhythm. However, there are many more physiological processes that cycle with a period of about 24 hours. Examples include body temperature, heart rate and blood pressure, respiratory rate, and the levels of various hormones. The concept of an endogenous clock or pacemaker originated from experiments showing that the mentioned circadian rhythms persisted in animals, even when they were placed in a continuously dark environment, with a constant temperature. These early experiments were soon replicated in humans, using long-term isolation experiments, where all external time cues (Zeitgebers) were eliminated. From these data, it became clear that the endogenous pacemaker in various animals does not have a period of exactly 24 hours. This so-called free-running period varies among different species. Importantly, external time cues, especially the light/dark cycle, can synchronize (entrain) the endogenous clock, so that the resulting behavioral and physiological rhythms do exactly follow a 24-hour period.

D. Homeostatic and Circadian Sleep Regulation

The sleep/wake cycle is not solely under circadian control. Homeostatic regulatory mechanisms pose another important influence on sleep-propensity. Sleep propensity clearly builds up when the time spent awake increases. Furthermore, an extended period of wakefulness (i.e., sleep loss) is followed by a compensatory increase of sleep afterward. Several experimental paradigms have been developed to disentangle the circadian and homeostatic contributions to sleep regulation. Examples include *constant routine* studies in which the influence of environmental and behavioral factors are kept as constant as possible over the experimental period, so that the 24-hour variation measured in a variable can be attributed mainly to the endogenous pacemaker. *Forced desynchrony* studies use a sleep/wake schedule with a period clearly different from 24 hours (e.g., 20 or 28 hours) that is forced upon the subjects, under constant dim-light conditions that do not entrain the circadian pacemaker. In this paradigm there is an increasing loss of synchronization between the rhythms imposed by the circadian pacemaker and the artificially induced sleep/wake cycle. This makes it possible to determine the influence of both circadian and homeostatic processes on a certain variable under study.

Based on data derived from experiments as just described, Borbély was the first to propose a mathematical model for the regulation of sleep, including the circadian and the homeostatic process (see Figure 26.1). This so-called two-process model still has an important place in our understanding of the sleep/wake cycle. More recently, an ultradian process governing the cycling of the different stages of sleep during the rest period has been added, resulting in a three-process model (Borbély & Achermann, 1999).

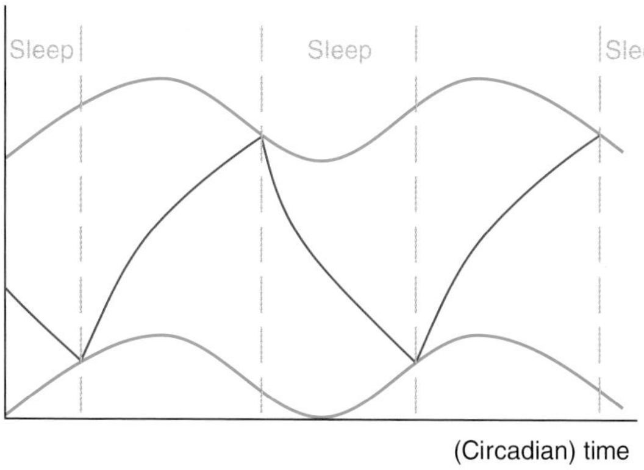

Figure 26.1 Schematic representation of the two-process model of sleep regulation. In this model, sleep timing is governed by the interaction between two processes. The sleep homeostat (process S, red line) represents the increasing need for sleep during the wake period and the decreasing need for sleep due to sleeping. Process C (green lines) represents the influence of the circadian clock on the boundary between sleep and wake: high during the day, low during the night. When S is high and the sleep boundary determined by C is lower, sleep results. Sleep continues until C begins to rise again and S has decreased sufficiently.

III. Sleep and Circadian Neurology

A. Characteristics of Normal Sleep

The study of sleep really took off with the invention of electroencephalography. Soon the idea that sleep is a passive bodily state was abandoned, and it became clear that "sleep is a dynamic behavior, not simply the absence of waking. Sleep is a special activity of the brain, controlled by elaborate and precise mechanisms" (Hobson, 1989). Nowadays, the architecture and stages of sleep are described using a combination of physiological variables (EEG, electro-oculography (EOG), and electromyography (EMG)). Still in use are the sleep stage criteria developed by Rechtschaffen and Kales in 1968 (Rechtschaffen & Kales, 1968). Using these measurements, sleep is broadly divided into two states with independent controls and functions: non-REM and REM sleep. NREM sleep is further divided into four stages with an increasing depth, mainly determined on the basis of EEG criteria (see Table 26.1).

After sleep onset, the different stages of sleep are encountered in a characteristic order. Sleep is entered in stage I non-REM sleep, which lasts for a few minutes. Sleep is easily interrupted during this stage. Besides a role in the initial wake-to-sleep transition, stage I may also occur as a transitional stage during the rest of the night. Stage II, heralded by the occurrence of sleep spindles and K-complexes in the EEG, follows, typically lasting about 30 minutes. The arousal threshold in this stage is increased compared to stage I. During the progression of stage II, there is a gradual appearance of high-voltage slow waves (delta activity) in the EEG. When delta waves are present for more than 20 percent of time, stage III is entered, soon followed by stage IV with more than 50 percent of slow waves. Stages III and IV (together called *slow wave sleep*) are the deepest stages of sleep, with the highest arousal threshold. When around 1.5 hours have passed since sleep-onset, the first episode of REM sleep is entered.

Table 26.1 Human Sleep Stages

	Wakefulness	Non-REM sleep				REM sleep
		Stage I	Stage II	Stage III	Stage IV	
EEG	Posterior alpha (8–13 Hz), mixed with frontocentral beta (>13 Hz)	Decrease of alpha activity (<50%)	Sleep spindles (14 Hz), K complexes	Delta waves (<4 Hz) 20–50%	Delta waves (<4 Hz) >50%	Mixed frequency, sawtooth waves
EMG (muscle tone)	Normal	Slightly diminished	Slightly diminished	Diminished	Diminished	Absent
EOG	Waking eye movements	Slow, rolling eye movements	Slow eye movements	Slow eye movements	Slow eye movements	Rapid eye movements

REM sleep is characterized by a mixed-frequency EEG that resembles the waking EEG, together with a striking reduction of muscle tone and the characteristic jerky eye movements. The arousal threshold during REM sleep is low, and it is this stage when most of our dreams occur. After about 15 minutes, a whole sleep cycle is started again, beginning with the lighter non-REM stages. When the progression through the different stages of sleep is plotted against time, a so-called *hypnogram* is derived (see Figure 26.2). From this, it becomes clear that a normal night of sleep consists of four to six sleep cycles as just described. The contents of the cycle change during the night: In the early hours, there is a large amount of slow wave sleep and relatively low REM sleep. Toward the morning, the amount of slow wave sleep diminishes, and REM sleep increases. The total amount of sleep varies greatly between persons, with a normal range as wide as five to 10 hours of sleep per night.

B. Symptoms of Sleep Disorders

Sleep disorders can be divided in several groups based on the main complaint. The first (and probably most common) complaint is too little sleep, or insomnia, which most often is psychophysiological. The opposite is hypersomnia, or too often and/or too long sleep. The amount and structure of sleep can be normal, but the timing disturbed—this is the hallmark of circadian disorders of sleep. Finally, there is a range of unwanted phenomena that can occur around and/or during sleep (such as restless legs, muscle jerks, sleepwalking, nightmares, etc.).

In every patient, one should inquire about the basic aspects of nighttime sleep, such as habitual sleep times, sleep hygiene, trouble falling asleep, frequent awakenings, and others. Not only the night should be evaluated—the daytime deserves attention as well. For example: A complaint of involuntary sleep episodes during the day, when there has been an opportunity to have enough nighttime sleep, always deserves further investigation. In addition to these general aspects, specific symptoms of sleep disorders should be evaluated. These include respiratory disturbances (snoring, witnessed apneas), abnormal motor behavior during sleep, and specific features of narcolepsy (such as cataplexy).

In addition to the patient's history, many sleep disorders require laboratory investigations for diagnosis. One of the mainstay tools is overnight polysomnography (PSG), which can be performed in a sleep clinic but also ambulatory. PSG consists of EEG, EMG, and EOG recordings to score sleep according to the Rechtschaffen and Kales criteria, together

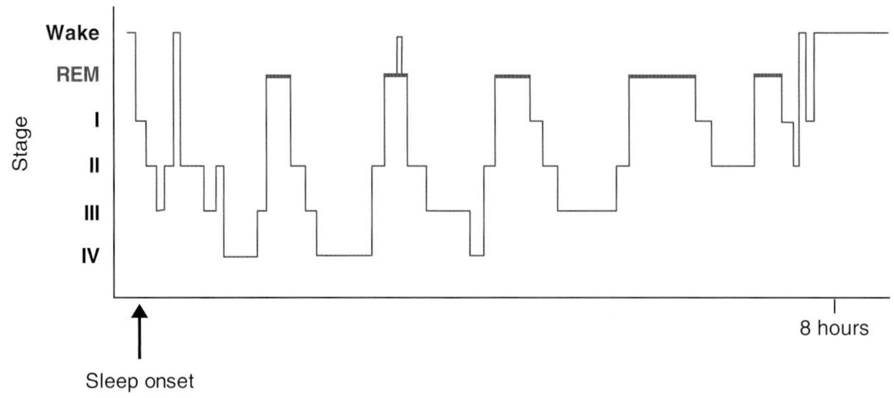

Figure 26.2 Schematic representation of a hypnogram in a healthy control subject, showing the different sleep stages as the night advances. In this example, there are 5 sleep cycles (from light NREM to slow-wave to REM sleep) during the night. Note that the amount of slow-wave sleep (stage III and IV) decreases during the night, whereas the amount of REM sleep increases.

with additional channels to record breathing effort, nasal and oral airflow, and limb movements. Excessive daytime sleepiness (EDS) can be objectively documented using a Multiple Sleep Latency Test (MSLT). During the MSLT, patients are asked to lie down and try to sleep during five 20-minute nap opportunities over the day. EDS is characterized by a mean sleep latency of less than 8 minutes. In addition, the premature occurrence of REM sleep during these short nap opportunities supports a diagnosis of narcolepsy. Besides keeping a sleep diary, wrist-worn actigraphy can be used to monitor rest/activity cycles over several days or weeks. Although rest/activity rhythms are not necessarily identical to sleep/wake rhythms, this can nevertheless give a fair indication of abnormal circadian timing (see Figure 26.9).

The range of sleep and circadian disorders has been brought together in a diagnostic and coding manual, the international classification of sleep disorders (ICSD, 2005). This diagnostic system provides a rational and systematic framework for an accurate and standardized diagnosis. This not only facilitates clinical practice, but is highly important for research studies as well.

C. Genetic Aspects of Sleep Disorders

Although many sleep disorders run in families, only a few have a known genetic basis. Like most other complex disorders, sleep disorders result from the coordinated contribution of genetic and environmental factors as well as gene–environment interactions. Evidence for genetic contribution is becoming available in several frequent sleep disorders. Family studies in restless legs syndrome reported linkage with 12q12-q21 and 14q13-q21, and an association study suggested that polymorphisms of monoamine oxidase-A confer susceptibility to this condition (Trenkwalder, Paulus & Walters, 2005). Human Leukocyte Antigen (HLA) polymorphisms are strongly associated with narcolepsy, but also with several other sleep disorders including sleepwalking, REM sleep behavior disorder, and Kleine-Levin syndrome (Tafti, Maret & Dauvilliers, 2005). One of the most challenging tasks is to find genetic susceptibility factors to the obstructive sleep apnea syndrome, the most prevalent hypersomnia. Preliminary quantitative genetic analyses suggest potential linkage with chromosomes 2p and 19p, once confounding effects such as obesity are accounted for (Patel, 2005). Also the probability of moderate to severe sleep disordered breathing is significantly higher in patients with apolipoprotein E4, independently of age, sex, body mass index, and ethnicity (Kadotani et al., 2001).

Up until now, only three diseases have been reported to result from single gene mutations: a rare form of primary insomnia, fatal familial insomnia, and familial advanced sleep phase syndrome. Fatal familial insomnia (FFI) is characterized by an increasing inability to initiate and maintain sleep (ultimately leading to complete agrypnia), together with autonomic and somatomotor disturbances. FFI is caused by a point mutation at codon 178 of the prion protein gene, which is responsible for the degeneration of specific thalamic nuclei (Medori et al., 1992). A missense mutation in a single patient with chronic insomnia was reported at position 192 in exon 6 of the gene coding GABA-A beta-3 subunit (Buhr et al., 2002). More detailed studies have been carried out in familial advanced sleep phase syndrome, a rare but striking circadian disorder. These findings will be described in detail later in the chapter.

IV. Neurobiology of Sleep and Circadian Rhythms

A. Circadian Biology

1. The Master Pacemaker

In the last three decades, compelling evidence has been collected that the endogenous master clock resides in a part of the anterior hypothalamus, the suprachiasmatic nucleus (SCN) (Buijs & Kalsbeek, 2001; Reppert & Weaver, 2002). Destruction of the SCN abolishes circadian rhythms in behavior, hormonal levels, metabolic rate, and body temperature (Moore & Eichler, 1972; Stephan & Zucker, 1972). When SCN tissue is isolated and kept in culture, a near 24-hour cycle of neuronal firing persists. When individual SCN neurons are kept in culture, circadian oscillations in firing rate persist as well, with small differences in period from cell to cell. The final piece of the puzzle was provided by experiments of Ralph et al. (1990): They transplanted fetal SCN material into the brains of (arrhythmic) SCN-lesioned hamsters, and showed this to restore circadian rhythmicity. Moreover, the resulting circadian period in the acceptor animal was determined by the donor.

As mentioned earlier, the SCN can function autonomously, without external time cues. An important aspect of circadian rhythms in intact animals however, is the fact that the circadian period is entrained by external influences, most notably the light/dark cycle. This entrainment is mediated by a specialized pathway providing photic input to the SNC: the retino-hypothalamic tract (Moore & Lenn, 1972). The retinal cells of the retinohypothalamic tract form a distinct population of photoreceptive cells involved in vision. The SCN has dense projections to various other parts of the hypothalamus, through which its influence on behavior and endocrine rhythms is mediated (Saper et al., 2005). Most SCN effects rely on this direct neuronal interaction, although in part paracrine signaling using diffusible factors may be involved (Silver et al., 1996).

2. Central and Peripheral Oscillators

Isolated SCN neurons all have slightly different periods and phases, and the average of these individual pacemakers makes up the final output. This concept is supported by studies using mouse chimeras in which the SCN is composed of a mixture of wild-type and *Clock* mutant neurons. The resulting circadian period depends on the proportion of wild-type and mutant cells in the SCN (Low-Zeddies & Takahashi, 2001). Several mechanisms have been proposed that underlie the coupling between individual SCN neurons, such as synaptic transmission, gap junctions, as well as paracrine mechanisms.

SCN neurons are not the only cells that express circadian oscillations. In fact, there are circadian oscillators throughout the body. Several genes making up the intracellular clock mechanism (see next) have been found to be rhythmically expressed in other brain areas as well as various peripheral organs (Buijs et al., 2001; Yamazaki et al., 2000). In contrast to SCN neurons, these slave oscillators sustain their 24-hour oscillations for a few days only when kept in culture, deprived of master clock input. It is thought that the collective SCN output synchronizes the timing of peripheral oscillators, which in turn regulate local rhythms in physiology and behavior. This hierarchical system confers precise period and phase control as well as stability of the widespread physiological systems that are regulated.

3. The Molecular Inner Workings of the Clock

Our conceptualization of the molecular mechanisms of the mammalian circadian clock has been greatly aided by the data available from the fruit fly *Drosophila melanogaster* (Panda et al., 2002). Most of the genes involved in the fly circadian clock have orthologs in mammals, and the general principles of interacting transcriptional feedback loops are similar. The specific differences between the molecular clock mechanisms in different species have been reviewed elsewhere (Young & Kay, 2001). Here, we will discuss the main parts of the mammalian (mouse model) clock. For details, see several recent reviews (Allada et al., 2001; Bromham & Penny,2003; Cermakian & Boivin, 2003; Lowrey & Takahashi, 2004; Panda et al., 2002; Reppert et al., 2002).

The intracellular clock mechanism is based on several interacting positive and negative transcriptional feedback loops that result in oscillating RNA and protein levels of key clock component (see Figure 26.3). The basic drive of the system forms the cycling transcriptional enhancement by two transcription factors, CLOCK and BMAL1. When CLOCK and BMAL1 heterodimerize, they can activate the transcription of three *period* genes (*Per1-3*) and two *cryptochrome* genes (*Cry1-2*) by binding to E-box elements in their promoters. The resulting protein products (PER1-3 and CRY1-2) constitute the basic negative feedback loop: PER and CRY proteins dimerize and are phosphorylated, after which they reenter the nucleus and directly repress the transcriptional activity of the CLOCK/BMAL1 complex, thereby inhibiting their own production (see Figure 26.3, red loop).

On top of this basic negative feedback loop (which in itself would be sufficient to generate oscillations) several other regulation loops are imposed. For example, a positive feedback loop is formed by another gene whose transcription is activated by the CLOCK/BMAL1 complex: *Rev-Erbα*. REV-ERBα then inhibits *Bmal1* transcription. As a result, when the CRY/PER complex represses the activity of CLOCK/BMAL1, this not only diminishes *Cry* and *Per* transcription, but the transcription of *Rev-Erb* α as well, thereby releasing the repressing effect of REV-ERB α and reactivating *Bmal1 and Clock* transcription (see Figure 26.3, green loop). The various transcriptional feedback loops result in the rhythmic regulation of the various genes, with *Bmal1* RNA levels peaking 12 hours out of phase relative to *Per* and *Cry* RNA. Further refinement and control of the period of oscillations is provided by post-translational processes, most notably phosphorylation, degradation, and regulated nuclear entry of the proteins in the various feedback loops. Key components in this part of the clock are the phosphorylation enzymes casein kinases 1 (CK1) δ and ε. CK1 δ/ε phosphorylate the PER, CRY, and BMAL1 proteins, targeting them for degradation. Furthermore, CK1 δ and ε form part of the PER/CRY multimeric complex.

There is partial compensation of function between various members of the clock gene family, as the clock continues to oscillate after single gene mutations in several components, although often the circadian period under constant conditions is altered (Reppert et al., 2002). However, PER and CRY outputs of the negative feedback loop are critical for maintaining a functional clock, as disruption of either the *Per1-2* genes together or both *Cry* genes result in acute and total behavioral arrhythmicity under constant conditions. Recent studies in monogenetic human circadian disorders have provided the first evidence for a functional homology of the human clock with that of the mouse (see later).

One of the mechanisms through which the synchronized molecular oscillation can be transduced into local rhythms is through the local clock-controlled genes (CCGs). These CCG for example can contain E-box enhancers that can be controlled directly by CLOCK/BMAL1 heterodimers. Using various high throughput DNA techniques (such as microarrays), many CCG currently are being identified. It is one of the challenges of the years to come, to identify the mechanisms through which the molecular clockwork in the SCN finally results in peripheral rhythms and ultimately behavior.

B. Sleep Biology

The study of the neural correlates underlying the different states of consciousness started in first half of the twentieth century. Soon a number of key structures were proposed. Based on his famous study of the *encephalitis lethargica* epidemic, Von Economo (1930) pointed to a crucial role for

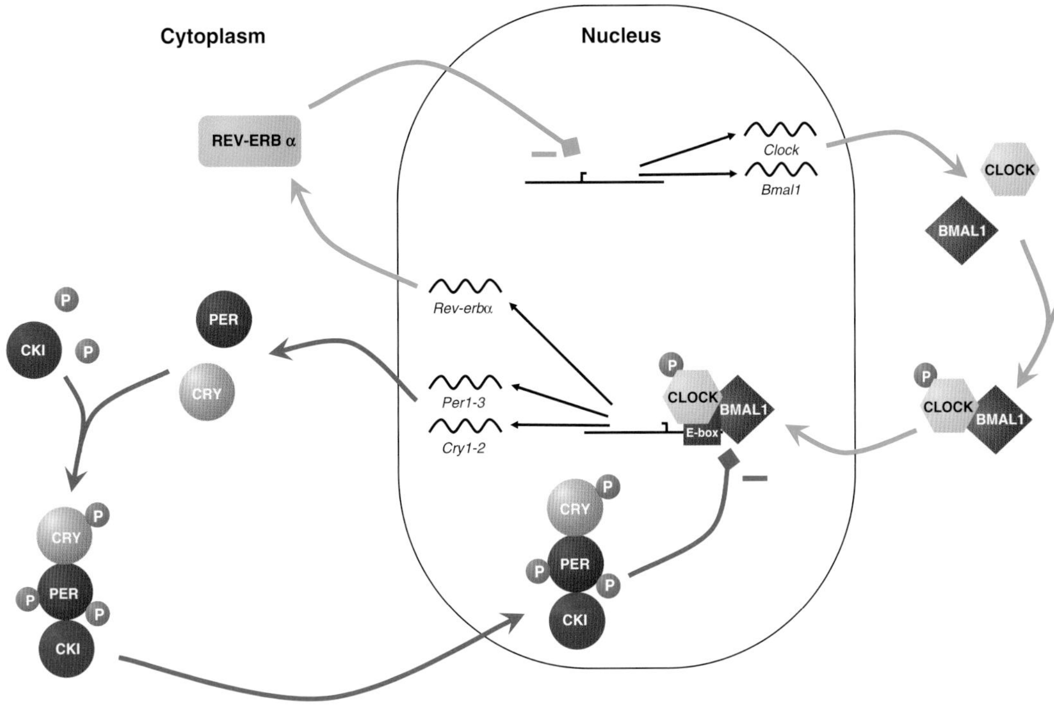

Figure 26.3 Schematic representation of the basic components of the mammalian molecular circadian clock. The core of the clock consists of two transcriptional feedback loops. The negative loop (shown in red) leads to decreased *Per/Cry* transcription (shown in red), by inhibiting CLOCK/BMAL1 mediated transcriptional activation. The positive loop (shown in green) results in increased *Per/Cry* transcription, as *Per/Cry* mediated inhibition of CLOCK/BMAL1 activity decreases REV/ERBα expression, and thus de-represses (increases) *Bmal1* transcription. Degradation and nuclear entrance of the various components is regulated in various ways, for example through CK1 mediated phosphorylation of the PER/CRY complex. Per: *Per* 1-3; Cry: *Cry* 1-2; CK1: *CK1* δ and ε.

the hypothalamus, as lesions in the posterior hypothalamus resulted in excessive sleepiness, whereas injury to the anterior hypothalamus resulted in severe insomnia. Moruzzi and Magoun (1949) focused on the hypothalamus/midbrain wake-promoting regions and showed that stimulation of the reticular formation in the brainstem induced waking. Several excellent reviews are available that give a comprehensive overview of the neuronal systems involved in the regulation of sleep and wakefulness (Espana & Scammell, 2004; Hobson & Pace-Schott, 2002; Pace-Schott & Hobson, 2002; Saper, Scammell & Lu, 2005). Here we will give a short overview of the most important systems and their interaction (see Figure 26.4).

Classical view regarded conscious states as a continuum from hypervigilance induced by stimulants going down to normal wakefulness, drowsiness, sleep, torpor, and hibernation, and ending with coma. This point of view was essentially based on the intensity of both external and internal somatosensory inputs thought to be essential for the excitation of the brain. Accordingly, wakefulness was considered to be the active state of the brain, and sleep was considered as a passive state following the decrease or the absence of sensory inputs. As mentioned before, this view is not correct: sleep is a vital and active state during which specific brain structures display high levels of neuronal activity comparable to those of wakefulness.

1. Wakefulness

Experiments in the 1930s using the so-called *Encéphale Isolé* and *Cerveau Isolé* in the cat model, indicated that the first preparation does not affect wakefulness whereas the second induces a state of continuous sleep. This was interpreted as the existence of a center for wakefulness between the two sections (i.e., the brainstem). Transsections at the level of hypothalamus further revealed that the anterior hypothalamus is implicated in sleep and the posterior hypothalamus is involved in wakefulness. The development of electrophysiology led to the reticular formation theory with the demonstration that the stimulation of the reticular formation neurons without stimulating sensorial afferents induces wakefulness (Moruzzi et al., 1949). The ascending activating reticular formation was therefore long considered as the active wakefulness system. However, a mediopontine pretrigeminal preparation resulted in a state of continuous wakefulness up to two to three days, and a chronic *Cerveau Isolé* preparation demonstrated that wakefulness could recover, suggesting that some rostral

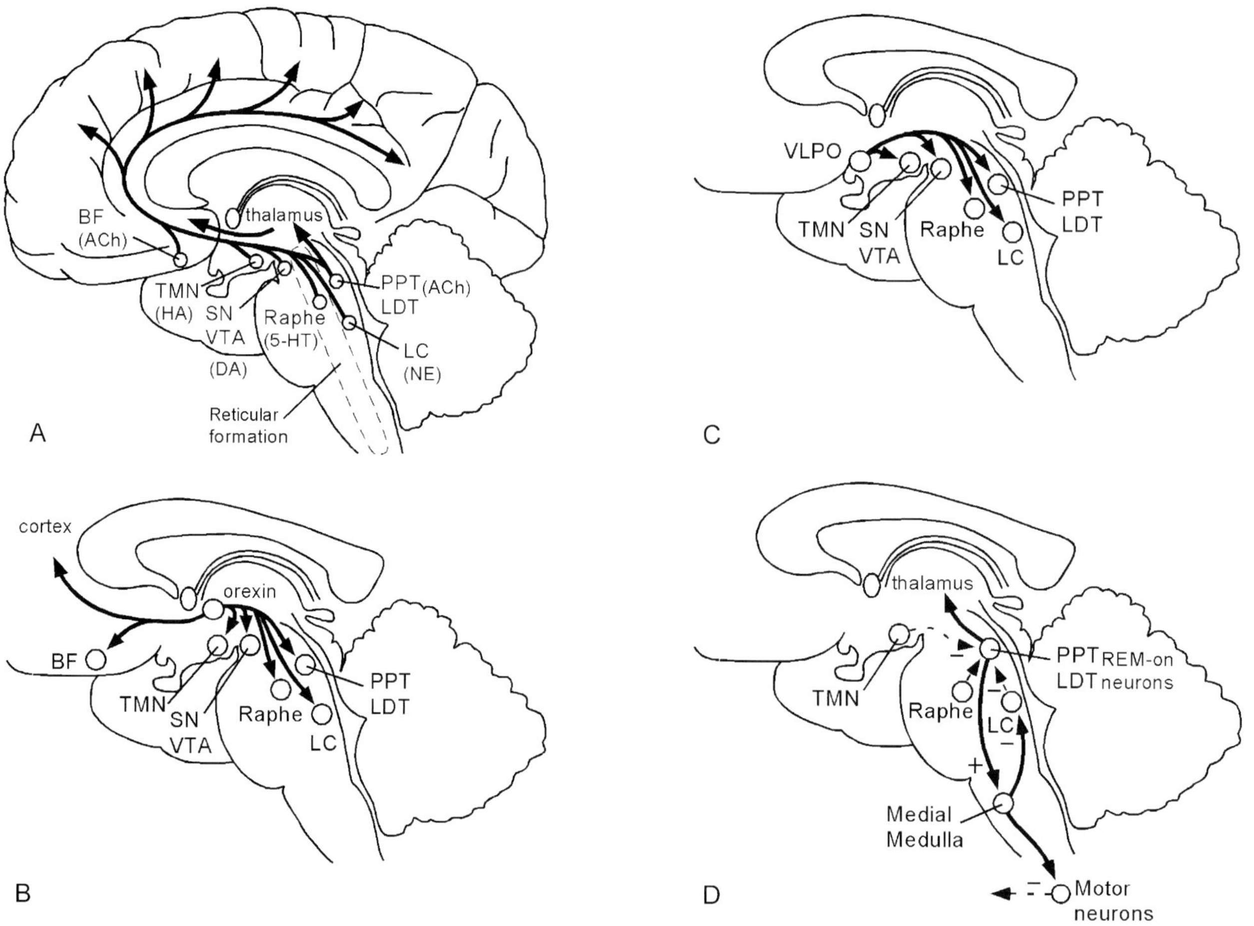

Figure 26.4 Schematic drawings of the various brain areas involved in the regulation of sleep and wakefulness, and in NREM/REM sleep regulation. **A**. Ascending arousal systems in the brainstem and posterior hypothalamus send projections throughout the forebrain. Cholinergic neurons in the pedunculopontine and laterodorsal tegmental areas (PPT/LDT) activate many forebrain targets, including the thalamus. Neurons in the locus coeruleus (LC), dorsal and median raphé, tuberomammillary nucleus (TMN), substantia nigra and ventral tegmental area (SN/VTA), and basal forebrain (BF) excite many cortical and subcortical targets. The reticular formation projects to the thalamus, hypothalamus, and basal forebrain. **B**. Hypocretin (Orexin) neurons in the lateral hypothalamic area innervate all of the ascending arousal systems, as well as the cerebral cortex. **C**. NREM sleep and **D**. REM sleep pathways. Neurons of the ventrolateral preoptic nucleus (VLPO) produce GABA and galanin, and inhibit all the arousal systems during NREM sleep. Many of these cells are active during REM sleep as well. REM sleep is driven by a distinct population of cholinergic PPT/LDT neurons. During wakefulness and NREM sleep, these cells are inhibited by norepinephrine, serotonin, and histamine, but during REM sleep, the aminergic neurons fall silent, thus disinhibiting the LDT/PPT REM-generating neurons. These cholinergic neurons also produce the atonia of REM sleep by activating the medial medulla, which inhibits motor neurons. The medial medulla also reduces excitatory signals from the LC that normally increase motor tone. ACh: acetylcholine; DA: dopamine; DR: dorsal raphé; HA: histamine; HCRT: hypocretin; NE: norepinephrine. Figure reprinted with permission from Espana et al. (2004).

structures are likely to be involved in the generation and maintenance of wakefulness (Villablanca, 1965).

The ascending reticular activating system originates from monoaminergic and cholinergic nuclei from rostral pons and the midbrain (see Figure 26.4A). Lesions of dopamine-containing neurons of the substantia nigra and the ventral tegmental area abolish the behavioral response despite a cortical desynchronization. Lesions of norepinephrine-containing neurons of the locus coeruleus have limited effect on wakefulness. However, cooling these neurons decreases wakefulness whereas their electrical stimulation induces cortical desynchronization and wakefulness. Note also that pharmacological inhibition of the catecholamine systems induces somnolence whereas stimulants (amphetamine or cocaine) induce hyperexcitability through increased catecholamine release.

Although acetylcholine seems necessary for the cortical desynchronization, lesions of both pontine and basal forebrain cholinergic neurons have limited effect on vigilance states despite a slowing of the cortical activity. However, almost all cholinergic neurons are found more active during wakefulness than during slow wave sleep and acetylcholine release parallels this activity. Acetylcholine producing neurons from the pedunculopontine and laterodorsal tegmentum nuclei (PPT/LDT) project to thalamic-relay and reticular nuclei and are essential in promoting wakefulness and inhibiting rhythmic pacemaker thalamic activity, which characterizes slow wave sleep (see Figure 26.4A).

The rostral structures implicated in wakefulness include the newly discovered hypocretin (orexin) neurons from the lateral hypothalamus, the histaminergic (tuberomammillary) nucleus from the posterior hypothalamus, and the cholinergic neurons from the basal forebrain with widespread excitatory projections to the cortex and caudally to the midbrain and the brainstem monoaminergic nuclei (see Figure 4A, B).

2. Non-REM (Slow Wave) Sleep

The EEG during slow wave sleep (SWS) is characterized by a widespread cortical synchronization with rhythmic activities in the delta (0.5–4 Hz) and sigma range (12–15 Hz). Electrical stimulation of solitary nucleus induces cortical synchrony, whereas its lesion induces a state of total insomnia, clearly indicating that this structure is involved in the generation of sleep. However, the fact that continuous slow waves are induced by the *Cerveau Isolé* preparation indicates that other rostral structures are involved as well. This is corroborated by observations in humans that lesions of the anterior hypothalamus produce total insomnia. It was also reported that electrical stimulation of the preoptic region can readily induce slow wave sleep.

More recent experiments using neurotoxic lesions of the preoptic area indicated a state of insomnia, which could be reversed by microinjections of muscimol into the posterior hypothalamus. It is noteworthy that a complete lesion of the thalamus suppresses spindle activity while delta activity persists. More recently it was demonstrated that SWS is generated by a reciprocal thalamocortical circuit, with a cortical pacemaker responsible for the slow component and a thalamic pacemaker (reticular thalamic nucleus) responsible for sigma activity (Llinas & Steriade, 2006). Therefore, some structures are permissive of SWS expression such as the solitary nucleus and the preoptic area, and some others are implicated in the generation of rhythmic oscillations that characterize this state such as the thalamus and thalamocortical network (see Figure 26.4C).

The serotonergic hypothesis of sleep had a widespread impact for over 10 years, and still exerts major influences in this field. The main discovery by Jouvet was that a total lesion of the raphé nucleus induces a complete insomnia. This was corroborated by the fact that parachlorophenylalanine (a tryptophane hydroxylase inhibitor) induces a complete insomnia, which can be reversed by serotonin injections. However, in the early 1980s, experimental evidence was provided indicating that neurotoxic lesions of serotonin neurons decreases sleep but is followed by a gradual recovery, and that cooling the dorsal raphé nuclei can induce sleep. Moreover, it was demonstrated that raphé neurons are electrically silent during sleep and serotonin release parallels this pattern of activity. However, local injection of serotonin into the preoptic area induces sleep with a delay, probably through the synthesis and accumulation of a yet unknown substance at this level.

Microinjection of adenosine into the preoptic area potently induces sleep (Portas et al., 1997), and it is well known that caffeine (inhibitor of the adenosine receptors) increases wakefulness (Yanik, Glaum & Radulovacki, 1987).

More recently the ventrolateral preoptic (VLPO) nucleus was identified as the major sleep-inducing structure (Sherin et al., 1996) containing specific sleep-active neurons (Gallopin et al., 2000). The major activity of the VLPO is inhibitory, through GABAergic (and probably galaninergic) projections to all previously mentioned hypothalamic and brainstem wake-promoting systems.

3. REM Sleep

In 1962, Jouvet demonstrated that a section rostral to the brainstem induces a complete disappearance of REM sleep, whereas after a section caudal to the brainstem all REM-related activities are still present. Further lesion experiments demonstrated the essential role of the pontine tegmentum in the generation of REM sleep (see Figure 26.4D). Electrophysiological and immunohistochemical investigations in the brainstem revealed the presence of REM-on cells in the gigantocellular and sublaterodorsal nuclei and REM-off cells in the locus coeruleus, the dorsal raphé, and the periaqueductal gray matter.

Neuronal networks generating the different aspects of REM sleep phenomenology are now well understood. Cortical desynchronization is under the control of mesopontine tegmentum neurons (which project to the intralaminar and reticular thalamic nuclei) and the magnocellular nucleus (which projects to the posterior hypothalamus, the thalamus, and the basal forebrain). More recently it was proposed that REM-on neurons from the peri-locus-coeruleus region send glutaminergic projection to the medial septum, and regulate theta rhythms during REM sleep (Lu et al., 2006). Muscle atonia is under the control of magnocellular neurons, which send inhibitory projections to the spinal motoneurons. These neurons receive afferents from the locus coeruleus. A few pontine tegmentum neurons directly project to the spinal motoneurons. Note that lesions of the peri-locus-coeruleus induce REM sleep without atonia.

Cholinergic stimulation at the level of the pontine reticular formation readily induces REM sleep. This effect is more specifically obtained if a cholinergic agonist (carbachol) is injected into the dorsal pontine reticular formation at the vicinity of the peri-locus coeruleus. In addition, as mentioned earlier, cholinergic neurons of the laterodorsal and the pedunculopontine tegmentum are REM-on cells (i.e., increase their activity prior and during REM sleep). A neurotoxic lesion of these nuclei produces a disappearance of REM sleep for up to three weeks.

4. Higher Order Regulation

By now, the overall picture of the different brain areas that are involved in the generation of the different states of consciousness is rather complete. But how do these different regions interact to generate the orchestrated changes between wakefulness and sleep, and between the different stages of sleep? Reciprocal interactions between sleep- and wake-promoting regions may be responsible for generating discrete behavioral states. More specifically, the sleep-promoting VLPO and the wake-promoting TMN/LC/DR have direct inhibitory projections to each other. This mutual inhibitory influence results in a bistable switching system, known in electrical engineering as a flip-flop (see Figure 26.5) (Saper, Chou & Scammell, 2001). This sleep switch results in self-reinforcing firing patterns, and avoids intermediate states between wakefulness and sleep. Recently, a similar model describing inhibitory interactions between REM-on and REM-off structures has been proposed in the regulation of the REM/NREM cycle (Lu et al., 2006).

One other property of the sleep flip-flop is its sensitivity to external perturbations. To function properly, a stabilizing factor is needed, for example keeping the system in the wake position during the day. A string of discoveries on the pathophysiology of the sleep disorder narcolepsy recently pinpointed the hypocretin system to have exactly such a function.

V. Disorders of Sleep and Circadian Rhythms

A. Narcolepsy as a Prototypical Sleep Disorder

1. Clinical Features

Narcolepsy is a primary sleep disorder, with an estimated prevalence of three to five per 10,000 in the Western population (Overeem et al., 2001). Excessive daytime sleepiness is the principal symptom. Patients may report a continuous feeling of sleepiness throughout the day, but more typically EDS is expressed by the occurrence of sudden episodes of irresistible sleepiness. EDS in narcolepsy is best described as an inability to stay awake for longer periods of time rather than a disorder with an increased need or amount of sleep (Overeem et al., 2001). Although patients fall asleep multiple times during the day, the total amount of sleep over 24 hours is not increased (Broughton et al., 1998). The second cardinal symptom of narcolepsy—and the most specific—is cataplexy: sudden attacks of skeletal muscle paralysis with preserved consciousness, triggered by emotions. In fact, narcolepsy patients may become literally "weak with laughter." All striated muscles may be involved during cataplexy, except the extraocular and respiratory musculature. Although laughter is the strongest trigger, various emotions may induce a cataplectic attack such as unexpectedly meeting an acquaintance, scoring a goal in sports, or merely thinking about the punch line of a joke.

Sleep paralysis, hypnagogic hallucinations, and automatic behavior often are reported by narcoleptic patients. More recently, it has become clear that disturbed nocturnal sleep is an integral feature of the disorder. This nocturnal sleep fragmentation can actually be described as an inability to stay asleep for longer periods of time, again pointing to a dysregulation of timing and maintenance of sleep and wakefulness as the primary pathophysiological mechanism. Nocturnal sleep recordings in narcoleptic patients show a short sleep onset, and frequent stage-shifts and arousals. Daytime recordings in a MSLT show a mean sleep onset latency of less than 8 minutes, and frequently two or more sleep onset REM periods (SOREMP: REM sleep occurring within the 20 minutes of a MSLT nap).

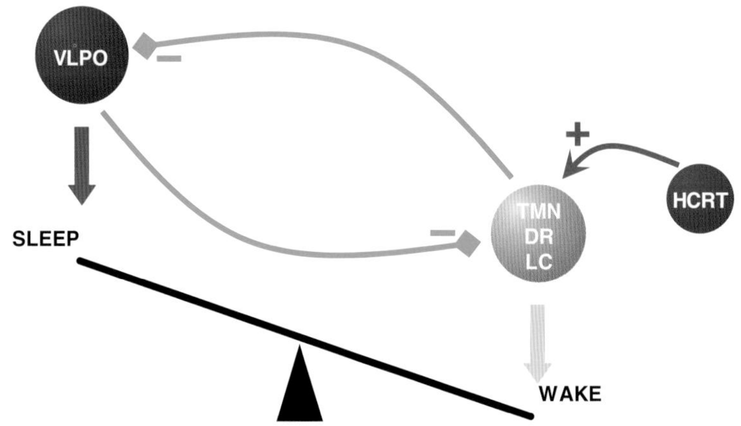

Figure 26.5 Schematic and simplified illustration of the hypothalamic sleep switch (Saper et al., 2001). Basically, the sleep and wake promoting areas in the brain have reciprocal inhibition connections, resulting in a bistable flip-flop switch. The hypocretin system stabilizes the switch in the wake position by excitatory projections to the wake areas.

2. Genetics of Human Narcolepsy

Human narcolepsy in its typical form is a sporadic disease. However, first degree relatives still have a 10 to 40 times increased risk of developing narcolepsy compared to the general population. True familial narcolepsy, with patients in multiple generations is rare, accounting for 1 to 2 percent of all narcoleptics. Around 20 monozygotic twin reports are available. Depending on how strict concordance criteria are applied, five to seven of those are concordant for the disease. In addition to genetic factors, most cases of narcolepsy require environmental factors for the disease to develop.

In the early 1980s the first reports emerged on the association of sporadic narcolepsy and specific HLA subtypes. At first an association was established with the (serologically determined) subtype DR2, but molecular typing at the DNA level pinpointed the HLA-DQ1 marker B1*0602 to be the main factor involved. Over 90 percent of sporadic narcolepsy patients are positive for HLA-DQB1*0602, compared to about 25 percent of the general population. As there are several autoimmune disorders closely associated with specific HLA types, these findings implied an autoimmune genesis for narcolepsy as well (see later). The increased risk of family members to develop narcolepsy is not fully explained by the HLA association. Furthermore, HLA-negative familial cases of narcolepsy have been described. Therefore, there may be other genetic factors involved in the pathophysiology of the disease. So far, there have been reports showing associations with polymorphisms in the tumor necrosis factor-α, and the catechol-O-methyltransferase genes. The latter may also modulate disease severity, and response to treatment with stimulants. Few genome-wide screenings studies have been performed up till now in familial narcolepsy, yielding linkage to two (large) genomic regions on chromosome 4p13-q21(Nakayama et al., 2000) and 21q (Dauvilliers et al., 2004; Kawashima et al., 2006). However, no further data is available on (candidate) genes from these regions, and their association with the disease.

3. Hypocretin Defects in Animal Narcolepsy

Toward the end of the 1970s, research into the pathophysiology of narcolepsy received an enormous boost with the discovery of a natural animal model in dogs. Canine narcolepsy is strikingly similar to the human condition, and cataplexy is the most prominent feature. In dogs, cataplexy is triggered easily by playing or when palatable food is presented (see Figure 26.6). Besides the clinical resemblance, there are striking similarities at the physiological and pharmacological levels (Baker, 1985; Foutz et al., 1979; Nishino & Mignot, 1997). W.C. Dement at Stanford University established the first colony with different breeds of narcoleptic animals. Although most donated dogs had a sporadic form of narcolepsy, it turned out that in some of these animals (Doberman Pinschers and Labrador Retrievers) the narcolepsy phenotype is transmitted as an autosomal recessive trait with full penetrance (Baker, 1985; Foutz et al., 1979).

Using the dog model, a cloning effort was initiated in the early 1990s to identify the canine narcolepsy gene, designated *canarc-1*. In 1999, a critical region was finally established, which contained only one known gene: *Hcrtr-2* (Lin et al., 1999). *Hcrtr-2* codes for the hypocretin receptor-2, one of the two receptors for the hypothalamic peptides hypocretin-1 and -2 (also known as orexin-A and -B). The hypocretin system was described in 1998, and first implicated in the regulation of feeding (de Lecea & Sutcliffe, 1999; Sakurai et al., 1998). In both narcoleptic Labradors and Dobermans, *Hcrtr-2* cDNAs were significantly shorter in affected

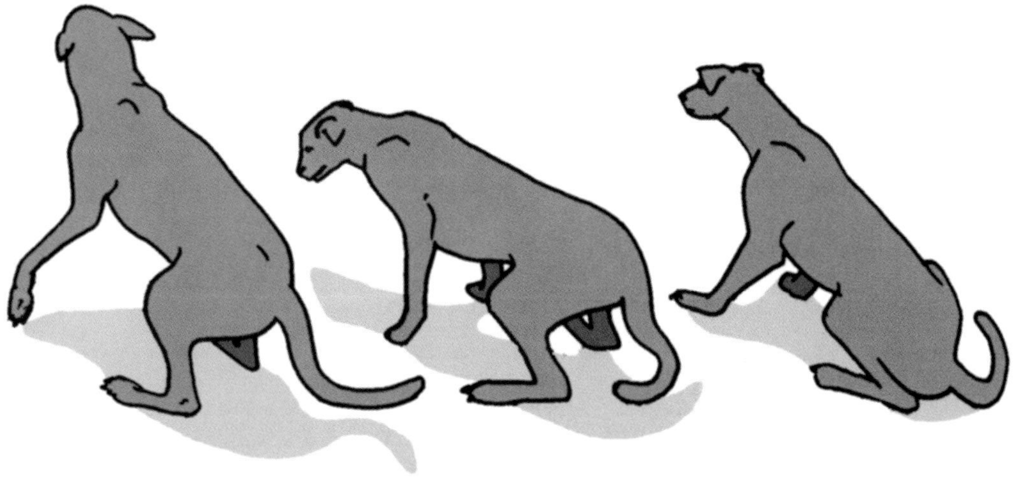

Figure 26.6 Attack of cataplexy, here elicited by a piece of palatable food, in a Doberman Pinscher with the heritable form of narcolepsy. Note the head drop, and the main involvement of the hind limbs.

dogs than in controls, suggesting a deletion in the transcripts. In the Dobermans, a so-called short interspersed nucleotide element (SINE) was found to cause aberrant splicing of the receptor mRNA, resulting in a truncated peptide (see Figure 26.7) (Lin et al., 1999). In the Labradors, a single base change in the 5 splice site consensus sequence results in skipping of exon 6 (see Figure 26.7) (Lin et al., 1999). Both mutations result in a receptor lacking the intracellular end, and therefore an absence of function.

Only two weeks after the publication of the canine narcolepsy gene, Yanagisiwa's group, which discovered the hypocretins a year before, reported on the phenotype of *preprohypocretin* knock-out mice (Chemelli et al., 1999). Infrared video recordings during the active period showed peculiar episodes of sudden behavioral arrest, which were short-lived and from which the animal abruptly recovered. Furthermore, it was found that the *hypocretin* knock-outs had several sleep abnormalities, including an increased but fragmented amount of sleep during the dark period, and a decreased REM latency as well as sleep-onset REM periods. Together, these data convincingly showed that mice lacking the hypocretin peptides develop narcolepsy (Chemelli et al., 1999).

4. Hypocretin Defects in Human Narcolepsy

Shortly after the implication of the hypocretin system in the pathogenesis of narcolepsy in animals, we were able to publish the first evidence of hypocretin involvement in the human disorder. We measured hypocretin-1 levels in the CSF of nine narcoleptic patients and eight control subjects. While in controls, a very stable concentration of hypocretin-1 was found (between 250 and 285 pg/ml); seven out of nine patients had no detectable hypocretin-1 in their CSF (Nishino et al., 2000). These findings quickly were followed by larger studies, and it is now clear that over 90 percent of patients with sporadic, HLA DQB1*0602 positive narcolepsy are hypocretin deficient (Nishino et al., 2001; Ripley et al., 2001). Furthermore, hypocretin-1 deficiency is highly specific for narcolepsy: Levels were normal in a large group of patients with various other sleep disorders (see Figure 26.8) (Mignot et al., 2002). In the recent revised version of the International Classification of Sleep Disorders, hypocretin-1 measurements therefore have been incorporated as a diagnostic tool for narcolepsy (ICSD, 2005).

By now, a number of neuropathological studies in postmortem brain tissue have been performed. Using both *in situ* hybridization in frozen brain tissue, and immunohistochemistry on fixed human brains, it was shown that there is no detectable hypocretin mRNA or peptide in the hypothalamus of narcoleptic subjects (Peyron et al., 2000; Thannickal et al., 2000). In addition, one of the studies found a significantly increased number of reactive astrocytes in the hypothalamus, suggesting neuronal degeneration. As two colocalizing markers in hypocretin neurons, dynorphin and NARP, also were shown to be absent in narcoleptic brains, it is likely that human narcolepsy results from a selective degeneration of hypocretin-producing neurons (Blouin et al., 2005; Crocker et al., 2005). The cause of such a neurodegenerative process remains unknown. Mainly based on the strong HLA association, the prevailing hypothesis states that an autoimmune reaction may be responsible, but direct evidence for this is lacking as of yet (Overeem et al., 2006).

5. From Hypocretin Defects to Symptomatology

Pharmacological data gathered over the last 30 years implicated a number of neurotransmitter systems in the various symptoms of narcolepsy. Most currently available stimulant medications (such as amphetamine-like compounds) primarily act by increasing dopaminergic neurotransmission

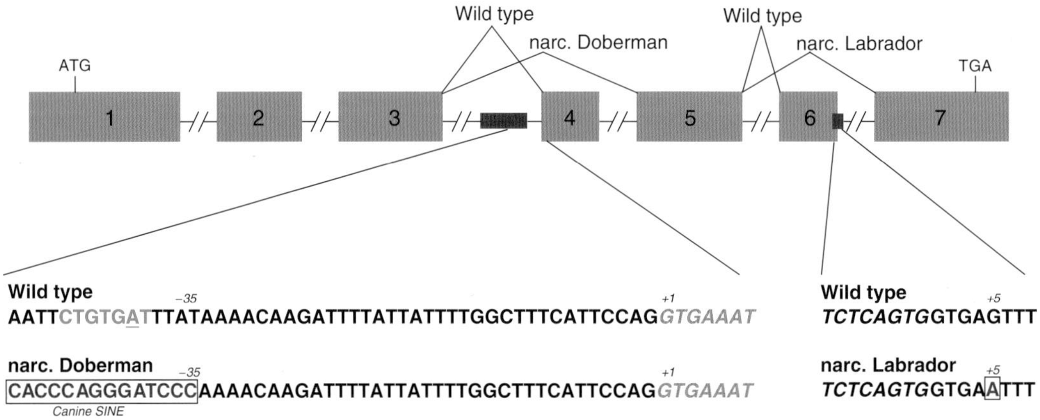

Figure 26.7 Schematic representation of the hypocretin receptor-2 gene in dogs. *Hcrtr-2* consists of 7 exons. Exon sequences are shown in italic. Mutations in both narcoleptic Dobermans and Labradors are boxed in red. In Dobermans, an insertion of a canine short interspersed nucleotide element (SINE) was found near the 5' splice site of exon 4, resulting in aberrant splicing and a premature stop codon at position 932. In Labradors, a G to A transition in the 5' splice junction consensus sequence results in deletion of exon 6 and a truncated receptor as well (Lin et al., 1999). Figure adapted from Lin et al. (1999) and Overeem et al. (2001).

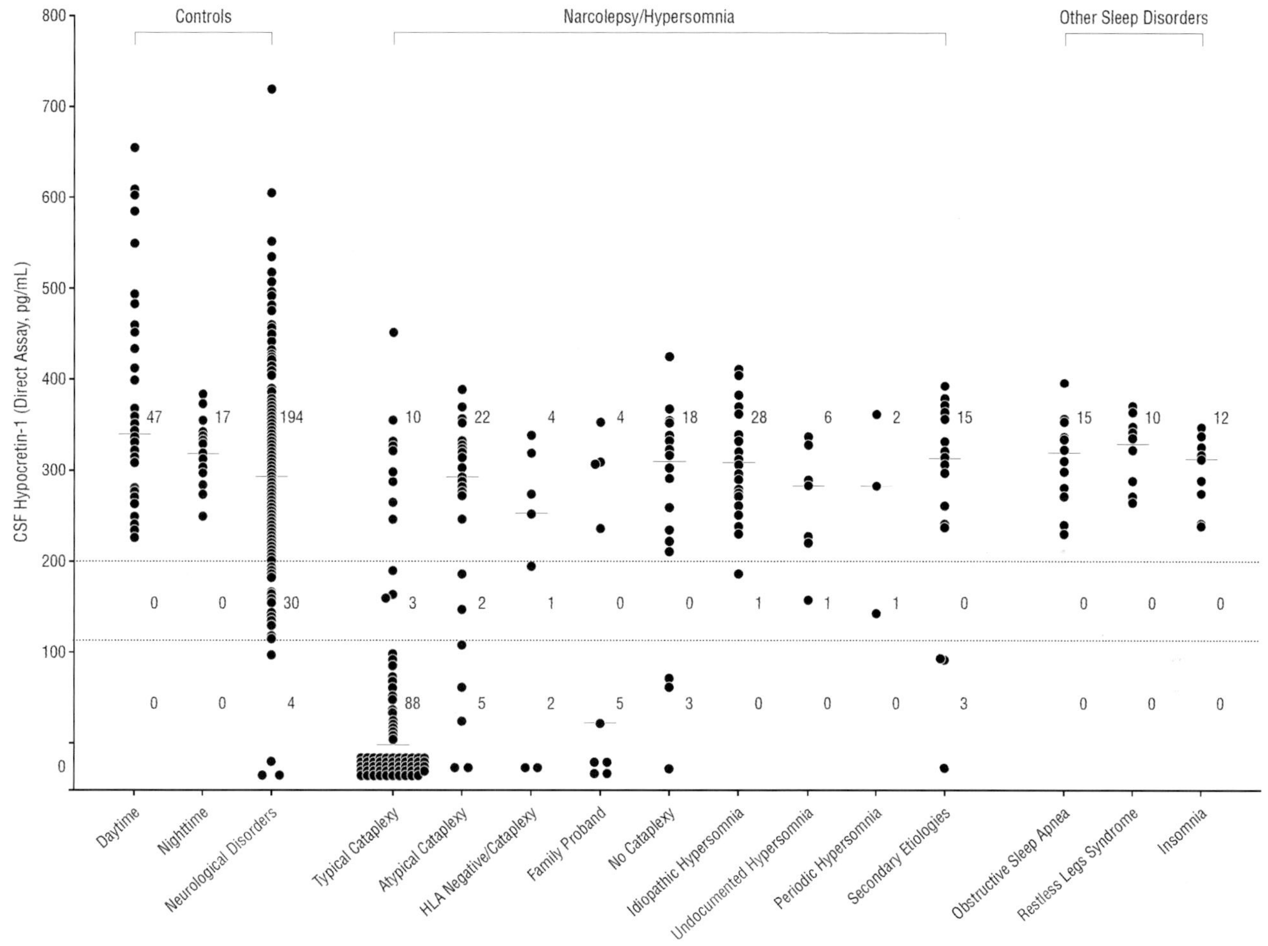

Figure 26.8 CSF hypocretin-1 levels in various categories of sleep disorders (Mignot et al., 2002). Each dot represents a single patient. Hypocretin-1 levels below 110 pg/ml were determined to be diagnostic for narcolepsy. Concentrations above 200 pg/ml best determine healthy controls. Levels below the detection limit of the assay are shown as 0. The number of subjects with values in each category is shown. Note that undetectable hypocretin levels are specific for narcolepsy, especially narcolepsy with typical cataplexy. In the group of neurological disorders, three patients had undetectable levels as well, all had Guillain Barré Syndrome (for details, see Ripley et al. (2001)). Figure reprinted with permission from Mignot et al. (2002).

(Nishino et al., 1997). More recent data also point to an important additional role of histaminergic transmission. Cataplexy is considered to result from an imbalance of cholinergic and monoaminergic neurotransmission at the brainstem level (Nishino et al., 1997). Compounds that increase monoaminergic (mainly adrenergic) tone are potent anticataplectics. All these data are based on the effects of *symptomatic* treatment however, and therefore do not give precise information on the underlying primary defects. For example, it is unlikely that narcolepsy is a state of continuous hypovigilance due to decreased dopaminergic signaling, although dopaminergic drugs are beneficial. If this would be the case, one would expect, for example, an increase in the amount of sleep over 24 hours, which is not the case.

Given the combined data in human patients and in several animal models, the hypocretin deficiency is likely to be the primary defect causing narcolepsy. Currently, models are developed to link the hypocretin deficiency to the clinical symptoms. For EDS, the sleep switch model of Saper and colleagues (2001) gives an excellent explanation. Although the flip-flop arrangement of wake- and sleep-promoting systems yields a network that results in discrete stages of consciousness and avoids intermediate states, it is inherently an unstable system, sensitive to slight disturbances. To function properly, it is necessary to stabilize the switch. For the waking position, the hypocretin system is situated exactly just to do that. The excitatory projections of hypocretin neurons to the several wake promoting systems (including the TMN, DR, and LC) form a "finger on the switch," keeping it in the wake position (see Figure 26.5) (Overeem et al., 2002; Saper et al., 2001). Loss of hypocretin transmission results in an unstable sleep switch, which nicely predicts the symptomatology of narcolepsy: in essence an unstable, fragmented sleep/wake pattern characterized by frequent jumps between wakefulness and sleep.

For cataplexy, the theoretical framework is less clear. Recently, Saper and colleagues proposed a flip-flop mechanism for the control of REM sleep as well (Lu et al., 2006). In this model, loss of hypocretin signaling is thought to result in a fragmentation of REM sleep, and possibly a dissociation of REM sleep leading to features such as atonia, which may result in cataplexy. However, the episodic character of cataplexy and more importantly its triggering by emotions remains unexplained. Therefore, other mechanisms may be involved. Based on the fact that strong emotions such as laughter can induce subclinical signs of motor inhibition even in healthy subjects (Overeem, Lammers & van Dijk, 1999), we have proposed the view that cataplexy may be an atavism of *tonic immobility* (Overeem, Lammers & van Dijk, 2002). Tonic immobility denotes a condition in which an animal is rendered immobile when faced with danger. In this view, this behavior may have been suppressed by the development of the hypocretin system, only to reemerge as cataplexy when hypocretin transmission is impaired.

Recent studies in which the activity of hypocretin neurons was measured in freely moving rats may support this view. The hypocretin neurons are relatively silent during most of the day, but are activated most when the animals are confronted with emotional stimuli, such as previously unencountered types of food, or a new environment (Lee, Hassani & Jones, 2005; Mileykovskiy, Kiyashchenko & Siegel, 2005). It was hypothesized that the hypocretin system is necessary to prevent the onset of motor inhibition during this kind of situation (Mileykovskiy et al., 2005). In the following years, we will have the opportunity to combine the vast amount of data gathered and create truly integrative and mechanistic models of the pathophysiology of narcolepsy.

B. ASPS as a Prototypical Circadian Disorder

1. Clinical Features

Circadian rhythm disorders are caused primarily by alterations in the endogenous circadian pacemaker or by a misalignment between the endogenous circadian rhythm and the externally imposed sleep/wake schedule (Reid & Zee, 2004). Several variants exist—endogenous disturbances, such as the delayed sleep phase type, the advanced sleep phase type, the nonentrained type; and exogenous disturbances, such as jet-lag or shift-work-induced sleep disorder.

The prevalence of advanced-sleep-phase syndrome (ASPS) in the general population is unknown, but considered to be very low. However, the prevalence may be underestimated as mild forms may not always be perceived as pathologic. Patients with ASPS have habitual sleep and wake times that are more than three hours earlier than societal means. A typical complaint is persistent and irresistible sleepiness in the late afternoon or early evening, interfering with desired activities during that time. Even when sleep onset is forcefully delayed, ASPS patients wake up very early in the morning. When social or professional obligations demand regular delayed sleep onset, this results in chronic sleep curtailment often with EDS. Interestingly, in contrast to healthy people who tend to delay their sleep/wake times when permitted (e.g., during weekends or holidays), ASPS patients even more profoundly advance their schedule under these circumstances (Jones et al., 1999). When sleep/activity periods are recorded over consecutive days, for example using actigraphy, ASPS patients show an advanced time of falling asleep at night and waking up in the morning during entrained conditions. Furthermore, during free-running conditions (without external time cues) they show a circadian period of less than 24 hours (see Figure 26.9).

2. Familial Forms of ASPS

Earlier reports of ASPS in younger patients already hinted at a genetic predisposition. Billiard described a 15-year-old girl with early evening sleepiness and early bedtimes together with early morning awakenings (Billiard et al., 1993). Symptoms had been present since childhood, and there was a report of similar symptoms in her mother and maternal grandfather. In 1999, Chris Jones and Louis Ptá ek described the first three families in which ASPS segregated as an autosomal dominant trait with high penetrance (Jones et al., 1999). For this, they developed strict diagnostic criteria to mark family members as affected, including specific bed and wake-up times, nocturnal sleep quality, and exclusion of other possible causes of sleep/wake disturbances. The affected individuals regularly fell asleep at about 7 or 8 P.M. and woke up at about 4 A.M. This was further reflected in significantly higher Horne-Östberg scores for "morningness" (Horne & Ostberg, 1976). In addition to the clinical assessment, part of the subjects was studied using laboratory polysomnography and actigraphy, which confirmed the clinical diagnosis of a strikingly advanced (>4 hours) sleep/wake cycle. Circadian phase estimations using core body temperature curves and plasma melatonin levels again showed similar phase advancement. Finally, one affected subject was studied for 18 days in time-isolation. The free-running period, determined both by rest/activity and body temperature, was significantly shortened to 23.3 hours (compared to the normal value of 24.2 hours). With this study, the existence of monogenic circadian rhythm variants in humans was clearly established, opening the way to determine the genetic substrates involved (see below). By now, multiple ASPS pedigrees have been described, in the United States as well as in Japan (Reid et al., 2001; Satoh et al., 2003).

3. Molecular Biology of Human ASPS

In the largest ASPS family described to date a linkage analysis was initiated, yielding a single marker on the telomere

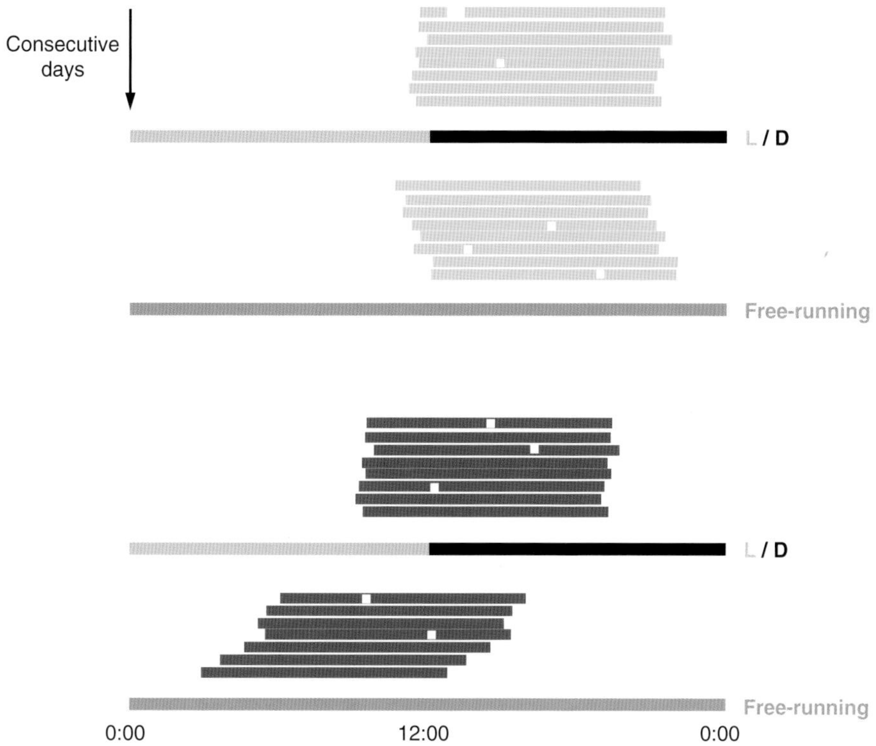

Figure 26.9 Schematic representation of rest/activity rhythms using actigraphy, over 7 consecutive days under light-dark conditions and free-running in the absence of time clues. Yellow and purple bars represent rest (sleep) periods. In the control subject (upper half) there is a sleep period at the usual hours. Furthermore, an endogenous rhythm close to 24 hours is maintained during free running conditions. A patient with advanced sleep phase syndrome shows a markedly advanced (earlier) moment of falling asleep and waking up during light/dark conditions. Additionally, during free running conditions, the endogenous circadian period is shorter than 24 hours (seen as a gradual shift in the sleep wake cycle). Figure adapted from Taheri and Mignot (2002).

of chromosome 2q (Toh et al., 2001). The critical region was defined further using additional marker sets. In the final critical region an attractive candidate gene was identified: *Per2*, a human ortholog of the Drosophila *period* gene. Based on the short period phenotypes of *period* mutants in flies and mice (Zheng et al., 1999), it was hypothesized that loss-of-function of h*Per2* could lead to a phase advance. Sequencing studies then demonstrated a base change resulting in a serine to glycine substitution at amino acid 662 (S662G) (Toh et al., 2001). In the same seminal study, it was further demonstrated that this amino acid change has functional consequences. First, it was shown that position 662 of hPER2 is within the CK1ε binding site. Second, it was shown that the mutation diminished the phosphorylation of PER2 by CK1ε. In the models for the mammalian clock, PER2 is a positive regulator of the *Bmal1* feedback loop (see Figure 26.3, through repression of REV/ERBα), so a phase-advance of h*Per2* could advance this whole loop. The exact mechanisms through which alterations in PER phosphorylation lead to this phase-advance is not known. It is thought that PER phosphorylation by CK1 leads to its degradation during the circadian cycle. Consequently, hypophosphorylation of the mutant h*Per2* could increase PER accumulation through impaired degradation, finally leading to an increased transcription of *Bmal1*, resulting in shortening of the circadian period and a sleep/wake phase advance. This study was the first to genetically dissect part of the human circadian clock, and to show its functional homology with other mammalian species.

Recently, another study showed the power of human circadian disorders due to single gene mutations, to further understand the clock function at the molecular level. Xu et al. (2005) performed a mutation screening of circadian candidate genes in another family with ASPS. They found a point mutation in the DNA sequence of *CK1δ* causing a threonine-to-alanine substitution in a region of the protein that is highly conserved across species. *In vitro* enzymatic assays showed that the mutant CK1δ had decreased activity in the phosphorylation of substrates, including PER1-3. Next, the mutant allele was introduced into *Drosophila*, and mutant flies were shown to have an elongated circadian period when compared to wild-type lines. In contrast to the effects of the CK1δ in flies, transgenic mice carrying the mutant gene had significantly shorter free-running periods under total dark conditions. Taken together, both *in vitro* and *in vivo* data directly show that CK1δ is a core clock component (see Figure 26.3). However, the findings also point to the complexity of the clock system, as the same mutation resulted in opposite effects on circadian period when introduced in flies and mice. Even though individual components are highly conserved across different species, the activity of these components and their combined network activity may be completely different.

VI. Conclusions

Over the last decades we have learned a tremendous amount about sleep, the intriguing behavior that we express during a third of our lifespan. The same holds true for the pervasive influence of our internal clock, and its interaction with the environment. The time has come to integrate the vast amount of data gathered into rational and functional models of higher order sleep/wake regulation. This will make it possible not only to understand the variety of sleep/wake and circadian disorders, but may also open up avenues toward new, rational therapies. Furthermore, they may help us to understand the elusive functions of sleep and to combat self-inflicted conditions such as jet-lag and the effects of shift work.

The advances in the field illustrate the power of basic (animal) research in general, and of molecular biology in particular. The narcolepsy–hypocretin success story is a prominent case in point. Within two years, the leap was made from the discovery of a new neurotransmitter system, the cloning of the hypocretin receptor mutation in narcoleptic dogs, to the demonstration that human narcolepsy is caused by hypocretin defects. Molecular genetics of human sleep disorders can also provide direct confirmation of basic research, as illustrated by the *Per* gene mutations in human ASPS. Moreover, such studies can *complement* animal data, as the *CK1*δ mutation in ASPS proved for the first time that CK1 is a core clock component. In the years to come, we will take on the more prevalent (but also more complex) disorders of sleep and waking, such as sleep apnea and the restless legs syndrome. Furthermore, we may further understand normal variants, such as the large differences in sleep need between individuals.

Of course, the ultimate goal is not only to understand human sleep disorders, but also to develop new, rational therapies. In order to fulfill this goal, the fields of pharmacogenetics and pharmacogenomics need further attention and effort.

For molecular biology in general, one of the future challenges will be to link genetic factors to complex phenotypes such as behavior. As the sleep/wake cycle and other 24-hour rhythms have such a clear behavioral phenotype, the circadian system forms a unique opportunity to gain such a direct understanding of the cellular and molecular mechanisms connecting genes to behavior.

Acknowledgments

Dr. S. Overeem is supported by a Veni grant from the Netherlands Organization for Scientific Research (grant no. 916.56.103).

References

Allada, R., Emery, P., Takahashi, J. S., Rosbash, M. (2001). Stopping time: The genetics of fly and mouse circadian clocks. *Annu Rev Neurosci* **24**, 1091–1119.

Aserinsky, E., Kleitman, N. (1953). Regularly occurring periods of eye motility, and concomitant phenomena during sleep. *Science* **118**, 273–274.

Baker, T. L. (1985). Sleep apnea disorders. Introduction to sleep and sleep disorders. *Med Clin North Am* **69**, 1123–1152.

Billiard, M., Verge, M., Aldaz, C., Carlander, B., Touchon, J., Besset, A. (1993). A case of advanced sleep phase syndrome (abstract). *Sleep Research* **109.**

Blouin, A. M., Thannickal, T. C., Worley, P. F., Baraban, J. M., Reti, I. M., Siegel, J. M. (2005). Narp immunostaining of human hypocretin (orexin) neurons: Loss in narcolepsy. *Neurology* **65**, 1189–1192.

Borbely, A. A., Achermann, P. (1999). Sleep homeostasis and models of sleep regulation. *J Biol Rhythms* **14**, 557–568.

Bromham, L., Penny, D. (2003). The modern molecular clock. *Nat Rev Genet* **4**, 216–224.

Broughton, R., Krupa, S., Boucher, B., Rivers, M., Mullington, J. (1998). Impaired circadian waking arousal in narcolepsy-cataplexy. *Sleep Res Online* **1**, 159–165.

Buhr, A., Bianchi, M. T., Baur, R., Courtet, P., Pignay, V., Boulenger, J. P. et al. (2002). Functional characterization of the new human GABA(A) receptor mutation beta 3 (R192H). *Hum Genet* **111**, 154–160.

Buijs, R. M., Kalsbeek, A. (2001). Hypothalamic integration of central and peripheral clocks. *Nat Rev Neurosci* **2**, 521–526.

Cermakian, N., Boivin, D. B. (2003). A molecular perspective of human circadian rhythm disorders. *Brain Res Brain Res Rev* **42**, 204–220.

Chemelli, R. M., Willie, J. T., Sinton, C. M., Elmquist, J. K., Scammell, T., Lee, C. et al. (1999). Narcolepsy in orexin knockout mice: Molecular genetics of sleep regulation. *Cell* **98**, 437–451.

Crocker, A., Espana, R. A., Papadopoulou, M., Saper, C. B., Faraco, J., Sakurai, T. et al. (2005). Concomitant loss of dynorphin, NARP, and orexin in narcolepsy. *Neurology* **65**, 1184–1188.

Dauvilliers, Y., Blouin, J. L., Neidhart, E., Carlander, B., Eliaou, J. F., Antonarakis, S. E. et al. (2004). A narcolepsy susceptibility locus maps to a 5 MB region of chromosome 21q. *Ann Neurol* **56**, 382–388.

de Lecea, L., Sutcliffe, J. G. (1999). The hypocretins/orexins: Novel hypothalamic neuropeptides involved in different physiological systems. *Cell Mol Life Sci* **56**, 473–480.

Dijk, D. J., Lockley, S. W. (2002). Integration of human sleep-wake regulation and circadian rhythmicity. *J Appl Physiol* **92**, 852–862.

Espana, R. A., Scammell, T. E. (2004). Sleep neurobiology for the clinician. *Sleep* **27**, 811–820.

Everson, C. A., Bergmann, B. M., Rechtschaffen, A. (1989). Sleep deprivation in the rat: III. Total sleep deprivation. *Sleep* **12**, 13–21.

Foutz, A. S., Mitler, M. M., Cavalli-Sforza, L. L., Dement, W. C. (1979). Genetic factors in canine narcolepsy. *Sleep* **1**, 413–421.

Gallopin, T., Fort, P., Eggermann, E., Cauli, B., Luppi, P. H., Rossier, J. et al. (2000). Identification of sleep-promoting neurons in vitro. *Nature* **404**, 992–995.

Hobson, J. A. (1989). *Sleep*. Scientific American Library, New York.

Hobson, J. A., Pace-Schott, E. F. (2002). The cognitive neuroscience of sleep: Neuronal systems, consciousness and learning. *Nat Rev Neurosci* **3**, 679–693.

Horne, J. A., Ostberg, O. (1976). A self-assessment questionnaire to determine morningness-eveningness in human circadian rhythms. *Int J Chronobiol* **4**, 97–110.

ICSD, 2005. *International Classification of Sleep Disorders: Second Edition*. American Academy of Sleep Medicine, Westchester, Il.

Jones, C. R., Campbell, S. S., Zone, S. E., Cooper, F., DeSano, A., Murphy, P. J. et al. (1999). Familial advanced sleep-phase syndrome: A short-period circadian rhythm variant in humans. *Nat Med* **5**, 1062–1065.

Kadotani, H., Kadotani, T., Young, T., Peppard, P. E., Finn, L., Colrain, I. M. et al. (2001). Association between apolipoprotein E epsilon 4 and sleep-disordered breathing in adults. *JAMA* **285**, 2888–2890.

Kawashima, M., Tamiya, G., Oka, A., Hohjoh, H., Juji, T., Ebisawa, T. et al. (2006). Genomewide association analysis of human narcolepsy and a new resistance gene. *Am J Hum Genet* **79**, 252–263.

Lee, M. G., Hassani, O. K., Jones, B. E. (2005). Discharge of identified orexin/hypocretin neurons across the sleep-waking cycle. *J Neurosci* **25**, 6716–6720.

Lin, L., Faraco, J., Li, R., Kadotani, H., Rogers, W., Lin, X. et al. (1999). The sleep disorder canine narcolepsy is caused by a mutation in the hypocretin (orexin) receptor 2 gene. *Cell* **98**, 365–376.

Llinas, R. R., Steriade, M. (2006). Bursting of thalamic neurons and states of vigilance. *J Neurophysiol* **95**, 3297–3308.

Low-Zeddies, S. S., Takahashi, J.S. (2001). Chimera analysis of the Clock mutation in mice shows that complex cellular integration determines circadian behavior. *Cell* **105**, 25–42.

Lowrey, P. L., Takahashi, J. S. (2004). Mammalian circadian biology: Elucidating genome-wide levels of temporal organization. *Annu Rev Genomics Hum Genet* **5**, 407–441.

Lu, J., Sherman, D., Devor, M., Saper, C. B. (2006). A putative flip-flop switch for control of REM sleep. *Nature* **441**, 589–594.

Medori, R., Tritschler, H. J., LeBlanc, A., Villare, F., Manetto, V., Chen, H. Y. et al. (1992). Fatal familial insomnia, a prion disease with a mutation at codon 178 of the prion protein gene. *N Engl J Med* **326**, 444–449.

Mignot, E., Lammers, G. J., Ripley, B., Okun, M., Nevsimalova, S., Overeem, S. et al. (2002). The role of cerebrospinal fluid hypocretin measurement in the diagnosis of narcolepsy and other hypersomnias. *Arch Neurol* **59**, 1553–1562.

Mileykovskiy, B. Y., Kiyashchenko, L. I., Siegel, J. M. (2005). Behavioral correlates of activity in identified hypocretin/orexin neurons. *Neuron* **46**, 787–798.

Moore, R.Y., Eichler, V. B. (1972). Loss of a circadian adrenal corticosterone rhythm following suprachiasmatic lesions in the rat. *Brain Res* **42**, 201–206.

Moore, R. Y., Lenn, N.J. (1972). A retinohypothalamic projection in the rat. *J Comp Neurol* **146**, 1–14.

Moruzzi, G., Magoun, H. W. (1949). Brain stem reticular formation and activation of the EEG. *Electroencephalogr Clin Neurophysiol* **1**, 455–473.

Nakayama, J., Miura, M., Honda, M., Miki, T., Honda, Y., Arinami, T. (2000). Linkage of human narcolepsy with HLA association to chromosome 4p13-q21. *Genomics* **65**, 84–86.

Nishino, S., Mignot, E. (1997). Pharmacological aspects of human and canine narcolepsy. *Prog Neurobiol* **52**, 27–78.

Nishino, S., Ripley, B., Overeem, S., Lammers, G. J., Mignot, E. (2000). Hypocretin (orexin) deficiency in human narcolepsy. *Lancet* **355**, 39–40.

Nishino, S., Ripley, B., Overeem, S., Nevsimalova, S., Lammers, G. J., Vankova, J. et al. (2001). Low cerebrospinal fluid hypocretin (orexin) and altered energy homeostasis in human narcolepsy. *Ann Neurol* **50**, 381–388.

Overeem, S., Lammers, G. J., van Dijk, J. G. (2002). Cataplexy: 'Tonic immobility' or 'REM-sleep atonia'? *Sleep Medicine* **3**, 471–477.

Overeem, S., Lammers, G. J., van Dijk, J. G. (1999). Weak with laughter. *Lancet* **354**, 838.

Overeem, S., Mignot, E., van Dijk, J. G., Lammers, G. J. (2001). Narcolepsy: clinical features, new pathophysiologic insights, and future perspectives. *J Clin Neurophysiol* **18**, 78–105.

Overeem, S., van Vliet, J. A., Lammers, G. J., Zitman, F. G., Swaab, D. F., Ferrari, M. D. (2002). The hypothalamus in episodic brain disorders. *Lancet Neurol* **1**, 437–444.

Overeem, S., Verschuuren, J. J., Fronczek, R., Schreurs, L., den, H. H., Hegeman-Kleinn, I. M. et al. (2006). Immunohistochemical screening for autoantibodies against lateral hypothalamic neurons in human narcolepsy. *J Neuroimmunol* **174**, 187–191.

Pace-Schott, E. F., Hobson, J. A. (2002). The neurobiology of sleep: Genetics, cellular physiology and subcortical networks. *Nat Rev Neurosci* **3**, 591–605.

Panda, S., Hogenesch, J. B., Kay, S. A. (2002). Circadian rhythms from flies to human. *Nature* **417**, 329–335.

Patel, S. R. (2005). Shared genetic risk factors for obstructive sleep apnea and obesity. *J Appl Physiol* **99**, 1600–1606.

Peyron, C., Faraco, J., Rogers, W., Ripley, B., Overeem, S., Charnay, Y. et al. (2000). A mutation in a case of early onset narcolepsy and a generalized absence of hypocretin peptides in human narcoleptic brains. *Nat Med* **6**, 991–997.

Portas, C. M., Thakkar, M., Rainnie, D. G., Greene, R. W., McCarley, R. W. (1997). Role of adenosine in behavioral state modulation: a microdialysis study in the freely moving cat. *Neuroscience* **79**, 225–235.

Ralph, M. R., Foster, R. G., Davis, F. C., Menaker, M. (1990). Transplanted suprachiasmatic nucleus determines circadian period. *Science* **247**, 975–978.

Rechtschaffen, A., Kales, A. (1968). A manual of standardized terminology, techniques and scoring system for sleep stages of human subjects. UCLA Brain Information Service/Brain Research Institute, Los Angeles.

Reid, K. J., Chang, A. M., Dubocovich, M. L., Turek, F. W., Takahashi, J. S., Zee, P. C. (2001). Familial advanced sleep phase syndrome. *Arch Neurol* **58**, 1089–1094.

Reid, K. J., Zee, P. C. (2004). Circadian rhythm disorders. *Semin Neurol* **24**, 315–325.

Reppert, S. M., Weaver, D. R. (2002). Coordination of circadian timing in mammals. *Nature* **418**, 935–941.

Ripley, B., Overeem, S., Fujiki, N., Nevsimalova, S., Uchino, M., Yesavage, J. et al. (2001). CSF hypocretin/orexin levels in narcolepsy and other neurological conditions. *Neurology* **57**, 2253–2258.

Sakurai, T., Amemiya, A., Ishii, M., Matsuzaki, I., Chemelli, R. M., Tanaka, H. et al. (1998). Orexins and orexin receptors: A family of hypothalamic neuropeptides and G protein-coupled receptors that regulate feeding behavior. *Cell* **92**, 573–585.

Saper, C. B., Chou, T. C., Scammell, T. E. (2001). The sleep switch: Hypothalamic control of sleep and wakefulness. *Trends Neurosci* **24**, 726–731.

Saper, C. B., Lu, J., Chou, T. C., Gooley, J. (2005). The hypothalamic integrator for circadian rhythms. *Trends Neurosci* **28**, 152–157.

Saper, C. B., Scammell, T. E., Lu, J. (2005). Hypothalamic regulation of sleep and circadian rhythms. *Nature* **437**, 1257–1263.

Satoh, K., Mishima, K., Inoue, Y., Ebisawa, T., Shimizu, T. (2003). Two pedigrees of familial advanced sleep phase syndrome in Japan. *Sleep* **26**, 416–417.

Schwartz, W. J. (1997). Understanding circadian clocks: from c-fos to fly balls. *Ann Neurol* **41**, 289–297.

Sherin, J. E., Shiromani, P. J., McCarley, R. W., Saper, C. B. (1996). Activation of ventrolateral preoptic neurons during sleep. *Science* **271**, 216–219.

Siegel, J. M. (2005). Clues to the functions of mammalian sleep. *Nature* **437**, 1264–1271.

Silver, R., LeSauter, J., Tresco, P. A., Lehman, M. N. (1996). A diffusible coupling signal from the transplanted suprachiasmatic nucleus controlling circadian locomotor rhythms. *Nature* **382**, 810–813.

Stephan, F. K., Zucker, I. (1972). Circadian rhythms in drinking behavior and locomotor activity of rats are eliminated by hypothalamic lesions. *Proc Natl Acad Sci U S A* **69**, 1583–1586.

Stickgold, R., Walker, M. P. (2005). Sleep and memory: The ongoing debate. *Sleep* **28**, 1225–1227.

Tafti, M., Maret, S., Dauvilliers, Y. (2005). Genes for normal sleep and sleep disorders. *Ann Med* **37**, 580–589.

Taheri, S., Mignot, E. (2002). The genetics of sleep disorders. *Lancet Neurol* **1**, 242–250.

Thannickal, T. C., Moore, R. Y., Nienhuis, R., Ramanathan, L., Gulyani, S., Aldrich, M. et al. (2000). Reduced number of hypocretin neurons in human narcolepsy. *Neuron* **27**, 469–474.

Toh, K. L., Jones, C. R., He, Y., Eide, E. J., Hinz, W. A., Virshup, D. M. et al. (2001). An hPer2 phosphorylation site mutation in familial advanced sleep phase syndrome. *Science* **291**, 1040–1043.

Trenkwalder, C., Paulus, W., Walters, A. S. (2005). The restless legs syndrome. *Lancet Neurol* **4**, 465–475.

Turek, F. W., van Reeth, O. (1996). Circadian rhythms. In Fregly, M. J., Blatteis, C. M., Eds., *Handbook of Physiology: Environmental Physiology*, 1329–1360. Oxford University Press, Oxford, UK.

Vertes, R. P., Siegel, J. M. (2005). Time for the sleep community to take a critical look at the purported role of sleep in memory processing. *Sleep* **28**, 1228–1229.

Villablanca, J. (1965). The electrocorticogram in the chronic cerveau isole cat. *Electroencephalogr Clin Neurophysiol* **19**, 576–586.

von Economo, C. (1930). Sleep as a problem of localization. *J Nerv Ment Dis* **71**, 249–259.

Xu, Y., Padiath, Q. S., Shapiro, R. E., Jones, C. R., Wu, S. C., Saigoh, N. et al. (2005). Functional consequences of a CKI delta mutation causing familial advanced sleep phase syndrome. *Nature* **434**, 640–644.

Yamazaki, S., Numano, R., Abe, M., Hida, A., Takahashi, R., Ueda, M. et al. (2000). Resetting central and peripheral circadian oscillators in transgenic rats. *Science* **288**, 682–685.

Yanik, G., Glaum, S., Radulovacki, M. (1987). The dose-response effects of caffeine on sleep in rats. *Brain Res* **403**, 177–180.

Young, M. W., Kay, S. A. (2001). Time zones: A comparative genetics of circadian clocks. *Nat Rev Genet* **2**, 702–715.

Zheng, B., Larkin, D. W., Albrecht, U., Sun, Z. S., Sage, M., Eichele, G. et al. (1999). The mPer2 gene encodes a functional component of the mammalian circadian clock. *Nature* **400**, 169–173.

27

Chronic Pain as a Molecular Disorder

John N. Wood and Stephen G. Waxman

The prevalence and unfortunate consequences of chronic pain states have become the focus of intense research interest. Using molecular genetics and pharmacology in rodents and a range of related techniques as well as functional imaging in man, many of the molecular events that underlie damage sensing and the eventual perception of pain have been identified. Although mechanisms of acute pain perception and the transduction of chemical and thermal stimuli are increasingly well understood, the mechanisms that underlie neuropathic pain (pain that results from direct damage to the nervous system) are still the subject of vigorous debate. Damage to, and altered electrical excitability of, peripheral damage-sensing neurons seems to play a significant role. A consensus is now emerging that there is also an important role for neuro-immune cell interactions in the early stages of the establishment of chronic pain. In this review we focus on the cells involved in the establishment of chronic pain states, the molecular mechanisms involved in altering pain thresholds, and the interplay between the immune system and neurons implicated in pain pathways.

I. Incidence of Chronic Pain

Disease-associated chronic pain states are extremely common. They include pain resulting from viral-induced damage, for example HIV-associated neuropathy and post-herpetic neuralgia, chronic inflammatory conditions of unknown etiology, and pain induced by damage to nerves (e.g., diabetic neuropathy, neuropathy associated with some anti-neoplastic drugs such as vincristine, or physical trauma). If we include conditions such as chronic back pain then the number of people suffering these conditions is remarkably large. A range of surveys of random population samples in European countries suggests that at any one time 1 in 12 of the population is suffering some form of chronic pain (Wood & Waxman, 2005). The majority of people are likely to experience an episode of chronic pain during their lifetime.

Drug treatment of chronic pain states is thus an extremely important area of clinical practice. Unfortunately, most analgesic drugs are associated with some side effects. Aspirin-like drugs that block arachidonic acid metabolism are highly effect anti-inflammatory agents, useful in chronic inflammatory conditions such as rheumatoid arthritis. However they may cause GI problems and stomach bleeding (Cox-1 inhibitors) or cardiovascular problems (Cox-2 inhibitors). Because these drugs are taken by such a high proportion of the population, the number of deaths attributable to side effects is high—in the range of 10,000 to 20,000 per annum in the United States.

Opioid analgesics are highly effective, but the associated problems of respiratory depression, constipation, and addiction limit the usefulness of these drugs for all but fairly severe pain conditions. In addition the underlying mechanisms responsible for chronic pain may be different from those that subserve the evolutionarily useful ability to detect acute or inflammatory pain. Spontaneous pain and mechanical allodynia (noxious response to light touch) are the hallmarks of human chronic pain, but these phenomena are rarely modeled in animals and as a result are incompletely understood. Identifying the molecular mechanisms that underlie chronic pain is thus of great importance for the design of new analgesics.

II. Neurons Involved in Damage Sensing

The skin, muscle, and viscera are all innervated with specialized sensory neurons that detect tissue-damaging stimuli and signal such events both to the central nervous system and to the local environment. The cell bodies are grouped together in ganglia containing 10,000 to 20,000 neurons that are found on either side of the spinal cord at every spinal level (dorsal root ganglia). The two trigeminal ganglia, similar in function to the dorsal root ganglia, are located in the head and are responsible for sensation in the face.

Probably more is known about DRG sensory neurons than any other neuronal subtype, because it is easy to culture the neurons and characterize them electrophysiologically, with histochemical markers, and in terms of their specialized sensory modality *in vivo*. Despite this, some confusion reigns in the field of nociceptive sensory neurons, partly because the cells themselves show a degree of plasticity in terms of their expression of markers and partly because of misconceptions about the characteristics of nociceptive neurons. It has become a conventional view that small diameter sensory neurons with unmyelinated fibers are involved in damage sensing, and light touch and proprioreceptive neurons have large diameter cell bodies and are myelinated with fast conduction velocities. In fact, a high proportion of myelinated fast conducting sensory neurons are nociceptive (Lawson, 2005), and express TrkA, the high affinity NGF receptor, as well as Nav1.8, a voltage-gated sodium channel particularly associated with damage-sensing neurons. About 20 percent of rat alpha/beta fast fibers are nociceptive (Djourhi & Lawson, 2005).

There is also increasing evidence that nonneuronal cells (e.g., keratinocytes and endothelial cells) may play a role as primary sensors of tissue damage, for example mechanical stress, indirectly signaling to sensory neurons through chemical mediators such as ATP. This kind of mechanism is particularly well documented in visceral pain and discussed further later. Other cell types may also release mediators in response to tissue damage that have a profound effect on pain thresholds and pain perception. Thus macrophages and microglia expressing P2X receptors have been implicated in the development of neuropathic pain (Tsuda et al., 2004). Similarly, the absence of major phenotypic effects of deleting sensory neuron thermosensitive TRP receptors in transgenic mice suggests that other heat sensitive channels present in keratinocytes may be signaling indirectly and/or in concert with receptors that are present on sensory neurons (Chung et al., 2003).

III. Damage Sensing

The specialized ionotrophic and metabotrophic receptors that are activated by tissue damage have been catalogued over the past several years. In addition, receptors for growth factors, cytokines, and inflammatory molecules such as interleukins that are expressed on neurons involved in pain pathways have been demonstrated to play a role in altering pain thresholds (Rutkowski et al., 2002). Now that whole genome sequence information is available, identifying transcripts encoding homologues of genes of known function (e.g., thermosensors) has become routine. However, determining function and ascribing physiological significance to new receptors remains a labor-intensive and time-consuming process. Efforts on this front have provided many insights into the molecules involved in nociceptive signal transduction and the regulation of peripheral pain thresholds.

A number of approaches to identifying genes involved in pain pathways have been used successfully. Expression cloning has been used to define receptors activated by chemical mediators known to elicit a sensation of pain. When expressed either in xenopus oocytes or mammalian cells, cDNA fractions that confer sensitivity to externally applied ligand (e.g., ATP or capsaicin) can be subfractionated until a single cDNA encoding the receptor of interest is identified (e.g., Caterina et al., 1997).

Homology cloning, where clones of related sequence are identified on the basis of their cross-hybridization to other unrelated cDNA, has been extremely productive, for example, in the identification of channels related to the Transient Receptor Potential receptor found to be gated by noxious heat, low pH, and capsaicin (e.g., Peier et al., 2002). This approach now has been overtaken by the bioinformatic analysis of related sequences in databases derived from whole genome sequencing projects. A complete description of all potential expressed ion channel encoding genes has been provided using this methodology (Yu & Catterall, 2004). However, homology cloning of mammalian counterparts of genes implicated in pain-like behavior in genetically amenable organisms such as *Drosophila* or *C. elegans* is still being used effectively to identify genes that may play a related role in higher organisms (Tobin & Bargman, 2004).

Another approach involves identifying genes that are selectively expressed in cell types involved in pain pathways. By subtracting transcripts present in other tissues from cDNA present in sensory neurons, a cohort of genes likely to have a role in nociceptor function can be identified (Akopian et al., 1999). Bioinformatic analysis of tissue-specific transcripts identified by microarray analysis has extended this approach to the identification of genes present in nociceptors.

Microarray technology also allows the scanning of expressed genes in normal and pathophysiological states to try to find altered expression of transcripts that may underlie pain pathology. Costigan et al. (2002) found effects similar to those reported by Wang et al. (2003) with about 240 genes dysregulated three days after axotomy. This powerful approach does have limitations, however. Small changes in gene expression are hard to detect. Splice variant differences, particularly involving small exon substitutions, are not detectable with presently available arrays. In addition, dramatic alterations in mRNA expression in a small subset of neurons may be swamped by background levels of the transcript in other tissues. Perhaps most importantly, transcriptional regulation may not be the principal site of gene dysfunction. For example, the redistribution of voltage-gated channels within the membrane of a damaged nerve may produce major excitability changes that are not a consequence of altered transcriptional regulation.

Classical gene mapping using inbred strains of mice and quantitative trait loci mapping or analysis of candidate genes in man also have identified important molecules in pain pathways (Yang et al., 2004). Studies in man are particularly significant. Some heritable pain insensitivity syndromes have mapped to neurotrophic factors (NGF-β) and their receptors (TrkA), demonstrating the important role of the neuronal cell types dependent on such trophic support for pain sensation. More recently ion channel mutations have been implicated in both chronic and inflammatory conditions and in heritable insensitivity to pain.

IV. Chemical Mediators of Nociception

It is relatively easy to characterize the locus of actions of endogenous mediators that elicit a sensation of pain using the genetic approaches described earlier. Damaged tissues release a range of molecules that have been shown to elicit a sensation of pain. Proteolytic cascades acting on soluble precursor molecules generate peptides involved in altering pain thresholds. Other mediators include lipids and nitric oxide that may signal between cells, as well as intracellular mediators that act downstream of algogenic compounds that also play a role in inducing pain or altering pain thresholds.

ATP is present in all cells at millimolar levels, and as a consequence is released into the extracellular environment on tissue damage. Both G-protein-coupled (GPCR) as well as ionotropic receptors on sensory neurons are activated by ATP. The ATP-gated cation channel P2X3 is expressed by nociceptive neurons, and has been assessed as an analgesic target by antisense studies, the generation of null mutant mice, and the development of specific pharmacological antagonists (North, 2003). There appears to be a strong case that this receptor plays a role in both inflammatory and neuropathic pain. Barclay et al. (2002) used antisense oligonucleotides administered intrathecally to functionally down-regulate P2X3 receptors. After seven days of treatment, P2X3 protein levels were reduced in the primary afferent terminals in the dorsal horn. After partial sciatic ligation, inhibition of the development of mechanical hyperalgesia as well as significant reversal of established hyperalgesia were observed within two days of antisense treatment. The time course of the reversal of hyperalgesia was consistent with down-regulation of P2X3 receptor protein and function. Despite these observations, there is no evidence that P2X3 receptors are up-regulated in neuropathic pain. There does in fact seem to be down-regulation of P2X3 following L5/L6 spinal nerve ligation in rats (Kage et al., 2002). A significantly reduced number of small diameter neurons exhibited a response to α, β–methyleneATP (a P2X3 selective agonist), but large diameter neurons and some small neurons retain their expression of functional P2X3 receptors. TNP-ATP is a potent antagonist of P2X3 receptors, but is metabolically unstable and also acts on P2X1-4 subtypes. Nevertheless TNP-ATP is capable of completely reversing tactile allodynia, albeit in a transient fashion over a period of about an hour (Tsuda et al., 1999).

More recently a potent stable antagonist of P2X3 and P2X2/3 heteromultimers has been developed. This compound, A317491 (Jarvis et al., 2002), reverses mechanical allodynia and thermal sensitivity in a rat neuropathic pain model. P2Y receptors may also play a regulatory role in neuropathic pain. Okada (2003) showed that intrathecal administration of P2Y receptor agonists UTP and UDP produced significant anti-allodynic effects in a rat sciatic nerve ligation model.

Proteolytic cascades give rise to *kinins*, blood-derived local-acting peptides that have broad effects mediated by the B1 and B2 G-protein-coupled bradykinin receptors. The kallikrein-kinin system controls blood circulation and kidney function, and promotes inflammatory pain, and wound healing in damaged tissues (Marceau & Regoli, 2004).

Lipid mediators, particularly prostaglandins, have long been known to play an important role in lowering pain thresholds. Their synthesis is blocked by anti-inflammatory drugs (NSAIDS) that inhibit the metabolism of arachidonic acid by cyclooxygenase enzymes. Many effects of prostanoids appear to mediate via GPCRs and the subsequent activation of protein kinases that alter the properties of voltage-gated channels. More short-lived lipids such as hydroperoxy-eicosatetranoic acids also derived from arachidonic acid can act directly on ion channels such as TrpV1 to depolarize sensory neurons (Hwang et al., 2000). A whole family of GPCRs have been found to be associated with sensory neurons in a fashion reminiscent of the expression of GPCRs associated with olfaction in the olfactory epithelia. As yet the range of ligands that activate these MAS-like receptors and their possible role in regulating nociceptor excitability are incompletely understood (Dong et al., 2001; Han et al., 2002).

Cannabinoids as well as opioids can inhibit pain pathways. CB1 receptors both on sensory neurons and within the CNS are known to be useful targets for agonists with analgesic activity in neuropathic pain (Fox et al., 1999). In the partial sciatic ligation model of neuropathic pain CB-selective agonists WIN55, 212–2, CP-55,940, and HU-210 produced complete reversal of mechanical hyperalgesia within three hours of subcutaneous administration. Zhang et al. (2003) showed that chronic pain models associated with peripheral nerve injury, but not peripheral inflammation, induce CB2 receptor expression in a highly restricted and specific manner within the lumbar spinal cord. Conventional opioid drugs have been shown unequivocally to be useful in treating acute inflammatory and certain neuropathic pain conditions, for example diabetic neuropathy (Rowbotham et al., 2003). More controversial is the role of the nociceptin/orphanin FQ system in regulating neuropathic pain. Initial reports suggested that nociceptin had analgesic effects in neuropathic pain models. In contrast, Mabuchi et al. (2001) have used a nociceptin/orphanin FQ antagonist, JTC-801, to demonstrate attenuation of thermal hyperalgesia in neuropathic pain models.

V. Mechanosensation

Mechanosensation is the least understood sensory modality in molecular terms. Early ideas of primary mechanosensors focused on the mammalian acid sensing ion channels (ASICs), which are members of a channel superfamily involved in mechanosensation in nematode worms (MEC-4 and MEC-10 mutants) and are highly expressed in sensory neurons (Waldmann & Lazdunski, 1999). There are four identified genes encoding ASIC subunits, ASIC1 to ASIC4, with two alternative splice variants of ASIC1 and ASIC2 taking the number of known subunits to six. Although protons are the only confirmed activator of ASICs, the homology between ASICs and MEC channels, coupled to high levels of expression of ASICs in sensory neurons, has led to the hypothesis that these channels function in mechanotransduction (Lewin & Stucky, 2000). ASIC subunits are found at appropriate sites to contribute to mechanosensation. However, studies show staining for ASIC subunits along the length of the fibers, not a specific enrichment at the terminals. Expression in sensory terminals is necessary for a role in the transduction of either acidic or mechanical stimuli. Moreover, the finding that the majority of αβ-fibers' sensory terminals are immunoreactive for ASICs is at odds with the long-known observation that low threshold mechanoreceptors are not activated by low pH (see Lewin & Stucky, 2000). Thus, Welsh et al. (2001) have proposed that ASICs may exist, like MEC-4 and MEC-10, in a multiprotein transduction complex that through an unknown mechanism masks the proton sensitivity of these channels.

Studies of knock-out mice do not support a role for ASICs as mechanotransducers in mammals. Using the neuronal cell body as a model of the sensory terminal mechanically activated currents in dorsal root ganglion (DRG) neurons have been characterized (Drew et al., 2004). Neurons from ASIC2 and ASIC3 null mutants were compared with wild-type controls. Neuronal subpopulations categorized by cell size, action potential duration, and isolectin B4 (IB4) binding generated distinct responses to mechanical stimulation consistent with their predicted *in vivo* phenotypes. In particular, there was a striking relationship between action potential duration and mechanosensitivity as has been observed *in vivo*. Putative low threshold mechanoreceptors exhibited rapidly adapting mechanically activated currents. Conversely, when nociceptors responded they displayed slowly or intermediately adapting currents that were smaller in amplitude than responses of low threshold mechanoreceptor neurons. No differences in current amplitude or kinetics were found between ASIC2 and/or ASIC3 null mutants and controls. Ruthenium red blocked mechanically activated currents in a voltage-dependent manner, with equal efficacy in wild-type and knock-out animals. Analysis of proton-gated currents revealed that in wild-type and ASIC2/3 double knock-out mice, the majority of low threshold mechanoreceptors did not exhibit ASIC-like currents but exhibited a persistent current in response to low pH. These findings are consistent with another ion channel type being important in DRG mechanotransduction. Lazdunski's group also investigated the effect of ASIC2 gene knock-out in mice on hearing, cutaneous mechanosensation, and visceral mechanonociception. Their data also failed to support a role of ASIC2 in mechanosensation (Roza et al., 2004).

In both *Drosophila* (NAN) and *C. elegans* (OSM-9) mutants, members of the transient receptor potential (TRP) family of channels have been implicated in mechanoreception. To date no channels with close homology to either NOMPC or NAN have been reported in mammals but TRPV4 shows moderate homology to OSM-9 (26% amino acid identity, 44% identity or conservative change; Liedtke et al., 2003). TRPV4 is widely expressed in rodents with the highest expression levels in the kidney and significant expression in liver, heart, testes, and brain. Interestingly, expression is also seen in cochlea, trigeminal ganglia, and Merkel cells, all of which are associated with mechanosensation, although the channel does not appear to be expressed at high levels by sensory neurons themselves. When heterologously expressed, TRPV4 is gated by hypotonicity and also by phorbol esters, lipids, and moderate temperatures. Gating by multiple stimuli also has been demonstrated for the related TRPV1 channel and has led to the suggestion that this channel acts as an integrator of multiple sensory stimuli. In the tail pressure behavioral assay, which measures nociceptive thresholds in response to compression of the tail, it was found that TRPV4 nulls had thresholds around twice those of controls, whereas von Frey withdrawal thresholds were unchanged. Overall, it is unclear if TRPV4 can be directly mechanically activated or if it participates in the detection of mechanical stimuli *in situ*; the striking phenotype reported by Suzuki et al. (2003) using electrophysiology is at odds with the relatively sparse expression of TRPV4 in DRG neurons.

In addition to TRPV4, a role for TRPV1 has been postulated in bladder mechanosensation and the polycystins, distantly related to TRP channels, may have a mechanosensory function (Nauli et al., 2003). Birder et al. (2002) demonstrated that, despite having apparently morphologically normal bladders, TRPV1 knock-out mice had deficits in voiding reflexes and spinal signaling of bladder volume. Distension of the bladder is known to evoke ATP release; however, the absence of TRPV1 caused a reduction in the amount of ATP released from both stretched whole bladders or from hypotonically swelled urothelial cells. Moreover, stimulation of cultured urothelial cells with capsaicin evoked ATP release, suggesting that TRPV1 activation is both necessary and sufficient to evoke ATP release. No group has reported gating of TRPV1 by mechanical stimuli and cutaneous mechanosensation is seemingly normal in TRPV1 nulls (Caterina et al., 2000). Hence the role of TRPV1 in this pathway remains to be determined; perhaps mechanical stimuli gate TRPV1 via a chemical (possibly a lipid) mediator, electrophysiological analysis of mechanically stimulated urothelial cells may be informative.

Polycystin 1 (PC1) regulates Ca^{2+} and K^+ channels via modulation of G-protein signaling pathways (Delmas et al., 2002) whereas PC2 is a Ca^{2+} permeable cation channel. Both have similar membrane topology to TRP channels. Mutations in either gene can cause polycystic kidney disease. Nauli et al. (2003) showed that the normal function of these proteins is pivotal to mechanosensation by the cilia of kidney epithelial cells. In animals lacking functional PC1 the normal increase in intracellular Ca^{2+} levels evoked by fluid stress of the cilium was either greatly reduced or absent. In wild-type cells removal of external Ca^{2+} inhibited such responses and the use of antibodies against the external domain of PC2 suggested that Ca^{2+} entry is via these channels. The authors postulate that PC1 (which has a large extracellular domain) may act as a mechanosensor that subsequently activates the tightly associated PC2 channel. Finally TrpA1 has been shown to be activated by mustard oil and involved in inflammatory responses, and responses to noxious cold although its role as a primary mechanosensor is not supported by studies of knock-out mice (Bautista et al., 2006).

Chemically mediated mechanosensation may also be a factor in noxious mechanosensation or allodynia. Endothelial cells release a number of factors, including nitric oxide, ATP, and substance P in response to changes in blood flow. Cockayne et al. (2000) showed that mice lacking the P2X3 receptor displayed marked bladder hyporeflexia, demonstrating reduced micturition frequency and increased bladder volume. They also showed that normally P2X3 receptors are present on sensory nerves innervating the bladder. Subsequent work by the same group showed that bladder distension evoked a graded release of ATP and the response of sensory fibers to bladder distension was attenuated in P2X3 knock-outs. Cook et al. (2002) showed that when keratinocytes or fibroblasts were mechanically lysed in the vicinity of sensory neurons, neurons were depolarized by ATP acting at P2X receptors. This raises the possibility that some noxious mechanical stimuli may activate nociceptors via damage to nearby cells and consequent ATP release. Nakamura and Strittmatter (1996) had previously proposed that P2Y1 purinergic receptors might contribute to touch-induced impulse generation. They identified this GPCR from an expression-cloning screen of *Xenopus* oocytes expressing DRG cRNAs; it was found that eggs expressing P2Y1 responded, via mechanically evoked ATP release, to a puff of external buffer with an inward current

VI. Thermoreception

Some TRP channels are also thermosensitive and seem to underlie the responses to noxious heat found in whole organisms as well as in sensory neurons in culture. These channels exhibit distinct thermal activation thresholds (>43° C for TRPV1, >52° C for TRPV2, >36° C for TRPV3, >27–35° C for TRPV4, <25–28° C for TRPM8, and <17° C for TRPA1), and are expressed in primary sensory neurons as well as other tissues.

Rather puzzlingly, behavioral responses to noxious heat do not seem to be substantially compromised in TRPV1 null mutants, but other TRPs are likely to play a cooperative role in temperature sensing in both sensory neurons and the skin (Peier et al., 2002; Woodbury et al., 2004).

VII. Voltage-gated Channels and the Transmission of Information to the Central Nervous System

Sodium channels comprise a family of 10 structurally related proteins that are expressed in spatially and temporally distinct patterns in the mammalian nervous system. As these channels underlie electrical signaling in nerve and muscle, the discovery that some sodium channels are selectively expressed in sensory neurons focused attention on these isoforms as potential analgesic drug targets. Sodium channel blockers that act as anesthetics at high doses are highly effective analgesics at lower concentrations (Strichartz, 2002). The sodium channels Nav1.8 and Nav1.9 are selectively expressed within the peripheral nervous system, predominantly in nociceptive sensory neurons, and Nav1.7 is found in both sympathetic and nociceptive sensory neurons, and has been shown to play a critical role in inflammatory pain using nociceptor-specific knock-out mice (Nassar et al., 2004). In addition, an embryonic channel Nav1.3 and a beta subunit, β-3, have been found to be upregulated in DRG neurons in some chronic neuropathic pain states.

Nav1.3, which is present in relatively high levels at embryonic stages, is normally present at low levels in the adult rat peripheral nervous system. Peripheral axotomy or other forms of nerve damage lead to the reexpression of Nav1.3 and the associated beta-3 subunit in sensory neurons, but not in primary motor neurons (Waxman et al., 1994). Nav1.3 is known to recover (reprime) rapidly from inactivation (Cummins et al., 2001). Peripheral axotomy has been shown to induce the expression of rapidly repriming TTX-sensitive sodium channels in damaged DRG neurons, and this event can also be reversed by the combined actions of GDNF and NGF (Boucher et al., 2000; Leffler et al., 2002). Concomitant with the reversal of Nav1.3 expression by GFNF, ectopic action potential generation is diminished and thermal and mechanical pain-related behavior in a rat CCI model is reversed (Boucher et al., 2000).

Nav1.3 is also up-regulated in nociceptive dorsal horn neurons following experimental spinal cord injury, and this up-regulation is associated with pain; antisense knockdown of Nav1.3 reduces levels of Nav1.3 in these dorsal horn neurons, attenuates their hyperexcitability, and ameliorates the pain behavior in spinal-cord injured animals (Hains et al., 2003). A similar up-regulation of Nav1.3 within dorsal horn neurons accompanies the development of allodynia and hyperalgesia in the chronic constriction injury model of neuropathic pain and these again are ameliorated by antisense knockdown of Nav1.3 (Hains et al., 2004). Moving further into the CNS, second-order dorsal horn neurons project to third-order neurons in the thalamus. After spinal cord contusion injuries at the T9 thoracic level, Nav 1.3 protein has been found to be up-regulated within thalamic neurons in ventroposterior lateral (VPL) and ventroposterior medial nuclei (Hains et al., 2005). Extracellular unit recordings showed increased spontaneous discharge, and enhanced responses to innocuous and noxious peripheral stimuli, as well as expanded peripheral receptive fields. The increased spontaneous discharge of these thalamic neurons continues after acute, high spinal cord transection, showing that thalamic hyperactivity has become autonomous and is not dependent on an increased ascending barrage from neurons at or close to the spinal cord contusion site. Intrathecal administration of antisense oligodeoxynucleotides directed against Nav1.3 caused a reduction in Nav1.3 expression in thalamic neurons and reversed the electrophysiological alterations that occurred after spinal cord contusion injury (Hains et al., 2006). Upregulation of Nav1.3 within thalamic neurons also occurs, and is accompanied by increased levels of background firing and evoked hyper-responsiveness, after chronic constriction injury of the sciatic nerve (Zhao et al., 2006a). Taken together, these data suggest that Nav1.3 reexpression both peripherally and centrally may play a significant role in increasing neuronal excitability, contributing to neuropathic pain after nerve and spinal cord injury.

Lindia et al. (2005) have questioned some aspects of these conclusions in a study where they used antisense oligonucleotides and reported them to be ineffective in ameliorating neuropathic pain in the spared nerve injury model, despite a 50 percent reduction in Nav1.3 immunoreactivity within DRG; they observed only a relatively small number of axotomized DRG neurons expressing Nav1.3 in this model. Their results are not directly comparable with those of Hains et al. (2003, 2004, 2005), however, due to differences in pain models and in Nav1.3 antisense sequences, and possible differences in tissue penetrability. Nonetheless, the results of Lindia et al. (2005) are important in demonstrating an analgesic role of low-dose systemic sodium channel blockers; the thesis that sodium channel over-activity underlies neuropathic pain is thus still supported, although their results do not demonstrate an essential role of any individual sodium channel isotype in the model that they studied.

Nav1.8 is expressed mainly in nociceptive neurons (Akopian et al., 1996; Djouhri et al., 2003). This channel contributes a majority of the sodium current underlying the depolarizing phase of the action potential in cells in which it is present (Blair & Bean, 2003; Renganathan et al., 2001). Functional expression of the channel is regulated by inflammatory mediators, including prostaglandins and NGF, and both antisense and knock-out studies support a role for the channel in contributing to inflammatory pain and noxious mechanosensation (e.g., Akopian et al., 1999). Antisense studies have also suggested a role for this protein in the development of neuropathic pain (Lai et al., 2002), and a deficit in ectopic action propagation has been described in the Nav1.8 null mutant mouse (Roza et al., 2003). However, neuropathic pain behavior at early time points seems to be normal in the Nav1.8 null mutant mouse (Kerr et al., 2000).

Recent studies have demonstrated that Nav1.8 is the sole functional voltage-gated sodium channel in sensory neurons at low temperatures, as it escapes the inactivation that disables other voltage-gated channels (Zimmerman et al., 2006). The ability to transmit information about tissue damage at low temperatures, shared with the sodium channels of cold-blooded animals, may explain the remarkably specific tissue expression of this evolutionarily ancient channel, which is found only in damage-sensing neurons. Another characteristic of this channel, however, seems to be an important factor in its ability to transmit information about tissue damage. This is the relatively positive potential at which the channel activates and inactivates, combined with its repriming characteristics, and its interplay with the voltage-gated sodium channel Nav1.7 with which it is frequently coexpressed in nociceptors (Rush et al., 2005). Nav1.7 is a critical player in inflammatory pain. Transgenic mice that no longer express Nav1.7 in nociceptive neurons show an almost complete inability to experience altered inflammatory pain thresholds (Nassar et al., 2004). Even more dramatically, gain of function mutations that result in enhanced Nav1.7 activity in man cause the dominant heritable condition erythromelalgia (erythermalgia), in which chronic episodic pain is evoked by even mild warmth. A number of different allelic mutants of Nav1.7 have been shown to cause this condition as a result of the altered biophysical properties of the channel that include hyperpolarizing shifts in the voltage-dependence of activation, slowed deactivation, and an increase in the channel's response to small, slow depolarizations, all of which contribute to hyperexcitability in sensory neurons (Dib-Hajj et al., 2005; Han et al., 2006; Rush et al., 2006; Waxman & Dib-Hajj, 2005).

Interestingly, related human mutations that show defective inactivation of Nav1.7 cause a related paroxysmal pain condition, where mechanical stimulation causes intense pain (Fertleman et al., 2006) Nav1.7 is thus a critical modulator of pain pathways, and to emphasize this, human loss of function Nav1.7 mutants are completely refractory to pain, while showing no other sensory deficits (Cox et al., 2006).

Insights from the study of human erythromelalgia mutants have provided some remarkable insights into the interplay between Nav1.7 and Nav1.8 in sensory signaling. Rush et al. (2006) showed that Nav1.7 mutants could render sympathetic neurons hypoexcitable, while rendering DRG neurons hyperexcitable. Thus the same mutations have directly opposite effects in different neuronal subtypes. The reason for this is that Nav1.8 is still able to sustain action potentials in the DRG neurons at the slightly depolarized potential caused by the Nav1.7 mutations, and the depolarizations cause inactivation of the TTXs channels that are found in sympathetic neurons that do not express Nav1.8.

Given our knowledge about the importance of particular sodium channel isoforms in pain pathways it is disappointing that isotype specific drugs have been so hard to develop. Conotoxins (small peptides from marine snails with channel blocking function) have been identified that block Nav1.8, and it may be that as with calcium channel blockers, natural products may provide useful sodium channel blocking analgesic drugs (Bulaj et al., 2006). Another possible route to analgesia is to block channel trafficking in a specific way.

Identification of annexin II/p11, which binds to Nav1.8 and facilitates the insertion of functional channels in the cell membrane (Okuse et al., 2002), may provide a target that can be used to modulate the expression of Nav1.8 and hence the level of Nav1.8 current in nociceptive neurons.

Nav1.9 is also expressed in nociceptive neurons (Dib-Hajj et al., 1998, 2002) and underlies a persistent sodium current with substantial overlap between activation and steady-state inactivation (Cummins et al., 1999) that has a probable role in setting thresholds of activation (Baker et al., 2003), suggesting that blockade of Nav1.9 might be useful for the treatment of pain. The phenotype of Nav1.9 knock-out mice shows few deficits in pain processing, however, and the significance of this channel as a drug target important in altering pain thresholds is still unclear. (Priest et al., 2005). Present evidence nonetheless makes a number of sodium channels highly attractive analgesic drug targets, with Nav1.7 clearly validated in both animal models and man.

A. Potassium Channels

Potassium channels also play an essential role in determining neuronal excitability. A variety of voltage-gated and nonvoltage-gated potassium channels present in sensory neurons have been found to alter in their expression after nerve injury. Of the voltage-gated K channels, Kv1.4 seems to be expressed specifically in small diameter mainly nociceptive sensory neurons, and the expression of the mRNA encoding this channel is lowered in a Chung model of neuropathic pain (Rasband et al., 1999). Knock-out studies have highlighted a significant role for TREK1—an osmosensitive, mechanically and thermally gated potassium channel present in nociceptive neurons (Alloui et al., 2005). In the absence of this channel, C-fibers and nociceptive sensory neurons are sensitized to noxious thermal and mechanical stimuli. In animal models of neuropathic pain, for example, a chronic constriction injury (CCI) model of neuropathic pain, the voltage-gated potassium channels Kv 1.2, 1.3, 1.4, 2.2, 4.2, and 4.3 mRNA levels in the ipsilateral DRG are reduced by two-thirds, compared to the contralateral side of the same animal No significant changes in Kv 1.5, 1.6, 2.1, 3.1, 3.2, 3.5, and 4.1 mRNA levels were detectable in the ipsilateral DRG. Passmore et al. (2003) have provided evidence that KCNQ potassium currents (responsible for the M-current) may also play a role in setting pain thresholds. Retigabine potentiates M-currents, and leads to a diminution of nociceptive input into the dorsal horn of the spinal cord in both neuropathic and inflammatory pain models in the rat. Finally the TASK-1, -2, and -3 channels (tandem of P domains in a weak inwardly rectifying K+ channel TWIK-related K+ channels) that are activated by low pH are also expressed in nociceptive neurons, although no knock-out data about their function are as yet available (Rau et al., 2006).

B. Calcium Channels and Transmitter Release

As voltage-gated calcium channels play a critical role in neuronal signaling through the regulation of neurotransmitter release, inhibition of these channels in electrically hyperexcitable neurons may lead to analgesia. Of the various calcium channel subtypes investigated, there is strong evidence of a role for Cav2.2 and the T-type calcium channel 3.2 in pain pathways. Mouse null mutants of N-type Cav2.2 calcium channels show dramatic diminution in neuropathic pain behavior in response to both mechanical and thermal stimuli (Kim et al., 2001). In addition, two highly effective analgesic drugs used in neuropathic pain conditions selectively target calcium channel subtypes. The conotoxin ziconotide blocks Cav2.2 alpha subunits, and the widely prescribed drug gabapentin binds with high affinity to α–2δ subunits of calcium channels (McGivern, 2006).

Ziconotide, a toxin derived from marine snails, blocks Cav2.2 channels with high affinity and has been found to have analgesic actions in animal models and man (McGivern, 2006). Intrathecal ziconotide blocks established heat hyperalgesia in a dose-dependent manner and causes a reversible blockade of established mechanical allodynia. Intrathecal ziconotide was found to be more potent, longer acting, and more specific in its actions than intrathecal morphine in this model of post-surgical pain, although with many side effects (Prommer 2006).

Voltage-gated calcium channels comprise a single alpha subunit and show structural homology with sodium channels, but the accessory subunits associated with these channels are more complex. The functional calcium channel complexes contain four proteins: α1 (170 kDa), α2 (150 kDa), β (52 kDa), δ (17–25 kDa), and γ (32 kDa). The $\alpha2\delta$ subunits are up-regulated severalfold in damaged dorsal root ganglia neurons, although the functional consequences of this increased expression remains unclear. $\alpha2\delta$-1 up-regulation in neuropathic pain correlates well with gabapentin sensitivity (Luo et al., 2002), suggesting that the $\alpha2\delta$-1 isoform is the most likely site of action of gabapentin. The up-regulation of α2-δ subunits occurs only in a subset of animal models of neuropathic pain that result in allodynia. Luo et al. (2002) compared DRG and spinal cord $\alpha2\delta$-1 subunit levels and gabapentin sensitivity in allodynic rats with mechanical nerve injuries (sciatic nerve chronic constriction injury, spinal nerve transection, or ligation), a metabolic disorder (diabetes), or chemical neuropathy (vincristine neurotoxicity). Allodynia occurred in all types of nerve injury investigated, but DRG and/or spinal cord $\alpha2\delta$-1 subunit up-regulation and gabapentin sensitivity coexisted only in mechanical and diabetic neuropathies, providing a possible explanation of why gabapentin is effective only in a subset of chronic pain states.

The use of knock-out mice has strengthened the case for calcium channels as useful drug targets in chronic and neuropathic pain conditions. Cav2.2 is very broadly expressed, but after global deletion of Cav2.2 it has proved possible to demonstrate major deficits in inflammatory and in particular neuropathic pain in this transgenic mouse using the Seltzer model (Kim et al., 2001). Thermal and mechanical thresholds are dramatically stabilized in the mutant mouse. A role for Cav2.2 in chronic pain is consistent with a known analgesic role for N-type calcium channel blockers.

Antisense oligonucleotides to Cav3.1, 3.2, and 3.2 showed that only knock-down of the T-type channel Cav3.2 had an effect on pain perception (Bourinet et al., 2005). The antisense treatment resulted in major anti-nociceptive, anti-hyperalgesic, and anti-allodynic effects, suggesting that Cav3.2 plays a major pronociceptive role in acute and chronic pain states. Taken together, the results provide direct evidence linking Cav3.2 T-type channels to pain perception and suggest that Cav3.2 may offer a useful new molecular target for the treatment of pain

VIII. Microglial Interactions and Chronic Pain

Peripheral inflammation and nerve damage were recognized to result in increased microglial activity in the dorsal horn innervated by the damaged nerves several years ago (Fu et al., 1999). Recently, the injection of activated microglia into the dorsal horn has been shown to induce neuropathic pain, while depletion of activated microglia has been found to block pain induction (Hains et al., 2006; Inoue, 2006; Frank et al., 2005; Tsuda et al., 2003). Intraspinal administration of microglia in which P2X4Rs had been induced and stimulated produced tactile allodynia in naive rats (Tsuda et al., 2003). These observations have heightened interest in the mediators that are responsible for microglial recruitment and activation, and the mechanism by which pain pathways are sensitized by these cells.

Candidate molecules for microglial recruitment and activation include fractalkine and various chemokines such as

CCL2 that could be released from damaged sensory neurons (Abbadie et al., 2003). Exogenous fractalkine causes increased responsiveness of lumbar wide dynamic range neurons to brush, pressure, and pinch applied to the hind paw. One day after spinal nerve ligation (SNL), minocycline attenuates after-discharge and responses to brush and pressure, presumably through block of microglial activation (Owolabi et al., 2006). Mice lacking the chemokine receptor chemotactic cytokine receptor 2 (CCR2) have a marked attenuation of monocyte recruitment in response to inflammatory stimuli. In acute pain tests, responses were equivalent in CCR2 knock-out and wild-type mice, and inflammatory pain was slightly diminished. Strikingly, the development of mechanical allodynia after neuropathic injury was totally abolished in CCR2 knock-out mice. Chronic pain resulted in the appearance of activated CCR2-positive microglia in the spinal cord. This recruitment and activation of macrophages and microglia both peripherally and centrally may contribute to inflammatory and neuropathic pain states.

There is also a requirement for extracellular ATP acting through P2X4 receptors on microglia for the establishment of neuropathic pain. Inoue and collaborators showed that the expression of the P2X4 receptor is enhanced in spinal microglia in a peripheral nerve injury model, and blocking P2X4 receptors produces a reduction of the neuropathic pain (Tsuda et al. 2003).

Cytokines such as interleukin-1 and-6 (IL-1 and IL-6) and tumor necrosis factor alpha (TNF-alpha) in the dorsal horn are increased after nerve lesion and, though mechanistic details are not yet fully understood, have been implicated

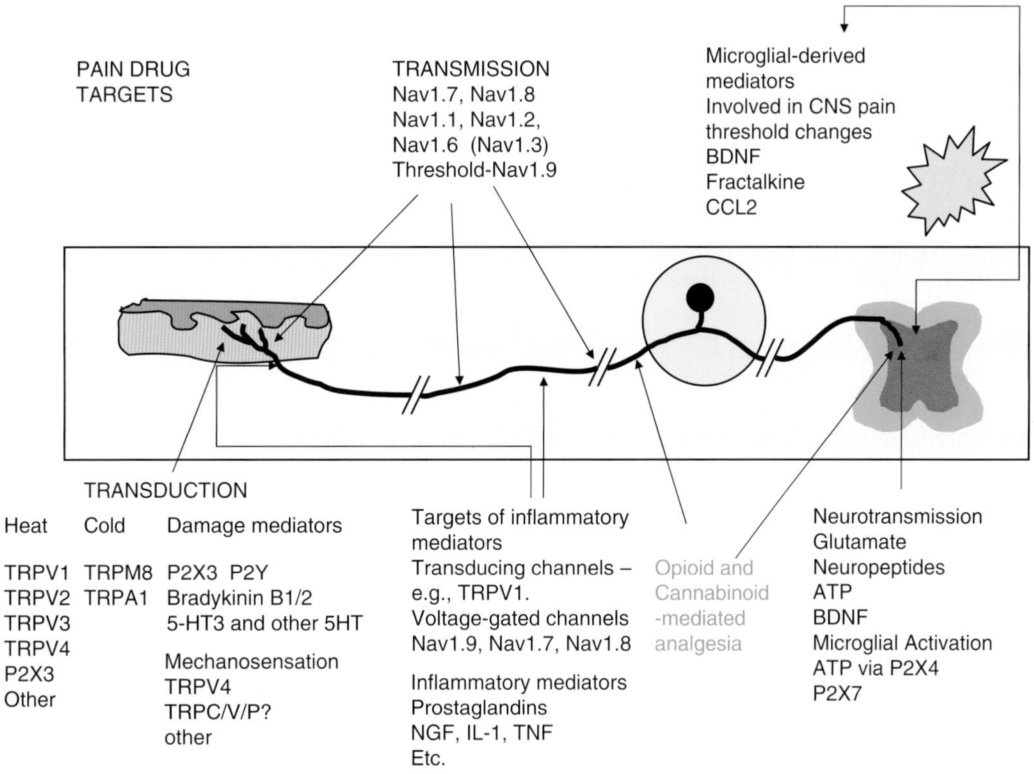

Figure 27.1 Mediators and receptors implicated in nociception and pain. Primary sensory neurons are activated by chemical thermal and mechanical stimuli at specialized terminals in the skin, nerve, and viscera. The biophysical properties of the receptors may be sensitized by inflammatory mediators released from damaged tissue. Voltage-gated sodium channels transmit information about tissue damage. Nav1.9 may be involved in setting inflammatory pain thresholds, and Nav1.7 is required for altered pain thresholds in inflammatory pain states. Nav1.8 is the only channel to function at low temperatures, and sustains action potential propagation at relatively depolarized potentials in nociceptive sensory neurons. Nav1.3 is up-regulated in neuropathic pain states. Endogenous opioid and cannabinoid receptors can mute nociceptive responses in the periphery and the CNS, and are the site of action of many effective analgesics. Neurotransmitters released in the spinal cord include glutamate, a variety of neuropeptides such as substance P and trophic factors such as BDNF. Inflamed states result in altered expression of pre- and post-synaptic receptors enhancing input into the CNS. Pathological input resulting from neuropathic pain states causes microglial activation, and the ATP-dependent release of chemokines and trophic factors such as BDNF that alter spinal cord physiology to cause chronic pain pathway activation.

in contributing to nerve-injury pain. They may be released on ATP binding to activated microglia.

Inflammatory cytokines such as IL-1, IL-6, and TNF are not only released from activated microglia, but they may also release the neurotrophin BDNF. Coull and collaborators have suggested that BDNF released on ATP-treatment of activated microglia may alter the properties of lamina 1 neurons to render GABA excitatory rather than inhibitory (Coull et al., 2004). Surprisingly BDNF released from nociceptors does not seem to play any role in neuropathic pain, although it has important pro-inflammatory actions (Zhou et al., 2006b).

Katsura et al. (1996) have made a case for Src-family kinases (SFKs) involvement within spinal cord microglia in mechanical hypersensitivity after peripheral nerve injury. Nerve injury induced an increase in SFK phosphorylation in the ipsilateral dorsal horn, and SFKs were activated only in microglia rather than neurons or astrocytes. Intrathecal administration of the Src-family tyrosine kinase inhibitor 4-amino-5-(4-chlorophenyl)-7-(t-butyl)pyrazolo[3,4-d] pyrimidine (PP2) suppressed nerve injury-induced mechanical hypersensitivity but not heat and cold hypersensitivity.

IX. Human Functional Imaging Studies

The central pathways involved in acute and chronic pain perception have been explored extensively using functional imaging (Apkarian et al., 2005). Although it has been a disappointment that functional imaging has failed to have any predictive power in terms of diagnosis, it has shed light on the complexity of central nervous system response to noxious input, and the widely distributed regions of the central nervous system that are activated in chronic pain states. fMRI and PET have allowed, for example, brain activation patterns to be compared between normal and chronic pain sufferers in response to an identical painful stimulus. In chronic pain patients, such stimuli lead to enhanced responses in pain processing areas, including the insular and cingulate cortices, suggesting that peripheral input is amplified somewhere along pain processing pathways (Dunley et al., 2005). This validates imaging techniques, but so far has provided little help in terms of defining new pain targets.

X. Conclusion

In Figure 27.1, we contrasted the role of ion channels in normal and injured primary sensory neurons, and the role of immune system cells within the central nervous system in modulating pain pathways and neuronal excitability.

We now understand the molecular basis of chemical and thermal activation of nociceptors, the channels involved in electrical signaling, and the mediators that change peripheral pain thresholds. Emphasis recently has shifted to the role of immune system cells in modulating chronic pain pathways; the development of therapeutic agents based on these observations is a time-consuming process, but there is good reason for optimism that new classes of analgesic drugs will be developed acting through molecular mechanisms defined over the past few years.

References

Abbadie, C., Lindia, J. A., Cumiskey, A. M., Peterson, L. B., Mudgett, J. S., Bayne, E.K. et al. (2003). Impaired neuropathic pain responses in mice lacking the chemokine receptor CCR2, *Proc Natl Acad Sci U S A* **100**, 7947–7952.

Akopian, A. N., Sivilotti, L., Wood, J. N. (1996). A tetrodotoxin-resistant voltage-gated sodium channel expressed by sensory neurons. *Nature* **379**, 257–262.

Akopian, A. N., Souslova, V., England, S., Okuse, K., Ogata, N., Ure, J. et al. (1999). The TTX-R sodium channel SNS has a specialized function in pain pathways, *Nature Neuroscience* **2**, 5481–5489.

Alloui, A., Zimmermann, K., Mamet, J., Duprat, F., Noel, J., Chemin, J. et al. (2006). TREK-1, a K+ channel involved in polymodal pain perception. *EMBO J* **25**, 2368–2376.

Apkarian, A. V., Bushnell, M. C., Treede, R. D., Zubieta, J. K. (2005). Human brain mechanisms of pain perception and regulation in health and disease. *Eur J Pain* **9**, 463–484.

Baker, M. D. et al. (2003). GTP-induced tetrodotoxin-resistant Na+ current regulates excitability in small diameter sensory neurones. *J Physiol* **548**, 373–382.

Barclay, J. et al. (2002). Functional downregulation of P2X3 receptor subunit in rat sensory neurons reveals a role in neuropathic and inflammatory pain. *J Neurosci* **22**, 8139–8147.

Bautista, D. M., Jordt, S. E., Nikai, T., Tsuruda, P. R., Read, A. J., Poblete, J. et al. (2006). TRPA1 mediates the inflammatory actions of environmental irritants and proalgesic agents. *Cell* **124**, 1269–1282.

Birder, L. A. et al. (2002). Altered urinary bladder function in mice lacking the vanilloid receptor TRPV1. *Nat Neurosci* **9**, 856–860.

Blair, N. T., Bean, B. P. (2003). Role of tetrodotoxin-resistant Na+ current slow inactivation in adaptation of action potential firing in small-diameter dorsal root ganglion neurons. *J Neurosci* **23**, 10338–10350.

Boucher, T. J. et al. (2000). Potent analgesic effects of GDNF in neuropathic pain states. *Science* **290**, 124–127.

Bourinet, E., Alloui, A., Monteil, A., Barrere, C., Couette, B., Poirot, O. et al. (2005). Silencing of the Cav3.2 T-type calcium channel gene in sensory neurons demonstrates its major role in nociception. *EMBO J* **24**, 315–324.

Brooks, J., Tracey, I. (2005). From nociception to pain perception: Imaging the spinal and supraspinal pathways. *J Anat* **207**, 19–33.

Bulaj, G., Zhang, M. M., Green, B. R., Fiedler, B., Layer, R. T., Wei, S. et al. (2006). Synthetic muO-conotoxin MrVIB blocks TTX-resistant sodium channel NaV1.8 and has a long-lasting analgesic activity. *Biochemistry* **45**, 7404–7414.

Caterina, M. J. et al. (1997). The capsaicin receptor: A heat-activated ion channel in the pain pathway. *Nature* **389**, 816–824.

Chung, M. K., Lee, H., Caterina, M. J. (2003). Warm temperatures activate TRPV4 in mouse 308 keratinocytes. *J Biol Chem* **278**, 32037–32046.

Cockayne, D. A. et al. (2000). Urinary bladder hyporeflexia and reduced pain-related behaviour in P2X3-deficient mice. *Nature* **407**, 1011–1015.

Cook, S. P., McCleskey, E. W. (2002). Cell damage excites nociceptors through release of cytosolic ATP. *Pain* **95**, 41–47.

Corey, D. P. et al. (2004). TrpA1 is a candidate for the mechanosensitive transduction channel of vertebrate hair cells. *Nature* **452**, 723–730.

Costigan, M. et al. (2002). Replicate rat genome oligonucleotide microarrays reveal hundreds of regulated genes in the DRG after peripheral nerve injury. *BMC Neurosci* **3**, 16.

Coull, J. A., Beggs, S., Boudreau, D., Boivin, D., Tsuda, M., Inoue, K. et al. (2005). BDNF from microglia causes the shift in neuronal anion gradient underlying neuropathic pain. *Nature* **438**, 1017–1021.

Cox, J. J. (2006). *SCN9A* loss of function mutations cause a congenital inability to experience pain. *Nature*, submitted.

Craner, M. J., Damarjian, T. G., Liu, S., Hains, B. C., Lo, A. C., Black, J. A. et al. (2005). Sodium channels contribute to microglia/macrophage activation and function in EAE and MS. *Glia* **49**, 220–229.

Cummins, T. R., Waxman, S. G. (1997). Down-regulation of tetrodotoxin-resistant sodium currents and up-regulation of a rapidly repriming tetrodotoxin-sensitive sodium current in small spinal sensory neurons following nerve injury. *J Neurosci* **17**, 3503–3504.

Cummins, T. R., Black, J. A., Dib-Hajj, S. D., Waxman, S. G. (2000). GDNF up-regulates expression of functional SNS and NaN sodium channels and their currents in axotomized DRG neurons. *J Neurosci* **20**, 8754–8761.

Cummins, T. R., Dib-Hajj, S. D., Black, J. A., Akopian, A. N., Wood, J. N., Waxman, S. G. (1999). A novel persistent tetrodotoxin-resistant sodium current in SNS-null and wild-type small primary sensory neurons. *J Neurosci*, **19**, RC 43, 1–6.

Delmas, P. (2004). Polycystins: From mechanosensation to gene regulation. *Cell* **118**, 145–148.

Dib-Hajj, S., Black, J. A., Cummins, T. R., Waxman, S. G. (2002). NaN/Nav1.9: A sodium channel with unique properties. *Trends Neurosci* **25**, 253–259.

Dib-Hajj, S. D., Tyrrell, L., Black, J. A., Waxman, S. G. (1998). NaN, a novel voltage-gated Na channel preferentially expressed in peripheral sensory neurons and down-regulated following axotomy. *Proc Natl Acad Sci* **95**, 8963–8968.

Dib-Hajj, S. D., Rush, A. M., Cummins, T. R., Hisama, F. M., Novella, S., Tyrrell, L, et al. (2005). Gain-of-function mutation in Nav1.7 in familial erythromelalgia induces bursting of sensory neurons. *Brain* **128**, 1847–1854.

Djouhri, L., Lawson, S. N. (2004). A beta-fiber nociceptive primary afferent neurons: A review of incidence and properties in relation to other afferent A-fiber neurons in mammals. *Brain Res Rev* **46**, 131–145.

Djouhri, L, et al. (2003). The sodium channel Nav1.8 (SNS/PN3): Expression and correlation with membrane properties in rat primary afferent neurons *J Physiol* **550**, 739–750.

Dong, X. et al. (2001). A diverse family of GPCRs expressed in specific subsets of nociceptive sensory neurons. *Cell* **106**, 619–632.

Drew, L. J. et al. (2004). ASIC2 and ASIC3 do not contribute to mechanically activated currents in mammalian sensory neurones. *J Physiol* **556**, 691–710.

Dunckley, P., Wise, R. G., Aziz, Q., Painter, D., Brooks, J., Tracey, I., Chang, L. (2005). Cortical processing of visceral and somatic stimulation: Differentiating pain intensity from unpleasantness. *Neuroscience* **133**, 533–542.

Fox, A. et al. (2001). The role of central and peripheral Cannabinoid-1 receptors in the antihyperalgesic activity of cannabinoids in a model of neuropathic pain. *Pain* **92**, 91–100.

Gong, H. C., Hang, J., Kohler, W., Li, L., Su, T. Z. (2001). Tissue-specific expression and gabapentin-binding properties of alpha-2 delta subunit subtypes. *J Membr Biol* **184**, 35–43.

Hains, B. C, Klein, J. P., Saab, C. Y., Craner, M. J., Black, J. A., Waxman, S. G. (2003). Upregulation of $Na_v1.3$ and involvement in neuronal hyper-excitability associated with neuropathic pain after spinal cord injury. *J Neurosci* **26**, 8881–8892.

Hains, B. C., Saab, C. Y., Klein, J. P., Craner, M. J., Waxman, S. G. (2004). Altered sodium channel expression in second-order spinal sensory neurons contributes to pain after peripheral nerve injury. *J Neurosci* **24**, 4832–4848.

Hains, B. C., Waxman, S. G. (2006). Activated microglia contribute to the maintenance of chronic pain after spinal cord injury. *J Neurosci* **26**, 4308–4317.

Hains, B. C., Saab, C. Y., Waxman, S. G. (2005). Changes in electrophysiological properties and sodium channel Nav1.3 expression in thalamic neurons after spinal cord injury. *Brain* **128**, 2359–2371.

Hains, B. C., Saab, C. Y., Waxman, S. G. (2006). Alterations in burst firing of thalamic VPL neurons and reversal by Na(v)1.3 antisense after spinal cord injury. *J Neurophysiol* **95**, 3343–3352.

Han, S. K. et al. (2002). Orphan G protein-coupled receptors MrgA1 and MrgC11 are distinctively activated by RF-amide-related peptides through the Galpha q/11 pathway. *Proc Natl Acad Sci* **99**, 14740–14745.

Han, C, Rush, A. M., Dib-Hajj, S. D., Li, S., Xu, Z., Wang, Y. et al. (2006). Sporadic onset of erythermalgia: A gain-of-function mutation in $Na_v1.7$. *Annals of Neurology* **59**, 553–558.

Hao, J. X., Xu, I. S., Wiesenfeld-Hallin, Z., Xu, X. J. (1998). Anti-hyperalgesic and anti-allodynic effects of intrathecal nociceptin/orphanin FQ in rats after spinal cord injury, peripheral nerve injury and inflammation. *Pain* **76**, 385–393.

Hwang, S. W. et al. (2000). Activation of capsaicin receptors by products of lipoxygenases: endogenous capsaicin-like substances. *Proc Natl Acad Sci* U S A **97**, 6155–6160.

Inoue, K. (2006). ATP receptors of microglia involved in pain. *Novartis Found Symp* **276**, 263–272.

Ishikawa, K. et al. (1999). Changes in expression of voltage-gated potassium channels in dorsal root ganglion neurons following axotomy. *Muscle & Nerve* **22**, 502–507.

Jarvis, M. F. et al. (2002). A-317491, a novel potent and selective non-nucleotide antagonist of P2X3 and P2X2/3 receptors, reduces chronic inflammatory and neuropathic pain in the rat. *Proc Natl Acad Sci U S A* **99**, 17179–17184.

Kage, K. et al. (2002). Alteration of DRG P2X3 receptor expression and function following spinal nerve ligation in the rat. *Exp Brain Res* **147**, 511–519.

Katsura, H., Obata, K., Mizushima, T., Sakurai, J., Kobayashi, K., Yamanaka, H. et al. (2006). Activation of Src-family kinases in spinal microglia contributes to mechanical hypersensitivity after nerve injury. *J Neurosci* **23; 26**, 8680–8690.

Kerr, B. J. et al. (2001). A role for the TTX-resistant sodium channel Nav 1.8 in NGF-induced hyperalgesia, but not neuropathic pain. *Neuroreport* **12**, 3077–3080.

Kim, C., Jun, K., Lee, T., Kim, S. S., McEnery, M. W., Chin, H. et al. (2001). Altered nociceptive response in mice deficient in the alpha(1B) subunit of the voltage-dependent calcium channel. *Mol Cell Neurosci* **18**, 235–245.

Lai, J. et al. (2002). Inhibition of neuropathic pain by decreased expression of the tetrodotoxin-resistant sodium channel, Nav1.8. *Pain* **95**, 143–152.

Lawson, S. N. (2002). Phenotype and function of somatic primary afferent nociceptive neurones with C-, Adelta- or Aalpha/beta-fibres. *Exp Physiol* **87**, 239–244.

Ledeboer, A., Sloane, E. M., Milligan, E. D., Frank, M. G., Mahony, J. H., Maier, S. F., Watkins, L. R. (2005). Minocycline attenuates mechanical allodynia and proinflammatory cytokine expression in rat models of pain facilitation. *Pain* **115**, 71–83.

Leffler, A., Cummins, T. R., Dib-Hajj, S. D., Hormuzdiar, W. N., Black, J. A., Waxman, S. G. (2002). Glial-derived neurotrophic factor and nerve growth factor reverse changes in repriming of TTX-sensitive Na^+ currents following axotomy of dorsal root ganglion neurons. *J Neurophysiol* **88**, 650–660.

Lewin, G., Stucky, C. (2000). Molecular basis of pain induction, Wood, J. N., Ed. Wiley, New York.

Liedtke, W., Friedman, J. M. (2003). Abnormal osmotic regulation in trpv4–/– mice. *Proc Natl Acad Sci U S A* **100**,13698–13703.

Lindia, J. A., Kohler, M. G., Martin, W. J., Abbadie, C. (2005). Relationship between sodium channel NaV1.3 expression and neuropathic pain behavior in rats. *Pain* **117**, 145–153.

Luo, Z. D. et al. (2002). Injury type-specific calcium channel alpha 2 delta-1 subunit up-regulation in rat neuropathic pain models correlates with antiallodynic effects of gabapentin. *J Pharmacol Exp Ther* **303**, 1199–1205.

Mabuchi, T. et al. (2003). Attenuation of neuropathic pain by the nociceptin/orphanin FQ antagonist JTC-801 is mediated by inhibition of nitric oxide production. *Eur J Neurosci* **17**, 1384–1392.

Marceau, F., Regoli, D. (2004). Bradykinin receptor ligands *Nat Rev Drug Discov* **10**, 845–852.

McGivern, J. G. (2006). Targeting N-type and T-type calcium channels for the treatment of pain. *Drug Discov Today* **11**, 245–253.

Nakamura, F., Strittmatter, S. M. (2004).(1996). P2Y1 purinergic receptors in sensory neurons: Contribution to touch-induced impulse generation. *Proc Natl Acad Sci U S A* **93**, 10465–10470.

Nassar, M. A. et al. (2004). Nociceptor-specific gene deletion reveals a major role for Na_v1.7 (PN1) in acute and inflammatory pain. *Proc Natl Acad Sci U S A* **101**, 12706–12711.

Nauli, S. M., Zhou, J. (2004). Polycystins and mechanosensation in renal and nodal cilia. *Bioessays* **8**, 844–856.

North, R. A. (2003). P2X3 receptors and peripheral pain mechanisms. *J Physiol.*

Okada, M., Nakagawa, T., Minami, M., Satoh, M. (2002). Analgesic effects of intrathecal administration of P2Y nucleotide receptor agonists UTP and UDP in normal and neuropathic pain model rats. *J Pharmacol Exp Ther* **303**, 66–73.

Okuse, K., Malik-Hall, M., Baker, M. D., Poon, W. Y. L., Kong, H., Chao, M. V., Wood, J. N. (2002). Annexin II light chain regulates sensory neuron-specific sodium channel expression. *Nature* **47**, 653–656.

Owolabi, S. A., Saab, C. Y. (2006). Fractalkine and minocycline alter neuronal activity in the spinal cord dorsal horn. *FEBS Lett* **580**, 4306–4310.

Passmore, G. M. et al. (2003). KCNQ/M currents in sensory neurons: Significance for pain therapy. *J Neurosci* **23**, 7227–7236.

Peier, A. M. et al. (2002). A heat-sensitive TRP channel expressed in keratinocytes. *Science* **296**, 2046–2049.

Priest, B. T., Murphy, B. A., Lindia, J. A., Diaz, C., Abbadie, C., Ritter, A. M. et al. (2005). Contribution of the tetrodotoxin-resistant voltage-gated sodium channel NaV1.9 to sensory transmission and nociceptive behavior. *Proc Natl Acad Sci U S A* **102**, 9382–9387.

Prommer, E. (2006). Ziconotide: A new option for refractory pain. *Drugs Today* (Barc) **42**, 369–378.

Rasband, M. N. et al. (2001). Distinct potassium channels on pain-sensing neurons. *Proc Natl Acad Sci U S A* **98**, 13373–13378.

Renganathan, M., Cummins, T. R., Waxman, S. G. (2001). Contribution of Na(v)1.8 sodium channels to action potential electrogenesis in DRG neurons. *J Neurophysiol* **86**, 629–640.

Roza, C. et al. (2004). Knockout of the ASIC2 channel in mice does not impair cutaneous mechanosensation, visceral mechanonociception and hearing. *J Physiol* **558**, 59–69.

Roza, C., Laird, J. M., Soslova, V., Wood, J. N., Cervero, F. (2003). The tetrodotoxin-resistant Na+ channel Nav1.8 is essential for the expression of spontaneous activity in damaged sensory axons of mice *J Physiol* **550**, 921–926.

Rowbotham, M. C. et al. (2003). Oral opioid therapy for chronic peripheral and central neuropathic pain. *N Engl J Medi* **348**, 1223–1232.

Rush, A. M., Dib-Hajj, S. D., Liu, S., Cummins, T. R., Black, J. A., Waxman, S. G. (2006). A single sodium channel mutation produces hyper- or hypoexcitability in different types of neurons. *Proc Natl Acad Sci U S A* **103**, 8245–8250.

Rutkowski, M. D., DeLeo, J. A. (2002). The role of cytokines in the initiation and maintenance of chronic pain. *Drug News Perspect* **15**, 626–632.

Saegusa, H., Matsuda, Y., Tanabe, T. (2002). Effects of ablation of N- and R-type Ca(2+) channels on pain transmission. *Neurosci Res* **43**, 1–7.

Strichartz, G. R. et al. (2002). Therapeutic concentrations of local anaesthetics unveil the potential role of sodium channels in neuropathic pain. *Novartis Found Symp* **241**, 189–201.

Suzuki, M., Mizuno, A., Kodaira, K., Imai, M. (2003). Impaired pressure sensation in mice lacking TRPV4. *J Biol Chem* **278**, 22664–22668.

Tobin, D. M., Bargmann, C. I. (2004). Invertebrate nociception: Behaviors, neurons and molecules. *J Neurobiol* **61**, 161–174.

Trang, T., Beggs, S., Salter, M. W. (2006). Purinoceptors in microglia and neuropathic pain. *Pflugers Arch* **452**, 645–652.

Tsuda, M. et al. (2003). P2X4 receptors induced in spinal microglia gate tactile allodynia after nerve injury. *Nature* **424**, 778–783.

Waldmann, R., Lazdunski, M. (1998). H(+)-gated cation channels: Neuronal acid sensors in the NaC/DEG family of ion channels. *Curr Opin Neurobiol* **8**, 418–424.

Wang, H. et al. (2003). Chronic neuropathic pain is accompanied by global changes in gene expression and shares pathobiology with neurodegenerative diseases. *J Peripher Nerv Syst* **2**, 128–133.

Watkins, L. R., Milligan, E. D., Maier, S. F. (2001). Glial activation: A driving force for pathological pain. *Trends Neurosci* **24**, 450–455.

Waxman, S. G. et al (1994). Type III sodium channel mRNA is expressed in embryonic spinal sensory neurons, and is re-expressed following axotomy. *J Neurophysiol* **72**, 466–471.

Waxman, S. G., Dib-Hajj, S. (2005). Erythermalgia: Molecular basis for an inherited pain syndrome. *Trends Mol Med* **11**, 555–562.

Welsh, M. J. (2003). Biochemical basis of touch perception: Mechanosensory function of degenerin/epithelial Na+ channels. *J Biol Chem* **277**, 2369–2372.

Woodbury, et al. (2004). Nociceptors lacking TRPV1 and TRPV2 have normal heat responses. *J Neurosci* **24**, 6410–6415.

Yang, Y. et al. (2004). Mutations in SCN9A, encoding a sodium channel alpha subunit, in patients with primary erythermalgia. *J Med Genet* **41**, 71–74.

Yu, F. H., Catterall, W. A. (2004). The VGL-chanome: A protein superfamily specialized for electrical signaling and ionic homeostasis. *Sci STKE* **253**, Re15.

Zhang, J. et al. (2003). Induction of CB2 receptor expression in the rat spinal cord of neuropathic but not inflammatory chronic pain models. *Eur J Neurosci* **12**, 2750–2754.

Zhao, J., Seereeram, A., Nassar, M. A., Levato, A., Pezet, S., Hathaway, G. et al. (2006b). London Pain Consortium. Nociceptor-derived brain-derived neurotrophic factor regulates acute and inflammatory but not neuropathic pain. *Mol Cell Neurosci* **31**, 539–548.

Zhao, P., Waxman, S. G., Hains, B. C. (2006b). Sodium channel expression in the ventral posterolateral nucleus of the thalamus after peripheral nerve injury. *Molecular Pain*, in press.

Zimmermann, K. et al. (2006). Nav1.8 is the molecular switch for cold pain transmission. Manuscript submitted.

28

Migraine as a Cerebral Ionopathy with Impaired Central Sensory Processing

Michel D. Ferrari, Arn M.J.M. van den Maagdenberg, Rune R. Frants, and Peter J. Goadsby

I. Migraine Is a Common Disabling Episodic Disorder

Migraine is a common primary headache disorder, typically characterized by disabling attacks of severe throbbing unilateral headache, accompanied by nausea, supersensitivity to sound and light, and head movement, lasting about a day. In one-third of patients there is a preceding aura that typically lasts 20 to 60 minutes. It usually consists of homonymous visual symptoms, such as flashing zigzag lights and visual loss, which begin paracentrally and slowly expand over minutes as a hemifield defect. Migraine auras may also include other transient focal neurological symptoms. Table 28.1A summarizes the diagnostic criteria for migraine according to the International Headache Society (Headache Classification Committee of The International Headache Society, 2004). The disease is broadly classified into:

- ▲ Migraine *with aura* (previously called classic(al) migraine), where at least some of the attacks are temporally associated with distinct transient focal neurological aura symptoms
- ▲ Migraine *without aura* (previously called common migraine), where there are no associated neurological symptoms of a focal nature (Silberstein et al., 2002).

Many patients believe that their attacks are precipitated by specific trigger factors as listed in Table 28.1B. However, when such patients are exposed to "their" trigger factor in a double-blind, placebo-controlled design, the outcome is seldom so clear, with the notable exception of nitric oxide donors, which are reliable triggers (Afridi et al., 2004; Iversen, 2001).

A. Epidemiology

Migraine may begin at any age, but rarely begins after the age of 50 years. The peak incidence in females is at age

Table 28.1 International Headache Society Features of Migraine (Headache Classification Committee of The International Headache Society, 2004)

A: Repeated episodic headache (4–72 hrs) with the following features:	
Any two of:	*Any one of*:
• Unilateral	• Nausea/vomiting
• Throbbing	• Photophobia and phonophobia
• Worsened by movement	
• Moderate or severe	
B. Triggers Believed to Precipitate Migraine Attacks (Lance & Goadsby, 2005)	
• Altered sleep patterns: becoming tired or oversleeping	
• Skipping meals	
• Overexertion	
• Weather change	
• Stress or relaxation from stress	
• Hormonal change, such as menstrual periods	
• Excess afferent stimulation: bright lights, strong smells	
• Chemicals: alcohol or nitrates	

12 to 13 years for migraine with aura and at age 14 to 17 years for migraine without aura. In males, the incidence of migraine peaks several years earlier: migraine with aura at age 5 years and migraine without aura at age 10 to 11 years (Haut et al., 2006). The overall prevalence of migraine in the general population is at least 12 percent, of which two-thirds is female. Peak prevalence is around age 40 years (Lipton et al., 2001; Scher et al., 1999). The median attack frequency is 18 migraine attacks per year; about 10 percent of migraine patients have attacks at least once weekly (Goadsby et al., 2002). Migraine is rated by WHO among the most disabling chronic disorders (Menken et al., 2000). Migraine has been estimated to be the most costly neurological disorder in the European community at more than €27 billion per year (Andlin-Sobocki et al., 2005) and costs the United States some $19.6 billion per year (Stewart et al., 2003).

B. Comorbidity

Migraine patients, especially those with migraine with aura, also have an increased risk (comorbidity) of a number of other episodic brain disorders (Goadsby et al., 2002). The highest and most consistently found increased risks are for:

▲ Epilepsy—two- to four-fold (Haut et al., 2006; Ludvigsson et al., 2006)

▲ Depression and anxiety disorders—two- to 10-fold (Breslau & Davis, 1993; Breslau et al., 2003; Radat & Swendsen, 2005)

▲ Patent Foramen Ovale—three-fold (Bousser & Welch, 2005)

▲ Stroke—three- to 14-fold increased risk depending on age and cofactors such as smoking and use of oral contraceptives (Bousser & Welch, 2005)

Migraineurs with a high attack frequency have a 16-fold increased risk of white matter and cerebellar lesions visible on MRI (Kruit et al., 2004). The increased risk for all these diseases is bidirectional, suggesting common underlying mechanisms, including a shared genetic background, increased excitability, neurovascular changes, and aberrations in the serotonin metabolism.

II. The Migraine Attack: Clinical Phases and Pathophysiology

Migraine attacks may consist of up to four distinct phases, although not every patient will experience all:

▲ Up to one-third of patients, at least sometimes, may experience premonitory symptoms for several hours before the aura or headache phase begins; these warning symptoms may include mood changes (e.g., depression or irritation), hyperactivation, fatigue, yawning, neck pain, smell disturbances, craving for particular food such as sweets or chocolate, and water retention resulting in swollen ankles and breasts (Giffin et al., 2003).

▲ Up to one-third of patients may have transient visual, sensory, motor, brainstem, or cognitive aura symptoms in at least some of their attacks; these focal neurological symptoms spread or march consecutively, usually last up to an hour, but sometimes may go on for several hours to days (Russell & Olesen, 1996).

▲ The headache phase, with headache and associated symptoms such as nausea, vomiting, and sensitivity to light, sound, and head movement; this phase may range from four to 72 hours but usually lasts for a day.

▲ The recovery phase, which may take several hours to sometimes several days (Giffin et al., 2005; Kelman, 2006).

The pathophysiology of the individual phases of the attack, once the attack has started, is now beginning to be well understood and will be discussed later (Table 28.2). What is essentially unknown is why and how migraine attacks are triggered (discussed in SectionVI).

III. The Premonitory Phase and the Hypothalamus

Very little is known about the pathogenesis of the prodromal warning symptoms. Patients report a distinctive collection of symptoms in the hours before an attack, known as premonitory symptoms (Giffin et al., 2003; Kelman, 2004). These are remarkably stereotyped and can be reproducibly triggered in some patients by nitroglycerin infusion (Afridi et al., 2004). In the rodent, *in vivo* D_2 receptor activation can lead to experimentally induced yawning (Mogilnicka & Klimek, 1977; Protais et al., 1983; Serra et al., 1986; Yamada et al., 1986). Similarly, apomorphine, a dopamine agonist, elicits yawning in migraineurs at doses that do not affect age-matched control groups (Blin et al., 1991). Apomorphine has also been reported to induce headache in 86 percent of migraine sufferers but none in age-matched control individuals (del Bene et al., 1994). Dopamine receptor agonist administration was reported to markedly worsen the headache in two patients with prolactinoma-associated headache (Levy et al., 2003). Conversely the dopamine receptor antagonist domperidone taken during the premonitory phase prevented the occurrence of migraine in uncontrolled trials (Amery & Waelkens, 1983; Waelkens, 1981, 1984). It recently has been reported that, similar to dorsal horn neurons (Levant & McCarson, 2001; Levey et al., 1993; van Dijken et al., 1996), there are dopamine receptors in the trigeminocervical complex of the rat (Bergerot & Goadsby, 2005). These are inhibitory in function (Bergerot et al., 2005). The only dopaminergic neurons known to innervate the spinal cord come from the hypothalamic nucleus A11 region (Skagerberg et al., 1982), with electrical stimulation suppressing the firing of spinal wide dynamic range neurons through D_2 receptors (Fleetwood-Walker et al., 1988). Taken together the available, albeit limited, data point to a possible dopaminergic/hypothalamic involvement in the premonitory phase of migraine.

IV. The Migraine Aura

A. Cortical Spreading Depression (CSD) in Experimental Animals

It is now well accepted that the migraine aura is not due to reactive vasoconstriction, as was previously believed for several decades, but rather is neurally driven and most likely caused by the human equivalent of the cortical spreading depression (CSD) of Leao (Haerter et al., 2005; Lauritzen, 1994). In experimental animals, CSD is a short-lasting (< 1 min) intense and steady depolarization of neuronal and glial cell membranes that spreads into contiguous areas of brain cortex at a rate of 2 to 5 mm/min, regardless of functional cortical divisions or arterial territories. It is accompanied by a transient total loss of neuroglial membrane integrity, a massive influx of Ca^{2+} and Na^+, and a massive efflux of K^+ causing highly elevated extracellular K^+ levels. This results in a spreading wave of brief excitation followed by a longer-lasting inhibition of spontaneous and evoked neuronal activity that traverses the cortex at a rate of about 3 to 5 mm/min.

In experimental animals, the electrophysiological changes are associated with characteristic triphasic cerebral blood flow (CBF) changes. Initially, there is a small and very brief reduction in CBF. This is followed by a profound increase of the CBF for several minutes. The third phase consists of a reduction of the CBF that may last for up to an hour and is accompanied by a loss of the cerebrovascular response to hypercapnia (Piper et al., 1991). CSD appears to be a self-defense mechanism of the brain to strong stimuli and can be triggered by electrical stimulation of brain tissue, cortical trauma, cerebral ischemia, or cortical application of high concentrations of K^+ or neuroexcitatory amino acids such as glutamate (Somjen, 2001).

B. CSD and the Migraine Aura

There is a considerable body of clinical evidence that CSD is the likely basis of migraine aura. Visual aura symptoms typically spread or march from the center of the visual field to the periphery at a speed of approximately 3 mm/min when translated to the visual cortex (Lashley, 1941). This is very similar to the propagation rate of CSD in experimental animals. The positive (e.g., scintillations, paraesthesias) and negative (e.g., scotomata, paresis) phenomena of the migraine aura could be well explained by the initial transient hyperexcitation front of CSD followed by neuronal depression. Most importantly however, functional neuroimaging studies in humans convincingly have demonstrated that the CBF changes that occur during migraine aura are very similar to those observed in experimental animals during CSD. Using functional MRI, Hadjikhani and colleagues (2001) found a focal increase in BOLD signal spreading into the occipital cortex at a rate of 3.5 mm/min. The cortical direction and speed of the spread were congruent with the visual experiences of the patient. The increased BOLD signal was followed by a decrease. This pattern would suggest an initial brief rise of CBF, followed by a longer lasting oligemia as seen in experimental CSD.

Although the evidence that CSD causes the migraine aura is mounting, there is much debate as to whether CSD may trigger the rest of the migraine attack as well through activation of the trigeminovascular system (see later). Although there is some evidence from animal experiments (Bolay et al., 2002), direct human evidence for this intriguing hypothesis is still lacking (Goadsby, 2001). We will discuss this issue at the end of the chapter, in the section describing the mechanisms for the migraine headache.

V. The Headache Phase

A. The Trigeminal Innervation of Pain-Producing Intracranial Structures

Surrounding the large cerebral vessels, pial vessels, large venous sinuses, and dura mater is a plexus of largely unmyelinated fibers that arise from the ophthalmic division of the trigeminal ganglion and in the posterior fossa from the upper cervical dorsal roots (McNaughton, 1938, 1966). Trigeminal fibers innervating cerebral vessels arise from neurons in the trigeminal ganglion that contain substance P (Edvinsson et al., 1983) and calcitonin gene-related peptide (CGRP) (Edvinsson et al., 1987), both of which can be released when the trigeminal ganglion is stimulated either in humans or cat (Goadsby et al., 1988). Stimulation of the cranial vessels, such as the superior sagittal sinus (SSS), is certainly painful in humans (Wolff, 1963). Human dural nerves that innervate the cranial vessels largely consist of small diameter myelinated and unmyelinated fibers that almost certainly subserve a nociceptive function.

B. Peripheral Connections: Plasma Protein Extravasation

Moskowitz and colleagues have provided a series of experiments to suggest that the pain of migraine may be a form of sterile neurogenic inflammation (Moskowitz & Cutrer, 1993). Although this is clinically unproven, the model system has been very helpful in understanding some aspects of trigeminovascular physiology and pharmacology (Moskowitz & Cutrer, 1993). Neurogenic plasma protein extravasation (PPE) can be seen during electrical stimulation of the trigeminal ganglion in the rat. PPE can be blocked by ergot alkaloids, indomethacin, acetylsalicylic acid, and serotonin-5$HT_{1B/1D}$ agonists (triptans) such as sumatriptan (Moskowitz & Cutrer, 1993). There are structural changes in the dura mater that are observed after trigeminal ganglion stimulation (Dimitriadou et al., 1991). These include mast cell degranulation and changes in post-capillary venules including platelet aggregation (Dimitriadou et al., 1992).

Table 28.2 Neuroanatomical Processing of Vascular Head Pain

	Structure	Comments
Target innervation:		
• Cranial vessels	Ophthalmic branch of	
• Dura mater	trigeminal nerve	
1st	Trigeminal ganglion	Middle cranial fossa
2nd	Trigeminal nucleus *(quintothalamic tract)*	Trigeminal n. caudalis & C_1/C_2 dorsal horns
3rd	Thalamus	Ventrobasal complex
		Medial n. of posterior group
		Intralaminar complex
Modulatory	Midbrain	Periaqueductal grey matter
	Hypothalamus	?
Final	Cortex	• Insulae
		• Frontal cortex
		• Anterior cingulate cortex
		• Basal ganglia

Although it is generally accepted that a sterile inflammatory response would cause pain, it is not clear whether such a response actually does occur in migraine and whether it is sufficient of itself, or requires other stimulators or promoters to be painful.

Although plasma extravasation in the retina, which is blocked by sumatriptan, can be seen after trigeminal ganglion stimulation in experimental animals, no such changes are seen with retinal angiography during acute attacks of migraine or cluster headache (May et al., 1998). A limitation of this study was the probable sampling of both retina and choroid elements in rat, given that choroidal vessels have fenestrated capillaries. More importantly however, although most established acute anti-migraine drugs proved effective in blocking experimental PPE (see earlier), the PPE model has proved *not* completely predictive of anti-migraine efficacy of putative anti-migraine drugs in humans. Despite showing high efficacy in blocking experimental PPE, substance P, neurokinin-1 antagonists, specific PPE blockers: CP122,288 and 4991w93, the endothelin antagonist Bosentan, and the neurosteroid ganaxolone all failed in clinical trials testing acute anti-migraine efficacy (May & Goadsby, 2001; Peroutka, 2005).

1. Sensitization and Migraine

Although it is as yet unclear whether there is a significant sterile inflammatory response in the dura mater during migraine, it is clear that some form of sensitization takes place during attacks, since allodynia is common. About two-thirds of patients complain of pain from nonnoxious stimuli, allodynia (Burstein et al., 2000; Selby & Lance, 1960). A particularly interesting aspect is the demonstration of allodynia in the upper limbs ipsilateral and contralateral to the pain. This finding is consistent with at least third-order neuronal sensitization, such as sensitization of thalamic neurons, and firmly places important parts of the pathophysiology of migraine within the central nervous system. Sensitization in migraine may be peripheral with local release of inflammatory markers, which would certainly activate trigeminal nociceptors. More likely in migraine there is a form of central sensitization, which may be classical central sensitization (Woolf, 1996), or a form of disinhibitory sensitization with dysfunction of descending modulatory pathways (Knight et al., 2002). Just as dihydroergotamine (DHE) can block trigeminovascular nociceptive transmission (Hoskin et al., 1996), probably at least by a local effect in the trigeminocervical complex, DHE can also block central sensitization associated with dural stimulation by an inflammatory soup (Pozo-Rosich & Oshinsky, 2005).

C. Neuropeptide Studies

Electrical stimulation of the trigeminal ganglion in both humans and the cat leads to increases in extracerebral blood flow and local release of both CGRP and SP. In the cat, trigeminal ganglion stimulation also increases cerebral blood flow by a pathway traversing the greater superficial petrosal branch of the facial nerve, again releasing a powerful vasodilator peptide, vasoactive intestinal polypeptide (VIP) (May & Goadsby, 1999). Interestingly, the VIP-ergic innervation of the cerebral vessels is predominantly anterior rather than posterior, and this may contribute to this region's vulnerability to CSD. In combination with the higher neuron/glial cell ratio, resulting in a lower glial K^+ reuptake buffer capacity, this may explain why the aura mostly commences posteriorly. Stimulation of the more specifically vascular pain-producing superior sagittal sinus (SSS) increases CBF and jugular vein CGRP levels (Zagami et al., 1990).

Human evidence that CGRP is elevated in the headache phase of migraine, both spontaneous (Gallai et al., 1995; Goadsby et al., 1990), and triggered attacks (Juhasz et al., 2003), although not in less severe attacks (Tvedskov et al., 2005), cluster headache (Fanciullacci et al., 1995; Goadsby & Edvinsson, 1994) and chronic paroxysmal hemicrania (Goadsby & Edvinsson, 1996). These data broadly support the view that the trigeminovascular system may be activated in a protective role in these conditions. Moreover, NO-donor triggered migraine, which is in essence typical migraine, also results in increases in CGRP that are blocked by sumatriptan, just as in spontaneous migraine (Juhasz et al., 2005). Recently, a specific nonpeptide CGRP antagonist, BIBN4096BS (Doods et al., 2000), demonstrated acute anti-migraine efficacy in a proof-of-concept trial (Olesen et al., 2004), firmly establishing blockade of the CGRP pathway as a novel and important new emerging treatment principle for acute migraine (Goadsby, 2005a). At the same time, the lack of any effect of CGRP blockers on PPE (Grant et al., 2005) suggests that this would not be the basis for the action of these new medicines.

D. Central Connections: The Trigeminocervical Complex

Fos immunohistochemistry is a method for looking at activated cells by plotting the expression of Fos protein. While after meningeal irritation with blood, Fos expression is noted only in the trigeminal nucleus caudalis, stimulation of the SSS in the cat (Kaube et al., 1993) and monkey (Goadsby & Hoskin, 1997) induces Fos-like immunoreactivity in the trigeminal nucleus caudalis and in the dorsal horn at the C_1 and C_2 levels. Similar activation profiles were obtained after SSS stimulation when using 2-deoxyglucose measurements (Goadsby & Zagami, 1991) and after stimulation of a branch of C_2, the greater occipital nerve, when measuring metabolic activity (Goadsby et al., 1997).

In experimental animals one can record directly from trigeminal neurons that have both supratentorial trigeminal input and input from the C_2 dorsal root via the greater occipital nerve. Stimulation of the greater occipital nerve for five minutes results in substantial increases in responses to supratentorial dural stimulation, which can last for over an hour (Bartsch & Goadsby, 2002). Conversely, stimulation of the middle meningeal artery dura mater with the C-fiber irritant mustard oil sensitizes responses to occipital muscle stimulation (Bartsch & Goadsby, 2003). Taken together these data suggest convergence of cervical and ophthalmic inputs at the level of the second order neuron (Bartsch & Goadsby, 2005). Moreover, stimulation of a lateralized structure, the middle meningeal artery, produces Fos expression bilaterally in both cat and monkey brain (Hoskin et al., 1999). This group of neurons from the superficial laminae of trigeminal nucleus caudalis and $C_{1/2}$ dorsal horns should be regarded functionally as the *trigeminocervical* complex (Bartsch & Goadsby, 2005).

These data demonstrate that trigeminovascular nociceptive information comes by way of the most caudal cells. This concept provides an anatomical explanation for the referral of pain to the back of the head in migraine. Moreover, experimental pharmacological evidence suggests that some abortive anti-migraine drugs, such as ergot derivatives, acetylsalicylic acid, and several triptans, can have actions at these second order neurons that reduce cell activity and suggest a further possible site for therapeutic intervention in migraine (Goadsby, 2005d). This action can be dissected out to involve each of the 5-HT_{1B}, 5-HT_{1D}, and 5-HT_{1F} receptor subtypes (Goadsby, 2004) and are consistent with the localization of these receptors on peptidergic nociceptors. Interestingly, triptans also influence the CGRP promoter (Durham et al., 1997), and regulate CGRP secretion from neurons in culture (Durham & Russo, 1999). Remarkably the effects of triptans may be activity-dependent in the sense that at least 5-HT_{1D} receptor expression on the cell surface depends on neuronal activation (Ahn & Basbaum, 2006). Furthermore, the demonstration that some part of this action is postsynaptic with either 5-HT_{1B} or 5-HT_{1D} receptors located *non*presynaptically (Maneesi et al., 2004) offers a prospect of highly anatomically localized treatment options (Goadsby, 2005b).

1. Higher Order Processing

Following transmission in the caudal brain stem and high cervical spinal cord information is relayed rostrally.

Thalamus. Processing of vascular nociceptive signals in the thalamus occurs in the ventroposteromedial (VPM) thalamus, medial nucleus of the posterior complex, and in the intralaminar thalamus (Zagami & Lambert, 1990). It has been shown by application of capsaicin to the SSS that trigeminal projections with a high degree of nociceptive input are processed in neurons,

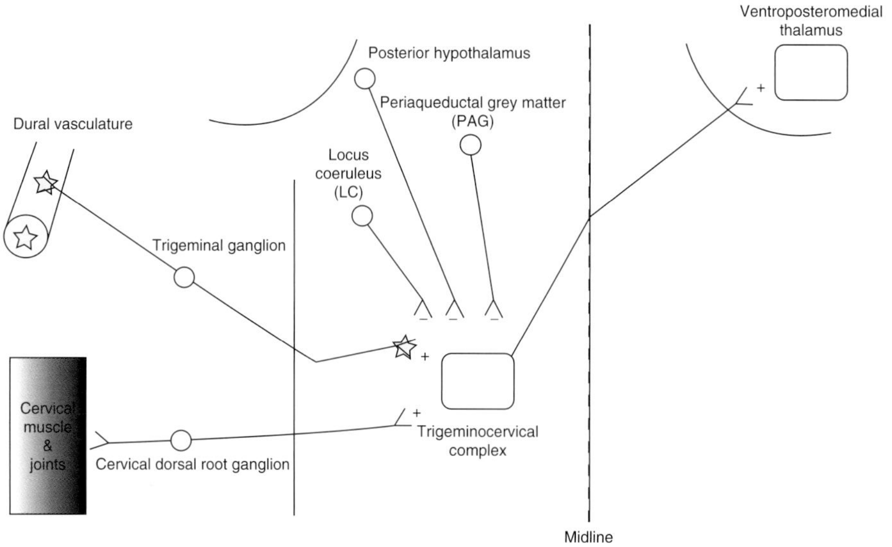

Figure 28.1 Illustration of some elements of migraine biology. Dural vessels and dura mater, and upper ($C_{1/2}$) cervical muscles and joints all project to neurons on the upper cervical spinal dorsal horn and trigeminal nucleus caudalis that may be labeled the trigeminocervical complex. The input is excitatory (+). This input then projects to trigeminovascular neurons in the ventroposteromedial thalamus (+) after crossing the midline. Descending brain systems from, for example, locus coeruleus, posterior hypothalamus (PH), and periaqueductal grey matter (PAG) have inhibitory influences (−) on trigeminocervical neurons. Although the sum of stimulating these structures is inhibitory it is noteworthy that individual neurons from PAG or PH can facilitate nociception (after Ferrari & Goadsby, 2006).

particularly in the VPM thalamus and in its ventral periphery (Zagami & Lambert, 1991). These neurons in the VPM can be modulated by activation of $GABA_A$ inhibitory receptors (Shields et al., 2003), and perhaps of more direct clinical relevance by propranolol though a β_1-adrenoceptor mechanism (Shields & Goadsby, 2005). Remarkably, triptans through $5\text{-HT}_{1B/1D}$ mechanisms can also inhibit VPM neurons locally, as demonstrated by micro-iontophoretic application, suggesting a hitherto unconsidered locus of action for triptans in acute migraine (see Figure 28.1) (Shields & Goadsby, 2006). Human imaging studies have confirmed activation of thalamus contralateral to pain in acute migraine, cluster headache, and in SUNCT (short-lasting unilateral neuralgiform headache with conjunctival injection and tearing) (Cohen & Goadsby, 2004).

Activation of modulatory regions. Stimulation of nociceptive afferents by stimulation of the SSS in the cat activates neurons in the ventrolateral periaqueductal grey matter (PAG) (Hoskin et al., 2001). PAG activation in turn feeds back to the trigeminocervical complex with an inhibitory influence (Knight & Goadsby, 2001). PAG is clearly included in the area of activation seen in positron emission tomography (PET) studies in migraineurs (Weiller et al., 1995). This typical negative feedback system will be further considered later as a possible mechanism for the symptomatic manifestations of migraine (Cohen & Goadsby, 2004).

Another potentially modulatory region activated by stimulation of nociceptive trigeminovascular input is the posterior hypothalamic grey. This area is crucially involved in several primary headaches, notably cluster headache, SUNCT, paroxysmal hemicrania, and hemicrania continua (Cohen & Goadsby, 2004; Matharu & Goadsby, 2005). Moreover, the clinical features of the premonitory phase, and other features of the disorder, suggest dopamine neuron involvement. Orexinergic neurons in the posterior hypothalamus (Beuckmann & Yanagisawa, 2002; Ebrahim et al., 2002) can be both pro- and anti-nociceptive (Bartsch et al., 2004), offering a further possible region whose dysfunction might involve the perception of head pain.

E. Central Modulation of Trigeminal Pain

1. Brain Imaging in Humans

Functional brain imaging studies with PET have demonstrated activation of the dorsal midbrain, including the periaqueductal grey (PAG), and the dorsal pons, near the locus coeruleus, in migraine without aura (Cohen & Goadsby, 2004). In spontaneous episodic (Afridi et al., 2005a) and chronic migraine (Matharu et al., 2004), and in nitroglycerin-triggered attacks (Afridi et al., 2005b), activation of the dorsolateral pons was seen (see Figure 28.2). These areas are active immediately after successful treatment of the headache but are not active interictally. The activation corresponds with the brain region that cause migraine-like headache when

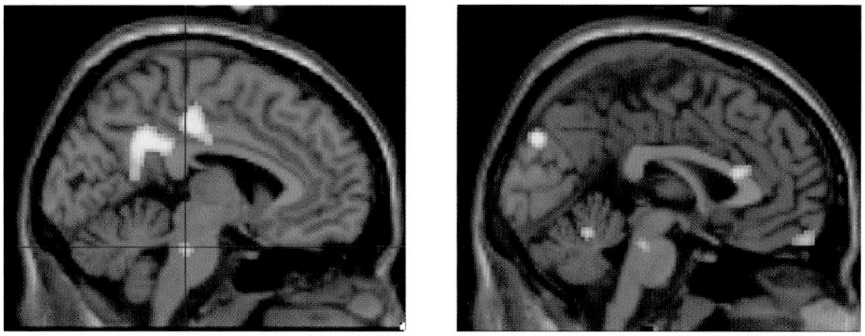

Figure 28.2 Positron emission tomography (PET) scans from groups of patients with episodic and chronic migraine studied during acute pain. Activation in the dorsal rostral pons is a constant finding in PET studies in migraine and may represent a crucial area that mediates important aspects of the pathophysiology of the disorder.

stimulated in patients with electrodes implanted for pain control (Raskin et al., 1987). Similarly, excess iron in the PAG of patients with episodic and chronic migraine (Welch et al., 2001), and chronic migraine can develop after a bleed into a cavernoma in the region of the PAG (Goadsby, 2002), or with a lesion of the pons (Obermann et al., 2006). What could dysfunction of these brain areas lead to?

2. Animal Experimental Studies of Sensory Modulation

Stimulation of nucleus locus coeruleus, the main central noradrenergic nucleus, reduces CBF through an α_2-adrenoceptor-linked mechanism in a frequency-dependent manner (Goadsby et al., 1982). This reduction is 25 percent overall but maximal in the occipital cortex. In parallel there is extracerebral vasodilatation. In addition, activation of the midbrain dorsal raphe nucleus, the main serotonin-containing nucleus in the brain stem, can increase CBF (Goadsby et al., 1985a, 1985b). Stimulation of PAG will inhibit SSS-evoked trigeminal neuronal activity in cats (Knight & Goadsby, 2001), whereas blockade of P/Q-type voltage-gated Ca^{2+} channels in the PAG facilitates trigeminovascular nociceptive processing (Knight et al., 2002), with the local GABAergic system in the PAG still intact (Knight et al., 2003).

3. Electrophysiology of Migraine in Humans

Welch (2005) summarized the arguments to support the hypothesis that migraine is due to a hyperexcitable brain state, which results in susceptibility to migraine attacks. The evidence comes from clinical studies applying a wide range of technologies, including psychophysical, cortical visual control, functional MRI, magnetoencephalography, evoked and event-related potentials, and transcranial magnetic stimulation. Furthermore, the high bidirectional comorbidity of migraine with epilepsy and the migraine-prophylactic efficacy of at least two anti-epileptic agents also seem to support the concept that the migraine brain is hyperexcitable. However, there is much discussion as to how to interpret the different findings (Kaube & Giffin, 2002). An attractive alternative explanation is provided by Schoenen and colleagues (Schoenen et al., 2003), who postulate that the essence of the migrainous brain is that it does not habituate to signals in a normal way. Contingent negative variation (CNV), an event-related potential, is abnormal in migraineurs compared to controls. Changes in CNV predict attacks and preventive therapies alter and normalize such changes. Further research into this area is clearly needed.

F. Can CSD Trigger the Mechanisms for the Headache Phase?

Another controversial area is whether CSD can initiate the migraine headache cascade. Haerter and colleagues (2005) reviewed the experimental animal evidence that CSD might activate the trigeminal sensory system, presumably by depolarizing perivascular trigeminal terminals at meningeal and dural blood vessels. KCl-induced CSD in the rat parietal cortex activates ipsilateral trigeminal nucleus caudalis neurons and CSD causes a long-lasting blood flow increase in the rat middle meningeal artery and a dural plasma protein leakage that can be inhibited by ipsilateral trigeminal nerve section (Bolay et al., 2002). Chronic daily, but not acute administration of migraine prophylactic drugs in rats, dose and duration dependently suppressed KCl-induced CSD frequency by 40 to 80 percent, and increased the trigger threshold for inducing CSD (Ayata et al., 2006), although there is one study to suggest topiramate can acutely inhibit CSD (Akerman & Goadsby, 2005).

Goadsby (2001), however, reviewed a number of mainly clinical arguments that go against the hypothesis that CSD may also trigger the headache mechanisms. The fact that migraine aura occurs in only up to one-third of migraine patients makes it difficult to explain how CSD might be involved in the majority of migraineurs in whom attacks are not associated with aura. The only explanation to rescue the "CSD headache trigger" hypothesis would be that CSD might occur in clinically silent subcortical areas of the brain without propagating through all six cortical laminae within the neocortex (Haerter et al., 2005). Direct human evidence for such events is difficult to obtain and, apart from possibly one case (Woods et al., 1994), is lacking. Other clinical observations that argue against the role of CSD in triggering the headache are that the headache may occur at the same side of the aura, instead of the expected opposite side if CSD would have triggered the trigeminovascular system, and that aura may sometimes occur after the headache has started. Furthermore, aura appears to be not confined to migraine; it is reported with attacks of cluster headaches (Bahra et al., 2002; Silberstein et al., 2000), paroxysmal hemicrania (Matharu & Goadsby, 2001), and hemicrania continua (Peres et al., 2002). Finally, intranasal ketamine could abort migraine aura without affecting the ensuing headache (Kaube et al., 2000). From this discussion, it is clear that the exact role of CSD in triggering migraine headache mechanisms in humans remains unclear and needs further investigation.

G. Migraine Symptoms in Relationship to the Pathophysiology

Migraine primarily seems an episodic brain disorder with impaired central sensory processing (Goadsby et al., 2002). Patients complain of pain in the head that is throbbing, but there is no reliable relationship between vessel diameter and the pain, or its treatment (Friberg et al., 1991; Limmroth et al., 1996). They complain of discomfort from normal lights and the unpleasantness of routine sounds. Some mention otherwise pleasant odors are unpleasant. Normal movement of the head causes pain, and many mention a sense of unsteadiness as if they have just stepped off a boat, having been nowhere near the water. All these features suggest that normal sensory input is processed abnormally.

The anatomical connections of, for example, the pain pathways are clear, the ophthalmic division of the trigeminal nerve subserves sensation within the cranium and explains why the top of the head is headache, and the maxillary division is *facial pain*. The convergence of cervical and trigeminal afferents explains why neck stiffness or pain is so common in primary headache. The genetics of ionopathies (see later) is opening up a plausible way to think about the episodic nature of migraine. However, where is the lesion, what is actually the pathology?

There is not a photon of extra light that migraine patients receive over others, nor do they hear more decibels of sound or do they smell more odor units. Accordingly, for photophobia, phonophobia, and osmophobia, the basis of the problem must be abnormal central processing of a normal signal. Perhaps electrophysiological changes in the brain have been mislabeled as *hyperexcitability*, whereas dyshabituation might be a simpler explanation (Goadsby, 2005c). If migraine was basically an attentional problem with changes in cortical synchronization (Niebur et al., 2002), *hypersynchronization*, all its manifestations could be accounted for in a single overarching pathophysiological hypothesis of a disturbance of subcortical sensory modulation systems. Though it seems likely that the trigeminovascular system and its cranial autonomic reflex connections, the trigeminal-autonomic reflex (May & Goadsby, 1999), act as a feed-forward system to facilitate the acute attack, the fundamental problem in migraine is in the brain. Unraveling its basis will deliver great benefits to patients and considerable understanding of some very fundamental neurobiological processes.

VI. The Migraine Trigger Threshold: Repeated Recurrence of Attacks

A. How Are Migraine Attacks Triggered?

The *disease* migraine rather arbitrarily is defined as having had at least five attacks of migraine without aura or at least two attacks of migraine with aura. One of the reasons for this is that it is said many people may experience a few sporadic attacks of migraine throughout life and no more. There are no data to argue this point either way. On this basis it has been suggested that the migraine attack is not abnormal; rather, the repeated occurrence of attacks is abnormal (Ferrari, 1998). In this respect, migraine is very similar to that other classical and often comorbid episodic brain disorder, epilepsy.

In order to understand the *disease* migraine, we must understand how migraine attacks are triggered and why patients get recurrent attacks. It is not unlikely that at least part of the answer is in an imbalance between the individual's trigger threshold ("defense") and the "attack" by migraine triggers. Here we discuss the growing evidence that the *disease* migraine might be due to a genetically determined reduced threshold for migraine triggers and that attacks may occur when migraine triggers are particularly strong or frequent; when there is a temporarily further reduction of the threshold due to endogenous factors such as menstruation, sleep deprivation, or (relaxation after) stress, that can facilitate the triggering of an attack; or when there is a temporal coincidence of both triggering and facilitating factors. Understanding the mechanisms involved in the triggering of migraine attacks will help to identify novel treatment targets for urgently needed specific prophylactic agents to prevent migraine attacks. An important initial step to achieve these goals is to unravel the genetic basis of the migraine threshold and to decipher the common pathways for triggering migraine attacks.

B. Genetic Epidemiology

Migraine often runs in families (Kors et al., 2004). Population-based studies have confirmed that the risk of migraine in first-degree relatives is 1.5- to 4-fold increased. The familial risk appeared greatest for patients with migraine with aura, with a young age at onset and a high attack severity and disease disability (Russell & Olesen, 1995; Stewart et al., 1997, 2006).

Some authors concluded, on the basis of different heritability estimates, that migraine with and without aura are different entities (Ludvigsson et al., 2006; Russell & Olesen, 1995; Russell et al., 2002). It seems very unlikely in view of a number of clinical and genetic arguments that migraine is different, rather the aura component may have some heritable biological distinction. These include the high comorbidity of attacks with and without aura within migraineurs and the remarkable intrapersonal variability of the disease presentation in the various stages of life; for example, with aura as a child, without aura as a young adult, and aura without headache after age 50. Furthermore, an Australian study, in over 6,000 twin pairs, identified disease subtypes (latent classes) by using so-called *latent class analysis* on the basis of the patterns and severity of the symptoms (Nyholt et al., 2004). The results did not support the hypothesis that migraine with and without aura are distinct disorders. A Finnish study of over 200 migraine families suggested that there is a continuum from pure migraine with aura at the neural end of the spectrum to pure migraine without aura at the headache end of the spectrum, and migraine with both with and without aura in between (Kallela et al., 2001). In conclusion, the different migraine subtypes appear to be different clinical expressions of the same disorder.

Studies of twin pairs are the classical method to investigate the relative importance of genetic and environmental factors. In twin pairs drawn from the general population, the pair-wise concordance rates for migraine were significantly higher among monozygotic than among dizygotic twin pairs, indicating that genetic factors are important in the susceptibility to migraine. However, as these concordance rates never reached 100 percent, environmental factors must be involved as well, making migraine a true multifactorial complex disorder (Gervil et al., 1999; Honkasalo et al., 1995; Mulder et al., 2003; Ulrich et al., 1999). The relative importance can be estimated from a very large population-based twin study investigating 30,000 twin pairs from six countries (Mulder et al., 2003). The heritability was 40 to 50 percent, and shared environmental factors were considered to have a minor effect on the susceptibility to migraine. This finding is in accordance with that of a comparison between twins raised together and raised apart (Svensson et al., 2003; Ziegler et al., 1998).

C. Finding Genes for Complex Disorders: The Migraine Gene Highway

The identification of genes for multifactorial disorders is hampered by a number of complicating factors. Multiple genes contribute to the susceptibility, each contributing gene displays a relatively low penetrance, and the phenotypic expression is modulated by variable endogenous and exogenous nongenetic factors. Furthermore, complex disorders usually are very prevalent and may start at older ages, complicating a reliable distinction between affected and nonaffected populations.

To identify genes for migraine, several genetic approaches have been applied. The first, and thus far most successful approach, has been the identification of genes in families with rare, monogenic subtypes of migraine. This has been done by using traditional linkage analyses (testing several hundreds to thousands of genetic markers spread over all chromosomes and selecting those markers, i.e., the chromosomal region best segregated with the disease), positional gene cloning techniques, and mutation analysis. This approach is based on the hypothesis that monogenic rare subtypes and multifactorial common types of migraine share common genes and related biochemical pathways for the trigger threshold and initiation mechanisms of attacks. Thus, the rare monogenic variant may serve as a genetic and/or functional model for the common complex types. In the latter case, the *functional* changes caused by the causative gene mutations are more relevant that the genes themselves, as they might hint at shared pathogenic pathways; the genes identified in the monogenic variant may not necessarily be involved in the common forms. So far, genes for three monogenic subtypes of migraine have been identified: Familial Hemiplegic Migraine (FHM), Sporadic Hemiplegic Migraine (SHM), and CADASIL. The details of these findings, and the implications for the common forms of migraine, will be explained later.

A second linkage analysis approach that is often used in complex traits is affected sib-pair analysis. With this approach, chromosomal areas that are shared by affected siblings with a probability higher than by chance alone are being identified. This is then followed by case-control association studies testing single nucleotide polymorphisms (SNPs) in candidate genes in the shared regions. The goal is to identify SNPs, and thus gene alleles, that statistically differ in frequency between cases and controls and cause increased susceptibility to the disease. A third, hypothesis-driven approach is direct testing of candidate genes in case-control association studies. An interesting new twist to this approach will be the possibility of nonhypothesis driven testing for genome-wide association by scanning hundreds of thousands of SNPs in extended and clinically homogenous populations (Hirschhorn & Daly, 2005).

D. Familial Hemiplegic Migraine (FHM) Genes

FHM is a rare, severe, monogenic subtype of migraine with aura, characterized by at least some degree of hemiparesis during the aura (Ferrari, 1998). The hemiparesis may last from minutes to several hours or even days. Patients are frequently initially misdiagnosed with epilepsy. Apart from the hemiparesis, the other headache and aura features of the FHM attack are identical to those of attacks of the common types of migraine. In addition to attacks with hemiparesis, the majority of FHM patients also experience attacks of "normal" migraine with or without aura (Ducros et al., 2001; Terwindt et al., 1998b).

As in the common forms of migraine, attacks of FHM may be triggered by mild head trauma. Thus, from a clinical point of view, FHM seems a valid model for the common forms of migraine (Ferrari, 1998). Major clinical differences, apart from the hemiparesis, include that FHM in 20 percent of the cases may also be associated with cerebellar ataxia and other neurological symptoms such as epilepsy, mental retardation, brain edema, and (fatal) coma. Thus far, three genes for FHM have been published, but based on unpublished linkage results in several families, there are more to come.

E. The FHM1 *CACNA1A* Gene

The first gene identified for FHM is the *CACNA1A* gene on chromosome 19p13. It is responsible for approximately 50 percent of all families with FHM (see Figure 28.3). The FHM1 gene encodes the ion-conducting, pore-forming α_{1A} subunit of $Ca_v2.1$ (P/Q-type), voltage-gated, neuronal calcium channels (Ophoff et al., 1996). The main function of neuronal P/Q-type calcium channels is to modulate release of neurotransmitters, both at peripheral neuromuscular junctions as well as central synapses, mainly within the cerebellum, brainstem, and cerebral cortex (Catterall, 1998). Over 50 *CACNA1A* mutations have been associated with a wide range of clinical phenotypes (Haan et al., 2005) (see also Figure 28.3). These include pure

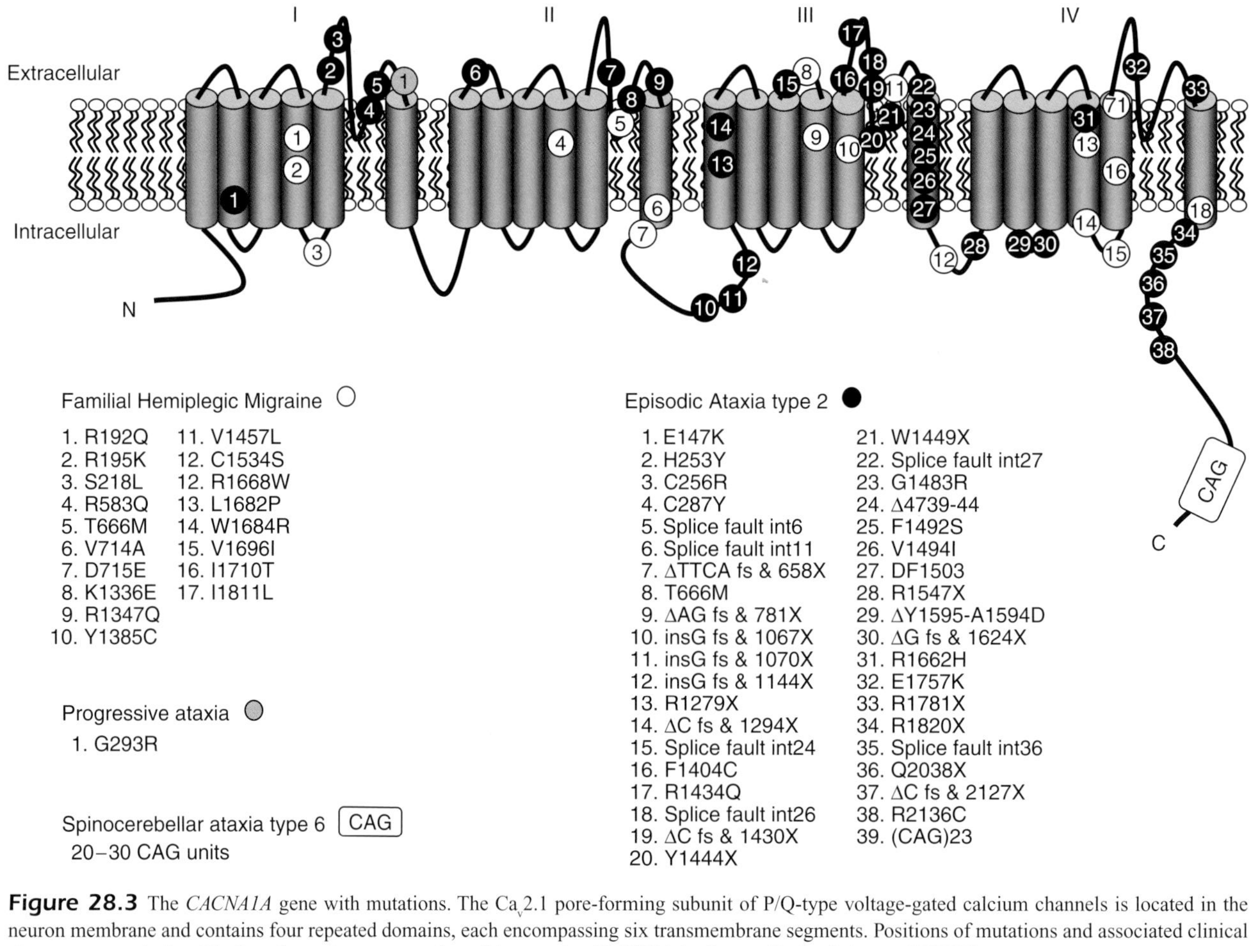

Figure 28.3 The *CACNA1A* gene with mutations. The $Ca_v2.1$ pore-forming subunit of P/Q-type voltage-gated calcium channels is located in the neuron membrane and contains four repeated domains, each encompassing six transmembrane segments. Positions of mutations and associated clinical phenotypes are depicted in the schematic representation of the protein. (*CACNA1A* ref. seq.: Genbank Ac. nr. X99897)

Extracellular

Intracellular

N

Familial Hemiplegic Migraine ○

1. E174K
2. T263M
3. G301R
4. T345A
5. T376M
6. R383H
7. R593W
8. A606T
9. G615R
10. V628M

11. Δ1804-1820 & ins TT
12. R689Q
13. E700K
14. D718N
15. M731T
16. R763H
17. L764P
18. P796R
19. M829R
20. R834Q

21. W887R
22. E902K
23. Δ935K-940S, ins I
24. R937P
25. Δnt2897-2898 > FS
26. P979L
27. X1021R

Basilar migraine ●

1. C515Y
2. R548H

Alternating Hemiplegia of Childhood ◐

1. T378N

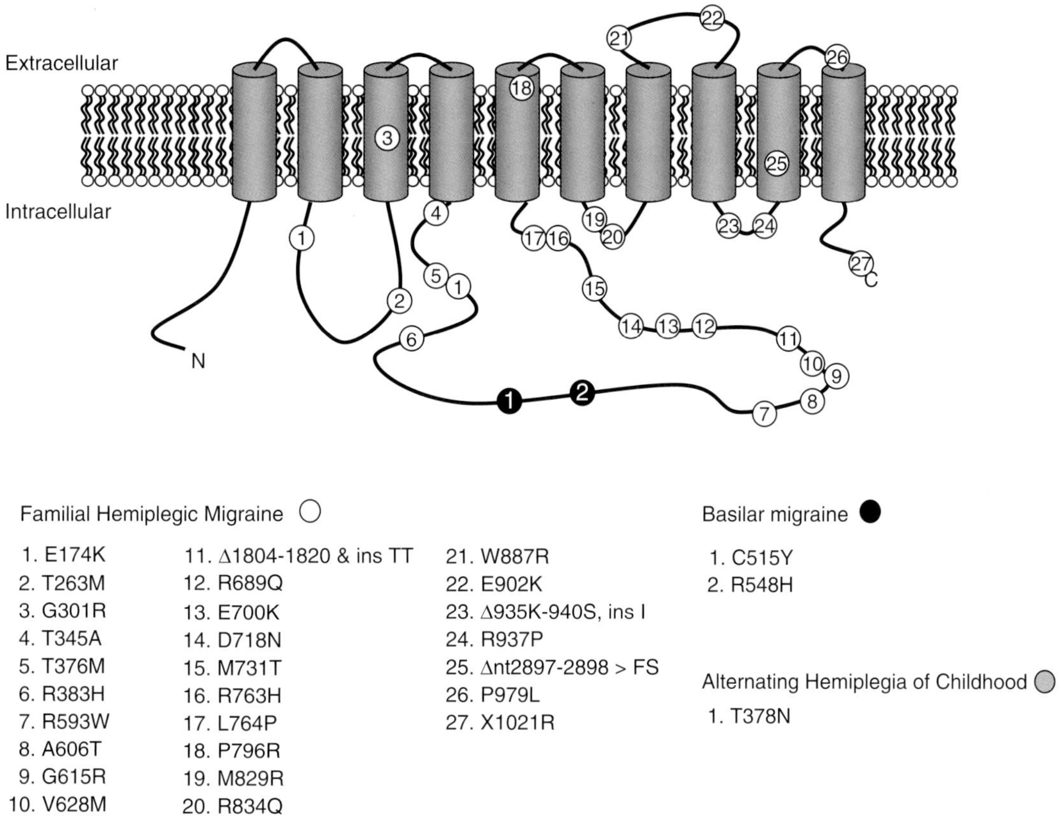

Figure 28.4 The *ATP1A2* gene with mutations. The alpha 2 subunit of sodium potassium pumps is located in the plasma membrane and contains 10 transmembrane segments. Positions of mutations and associated clinical phenotypes are indicated in the schematic drawing of the protein. (*ATP1A2* ref. seq.: Genbank Ac.nr. NM_000702)

forms of FHM (Ophoff et al., 1996), combinations of FHM with various degrees of cerebellar ataxia (Ducros et al., 2001; Ophoff et al., 1996), or fatal coma due to excessive cerebral edema (Kors et al., 2001), and disorders not associated with FHM such as episodic ataxia type 2 (Jen et al., 2004; Ophoff et al., 1996), progressive ataxia (Yue et al., 1997), spinocerebellar ataxia type 6 (Zhuchenko et al., 1997), and absence (Imbrici et al., 2004) and generalized epilepsy (Haan et al., 2005; Jouvenceau et al., 2001).

Interestingly, in several FHM families, FHM1 *CACNA1A* mutations also were found in family members who had only "normal" nonparetic migraine but no FHM. This suggests that gene mutations for FHM may also be responsible for the common forms of migraine, probably due to different genetic and nongenetic modulating factors.

F. The FHM2 *ATP1A2* Gene

The *ATP1A2* FHM2 gene on chromosome 1q23 encodes the α_2 subunit of a Na$^+$,K$^+$ pump ATPase (De Fusco et al., 2003; Marconi et al., 2003). This catalytic subunit binds Na$^+$, K$^+$, and ATP, and utilizes ATP hydrolysis to exchange Na$^+$ ions out of the cell for K$^+$ ions into the cell. Na$^+$ pumping provides

the steep Na$^+$ gradient essential for the transport of glutamate and Ca^{2+}. The gene is predominantly expressed in neurons at neonatal age and in glial cells at adult age (De Fusco et al., 2003; Vanmolkot et al., 2003). In adults, an important function of this specific ATPase is to modulate the reuptake of potassium and glutamate from the synaptic cleft into the glial cell. Mutations in the *ATP1A2* gene are responsible for at least 20 percent of FHM cases (see Figure 28.4) and have been associated with pure FHM (De Fusco et al., 2003; Riant et al., 2005; Vanmolkot et al., 2006) and FHM in combinations with cerebellar ataxia (Spadaro et al., 2004), alternating hemiplegia of childhood (Bassi et al., 2004; Swoboda et al., 2004), benign focal infantile convulsions (Vanmolkot et al., 2003), and other forms of epilepsy (Haan et al., 2005).

In an Italian family a variant in the *ATP1A2* gene segregated with basilar migraine, a subtype of migraine with aura characterized by aura symptoms attributable to the brainstem and both occipital lobes (Ambrosini et al., 2005). Unfortunately, no functional studies were reported, precluding a definite conclusion as to whether this gene variation is also *causally* linked to basilar migraine. Of note, in two non-FHM migraine families *ATP1A2* variants were identified,

Familial Hemiplegic Migraine ○

1. Q1489K
2. L1649Q

Inactivation gate (Box A) ●

1. S1505X (SMEI)
2. N1509 fs 1511X (SMEI)

S4 voltage sensor region in Domain IV (Box B) ●

1. R1635X 3. R1648C (SMEI) 5. R1657C (GEFS+)
2. R1645C (SMEI) 4. R1648H (GEFS+)

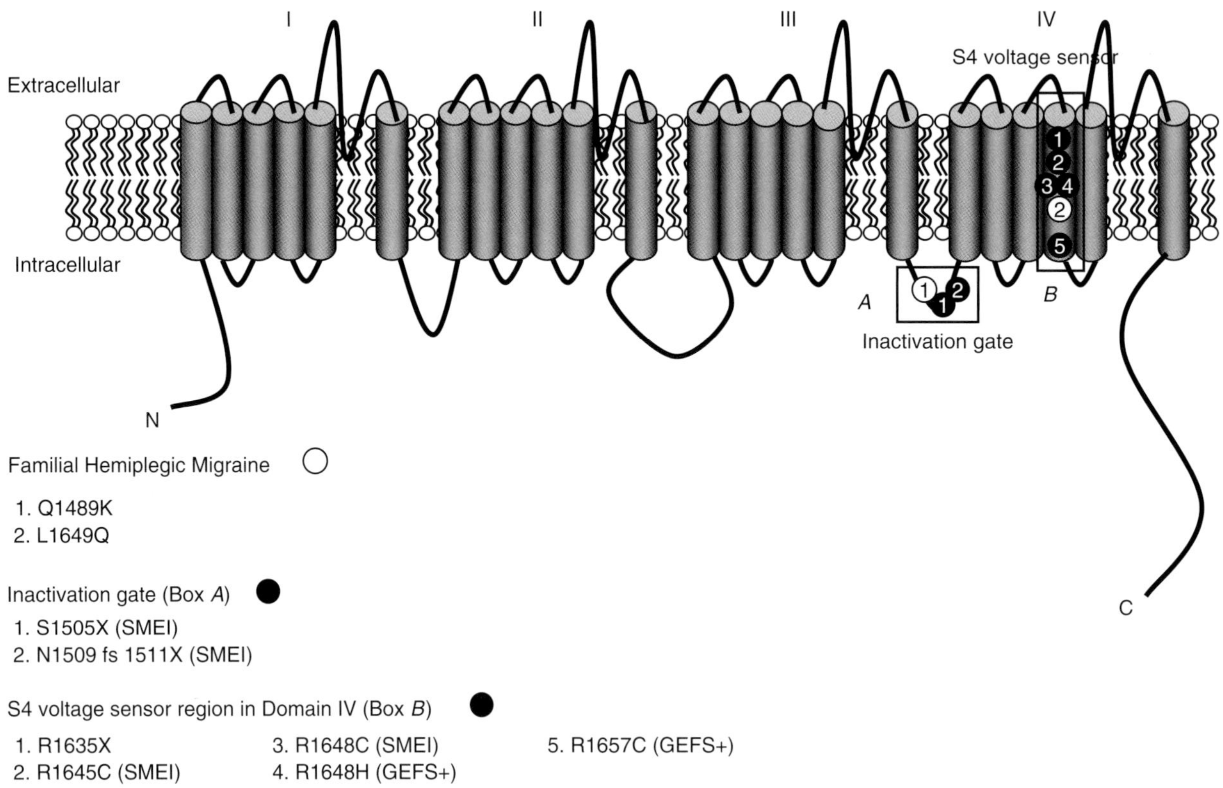

Figure 28.5 The *SCN1A* gene with mutations. The pore-forming subunit of Na$_v$1.1 sodium channels is located in the neuron membrane and contains four repeated domains, each encompassing six transmembrane segments. Over 150 mutations have been identified for severe myoclonic epilepsy in infancy (SMEI) or generalized epilepsy with febrile seizures plus (GEFS+) (reviewed in Kanai et al., 2004; Meisler & Kearney, 2005). For clarity only SMEI and GEFS+ mutations are shown, which reside in two functional domains (e.g., inactivation gate and D4/S4 voltage sensor), where familial hemiplegic migraine mutations were identified. Positions of mutations and associated clinical phenotypes are depicted in the schematic representation of the protein. (*SCN1A* ref. seq.: Genbank Ac.nr. NM_006920)

suggesting that this gene may be involved in the susceptibility to common forms of migraine (Todt et al., 2005).

G. The FHM3 *SCNA1* Gene

The *SCNA1* gene on chromosome 2q24 encodes the alpha subunit of a neuronal voltage-gated sodium (Na$_v$1.1) channel. The Na$_v$1.1 channel is mainly responsible for the generation and propagation of neuronal action potentials. Different mutations in this gene are known to be associated with epilepsy and febrile seizures (for review, Meisler & Kearney, 2005). Recently, Dichgans and colleagues (2005) found a novel Q1489K mutation in three German FHM families of common ancestry. Another new *SCNA1* mutation was found in a North-American FHM family confirming the relationship between *SCNA1* and FHM3 (Vanmolkot et al., unpublished observations; also see Figure 28.5).

H. Sporadic Hemiplegic Migraine (SHM)

Hemiplegic migraine patients are not always clustered in families. Sporadic patients, without affected family members, often are seen and may sometimes represent the first "FHM patient" (*de novo* mutation) in a family (Thomsen et al., 2003a). Apart from sharing the clinical phenotype with the common forms of migraine, similar to FHM, SHM and normal migraine also show a remarkable genetic epidemiological relationship. SHM patients have a highly increased risk of also suffering from typical migraine with aura and their first-degree relatives have a highly increased risk of both migraine with and without aura (Thomsen et al., 2003b). Although in an initial study, *CACNA1A* mutations were found in only 2 of 27 SHM patients (Terwindt et al., 2002), a recent, much larger study (de Vries et al., unpublished data) did find mutations in the FHM genes in a higher proportion of SHM patients, confirming a genetic relationship between FHM and SHM.

I. Migraine as a Cerebral Ionopathy?

The next question we need to address is, what is the evidence that a dysfunction of ion flux or transportation is involved in the pathogenesis of the common forms of migraine? At this point in time, the evidence is primarily circumstantial, but growing and ranging from clinical arguments to genetic, clinical neurophysiological, and neuropharmacological evidence.

First, migraine shares strikingly similar clinical characteristics with established channelopathies such as FHM and SHM (see earlier), but also with episodic neuromuscular disorders such as myotonia and periodic paralysis (Ferrari & Goadsby, 2006). These include the episodic presentation of the symptoms; a similar distribution for duration and frequency of the attacks; similar trigger factors for attacks such as emotion, stress, food, alcohol and weather changes; and a similar gender-related expression with an onset of attacks mostly around puberty and amelioration after age 40.

Second, FHM1 *CACNA1A* and FHM2 *ATP1A2* gene mutations have been found in patients with only the common forms of migraine, without the FHM phenotype. Furthermore, some (but not all) linkage and association studies do suggest a role of FHM genes in normal migraine (Haan et al., 2005). Size and homogeneity of the study populations in such studies are clearly important complicating factors.

A third line of evidence comes from clinical neurophysiologal studies in migraineurs. Ambrosini and colleagues (2005) found single fiber abnormalities, suggesting an altered release of acetylcholine at the neuromuscular junction, which is mainly controlled by P/Q-type $Ca_v2.1$ channels. Terwindt and colleagues (2004), however, could not replicate these findings in FHM1 patients with demonstrated gene mutations. Sandor and colleagues (2001) found evidence of subclinical cerebellar dysfunction. A higher proportion of migraine patients than healthy controls showed a systematic horizontal deviation when subjected to an automated sensitive test for cerebellar coordination. As P/Q-type $Ca_v2.1$ channels are highly expressed in cerebellar Purkinje cells, this finding seems to suggest a dysfunction of $Ca_v2.1$ channels in Purkinje cells of migraineurs, although again eye movement problems have not been replicated by others (Wilkinson et al., 2006).

Finally, neuropharmacological animal experiments suggest that application of selective blockers of P/Q-type $Ca_v2.1$ channels within areas of the brainstem that have been associated with important migraine mechanisms (see first part of this chapter), can modulate these mechanisms. These include inhibition of the release of CGRP and neurogenic inflammation (Asakura et al., 2000), facilitation of trigeminal firing (Knight et al., 2002), and modulation of nociceptive transmission in the trigeminocervical complex (Shields et al., 2005).

J. Functional Consequences of FHM Gene Mutations

Understanding the functional consequences of gene mutations is crucial to the understanding of the disease pathways. For the FHM genes, this has been studied in cellular models, in knock-out mouse models, and in transgenic knock-in mouse models carrying a human pathogenic mutation.

1. Functional Studies in Cellular Models for FHM1 *CACNA1A* Mutations

Several FHM and episodic ataxia type 2 mutations have been analyzed with electrophysiological techniques in neuronal and nonneuronal cell models (Cao et al., 2004; Hans et al., 1999; Imbrici et al., 2004; Jouvenceau et al., 2001; Kraus et al., 1998, 2000; Tottene et al., 2002). Although episodic ataxia type 2 *CACNA1A* mutations all show a dramatic decrease or even complete loss of current density (Guida et al., 2001; Imbrici et al., 2004; Jen et al., 2001; Jeng et al., 2006; Jouvenceau et al., 2001; Spacey et al., 2004; Wan et al., 2005; Wappl et al., 2002), FHM1 mutations cause different effects on channel conductance, kinetics, and/or expression in transfected cells (Cao et al., 2004; Hans et al., 1999; Kraus et al., 1998, 2000; Tottene et al., 2002, 2005).

The most consistent change found with FHM1 mutations, when tested in a single channel configuration, was a hyperpolarizing shift of about 10 mV of the activation voltage (Hans et al., 1999; Kraus et al., 1998; Tottene et al., 2005). The change in calcium influx, however, could alter during high neuronal activity. Mutant T666M and V714A channels have a low conductance mode that may sometimes switch to the wild-type state (Hans et al., 1999). For other FHM mutations, like R583Q and D715E, accumulation of inactivated channels was observed during repetitive stimulation (Kraus et al., 2000). Such phenomena could contribute to the paroxysmal presentation of symptoms.

The gain-of-function effects found in single channel test models, *in theory*, will lead to an easier opening of channels in neurons. The overall change in calcium influx, however, is difficult to predict and seems to depend, at least partly, on the model that is being used. For instance, Cao and colleagues (2004) found evidence for *reduced* calcium influx at the whole cell level in transfections of cultured mouse hippocampal neurons. What will happen in the mutant brain will be determined by the delicate interplay between the functional effects of a particular mutation, the different channel properties and density, the different channel subunits, and the direct and indirect cellular environment. The observation that different auxillilary beta-subunits of calcium channels can modulate the consequence of FHM mutations certainly adds to the complexity of predicting calcium channel functioning (Mullner et al., 2004). It seems, therefore, more appropriate to study the functional consequences of gene mutations in knock-in mouse models carrying human pathogenic mutat ions (see next).

2. Functional Studies in Naturally Occurring *Cacna1a* Mouse Mutants

Naturally occurring mouse mutants with *Cacna1a* missense mutations (Tottering, Rocker, Rolling Nagoya) or

Cacna1a truncation mutations (Leaner) display different combinations and severities of various types of epilepsy and ataxia (Pietrobon, 2005). A reduction in calcium current density appears to be the main effect of these mutated P/Q-type channels (Dove et al., 1998; Lorenzon et al., 1998; Mori et al., 2000; Wakamori et al., 1998), with a change in channel kinetics for the Leaner and Rolling Nagoya mutants (Dove et al., 1998; Lorenzon et al., 1998; Mori et al., 2000). Two *Cacna1a*-null (knock-out) mouse models were generated showing ataxia, dystonia, and lethality at a young age (Fletcher et al., 2001; Jun et al., 1999). The total Ca^{2+} influx in cerebellar cells and neurotransmission at the neuromuscular junction was reduced in these mice. Loss of P/Q type channels could be (partly) compensated for by N-, R- and L-type channels. Moreover, leaner mice showed an increased threshold for CSD and a reduced release of cortical glutamate (Ayata et al., 2000).

3. Functional Studies in FHM1 *Cacna1a* Knock-in Mouse Mutants

Very recently, a knock-in mouse model was generated, carrying the human FHM1 R192Q mutation (van den Maagdenberg et al., 2004). Unlike the natural *Cacna1a* mutant mouse models, transgenic R192Q mice exhibit no overt clinical phenotype or structural abnormalities. This is very similar to the human situation: the R192Q mutation causes only mild FHM attacks, very similar to the common forms of migraine, albeit with hemiparesis, without other neurological symptoms. Extensive functional analysis revealed multiple gain-of-function effects. These include increased Ca^{2+} influx in cerebellar neurons; increased release of neurotransmitters at the neuromuscular junction, both spontaneous and upon stimulation at low Ca^{2+}; and in the intact animal, a reduced trigger-threshold for CSD that propagates with increased velocity. It seems that whole-animal studies may be better suited to dissect the effects of mutations and to understand the integrated physiology of the disease. Other studies aiming at functional changes within the brainstem are under way and will shed more light on the important question whether migraine-related mechanisms within the trigeminocervical complex are also affected.

4. Functional Studies in Cellular Models for FHM2 *ATP1A2* Mutations

Functional analysis of mutated proteins revealed inhibition of pump activity, decreased affinity for K^+, and decreased catalytic turnover resulting in reduced reuptake of K^+ and glutamate into glial cells (De Fusco et al., 2003; Segall et al., 2005).

5. Functional Studies in *Atp1a2* Knock-out Mouse Models

Two groups have generated α2-subunit deficient (*Atp1a2*-null) mice (Ikeda et al., 2004; James et al., 1999). These mice died immediately after birth because of severe motor deficits and absent respiration (Ikeda et al., 2003; James et al., 1999). *Atp1a2*-null fetuses of 18.5 days revealed selective neuronal apoptosis in the amygdala and piriform cortex in response to neural hyperactivity (Ikeda et al., 2003). *Atp1a2*-null mice on 129sv genetic background displayed frequent generalized seizures and died within 24 hours after birth (Ikeda et al., 2004). Epilepsy is also a feature of the clinical phenotype in humans with *ATP1A2* mutations. Heterozygous *Atp1a2+/−* mice are viable. Their heart shows a hypercontractile state with positive inotropic response and resembles what typically is seen after the administration of cardiac glycosides (James et al., 1999). In addition, they revealed enhanced fear and anxiety behaviors after conditioned fear stimuli; this is probably due to neuronal hyperactivity in the amygdala and piriform cortex (Ikeda et al., 2003). There are no *Atp1a2* knock-in models available.

6. Functional Studies in Cellular Models for the FHM3 Mutation

Functional consequences of the Q1489K *SCN1A* mutation have been analyzed in the highly homologous SCN5A protein. This missense mutation is located within a domain critical for fast inactivation of the sodium channel. It causes a two- to four-fold increased acceleration of recovery from fast inactivation (Dichgans et al., 2005). This would predict enhanced neuronal excitation and release of neurotransmitters. Unfortunately, no knock-out or knock-in mouse mutants are available for this gene.

K. A Common Mechanism for Triggering FHM Attacks

Three different genes encoding three very different proteins with apparent different mechanisms now have been associated with FHM. How can we fit these apparently very diverting mechanisms into one (final) common pathway? There is one obvious candidate, cortical spreading depression (Moskowitz et al., 2004). Mutations in the FHM1 calcium gene cause increased neuronal release of neurotransmitters, and in the cortex, more specifically the neuroexcitatory amino acid glutamate (Pietrobon et al., unpublished data) that can induce, maintain, and propagate CSD. Mutations in the FHM2 sodium, potassium pump gene cause reduced reuptake of K^+ and glutamate from the synaptic cleft into the glia cell. Mutations in the FHM3 sodium channel gene result in hyperexcitability and most likely increased release of neurotransmitters in the synaptic cleft. The overall result is increased levels of glutamate and K^+ in the synaptic cleft resulting in an increased propensity for CSD. This would easily explain the aura of FHM attacks. More controversial, however, is whether the enhanced tendency for CSD might also be responsible for triggering the headache phase, for example, by activation of the trigeminovascular system (for discussion see earlier).

Lastly, the most important question that remains to be answered is whether the same mechanisms also are involved in the common forms of migraine with and without aura. Interestingly, although drugs with migraine prophylactic activity belong to a wide range of pharmacological classes (e.g., anti-epiletics and blockers of calcium channels and serotinergic, beta-adrenergic, and histaminergic receptors), they all seem to share anti-CSD activity (Akerman & Goadsby, 2005; Ayata et al., 2006). CSD inhibition may thus be a promising model system to contribute to the development of preventive medicines.

An interesting compound to watch in this respect will be Tonabersat, which has entered clinical trials in migraine. Tonabersat (SB-220453) inhibits CSD, CSD-induced nitric oxide (NO) release, and cerebral vasodilation. It does not constrict isolated human blood vessels, but does inhibit trigeminally induced craniovascular effects (Goadsby, 2005b). Tonabersat is inactive in the human NO-model of migraine, as is propranolol (Tvedskov et al., 2004b), although valproate showed some activity in that model (Tvedskov et al., 2004a). If proven effective, it would be the first migraine prophylactic agent developed on the basis of the CSD hypothesis.

L. Cadasil

Cerebral Autosomal Dominant Arteriopathy with Subcortical Infarcts and Leucoencephalopathy (CADASIL) is a severe arteriopathy caused by mutations in the *Notch3* gene on chromosome 19p13 (Joutel et al., 1996). CADASIL is clinically characterized by recurrent subcortical infarcts, white matter lesions on MRI, dementia and other neurospychiatric symptoms, and most relevant here, migraine with aura in 40 percent of patients (Joutel et al., 1996; Tournier-Lasserve et al., 1993). Migraine is usually the presenting symptom, approximately 10 years before the other symptoms become apparent. The *Notch3* gene is involved primarily in the regulation of arterial differentiation and maturation of vascular smooth muscle cells (Domenga et al., 2004). Neuronal effects have not been found. Why mutations in this gene would cause migraine is unclear. Because of the apparent primarily vascular effects of the gene, the CADASIL-migraine relationship may be seen as a support for an important contribution of blood vessels to migraine pathogenesis, although there is no direct evidence for this hypothesis. Alternatively, cerebral ischaemia may serve as a focus for CSD, which would then set off migraine aura and migraine attacks.

Notch3 CADASIL mouse models are available, but do not seem to express a CADASIL-like cerebral phenotype. This is true for both a conventional transgenic mouse model, over-expressing mutant Notch3 in vascular smooth muscle cells (Ruchoux et al., 2003), and a R142C knock-in mouse model (Lundkvist et al., 2005). The former model showed compromised cerebrovascular reactivity and impaired CBF autoregulation probably due to decreased relaxation or increased resistance of cerebral vessels. Studying both neuronal and vascular changes in these mice may prove invaluable in further dissection of the triggering mechanisms for migraine attacks.

M. Vascular Retinopathy and Migraine

A rare syndrome, clinically characterized by a combination of cerebroretinal vasculopathy, Raynaud phenomenon, migraine, pseudotumour cerebri, and variable other forms of vascular dysfunction, has been linked to chromosome 3p21 in three families (Ophoff et al., 2001; Terwindt et al., 1998a). Migraine is clearly part of the syndrome (Hottenga et al., 2005). Identification of the responsible gene for this neurovascular syndrome will evidently be important for a wide range of vascular disorders, including the pathogenesis of migraine.

N. Other Candidate Genes and Loci for Migraine

1. Linkage Studies

A number of genome-wide linkage studies have found significant or suggestive linkage for migraine and non-FHM loci, two of which have been replicated in independent samples. Linkage to chromosome 6p12.2-p21.1 in a large family with migraine with and without aura from northern Sweden (Carlsson et al., 2002) was confirmed in Australian patients (Nyholt et al., 2005), albeit with low evidence for linkage. Linkage to 4q24 in 50 Finnish families with migraine with aura (Wessman et al., 2002) was confirmed in a study in 289 Icelandic patients with migraine without aura (Bjornsson et al., 2003); both the Finnish and Icelandic populations are considered genetic isolates. Other migraine loci that have been found, but not yet replicated, are 1q31 (Gardner et al., 1997), 11q24 (Cader et al., 2003), 14q21.1-q22.3 (Soragna et al., 2003), 15q11-q13 (Russo et al., 2005), and Xq24-28 (Nyholt et al., 1998). The variety in loci reported is probably a reflection of the genetic heterogeneity of migraine.

2. Linkage Studies Using Quantitative Trait and Trait Component Analyses

Nyholt and colleagues (Lea et al., 2005a; Nyholt et al., 2005) used a quantitative trait analysis in 790 independent sib-pairs, selected from a large Australian sample of 12,245 twins, that were concordant for LCA migraine class. They found significant linkage on chromosome 5q21 for a severe migraine phenotype with pulsating headache. Interestingly, they also found certain loci for specific migraine characteristics such as phonophobia, photophobia, nausea/vomiting, and pulsating quality of the headache, again with low evidence for linkage.

A Finnish study used the individual clinical symptoms of migraine (trait component analysis) to determine affection status in genome-wide linkage analyses of 50 migraine families (Anttila et al., 2006). The previously identified chromosome 4q24 locus (Wessman et al., 2002) now was found to link to several traits. Novel loci were identified for pulsation trait on 17p13, an age at onset trait on 4q28, and a trait combination phenotype (International Headache Society full criteria) on 18q12. Furthermore, suggestive or nearly suggestive evidence of linkage was observed for phonophobia and aggravation by physical exercise. The use of symptom components of migraine rather than the full end diagnosis is a promising novel approach to stratify samples for genetic studies.

How relevant linkage findings for individual symptoms are, however, remains to be proven. It is reasonable to suggest that such findings might reflect only general sensitivities rather than migraine-specific relationships. That is, patients showing linkage, for example to the photophobia gene locus might show a tendency for photophobia under a variety of conditions (e.g., influenza or stomach pain). Another example is, how relevant would it be for the understanding of the pathogenesis of myocardial infarction and the development of preventive treatments for this disease, to find the gene for pain irradiating into the left arm? Another critical point is that migraine patients usually show a variable and changing pattern of symptoms over their lifetime. For instance, they may have severe nausea and vomiting together with aura as part of their migraine attacks at young age, may "lose" the aura and vomiting in their twenties to have attacks of migraine without aura, to end up with attacks of isolated aura's without headache or other associated symptoms. It seems that the presence of the individual clinical characteristics are time-locked rather than gene-locked, and that asking the patient for their symptoms may give different answers when asked at different stages in life.

3. Association Studies

In complex diseases, multiple genes are expected to contribute to the phenotype. Each gene has only a limited contribution. Finding such genes by using the classical linkage in family material with common forms of migraine may therefore be difficult. Association studies are considered a powerful alternative to detect genes if they confer moderate to high increased susceptibility to disease. However, there are a number of important pitfalls when conducting such studies. Many association studies in migraine have, for example, been conducted with insufficient sample sizes, inadequate definition of patients, inadequate control samples, and most importantly, replication in a separate study. Detailed overview of the numerous association studies in migraine have been published (Montagna et al., 2005). The latter review (Montagna et al., 2005) also provides an excellent overview of the many association studies that have been done on the relationship between migraine and genes for dopamine receptors and genes involved in the metabolism and transportation of serotonin (5-HT). In brief, none of the associations have been convincingly replicated. Here, we shall discuss only briefly those associations that have been replicated at least once.

The enzyme 5,10-methylenetetrahydrofolate reductase gene (MTHFR) plays a role in maintaining homocysteine levels. An association between the C677T variant in MTHFR and migraine with aura has been found in several clinic-based (and therefore selected) study populations (Kara et al., 2003; Kowa et al., 2000; Lea et al., 2004; Oterino et al., 2004) but also, and most importantly, in a large sample taken from the general population (Scher et al., 2006). This makes MTHFR the first migraine risk gene at the population level. The association was found to be enhanced in the presence of another variant (A1298C) in the same gene (Kara et al., 2003), and in combination with an angiotensin I-converting enzyme (ACE) DD/ID genotype (Lea et al., 2005b). If replicated, this would indicate also the first gene–gene interaction to be involved in modulating the risk for migraine. Other replicated associations, but only in selected clinic-based samples, include associations with a progesterone receptor (PGR) *Alu* insertion in two independent populations (Colson et al., 2005); the estrogen receptor 1 (ESR1) in two independent populations for the G594A polymorphism (Colson et al., 2005), although Oterino and colleagues (2006) found an association only for the G325C polymorphism (3-fold increased risk) but not for the G594A polymorphism; the tumor necrosis factor gene in two separate studies (Rainero et al., 2004; Trabace et al., 2002); and variants in the ACE (Kowa et al., 2005; Lea et al., 2005b; Paterna et al., 2000). Remarkably, in one study the ACE-DD variant seemed to have a slight protective effect against migraine in male patients (Lin et al., 2005). Also for ESR1 and PGR variants a synergistic effect increasing the risk for migraine has been observed (Colson et al., 2005).

VII. Conclusions

Migraine is a highly prevalent, multifactorial, episodic disorder of the brain, with high impact on patients and society. Changes within the trigeminocervical complex and trigeminovascular system, mainly leading to abnormal central sensory processing, are crucial pathophysiological mechanisms of the migraine attack. Interfering with these neurovascular mechanisms offers new avenues for novel specific *acute* treatments of the migraine attack, hopefully not associated with potential cardiovascular complications. Effective and well-tolerated treatments to *prevent* attacks, rather than abort once started, are dearly needed. New insights into the genetics and molecular biology of the migraine trigger threshold suggest that migraine might be a cerebral ionopathy resulting in enhanced propensity for CSD as a potential triggering mechanism for migraine aura and possibly other symptoms of the attack. Similar mechanisms might be involved in changing the modulatory role of the trigeminocervical complex on central trigeminal pain and other sensory signal transmission. Pharmacological interventions aimed at normalizing the disturbed ion homeostasis might offer new avenues for the development of novel specific migraine prophylactic treatments.

References

Afridi, S., Giffin, N. J., Kaube, H., Friston, K. J., Ward, N. S., Frackowiak, R. S. J., and Goadsby, P. J. (2005a). A PET study in spontaneous migraine. *Archives of Neurology* **62**, 1270–1275.

Afridi, S., Kaube, H., and Goadsby, P. J. (2004). Glyceryl trinitrate triggers premonitory symptoms in migraineurs. *Pain* **110**, 675–680.

Afridi, S., Matharu, M. S., Lee, L., Kaube, H., Friston, K. J., Frackowiak, R. S. J., and Goadsby, P. J. (2005b). A PET study exploring the laterality of brainstem activation in migraine using glyceryl trinitrate. *Brain* **128**, 932–939.

Ahn, A. H. and Basbaum, A. I. (2006). Tissue injury regulates serotonin 1D receptor expression: Implications for the control of migraine and inflammatory pain. *J Neurosci* **26**, 8332–8338.

Akerman, S. and Goadsby, P. J. (2005). Topiramate inhibits cortical spreading depression in rat and cat: impact in migraine aura. *NeuroReport* **16**, 1383–1387.

Ambrosini, A., D'Onofrio, M., Grieco, G. S., Di Mambro, A., Fortini, D., Nicoletti, F. et al. (2005). A new mutation on the ATPA2 gene in one Italian family with basilar-type migraine linked to the FHM2 locus. *Neurology* **64**, A132.

Amery, W. K. and Waelkens, J. (1983). Prevention of the last chance: An alternative pharmacological treatment of migraine. *Headache* **23**, 37–38.

Andlin-Sobocki, P., Jonsson, B., Wittchen, H. U., and Olesen, J. (2005). Cost of disorders of the brain in Europe. *Eur J Neurol* **12**, 1–27.

Anttila, V., Kallela, M., Oswell, G., Kaunisto, M. A., Nyholt, D. R., Hamalainen, E. et al. (2006). Trait components provide tools to dissect the genetic susceptibility of migraine. *Am J Hum Genet* **79**, 85–99.

Asakura, K., Kanemasa, T., Minagawa, K., Kagawa, K., Yagami, T., Nakajima, M., and Ninomiya, M. (2000). α-Eudesmol, a P/Q-type Ca^{2+} channel blocker, inhibits neurogenic vasodilation and extravasation following electrical stimulation of trigeminal ganglion. *Brain Research* **873**, 94–101.

Ayata, C., Jin, H., Kudo, C., Dalkara, T., and Moskowitz, M. A. (2006). Suppression of cortical spreading depression in migraine prophylaxis. *Ann Neurol* **59**, 652–661.

Ayata, C., Shimizu-Sasamata, M., Lo, E. H., Noebels, J. L., and Moskowitz, M. A. (2000). Impaired neurotransmitter release and elevated threshold for cortical spreading depression in mice with mutations in the alpha 1A subunit of P/Q type calcium channels. *Neuroscience* **95**, 639–645.

Bahra, A., May, A., and Goadsby, P. J. (2002). Cluster headache: A prospective clinical study in 230 patients with diagnostic implications. *Neurology* **58**, 354–361.

Bartsch, T. and Goadsby, P. J. (2002). Stimulation of the greater occipital nerve induces increased central excitability of dural afferent input. *Brain* **125**, 1496–1509.

Bartsch, T. and Goadsby, P. J. (2003). Increased responses in trigeminocervical nociceptive neurones to cervical input after stimulation of the dura mater. *Brain* **126**, 1801–1813.

Bartsch, T. and Goadsby, P. J. (2005). Anatomy and physiology of pain referral in primary and cervicogenic headache disorders. *Headache Currents* **2**, 42–48.

Bartsch, T., Levy, M. J., Knight, Y. E., and Goadsby, P. J. (2004). Differential modulation of nociceptive dural input to [hypocretin] Orexin A and B receptor activation in the posterior hypothalamic area. *Pain* **109**, 367–378.

Bassi, M. T., Bresolin, N., Tonelli, A., Nazos, K., Crippa, F., Baschirotto, C. et al. (2004). A novel mutation in the ATP1A2 gene causes alternating hemiplegia of childhood. *J Med Genet* **41**, 621–628.

Bergerot, A. and Goadsby, P. J. (2005). Distribution of the dopamine D1 and D2 receptors in the rat trigeminocervical complex. *Cephalalgia* **25**, 872.

Bergerot, A., Storer, R. J., and Goadsby, P. J. (2005). Dopamine inhibits trigeminovascular transmission in the rat. *Cephalalgia* **25**, 862.

Beuckmann, C. T. and Yanagisawa, M. (2002). Orexins: From neuropeptides to energy homeostasis and sleep/wake regulation. *Journal of Molecular Medicine* **80**, 329–342.

Bjornsson, A., Gudmundsson, G., Gudfinnsson, E., Hrafnsdottir, M., Benedikz, J., Skuladottir, S. et al. (2003). Localization of a gene for migraine without aura to chromosome 4q21. *Am J Hum Genet* **73**, 986–993.

Blin, O., Azulay, J., Masson, G., Aubrespey, G., and Serratrice, G. (1991). Apomorphine-induced yawning in migraine patients: Enhanced responsiveness. *Clin Neuropharmacol* **14**, 91–95.

Bolay, H., Reuter, U., Dunn, A. K., Huang, Z., Boas, D. A., and Moskowitz, M. A. (2002). Intrinsic brain activity triggers trigeminal meningeal afferents in a migraine model. *Nature Medicine* **8**, 136–142.

Bousser, M. G. and Welch, K. M. (2005). Relation between migraine and stroke. *Lancet Neurol* **4**, 533–542.

Breslau, N. and Davis, G. C. (1993). Migraine, physical health and psychiatric disorder: A prospective epidemiological study in young adults. *Journal of Psychiatric Research* **27**, 211–221.

Breslau, N., Lipton, R. B., Stewart, W. F., Schultz, L. R., and Welch, K. M. (2003). Comorbidity of migraine and depression: Investigating potential etiology and prognosis. *Neurology* **60**, 1308–1312.

Burstein, R., Yarnitsky, D., Goor-Aryeh, I., Ransil, B. J., and Bajwa, Z. H. (2000). An association between migraine and cutaneous allodynia. *Annals of Neurology* **47**, 614–624.

Cader, Z. M., Noble-Topham, S., Dyment, D. A., Cherny, S. S., Brown, J. D., Rice, G. P., and Ebers, G. C. (2003). Significant linkage to migraine with aura on chromosome 11q24. *Hum Mol Genet* **12**, 2511–2517.

Cao, Y. Q., Piedras-Renteria, E. S., Smith, G. B., Chen, G., Harata, N. C., and Tsien, R. W. (2004). Presynaptic Ca^{2+} channels compete for channel type-preferring slots in altered neurotransmission arising from Ca^{2+} channelopathy. *Neuron* **43**, 387–400.

Carlsson, A., Forsgren, L., Nylander, P. O., Hellman, U., Forsman-Semb, K., Holmgren, G. et al. (2002). Identification of a susceptibility locus for migraine with and without aura on 6p12.2-p21.1. *Neurology* **59**, 1804–1807.

Catterall, W. A. (1998). Structure and function of neuronal Ca2+ channels and their role in neurotransmitter release. *Cell Calcium* **24**, 307–323.

Cohen, A. S. and Goadsby, P. J. (2004). Functional neuroimaging of primary headache disorders. *Current Neurology and Neuroscience Reports* **4**, 105–110.

Colson, N. J., Lea, R. A., Quinlan, S., MacMillan, J., and Griffiths, L. R. (2005). Investigation of hormone receptor genes in migraine. *Neurogenetics* **6**, 17–23.

De Fusco, M., Marconi, R., Silvestri, L., Atorino, L., Rampoldi, L., Morgante, L. et al. (2003). Haploinsufficiency of ATP1A2 encoding the Na^+/K^+ pump α2 subunit associated with familial hemiplegic migraine type 2. *Nature Genetics* **33**, 192–196.

del Bene, E., Poggonioni, M., and de Tommasi, F. (1994). Video assessment of yawning induced by sublingual apomorphine in migraine. *Headache* **34**, 536–538.

Dichgans, M., Freilinger, T., Eckstein, G., Babini, E., Lorenz-Depiereux, B., Biskup, S. et al. (2005). Mutation in the neuronal voltage-gated sodium channel *SCN1A* causes familial hemiplegic migraine. *The Lancet* **366**, 371–377.

Dimitriadou, V., Buzzi, M. G., Moskowitz, M. A., and Theoharides, T. C. (1991). Trigeminal sensory fiber stimulation induces morphological changes reflecting secretion in rat dura mater mast cells. *Neuroscience* **44**, 97–112.

Dimitriadou, V., Buzzi, M. G., TheoharidTes, T. C., and Moskowitz, M. A. (1992). Ultrastructural evidence for neurogenically mediated changes in blood vessels of the rat dura mater and tongue following antidromic trigeminal stimulation. *Neuroscience* **48**, 187–203.

Domenga, V., Fardoux, P., Lacombe, P., Monet, M., Maciazek, J., Krebs, L. T. et al. (2004). Notch3 is required for arterial identity and maturation of vascular smooth muscle cells. *Genes Dev* **18**, 2730–2735.

Doods, H., Hallermayer, G., Wu, D., Entzeroth, M., Rudolf, K., Engel, W., and Eberlein, W. (2000). Pharmacological profile of BIBN4096BS, the first selective small molecule CGRP antagonist. *British Journal of Pharmacology* **129**, 420–423.

Dove, L. S., Abbott, L. C., and Griffit, W. H. (1998). Whole-cell and single-channel analysis of P-type calcium currents in cerebellar purkinje cells of leaner mutant mice. *Journal of Neuroscience* **18**, 7687–7699.

Ducros, A., Denier, C., Joutel, A., Cecillon, M., Lescoat, C., Vahedi, K. et al. (2001). The clinical spectrum of familial hemiplegic migraine associated with mutations in a neuronal calcium channel. *New England Journal of Medicine* **345**, 17–24.

Durham, P. L. and Russo, A. F. (1999). Regulation of calcitonin gene-related peptide secretion by a serotonergic antimigraine drug. *Journal of Neuroscience* **19**, 3423–3429.

Durham, P. L., Sharma, R. V., and Russo, A. F. (1997). Repression of the calcitonin gene-related peptide promoter by 5-HT1 receptor activation. *Journal of Neuroscience* **17**, 9545–9553.

Ebrahim, I. O., Howard, R. S., Kopelman, M. D., Sharief, M. K., and Williams, A. J. (2002). The hypocretin/orexin system. *Journal of the Royal Society of Medicine* **95**, 227–230.

Edvinsson, L., Ekman, R., Jansen, I., McCulloch, J., and Uddman, R. (1987). Calcitonin gene-related peptide and cerebral blood vessels: Distribution and vasomotor effects. *Journal of Cerebral Blood Flow and Metabolism* **7**, 720–728.

Edvinsson, L., Rosendahl-Helgesen, S., and Uddman, R. (1983). Substance P: Localization, concentration and release in cerebral arteries, choroid plexus and dura mater. *Cell and Tissue Research* **234**, 1–7.

Fanciullacci, M., Alessandri, M., Figini, M., Geppetti, P., and Michelacci, S. (1995). Increase in plasma calcitonin gene-related peptide from extracerebral circulation during nitroglycerin-induced cluster headache attack. *Pain* **60**, 119–123.

Ferrari, M. D. (1998). Migraine. *The Lancet* **351**, 1043–1051.

Ferrari, M. D. and Goadsby, P. J. (2006). Migraine as a cerebral ionopathy with abnormal central sensory processing. In S. Gilman and T. Pedley, Eds., *Neurobiology of disease*, in press. Elsevier, New York.

Fleetwood-Walker, S. M., Hope, P. J., and Mitchell, R. (1988). Antinociceptive actions of descending dopaminergic tracts on cat and rat dorsal horn somatosensory neurones. *J Physiol* **399**, 335–348.

Fletcher, C. F., Tottene, A., Lennon, V. A., Wilson, S. M., Dubel, S. J., Paylor, R. et al. (2001). Dystonia and cerebellar atrophy in CACNA1A null mice lacking P/Q calcium channel activity. *FASEB J* **15**, 1288–1290.

Friberg, L., Olesen, J., Iversen, H. K., and Sperling, B. (1991). Migraine pain associated with middle cerebral artery dilatation: Reversal by sumatriptan. *Lancet* **338**, 13–17.

Gallai, V., Sarchielli, P., Floridi, A., Franceschini, M., Codini, M., Trequattrini, A., and Palumbo, R. (1995). Vasoactive peptide levels in the plasma of young migraine patients with and without aura assessed both interictally and ictally. *Cephalalgia* **15**, 384–390.

Gardner, K., Barmada, M., Ptacek, L. J., and Hoffman, E. P. (1997). A new locus for hemiplegic migraine maps to chromosome 1q31. *Neurology* **49**, 1231–1238.

Gervil, M., Ulrich, V., Kaprio, J., Olesen, J., and Russell, M. B. (1999). The relative role of genetic and environmental factors in migraine without aura. *Neurology* **53**, 995–999.

Giffin, N. J., Lipton, R. B., Silberstein, S. D., Tvedskov, J. F., Olesen, J., and Goadsby, P. J. (2005). The migraine postdrome: An electronic diary study. *Cephalalgia* **25**, 958.

Giffin, N. J., Ruggiero, L., Lipton, R. B., Silberstein, S., Tvedskov, J. F., Olesen, J. et al. (2003). Premonitory symptoms in migraine: An electronic diary study. *Neurology* **60**, 935–940.

Goadsby, P. J. (2001). Migraine, aura and cortical spreading depression: Why are we still talking about it? *Annals of Neurology* **49**, 4–6.

Goadsby, P. J. (2002). Neurovascular headache and a midbrain vascular malformation—Evidence for a role of the brainstem in chronic migraine. *Cephalalgia* **22**, 107–111.

Goadsby, P. J. (2004). Prejunctional and presynaptic trigeminovascular targets: What preclinical evidence is there? *Headache Currents* **1**, 1–6.

Goadsby, P. J. (2005a). Calcitonin gene-related peptide antagonists as treatments of migraine and other primary headaches. *Drugs* **65**, 2557–2567.

Goadsby, P. J. (2005b). Can we develop neurally-acting drugs for the treatment of migraine? *Nature Reviews Drug Discovery* **4**, 741–750.

Goadsby, P. J. (2005c). Migraine pathophysiology. *Headache* **45**, S14–S24.

Goadsby, P. J. (2005d). New targets in the acute treatment of headache. *Current Opinion in Neurology* **18**, 283–288.

Goadsby, P. J. and Edvinsson, L. (1994). Human *in vivo* evidence for trigeminovascular activation in cluster headache. *Brain* **117**, 427–434.

Goadsby, P. J. and Edvinsson, L. (1996). Neuropeptide changes in a case of chronic paroxysmal hemicrania—Evidence for trigemino-parasympathetic activation. *Cephalalgia* **16**, 448–450.

Goadsby, P. J., Edvinsson, L., and Ekman, R. (1988). Release of vasoactive peptides in the extracerebral circulation of man and the cat during activation of the trigeminovascular system. *Annals of Neurology* **23**, 193–196.

Goadsby, P. J., Edvinsson, L., and Ekman, R. (1990). Vasoactive peptide release in the extracerebral circulation of humans during migraine headache. *Annals of Neurology* **28**, 183–187.

Goadsby, P. J. and Hoskin, K. L. (1997). The distribution of trigeminovascular afferents in the nonhuman primate brain *Macaca nemestrina*: A c-fos immunocytochemical study. *Journal of Anatomy* **190**, 367–375.

Goadsby, P. J., Hoskin, K. L., and Knight, Y. E. (1997). Stimulation of the greater occipital nerve increases metabolic activity in the trigeminal nucleus caudalis and cervical dorsal horn of the cat. *Pain* **73**, 23–28.

Goadsby, P. J., Lambert, G. A., and Lance, J. W. (1982). Differential effects on the internal and external carotid circulation of the monkey evoked by locus coeruleus stimulation. *Brain Research* **249**, 247–254.

Goadsby, P. J., Lipton, R. B., and Ferrari, M. D. (2002). Migraine-current understanding and treatment. *New England Journal of Medicine* **346**, 257–270.

Goadsby, P. J., Piper, R. D., Lambert, G. A., and Lance, J. W. (1985a). The effect of activation of the nucleus raphe dorsalis (DRN) on carotid blood flow. I: The Monkey. *American Journal of Physiology* **248**, R257–R262.

Goadsby, P. J., Piper, R. D., Lambert, G. A., and Lance, J. W. (1985b). The effect of activation of the nucleus raphe dorsalis (DRN) on carotid blood flow. II: The Cat. *American Journal of Physiology* **248**, R263–R269.

Goadsby, P. J. and Zagami, A. S. (1991). Stimulation of the superior sagittal sinus increases metabolic activity and blood flow in certain regions of the brainstem and upper cervical spinal cord of the cat. *Brain* **114**, 1001–1011.

Grant, A. D., Pinter, E., Salmon, A. M., and Brain, S. D. (2005). An examination of neurogenic mechanisms involved in mustard oil-induced inflammation in the mouse. *Eur J Pharmacol* **507**, 273–280.

Guida, S., Trettel, F., Pagnutti, S., Mantuano, E., Tottene, A., Veneziano, L. et al. (2001). Complete loss of P/Q calcium channel activity caused by a CACNA1A missense mutation carried by patients with episodic ataxia type 2. *Am J Hum Genet* **68**, 759–764.

Haan, J., Kors, E. E., Vanmolkot, K. R., van den Maagdenberg, A. M., Frants, R. R., and Ferrari, M. D. (2005). Migraine genetics: an update. *Curr Pain Headache Rep* **9**, 213–220.

Hadjikhani, N., Sanchez del Rio, M., Wu, O., Schwartz, D., Bakker, D., Fischl, B. et al. (2001). Mechanisms of migraine aura revealed by functional MRI in human visual cortex. *Proceedings of the National Academy of Sciences (USA)* **98**, 4687–4692.

Haerter, K., Ayata, C., and Moskowitz, M. A. (2005). Cortical spreading depression: A model for understanding migraine biology and future drug targets. *Headache Currents* **2**, 97–103.

Hans, M., Luvisetto, S., Williams, M. E., Spagnolo, M., Urrutia, A., Tottene, A. et al. (1999). Functional consequences of mutations in the human alpha(1A) calcium channel subunit linked to familial hemiplegic migraine. *Journal of Neuroscience* **19**, 1610–1619.

Haut, S. R., Bigal, M. E., and Lipton, R. B. (2006). Chronic disorders with episodic manifestations: Focus on epilepsy and migraine. *Lancet Neurol* **5**, 148–157.

Headache Classification Committee of The International Headache Society. (2004). The International Classification of Headache Disorders (second edition). *Cephalalgia* **24**, 1–160.

Hirschhorn, J. N. and Daly, M. J. (2005). Genome-wide association studies for common diseases and complex traits. *Nat Rev Genet* **6**, 95–108.

Honkasalo, M.-L., Kaprio, J., Winter, T., Heikkilä, K., Sillanpää, M., and Koskenvuo, M. (1995). Migraine and concomitant symptoms among 8167 adult twin pairs. *Headache* **35**, 70–78.

Hoskin, K. L., Bulmer, D. C. E., Lasalandra, M., Jonkman, A., and Goadsby, P. J. (2001). Fos expression in the midbrain periaqueductal grey after trigeminovascular stimulation. *Journal of Anatomy* **197**, 29–35.

Hoskin, K. L., Kaube, H., and Goadsby, P. J. (1996). Central activation of the trigeminovascular pathway in the cat is inhibited by dihydroergotamine. A c-Fos and electrophysiology study. *Brain* **119**, 249–256.

Hoskin, K. L., Zagami, A., and Goadsby, P. J. (1999). Stimulation of the middle meningeal artery leads to Fos expression in the trigeminocervical nucleus: A comparative study of monkey and cat. *Journal of Anatomy* **194**, 579–588.

Hottenga, J. J., Vanmolkot, K. R., Kors, E. E., Kheradmand Kia, S., de Jong, P. T., Haan, J. et al. (2005). The 3p21.1-p21.3 hereditary vascular retinopathy locus increases the risk for Raynaud's phenomenon and migraine. *Cephalalgia* **25**, 1168–1172.

Ikeda, K., Onaka, T., Yamakado, M., Nakai, J., Ishikawa, T. O., Taketo, M. M., and Kawakami, K. (2003). Degeneration of the amygdala/piriform cortex and enhanced fear/anxiety behaviors in sodium pump alpha2 subunit (Atp1a2)-deficient mice. *J Neurosci* **23**, 4667–4676.

Ikeda, K., Onimaru, H., Yamada, J., Inoue, K., Ueno, S., Onaka, T. et al. (2004). Malfunction of respiratory-related neuronal activity in Na+, K+-ATPase alpha2 subunit-deficient mice is attributable to abnormal Cl- homeostasis in brainstem neurons. *J Neurosci* **24**, 10693–10701.

Imbrici, P., Jaffe, S. L., Eunson, L. H., Davies, N. P., Herd, C., Robertson, R. et al. (2004). Dysfunction of the brain calcium channel $Ca_V2.1$ in absence epilepsy and episodic ataxia. *Brain* **127**, 2682–2692.

Iversen, H. (2001). Human migraine models. *Cephalalgia* **21**, 781–785.

James, P. F., Grupp, I. L., Grupp, G., Woo, A. L., Askew, G. R., Croyle, M. L. et al. (1999). Identification of a specific role for the Na,K-ATPase alpha 2 isoform as a regulator of calcium in the heart. *Mol Cell* **3**, 555–563.

Jen, J., Kim, G. W., and Baloh, R. W. (2004). Clinical spectrum of episodic ataxia type 2. *Neurology* **62**, 17–22.

Jen, J., Wan, J., Graves, M., Yu, H., Mock, A. F., Coulin, C. J. et al. (2001). Loss-of-function EA2 mutations are associated with impaired neuromuscular transmission. *Neurology* **57**, 1843–1848.

Jeng, C. J., Chen, Y. T., Chen, Y. W., and Tang, C. Y. (2006). Dominant-negative effects of human P/Q-type Ca2+ channel mutations associated with episodic ataxia type 2. *Am J Physiol Cell Physiol* **290**, 1209–1220.

Joutel, A., Corpechot, C., Ducros, A., Vahedi, K., Chabriat, H., Mouton, P. et al. (1996). Notch3 mutations in CADASIL, a hereditary adult-onset condition causing stroke and dementia. *Nature* **383**, 707–710.

Jouvenceau, A., Eunson, L. H., Spauschus, A., Ramesh, V., Zuberi, S. M., Kullmann, D. M., and Hanna, M. G. (2001). Human epilepsy associated with dysfunction of the brain P/Q-type calcium channel. *Lancet* **358**, 801–807.

Juhasz, G., Zsombok, T., Jakab, B., Nemeth, J., Szolcsanyi, J., and Bagdy, G. (2005). Sumatriptan causes parallel decrease in plasma calcitonin gene-related peptide (CGRP) concentration and migraine headache during nitroglycerin induced migraine attack. *Cephalalgia* **25**, 179–183.

Juhasz, G., Zsombok, T., Modos, E. A., Olajos, S., Jakab, B., Nemeth, J. et al. (2003). NO-induced migraine attack: Strong increase in plasma calcitonin gene-related peptide (CGRP) concentration and negative correlation with platelet serotonin release. *Pain* **106**, 461–470.

Jun, K., Piedras-Renteria, E. S., Smith, S. M., Wheeler, D. B., Lee, S. B., Lee, T. G. et al. (1999). Ablation of P/Q-Type Ca(2+) channel currents, altered synaptic transmission, and progressive ataxia in mice lacking the alpha(1A)- subunit. *Proceedings of the National Academy of Science USA* **96**, 15245–15250.

Kallela, M., Wessman, M., Havanka, H., Palotie, A., and Farkkila, M. (2001). Familial migraine with and without aura: Clinical charecteristics and co-occurence. *European Journal of Neurology* **8**, 441–449.

Kanai, K., Hirose, S., Oguni, H., Fukuma, G., Shirasaka, Y., Miyajima, T. et al. (2004). Effect of localization of missense mutations in SCN1A on epilepsy phenotype severity. *Neurology* **63**, 329–334.

Kara, I., Sazci, A., Ergul, E., Kaya, G., and Kilic, G. (2003). Association of the C677T and A1298C polymorphisms in the 5,10 methylenetetrahydrofolate reductase gene in patients with migraine risk. *Brain Res Mol Brain Res* **111**, 84–90.

Kaube, H. and Giffin, N. J. (2002). The electrophysiology of migraine. *Current Opinion in Neurology* **15**, 303–309.

Kaube, H., Herzog, J., Kaufer, T., Dichgans, M., and Diener, H. C. (2000). Aura in some patients with familial hemiplegic migraine can be stopped by intranasal ketamine. *Neurology* **55**, 139–141.

Kaube, H., Keay, K. A., Hoskin, K. L., Bandler, R., and Goadsby, P. J. (1993). Expression of c-*Fos*-like immunoreactivity in the caudal medulla and upper cervical cord following stimulation of the superior sagittal sinus in the cat. *Brain Research* **629**, 95–102.

Kelman, L. (2004). The premonitory symptoms (prodrome): A tertiary care study of 893 migraineurs. *Headache* **44**, 865–872.

Kelman, L. (2006). The postdrome of the acute migraine attack. *Cephalalgia* **26**, 214–220.

Knight, Y. E., Bartsch, T., and Goadsby, P. J. (2003). Trigeminal antinociception induced by bicuculline in the periaqueductal grey (PAG) is not affected by PAG P/Q-type calcium channel blockade in rat. *Neuroscience Letters* **336**, 113–116.

Knight, Y. E., Bartsch, T., Kaube, H., and Goadsby, P. J. (2002). P/Q-type calcium channel blockade in the PAG facilitates trigeminal nociception: A functional genetic link for migraine? *Journal of Neuroscience* **22**, 1–6.

Knight, Y. E. and Goadsby, P. J. (2001). The periaqueductal gray matter modulates trigeminovascular input: A role in migraine? *Neuroscience* **106**, 793–800.

Kors, E. E., Terwindt, G. M., Vermeulen, F. L. M. G., Fitzsimons, R. B., Jardine, P. E., Heywood, P. et al. (2001). Delayed cerebral edema and fatal coma after minor head trauma: Role of CACNA1A calcium channel subunit gene and relationship with familial hemiplegic migraine. *Annals of Neurology* **49**, 753–760.

Kors, E. E., Vanmolkot, K. R., Haan, J., Frants, R. R., van den Maagdenberg, A. M., and Ferrari, M. D. (2004). Recent findings in headache genetics. *Curr Opin Neurol* **17**, 283–288.

Kowa, H., Fusayasu, E., Ijiri, T., Ishizaki, K., Yasui, K., Nakaso, K. et al. (2005). Association of the insertion/deletion polymorphism of the angiotensin I-converting enzyme gene in patients of migraine with aura. *Neurosci Lett* **374**, 129–131.

Kowa, H., Yasui, K., Takeshima, T., Urakami, K., Sakai, F., and Nakashima, K. (2000). The homozygous C677T mutation in the methylenetetrahydrofolate reductase gene is a genetic risk factor for migraine. *Am J Med Genet* **96**, 762–764.

Kraus, R. L., Sinnegger, M. J., Glossmann, H., Hering, S., and Striessnig, J. (1998). Familial hemiplegic migraine mutations change alpha(1A) Ca^{2+} channel kinetics. *Journal of Biological Chemistry* **273**, 5586–5590.

Kraus, R. L., Sinnegger, M. J., Koschak, A., Glossmann, H., Stenirri, S., Carrera, P., and Striessnig, J. (2000). Three new familial hemiplegic migraine mutants affect P/Q-type Ca(2+) channel kinetics. *Journal of Biological Chemistry* **275**, 9239–9243.

Kruit, M. C., van Buchem, M. A., Hofman, P. A., Bakkers, J. T., Terwindt, G. M., Ferrari, M. D., and Launer, L. J. (2004). Migraine as a risk factor for subclinical brain lesions. *Journal of the American Medical Association* **291**, 427–434.

Lance, J. W. and Goadsby, P. J. (2005). *Mechanism and management of headache*. Elsevier, New York.

Lashley, K. S. (1941). Patterns of cerebral integration indicated by the scotomas of migraine. *Archives of Neurology and Psychiatry* **46**, 331–339.

Lauritzen, M. (1994). Pathophysiology of the migraine aura. The spreading depression theory. *Brain* **117**, 199–210.

Lea, R. A., Nyholt, D. R., Curtain, R. P., Ovcaric, M., Sciascia, R., Bellis, C. et al. (2005a). A genome-wide scan provides evidence for loci influencing a severe heritable form of common migraine. *Neurogenetics* **6**, 67–72.

Lea, R. A., Ovcaric, M., Sundholm, J., MacMillan, J., and Griffiths, L. R. (2004). The methylenetetrahydrofolate reductase gene variant C677T influences susceptibility to migraine with aura. *BMC Med* **2**, 3.

Lea, R. A., Ovcaric, M., Sundholm, J., Solyom, L., Macmillan, J., and Griffiths, L. R. (2005b). Genetic variants of angiotensin converting enzyme and methylenetetrahydrofolate reductase may act in combination to increase migraine susceptibility. *Brain Res Mol Brain Res* **136**, 112–117.

Levant, B. and McCarson, K. E. (2001). D(3) dopamine receptors in rat spinal cord: implications for sensory and motor function. *Neurosci Lett* **303**, 9–12.

Levey, A. I., Hersch, S. M., Rye, D. B., Sunahara, R. K., Niznik, H. B., Kitt, C. A. et al. (1993). Localization of D1 and D2 dopamine receptors in brain with subtype-specific antibodies. *Proc Natl Acad Sci U S A* **90**, 8861–8865.

Levy, M. J., Matharu, M. S., and Goadsby, P. J. (2003). Prolactinomas, dopamine agonist and headache: Two case reports. *European Journal of Neurology* **10**, 169–174.

Limmroth, V., May, A., Auerbach, P., Wosnitza, G., Eppe, T., and Diener, H. C. (1996). Changes in cerebral blood flow velocity after treatment with sumatriptan or placebo and implications for the pathophysiology of migraine. *Journal of Neurological Sciences* **138**, 60–65.

Lin, J. J., Wang, P. J., Chen, C. H., Yueh, K. C., Lin, S. Z., and Harn, H. J. (2005). Homozygous deletion genotype of angiotensin converting enzyme confers protection against migraine in man. *Acta Neurol Taiwan* **14**, 120–125.

Lipton, R. B., Stewart, W. F., Diamond, S., Diamond, M. L., and Reed, M. (2001). Prevalence and burden of migraine in the United States: Data from the American Migraine Study II. *Headache* **41**, 646–657.

Lorenzon, N. M., Lutz, C. M., Frankel, W. N., and Beam, K. G. (1998). Altered calcium channel currents in purkinje cells of neurological mutant mouse leaner. *Journal of Neuroscience* **18**, 4482–4489.

Ludvigsson, P., Hesdorffer, D., Olafsson, E., Kjartansson, O., and Hauser, W. A. (2006). Migraine with aura is a risk factor for unprovoked seizures in children. *Ann Neurol* **59**, 210–213.

Lundkvist, J., Zhu, S., Hansson, E. M., Schweinhardt, P., Miao, Q., Beatus, P. et al. (2005). Mice carrying a R142C Notch 3 knock-in mutation do not develop a CADASIL-like phenotype. *Genesis* **41**, 13–22.

Maneesi, S., Akerman, S., Lasalandra, M. P., Classey, J. D., and Goadsby, P. J. (2004). Electron microsopic demonstration of pre- and postsynaptic 5-HT_{1D} and 5-HT_{1F} receptor immunoreactivity (IR) in the rat trigeminocervical complex (TCC) new therapeutic possibilities for the triptans. *Cephalalgia* **24**, 148.

Marconi, R., De Fusco, M., Aridon, P., Plewnia, K., Rossi, M., Carapelli, S. et al. (2003). Familial hemiplegic migraine type 2 is linked to 0.9Mb region on chromosome 1q23. *Ann Neurol* **53**, 376–381.

Matharu, M. S., Bartsch, T., Ward, N., Frackowiak, R. S. J., Weiner, R. L., and Goadsby, P. J. (2004). Central neuromodulation in chronic migraine patients with suboccipital stimulators: A PET study. *Brain* **127**, 220–230.

Matharu, M. S. and Goadsby, P. J. (2001). Post-traumatic chronic paroxysmal hemicrania (CPH) with aura. *Neurology* **56**, 273–275.

Matharu, M. S. and Goadsby, P. J. (2005). Functional brain imaging in hemicrania continua: Implications for nosology and pathophysiology. *Current Pain and Headache Reports* **9**, 281–288.

May, A. and Goadsby, P. J. (1999). The trigeminovascular system in humans: Pathophysiological implications for primary headache syndromes of the neural influences on the cerebral circulation. *Journal of Cerebral Blood Flow and Metabolism* **19**, 115–127.

May, A. and Goadsby, P. J. (2001). Substance P receptor antagonists in the therapy of migraine. *Expert Opinion in Investigational Drugs* **10**, 1–6.

May, A., Shepheard, S., Wessing, A., Hargreaves, R. J., Goadsby, P. J., and Diener, H. C. (1998). Retinal plasma extravasation can be evoked by trigeminal stimulation in rat but does not occur during migraine attacks. *Brain* **121**, 1231–1237.

McNaughton, F. L. (1938). The innervation of the intracranial blood vessels and dural sinuses. *Proceedings of the Association for Research in Nervous & Mental Diseases* **18**, 178–200.

McNaughton, F. L. (1966). The innervation of the intracranial blood vessels and the dural sinuses. In S. Cobb, A. M. Frantz, W. Penfield, and H. A. Riley, Eds., *The circulation of the brain and spinal cord*, 178–200. Hafner Publishing Co. Inc, New York.

Meisler, M. H. and Kearney, J. A. (2005). Sodium channel mutations in epilepsy and other neurological disorders. *J Clin Invest* **115**, 2010–2017.

Menken, M., Munsat, T. L., and Toole, J. F. (2000). The global burden of disease study—Implications for neurology. *Archives of Neurology* **57**, 418–420.

Mogilnicka, E. and Klimek, V. (1977). Drugs affecting dopamine neurons and yawning behavior. *Pharmacol Biochem Behav* **7**, 303–305.

Montagna, P., Pierangeli, G., Cevoli, S., Mochi, M., and Cortelli, P. (2005). Pharmacogenetics of headache treatment. *Neurological Sciences* **26**, S143–S147.

Mori, Y., Wakamori, M., Oda, S., Fletcher, C. F., Sekiguchi, N., Mori, E. et al. (2000). Reduced voltage sensitivity of activation of P/Q-type Ca^{2+} channels is associated with the ataxic mouse mutation rolling Nagoya (tg(rol)). *Journal of Neuroscience* **20**, 5654–5662.

Moskowitz, M. A., Bolay, H., and Dalkara, T. (2004). Deciphering migraine mechanisms: Clues from familial hemiplegic migraine genotypes. *Ann Neurol* **55**, 276–280.

Moskowitz, M. A. and Cutrer, F. M. (1993). SUMATRIPTAN: A receptor-targeted treatment for migraine. *Annual Review of Medicine* **44**, 145–154.

Mulder, E. J., Van Baal, C., Gaist, D., Kallela, M., Kaprio, J., Svensson, D. A. et al. (2003). Genetic and environmental influences on migraine: A twin study across six countries. *Twin Research* **6**, 422–431.

Mullner, C., Broos, L. A., van den Maagdenberg, A. M., and Striessnig, J. (2004). Familial hemiplegic migraine type 1 mutations K1336E, W1684R, and V1696I alter Cav2.1 Ca2+ channel gating: Evidence for beta-subunit isoform-specific effects. *J Biol Chem* **279**, 51844–51850.

Niebur, E., Hsiao, S. S., and Johnson, K. O. (2002). Synchrony: A neural mechanism for attentional selection? *Current Opinion in Neurobiology* **12**, 190–194.

Nyholt, D. R., Dawkins, J. L., Brimage, P. J., Goadsby, P. J., Nicholson, G. A., and Griffiths, L. R. (1998). Evidence for an X-linked genetic component in familial typical migraine. *Human Molecular Genetics* **7**, 459–463.

Nyholt, D. R., Gillespie, N. G., Heath, A. C., Merikangas, K. R., Duffy, D. L., and Martin, N. G. (2004). Latent class and genetic analysis does not support migraine with aura and migraine without aura as separate entities. *Genet Epidemiol* **26**, 231–244.

Nyholt, D. R., Morley, K. I., Ferreira, M. A., Medland, S. E., Boomsma, D. I., Heath, A. C. et al. (2005). Genomewide significant linkage to migrainous headache on chromosome 5q21. *Am J Hum Genet* **77**, 500–512.

Obermann, M., Gizewski, E. R., Limmroth, V., Diener, H.-C., and Katsarava, Z. (2006). Symptomatic migraine and pontine vascular malformation: Evidence for a key role of the brainstem in the pathophysiology of chronic migraine. *Cephalalgia* **26**, in press.

Olesen, J., Diener, H.-C., Husstedt, I.-W., Goadsby, P. J., Hall, D., Meier, U. et al. (2004). Calcitonin gene-related peptide (CGRP) receptor antagonist BIBN4096BS is effective in the treatment of migraine attacks. *New England Journal of Medicine* **350**, 1104–1110.

Ophoff, R. A., DeYoung, J., Service, S. K., Joosse, M., Caffo, N. A., Sandkuijl, L. A. et al. (2001). Hereditary vascular retinopathy, cerebroretinal vasculopathy, and hereditary endotheliopathy with retinopathy, nephropathy and stroke map to a single locus on chromosome 3p21.1-p21.3. *Am J Hum Genet* **69**, 447–453.

Ophoff, R. A., Terwindt, G. M., Vergouwe, M. N., van Eijk, R., Oefner, P. J., Hoffman, S. M. G. et al. (1996). Familial hemiplegic migraine and episodic ataxia type-2 are caused by mutations in the Ca^{2+} channel gene CACNL1A4. *Cell* **87**, 543–552.

Oterino, A., Pascual, J., Ruiz de Alegria, C., Valle, N., Castillo, J., Bravo, Y. et al. (2006). Association of migraine and ESR1 G325C polymorphism. *Neuroreport* **17**, 61–64.

Oterino, A., Valle, N., Bravo, Y., Munoz, P., Sanchez-Velasco, P., Ruiz-Alegria, C. et al. (2004). MTHFR T677 homozygosis influences the presence of aura in migraineurs. *Cephalalgia* **24**, 491–494.

Paterna, S., Di Pasquale, P., D'Angelo, A., Seidita, G., Tuttolomondo, A., Cardinale, A. et al. (2000). Angiotensin-converting enzyme gene deletion polymorphism determines an increase in frequency of migraine attacks in patients suffering from migraine without aura. *Eur Neurol* **43**, 133–136.

Peres, M. F. P., Siow, H. C., and Rozen, T. D. (2002). Hemicrania continua with aura. *Cephalalgia* **22**, 246–248.

Peroutka, S. J. (2005). Neurogenic inflammation and migraine: Implications for therapeutics. *Molecular Interventions* **5**, 306–313.

Pietrobon, D. (2005). Function and dysfunction of synaptic calcium channels: Insights from mouse models. *Curr Opin Neurobiol* **15**, 257–265.

Piper, R. D., Lambert, G. A., and Duckworth, J. W. (1991). Cortical blood flow changes during spreading depression in cats. *American Journal of Physiology* **261**, H96–H102.

Pozo-Rosich, P. and Oshinsky, M. (2005). Effect of dihydroergotamine (DHE) on central sensitisation of neurons in the trigeminal nucleus caudalis. *Neurology* **64**, A151.

Protais, P., Dubuc, I., and Costentin, J. (1983). Pharmacological characteristics of dopamine receptors involved in the dual effect of dopamine agonists on yawning behaviour in rats. *Eur J Pharmacol* **94**, 271–280.

Radat, F. and Swendsen, J. (2005). Psychiatric comorbidity in migraine: A review. *Cephalalgia* **25**, 165–178.

Rainero, I., Grimaldi, L. M., Salani, G., Valfre, W., Rivoiro, C., Savi, L., and Pinessi, L. (2004). Association between the tumor necrosis factor-alpha -08 G/A gene polymorphism and migraine. *Neurology* **62**, 141–143.

Raskin, N. H., Hosobuchi, Y., and Lamb, S. (1987). Headache may arise from perturbation of brain. *Headache* **27**, 416–420.

Riant, F., De Fusco, M., Aridon, P., Ducros, A., Ploton, C., Marchelli, F. et al. (2005). ATP1A2 mutations in 11 families with familial hemiplegic migraine. *Hum Mutat* **26**, 281.

Ruchoux, M. M., Domenga, V., Brulin, P., Maciazek, J., Limol, S., Tournier-Lasserve, E., and Joutel, A. (2003). Transgenic mice expressing mutant Notch3 develop vascular alterations characteristic of cerebral autosomal dominant arteriopathy with subcortical infarcts and leukoencephalopathy. *Am J Pathol* **162**, 329–342.

Russell, M. B. and Olesen, J. (1995). Increased familial risk and evidence of genetic factor in migraine. *British Medical Journal* **311**, 541–544.

Russell, M. B. and Olesen, J. (1996). A nosographic analysis of the migraine aura in a general population. *Brain* **119**, 355–361.

Russell, M. B., Ulrich, V., Gervil, M., and Olesen, J. (2002). Migraine without aura and migraine with aura are distinct disorders. A population-based twin survey. *Headache* **42**, 332–336.

Russo, L., Mariotti, P., Sangiorgi, E., Giordano, T., Ricci, I., Lupi, F. et al. (2005). A new susceptibility locus for migraine with aura in the 15q11-q13 genomic region containing three GABA-A receptor genes. *Am J Hum Genet* **76**, 327–333.

Sandor, P. S., Mascia, A., Seidel, L., de Pasqua, V., and Schoenen, J. (2001). Subclinical cerebellar impairment in the common types of migraine: A three-dimensional analysis of reaching movements. *Annals of Neurology* **49**, 668–672.

Scher, A. I., Stewart, W. F., and Lipton, R. B. (1999). Migraine and headache: A meta-analytic approach. In I. K. Crombie, Ed., *Epidemiology of Pain*, 159–170. IASP Press, Seattle, Washington.

Scher, A. I., Terwindt, G. M., Verschuren, W. M., Kruit, M. C., Blom, H. J., Kowa, H. et al. (2006). Migraine and MTHFR C677T genotype in a population-based sample. *Ann Neurol* **59**, 372–375.

Schoenen, J., Ambrosini, A., Sandor, P. S., and Maertens de Noordhout, A. (2003). Evoked potentials and transcranial magnetic stimulation in migraine: Published data and viewpoint on their pathophysiologic significance. *Clinical Neurophysiology* **114**, 955–972.

Segall, L., Mezzetti, A., Scanzano, R., Gargus, J. J., Purisima, E., and Blostein, R. (2005). Alterations in the alpha 2 isoform of Na,K-ATPase associated with familial hemiplegic migraine type 2. *Proc Natl Acad Sci U S A* **102**, 11106–11111.

Selby, G. and Lance, J. W. (1960). Observations on 500 cases of migraine and allied vascular headache. *Journal of Neurology, Neurosurgery and Psychiatry* **23**, 23–32.

Serra, G., Collu, M., and Gessa, G. L. (1986). Dopamine receptors mediating yawning: Are they autoreceptors? *European Journal of Pharmacology* **120**, 187–192.

Shields, K. G. and Goadsby, P. J. (2005). Propranolol modulates trigeminovascular responses in thalamic ventroposteromedial nucleus: A role in migraine? *Brain* **128**, 86–97.

Shields, K. G. and Goadsby, P. J. (2006). Serotonin receptors modulate trigeminovascular responses in ventroposteromedial nucleus of thalamus: A migraine target? *Neurobiology of Disease*, in press.

Shields, K. G., Kaube, H., and Goadsby, P. J. (2003). GABA receptors modulate trigeminovascular nociceptive transmission in the ventroposteromedial (VPM) thalamic nucleus of the rat. *Cephalalgia* **23**, 728.

Shields, K. G., Storer, R. J., Akerman, S., and Goadsby, P. J. (2005). Calcium channels modulate nociceptive transmission in the trigeminal nucleus of the cat. *Neuroscience* **135**, 203–212.

Silberstein, S. D., Lipton, R. B., and Goadsby, P. J. (2002). *Headache in Clinical Practice*. Martin Dunitz, London.

Silberstein, S. D., Niknam, R., Rozen, T. D., and Young, W. B. (2000). Cluster headache with aura. *Neurology* **54**, 219–221.

Skagerberg, G., Bjorklund, A., Lindvall, O., and Schmidt, R. H. (1982). Origin and termination of the diencephalo-spinal dopamine system in the rat. *Brain Res Bull* **9**, 237–244.

Somjen, G. G. (2001). Mechanisms of spreading depression and hypoxic spreading depression-like depolarization. *Physiol. Rev.* **81**, 1065–1096.

Soragna, D., Vettori, A., Carraro, G., Marchioni, E., Vazza, G., Bellini, S., R., T. et al. (2003). A locus for migraine without aura maps on chromosome 14q21.2-q22.3. *Am J Hum Genet* **72**, 161–167.

Spacey, S. D., Hildebrand, M. E., Materek, L. A., Bird, T. D., and Snutch, T. P. (2004). Functional implications of a novel EA2 mutation in the P/Q-type calcium channel. *Ann Neurol* **56**, 213–220.

Spadaro, M., Ursu, S., Lehmann-Horn, F., Veneziano, L., Antonini, G., Giunti, P. et al. (2004). A G301R Na+/K+-ATPase mutation causes familial hemiplegic migraine type 2 with cerebellar signs. *Neurogenetics* **5**, 177–185.

Stewart, W. F., Bigal, M. E., Kolodner, K., Dowson, A., Liberman, J. N., and Lipton, R. B. (2006). Familial risk of migraine: Variation by proband age at onset and headache severity. *Neurology* **66**, 344–348.

Stewart, W. F., Ricci, J. A., Chee, E., Morganstein, D., and Lipton, R. (2003). Lost productive time and cost due to common pain conditions in the US workforce. *JAMA* **290**, 2443–2454.

Stewart, W. F., Staffa, J., Lipton, R. B., and Ottman, R. (1997). Familial risk of migraine: A population-based study. *Annals of Neurology* **41**, 166–172.

Svensson, D. A., Larsson, B., Waldenlind, E., and Pedersen, N. L. (2003). Shared rearing environment in migraine: Results from twins reared apart and twins reared together. *Headache* **43**, 235–244.

Swoboda, K. J., Kanavakis, E., Xaidara, A., Johnson, J. E., Leppert, M. F., Schlesinger-Massart, M. B. et al. (2004). Alternating hemiplegia of childhood or familial hemiplegic migraine?: A novel ATP1A2 mutation. *Annals of Neurology* **55**, 884–887.

Terwindt, G., Kors, E., Haan, J., Vermeulen, F., Van den Maagdenberg, A., Frants, R., and Ferrari, M. (2002). Mutation analysis of the CACNA1A calcium channel subunit gene in 27 patients with sporadic hemiplegic migraine. *Arch Neurol* **59**, 1016–1018.

Terwindt, G. M., Haan, J., Ophoff, R. A., Groenen, S. M. A., Storimans, C., Lanser, J. B. K. et al. (1998a). Clinical and genetic analysis of a large Dutch family with autosomal dominant vascular retinopathy, migraine and Raynaud's phenomenon. *Brain* **121**, 303–316.

Terwindt, G. M., Kors, E. E., Vein, A. A., Ferrari, M. D., and van Dijk, J. G. (2004). Single-fiber EMG in familial hemiplegic migraine. *Neurology* **63**, 1942–1943.

Terwindt, G. M., Ophoff, R. A., Haan, J., Vergouwe, M. N., van Eijk, R., Frants, R. R., and Ferrari, M. D. (1998b). Variable clinical expression of mutations in the P/Q-type calcium channel gene in familial hemiplegic migraine. *Neurology* **50**, 1105–1110.

Thomsen, L. L., Ostergaard, E., Olesen, J., and Russell, M. B. (2003a). Evidence for a separate type of migraine with aura: Sporadic hemiplegic migraine. *Neurology* **60**, 595–601.

Thomsen, L. L., Ostergaard, E., Romer, S. F., Andersen, I., Eriksen, M. K., Olesen, J., and Russell, M. B. (2003b). Sporadic hemiplegic migraine is an aetiologically heterogeneous disorder. *Cephalalgia* **23**, 921–928.

Todt, U., Dichgans, M., Jurkat-Rott, K., Heinze, A., Zifarelli, G., Koenderink, J. B. et al. (2005). Rare missense variants in ATP1A2 in families with clustering of common forms of migraine. *Hum Mutat* **26**, 315–321.

Tottene, A., Pivotto, F., Fellin, T., Cesetti, T., van den Maagdenberg, A. M., and Pietrobon, D. (2005). Specific kinetic alterations of human CaV2.1 calcium channels produced by mutation S218L causing familial hemiplegic migraine and delayed cerebral edema and coma after minor head trauma. *J Biol Chem* **280**, 17678–17686.

Tottene, A., Tottene, A., Fellin, T., Pagnutti, S., Luvisetto, S., Striessnig, J. et al. (2002). Familial hemiplegic migraine mutations increase Ca^{2+} influx through single human CaV2.1 channels and decrease maximal $Ca_V2.1$ current density in neurons. *Proceedings of the National Academy of Science USA* **99**, 13284–13289.

Tournier-Lasserve, E., Joutel, A., Melki, J., Weissenbach, J., Lathrop, G. M., Chabriat, H. et al. (1993). Cerebral autosomal dominant arteriopathy with subcortical infarcts and leukoencephalopathy maps to chromosome 19p12. *Nature Genetics* **3**, 256–259.

Trabace, S., Brioli, G., Lulli, P., Morellini, M., Giacovazzo, M., Cicciarelli, G., and Martelletti, P. (2002). Tumor necrosis factor gene polymorphism in migraine. *Headache* **42**, 341–345.

Tvedskov, J. F., Lipka, K., Ashina, M., Iversen, H. K., Schifter, S., and Olesen, J. (2005). No increase of calcitonin gene-related peptide in jugular blood during migraine. *Ann Neurol* **58**, 561–568.

Tvedskov, J. F., Thomsen, L. L., Iversen, H. K., Gibson, A., Wiliams, P., and Olesen, J. (2004a). The prophylactic effect of valproate on glyceryl trinitrate induced migraine. *Cephalalgia* **24**, 576–585.

Tvedskov, J. F., Thomsen, L. L., Iversen, H. K., Williams, P., Gibson, A., Jenkins, K. et al. (2004b). The effect of propranolol on glyceryl trinitrate-induced headache and arterial response. *Cephalalgia* **24**, 1076–1087.

Ulrich, V., Gervil, M., Kyvik, K. O., Olesen, J., and Russell, M. B. (1999). Evidence of a genetic factor in migraine with aura: A population based Danish twin study. *Annals of Neurology* **45**, 242–246.

van den Maagdenberg, A. M. J. M., Pietrobon, D., Pizzorusso, T., Kaja, S., Broos, L. A. M., Cesetti, T. et al. (2004). A Cacna1a knock-in migraine mouse model with increased susceptibility to cortical spreading depression. *Neuron* **41**, 701–710.

van Dijken, H., Dijk, J., Voom, P., and Holstege, J. C. (1996). Localization of dopamine D2 receptor in rat spinal cord identified with immunocytochemistry and in situ hybridization. *Eur J Neurosci* **8**, 621–628.

Vanmolkot, K. R., Kors, E. E., Turk, U., Turkdogan, D., Keyser, A., Broos, L. A. et al. (2006). Two de novo mutations in the Na,K-ATPase gene ATP1A2 associated with pure familial hemiplegic migraine. *Eur J Hum Genet* **14**, 555–560.

Vanmolkot, K. R. J., Kors, E. E., Hottenga, J. J., Terwindt, G. M., Haan, J., Hoefnagels, W. A. J. et al. (2003). Novel mutations in the Na^+,K^+-ATPase pump gene ATP1A2 associated with Familial Hemiplegic Migraine and Benign Familial Infantile Convulsions. *Annals of Neurology* **54**, 360–366.

Waelkens, J. (1981). Domperidone in the prevention of complete classical migraine. *British Medical Journal* **284**, 944.

Waelkens, J. (1984). Dopamine blockade with domperidone: Bridge between prophylactic and abortive treatment of migraine? A dose-finding study. *Cephalalgia* **4**, 85–90.

Wakamori, M., Yamazaki, K., Matsunodaira, H., Teramoto, T., Tanaka, I., Niidome, T. et al. (1998). Single tottering mutations responsible for the neuropathic phenotype of the P-Type calcium channel. *Journal of Biological Chemistry* **273**, 34857–34867.

Wan, J., Khanna, R., Sandusky, M., Papazian, D. M., Jen, J. C., and Baloh, R. W. (2005). CACNA1A mutations causing episodic and progressive ataxia alter channel trafficking and kinetics. *Neurology* **64**, 2090–2097.

Wappl, E., Koschak, A., Poteser, M., Sinnegger, M. J., Walter, D., Eberhart, A. et al. (2002). Functional consequences of P/Q-type Ca2+ channel Cav2.1 missense mutations associated with episodic ataxia type 2 and progressive ataxia. *J Biol Chem* **277**, 6960–6966.

Weiller, C., May, A., Limmroth, V., Juptner, M., Kaube, H., Schayck, R. V. et al. (1995). Brain stem activation in spontaneous human migraine attacks. *Nature Medicine* **1**, 658–660.

Welch, K. M., Nagesh, V., Aurora, S., and Gelman, N. (2001). Periaqueductal grey matter dysfunction in migraine: Cause or the burden of illness? *Headache* **41**, 629–637.

Welch, K. M. A. (2005). Brain hyperexcitability: The basis for antiepileptic drugs in migraine prevention. *Headache* **45**, S25–S32.

Wessman, M., Kallela, M., Kaunisto, M. A., Marttila, P., Sobel, E., Hartiala, J. et al. (2002). A susceptibility locus for migraine with aura, on chromosome 4q24. *American Journal of Human Genetics* **70**, 652–662.

Wilkinson, F., Karanovic, O., Ross, E. C., Lillakas, L., and Steinbach, M. J. (2006). Ocular motor measures in migraine with and without aura. *Cephalalgia* **26**, 660–671.

Wolff, H. G. (1963). *Headache and Other Head Pain.* Oxford University Press, New York.

Woods, R. P., Iacoboni, M., and Mazziotta, J. C. (1994). Bilateral spreading cerebral hypoperfusion during spontaneous migraine headache. *New England Journal of Medicine* **331**, 1689–1692.

Woolf, C. J. (1996). Windup and central sensitization are not equivalent. *Pain* **66**, 105–108.

Yamada, K., Tanaka, M., Shibata, K., and Furukawa, T. (1986). Involvement of septal and striatal dopamine D-2 receptors in yawning behavior in rats. *Psychopharmacology (Berl)* **90**, 9–13.

Yue, Q., Jen, J. C., Nelson, S. F., and Baloh, R. W. (1997). Progressive ataxia due to a missense mutation in a calcium-channel gene. *Am J Hum Genet* **61**, 1078–1087.

Zagami, A. S., Goadsby, P. J., and Edvinsson, L. (1990). Stimulation of the superior sagittal sinus in the cat causes release of vasoactive peptides. *Neuropeptides* **16**, 69–75.

Zagami, A. S. and Lambert, G. A. (1990). Stimulation of cranial vessels excites nociceptive neurones in several thalamic nuclei of the cat. *Experimental Brain Research* **81**, 552–566.

Zagami, A. S. and Lambert, G. A. (1991). Craniovascular application of capsaicin activates nociceptive thalamic neurons in the cat. *Neuroscience Letters* **121**, 187–190.

Zhuchenko, O., Bailey, J., Donnen, P., Ashizawa, T., Stockton, D. W., Amos, C. et al. (1997). Autosomal dominant cerebellar ataxia (SCA6) associated with small polyglutamine expansions in the α1A-voltage-dependent calcium channel. *Nature Genetics* **15**, 62–69.

Ziegler, D. K., Hur, Y. M., Bouchard Jr, T. J., Hassanein, R. S., and R., B. (1998). Migraine in twins raised together and apart. *Headache* **38**, 417–422.

29

Myelin Protein Zero and CMT1B: A Tale of Two Phenotypes

John A. Kamholz, Michelle Brucal, Jun Li, and Michael Shy

I. Introduction

The inherited peripheral neuropathies, now collectively called Charcot-Marie-Tooth (CMT) disease, were first described in 1886 by Charcot and Marie in France and independently by Tooth in England (Charcot & Marie, 1886; Tooth, 1886). These neurologists recognized that individuals with this disease shared a similar clinical phenotype, including weakness and atrophy of muscles innervated by the peroneal nerve and a characteristic foot deformity. In addition, they also noted that the disease clustered within families, suggesting that it had a genetic etiology. Dejerine and Sottas (1893) further extended the phenotype of CMT disease to include patients with particularly severe motor and sensory deficits with onset in infancy, and Roussy and Levy (1926) identified cases of CMT in which the patients had both tremor and ataxia.

With the advent of reliable techniques to measure the conduction velocity of human peripheral nerves, Dyck and Lambert (1968) and Harding and Thomas (1980a,b) demonstrated that patients with CMT could be divided into two clinical groups: one with relatively slow nerve conduction velocities and pathological evidence of demyelination, which they called CMT type 1; and a second with relatively normal nerve conduction velocities and pathologic evidence of axonal degeneration, which they designated CMT type 2. Patients in both groups, however, had atrophy, weakness, and sensory loss, predominantly in the distal legs, which had developed during the first two decades of life, and which segregated predominantly as an autosomal dominant trait.

Linkage studies using minor blood group antigen polymorphisms to identify the genetic etiology of CMT were first carried out by Bird and coworkers (1982), who showed that a gene causing CMT type 1 in several large families was linked to the Duffy blood group locus on chromosome 1. Interestingly, linkage studies using DNA polymorphisms (Lupski et al., 1991; Raeymaekers et al., 1991) demonstrated that the most common form of CMT type 1, renamed CMT1A, was caused by a 1.4 Mb duplication on the short arm of chromosome 17, which included a region containing the myelin structural gene, *PMP22*. Deletion of this same region, in contrast, was found to cause hereditary neuropathy with liability to pressure palsies (HNPP) (Chance et al., 1993), an inherited condition in which individuals are susceptible to episodes of focal weakness and sensory loss due to small areas of demyelination from minor trauma. An inherited neuropathy caused by a genetic defect in the major myelin structural protein, myelin protein zero (MPZ), which had been mapped to the short arm of chromosome 1 also was soon identified (Hayasaka et al., 1993) and renamed CMT1B. In fact, this same genetic defect also was found in the families initially described by Bird and colleagues in which the disease gene had been linked to chromosome 1 and the Duffy blood group locus. Finally, the gene locus of an X-linked form of CMT, now renamed CMTX1, was mapped, and mutations identified in the *GJ1* gene encoding the gap junction protein, connexin 32 (Cx32) (Bergoffen et al., 1993).

Over the last 10 years, more than 30 genes have been identified that cause inherited neuropathy (www.molgen.ua.ac.be/CMTMutations/), most of which, surprisingly, are not uniquely expressed in peripheral nerve. These include genes involved in myelination, genes involved in vesicular transport, genes involved in mitochondrial function, and genes involved in axonal transport, suggesting that the peripheral nervous system is particularly vulnerable to a wide range of genetic insults. Although the reasons for this vulnerability are not known, they are likely a result of the set of unique biological and physiological processes in the peripheral nervous system. The biological background relevant to the function of the peripheral nervous system is thus reviewed in the following section.

II. Biological Background

Most peripheral nerves contain both motor and sensory axons enclosed within a single nerve sheath or epineurium. Within this sheath, however, groups of motor and sensory axons are further enclosed within smaller diameter tubular structures, called endoneurium. During development, Schwann cell precursors from the neural crest migrate out and contact the developing peripheral axons (Harrison, 1924; Le Douarin & Dupin, 1993) where they ensheath them in bundles. These Schwann cells further differentiate into myelinating or nonmyelinating Schwann cells. During this process of "radial sorting" (Webster, 1993), cells destined to become myelinating Schwann cells establish a one-to-one association with axons, the *promyelinating* stage of Schwann cell development, and then initiate the program of myelination. This includes concentric wrapping of the axon by the Schwann cell membrane and transcription of a set of myelin-specific genes (Scherer, 1997; Webster, 1993). In contrast, cells destined to become nonmyelinating Schwann cells do not establish a one-to-one relationship with axons, continue to ensheath bundles of axons, and do not activate this set of myelin-specific genes (Mirsky & Jessen, 1996; Webster, 1993).

Interestingly, the choice whether or not to become a myelinating Schwann cell is directed by the axon, at least in part, by means of a signal-transduction cascade activated on the Schwann cell surface by an axon-associated form of neuregulin-1 (Nave, 2006; Taveggia et al., 2005). For this reason all immature Schwann cells are essentially equipotent, and have the potential to become either myelinating or nonmyelinating cells, depending on the input from the interacting axon. In addition, interruption of Schwann cell-axonal contact, as during the process of Wallerian degeneration, can reverse the differentiation process, returning the Schwann to an immature phenotype (Scherer et al., 1997).

One of the major functions of the myelin sheath is to increase axonal conduction velocity without a significant increase in axonal diameter. This is accomplished by saltatory conduction (from the Latin, *saltare*, to hop or dance). During this process nerve impulses jump between the electrically excitable regions of the axon, called nodes of Ranvier, where a high density of voltage-sensitive sodium channels are clustered between the electrically insulated areas ensheathed by myelinating Schwann cells. Conduction velocity of an axon is determined both by the length of the average myelin internodal segment and the diameter of the myelinated fiber. Because most peripheral nerves are mixed, carrying both sensory and motor fibers, their axon populations contain both large and small diameter myelinated axons with differing conduction velocities. The average conduction velocity for such a mixed nerve is thus determined by the speed of its largest diameter, fastest conducting myelinated fiber population (Siegel, 2006).

Recent investigations of myelinated axons and their nodes of Ranvier have demonstrated a surprising structural complexity (Arroyo & Scherer, 2000; Kazarinova-Noyes & Shrager, 2002; Rasband et al., 2006; Salzer, 2003). As can be seen in Figures 29.1, 29.2, and 29.3, the myelin sheath consists of two separate subdomains, one compact, the other noncompact, each of which contains a unique, nonoverlapping set of protein constituents. In the compact region the myelin structural proteins MPZ, PMP22, and MBP are localized to either the major dense- or intraperiod-lines of the two dimensional repeating myelin structure. The adjacent noncompact region of myelin is also composed of two subdomains, called the paranode and the juxtaparanode. The paranodal region consists of loops of Schwann cell membrane attached to the underlying axonal surface by tight junctions (sometimes called axo-glial junctions) and to themselves by reflexive gap junctions, which together act to electrically insulate the axon.

The paranodal region of the Schwann cell membrane contains Cx32, which makes up the reflexive gap junctions, Neurofascin 155, which is part of the Schwann cell portion of the axo-glial junction, and MAG; the underlying axonal surface, in contrast, contains Neurofascin 186, Caspr, Contactin and βIVspectrin, all of which are involved in the structure of

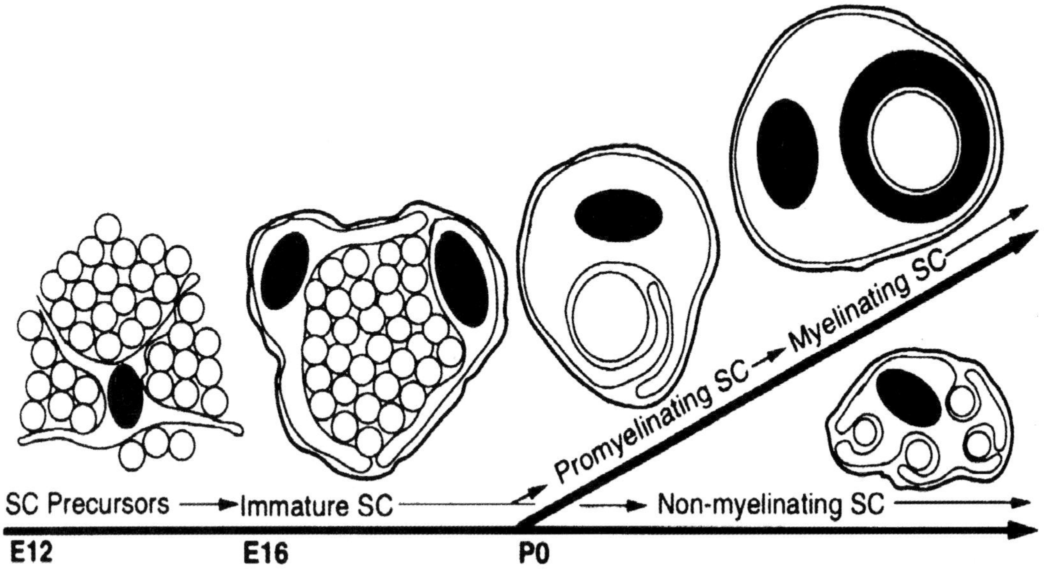

Figure 29.1 Graphic representation of Schwann cell differentiation into myelinating and nonmyelinating Schwann cells. Myelinating Schwann cells establish one-to-one relationships with axons, concentrically wrapping a single axon. Nonmyelinating Schwann cells, however, do not form a myelin sheath and surround bundles of axons.

the axonal side of the axo-glial junction. The juxtaparanodal region, the portion of Schwann cell and underlying axonal membrane adjacent to the paranode, contains potassium channels and Caspr 2, both expressed on the axonal surface. The organization of the nodal region, at least in the peripheral nervous system, is clearly regulated by Schwann cell contact with the axonal surface, probably through the Schwann cell protein gliomedin interacting with the axonal protein Neurofascin 186 (Eshed et al., 2005). These complex cellular structures formed by myelinating Schwann cells and their axons are analogous in many respects to the neuromuscular junction formed between motor axons and muscle cells; both are highly ordered multicomponent systems formed by the interaction of two distinct cell types in order to carry out a specific biological function related to nerve transmission.

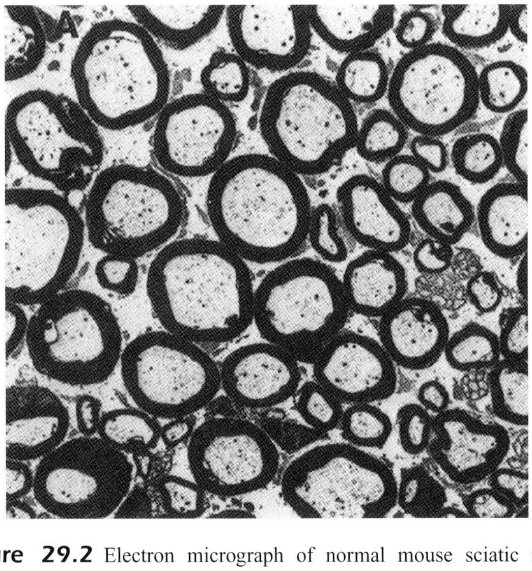

Figure 29.2 Electron micrograph of normal mouse sciatic nerve. Note the different axonal populations within the nerve, whose composition reflects the presence of large and small diameter myelinated axons. (With permission from de Waegh and Brady, 1990.)

Nerve development and associated Schwann cell differentiation, myelination, and its maintenance, and the establishment of an electrically insulated node of Ranvier capable of saltatory conduction provide at least in part, a biological framework for understanding the pathogenesis of all types of peripheral neuropathy, including CMT. In fact, one might anticipate that inherited peripheral neuropathies would be caused by mutations that alter crucial aspects of this biological process, such as the critical interactions between Schwann cells and their axons at the paranodal region, or the process of myelin compaction. For the purpose of developing rational treatments, each of these sites should be considered potential targets of therapy.

The maintenance and regulation of axonal transport is also important for understanding the pathogenesis of peripheral neuropathy. Axonal transport is necessary to support neuronal and axonal energy metabolism and thus to maintain and regulate axonal membrane voltage for saltatory conduction. Although neurons are polarized cells, some more than a meter long, metabolic reactions occur mainly in their cell body, so that nutrients and their byproducts must be transported down to the distal portion of the cell and back by the energy-dependent activity of axonal transport. Disruption of axonal transport thus could lead to neuronal dysfunction, axonal damage, demyelination, and/or neuropathy (Chevalier-Larsen & Holzbaur, 2006; Roy et al., 2005).

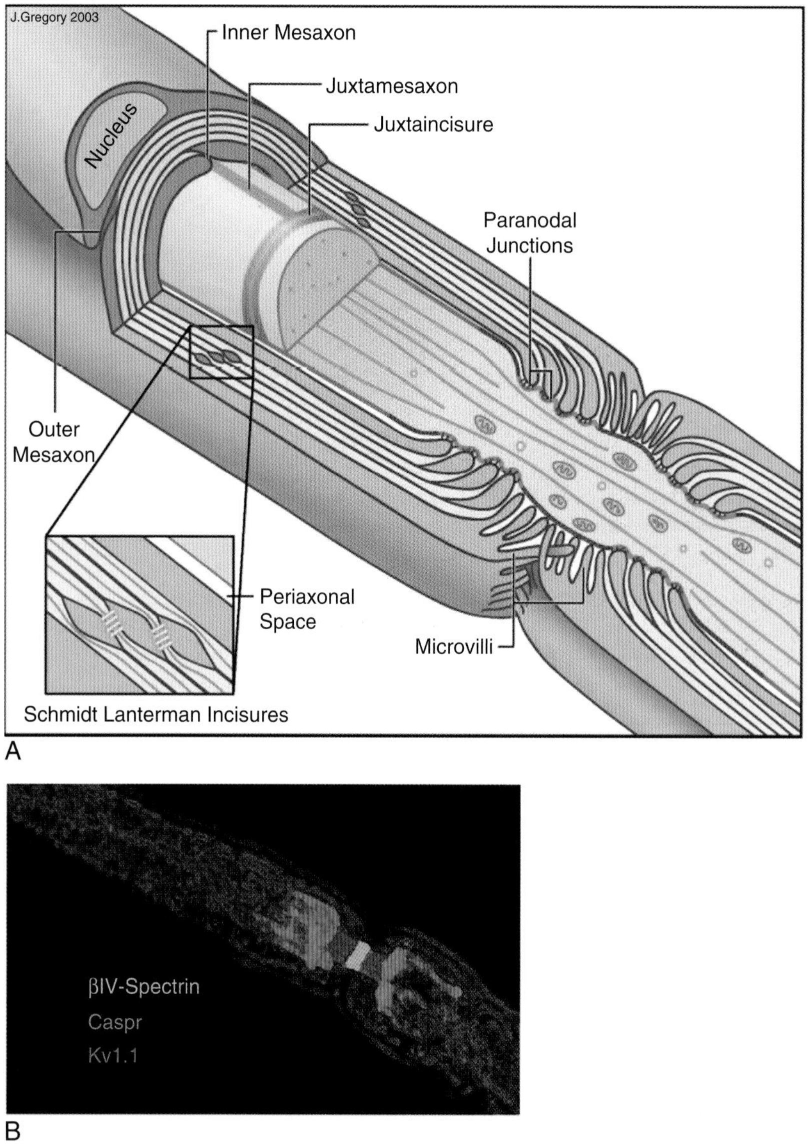

Figure 29.3 Structural organization of myelinated PNS axons. **A**. Graphic representation of a myelinated PNS axon highlighting the domains of the nerve in cross section. **B**. Teased sciatic nerve fiber viewed by confocal immunofluorescence depicting axonal domains. The node is stained for βIV-Spectrin (green), the paranodes for Caspr (blue), and the juxtaparanodes for Kv1.1 (red). (With permission from Salzer, 2003.)

During the transport process neurons move a wide variety of cellular constituents including proteins, protein complexes, cytoskeletal elements, lipids, membrane vesicles, including synaptic vesicle precursors, endosomes, and lysosomes, and organelles such as mitochondria and ribosomes. The transport process itself is mediated by the protein motors kinesin and cytoplasmic dynein, and takes place on a system of microtubular cables, both in the antegrade and retrograde directions (Duncan & Goldstein, 2006). Disruption of axonal transport can cause axonal metabolic and energy deficits, which can lead to axonal damage and dysfunction, producing neurological signs and symptoms (Chevalier-Larsen & Holzbaur, 2006; Roy et al., 2005). One of the hereditary spastic paraplegias, for example, SPG10, in which there is dysfunction the long motor tracts to the legs, is caused by a point mutation in the kinesin KIF5A heavy chain gene, demonstrating that a disruption in axonal transport is the cause of this syndrome (Blair et al., 2006: Fichera et al., 2004; Reid et al., 2002). In addition, a form of hereditary motor neuron disease has recently been identified due to mutation in the dynactin-1 gene, encoding one of the proteins that is necessary for cytoplasmic dynein to interact with microtubules, again implicating alteration of an axonal transport motor in the pathogenesis of this neurodegenerative disease (Jablonka et al., 2004; Puls et al., 2005).

Importantly, axonal transport also is altered in both Trembler mice and patients with CMT1A (Brady et al., 1999; Sahenk, 1999; Watson et al., 1994), so that an axonal transport defect may contribute to axonal degeneration even in demyelinating neuropathies. Consistent with this notion, Krajewski and coworkers have found that loss of axons rather than slowed conduction velocities is the major cause of neurological disability in patients with CMT1A (Krajewski et al., 2000).

The mechanisms underlying the regulation of axonal energy metabolism are also important for understanding the pathogenesis of CMT2. Zuchner and colleagues, for example, have recently found that the inherited axonal neuropathy CMT2A is caused by mutations in the gene encoding mitofusin 2 (MFN2), a protein involved in the regulation of mitochondrial fusion (Zuchner et al., 2004; Zuchner & Vance, 2006). Mitochondria are known to fuse into syncicial chains, presumably in order to redistribute their metabolic components such as tRNAs and rRNAs. Mutations in MFN2 disrupt mitochondrial fusion, and also lead to a reduction in mitochondrial ATP synthesis. Interestingly, MFN2 mutations may also alter axonal transport of mitochondria, suggesting that a mitochondrial transport defect may contribute to the pathogenesis of CMT2A.

Abnormalities of mitochondrial function also have been implicated in several other forms of inherited neuropathy. Mutations in GDAP1, for example, cause CMT4A, an autosomal recessive demyelinating neuropathy. GDAP1 is a putative glutathione transferase, predominantly expressed in neurons and localized to mitochondria, suggesting that abnormalities of mitochondrial function are involved in its pathogenesis (Pedrola et al., 2005). In addition, mutations in the small heat shock proteins HSP27 and HSP22 have both been found to cause autosomal dominant forms of CMT2, designated CMT2F and CMT2L. Small HSPs like HSP27 and HSP22 have been shown to protect against H_2O_2-mediated cell death (Mehlen et al., 1995), and neuronal cell lines transfected with mutant HSP22 or HSP27 showed a reduced viability (Irobi et al., 2004). Also, Hsp27 has been shown to be directly responsible for a stable mitochondrial membrane potential through an increase and maintenance of the reduced form of the redox modulator glutathione. Taken together these data suggest that maintenance of axonal energy metabolism is important for axonal homeostasis, and that alterations in mitochondrial function and/or axonal transport of mitochondria can lead to axonal damage and neurodegeneration. Manipulating the function of mitochondria and/or axonal transport are thus areas of potential therapeutic importance for patients with CMT.

Defects in axonal transport and mitochondrial function are not only important for understanding inherited peripheral neuropathy, but also have been implicated in the pathogenesis of a number of other neurodegenerative disorders, including Alzheimer disease (AD), Parkinson disease (PD), Huntington's disease (HD), Friedreich's ataxia (FA), and amyotrophic lateral sclerosis (ALS) (Chevalier-Larsen & Holzbaur, 2006; Roy et al., 2005). Mutations in the gene encoding superoxide dismutase 1 (SOD1), for example, the cause the most common type of familial ALS (FALS), produce an altered SOD1 protein that is abnormally localized to mitochondria and decreases its respiratory function, suggesting that altered axonal energy metabolism is involved in the pathogenesis of FALS (Pasinelli et al., 2004). The defect in FA is in a mitochondrial protein, involved in both iron homeostasis and respiration, and the FA mutations can cause oxidative stress in affected neurons, leading to axonal degeneration and apoptosis (Lodi et al., 2006). In addition, several of the known mutations causing familial PD have been shown to alter the function of complex I of the respiratory chain, implicating a defect in oxidative metabolism in the pathogenesis of PD (Savitt et al., 2006). Maintenance of axonal transport and axonal energy metabolism is thus likely important for preventing many of the most common neurodegenerative diseases, as well as CMT. A schematic drawing of a myelinated internode is shown in Figure 29.4, and the locations where alteration in cellular function could cause neuropathy are indicated.

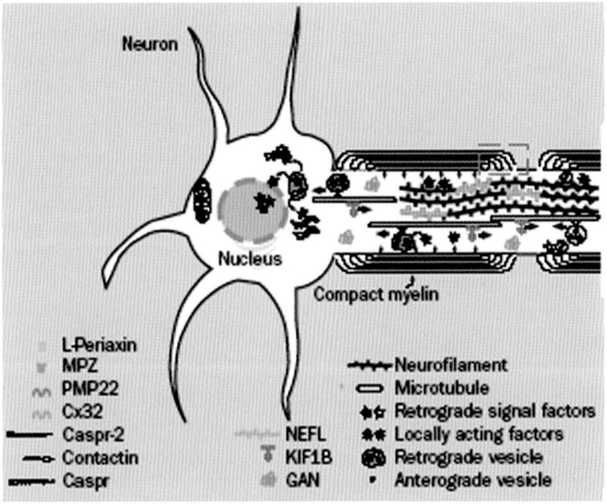

Figure 29.4 Graphic representation of a myelinated axon. Those proteins associated with neuropathy, for example, MPZ, PMP22, and Cx32, are depicted in their respective intracellular locations.

III. The Clinical Phenotypes of CMT1B

Mutations in the gene encoding the major PNS myelin protein, myelin protein zero (MPZ), cause the inherited demyelinating neuropathy designated CMT1B. To date there are over 90 different mutations in MPZ known to cause peripheral neuropathy in patients. An initial review of the phenotypic spectrum of CMT1B suggested that mutations in the MPZ gene could produce a wide variety of clinical phenotypes, including congenital demyelinating neuropathy, Dejerine-Sotas syndrome (a severe early onset demyelinating neuropathy), and so-called CMT1B, an adult onset disease similar to CMT1A. In several recent and more detailed studies of the clinical phenotypes of patients with MPZ mutations, however, these individuals could be divided into two distinct phenotypic groups: one with slow nerve conduction velocities and onset of symptoms during the period of motor development, and a second with essentially normal nerve conduction velocities and the onset of symptoms as adults (Hattori et al., 2003; Shy et al., 2004). There was essentially no clinical overlap between the early onset and late onset groups, and both were clinically and physiologically distinct from patients with CMT1A caused by a duplication of the PMP22 gene region. This is graphically demonstrated in Figure 29.5, in which the nerve conduction velocities of a group of CMT1B patients are plotted along with those of a cohort of patients with CMT1A.

Although the molecular mechanisms causing these two forms of CMT1B are not yet known, these data are important since they suggest that MPZ mutations can also be divided into two pathogenic groups: one that affects the process of myelination and/or myelin development, causing delayed motor development and slowed nerve conduction velocities; and a second, which is associated with clinically normal myelin development and normal nerve conduction velocities, but which causes axonal degeneration and weakness in later life. The identification of these two distinct clinical presentations of neuropathy in patients with MPZ mutations thus has clear implications for the molecular and cellular pathogenesis of the disease process, as well as for the function of MPZ in myelination.

IV. Myelin Protein Zero (MPZ)

MPZ, a transmembrane protein of 219 amino acids, is a member of the immunoglobulin supergene family. It has a single immunoglobulin-like extracellular domain of 124 amino acids, a single transmembrane domain of 25 amino acids, and a single cytoplasmic domain of 69 amino acids (Lemke & Axel, 1985; Uyemura et al., 1995). MPZ is also post-translationally modified by the addition of an N-linked oligosaccharide at a single asparagine residue in the extracellular domain, as well by the addition of sulfate, acyl, and phosphate groups (D'Urso et al., 1990; Eichberg & Iyer, 1996).

MPZ, like other members of the immunoglobulin superfamily, is a homophilic adhesion molecule (Filbin et al., 1990). Heterologous cells expressing MPZ adhere to each other in an *in vitro* cell interaction assay (Xu et al., 2001); absence of MPZ expression *in vivo* in *MPZ* knockout mice produces poorly compacted myelin sheaths (Giese et al., 1992). Interestingly, overexpression of MPZ also disrupts myelination by inhibiting Schwann cell wrapping of axons, also consistent with an adhesive function for the protein. Glycosylation and acylation of MPZ are also necessary for adhesion, since acylation at cysteine residue 153 (Gao et al.,

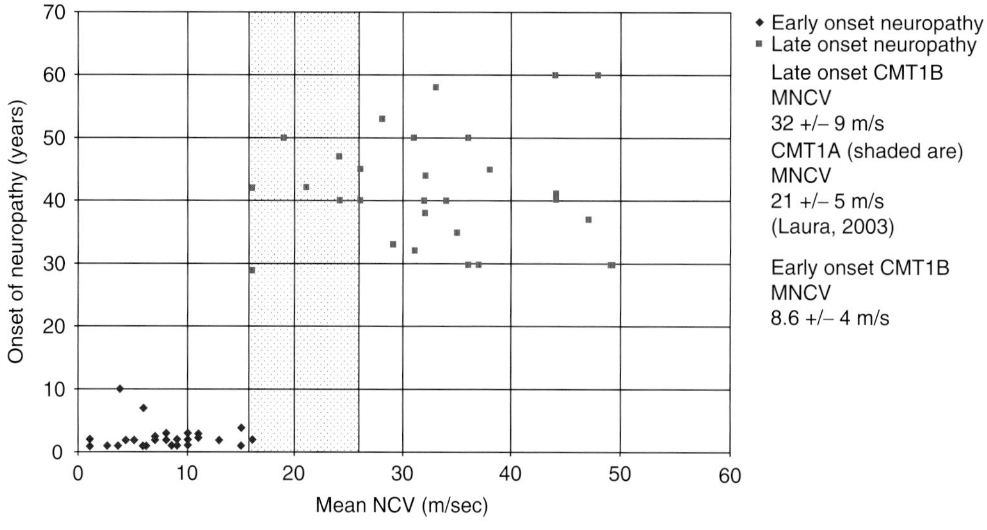

Figure 29.5 Correlation between nerve conduction velocity (NCV) and early or late onset phenotypes for patients with MPZ mutations. Areas in gray represent the mean NCV ± SD for CMTIA patients described by Laura et al., 2003.

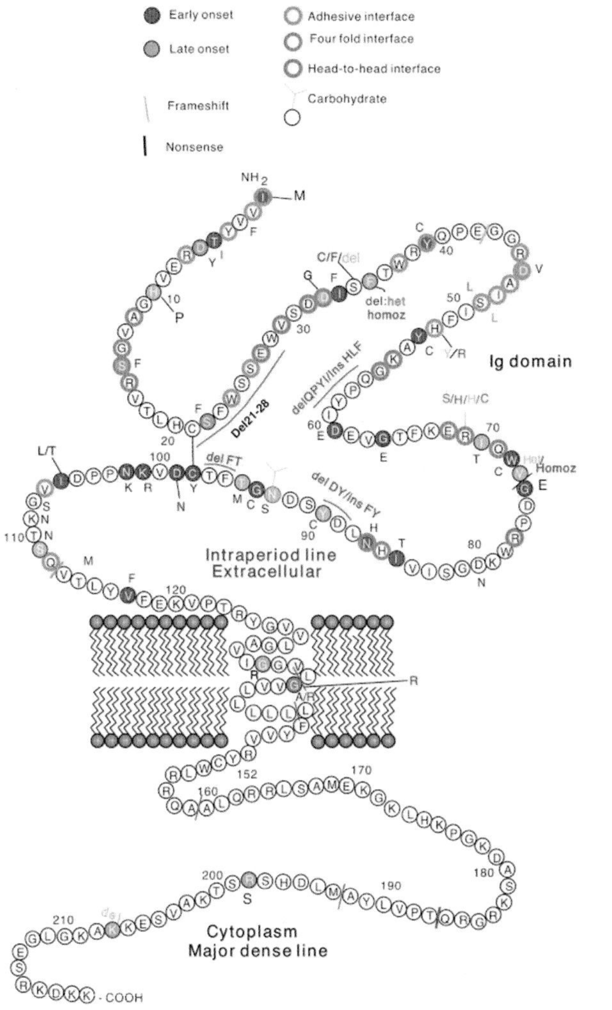

Figure 29.6 Graphic representation of the secondary structure of MPZ highlighting those residues associated with early (red) and late (blue) onset neuropathy. Note the numbering system for the mutations does not include the 29 amino acid leader peptide that is cleaved before insertion into the myelin sheath.

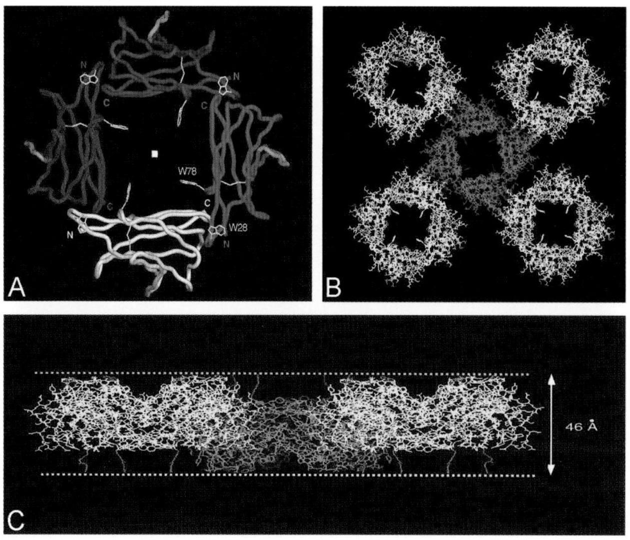

Figure 29.7 Tetrameric assembly of MPZ. The extracellular domains of MPZ form homotetramers in *cis* (**A**) and each homotetramer interacts in *trans* with another homotetramer on the opposing membrane surface (**B**, perpendicular view; **C**, side view). (With permission from Shapiro et al., 1996.)

2000), formation of the C21-C98 disulfide bond (Zhang & Filbin, 1994), and glycosylation at asparagine residue 93 (Filbin & Tennekoon, 1993) are each necessary for MPZ-mediated adhesion *in vitro*. In addition, alteration of glycosylation at asparagine 93 also causes peripheral neuropathy in patients. Taken together, these data demonstrate that MPZ plays an essential role in myelination, probably by holding together adjacent wraps of myelin membrane through MPZ-mediated homotypic interactions. A schematic diagram of MPZ and the location of a number of its known mutations is shown in Figure 29.6.

Consistent with its role as an adhesion molecule, crystallographic analysis of the extracellular domain of rat MPZ demonstrates that it forms a compact sandwich of beta-sheets held together by a disulfide bridge, similar to that of other members of the Ig-superfamily. In the crystal structure each Ig-domain monomer interacts by way of a four-fold interface to form a homotetramer, a doughnut-like structure with a large central hole, as well as by way of a separate adhesive interface. This structure suggests that MPZ monomers interact within the plane of the Schwann cell membrane to form a lattice of homotetramers, which in turn could interact with similar structures on the opposing membrane surface to mediate myelin compaction (Shapiro et al., 1996) as is shown in Figure 29.7.

This model of MPZ structure thus suggests that mutations in MPZ could cause neuropathy by altering protein–protein interactions at one of these two major crystallographic interfaces, thereby inhibiting homotypic adhesion. Supporting this notion, some MPZ mutations in critical residues cause either early or late onset neuropathy, as shown in Table 29.1. In addition, mutations that are known to alter glycosylation, such as N93S or T95M (Blanquet-Grossard et al., 1996) and homotypic adhesion cause peripheral neuropathy. There is little correlation, however, between the clinical phenotype of patients and the location of the MPZ mutation. In addition, mutations in the cytoplasmic domain of the protein can also abolish MPZ-mediated adhesion *in vitro*, and can cause neuropathy in patients (Xu et al., 2001). The simple structural model of MPZ-mediated adhesion derived from the crystal structure of the extracellular domain of the protein is thus probably incomplete.

Consistent with a more complex model of MPZ structure and function, the cytoplasmic domain of MPZ is also

Table 29.1 Patient Mutations Disrupting cis/trans Homotypic Interactions

A) Mutations Affecting The Four-Fold Interface (*cis*)
Early Onset: R69C, R69H, R69S
Late Onset: S15F, E68V, R69H
B) Mutations Affecting The Adhesive Interface (*trans*)
Early Onset: S49L, H52R
Late Onset: D46V, A47V, H52R

necessary for MPZ-mediated homotypic adhesion. Deletion of 28 amino acids from the carboxy-terminus of the protein abolishes adhesion *in vitro* (Filbin et al., 1999), and nonsense mutations within the cytoplasmic domain cause particularly severe forms of demyelinating peripheral neuropathy in patients (Mandich et al., 1999). Interestingly, coexpression of an MPZ truncated within the cytoplasmic domain with a wild-type MPZ in heterologous cells also inhibits MPZ-mediated adhesion, suggesting that the cytoplasmic domains can interact or can influence the structure and function of the extracellular domain. A PKC substrate motif, RSTK, located between amino acids 198 and 201 of the cytoplasmic domain, is also necessary for MPZ-mediated adhesion *in vitro*, and mutation of this residue in a patient causes peripheral neuropathy (Xu et al., 2001).

In addition, we have also found that both RACK1, an activated PKC binding protein, and a 65 kDa protein identified through a yeast two-hybrid screen (p65), interact with the cytoplasmic domain of MPZ (Hrstka et al., 2004; In press, 2007) to phosphorylate MPZ at its PKC substrate motif. Phosphorylation is necessary for MPZ-mediated adhesion *in vitro*. Although the mechanisms by which the cytoplasmic domain regulates homotypic adhesion are not known, MPZ clearly participates in an adhesion-mediated signal transduction cascade, similar to that of other adhesion molecules such as the cadherins and the integrins by interacting with the cell cytoskeleton. A model of the protein interactions within the cytoplasmic domain of MPZ are shown schematically in Figure 29.8.

The absence of MPZ expression not only disrupts myelin compaction, but it also causes dysregulation of myelin-specific gene expression and abnormalities of myelin protein localization in Schwann cells (Menichella et al., 2001; Xu et al., 2000). MPZ thus has two separate functions in the peripheral nervous system. The first, a predominantly structural function, is to hold together adjacent wraps of myelin membrane through MPZ-mediated homotypic interactions. The second, a regulatory function, perhaps mediated by a MPZ signal transduction cascade, is to modulate the process of myelin assembly and/or myelin maintenance. The clinical phenotype of patients with mutations in MPZ thus also might be expected to reflect these two separate roles of protein function.

V. Cellular and Molecular Mechanisms of Neuropathy Caused by MPZ Mutations

The cellular and molecular mechanisms underlying the clinical phenotypes of CMT1B currently are not well understood, but there are a few reasonable possibilities. The late onset MPZ mutants could exert their cellular effects by altering the process of axonal transport and/or axonal energy metabolism. Individuals with late onset mutations have essentially normal nerve conduction velocities, so that the structure of their nodes of Ranvier must be relatively intact, and their saltatory conduction must occur normally. Interactions between their Schwann cells and axons, however, are likely disturbed, since patients develop a length-dependent axonal neuropathy, in spite of the fact that MPZ is expressed only in myelinating Schwann cells. This could be a direct consequence of altered myelin structure on myelinated axons; it also could be, however, an indirect effect due to changes in the MPZ-mediated signal transduction cascade, altered myelin turnover, or abnormalities of myelin remodeling or repair. Consistent with this interpretation, we have identified a patient with an R198S mutation, located within the intracellular PKC recognition site of MPZ that alters both MPZ-mediated signal transduction and homotypic adhesion, and causes a mild, late onset neuropathy.

In contrast, early onset mutants probably interfere with the process of myelin development. This could be due to more severe abnormalities of myelin structure, leading to dysmyelination, demyelination or Schwann cell damage. In addition, early onset mutants could have dominant negative or gain of function properties due to protein misfolding or altered protein interactions. Wrabetz and coworkers, for

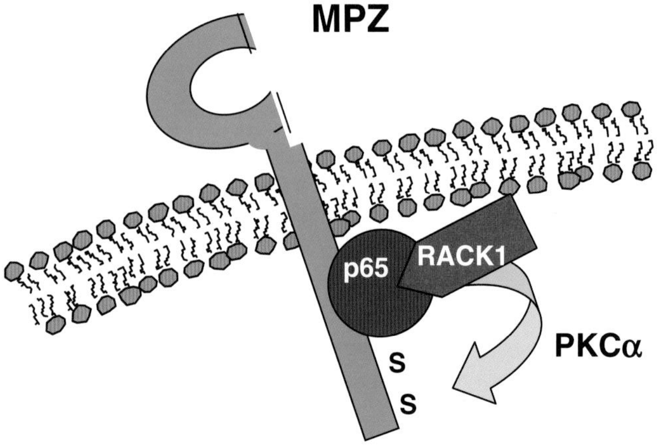

Figure 29.8 Graphic representation of the MPZ mediated signal transduction cascade. The cytoplasmic domain of MPZ facilitates homotypic adhesion through interactions with p65, RACK 1, and PKC-α, ultimately resulting in the phosphorylation of two serine residues, 199 and 204. (Schematic courtesy of J. Lilien and J. Balsamo, University of Iowa.)

example, have demonstrated that mice expressing the S34C mutation, causing early onset CMT1B in patients, have compact myelin that likely incorporates the mutant protein, but have abnormal myelin packing compared to wild-type animals (Wrabetz et al., 2006). In addition, Kirschner and coworkers have found myelin packing abnormalities in nerve biopsies from patients with both early and late onset forms of CMT1B (Kirschner et al., 1996; Kirschner et al., submitted), suggesting that mutant forms of MPZ are incorporated into the myelin sheath and alter myelin structure in patients with both clinical phenotypes.

As noted earlier, the pathogenesis of an MPZ mutation depends, at least in part, on its ability to be transported to the cell surface and incorporated into the myelin sheath. In order to identify the ability of mutant MPZs to be transported to the cell surface, we and others have analyzed MPZ expression in a heterologous cell transfection assay. cDNAs encoding the mutant proteins or controls were fused at the carboxy-terminus to a green fluorescent protein (GFP) cDNA and the construct transfected into Cos7 or HeLa cells. The location of the mutant fusion protein was then identified by fluorescence microscopy. As can be seen in Figure 29.9, some mutant proteins, such as R69C associated with an early onset phenotype, are retained within the cell, probably within the ER. Some mutant proteins, however, can be transported to the cell surface in this assay, and thus could be incorporated into the myelin sheath. These include one from the early onset group, H52R, and two from the late onset group, H10P and T95M. Interestingly, expression of the mutant proteins does not alter the transport of the wild-type proteins in this assay, suggesting that interactions between wild-type and mutant MPZ within the cell are unlikely to play a role in disease pathogenesis.

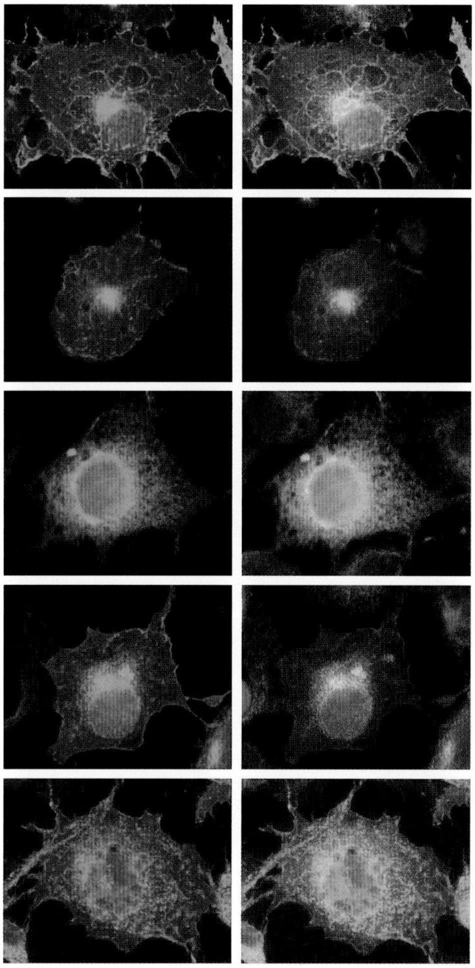

Figure 29.9 Cell surface expression of mutant MPZ proteins. Cos7 cells transiently transfected with MPZ-fluorescent fusion proteins (green) and stained with anti-calnexin antibody (red). Wild-type MPZ and 3 out of 4 mutants (one early onset, H52R, and two late onset, H10P and T95M) are transported to the cell surface, whereas the early onset mutant, R69C, is not. Colocalization of R69C with calnexin indicates ER retention (yellow). (Unpublished data, J. Kamholz, Wayne State University.)

VI. R69C Mutation

Analysis of a nerve biopsy specimen from a patient with early onset neuropathy caused by the R69C mutation, shown in Figure 29.10, demonstrates absence of the large myelinated fiber population, uniformly myelinated internodes that are shorter than normal, but also with significant evidence of demyelination and/or remyelination. Nerve conduction velocities in this patient are very slow, and motor and sensory amplitudes are reduced (Bai et al., 2006). These findings strongly suggest that the R69C mutation has a major cellular effect on myelin development, rather than on myelin maintenance. Since the R69C protein is not transported to the cell surface in our transfection assay, its effect on myelination may be due to a gain of function mechanism. Interestingly, the myelin sheath looks morphologically normal in smaller myelinated axons from this same patient, and immunoelectron microscopy of MPZ expression in these Schwann cells is also normal. Perhaps the level of R69C protein expression is important for its pathogenesis: high levels of expression in large myelinated fibers cause Schwann cell apoptosis; lower levels of expression in smaller myelinated fibers can be incorporated into the myelin sheath, which looks morphologically normal, although shorter than wild-type.

VII. H10P Mutation

Recently we have obtained peripheral nerve samples at autopsy from a patient with an H10P mutation (see Figure 29.11). This mutation causes a late onset axonal degeneration, but without significant effect on nerve conduction velocities. Consistent with the clinical and

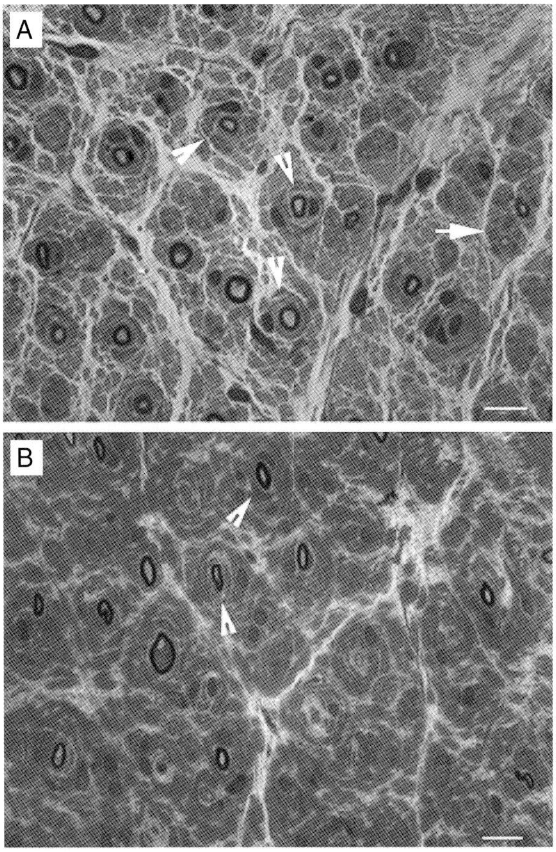

Figure 29.10 Nerve biopsy results from patient with R69C mutation. Toluidine blue staining of a 20-year-old sural nerve biopsy demonstrates the formation of onion bulbs, indicating chronic demyelination and remyelination (**A**). Similar staining of a recent biopsy sample from the other sural nerve demonstrates analogous axonal loss and pathology (**B**). (With permission from Bai et al., 2006.)

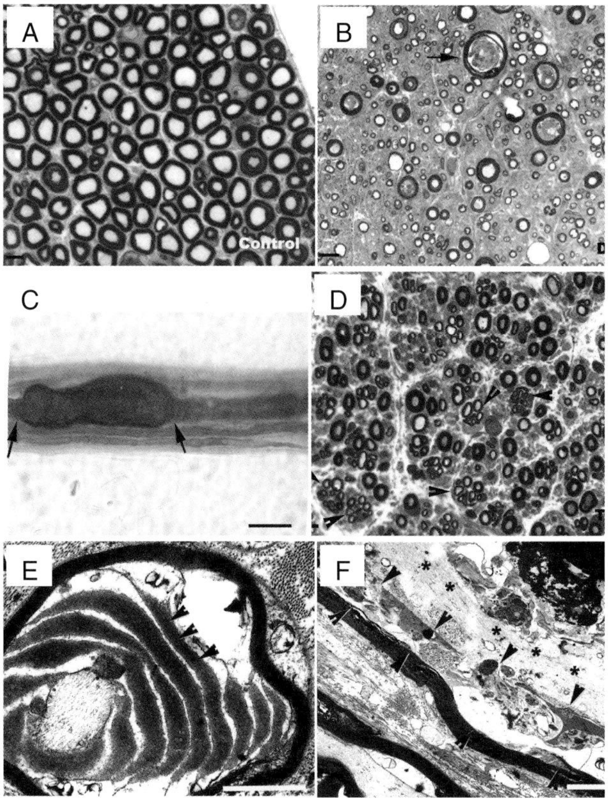

Figure 29.11 Autopsy results from peripheral nerve samples of patient with H10P mutation. Toluidine blue staining of lumbar dorsal root (**B**) and tibial nerve (**D**) of H10P patient indicate a reduction in density of myelinated nerve fibers in addition to several enlarged nerve fibers (**B**) and formation of regenerative clusters (**D**), compared to control sciatic nerve (**A**). Teased fiber analysis of H10P nerve demonstrates the segmental nature of the enlargement (**C**). These enlargements were found by EM to be associated with an accumulation of amorphous material in the intralaminar (**E**) and periaxonal spaces (**F**). (With permission from Li et al., 2006.)

electrophysiological phenotype, the peripheral nerve myelin from this patient appears normal at the light microscopic level. There are, however, numerous large axonal swellings, mainly at the paranodal regions, as well as increased numbers of Schmidt-Lantermann inscisures (Li et al., 2006). In addition, a proteinaceous material that contains MPZ has accumulated between the abaxonal surface of the Schwann cell and the axon, suggesting an abnormality of myelin maintenance or MPZ turnover. Interestingly, antibodies raised to an N-terminal peptide with an H10P mutation recognize the mutant peptide but not its wild-type counterpart (unpublished data), suggesting that the H10P mutation alters the structure of the N-terminus of MPZ. This mutant structure could lead to abnormal MPZ interactions, a change in myelin packing, and altered myelin protein turnover. Neither how the cell recognizes this altered myelin structure nor how it subsequently causes axonal degeneration is currently understood.

VIII. Summary and Conclusions

In this chapter we have summarized the history and biological background of inherited demyelinating peripheral neuropathy, and described what is currently known of the pathogenesis of CMT1B, caused by mutations in myelin protein zero (MPZ). From this analysis, we believe there are four key findings currently relevant to understanding the pathogenesis of CMT1B. The first is the molecular description of the developmental pathway of Schwann cell differentiation and myelination, and the identification of the key molecules involved in the regulation of this process. The second is the description of the crystal structure of the extracellular domain of MPZ, and the demonstration that MPZ is a homotypic adhesion molecule. The third is the identification of a large cohort of more than 90 patients with MPZ mutations. The fourth, and perhaps most important, is the finding that patients with MPZ mutations can be divided into two phenotypic classes, early and late onset, suggesting that an abnormality of either myelin development or axonal/glial interactions is the cause of CMT1B neuropathy.

Of these four key findings, two focus on a detailed analysis of the molecular biology of MPZ and nerve development, and the other two focus on the identification of patients with MPZ mutations and their careful clinical and physiological evaluation. Elucidation of the molecular pathogenesis of CMT1B, and probably most inherited diseases, thus requires advances in the molecular and cellular biology of the affected gene or protein, as well as in the clinical description and natural history of the disease. This amalgamation of hypothesis-driven cellular and molecular biology research with the more descriptive clinical evaluation of patients is not only important for understanding the pathogenesis of the disease process, but also, more importantly, for monitoring the results of new treatments for the disease. A truly workable molecular medicine enterprise for the future will thus require a close collaboration of molecular biologists with an interest in disease mechanisms and clinicians with an interest in molecular and cellular biology.

References

Arroyo, E. J., Scherer, S. S. (2000). On the molecular architecture of myelinated fibers. *Histochem Cell Biol* **113**, 1–18.

Bai, Y., Ianokova, E. et al. (2006). Effect of an R69C mutation in the myelin protein zero gene on myelination and ion channel subtypes. *Arch Neurol* **63**, 1787–1794.

Bergoffen, J., Scherer, S. S. et al. (1993). Connexin mutations in X-linked Charcot-Marie-Tooth disease. *Science* **262**, 2039–2042.

Bird, T. D., Ott, J., Giblett, E. R. (1982). Evidence for linkage of Charcot-Marie-Tooth neuropathy to the Duffy locus on chromosome 1. *Am J Hum Genet* **34**, 388–394.

Blair, M. A., Ma, S. et al. (2006). Mutation in KIF5A can also cause adult-onset hereditary spastic paraplegia. *Neurogenetics* **7**, 47–50.

Brady, S. T., Witt, A. S., Kirkpatrick, L. L. et al. (1999). Formation of compact myelin is required for maturation of the axonal cytoskeleton. *J Neurosci* **19**, 7278–7288.

Charcot, J., Marie, P. (1886). Sue une forme particulaire d'atrophie musculaire progressive souvent familial debutant par les pieds et les jamber et atteingnant plus tard les mains. *Rev Med* **6**, 97–138.

Chevalier-Larsen, E., Holzbaur E. L. (2006). Axonal transport and neurodegenerative disease. *Biochim Biophys Acta* **1762**, 1094–1098.

Déjérine, H., Sottas, J. (1893). Sur la nevritte interstitielle, hypertrophique et progressive de l'enfance. *CR Soc Biol Paris* **45**, 63–96.

Duncan, J. E., Goldstein L. S. (2006). The genetics of axonal transport and axonal transport disorders. *PLoS Genet* **2**, e124.

D'Urso, D., Brophy, P. J., Staugaitis, S. M., Gillespie, C. S., Frey, A. B., Stempak, J. G., Colman, D. R. (1990). Protein zero of peripheral nerve myelin: Biosynthesis, membrane insertion, and evidence for homotypic interaction. *Neuron* **4**, 449–460.

Dyck, P. J., Lambert, E. H. (1968). Lower motor and primary sensory neuron diseases with peroneal muscular atrophy: I, neurologic, genetic, and electrophysiologic findings in hereditary polyneuropathies. *Arch Neurol* **18**, 603–618.

Dyck, P. J., Lambert, E. H. (1968). Lower motor and primary sensory neuron diseases with peroneal muscular atrophy: II, neurologic, genetic, and electrophysiologic findings in various neuronal degenerations. *Arch Neurol* **18**, 619–625.

Eichberg, J., Iyer, S. (1996). Phosphorylation of myelin protein: Recent advances. *Neurochem Res* **21**, 527–535.

Eshed, Y., Feinberg, K. et al. (2005). Gliomedin mediates Schwann cell-axon interaction and the molecular assembly of the nodes of Ranvier. *Neuron* **47**, 215–229.

Fichera, M., Lo Giudice, M. et al. (2004). Evidence of kinesin heavy chain (KIF5A) involvement in pure hereditary spastic paraplegia. *Neurology* **63**, 1108–1110.

Filbin, M. T., Tennekoon, G. I. (1993). Homophilic adhesion of the myelin P0 protein requires glycosylation of both molecules in the homophilic pair. *J Cell Biol* **122(2)**, 451–459.

Filbin, M. T., Walsh, F. S., Trapp, B. D., Pizzey, J. A., Tennekoon, G. I. (1990). Role of myelin Po protein as a homophilic adhesion molecule. *Nature* **344**, 871–872.

Filbin, M. T., Zhang, K., Li, W., Gao, Y. (1999). Characterization of the effect on adhesion of different mutations in myelin P0 protein. *Ann N Y Acad Sci* **883**, 160–167.

Giese, K. P., Martini, R., Lemke, G., Soriano, P., Schachner, M. (1992). Mouse P0 gene disruption leads to hypomyelination, abnormal expression of recognition molecules, and degeneration of myelin and axons. *Cell* **71**, 565–576.

Harding, A. E., Thomas, P. K. (1980a). The clinical features of hereditary motor and sensory neuropathy types I and II. *Brain* **103**, 259–280.

Harding, A. E., Thomas, P. K. (1980b). Genetic aspects of hereditary motor and sensory neuropathy (types I and II). *J Med Genet* **17**, 329–336.

Harrison, R. G. (1924). Neuroblast versus sheath cell in the development of peripheral nerves. *J Comp Neurol* **37**, 123–194.

Hattori, N., Yamamoto, M. et al. (2003). Demyelinating and axonal features of Charcot-Marie-Tooth disease with mutations of myelin-related proteins (PMP22, MPZ and Cx32): A clinicopathological study of 205 Japanese patients. *Brain* **126**, 134–151.

Hayasaka, K., Himoro, M. et al. (1993). Structure and chromosomal localization of the gene encoding the human myelin protein zero (MPZ). *Genomics* **17(3)**, 755–758.

Irobi, J., Van Impe, K. et al. (2004). Hot-spot residue in small heat-shock protein 22 causes distal motor neuropathy. *Nat Genet* **36(6)**, 597–601.

Jablonka, S., Wiese, S. et al. (2004). Axonal defects in mouse models of motoneuron disease. *J Neurobiol* **58**, 272–286.

Kazarinova-Noyes, K., Shrager, P. (2002). Molecular constituents of the node of Ranvier. *Mol Neurobiol* **26**, 167–182.

Kirschner, D. A., Szumowski, K. et al. (1996). Inherited demyelinating peripheral neuropathies: Relating myelin packing abnormalities to P0 molecular defects. *J Neurosci Res* **46**, 502–508.

Krajewski, K. M., Lewis, R. A. et al. (2000). Neurological dysfunction and axonal degeneration in Charcot-Marie-Tooth disease type 1A. *Brain* **123**, 1516–1527.

Le Douarin, N. M., Dupin, E. (1993). Cell lineage analysis in neural crest ontogeny. *J Neurobiol* **24**, 146–161.

Lemke, G., Axel, R. (1985). Isolation and sequence of a cDNA encoding the major structural protein of peripheral myelin. *Cell* **40**, 501–508.

Li, J., Bai, Y. et al. (2006). Major myelin protein gene (P0) mutation causes a novel form of axonal degeneration. *J Comp Neurol* **498**, 252–265.

Lodi, R., Tonon, C. et al. (2006). Friedreich's ataxia: From disease mechanisms to therapeutic interventions. *Antioxid Redox Signal* **8**, 438–443.

Lupski, J. R., de Oca-Luna, R. M., Slaugenhaupt, S. et al. (1991). DNA duplication associated with Charcot-Marie-Tooth disease type 1A. *Cell* **66**, 219–232.

Mandich, P., Mancardi, G. L., Varese, A., Soriani, S., Di Maria, E., Bellone, E. et al. (1999). Congenital hypomyelination due to myelin protein zero Q215X mutation. *Ann Neurol* **45**, 676–678.

Mehlen, P., Preville, X. et al. (1995). Constitutive expression of human hsp27, Drosophila hsp27, or human alpha B-crystallin confers resistance to TNF- and oxidative stress-induced cytotoxicity in stably transfected murine L929 fibroblasts. *J Immunol* **154(1)**, 363–374.

Menichella, D. M., Arroyo, E. J., Awatramani, R., Xu, T., Baron, P., Vallat, J. M. et al. (2001). Protein zero is necessary for E-cadherin-mediated adherens junction formation in Schwann cells. *Mol Cell Neurosci* **18**, 606–618.

Mirsky, R., Jessen, K. R. (1996). Schwann cell development, differentiation and myelination. *Curr Opin Neurobiol* **6**, 89–96.

Nave, K. A., Salzer, J. L. (2006). Axonal regulation of myelination by neuregulin 1. *Curr Opin Neurobiol* **16**, 492–500.

Pasinelli, P., Belford, M. E. et al. (2004). Amyotrophic lateral sclerosis-associated SOD1 mutant proteins bind and aggregate with Bcl-2 in spinal cord mitochondria. *Neuron* **43**, 19–30.

Pedrola, L., Espert, A. et al. (2005). GDAP1, the protein causing Charcot-Marie-Tooth disease type 4A, is expressed in neurons and is associated with mitochondria. *Hum Mol Genet* **14**, 1087–1094.

Puls, I., Oh, S. J. et al. (2005). Distal spinal and bulbar muscular atrophy caused by dynactin mutation. *Ann Neurol* **57**, 687–694.

Raeymaekers, P., Timmerman, V., Nelis, E. et al. (1991). Duplication in chromosome 17p11.2 in Charcot-Marie-Tooth neuropathy type 1a (CMT 1a). *Neuromuscul Disord* **1**, 93–97.

Rasband, M. N. (2006). Neuron-glia interactions at the node of Ranvier. *Results Probl Cell Differ* **43**, 129–149.

Reid, E., Kloos, M. et al. (2002). A kinesin heavy chain (KIF5A) mutation in hereditary spastic paraplegia (SPG10). *Am J Hum Genet* **71**, 1189–1194.

Roussy, G., Levy, G. (1926). A sept cas d'une maladie familiale particulaire. Rev Neurol **33**, 427–450.

Roy, S., Zhang, B. et al. (2005). Axonal transport defects: A common theme in neurodegenerative diseases. *Acta Neuropathol (Berl)* **109**, 5–13.

Sahenk, Z. (1999). Abnormal Schwann cell-axon interactions in CMT neuropathies. The effects of mutant Schwann cells on the axonal cytoskeleton and regeneration-associated myelination. *Ann N Y Acad Sci* **883**, 415–426.

Salzer, J. L. (2003). Polarized domains of myelinated axons. *Neuron* **40**, 297–318.

Savitt, J. M., Dawson, V. L. et al. (2006). Diagnosis and treatment of Parkinson disease: Molecules to medicine. *J Clin Invest* **116**, 1744–1754.

Scherer, S. S. (1997). The biology and pathobiology of Schwann cells. *Curr Opin Neurol* **10**, 386–397.

Shapiro, L., Doyle, J. P., Hensley, P., Colman, D. R., Hendrickson, W. A. (1996). Crystal structure of the extracellular domain from P0, the major structural protein of peripheral nerve myelin. *Neuron* **17**, 435–449.

Shy, M. E., Jani, A. et al. (2004). Phenotypic clustering in MPZ mutations. *Brain* **127**, 371–384.

Siegel, G. (2006). *Basic Neurochemistry: Molecular, Cellular and Medical Aspects, 7th ed.*, 51–52. Elsevier Academic Press, Amsterdam, Netherlands..

Taveggia, C., Zanazzi, G. et al. (2005). Neuregulin-1 type III determines the ensheathment fate of axons. *Neuron* **47**, 681–694.

Tooth, H. (1886). *The Peroneal Type of Progressive Muscular Atrophy*. Lewis, London.

Uyemura, K., Asou, H., Takeda, Y. (1995). Structure and function of peripheral nerve myelin proteins. *Prog Brain Res* **105**, 311–318.

Watson, D. F., Nachtman, F. N. et al. (1994). Altered neurofilament phosphorylation and beta tubulin isotypes in Charcot-Marie-Tooth disease type 1. *Neurology* **44**, 2383–2387.

Webster, H. (1993). Development of peripheral nerve fibers. In Dyck, P. K. T., Low, P. A., Poduslo, J. F., Eds., *Peripheral Neuropathy*, 243–266. WB Saunders, Philadelphia.

Wrabetz, L., D'Antonio, M. et al. (2006). Different intracellular pathomechanisms produce diverse Myelin Protein Zero Neuropathies in transgenic mice. *J Neurosci* **26(8)**, 2358–2368.

Xu, W., Manichella, D., Jiang, H., Vallat, J. M., Lilien, J., Baron, P. et al. (2000). Absence of P0 leads to the dysregulation of myelin gene expression and myelin morphogenesis. *J Neurosci Res* **60**, 714–724.

Xu, W., Shy, M., Kamholz, J., Elferink, L., Xu, G., Lilien, J., Balsamo, J. (2001). Mutations in the cytoplasmic domain of P0 reveal a role for PKC-mediated phosphorylation in adhesion and myelination. *J Cell Biol* **155**, 439–446.

Zuchner, S., Mersiyanova, I. V. et al. (2004). Mutations in the mitochondrial GTPase mitofusin 2 cause Charcot-Marie-Tooth neuropathy type 2A. *Nat Genet* **36**, 449–451.

Zhang, K., Filbin, M. T. (1994). Formation of a disulfide bond in the immunoglobulin domain of the myelin P0 protein is essential for its adhesion. *J Neurochem* **63**, 367–370.

Zuchner, S., Vance, J. M. (2006). Molecular genetics of autosomal-dominant axonal Charcot-Marie-Tooth disease. *Neuromolecular Med* **8**, 63–74.

30

Demyelinating Diseases: Immunological Mechanisms in the Pathogenesis of Multiple Sclerosis

Hartmut Wekerle and Alexander Flügel

I. Introduction

Multiple sclerosis, the most frequent inflammatory demyelination disease, provides an extremely complex clinical and histological picture. The variegated neurological defects are caused by circumscript lesions, which are distributed throughout the white matter of the central nervous system (CNS), and which are characterized by inflammatory and degenerative changes. There is prominent inflammation with round cell infiltrates, local edema formation, and glial activation. At the same time the MS lesion presents with destruction of myelin sheaths and myelin forming oligodendrocytes, as well as with degeneration of the local axons.

There is evidence that these seemingly incoherent changes ultimately are caused by an autoimmune attack, where brain-specific T cells are led to attack the body's own brain. This autoimmune pathogenesis raises numerous questions, most of them directly relevant for diagnosis and treatment of the disease. For example, which is the origin of the autoimmune T cells? Where do they come from, and in which circumstances are they entering the brain? How can they mount an immune attack in the CNS, which is reputed to suppress immune responses within its own confines? How do the autoaggressive T cells interact with local CNS cells, and how do they finally trigger a pathogenic cascade, which ends up in degeneration of myelin structures and neuronal processes? And, most importantly, will it be possible to interrupt therapeutically the autoimmune attack specifically and at an early stage?

This chapter attempts to summarize our present insight in the immune mechanism of MS and to answer the questions raised as far as presently possible.

II. Autoimmunity versus Self-Tolerance

If we weigh the possibility that MS ultimately is caused and driven by an autoimmune reaction, it is crucial to know the mechanisms by which the immune system normally tolerates self tissues, and in which circumstances it initiates an attack against the own organisms. Indeed, the capacity of distinguishing between self (i.e., components of the organism) and potentially dangerous nonself, is a hallmark of multicellular organisms. All along the evolutionary phyla self–nonself discrimination is found in different versions. In vertebrates, it is the immune system, which must distinguish most sharply between self, desirable components of the own body, and nonself structures. Immune responses react against foreign, potentially dangerous invaders with the ultimate aim to destroy and eliminate them; at the same time, however, they tolerate the organism's own cells and tissues.

The immune system learns to distinguish between self and potentially dangerous nonself structure in primary immune organs, the site, where the mature immune repertoires are generated. T cells learn self-tolerance in the thymus, B cells, instead, in the bone marrow. During T and B cell differentiation, the maturing cells go through cellular milieus, where they are confronted with stromal cells, which present a spectrum of organ-specific self antigens, which grossly represent the universe of tissue specific self-antigens found on the body's different organs and tissues. Each encounter between a young T cell and these stromal cells contributes to immunological self-tolerance.

Although in the case of B cell tolerization these processes remain still largely obscure, there is an emerging picture of intrathymic tolerization of T cells (Kyewski & Klein, 2006). The thymus is the site where the diverse T cell receptor repertoire is created, and, implicitly, where self-tolerance is generated. Both processes are the results of contact interactions between the T cell on the one side, and the stromal cells—thymus epithelium and thymic dendritic cells (DCs)—on the other (Ladi et al., 2006). In a first step, T cells develop in the cortex to reach a stage close to maturity. At this point the "pre-T cell" population includes many self-reactive clones, cells that have receptors reacting against structures of regular tissues. Most of these self-reactive T cells are purged out in a subsequent step of maturation, when they leave the cortical milieus and enter into the thymic medulla. In this new milieu, the fresh T cells encounter now medullary epithelial cells and DCs, which again, like cortical epithelium, expose on their surface self antigens. But in the medulla, exposure of T cells to self antigens results in physical elimination of the self-recognizing T cells.

At this point, the emerging T cell repertoire is purged of potentially hazardous self-reactive T cells. However, where would the many autoantigens come from? Medullary epithelial cells form mosaics composed of cell groups that produce genuine organ-specific self-antigens, thus mimicking pancreatic islet cells, liver, and many others. Interestingly, the self antigens presented within the thymus medulla include most, if not all determinants of CNS and testis, the classical organs enjoying immune privilege (Kyewski & Klein, 2006). Both organs are secluded from blood circulation by a tight endothelial blood-tissue barrier, and thus they would be out of reach to T cells irrespective of their specificity for self or nonself antigen.

In general, thymic depletion of autoreactive T cell clone works quite reliably, but still, the mechanism is leaky. Especially with regard to CNS self-antigens, thymic elimination of complementary, CNS autoreactive T cells is far from complete. In fact, the immune repertoire of healthy rodents and humans abounds with myelin-specific T cell clones. Thus, organ-specific autoreactive T cell clones constitute a sizeable proportion of the circulating mature immune cell compartment.

The leakiness of central thymic tolerance mechanisms may appear frightening. Autoreactive T cells sneaking through negative selection could attack at any time specific tissues. Fortunately, however, the central self-tolerance is complemented by tolerogenic mechanisms acting in the peripheral immune system, the lymph nodes, and spleen. There, T cells are silenced and in some cases even eliminated, when confronted to their (self-) antigen in the absence of adequate costimulatory molecules. These structures are up-regulated in antigen presenting cells in response to danger signals, for example, pathogen-associated molecular patterns (PAMPs). In addition, autoreactive T cells are kept in check by coinhibitory signals and specialized immune regulatory immune cells.

There are several scenarios, however, to break tolerance, the most popular one via *antigenic mimicry*, a concept referring to sequence similarities between peptides from microbes and autoantigens. An autoreactive T cell can be activated erroneously by a microbial peptide, which shares crucial structural features with a self-antigen. Upon such activation, the cross-reactive T cell is primed to attack the tissues producing and presenting the self-antigen in question. Alternatively, autoreactive T cells can be activated by microbial superantigens or by antigen-nonspecific proinflammatory factors of the innate immune system, which are abundantly expressed in the inflammatory milieu. Finally, loss of tolerance might occur by weakening of the regulatory mechanisms. However, as we will discuss later, the brain tissue might make use of another tolerogenic mechanism, *immune ignorance*, that is, the inability of autoreactive T cells to get in contact with body antigens due inaccessibility of the organ.

III. Immune Reactivity in the CNS

A. Immune Privilege

Healthy CNS tissues provide a barren ground for the unfolding immune reactions. Indeed, more than half a century ago, Sir Peter Medawar noted that tissues grafted on brain cortex of normal rabbits are much less readily rejected than similar grafts transplanted on a skin bed (Medawar, 1948). Brain-grafted skin was not attacked by the recipient's immune system, but tolerated; the graft enjoyed a privileged situation. Medawar's school extended this type of immunological nonreactivity to other organs, coining the term *immune privilege* (Barker & Billingham, 1977).

The immune privileged status of the central nervous system (CNS) is maintained by a number of structural peculiarities, which distinguish the organ from other tissues, and which stand against successful immune reactions. First, brain and spinal cord parenchyma are secluded from the rest of the body. Its blood vessels are lined by specialized endothelial cells interlaced by tight junctions forming a blood–brain barrier (BBB) that holds back most blood-borne macromolecules and cells. Furthermore, in contrast to most other tissues in the body, the CNS lacks a fully organized lymphatic vasculature, which would drain CNS interstitial fluid into the periphery and allow the exit of antigen laden antigen presenting cells into lymph nodes and spleen.

Second, and most important, the normally functioning CNS tissues form a milieu unfavorable to an unfolding immune response. The CNS tissue fails to contain professional antigen presenting cells, dendritic cell (DC) population, which would be able to activate naïve, antigen unexperienced T cells. There may be some DCs in the leptomeninges or in the choroid plexus (Matyszak & Perry, 1996; Serafini et al., 2000), but in the CNS parenchyma it is the microglia, specialized phagocytes derived from bone-marrow progenitors, that serves as the brain's first-line police (Kreutzberg, 1996). Microglial cells are able to spot intruded microbes, and also respond to other potential threats, such as those surrounding neuronal degeneration (vide infra). However, in the healthy CNS, microglial cells are not efficient antigen presenting cells. They are armored with pattern recognition receptors of the innate immune system, receptors that recognize microbial structures, along with extruded nucleic acids. But in health, resting microglial cells rarely produce MHC antigens and costimulatory molecules in levels sufficient to allow antigen presentation and activation of T lymphocytes. In general, intact CNS tissues lack most of the structures that are required by immune cells to survive and to properly function in order to achieve a productive immune response. This deficit includes, besides MHC and costimulatory molecules, cell adhesion molecules, cytokines, and chemokines.

The conspicuous lack of immune molecules in healthy CNS tissues is by no means due to the incapacity of neural cells of producing these structures, but to the active repression of their production by electrically active neurons. In fact, neuronal activity not only represses immune-relevant gene expression on local glia cells, but on the neuronal membrane itself.

Tissue culture studies indicated that mature, functional neurons fail to express MHC class I proteins. These can, however, be induced readily by paralyzing the neurons, for example using the sodium channel blocker tetrodotoxin and simultaneous stimulation with pro-inflammatory cytokine interferon-γ(IFN). Such treatment leads to the graded induction of all components required for the formation of MHC class I molecules: These include, besides class I heavy and light chains (β_2-microglobulin), the TAP family of peptide transporters (Neumann et al., 1995). Once these gene elements are derepressed, the neurons expose structurally intact MHC class I protein on their membrane, and are able to present class-I-bound peptides to specific $CD8^+$ T cells. As a consequence, class I induced neurons presenting viral peptide are recognized and attacked by virus-specific cytotoxic T cells (Medana et al., 2000).

Apart from this classically immunological role, MHC class I expression may have nonimmune, developmental functions. In one study, MHC class I expression was detected in certain neurons, especially in the developing CNS (Huh et al., 2000), whereas another group related MHC class I expression to synaptic plasticity (Oliveira et al., 2004).

By which mechanisms could neuronal activity repress MHC repression? It is known that neuronal membrane activity affects gene expression in several complex ways. Calcium fluxes have a central role in these signaling cascades that involve transcription factors like the cyclic-AMP-response-element-binding protein (CREB), c-fos and NF-AT (nuclear factor of activated T cells) and NF-κB (nuclear factor-κB), and that initiate diverse programs of gene expression (West et al., 2002). Electrically active neurons intensively interact with surrounding glial cells. The signals used include soluble mediators, molecules involved in surface contacts, but also ion fluxes that act on neighboring cells. But the signaling is by no means unidirectional; the glial cells themselves talk back to the neuron modifying neuronal function (Fields & Stevens-Graham, 2002).

Anti-inflammatory mediators come in manifold guise. These agents include anti-inflammatory cytokines, receptor antagonists, and neurotransmitters. Previous studies established that neurotrophins are particularly efficient mediators of MHC class I repression. In organotypic explant cultures of neonatal hippocampus tissue, addition of antibodies neutralizing neurotrophins facilitates induction of MHC class II proteins on microglia cells, and addition of exogenous neurotrophins reverses their expression (Neumann et al., 1998). Interestingly, several neurotrophic factors act also *in vivo* as anti-inflammatory agents. Thus, infusion of soluble NGF (Villoslada et al., 2000), or local deposition by engineered, NGF secreting T cells (Flügel et al., 2001b) reduced or suppressed Experimental Autoimmune Encephalomyelitis (EAE) in rodents or primates.

B. Inflammation—Degeneration

The brain's immune-privileged status is efficient, but it is conditional. Although immune privilege in the CNS is reliably upheld in health, there are diverse pathological conditions that facilitate *de novo* induction of immune genes. These conditions include autoimmune inflammation, such as in MS and in animal models, EAE. The diseases are created by autoimmune T cells that enter the CNS, release there pro-inflammatory mediators powerful enough to overcome the anti-inflammatory regulation by neurons. Consequently, the immune-privileged CNS milieu is mutated to a milieu permissive and supportive of immune reactions (Huang et al., 2000). Similar changes are noted in CNS responses to microbial infections. The infected tissue displays expression of MHC antigens as well as cytokines and cell adhesion molecules required for immune reactions. These genes, however, are activated mainly via responses knitting together innate and adaptive immune reactivity. Microglia cells are the first-line sentinels channeling this response, sensing intruded foreign structures, toll-like receptors, and other pattern recognition receptors (Nguyen et al., 2002).

The most intriguing pathological condition, which furthers local immune reactivity in the CNS tissue, revolves around neuronal degeneration. Indeed, neural cell degeneration and inflammation are tightly connected, and their relationship is reciprocal. Inflammatory responses go along with endogenous neurodegenerative processes, as they underlie for example Alzheimer's and Parkinson disease, amyotrophic lateral sclerosis, and adrenoleukodystrophy, but also with neurodegeneration following exogenous noxes, such as trauma, ischemia. As pointed out earlier, it seems that in all these situations, the weakening of neuronal function relaxes repression of immune genes, and thus creates a milieu that supports inflammation and immune reactivity.

Much of our insights into paradegenerative inflammation come from rodent models based on the disruption (axotomy) of peripheral nerves. In the facial nerve model developed in the group of Georg Kreutzberg, the facial nerve, a purely motor nerve, is clipped or crushed in its peripheral part. As a response to this mechanical lesion, degenerative processes migrate back into the nerves central origin, the facial nucleus, which is embedded in the brainstem. Dependent on the circumstances, neuronal degeneration can result in apoptotic death, or remain sublethal. As a hallmark of the degenerative response, the synapses touching the neuronal somata are removed, "stripped" from the neuronal membrane by activated microglia cells, which in the end form a tight sheath surrounding the neuron (Raivich et al., 1999).

Activation is not the only change in the facial nucleus triggered by the peripheral nerve lesion. Neuronal degeneration kicks off a large-scale remodeling of the local microenvironment. In particular, expression of genes relevant for immune responses are induced following neurodegeneration (Moran & Graeber, 2004). As a result, the immune privileged CNS tissue turns hospitable for immune reactions. Indeed, in the facial nerve model, the lesioned nucleus spontaneously attracts immune cells in untreated rodents (Raivich et al., 1998). In animals with ongoing EAE, the degenerating facial

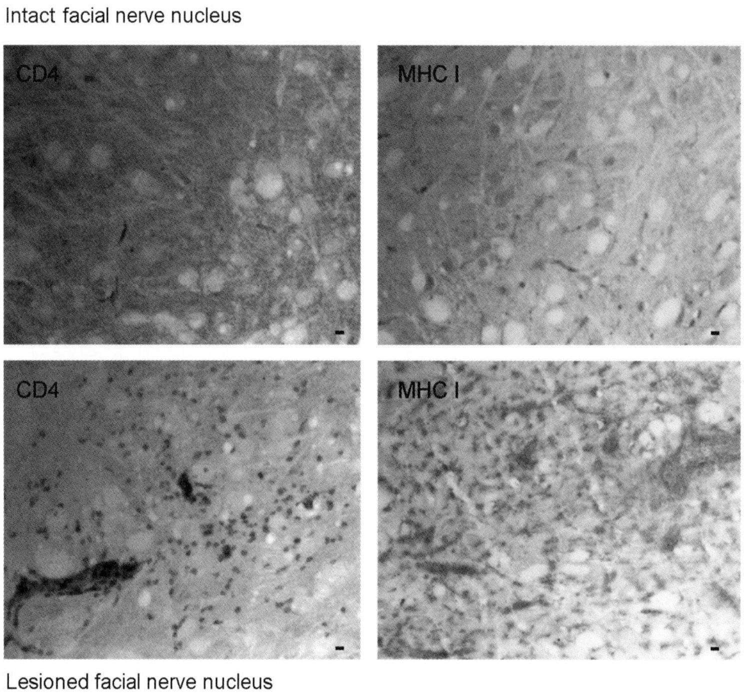

Figure 30.1 Immune cell ($CD4^+$ cells, left) infiltration and MHC class I (right) upregulation in the degenerating facial nerve nucleus of rats 3 days following facial nerve axotomy and tEAE induction by transfer of 3×10^6 MBP-specific T cells. Upper panels: counter-lateral nonlesioned facial nerve nucleus. Lower panels lesioned nucleus. Immunohistochemistry. Magnification bar: 10 μm.

nucleus becomes densely infiltrated, whereas the counterlateral intact nucleus remains free of inflammatory cells (see Figure 30.1) (Maehlen et al., 1989). Conversely, focal CNS inflammatory reactions can cause neurodegeneration with disruption of local axons. Injection of pro-inflammatory cytokines into corticospinal tracts of rat subclinically sensitized against myelin oligodendrocyte glycoprotein (MOG) creates local infiltrations with destruction of myelin and axons (Kerschensteiner et al., 2004).

These experimental models describe the behavior of autoimmune, pathogenic T cells, and it should be kept in mind that autoimmunity represents a pathogenic exaggeration of physiological immune responses. Thus, the behavior of autoimmune T cells within the CNS stands for the behavior of T cells participating in a fresh immune response against foreign antigen.

IV. MS—Immune Pathogenesis?

Many neurologists and immunologists are convinced that MS has an autoimmune pathogenesis, a concept that leads them to design therapies directed at autoimmune processes. But, what is the evidence that supports the autoimmune hypothesis? Further, if indeed there are autoimmune responses driving the pathogenesis of MS, are they primary or are they merely secondary consequences of other pathogenic processes?

These are not easy questions. In animal models, ultimate proof of T cell mediated autoimmune diseases demands transfer of the disease. For example, CNS autoreactive T cells have been isolated from rodents with actively induced EAE. The T cells were propagated in cultures as pure lines, and, when transferred into naïve animals, they mediated classical acute EAE (Ben-Nun et al., 1981). Obviously, transfer of suspected autoimmune effector cells from MS patients to healthy recipients is impossible, which foils attempts to formally prove a putative autoimmune pathogenesis of MS.

Understanding the pathogenesis of MS is further complicated by the enormously complex and varied nature of the disease. Course and clinical picture are rarely the same in two patients. To add another dimension of complexity, lesion patterns can be divided in several distinct archetypes, each one suggesting a distinct inflammatory effector pathway. The following lines of evidence support the autoimmune concept of MS.

A. Genetic Control of Disease Susceptibility

Clearly, susceptibility to MS is strongly influenced by genetic factors. Consider, for example, disease concordance in monozygotic (identical) and dizygotic twin pairs. In monozygotic twins the probability that both siblings develop the disease is high, 1:3. In sharp contrast, in dizygotic twins, whose genetic similarity corresponds to nontwin brothers and sisters, the probability of double affection is much lower, 1:25 (Willer et al., 2003). The high concordance in monozygotic twins points to the importance of genes. However, the fact that in these identical twins concordance is lower than 100 percent leaves room for additional, nongenetic factors, as contributed by the environment. Large scale genomic screenings have established that most, probably a large number of genes can affect the development of MS, but is has been extremely difficult to identify such factors beyond doubt (Dyment et al., 2004).

Numerous gene polymorphisms have been proposed as "MS genes," but few if any have stood the test of time. The only class of genes that contribute to MS receptiveness without any doubt are class II genes of the human Major Histocompatibility Complex, HLA. Most importantly, a firm association of HLA DR2 and MS was discovered as early as in 1973, (Jersild et al., 1973), and confirmed later in numerous populations. DR2 is an "immune" gene par excellence, which controls and directs T cell immunity against self and foreign antigens on several levels. As a class II gene, DR2 influences the clonal composition of $CD4^+$ T cell repertoire. In the mature immune response, DR2 binds a large, though limited spectrum of (auto)antigenic peptides and presents them to the antigen receptors of specific $CD4^+$ T cells. Indeed, DR2 analogs strongly control autoimmune reactions against brain structures in experimental models, both in rodents as in primates ('t Hart et al., 2001). Considering these functions, the association of DR2 with MS suggests a participation of immune reactivity in the pathogenesis of the disease.

B. Myelin-specific Antibodies and T Cells in MS Patients and Humanized Mice

Having studied a large panel of MS plaque samples, biopsies as well as necropsies, Lassmann and Lucchinetti distinguished four particular patterns of structural changes (Lassmann et al., 2001). The first pattern is dominated by round cell infiltrates and activated macrophages and microglia cells, suggesting a T cell driven, macrophage executed pathogenesis, such as seen in classical rodent models of EAE. Although this observation per se by no means *proves* autoimmune T cells as the actual pathogens, there is additional evidence of such a role. Oksenberg et al. (1993) cloned the genes of TCR sequence out of MS lesions and found sequences similar or even identical with TCR β chains of myelinbasic protein (MBP)-specific T cell clones isolated from the peripheral blood of MS patients. Later, Fugger and colleagues introduced both receptor chains of an MS-derived, MBP-specific T cell into the genome of a mouse transgenic for human DR2. A proportion of these double-transgenic, "humanized" mice developed EAE, with CNS lesions that recapitulated at least part of the active MS lesion (Madsen et al., 1999). These findings support, but don't prove, autoimmune attacks in the development of at least some MS lesions.

Autoimmune B cells and their autoantibody products, especially those binding MOG, could have a role in the development of another class of MS lesions, Lassmann's pattern II. Pattern II lesions show large-scale demyelination with myelin debris often decorated with immunoglobulin and activated complement (Storch et al., 1998a). These changes strongly resemble the lesions induced in monkeys by immunization with recombinant MOG protein, where autoantibodies are known to have an active demyelinating function (Genain et al., 1995a). In line with these reports, soluble anti-MOG autoantibodies have been extracted recently from CNS tissue from MS patients (O'Connor et al., 2005). Possibly the most significant argument in favor of pathogenic autoantibodies comes from a recent study of rare cases, where pattern II lesions were identified in biopsies. In these patients, removal of circulating autoantibodies led to alleviation of neurological defects (Keegan et al., 2005).

C. Immunomodulatory Therapies

The ultimate argument for autoimmunity in MS would be a successful therapy specifically removing the autoimmune effector cells. Such a radical therapy is not available today, but several of the drugs currently used for treatment of MS are deemed to act by modifying immune cells. Interferon-β and copaxone, for example, have multiple effects, such as the ability to drive putatively pathogenic Th1 T cells into the beneficial Th2 lineage (Neuhaus et al., 2001). Th1 to Th2 conversion also appears to be the mechanism underlying vaccination with altered peptide ligands, APL. These are synthetic peptides analogous to major autoantigenic peptide epitopes on myelin proteins, which act as partial agonists. They do activate myelin-specific T cells but instead of driving them to the Th1 lineage, they push the T cells to the regulatory Th2 side. Vaccination with MBP-analog APL were beneficial in some, but not in all patients (Bielekova et al., 2000; Kappos et al., 2000). Then, blockade of migration (Miller et al., 2003) and physical deletion of T cells (Coles et al., 1999) are associated with a reduction of MS relapses and the lesion load in the CNS.

D. The Active MS Plaque Resembles Lesions of EAE Models

Although as mentioned, MS lesions are highly diverse, and have been separated into several archetypes, each one representing the result of distinct pathogenic pathways, there seems to be a common denominator. The early, florid MS lesion commonly is dominated by a stereotypic set of changes. Fresh MS plaques are dominated by conspicuous inflammatory infiltrates, mainly arranged around small postcapillary venules, but also disseminating into the surrounding parenchymal areas. These infiltrates are composed of mononuclear cells, including $CD8^+$ T cells and monocytes/macrophages. The normally tight BBB is ruptured, causing formation of local edemas. Then, one finds activation of glial cells (astrocytes as well as microglial cells). Finally, axonal myelin sheaths and myelin-forming oligodendrocytes are destroyed, and there is marked disruption of local axonal processes. All these changes are seen also in some, though not all, EAE models, which are undisputedly caused by a T cell autoimmune attack (Storch et al., 1998b).

Thus, it appears fair to state that so far the autoimmune concept of MS still waits to be proven, but that there is impressive indirect evidence in support of this concept. Obviously, there is a strong need to develop technologies that allow the identification and characterization of putative immunological effector cells of their target structures in CNS and immune system, and which will be of help in using this knowledge for the development of more specific and efficient therapies.

V. Animal Models of MS—Experimental Autoimmune Encephalomyelitis (EAE)

Research into the pathogenesis of human diseases is tightly restricted by ethical and biological limitations. In the case of autoimmune diseases, for example, identification of a putative pathogenic autoimmune agent cannot be done by its transfer to another healthy human recipient. Instead we depend on suitable experimental animal models. This need is particularly obvious for diseases of the CNS, which are especially complex, and whose lesions cannot be routinely biopsied for recovery and study of the changed tissue.

Animal models should represent the human disease as closely as possible and should allow approaches that are impossible in human research, namely the direct study of the lesioned tissue and the evaluation of the pathogenic function of cellular and molecular processes over time.

Experimental Autoimmune Encephalomyelitis (EAE) in its multiple variants is used commonly as a model for MS. Although no single EAE variant represents human MS in all aspects and in its entire complexity, there are different EAE versions that quite faithfully represent particular features of the human disease. This chapter will describe some of the most representative EAE variants.

A. Induction of EAE and Its Target Autoantigens

Rivers and colleagues often are quoted as the pioneers of EAE. They induced demyelinating encephalomyelitis in rhesus monkeys after repeated (>50 times) inoculation with rabbit CNS (Rivers et al., 1933). Based on this observation, subsequent research extended active EAE induction to a large number of additional species, especially rodents (see Table 30.1), and thus opened a new and enormously productive field of research. Among many other accomplishments, EAE studies led to the formulation of complete Freund's adjuvant mycobacteria in mineral oil/water emulsion of overriding immunogenicity (Freund et al., 1947).

Table 30.1 Selected Rodent and Primate EAE Models

Species Strain	Antigen, Reference	Mode of Induction	Disease Pattern	MS Features and Pathology	References	MHC Restriction
MOUSE-EAE						
C57BL6	***MOG (aa 35-55)***	aEAE, pEAE	Chronic-progressive	Demyelination, acute inflammation, axonal damage	Mendel et al. (1995); Bernard et al. (1997)	**H-2^{b}**
C3H.SW	**MOG** (**aa 35–55**)	aEAE	Chronic	Inflammation, demyelination	Mendel et al. (1995)	**H-2^{b}**
SJL	***Spinal cord homogenate***	aEAE	Relapsing-remitting	Demyelination, inflammation	Lublin et al. (1981); Brown et al. (1982)	**H-2^{s}**
	MBP	aEAE	Relapsing-remitting	Demyelination, inflammation	Pettinelli et al. (1982); Lublin (1985); Merrill et al. (1992)	
	PLP	aEAE, pEAE	Relapsing-remitting	Demyelination, inflammation	Tuohy et al. (1988); Sobel et al. (1990); Kuchroo et al. (1992)	
	MOG (aa 92–106)	aEAE	Severe acute	Inflammation, demyelination	Amor et al. (1994); Tsunoda et al. (2000)	
	MOBP (myelin-associated/ oligodendrocyte basic protein) (aa 37–60)	aEAE	Acute	Inflammation	Holz et al. (2000)	
	CNPase (2′,3′-cyclic nucleotide 3′-phosphodiesterase)	aEAE	Acute	Inflammation	Morris-Downes et al. (2002)	
	Oligodendrocyte-specific glycoprotein (OSP)	aEAE	Acute	Inflammation	Morris-Downes et al. (2002)	
B10.S	***CNS homogenate***	aEAE	Resistent		Linthicum & Frelinger (1982)	**H-2^{s}**
	MOG	aEAE	Resistent		Maron et al. (1999)	
ASW	***MOG (aa 92–106)***	aEAE	Depending on Bordetella pertussis application: + secondary progressive (SP)-primary progressive (PP)	Acute inflammation, demyelination	Tsunoda et al. (2000)	**H-2^{s}**
Biozzi	***Spinal cord homogenate***	aEAE	Relapsing-remitting	Inflammation, demyelination in relapse phases	Baker et al. (1990)	**H-2^{g7}**
	MBP (aa 12–26; 21–35)	aEAE	Native MBP: resistant, MBP 12–26 and MBP 21–35: mild acute	Inflammation	Amor et al. (1996)	
	MOG (aa 1–22; 43–57; 134–148)	aEAE	Chronic relapsing (MOG 1–22) Chronic acute (MOG 35–55)	Inflammation, no demyelination Native 'MOG: demyelination	Amor et al. (1994); Smith et al. (2005)	

(Continued)

Table 30.1 Selected Rodent and Primate EAE Models—Cont'd

Species Strain	Antigen, Reference	Mode of Induction	Disease Pattern	MS Features and Pathology	References	MHC Restriction
	PLP (aa 56–70)	aEAE	Chronic relapsing	Inflammation, demyelination	Amor et al. (1993)	
	Ab-crystallin (aa 1–16)	aEAE	Acute mild	Inflammation	Thoua et al. (2000)	
	Glial fibrillary acidic protein GFAP	aEAE	Acute severe	Inflammation	Amor et al. (2005)	
NOD	***MOG (aa 35–55)***	aEAE	Chronic relapsing remitting resistant	Inflammation, demyelination	Bernard et al. (1997)	**H-2^{g7}**
BALB/c	**Spinal cord homogenate**	aEAE			Levine and Sowinski (1974); Teuscher et al. (1987); Linthicum & Frelinger (1982)	**H-2^{d}**
	MBP (aa 59–76)	aEAE, pEAE	aEAE resistant, pEAE: acute	Inflammation, demyelination	Abromson-Leeman et al. (1995)	
	MOG (aa 35–55)	aEAE	Partially resistant, mild acute	Inflammation	Zhu et al. (2006)	
	PLP (aa 40–59)	aEAE	Resistant		Hurwitz et al. (2002)	
PL/J	***MBP (aa 1–37 and 89–169)***	aEAE	Chronic relapsing	Demyelination, acute inflammation	Zamvil et al. (1985a); Zamvil et al. (1985b)	**H-2^{u}**
	MOG (aa 35–55)	aEAE	Relapsing-remitting	Demyelination, acute inflammation	Kerlero de Rosbo et al. (1995)	
	PLP (aa 43–64)	aEAE, pEAE	Chronic relapsing	Demyelination, acute inflammation	Kerlero de Rosbo et al. (1995)	
RAT EAE Lewis	**MBP (aa 68–88)**	aEAE, pEAE	Acute, relapses after cyclosporine	Inflammation	Ben-Nun et al. (1981)	**RT1-l**
	PLP	aEAE pEAE	Weak acute	Inflammation	Chalk et al. (1994); Stepaniak et al. (1995); Yamamura et al. (1986)	
	MOG (an 35-55)	aEAE, pEAE	Weak acute, after native MOG chronic progressive	Inflammation, demyelination with native MOG	Adelmann et al. (1995); Johns et al. (1995); Linington et al. (1993)	
	S100β (αα 76–91)	Passive	Weak acute	Inflammation	Kojima et al. (1994)	
	GFAP	pEAE	Weak acute	Inflammation	Berger et al. (1997); Wekerle et al. (1994)	
Dark Agouti (DA)	**Spinal cord homogenate**	aEAE in-complete FA	Chronic relapsing EAE	Inflammation, demyelination	Lorentzen et al. (1995)	**RT1-av1**
	MBP (aa 62-75)	EAE, pEAE	Acute	Inflammation	Miyakoshi et al. (2003); Stepaniak et al. (1995); Stepaniak et al. (1997); Smeltz et al. (1998)	

		aEAE, pEAE	Chronic, Relapsing-remitting	Inflammation, demyelination	Papadopoulos et al. (2006); Stefferl et al. (2000); Storch et al. (1998b)	
	BTN Butyrophilin (aa 74–90)	aEAE	No disease	Inflammation	Stefferl et al. (2000)	
Brown Norway (BN)	**MBP**	pEAE	Acute	Inflammation	Happ et al. (1988)	**RT1-n**
	Rat spinal cord	aEAE	Acute	Inflammation	Levine & Sowinski (1975)	
	MOG	aEAE	Chronic progressive	Inflammation, demyelination	Stefferl et al. (1999)	
PVG	**MBP**	Active resistant, pEAE susceptible	Acute	Inflammation	Ben-Nun et al. (1982)	**RT1-c**
NONHUMAN-PRIMATE-EAE						
Macaca mulatta (Rhesus monkey)	**Rabbit brain extract**	aEAE	Acute and chronic progressive	Inflammation, demyelination	Rivers et al. (1933); Kabat et al. (1947)	Mamu-DPB1*01, outbred
	Brain extract with adjuvant	aEAE	Acute and chronic	Inflammation with necrosis		
	MOG (aa 4–20, 35–50, 94–116)	aEAE	Acute and chronic	Inflammation with necrosis	Brok et al. (2001); Kerlero de Rosbo et al. (2000)	
	MBP (aa 61–82, 80–105, 170–186)	aEAE, pEAE	Acute and chronic	Inflammation with necrosis	Rose et al. (1994); 't Hart et al. (2005); Meinl et al. (1997)	
Macaca fascicularis (Cynomolgus monkey)	**Brain homogenate**	aEAE	Acute and chronic relapsing	Inflammation, necrotic lesions	Stewart et al. (1991); Massacesi et al. (1992)	Outbred
	MBP	aEAE	Acute and chronic relapsing	Inflammation necrotic lesions	Massacesi et al. (1992); Stewart et al. (1991)	
Macaca nemestrina	**Brain homogenate and MBP**	aEAE	Relatively resistant		Brok et al. (2001)	
Common marmoset Calithrix jacchus	**Myelin homogenate**	aEAE	Acute – chronic progressive	Inflammation, demyelination, necrotic lesions	't Hart et al. (1998); Mancardi et al. (2001)	Caja-DRB*W1201, outbred
	MBP	aEAE, pEAE	Mild acute	Inflammation, no demyelination	Genain et al. (1995a); Genain et al. (1995b); Mancardi et al. (2001)	
	PLP	active	Mild acute	Inflammation, no demyelination	Genain & Hauser (1997)	
	MOG (aa 1–124, 14–36)	Active	relapsing remitting, secondary progressive	Inflammation, demyelination	't Hart et al. (1998); Genain & Hauser (2001)	

The clinical character of active EAE—severity and neurological defects—is variably between individual models; it depends strongly on the species/strain of the experimental organism, the neural autoantigen, and the type of adjuvant (see Table 30.1).

Most EAE models are driven by T cells and macrophages, but less so by B cells. This was formally proven by the isolation of $CD4^+$ T cells reactive against defined brain autoantigens and their transfer of EAE to naïve recipients (Ben-Nun et al., 1981; Mokhtarian et al., 1984). The T cell transfer approach demonstrated that encephalitogenic potential is by no means limited to myelin-specific T cells, but that many, if not all CNS proteins can act as targets of autoimmune responses (see Table 30.1) (Wekerle et al., 1994).

Naturally occurring MS-like autoimmune CNS diseases in animals, such as comparable to autoimmune diabetes mellitus in the NOD mouse, are not known. However, there are several T cell receptor (TCR) transgenic mice (i.e., mice with defined autoimmune TCR α and β chains inserted into their germline) that develop spontaneous CNS inflammation (see Table 30.2). The immune repertoires of these mice are dominated by T cells responsive to defined antigens. The first myelin-specific TCR-transgenic mice had T cells that recognize MBP Ac1-10 in MHC class II I-A^u context (Goverman et al., 1993). When kept in a conventional, "dirty" environment, some of these mice (14–40%) developed spontaneous EAE. Under SPF conditions, the mice remained healthy, which may point to environmental factors contributing to the trigger of the autoimmune attack. Spontaneous EAE also was noted in other TCR transgenic mice, specific for MBP, proteolipid protein (PLP), and MOG. Lafaille et al. detected that EAE incidence rises to 100 percent in MBP reactive TCR transgenic, RAG-2 knock-out that lack an endogenous adaptive immune system (Lafaille et al., 1994). In the double-transgenic mice, EAE development was suppressed by transfers of CD4+CD25+ regulatory T cells (Furtado et al., 2001). Another interesting set of transgenic mice with spontaneous EAE carry human myelin-specific TCRs derived from peripheral blood lymphocytes of patients with MS. Disease incidence in this humanized TCR transgenic mice also was increased in RAG-deficient mice (Madsen et al., 1999). Most recently, two groups described double-transgenic mice with MOG-specific T and B cells, which develop at a high rate spontaneous optico-spinal EAE in the presence of an intact immune system (Krishnamoorthy et al., 2006), and (Bettelli et al., 2006a) a model, which underlines the potential importance of T-B interactions in triggering and shaping autoimmune responses (see Table 30.2).

B. The EAE Lesion: Pathology and Distribution

The hallmarks of active MS lesion comprise acute infiltration by inflammatory cells, gliosis (astrocytic activation) demyelination (sometimes coincident with some remyelination), axonal degeneration. Most EAE models just partially reflect part of, but not the complete spectrum of this pathology (see Table 30.2). Thus, classical MBP-induced EAE lesions are characterized by acute inflammation of the CNS with breakdown of the BBB and gliosis, but not large-scale demyelination (see Table 30.2). Like in MS, EAE infiltrates are formed mainly by T cells and activated macrophages and preferentially arranged around small venules and the surrounding parenchyma. In addition, the autoaggressive cells infiltrate deeply into the parenchyma of CNS white and gray matter, as revealed in studies using encephalitogenic T cells labeled with fluorescent proteins (green fluorescent protein, GFP; see Figure 30.2).

Neuronal degeneration seems to be a less prominent feature of acute EAE models and axonal pathology is mostly transient (Aboul-Enein et al., 2006; Smith et al., 2000). Overt destruction of axons with degenerative response, however, is seen regularly in chronic EAE (Gold et al., 2006; Kornek et al., 2000) (see Table 30.2).

Recently, gray matter (cortical) lesions and diffuse axonal damage in "normal appearing white matter" (the tissue surrounding demyelinating plaques) during disease progression came into the focus of MS research (Kutzelnigg et al., 2005). Clearly, gray matter lesions are noted in models of chronic EAE (e.g., MOG EAE in marmosets and in LEWI.W and LEWI.ARI rats (Gold et al., 2006; Merkler et al., 2006; Pomeroy et al., 2005)), and can be induced experimentally by deposition of inflammatory stimuli in the gray matter of primed

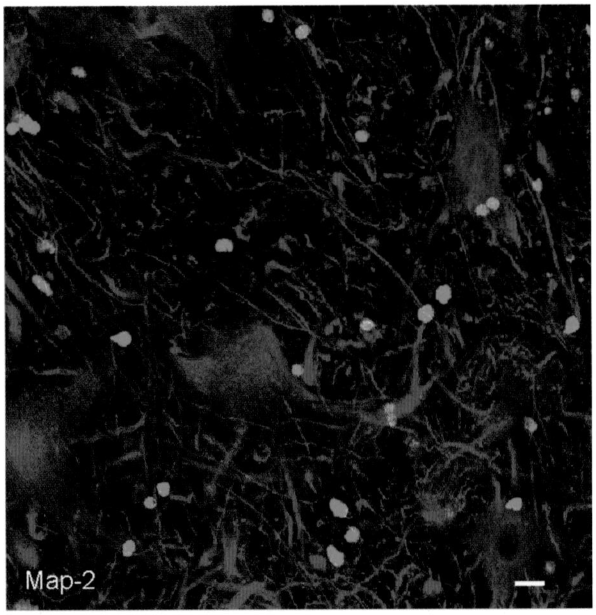

Figure 30.2 GFP^+ MBP-specific T cells in spinal cord gray matter 4 days after induction of tEAE by intravenous transfer of 3×10^6 cells. Magnification bar: 10 μm. Green: MBP-specific T cells. Red: Map-2^+ neuronal cells and processes.

animals (Kerschensteiner et al., 2004; Phillips et al., 1995; Sun et al., 2004). However, none of the currently available models presents the global neural degenerative changes in normal appearing white matter seen in MS (see Table 30.2).

The distribution of the pathological lesions in conventional EAE differs significantly from the one in MS. The predilection sites of MS, brainstem, cerebellum, optic nerve, and periventricular area of the forebrain typically are spared in most EAE variants. Instead, classical EAE lesions (e.g., after transfer of MBP specific T cells) predominantly start up in the spinal cord, forming a stereotypic caudal-cranial gradient. However, other distribution patterns can be pro-

Table 30.2 Representation of MS Hallmarks by EAE Models

MS Course	MS Clinic and Lesion Distribution	MS Pathology	MS Pathomechanisms
Acute demyelinating encephalmyelitis (ADEM): "classical" EAE models: Mouse: SJL mice: MOG-aEAE/tEAE Rat: Lewis rat: MBP-aEAE/tEAE Non-human primate: Rhesus monkey: Myelin, MBP, MOG aEAE	**Spinal cord lesions** (pareses, vegetative signs): Most mouse and rat EAE models e.g.: Mouse: C57BL6 MOG aEAE (Mendel et al., 1995; Bernard et al., 1997) Rat: Lewis rat EAE (Ben-Nun et al., 1981)	**Acute inflammation with perivenous round cell infiltrates:** All EAE models	**CD4+ T cells:** Underlie pathogenesis of most "classical" EAE models: Proof of concept: Adoptive transfer EAE (Ben-Nun et al., 1981; Huitinga et al., 1990) Depletion of CD4+ T cells abrogates EAE Protective role of CD4+ "regulatory" T cells:
Relapsing remitting (RR): Mice: SJL: MBP, PLP-aEAE (Pettinelli et al., 1982; Lublin, 1985; Merrill et al., 1992; Tuohy et al., 1988; Sobel et al., 1990; Kuchroo et al., 1992) (B6 × SJL) F1: MOG-aEAE (Skundric et al., 2003; Zhang et al., 1997) PL/J: MOG aEAE (Kerlero de Rosbo et al., 1995) Rat: DA: MOG (Papadopoulos et al., 2006; Stefferl et al., 2000; Storch et al., 1998b) Nonhuman primate: Marmoset: Myelin, MOG ('t Hart et al., 1998)	**Cerebral lesions:** brain stem, cerebellar involvement (ataxia, spinning): Mouse: C3/HEJ PLP (aa 190–209, 215–232), TcR transgenic IFN-γ deficient mice (Wensky et al., 2005) Rat: Panencephalitis in S100β-EAE of Lewis rats (Kojima et al., 1994) Nonhuman primate: Most EAE models in rhesus and common marmosets ('t Hart et al., 2005)	**Demyelination/remyelination:** T cell- and macrophage-mediated: chronic EAE models in mice, e.g. MOG peptide aa 35-55-induced EAE in C57/BL6 mice Antibody-contribution: aEAE with MOG (extracellular part, refolded protein) in susceptible mice, rats and non-human primates (Iglesias et al., 2001; 't Hart et al., 2005)	**CD8+ T cells:** tEAE with myelin-specific CD8+ T cells (Huseby et al., 2001; Sun et al., 2001; Cabarrocas et al., 2003; Ford & Evavold, 2005) CD8+ T cell mediated axonal transaction (Medana et al., 2001) Protective role of CD8+ T cells: (Najafian et al., 2003; Sun et al., 1988; Abdul-Majid et al., 2003; Hu et al., 2004; Koh et al., 1992)
Primary/secondary progressive (PP, CREAE): Mice: C57BL6 MOG aEAE (Mendel et al., 1995; Bernard et al., 1997) Rat: DA: MOG (Papadopoulos et al., 2006; Stefferl et al., 2000; Storch et al., 1998b) Nonhuman primate: Common Marmoset, MOG (aa 1-125) ('t Hart et al., 1998; Genain & Hauser, 2001)	**Optic neuritis:** Mice: MOG EAE in C57BL6 mice (Shao et al., 2004), MOBP-EAE in SJL/J mice Kerlero de Rosbo, 2004 38192 /id} T cell receptor transgenic mice (Bettelli et al., 2003; Krishnamoorthy et al., 2006; Waldner et al., 2000) Rat: MOG-EAE in DA and Lewis rats (Optic neuritis) (Stefferl et al., 2000; Storch et al., 1998b; Berger et al., 1997; Sakuma et al., 2004) Devic's disease (Optic neuritis and myelitis): Combined T cell transgenic and B-cell transgenic (both anti-MOG) C57BL6 mice, spontaneous disease (Krishnamoorthy et al., 2006)	**Axonal pathology, neurodegeneration:** Chronic EAE in mice and rats (Kornek et al., 2000; Gold et al., 2006) Marmoset, myelin homogenate aEAE ('t Hart et al., 1998; Mancardi et al., 2001)	**B-cells and anti-myelin antibodies:** Local depositions of antibodies and complement in EAE lesions Application of demyelinating antibodies (Linington et al., 1988) Generation of B cell-transgenic (anti-MOG) mice (Litzenburger et al., 1998)
Spontaneous autoimmune encephalomyelitis: Mouse: T cell receptor transgenic mice (Bettelli et al., 2003; Brabb et al., 1997; Ellmerich et al., 2005; Goverman et al., 1993; Krishnamoorthy et al., 2006; Lafaille et al., 1994; Waldner et al., 2000; Wensky et al., 2005)	**Cortical lesions**: Nonhuman primate: MOG-EAE in marmosets (Merkler et al., 2006; Pomeroy et al., 2005) Mice and rats: MOG EAE in LEWI.W and LEWI.ARI rats (Gold et al., 2006), EAE induction after focal lesioning: (Kerschensteiner et al., 2004; Phillips et al., 1995; Sun et al., 2004)	**Diffuse inflammation and axonal injury in the normal appearing white matter**: No model available (Gold et al., 2006)	**Degenerative component of demyelination and axonal pathology** (Hobom et al., 2004; Lev et al., 2004; Linker et al., 2002; Offen et al., 2000)

duced even within the same strain by changing the inducing conditions. Transfer of MOG or S100b-specific T cells into Lewis rats leads to much more widespread inflammation of the CNS, involving forebrain and optic nerves. The factors that determine the distribution of EAE lesions remain unknown. Besides the nature and distribution of the autoantigen, genetic variations and the cytokine response have been named as essential factors (Abromson-Leeman et al., 2004), C3H/HeJ; PLP (Krishnamoorthy et al., 2006; Muller et al., 2005; Wensky et al., 2005). Thus, "atypical" EAE arising in IFN-γ deficient, RAG-2 k.o. TCR transgenic mice show a predilection of pathological lesions relocated into the brainstem and cerebellum (see Table 30.2) (Wensky et al., 2005). Isolated optic neuritis is observed in MOG-TCR transgenic mice of C57BL6 H-2^b background (Bettelli et al., 2003). These mice crossed with MOG B cell transgenic mice resulted in an optico-spinal pathology affecting the optic nerve and spinal cord (Krishnamoorthy et al., 2006).

Degenerative changes of the CNS create local milieus particularly attractive for autoimmune inflammation. This has been most impressively demonstrated in the facial nerve axotomy model (Kreutzberg, 1996). Transection of the peripheral facial nerve combined with induction of EAE led to the infiltration of the facial nerve nucleus located within the brainstem, a CNS area that usually is spared during classical EAE. Importantly, inflammation occurred selectively in the lesioned but not in the contralateral healthy facial nerve nucleus (see Figure 30.1) (Flügel et al., 2000; Konno et al., 1990; Maehlen et al., 1989). Peripheral axotomy triggers retrograde neuronal degeneration, which in the facial nucleus allows production of chemotactic factors (Flügel et al., 2001a) and a host of MHC determinants (see Figure 30.1), cell adhesion molecules and costimulatory proteins (Moran & Graeber, 2004), which together create an "immune-friendly" islet within the generally "immune-hostile" CNS tissue.

C. Clinical Courses in EAE

MS is a chronic disease often with lifelong course. Classical active and adoptive transfer EAE models (such as MBP-induced EAE in the Lewis rat) are, however, monophasic diseases (see Table 30.1). The animals develop clinical disease, which arises after a disease-free prodromal period, and after recovery the animals are largely protected against reinduction of EAE.

Only a few models show a relapsing remitting (RR) disease course, such as observed in many MS patients. The RR course is represented in SJL/J or Biozzi mice immunized with PLP or MBP, respectively (McRae et al., 1995; Sobel & Kuchroo, 1992) (see Table 30.2). The cause of relapses in these models was related to *determinant spreading*, a consecutive expansion of brain target antigens recognized by encephalitogenic T cells following individual bouts of EAE (Lehmann et al., 1992; McRae et al., 1995; Yu et al., 1996).

MOG-induced EAE in C57BL/6 mice, the most widely used murine model, takes a primary chronic disease course (Mendel et al., 1995). A secondary progressive disease course was achieved by adding *Bordetella pertussis* to MOG92-106 immunization in A.SW mice (H-2^s) (see Tables 30.1, 30.2).

D. EAE Genetics

There is no doubt that susceptibility to MS is strongly influenced by genetic factors, and this is true for inducibility of EAE as well. But, apart from the disease favoring HLA haplotype (HLA-DRB1*1501 allele), neither linkage analyses nor association studies were successful in indisputably identifying additional candidate genes.

The susceptibility for EAE induction varies strongly between different strains and, like in MS, is strongly influenced by genes within the inside and outside of the MHC. The contribution of MHC-associated genes was studied in inbred rodent strains congenic for either the entire or parts of the MHC. The importance of the MHC has been demonstrated in congenic Lewis rats with different MHC haplotypes but identical non-MHC genomic background. Lewis rats carrying the RT1.N haplotype are super-responsive, the same rat strain with the RT1.L haplotype is a poor responder, and the middle position is taken by Lewis rats of the RT1.AV1 haplotype (Becanovic et al., 2004). The influence of non-MHC genes, however, became clear in another set of congenic rats: DA, LEW.1AV1, PVG.1AV1, and ACI rats all carry the identical RT1.AV1 haplotype, and, yet, their susceptibility to MOG-induced EAE varies considerably. ACI rats are poor responders, PVG.1AV1 and LE.1AV1 are moderately affected, and DA rats develop severe disease (see Table 30.1).

A most effective technique to evaluate the importance of distinct genes in EAE pathogenesis relies on transgenic manipulation of the rodent germline. Introduction of genes via injection of DNA fragments into a fertilized egg, or transgene transfer via lentiviral vectors has been achieved in most species used for EAE studies. However, knock-out technology (i.e., the targeted elimination or disruption of defined genes) is still limited to mice due to lack of suitable embryonic stem cells in other species. These technologies have provided a wealth of important new information about the role of disease-modifying genes including cytokines, adhesion molecules, costimulatory factors, proteases (Gran et al., 2004; Grewal & Flavell, 1996; Linker et al., 2005; Owens et al., 2001; Wong et al., 1999).

E. Pathogenesis of EAE: Autoimmune T and B Cells

For an extensive period of time, there was a general consensus that αβTCR+ myelin-reactive $CD4^+$ T cells, the culprits of classical EAE, are typical TH1 lymphocytes producing the pro-inflammatory cytokines IFNγ and IL-2 as markers. This was based on the observation that indeed $CD4^+$ T cell populations that transfer EAE contain lymphocytes expressing the TH1 phenotype (Ben-Nun et al., 1981; Kuchroo et al., 1992; Pettinelli & McFarlin, 1981; Zamvil et al., 1985b). This concept conflicted, however, with studies that neutralized IFN-γ with antibodies (Billiau et al., 1988) or induced EAE in IFN-γ-deficient mice (Krakowski & Owens, 1996; Willenborg et al., 1996); in both situations absence of IFN-γ had a paradoxical, enhancing effect on EAE. Recent work resolved this seeming contradiction. It led to the discovery of a newly recognized subset of $CD4^+$ T cells (TH-17 cells), which produce IL-17 as marker cytokine, rather than IFN-γ (Langrish et al., 2005; Park et al., 2005), and that these cells require a mix of TGF-β, and IL-6 for lineage decision, instead of IL-12 (Bettelli et al., 2006b; Veldhoen et al., 2006).

TH2 cells, the $CD4^+$ lymphocytes producing IL-4 and IL-10, which are traditionally credited with a function as beneficial regulators in autoimmunity, may also have a pathogenic role, though under special conditions. MBP Ac1-11-specific TH2 $CD4^+$ T cells of B10.PL mice ($H\text{-}2^u$) were capable of transfering EAE to immune-compromised, RAG-deficient mice (Lafaille et al., 1997). Also, Th2-like T cells may be involved in allergic shock responses by mice primed to myelin peptides in the absence of full-blown clinical EAE (Pedotti et al., 2001).

As mentioned before, in MS brain lesions, it is $CD8^+$ rather than $CD4^+$ T cells that dominate the inflammatory infiltrates and expand there clonally (Babbe et al., 2000). In EAE, CD8 T cells rarely were considered as pathogenic effectors, but were described repeatedly in the function regulatory cells (Abdul-Majid et al., 2003; Hu et al., 2004; Koh et al., 1992; Najafian et al., 2003; Sun et al., 1988). Only recently were myelin autoreactive $CD8^+$ T cells identified as mediators of autoimmune brain inflammation. Independent groups characterized MBP (aa 79-87) specific $CD8^+$ T cells in C3H mice (Huseby et al., 2001) and MOG (aa 40-55) (Sun et al., 2001) or 37-46 (Ford & Evavold, 2005) specific $CD8^+$ T cells in C57BL6 mice.

Although T cells are the specific effectors in controlling location, character, and course of the autoimmune response in the CNS, they are not the only inflammatory cells required to bring about EAE. They are supported by members of the innate immune system, well-known makers of proinflammatory and cytotoxic mediators such as cytokines, proteases, reactive oxygen species. Indeed, depletion of macrophages, which are abundantly found in EAE lesions, mitigates or prevents clinical EAE (Huitinga et al., 1990). Also mast cells contribute to EAE during T cell priming in the periphery as well as during the CNS inflammation as demonstrated in mast cell deficient mice, which reveal reduced pathogenesis in active and adoptive transfer EAE (Brenner et al., 1994; Gregory et al., 2005; Secor et al., 2000).

Mature B cells are clearly not required for the induction of purely "cellular" EAE. B cell deficient mice immunized with antigen or receiving transfers of encephalitogenic T cells readily develop acute and chronic EAE (Dittel et al., 2000; Fillatreau et al., 2002; Hjelmström et al., 1998; Svensson et al., 2002; Wolf et al., 1996). Yet, autoantigen-specific B cells may substantially contribute to the pathogenesis of disease, for example, by releasing myelin-specific B cell antibodies binding to myelin, and thus initiate myelin destruction. This has been convincingly shown in the case of antibodies directed against MOG, a prime target autoantigen located at the outer myelin sheath. Anti-myelin antibodies may act via opsonization of the myelin sheaths or by a direct toxic effect on oligodendrocytes (Marta et al., 2005). The importance of the antibody component in EAE pathogenesis is undermined by the fact that putatively "resistant" EAE models (e.g., BN rats) become susceptible and develop chronic demyelinating EAE after immunization against structurally intact MOG, which induces a demyelinating antibody response (Stefferl et al., 1999).

There may be additional ways that myelin-specific B cells amplify the pathogenic potential of T cells. Recently, two groups described double-transgenic mice that develop spontaneous EAE at young age. The mice have immune repertoire containing large proportions of MOG specific T and B cells (Bettelli et al., 2006a; Krishnamoorthy et al., 2006). It appears that both lymphocyte populations interact on different levels—privileged presentation of MOG, release of cytokines, production of demyelinating antibodies—to achieve the MS-like disease.

VI. B Cell Autoimmunity in Multiple Sclerosis

One prominent diagnostic feature of MS is the presence of oligoclonal bands of immunoglobulins (OCBs) in the cerebrospinal fluid (CSF). Due to their distinct electric charge they can be separated by electrophoresis. If these have been produced by a limited number of B cell clones, electrophoresis separates them into discrete individual bands. In contrast, in polyclonal mixtures, the single bands overlap and form together a broad coherent smear, as is the case in blood immunoglobulins.

In MS, OCBs are produced by a few immunoglobulin secreting B cell clones, located either in the leptomeningeal membranes surrounding the CNS tissues, or within the tissue itself. These are by no means exclusive to MS, but are noted also in other CNS conditions, mainly infections, and most MS patients do display OCBs, which makes them a useful diagnostic tool. Sequence analyses established genealogies of clonal B cells suggesting an antigen-driven immune response as the origin, but the target antigens (or autoantigens) remain to be established. Additional hints for B cell participation in

MS comes from histology of the lesion and from therapy. B cells have been described from the beginning of modern immunocytochemistry, mainly within fresh plaques (Esiri, 1980). Like in CSF B cells, parenchymal infiltrate B cells show evidence of antigen driven activation.

The opticospinal variant of MS, Devic's disease, poses a particular enigma. A major proportion of patients have serum antibodies binding to aquaporin, a water channel involved in water homeostasis, and expressed, besides renal epithelia, also in astrocyte endfeet (Lennon et al., 2005). Autoantigen turned out to be aquaporin (Lennon et al., 2005). A highly visible study of MS patients correlated autoantibodies binding recombinant MOG protein with a more violent course (Berger et al., 2003), a trial that has been reproduced in some, but not all succeeding studies.

Finally, and particularly promising, there seems to be a subset of MS patients who respond well to plasmapheresis. A recent retrospective study indicates that the responders have lesions of a type suggesting a pathogenesis comprising humoral autoantibodies as demyelinating agents (Keegan et al., 2005).

VII. Immune Demyelination

Destruction of periaxonal myelin sheaths is the hallmark of the MS lesion. It can be brought about either by inflammatory factors damaging the myelin sheath directly or more indirectly by injuring the myelin-forming glia cells, the oligodendrocyte. There are several mechanisms potentially contributing to either mode.

The most selective and possibly most efficient way to destroy central myelin is via humoral autoantibodies that bind to surface structures of the sheath. Fixation of the antibodies causes conformational changes in the immunoglobulin constant Fc part, which allows them to recruit and activate complement, and/or macrophages. These then attack myelin mostly by perforating myelin membrane and eventually cause cell death by apoptosis or necrosis.

Obviously, humoral autoantibodies can bind only to antigenic structures, which stand out of the surface membrane. However, only a limited number of such targets seem to be accessible to autoantibodies. The most prominent of these is MOG, a member of the immunoglobulin gene superfamily. In quantitative terms, MOG represents only a minor component among all myelin proteins, but its importance as a potential autoantigen rests on its strategic positioning on the myelin surface. MOG is located selectively on the outer surface of central myelin, protruding through the glycocalix that envelopes the membrane. MOG is not found in the inner wraps of myelin, and it is highly selective for the CNS. It is not produced by Schwann cells, the myelin-forming cell of peripheral nerves.

In addition to MOG, galactoside cerebroside (GalC), a glycolipid is a target for autoantibodies at least in culture and *in vivo* systems of experimental demyelination. Finally, it is possible that transmembrane PLP exhibits surface epitopes that could be reached by humoral autoantibodies (Jung et al., 1996), although, to our knowledge, direct anti-myelin cytotoxic effects of anti-PLP autoantibodies have not been reported.

As mentioned, the demyelinating potential of an anti-MOG antibody has been shown first in two studies of mouse and rat EAE. The investigators induced classical EAE in rodent either by active immunization with EAE (Schluesener et al., 1987) or by transfer of activated encephalitogenic T cells (Linington et al., 1988), and then shortly before or during clinical EAE infused a MAb 8-18C5 (Linington et al., 1984), an antibody specific for a conformational epitope of MOG (Brehm et al., 1999). In all cases, MOG binding autoantibody strikingly enhanced the clinical deficiencies, and histologically, produced large, confluent demyelinating plaques surrounding hyperacute inflammatory foci.

Similar changes were seen after active immunization of rodents (Adelmann et al., 1995; Amor et al., 1994) or primates (Genain et al., 1995a) with recombinant MOG, but not with MOG peptides. Although immunization of MOG peptide causes EAE via activation of encephalitogenic T cells, it does not induce demyelinating autoantibody responses. Demyelinating autoantibodies are primarily binding to conformational determinants on MOG, are, however induced by immunization against recombinant (i.e., correctly folded) MOG protein (Brehm et al., 1999; von Büdingen et al., 2002).

Finally, Litzenburger et al. (1998) created a transgenic knock-in mouse, whose germline J(H) locus was replaced by the rearranged immunoglobulin H chain V gene of Linington's classic pathogenic MOG-specific MAb 8-18C5. These mice produce high titers of MOG binding autoantibodies, but do not develop spontaneous EAE. If, however, challenged by immunization against a strong T cell encephalitogen (e.g., PLP in complete Freund's adjuvant), the animals produce a more intensive clinical EAE with larger demyelinating lesions than their wild-type counterparts.

The observations made in experimental models of autoimmune demyelination may well hold true for human MS as well. There is a particular pattern of MS plaque changes, which has been classified by Lassmann and Lucchinetti as antibody-mediated (Lassmann et al., 2001). Like in autoantibody-enhanced EAE, in type II lesions, decaying myelin is decorated by immunoglobulin and activated complement (Lucchinetti et al., 2000). The bound immunoglobulin may include anti-MOG autoantibodies, as suggested by direct antigen binding studies (Genain et al., 1999) and by immunoglobulin elution (O'Connor et al., 2005). Most importantly, as indicated by a recent limited trial, patients with type II plaques seem to respond more vigorously to removal of humoral (auto)antibodies by plasmapheresis than patients with other lesion patterns (Keegan et al., 2005), a finding that plausibly argues in favor of an active role of humoral autoantibodies in at least a subset of MS patients.

Anti-MOG reactive antibodies are found commonly in the serum of MS patients, but also in patients with other CNS disorders, or even in healthy volunteers (Lindert et al., 1999; Xiao et al., 1991). Of note, however, antibodies binding to contiguous peptide epitopes are much less tightly restricted to MS than are the much rarer anticonformational antibodies, the ones resembling demyelinating autoantibodies in EAR models (Haase et al., 2001). One study linked anti-MOG antibody titers to a particularly progressive course (Berger et al., 2003), an observation, which, however seems to have been diluted by subsequent trials (Rauer et al., 2006).

Type I lesions of human MS as well as most T cell mediated versions of EAE do not present evidence of humoral demyelination, but seem to be mediated purely by cellular mechanisms. Macrophage and T cells seem to attack myelin structures via inflammatory factors including pro-inflammatory cytokines, contact dependent "death signals," and free radicals.

The confluent nature of demyelinating plaques suggests diffusable factors as mediators of myelin destruction. One of the first cytokines suspected to contribute to inflammatory demyelination counts tumor necrosis factor-α (TNF)-α. (TNF)-α was reported to lyse myelinating CNS cell cultures (Selmaj & Raine, 1988). *In vivo*, large scale primary demyelination was noted in transgenic mice releasing large quantities of (TNF)-α into brain white matter (Probert et al., 1995), an effect that could be reversed by treatment with neutralizing anti-cytokine MAbs, and that is not dependent on the presence of immunocompetent lymphocytes (Kassiotis et al., 1999).

However, it is important to keep in mind that TNF-α has diverse functions and activities in the CNS inflammatory disease. It affects invasion of the CNS tissue by autoimmune T cells by controlling their transition from the perivascular cuff area into the parenchyma (Riminton et al., 1998), possibly by regulating proteases required to cross the parenchymal basement membrane (Agrawal et al., 2006).

Furthermore, dependent on its concentration level in the tissue, TNF-α can assume a neuroprotective activity. In neuronal cultures, the cytokine protects neurons from excitatory death (Cheng et al., 1994), or from the damaging effect of amyloid beta-peptide (Barger et al., 1995). In addition, *in vivo* experiments using mice with deficient TNF-α signaling chains argued in favor of TNF-α's neuroprotective action in post-traumatic neuronal degeneration (Diem et al., 2001; Sullivan et al., 1999). The ambiguous action of TNF-α in CNS inflammatory disease may explain why a therapeutic trial testing a TNF-α inhibitor, a therapy with glorious success in rheumatoid arthritis (Feldmann & Maini, 2001), failed in the treatment of MS (The Lenercept MS Study Group and The UBC MS/MRI Analysis Group, 1999).

Like TNF-α, IFN-γ, the other main pro-inflammatory cytokine, appears to have a complex role in CNS inflammatory demyelination. Surprisingly, transgenic mice deficient of IFN-γ or its receptor, show an enhanced sensibility to induction of EAE. Heremans and Billiau were the first to show that neutralization of IFN-γ paradoxically enhances EAE in some, though not in other strains of mice (Billiau et al., 1988). Along this line, susceptibility of IFN-γ or IFNγ receptor deficient mice is equal or even higher than in their wild-type littermates (Ferber et al., 1996; Krakowski & Owens, 1996).

Although there is evidence that the paradoxically protective effect of IFN-γ in -CFA induced EAE models may involve primarily the peripheral immunization phase, and the innate response to the mycobacterial adjuvant component (Matthys et al., 2000), there may be an additional effect on the myelin sheath.

Recently, IFN-γ emerged as a pro-inflammatory cytokine with a possible myelinotoxic activity. One study reported that in the absence of an intact IFN-γ signaling chain, toxic myelin damage is alleviated, and there is evidence that the cytokine might act on microglia/macrophage activation and mobilization of oligodendrocyte progenitors (Maña et al., 2006). In a similar system, administration of IFN-γ delays remyelination acting directly on the myelinating glia (Lin et al., 2006).

Proteases are tightly associated with the unfolding and lasting inflammatory anti-myelin response. Proteases are commonly present in the cerebrospinal fluid both of animals with EAE (Gijbels et al., 1993) as well as in patients during active phases of MS (Leppert et al., 1998). They have numerous and diverse effects in the development of inflammatory demyelination. Proteases (e.g., members of the MMP family) may contribute to the passage of inflammatory cells through the endothelial blood–brain barrier and its associated basal laminas (Agrawal et al., 2006). Other proteases, released by activated macrophages (Hendriks et al., 2005), may directly act from outside on the myelin sheath (Scarisbrick et al., 2002). This is in contrast with caspases, proteases located within the cell, which drive apoptosis of myelinating oligodendrocytes (Hisahara et al., 2001). Interestingly, there are particular proteases that degrade proteins that *inhibit de*myelination and thus support *re*myelination in demyelinated plaques (Larsen et al., 2003).

A final, as-yet unproven but attractive mechanism of myelin destruction involves direct lysis of class I induced axons by cytotoxic $CD8^+$ T cell axons. This mechanism is speculative, indeed, but it rests on indirect, but complementary pieces of evidence. Firstly, it is important to know that, in contrast to T cell mediated EAE models, the active MS lesion indeed is dominated by $CD8^+$ T cells, potential cytotoxic killer cells (Friese & Fugger, 2005). These lymphocytes invade the affected parenchyma, and they seem to expand there (Babbe et al., 2000; Skulina et al., 2004). Direct cytotoxic effects of CNS infiltrating

CD8⁺ T cells have not been shown to date, but in another putative CD8⁺ mediated CNS disease, paraneoplastic Rasmussen's encephalomyelitis, CD8⁺ T cells were shown apposed to class I expressing neurons, directing their granzyme B vesicular apparatus toward their targets (Bien et al., 2002).

That indeed class I induced neurons can be attacked directly by CD8⁺ killer T cells, has been shown in a rodent culture system. Rat hippocampus neurons paralyzed by tetrodotoxin were induced by interferon-γ treatment to express intact class I protein (Neumann et al., 1995), then exposed to a viral antigenic peptide. These neurons immediately were attacked by virus-specific CD8⁺ T cells, when cultured together. Interestingly, the killer T cells first attacked axons and disrupted them within half an hour in a perforin dependent fashion (Medana et al., 2001).

Obviously, lysis of cultured neurons by virus-specific T cells is a situation far remote from the pathogenesis of human MS, where a cytotoxic action of myelin-specific autoimmune killer T cells on oligodendrocytes would be predicted. Again, such interaction has not been shown formally, but in principle, such a mechanism appears feasible. Under regular conditions, oligodendrocytes do not expose MHC class I nor class II proteins. In fact, developmentally enforced production of MHC class I determinants either by targeting the polymorphic class I H chain gene into oligodendrocytes (Power et al., 1996; Turnley et al., 1991; Yoshioka et al., 1991), or the class I-inducing IFN-γ gene (Corbin et al., 1996; Horwitz et al., 1997) seems to interfere with regular myelination of the CNS in transgenic mice (Corbin et al., 1996).

However, expression of class I antigens can be induced in mature oligodendrocytes, and such class I positive oligodendrocytes qualify as potential targets of class I-restricted CD8⁺ killer T cells. Oligodendrocytes express *de novo* class I in response to virus infections in rodents (Redwine et al., 2001; Suzumura et al., 1986), and also in humans with viral (e.g., JC virus) encephalomyelitis (Achim & Wiley, 1992) and MS plaques (Höftberger et al., 2004).

Myelin-specific CD8⁺ T cells have been described by several groups recently. The first report by Huseby et al. (1999) described the isolation of MBP-specific CD8⁺ T cells from shiverer mutant mice lacking MBP (see Table 30.2). These T cells transferred to naïve recipients severe EAE with unconventional neurological defects (Huseby et al., 2001). MOG-specific CD8⁺ T cells were derived from wild-type mice immunized against a MOG peptide. These T cells were also encephalitogenic, when transferred to hosts expressing MHC class I, but not to class I knock-out mice (Sun et al., 2001). None of these T cells, however, have been directly monitored for direct cytotoxic killing of oligodendrocytes *in vitro* or *in vivo*.

To date, there are good reasons to search for myelin-specific CD8 T cells, which could have a pathogenic function in MS or its experimental models, but such evidence is still standing out.

VIII. Beneficial Brain Autoimmunity?

Nature rarely, if ever, produces nonsense. So, is there a good reason for the existence of so many autoreactive T cells in the healthy immune system? Do we have to consider these cells solely as potential time bombs, threatening to attack ourselves throughout our lives? Or could they have a positive, beneficial function?

Irun Cohen was presumably the first to comprehend the preformed immune repertoire not only as a shield protecting us against intruded exogenous microbes, but as a tool supporting tissue repair and regeneration. Impressed by the abundance of auto-reactive T cells complementary to essentially all organ-"specific" self-determinants, he proposed the concept of the *Immunological Homunculus* (Cohen, 1992). This concept implies that the immune system has a positive image of the body's antigenic universe. The homunculus concept derives from neurobiology, where the neurological homunculus describes the contiguous projection of the individual parts of the body to correspondingly located regions in motor and sensible brain areas. Accordingly, each organ with its specific, antigenic structures is projected in the immune repertoire, where it forms a complementary immunological homunculus. This representative internal immune image helps immune cells to scan the body's tissues, for integrity and its loss, and if so required, to contribute to tissue restoration.

The homunculus concept of beneficial immunity was verified experimentally by several groups, most consequently by the group of Michal Schwartz (Schwartz et al., 1999). Studying the effect of inflammation on regeneration of spinal cord trauma, the workers first explored transfers of macrophages into lesions. Blood-derived autologous macrophages instilled into the gap between two separated axonal stumps, significantly increased axon regeneration as assessed by histology and electrophysiology (Rapalino et al., 1998).

Later the investigators extended their studies to "true" autoimmune T cells, namely MBP-specific T cell line cells. These T cells transfer EAE in intact hosts, but at the same time they exert a beneficial effect when infused into rodents suffering from traumatic neuron degeneration, such as a partially crushed optic nerve (Moalem et al., 1999). The autoimmune therapy strategy was reproduced by the group in a variety of different models of neuronal degeneration with a variety of brain autoantigens (or modifications thereof) as targets (Schwartz & Moalem, 2001). Some of these results have been repeated by some (Benner et al., 2004), but not other groups (Jones et al., 2004).

Therapeutic use of *de facto* autoimmune responses has been made recently in animal models of Alzheimer's disease, where accumulation of amyloid-derived complexes plays a pivotal role in the neurodegenerative process. Immunization of transgenic mice against the Aß peptide led to the resolution of amyloid deposits (Schenk et al., 1999). Vaccination of people with Alzheimer's disease against amyloid peptides came to a mixed result. A number of the treated patients showed beneficial effects with a slowed decline of cognitive capacity, but other patients developed meningoencephalitis (Hock et al., 2003). The positive treatment effects may be mediated by humoral antibodies that bind to aggregated amyloid in the CNS and initiate their clearance. Hence, current efforts are aimed at developing protocols that lead to the production of such antibodies, while avoiding a cellular autoimmune attack.

Another vaccination approach using humoral antibodies is emerging in the field of post-traumatic axonal regeneration. Historically, a monoclonal antibody, which neutralized *in vitro* suppression of axonal migration by CNS myelin marked the beginning of this field (Caroni & Schwab, 1989). After years of search, the target of this MAb was identified as Nogo A simultaneously by three groups (Chen et al., 2000; GrandPré et al., 2000; Prinjha et al., 2000).

Vaccination with recombinant protein or Nogo-encoding DNA is thought to neutralize the inhibitory effect on CNS myelin. A problem is that CNS expression may result in T cell mediated autoimmunity. Expression outside of CNS may undermine CNS targeted effect. Finally, Nogo is not the only axonal inhibitor. Myelin-associated glycoprotein (MAG) and oligodendrocyte-myelin glycoprotein (OMgp) similarly inhibit neuronal outgrowth, interestingly as Nogo signaling via Nogo 66 receptor. Obviously, vaccination using active or passive immunization appears as a useful strategy to support axonal regeneration after traumatic neuronal lesions, but like immunotherapy of Alzheimer's disease, there is a risk of inflammatory side effects that have to be contained.

References

't Hart, B. A., Bauer, J., Brok, H. P. M., Amor, S. (2005). Non-human primate models of experimental autoimmune encephalomyelitis: Variations on a theme. *J Neuroimmunol* **168**, 1–12.

't Hart, B. A., Bauer, J., Muller, H. J., Melchers, B., Nicolay, K., Brok, H. et al. (1998). Histopathological characterization of magnetic resonance imaging detectable brain white matter lesions in a primate model of multiple sclerosis: A correlative study in the experimental autoimmune encephalomyelitis model in common marmosets (*Callithrix jacchus*). *Am J Pathol* **153**, 649–663.

't Hart, B. A., Brok, H. P. M., Amor, S., Bontrop, R. E. (2001). The major histocompatibility complex influences the etiopathogenesis of MS-like disease in primates at multiple levels. *Hum Immunol* **62,** 1371–1381.

Abdul-Majid, K. B., Wefer, J., Stadelmann, C., Stefferl, A., Lassmann, H., Olsson, T., Harris, R. A. (2003). Comparing the pathogenesis of experimental autoimmune encephalomyelitis in CD4(−/−) and CD8(−/−) DBA/1 mice defines qualitative roles of different T cell subsets. *J Neuroimmunol* **141**, 10–19.

Aboul-Enein, F., Weiser, P., Höftberger, R., Lassmann, H., Bradl, M. (2006). Transient axonal injury in the absence of demyelination: A correlate of clinical disease in acute experimental autoimmune encephalomyelitis. *Acta Neuropathol* **111**, 539–547.

Abromson-Leeman, S., Alexander, J., Bronson, R., Carroll, J., Southwood, S., Dorf, M. (1995). Experimental autoimmune encephalomyelitis-resistant mice have highly encephalitogenic myelin basic protein (MBP)-specific T cell clones that recognize a MBP peptide with high affinity for MHC class II. *J Immunol* **154**, 388–398.

Abromson-Leeman, S., Bronson, R., Luo, Y., Berman, M., Leeman, R., Leeman, J., Dorf, M. (2004). T-cell properties determine disease site, clinical presentation, and cellular pathology of experimental autoimmune encephalomyelitis. *Am J Pathol* **165**, 1519–1533.

Achim, C. L., Wiley, C. A. (1992). Expression of major histocompatibility complex antigens in the brains of patients with progressive multifocal leukoencephalopathy. *J Neuropathol Exp Neurol* **51**, 257–263.

Adelmann, M., Wood, J., Benzel, I., Fiori, P., Lassmann, H., Matthieu, J-M. et al. (1995). The N-terminal domain of the myelin oligodendrocyte glycoprotein (MOG) induces acute demyelinating experimental autoimmune encephalomyelitis in the Lewis rat. *J Neuroimmunol* **63**, 17–27.

Agrawal, S., Anderson, P., Durbeej, M., Van Rooijen, N., Ivars, F., Opdenakker, G., Sorokin, L. M. (2006). Dystroglycan is selectively cleaved at the parenchymal basement membrane at sites of leukocyte extravasation in experimental autoimmune encephalomyelitis. *J Exp Med* **203**, 1007–1019.

Amor, S., Baker, D., Groome, N., Turk, J. L. (1993). Identification of a major encephalitogenic epitope of proteolipid protein (residues 56–70) for the induction of experimental allergic encephalomyelitis in Biozzi AB/H and nonobese diabetic mice. *J Immunol* **150**, 5666–5672.

Amor, S., Groome, N., Linington, C., Morris, M. M., Dornmair, K., Gardinier, M. V. et al. (1994). Identification of epitopes of myelin oligodendrocyte glycoprotein for the induction of experimental allergic encephalomyelitis in SJL and Biozzi AB/H mice. *J Immunol* **153**, 4349–4356.

Amor, S., O'Neill, J. K., Morris, M. M., Smith, R. M., Wraith, D. C., Groome, N. et al. (1996).Encephalitogenic epitopes of myelin basic protein, proteolipid protein, and myelin oligodendrocyte glycoprotein for experimental allergic encephalomyelitis induction in Biozzi ABH (H-2^{Ag7}) mice share an amino acid motif. *J Immunol* **156**, 3000–3008.

Amor, S., Smith, P. A., 't Hart, B. A., Baker, D. (2005). Biozzi mice: Of mice and human neurological diseases. *J Neuroimmunol* **165**, 1–10.

Babbe, H., Roers, A., Waisman, A., Lassmann, H., Goebels, N., Hohlfeld, R. et al. (2000). Clonal expansion of $CD8^+$ T cells dominate the T cell infiltrate in active multiple sclerosis lesions shown by micromanipulation and single cell polymerase chain reaction. *J Exp Med* **192**, 393–404.

Baker, D., O'Neill, J. K., Gschmeissner, S. E., Wilcox, C. E., Butter, C., Turk, J. L. (1990). Induction of chronic experimental allergic encephalomyelitis in Biozzi mice. *J Neuroimmunol* **28**, 261–270.

Barger, S. W., Hörster, D., Furukawa, K., Goodman, Y., Krieglstein, J., Mattson, M. P. (1995). Tumor necrosis factors α and β protect neurons against amyloid β-peptide toxicity: Evidence for involvement of a κB-binding factor and attenuation of peroxide and Ca^{2+} accumulation. *Proc Natl Acad Sci USA* **92**, 9328–9332.

Barker, C. F., Billingham, R. E. (1977). Immunologically privileged sites. *Adv Immunol* **25**, 1–54.

Becanovic, K., Jagodic, M., Wallstrom, E., Olsson, T. (2004). Current gene-mapping strategies in experimental models of multiple sclerosis. *Scand J Immunol* **60**, 39–51.

Ben-Nun, A., Eisenstein, S., Cohen, I. R. (1982). Experimental autoimmune encephalomyelitis (EAE) in genetically resistant rats: PVG rats resist active induction of EAE but are susceptible to and can generate EAE effector T cell lines. *J Immunol* **129**, 918–919.

Ben-Nun, A., Wekerle, H., Cohen, I. R. (1981). The rapid isolation of clonable antigen-specific T lymphocyte lines capable of mediating autoimmune encephalomyelitis. *Eur J Immunol* **11**, 195–199.

Benner, E. J., Mosley, R. L., Destache, C. J., Lewis, T., Jackson-Lewis, V., Gorantla, S. et al. (2004). Therapeutic immunization protects dopaminergic neurons in a mouse model of Parkinson's disease. *Proc Natl Acad Sci USA* **101**, 9435–9440.

Berger, T., Rubner, P., Schautzer, F., Egg, R., Ulmer, H., Mayringer, I. et al.,. (2003). Antimyelin antibodies as a predictor of clinically definite multiple sclerosis after a first demyelinating event. *N Engl J Med* **349**, 139–145.

Berger, T., Weerth, S., Kojima, K., Linington, C., Wekerle, H., Lassmann, H. (1997). Experimental autoimmune encephalomyelitis: The antigen specificity of T-lymphocytes determines the topography of lesions in the central and peripheral nervous system. *Lab Invest* **76**, 355–364.

Bernard, C. C. A., Johns, T. G., Slavin, A., Ichikawa, M., Ewing, C., Liu, J., Bettapura, J. (1997). Myelin oligodendrocyte glycoprotein: A novel candidate autoantigen in multiple sclerosis. *J Mol Med* **75**, 77–88.

Bettelli, E., Baeten, D., Jäger, A., Sobel, R. A., Kuchroo, V. K. (2006a). Myelin oligodendrocyte glycoprotein-specific T and B cells cooperate to induce a Devic-like disease in mice. *J Clin Invest* **116**, 2393–2402.

Bettelli, E., Carrier, Y., Gao, W., Korn, T., Strom, T. B., Oukka, M. et al. (2006b). Reciprocal developmental pathways for the generation of pathogenic effector TH17 and regulatory T cells. *Nature* **441**, 235–238.

Bettelli, E., Pagany, M., Weiner, H. L., Linington, C., Sobel, R. A., Kuchroo, V. K. (2003). Myelin oligodendrocyte glycoprotein-specific T cell receptor transgenic mice develop spontaneous autoimmune optic neuritis. *J Exp Med* **197**, 1073–1081.

Bielekova, B., Goodwin, B., Richert, N., Cortese, I., Kondo, T., Afshar, G. et al., (2000). Encephalitogenic potential of the myelin basic protein peptide (amino acids 83-99) in multiple sclerosis: Results of a phase II clinical trial with an altered peptide ligand. *Nature Med* **6**, 1167–1175.

Bien, C. G., Bauer, J., Deckwerth, T. L., Wiendl, H., Deckert, M., Wiestler, O. D. et al. (2002). Destruction of neurons by cytotoxic T cells: A new pathogenic mechanism in Rasmussen's encephalitis. *Ann Neurol* **51**, 311–318.

Billiau, A., Heremans, H., Vandekerckhove, F., Dijkmans, R., Sobis, H., Meulepas, E., Carton, H. (1988). Enhancement of experimental allergic encephalomyelitis in mice by antibodies against IFN-γ. *J Immunol* **140**, 1506–1510.

Brabb, T., Goldrath, A. W., Von Dassow, P., Paez, A., Liggitt, H. D., Goverman, J. et al. (1997). Triggers for autoimmune disease in a murine TCR transgenic model for multiple sclerosis. *J Immunol* **159**, 497–507.

Brehm, U., Piddlesden, S. J., Gardinier, M. V., Linington, C. (1999). Epitope specificity of demyelinating monoclonal autoantibodies directed against the human myelin oligodendrocyte glycoprotein. *J Neuroimmunol* **97**, 9–15.

Brenner, T., Soffer, D., Shalit, M., Levi-Schaffer, F. (1994). Mast cells in experimental allergic encephalomyelitis: Characterization, distribution in the CNS and in vitro activation by myelin basic protein and neuropeptides. *J Neurol Sci* **122**, 210–213.

Brok, H. P. M., Bauer, J., Jonker, M., Blezer, E., Amor, S., Bontrop, R. E. et al. (2001). Non-human primate models of multiple sclerosis. *Immunol Rev* **183**, 173–185.

Brown, A., McFarlin, D. E., Raine, C. S. (1982). Chronologic neuropathology of relapsing experimental allergic encephalomyelitis in the mouse. *Lab Invest* **46**, 171–185.

Cabarrocas, J., Bauer, J., Piaggio, E., Liblau, R., Lassmann, H. (2003). Effective and selective immune surveillance of the brain by MHC class I-restricted cytotoxic T lymphocytes. *Eur J Immunol* **33**, 1174–1182.

Caroni, P., Schwab, M. E. (1989). Codistribution of neurite growth inhibitors and oligodendrocytes in rat CNS: Appearance follows nerve fiber growth and precedes myelination. *Dev Biol* **136**, 287–295.

Chalk, J. B., McCombe, P. A., Smith, R., Pender, M. P. (1994). Clinical and histological findings in proteolipid protein-induced experimental autoimmune encephalomyelitis (EAE) in the Lewis rat. Distribution of demyelination differs from that in EAE induced by other antigens. *J Neurol Sci* **123**, 154–161.

Chen, M. S., Huber, A. B., Van der Haar, M. E., Frank, M., Schnell, L., Spillmann, A. A. et al. (2000). Nogo-A is a myelin-associated neurite outgrowth inhibitor and an antigen for monoclonal antibody IN-1. *Nature* **403**, 434–439.

Cheng, B., Christakos, S., Mattson, M. P. (1994). Tumor necrosis factors protect neurons against metabolic-excitotoxic insults and promote maintenance of calcium homeostasis. *Neuron* **12**, 139–153.

Cohen, I. R. (1992). The cognitive paradigm and the immunological homunculus. *Immunol Today* **13**, 490–494.

Coles, A. J., Wing, M. G., Molyneux, P. D., Paolillo, A., Davie, C. M., Hale, G. et al. (1999). Monoclonal antibody treatment exposes three mechanisms underlying the clinical course of multiple sclerosis. *Ann Neurol* **46**, 296–304.

Corbin, J. G., Kelly, D., Rath, E. M., Baerwald, K. D., Suzuki, K., Popko, B. (1996). Targeted CNS expression of interferon-γ in transgenic mice leads to hypomyelination, reactive gliosis, and abnormal cerebellar development. *Mol Cell Neurosci* **7**, 354–370.

Diem, R., Meyer, R., Weishaupt, J. H., Bähr, M. (2001). Reduction of potassium currents and phosphatidylinositol 3-kinase-dependent Akt phosphorylation by tumor necrosis factor-α rescues axotomized retinal ganglion cells from retrograde cell death in vivo. *J Neurosci* **21**, 2058–2066.

Dittel, B. N., Urbania, T. H., Janeway, C. A. (2000). Relapsing and remitting experimental autoimmune encephalomyelitis in B cell deficient mice. *J Autoimmunity* **14**, 311–318.

Dyment, D. A., Ebers, G. C., Sadovnick, A. D. (2004). Genetics of multiple sclerosis. *Lancet Neurology* **3**, 104–110.

Ellmerich, S., Mycko, M., Takács, K., Waldner, H., Wahid, F. N., Boyton, R. J. et al. (2005). High incidence of spontaneous disease in an HLA-DR15 and TCR transgenic multiple sclerosis model. *J Immunol* **174**, 1938–1946.

Esiri, M. M. (1980). Multiple sclerosis: A quantitative and qualitative study of immunoglobulin-containing cells in the central nervous system. *Neuropathol Appl Neurobiol* **6**, 9–21.

Feldmann, M., Maini, R. N. (2001). Anti-TNF-α therapy of rheumatoid arthritis: What have we learned? *Annu Rev Immunol* **19**, 163–196.

Ferber I, Brocke S, Taylor-Edwards C, Ridgway W, Dinisco C, Steinman L et al. (1996). Mice with a disrupted IFN-γ gene are susceptible to the induction of experimental autoimmune encephalomyelitis (EAE). *J Immunol* **156**, 5–7.

Fields, R. D., Stevens-Graham, B. (2002). New insights into neuron-glia communication. *Science* **298**, 556–562.

Fillatreau, S., Sweenie, C. H., McGeachy, M. J., Gray, D., Anderton, S. M. (2002). B cells regulate autoimmunity by provision of IL-10. *Nature Immunol* **3**, 944–950.

Flügel, A., Hager, G., Horvat, A., Spitzer, C., Singer, G. M. A., Graeber, M. B. et al. (2001a). Neuronal MCP-1 expression in response to remote nerve injury. *J Cereb Blood Flow Metab* **21**, 69–76.

Flügel, A., Matsumuro, K., Neumann, H., Klinkert, W. E. F., Birnbacher, R., Lassmann, H. et al. (2001b). Anti-inflammatory activity of nerve growth factor in experimental autoimmune encephalomyelitis: Inhibition of monocyte transendothelial migration. *Eur J Immunol* **31**, 11–22.

Flügel, A., Schwaiger, F-W., Neumann, H., Medana, I., Willem, M., Wekerle, H. et al. (2000). Neuronal FasL induces cell death of encephalitogenic T lymphocytes. *Brain Pathol* **10**, 353–364.

Ford, M. L., Evavold, B. D. (2005). Specificity, magnitude, and kinetics of MOG-specific $CD8^+$ T cell responses during experimental autoimmune encephalomyelitis. *Eur J Immunol* **35**, 76–85.

Freund, J., Stern, E. R., Pisani, T. M. (1947). Isoallergic encephalomyelitis and radiculitis in Guinea pigs after one injection of brain and mycobacteria in water-in-oil emulsion. *J Immunol* **57**, 179–195.

Friese, M. A., Fugger, L. (2005). Autoreactive $CD8^+$ T cells in multiple sclerosis: A new target for therapy? *Brain* **128**, 1747–1763.

Furtado, G. D., Olivares-Villagomez, D., de Lafaille, M. A. C., Wensky, A. K., Latkowski, J. A., Lafaille, J. J. (2001). Regulatory T cells in spontaneous autoimmune encephalomyelitis. *Immunol Rev* **182**, 122–134.

Genain, C. P., Cannella, B., Hauser, S. L., Raine, C. S. (1999). Identification of autoantibodies associated with myelin damage in multiple sclerosis. *Nature Med* **5**, 170–175.

Genain, C. P., Hauser, S. L. (1997). Creation of a model for multiple sclerosis in *Callithrix jacchus* marmosets. *J Mol Med* **75**, 187–197.

Genain, C. P., Hauser, S. L. (2001). Experimental allergic encephalomyelitis in the New World monkey *Callithrix jacchus*. *Immunol Rev* **183**, 159–172.

Genain, C. P., Nguyen, M-H., Letvin, N. L., Pearl, R., Davis, R. L., Adelmann, M. et al. (1995a). Antibody facilitation of multiple sclerosis-like lesions in a nonhuman primate. *J Clin Invest* **96**, 2966–2974.

Genain, C. P., Roberts, T., Davis, R. L., Nguyen, M-H., Uccelli, A., Faulds, D. et al. (1995b). Prevention of autoimmune demyelination in non-human primates by a cAMP-specific phosphodiesterase inhibitor. *Proc Natl Acad Sci USA* **92**, 3601–3605.

Gijbels, K., Proost, P., Masure, S., Carton, H., Billiau, A., Opdenakker, G. (1993). Gelatinase B is present in the cerebrospinal fluid during experimental autoimmune encephalomyelitis and cleaves myelin basic protein. *J Neurosci Res* **36**, 432–440.

Gold, R., Linington, C., Lassmann, H. (2006). Understanding pathogenesis and therapy of multiple sclerosis via animal models: 70 years of merits and culprits in experimental autoimmune encephalomyelitis research. *Brain* **129**, 1953–1971.

Goverman, J., Woods, A., Larson, L., Weiner, L. P., Hood, L., Zaller, D. M. (1993). Transgenic mice that express a myelin basic protein-specific T cell receptor develop spontaneous autoimmunity. *Cell* **72**, 551–560.

Gran, B., Zhang, G. X., Rostami, A. (2004). Role of the IL-12/IL-23 system in the regulation of T-cell responses in central nervous system inflammatory demyelination. *CRC Crit Rev Immunol* **24**, 111–128.

GrandPré, T., Nakamura, F., Vartanian, T., Strittmatter, S. M. (2000). Identification of the Nogo inhibitor of axon regeneration as a reticulon protein. *Nature* **403**, 439–443.

Gregory, G. D., Robbie-Ryan, M., Secor, V. H., Sabatino, J. J., Brown, M. A. (2005). Mast cells are required for optimal autoreactive T cell responses in a murine model of multiple sclerosis. *Eur J Immunol* **35**, 3478–3486.

Grewal, I. S., Flavell, R. A. (1996). A central role of CD40 ligand in the regulation of $CD4^+$ T cell responses. *Immunol Today* **17**, 410–414.

Haase, C. G., Guggenmos, J., Brehm, U., Andersson, M., Olsson, T., Reindl, M. et al. (2001). The fine specificity of the myelin oligodendrocyte glycoprotein autoantibody response in patients with multiple sclerosis and normal healthy controls. *J Neuroimmunol* **114**, 220–225.

Happ, M. P., Wettstein, P., Dietzschold, B., Heber-Katz, E. (1998). Genetic control of the development of experimental allergic encephalomyelitis in rats: Separation of MHC and non-MHC gene effects. *J Immunol* **141**, 1489–1494.

Hendriks, J. J. A., Teunissen, C. E., De Vries, H. E., Dijkstra, C. D. (2005). Macrophages and neurodegeneration. *Brain Res Rev* **48**, 185–195.

Hisahara, S., Yuan, J., Momoi, T., Okano, H., Miura, M. (2001). Caspase-11 mediates oligodendrocyte cell death and pathogenesis of autoimmune mediated demyelination. *J Exp Med* **193**, 111–122.

Hjelmström, P., Juedes, A. E., Fjell, J., Ruddle, N. H. (1998). B cell deficient mice develop experimental allergic encephalomyelitis with demyelination after myelin oligodendrocyte glycoprotein immunization. *J Immunol* **161**, 4480–4483.

Hobom, M., Storch, M. K., Weissert, R., Maier, K., Radhakrishnan, A., Kramer, B. et al. (2004). Mechanisms and time course of neuronal degeneration in experimental autoimmune encephalomyelitis. *Brain Pathol* **14**, 148–157.

Hock, C., Konietzko, U., Streffer, J. R., Tracy, J., Signorell, A., Müller-Tillmanns, B. et al. (2003). Antibodies against β-amyloid slow cognitive decline in Alzheimer's disease. *Neuron* **38**, 547–554.

Höftberger, R., Aboul-Enein, F., Brück, W., Lucchinetti, C., Rodriguez, M., Schmidbauer, M. et al. (2004). Expression of major histocompatibility complex class I molecules on the different cell types in multiple sclerosis lesions. *Brain Pathol* **14**, 43–50.

Holz, A., Bielekova, B., Martin, R., Oldstone, M. B. A. (2000). Myelin-associated oligodendrocytic basic protein: Identification of an encephalitogenic epitope and association with multiple sclerosis. *J Immunol* **164**, 1103–1109.

Horwitz, M. S., Evans, C. F., McGavern, D. B., Rodriguez, M., Oldstone, M. B. A. (1997). Primary demyelination in transgenic mice expressing interferon-γ. *Nature Med* **3**, 1037–1041.

Hu, D., Ikizawa, K., Lu, L. R., Sanchirico, M. E., Shinohara, M. L., Cantor, H. (2004). Analysis of regulatory CD8 T cells in Qa-1-deficient mice. *Nature Immunol* **5**, 516–523.

Huang, D. R., Han, Y. L., Rani, M. R. S., Glabinski, A., Trebst, C., Sorensen, T. et al. (2000). Chemokines and chemokine receptors in inflammation of the nervous system: Manifold roles and exquisite regulation. *Immunol Rev* **177**, 52–67.

Huh, G. S., Boulanger, L. M., Du, H., Riquelme, P. A., Brotz, T. M., Shatz, C. J. (2000). Functional requirement for class I MHC in CNS development and plasticity. *Science* **290**, 2155–2159.

Huitinga, I., Van Rooijen, N., De Groot, C. J. A., Uitdehaag, B. M. J., Dijkstra, C. D. (1990). Suppression of experimental allergic encephalomyelitis in Lewis rats after elimination of macrophages. *J Exp Med* **172**, 1025–1033.

Hurwitz, A. A., Sullivan, T. J., Sobel, R. A., Allison, J. P. (2002). Cytotoxic T lymphocyte antigen-4 (CTLA-4) limits the expansion of encephalitogenic T cells in experimental autoimmune encephalomyelitis (EAE)-resistant BALB/c mice. *Proc Natl Acad Sci USA* **99**, 3013–3017.

Huseby, E. S., Liggitt, D., Brabb, T., Schnabel, B., Öhlén, C., Goverman, J. (2001). A pathogenic role for myelin specific $CD8^+$ T cells in a model for multiple sclerosis. *J Exp Med* **194**, 669–676.

Huseby, E. S., Öhlén, C., Goverman, J. (1999). Myelin basic protein specific cytotoxic T cell tolerance is maintained in vivo by a single dominant epitope in H-2^k mice. *J Immunol* **163**, 1115–1118.

Iglesias, A., Bauer, J., Litzenburger, T., Schubart, A., Linington, C. (2001). T- and B-cell responses to myelin oligodendrocyte glycoprotein in experimental autoimmune encephalomyelitis and multiple sclerosis. *Glia* **36**, 220–234.

Jersild, C., Fog, T., Hansen, G. S., Thomsen, M., Svejgaard, A. (1973). Histocompatibility determinants in multiple sclerosis, with special reference to clinical course. *Lancet* **ii**, 1221–1225.

Johns, T. G., Kerlero de Rosbo, N., Menon, K. K., Abo, S., Gonzales, M. F., Bernard, C. C. A. (1995). Myelin oligodendrocyte glycoprotein induces a demyelinating encephalomyelitis resembling multiple sclerosis. *J Immunol* **154**, 5536–5541.

Jones, T. B., Ankeny, D. P., Guan, Z., McGaughy, V., Fisher, L. C., Basso, D. M., Popovich, P. G. (2004). Passive or active immunization with myelin basic protein impairs neurological function and exacerbates neuropathology after spinal cord injury in rats. *J Neurosci* **24**, 3752–3761.

Jung, M., Sommer, I., Schachner, M., Nave, K-A. (1996). Monoclonal antibody O10 defines a conformationally sensitive cell surface epitope of proteolipid protein (PLP): Evidence that PLP misfolding underlies dysmyelination in mutant mice. *J Neurosci* **16**, 7920–7929.

Kabat, E. A., Wolf, A., Bezer, A. E. (1947). The rapid production of acute disseminated encephalomyelitis in rhesus monkeys by injection of heterozygous and homologous brain tissue. *J Exp Med* **85**, 117–129.

Kappos, L., Comi, G., Panitch, H., Oger, J., Antel, J., Conlon, P., Steinman, L. (2000). The APL in Relapsing MS Study Group. Induction of a non-encephalitogenic Th2 autoimmune response in multiple sclerosis after administration of an altered peptide ligand in a placebo controlled, randomized phase II trial. *Nature Med* 1176–1182.

Kassiotis, G., Bauer, J., Akassoglou, K., Lassmann, H., Kollias, G., Probert, L. (1999). A tumor necrosis factor-induced model of human primary demyelinating diseases develops in immunodeficient mice. *Eur J Immunol* **29**, 912–917.

Keegan, M., König, F., McClelland, R., Brück, W., Morales, Y., Bitsch, A. et al. (2005). Relation between humoral pathological changes in multiple sclerosis and response to therapeutic plasma exchange. *Lancet* **366**, 579–582.

Kerlero de Rosbo, N., Brok, H. P. M., Bauer, J., Kaye, J. F., 't Hart, B. A., Ben-Nun, A. (2000). Rhesus monkeys are highly susceptible to experimental autoimmune encephalomyelitis induced by myelin oligodendrocyte glycoprotein: Characterization of immunodominant T- and B-cell epitopes. *J Neuroimmunol* **110**, 83–96.

Kerlero de Rosbo, N., Mendel, I., Ben-Nun, A. (1995). Chronic relapsing experimental autoimmune encephalomyelitis with a delayed onset and an atypical course, induced on PL/J mice by myelin oligodendrocyte glycoprotein (MOG)-derived peptide: Preliminary analysis of MOG T cell epitopes. *Eur J Immunol* **25**, 985–993.

Kerschensteiner, M., Stadelmann, C., Buddeberg, B. S., Merkler, D., Bareyre, F. M., Anthony, D. C. et al. (2004). Targeting experimental autoimmune encephalomyelitis lesions to a predetermined axonal tract system allows for refined behavioral testing in an animal model of multiple sclerosis. *Am J Pathol* **164**, 1455–1469.

Koh, D-R., Fung-Leung, W-P., Ho, A., Gray, D., Acha-Orbea, H., Mak, T. W. (1992). Less mortality but more relapses in experimental allergic encephalomyelitis in $CD8^{-/-}$ mice. *Science* **256**, 1210–1213.

Kojima, K., Berger, T., Lassmann, H., Hinze-Selch, D., Zhang, Y., Gehrmann, J. et al. (1994). Experimental autoimmune panencephalitis and uveoretinitis in the Lewis rat transferred by T lymphocytes specific for the S100 β molecule, a calcium binding protein of astroglia. *J Exp Med* **180**, 817–829.

Konno, H., Yamamoto, T., Suzuki, H., Yamamoto, H., Iwasaki, Y., Ohara, Y. et al. (1990). Targeting of adoptively transferred experimental allergic encephalomyelitis lesion at the site of Wallerian degeneration. *Acta Neuropathol* **80**, 521–526.

Kornek, B., Storch, M. K., Weissert, R., Wallström, E., Stefferl, A., Olsson, T. et al. Multiple sclerosis and chronic autoimmune encephalomyelitis—A comparative quantitative study of axonal injury in active, inactive, and remyelinated lesions. *Am J Pathol* **157**, 267–276.

Krakowski, M., Owens, T. (1996). Interferon-γ confers resistance to experimental allergic encephalomyelitis. *Eur J Immunol* **26**, 1641–1646.

Kreutzberg, G. W. (1996). Microglia: A sensor for pathological events in the CNS. *Trends Neurosci* **19**, 312–318.

Krishnamoorthy, G., Lassmann, H., Wekerle, H., Holz, A. (2006). Spontaneous opticospinal encephalomyelitis in a double-transgenic mouse model of autoimmune T cell/B cell cooperation. *J Clin Invest* **116**, 2385–2392.

Kuchroo, V. K., Sobel, R. A., Laning, J. C., Martin, C. A., Greenfield, E., Dorf, M. E., Lees, M. B. (1992). Experimental allergic encephalomyelitis mediated by cloned T cells specific for a synthetic peptide of myelin proteolipid protein. Fine specificity and T cell receptor Vβ usage. *J Immunol* **148**, 3776–3782.

Kutzelnigg, A., Lucchinetti, C. F., Stadelmann, C., Brück, W., Rauschka, H., Bergmann, M. et al. (2005). Cortical demyelination and diffuse white matter injury in multiple sclerosis. *Brain* **128**, 2705–2712.

Kyewski, B., Klein, L. (2006). Central role for central tolerance. *Annu Rev Immunol* **24**, 571–606.

Ladi, E., Yin, X. Y., Chtanova, T., Robey, E. A. (2006). Thymic microenvironments for T cell differentiation and selection. *Nature Immunol* **7**, 338–343.

Lafaille, J., Nagashima, K., Katsuki, M., Tonegawa, S. (1994). High incidence of spontaneous autoimmune encephalomyelitis in immunodeficient anti-myelin basic protein T cell receptor mice. *Cell* **78**, 399–408.

Lafaille, J. J., Van de Keere, F., Hsu, A. L., Baron, J. L., Haas, W., Raine, C. S., Tonegawa, S. (1997). Myelin basic protein-specific T helper 2 (Th2) cells cause experimental autoimmune encephalomyelitis in immunodeficient hosts rather than protect them from the disease. *J Exp Med* **186**, 307–312.

Langrish, C. L., Chen, Y., Blumenschein, W. M., Mattson, J., Basham, B., Sedgwick, J. D. et al. (2005). IL-23 drives a pathogenic T cell population that induces autoimmune inflammation. *J Exp Med* **201**, 233–240.

Larsen, P. H., Wells, J. E., Stallcup, W. B., Opdenakker, G., Yong, V. W. (2003). Matrix metalloproteinase-9 facilitates remyelination in part by processing the inhibitory NG2 proteoglycan. *J Neurosci* **23**, 11127–11135.

Lassmann, H., Brück, W., Lucchinetti, C. (2001). Heterogeneity of multiple sclerosis pathogenesis: Implications for diagnosis and therapy. *Trends Mol Med* **7**, 115–121.

Lehmann, P. V., Forsthuber, T., Miller, A., Sercarz, E. E. (1992). Spreading of T-cell autoimmunity to cryptic determinants of an autoantigen. *Nature* **358**, 155–157.

Lennon, V. A., Kryzer, T. J., Pittock, S. J., Verkman, A. S., Hinson, S. R. (2005). IgG marker of optic-spinal multiple sclerosis binds to the aquaporin-4 water channel. *J Exp Med* **202**, 473–477.

Leppert, D., Ford, J., Stabler, G., Grygar, C., Lienert, C., Huber, S. et al. (1998). Matrix metalloproteinase-9 (gelatinase B) is selectively elevated in CSF during relapses and stable phases of multiple sclerosis. *Brain* **121**, 2334.

Lev, N., Barhum, Y., Melamed, E., Offen, D. (2004). Bax-ablation attenuates experimental autoimmune encephalomyelitis in mice. *Neurosci Lett* **359**, 139–142.

Levine, S., Sowinski, R. (1974). Experimental allergic encephalomyelitis in congenic strains of mice. *Immunogenetics* **1**, 352–356.

Levine, S., Sowinski, R. (1975). Allergic encephalomyelitis in the reputedly resistant Brown Norway strain of rats. *J Immunol* **114**, 597–601.

Lin, W. S., Kemper, A., Dupree, J. L., Harding, H. P., Ron, D., Popko, B. (2006). Interferon-γ inhibits central nervous system remyelination through a process modulated by endoplasmic reticulum stress. *Brain* **129**, 1306–1318.

Lindert, R-B., Haase, C. G., Brehm, U., Linington, C., Wekerle, H., Hohlfeld, R. (1999). Multiple sclerosis: B- and T-cell responses to the extracellular domain of the myelin oligodendrocyte glycoprotein. *Brain* **122**, 2089–2099.

Linington, C., Berger, T., Perry, L., Weerth, S., Hinze-Selch, D., Zhang, Y. et al. (1993). T cells specific for the myelin oligodendrocyte glycoprotein (MOG) mediate an unusual autoimmune inflammatory response in the central nervous system. *Eur J Immunol* **23**, 1364–1372.

Linington, C., Bradl, M., Lassmann, H., Brunner, C., Vass, K. (1988). Augmentation of demyelination in rat acute allergic encephalomyelitis by circulating mouse monoclonal antibodies directed against a myelin/oligodendrocyte glycoprotein. *Am J Pathol* **130**, 443–454.

Linington, C., Webb, M., Woodhams, P. L. (1984). A novel myelin-associated glycoprotein defined by a mouse monoclonal antibody. *J Neuroimmunol* **6**, 387–396.

Linker, R. A., Mäurer, M., Gaupp, S., Martini, R., Holtmann, B., Giess, R. et al., (2002). CNTF is a major protective factor in demyelinating CNS disease: A neurotrophic cytokine as modulator in neuroinflammation. *Nature Med* **8**, 620–624.

Linker, R. A., Sendtner, M., Gold, R. (2005). Mechanisms of axonal degeneration in EAE—Lessons from CNTF and MHC I knockout mice. *J Neurol Sci* **233**, 167–172.

Linthicum, D. S., Frelinger, J. A. (1982). Acute autoimmune encephalomyelitis in mice. II. Susceptibility is controlled by the combination of H-2 and histamine sensitization genes. *J Exp Med* **156**, 31–40.

Litzenburger, T., Fässler, R., Bauer, J., Lassmann, H., Linington, C., Wekerle, H., Iglesias, A. (1988). B lymphocytes producing demyelinating autoantibodies: Development and function in gene-targeted transgenic mice. *J Exp Med* **188**, 169–180.

Lorentzen, J. C., Issazadeh, S., Storch, M., Mustafa, M. I., Lassmann, H., Linington, C. et al. (1995). Protracted, relapsing and demyelinating experimental autoimmune encephalomyelitis in DA rats immunized with syngeneic spinal cord and incomplete Freund's adjuvant. *J Neuroimmunol* **63**, 193–205.

Lublin, F. D. (1985). Adoptive transfer of murine relapsing experimental allergic encephalomyelitis. *Ann Neurol* **17**, 188–190.

Lublin, F. D., Maurer, P. H., Berry, R. G., Tippett, D. (1981). Delayed relapsing experimental allergic encephalomyelitis in mice. *J Immunol* **126**, 819–822.

Lucchinetti, C. F., Brück, W., Parisi, J., Scheithauer, B., Rodriguez, M., Lassmann, H. (2000). Heterogeneity of multiple sclerosis lesions: Implications for the pathogenesis of multiple sclerosis. *Ann Neurol* **47**, 707–717.

Madsen, L. S., Andersson, E. C., Jansson, L., Krogsgaard, M., Andersen, C. B., Engsberg, J. et al. (1999). A humanized model for multiple sclerosis using HLA DR2 and a human T cell receptor. *Nature Genet* **23**, 343–347.

Maehlen, J., Olsson, T., Zachau, A., Klareskog, L. (1989). Local enhancement of major histocompatibility complex (MHC) class I and class II expression and cell infiltration in experimental allergic encephalomyelitis around axotomized motor neurons. *J Neuroimmunol* **23**, 125–132.

Maña, P., Liñares, D., Foordham, S., Staykova, M., Willenborg, D. O. (2006). Deleterious role of IFN-γ in a toxic model of central nervous system demyelination. *Am J Pathol* **168**, 1464–1473.

Mancardi, G. L., 't Hart, B. A., Roccatagliata, L., Brok, H., Giunti, D., Bontrop, R. et al. (2001). Demyelination and axonal damage in a non-human primate model of multiple sclerosis. *J Neurol Sci* **184**, 41–49.

Maron, R., Hancock, W. W., Slavin, A., Hattori, M., Kuchroo, V. K., Weiner, H. L. (1999). Genetic susceptibility or resistance to autoimmune encephalomyelitis in MHC congenic mice is associated with differential production of pro- and anti-inflammatory cytokines. *Int Immunol* **11**, 1573–1580.

Marta, C. B., Oliver, A. R., Sweet, R. A., Pfeiffer, S. E., Ruddle, N. H. (2005). Pathogenic myelin oligodendrocyte glycoprotein antibodies recognize glycosylated epitopes and perturb oligodendrocyte physiology. *Proc Natl Acad Sci USA* **102**, 13992–13997.

Massacesi, L., Joshi, N., Lee-Parritz, D., Rombos, A., Letvin, N. L., Hauser, S. L. (1992). Experimental allergic encephalomyelitis in *Cynomolgus* monkeys. Quantitation of T cell responses in peripheral blood. *J Clin Invest* **90**, 399–404.

Matthys, P., Vermeire, K., Heremans, H., Billiau, A. (2000). The protective effect of IFN-γ in experimental autoimmune diseases: A central role of mycobacterial adjuvant induced myelopoiesis. *J Leukocyte Biol* **68**, 447–454.

Matyszak, M. K., Perry, V. H. (1996). The potential role of dendritic cells in immune-mediated inflammatory diseases in the central nervous system. *Neuroscience* **74**, 599–608.

McRae, B. L., Vanderlugt, C. L., Dal Canto, M. C., Miller, S. D. (1995). Functional evidence for epitope spreading in the relapsing pathology of experimental autoimmune encephalomyelitis. *J Exp Med* **182**, 75–85.

Medana, I., Martinic, M. M. A., Wekerle, H., Neumann, H. (2001). Transection of MHC class I-induced neurites by cytotoxic T lymphocytes. *Am J Pathol* **159**, 809–815.

Medana, I. M., Gallimore, A., Oxenius, A., Martinic, M. M. A., Wekerle, H., Neumann, H. (2000). MHC class I-restricted killing of neurons by virus specific $CD8^+$ T lymphocytes is effected through the Fas/FasL, but not the perforin pathway. *Eur J Immunol* **30**, 3623–3633.

Medawar, P. B. (1948). Immunity to homologous grafted skin. III. The fate of skin homografts transplanted to the brain, to subcutaneous tissue, and to anterior chamber of the eye. *Br J Exp Pathol* **29**, 58–69.

Meinl, E., Hoch, R. M., Dornmair, K., De Waal Malefijt, R., Bontrop, R. E., Jonker, M. et al. (1997). Encephalitogenic potential of myelin basic protein-specific T cells isolated from normal rhesus macaques. *Am J Pathol* **150**, 445–453.

Mendel, I., Kerlero de Rosbo, N., Ben-Nun, A. (1995). A myelin oligodendrocyte glycoprotein peptide induces typical chronic experimental autoimmune encephalomyelitis in $H\text{-}2^b$ mice: Fine specificity and T cell receptor Vβ expression of encephalitogenic T cells. *Eur J Immunol* **25**, 1951–1959.

Merkler, D., Boscke, R., Schmelting, B., Czeh, B., Fuchs, E., Bruck, W., Stadelmann, C. (2006). Differential macrophage/microglia activation in neocortical EAE lesions in the marmoset monkey. *Brain Pathol* **16**, 117–123.

Merrill, J. E., Kono, D. H., Clayton, J., Ando, D. G., Hinton, D. R., Hofman, F. M. (1992). Inflammatory leukocytes and cytokines in the peptide-induced disease of experimental allergic encephalomyelitis in SJL and B10.PL mice. *Proc Natl Acad Sci USA* **89**, 574–578.

Miller, D. H., Khan, O. A., Sheremata, W. A., Blumhardt, L. D., Rice, G. P. A., Libonati, M. A. et al. (2003). A controlled trial of Natalizumab for relapsing multiple sclerosis. *N Engl J Med* **348**, 15–23.

Miyakoshi, A., Yoon, W. K., Jee, Y., Matsumoto, Y. (2003). Characterization of the antigen specificity and TCR repertoire, and TCR-based DNA vaccine therapy in myelin basic protein-induced autoimmune encephalomyelitis in DA rats. *J Immunol* **170**, 6371–6378.

Moalem, G., Leibowitz-Amit, R., Yoles, E., Mor, F., Cohen, I. R., Schwartz, M. (1999). Autoimmune T cells protect neurons from secondary degeneration after central nervous system axotomy. *Nature Med* **5**, 49–55.

Mokhtarian, F., McFarlin, D. E., Raine, C. S. (1984). Adoptive transfer of myelin basic protein-sensitized T cells produces chronic relapsing demyelinating disease in mice. *Nature* **309**, 356–358.

Moran, L. B., Graeber, M. B. (2004). The facial nerve axotomy model. *Brain Res Rev* **44**, 154–178.

Morris-Downes, M. M., McCormack, K., Baker, D., Sivaprasad, D., Natkunarajah, J., Amor, S. (2002). Encephalitogenic and immunogenic potential of myelin-associated glycoprotein (MAG), oligodendrocyte-specific glycoprotein (OSP) and 2′,3′-cyclic nucleotide 3′-phosphodiesterase (CNPase) in ABH and SJL mice. *J Neuroimmunol* **122**, 20–33.

Muller, D. M., Pender, M. P., Greer, J. M. (2005). Blood-brain barrier disruption and lesion localisation in experimental autoimmune encephalomyelitis with predominant cerebellar and brainstem involvement. *J Neuroimmunol* **160**, 162–169.

Najafian, N., Chitnis, T., Salama, A. D., Zhu, B., Benou, C., Yuan, X. et al. (2003). Regulatory functions of $CD8^+CD28^-$ T cells in an autoimmune disease model. *J Clin Invest* **112**, 1037–1048.

Neuhaus, O., Farina, C., Wekerle, H., Hohlfeld, R. (2001). Mechanisms of action of glatiramer acetate in multiple sclerosis. *Neurol* **56**, 702–706.

Neumann, H., Cavalié, A., Jenne, D. E., Wekerle, H. (1995). Induction of MHC class I genes in neurons. *Science* **269**, 549–552.

Neumann, H., Misgeld, T., Matsumuro, K., Wekerle, H. (1998). Neurotrophins inhibit major histocompatibility class II inducibility of microglia: Involvement of the p75 neurotrophin receptor. *Proc Natl Acad Sci USA* **95**, 5779–5784.

Nguyen, M. D., Julien, J. P., Rivest, S. (2002). Innate immunity: The missing link in neuroprotection and neurodegeneration? *Nature Rev Neurosci* **3**, 216–227.

O'Connor, K. C., Appel, H., Bregoli, L., Call, M. E., Catz, I., Chan, J. A. et al. (2005). Antibodies from inflamed central nervous system tissue recognize myelin oligodendrocyte glycoprotein. *J Immunol* **175**, 1974–1982.

Offen, D., Kaye, J. F., Bernard, O., Merims, D., Coire, C. I., Panet, H. et al., (2000). Mice overexpressing bcl-2 in their neurons are resistant to myelin oligodendrocyte glycoprotein (MOG)-induced experimental autoimmune encephalomyelitis (EAE). *J Mol Neurosci* **15**, 167–176.

Oksenberg, J. R., Panzara, M. A., Begovich, A. B., Mitchell, D., Erlich, H. A., Murray, R. S. et al. (1993). Selection for T-cell receptor Vβ-Dβ-Jβ gene rearrangements with specificity for a myelin basic protein peptide in brain lesions of multiple sclerosis. *Nature* **362**, 68–70.

Oliveira, A. L. R., Thams, S., Lidman, O., Piehl, F., Hökfelt, T., Kärre, K. et al.,. (2004). A role for MHC class I molecules in synaptic plasticity and regeneration of neurons after axotomy. *Proc Natl Acad Sci USA* **101**, 17843–17848.

Owens, T., Wekerle, H., Antel, J. (2001). Genetic models of CNS inflammation. *Nature Med* **7**, 161–165.

Papadopoulos, D., Pham-Dinh, D., Reynolds, R. (2006). Axon loss is responsible for chronic neurological deficit following inflammatory demyelination in the rat. *Exp Neurol* **197**, 373–385.

Park, H., Li, Z. X., Yang, X. X. O., Chang, S. H., Nurieva, R., Wang, Y. et al. (2005). A distinct lineage of CD4 T cells regulates tissue inflammation by producing interleukin 17. *Nature Immunol* **6**, 1133–1141.

Pedotti, R., Mitchell, D., Wedemeyer, J., Karpuj, M., Chabas, D., Hattab, E. M. et al. (2001). An unexpected version of horror autotoxicus: Anaphylactic shock to a self peptide. *Nature Med* **2**, 216–222.

Pettinelli, C. B., Fritz, R. B., Chou, C-H. J., McFarlin, D. E. (1982). Encephalitogenic activity of guinea pig myelin basic protein. *J Immunol* **129**, 1209–1211.

Pettinelli, C. B., McFarlin, D. E. (1981). Adoptive transfer of experimental allergic encephalomyelitis in SJL/J mice after in vitro activation of lymph node cells with myelin basic protein: Requirement for Lyt-1+2- T lymphocytes. *J Immunol* **127**, 1420–1423.

Phillips, M. J., Weller, R. O., Iannotti, F. (1995). Focal brain damage enhances experimental allergic encephalomyelitis in brain and spinal cord. *Neuropathol Appl Neurobiol* **21**, 189–200.

Pomeroy, I. M., Matthews, P. M., Frank, J. A., Jordan, E. K., Esiri, M. M. (2005). Demyelinated neocortical lesions in marmoset autoimmune encephalomyelitis mimic those in multiple sclerosis. *Brain* **128**, 2713–2721.

Power, C., Kong, P-A., Trapp, B. D. (1996). Major histocompatibility complex class I expression in oligodendrocytes induces hypomyelination in transgenic mice. *J Neurosci Res* **44**, 165–173.

Prinjha, R., Moore, S. E., Vinson, M., Blake, S., Morrow, R., Christie, G. et al. (2000). Inhibitor of neurite outgrowth in humans. *Nature* **403**, 383–384.

Probert, L., Akassoglou, K., Pasparakis, M., Kontogeorgos, G., Kollias, G. (1995). Spontaneous inflammatory demyelinating disease in transgenic mice showing central nervous system-specific expression of tumor necrosis factor α. *Proc Natl Acad Sci USA* **92**, 11294–11298.

Raivich, G., Bohatschek, M., Kloss, C. U. A., Werner, A., Jones, L. L., Kreutzberg, G. W. (1999). Neuroglial activation repertoire in the injured brain: Graded response, molecular mechanisms and cues to physiological function. *Brain Res Rev* **30**, 77–105.

Raivich, G., Jones, L. L., Kloss, C. U. A., Werner, A., Neumann, H., Kreutzberg, G. W. (1998). Immune surveillance in the injured nervous system: T lymphocytes invade the axotomized mouse facial motor nucleus and aggregate around sites of neuronal degeneration. *J Neurosci* **18**, 5804–5816.

Rapalino, O., Lazarov-Spiegler, O., Agranov, E., Velan, G. J., Yoles, E., Fraidakis, M. et al. (1998). Implantation of stimulated homologous macrophages results in partial recovery of paraplegic rats. *Nature Med* **4**, 814–821.

Rauer, S., Euler, B., Reindl, M., Berger, T. (2006). Antimyelin antibodies and the risk of relapse in patients with a primary demyelinating event. *J Neurol Neurosurg Psych* **77**, 739–742.

Redwine, J. M., Buchmeier, M. J., Evans, C. F. (2001). In vivo expression of major histocompatibility complex molecules on oligodendrocytes and neurons during viral infection. *Am J Pathol* **159**, 1219–1224.

Riminton, D. S., Körner, H., Strickland, D. H., Lemckert, F. A., Pollard, J. D., Sedgwick, J. D. (1998). Challenging cytokine redundancy: Inflammatory cell movement and clinical course in experimental autoimmune encephalomyelitis are normal in lymphotoxin-deficient, but not tumor necrosis factor-deficient mice. *J Exp Med* **187**, 1517–1528.

Rivers, T. M., Sprunt, D. H., Berry, G. P. (1933). Observations on attempts to produce acute disseminated encephalomyelitis in monkeys. *J Exp Med* **58**, 39–53.

Rose, L. M., Richards, T., Alvord, E. C. (1994). Experimental allergic encephalomyelitis (EAE) in nonhuman primates: A model of multiple sclerosis. *Lab Animal Sci* **44**, 508–512.

Sakuma, H., Kohyama, K., Park, I-K., Miyakoshi, A., Tanuma, N., Matsumoto, Y. (2004). Clinicopathological study of a myelin oligodendrocyte glycoprotein-induced demyelinating disease in LEW.1AV1 rats. *Brain* **127**, 2201–2213.

Scarisbrick, I. A., Blaber, S. I., Lucchinetti, C. F., Genain, C. P., Blaber, M., Rodriguez, M. (2002). Activity of a newly identified serine protease in CNS demyelination. *Brain* **125**, 1283–1296.

Schenk, D., Barbour, R., Dunn, W., Gordon, G., Grajeda, H., Guido, T. et al. (1999). Immunization with amyloid-β attenuates Alzheimer disease-like pathology in the PDAPP mouse. *Nature* **400**, 173–177.

Schluesener, H. J., Sobel, R. A., Linington, C., Weiner, H. L. (1987). A monoclonal antibody against a myelin oligodendrocyte glycoprotein induces relapses and demyelination in central nervous system autoimmune disease. *J Immunol* **139**, 4016–4021.

Schwartz, M., Moalem, G. (2001). Beneficial immune activity after CNS injury: Prospects for vaccination. *J Neuroimmunol* **113**, 192–195.

Schwartz, M., Moalem, G., Leibowitz-Amit, R., Cohen, I. R. (1999). Innate and adaptive immune responses can be beneficial for CNS repair. *Trends Neurosci* **22**, 295–299.

Secor, V. H., Secor, W. E., Gutekunst, C-A., Brown, M. A. (2000). Mast cells are essential for early onset and severe disease in a murine model of multiple sclerosis. *J Exp Med* **191**, 813–821.

Selmaj, K. W., Raine, C. S. (1988). Tumor necrosis factor mediates myelin and oligodendrocyte damage in vitro. *Ann Neurol* **23**, 339–346.

Serafini, B., Columba-Cabezas, S., Di Rosa, F., Aloisi, F. (2000). Intracerebral recruitment and maturation of dendritic cells in the onset and progression of experimental autoimmune encephalomyelitis. *Am J Pathol* **157**, 1991–2002.

Shao, H., Huang, Z. G., Sun, S. L., Kaplan, H. J., Sun, D. M. (2004). Myelin/oligodendrocyte glycoprotein-specific T-cells induce severe optic neuritis in the C57Bl/6 mouse. *Invest Ophthalmol Vis Sci* **45**, 4060–4065.

Skulina, C., Schmidt, S., Dornmair, K., Babbe, H., Roers, A., Rajewsky, K. et al. (2004). Multiple sclerosis: Brain-infiltrating $CD8^+$ T cells persist as clonal expansions in the cerebrospinal fluid and blood. *Proc Natl Acad Sci USA* **101**, 2428–2433.

Skundric, D. S., Zakarian, V., Dai, R. J., Lisak, R. P., Tse, H. Y., James, J. (2003). Distinct immune regulation of the response to $H\text{-}2^b$ restricted epitope of MOG causes relapsing-remitting EAE in $H\text{-}2^{b/s}$ mice. *J Neuroimmunol* **136**, 34–45.

Smeltz, R. B., Wolf, N. A., Swanborg, R. H. (1998). Delineation of two encephalitogenic myelin basic protein epitopes for DA rats. *J Neuroimmunol* **87**, 43–48.

Smith, P. A., Heijmans, N., Ouwerling, B., Breij, E. C., Evans, N., Van Noort, J. M. et al. (2005). Native myelin oligodendrocyte glycoprotein promotes severe chronic neurological disease and demyelination in Biozzi ABH mice. *Eur J Immunol* **35**, 1311–1319.

Smith, T., Groome, A., Zhu, B., Turski, L. (2000). Autoimmune encephalomyelitis ameliorated by AMPA antagonists. *Nature Med* **6**, 62–66.

Sobel, R. A., Kuchroo, V. K. (1992). The immunopathology of acute experimental allergic encephalomyelitis induced with myelin proteolipid protein. *J Immunol* **149**, 1444–1451.

Sobel, R. A., Tuohy, V. K., Lu, Z., Laursen, R. A., Lees, M. B. (1990). Acute experimental allergic encephalomyelitis in SJL/J mice induced by a synthetic peptide of myelin proteolipid protein. *J Neuropathol Exp Neurol* **49**, 468–479.

Stefferl, A., Linington, C., Holsboer, F., Reul, J. M. H. M. (1999). Susceptibility and resistance to experimental allergic encephalomyelitis: Relationship with hypothalamic-pituitary-adrenocortical axis responsiveness in the rat. *Endocrinology* **140**, 4932–4938.

Stefferl, A., Schubart, A., Storch, M., Amini, A., Mather, I. H., Lassmann, H., Linington, C. (2000). Butyrophilin, a milk protein, modulates the encephalitogenic T cell response to myelin oligodendrocyte glycoprotein in experimental autoimmune encephalomyelitis. *J Immunol* **165**, 2859–2865.

Stepaniak, J. A., Gould, K. E., Sun, D., Swanborg, R. H. (1995). A comparative study of experimental autoimmune encephalomyelitis in Lewis and DA rats. *J Immunol* **155**, 2762–2769.

Stepaniak, J. A., Wolf, N. A., Sun, D., Swanborg, R. H. (1997). Interstrain variability of autoimmune encephalomyelitis in rats: Multiple encephalitogenic myelin basic protein epitopes for DA rats. *J Neuroimmunol* **78**, 79–85.

Stewart, W. A., Alvord, E. C., Hruby, S., Hall, L. D., Paty, D. W. (1991). Magnetic resonance imaging of experimental allergic encephalomyelitis in primates. *Brain* **114**, 1069–1096.

Storch, M. K., Piddlesden, S., Haltia, M., Iivanainen, M., Morgan, P., Lassmann, H. (1998a). Multiple sclerosis: In situ evidence for antibody- and complement-mediated demyelination. *Ann Neurol* **43**, 465–471.

Storch, M. K., Stefferl, A., Brehm, U., Weissert, R., Wallström, E., Kerschensteiner, M. et al. (1998b). Autoimmunity to myelin oligodendrocyte glycoprotein in rats mimics the spectrum of multiple sclerosis pathology. *Brain Pathol* **8**, 681–694.

Sullivan, P. G., Bruce-Keller, A. J., Rabchevsky, A. G., Christakos, S., St.Clair, D. K., Mattson, M. P., Scheff, S. W. (1999). Exacerbation of damage and altered NF-κB activation in mice lacking tumor necrosis factor receptors after traumatic brain injury. *J Neurosci* **19**, 6248–6256.

Sun, D., Newman, T. A., Perry, V. H., Weller, R. O. (2004). Cytokine-induced enhancement of autoimmune inflammation in the brain and spinal cord: Implications for multiple sclerosis. *Neuropathol Appl Neurobiol* **30**, 374–384.

Sun, D., Qin, Y., Chluba, J., Epplen, J. T., Wekerle, H. (1988). Suppression of experimentally induced autoimmune encephalomyelitis by cytolytic T-T- cell interactions. *Nature* **332**, 843–845.

Sun, D. M., Whitaker, J. N., Huang, Z. G., Liu, D., Coleclough, C., Wekerle, H., Raine, C. S. (2001). Myelin antigen-specific $CD8^+$ T cells are encephalitogenic and produce severe disease in C57BL/6 mice. *J Immunol* **166**, 7579–7587.

Suzumura, A., Lavi, E., Weiss, S. R., Silberberg, D. H. (1986). Coronavirus infection induces H-2 antigen expression on oligodendrocytes and astrocytes. *Science* **232**, 991–993.

Svensson, L., Abdul-Majid, K. B., Bauer, J., Lassmann, H., Harris, R. A., Holmdahl, R. (2002). A comparative analysis of B cell-mediated myelin oligodendrocyte glycoprotein-experimental autoimmune encephalomyelitis pathogenesis in B cell-deficient mice reveals an effect on demyelination. *Eur J Immunol* **32**, 1939–1946.

Teuscher, C., Blankenhorn, E. P., Hickey, W. F. (1987). Differential susceptibility to actively induced experimental allergic encephalomyelitis and experimental allergic orchitis among BALB/c substrains. *Cell Immunol* **110**, 294–304.

The Lenercept MS Study Group, The UBC MS/MRI Analysis Group. (1999). TNF neutralization in MS. Results of a randomized placebo controlled multicenter trial. *Neurol* **53**, 457–465.

Thoua, N. M., Van Noort, J. M., Baker, D., Bose, A., Van Sechel, A. C., Van Stipdonk, M. J. B. et al. (2000). Encephalitogenic and immunogenic potential of the stress protein β-crystallin in Biozzi ABH ($H\text{-}2A^{g7}$). *J Neuroimmunol* **104**, 47–57.

Tsunoda, I., Kuang, L. Q., Theil, D. J., Fujinami, R. S. (2000). Antibody association with a novel model for primary progressive multiple sclerosis: Induction of relapsing-remitting and progressive forms of EAE in $H2^s$ mouse strains. *Brain Pathol* **10**, 402–418.

Tuohy, V. K., Sobel, R. A., Lees, M. B. (1988). Myelin proteolipid protein-induced experimental allergic encephalomyelitis: Variation of disease expression in different strains of mice. *J Immunol* **140**, 1868–1873.

Turnley, A. M., Morahan, G., Okano, H., Bernard, O., Mikoshiba, K., Allison, J. et al. (1991). Dysmyelination in transgenic mice resulting from expression of class I histocompatibility molecules in oligodendrocytes. *Nature* **353**, 566–569.

Veldhoen, M., Hocking, R. J., Flavell, R. A., Stockinger, B. (2006). Signals mediated by transforming growth factor-β initiate autoimmune encephalomyelitis, but chronic inflammation is needed to sustain disease. *Nature Immunol* **7**, 1151–1156.

Villoslada, P., Hauser, S. L., Bartke, I., Unger, J., Heald, N., Rosenberg, D. et al. (2000). Human nerve growth factor protects common marmosets against autoimmune encephalomyelitis by switching the balance of T helper cell type 1 and 2 cytokines within the central nervous system. *J Exp Med* **191**, 1799–1806.

von Büdingen, H. C., Hauser, S. L., Fuhrmann, A., Nabavi, C. B., Lee, J. I., Genain, C. P. (2002). Molecular characterization of antibody specificities against myelin/oligodendrocyte glycoprotein in autoimmune demyelination. *Proc Natl Acad Sci USA* **99**, 8207–8212.

Waldner, H., Whitters, M. J., Sobel, R. A., Collins, M., Kuchroo, V. K. (2000). Fulminant spontaneous autoimmunity of the central nervous system in mice transgenic for the myelin proteolipid protein-specific T cell receptor. *Proc Natl Acad Sci USA* **97**, 3412–3417.

Wekerle, H., Kojima, K., Lannes-Vieira, J., Lassmann, H., Linington, C. (1994). Animal models. *Ann Neurol* **36**, S47–S53.

Wensky, A. K., Furtado, G. C., Garibaldi Marcondes, M. C., Chen, S. H., Manfra, D., Lira, S. A. et al. (2005). IFN-γ determines distinct clinical outcomes in autoimmune encephalomyelitis. *J Immunol* **174**, 1416–1423.

West, A. E., Griffith, E. C., Greenberg, M. E. (2002). Regulation of transcription factors by neuronal activity. *Nature Rev Neurosci* **3**, 921–931.

Willenborg, D. O., Fordham, S. A., Bernard, C. C. A., Cowden, W. B., Ramshaw, I. A. (1996). IFN-γ plays a critical down-regulatory role in the induction and effector phase of myelin oligodendrocyte glycoprotein-induced autoimmune encephalomyelitis. *J Immunol* **157**, 3223–3227.

Willer, C. J., Dyment, D. A., Risch, N. J., Sadovnick, A. D., Ebers, G. C. (2003). Twin concordance and sibling recurrence rates in multiple sclerosis. *Proc Natl Acad Sci USA* **100**, 12877–12882.

Wolf, S. D., Dittel, B. N., Hardardottir, F., Janeway, C. A. (1996). Experimental autoimmune encephalomyelitis induction in genetically B cell-deficient mice. *J Exp Med* **184**, 2271–2278.

Wong, F. S., Dittel, B. N., Janeway, C. A. (1999). Transgenes and knock-out mutations in animal models of type 1 diabetes and multiple sclerosis. *Immunol Rev* **169**, 93–106.

Xiao, B., Linington, C., Link, H. (1991). Antibodies to myelin-oligodendrocyte glycoprotein in cerebrospinal fluid from patients with multiple sclerosis and controls. *J Neuroimmunol* **31**, 91–96.

Yamamura, T., Namikawa, T., Endoh, M., Kunishita, T., Tabira, T. (1986). Experimental allergic encephalomyelitis induced by proteolipid apoprotein in Lewis rats. *J Neuroimmunol* **12**, 143–153.

Yoshioka, T., Feigenbaum, L., Jay, G. (1991). Transgenic mouse model for demyelination. *Mol Cell Biol* **11**, 5479–5486.

Yu, M., Nishiyama, A., Trapp, B. D., Tuohy, V. K. (1996). Interferon-β inhibits progression of relapsing-remitting experimental autoimmune encephalomyelitis. *J Neuroimmunol* **64**, 91–100.

Zamvil, S. S., Nelson, P. A., Mitchell, D. J., Knobler, R. L., Fritz, R. B., Steinman, L. (1985a). Encephalitogenic T cell clones specific for myelin basic protein: An unusual bias in antigen recognition. *J Exp Med* **162**, 2107–2124.

Zamvil, S. S., Nelson, P. A., Trotter, J., Mitchell, D. J., Knobler, R. L., Fritz, R. B., Steinman, L. (1985b). T-cell clones specific for myelin basic protein induce chronic relapsing paralysis and demyelination. *Nature* **317**, 355–358.

Zhang, B., Yamamura, T., Kondo, T., Fujiwara, M., Tabira, T. (1997). Regulation of experimental autoimmune encephalomyelitis by natural killer cells. *J Exp Med* **186**, 1677–1687.

Zhu, B., Guleria, I., Khosroshani, A., Chitnis, T., Imitola, J., Azuma, M. et al. (2006). Differential role of programmed death-1 ligand and programmed death-2 ligand in regulating the susceptibility and chronic progression of experimental autoimmune encephalomyelitis. *J Immunol* **176**, 3480–3489.

31

Autoimmune and Genetic Disorders of the Neuromuscular Junction and Motor Nerve Terminal

Angela Vincent

I. Introduction

A. The Neuromuscular Junction

Despite the considerable knowledge that has accrued concerning central synapses, the neuromuscular junction (NMJ) remains a protypic synapse, although its structure is rather different from those of the CNS. The unmyelinated motor nerve terminals are separated by a 500 A synaptic cleft from the postsynaptic muscle membrane (see Figure 31.1). The synaptic cleft contains a basal lamina that includes many of the proteins that are found elsewhere in extracellular matrices, such as collagens, laminins, fibronectin, and perlecan, but some of the isoforms (e.g., collagen IV) are specific (Sanes & Lichtman, 1999, 2001). These proteins are not only important structural elements, but they help anchor some of the key elements involved in NMJ development and function. For example, perlecan anchors acetylcholine esterase (AChE) via ColQ, a collagen like molecule, and agrin and neuregulins, that are secreted from the nerve terminal, concentrate in the basal lamina where they can interact with their targets on the postsynaptic membrane. Some of these interactions are important for the location of membrane proteins such as voltage-gated calcium channels presynaptically (Nishimune et al., 2004) and the dystroglycans post-synaptically (Patton, 2003).

The postsynaptic membrane at the neuromuscular junction forms a series of folds, and these are particularly deep and complex in human muscle (Slater, 2003; Wood & Slater, 2001). The acetylcholine receptors are found at the top one-third of these folds, whereas the voltage-gated sodium channels are anchored at the bottom of the folds (diagrammatically illustrated in Figure 31.1). The developmental neurobiology of the neuromuscular junction has been an area of very active research (Sanes & Lichtman, 2001), and many of the key proteins are relevant to the diseases to be discussed.

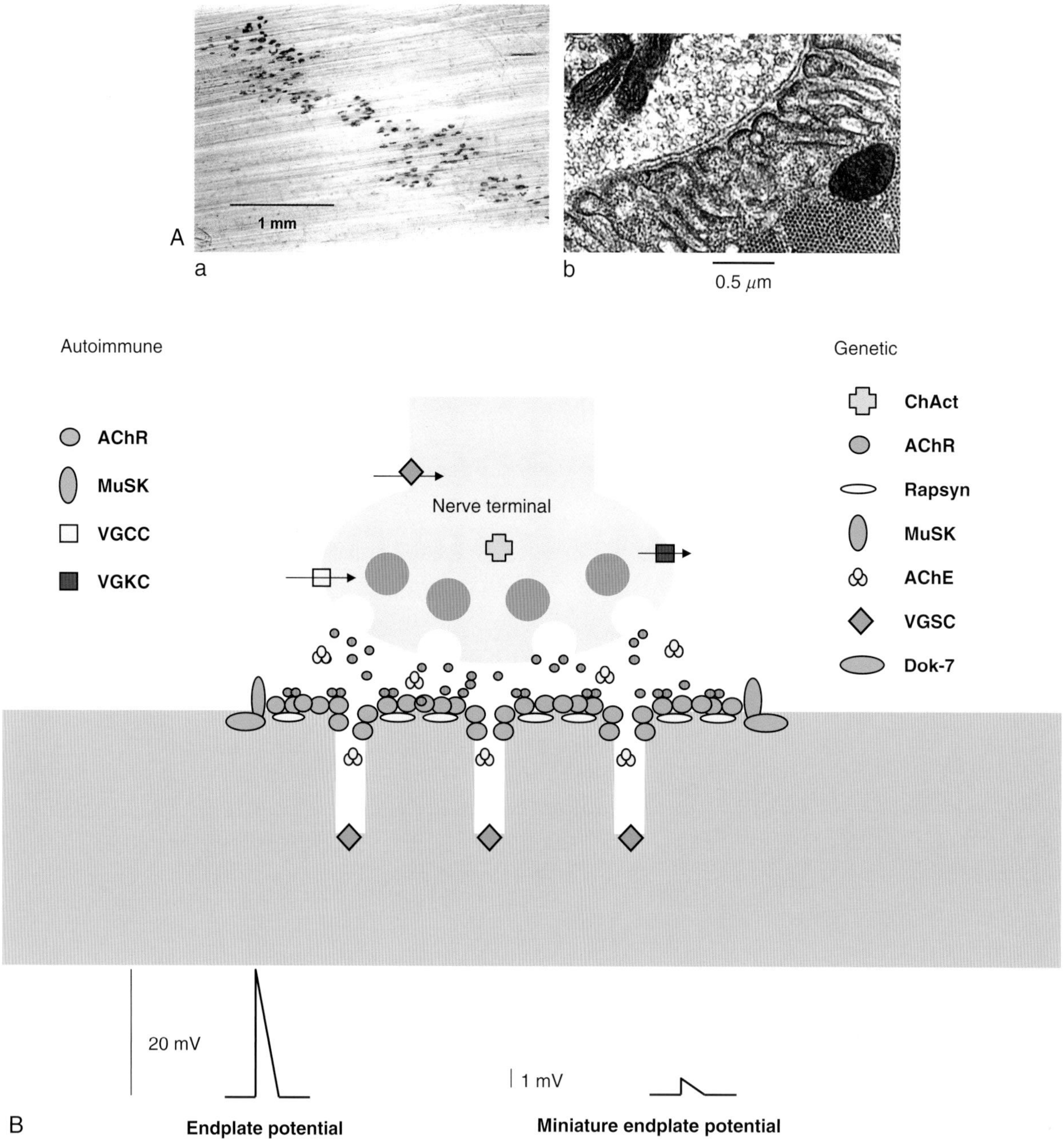

Figure 31.1 The neuromuscular junction and acetylcholine receptors. (**A**) Electronmicroscopic image of the human neuromuscular junction. The nerve terminal contains a mitochondrion and many synaptic vesicles. The basal lamina can be seen as a thin line between the pre- and post-synaptic membranes. The tops of the post-synaptic folds are electron dense due to the high density of AChRs. The folds are very long and below them the muscle fibrils can be seen in cross-section (courtesy of Prof. Clarke Slater). (**B**) Diagrammatic representation of the NMJ with the ion channels and other molecules that are essential for its normal function and that are targets for autoimmune and genetic diseases.

When the nerve impulse reaches the motor nerve terminal, the depolarization opens voltage-gated calcium channels (VGCCs) that are located in active zones along the terminal membrane where it abuts the muscle (see Figure 31.1). The resulting transient, and highly localized, influx of calcium leads to the release of individual packets or "quanta" of acetylcholine (ACh) into the synaptic cleft. About 10,000 molecules of ACh are stored in each packet and many of these succeed in avoiding immediate hydrolysis by AChE. This is probably because the density of AChE molecules in the primary synaptic cleft is lower than that of the AChRs on the postsynaptic membrane, and the hydrolysis of each molecule of ACh is relatively slow. Thus the ACh can bind, instead, to the acetylcholine receptors (AChRs) on the postsynaptic membrane. This leads to the opening of the AChR associated ion channels and depolarization of the motor endplate, which can be measured by an intracellular electrode as an endplate potential (EPP) or, under voltage clamp conditions, as an endplate current (EPC; see Figure 31.1). The EPP in normal human muscles is around 20 mV, representing the release of about 30 packets of ACh. Miniature EPPs (MEPPs) are the much smaller depolarizations that occur when a packet of ACh is spontaneously released. Their amplitude and duration are useful indicators of AChR function in disease states since they reflect the density of AChRs and the kinetics of the individual AChRs.

Under normal conditions, the EPP rapidly depolarizes the postsynaptic membrane from its resting potential of around −70 mV. When it reaches a critical firing threshold, the voltage-gated sodium channels open and an action potential is initiated that propagates along the muscle fiber, activating the contractile apparatus. The AChRs close spontaneously, ACh unbinds, and ACh is hydrolyzed by AChE, thus limiting the duration of the response. Presynaptically, the calcium channels close spontaneously, the resting membrane potential is restored by efflux of potassium through voltage-gated potassium channels and the electrochemical gradient maintained by the sodium/potassium ATPase. The extent to which the EPP exceeds that necessary to initiate the action potential usually is called the safety factor for neuromuscular transmission (see Wood & Slater, 2001). Interestingly, as will be discussed, the pre- and post-synaptic membranes communicate with each other, and changes in function on one side of the synapse may be reflected by compensatory changes on the other.

The neuromuscular junction also has provided a model for different disease processes. It is particularly vulnerable to circulating factors because it has no blood–brain barrier. In various parts of the world, envenomation (by, for instance, snakes, spiders, scorpions) causes neuromuscular junction paralysis or hyperexcitability (see Hodgson et al., 2002; Senanayake et al., 1992 for reviews). Botulism is still a major problem in some countries. Poisoning by environmental or self-administered insecticides that block AChE is common, and there are a variety of plant extracts that also interfere with neuromuscular transmission. By contrast, the most common disorders in the western world are caused by autoantibodies to the ion channels on the pre- and post-synaptic membranes, or associated proteins. This review will concentrate on these conditions and disease mechanisms but will not discuss in any detail the immunological aspects of these disorders.

Throughout this review, emphasis will be placed on the general concepts regarding disease specificity, mechanisms, phenotypic variability, and compensatory mechanisms. Understanding these still represents a challenge.

II. Autoimmune Disorders

A. Myasthenia Gravis

1. Clinical Features and History

Myasthenia gravis usually presents in young adult or adult life as muscle weakness and fatigue, which can be generalized or limited in distribution. Fatigue can occur during any task that requires repetitive movements. Typically the extraocular muscles of the eye are involved, causing double vision, and ptosis may result from weakness of eye-lid elevation. Involvement of the facial and bulbar muscles can result in loss of smile, poor speech, and choking, and this and respiratory muscle weakness can be life-threatening. Cholinesterase inhibitors, by prolonging the action of ACh, tend to lead to clinical improvement (for reviews of the clinical features of MG, see Drachman, 1994; Vincent et al., 2000).

There is a long history of hypotheses regarding the molecular basis of myasthenia gravis (MG). At the beginning of the twentieth century it was proposed that the weakness might be due to an autotoxic circulating factor that interfered with neuromuscular transmission (see Vincent 2002 for a review). It was only when the quantal nature of neuromuscular transmission was demonstrated in the 1950s by Katz and his colleagues that better understanding of the disease could be achieved. First it was found that the miniature endplate potentials are reduced substantially in muscle biopsies from MG patients (Elmqvist et al., 1964). This was interpreted as due either to a reduction in the amount of ACh in each quanta or to a reduction in the postsynaptic sensitivity to ACh. At that time the existence of receptors for ACh was still hypothetical, since there was no direct way of measuring them. Quite independently, it was found that the snake toxin, α-bungarotoxin, from the Taiwan banded Krait, bound irreversibly to the neuromuscular junction, and that it bound highly specifically to receptors for ACh (see Chu, 2005). It was then possible to show that the number of AChRs was reduced at the neuromuscular junctions of MG patients using ^{125}I-α-bungarotoxin to measure the AChRs (Fambrough, Drachman & Satyamurti, 1973).

The electric organs of eels and rays are derived from branchial arch myotubes and related to muscle (see Keesey, 2005). Each organ is made up of stacks of flat cylindrical plates whose ventral surface is studded with AChRs. The AChRs were purified from detergent extracts of these tissues using Krait neurotoxins, such as α-cobratoxin, for affinity-chromatography. Patrick and Lindstrom (1973) found that rabbits, and subsequently other species, developed muscle weakness when immunized with purified Torpedo or eel AChR and their weakness responded to cholinesterase inhibitors (Patrick & Lindstrom, 1973). Following this seminal observation, Lindstrom and colleagues (1976) established a radioimmunoprecipitation assay for AChR antibodies and showed its specificity for the diagnosis of MG.

It was not until somewhat later that the subunit structure and N-terminal sequences of the AChR subunits could be identified, leading eventually to the cloning and expression of many species of AChR. The AChR is an oligomeric membrane protein consisting of five subunits (see Figure 31.2): α^2, β, γ, and δ in embryonic or denervated muscle (fetal form), and α^2, β, δ, and ε at the adult endplate (adult form). The antibodies are measured by immunoprecipitation of ^{125}I α-BuTx labeled AChR extracted from ischemic human muscle, which consists mainly of the fetal form (because ischemia causes denervation) or more recently from cell lines expressing adult and fetal AChRs, respectively (Beeson et al., 1996). AChR antibody titers vary widely between patients, ranging from 0 to more than 1000 nm/l, with little correlation with clinical severity between individuals. The AChR antibodies are very heterogeneous between and within individuals, as demonstrated by studies on their light chains, IgG subclasses, and ability to bind to different regions on the AChR (Vincent et al., 1987). Nevertheless, within an individual the levels correlate well with clinical response to treatments such as plasma exchange (Newsom-Davis et al., 1978) and immunosuppression with steroids. Most importantly, the role of serum antibodies can be demonstrated by passive transfer of purified IgG from MG patients to mice; this results in reduced numbers of AChRs and reduced amplitude of the miniature endplate potentials in the mouse muscle (like those in the patients' muscles; Toyka et al., 1975).

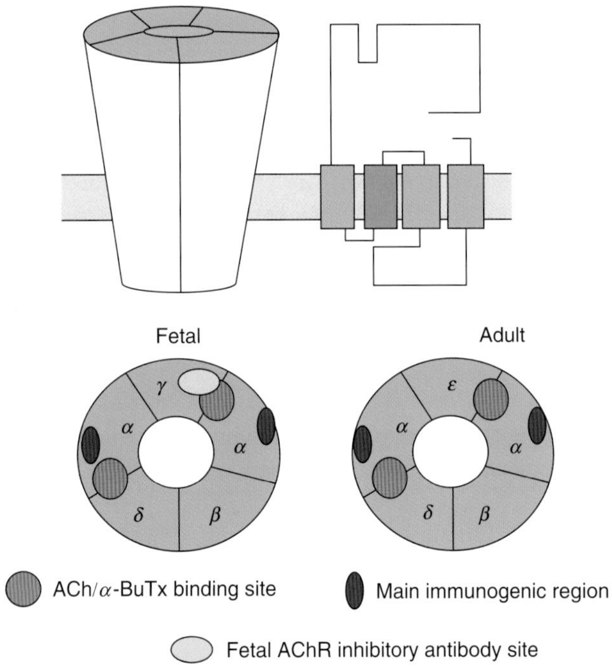

Figure 31.2 The two forms of the acetylcholine receptor. The acetylcholine receptor is a pentameric membrane protein that occurs in an adult and fetal isoform. Acetylcholine (ACh) and alpha-bungarotoxin (α-bungarotoxin) bind to sites on the interfaces between the α and adjacent subunits. Many of the antibodies in myasthenia bind to a main immunogenic region on the two α subunits. Other antibodies are specific to the γ subunit and if present in mothers can cause fetal paralysis by inhibiting fetal AChR function.

2. Epidemiology and Clinical Heterogeneity

Ocular MG MG often presents with extraocular muscle weakness (leading to diplopia) and may never spread to other muscles. It is still not clear why the extraocular muscles are so susceptible, although it is notable that they are often the first to be affected during botulism or following envenomation with snake neurotoxins. In general they should be resistant to fatigue with high blood flow, mitochondria content, and metabolic rate (Yu Wai Man et al., 2005). However, the motor unit sizes are small and the firing frequencies high, and the extraocular muscles consist of twitch and tonic muscle fibers, some of which have multiple neuromuscular junctions rather than a single one in each fiber. In these fibers there is no action potential generated; rather the endplate potential itself is responsible for activating the contractile apparatus. Thus any reduction in the endplate potential, as would occur when AChRs are lost, could have a direct effect on the strength of muscle contraction. The multiple-innervated endplates may also contain fetal-type AChRs (Kaminski & Ruff, 1997), which could make them susceptible to the action of antibodies specific for fetal AChRs, but this is unlikely to be a significant factor in their susceptibility. In fact, overall the extraocular muscles are a rich source of adult-type AChRs (McLennan et al., 1997).

One factor that might make the extraocular muscle more susceptible in MG is the low expression of complement regulators, which are now known to be expressed at higher concentration in other muscles (Kaminski et al., 2004); this would make them more vulnerable to complement-mediated damage. Interestingly in the Lambert Eaton myasthenic syndrome (LEMS, see later), in which complement-mediated damage is not thou ght to occur, ocular muscle weakness is uncommon.

3. Generalized AChR

In Eastern countries, particularly Chinese populations, ocular MG is common and frequently occurs in children (Zhang et al., 2007). Most patients in the Western world, however, progress to generalized weakness, at least if not treated promptly. These patients can be divided into early onset, late onset, and thymoma-MG, depending on the presence of AChR antibodies, age at onset, HLA association, and thymic pathology (Compston et al., 1980; see Table 31.1). The association of early onset MG with HLA B8DR3 haplotype suggests genetic susceptibility to development of MG in this age group (Compston et al., 1980). This may relate to polymorphic variants in the AChR genes (Djabiri et al., 1997) as well as in the major histocompatibility, HLA, genes (Giraud et al., 2004; Vandiedonck et al., 2004). However, there is no evidence that the AChR antibody characteristics or pathogenic mechanisms characteristically differ between these three groups, suggesting that the antibodies are the common final pathway that can be reached by a number of different etiological routes. In the older patients, the HLA association is less clear (but not B8DR3) and the thymus is essentially normal for age. Importantly, these patients are being recognized much more frequently in the population, and may well be underdiagnosed or misdiagnosed as stroke or motor neuron disease (Vincent et al., 2004).

4. Neonatal Myasthenia Gravis and Arthrogryposis Multiplex Congenita

A proportion of babies born to MG mothers have transient respiratory and feeding difficulties, owing to transplacental transfer of maternal AChR antibodies. According to the only available study (Hesselmans et al., 1993), the adult AChR is expressed at many NMJs in the fetus by 33 weeks gestation. The effects on the fetus, therefore, will depend not only on the maternal antibody levels but also on the efficiency of transfer at different stages of the pregnancy and the specificity of the antibodies for fetal or adult AChR. In a few reported cases, mothers with MG have given birth to babies with arthrogryposis multiplex congenita, a relatively common condition that arises within any situation that reduces fetal movement *in utero*, and involves fixed joint contractures associated with inadequate development of the lungs. When associated with maternal antibodies, it is not uncommon for several consecutive pregnancies to be affected. The mother's antibodies completely inhibit the function of fetal AChR, but do not necessarily have an effect on adult AChR function; this explains the fetal paralysis and development of deformities (Polizzi et al., 2002). In some cases, the maternal antibodies are relatively specific for the fetal form and the mother herself is asymptomatic (Vincent et al., 1995). An animal model of this condition was induced by injecting the maternal plasmas into pregnant mice; the offspring were found to be paralyzed and to exhibit fixed joint contractures (Jacobson et al., 1999). Maternal antibodies to other antigens are beginning to be considered as a possible cause of other developmental disorders (e.g., Rothenberg et al., 2004).

5. Etiology

Despite several decades of research, the etiology of MG is unknown. An understanding of the etiology should be derived from a full characterization of the AChR antibodies since they should reflect the precipitating event. All studies have shown that the AChR antibodies are heterogeneous and bind variably to several sites on the AChR. They are mainly IgG1 and IgG3 subclasses, which bind and activate complement effectively (Vincent et al., 1987). The antibodies, at least those that are measured in the radioimmunoprecipitation assays used routinely for diagnosis, bind with high affinity (or avidity) to detergent-extracted human AChR, and do not bind well to denatured or recombinant subunits on western blots, and very variably to AChR extracted from muscles derived from other mammalian species. The binding sites on the AChR were therefore defined by competition with monoclonal antibodies raised against purified electric fish AChRs (see Tzartos et al., 1998). The monoclonal antibodies bound mainly to a "main immunogenic region" (MIR) on the two AChR alpha subunits (Tzartos et al., 1981), and competed with a high proportion of MG antibodies. A recent publication, however, demonstrates that antibodies to the MIR are not necessarily predominant in all patients (Fostieri et al., 2005; although these antibodies may be the most pathogenic because of their ability to bind divalently to the AChRs). Moreover, immunization against purified fetal-type AChR produced monoclonal antibodies that bound to alpha, beta, delta, and gamma subunits (Whiting et al., 1986), and competed more variably with antibodies in different patients for binding to human AChR.

Overall, the characteristics of the antibodies suggest that the human AChR is the immunogen in MG, and that the antibodies do not arise as the result of a cross-reaction with microbial antigens (e.g., Schimmbeck et al., 1981; Stefansson et al., 1985) or as the result of an imbalance in an idiotypic network (Dwyer et al., 1986). An alternative possibility, however, is that an event that generates a low-affinity

Table 31.1 Subtypes of Myasthenia Gravis

	Age at Onset	HLA Association	Thymic Pathology	AChR Antibodies
AChR-MG				
Early onset	<40 years	DR3 B8	Hyperplasia	High
Late onset	>40 years	DR2B7 but not very strong	Normal for age	Low
Thymoma	Variable	None known	Thymic tumor	Moderate
Ocular	Variable	None known	Not known	Low or negative
AChR/MuSK negative MG	Variable	None known	Hyperplasia in some	Negative
MuSK-MG	Variable	DR14DQ5	Normal for age	Negative

or cross-reactive antibody may lead, by a process of determinant spreading and somatic mutation, to a high affinity reaction against the AChR (Vincent et al., 1994; and see Vincent et al., 1998 for discussion).

6. Pathophysiology

Early studies on neuromuscular junctions in patients' muscle biopsies showed that the main defect was a reduction in the number of functional AChRs (Fambrough et al., 1973), and that this was sufficient to explain the reduced amplitude of miniature endplate potentials (Elmqvist et al., 1964) and the patients' weakness. Indeed there was a correlation between miniature endplate potential amplitudes and α-bungarotoxin binding to NMJs (Ito et al., 1978).

The loss of AChRs results from several immunologically mediated processes. First some antibodies might produce a pharmacological blockade of ACh-induced ion channel function. Antibodies to the ACh/α-BuTx binding site are found in many MG patients (Vincent & Newsom-Davis, 1982), but they usually comprise only a small proportion of the total AChR antibody population. Some MG sera block function substantially in muscle cell cultures (e.g., Weinberg & Hall, 1979), and sometimes the effect is transient (Bufler et al., 1998), but inhibition of AChR function at neuromuscular junction has been shown only in a few studies (e.g., Burges et al., 1990). However, after an overnight incubation in MG sera, a reduction in AChRs can be demonstrated in cell lines. Several authors clearly demonstrated that this was the result of cross-linking of AChRs by divalent antibodies (e.g., Appel et al., 1977; Heineman et al., 1977; Heineman et al., 1978). These studies looked at the degradation of ^{125}I-bungarotoxin-labeled AChRs, which are normally internalized and degraded with release of ^{125}I into the medium. In cultured cells this process has a half-life of around 10 to 16 hours, but in the presence of AChR antibodies, the half-life fell to less than 10 hours.

Similar approaches were used in the mouse passive transfer model. ^{125}I-bungarotoxin was injected in to the thoracic cavity in order that it could bind to the diaphragm neuromuscular junctions before injection of the MG IgG preparation. The ^{125}I-bungarotoxin labeling decreased faster compared to control treated animals, and the half-life of the AChRs was reduced from about 10 days to less than 5 days (Stanley et al., 1978; Wilson et al., 1982a). Moreover, the appearance of new AChRs could be analyzed by pretreating the AChRs with cold bungarotoxin, to block specific binding of ^{125}I-bungarotoxin, and then measuring *ex vivo* the appearance of new ^{125}I-bungarotoxin binding sites (AChRs). These experiments showed that the MG antibodies could increase the synthesis rate of new AChRs (Wilson et al., 1982b), as well as the degradation rate. Since then, analysis of muscles from MG patients has confirmed an increase in the expression of AChR subunits (Guyon et al., 1998). Thus from these studies, the amount of AChR at the NMJ will be a balance between the degradation induced by antibodies and the compensatory increase in AChR synthesis.

A major pathogenic mechanism at the neuromuscular junction is complement-mediated destruction of the post-synaptic membrane. This was studied mainly by Engel and his colleagues (reviewed in Engel 1984) in MG as well as its animal model, EAMG, in rats. They found IgG localized to the NMJ (Engel et al., 1977; Sahashi et al., 1977) and the distribution tended to correspond to the distribution of remaining AChRs (determined by peroxidase-bungarotoxin binding). There were deposits of complement components C3 and C9, and of the membrane attack complex (MAC), and the synaptic cleft was widened and contained debris that also immunostained for IgG and complement (Sahashi et al., 1980). Interestingly, the amount of the MAC appeared to correlate inversely with the number of AChRs, suggesting that the terminal membrane attack complex is directly responsible for AChR loss (Engel & Arahata, 1987). The results of this process would not only lead to loss of AChR-containing membrane but also to loss of the post-synaptic folds. Most NMJs from MG patients have marked simplification of the post-synaptic folds even in the absence of other damage. Since the voltage-gated sodium channels are thought to be situated at the bottom of these folds, their loss would have the effect of increasing the threshold for neuromuscular transmission, and reducing the safety factor. This was demonstrated both in MG and experimental MG by Ruff and Lennon (1998). The amount of current that needed to be injected post-synaptically (simulating an EPP) in order to initiate an action potential was increased at the NMJs, presumably because the loss of post-synaptic folds had reduced the number of voltage-gated sodium channels.

The three main mechanisms considered earlier probably vary to some extent between patients, although it is difficult to study this since few biopsies are available. It seems very likely that there are differences in the degree of complement mediated attack according to the specificity and IgG subclass of the antibodies, the patient's complement levels, and the ability of the NMJ to repair itself during this process. In addition, complement regulatory proteins have been identified at the NMJ (Kaminski et al., 2003) suggesting another disease modifying factor that might be important in individual patients.

In addition, like most biological systems, the neuromuscular junction tries to compensate for the loss of AChRs that causes interference with neuromuscular transmission, by increasing AChR synthesis, as mentioned previously. In addition, the motor nerve also seems to recognize the impaired neuromuscular transmission. An increase in the number of ACh packets released, which correlated inversely with the amplitude of the MEPPs at individual endplates (Plomp et al., 1994), indicates that retrograde signaling from the postsynaptic to presynaptic components can lead to compensatory changes in ACh release. How this occurs and whether it involves modulation of existing nerve terminal proteins or synthesis of new proteins, either in the motor neuron cell body or at the motor nerve terminal, is still unknown.

Overall, whatever the combination of mechanisms involved, the antibodies cause a reduction in the endplate potential so that it does not reach the critical firing threshold—either initially or during repeated effort (when the release of ACh naturally decreases a little). As a result of these changes, there is weakness at rest and/or increasing fatigue during sustained efforts.

7. The Thymus in MG

There are two main thymic pathologies in MG patients (see Table 31.1). In younger patients, usually under the age of 50, the thymus often is enlarged with frequent lymphocytic infiltrates and germinal centers, similar to those found in lymph nodes. Normally the thymus does not contain many B cells or antibody-producing plasma cells, but it does contain muscle-like myoid cells that lie sparsely in the thymic medulla and express fetal AChR (Schluepp et al., 1987) and other muscle proteins (Mesnard Rouiller et al., 2004). In MG these myoid cells may be found at the edge of the germinal centers, suggesting that their AChR may stimulate the aberrant immune response (Roxanis et al., 2002). It is possible that the epithelium of the thymic medulla may also express individual AChR subunits, as shown by lacZ reporter gene expression in cultured epithelial cells (Salmon et al., 1998). The germinal centers contain T and B cells specific for AChR (see Roxanis et al., 2003) and AChR antibodies are synthesized by cultured thymic lymphocytes *ex vivo* (Scadding et al., 1981). Removal of the thymus results in a moderate fall in AChR antibody with clinical improvement in a proportion of patients (Vincent et al., 1993), although most also require long-term immunosuppressive treatments.

A thymic tumor, thymoma, is found in about 10 percent of MG patients. It usually presents in middle age and may be associated with other autoimmune disorders, including acquired neuromyotonia (see later). The relationship between the tumor and the disease is not clear. The thymoma itself does not contain myoid cells, and AChR subunits are present at very low levels, with the exception of the adult-specific epsilon subunit (McLennan et al., submitted), although epitopes of AChRs may be present in the epithelial cells (Marx et al., 1989) as suggested by the study mentioned earlier. However, the adjacent normal thymus will contain AChR-expressing myoid cells and it is possible that abnormal sensitization of T cells to AChR epitopes within the thymoma initiates an immune response against the epsilon subunit, which then leads, via determinant spreading, to the immune response against the whole AChR that typifies MG (discussed in Vincent et al., 1998). Thymomas appear to generate mature CD4 and CD8 T cells (Buckley et al., 2001) or, in another study using different techniques, naïve T cells (Strobel et al., 2002) that may lack regulatory capacity (Strobel et al., 2004). Nevertheless, thymectomy generally does not lead to clinical improvement and most patients require immunosuppressive treatments.

8. Myasthenia Gravis without AChR Antibodies

About 10 to 15 percent of all MG patients with generalized symptoms do not have detectable AChR antibodies by current laboratory methods. Their symptoms are similar to those of other MG patients, although bulbar symptoms are more common (see later), and they respond well to plasma exchange, indicating an antibody-mediated condition. In addition, their immunoglobulins passively transfer a defect in neuromuscular transmission to mice (Burges et al., 1994; Mossman et al., 1986).

Earlier studies indicated that the plasma from a proportion of theses patients inhibited $^{23}Na+$ ion flux through the AChR expressed by various muscle-like cell lines (Yamamoto et al., 1991). The plasma factor involved, however, appeared to be an IgM rather than IgG antibody and acted directly on ACh induced ion currents, unlike the antibodies in most MG patients that do not directly block AChR function (see earlier). A more recent study suggests that the non-IgG fraction acts by increasing the rate of desensitization that occurs during agonist application *in vitro* (Spreadbury et al., 2005); whether this is relevant to NMJ function *in vivo* is difficult to say. Since the effects seen can be reversed by washing (see also Bufler et al., 1998), this has led to the hypothesis that the antibodies are low-affinity and directed against the AChR. In some patients, at least, this appears to be the case (Leite et al., submitted).

9. Myasthenia Gravis with MuSK Antibodies (MuSK-MG)

In other patients without AChR antibodies, another IgG antibody is present. MuSK is a transmembrane protein restricted to the neuromuscular junction in adult muscle. During development it plays an essential role in orchestrating the clustering of AChRs under the motor nerve terminal during formation of the NMJ (see Liyanage et al., 2002; Sanes & Lichtman, 1999). Its role in adult muscle is not clear, but a recent study showed that treatment of normal rodent muscle with sRNAi inhibiting MuSK synthesis led to a slow dispersal of AChRs from the NMJ (Kong et al., 2003) and suggests that MuSK is important for maintaining postsynaptic structure (see also Lichtman & Sanes, 1998, for a recent review).

MuSK therefore was an attractive candidate antigen for patients with MG. Antibodies to MuSK were first identified by Hoch et al. (2001). The antibodies bound to transfected cell lines, immunoprecipitated MuSK from detergent-extracted cells, and bound to recombinant MuSK on ELISA. A subsequent study also demonstrated the antibodies by immunoprecipitation from a muscle cell line (Scuderi et al., 2003). Although originally found in a high proportion of the patients who are negative for AChR antibodies, and highly specific for this form of MG (Evoli et al., 2003; McConville et al., 2004), it appears that their incidence is variable throughout the world and in some countries there are few or no MuSK antibody positive patients (see Vincent & Leite, 2005; Vincent et al., in preparation). Interestingly, the thymus in MuSK antibody positive MG patients is very similar to control thymi and differs from that in patients with AChR antibodies, or even those with neither antibody (Leite et al., 2005).

MuSK-MG is associated with HLA DR14-DQ5 in one study (Niks et al., 2006). In contrast to the IgG1 and IgG3 antibodies in typical MG, the MuSK antibodies are mainly IgG4 (McConville et al., 2004). Since IgG4 is not a strong activator of complement, this would tend to rule out a role for this mechanism, and complement deposition was not found at MuSK NMJs (Shiraishi et al., 2005). In addition, there was little evidence of AChR loss, again contrasting with the findings in MG. Thus it is not clear how the MuSK antibodies cause MG and the downstream effects of interfering with MuSK function are not fully understood. Interestingly, a recent publication has identified a new intracellular binding partner of MuSK, Dok7 (Otago et al., 2006). Knock-out of this protein in animal studies leads to defective neuromuscular junction formation and inherited defects in patients with a limb-girdle type of myasthenic syndrome in whom there are abnormally small NMJs with normal AChR density but decreased total numbers (Beeson et al., 2006; Slater et al., 2006). Although the NMJs in these genetic disorders are abnormally small, the relative preservation of AChR numbers and the distribution of muscle weakness are not altogether unlike those found in MuSK patients.

One additional factor that needs to be considered, however, is that the limb muscles that were studied by Shiraishi et al. (2005) are, in general, relatively resistant to the effects of MuSK antibodies. Limb muscle electrophysiology (Nemoto et al., 2005) may be normal in MuSK-MG whereas facial muscles are abnormal (Farrugia et al., 2006a), suggesting that the MuSK antibodies are most pathogenic in facial muscles and that these would be the best muscles in which to investigate pathogenic mechanisms. This is also evident clinically, since ocular, facial, bulbar, neck, and respiratory muscles are the most affected in the majority of MuSK-MG patients (Evoli et al., 2003). This has been confirmed in magnetic resonance imaging (MRI) studies, which showed muscle atrophy in tongue and facial muscles in up to 50 percent of patients (Farrugia et al., 2006; see Figure 31.3). This atrophy was greater than that found in patients with AChR antibodies and similar clinical histories, suggesting that the MuSK antibodies do have direct effects on muscle itself. There is also preliminary evidence (Benveniste et al., 2005) that the MuSK antibodies may up-regulate a ring-finger ligase, MURF-1, that is involved in induction of muscle atrophy (Glass et al., 2005). Animal studies are required to determine the role of MuSK in adult muscle, and to look for differences between facial and limb NMJs.

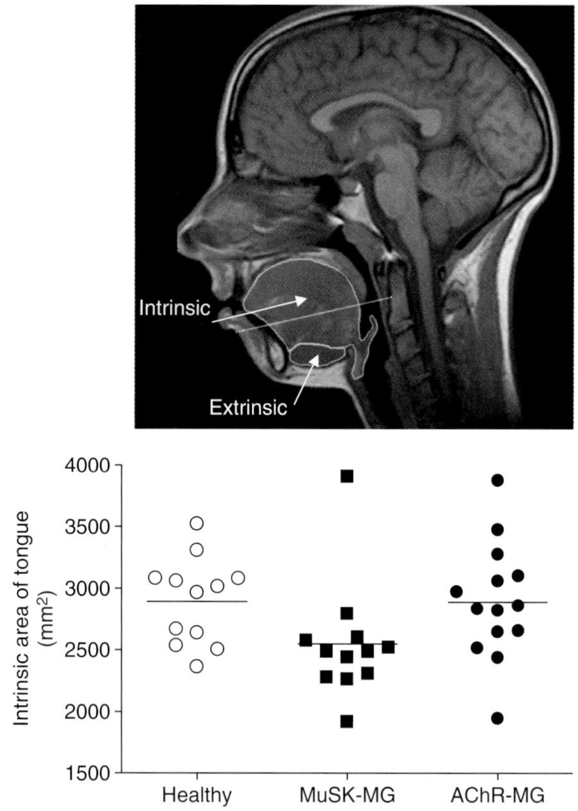

Figure 31.3 Tongue atrophy in MuSK-MG. The intrinsic area of the tongue was measured on the midline sagittal T_1W sequences, and the areas are plotted for individual patients. The areas in the MuSK-MG patients are significantly lower than those in the healthy individual controls ($p = 0.012$). The results in the AChR-MG patients are not different. Data and figure revised from Farrugia et al., *Brain,* 2006 with permission of OUP.

B. The Lambert-Eaton Myasthenic Syndrome

1. Clinical Features and History

The Lambert Eaton myasthenic syndrome was first described by Lambert and Eaton and differs clinically in several ways from MG. Weakness more commonly involves the trunk and legs with ocular involvement uncommon. Weakness improves during sustained effort as can be shown by electromyography *in vivo*; the compound muscle action potential increases during high frequency stimulation or following a brief period of voluntary contraction. Other defining features of LEMS are that the reflexes are absent or depressed but may become stronger after voluntary contraction, and that autonomic symptoms (dry mouth, constipation, male impotence) are frequent, suggesting that the target antigen may be common to certain autonomic systems (reviewed by O'Neill et al., 1993).

2. Etiology

LEMS is one of the few autoimmune diseases in which there is a clear etiological factor. Small cell lung cancer (SCLC) is present in about 50 percent of patients, and the evidence strongly suggests that the immune response is directed primarily against the tumor in these cases (see later). This is analogous to other paraneoplastic autoimmune conditions in which an immune response against a tumor, often SCLC or gynecological, leads indirectly to neurological damage (see Chapter 32). In the remaining 50 percent of patients, however, the cause is unknown. Interestingly, like early-onset MG patients, the noncancer subgroup is associated with HLA B8DR3, suggesting a genetic predisposition to autoimmunity, but the triggering factor(s) are unknown (Wierz et al., 2003). These patients often have other autoimmune disorders, such as thyroid disease, vitiligo, pernicious anemia, celiac disease, and juvenile onset diabetes mellitus (see O'Neill et al., 1993).

3. Pathophysiology

The physiology differs from that in MG. Recordings from LEMS muscle biopsies *in vitro* show that the miniature endplate potentials are normal in amplitude but that the endplate potentials are very small and the number of packets or quanta of ACh released from the NMJ are markedly reduced from around 20 per nerve impulse to less than 5 (Elmqvist & Lambert 1968, 1971). Repetitive stimulation showed that the endplate potential increased during repetitive stimulation and also when extracellular calcium was raised. These results suggested a possible defect in the presynaptic voltage gated calcium channels (VGCC).

Pioneering work by Andrew Engel and colleagues in the 1980s used freeze fracture electron microscopic studies of motor nerve terminals from healthy individuals and LEMS patients to study these channels. Freeze fracture of the NMJ reveals double parallel rows of intramembranous particles (each about 10–20 nm in diameter), which are present at the regions of the motor nerve terminal known as active zones, and are thought to represent the VGCCs. In LEMS, there was a marked reduction in the number of active zone particles, in the number of particles per active zone, and a disruption of their organization (reviewed by Engel, 1991).

Independently, it was shown that LEMS is an autoimmune disease. Plasma exchange was found to lead to clinical improvement associated with an increase in CMAP amplitudes (Lang et al., 1981) and patients also benefit from long-term immunosuppressive drugs or intravenous immunoglobulin therapy. Crucially, daily injection of LEMS plasmas, or IgG fractions, into mice reproduced the principal neurophysiologic changes of LEMS with reduced endplate potential amplitudes and reduced quantal content (Lang et al., 1981). The active zone particles were reduced in number and dispersed, as in the human studies (Fukunaga et al., 1983), and strikingly this process was preceded by a reduction in the distance between particles, suggesting that the divalent antibodies were acting by cross-linking the extracellular domain of the VGCCs (Fukuoka et al., 1987a). Moreover, IgG could be demonstrated at the presynaptic active zones of mice that had received multiple intraperitoneal doses of LEMS IgG (Fukunaga et al., 1987b). Importantly, divalent $F(ab)_2$ IgG molecules, but not monovalent F(ab)s, were able to cause the neurophysiological changes, indicating that divalency of the antibody is essential (Peers et al., 1993). F(ab)2 IgGs do not fix complement—therefore, complement activation is not required, confirming passive transfer experiments in complement-dependent mice (Lang et al., 1983). These observations implicate cross-linking and internalization as the main mechanisms for VGCC loss in LEMS; why the antibodies do not activate complement is not yet clear as the IgG subclasses have not been identified.

VGCC subtypes are transmembrane proteins comprising $\alpha1$, β, and $\alpha2/\delta$ subunits. There are more than eight different $\alpha1$ subunits that contain the Ca^{2+} conducting channel, and are the principal determinants of the functional properties. Different drugs and neurotoxins can be used to distinguish the subtypes. In particular, the cone snail derived toxins inhibit the function of VGCCs in neurons. ω-Conotoxin (ω-CmTx) MVIIC reduces transmitter release at the mouse neuromuscular junction, indicating that the $\alpha1A$ (P/Q-type) VGCC are involved (Uchitel et al., 1992) and VGCCs extracted from human or mammalian cerebellum and prelabeled with ^{125}1-ω-Conotoxin (ω-CmTx) MVIIC can be immunoprecipitated by LEMS IgG (Lennon et al., 1995; Motomura et al., 1995). LEMS IgG reduces VGCC channel activity in cultured human embryonic kidney cells that have been engineered to express the $\alpha1A$ VGCCs (Pinto et al., 1998) but not those expressing the $\alpha1B$ channels. Interestingly, in cultured cerebellar granular neurons there was a decrease in $\alpha1A$ VGCC channel currents accompanied by an up-regulation of other VGCC subtypes. This up-regulation (analogous to the increased AChR synthesis in NG mentioned earlier) was also demonstrated at the mouse NMJ after passive transfer of LEMS IgG, illustrating, again, the ability of the NMJ to compensate for disease-induced changes (Giovannini et al., 2002).

At autonomic synapses between post-gangionic nerves and smooth muscle, multiple subtypes of VGCC are involved in the release of neurotransmitter, as can be shown by applying sequentially different neurotoxins to block each subtype. The muscle tension generated at different stimulation frequencies was reduced in mice injected with LEMS IgG, mainly due to reduced activity through the $\alpha1A$ channels (Waterman et al., 1997). Thus the effects of LEMS IgG on autonomic synapses are selective for the α1A channels; why the patients' symptoms can be so marked is not clear but perhaps there are a higher proportion of P-type VGCCs in human autonomic nerve endings.

C. Acquired Neuromyotonia

1. Clinical Features and History

Acquired neuromyotonia (NMT), or Isaac's syndrome, differs from other myasthenia syndromes in being associated mainly with muscle hyperactivity. The patients usually present between 20 and 60 years of age, and clinical features include muscle stiffness, cramps, myokymia (visible undulation of the muscle), pseudomyotonia (slow relaxation after contraction), and weakness, most prominent in the limbs and trunk but found sometimes in isolated muscle groups including the face. Typically the myokymia continues during sleep, indicating that it is generated peripherally. Increased sweating is common. On electromyography, there are spontaneous motor unit discharges occurring in distinctive doublets, triplets, or longer runs with high intraburst frequency (see Newsom-Davis & Mills, 1993). The muscle hyperactivity is due to hyperexcitability of the motor nerves, generated mainly distally, perhaps at the NMJ itself, but in some cases more proximally. Cramp fasciculation syndrome may form part of the same clinical spectrum (Hart et al., 2002).

2. Etiology

The most common precipitating factor is a thymic tumor. About 20 percent of patients with NMT have a thymoma that can predate the symptoms or first be identified afterward. Some of the thymoma patients have MG as well as neuromyotonia (Hart et al., 2002), and the range of autoimmune diseases that can occur with this tumor is remarkable. In addition, there are anecdotal reports of associated infections, and some patients without thymomas appear to have a monophasic illness that recovers spontaneously within one to two years. Thus it is likely that some cases occur secondarily to infection with a microbe that is either cross-reactive with the immune target or nonspecifically stimulates autoimmunity.

3. Pathophysiology

There have been no reports of muscle biopsy studies and the pathogenic mechanisms are largely inferred from *in vitro* studies (see later). Since NMT may be associated with thymomas and other autoimmune diseases or other autoantibodies, and the cerebrospinal fluid often contains oligoclonal bands, it was proposed that it was autoimmune (Sinha et al., 1991). This became clear when it was shown that some patients responded to plasma exchange and that passive transfer of their IgG to mice resulted in curare resistance of neuromuscular transmission (Sinha et al., 1991). Further work indicated that this was due to an increase in the number of packets of ACh released, and that the IgG also caused prolonged action potentials in the mouse sensory nerves and repetitive activity in dorsal root ganglion cultures (Shillito et al., 1995). All these findings were similar to those of normal tissue in the presence of a low concentration of a voltage-gated potassium channel blocker, 4-aminopyridine. This suggested that loss of voltage-gated potassium channels may underlie the electrophysiological findings in patients.

VGKC represent a subgroup of a large family of potassium channels. The shaker-type VGKC are targets for the snake toxin, α-dendrotoxin, that blocks VGKC Kv1.1, 1.2, and 1.6. Each VGKC consists of four transmembrane α subunits that combine as homomultimeric and heteromultimeric tetramers. Kv1.1 and 1.2 are highly expressed in the peripheral nervous system, particularly in the juxtaparanodal region of the nodes of Ranvier (Devaux et al., 2004). The subtype expressed at the motor nerve terminal, however, is not known, and α-dendrotoxin has relatively small effects on neuromuscular transmission *in vitro* although it can cause repetitive EPPs (Harvey, 2001).

Antibodies to VGKCs can be detected by immunostaining of xenopus oocytes or cell lines engineered to express the different Kv subtypes individually (Hart et al., 1997; Kleopa et al., 2006). In practice, however, they are usually measured by immunoprecipitation of ^{125}I-α-dendrotoxin-labeled VGKCs extracted from human frontal cortex (Hart et al., 1997; Shillito et al., 1995). Since all three Kv subtypes bind dendrotoxin, the specificity of the antibodies for each subtype is not yet clear, and any analysis is likely to be confounded by presence of heterooligomers of different Kv subunits that are present in the brain extract used as a source of the Kvs. Antibodies to Kv1.2 probably dominate in most patients, but antibodies to Kv1.6 may be important in some sera (Kleopa et al., 2006). There are several reports that the antibodies, or IgG fractions, can reduce potassium channel currents in neuroblastoma cells, Nb-1 (Sonoda et al., 1996), and in cells transfected with the Kv1.1 or 1.6 subtypes. Moreover, this does not require complement but does require divalent (Fab)2 antibodies (Tomimitsu et al., 2004). So the immunological mechanisms are likely to be similar to those in LEMS. For a recent review, see Arimura et al. (2002).

The relationship between thymomas and these antibodies is intriguing. Thymoma epithelial cells express many ion channels including sodium and potassium channels (Marx et al., 1991). Figure 31.4 illustrates a patient with thymoma, with neuromyotonia and Kv1.6 antibodies; a second antibody (to glutamic acid decarboxylase) developed at a time when the patient had improved clinically but was still undergoing immunoabsorption treatments. The association at different time points with more than one antibody is typical of thymoma-related disorders.

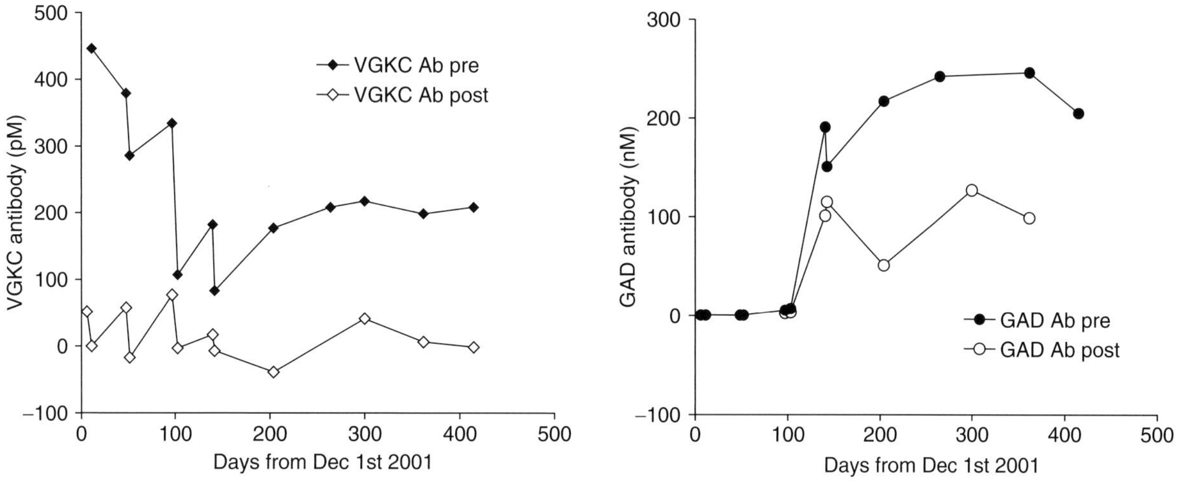

Figure 31.4 Antibodies to VGKC and to glutamic acid decarboxylase in a patient with thymoma. Note that the GAD antibodies show a different time course to the VGKC antibodies and only first appear at a time when the VGKC antibodies are being reduced by immunoabsorption treatments. Taken with permission from Antozzi et al. 2005.

4. Central Nervous System Disease with VGKC Antibodies

Some NMT patients have sensory symptoms, and CNS symptoms such as insomnia, hallucinations, delusions, and personality change are not infrequent (Hart et al., 2002). Thus the spectrum of disease is wider than originally realized, which correlates well with the important role of VGKCs in all forms of neuronal function. Relatively few patients have dominant CNS symptoms, which usually are referred to as Morvan's syndrome, and includes insomnia, memory loss, autonomic dysfunction, and neuromyotonia (e.g., Liguori et al., 2001). Recently, other patients with VGKC antibodies have been found to have mainly or entirely CNS symptoms, often presenting as a nonparaneoplastic form of limbic encephalitis with memory loss, seizures, and disorientation (Thieben et al., 2004; Vincent et al., 2004). In these patients, Kv1.1 antibodies predominate in one study (Kleopa et al., 2006), but in other respects it is not clear what determines the clinical phenotype.

D. Other Autoimmune Ion Channel Diseases

There are increasing numbers of diseases associated with autoimmunity to various ion channels or receptors in a variety of disciplines. Autoantibodies to ganglionic AChR, which are present in both the sympathetic and parasympathetic nervous systems, have been identified in some patients with idiopathic or paraneoplastic autonomic neuropathies and appear to be pathogenic, although there is relatively little data as yet concerning the pathogenic mechanisms or response of the patients to immunotherapies (Vernino & Lennon, 2003). Antibodies to GluR3 glutamate receptors were demonstrated in Rasmussen's encephalitis (Rogers et al., 1994), but may also be found in other patients with epilepsy (Mantegazza et al., 2002), although their presence is still controversial (Watson et al., 2004) and two patients have been found to have antibodies instead to the alpha7 nicotinic AChR (Watson et al., 2005). Recently antibodies to the NR2 glutamate receptors were identified in patients with systemic lupus erythematosus and neuropsychiatric symptoms (DeGeorgio et al., 2004). These and other recent findings indicate the likelihood of further diseases to be identified in the future.

III. Genetic Disorders

A. Congenital Myasthenic Syndromes

In contrast to the autoimmune conditions already discussed, the congenital myasthenic syndromes (CMS) are rare inherited disorders that result from mutations in different key proteins at the neuromuscular junction (see Figure 31.1B and Table 31.2). The mutations and their pathogenic mechanisms will not be discussed here, but clinically they can be difficult to distinguish from the autoimmune disorders, except that they usually, but not always, present around birth or in early childhood and do not respond to immunotherapies (see Beeson et al., 2006b, for a review).

The most common disorder is the AChR deficiency syndrome. This can be caused either by mutations in the AChR genes (usually ε) or in the gene encoding rapsyn that anchors the AChR at the postsynaptic membrane (see Figure 31.1). In both situations the resulting loss of adult-type AChRs from the postsynaptic membrane causes a myasthenic syndrome that is similar to MG. The fact that the babies do not die at birth and many patients survive well into adult life is

Table 31.2 Key Proteins at the Neuromuscular Junction and Their Involvement in Autoimmune and Genetic Disorders

Location	Protein	Autoimmune Disease	Genetic Disease
Membrane or extracellular matrix	Acetylcholine receptor	Yes	Yes
	Muscle-specific kinase	Yes	One case of congenital myasthenic syndrome
	Voltage-gated calcium channel	Yes	Not yet identified in congenital myasthenic syndromes
	Voltage-gated potassium channel	Yes	Not yet identified in congenital myasthenic syndromes
	Voltage-gated sodium channels	None known	One case of congenital myasthenic syndrome
	Acetylcholine esterase	None known	Yes
Cytoplasmic	Choline acetylase	No	Yes
	Rapsyn	No	Yes
	Dok-7	No	Yes

Data are found in this chapter and in Beeson et al. (2005).

because the γ subunit of the AChR (see Figure 31.2) continues to be synthesized in human muscle and can continue to contribute to AChRs at the NMJ. Because some of the patients don't present clinically until adult life, particularly a subset of those with rapsyn mutations, the condition can be misdiagnosed as autoimmune MG. The limb girdle type of CMS associated with Dok-7 mutations (Beeson et al., 2006a; Slater et al., 2006) may also present in adult life.

IV. General Concepts

The observations on myasthenia gravis and the other autoimmune conditions described here provide a basis for understanding the disease processes in general terms and demonstrate unequivocally how different types of disease process, autoimmune, genetic, toxic, can affect a relatively "simple" synapse (see Table 31.2). The main differences between clinical and experimental observations on autoimmune and genetic disorders are summarized in Table 31.3; these are not as obvious as one might expect. Several of the CMS do not present at birth or during infancy, and when they do may be episodic. Some of the autoimmune diseases present in childhood, although this is rare in Caucasians. Thus there can be difficulties in determining whether a patient who presents in adult life with a neuromuscular problem has an autoimmune disease or a delayed onset CMS. If the patient does not respond to immunosuppressive therapies, a genetic defect should be considered.

A major distinction between the genetic and autoimmune diseases, however, relates to the localization of the target molecule and the response to treatments. Whereas genetic

Table 31.3 Differences between Autoimmune and Genetic Diseases

	Genetic	Autoimmune
Age at onset	Often at birth or during infancy but can be later or adulthood	Usually but not always adult; occasionally within first five years
Presentation	Usually gradual if not present at birth but can be episodic	Often subacute and rapidly progressive
Course	Usually stable or slowly progressive	Often fluctuating
Target	Any important functional protein	Extracellular or membrane proteins accessible to antibodies
Response to immuno-suppressive treatments	None	Usually good

disorders can involve functional molecules within the cytoplasm, such as acetylcholine transferase and rapsyn, as well as cell membrane proteins, in general targets for autoimmunity are membrane proteins with sufficient extracellular domains to provide a target for circulating antibodies to bind.

In all these conditions there is considerable phenotypic variability between patients, and in the autoimmune conditions particularly there can also be marked variability between patients and from day to day. How can one explain these features? Some of the phenotypic differences can be due to features of the immune response, such as antibody specificity and complement activity, which could modify the pathogenic mechanisms. It is also likely that other factors affect the ability of the antibodies to reduce neuromuscular transmission. The neuromuscular synaptic space can be thought of as a tiny disk-like space (30 μm diameter, 0.05 μm depth) with, on one side, a very high density of AChRs (total number approximately 2×10^7 per NMJ). For these AChRs to be saturated by antibodies requires a considerable flow of extracellular fluid through the synaptic space. Meanwhile, new AChRs are being synthesized, probably at an increased rate, and the balance between antibody-mediated degradation and synthesis will also depend on blood flow into the muscle and metabolic activity. If one appreciates that different individuals will have different antibody levels and affinities and different rates of AChR synthesis and complement levels, and that different muscles may have different metabolism and blood flow, it is easy to see how difficult it is to predict which muscles are involved and how severely.

Additional factors may be the other proteins at the neuromuscular junction. Retrograde signaling must be occurring in order to allow the presynaptic release of ACh to increase in MG, but how does this occur and will genetic factors underlie the patient's ability to compensate? Structural features similarly may well be genetically determined. Rapsyn is an intracellular protein that anchors the AChRs (see Figure 31.1). Experimental autoimmune MG in rats is less easy to demonstrate in older animals who are protected apparently by the higher concentration of rapsyn at the neuromuscular junction. Indeed transfection of leg muscles *in vivo* with DNA for rapsyn protected the NMJs from passively transferred experimental MG (Losen et al., 2005). Therefore differences in rapsyn expression in different muscles or in different individuals might be another factor that determines susceptibility to antibody-mediated degradation. In the LEMS and NMT, analogous considerations may apply and there is already evidence that compensatory increases in the expression of other VGCCs or VGKCs probably may take place.

As mentioned in the introduction, neuromuscular transmission is dependent on the EPP reaching a threshold, and is therefore (except in the ocular and some other small muscles) an all or none process. Anything that increases the EPPs sufficiently for threshold to be reached in a number of fibers that were previously inactive will produce a substantial benefit. Conversely, if the EPP is only just above threshold, although weakness will not be present, any further small decrease in EPP amplitude will lead to transmission failure at a significant number of NMJs. It may be that, if one can identify more of the determinants of the phenotypic variability and compensatory mechanisms in these conditions, it will help to define new approaches to treatments.

References

Antozzi, C., Frassoni, C., Vincent, A., Regondi, M.C., Andreetta, F., Bernasconi, P. et al. (2005). Sequential antibodies to potassium channels and glutamic acid decarboxylase in neuromyotonia. *Neurology* **64**, 1290–1293.

Appel, S.H., Anwyl, R., McAdams, M.W., and Elias, S. (1977). Accelerated degradation of acetylcholine receptor from cultured rat myotubes with myasthenia gravis sera and globulins. *Proc Natl Acad Sci U S A* **74**, 2130–2134.

Arimura, K., Sonoda, Y., Watanabe, O., Nagado, T., Kurono, A., Tomimitsu, H. et al. (2002). Isaacs' syndrome as a potassium channelopathy of the nerve. *Muscle Nerve* **Suppl 11**, S55–58.

Beeson, D., Hantai, D., Lochmuller, H., and Engel, A.G. (2005). 126th International Workshop: Congenital myasthenic syndromes, 24–26 September 2004, Naarden, the Netherlands. *Neuromuscul Disord* **15**, 498–512.

Beeson, D., Higuchi, O., Palace, J., Cossins, J., Spearman, H., Maxwell, S. et al. (2006). Dok-7 Mutations Underlie a Neuromuscular Junction Synaptopathy. *Science*.

Beeson, D., Jacobson, L., Newsom-Davis, J., and Vincent, A. (1996). A transfected human muscle cell line expressing the adult subtype of the human muscle acetylcholine receptor for diagnostic assays in myasthenia gravis. *Neurology* **47**, 1552–1555.

Benveniste, O., Jacobson, L., Farrugia, M.E., Clover, L., and Vincent, A. (2005). MuSK antibody positive myasthenia gravis plasma modifies MURF-1 expression in C2C12 cultures and mouse muscle in vivo. *J Neuroimmunol* **170**, 41–48.

Buckley, C., Douek, D., Newsom-Davis, J., Vincent, A., and Willcox, N. (2001). Mature, long-lived CD4+ and CD8+ T cells are generated by the thymoma in myasthenia gravis. *Ann Neurol* **50**, 64–72.

Buckley, C., Oger, J., Clover, L., Tuzun, E., Carpenter, K., Jackson, M., and Vincent, A. (2001). Potassium channel antibodies in two patients with reversible limbic encephalitis. *Ann Neurol* **50**, 73–78.

Bufler, J., Pitz, R., Czep, M., Wick, M., and Franke, C. (1998). Purified IgG from seropositive and seronegative patients with mysasthenia gravis reversibly blocks currents through nicotinic acetylcholine receptor channels. *Ann Neurol* **43**, 458–464.

Burges, J., Vincent, A., Molenaar, P.C., Newsom-Davis, J., Peers, C., and Wray, D. (1994). Passive transfer of seronegative myasthenia gravis to mice. *Muscle Nerve* **17**, 1393–1400.

Burges, J., Wray, D.W., Pizzighella, S., Hall, Z., and Vincent, A. (1990). A myasthenia gravis plasma immunoglobulin reduces miniature endplate potentials at human endplates in vitro. *Muscle Nerve* **13**, 407–413.

Burke, G., Cossins, J., Maxwell, S., Owens, G., Vincent, A., Robb, S. et al. (2003). Rapsyn mutations in hereditary myasthenia: Distinct early- and late-onset phenotypes. *Neurology* **61**, 826–828.

Chevessier, F., Faraut, B., Ravel-Chapuis, A., Richard, P., Gaudon, K., Bauche, S. et al. (2004). MUSK, a new target for mutations causing congenital myasthenic syndrome. *Hum Mol Genet* **13**, 3229–3240.

Chu, N.S. (2005). Contribution of a snake venom toxin to myasthenia gravis: The discovery of alpha-bungarotoxin in Taiwan. *J Hist Neurosci* **14**, 138–148.

Compston, D.A., Vincent, A., Newsom-Davis, J., and Batchelor, J.R. (1980). Clinical, pathological, HLA antigen and immunological evidence for disease heterogeneity in myasthenia gravis. *Brain* **103**, 579–601.

DeGiorgio, L.A., Konstantinov, K.N., Lee, S.C., Hardin, J.A., Volpe, B.T., and Diamond, B. (2001). A subset of lupus anti-DNA antibodies cross-reacts with the NR2 glutamate receptor in systemic lupus erythematosus. *Nat Med* **7**, 1189–1193.

Devaux, J.J., Kleopa, K.A., Cooper, E.C., and Scherer, S.S. (2004). KCNQ2 is a nodal K+ channel. *J Neurosci* **24**, 1236–1244.

Djabiri, F., Caillat-Zucman, S., Gajdos, P., Jais, J.P., Gomez, L., Khalil, I. et al. (1997). Association of the AChRalpha-subunit gene (CHRNA), DQA1*0101, and the DR3 haplotype in myasthenia gravis. Evidence for a three-gene disease model in a subgroup of patients. *J Autoimmun* **10**, 407–413.

Drachman, D.B. (1994). Myasthenia gravis. *N Engl J Med* **330**, 1797–1810.

Drachman, D.B., Angus, C.W., Adams, R.N., Michelson, J.D., and Hoffman, G.J. (1978). Myasthenic antibodies cross-link acetylcholine receptors to accelerate degradation. *N Engl J Med* **298**, 1116–1122.

Dwyer, D.S., Vakil, M., and Kearney, J.F. (1986). Idiotypic network connectivity and a possible cause of myasthenia gravis. *J Exp Med* **164**, 1310–1318.

Elmqvist, D., Hofmann, W.W., Kugelberg, J., and Quastel, D.M. (1964). An electrophysiological investigation of neuromuscular transmission in myasthenia gravis. *J Physiol* **174**, 417–434.

Elmqvist, D. and Lambert, E.H. (1968). Detailed analysis of neuromuscular transmission in a patient with the myasthenic syndrome sometimes associated with bronchogenic carcinoma. *Mayo Clin Proc* **43**, 689–713.

Engel, A.G. (1984). Myasthenia gravis and myasthenic syndromes. *Ann Neurol* **16**, 519–534.

Engel, A.G. (1991). Review of evidence for loss of motor nerve terminal calcium channels in Lambert-Eaton myasthenic syndrome. *Ann N Y Acad Sci* **635**, 246–258.

Engel, A.G. and Arahata, K. (1987). The membrane attack complex of complement at the endplate in myasthenia gravis. *Ann N Y Acad Sci* **505**, 326–332.

Engel, A.G., Lambert, E.H., and Howard, F.M. (1977). Immune complexes (IgG and C3) at the motor end-plate in myasthenia gravis: Ultrastructural and light microscopic localization and electrophysiologic correlations. *Mayo Clin Proc* **52**, 267–280.

Evoli, A., Tonali, P.A., Padua, L., Monaco, M.L., Scuderi, F., Batocchi, A.P. et al. (2003). Clinical correlates with anti-MuSK antibodies in generalized seronegative myasthenia gravis. *Brain* **126**, 2304–2311.

Fambrough, D.M., Drachman, D.B., and Satyamurti, S. (1973). Neuromuscular junction in myasthenia gravis: Decreased acetylcholine receptors. *Science* **182**, 293–295.

Farrugia, M.E., Kennett, R.P., Newsom-Davis, J., Hilton-Jones, D., and Vincent, A. (2006). Single-fiber electromyography in limb and facial muscles in muscle-specific kinase antibody and acetylcholine receptor antibody myasthenia gravis. *Muscle Nerve* **33**, 568–570.

Farrugia, M.E., Robson, M.D., Clover, L., Anslow, P., Newsom-Davis, J., Kennett, R. et al. (2006). MRI and clinical studies of facial and bulbar muscle involvement in MuSK antibody-associated myasthenia gravis. *Brain* **129**, 1481–1492.

Fostieri, E., Tzartos, S.J., Berrih-Aknin, S., Beeson, D., and Mamalaki, A. (2005). Isolation of potent human Fab fragments against a novel highly immunogenic region on human muscle acetylcholine receptor which protect the receptor from myasthenic autoantibodies. *Eur J Immunol* **35**, 632–643.

Fukunaga, H., Engel, A.G., Lang, B., Newsom-Davis, J., and Vincent, A. (1983). Passive transfer of Lambert-Eaton myasthenic syndrome with IgG from man to mouse depletes the presynaptic membrane active zones. *Proc Natl Acad Sci U S A* **80**, 7636–7640.

Fukuoka, T., Engel, A.G., Lang, B., Newsom-Davis, J., Prior, C., and Wray, D.W. (1987a). Lambert-Eaton myasthenic syndrome: I. Early morphological effects of IgG on the presynaptic membrane active zones. *Ann Neurol* **22**, 193–199.

Fukuoka, T., Engel, A.G., Lang, B., Newsom-Davis, J., and Vincent, A. (1987b). Lambert-Eaton myasthenic syndrome: II. Immunoelectron microscopy localization of IgG at the mouse motor end-plate. *Ann Neurol* **22**, 200–211.

Giovannini, F., Sher, E., Webster, R., Boot, J., and Lang, B. (2002). Calcium channel subtypes contributing to acetylcholine release from normal, 4-aminopyridine-treated and myasthenic syndrome auto-antibodies-affected neuromuscular junctions. *Br J Pharmacol* **136**, 1135–1145.

Giraud, M., Beaurain, G., Eymard, B., Tranchant, C., Gajdos, P., and Garchon, H.J. (2004). Genetic control of autoantibody expression in autoimmune myasthenia gravis: Role of the self-antigen and of HLA-linked loci. *Genes Immun* **5**, 398–404.

Glass, D.J. (2005). Skeletal muscle hypertrophy and atrophy signaling pathways. *Int J Biochem Cell Biol* **37**, 1974–1984.

Guyon, T., Wakkach, A., Poea, S., Mouly, V., Klingel-Schmitt, I., Levasseur, P. et al. (1998). Regulation of acetylcholine receptor gene expression in human myasthenia gravis muscles. Evidences for a compensatory mechanism triggered by receptor loss. *J Clin Invest* **102**, 249–263.

Harper, C.M. (2004). Congenital myasthenic syndromes. *Semin Neurol* **24**, 111–123.

Hart, I.K., Maddison, P., Newsom-Davis, J., Vincent, A., and Mills, K.R. (2002). Phenotypic variants of autoimmune peripheral nerve hyperexcitability. *Brain* **125**, 1887–1895.

Hart, I.K., Waters, C., Vincent, A., Newland, C., Beeson, D., Pongs, O. et al. (1997). Autoantibodies detected to expressed K+ channels are implicated in neuromyotonia. *Ann Neurol* **41**, 238–246.

Harvey, A.L. (2001). Twenty years of dendrotoxins. *Toxicon* **39**, 15–26.

Heinemann, S., Bevan, S., Kullberg, R., Lindstrom, J., and Rice, J. (1977). Modulation of acetylcholine receptor by antibody against the receptor. *Proc Natl Acad Sci U S A* **74**, 3090–3094.

Heinemann, S., Merlie, J., and Lindstrom, J. (1978). Modulation of acetylcholine receptor in rat diaphragm by anti-receptor sera. *Nature* **274**, 65–68.

Hesselmans, L.F., Jennekens, F.G., Van den Oord, C.J., Veldman, H., and Vincent, A. (1993). Development of innervation of skeletal muscle fibers in man: Relation to acetylcholine receptors. *Anat Rec* **236**, 553–562.

Hoch, W., McConville, J., Helms, S., Newsom-Davis, J., Melms, A., and Vincent, A. (2001). Auto-antibodies to the receptor tyrosine kinase MuSK in patients with myasthenia gravis without acetylcholine receptor antibodies. *Nat Med* **7**, 365–368.

Hodgson, W.C. and Wickramaratna, J.C. (2002). In vitro neuromuscular activity of snake venoms. *Clin Exp Pharmacol Physiol* **29**, 807–814.

Hoffmann, K., Muller, J.S., Stricker, S., Megarbane, A., Rajab, A., Lindner, T.H. et al. (2006). Escobar syndrome is a prenatal myasthenia caused by disruption of the acetylcholine receptor fetal gamma subunit. *Am J Hum Genet* **79**, 303–312.

Ito, Y., Miledi, R., Vincent, A., and Newsom-Davis, J. (1978). Acetylcholine receptors and end-plate electrophysiology in myasthenia gravis. *Brain* **101**, 345–368.

Jacobson, L., Polizzi, A., Morriss-Kay, G., and Vincent, A. (1999). Plasma from human mothers of fetuses with severe arthrogryposis multiplex congenita causes deformities in mice. *J Clin Invest* **103**, 1031–1038.

Kaminski, H.J., Li, Z., Richmonds, C., Lin, F., and Medof, M.E. (2004). Complement regulators in extraocular muscle and experimental autoimmune myasthenia gravis. *Exp Neurol* **189**, 333–342.

Kaminski, H.J., Li, Z., Richmonds, C., Ruff, R.L., and Kusner, L. (2003). Susceptibility of ocular tissues to autoimmune diseases. *Ann N Y Acad Sci* **998**, 362–374.

Kaminski, H.J. and Ruff, R.L. (1997). Ocular muscle involvement by myasthenia gravis. *Ann Neurol* **41**, 419–420.

Keesey, J. (2005). How electric fish became sources of acetylcholine receptor. *J Hist Neurosci* **14**, 149–164.

Kleopa, K.A., Elman, L.B., Lang, B., Vincent, A., and Scherer, S.S. (2006). Neuromyotonia and limbic encephalitis sera target mature Shaker-type K+ channels: Subunit specificity correlates with clinical manifestations. *Brain* **129**, 1570–1584.

Kong, X.C., Barzaghi, P., and Ruegg, M.A. (2004). Inhibition of synapse assembly in mammalian muscle in vivo by RNA interference. *EMBO Rep* **5**, 183–188.

Lambert, E.H. and Elmqvist, D. (1971). Quantal components of end-plate potentials in the myasthenic syndrome. *Ann N Y Acad Sci* **183**, 183–199.

Lang, B., Newsom-Davis, J., Wray, D., Vincent, A., and Murray, N. (1981). Autoimmune aetiology for myasthenic (Eaton-Lambert) syndrome. *Lancet* **2**, 224–226.

Leite, M.I., Strobel, P., Jones, M., Micklem, K., Moritz, R., Gold, R. et al. (2005). Fewer thymic changes in MuSK antibody-positive than in MuSK antibody-negative MG. *Ann Neurol* **57**, 444–448.

Lennon, V.A., Kryzer, T.J., Griesmann, G.E., O'Suilleabhain, P.E., Windebank, A.J., Woppmann, A. et al. (1995). Calcium-channel antibodies in the Lambert-Eaton syndrome and other paraneoplastic syndromes. *N Engl J Med* **332**, 1467–1474.

Lichtman, J.W. and Sanes, J.R. (2003). Watching the neuromuscular junction. *J Neurocytol* **32**, 767–775.

Liguori, R., Vincent, A., Clover, L., Avoni, P., Plazzi, G., Cortelli, P. et al. (2001). Morvan's syndrome: Peripheral and central nervous system and cardiac involvement with antibodies to voltage-gated potassium channels. *Brain* **124**, 2417–2426.

Lindstrom, J.M., Seybold, M.E., Lennon, V.A., Whittingham, S., and Duane, D.D. (1976). Antibody to acetylcholine receptor in myasthenia gravis. Prevalence, clinical correlates, and diagnostic value. *Neurology* **26**, 1054–1059.

Liyanage, Y., Hoch, W., Beeson, D., and Vincent, A. (2002). The agrin/muscle-specific kinase pathway: New targets for autoimmune and genetic disorders at the neuromuscular junction. *Muscle Nerve* **25**, 4–16.

Losen, M., Stassen, M.H., Martinez-Martinez, P., Machiels, B.M., Duimel, H., Frederik, P. et al. (2005). Increased expression of rapsyn in muscles prevents acetylcholine receptor loss in experimental autoimmune myasthenia gravis. *Brain* **128**, 2327–2337.

MacLennan, C., Beeson, D., Buijs, A.M., Vincent, A., and Newsom-Davis, J. (1997). Acetylcholine receptor expression in human extraocular muscles and their susceptibility to myasthenia gravis. *Ann Neurol* **41**, 423–431.

Mantegazza, R., Bernasconi, P., Baggi, F., Spreafico, R., Ragona, F., Antozzi, C. et al. (2002). Antibodies against GluR3 peptides are not specific for Rasmussen's encephalitis but are also present in epilepsy patients with severe, early onset disease and intractable seizures. *J Neuroimmunol* **131**, 179–185.

Marx, A., Kirchner, T., Hoppe, F., O'Connor, R., Schalke, B., Tzartos, S., and Muller-Hermelink, H.K. (1989). Proteins with epitopes of the acetylcholine receptor in epithelial cell cultures of thymomas in myasthenia gravis. *Am J Pathol* **134**, 865–877.

Marx, A., Siara, J., and Rudel, R. (1991). Sodium and potassium channels in epithelial cells from thymus glands and thymomas of myasthenia gravis patients. *Pflugers Arch* **417**, 537–539.

McConville, J., Farrugia, M.E., Beeson, D., Kishore, U., Metcalfe, R., Newsom-Davis, J., and Vincent, A. (2004). Detection and characterization of MuSK antibodies in seronegative myasthenia gravis. *Ann Neurol* **55**, 580–584.

Mesnard-Rouiller, L., Bismuth, J., Wakkach, A., Poea-Guyon, S., and Berrih-Aknin, S. (2004). Thymic myoid cells express high levels of muscle genes. *J Neuroimmunol* **148**, 97–105.

Mossman, S., Vincent, A., and Newsom-Davis, J. (1986). Myasthenia gravis without acetylcholine-receptor antibody: A distinct disease entity. *Lancet* **1**, 116–119.

Motomura, M., Johnston, I., Lang, B., Vincent, A., and Newsom-Davis, J. (1995). An improved diagnostic assay for Lambert-Eaton myasthenic syndrome. *J Neurol Neurosurg Psychiatry* **58**, 85–87.

Nagado, T., Arimura, K., Sonoda, Y., Kurono, A., Horikiri, Y., Kameyama, A. et al. (1999). Potassium current suppression in patients with peripheral nerve hyperexcitability. *Brain* **122 (Pt 11)**, 2057–2066.

Nemoto, Y., Kuwabara, S., Misawa, S., Kawaguchi, N., Hattori, T., Takamori, M., and Vincent, A. (2005). Patterns and severity of neuromuscular transmission failure in seronegative myasthenia gravis. *J Neurol Neurosurg Psychiatry* **76**, 714–718.

Newsom-Davis, J., Buckley, C., Clover, L., Hart, I., Maddison, P., Tuzum, E., and Vincent, A. (2003). Autoimmune disorders of neuronal potassium channels. *Ann N Y Acad Sci* **998**, 202–210.

Newsom-Davis, J. and Mills, K.R. (1993). Immunological associations of acquired neuromyotonia (Isaacs' syndrome). Report of five cases and literature review. *Brain* **116 (Pt 2)**, 453–469.

Newsom-Davis, J., Pinching, A.J., Vincent, A., and Wilson, S.G. (1978). Function of circulating antibody to acetylcholine receptor in myasthenia gravis: Investigation by plasma exchange. *Neurology* **28**, 266–272.

Niks, E.H., Kuks, J.B., Roep, B.O., Haasnoot, G.W., Verduijn, W., Ballieux, B.E. et al. (2006). Strong association of MuSK antibody-positive myasthenia gravis and HLA-DR14-DQ5. *Neurology* **66**, 1772–1774.

Nishimune, H., Sanes, J.R., and Carlson, S.S. (2004). A synaptic laminin-calcium channel interaction organizes active zones in motor nerve terminals. *Nature* **432**, 580–587.

Okada, K., Inoue, A., Okada, M., Murata, Y., Kakuta, S., Jigami, T. et al. (2006). The muscle protein Dok-7 is essential for neuromuscular synaptogenesis. *Science* **312**, 1802–1805.

O'Neill, J.H., Murray, N.M., and Newsom-Davis, J. (1988). The Lambert-Eaton myasthenic syndrome. A review of 50 cases. *Brain* **111 (Pt 3)**, 577–596.

Papanastasiou, D., Poulas, K., Kokla, A., and Tzartos, S.J. (2000). Prevention of passively transferred experimental autoimmune myasthenia gravis by Fab fragments of monoclonal antibodies directed against the main immunogenic region of the acetylcholine receptor. *J Neuroimmunol* **104**, 124–132.

Patrick, J. and Lindstrom, J. (1973). Autoimmune response to acetylcholine receptor. *Science* **180**, 871–872.

Patton, B.L. (2003). Basal lamina and the organization of neuromuscular synapses. *J Neurocytol* **32**, 883–903.

Peers, C., Johnston, I., Lang, B., and Wray, D. (1993). Cross-linking of presynaptic calcium channels: A mechanism of action for Lambert-Eaton myasthenic syndrome antibodies at the mouse neuromuscular junction. *Neurosci Lett* **153**, 45–48.

Pinto, A., Gillard, S., Moss, F., Whyte, K., Brust, P., Williams, M. et al. (1998). Human autoantibodies specific for the alpha1A calcium channel subunit reduce both P-type and Q-type calcium currents in cerebellar neurons. *Proc Natl Acad Sci U S A* **95**, 8328–8333.

Pinto, A., Moss, F., Lang, B., Boot, J., Brust, P., Williams, M. et al. (1998). Differential effect of Lambert-Eaton myasthenic syndrome immunoglobulin on cloned neuronal voltage-gated calcium channels. *Ann N Y Acad Sci* **841**, 687–690.

Plomp, J.J., Van Kempen, G.T., De Baets, M.B., Graus, Y.M., Kuks, J.B., and Molenaar, P.C. (1995). Acetylcholine release in myasthenia gravis: Regulation at single end-plate level. *Ann Neurol* **37**, 627–636.

Polizzi, A., Huson, S.M., and Vincent, A. (2000). Teratogen update: Maternal myasthenia gravis as a cause of congenital arthrogryposis. *Teratology* **62**, 332–341.

Roberts, A., Perera, S., Lang, B., Vincent, A., and Newsom-Davis, J. (1985). Paraneoplastic myasthenic syndrome IgG inhibits 45Ca2+ flux in a human small cell carcinoma line. *Nature* **317**, 737–739.

Rogers, S.W., Andrews, P.I., Gahring, L.C., Whisenand, T., Cauley, K., Crain, B. et al. (1994). Autoantibodies to glutamate receptor GluR3 in Rasmussen's encephalitis. *Science* **265**, 648–651.

Rothenberg, S.P., da Costa, M.P., Sequeira, J.M., Cracco, J., Roberts, J.L., Weedon, J., and Quadros, E.V. (2004). Autoantibodies against folate receptors in women with a pregnancy complicated by a neural-tube defect. *N Engl J Med* **350**, 134–142.

Roxanis, I., Micklem, K., McConville, J., Newsom-Davis, J., and Willcox, N. (2002). Thymic myoid cells and germinal center formation in myasthenia gravis; possible roles in pathogenesis. *J Neuroimmunol* **125**, 185–197.

Roxanis, I., Micklem, K., and Willcox, N. (2001). True epithelial hyperplasia in the thymus of early-onset myasthenia gravis patients: Implications for immunopathogenesis. *J Neuroimmunol* **112**, 163–173.

Ruff, R.L. and Lennon, V.A. (1998). End-plate voltage-gated sodium channels are lost in clinical and experimental myasthenia gravis. *Ann Neurol* **43**, 370–379.

Sahashi, K., Engel, A.G., Lambert, E.H., and Howard, F.M., Jr. (1980). Ultrastructural localization of the terminal and lytic ninth complement component (C9) at the motor end-plate in myasthenia gravis. *J Neuropathol Exp Neurol* **39**, 160–172.

Sahashi, K., Engel, A.G., Linstrom, J.M., Lambert, E.H., and Lennon, V.A. (1978). Ultrastructural localization of immune complexes (IgG and C3) at the end-plate in experimental autoimmune myasthenia gravis. *J Neuropathol Exp Neurol* **37**, 212–223.

Salmon, A.M., Bruand, C., Cardona, A., Changeux, J.P., and Berrih-Aknin, S. (1998). An acetylcholine receptor alpha subunit promoter confers intrathymic expression in transgenic mice. Implications for tolerance of a transgenic self-antigen and for autoreactivity in myasthenia gravis. *J Clin Invest* **101**, 2340–2350.

Sanes, J.R. and Lichtman, J.W. (1999). Development of the vertebrate neuromuscular junction. *Annu Rev Neurosci* **22**, 389–442.

Sanes, J.R. and Lichtman, J.W. (2001). Induction, assembly, maturation and maintenance of a postsynaptic apparatus. *Nat Rev Neurosci* **2**, 791–805.

Scadding, G.K., Vincent, A., Newsom-Davis, J., and Henry, K. (1981). Acetylcholine receptor antibody synthesis by thymic lymphocytes: Correlation with thymic histology. *Neurology* **31**, 935–943.

Schluep, M., Willcox, N., Vincent, A., Dhoot, G.K., and Newsom-Davis, J. (1987). Acetylcholine receptors in human thymic myoid cells in situ: An immunohistological study. *Ann Neurol* **22**, 212–222.

Schwimmbeck, P.L., Dyrberg, T., Drachman, D.B., and Oldstone, M.B. (1989). Molecular mimicry and myasthenia gravis. An autoantigenic site of the acetylcholine receptor alpha-subunit that has biologic activity and reacts immunochemically with herpes simplex virus. *J Clin Invest* **84**, 1174–1180.

Scuderi, F., Marino, M., Colonna, L., Mannella, F., Evoli, A., Provenzano, C., and Bartoccioni, E. (2002). Anti-p110 autoantibodies identify a subtype of "seronegative" myasthenia gravis with prominent oculobulbar involvement. *Lab Invest* **82**, 1139–1146.

Senanayake, N. and Roman, G.C. (1992). Disorders of neuromuscular transmission due to natural environmental toxins. *J Neurol Sci* **107**, 1–13.

Shillito, P., Molenaar, P.C., Vincent, A., Leys, K., Zheng, W., van den Berg, R.J. et al. (1995). Acquired neuromyotonia: Evidence for autoantibodies directed against K+ channels of peripheral nerves. *Ann Neurol* **38**, 714–722.

Shiono, H., Roxanis, I., Zhang, W., Sims, G.P., Meager, A., Jacobson, L.W. et al. (2003). Scenarios for autoimmunization of T and B cells in myasthenia gravis. *Ann N Y Acad Sci* **998**, 237–256.

Shiraishi, H., Motomura, M., Yoshimura, T., Fukudome, T., Fukuda, T., Nakao, Y. et al. (2005). Acetylcholine receptors loss and postsynaptic damage in MuSK antibody-positive myasthenia gravis. *Ann Neurol* **57**, 289–293.

Sinha, S., Newsom-Davis, J., Mills, K., Byrne, N., Lang, B., and Vincent, A. (1991). Autoimmune aetiology for acquired neuromyotonia (Isaacs' syndrome). *Lancet* **338**, 75–77.

Slater, C.R. (2003). Structural determinants of the reliability of synaptic transmission at the vertebrate neuromuscular junction. *J Neurocytol* **32**, 505–522.

Slater, C.R., Fawcett, P.R., Walls, T.J., Lyons, P.R., Bailey, S.J., Beeson, D. et al. (2006). Pre- and post-synaptic abnormalities associated with impaired neuromuscular transmission in a group of patients with 'limb-girdle myasthenia.' *Brain* **129**, 2061–2076.

Sonoda, Y., Arimura, K., Kurono, A., Suehara, M., Kameyama, M., Minato, S. et al. (1996). Serum of Isaacs' syndrome suppresses potassium channels in PC-12 cell lines. *Muscle Nerve* **19**, 1439–1446.

Spreadbury, I., Kishore, U., Beeson, D., and Vincent, A. (2005). Inhibition of acetylcholine receptor function by seronegative myasthenia gravis non-IgG factor correlates with desensitisation. *J Neuroimmunol* **162**, 149–156.

Stanley, E.F. and Drachman, D.B. (1978). Effect of myasthenic immunoglobulin on acetylcholine receptors of intact mammalian neuromuscular junctions. *Science* **200**, 1285–1287.

Stefansson, K., Dieperink, M.E., Richman, D.P., Gomez, C.M., and Marton, L.S. (1985). Sharing of antigenic determinants between the nicotinic acetylcholine receptor and proteins in Escherichia coli, Proteus vulgaris, and Klebsiella pneumoniae. Possible role in the pathogenesis of myasthenia gravis. *N Engl J Med* **312**, 221–225.

Strobel, P., Helmreich, M., Menioudakis, G., Lewin, S.R., Rudiger, T., Bauer, A. et al. (2002). Paraneoplastic myasthenia gravis correlates with generation of mature naive CD4(+) T cells in thymomas. *Blood* **100**, 159–166.

Strobel, P., Rosenwald, A., Beyersdorf, N., Kerkau, T., Elert, O., Murumagi, A. et al. (2004). Selective loss of regulatory T cells in thymomas. *Ann Neurol* **56**, 901–904.

Thieben, M.J., Lennon, V.A., Boeve, B.F., Aksamit, A.J., Keegan, M., and Vernino, S. (2004). Potentially reversible autoimmune limbic encephalitis with neuronal potassium channel antibody. *Neurology* **62**, 1177–1182.

Tomimitsu, H., Arimura, K., Nagado, T., Watanabe, O., Otsuka, R., Kurono, A. et al. (2004). Mechanism of action of voltage-gated K+ channel antibodies in acquired neuromyotonia. *Ann Neurol* **56**, 440–444.

Toyka, K.V., Brachman, D.B., Pestronk, A., and Kao, I. (1975). Myasthenia gravis: Passive transfer from man to mouse. *Science* **190**, 397–399.

Tzartos, S.J., Barkas, T., Cung, M.T., Mamalaki, A., Marraud, M., Orlewski, P. et al. (1998). Anatomy of the antigenic structure of a large membrane autoantigen, the muscle-type nicotinic acetylcholine receptor. *Immunol Rev* **163**, 89–120.

Tzartos, S.J., Cung, M.T., Demange, P., Loutrari, H., Mamalaki, A., Marraud, M. et al. (1991). The main immunogenic region (MIR) of the nicotinic acetylcholine receptor and the anti-MIR antibodies. *Mol Neurobiol* **5**, 1–29.

Tzartos, S.J., Rand, D.E., Einarson, B.L., and Lindstrom, J.M. (1981). Mapping of surface structures of electrophorus acetylcholine receptor using monoclonal antibodies. *J Biol Chem* **256**, 8635–8645.

Uchitel, O.D., Protti, D.A., Sanchez, V., Cherksey, B.D., Sugimori, M., and Llinas, R. (1992). P-type voltage-dependent calcium channel mediates presynaptic calcium influx and transmitter release in mammalian synapses. *Proc Natl Acad Sci U S A* **89**, 3330–3333.

Vandiedonck, C., Beaurain, G., Giraud, M., Hue-Beauvais, C., Eymard, B., Tranchant, C. et al. (2004). Pleiotropic effects of the 8.1 HLA haplotype in patients with autoimmune myasthenia gravis and thymus hyperplasia. *Proc Natl Acad Sci U S A* **101**, 15464–15469.

Vernino, S. and Lennon, V.A. (2003). Neuronal ganglionic acetylcholine receptor autoimmunity. *Ann N Y Acad Sci* **998**, 211–214.

Vernino, S., Low, P.A., and Lennon, V.A. (2003). Experimental autoimmune autonomic neuropathy. *J Neurophysiol* **90**, 2053–2059.

Vincent, A. (2002). Unravelling the pathogenesis of myasthenia gravis. *Nat Rev Immunol* **2**, 797–804.

Vincent, A., Bowen, J., Newsom-Davis, J., and McConville, J. (2003). Seronegative generalised myasthenia gravis: Clinical features, antibodies, and their targets. *Lancet Neurol* **2**, 99–106.

Vincent, A., Buckley, C., Schott, J.M., Baker, I., Dewar, B.K., Detert, N. et al. (2004). Potassium channel antibody-associated encephalopathy: A potentially immunotherapy-responsive form of limbic encephalitis. *Brain* **127**, 701–712.

Vincent, A., Clover, L., Buckley, C., Grimley Evans, J., and Rothwell, P.M. (2003). Evidence of underdiagnosis of myasthenia gravis in older people. *J Neurol Neurosurg Psychiatry* **74**, 1105–1108.

Vincent, A., Jacobson, L., and Shillito, P. (1994). Response to human acetylcholine receptor alpha 138-199: Determinant spreading initiates autoimmunity to self-antigen in rabbits. *Immunol Lett* **39**, 269–275.

Vincent, A. and Leite, M.I. (2005). Neuromuscular junction autoimmune disease: Muscle specific kinase antibodies and treatments for myasthenia gravis. *Curr Opin Neurol* **18**, 519–525.

Vincent, A., Newland, C., Brueton, L., Beeson, D., Riemersma, S., Huson, S.M., and Newsom-Davis, J. (1995). Arthrogryposis multiplex congenita with maternal autoantibodies specific for a fetal antigen. *Lancet* **346**, 24–25.

Vincent, A. and Newsom-Davis, J. (1982). Acetylcholine receptor antibody characteristics in myasthenia gravis. I. Patients with generalized myasthenia or disease restricted to ocular muscles. *Clin Exp Immunol* **49**, 257–265.

Vincent, A., Newsom-Davis, J., Newton, P., and Beck, N. (1983). Acetylcholine receptor antibody and clinical response to thymectomy in myasthenia gravis. *Neurology* **33**, 1276–1282.

Vincent, A., Palace, J., and Hilton-Jones, D. (2001). Myasthenia gravis. *Lancet* **357**, 2122–2128.

Vincent, A., Scadding, G.K., Thomas, H.C., and Newsom-Davis, J. (1978). In-vitro synthesis of anti-acetylcholine-receptor antibody by thymic lymphocytes in myasthenia gravis. *Lancet* **1**, 305–307.

Vincent, A., Whiting, P.J., Schluep, M., Heidenreich, F., Lang, B., Roberts, A. et al. (1987). Antibody heterogeneity and specificity in myasthenia gravis. *Ann N Y Acad Sci* **505**, 106–120.

Vincent, A., Willcox, N., Hill, M., Curnow, J., MacLennan, C., and Beeson, D. (1998). Determinant spreading and immune responses to acetylcholine receptors in myasthenia gravis. *Immunol Rev* **164**, 157–168.

Waterman, S.A., Lang, B., and Newsom-Davis, J. (1997). Effect of Lambert-Eaton myasthenic syndrome antibodies on autonomic neurons in the mouse. *Ann Neurol* **42**, 147–156.

Watson, R., Jepson, J.E., Bermudez, I., Alexander, S., Hart, Y., McKnight, K. et al. (2005). Alpha7-acetylcholine receptor antibodies in two patients with Rasmussen encephalitis. *Neurology* **65**, 1802–1804.

Watson, R., Jiang, Y., Bermudez, I., Houlihan, L., Clover, L., McKnight, K. et al. (2004). Absence of antibodies to glutamate receptor type 3 (GluR3) in Rasmussen encephalitis. *Neurology* **63**, 43–50.

Weinberg, C.B. and Hall, Z.W. (1979). Antibodies from patients with myasthenia gravis recognize determinants unique to extrajunctional acetylcholine receptors. *Proc Natl Acad Sci U S A* **76**, 504–508.

Whiting, P.J., Vincent, A., Schluep, M., and Newsom-Davis, J. (1986). Monoclonal antibodies that distinguish between normal and denervated human acetylcholine receptor. *J Neuroimmunol* **11**, 223–235.

Whitney, K.D. and McNamara, J.O. (2000). GluR3 autoantibodies destroy neural cells in a complement-dependent manner modulated by complement regulatory proteins. *J Neurosci* **20**, 7307–7316.

Wilson, S., Vincent, A., and Newsom-Davis, J. (1983). Acetylcholine receptor turnover in mice with passively transferred myasthenia gravis. I. Receptor degradation. *J Neurol Neurosurg Psychiatry* **46**, 377–382.

Wilson, S., Vincent, A., and Newsom-Davis, J. (1983). Acetylcholine receptor turnover in mice with passively transferred myasthenia gravis. II. Receptor synthesis. *J Neurol Neurosurg Psychiatry* **46**, 383–387.

Wirtz, P.W., Willcox, N., Roep, B.O., Lang, B., Wintzen, A.R., Newsom-Davis, J., and Verschuuren, J.J. (2003). HLA-B8 in patients with the Lambert-Eaton myasthenic syndrome reduces likelihood of associated small cell lung carcinoma. *Ann N Y Acad Sci* **998**, 200–201.

Wood, S.J. and Slater, C.R. (1997). The contribution of postsynaptic folds to the safety factor for neuromuscular transmission in rat fast- and slow-twitch muscles. *J Physiol* **500 (Pt 1)**, 165–176.

Wood, S.J. and Slater, C.R. (2001). Safety factor at the neuromuscular junction. *Prog Neurobiol* **64**, 393–429.

Yamamoto, T., Vincent, A., Ciulla, T.A., Lang, B., Johnston, I., and Newsom-Davis, J. (1991). Seronegative myasthenia gravis: A plasma factor inhibiting agonist-induced acetylcholine receptor function copurifies with IgM. *Ann Neurol* **30**, 550–557.

Yu Wai Man, C.Y., Chinnery, P.F., and Griffiths, P.G. (2005). Extraocular muscles have fundamentally distinct properties that make them selectively vulnerable to certain disorders. *Neuromuscul Disord* **15**, 17–23.

Zhang X, Yang M, Xu J, Zhang M, Lang B, Wang W, Vincent A. (2007). Clinical and serological study of myasthenia gravis in HuBei province, China. J Neurol Neurosurg Psychiatry, in press.

32

Paraneoplastic Neurologic Syndromes

Josep Dalmau

I. Definition and Frequency

Among the many neurological complications that patients with cancer develop, there is a group of disorders, called paraneoplastic neurologic syndromes (PNS), in which the neurologic dysfunction is immune-mediated. The main targets of these immune responses are the neurons and peripheral nerves, but any part of the central or peripheral nervous system, including retina and muscles, can be involved (Bataller & Dalmau, 2004). The trigger of the immune responses is the tumor, which by diverse mechanisms breaks immune tolerance to proteins normally expressed in the nervous system, or disrupts the function of neurons or nerve and muscle cells by inflammatory or cytokine-related processes. In some PNS, one area of the brain or a subset of neurons is involved, whereas in others, multiple regions of the nervous system can be affected. The resulting syndromes are listed in Table 32.1.

In approximately 70 percent of patients, the development of paraneoplastic neurologic symptoms occurs at early stages of the cancer, months or sometimes years before the tumor is detectable. Therefore, recognition of these disorders is not only important for early diagnosis and treatment of the tumor, but also for prompt use of immunotherapy before the neuronal dysfunction is irreversible. In addition to these practical implications, PNS are interesting models of antitumor immune responses that progress to become autoimmune disorders capable of reaching antigens behind the blood–brain barrier.

The tumors more frequently involved in PNS are those that express neuroendocrine proteins such as small-cell lung cancer (SCLC) or neuroblastoma, affect organs with immunoregulatory functions (thymoma), or derive from cells that produce immunoglobulins (plasma cell dyscrasias, B-cell lymphomas). About 3 to 5 percent of patients with SCLC, 15 to 20 percent with thymomas, and 3 to 10 percent with B-cell neoplasms develop PNS (Elrington et al., 1991; Ropper & Gorson, 1998; Vernino & Lennon, 2004). For other tumors, the frequency of PNS is well below 1 percent; the more frequent neoplasms include ovary, breast, germ-cell tumors of the testis, and teratomas.

Table 32.1 Paraneoplastic Syndromes of the Nervous System

Area Involved	Classical Syndromes	Nonclassical Syndromes
CNS	Cerebellar degeneration Limbic encephalitis Opsoclonus-myoclonus Encephalomyelitis	Brainstem encephalitis Stiff-person syndrome Necrotizing myelopathy Motor neuron disease
Dorsal root ganglia or peripheral nerves	Subacute sensory neuronopathy Gastrointestinal paresis or pseudo-obstruction	Acute sensorimotor neuropathy (Guillain-Barré syndrome, plexitis) Subacute and chronic sensorimotor neuropathies Neuropathy of plasma cell dyscrasias and lymphoma Vasculitis of the nerve and muscle Pure autonomic neuropathy Acquired neuromyotonia
Multiple levels	Encephalomyelitis	
Neuromuscular junction	LEMS	Myasthenia gravis
Muscle	Dermatomyositis	Acute necrotizing myopathy Polymyositis
Eye and retina	Cancer-associated retinopathy Melanoma-associated retinopathy	Optic neuritis

II. Immune-mediated Pathogenic Mechanisms

The best evidence that many PNS are immune-mediated comes from the demonstration of antineuronal antibodies in the serum and cerebrospinal fluid (CSF) of patients (see Table 32.2). These antibodies react with proteins expressed by normal neurons and the patient's tumor (onconeuronal antigens). For clinicians, the discovery of these antibodies has been extremely useful (Graus et al., 2004), allowing the development of specific diagnostic tests (discussed later). However, only a few antibodies appear to have a direct pathogenic role in causing the neurologic dysfunction. The rest of the antibodies occur in association with cytotoxic T-cell responses that are the main effectors of the neuronal degeneration.

A. Antibody-mediated PNS

Antibodies with a direct pathogenic effect are those that target cell-surface antigens usually expressed in the peripheral nerve or neuromuscular junction. These include antibodies to P/Q-type voltage-gated calcium channels (VGCC) in patients with the Lambert-Eaton myasthenic syndrome (LEMS) (Motomura et al., 1995), antibodies to the acetylcholine receptor (AChR) in patients with myasthenia gravis, and antibodies to voltage-gated potassium channels (VGKC) in patients with neuromyotonia (Hart et al., 2002). In these disorders removal of the antibodies with plasma exchange or modulation of the immune response with intravenous IgG (IV-Ig) often results in neurologic improvement. The pathogenic role of these antibodies has been demonstrated by modeling the clinical or neurophysiologic abnormalities with passive transfer of serum or IgG to animals (Lang et al., 2003; Newsom-Davis et al., 2003).

Recent studies indicate that antibodies to VGKC also associate with disorders of the central nervous system (CNS), such as limbic encephalitis (LE) or Morvan's syndrome (a disorder that includes symptoms of neuromyotonia and encephalitis) (Vincent et al., 2004). Related clinical syndromes recently have been described in association with antibodies to uncharacterized antigens that are expressed in the cell surface of neurons of the limbic system (hippocampal neuropil antigens). These novel antibodies appear to be directed to diverse cell surface antigens and occur in patients with LE without VGKC antibodies (Ances et al., 2005). Because disorders associated with these antibodies usually respond to immunotherapy and IgG-depleting strategies, and the titers correlate with the symptoms, a direct pathogenic role of the antibodies has been implied.

B. T-Cell mediated PNS

In contrast to the previously discussed disorders, the majority of PNS of the central nervous system appear to be mediated by cytotoxic T-cell responses against intracellular onconeuronal antigens. These immune responses usually are accompanied by antibodies against the same onconeuronal antigens. Attempts to reproduce the neurologic disorders by passive transfer of antibodies or animal immunization with the recombinant antigens have been unsuccessful (Graus et al., 1991; Sillevis- Smitt et al., 1995; Tanaka et al., 1994, 1995).

Table 32.2 Antibodies, Paraneoplastic Syndromes, and Associated Cancers

Antibody	Syndrome	Associated Cancers
Well-characterized Paraneoplastic Antibodies*		
Anti-Hu (ANNA-1)	PEM including cortical, limbic, brainstem encephalitis, PCD, myelitis; PSN, autonomic dysfunction	SCLC, other
Anti-Yo (PCA-1)	PCD	Gynecological, breast
Anti-Ri (ANNA-2)	PCD, brainstem encephalitis, opsoclonus-myoclonus	Breast, gynecological, SCLC
Anti-CV2/CRMP5	PEM, PCD, chorea, peripheral neuropathy	SCLC, thymoma, other
Anti-Ma proteins**	Limbic, hypothalamic, brainstem encephalitis (infrequently PCD)	Germ-cell tumors of testis, other solid tumors
Anti-amphiphysin	Stiff-man syndrome, PEM	Breast
Anti-recoverin#	Cancer-associated retinopathy (CAR)	SCLC
Partially-characterized Paraneoplastic Antibodies*		
Anti-Tr	PCD	Hodgkin's lymphoma
Anti-Zic4	PCD	SCLC
Hippocampal neuropil	LE	Teratoma of the ovary
mGluR1	PCD	Hodgkin's lymphoma
ANNA3	Various PNS of the CNS	SCLC
PCA2	Various PNS of the CNS	SCLC
Anti-bipolar cells of the retina	Melanoma-associated retinopathy (MAR)	Melanoma
Antibodies that Occur with and Without Cancer Association		
Anti-VGCC	LEMS, PCD	SCLC
Anti-AChR	Myasthenia gravis	Thymoma
Anti-VGKC	Neuromyotonia, LE	Thymoma, others
Anti-nAChR	Subacute pandysautonomia	SCLC, others

*Well-characterized antibodies are those directed against antigens whose molecular identity is known, or that have been identified by several investigators.

**Patients with antibodies to Ma2 are usually men with testicular cancer. Patients with additional antibodies to other Ma proteins are men or women with a variety of solid tumors.

#Other antibodies reported in a few or isolated cases include antibodies to tubby-like protein and the photoreceptor-specific nuclear receptor.

PEM: paraneoplastic encephalomyelitis; PCD: paraneoplastic cerebellar degeneration; LE: limbic encephalitis; SCLC: small-cell lung cancer; VGCC: voltage-gated calcium channels; VGKC: voltage-gated potassium channels; nAChR: neuronal (or ganglionic) acetylcholine receptor.

Furthermore, these patients usually do not improve with IgG-depleting strategies and the response to immunotherapy is generally poor. Studies of the CNS of these patients show extensive infiltrates of T-cells, neuronophagic nodules, gliosis, and microglial activation. The T-cell infiltrates are composed of CD4 lymphocytes that predominate in perivascular and interstitial regions, and CD8 lymphocytes that usually form neuronophagic nodules (Bernal et al., 2002; Jean et al., 1994); these T-cells surround neurons and use perforin and granzyme B, two membranolytic proteins that induce apoptosis (see Figure 32.1; T-cells IgG) (Blumenthal et al., 2006).

These findings have led to the proposal that T-cells are the main effectors of these disorders, although a complementary role of the antibodies cannot be completely ruled out. For example, in addition to the T-cell infiltrates, deposits of IgG and onconeuronal antibodies frequently are encountered in the same brain regions, and most patients have intrathecal synthesis of antibodies. Analysis of the T-cell receptor usage and detection of oligoclonal expansion of T-cells have suggested that these lymphocytes are driven by antigens restricted to the tumor and nervous system (Voltz et al., 1998). Although the onconeuronal antigen-specificity of the brain infiltrating T-cells has not been established, this specificity has been demonstrated in studies using lymphocytes from peripheral blood of patients, including classical T-cell proliferation assays (Benyahia et al., 1999), analysis of dendritic cells as antigen-presenting cells (Albert et al., 1998; Tanaka et al., 1998), and experiments with fibroblasts manipulated to express onconeuronal antigens (Tanaka et al., 1999).

The molecular identity of most target antigens has been identified by screening diverse cDNA expression libraries with patients' sera. The function of some of these proteins in normal neurons is partially known, but their role in the tumor cells remains unclear. The main paraneoplastic antigens can

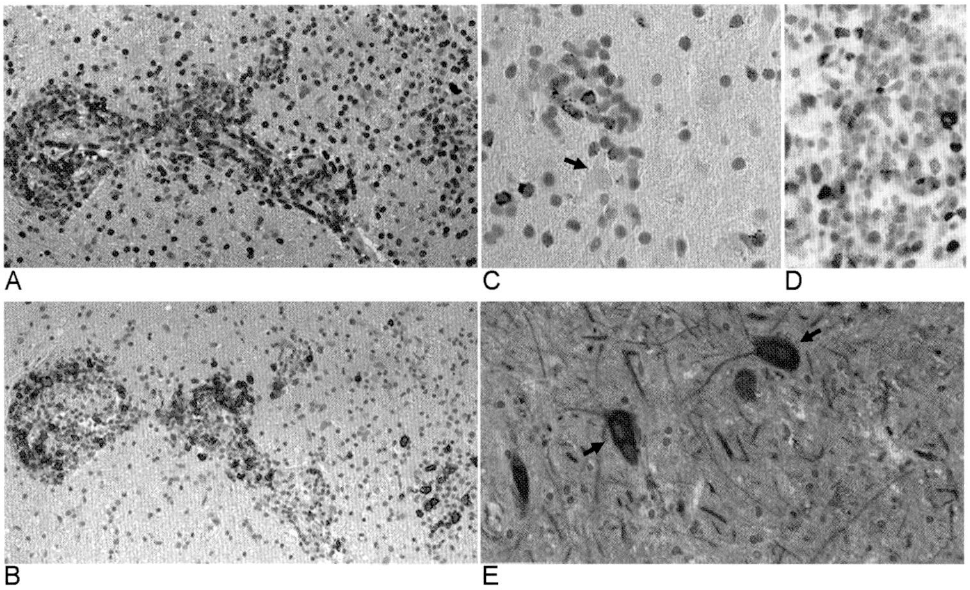

Figure 32.1 Inflammatory infiltrates in the hippocampus of a patient with paraneoplastic anti-Ma2-associated encephalitis. **A, B.** Consecutive tissue sections immunolabeled with CD3, a pan-T-cell marker (**A**), and CD20, a marker of B-cells (**B**). Note that the predominant infiltrates of T-cells are accompanied by a significant number of B-cells. (×100, counterstained with hematoxylin). **C, D.** Infiltrates of T-cells showing expression of TIA-1, a cytotoxic granule-associated protein; arrow points to a neuron in close contact with TIA-1 expressing T-cells. Most of these cells use granzyme-B (**D**) and less frequently perforin (not shown), two membranolytic proteins that induce target cell apoptosis (×400, counterstained with hematoxylin). **E.** Deposits of IgG in neurons (arrows); this type of IgG immunolabeling was identified in approximately 10% of neurons of a small biopsy specimen; no IgM was identified. Glial cells did not contain IgG or IgM (×400, counterstained with hematoxylin). Adapted from Blumenthal et al., 2006.

be divided in several groups (see Table 32.3) (reviewed by Musunuru & Darnell, 2001; Rosenfeld & Dalmau, 2000):

▲ RNA binding proteins, including the Hu proteins (HuC, HuD, HelN1) that are targets of Hu antibodies, and Nova proteins that are targets of Ri antibodies

▲ Antigens involved in axonal outgrowth and guidance (CRMP5) that are targets of CV2 antibodies

▲ The Myc-interacting protein antigen (cdr2) that is the target of anti-Yo antibodies

▲ Antigens contained in nuclear bodies (Ma proteins) that are targets of anti-Ma1 and anti-Ma2 (also called Ta) antibodies (Rosenfeld et al., 2001)

▲ The synaptic-endocytic protein, amphiphysin I, that is the target of amphiphysin-antibodies

C. The Development of PNS

The central concept in the pathogenesis of most PNS is that the ectopic expression of neuronal proteins by a tumor triggers the immune response that eventually becomes an autoimmune disease causing neuronal degeneration. There are probably several mechanisms whereby the involved tumors trigger the immune response. There is data indicating that apoptotic tumor cells are captured and presented to the immunological system by dendritic cells (professional antigen presenting cells) in the regional lymph nodes (Albert et al., 1998). This type of antigen presentation correlates with the frequent clinical identification of the tumor at the organ-draining lymph nodes; for example, most patients with anti-Hu associated encephalomyelitis have tumor localized in the mediastinal lymph nodes and in many instances this is the only site where the tumor can be detected (Dalmau et al., 1992; Graus et al., 2001). A similar mechanism of antigen presentation is clinically suggested for patients with PNS associated with breast and ovarian cancer (Rojas et al., 2000). Compelling examples are the patients with anti-Yo associated cerebellar degeneration in whom neoplastic cells with immunohistochemical features of breast cancer (i.e., ErbB2 or Her2/Neu) can be detected only in the axillary lymph nodes (see Figure 32.2; breast PET).

However, the presentation of onconeuronal antigens to the immunological system does not always occur at the regional lymph nodes. The best example of an alternative mechanism of antigen-presentation comes from patients with anti-Ma2-associated encephalitis. In young men this paraneoplastic disorder usually associates with testicular

Table 32.3 Main Paraneoplastic Antigens, Molecular Features, and Function

Antigens	Molecular Features and Function
HuC, HuD, HelN-1, HuR (targets of "Hu antibodies") (Chung et al., 1996)	Homologous to the Drosophila proteins Elav, RBp9, and Sex-lethal. Elav: essential for development and maintenance of the nervous system Hu proteins contain 3 RNA recognition motifs (RRM); they bind to AU rich elements in the 3′ untranslated regions of several growth-related mRNAs (c-myc, c-fos, GM-CSF, among others) Hu proteins are involved in regulating mRNA stability, localization, and translation The homology to Elav, the very early expression during mammalian neuronal development, and the RNA binding properties, support a role of the Hu proteins in neuronal growth and differentiation. Hu antibodies react with the first 2 RNA binding domains of HuD, HuC, and HelN-1
Nova-1 and -2 (targets of "Ri antibodies") (Dredge et al., 2005)	Contain 3 RNA binding motifs homologous to the KH motifs found in hnRNP-K protein Nova-1 predominantly expressed in hindbrain and ventral spinal cord; Nova 2 is expressed in neocortex and hippocampus Nova-1 regulates alternative splicing of pre-mRNA encoding GABA(A)Rgamma2, GlyRalpha2, and Nova-1 expression itself Nova-1 null mice die post-natally from apoptotic death of spinal and brainstem neurons Ri antibodies bind to the 3rd KH domain of Nova-1 and block the Nova-1 RNA interactions
CRMP5 (targets of "CV2 antibodies") (Bretin et al., 2005)	Involved in signaling function in axon outgrowth and guidance
Cdr2 (or cdr62) (target of Yo antibodies) (Okano et al., 1999)	Contains a zinc finger domain and helix-leucine zipper (HLZ) motif typical of proteins that bind to DNA as hetero- or homodimers cdr2 interact specifically with the HLZ of c-Myc (both proteins co-localize in the cytoplasm of Purkinje cells) Yo antibodies block the interaction between cdr2 and c-Myc
Ma proteins (target of Ma1, Ma2 (or Ta) antibodies) (Rosenfeld et al., 2001)	Localized to nuclear structures resembling interchromatin granule clusters or "speckles" Ma2 contains a polypyrimidine tract, a common feature of ribosomal proteins Possibly involved in the coupling of transcription and pre-mRNA processing in neurons and testis
Amphiphysin I (Farsad et al., 2003)	Brain specific protein enriched at the synapse Major binding partner of several components of the clathrin-mediated endocytic machinery (interacts with dynamin) Mutant amphiphysin with amino acid substitutions in the SH3 domain, induces accumulation of intracellular aggregates (amphiphysin, clathrin, AP-2) and block endocytosis Amphiphysin antibodies from patients with stiff-man syndrome bind to the C-terminus of the protein that contains SH3

tumors (Dalmau et al., 2004). Recent studies demonstrate that the paraneoplastic anti-Ma2 immune response can be triggered by microscopic tumors confined to the seminiferous tubules (intratubular germ-cell neoplasms, or *carcinoma in situ* of the testis) (Mathew et al., 2006). These tumors often are associated with extensive inflammatory infiltrates (see Figure 32.3). Since testis is an immunoprivileged organ, antigen presentation probably occurs at the tumor site; there is evidence that subpopulations of testicular macrophages are efficient antigen presenting cells in humoral and cell-mediated immune responses (Bryniarski et al., 2004), and that damaged seminiferous tubules express increased amounts of cytokines promoting immune responses (Sundstrom et al., 1999).

The next critical step in the immunopathogenesis of PNS relates to how the onconeuronal cytotoxic T-cells recognize antigens that are intracellular and expressed at immunoprivileged cells (i.e., neurons) and sites (i.e., behind the blood–brain barrier). The involved mechanisms are unclear, but the associated antibodies may play a complementary role that could be critical at initial stages of the disorder. For example, internalization of antibodies causing disruption of the target antigens could lead to neuronal degeneration and apoptosis releasing antigens to microglia, which are the main antigen-presenting cells in the CNS. Activated antigen-specific T-cells (able to cross the blood–brain barrier) would then react with microglial-expressing onconeuronal antigens and release proinflammatory cytokines. In this microenvironment neurons are able to express class I MHC allowing direct antigen presentation to the cytotoxic T cells (Neumann et al., 1997).

This model would fit with the indicated immunopathological findings in the brain of these patients, including intraneuronal deposits of IgG and cytotoxic T-cells surrounding and indenting neurons undergoing degeneration (see Figure 32.1) (Blumenthal et al., 2006). The high titers of onconeuronal antibodies usually present in patients' sera (allowing some IgG to cross the blood–brain barrier) and the frequent detection of intrathecal synthesis of antibodies would support this model in which the antibodies play a critical role in the initial stages of immune-mediated neuronal degeneration.

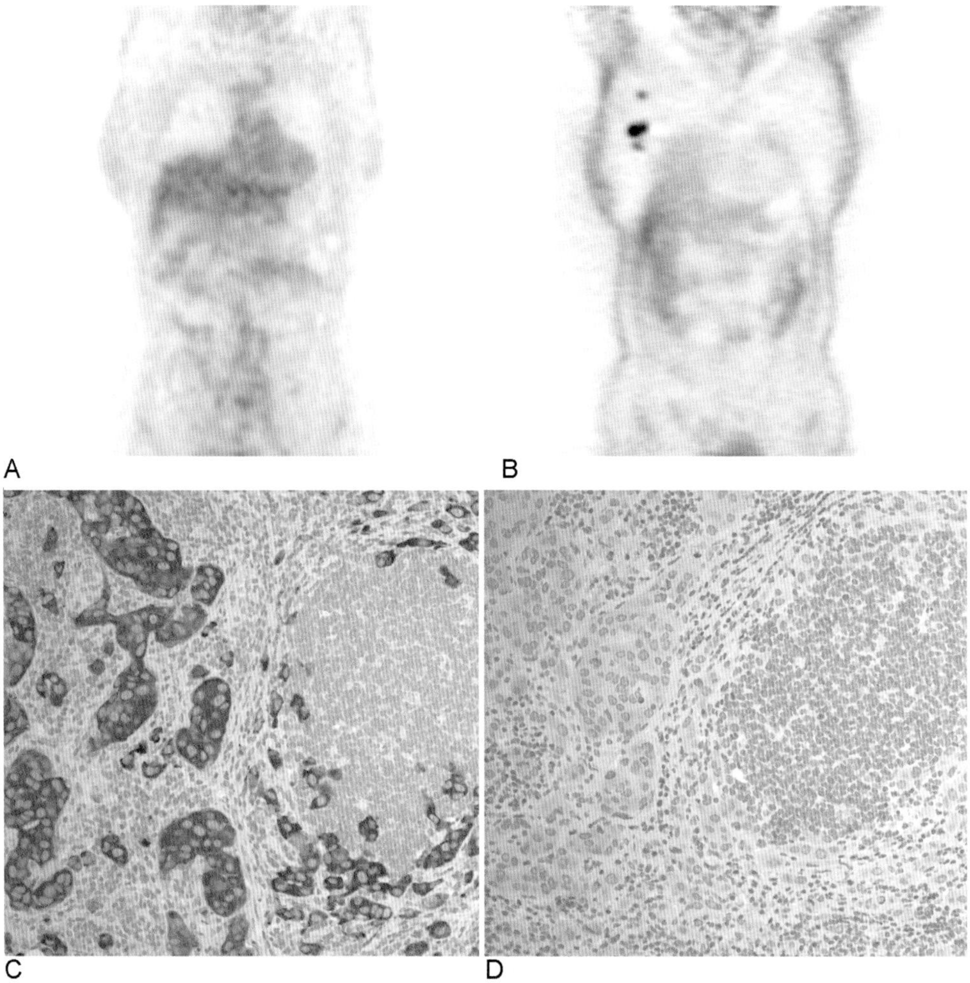

Figure 32.2 Demonstration of a cdr2-expressing tumor in a patient whose paraneoplastic cerebellar degeneration preceded by 5 years the detection of the cancer (breast cancer detected only in the axillary lymph nodes). **A, B.** The body FDG-PET obtained in 2005 (**B**) demonstrates FDG hyperactive abnormalities in the right axillary region that were not present in 2003 (**A**). **C, D.** Sections of the involved axillary lymph nodes (containing tumor cells) incubated with patient's Yo antibodies. The reactivity of the tumor cells in **C** (brown cells) was abrogated when the antibodies were preabsorbed with the Yo antigen or cdr2 (**D**), confirming that the tumor reactivity was cdr2-specific (×400, counterstained with hematoxylin). Adapted from Mathew et al., 2006, *J Neural Sci,* 250, 153–155.

D. Other Immune-mediated Mechanisms

In contrast to the previous disorders in which the tumor cells express neuronal proteins that trigger the immune response, there are other PNS in which the neoplastic cells themselves (myeloma, Waldenstrom's macroglobulinemia) produce immunoglobulins or immunoglobulin fragments that affect the function of peripheral nerves (Dimopoulos et al., 2000; Ropper & Gorson, 1998). These abnormal immunoglobulins may have specific antibody activity against nerve antigens, such as myelin-associated glycoprotein (MAG) and gangliosides (Ilyas et al., 1992). Pathological studies show widening of the myelin lamellae that correlate with the clinical and electrophysiological features of a predominantly demyelinating sensory neuropathy. In addition, there are abnormal immunoglobulins synthesized by the neoplastic cells that do not have specific antibody activity but cause nerve damage through deposition of immunoglobulin fragments (amyloid) or immunocomplexes.

In other PNS the occurrence of immunological mechanisms is evident, but the molecular identity of the antigens is not known. These disorders include, among others, vasculitis of the nerve and muscle (T-cell infiltrates), dermatomyositis (inflammatory infiltrates of CD4 cells and deposits of IgG and complement in vessels leading to muscle ischemia and

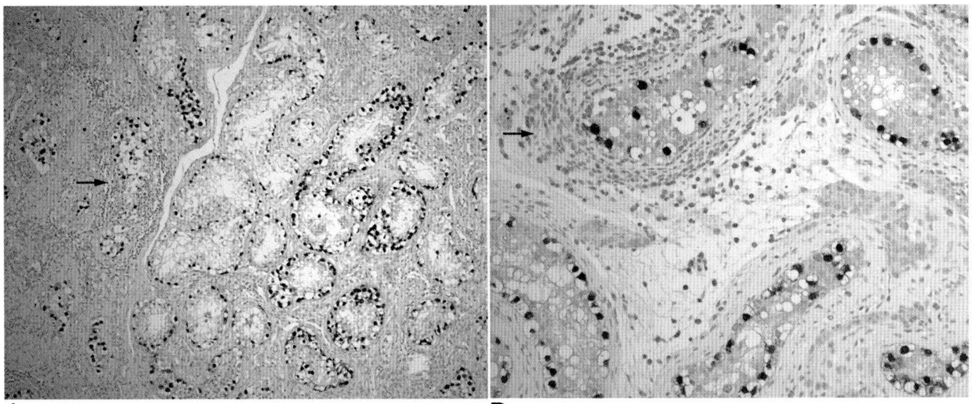

A B

Figure 32.3 Intratubular germ-cell neoplasm or *carcinoma in situ* of the testis in a patient with paraneoplastic anti-Ma2-associated encephalitis. Note the intratubular location of the neoplastic cells (brown staining) demonstrated with Oct4, a specific germ-cell tumor marker. **A.** Inflammatory infiltrates are shown with an arrow. **B.** Demonstrates, at higher magnification, the tumor cells and inflammatory infiltrates (arrow) (**A** ×100, **B** ×400, counterstained with hematoxylin).

perifascicular atrophy), and polymyositis (infiltrates of CD8 cells in muscle) (Rudnicki & Dalmau, 2000). High levels of circulating cytokines (IL-6, VEGF) have been associated with a neuropathy resembling a chronic inflammatory demyelinating neuropathy in patients with the POEMS syndrome (polyneuropathy, organomegaly, endocrinopathy, M protein, and skin changes) (Gherardi et al., 1996; Scarlato et al., 2005). Patients with the POEMS syndrome usually have an underlying sclerotic myeloma, or Castleman's disease (angiofollicular lymphoid hyperplasia) (Dispenzieri & Gertz, 2005).

III. The Immune Response as an Aid to the Diagnosis of PNS

Since PNS usually precede the diagnosis of the cancer and similar syndromes can occur without cancer, the recognition of PNS can be difficult. For that reason, the discovery of antibodies specifically associated with PNS was a breakthrough in the diagnosis of these disorders (Bataller & Dalmau, 2004). Among many possible scenarios, the most frequent presentation is a patient who in a matter of days or a few weeks develops neurologic symptoms without an apparent cause. Initial screening including blood tests and CSF analysis may be normal or show inflammatory (no specific) abnormalities in the CSF. Radiologic and other imaging studies including chest XR; CT of chest, abdomen, and pelvis; and ultrasound may suggest a tumor, but in many instances (~30–50%) can be normal. In this setting, the discovery that the patient harbors a paraneoplastic antibody, confirms the paraneoplastic origin of the symptoms and focuses the search of the tumor to a few organs. Depending on the patient's gender and type of antibody, the search may then include additional studies, such as body fluorodeoxyglucose-PET (FDG-PET), MRI of the breast and axillary nodes, or ultrasound of the testis, among others (Younes-Mhenni et al., 2004). The identification of mild or questionable findings with any test should be viewed with high suspicion for an occult tumor. For example, detection of microcalcifications or subtle changes in the testicular ultrasound of a young man with Ma2 antibodies is indication for orchiectomy unless another Ma2-expressing tumor is identified (Mathew et al., 2006).

Most paraneoplastic antibodies are detected by immunohistochemical and immunoblot techniques, using rat brain, human neuronal proteins, or recombinant paraneoplastic antigens (see Figure 32.4). Not all paraneoplastic antibodies have the same syndrome specificity; for example, some antibodies (i.e., Hu, CV2/CRMP5) can be identified at low titers in the serum of cancer patients without PNS (Bataller et al., 2004), whereas other antibodies (i.e., Ma2) always associate with PNS. The appropriate interpretation of the presence or absence of paraneoplastic antibodies in the diagnosis of PNS requires the following considerations:

1. Detection of any of the "well-characterized paraneoplastic antibodies" (see Table 32.2) in association with neurological symptoms of unknown etiology is diagnostic of PNS. However, follow-up antibody titers are not useful for most antibodies (Llado et al., 2004); two possible exceptions are anti-Tr and Ma2 antibodies for which titers appear to correlate with the clinical course (Bernal et al., 2003).
2. Approximately 40 percent of patients with PNS of the CNS do not have detectable antibodies. In these patients the diagnosis of PNS relies on the exclusion of other etiologies and demonstration of the cancer.

3. The majority of PNS of the peripheral nerves, neuromuscular junction, and muscle do not associate with paraneoplastic antibodies. The only exceptions are anti-Hu, which associates with sensory neuronopathy (involving dorsal root ganglia and nerve roots) and anti-CV2/CRMP5, which associates with mixed sensorimotor axonal neuropathy (Antoine et al., 2001).

4. Paraneoplastic antibodies can assist in determining if a patient's tumor is involved in the PNS. For example, the detection of a tumor that does not usually associate with a specific immunity should raise suspicion of a second primary neoplasm. In these patients if the atypical tumor does not react with the paraneoplastic antibodies, the search for a second neoplasm should be strongly considered.

5. There are antibodies of high diagnostic value that associate with syndromes of the peripheral nervous system whether they are paraneoplastic or not. These antibodies, included in Table 32.2, should not be considered paraneoplastic markers. For example, antibodies to P/Q-type VGCC occur in LEMS whether the disorder is associated with cancer (about 60% of cases) or not (Motomura et al., 1997).

IV. Diagnostic Criteria of PNS

Since paraneoplastic antibodies are not always present in patients with PNS, and some antibodies can occur in cancer patients without PNS, a set of diagnostic guidelines were recently proposed by a group of investigators (Graus et al., 2004). These guidelines take into account:

- ▲ Whether the syndrome is typical or not (see Table 32.1). A syndrome is considered typical when the etiology is frequently paraneoplastic such as LE or opsoclonus-myoclonus, among others. Syndromes that usually occur without a cancer association are considered atypical (i.e., parkinsonism or brainstem encephalopathy).
- ▲ The type of paraneoplastic antibody; well-characterized antibodies are those whose target antigens are known, and for which there is extensive clinical experience. Other antibodies, for which the experience is limited or the target antigens are unknown, are considered "partially characterized" antibodies (see Table 32.2).
- ▲ The presence or absence of cancer.

Based on these criteria and the time period between the development of the neurologic symptoms and detection of

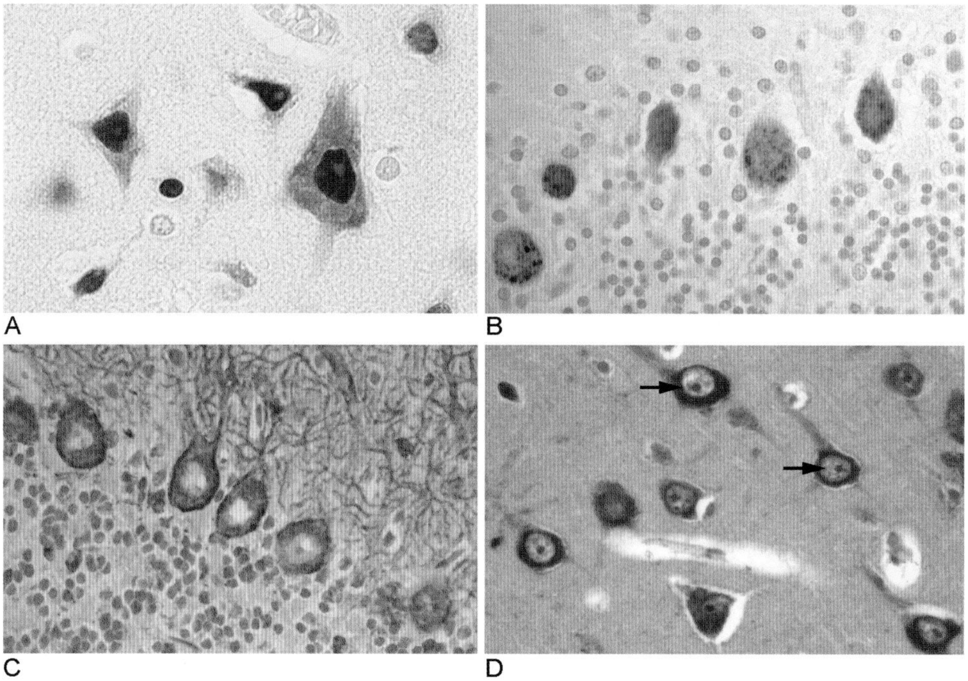

Figure 32.4 Classical paraneoplastic antibodies. **A.** Reactivity of anti-Hu with human pyramidal cortical neurons (tissue obtained from autopsy of a neurologically normal individual). There is predominant reactivity with the nuclei of the neurons, sparing the nucleoli. **B.** Reactivity of anti-Yo with human cerebellar cortex. Note the predominant reactivity with the cytoplasm of Purkinje cells. **C.** Reactivity of CRMP5/CV2 antibodies with rat brain. There is intense reactivity with Purkinje cells and dentrites in the molecular layer. **D.** Reactivity of anti-Ma2 antibody with neurons from human hippocampus. There is reactivity with the cytoplasm of neurons and intense immunolabeling of nuclear speckles (arrows) (All panels ×400, counterstained with hematoxylin).

tthe tumor, two levels of evidence were proposed: definite PNS or possible PNS (see Table 32.4).

V. The Clinical Manifestations of Paraneoplastic Immunity: Effects on the Nervous System

Except for a few autoantigens, such as CDR2 and Tr that are predominantly expressed by Purkinje cells of the cerebellum (Furneaux et al., 1990; Graus et al., 1998), and the novel hippocampal neuronal antigens, the rest of the known paraneoplastic antigens are expressed throughout the nervous system. Therefore, it is unclear why the immune responses to these antigens (such as Hu, CRMP5, or Ma2) preferentially associate with specific syndromes and not with more widespread clinical features. These syndromes include among others, paraneoplastic cerebellar degeneration (PCD), LE, opsoclonus-myoclonus, sensory neuronopathy, and LEMS.

Two clinical features shared by most syndromes of the CNS are the rapid development of symptoms and the frequent presence of inflammatory findings in the CSF analysis, including moderate lymphocytic pleocytosis, increased protein concentration, elevated IgG index, and oligoclonal bands (Posner, 1995). The clinical features and immunological findings of classical PNS are reviewed here.

A. Paraneoplastic Cerebellar Degeneration (PCD)

This disorder usually presents with dizziness, gait unsteadiness, oscillopsia, and evolves in a few days or weeks to severe cerebellar dysfunction with ataxia of gait and extremities. Eventually most patients become wheelchair bound, with dysarthria, dysphagia, blurry vision or diplopia, but preserved cognitive functions, and absent or very mild impairment of sensation and reflexes. At the early stages of PCD the brain MRI is usually normal, and as the disease progresses, cerebellar atrophy develops.

The pathological hallmark of PCD is loss of the Purkinje cells of the cerebellum as a result of an immune attack that usually associates with inflammatory infiltrates in the cerebellar cortex (rarely present at the time of autopsy), deep cerebellar nuclei, and inferior olivary nuclei. Almost all well-characterized antineuronal antibodies have been reported in association with PCD (see Table 32.2). Serological markers that associate with "pure" PCD include Yo (Peterson et al., 1992), Tr (Bernal et al., 2003), VGCC (Graus et al., 2002), and infrequently Zic4 and Ma2 antibodies (Bataller et al., 2004). Patients with VGCC may have associated symptoms of LEMS that can be overlooked if they have severe cerebellar dysfunction (Mason et al., 1997). Careful clinical evaluation is important as LEMS usually responds to treatment of the tumor and immunosuppressants but PCD does not.

PCD can be the presentation of a disorder that progresses to involve multiple areas of the CNS, called paraneoplastic encephalomyelitis. Patients with encephalomyelitis usually develop anti-Hu or anti-CV2/CRMP5 antibodies. Between 30 and 40 percent of patients with PCD do not have detectable paraneoplastic antibodies; in these patients the diagnosis relies on the exclusion of other etiologies and demonstration of the cancer.

PCD rarely responds to treatment. An exception is the group of patients with anti-Tr antibodies and Hodgkin's lymphoma; approximately 20 percent show improvement after treating the tumor and using corticosteroids, IV-Ig, or plasma exchange.

B. Paraneoplastic Limbic Encephalitis (LE)

Patients with paraneoplastic LE present with anxiety, depression, confusion, delirium, hallucinations, seizures, or short-term memory loss (Gultekin et al., 2000).

Table 32.4 Diagnostic Criteria of PNS of the CNS

Definite PNS
1) Classical syndrome *with* cancer diagnosed within 5 years of neurologic symptom development.
2) Nonclassical syndrome that resolves or significantly improves after cancer treatment.
3) Nonclassical syndrome *with* cancer diagnosed within 5 years of neurologic symptom development *and* positive antineuronal antibodies.
4) Neurological syndrome (classical or not) *without* cancer and *with* well-characterized antineuronal antibodies.
Possible PNS
1) Classical syndrome *with* high risk of cancer, *without* antineuronal antibodies.
2) Neurological syndrome (classical or not) *without* cancer and *with* partially-characterized antineuronal antibodies.
3) Nonclassical syndrome *with* cancer diagnosed within 2 years of neurologic symptom development, *without* antineuronal antibodies.

Recommended diagnostic criteria for paraneoplastic neurologic syndromes (guidelines from the Paraneoplastic Neurological Syndrome Euronetwork (Graus et al. 2004); www.pnseuronet.org/about/).

In approximately 80 percent of patients the MRI T2 and FLAIR sequences show hyperintense symmetric or asymmetric abnormalities involving the medial temporal lobes (see Figure 32.5). The brain FDG-PET frequently shows uptake of FDG in the medial temporal lobes, even when the MRI is normal (Fakhoury et al., 1999). Almost all patients have an abnormal EEG that includes uni- or bilateral temporal lobe epileptic discharges, or slow background activity. The combination of clinical, MRI, EEG, and CSF findings, along with antineuronal antibody testing, identify most cases of paraneoplastic LE (Lawn et al., 2003). Recent studies have shown three groups of immune-mediated LE, each likely including several subphenotypes (Ances et al., 2005):

1. **LE associated with antibodies to intracellular paraneoplastic antigens (Hu, Ma2, CV2/ CRMP5).** This group of disorders is mediated by cytotoxic T-cell responses and, in general, they are poorly responsive to treatment. An exception is the subgroup of patients with Ma2 antibodies, in which 30 percent respond to treatment of the tumor and immunotherapy (corticosteroids and IVIg). In addition to LE, each autoimmunity may associate with other syndromes:

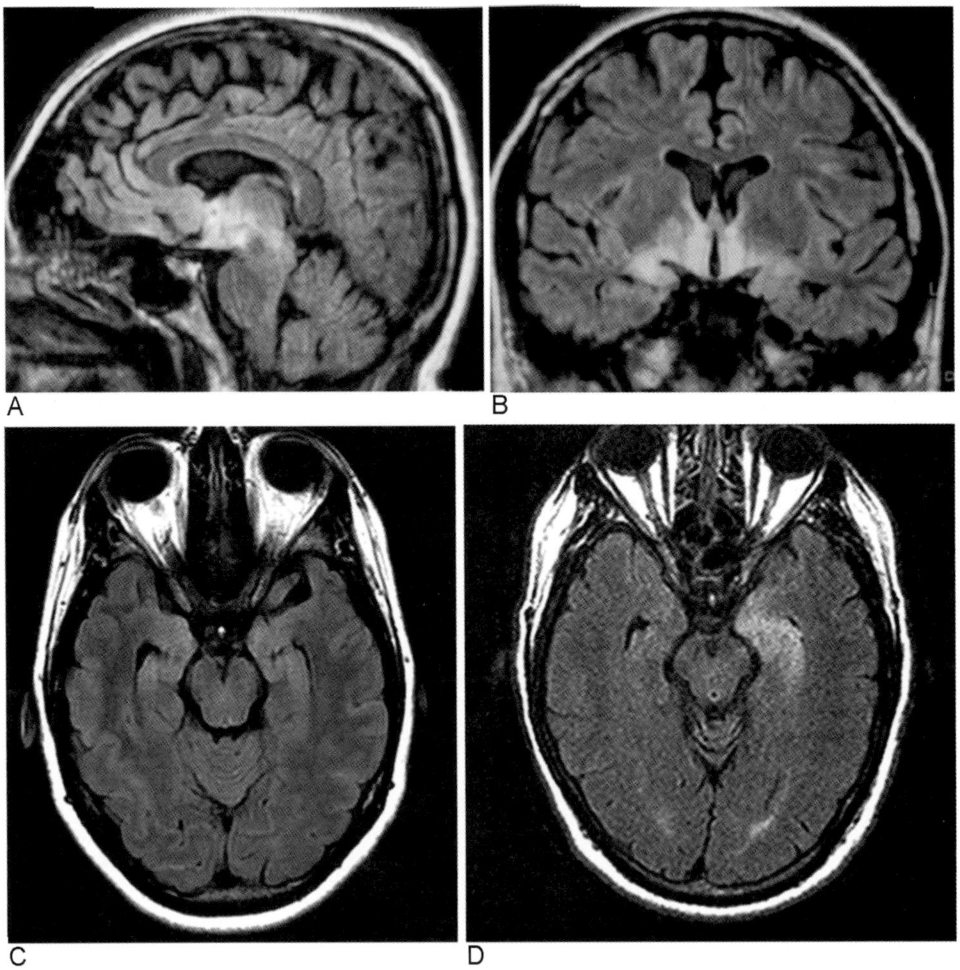

Figure 32.5 MRI of three patients with antibody-associated LE. **A, B.** Sagittal and coronal MRI FLAIR sequences of a patient with paraneoplastic anti-Ma2-associated encephalitis and a germ-cell tumor of the testis. Note that the MRI hyperintense abnormalities predominantly involve the medial temporal lobes, diencephalon (hypothalamus), and upper brainstem. These findings are typical of this disorder. **C.** Axial MRI FLAIR sequence of a patient with LE, ovarian teratoma, and hippocampal neuropil antibodies. There is mild bilateral FLAIR hyperintensity in the medial temporal lobes. **D.** Patient with VGKC antibodies and nonparaneoplastic LE. The MRI shows asymmetric involvement of the medial temporal lobes.

a. *Anti-Hu*. Patients with this immunity may develop dysfunction of any part of the nervous system resulting in multifocal symptoms (encephalomyelitis). Other than the limbic system, areas frequently affected include brainstem, cerebellum, spinal cord, dorsal root ganglia (sensory neuronopathy), and autonomic centers and nerves, resulting in life-threatening gastrointestinal paresis, orthostatic hypotension, or cardiac dysrhythmias. The most frequently associated tumor is SCLC (Graus et al., 2001).

b. *Anti-Ma2*. About 90 percent of patients with this immunity develop dysfunction of the limbic system, diencephalon, or upper brainstem (Rosenfeld et al., 2001). Therefore, in addition to the clinical picture of LE described earlier, patients may develop hypothalamic symptoms (excessive day time sleepiness with low CSF hypocretin levels, and hormonal deficits), complex supranuclear and nuclear ophthalmoparesis that usually presents with vertical gaze limitation, and a syndrome characterized by severe rigidity and hypokinesis. A few patients develop opsoclonus or cerebellar ataxia (Dalmau et al., 2004). The MRI findings often correlate with the clinical syndrome, and may show contrast enhancement. In men younger than 50 years, the tumor is almost always in the testis (germ-cell tumor); in other patients, the tumor association is diverse (cancer of the lung, breast, ovary, among others).

c. *Anti-CV2/CRMP5*. These patients may develop LE, but more frequently show cerebellar ataxia, axonal sensorimotor neuropathy, chorea, uveitis, and optic neuritis (Antoine et al., 2001; Vernino et al., 2002; Yu et al., 2001). For this reason, the MRI studies may show abnormalities involving the medial temporal lobes combined with FLAIR and T2 abnormalities in others areas of the CNS; those with chorea frequently show hyperintensities in the striatum and caudate. The tumors more frequently involved are SCLC and thymoma (Antoine et al., 2001). In patients with SCLC, CV2/CRMP5 antibodies often occur in association with anti-Hu antibodies.

2. **LE associated with antibodies to VGKC**. These patients may present with the typical features of LE, but when compared with other immunotypes, they are more likely to develop hyponatremia, and less likely to have CSF pleocytosis, elevated IgG index, or intrathecal synthesis of specific antibodies (Ances et al., 2005; Thieben et al., 2004; Vincent et al., 2004). Only 20 percent of patients with VGKC antibodies have an underlying tumor (usually SCLC or thymoma); for the other patients the trigger of the immune response is unknown. The brain MRI findings are similar to other types of LE; in some instances the radiologic involvement appears unilateral or can affect basal ganglia (Thieben et al., 2004).

Approximately 80 percent of patients with LE and VGKC antibodies respond to treatment, including corticosteroids, plasma exchange, or IV-Ig. Some patients have spontaneous improvement of symptoms. Other than LE, patients with VGKC can develop peripheral nerve hyperexcitability, autonomic dysfunction, hyperhydrosis, rapid eye movement sleep behavior abnormalities, and seizures (Iranzo et al., 2005; Liguori et al., 2001; Vincent et al., 2004).

3. **Antibodies to novel hippoccampal neuropil antigens**. This is a heterogeneous group of disorders that may encompass multiple antigens. The common clinical phenotype includes predominant behavioral and psychiatric symptoms (that may obscure short-term memory deficits), seizures, and brain MRI abnormalities that are less frequently restricted to the hippocampus than in classical LE (Ances et al., 2005). A subphenotype that includes young women with ovarian teratoma frequently develops central hypoventilation. FDG-PET may reveal foci of hypermetabolism in the frontotemporal lobes, brainstem, or cerebellum. Combining MRI and FDG-PET studies, the temporal lobes are preferentially affected. Patients harboring these novel antibodies are more likely to have intrathecal IgG synthesis, CSF pleocytosis, and systemic tumors (thymoma, ovarian teratoma) than those with VGKC antibodies, and do not develop hyponatremia (Vitaliani et al., 2005). The molecular identity of the antigens is unknown, but by using immunohistochemical techniques they are highly enriched, with specific patterns of expression in the hippocampus and sometimes, molecular layer of the cerebellum (see Figure 32.6).

In contrast to patients with antibodies to intracellular antigens, the encephalitis of patients with hippocampal neuropil antibodies improves with immunotherapy and, if present, treatment of the associated tumor. As occurs with patients with LE and VGKC antibodies, the clinical improvement associates with improvement of MRI and FDG-PET abnormalities and a decrease of antibody titers (Ances et al., 2005).

C. Paraneoplastic Sensory Neuronopathy

Patients with this disorder develop pain, numbness, and sensory deficits that can affect limbs, trunk, and cranial nerves, including hearing loss (Horwich et al., 1977). In addition to the sensory loss, the resulting syndrome includes sensory ataxia and decreased or absent reflexes. Nerve conduction studies show decreased or absent sensory nerve action potentials with normal or near-normal motor-conduction velocities (Camdessanche et al., 2002). Paraneoplastic sensory neuronopathy results from an immune attack against the neurons of the dorsal root ganglia, which frequently is triggered by a SCLC. Pathological studies show prominent infiltrates of T-cells and neuronal degeneration. This disorder

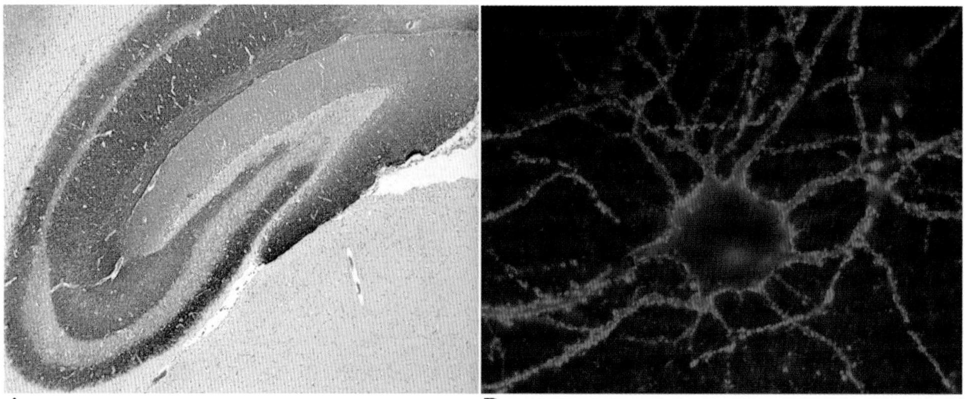

Figure 32.6 Immunohistochemical features of hippocampal neuropil antibodies from a patient with LE and carcinoma of the thymus. **A.** Saggital section of rat brain reacted with the CSF of the patient. Note the intense reactivity (brown staining) predominantly involving the neuropil (dendrites, synaptic rich regions) of the hippocampus. Different from classical paraneoplastic antibodies that usually react with intracellular antigens (see Figure 32.4), the target antigens of the hippocampal neuropil antibodies are located on the cell surface. **B.** Reactivity of the patient's CSF antibodies with live rat hippocampal neurons. There is intense reactivity with the cell surface of the neuronal processes (**A** ×5 counterstained with hematoxylin; **B** ×800 immunofluorescence, oil).

may occur in isolation, but often precedes or coincides with the development of encephalomyelitis, suggesting a common pathogenic mechanism. In both instances, detection of anti-Hu antibodies is frequent (Molinuevo et al., 1998). Sensorimotor neuropathies associated with anti-Hu antibodies often result from mixed involvement of dorsal root ganglia and peripheral nerves (Oh et al., 2005). In the latter, serum CV2/CRMP5 antibodies can also be present (Camdessanche et al., 2002). The anti-Hu associated sensory neuronopathy may stabilize or improve with prompt treatment of the tumor and corticosteroids (Oh et al.,1997).

D. Paraneoplastic Opsoclonus-Myoclonus

Opsoclonus is an eye movement disorder characterized by chaotic, conjugate, arrhythmic, and multidirectional saccades. These symptoms often associate with myoclonus and truncal ataxia. The tumors more frequently involved are SCLC, and breast and gynecological cancers in adults, and neuroblastoma in children. The majority of patients do not harbor specific paraneoplastic antibodies, but studies (immunohistochemistry, flow cytometry, cDNA library screening) show that many patients develop antibodies to cell surface and intracellular neuronal proteins of unknown identity (Antunes et al., 2000; Bataller et al., 2003; Blaes et al., 2005).

The best characterized antibody associated with opsoclonus-myoclonus is anti-Ri. Patients with anti-Ri constitute a minority (10%) of all cases of paraneoplastic opsoclonus-myoclonus. Anti-Ri usually associates with cancer of the breast, gynecological tumors, and SCLC (Luque et al., 1991). The target antigens are Nova 1 and 2 (Buckanovich et al., 1993) (see Table 32.3). Paraneoplastic opsoclonus may respond to immunotherapy, but improvement depends on tumor control (Bataller et al., 2001). The neuroblastic tumors of children with opsoclonus-myoclonus have a better prognosis than those without paraneoplastic syndromes; this has been attributed to the anti-tumor immune response (Korfei et al., 2005) and tumor-related genetic factors (Gambini et al., 2003).

E. Lambert-Eaton Myasthenic Syndrome (LEMS)

LEMS is one of the most characteristic PNS of the peripheral nervous system. The disorder results from antibodies against the presynaptic P/Q-type VGCC, which block the entry of calcium and interfere with the release of acetylcholine vesicles. The transfer of IgG from patients with LEMS into mice reproduces the clinical and electrophysiologic features of the disease (Fukunaga et al., 1983; Lang et al., 1987). The detection of antibodies to P/Q-type VGCC is used as a highly sensitive and specific serologic test for LEMS (Motomura et al., 1995).

Patients with LEMS develop proximal muscle weakness and mild involvement of cranial nerves that may result in transient ptosis and diplopia. The onset of symptoms is usually gradual over months, but occasionally can be more acute. Patients complain of fatigue, leg weakness, muscle aches, and vague paresthesias. Reflexes usually are decreased or abolished and enhance after brief muscle contraction (O'Neill, Murray & Newsom-Davis, 1988). Similarly, after brief exercise the strength may improve. Dry mouth and other symptoms of autonomic dysfunction occur in 90 percent of the patients.

Approximately 60 percent of patients with LEMS have an underlying neoplasm, usually SCLC, or rarely other tumors such as lymphoma (Wirtz et al., 2002). Paraneoplastic LEMS may associate with PCD and paraneoplastic encephalomyelitis (Mason et al., 1997). Treatment of the tumor and medication that enhances acetylcholine release (3,4-diaminopyridine, or combination of pyridostigmine and guanidine) are usually effective (Sanders et al., 2000). IV-Ig and plasma exchange improve symptoms within two to four weeks, but the benefit is transient. Long-term immunotherapy with prednisone or azathioprine is an alternative for patients who do not improve with 3,4-diaminopyridine (Newsom-Davis, 2003).

VI. The Clinical Manifestations of Paraneoplastic Immunity: Effects on the Tumor

The concept that most PNS result from immune responses triggered by the tumor expression of neuronal proteins carries the implication that the immune response also attacks the tumor cells. Furthermore, the small size and limited metastatic burden of tumors associated with PNS have suggested that the anti-tumor immune response is clinically effective in preventing tumor growth (Darnell & Posner, 2003). However, the data supporting the efficacy of the immune response is scarce and controversial. There are only two clinical series, both in patients with SCLC, supporting this hypothesis (Graus et al., 1997; Maddison et al., 2001). One study was conducted in patients without PNS, and showed that those with low titers of anti-Hu antibodies were more likely to have limited tumor stage and better response to chemotherapy than patients without antibodies (Graus et al., 1997). In this study the strongest association was between the presence of anti-Hu antibodies and increased response to chemotherapy; the mechanisms underlying the chemosensitivity were not examined. The second study compared SCLC patients with and without LEMS, and found that patients with LEMS had longer survival (Maddison et al., 2001). Because most of these patients developed the neurologic disorder before the tumor diagnosis, a lead-time bias could not be ruled out.

Other data supporting clinically effective anti-tumor immunity derives from cases of spontaneous tumor regression (Byrne et al., 1997; Darnell & DeAngelis, 1993; Gill et al., 2003; Zaheer et al., 1993). However, some of these patients received oncologic treatment and therefore the tumor regression was not truly spontaneous (Darnell & DeAngelis, 1993), others lacked pathological demonstration of a tumor (Byrne et al., 1997), and some did not have detectable immune responses or long-term follow-up (Gill et al., 2003; Zaheer et al., 1993).

Rather than effective antitumor immunity, most clinical series of PNS suggest that the immune response is triggered at early stages of cancer development and that the neurologic symptoms forewarn of the presence of a tumor that otherwise would have manifested clinically months or years later. In these studies most patients developed PNS before the tumor diagnosis, and none had spontaneous tumor regression (Dalmau et al., 1992; Graus et al., 2001; Hetzel et al., 1990; Lucchinetti et al., 1998; Peterson et al., 1992; Rojas et al., 2000; Rojas-Marcos et al., 2003; Sillevis et al., 2002). Furthermore, two series of patients with PNS that examined the effect of immunotherapy on tumor growth showed that it did not favor tumor progression as one would expect if the antitumor immune response was clinically effective (Keime-Guibert et al., 1999; Vernino et al., 2004).

A recent study of patients with paraneoplastic anti-Ma2 associated encephalitis suggests a model of what probably occurs in most PNS triggered by the tumor expression of neuronal proteins. This study examined 25 men with anti-Ma2-associated encephalitis younger than 50 years, and found that 19 had testicular germ-cell tumors (Mathew et al., 2006). The other six patients underwent extensive cancer screening but no tumor was found. Due to relentless progression of neurological symptoms, detection of Ma2 antibodies, and development of subtle changes (microcalcifications) in the ultrasound of the testes, these six patients eventually had orchiectomy, revealing in all instances a microscopic carcinoma *in situ* or intratubular germ-cell neoplasm of the testis. This type of neoplasm is considered a common precursor of most testicular cancers, and takes approximately five years to become invasive (Montironi, 2002). The orchiectomy specimens also showed inflammatory infiltrates and tubular fibrosis, resulting in areas of scar tissue, resembling a "burnt out" tumor. Findings from this study, facilitated by the location of the tumor in an easily accessible organ, suggest (1) that the paraneoplastic immune response can be triggered at the microscopic, preinvasive stage of tumor development, (2) the antitumor immunity may be partially effective, and (3) in most patients the antitumor effects are not sustained enough to prevent tumor growth.

VII. Treatment of Paraneoplastic Syndromes

The treatment of most PNS of the CNS is complicated by:

- ▲ The lack of animal models for all classical syndromes; only two disorders have been modeled: a rare subphenotype of cerebellar degeneration (Sillevis et al., 2000), and stiff-person syndrome (Sommer et al., 2005)
- ▲ The involvement of cytotoxic T-cell mechanisms that make these disorders resistant to strategies used for antibody-mediated disorders

▲ The rapid progression of symptoms and narrow window for intervention to prevent irreversible neurologic deficits

▲ The logistics of combining oncologic and immunosuppressive therapies in patients with poor clinical condition.

These limitations contrast with the frequent response to treatment of PNS of the peripheral nervous system, such as LEMS (as well as myasthenia gravis and neuromyotonia, not included in this review), in which the antigens are located on the cell surface and the disorder is directly mediated by antibodies.

Keeping these considerations in mind, the general treatment approach for PNS of the CNS relies on prompt diagnosis and treatment of the tumor. Because the simultaneous use of chemotherapy and some immunosuppressants may result in significant toxicity, two levels of immunologic intervention are suggested. Patients with progressive PNS who are receiving chemotherapy should be considered for immunosuppression or immunomodulation that may include oral or intravenous corticosteroids, IV-Ig, or plasma exchange. Patients with progressive PNS, who are not receiving chemotherapy, should be considered for more aggressive immunosuppression that may include oral or intravenous cyclophosphamide, tacrolimus, cyclosporine, or rituximab (Keime-Guibert et al., 2000; Shams'ili et al., 2006; Vernino et al., 2004). Although there is no compelling evidence than any of these immunosuppressants is better than another for patients with PNS, we favor the use of corticosteroids, IV-Ig, and cyclophosphamide.

Considering the critical role of the cytotoxic T-cell response in many PNS of the CNS and that the immune response is triggered systemically by the tumor, any strategy that prevents T-cells from endothelial transmigration into the CNS (i.e., antibodies to α-4 integrins or VLA-4) should be considered in future clinical trials.

References

Albert, M. L., Darnell, J. C., Bender, A., Francisco, L. M., Bhardwaj, N., and Darnell, R. B. (1998). Tumor-specific killer cells in paraneoplastic cerebellar degeneration. *Nat Med* **11**, 1321–1324.

Ances, B. M., Vitaliani, R., Taylor, R. A., Liebeskind, D. S., Voloschin, A., Houghton, D. J. et al. (2005). Treatment-responsive limbic encephalitis identified by neuropil antibodies: MRI and PET correlates. *Brain* **128**, 1764–1777.

Antoine, J. C., Honnorat, J., Camdessanche, J. P., Magistris, M., Absi, L., Mosnier, J. F. et al. (2001). Paraneoplastic anti-CV2 antibodies react with peripheral nerve and are associated with a mixed axonal and demyelinating peripheral neuropathy. *Ann Neurol* **49**, 214–221.

Antunes, N. L., Khakoo, Y., Matthay, K. K., Seeger, R. C., Stram, D. O., Gerstner, E. et al. (2000). Antineuronal antibodies in patients with neuroblastoma and paraneoplastic opsoclonus-myoclonus. *J Pediatr Hematol Oncol* **22**, 315–320.

Bataller, L. and Dalmau, J. O. (2004). Paraneoplastic disorders of the central nervous system: Update on diagnostic criteria and treatment. *Semin Neurol* **24**, 461–471.

Bataller, L., Graus, F., Saiz, A., and Vilchez, J. J. (2001). Clinical outcome in adult onset idiopathic or paraneoplastic opsoclonus-myoclonus. *Brain* **124**, 437–443.

Bataller, L., Rosenfeld, M. R., Graus, F., Vilchez, J. J., Cheung, N. K., and Dalmau, J. (2003). Autoantigen diversity in the opsoclonus-myoclonus syndrome. *Ann Neurol* **53**, 347–353.

Bataller, L., Wade, D. F., Graus, F., Stacey, H. D., Rosenfeld, M. R., and Dalmau, J. (2004). Antibodies to Zic4 in paraneoplastic neurologic disorders and small-cell lung cancer. *Neurology* **62**, 778–782.

Benyahia, B., Liblau, R., Merle-Beral, H., Tourani, J. M., Dalmau, J., and Delattre, J. Y. (1999). Cell-mediated autoimmunity in paraneoplastic neurological syndromes with anti-Hu antibodies. *Ann Neurol* **45**, 162–167.

Bernal, F., Graus, F., Pifarre, A., Saiz, A., Benyahia, B., and Ribalta, T. (2002). Immunohistochemical analysis of anti-Hu-associated paraneoplastic encephalomyelitis. *Acta Neuropathol (Berl)* **103**, 509–515.

Bernal, F., Shams'ili, S., Rojas, I., Sanchez-Valle, R., Saiz, A., Dalmau, J. et al. (2003). Anti-Tr antibodies as markers of paraneoplastic cerebellar degeneration and Hodgkin's disease. *Neurology* **60**, 230–234.

Blaes, F., Fuhlhuber, V., Korfei, M., Tschernatsch, M., Behnisch,W., Rostasy, K. et al. (2005). Surface-binding autoantibodies to cerebellar neurons in opsoclonus syndrome. *Ann Neurol* **58**, 313–317.

Blumenthal, D. T., Salzman, K., Digre, K. B., Jensen, R. L., Dunson, W., and Dalmau, J. (2006). Early pathological findings and long-term improvement in anti-Ma2 associated encephalitis. *Neurology,* **67**, 146–149.

Bretin, S., Reibel, S., Charrier, E., Maus-Moatti, M., Auvergnon, N., Thevenoux, A. et al. (2005). Differential expression of CRMP1, CRMP2A, CRMP2B, and CRMP5 in axons or dendrites of distinct neurons in the mouse brain. *J Comp Neurol* **486**, 1–17.

Bryniarski, K., Szczepanik, M., Maresz, K., Ptak, M., and Ptak, W. (2004). Subpopulations of mouse testicular macrophages and their immunoregulatory function. *Am J Reprod Immunol* **52**, 27–35.

Buckanovich, R. J., Posner, J. B., and Darnell, R. B. (1993). Nova, the paraneoplastic Ri antigen, is homologous to an RNA-binding protein and is specifically expressed in the developing motor system. *Neuron* **11**, 657–672.

Byrne, T., Mason, W. P., Posner, J. B., and Dalmau, J. (1997). Spontaneous neurological improvement in anti-Hu associated encephalomyelitis. *J Neurol Neurosurg Psychiatry* **62**, 276–278.

Camdessanche, J. P., Antoine, J. C., Honnorat, J., Vial, C., Petiot, P., Convers, P., and Michel, D. (2002). Paraneoplastic peripheral neuropathy associated with anti-Hu antibodies. A clinical and electrophysiological study of 20 patients. *Brain* **125**, 166–175.

Chung, S. M., Jiang, L., Cheng, S., and Furneaux, H. (1996). Purification and properties of HuD, a neuronal RNA-binding protein. *J Biol Chem* **271**, 11518–11524.

Dalmau, J., Graus, F., Rosenblum, M. K., and Posner, J. B. (1992). Anti-Hu–associated paraneoplastic encephalomyelitis/sensory neuronopathy. A clinical study of 71 patients. *Medicine (Baltimore)* **71**, 59–72.

Dalmau, J., Graus, F., Villarejo, A., Posner, J. B., Blumenthal, D., Thiessen, B. et al. (2004). Clinical analysis of anti-Ma2-associated encephalitis. *Brain* **127**, 1831–1844.

Darnell, R. B. and DeAngelis, L. M. (1993). Regression of small-cell lung carcinoma in patients with paraneoplastic neuronal antibodies. *Lancet* **341**, 21–22.

Darnell, R. B. and Posner, J. B. (2003). Observing the invisible: Successful tumor immunity in humans. *Nat Immunol* **4**, 201.

Dimopoulos, M. A., Panayiotidis, P., Moulopoulos, L. A., Sfikakis, P., and Dalakas, M. (2000). Waldenstrom's macroglobulinemia: Clinical features, complications, and management. *J Clin Oncol* **18**, 214.

Dispenzieri, A. and Gertz, M. A. (2005). Treatment of Castleman's disease. *Curr Treat Options Oncol* **6**, 255–266.

Dredge, B. K., Stefani, G., Engelhard, C. C., and Darnell, R. B. (2005). Nova autoregulation reveals dual functions in neuronal splicing. *EMBO J* **24**, 1608–1620.

Elrington, G. M., Murray, N. M., Spiro, S. G., and Newsom-Davis, J. (1991). Neurological paraneoplastic syndromes in patients with small cell lung cancer: A prospective survey of 150 patients. *J Neurol Neurosurg Psychiatry* **54**, 764–767.

Fakhoury, T., Abou-Khalil, B., and Kessler, R. M. (1999). Limbic encephalitis and hyperactive foci on PET scan. *Seizure* **8**, 427–431.

Farsad, K., Slepnev, V., Ochoa, G., Daniell, L., Haucke, V., and De Camilli, P. (2003). A putative role for intramolecular regulatory mechanisms in the adaptor function of amphiphysin in endocytosis. *Neuropharmacology* **45**, 87–796.

Fukunaga, H., Engel, A. G., Lang, B., Newsom-Davis, J., and Vincent, A. (1983). Passive transfer of Lambert-Eaton myasthenic syndrome with IgG from man to mouse depletes the presynaptic membrane active zones. *Proc Natl Acad Sci* **80**, 7636–7640.

Furneaux, H. M., Rosenblum, M. K., Dalmau, J., Wong, E., Woodruff, P., Graus, F., and Posner, J. B. (1990). Selective expression of Purkinje-cell antigens in tumor tissue from patients with paraneoplastic cerebellar degeneration. *N Engl J Med* **322**, 1844–1851.

Gambini, C., Conte, M., Bernini, G., Angelini, P., Pession, A., Paolucci, P. et al. (2003). Neuroblastic tumors associated with opsoclonus-myoclonus syndrome: Histological, immunohistochemical and molecular features of 15 Italian cases. *Virchows Arch* **442**, 555–562.

Gherardi, R. K., Belec, L., Soubrier, M., Malapert, D., Zuber, M., Viard, J. P. et al. (1996). Overproduction of proinflammatory cytokines imbalanced by their antagonists in POEMS syndrome. *Blood* **87**, 1458–1465.

Gill, S., Murray, N., Dalmau, J., and Thiessen, B. (2003). Paraneoplastic sensory neuronopathy and spontaneous regression of small cell lung cancer. *Can J Neurol Sci* **30**, 269–271.

Graus, F., Dalmau, J., Rene, R., Tora, M., Malats, N., Verschuuren, J. J. et al. (1997). Anti-Hu antibodies in patients with small-cell lung cancer: Association with complete response to therapy and improved survival. *J Clin Oncol* **15**, 2866–2872.

Graus, F., Delattre, J. Y., Antoine, J. C., Dalmau, J., Giometto, B., Grisold, W. et al. (2004). Recommended diagnostic criteria for paraneoplastic neurological syndromes. *J Neurol Neurosurg Psychiatry* **75**, 1135–1140.

Graus, F., Gultekin, S. H., Ferrer, I., Reiriz, J., Alberch, J., and Dalmau, J. (1998). Localization of the neuronal antigen recognized by anti-Tr antibodies from patients with paraneoplastic cerebellar degeneration and Hodgkin's disease in the rat nervous system. *Acta Neuropathol (Berl)* **96**, 1–7.

Graus, F., Illa, I., Agusti, M., Ribalta, T., Cruz-Sanchez, F., and Juarez, C. (1991). Effect of intraventricular injection of an anti-Purkinje cell antibody (anti-Yo) in a guinea pig model. *J Neurol Sci* **106**, 82–87.

Graus, F., Keime-Guibert, F., Rene, R., Benyahia, B., Ribalta, T., Ascaso, C. et al. (2001). Anti-Hu-associated paraneoplastic encephalomyelitis: Analysis of 200 patients. *Brain* **124**, 1138–1148.

Graus, F., Lang, B., Pozo-Rosich, P., Saiz, A., Casamitjana, R., and Vincent, A. (2002). P/Q type calcium-channel antibodies in paraneoplastic cerebellar degeneration with lung cancer. *Neurology* **59**, 764–766.

Gultekin, S. H., Rosenfeld, M. R., Voltz, R., Eichen, J., Posner, J. B., and Dalmau, J. (2000). Paraneoplastic limbic encephalitis: Neurological symptoms, immunological findings and tumour association in 50 patients. *Brain* **123**, 1481–1494.

Hart, I. K., Maddison, P., Newsom-Davis, J., Vincent, A., and Mills, K. R. (2002). Phenotypic variants of autoimmune peripheral nerve hyperexcitability. *Brain* **125**, 1887–1895.

Hetzel, D. J., Stanhope, C. R., O'Neill, B. P., and Lennon, V. A. (1990). Gynecologic cancer in patients with subacute cerebellar degeneration predicted by anti-Purkinje cell antibodies and limited in metastatic volume. *Mayo Clin Proc* **65**, 1558–1563.

Horwich, M. S., Cho, L., Porro, R. S., and Posner, J. B. (1977). Subacute sensory neuropathy: A remote effect of carcinoma. *Ann Neurol* **2**, 7–19.

Ilyas, A. A., Cook, S. D., Dalakas, M. C., and Mithen, F. A. (1992). Anti-MAG IgM paraproteins from some patients with polyneuropathy associated with IgM paraproteinemia also react with sulfatide. *J Neuroimmunol* **37**, 85–92.

Iranzo, A., Graus, F., Clover, L., Morera, J., Bruna, J., Vilar, C. et al. (2005). Rapid eye movement sleep behavior disorder and potassium channel antibody-associated limbic encephalitis. *Ann Neurol.*

Jean, W. C., Dalmau, J., Ho, A., and Posner, J. B. (1994). Analysis of the IgG subclass distribution and inflammatory infiltrates in patients with anti-Hu-associated paraneoplastic encephalomyelitis. *Neurology* **44**, 140–147.

Keime-Guibert, F., Graus, F., Broet, P., Rene, R., Molinuevo, J. L., Ascaso, C., and Delattre, J. Y. (1999). Clinical outcome of patients with anti-Hu-associated encephalomyelitis after treatment of the tumor. *Neurology* **53**, 1719–1723.

Keime-Guibert, F., Graus, F., Fleury, A., Rene, R., Honnorat, J., Broet, P., and Delattre, J. Y. (2000). Treatment of paraneoplastic neurological syndromes with antineuronal antibodies (anti-Hu, anti-Yo) with a combination of immunoglobulins, cyclophosphamide, and methylprednisolone. *J Neurol Neurosurg Psychiatry* **68**, 479–482.

Korfei, M., Fuhlhuber, V., Schmidt-Woll, T., Kaps, M., Preissner, K. T., and Blaes, F. (2005). Functional characterization of autoantibodies from patients with pediatric opsoclonus-myoclonus syndrome. *J Neuroimmunol* **170**, 150–157.

Lang, B., Newsom-Davis, J., Peers, C., Prior, C., and Wray, D. W. (1987). The effect of myasthenic syndrome antibody on presynaptic calcium channels in the mouse. *J Physiol* **390**, 257–270.

Lang, B., Pinto, A., Giovannini, F., Newsom-Davis, J., and Vincent, A. (2003). Pathogenic autoantibodies in the Lambert-Eaton myasthenic syndrome. *Ann N Y Acad Sci* **998**, 187–195.

Lawn, N. D., Westmoreland, B. F., Kiely, M. J., Lennon, V. A., and Vernino, S. (2003). Clinical, magnetic resonance imaging, and electroencephalographic findings in paraneoplastic limbic encephalitis. *Mayo Clin Proc* **78**, 1363–1368.

Liguori, R., Vincent, A., Clover, L., Avoni, P., Plazzi, G., Cortelli, P. et al. (2001). Morvan's syndrome: Peripheral and central nervous system and cardiac involvement with antibodies to voltage-gated potassium channels. *Brain* **124**, 2417–2426.

Llado, A., Mannucci, P., Carpentier, A. F., Paris, S., Blanco, Y., Saiz, A. et al. (2004). Value of Hu antibody determinations in the follow-up of paraneoplastic neurologic syndromes. *Neurology* **63**, 1947–1949.

Lucchinetti, C. F., Kimmel, D. W., and Lennon, V. A. (1998). Paraneoplastic and oncologic profiles of patients seropositive for type 1 antineuronal nuclear autoantibodies. *Neurology* **50**, 652–657.

Luque, F. A., Furneaux, H. M., Ferziger, R., Rosenblum, M. K., Wray, S. H., Schold, S. C. et al. (1991). Anti-Ri: An antibody associated with paraneoplastic opsoclonus and breast cancer. *Ann Neurol* **29**, 241–251.

Maddison, P., Lang, B., Mills, K., and Newsom-Davis, J. (2001). Long term outcome in Lambert-Eaton myasthenic syndrome without lung cancer. *J Neurol Neurosurg Psychiatry* **70**, 212–217.

Mason, W. P., Graus, F., Lang, B., Honnorat, J., Delattre, J. Y., Valldeoriola, F. et al. (1997). Small-cell lung cancer, paraneoplastic cerebellar degeneration and the Lambert-Eaton myasthenic syndrome. *Brain* **120**, 1279–1300.

Mathew, R. M., Vandenberghe, R., Garcia-Merino, A., Yamamoto, T., Landolfi, J. C., Rosenfeld, M. R., et al. (2007). Orchiectomy for suspected microscopic tumor in patients with anti-Ma2-associated encephalitis. *Neurology* **68**, 900–905.

Molinuevo, J. L., Graus, F., Serrano, C., Rene, R., Guerrero, A., and Illa, I. (1998). Utility of anti-Hu antibodies in the diagnosis of paraneoplastic sensory neuropathy. *Ann Neurol* **44**, 976–980.

Montironi, R. (2002). Intratubular germ cell neoplasia of the testis: Testicular intraepithelial neoplasia. *Eur Urol* **41**, 651–654.

Motomura, M., Johnston, I., Lang, B., Vincent, A., and Newsom-Davis, J. (1995). An improved diagnostic assay for Lambert-Eaton myasthenic syndrome. *J Neurol Neurosurg Psychiatry* **58**, 85–87.

Motomura, M., Lang, B., Johnston, I., Palace, J., Vincent, A., and Newsom-Davis, J. (1997). Incidence of serum anti-P/O-type and anti-N-type calcium channel autoantibodies in the Lambert-Eaton myasthenic syndrome. *J Neurol Sci* **147**, 35–42.

Musunuru, K. and Darnell, R. B. (2001). Paraneoplastic neurologic disease antigens: RNA-binding proteins and signaling proteins in neuronal degeneration. *Annu Rev Neurosci* **24**, 239–262.

Neumann, H., Schmidt, H., Cavalie, A., Jenne, D., and Wekerle, H. (1997). Major histocompatibility complex (MHC) class I gene expression in single neurons of the central nervous system: Differential regulation by interferon (IFN)-gamma and tumor necrosis factor (TNF)-alpha. *J Exp Med* **185**, 305–316.

Newsom-Davis, J. (2003). Therapy in myasthenia gravis and Lambert-Eaton myasthenic syndrome. *Semin Neurol* **23**, 191–198.

Newsom-Davis, J., Buckley, C., Clover, L., Hart, I., Maddison, P., Tuzum, E., and Vincent, A. (2003). Autoimmune disorders of neuronal potassium channels. *Ann N Y Acad Sci* **998**, 202–210.

O'Neill, J. H., Murray, N. M., and Newsom-Davis, J. (1988). The Lambert-Eaton myasthenic syndrome. A review of 50 cases. *Brain* **111**, 577–596.

Oh, S. J., Dropcho, E. J., and Claussen, G. C. (1997). Anti-Hu-associated paraneoplastic sensory neuropathy responding to early aggressive immunotherapy: Report of two cases and review of literature. *Muscle Nerve* **20**, 1576–1582.

Oh, S. J., Gurtekin, Y., Dropcho, E. J., King, P., and Claussen, G. C. (2005). Anti-Hu antibody neuropathy: A clinical, electrophysiological, and pathological study. *Clin Neurophysiol* **116**, 28–34.

Okano, H. J., Park, W. Y., Corradi, J. P., and Darnell, R. B. (1999). The cytoplasmic Purkinje onconeural antigen cdr2 down-regulates c-Myc function: Implications for neuronal and tumor cell survival. *Genes Dev* **13**, 2087–2097.

Peterson, K., Rosenblum, M. K., Kotanides, H., and Posner, J. B. (1992). Paraneoplastic cerebellar degeneration. I. A clinical analysis of 55 anti-Yo antibody-positive patients. *Neurology* **42**, 1931–1937.

Posner, J. B. (1995). Neurologic Complications of Cancer. F. A. Davis Company, Philadelphia.

Rojas, I., Graus, F., Keime-Guibert, F., Rene, R., Delattre, J. Y., Ramon, J. M. et al. (2000). Long-term clinical outcome of paraneoplastic cerebellar degeneration and anti-Yo antibodies. *Neurology* **55**, 713–715.

Rojas-Marcos, I., Rousseau, A., Keime-Guibert, F., Rene, R., Cartalat-Carel, S., Delattre, J. Y., and Graus, F. (2003). Spectrum of paraneoplastic neurologic disorders in women with breast and gynecologic cancer. *Medicine (Baltimore)* **82**, 216–223.

Ropper, A. H. and Gorson, K. C. (1998). Neuropathies associated with paraproteinemia. *N Engl J Med* **338**, 1601–1607.

Rosenfeld, M. R. and Dalmau, J. (2000). Paraneoplastic neurologic disorders and onconeuronal antigens. In Pfaff, D. W., Berrettini,W. H., Joh, T. H., and Maxson, S. C., Eds., *Genetic Influences on Neural and Behavioral Functions*, 217–230. CRC Press, Boca Raton.

Rosenfeld, M. R., Eichen, J. G., Wade, D. F., Posner, J. B., and Dalmau, J. (2001). Molecular and clinical diversity in paraneoplastic immunity to Ma proteins. *Ann Neurol* **50**, 339–348.

Rudnicki, S. A. and Dalmau, J. (2000). Paraneoplastic syndromes of the spinal cord, nerve, and muscle. *Muscle Nerve* **23**, 1800–1818.

Sanders, D. B., Massey, J. M., Sanders, L. L., and Edwards, L. J. (2000). A randomized trial of 3,4-diaminopyridine in Lambert-Eaton myasthenic syndrome. *Neurology* **54**, 603–607.

Scarlato, M., Previtali, S. C., Carpo, M., Pareyson, D., Briani, C., Del Bo, R. et al. (2005). Polyneuropathy in POEMS syndrome: Role of angiogenic factors in the pathogenesis. *Brain* **128**, 1911–1920.

Shams'ili, S., de Beukelaar, J., Gratama, J. W., Hooijkaas, H., Van Den, B. M., Van't Veer, M., and Sillevis, S. P. (2006). An uncontrolled trial of rituximab for antibody associated paraneoplastic neurological syndromes. *J Neurol* **253**, 16–20.

Sillevis Smitt, P. A., Manley, G. T., and Posner, J. B. (1995). Immunization with the paraneoplastic encephalomyelitis antigen HuD does not cause neurologic disease in mice. *Neurology* **45**, 1873–1878.

Sillevis, S. P., Grefkens, J., De Leeuw, B., Van Den, B. M., van Putten, W., Hooijkaas, H., and Vecht, C. (2002). Survival and outcome in 73 anti-Hu positive patients with paraneoplastic encephalomyelitis/sensory neuronopathy. *J Neurol* **249**, 745–753.

Sillevis, S. P., Kinoshita, A., De Leeuw, B., Moll, W., Coesmans, M., Jaarsma, D. et al. (2000). Paraneoplastic cerebellar ataxia due to autoantibodies against a glutamate receptor. *N Engl J Med* **342**, 21–27.

Sommer, C., Weishaupt, A., Brinkhoff, J., Biko, L., Wessig, C., Gold, R., and Toyka, K. V. (2005). Paraneoplastic stiff-person syndrome: Passive transfer to rats by means of IgG antibodies to amphiphysin. *Lancet* **365**, 1406–1411.

Sundstrom, J., Verajnkorva, E., Salminen, E., Pelliniemi, L. J., and Pollanen, P. (1999). Experimental testicular teratoma promotes formation of humoral immune responses in the host testis. *J Reprod Immunol* **42**, 107–126.

Tanaka, K., Tanaka, M., Inuzuka, T., Nakano, R., and Tsuji, S. (1999). Cytotoxic T lymphocyte-mediated cell death in paraneoplastic sensory neuronopathy with anti-Hu antibody. *J Neurol Sci* **163**, 159–162.

Tanaka, K., Tanaka, M., Onodera, O., Igarashi, S., Miyatake, T., and Tsuji, S. (1994). Passive transfer and active immunization with the recombinant leucine-zipper (Yo) protein as an attempt to establish an animal model of paraneoplastic cerebellar degeneration. *J Neurol Sciences* **127**, 153–158.

Tanaka, M., Tanaka, K., Onodera, O., and Tsuji, S. (1995). Trial to establish an animal model of paraneoplastic cerebellar degeneration with anti-Yo antibody. 1. Mouse strains bearing different MHC molecules produce antibodies on immunization with recombinant Yo protein, but do not cause Purkinje cell loss. *Clin Neurol Neurosurg* **97**, 95–100.

Tanaka, M., Tanaka, K., Shinozawa, K., Idezuka, J., and Tsuji, S. (1998). Cytotoxic T cells react with recombinant Yo protein from a patient with paraneoplastic cerebellar degeneration and anti-Yo antibody. *J Neurol Sci* **161**, 88–90.

Thieben, M. J., Lennon, V. A., Boeve, B. F., Aksamit, A. J., Keegan, M., and Vernino, S. (2004). Potentially reversible autoimmune limbic encephalitis with neuronal potassium channel antibody. *Neurology* **62**, 1177–1182.

Vernino, S. and Lennon, V. A. (2004). Autoantibody profiles and neurological correlations of thymoma. *Clin Cancer Res* **10**, 7270–7275.

Vernino, S., O'Neill, B. P., Marks, R. S., O'Fallon, J. R., and Kimmel, D. W. (2004). Immunomodulatory treatment trial for paraneoplastic neurological disorders. *Neurooncol* **6**, 55–62.

Vernino, S., Tuite, P., Adler, C. H., Meschia, J. F., Boeve, B. F., Boasberg, P. et al. (2002). Paraneoplastic chorea associated with CRMP-5 neuronal antibody and lung carcinoma. *Ann Neurol* **51**, 625–630.

Vincent, A., Buckley, C., Schott, J. M., Baker, I., Dewar, B. K., Detert, N. et al. (2004). Potassium channel antibody-associated encephalopathy: A potentially immunotherapy-responsive form of limbic encephalitis. *Brain* **127**, 701–712.

Vitaliani, R., Mason, W., Ances, B., Zwerdling, T., Jiang, Z., and Dalmau, J. (2005). Paraneoplastic encephalitis, psychiatric symptoms, and hypoventilation in ovarian teratoma. *Ann Neurol* **58**, 594–604.

Voltz, R., Dalmau, J., Posner, J. B., and Rosenfeld, M. R. (1998). T-cell receptor analysis in anti-Hu associated paraneoplastic encephalomyelitis. *Neurology* **51**, 1146–1150.

Wirtz, P. W., Smallegange, T. M., Wintzen, A. R., and Verschuuren, J. J. (2002). Differences in clinical features between the Lambert-Eaton myasthenic syndrome with and without cancer: An analysis of 227 published cases. *Clin Neurol Neurosurg* **104**, 359–363.

Younes-Mhenni, S., Janier, M. F., Cinotti, L., Antoine, J. C., Tronc, F., Cottin, V. et al. (2004). FDG-PET improves tumour detection in patients with paraneoplastic neurological syndromes. *Brain* **127**, 2331–2338.

Yu, Z., Kryzer, T. J., Griesmann, G. E., Kim, K.-K., Benarroch, E. E., and Lennon, V. A. (2001). CRMP-5 neuronal autoantibody: Marker of lung cancer and thymoma-related autoimmunity. *Ann Neurol* **49**, 146–154.

Zaheer, W., Friedland, M. L., Cooper, E. B., DoRosario, A., Burd, R. M., Gagliardi, J., and Torstenson, G. (1993). Spontaneous regression of small cell carcinoma of lung associated with severe neuropathy. *Cancer Invest* **11**, 306–309.

33

Mitochondrial Disorders

Salvatore DiMauro and Eric A. Schon

Even making allowance for the biased point of view of the authors, it is still fair to say that much of contemporary neurology is *mitochondrial neurology*, not only because of the extraordinary development in the past two decades of primary mitochondrial encephalomyopathies, that is, neurological disorders due to molecular defects impairing mitochondrial function directly, but also because mitochondrial dysfunction is involved in the pathogenic mechanism of most, if not all, neurodegenerative disorders.

It is also fair to say that few areas of medicine have seen as many conceptual innovations as mitochondrial medicine, first and foremost the concept of endosymbiosis, which was proposed at the beginning of last century but was popularized in 1967 by Lynn Sagan Margulis (Sagan, 1967). Endosymbiosis occurred about two billion years ago, when the atmosphere of the earth, until then composed of hydrogen, ammonia, and methane, became increasingly rich in oxygen and hydrogen, toxic compounds to most organisms. Early eukaryotic cells, threatened by this poisonous environment, were invaded—and rescued—by bacteria, which had not only adapted to the oxygen-rich atmosphere but had developed means by which oxygen became the terminal electron acceptor of a pathway (the respiratory chain) that "detoxified" it to produce water and generated much more energy than anaerobic glycolysis. This biological event is oddly reminiscent of Prometheus' mythical gift of fire to humans.

A corollary of endosymbiosis is that all eukaryotic cells, including ours, still contain, in addition to their original nuclear DNA (nDNA), a genetic "relic" of the protobacterial invaders (a.k.a. mitochondria), namely mitochondrial DNA (mtDNA). As two genomes in one cell cannot function any better than two governments in one country, it is hardly surprising that, in the course of evolution, mtDNA has given up much of its autonomy (in fact, most of its genes have actually been transferred to the nuclear genome) and it has become the slave of nDNA. What is more surprising is that clinical scientists ignored mtDNA—whose presence was documented in the early 1960s (Nass & Nass, 1963; Schatz et al., 1964)—until 1988, when the first mutations in mtDNA were associated with human disease (Holt et al., 1988; Wallace et al., 1988).

Because mtDNA is a relic, not a fossil, it encodes 13 functional proteins, all components of the respiratory chain. However, mitochondria contain almost 1,300 proteins (and counting), essentially all encoded by nDNA, which subserve a wide variety of function besides the provision of energy. About half of them have "housekeeping" functions, such as protein import and sorting, protein translation and stability, and nucleic acid metabolism. The rest of the mitochondrial proteins perform specialized functions beyond ATP production, including lipid metabolism, the TCA cycle, the urea cycle reactions, stress response, and one version of programmed cell death (apoptosis) (see Figure 33.1).

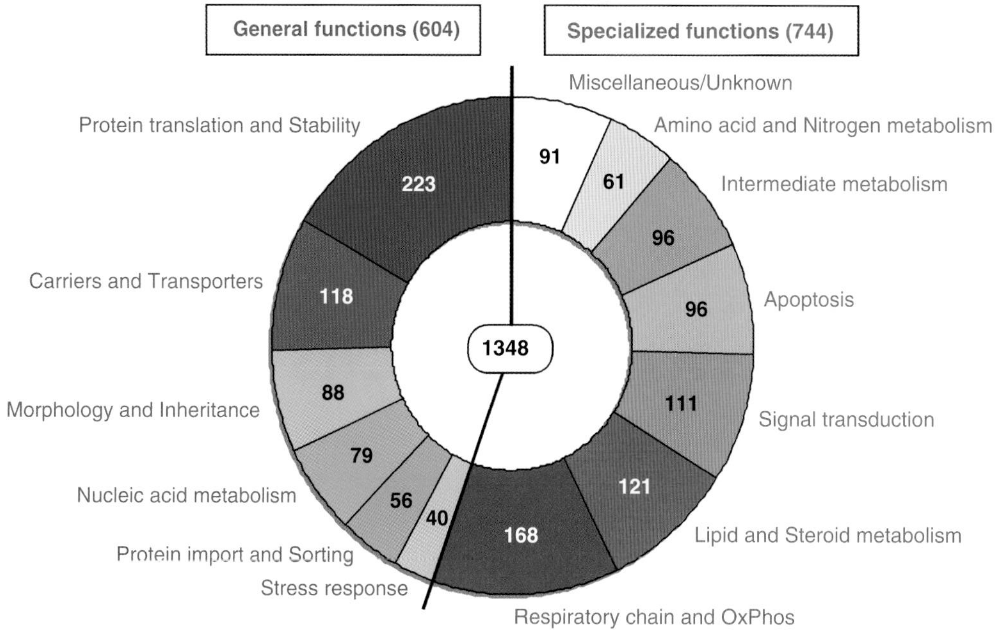

Figure 33.1 Pie chart of mitochondrial proteins divided into functional categories subserving general or specialized functions.

Thus, a neurological classification of mitochondrial diseases might include numerous categories (see Table 33.1):

1. Defects of the respiratory chain, that are, by convention, what we refer to under the term *mitochondrial encephalomyopathies*: they have taken center stage since 1988 and will keep center stage in this review.
2. Defects of other metabolic pathways, such as the pyruvate dehydrogenase complex (PDHC), β-oxidation, or the Krebs cycle.
3. Defects of nDNA-encoded protein transport, a complex machinery that involves mitochondrial targeting, chaperone-assisted protein unfolding and refolding, and energy-dependent crossing of the outer and inner mitochondrial membranes (OMM and IMM).
4. Defects of ancient but long-neglected functions of mitochondria—motility, fusion, and fission—are emerging as important causes of neurological diseases and will be discussed in more detail later.
5. Neurodegenerative disorders are being associated with mutations in increasing numbers of mitochondrial proteins, including Friedreich ataxia (frataxin), Parkinson disease (parkin, Pink1), and amyotrophic lateral sclerosis (SOD1). Although the precise functions of these proteins is still controversial, their pathogenic roles seem to involve iron homeostasis, calcium handling, oxidative stress, and eventually, mitochondrially mediated apoptosis (Tieu & Przedborski, 2006).
6. To quote the late Anita Harding, "is growing old the most common mitochondrial disease of all?" (Harding, 1992). Almost certainly yes, although it is still debated how much of normal aging is due to mtDNA-related versus nDNA-related mechanisms (DiMauro et al., 2002).

Let us now return to what is arguably the most important—and certainly the best known—function of mitochondria, energy generation. Oxidative phosphorylation (OXPHOS) burns the hydrogen of carbohydrates and fats in our diet with the oxygen that we breathe to generate energy (ATP and heat) and water (see Figure 33.2). After pyruvate has been oxidized by PDHC and fatty acyl-CoAs have been oxidized by the reactions of the β-oxidation spiral, the common resulting metabolite, acetyl-CoA, is further oxidized in the Krebs (or

Table 33.1 Mitochondrial Diseases: A Neurological Classification

- **Defects of the respiratory chain**
  - Mutations in mtDNA
  - Mutations in nDNA
- **Defects of other metabolic pathways**
  - Defects in PDHC
  - Defects in β-oxidation
  - Defects of the Krebs cycle
- **Defects of mitochondrial protein importation**
  - Mohr-Tranebjaerg syndrome
  - HSP
- **Defects in mitochondrial motility, fusion, and fission**
  - Autosomal dominant optic atrophy
  - CMT-2A
  - CMT-4A
- **Mutations of mitochondrial protein in neurodegenerative diseases**
  - Frataxin (FA), parkin (PD), Pink1 (PD), SOD1 (ALS)
- **Aging**

TCA) cycle. The reducing equivalents produced by the Krebs cycle and by β-oxidation are transferred to either NAD^+, generating NADH, or to FAD, generating $FADH_2$. NADH transfers electrons to complex I of the electron-transport chain, and $FADH_2$ cedes electrons, either from succinate (in the Krebs cycle) to complex II or from reduced ETF (electron transfer protein, at the end of β-oxidation), to coenzyme Q10 (CoQ10). The electron transport chain consists of four multimeric complexes (complexes I to IV) and two small electron carriers (CoQ10 and cytochrome *c*). The energy generated by these reactions pumps protons from the mitochondrial matrix (the space within the IMM) to the intermembrane space (IMS)—the space between IMM and OMM—at three sites (complex I, complex III, and complex IV). This builds up a proton gradient across the IMM, so that mitochondria are veritable biological capacitors (see Figure 33.2).

In considering disorders due to defects of the respiratory chain (mitochondrial encephalomyopathies *sensu stricto*), another unique molecular concept emerges, which has important implications for the correct diagnosis, sound genetic counseling, and gleaning of physiopathologic mechanisms and phenotypic expression, namely, the respiratory chain is the only biochemical pathway in the cell dependent on two genomes. Of the approximately 90 polypeptides that compose

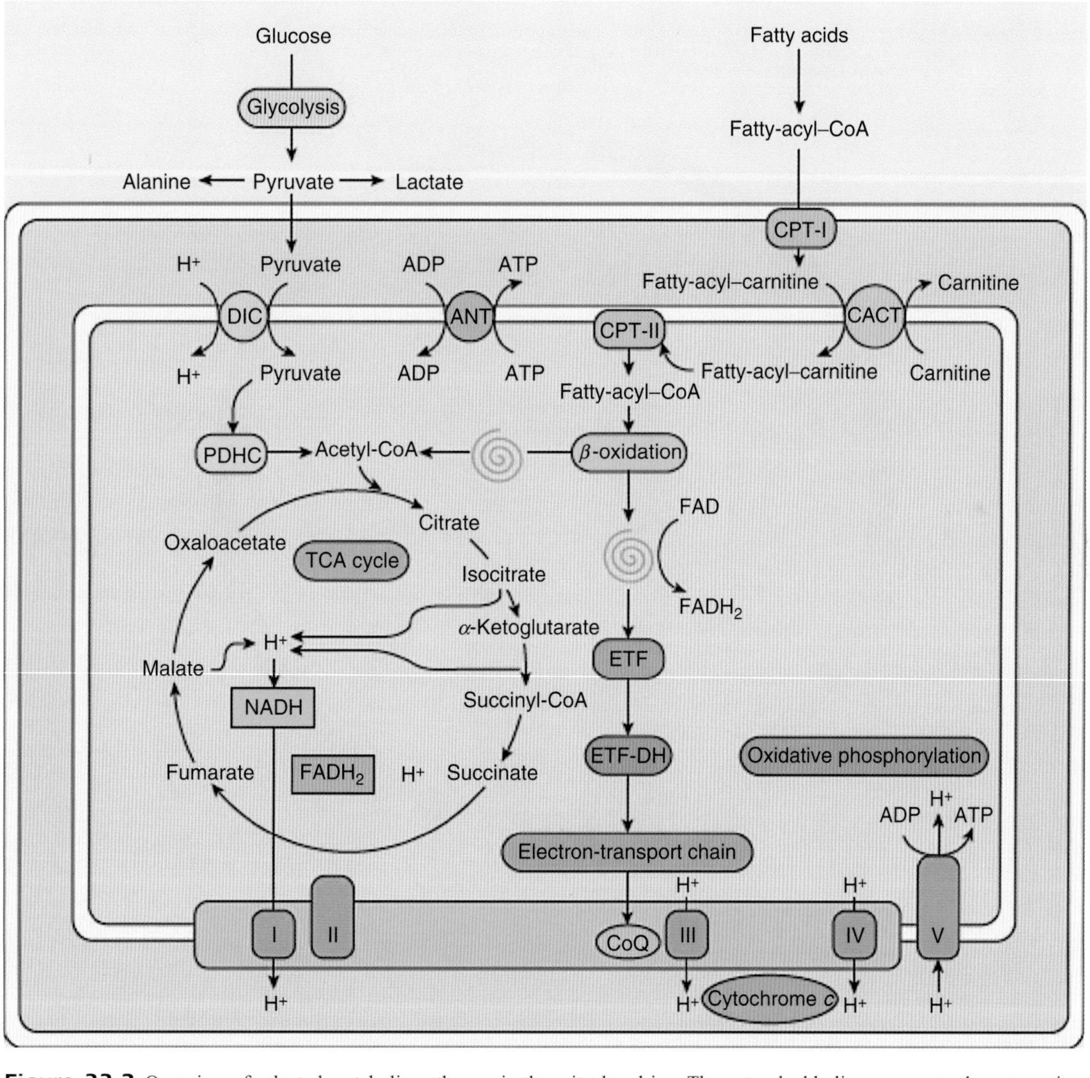

Figure 33.2 Overview of selected metabolic pathways in the mitochondrion. The outer double line represents the outer mitochondrial membrane (OMM) and the inner double line, the inner mitochondrial membrane (IMM). The spirals represent the sequential reactions of the β-oxidation pathway, resulting in the liberation of acetyl-coenzyme A (CoA) and the reduction of flavoprotein. Abbreviations: ADP, adenosine diphosphate; ATP, adenosine triphosphate; ANT, adenine nucleotide translocator; CACT, carnitine-acylcarnitine translocase; CoQ, coenzyme Q10; CPT, carnitine palmitoyltransferase; DIC, dicarboxylate carrier; ETF, electron transfer flavoprotein; ETFDH, electron-transfer dehydrogenase; FAD, flavin adenine dinucleotide; FADH2, reduced FAD; NADH, reduced nicotinamide adenine dinucleotide; PDHC, pyruvate dehydrogenase complex; TCA, tricarboxylic acid; I, complex I; II, complex II; III, complex III; IV, complex IV; V, complex V of the respiratory chain.

the respiratory chain (see Figure 33.2), 13 are encoded by mtDNA and all others by nDNA: After being synthesized in the cytoplasm, they are imported into mitochondria, where they are assembled, together with their mtDNA-encoded counterparts, into the respective holoenzymes in the mitochondrial inner membrane.

Although the small circle of mtDNA (16,569 base-pairs) has proven to be a veritable Pandora's box of diseases, currently known to harbor 192 pathogenic point mutations (see Figure 33.3) (Schon, 2006), mutations in nuclear genes ought to be even more numerous. The predominance of Mendelian mitochondrial diseases is predicated not only on the greater number of nDNA- than mtDNA-encoded respiratory chain subunits, but, as mentioned earlier, on the many nuclear factors needed for the proper assembly and functioning of the respiratory chain, including components of the mitochondrial protein import machinery, assembly factors for the individual complexes, enzymes involved in phospholipid synthesis, and enzymes necessary for the replication and integrity of mtDNA itself.

Discovery of pathogenic mutations in "the other genome" has brought a new type of genetics, mitochondrial genetics, to the attention of practicing physicians. The distinctive rules of mitochondrial genetics include maternal inheritance, heteroplasmy and the threshold effect, and mitotic segregation. Briefly, *maternal inheritance* refers to the fact that all

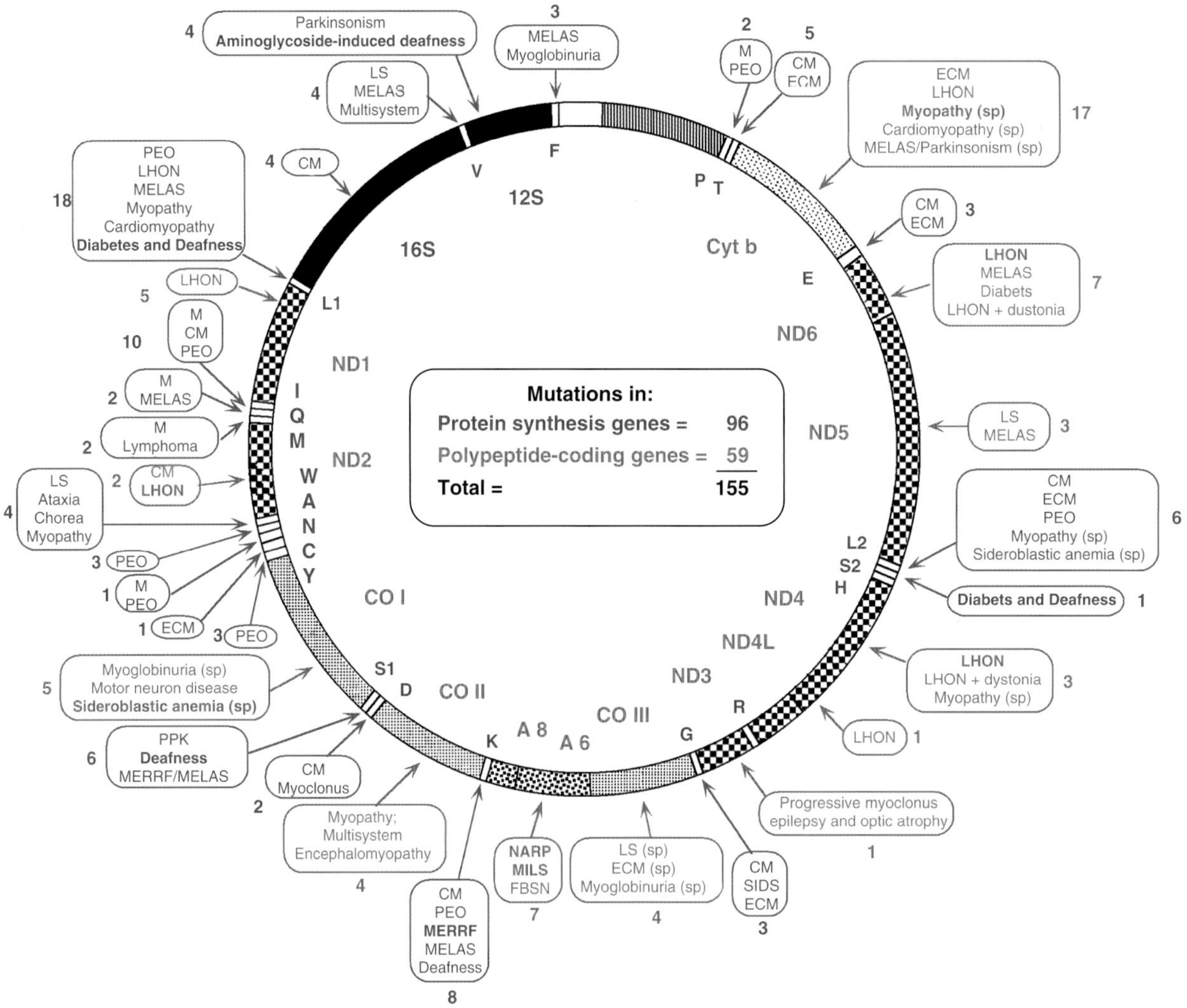

Figure 33.3 Morbidity map of the human mitochondrial genome. The differently colored areas of the 16,560-bp mtDNA represent protein-coding genes for the seven subunits of complex I (ND), the three subunits of cytochrome c oxidase (CO); cytochrome b (cyt b), and the two subunits of ATP synthase (A6 and A8), the 12S and 16S rRNAs (12S, 16S), and the 22 tRNAs identified by one-letter codes for the corresponding amino acids. Diseases due to mutations that impair mitochondrial protein synthesis are shown in blue. Diseases due to mutations in protein-coding genes are shown in red. Abbreviations: ECM, encephalomyopathy; FBSN, familial bilateral striatal necrosis; LHON, Leber's hereditary optic neuropathy; LS, Leigh syndrome; MELAS, mitochondrial encephalomyopathy, lactic acidosis, and strokelike episodes; MERRF, myoclonus epilepsy and ragged-red fibers; MILS, maternally inherited Leigh syndrome; NARP, neuropathy, ataxia, retinitis pigmentosa; PEO, progressive external ophthalmoplegia; PPK, palmoplantar keratoderma; SIDS, sudden infant death syndrome.

mtDNA comes to the zygote only from the oocyte: thus, as a rule, pathogenic mutations of mtDNA (and related diseases) are transmitted from a woman to all her children, but only her daughters will pass it on to their progeny, with no evidence of male-to-child transmission. *Heteroplasmy* refers to the coexistence of mutant and wild-type mtDNAs in the same cell, tissue, or individual, and is predicated on the notion that mtDNA is present in hundreds or thousands of copies in each cell (*polyplasmy*). A corollary of heteroplasmy is the *threshold effect*: a certain minimum number of mutant mtDNAs will be needed to impair oxidative phosphorylation and cause symptoms, and the threshold will be lower in tissues that are more dependent on oxidative metabolism. Because mitochondrial division and mtDNA replication are essentially unrelated to the cell cycle, the proportion of mutant mtDNAs in daughter cells following cell division may shift (due to random drift) and the clinical phenotype may change accordingly (*mitotic segregation*) (DiMauro & Schon, 2003).

From the genetic point of view, mitochondrial diseases can be classified into two major groups, those due to mutations in mtDNA, whose transmission is governed by the loose rules of mitochondrial genetics, and those due to mutations in nuclear DNA, whose transmission is governed by the stricter rules of Mendelian genetics. Each group comes in different "flavors," which are, as expected, more numerous for nDNA mutations (see Table 33.2).

Before going any further, we should dispel the notion that mitochondrial diseases are of purely academic interest, as recent epidemiological data show that they are, in fact, among the most common genetic disorders and a major burden to society (Schaefer et al., 2004). When studies in children and adults are combined and both nDNA and mtDNA mutations are considered, the minimum prevalence is at least 1 in 5,000. A study of adults with mtDNA mutations in Northern England has shown an overall prevalence of 6.57/100,000 (Chinnery et al., 2000), with an especially high prevalence of Leber's hereditary optic atrophy (LHON) (3.22/100,000) (Man et al., 2003).

I. Disorders Due to Mutations in mtDNA

Table 33.3 was drawn several years ago but still serves the original purpose of summarizing the clinical features of six common syndromes to illustrate general principles. Detailed clinical descriptions of mitochondrial syndromes affecting different systems can be found in *Mitochondrial Medicine*, the first textbook devoted to this topic (DiMauro et al., 2006c). The rules of mitochondrial genetics have gone a long way to explain the protean clinical features of mtDNA-related diseases, although many questions remain unanswered, especially those concerning pathogenesis.

Table 33.3 reveals an apparent exception to the maternal inheritance rule of mitochondrial genetics: single deletions of mtDNA are neither inherited from the mother nor transmitted to the progeny, and disorders due to mtDNA deletions, such as Kearns-Sayre syndrome (KSS), progressive external ophthalmoplegia (PEO), and Pearson syndrome (PS), are almost always sporadic. This is attributed to the "bottleneck" between the oöcyte/oögonium (where the giant deletion probably arises) and the embryo, such that only a small minority of maternal DNA populates the fetus (Thorburn, 2006). However, a few cases of maternal transmission of single deletions have been reported, raising the issue of how reassuring we can be in counseling carriers of single mtDNA deletions. A multicenter retrospective study of 226 families has confirmed that the risk of a carrier woman to have affected children is very small, but still finite (1 in 24 births) (Chinnery et al., 2004).

The involvement of multiple unrelated tissues, also illustrated in Table 33.3, which is at once bewildering and diagnostically useful, is due, of course, to the ubiquitous

Table 33.2 A Genetic Classification of the Mitochondrial Encephalomyopathies

Genetic Classification	
• **Defects of mtDNA**	• **Defects of nDNA**
— Mutations in protein synthesis genes	— Mutations in respiratory chain subunits
• tRNA, rRNA, rearrangements	• Complex I, Complex II, CoQ10
— Mutations in protein-coding genes	— Mutations in ancillary proteins
• Multisystemic (LHON, NARP/MILS)	• Complex III, Complex IV, Complex V
• Tissue-specific	— Defects of intergenomic signaling
	• AR-PEO with multiple Δ-mtDNA
	• mtDNA depletion
	• Defective mtDNA translation
	— Defects of protein transport
	• Mohr-Tranebjaerg syndrome
	— Defects of the lipid milieu
	• Barth syndrome

Table 33.3 Clinical Features of mtDNA-related Disorders. Boxes Highlight Typical Features of Specific Syndromes (Except Leigh Syndrome, Which Is Defined by Neuroradiological or Neuropathological Alterations)

		Δ-mtDNA		tRNA		ATPase	
Tissue	Symptom/Sign	KSS	Pearson	MERRF	MELAS	NARP	MILS
CNS	Seizures	−	−	+	+	−	+
	Ataxia	+	−	+	+	+	±
	Myoclonus	−	−	−	−	−	−
	Psychomotor retardation	−	−	−	−	−	+
	Psychomotor regression	+	−	±	+	−	−
	Hemiparesis/hemianopia	−	−	−	−	−	−
	Cortical blindness	−	−	−	+	−	−
	Migraine-like headaches	−	−	−	+	−	−
	Dystonia	−	−	−	−	−	−
PNS	Peripheral neuropathy	±	−	±	±	+	−
Muscle	Weakness	+	−	+	+	+	+
	Ophthalmoplegia	−	±	−	−	−	−
	Ptosis	+	−	−	−	−	−
Eye	Pigmentary retinopathy	−	−	−	−	+	±
	Optic atrophy	−	−	−	−	−	−
	Cataracts	−	−	−	−	−	−
Blood Endocrine	Sideroblastic anemia	±	+	−	−	−	−
	Diabetes mellitus	−	−	−	−	−	−
	Short stature	+	−	+	+	−	−
	Hypoparathyroidism	±	−	−	−	−	−
Heart	Conduction block	−	−	−	−	−	−
	Cardiomyopathy	±	−	−	±	−	±
GI	Exocrine pancreatic dysfunction	−	−	−	−	−	−
	Intestinal pseudoobstruction	−	−	−	−	−	−
ENT	Sensorineural hearing loss	−	−	+	+	±	−
Kidney	Fanconi's syndrome	±	±	−	±	−	−
Laboratory	Lactic acidosis	−	−	−	−	−	−
	Muscle biopsy: RRF	+	±	+	+	−	−
Inheritance	Maternal	−	−	+	+	+	+
	Sporadic	−	−	−	−	−	−

Abbreviations: CNS, central nervous system; PNS, peripheral nervous system; GI, gastrointestinal system; ENT, ear, nose, throat; Δ-mtDNA, deleted mtDNA.

nature of mtDNA. The apparent randomness of affected tissues is often explained by their dependence on oxidative metabolism, although a strict hierarchy of vulnerability to OXPHOS defects should not be expected.

The extreme variability of clinical severity and tissue involvement among maternal members of the same family most often is explained by different degrees of heteroplasmy in the same or different tissues, and is best exemplified by two clinical presentations of the same mutation, T8993G in the ATPase 6 gene of mtDNA. When the proportion of mutant mtDNA is around 70 percent, the disease manifests in young adults as NARP (neuropathy, ataxia, retinitis pigmentosa), but when the proportion surpasses 90 percent, the disease manifests in infants as maternally inherited Leigh syndrome (MILS) (Holt et al., 1990; Tatuch et al., 1992). The frequent occurrence of oligosymptomatic carriers of a mtDNA mutation, such as the A3233G change in $tRNA^{Leu(UUR)}$ most often associated with MELAS (mitochondrial encephalomyopathy, lactic acidosis, and stroke-like episodes) should serve as a warning to practitioners that isolated or "soft" signs (short stature, migraine, hearing loss, diabetes, exercise intolerance) should not been ignored but, in fact, should be carefully noted to reveal a hidden pattern of maternal inheritance. Another practical tip regards the best tissue in which to document the mutation in asymptomatic or oligosymptomatic carriers: Although blood is most commonly used, urinary sediment is both more easily accessible and more sensitive (McDonnell et al., 2004; Shanske et al., 2004).

The typical histochemical picture of the muscle biopsy (still a crucial diagnostic assay) is a mosaic pattern of ragged-red fibers (RRF) with the modified Gomori trichrome stain, or of "ragged-blue" fibers with the succinate dehydrogenase (SDH) stain (see Figure 33.4A, B), both reflecting massive proliferation of mitochondria in mutation-rich segments of individual fibers (hence the mosaic appearance in cross sections). When mutations impair mitochondrial protein synthesis as a whole (the Δ-mtDNA and Trna columns in Table 33.3), RRF are negative for the cytochrome *c* oxidase (COX) histochemical reaction (see Figure 33.4C): Partial deficits of COX are best revealed by the superimposed SDH and COX reactions in the same muscle section (see Figure 33.4D). In biopsies from patients with mutations in mtDNA protein-coding genes (ATPase column in Table 33.3), muscle histochemistry is more variable. As a rule, no RRF are seen in patients with NARP/MILS or with Leber's hereditary optic neuropathy and mutations in complex I (NADH dehydrogenase, ND) genes, whereas patients with myopathy and mutations in *cyt b* or ND genes have RRF, which are, as expected, COX-positive (see Figure 33.5).

Because mtDNA mutates spontaneously at a high rate, and most changes are neutral polymorphisms—a situation exploited in anthropology (Wallace, 2005) and forensic medicine (Anslinger et al., 2001; Gill et al., 1994)—a set of canonical rules have been established to prove the pathogenicity of a novel mtDNA mutation. First, the mutation should not be present in normal individuals of the same ethnic group. Second, it should alter a site conserved in evolution and, therefore, be functionally important. Third, it should cause single or multiple respiratory chain enzyme deficiencies in affected tissues or defects of mitochondrial protein synthesis and respiration demonstrable in cybrid cell lines; that is, immortalized human cell lines that are first depleted of their own mtDNA and then repopulated with varying proportions of mutant mtDNA (King & Attardi, 1989). Fourth, there should be a correlation between degree of heteroplasmy and clinical severity as well as a correlation between degree of heteroplasmy and cell pathology (i.e., COX-negative RRF), best documented by single fiber PCR (Sciacco et al., 1994).

As the last criterion implies, a corollary of these guidelines is that pathogenic mutations are usually heteroplasmic, whereas neutral polymorphisms are homoplasmic. In general

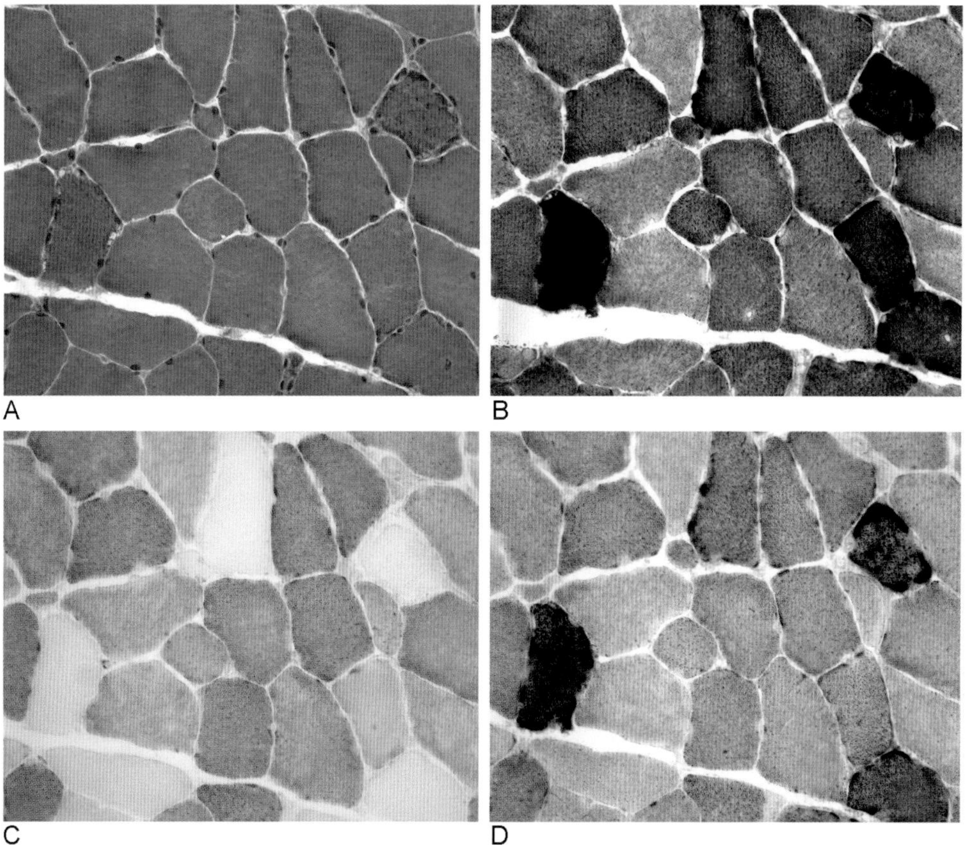

Figure 33.4 Serial cross-sections of a muscle biopsy from a patient with Kearns-Sayre syndrome, to illustrate the typical histochemical changes seen in mtDNA-related disorders that impair protein synthesis in toto. **A.** Modified Gomori trichrome stain, showing ragged-red fibers (RRF); **B.** Succinate dehydrogenase (SDH) stain, showing mitochondrial proliferation ("ragged-blue" fibers); **C.** Cytochrome c oxidase (COX) stain, showing that some RRF or ragged-blue fibers are blank for COX (COX-negative fibers); **D.** Combined SDH + COX stain facilitates detection of COX-deficient fibers, which appear blue (courtesy of Dr. Eduardo Bonilla, Columbia University Medical Center).

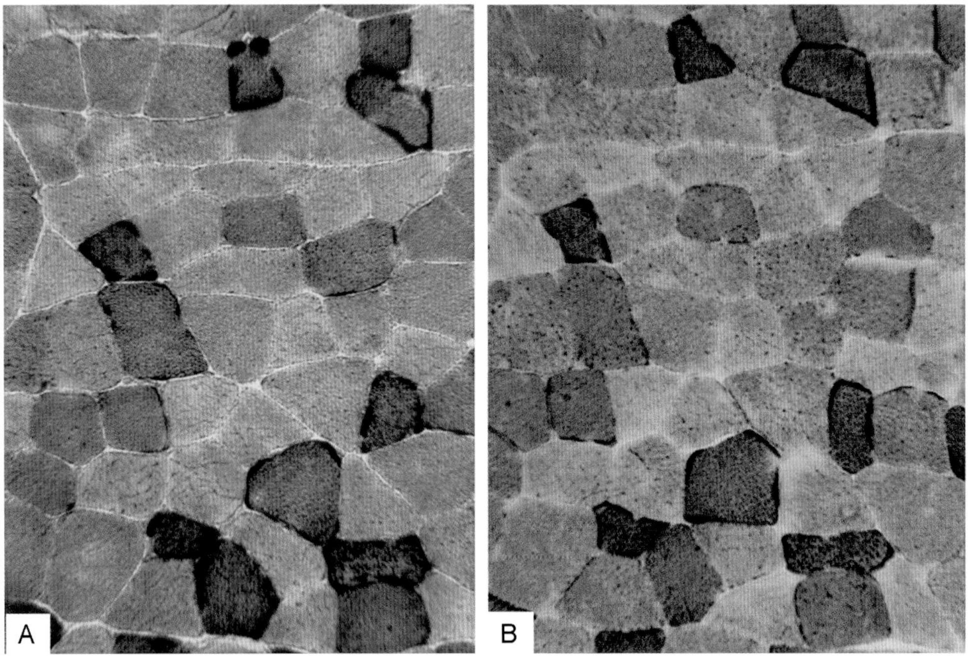

Figure 33.5 Serial cross-section of a muscle biopsy from a patient with complex III deficiency and a pathogenic mutation in the cyt b gene. **A.** SDH stain, showing a mosaic pattern of ragged-blue fibers; **B.** COX stain, showing that the ragged-blue fibers stain intensely for COX (COX-positive fibers) (courtesy of Dr. Eduardo Bonilla, Columbia University Medical Center).

this is true, but there are many exceptions and an increasing awareness of the possible or documented pathogenicity of homoplasmic mutations. In fact, the first point mutation associated with a human disease, LHON (G11778A in *ND4*), was homoplasmic (Wallace et al., 1988), as are other mutations causing LHON (Carelli et al., 2006b). Similarly, most nonsyndromic forms of deafness are due to homoplasmic mutations, including A15555G in the 12S rRNA gene (Prezant et al., 1993) and two mutations in the $tRNA^{Ser(UCN)}$ gene (Li et al., 2004; Sue et al., 1999). Several other examples of homoplasmic mutations clearly associated with various clinical presentations have been published (for review, see DiMauro & Davidzon, 2005).

A "cause celèbre" of homoplasmy regarded a large pedigree in which the *metabolic syndrome* (syndrome X or dyslipidemic hypertension), which includes various combinations of central obesity, atherogenic dyslipidemia, hypertension, hypomagnesemia, and insulin resistance, was transmitted maternally and attributed to a homoplasmic mutation (T4291C) in the $tRNA^{Ile}$ (Wilson et al., 2004). Because the metabolic syndrome afflicts about one fourth of the U.S. population, this finding did not go unnoticed (Hampton, 2004). The challenge with this, as with other homoplasmic mutations, is to go beyond the association and to document a deleterious functional effect. Nuclear magnetic resonance spectroscopy (MRS) and muscle biopsy were performed in a single 55-year-old member of the family with the metabolic syndrome. MRS showed decreased ATP production and a few fibers with mitochondrial proliferation were seen in the biopsy, rather meager evidence of mitochondrial dysfunction, especially considering the age of the patient studied.

Most canonical criteria for pathogenicity listed earlier do not apply to homoplasmic mutations, except for biochemical evidence of respiratory chain dysfunction or direct evidence that the expression of mutated genes is impaired. Defective ATP production due to distinct respiratory chain dysfunction has been documented in cybrid cells harboring different LHON mutations (Baracca et al., 2005), and high resolution Northern blots have shown decreased steady-state levels of the relevant tRNAs in affected tissues of patients with homoplasmic mutations causing cardiopathy (Taylor et al., 2003) or myopathy (McFarland et al., 2004). Similar studies are necessary to document a pathogenic role of the T4291C mutation in patients with the metabolic syndrome.

Another major question underlying the pathogenic mechanism of homoplasmic mtDNA mutations is why they are expressed in some family members but not in others, and why they can result in different clinical phenotypes. At least four factors can influence phenotypic expression: environmental factors, nuclear DNA background, tissue-specific expression of interacting genes, and mtDNA haplotype. The importance of environmental factors is exemplified by the deleterious effect of aminoglycoside exposure in triggering deafness in carriers of the A1555G mutation (Estivill et al., 1998; Prezant et al., 1993). The importance of the nuclear background is illustrated by the influence of the X-chromosome haplotype on the penetrance and gender bias of LHON mutations (Hudson et al., 2005).

We cannot discuss the pathogenicity of homoplasmic mtDNA mutations without considering the importance of mtDNA haplotypes. In their migration out of Africa, human beings have accumulated distinctive variations from the mtDNA of our ancestral "mitochondrial Eve," resulting in several haplotypes characteristic of different ethnic groups (Wallace et al., 1999). It has been suggested that different mtDNA haplotypes may modulate oxidative phosphorylation, thus influencing the overall physiology of individuals and predisposing them to—or protecting them from—certain diseases (Carelli et al., 2006a). Clearly, much work remains to be done to better define both the pathogenic role of homoplasmic mutations and the modulatory role of haplotypes in health and disease.

Eighteen years after the discovery of pathogenic mutations in mtDNA, it is sobering to note that our understanding of how different mtDNA mutations cause different syndromes is, at best, rudimentary. In fact, it is counterintuitive that mtDNA mutations should cause different syndromes in the first place, if—as conventional wisdom dictates—mtDNA rearrangements and point mutations in tRNA or tRNA genes impair mitochondrial protein synthesis and ATP production. This common final pathway would predict an ill-defined common set of symptoms and signs with many overlaps and few distinctive features. In fact, the reverse is true: Although some clinical overlap does occur in mtDNA-related diseases, most major mutations result in well-defined and clinically identifiable syndromes.

One major obstacle to functional studies of mtDNA mutations is the lack of animal models due to the still formidable problem of introducing mtDNA into the mitochondria of mammalian cells. An alternative approach has been to use cybrid cell lines harboring various proportions of mutant mtDNA (see earlier). This technique has largely confirmed the prediction that single deletions or mutations in tRNA genes impair respiration and protein synthesis and decrease ATP production (Pallotti et al., 2004). Elegant as it is, the *in vitro* cybrid system cannot replace animal models in understanding clinical expression. To be sure, two transmitochondrial mice or "mito-mice" were obtained through clever, if circuitous, stratagems, one harboring mtDNA deletions (Inoue et al., 2000), and the other harboring a point mutation for chloramphenicol resistance (Sligh et al., 2000). Although these animal prototypes are important proofs of principle (Hirano, 2001), there has been no recent progress in this area.

In an attempt to explain the distinctive brain symptoms in patients with KSS, MERRF, and MELAS, the different mutations have been "mapped" indirectly through immunohistochemical techniques. Consistent with clinical symptomatology and laboratory data, immunocytochemical evidence suggests that the 3243-MELAS mutation is abundant in the walls of cerebral arterioles (Betts et al., 2006), the 8344-MERRF mutation is abundant in the dentate nucleus of the cerebellum (Tanji et al., 2001), and the MELAS mutation abounds in the choroid plexus (see Figure 33.6, top panel) and in pial vessels (see Figure 33.6, lower panel) (Tanji et al., 2000). However, these data do not explain what directs each mutation to a particular area of the brain.

Finally, mutations in different tRNA genes may have different mechanisms of action, as suggested by the selective tissue vulnerability associated with mutations in certain tRNA: for example, cardiopathy often is associated with mutations in $tRNA^{Ile}$; diabetes is a frequent manifestation of the T14709C mutation in $tRNA^{Glu}$; and multiple lipomas have been reported only in patients with mutations in $tRNA^{Lys}$. But, again, these are associations, not explanations. It is fair to conclude that the pathogenesis of mtDNA-related disorders is still largely *terra incognita*.

II. Disorders Due to Mutations in nDNA

Table 33.4 summarizes the three major categories of mitochondrial diseases due to nDNA mutations. In two of these, the respiratory chain is affected directly or indirectly, whereas in the third group the respiratory chain is affected only indirectly, if at all.

A. Mutations Affecting the Respiratory Chain

Conceptually, the simplest group comprises diseases due to mutations in genes encoding subunits of respiratory chain complexes (direct hits). Interestingly, thus far direct hits have been found only for subunits of the largest complex (complex I) and of the smallest (complex II). This suggests that deleterious mutations in the terminal complexes of the respiratory chain are either extremely rare or incompatible with life. One explanation may be that complexes I and II are "in parallel," allowing for some residual electron transport even when only one complex is functioning, whereas complexes III, IV, and V are "in series." Direct hits do occur in the mtDNA-encoded subunits of complexes III (cytochrome b), IV (COX I, II, or III), and V (ATPase 6), but one could argue that the heteroplasmic nature of these mutations allows for some residual activity. More difficult is to explain why severe defects of COX due to mutations in assembly proteins (for example SCO2) are still compatible with life, albeit a very short one. Most mutations in complex I or in complex II subunits cause Leigh syndrome (LS), a devastating neurodegenerative disorder of infancy or childhood, whose signature neuropathological lesions (bilaterally symmetrical foci of cystic cavitation, vascular proliferation, neuronal loss, and demyelination in the basal ganglia, brainstem, and posterior columns of the spinal cord) seem to reflect the stereotypical

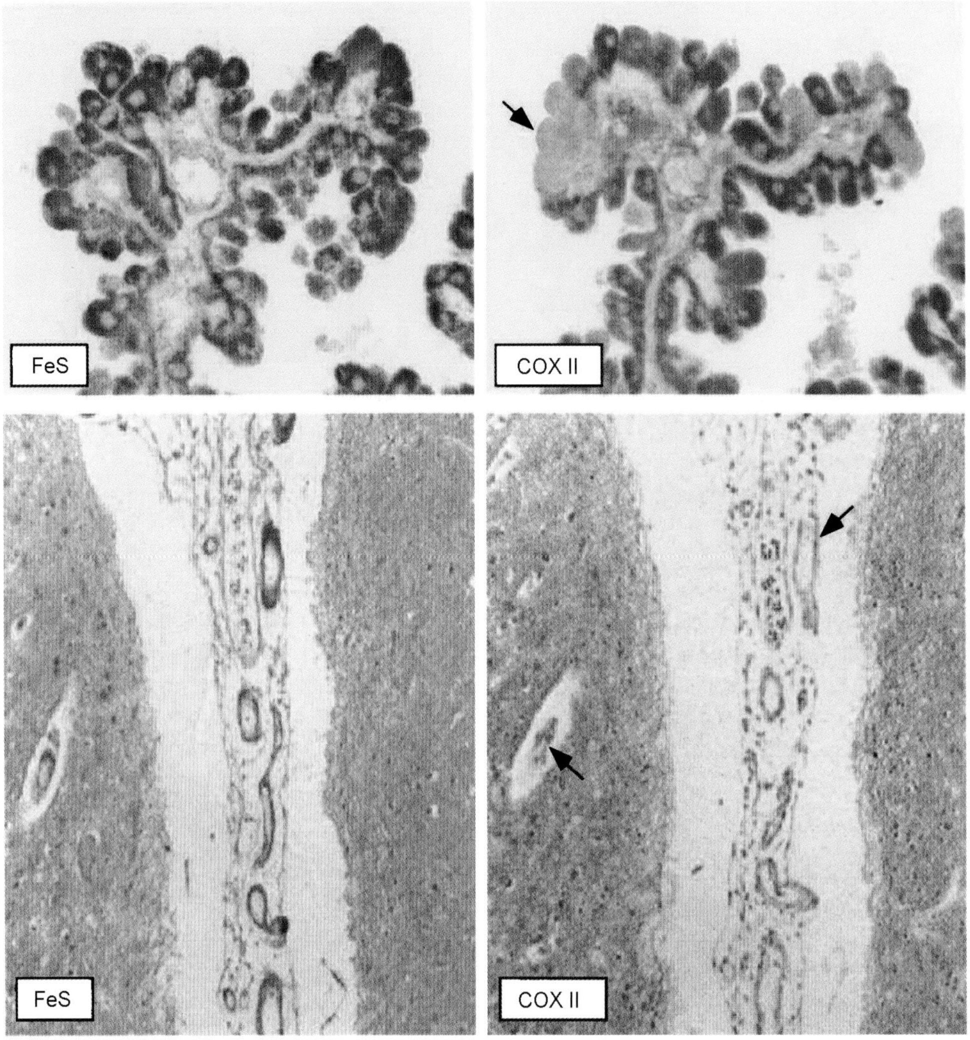

Figure 33.6 Neuropathological images from the postmortem brain of a MELAS patient. The upper panel shows adjacent sections of choroids plexus illustrating immunodeficiency (arrow) for the mtDNA-encoded COX II subunit in scattered epithelial cells, but normal immunoreaction for the nDNA-encoded nonheme iron sulfur protein (FeS). The lower panel shows serial sections of frontal cortex, illustrating COX II immunodeficiency (arrows) and normal immunoreaction for FeS in pial and intracortical arterioles (from Tanji et al. 2001, with permission).

ravages caused by defective oxidative metabolism on the developing nervous system.

The second group of disorders is due to mutations in ancillary proteins (indirect hits); that is, proteins that are not part of any complex, but are needed to direct the proper assembly of the various nDNA- and mtDNA-encoded subunits together with their prosthetic groups. Important clues to the molecular etiology of these disorders, and especially COX deficiency, came from yeast genetics, because most assembly genes needed for COX assembly in yeast have human homologues. A further shortcut to find mutant genes without resorting to the tedious sequencing of multiple candidate COX-assembly genes was to use monochromosomal hybrid fusion or microcell-mediated chromosome transfer to COX-deficient cultured cells from patients, looking for complementation. This is how the most common gene responsible for COX-deficient LS was discovered: When complementation occurred after transferring chromosome 9, fine mapping with microsatellite markers allowed two groups of investigators to zero in on a good suspect, the *SURF1* gene (Tiranti et al., 1998; Zhu et al., 1998). Sequencing of *SURF1* soon revealed more than 30 distinct mutations in children with LS and COX deficiency (Pequignot et al., 2001).

Another interesting way to discovering mutant genes is *integrative genomics*, which is based on combined information derived from studies of DNA, mRNA, and proteomics. The example here is another form of COX deficient LS associated with liver diseases and prevalent in the Saguenay-Lac

Table 33.4 Classification of Mitochondrial Disorders Due to Nuclear DNA Mutations

	Protein or Function	Clinical Features	L.A.	RRF	Muscle Biochemistry	Molecular Genetics
Mutations affecting the RC	RC subunits	LS	+	–	I; II	*NDUFS; NDUFV SDHA*
	Assembly proteins	LS; LSFC; EE; GRACILE; Leuko encephalopathy; EM; nephrosis	+	–	I; III; IV; V	*B17.2L; BCS1L; SURF1 SCO2; SCO1; COX10; COX15; LRPPRC; ETHE1; COQ1; COQ2*
Defects of intergenomic signaling	Multiple mtDNA deletions	AD-PEO; AR-PEO; ARCO; MNGIE; SANDO	+	COX-	I+III+IV	*ANT1; POLG; TWINKLE; TP*
	mtDNA depletion	Hepatocerebral; myopathic; Alpers	+	COX-	I+III+IV	*dGK; TK2; POLG*
	Defects of mtDNA translation	Hepatocerebral; generalized; MLASA	+	COX-	I+III+IV	*EFG1; MRPS16; PUS1*
Other nDNA mutations	Protein transport	Mohr-Tranebjaerg syndrome; HSP	–	–	?	*TIMM8A; HSP60*
	Fusion/fission/motility	AD-Optic atrophy; CMT2A; HSP	+/–	–	?	*OPA1; MNF2 KIF5A*
	Lipid milieu	Barth syndrome	–	+	IV	*G4.5*
	Late-onset Neurodegenerative disorders	FA; PD; ALS	–	–	I (PD)	*Frataxin; Parkin; PINK1; SOD1*

Abbreviations: L.A., lactic acidosis; LS, leigh syndrome; LSFC, leigh syndrome, french canadian; EE, ethylmalonic encephalopathy; GRACILE, growth retardation, aminoaciduria, lactic Acidosis, cholestasis, iron overload, early death; EM, encephalomyopathy; AD-PEO, autosomal dominant progressive external Ophtalmoplegia; AR-PEO, Autosomal Recessive PEO; ARCO, Autosomal Recessive Ophthalmoplegia and cardiopathy; MNGIE, mitochondrial neurogastrointestinal encephalomyopathy; SANDO, sensory ataxic neuropathy, dysarthria, and ophthalmoplegia; MLASA, mitochondrial myopathy, lactic acidosis, and sideroblastic anemia; HSP, hereditary spastic paraplegia; CMT, charcot-marie-tooth; FA, friedreich ataxia; PD, parkinson disease; ALS, amyotrophic lateral sclerosis. For the gene symbols, see text.

Saint-Jean region of Quebec, called LS-French-Canadian type (LSFC) (Morin et al., 1993). The responsible gene locus, 2p16-21, was known from genetic linkage analysis; RNA expression data sets were used to score human genes for similarity to known mitochondrial genes; and organellar proteomics served to score human genes for their likelihood to produce mitochondrial proteins. When these three data sets were intersected, a single candidate gene emerged as the culprit, *LRPPRC*, and proof of guilt was provided by the identification of deleterious mutations in patients with LSFC (Mootha et al., 2003). A bioinformatics approach was also used to identify the first mutant assembly gene (*B17.2L*) responsible for complex I deficiency in a child with severe cavitating leukoencephalopathy (Ogilvie et al., 2005).

A special subgroup of respiratory chain disorders due to mutations in ancillary proteins is characterized by deficiency of coenzyme Q10 (CoQ10), a small carrier that transfers electrons from complexes I and II to complex III, and receives electrons from the β-oxidation pathway via the electron tranfer flavoprotein dehydrogenase (ETFDH) (see Figure 33.2). Mutations in two CoQ10 biosynthetic enzymes have been identified in infants with encephalomyopathy (one of them had LS) and nephrotic syndrome (Lopez et al., 2006; Quinzii et al., 2006). Secondary CoQ10 deficiency occurs also in patients with the syndrome of ataxia and oculomotor apraxia (AOA1), but the relationship between mutations in the *APTX* gene (encoding a protein involved in DNA single strand repair) and CoQ10 deficiency remains unclear (Quinzii et al., 2005).

Aside from their heuristic value, knowledge of the molecular defects in these disorders causing fatal infantile syndromes offers young parents who have lost one child the option of prenatal diagnosis.

B. Defects of Intergenomic Signaling

Here, we return to alterations of mtDNA, except that these are not due to primary mutations of the mitochondrial genome but are the result of garbled messages from the nuclear genome, which controls mtDNA replication, maintenance, and translation. The resulting Mendelian disorders are characterized by qualitative (multiple deletions) or quantitative (depletion) alterations of mtDNA, or by defective translation of mtDNA-encoded respiratory chain components.

1. Multiple mtDNA Deletions

From the clinical point of view, multiple mtDNA deletion syndromes share the cardinal feature of ocular and limb myopathy (PEO, ptosis, proximal weakness), which are almost invariably associated with involvement of extramuscular systems, including peripheral nerves (sensorimotor neuropathy), the brain (ataxia, dementia, psychoses), the ear (sensorineural hearing loss), and the eye (cataracts). Mutations in several genes, all involved in the homeostasis of the mitochondrial nucleotide pool, have been associated with PEO and multiple mtDNA deletions. These include *ANT1*, which encodes the adenosine nucleoside translocator; *PEO1*, which encodes a helicase called Twinkle; *TP*, which encodes the cytosolic enzyme thymidine phosphorylase; *POLG1*, which encodes the mitochondrial polymerase γ; and *POLG2*, which encodes the dimeric accessory subunit of *POLG1* (Spinazzola & Zeviani, 2005). Two of these disorders deserve special attention because of their molecular and clinical pleomorphism.

The first is MNGIE (mitochondrial neurogastrointestinal encephalomyopathy), a devastating autosomal recessive multisystem disease of young adults due to mutations in *TP* (Nishino et al., 1999) and characterized clinically by PEO, neuropathy, leukoencephalopathy, and intestinal dysmotility leading to cachexia and early death. The virtual lack of TP activity wrecks mtDNA synthesis, causing not only multiple deletions, but also depletion and point mutations, which are evident in skeletal muscle, although this tissue does not express TP (Hirano et al., 2005). This muscle paradox suggests that TP deficiency acts through a toxic intermediate. In fact, two toxic intermediates, thymidine and deoxyuridine, accumulate massively in the blood of MNGIE patients. Hemodialysis, an obvious therapeutic approach, had only transient effects, but allogeneic bone marrow transplantation in one patient restored TP activity in buffy coat cells and normalized blood levels of thymidine and deoxyuridine. Although the patient was stable a year after the procedure, clinical efficacy remains to be documented (Hirano et al., 2006).

The other disorder, in fact, disorders, are associated with mutations in *POLG1*, and can be inherited as autosomal recessive or autosomal dominant traits. Both forms of inheritance occur in adults with PEO (but also in some without PEO), who present with a wide spectrum of associated symptoms and signs, including ataxia, peripheral neuropathy, parkinsonism, psychiatric disorders, myoclonus epilepsy mimicking MERRF, or gastrointestinal symptoms mimicking MNGIE (DiMauro et al., 2006a). These patients have multiple mtDNA deletions in muscle. Autosomal recessive inheritance is the rule in children with Alpers syndrome and mutations in *POLG1*, a severe hepatocerebral disease associated with mtDNA depletion and extreme vulnerability to valproate administration (Naviaux & Nguyen, 2004). This clinical heterogeneity, in part at least, can be attributed to the site of the mutation in the catalytic subunit, which has a polymerase (i.e., replicating) domain and an exonuclease (i.e., proofreading) domain joined by a "linker" region: Most patients with Alpers syndrome have at least one mutation in the linker region and the other in the polymerase domain, whereas adults with PEO-plus tend to have mutations in the polymerase domain. To complicate things even further, mutations in the dimeric accessory subunit POLG2, which is responsible for processive DNA synthesis and tight binding of the POLG complex to DNA, can also cause autosomal dominant PEO (Longley et al., 2006).

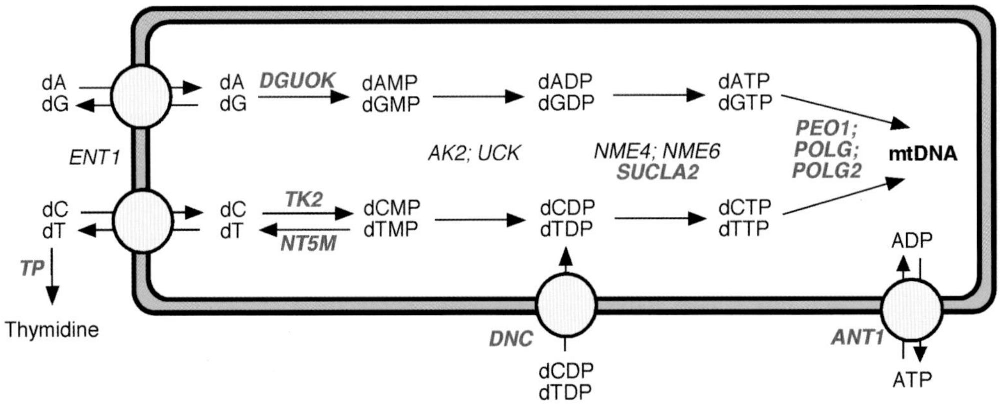

Figure 33.7 Schematic representation of the mitochondrial nucleotide pool, showing in red the genes with mutations that have been associated with depletion of mtDNA (*SUCLA2*, *TK2*, *DGUOK*, *POLG*, *TP*), multiple deletions of mtDNA (*PEO1*, *ANT1*, *POLG*), or both. Mutations in *DNC* have been associated with Amish microencephaly. *ANT1*, adenosine nucleotide translocator; *DGUOK*, doeoxyguanosine kinase; *DNC*, dinucleotide carrier; *ENT1*, equilibrative nucleoside transporter 1; *NT5M*, 5′,3′-nucleotidase, mitochondrial; *POLG*, polymerase γ; *SUCLA2*, succinyl-CoA synthetase/ligase; *TP*, thymidine phosphorylase (reproduced from DiMauro et al., 2006, with permission).

2. Depletion of mtDNA

We have seen how mutations in some proteins controlling the mitochondrial nucleotide pool, such as TP, can cause both multiple mtDNA deletions and mtDNA depletion, whereas some mutations in POLG cause predominantly mtDNA depletion and result in a severe infantile hepatocerebral disorder (Alpers syndrome). Other members of this same family of proteins, when mutated, also cause predominantly mtDNA depletion (see Figure 33.7). For reasons that are not completely clear, the degree of depletion varies in different tissues, but two major syndromes have emerged thus far: (1) hepatocerebral syndrome, caused by mutation either in *POLG1* or in *DGUOK*, which encodes the enzyme deoxyguanosine kinase (dGK); and (2) a purely or predominantly myopathic syndrome associated with mutations in *TK2*, which encodes the enzyme thymidine kinase, or with mutations in *SUCLA2* (Elpeleg et al. 2005), encoding the β subunit of the mitochondrial matrix enzyme succinyl coenzyme A synthetase (Spinazzola & Zeviani, 2005).

It soon became clear, however, that known mutant genes did not explain all cases of mtDNA depletion and, somewhat more surprisingly, mtDNA depletion was not always due to altered control of mitochondrial nucleotide homeostasis. The first example of a new, albeit still unclear, pathogenic mechanism came with the discovery of mutations in *MPV17*, which encodes an IMM protein of unknown function, in children with hepatocerebral syndrome (Spinazzola et al., 2006). The importance of this gene was bolstered by the finding that the same homozygote mutation encountered in a Southern Italian family is the cause of a baffling disease endemic in the Navajo population of the Southwestern United States, called Navajo neurohepatopathy (NNH), to stress that neuropathy rather than encephalopathy accompanies liver dysfunction in this condition (Karadimas et al., 2006). The identity of a recessive mutation in Southern Italian and Navajo families raises the seemingly outlandish possibility of a genetic founder effect. A closer look at history, however, reveals that Francisco Coronado led an army of *conquistadores* from Mexico into present-day Arizona and New Mexico in the early sixteenth century, and Italians were serving in the Spanish army. It will be interesting to determine if the identical mutation is a result of history or hazard: More importantly, we now have the genetic tool to provide sound genetic counseling in the hopes of eradicating this dreadful disease.

3. Defects of mtDNA Translation

Faithful translation of the 13 mtDNA-encoded subunits of the respiratory chain requires not only intact mtDNA, a trustworthy POLG, and the availability of nucleotide building blocks, but also several initiation, elongation, and termination factors, all encoded by nDNA. Defects in translation of mtDNA result in severe combined defects of all respiratory chain complexes containing mtDNA-encoded subunits, a biochemical pattern indistinguishable from that of mtDNA depletion (the biochemical defects in patients with single or multiple mtDNA deletions or tRNA mutations are similar but not as severe). It is important to think of this pathogenic mechanism in infants or children with hepatocerebral syndrome, encephalopathy, or cardiomyopathy and multiple respiratory chain defects but without any evidence of mtDNA depletion. Thus far, three molecular defects have been described, one involving *EFG1*, which encodes one of four ribosomal elongation factors (Coenen et al., 2004), the second involving the mitochondrial ribosomal protein subunit 16 (*MRPS16*) (Miller et al., 2004), and the third involving *TSFM*, which encodes the mitochondrial elongation factor EFTs (Smeitink et al., 2006). A third syndrome is due to defective pseudouridylation of mitochondrial tRNA genes and is characterized by myopathy, lactic acidosis, and sideroblastic anemia (MLASA): Mutations in the gene *PUS1*, which encodes the +RNA modified enzyme pseudouridine synthase 1, have been identified in three families (Bykhovskaya et al., 2004; Fernandez-Vizarra et al., 2006).

III. Other nDNA Mutations

As we mentioned at the outset, a number of Mendelian disorders affect the mitochondrial respiratory chain in an indirect way, if they affect it at all. We will consider here only three pathogenic mechanisms, because of their conceptual interest:

- ▲ Defects of the lipid milieu
- ▲ Defects of mitochondrial protein importation
- ▲ Defects of mitochondrial motility, fusion, and fission

A. Defects of the Lipid Milieu

The complexes of the respiratory chain are embedded in the lipid milieu of the IMM, whose major component is cardiolipin, an acidic phospholipid. It has been documented that cardiolipin does not have merely a structural, scaffolding function, but participates in the formation of supercomplexes (stoichiometric assemblies of individual respiratory chain complexes into functional units) (Zhang et al., 2005) and interacts directly with COX (Sedlak et al., 2006); conversely, intact respiratory chain function is essential for cardiolipin biosynthesis (Gohil et al., 2004). It stands to reason, therefore, that genetic abnormalities of cardiolipin could impair respiratory chain function in humans. Barth syndrome, an X-linked recessive disorder causing mitochondrial myopathy, cardiopathy, and growth retardation, is such a disease, due to mutations in the gene (unimaginatively first called *G4.5*, but now known as *TAZ*), which encodes a phospholipid acyltransferase more imaginatively called tafazzin (Schlame & Ren, 2006). Tafazzin promotes structural uniformity and molecular symmetry among cardiolipin molecular species, and mutations in *TAZ* alter the concentration and composition of cardiolipin, leading to altered mitochondrial architecture and function. Some TAZ mutations cause mislocalization of cardiolipin from MOM and MIM to the mitochondrial matrix (Claypool et al., 2006).

B. Defects of Mitochondrial Protein Importation

After getting settled comfortably in its new eukaryotic "home," the human protobacterium-turned-mitochondrion has been content to keep only 13 original genes. As a consequence, almost 1,300 mitochondrial proteins are now encoded by nDNA genes, synthesized in the cytoplasm, and imported into mitochondria. How? Mitochondrial importation is a complex process, with different pathways for the targeting and sorting of proteins to one of the four mitochondrial compartments (OMM, IMM, the intermembrane space, and the matrix enveloped by the IMM). Key components of the import machinery are members of the "heat shock" protein (HSP) family, chaperones that are needed for the unfolding and refolding of mitochondrially targeted protein as they transit through the import receptors (the "toll booths" of mitochondria) and are directed to appropriate compartments. Most, but not all, mitochondrial proteins (especially those destined to the IMM or the matrix) have "targeting signals," located (usually, but not always) at the N-terminus of the protein. Once inside the mitochondrion, the mitochondrial targeting signal (MTS, or leader peptide) is cleaved to release the mature peptide. The importation machinery consists of polymeric translocases located in the outer membrane (TOM) or the inner membrane (TIM). In collaboration with sorting and assembly machinery (SAM), a presequence translocation-associated motor (PAM), plus a mitochondrial import and assembly (MIA) pathway specific for a subset of intermembrane space proteins (Gabriel et al., 2006), TOM and TIM sort out incoming polypeptides to the proper compartments (Chacinska & Rehling, 2004).

A few mutations in leader peptides have been associated with specific enzyme defects, such as methylmalonic acidemia (Ledley et al., 1990) or PDHC deficiency (Takakubo et al., 1995), but remarkably few human diseases have been attributed to genetic defects of the general importation machinery. One of these is an X-linked recessive deafness-dystonia syndrome (Mohr-Tranebjaerg syndrome) due to mutations in the gene (*TIMM8A*) encoding the deafness/dystonia protein (DDP), an MIA pathway protein located in the intermembrane space (Roesch et al., 2002). Another is an autosomal dominant form of hereditary spastic paraplegia (HSP) due to mutations in the import chaperonin HSP60 (Hansen et al., 2002).

Unless most disorders due to disruption of the general importation machinery are incompatible with life, as suggested long ago by Fenton (1995), we can expect many more such disorders to be identified in the near future.

C. Defects of Mitochondrial Motility, Fusion, and Fission

This is a relatively new area of interest for clinical scientists, but it is an area that has already yielded exciting results and is sure to yield many more in the years to come. Remembering their bacterial origin, mitochondria move, fuse, and divide within cells, where they often form tubular networks, which may favor the delivery of organelles to areas of high energy demand (Bossy-Wetzel et al., 2003). The need for mitochondrial motility is nowhere more evident than in motor neurons of the anterior horn cells, where mitochondria have to travel a very long way from the cell soma to the neuromuscular junction. Mitochondria travel on microtubular rails, propelled by motor proteins, usually GTPases, called kinesins (when mitochondria travel downstream) or dyneins (when they travel upstream). The first defect of mitochondrial motility has been identified in a family with autosomal dominant hereditary spastic paraplegia and mutations in a gene (*KIF5A*) encoding one of the kinesins. Interestingly, the mutation affects a region of the protein involved in microtubule binding (Fichera et al., 2004).

In yeast, at least three proteins are required for mitochondrial fission: Dnm1p (dynamin-related protein), Fis1p (fission-related protein), and Mdv1p (mitochondrial division protein). Two other proteins are required for mitochondrial fusion: Fzo1p (the yeast homolog of the *Drosophila* "fuzzy onion" protein) and Ugo1p (*ugo* is Japanese for "fusion"). Because all five proteins are located in the OMM, for fission to occur, OMM and IMM have to establish contact sites, apparently through the action of yet another protein called Mgm1p (mitochondrial genome maintenance protein). Mutations in the human orthologs of Mgm1p (OPA1) and Fzo1p (mitofusin 2 or MFN2) have been associated with human diseases.

Mutations in *OPA1* cause autosomal dominant optic atrophy, the Mendelian counterpart, as it were, of LHON (Alexander et al., 2000; Delettre et al., 2000). Mutations in *MFN2* cause an autosomal dominant axonal variant of Charcot-Marie-Tooth (CMT) disease (Lawson et al., 2005; Zuchner et al., 2004). A recent review of 62 unrelated axonal CMT families revealed *MFN2* mutations in 26 patients from 15 families, suggesting that this is a major cause of axonal CMT (Chung et al., 2006). Also, and interestingly, some patients had variable involvement of the central nervous system, and some had early onset and optic atrophy. Also, mutations in GDAP1, the gene encoding ganglioside-induced differentiation protein 1, which is located in the MOM and regulates the mitochondrial network (Niemann et al., 2005), cause CMT type 4A, an autosomal recessive, severe, early-onset form of either demyelinating or axonal neuropathy (Pedrola et al., 2005).

It is clear that these diseases aren't but the proverbial tip of what will be an iceberg of human neurodegenerative disorders directly or indirectly linked to abnormal mitochondrial motility, fusion, or fission.

IV. Conclusions

The molecular complexity of mitochondrial neurology is staggering, as we hope this review has made clear. Nor does the story end here, as we have largely confined our discussion to molecular mechanisms impairing, directly or indirectly, the respiratory chain. The involvement of mitochondria in late-onset neurodegenerative disorders and aging requires separate chapters (see, for example, Tieu & Przedborski, 2006). Finally, although the therapy of mitochondrial disorders is still woefully inadequate, there is a fervor of research aimed at developing novel therapeutic strategies involving fascinating molecular concepts, some of which have been reviewed recently (DiMauro et al., 2006b).

Acknowledgments

This work has been supported by grants from the National institutes of Health (NINDS 11766 and HD32062), from the Muscular Dystrophy Association, and from the Marriott Mitochondrial Disorder Clinical Research Fund (MMDCRF).

References

Alexander, C., Votruba, M., Pesch, U. E. A., Thiselton, D. L., Mayer, S., Moore, A. et al. (2000). OPA1, encoding a dynamin-related GTPase, is mutated in autosomal dominant optic atrophy linked to chromosome 3q28. *Nature Genet* **26**, 211–215.

Anslinger, K., Weichhold, G., Keil, W., Bayer, B. (2001). Identification of the skeletal remains of Martin Bormann by mtDNA analysis. *Int J Legal Med* **114**, 194–196.

Baracca, A., Solaini, G., Sgarbi, G., Lenaz, G., Baruzzi, A., Schapira, A. H. V. et al. (2005). Severe impairment of complex I-driven ATP synthesis in Leber's hereditary optic neuropathy cybrids. *Arch Neurol* **62**, 730–736.

Betts, J., Jaros, E., Perry, R. H., Schaefer, A. M., Taylor, R. W., Abdel-All, Z. et al. (2006). Molecular neuropathology of MELAS: Level of heteroplasmy in individual meurones and evidence of extensive vascular involvement. *Neuropath Appl Neurobiol* **32**, 359–373.

Bossy-Wetzel, E., Barsoum, M. J., Godzik, A., Schwartzenbacher, R., Lipton, S. A. (2003). Mitochondrial fission in apoptosis, neurodegeneration and aging. *Curr Opin Cell Biol* **15**, 706–716.

Bykhovskaya, Y., Casas, K. A., Mengesha, E., Inbal, A., Fischel-Ghodsian, N. (2004). Missense mutation in pseudouridine synthase 1 (*PUS1*) causes mitochondrial myopathy and sideroblastic anemia (MLASA). *Am J Hum Genet* **74**, 1303–1308.

Carelli, V., Achilli, A., Valentino, M. L., Rengo, C., Semino, O., Pala, M. et al. (2006a). Haplogroup effect and recombination of mitochondrial DNA: Novel clues from the analysis of Leber hereditary optic neuropathy pedigrees. *Am J Hum Genet* **78**, 564–574.

Carelli, V., Barboni, P., Sadun, A. A. (2006b). Mitochondrial ophthalmology. In DiMauro, S., Hirano, M., Schon, E. A., Eds., *Mitochondrial Medicine,* 105–142. Informa Healthcare, London.

Chacinska, A., Rehling, P. (2004). Moving proteins from the cytosol into mitochondria. *Biochem Soc Trans* **32**, 774–776.

Chinnery, P. F., DiMauro, S., Shanske, S., Schon, E. A., Zeviani, M., Mariotti, C. et al. (2004). Risk of developing a mitochondrial DNA deletion disorder. *Lancet* **364**, 592–595.

Chinnery, P. F., Wardell, T. M., Singh-Kler, R., Hayes, C., Johnson, M. A., Taylor, R. W. et al. (2000). The epidemiology of pathogenic mitochondrial DNA mutations. *Ann Neurol* **48**, 188–193.

Chung, K. W., Kim, S. B., Park, K. D., Choi, K. G., Lee, J. H., Eun, H. W. et al. (2006). Early onset severe and late-onset mild Charcot-Marie-Tooth disease with mitofusin 2 (MFN2) mutations. *Brain* **129**, 2103–2118.

Claypool, S. M., McCaffrey, J. M., Koehler, C. M. (2006). Mitochondrial mislocalization and altered assembly of a cluster of Barth syndrome mutant tafazzins. *J Cell Biol* **174**, 379–390.

Coenen, M. J. H., Antonicka, H., Ugalde, C., Sasarman, F., Rossi, P., Heister, J. G. A. M. A. et al. (2004). Mutant mitochondrial elongation factor G1 and combined oxidative phosphorylation deficiency. *N Engl J Med* **351**, 2080–2086.

Delettre, C., Lenaers, G., Griffoin, J-M., Gigarel, N., Lorenzo, C., Belenguer, P. et al. (2000). Nuclear gene OPA1, encoding a mitochondrial dynamin-related protein, is mutated in dominant optic atrophy. *Nature Genet* **26**, 207–210.

DiMauro, S., Davidzon, G. (2005). Mitochondrial DNA and disease. *Ann Med* **37**, 222–232.

DiMauro, S., Davidzon, G., Hirano, M. (2006a). A polymorphic polymerase. *Brain* **126**, 1637–1639.

DiMauro, S., Hirano, M., Schon, E. A. (2006b). Approaches to the treatment of mitochondrial diseases. *Muscle and Nerve* **34**, 265–283.

DiMauro, S., Hirano, M., Schon, E. A., Eds. (2006c). *Mitochondrial Medicine*. Informa Healthcare, London.

DiMauro, S., Schon, E. A. (2003). Mitochondrial respiratory-chain diseases. *N Engl J Med* **348**, 2656–2668.

DiMauro, S., Tanji, K., Bonilla, E., Pallotti, F., Schon, E. A. (2002). Mitochondrial abnormalities in muscle and other aging cells: Classification, causes, and effects. *Muscle & Nerve* **26**, 597–607.

Elpeleg, O., Miller, C., Hershkovitz, E., Bitner-Glindzicz, M., Bondi-Rubinstein, G., Rahman, S. et al. (2005). Deficiency of the ADP-forming succinyl-CoA synthase activity is associated with encephalomyopathy and mitochondrial DNA depletion. *Am J Hum Genet* **76**, 1081–1086.

Estivill, X., Govea, N., Barcelo, A., Perello, E., Badenas, C., Romero, E. et al. (1998). Familial progressive sensorineural deafness is mainly due to the mtDNA A1555G mutation and is enhanced by treatment with aminoglycosides. *Am J Hum Genet* **62**, 27–35.

Fenton, W. A. (1995). Mitochondrial protein transport—A system in search of mutations. *Am J Hum Genet* **57**, 235–238.

Fernandez-Vizarra, E., Berardinelli, A., Valente, L., Tiranti, V., Zeviani, M. (2006). Nonsense mutation in pseudouridylate synthase 1 (PUS1) in two brothers affected by myopathy, lactic acidosis and sideroblastic anemia (MLASA). *J Med Genet*, in press.

Fichera, M., Lo Giudice, M., Falco, M., Sturnio, M., Amata, A., Calabrese, O. et al. (2004). Evidence of kinesin heavy chain (*KIF5A*) involvement in pure hereditary spastic paraplegia. *Neurology* **63**, 1108–1110.

Gabriel, K., Milenkovic, D., Chacinska, A., Muller, J., Guiard, B., Pfanner, N., Meisinger, C. (2006). Movel mitochondrial intermembrane space proteins as substrates of the MIA import pathway. *J Mol Biol*, in press.

Gill, P., Ivanov, P. L., Kimpton, C., Piercy, R., Benson, N., Tully, G. et al. (1994). Identification of the remains of the Romanov family by DNA analysis. *Nature Genet* **6**, 130–135.

Gohil, V. M., Hayes, P., Matsuyama, S., Schagger, H., Schlame, M., Greenberg, M. L. (2004). Cardiolipin biosynthesis and mitochondrial respiratory chain function are interdependent. *J Biol Chem* **279**, 42612–42618.

Hampton, T. (2004). Mitochondrial defects may play a role in the metabolic syndrome. *J Amer Med Assoc* **292**, 2823–2824.
Hansen, J. J., Durr, A., Cournu-Rebeix, I., Georgopoulos, C., Ang, D., Nyholm Nielsen, M. et al. (2002). Hereditary spastic paraplegia SPG13 is associated with a mutation in the gene encoding the mitochondrial chaperonin Hsp60. *Am J Hum Genet* **70**, in press.
Harding, A. E. (1992). Growing old: the most common mitochondrial disease of all? *Nature Genet* **2**, 251–252.
Hirano, M. (2001). Transmitochondrial mice: Proof of principle and promises. *Proc Natl Acad Sci USA* **98**, 401–403.
Hirano, M., Lagier-Tourenne, C., Valentino, M. L., Marti, R., Nishigaki, Y. (2005). Thymidine phosphorylase mutations cause instability of mitochondrial DNA. *Gene* **354**, 152–156.
Hirano, M., Marti, R., Casali, C., Tadesse, B. S., Uldrick, T., Fine, B. et al. (2006). Allogeneic stem cell transplantation corrects biochemical derangements in MNGIE. *Neurology* **67**, 1458–1460.
Holt, I. J., Cooper, J. M., Morgan Hughes, J. A., Harding, A. E. (1988). Deletions of muscle mitochondrial DNA. *Lancet* **1**, 1462–1462.
Holt, I. J., Harding, A. E., Petty, R. K., Morgan Hughes, J. A. (1990). A new mitochondrial disease associated with mitochondrial DNA heteroplasmy. *Am J Hum Genet* **46**, 428–433.
Hudson, G., Keers, S., Man, P. Y. W., Griffiths, P., Huoponen, K., Savontaus, M-L. et al. (2005). Identification of an X-chromosomal locus and haplotype modulating the phenotype of a mitochondrial DNA disorder. *Am J Hum Genet* **77**, 1086–1091.
Inoue, K., Nakada, K., Ogura, A., Isobe, K., Goto, Y-i., Nonaka, I., Hayashi, J-I. (2000). Generation of mice with mitochondrial dysfunction by introducing mouse mtDNA carrying a deletion into zygotes. *Nature Genet* **26**, 176–181.
Karadimas, C. L., Vu, T. H., Holve, S. A., Quinzii, C., Tanji, K., Bonilla, E. et al. (2006). Navajo neurohepatopathy is caused by a mutation in the MPV17 gene. *Am J Hum Genet* **79**, 544–548.
King, M. P., Attardi, G. (1989). Human cells lacking mtDNA: Repopulation with exogenous mitochondria by complementation. *Science* **246**, 500–503.
Lawson, V. H., Graham, B. V., Flanigan, K. M. (2005). Clinical and electrophysiologic features of CMT2A with mutations in the mitofusin 2 gene. *Neurology* **65**, 197–204.
Ledley, F. D., Jansen, R., Nham, S. U., Fenton, W. A., Rosenberg, L. E. (1990). Mutation eliminating mitochondrial leader sequence of methylmalonyl-CoA mutase causes mut0 methylmalonic acidemia. *Proc Natl Acad Sci USA* **87**, 3147–3150.
Li, X., Fischel-Ghodsian, N., Schwartz, F., Yan, Q., Friedman, R. A., Guan, M. X. (2004). Biochemical characterization of the mitochondrial tRNASer(UCN) T7511C mutation associated with nonsyndromic deafness. *Nucleic Acids Res* **32**, 867–877.
Longley, M. J., Clark, S., Man, C. Y. W., Hudson, G., Durham, S. E., Taylor, R. W. et al. (2006). Mutant POLG2 disrupts DNA polymerase gamma subunits and causes progressive external ophthalmoplegia. *Am J Hum Genet* **78**, 1026–1034.
Lopez, L. C., Quinzii, C., Schuelke, M., Kanki, T., Naini, A., DiMauro, S., Hirano, M. (2006). Leigh syndrome with nephropathy and CoQ10 deficiency due to decaproneyl diphosphate synthase subunit 2 (PDSS2) mutations. *Am J Hum Genet*, in press.
Man, P. Y., Griffiths, P. G., Brown, D. T., Howell, N., Turnbull, D. M., Chinnery, P. F. (2003). The epidemiology of Leber hereditary optic neuropathy in the Northeast of England. *Am J Hum Genet* **72**, 333–339.
McDonnell, M. T., Schaefer, A. M., Blakely, E. L., McFarland, R., Chinnery, P. F., Turnbull, D. M., R. W. T (2004). Noninvasive diagnosis of the 3243A>G mitochondrial DNA mutation using urinary epithelial cells. *Eur J Hum Genet* **12**, 778–781.
McFarland, R., Schaefer, A. M., Gardner, A., Lynn, S., Hayes, C. M., Barron, M. J. et al. (2004). Familial myopathy: New insights into the T14709C mitochondrial tRNA mutation. *Ann Neurol* **55**, 478–484.
Miller, C., Saada, A., Shaul, N., Shabtai, N., Ben-Shalom, E., Shaag, E. et al. (2004). Defective mitochondrial translation caused by a ribosomal protein (MRPS16) mutation. *Ann Neurol* **56**, 734–738.
Mootha, V. K., Lepage, P., Miller, K., Bunkenborg, J., Reich, M., Hjerrild, M. et al. (2003). Identification of a gene causing human cytochrome *c* oxidase deficiency by integrative genomics. *Proc Natl Acad Sci USA* **100**, 605–610.
Morin, C., Mitchell, G., Larochelle, J., Lambert, M., Ogier, H., Robinson, B. H., DeBraekeleer, M. (1993). Clinical, metabolic, and genetic aspects of cytochrome c oxidase deficiency in Saguenay-Lac-Saint-Jean. *Am J Hum Genet* **53**, 488–496.
Nass, S., Nass, M. (1963). Intramitochondrial fibers with DNA characteristics. *J Cell Biol* **19**, 593–629.
Naviaux, R. K., Nguyen, K. V. (2004). *POLG* mutations associated with Alpers' syndrome and mitochondrial DNA depletion. *Ann Neurol* **55**, 706–712.
Niemann, A., Ruegg, M., La Padula, V., Schenone, A., Suter, U. (2005). Ganglioside-induced differentiation associated protein 1 is a regulator of the mitochondrial network: New implications for Charcot-Marie-Tooth disease. *J Cell Biol* **170**, 1067–1078.
Nishino, I., Spinazzola, A., Hirano, M. (1999). Thymidine phosphorylase gene mutations in MNGIE, a human mitochondrial disorder. *Science* **283**, 689–692.
Ogilvie, I., Kennaway, N. G., Shoubridge, E. A. (2005). A molecular chaperone for mitochondrial complex I assembly is mutated in a progressive encephalopathy. *J Clin Invest* **115**, 2784–2792.
Pallotti, F., Baracca, A., Hernandez-Rosa, E., Walker, W. F., Solaini, G., Lenaz, G. et al. (2004). Biochemical analysis of respiratory function in cybrid cell lines harbouring mitochondrial DNA mutations. *Biochem J* **384**, 287–293.
Pedrola, L., Espert, A., Wu, X., Claramunt, R., Shy, M. E., Palau, F. (2005). GDAP1, the protein causing Charcot-Marie-Tooth disease type 4A, is expressed in neurons and is associated with mitochondria. *Hum Mol Genet* **14**, 1087–1094.
Pequignot, M. O., Dey, R., Zeviani, M., Tiranti, V., Godinot, C., Poyau, A. et al. (2001). Mutations in the *SURF1* gene associated with Leigh syndrome and cytochrome *c* oxidase deficiency. *Hum Mut* **17**, 374–381.
Prezant, T. R., Agapian, J. V., Bohlman, M. C., Bu, X., Oztas, S., Qiu, W-Q. et al. (1993). Mitochondrial ribosomal RNA mutation associated with both antibiotic-induced and non-syndromic deafness. *Nature Genet* **4**, 289–293.
Quinzii, C., Kattah, A. G., Naini, A., Akman, H. O., Mootha, V. K., DiMauro, S., Hirano, M. (2005). Coenzyme Q deficiency and cerebellar ataxia associated with an aprataxin mutation. *Neurology* **64**, 539–541.
Quinzii, C., Naini, A., Salviati, L., Trevisson, E., Navas, P., DiMauro, S., Hirano, M. (2006). A mutation in para-hydoxybenzoate-polyprenyl transferase (COQ2) causes primary coenzyme Q10 deficiency. *Am J Hum Genet* **78**, 345–349.
Roesch, K., Curran, S. P., Tranebjaerg, L., Koehler, C. M. (2002). Human deafness dystonia syndrome is caused by a defect in assembly of the DDP1/TIMM8a-TIMM13 complex. *Hum Mol Genet* **11**, 477–486.
Sagan, L. (1967). On the origin of mitosing cells. *J Theor Biol* **14**, 255–274.
Schaefer, A. M., Taylor, R. W., Turnbull, D. M., Chinnery, P. F. (2004). The epidemiology of mitochondrial disorders—Past, present and future. *Biochim Biophys Acta* **1659**, 115–120.
Schatz, G., Haslbrunner, E., Tuppy, H. (1964). Deoxyribonucleic acid associated with yeast mitochondria. *Biochem Biophys Res Comm* **15**, 127–132.
Schlame, M., Ren, M. (2006). Barth syndrome, a human disorder of cardiolipin metabolism. *FEBS Lett* **580**, 5450–5455.
Schon, E. A. (2006). Appendix. In DiMauro, S., Hirano, M., Schon, E. A., Eds., *Mitochondrial Medicine*, 329–335. Taylor & Francis, London.
Sciacco, M., Bonilla, E., Schon, E. A., DiMauro, S., Moraes, C. T. (1994). Distribution of wild-type and common deletion forms of mtDNA in normal and respiration-deficient muscle fibers from patients with mitochondrial myopathy. *Hum Mol Genet* **3**, 13–19.

Sedlak, E., Panda, M., Dale, M. P., Weintraub, S. T., Robinson, N. C. (2006). Photolabeling of cardiolipin binding subunits within bovine heart cytochrome c oxidase. *Biochem* **45**, 746–754.

Shanske, S., Pancrudo, J., Kaufmann, P., Engelstad, K., Jhung, S., Lu, J. et al. (2004). Varying loads of the mitochondrial DNA A3243G mutation in different tissues: Implications for diagnosis. *Am J Med Genet* **130A**, 134–137.

Sligh, J. E., Levy, S. E., Waymire, K. G., Allard, P., Dillehay, D. L., Nusinowitz, S. et al. (2000). Maternal germ-line transmission of mutant mtDNAs from embryonic stem cell-derived chimeric mice. *Proc Natl Acad Sci USA* **97**, 14461–14466.

Smeitink, J. A. M., Elpeleg, O., Antonicka, H., Diepstra, H., Saada, A., Smits, P. et al. (2006). Distinct clinical phenotypes associated with a mutation in the mitochondrial translation elongation factor EFTs. *Am J Hum Genet* **79**, 869–877.

Spinazzola, A., Viscomi, C., Fernandez-Vizarra, E., Carrara, F., D'Adamo, P., Calvo, S. et al. (2006). MPV17 encodes an inner mitochondrial membrane protein and is mutated in infantile hepatic mitochondrial DNA depletion. *Nature Genet* **38**, 570–575.

Spinazzola, A., Zeviani, M. (2005). Disorders of nuclear-mitochondrial intergenomic signaling. *Gene* **354**, 162–168.

Sue, C. M., Tanji, K., Hadjigeorgiou, G., Andreu, A. L., Nishino, I., Krishna, S. et al. (1999). Maternally inherited hearing loss in a large kindred with a novel mutation in the mitochondrial DNA $tRNA^{Ser(UCN)}$ gene. *Neurology* **52**, 1905–1908.

Takakubo, F., Cartwright, P., Hoogenraad, N., Thorburn, D. R., Collins, F., Lithgow, T., Dahl, H. (1995). An amino acid substitution in the pyruvate dehydrogenase E1α gene, affecting mitochondrial import of the precursor protein. *Am J Hum Genet* **57**, 772–780.

Tanji, K., Kunimatsu, T., Vu, T. H., Bonilla, E. (2001). Neuropathological features of mitochondrial disorders. *Cell & Develop Biol* **12**, 429–439.

Tanji, K., Schon, E. A., DiMauro, S., Bonilla, E. (2000). Kearns-Sayre syndrome: Oncocytic transformation of choroid plexus epithelium. *J Neurol Sci* **178**, 29–36.

Tatuch, Y., Christodoulou, J., Feigenbaum, A., Clarke, J., Wherret, J. , Smith, C., Rudd, N., Petrova-Benedict, R., Robinson, B. H. (1992). Heteroplasmic mtDNA mutation (T>G) at 8993 can cause Leigh disease when the percentage of abnormal mtDNA is high. *Am J Hum Genet* **50**, 852–858.

Taylor, R. W., Giordano, C., Davidson, M. M., d'Amati, G., Bain, H., Hayes, C. M. et al. (2003). A homoplasmic mitochondrial transfer ribonucleic acid mutation as a cause of maternally inherited cardiomyopathy. *J Am Coll Cardiol* **41**, 1786–1796.

Thorburn, D. R. (2006). Mitochondrial reproductive medicine. In DiMauro, S., Hirano, M., Schon, E. A., Eds., *Mitochondrial Medicine*, 241–259. Informa Healthcare, London.

Tieu, K., Przedborski, S. (2006). Mitochondrial dysfunction and neurodegenerative disorders. In DiMauro, S., Hirano, M., Schon, E. A., Eds., *Mitochondrial Medicine*, 279–307. Informa Healthcare, London.

Tiranti, V., Hoertnagel, K., Carrozzo, R., Galimberti, C., Munaro, M., Granatiero, M. et al. (1998). Mutations of SURF-1 in Leigh disease associated with cytochrome c oxidase deficiency. *Am J Hum Genet* **63**, 1609–1621.

Wallace, D. C. (2005). The mitochondrial genome in human adaptive radiation and disease: On the road to therapeutics and performance enhancement. *Gene* **354**, 169–180.

Wallace, D. C., Brown, M. D., Lott, M. T. (1999). Mitochondrial DNA variation in human evolution and disease. *Gene* **238**, 211–230.

Wallace, D. C., Singh, G., Lott, M. T., Hodge, J. A., Schurr, T. G., Lezza, A., Elsas, L. J., Nikoskelainen, E. K. (1988). Mitochondrial DNA mutation associated with Leber's hereditary optic neuropathy. *Science* **242**, 1427–1430.

Wilson, F. H., Hariri, A., Farhi, A., Zhao, H., Petersen, K. F., Toka, H. R. et al. (2004). A cluster of metabolic defects caused by mutation in a mitochondrial tRNA. *Science* **306**, 1190–1194.

Zhang, M., Mileykovskaya, E., Dowhan, W. (2005). Cardiolipin is essential for organization of complexes III and IV into a supercomplex in intact yeast mitochondria. *J Biol Chem* **280**, 29403–29408.

Zhu, Z., Yao, J., Johns, T., Fu, K., De Bie, I., Macmillan, C. et al. (1998). SURF1, encoding a factor involved in the biogenesis of cytochrome c oxidase, is mutated in Leigh syndrome. *Nature Genet* **20**, 337–343.

Zuchner, S., Mersiyanova, I. V., Muglia, M., Bissar-Tadmouri, N., Rochelle, J., Dadali, E. L. et al. (2004). Mutations in the mitochondrial GTPase mitofusin 2 cause Charcot-Marie-Tooth neuropathy type 2A. *Nature Genet* **36**, 449–451.

34

Molecular Neurology of HIV-1 Infection and AIDS

Marcus Kaul, and Stuart A. Lipton

I. Introduction

The worldwide development of disease associated with infection by the human immunodeficiency virus-1 (HIV-1) and acquired immunodeficiency syndrome (AIDS) is alarming, with estimated numbers having grown from more than 35 million existing infections in 2001 to more than 40 million in 2005, and over 20 million deaths since 1981 (UNAIDS, 2005). HIV-1 destroys the immune system of its host and eventually leads to AIDS, and also provokes a variety of neurological problems that can culminate in frank dementia. AIDS-related opportunistic infections may affect the central nervous system (CNS) more often in the absence of treatment than in the presence of medication, but HIV infection itself can also induce a number of neurological syndromes (Petito et al., 1986). Neuropathological conditions directly triggered by HIV-1 include peripheral neuropathies, vacuolar myelopathy, and a clinical syndrome of cognitive and motor dysfunction that has been designated HIV-associated dementia (HAD) (Gendelman et al., 2005; Glass et al., 1993; Kaul et al., 2001; Power et al., 2002). A mild form of HAD is termed minor cognitive motor disorder (MCMD) (Ellis et al., 1997; Gendelman et al., 2005; Kaul et al., 2001).

The mechanisms contributing to the development of MCMD and HAD remain incompletely understood, but interestingly anemia in HIV-1 infection is a major risk factor for the development of neuropsychological impairment (McArthur et al., 1993). On the other hand, the discovery in the brain of cellular binding sites for HIV-1, namely chemokine receptors, and further progress in understanding neuroinflammation and neural stem cell biology continue to provide new and surprising insights into HIV-mediated neurological syndromes (Gonzalez-Scarano & Martin-Garcia, 2005; Jones & Power, 2006; Kaul et al., 2001, 2005; Kramer-Hammerle et al., 2005; Lavi et al., 1998; Miller & Meucci, 1999; Minghetti, 2005). The present chapter will discuss recent developments regarding the understanding of the neurotoxicity of HIV-1 in the central nervous system (CNS) and potential approaches for therapy and prevention of HAD.

II. The Influence of Highly-Active Antiretroviral Therapy on the Epidemiology of HIV-Associated Dementia

In the early years when AIDS was first recognized as a disease, the majority of severe neurological symptoms occurred in advanced stages of systemic HIV-1 infection, and the prevalence of HAD was estimated to be 20 to 30 percent in individuals with low CD4 T-cell counts (McArthur et al., 1993). Additionally, as mentioned earlier, anemia was shown to be a risk factor for neuropsychological impairment (McArthur et al., 1993). The introduction of highly active antiretroviral therapy (HAART) has increased the life expectancy of people infected with HIV-1 and resulted in at least temporary decrease in the incidence of HAD to as low as 10.5 percent (McArthur et al., 2003). This effect attests to the point that the effects of HIV-1 infection in the brain should always be considered in conjunction with systemic conditions, and it is now widely understood that peripheral infection and associated immune response or inflammatory processes can influence cells in the CNS (Chakravarty & Herkenham, 2005; Turrin & Rivest, 2004). Indeed, improved control of peripheral viral replication and the treatment of opportunistic infections continue to extend survival times, but HAART fails to provide protection from MCMD or HAD, or to reverse the disease in most cases (Cunningham et al., 2000).

Although MCMD may be more prevalent than frank dementia in the HAART era, HAD continues to be a significant independent risk factor for death due to AIDS, and it is assumed to be the most common cause of dementia worldwide among people of age 40 or younger (Ellis et al., 1997; Lipton & Gendelman, 1995). Moreover, the proportion of new cases of HAD displaying a CD4 cell count greater then $200\,\mu L^{-1}$ is growing (McArthur et al., 2003), and another recent study found that in a group of 669 HIV patients who died between 1996 and 2001, more than 90 percent had been diagnosed with HAD as an AIDS-defining condition within the last 12 month of life (Welch & Morse, 2002). This situation at least in part might be due to poor penetration into the CNS of the treatment drugs, including HIV protease inhibitors and several of the nucleoside analogues, and in fact distinct patterns of viral drug resistance have been observed in plasma and cerebrospinal fluid (CSF) compartments (Cunningham et al., 2000; Kaul et al., 2005; Kramer-Hammerle et al., 2005).

Although HIV seems to penetrate into the CNS soon after infection of the periphery, and then resides primarily in perivascular macrophages and microglia (Gartner, 2000; Gonzalez-Scarano & Martin-Garcia, 2005; Ho et al., 1985), current therapeutic guidelines for AIDS suggest to start HAART only once the number of $CD4^{+}$ T-cells begin to decline. Since this might occur up to some years after peripheral infection, HAART is unlikely to prevent the entry of HIV-1 into the CNS (Kramer-Hammerle et al., 2005). Consequently, as people live longer with HIV-1 and AIDS, the prevalence of dementia is expected to rise, and in recent years the incidence of HAD as an AIDS-defining illness has also increased (Cunningham et al., 2000; Jones & Power, 2006; Kaul et al., 2001, 2005; Kramer-Hammerle et al., 2005; Lipton, 1997b; McArthur et al., 2003). Therefore, a better understanding of the pathogenesis of HAD, including viral and host factors, is urgently required in order to identify additional therapeutic targets for the prevention and treatment of this neurodegenerative disease since its prevalence, at least in the MCMD form, appears to be increasing.

III. Neuropathology of HIV Infection and Pathogenesis of HIV-Associated Dementia

The neuropathological hallmarks of HIV infection in the brain are termed HIV encephalitis and include widespread reactive astrocytosis, myelin pallor, microglial nodules, activated resident microglia, multinucleated giant cells, and infiltration by peripheral blood-derived cells, including macrophages (Budka, 1991).

Histological studies in specimens from HIV-1 infected humans and SIV infected rhesus macaques found that both lymphocytes and monocytes infiltrate the brain (Kalams & Walker, 1995; Prospero-Garcia et al., 1996). The pathophysiological relevance of CNS invading lymphocytes in HAD is not clearly established (Kalams & Walker, 1995; Mennicken et al., 1999). However, in one study invasive lymphocytes and activated microglia in brains with HIV-1 encephalitis showed strong immunoreactivity for IL-16, a natural ligand of CD4. Since this cytokine inhibits HIV-1 propagation, lymphocytes might contribute to an innate antiviral immune response in the CNS in addition to microglia (Zhao et al., 2004b).

Surprisingly, measures of cognitive function do not correlate well with numbers of HIV-infected cells, multinucleated giant cells, or viral antigens in CNS tissue (Glass et al., 1995; Masliah et al., 1997). In contrast, increased numbers of microglia (Glass et al., 1995), elevated tumor necrosis factor-alpha (TNF-α) mRNA in microglia and astrocytes (Wesselingh et al., 1997), evidence of excitotoxins (Giulian et al., 1996; Heyes et al., 1991; Jiang et al., 2001), decreased synaptic and dendritic density (Everall et al., 1999; Masliah et al., 1997), and selective neuronal loss (Fox et al., 1997; Masliah et al., 1992) constitute the pathologic features most closely associated with the clinical signs of HAD. Furthermore, signs of neuronal apoptosis have been linked to HAD (Adle-Biassette et al., 1995; Gelbard et al., 1995; Petito & Roberts, 1995), although this finding is not clearly associated with viral burden (Adle-Biassette et al., 1995) or a history of dementia (Adle-Biassette et al., 1999). The localization of apoptotic neurons is correlated with evidence of structural atrophy and closely associated with signs of microglial activation, especially within subcortical deep gray structures (Adle-Biassette et al., 1999), which may show a predilection for atrophy in HAD.

The neuropathology observed in postmortem specimens from HAD patients in combination with extensive studies using both *in vitro* and animal models of HIV-induced neurodegeneration have led to a fairly complex model for the pathogenesis of HAD. The available information strongly suggests that the pathogenesis of HAD might be most effectively explained when viewed as similar to the multi-hit model of oncogenesis. Figure 34.1 presents a model of potential intercellular interactions and alterations of normal cell functions that can lead to neuronal injury and death in the setting of HIV infection (Kaul et al., 2001). Macrophages and microglia can be infected by HIV-1, but they can also be activated by factors released from infected cells. These factors include cytokines and shed viral proteins such as gp120. Variations of the HIV-1 envelope protein gp120, in particular in its V1, V2, and V3 loop sequences, have been implicated in modulating the activation of macrophages and microglia (Power et al., 1998). Factors released by activated microglia affect all cell types in the CNS, resulting in induction or upregulation of cytokines, chemokines, and endothelial adhesion molecules (Gartner, 2000; Kaul et al., 2001; Lipton & Gendelman, 1995).

Some of these factors may directly or indirectly contribute to neuronal damage and apoptosis. Directly neurotoxic factors released from activated microglia include excitatory amino acids (EAAs) and related substances, such as quinolinate, cysteine, and a not completely characterized amine compound named Ntox (Brew et al., 1995; Giulian et al., 1990, 1993; Jiang et al., 2001; Lipton et al., 1991; Yeh et al., 2000; Zhao et al., 2004a). EAAs induce neuronal apoptosis through a process known as excitotoxicity. This detrimental process engenders excessive Ca^{2+} influx and free radical (nitric oxide and superoxide anion) formation by overstimulation of glutamate receptors (Bonfoco et al., 1995; Lipton et al., 1991). Certain HIV proteins, such as gp120 and Tat, also have been reported to be *directly* neurotoxic, although high concentrations of viral protein may be needed, or neurons may have to be cultured in isolation to see these direct effects (Meucci et al., 1998; Liu et al., 2000). It is important to note that toxic viral proteins, among other factors released from microglia, and glutamate, released by astrocytes, may act in concert to promote neurodegeneration even in the absence of extensive viral invasion of the CNS.

IV. HIV Entry into the Brain and Development of Minor Cognitive-Motor Disorder and HIV-Associated Dementia

HIV-1 productively infects macrophages and lymphocytes, first in the periphery and then in the brain, after binding of the viral envelope protein gp120 to one of several possible chemokine receptors in conjunction with the cell surface protein CD4. Depending on the primary sequence of their gp120, different HIV-1 strains may use CCR5 (CD195), CCR3, CXCR4 (CD184), or a combination of these chemokine receptors to enter target cells (Dragic et al., 1996; He et al., 1997; Oberlin et al., 1996).

Since most transmitted viruses use CCR5, genetic mutation of this receptor molecule (Δ32-CCR5) can provide substantial protection against HIV-1 infection (Liu et al., 1996). Some individuals who become infected but remain asymptomatic long term and do not progress to AIDS have been found to express high levels of certain CCR5-binding β-chemokines (Paxton et al., 1996). Again a few people never show seroconversion and seem to mount an unconventional, very effective humoral immune response that includes IgA antibodies against viral glycoprotein 41 (gp41) and IgG recognizing a CD4-gp120 complex (Lopalco et al., 2005).

Soon after infection in the periphery, HIV penetrates into the CNS where the virus primarily resides in microglia and macrophages (Gartner, 2000; Koenig et al., 1986). Viral load in brain can be measured by quantitative PCR, and the highest concentrations of virus are detected in those structures of the CNS most often affected in patients with HAD (McArthur et al., 1997; Wiley et al., 1998). However, infection of macrophages and microglia alone does not seem to initiate neurodegeneration, and it has therefore been proposed that additional factors associated with advanced HIV infection in the periphery, thus outside the CNS, provide important triggers for events leading to dementia (Gartner, 2000). An elevated number of circulating monocytes that express CD16 and CD69 could constitute one such factor. These cells are activated and tend to adhere to and transmigrate through the normal endothelium

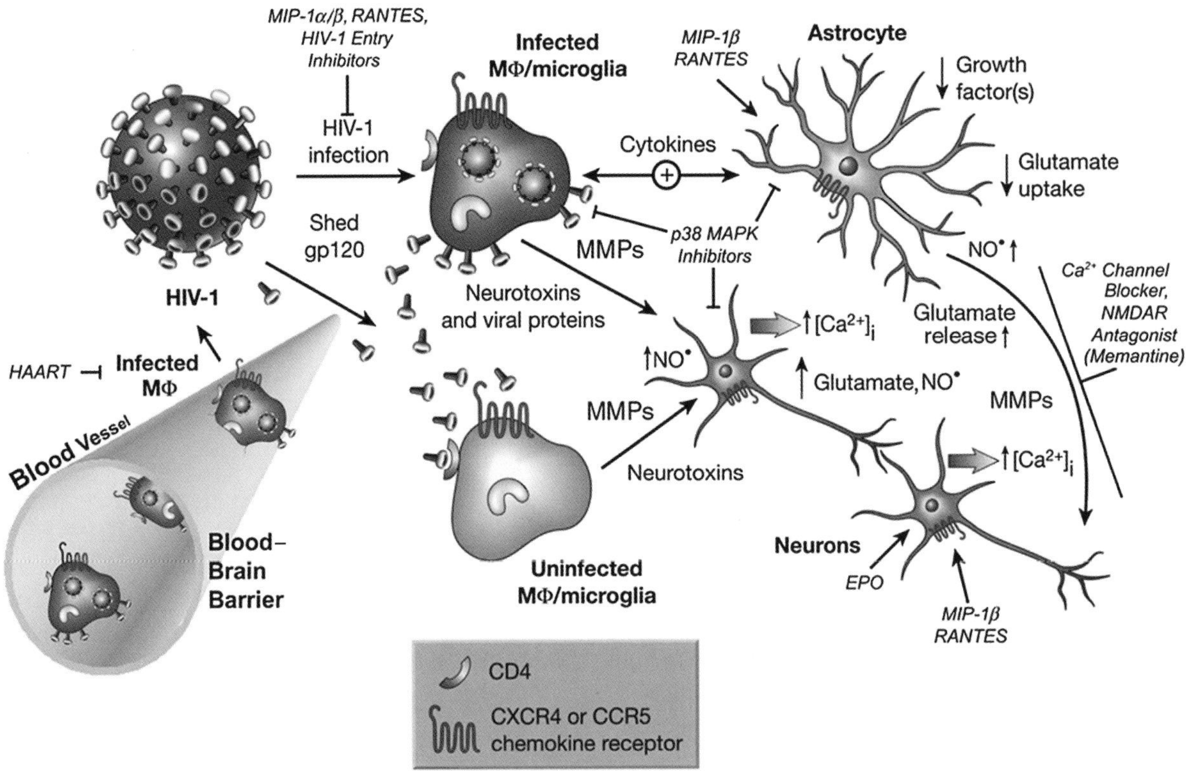

Figure 34.1 Current model of HIV-1 neuropathology indicating presumably toxic or protective factors and potential sites for therapeutic intervention (*protective factors are shown in italic*). Neuronal injury and death induced by HIV-1 infection: Immune-activated and HIV-infected, brain-infiltrating macrophages (MΦ) and microglia release potentially neurotoxic substances. These substances include quinolinic acid and other excitatory amino acids, such as glutamate and L-cysteine, arachidonic acid, PAF, NTox, free radicals, TNF-α, and probably others. These factors from MΦ/microglia and also possibly from reactive astrocytes contribute to neuronal injury, dendritic and synaptic damage, and apoptosis as well as to astrocytosis. Entry of HIV-1 into MΦ/microglia occurs via gp120 binding, and therefore it is not surprising that gp120 (or a fragment thereof) is capable of activating uninfected MΦ/microglia to release similar factors to those secreted in response to productive HIV infection. MΦ/microglia express CCR5 and CXCR4 chemokine receptors on their surface in addition to CD4 and viral gp120 binds via these receptors. Some populations of neurons and astrocytes have been reported to also possess CXCR4 and CCR5 receptors on their surface, raising the possibility of direct interaction with gp120. MΦ/microglia and astrocytes have mutual feedback loops (bidirectional arrow). Cytokines participate in this multicellular network in several ways. For example, HIV-infection or gp120-stimulation of MΦ/microglia enhances their production of TNF-α and IL-1ß (cytokines—arrow). The TNF-α and IL-1ß produced by MΦ/microglia stimulate astrocytosis. Arachidonate released from MΦ/microglia impairs astrocyte clearing of the neurotransmitter glutamate and thus contributes to excitotoxicity. In conjunction with cytokines, the α-chemokine SDF-1 stimulates reactive astrocytes to release glutamate in addition to the free radical nitric oxide [$NO^{\bullet}$], which in turn may react with superoxide ($O_2^{\bullet -}$) to form the neurotoxic molecule peroxynitrite ($ONOO^-$). NO might also activate extracellular matrix metalloproteinases (MMPs), which can then proteolytically affect neurons, and also cleave membrane-anchored fractalkine (Kaul et al., 2005). Neuronal injury is mediated primarily by overactivation of NMDARs with resultant excessive influx of Ca^{2+}. This, in turn, leads to overactivation of a variety of potentially harmful signaling systems, the formation of free radicals, and release of additional neurotransmitter glutamate. Glutamate subsequently overstimulates NMDARs on neighboring neurons, resulting in further injury. This final common pathway of neurotoxic action can be blocked by NMDAR antagonists. For certain neurons, depending on their exact repertoire of ionic channels, this form of damage can also be ameliorated to some degree by calcium channel antagonists or non-NMDAR antagonists. Additionally, MIP-1β and RANTES, agonists of β- chemokine receptors, which are present in the CNS on neurons, astrocytes, and microglia, can confer partial protection against neuronal apoptosis induced by HIV/gp120 or NMDA. Modified from Kaul et al. (2001) and Kaul et al. (2005).

of the brain microvasculature. Once localized in the perivascular space, those monocytic cells might subsequently initiate processes deleterious to neurons (Gartner, 2000).

The blood–brain barrier (BBB) also plays a crucial role in HIV infection of the CNS (Asensio & Campbell, 1999; Gartner, 2000; Nottet et al., 1996; Persidsky et al., 1997). Microglia and astrocytes produce chemokines—cell migration/chemotaxis inducing cytokines—such as monocyte chemoattractant protein (MCP)-1, which appear to regulate migration of peripheral blood mononuclear cells through the BBB (Asensio & Campbell, 1999). In fact, a mutant MCP- 1 allele that causes increased infiltration of mononuclear phagocytes into tissues recently has been implicated in an increased risk of HAD (Gonzalez et al., 2002).

Cell migration also engages adhesion molecules, and increased expression of vascular cell adhesion molecule-1

(VCAM-1) has been implicated in mononuclear cell migration into the brain during HIV and SIV infection (Nottet et al., 1996; Persidsky et al., 1997; Sasseville et al., 1994). As an alternative to entry via infected macrophages, it has been suggested that the inflammatory cytokine, TNF-α, promotes a paracellular route for HIV-1 across the BBB (Fiala et al., 1997). Interestingly, alterations in the BBB occur even in the absence of intact virus in transgenic mice expressing the HIV envelope protein gp120 in a form that circulates in plasma (Marshall et al., 1998). This finding suggests that circulating virus or envelope proteins may provoke BBB dysfunction during the viremic phase of primary infection. On the part of the host, a vicious cycle of immune dysregulation and BBB dysfunction might be required to achieve sufficient entry of infected or activated immune cells into the brain to cause neuronal injury (Bazan et al., 1997; Kaul et al., 2001). On the side of the virus, variations of the envelope protein gp120 might also influence the timing and extent of events allowing viral entry into the CNS and leading to neuronal injury (Power et al., 1998).

V. The Role of Chemokine Receptors in HIV-1 Infection and HIV-Associated Dementia

Chemokine receptors are seven transmembrane-spanning domain, G-protein coupled receptors, and as such trigger intracellular signaling events. While chemokines and their receptors originally were shown to mediate leukocyte trafficking and to contribute intimately to the organization of inflammatory responses of the immune system, they are now known to contribute to far more physiological and pathological processes (Bazan et al., 1997; Oberlin et al., 1996; Tran & Miller, 2003). The additional functions include the intricate control of organogenesis, including hematopoiesis, angiogenesis, and development of heart and brain (Locati & Murphy, 1999; Ma et al., 1998; Tachibana et al., 1998; Zou et al., 1998). Furthermore, chemokines and their receptors are essential for maintenance, maturation, and migration of hematopoietic and neural stem cells (Lapidot & Petit, 2002; Tran & Miller, 2003). However, the most prominent pathological function of certain chemokine receptors seems to be the mediation of HIV-1 infection (Alkhatib et al., 1996; Bleul et al., 1996; Locati & Murphy, 1999).

Infection of macrophages and lymphocytes by HIV-1 can occur after binding of the viral envelope protein gp120 to one of several possible chemokine receptors in conjunction with CD4. Generally, lymphocytes are infected via the α-chemokine receptor CXCR4 and/or the β-chemokine receptor CCR5. In contrast, macrophages and microglia primarily are infected via the β-chemokine receptor CCR5 or CCR3, but the α-chemokine receptor CXCR4 may also be involved (Chen et al., 2002; He et al., 1997; Michael & Moore, 1999; Ohagen et al., 1999). The HIV co-receptors CCR5 and CXCR4, among other chemokine receptors, are also present on neurons and astrocytes (Kaul et al., 2006; Rottman et al., 1997; Zhang et al., 1998), although these cells are not thought to harbor productive infection. Several *in vitro* studies strongly suggest that CXCR4 is directly involved in HIV-associated neuronal damage, and that CCR5 may additionally serve a protective role (Hesselgesser et al., 1998; Kaul & Lipton, 1999; Kaul et al., 2006; Meucci et al., 1998).

In cerebrocortical neurons and neuronal cell lines from humans and rodents, picomolar concentrations of HIV-1 gp120, as well as intact virus, can induce neuronal death via CXCR4 receptors (Chen et al., 2002; Garden et al., 2004; Hesselgesser et al., 1998; Kaul & Lipton, 1999; Kaul et al., 2006; Ohagen et al., 1999). In mixed neuronal/glial cerebrocortical cultures that mimic the cellular composition of the intact brain, this apoptotic death appears to be mediated predominantly via the release of microglial toxins rather than by direct neuronal damage (Chen et al., 2002; Garden et al., 2004; Kaul & Lipton, 1999). However, nanomolar concentrations of SDF-1α/β interacting with CXCR4 can induce apoptotic death of neurons in the absence of microglia, suggesting a possible direct interaction with neurons while interaction with astrocytes can also occur (Bezzi et al., 2001; Kaul & Lipton, 1999; Zheng et al., 1999). In contrast to these findings, it has been reported that somewhat higher concentrations of SDF-1α provide neuroprotection from CXCR4 (X4)-preferring gp120-induced damage of isolated hippocampal neurons (Meucci et al., 1998).

Using mixed neuronal/glial cerebrocortical cultures from rat and mouse, we have further investigated the role of chemokine receptors in the neurotoxicity of gp120. We found that gp120 from CXCR4 (X4)-preferring as well as CCR5 (R5)-preferring and dual tropic HIV-1 strains all were able to trigger neuronal death. Although, as expected, gp120 from at least some X4-preferring HIV-1 strains did not exhibit neurotoxicity in CXCR4-deficient cerebrocortical cultures, dual tropic $gp120_{SF2}$ surprisingly displayed even greater neurotoxicity in CCR5 knockout cultures compared to wild-type or CXCR4-deficient cultures (Kaul, 2002; Kaul et al., 2006). These findings are consistent with a primarily neurotoxic effect of CXCR4 activation by gp120. In contrast, activation of CCR5 at least in part might be neuroprotective, depending on the HIV-1 strain from which a given gp120 originated. Along these lines, we had observed earlier that the CCR5 ligands MIP-1β and RANTES could protect neurons against gp120-induced toxicity (Kaul & Lipton, 1999). The protective mechanism of these β-chemokines involves competition for receptor binding in the case of CCR5-preferring gp120 and heterologous desensitization of receptor signaling if the viral envelope protein utilizes CXCR4 (Kaul et al., 2006).

Since *in vitro* inhibition of microglial activation is sufficient to prevent neuronal death after gp120 exposure, it seems likely that stimulation of CXCR4 in macrophages/

microglia is a prerequisite for the neurotoxicity of gp120 (Kaul & Lipton, 1999; Ohagen et al., 1999). In contrast, SDF-1 might directly activate CXCR4 in astrocytes and neurons to trigger neuronal death, for example, by reversing glutamate uptake in astrocytes (Bezzi et al., 2001; Hesselgesser et al., 1998; Kaul & Lipton, 1999; Kaul et al., 2001). SDF-1 is produced by astrocytes, macrophages, neurons, and Schwann cells (Gleichmann et al., 2000; McGrath et al., 1999; Stumm et al., 2002; Zheng et al., 1999). An increase in SDF-1 mRNA has been detected in HIV encephalitis (Zhang et al., 1998), and protein expression of SDF-1 also appears to be elevated in the brains of HIV patients (Langford et al., 2002). To what degree the increased expression of SDF-1 aggravates neuronal damage by HIV-1 remains to be shown. We had reported previously that intact SDF-1 can be toxic to mature neurons in a CXCR4-dependent manner, at least in culture (Kaul, 2002; Kaul & Lipton, 1999; Zheng et al., 1999). Additionally, it was recently reported that cleavage of SDF-1 by matrix metalloproteinases (MMPs) may contribute to neuronal injury and thus HAD via a non-CXCR4-mediated mechanism (Zhang et al., 2003). Importantly, increased expression and activation of MMPs, including MMP-2 and MMP-9, were detected in HIV-infected macrophages and also in postmortem brain specimens from AIDS patients compared with uninfected controls (Johnston et al., 2000). As elegantly shown by Power and colleagues, MMP-2 released from HIV-infected macrophages is able to proteolytically remove four amino acids from the N-terminus of SDF-1. This truncated form of SDF-1 no longer binds CXCR4 and appears to be an even more powerful neurotoxin than full length SDF-1 (Zhang et al., 2003).

VI. Chemokines and HIV/gp120 Influence Neural Stem and Progenitor Cells

The CXCR4-SDF-1 receptor-ligand axis plays an important role in the physiological function of hematopoietic and neural stem cells (Asensio & Campbell, 1999; Tran & Miller, 2003). This fact suggests the potential for HIV-1 and its envelope protein to directly interfere with biological functions of neural stem and progenitor cells.

In cultures of primary mouse and human neural progenitor cells obtained from fetal tissue, cells stain positively for the neural stem cell marker nestin and readily undergo cell division. After several rounds of proliferation, the progenitors exit the cell cycle and express neuronal markers such as βIII-tubulin (TuJ1). Our immunocytochemical studies showed that these progenitors expressed CXCR4 and CCR5 chemokine receptors. We observed that exposure to HIV-1/gp120 reduced the number of progenitors and differentiating neurons. Accounting for these observations, we found that gp120 inhibited proliferation of neural progenitor cells without producing apoptosis. The resulting decrease in neural stem cell proliferation engendered by gp120 also meant that there were fewer progenitor cells present to differentiate into neurons, thus impairing neurogenesis (Okamoto, McKercher, Kaul, Lipton, unpublished).

These findings were complemented and extended by others using commercially generated human neural progenitor cells (Krathwohl & Kaiser, 2004a; Krathwohl & Kaiser, 2004b). In those experiments, chemokines promoted the quiescence and survival of human neural progenitor cells via stimulation of CXCR4 and CCR3 and a mechanism that involves downregulation of extracellularly regulated kinase-1 and -2 (ERK-1/2) with simultaneous upregulation of the neuronal glycoprotein Reelin (Krathwohl & Kaiser, 2004a). Exposure to HIV-1 caused quiescence of neural progenitors, again through engagement of CXCR4 and CCR3. The coat protein HIV-1/gp120 reportedly downregulated ERK-1/2 but had no effect on Reelin (Krathwohl & Kaiser, 2004b). Interestingly, the effects of both the chemokines and HIV-1/gp120 were reversible and could be inhibited with recombinant Apolipoprotein E3 (ApoE3), but not ApoE4. Although it is widely accepted that HIV-1 fails to productively infect neurons, it has been reported that neural progenitor cells are permissive to the virus (Mattson et al., 2005). The apparent ability of HIV-1/gp120 to interfere with the normal function of neural progenitor cells suggested the possibility that HAD might develop as a consequence not only of injury and death of existing neurons but also due to virus-induced disturbance of potential repair mechanisms in the CNS (see Figure 34.2).

VII. The Role of Macrophages and Microglia in HIV-Induced Neuronal Injury and HIV-Associated Dementia

Macrophages and microglia play a pivotal role, although somewhat paradoxical, in the pathobiology of HAD (Kaul et al., 2001; Luo et al., 2003; Milligan et al., 1991). Under steady-state conditions, mononuclear phagocytes, macrophages, and microglia act as scavengers and sentinel cells eliminating foreign material, and secreting trophic factors critical for maintenance of homeostasis within the CNS microenvironment (Elkabes et al., 1996; Gras et al., 2003; Lazarov-Spiegler et al., 1996; Rapalino et al., 1998; Zheng et al., 1999). A number of neurotrophins are secreted by macrophages (Robinson et al., 1986). These factors include but are not limited to, brain-derived neurotrophic factor (BDNF) (Miwa et al., 1997), insulin-like growth factor-2 (IGF-2) (Nicholas et al., 2002), β-nerve growth factor (βNGF) (Grace et al., 1999), transforming growth factor beta (TGF-β) (Chao et al., 1995), neurotrophin-3 (NT3) (Kullander et al., 1997) and glial-derived neurotrophic factor (GDNF) (Batchelor et al., 1999). During disease, however, a

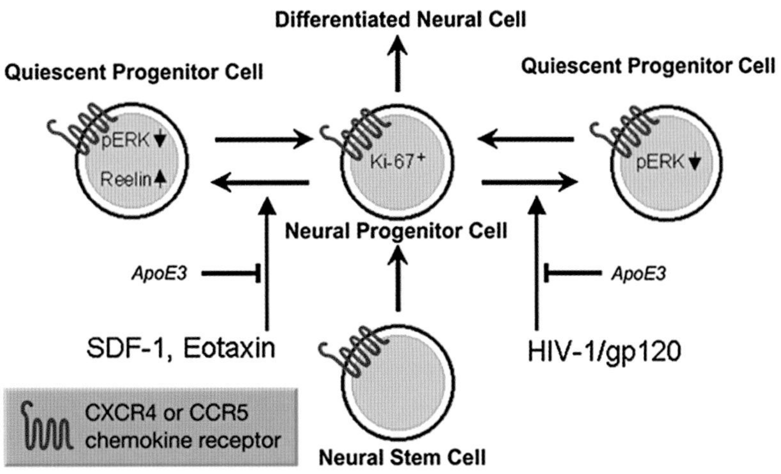

Figure 34.2 Current model of HIV-1 interference with the function of neural progenitor cells and potential sites for therapeutic intervention (*protective factors are shown in italics*): Exposure to chemokines, SDF-1 and Eotaxin, or HIV-1/gp120 of mouse or human neural progenitor cells (NPCs) reduces proliferation and promotes quiescence. ApoE3 inhibits these effects on NPCs. NPCs express nestin and show decreased proliferation as judged by decreased BrdU incorporation. However, NPCs do not undergo apoptosis, as evidenced by lack of TUNEL staining and nuclear condensation under the same conditions (Krathwohl & Kaiser, 2004a, 2004b; S. Okamoto, S. McKercher, M. Kaul & S. A. Lipton, unpublished observations). Modified from Kaul et al. (2001) and Kaul et al. (2005).

dysregulation of macrophage neurotrophic factors by viral infection and/or immune activation may occur and protective function may cease in favor of destructive ones. Also, such dysregulation may be as important as the production of neurotoxins for eliciting neuronal damage. Additionally, some neurotrophic factors are regulated by cytokines. For example, TNF-α (a candidate HIV-1-induced neurotoxin) produced by immune competent microglia can play a neurotrophic role by inducing biologically active TGF-β (Chao et al., 1995). TGF-β is a protective cytokine for mammalian neurons, particularly in protection against glutamate neurotoxicity, hypoxia and gp120-mediated neural injury (Meucci & Miller, 1996). This cytokine also affects long-term synaptic facilitation (Milligan et al., 1991).

HIV establishes a latent and persistent infection within macrophages (Koenig et al., 1986). The majority of HIV within the CNS appears to be localized within perivascular and blood-derived parenchymal brain macrophages and microglia (Koenig et al., 1986). Astrocytes, oligodendrocytes, and brain endothelial cells are rarely infected, if at all (Tornatore et al., 1994). As a result of viral infection and resultant immune activation macrophages produce and release a variety of neurotoxins within the brain (Gendelman et al., 1997; Kaul et al., 2001; Nath, 1999). These products comprise not only viral proteins, such as gp120 (Brenneman et al., 1988), gp41 (Adamson et al., 1996), and Tat (Nath et al., 1996), but also host cell-encoded products including platelet activating factor (PAF) (Gelbard et al., 1994), glutamate (Jiang et al., 2001), arachidonic acid and its metabolites (Nottet et al., 1995), pro-inflammatory cytokines, such as interleukin-1 beta (IL-1β), tumor necrosis factor alpha (TNF-α), TNF-related apoptosis inducing ligand (TRAIL) (Gelbard et al., 1993; Ryan et al., 2001), quinolinic acid (Heyes et al., 1991; Kerr et al., 1997), NTox (Giulian et al., 1996), and indirectly nitric oxide (Adamson et al., 1996), among others. Interestingly, in any case it seems that activation of p53 in microglia plays a crucial role for neurotoxicity to occur upon exposure of the cells to HIV-1/gp120 (Garden et al., 2004). In this manner macrophages, which were once protective constituents of the immune system, are now responsible for tissue damage, though it is still unclear how macrophages evolve from producing neurotrophins to producing neurotoxins. Perhaps HIV-1 infection and immune activation induces a transition between neurotrophic and neurotoxic activities.

VIII. Molecular Mechanisms of Neuronal Injury and Death in HIV-Associated Dementia

A. Physiological and Pathological Roles of Ionotropic Glutamate Receptors: The Bright and Dark Sides of N-Methyl-D-Aspartate Receptors (NMDARs)

A recurring question has been whether HIV-1 or its component proteins induce neuronal damage predominantly by an indirect route (e.g., via toxins produced by infected or immune-stimulated macrophages and/or astrocytes), or by a direct route (e.g., via binding to neuronal receptors) (Kaul & Lipton, 1999; Kaul et al., 2001; Lipton, 1997a; Lipton & Gendelman, 1995). Several lines of evidence suggest that HIV-associated neuronal injury involves predominantly an indirect route from macrophage and astrocyte toxins resulting in excessive activation of NMDARs and Ca^{2+} influx with consequent excitotoxicity (see Figure 34.1) (Doble, 1999; Dreyer et al., 1990; Giulian et al., 1990; Kaul & Lipton, 1999; Kaul et al., 2001; Lipton et al., 1991; Olney, 1969; Sardar et al., 1999).

Under physiological conditions, activation of ionotropic glutamate receptors in neurons initiates transient depolarization and excitation. AMPARs mediate a fast component of excitatory postsynaptic potentials, and NMDARs underlie a slower component. Presynaptic release of glutamate and consequent depolarization of the postsynaptic neuronal membrane via AMPAR-coupled channels relieve the Mg^{2+} block of the NMDAR-associated ion channel that occurs under resting conditions. This effect allows subsequent controlled Ca^{2+} influx through the NMDAR-coupled ion channel. This voltage-dependent modulation of the NMDAR results in activity-driven synaptic modulation (Bigge, 1999; Doble, 1999). However, extended and/or excessive NMDAR activation, particularly extrasynaptic receptors, and consequent excitotoxicity is triggered by sustained elevation of the intracellular Ca^{2+} concentration, compromised cellular energy metabolism, and resultant free radical formation (Doble, 1999; Lipton & Rosenberg, 1994; Olney, 1969).

A role for excitotoxicity in brain disorders was first suggested by the work of Olney following the pioneering work in 1957 of Lucas and Newhouse in the retina (Olney, 1969; Olney & Sharpe, 1969). Subsequently, several lines of evidence indicated that excessive stimulation of glutamate receptors contributes to the neuropathological process in a large number of disorders, including stroke, head and spinal cord injury, seizures, Huntington's disease, Parkinson disease, possibly Alzheimer's disease, amyotrophic lateral sclerosis, multiple sclerosis, glaucoma, and HIV-1 associated dementia (Brauner-Osborne et al., 2000; Doble, 1999; Lipton & Gendelman, 1995). Indeed, excitotoxicity seems to represent a common final pathway of neuronal injury and death in a wide variety of neurodegenerative disorders (Lipton & Rosenberg, 1994).

The NMDAR has attracted particular interest as a major player in excitotoxicity because this receptor, in contrast to most non-NMDARs (AMPA and KA receptors), is highly permeable to Ca^{2+}, and excessive Ca^{2+} influx can trigger excitotoxic neuronal injury (Choi, 1988; Weiss & Sensi, 2000). In addition, NMDAR antagonists effectively prevent some forms of glutamate neurotoxicity, both *in vitro* and *in vivo* in animal studies (Bigge, 1999; Choi et al., 1988a; Doble, 1999). This potential as a therapeutic agent was recently borne out in human phase II and III clinical trials with the NMDAR open-channel blocker, memantine, based upon pioneering work by our group (see later). However, AMPA and KA receptors can also mediate excitotoxicity and contribute to neuronal damage under certain conditions (Bigge, 1999; Doble, 1999). For example, a subpopulation of Ca^{2+}- or Zn^{2+}-permeable AMPA receptor-coupled channels have been implicated in selective neurodegenerative disorders, such as ischemia, epilepsy, Alzheimer's disease, and amyotrophic lateral sclerosis (Weiss & Sensi, 2000). Also transgenic mice overexpressing AMPARs display increased damage subsequent to ischemia when compared to control animals (Le et al., 1997).

B. Toxic Signaling Pathways Downstream from NMDARs

Excessive stimulation of the NMDAR induces several detrimental intracellular signals that contribute to neuronal cell death by apoptosis or necrosis, depending on the intensity of the initial insult (Nicotera et al., 1997). For example, excessive Ca^{2+} influx through NMDAR-coupled ion channels leads to an elevation of the intracellular free Ca^{2+} concentration to a point that results in Ca^{2+} overload of mitochondria, depolarization of the mitochondrial membrane potential, and a decrease in ATP synthesis. The scaffolding protein PSD-95 (postsynaptic density-95) links the principal subunit of the NMDAR (NR1) with neuronal nitric oxide synthase (nNOS), a Ca^{2+}-activated enzyme, and thus brings nNOS into close proximity to Ca^{2+} via the NMDAR-operated ion channel (Sattler et al., 1999). Excessive intracellular Ca^{2+} overstimulates nNOS and protein kinase cascades with consequent generation of deleterious levels of free radicals, including reactive oxygen species (ROS) and nitric oxide (NO) (Nicotera et al., 1997). NO can react with ROS to form cytotoxic peroxynitrite ($ONOO^-$) (Nicotera et al., 1997). However, in alternative redox states, NO can inhibit NMDARs (Lei et al., 1992; Lipton et al., 1993), activate p21ras (Gonzalez-Zulueta et al., 2000), and inhibit caspases (Tenneti et al., 1997) via S-nitrosylation (transfer of the NO group to critical cysteine thiols), thereby attenuating apoptosis in cerebrocortical neurons.

Importantly, excessive Ca^{2+} influx and free radicals also activate stress-related p38 mitogen-activated protein kinase (p38 MAPK) and, via c-Jun N-terminal kinase (JNK), c-Jun in cerebrocortical or hippocampal neurons. In turn, p38 MAPK also phosphorylates/activates transcription factors, including direct activation of myocyte enhancer factor 2 (MEF2). Activation of these pathways has been implicated in neuronal apoptosis, probably in conjunction with caspase activation (Kaul & Lipton, 1999; Mukherjee et al., 1999). As stated earlier, excessive intracellular Ca^{2+} accumulation after NMDAR stimulation leads to depolarization of the mitochondrial membrane potential ($\Delta\psi m$) and a drop in the cellular ATP concentration. If the initial excitotoxic insult is fulminant, the neuronal cells do not recover their ATP levels and die at this point because of the loss of ionic homeostasis, resulting in acute swelling and lysis (necrosis). If the insult is more mild, ATP levels recover, and the neuronal cells enter a delayed death pathway requiring energy, known as apoptosis (Nicotera et al., 1997). Interestingly, Zn^{2+} can substitute for Ca^{2+} and lead to neuronal death by these and other pathways (Aizenman et al., 2000; Choi et al., 1988b; Weiss & Sensi, 2000).

It has been reported that NMDAR-mediated excitotoxicity leading to neuronal apoptosis also involves activation of the Ca^{2+}/calmodulin-regulated protein phosphatase calcineurin (Nicotera et al., 1997), mitochondrial permeability transition, release of cytochrome c from mitochondria (Budd et al., 2000), activation of caspase-3 (Tenneti et al., 1998), lipid peroxidation (Tenneti et al., 1998), and cytoskeletal breakdown (Nicotera et al., 1997). Inhibition of calcineurin and caspase-3 by FK506 and caspase inhibitors, respectively, can attenuate this form of excitotoxicity (Nicotera et al., 1997; Tenneti et al., 1998). It has been proposed that the adenine nucleotide translocator (ANT) is a part of the mitochondrial permeability transition pore (PTP) and participates in mitochondrial depolarization. Indeed, our group found that pharmacologic blockade of the ANT with bongkrekic acid prevented collapse of the mitochondrial membrane potential ($\Delta\psi m$), as well as subsequent caspase-3 activation and NMDA-induced neuronal apoptosis. However, treatment with bongkrekic acid failed to inhibit the transient drop in ATP concentration (although it hastened the recovery of ATP levels) and did not prevent the liberation of cytochrome c into the cytosol. Thus, initiation of caspase-3 activation and resultant neuronal apoptosis after NMDAR activation require a factor(s) in addition to cytochrome c release (Budd et al., 2000).

C. HIV-1 Infection of the Brain and Activation of NMDARs

Analysis of specimens from AIDS patients (Sardar et al., 1999) as well as *in vivo* and *in vitro* experiments indicate that HIV-1 infection creates excitotoxic conditions, predominantly via an indirect route. HIV-1 infection induces soluble factors in macrophage/microglia and/or astrocytes, such as glutamate and glutamate-like molecules, viral proteins, cytokines, chemokines, and arachidonic acid metabolites (Bezzi et al., 2001; Kaul et al., 2001; Lipton, 1997a, 1998; Lipton & Gendelman, 1995).

However, it has also been suggested that HIV-1 or its protein components can directly interact with neurons and modulate NMDAR function, at least under some conditions (Meucci et al., 1998; Savio & Levi, 1993). Picomolar concentrations of soluble HIV/gp120 induce injury and apoptosis in primary rodent and human neurons both *in vitro* and *in vivo* (Brenneman et al., 1988; Lannuzel et al., 1995). Additionally, our group and subsequently several others have shown that gp120 contributes to NMDAR-mediated neurotoxicity (Lipton et al., 1991). Both voltage-gated Ca^{2+} channel blockers and NMDAR antagonists can ameliorate gp120-induced neuronal cell death *in vitro* (Dreyer et al., 1990; Lipton et al., 1991). Transgenic mice expressing gp120 manifest neuropathological features that are similar to the findings in brains of AIDS patients, and in these mice neuronal damage is ameliorated by the NMDAR antagonist memantine (Toggas et al., 1994; Toggas et al., 1996) (see later). It is also conceivable that other glutamate receptors in addition to NMDARs influence HIV-associated neuronal damage. Interestingly, stimulation of specific subtypes of the G protein-coupled mGluRs interferes with excitotoxic NMDAR-mediated activation of MAPKs and can attenuate subsequent neuronal cell death (Mukherjee et al., 1999).

In the case of HAD, macrophages and microglia play a crucial role because they are the predominant cells productively infected with HIV-1 in the brain (Lipton & Gendelman, 1995) (see Figure 34.1), although infection of astrocytes has been observed in pediatric cases (reviewed in Brack-Werner & Bell, 1999). In accordance with the report that the presence of macrophages/microglia correlates with the severity of HAD (Glass et al., 1995), in our hands, the predominant mode of neurotoxicity of HIV-1 or gp120 requires the presence and activation of macrophages/microglia (Kaul & Lipton, 1999; Lipton, 1992c, 1994). Moreover, HIV-1-infected or gp120-stimulated mononuclear phagocytes have been shown to release neurotoxins that directly stimulate the NMDAR (Giulian et al., 1990; Kaul & Lipton, 1999; Lipton et al., 1991). Those macrophage toxic factors include molecules that directly or indirectly act as NMDAR agonists, such as quinolinic acid, cysteine, platelet-activating factor (PAF), and a low-molecular weight compound designated NTox (Lipton, 1998; Lipton & Gendelman, 1995; Yeh et al., 2000).

Additionally, HIV-infected or immune-activated macrophages/microglia and possibly astrocytes produce inflammatory cytokines, including TNF-α and IL-1β, arachidonic acid metabolites, and free radicals (ROS and NO) that may indirectly contribute to excitotoxic neuronal damage (see Figure 34.1) (Bezzi et al., 2001; Lipton, 1998; Lipton & Gendelman, 1995). TNF-α and IL-1β may amplify neurotoxin production by stimulating adjacent glial cells and by increasing immunologic NOS activity (Adamson et al., 1996; Lipton, 1998).

In contrast to these indirect neurotoxic pathways, it has been reported that gp120 can directly interact with neurons in the absence of glial cells. Recently, gp120 was found to act at chemokine receptors directly on isolated neurons in culture to induce their death (Meucci et al., 1998). Additionally, higher nanomolar concentrations of gp120 have been reported to interact with the glycine binding site of the NMDAR (Fontana et al., 1997). Furthermore, gp120 may produce a direct excitotoxic influence via NMDAR-mediated Ca^{2+} oscillations in rat hippocampal neurons (Lo et al., 1992), and may bind to noradrenergic axon terminals in neocortex, where it possibly potentiates NMDA-evoked noradrenaline release (Pittaluga et al., 1996). Nonetheless, many if not all of these direct effects on neurons were observed *in vitro* in the absence of glial cells. Since glial cells are known to modify these death pathways, we feel that under *in vivo* conditions, the indirect route to neuronal injury is the predominant one, based on studies in mixed neuronal-glial cultures and on work in a gp120-transgenic mouse (see later and Figure 34.1).

Along these lines, gp120 has been found to aggravate excitotoxic conditions by impairing astrocyte uptake of glutamate via arachidonic acid that is released from activated macrophages/microglia (Dreyer & Lipton, 1995; Lipton, 1997a). The α-chemokine SDF-1, the cytokine TNF-α, and metabolites of arachidonic acid, such as prostaglandins, also stimulate a Ca^{2+}-dependent release of glutamate by astrocytes (Bezzi et al., 1998, 2001). Moreover, HIV-1 can induce astrocytic expression of the β–chemokine known as macrophage chemotactic protein-1 (MCP-1). This β–chemokine in turn attracts additional mononuclear phagocytes and microglia to further enhance the potential for indirect neuronal injury via the release of macrophage toxins (Conant et al., 1998).

In our view, therefore, HIV-1 infection and its associated neurological dysfunction involve both chemokine receptors and NMDAR-mediated excitotoxicity. This dual receptor involvement raises the question of whether G protein-coupled chemokine receptors and ionotropic glutamate receptors might influence each other's activity. Indeed, the β–chemokine known as "regulated and activated normal T cell expressed and secreted" (RANTES), which binds to chemokine receptors CCR1, CCR3, and CCR5, can abrogate neurotoxicity induced by gp120 (Kaul & Lipton, 1999) or by excessive NMDAR stimulation (Bruno et al., 2000). In turn, excitotoxic stimulation can enhance expression of CCR5 (Galasso et al., 1998). Whether or not these findings reflect a mechanism of feedback or crosstalk in which chemokines indirectly antagonize the stimulation of the NMDAR awaits to be elucidated.

IX. Prevention and Therapy of HIV-Associated Dementia: Previous and Potential Future Strategies

A. Previous Approaches to Treatment of HAD

A truly effective pharmacotherapy for HAD has yet to be developed. Previous approaches to cope with HAD reflect the challenging complexity inherent in the treatment of patients with AIDS (reviewed by Melton et al., 1997 and Clifford, 1999). Previous and current therapeutic approaches include various antiretroviral compounds, alone or in combination: (1) reverse transcriptase inhibitors, including Zidovudine, Didanosine, Zalcitabine, Stavudine, and Lamivudine; and (2) protease inhibitors, such as Saquinavir, Ritonavir, and Indinavir. Of these only Zidovudine has been shown to cross the blood–brain barrier to some extent, and Zidovudine has a beneficial effect on HAD but the effect is not longlasting. The other antiretroviral drugs may not penetrate the brain sufficiently to eradicate the virus in the CNS. Thus an adjunctive treatment besides antiretroviral drugs is needed.

B. Current and Future Potential Therapeutic Strategies Targeting Receptors for Glutamate, Chemokines, and Erythropoietin

In past therapeutic attempts, Pentoxyfylline, an inhibitor of production and action of TNF-α, and the neurotrophic peptide T were tested as investigational agents (Melton et al., 1997), but clinical studies assessing their therapeutic potential did not prove substantial benefits. Previous, small clinical trials of the voltage-activated (L-type) calcium channel blocker, nimodipine, and a PAF inhibitor suggested some therapeutic benefit but were not conclusive (Clifford, 1999; Schifitto et al., 1999; Navia et al., 1998).

From the pathogenesis of HAD as described earlier, several potential therapeutic strategies appear viable (see Figure 34.1). NMDAR antagonists are among the agents under consideration. Others include certain chemokines and cytokines, and antagonists for their receptors, p38 MAPK inhibitors, caspase inhibitors, and antioxidants (free radical scavengers or other inhibitors of excessive nitric oxide or reactive oxygen species) (Clifford, 1999; Digicaylioglu et al., 2004b; Kaul et al., 2001; Lipton, 2004; Turchan et al., 2003).

Chemokine receptors allow HIV-1 to enter cells and as such are major potential therapeutic targets in the fight against AIDS in general (Michael & Moore, 1999). Antagonists of CXCR4 and CCR5 inhibit HIV-1 entry and are being assessed in clinical trials (Michael & Moore, 1999). However, the benefit of inhibitors of chemokine receptors for HIV-associated neurological complications awaits study (Gartner, 2000; Kaul et al., 2001, 2005). Interestingly, certain chemokines have been shown to protect neurons from injury, even though the virus does not productively infect neurons. In particular, β-chemokines and fractalkine prevent gp120-induced neuronal apoptosis *in vitro* (Bruno et al., 2000; Kaul & Lipton, 1999), and, similarly, some β-chemokines (i.e., CCR5 agonists) can ameliorate NMDAR-mediated neurotoxicity (Bruno et al., 2000; Kaul & Lipton, 2001). Additionally, the CCR5 ligands MIP-1α, MIP-1β, and RANTES are able to suppress HIV-1 infection in the periphery and are highly expressed in long-term HIV-1 infected individuals who do not, or only very slowly, progress to AIDS (Cocchi et al., 1995; Paxton et al., 1998; Scala et al., 1997; Zagury et al., 1998). HIV-infected patients with relatively higher

CSF concentrations of MIP-1α/β and RANTES performed better on neuropsychological measures then those with low or undetectable levels (Letendre et al., 1999). These findings support the hypothesis that selected β-chemokines may represent a potential treatment modality for AIDS and HAD. One of the efforts underway aims at modification of natural CCR5 ligands in order to avoid adverse inflammatory side effects upon application (De Clercq, 2004; Pierson et al., 2004; Verani & Lusso, 2002).

Previously, we have shown that the cytokine erythropoietin (EPO) may not only be effective in treating anemia but also for protecting neurons, since it prevents NMDAR-mediated and HIV-1/gp120-induced neuronal death in mixed cerebrocortical cultures (Digicaylioglu & Lipton, 2001; Digicaylioglu et al., 2004b). Since EPO is already clinically approved for the treatment of anemia, human trials of EPO as a neuroprotectant from HIV-associated dementia may be expedited (Lipton, 2004). Additionally, EPO plus insulin-like growth factor-1 act synergistically as neuroprotectants by activating the PI3K/Akt pathway (Digicaylioglu et al., 2004a), so the use of these two cytokines in conjunction has been advocated for clinical trials (Lipton, 2004).

NMDAR antagonists have been shown to attenuate neuronal damage due to either HIV-infected macrophages or HIV/gp120, both *in vitro* and *in vivo* (Chen et al., 2002; Dreyer et al., 1990; Lipton, 1992a; Toggas et al., 1996). Both voltage-gated Ca^{2+} channel blockers and NMDAR antagonists can ameliorate gp120-induced neuronal cell death *in vitro* (Dreyer et al., 1990; Lipton et al., 1991). Transgenic mice expressing gp120 in their CNS manifest neuropathological features that are similar to the findings in brains of AIDS patients, and in these mice neuronal damage is ameliorated by the NMDAR antagonist memantine (Toggas et al., 1994, 1996). Memantine-treated gp120 transgenics and nontransgenic control mice retain a density of presynaptic terminals and dendrites that is similar to untreated non-tg/wild-type controls but significantly higher than in untreated gp120 transgenic animals (Toggas et al., 1996). This finding supports the hypothesis that the HIV-1 surface glycoprotein is sufficient to initiate downstream of chemokine receptor activation excitotoxic neuronal injury and death. It also shows that an antagonist of NMDAR overstimulation can ameliorate HIV-associated neuronal damage *in vivo*, an observation that another group recently confirmed (Anderson et al., 2004).

However, the majority of NMDAR antagonists have unacceptable psychotomimetic side effects in humans, and this problem and its solution is discussed next.

NMDAR antagonist drugs with fewer adverse effects were thought to include the glycine site antagonists, but these can cause dizziness and sedation in healthy human volunteers (Lees, 1997). In general, competitive antagonists for the glutamate or glycine coagonist sites may be doomed to failure because they inhibit normal brain function (which occurs at lower levels of agonist) before they block pathological actions (which occur at higher levels of agonist); hence, normal brain areas are affected in an adverse manner prior to the drugs becoming effective in pathologically injured brain regions. Thus, in our view, uncompetitive open-channel blockers have the best chance of emerging as acceptable agents in clinical practice, as discussed next.

As alluded to earlier, many NMDAR antagonists are not clinically tolerated, whereas some others appear to be tolerated by humans at concentrations that are effective neuroprotectants (Lipton, 1993; Lipton & Rosenberg, 1994; Parsons et al., 1999). Several NMDAR antagonists prevent neuronal injury in animal models of a variety of neurological disorders, including HIV-associated dementia, focal stroke, Parkinson disease, Huntington's disease, Alzheimer's disease, amyotrophic lateral sclerosis, neuropathic pain, glaucoma, and others (Choi, 1988; Doble, 1999; Lipton & Rosenberg, 1994; Parsons et al., 1999). Of these drugs, two of the most promising, because of their long experience in patients with other diseases, are memantine (Bormann, 1989; Lipton 2006; Parsons et al., 1999) and nitroglycerin (Lipton, 1993; Lipton & Gendelman, 1995; Lipton & Rosenberg, 1994), as well as new combinatorial agents combining features of both of these drugs (Lipton & Kieburtz, 1998).

Our group was the first to show that memantine blocks the NMDAR-associated ion channel, preferentially when it is open for prolonged (pathological) periods of time; conversely, we showed that during normal neurotransmission, when there is less NMDAR-operated channel activity, memantine has relatively little effect on this activity (Chen & Lipton, 1997; Chen et al., 1992, 1998; Kaul et al., 2001; Le & Lipton, 2001; Lipton, 1992a, 1992b, 1993, 1998; Lipton & Gendelman, 1995; Lipton & Kieburtz, 1998; Lipton & Rosenberg, 1994; Pellegrini & Lipton, 1993; Stieg et al., 1999). We found that unlike other NMDAR open-channel blockers, such as dizocilpine (MK-801), memantine does not remain in the channel for an excessively long time, and hence we discovered that this short dwell time (or relatively fast off rate) is a key factor to memantine's lack of clinical side effects. For example, we found that the relatively short dwell time accounts for the fact that neuroprotective concentrations of memantine manifest little or no effect on the NMDAR component of excitatory post-synaptic potentials (EPSPs), on long-term potentiation (LTP), and on performance in the Morris water maze behavioral task (Chen & Lipton, 1997; Chen et al., 1992, 1998).

Interestingly, although Mg^{2+} has an even shorter dwell time in the channel than memantine, the Mg^{2+} effect is so short-lived that it does not effectively block the NMDAR-associated channel during insult, and thus does not afford significant neuroprotection under most conditions. In contrast, MK-801 can afford neuroprotection by effectively blocking the NMDAR-associated channel, but is not clinically tolerated because its block is too prolonged, contributing to its very high affinity of action, and thus MK-801 blocks all normal physiological activity (Chen & Lipton, 1997; Chen et al., 1992, 1998). Thus, it has been possible to use memantine safely in humans for over 20 years in Europe as a treatment for Parkinson disease and spasticity (Chen & Lipton, 1997; Chen et al., 1992, 1998; Parsons et al., 1999). Our fortuitous breakthrough was the realization that a low-affinity agent such as memantine can afford significant neuroprotection while leaving normal physiological function relatively unaffected.

The affinity of a channel-blocking drug is related to the ratio of its on-time to its off-rate (the latter representing the inverse of its dwell time in the channel). Importantly, the on-time is influenced by the concentration of the drug, but the off-rate (or dwell time) is not concentration related and instead is purely an intrinsic property of the antagonist. Hence, we realized that the off-rate (and hence the dwell time) was a key property of an open-channel blocking drug that contributes to its affinity, to its efficacy as a neuroprotectant, and to its safety or tolerability in the brain. Another important realization in our work was that memantine was selective for NMDARs at a neuroprotective concentration despite its relatively low affinity ($IC_{50} = \sim 1\ \mu M$), and, in fact, a high-affinity agent, such as MK-801, would be toxic. In other words, one does not need high affinity in a drug in order to have high selectivity for its target. Quite the opposite is desired in the brain: A low-affinity agent such as memantine is preferred because this results in relative sparing of normal neurotransmission; however, the drug also needs to be selective for its target receptor in order to avoid side effects stemming from interactions with unwanted targets.

We realized that there were benefits of such a low-affinity agent. For example, increasing concentrations of glutamate/glycine or other NMDA agonists cause NMDAR channels to remain open on average for a greater fraction of time. Under pathological conditions of increased glutamate (and glycine), we discovered that the open-channel blocking drug, memantine, has a better chance to enter the channel and block it (after all, the drug can get into the channel only when it is open, and statistically, more drug will get into the channel when, on average, the channel is open longer). It is because of this mechanism of action that the destructive effects of greater (pathological) concentrations of glutamate are prevented to a greater extent than the effects of lower (physiological) concentrations, which are relatively spared (Chen & Lipton, 1997; Chen et al., 1998; Lipton, 1993, 2006; Lipton & Rosenberg, 1994). This mechanism of inhibition is termed uncompetitive antagonism, defined as the action of the antagonist being contingent upon prior activation of the receptor by the agonist. Moreover, in animal model systems, clinically tolerated concentrations of memantine can ameliorate neuronal injury associated with either focal cerebral ischemia or HIV-1 proteins, both *in vitro* and *in vivo* (Chen et al., 1992; Erdo & Schafer, 1991; Keilhoff & Wolf, 1992; Lipton, 1992a; Lipton & Jensen, 1992; Muller et al., 1992; Osborne & Quack, 1992; Pellegrini & Lipton, 1993; Sathi et al., 1993; Seif el Nasr et al., 1990; Toggas et al., 1996). Thus, because of its lack of major adverse effects under pathological conditions and recent evidence of efficacy in human clinical trials, memantine was the first NMDAR antagonist that was clinically successful, being approved recently for the treatment of moderate-to-severe Alzheimer's disease by both the European Union and the FDA.

Our group also has shown that another potentially clinically useful modulatory agent of the NMDAR is nitroglycerin, which produces nitric oxide-related molecules. Nitric oxide ($NO^{\bullet}$, where the dot represents one unpaired electron in the outer molecular orbital) can contribute to neuronal damage. One of the pathways to neurotoxicity involves the reaction of $NO^{\bullet}$ with $O_2^{\bullet -}$ to form peroxynitrite ($ONOO^-$) (Beckman et al., 1990; Dawson et al., 1991, 1993; Lipton et al., 1993). In contrast, $NO^{\bullet}$ can be converted to a chemical state that has just the opposite effect, that is, one that protects neurons from injury due to NMDA receptor-mediated overstimulation. The change in chemical state is dependent on the removal or addition of an electron to $NO^{\bullet}$. We and our colleagues have demonstrated that with one less electron, $NO^{\bullet}$ acts like nitrosonium ion (NO^+), which facilitates reaction with critical thiol group(s) (R-SH or, more properly, thiolate anion, $R\text{-}S^-$) comprising a redox modulatory site(s) on the NMDA receptor-channel complex, which decreases channel activity (Choi et al., 2000, 2001; Kim et al., 1999; Lei et al., 1992; Lipton et al., 1993). Our group has further shown that this reaction can afford neuronal protection from overstimulation of NMDA receptors, as well as other reactions, which would otherwise result in excessive Ca^{2+} influx (Lipton & Stamler, 1994).

One such drug that can react with NMDA receptors in this redox-related manner is the common vasodilator nitroglycerin (Lei et al., 1992; Lipton, 1993; Lipton & Rosenberg, 1994; Lipton et al., 1993). Chronic use of nitroglycerin induces tolerance to the drug's effects on the cardiovascular system, thus avoiding systemic adverse effects such as hypotension. However, during chronic use, nitroglycerin still appears to work in the brain to attenuate NMDA receptor-mediated neurotoxicity (Lipton, 1993). Nonetheless, the exact dosing regimen has yet to be worked out for the neuroprotective effects of nitroglycerin in the brain; therefore, caution has to be exercised before attempting to implement this form of therapy. In preliminary experiments, including those using animal models of focal ischemia, high concentrations of nitroglycerin were neuroprotective during various NMDA receptor-mediated insults (Lipton & Wang, 1996; Stieg et al., 1999). Our *in vivo* data suggest that this effect of nitroglycerin may, at least in part, be due to a direct effect on neurons, consistent with an action at the NMDAR redox modulatory site(s) (Lipton & Wang, 1996; Stieg et al., 1999).

As the structural basis for redox modulation of the NMDA receptor recently has been further elucidated (Choi et al., 2000, 2001; Das et al., 1998; Kim et al., 1999; Kohr et al., 1994; Sullivan et al., 1994), it has become possible to design even better redox reactive reagents of clinical value, for example, with the NO group in appropriate redox state, targeted specifically to the NMDAR. This targeting strategy avoids hypotensive and other adverse effects of acute systemic administration of NO-related drugs. We and our colleagues have accomplished this goal by synthesizing a series of nitro-memantine compounds; that is, using the NMDA channel blockade by memantine to target the NO group to the NMDA receptor (Lipton, 2006). However, for this to work in an efficient manner, one would want to know the correct length of the "arm" or "bridge" that chemically links memantine with an NO group. To determine the length of this bridge, one should ideally know the location of both the ion pore and the redox site(s) on the NMDA receptor/channel complex.

In recent years, the channel pore has been localized to the second membrane loop of NMDA receptor subunits. Our laboratory has now characterized the redox modulatory sites of NMDAR at a molecular level using site-directed mutagenesis of recombinant NMDAR subunits (NR1; NR2A-D; NR3A, B) as well as crystallographic modeling techniques (Choi et al., 2000, 2001; Das et al., 1998; Kim et al., 1999; Kohr et al., 1994; Lipton et al., 2002; Sullivan et al., 1994). These approaches have facilitated the design strategy of NO-group targeting to the NMDA receptor. Unlike many, if not all, of the other drugs currently under investigation, memantine, nitroglycerin, and combinatorial nitro-memantine compounds have a high degree of clinical tolerability at neuroprotective doses. These facts should expedite clinical studies for the use of these drugs in patients with a variety of neurological disorders mediated, at least in part, by excessive NMDA receptor activity (Lipton, 2006; Lipton & Rosenberg, 1994).

Because of the apparent clinical safety of memantine, nitroglycerin, and combinatorial nitro-memantine compounds, they have the potential for expeditious trials in humans. In fact, memantine was proven effective in a phase III multicenter clinical trial in patients with severe Alzheimer's disease in the United States, and was also shown to hold promise for HAD in a recent phase II clinical trial in the United States (Jain, 2000; Susman, 2001). Memantine revealed a trend toward improvement on neuropsychological test scores above the control group, and significant improvement in a last-observation-carried-forward (LOCF) analysis. Concomitantly, magnetic resonance spectroscopy (MRS) values for the N-acetylaspartate (NAA) to choline ratio were also significantly improved, suggesting neuronal protection.

Finally, p38 MAPK inhibitors have been shown to reduce or abrogate neuronal apoptosis due to exposure to HIV/gp120 or SDF-1, or excitotoxicity (Kaul & Lipton, 1999; Kikuchi et al., 2000). The pharmaceutical industry is currently developing p38 inhibitors for a variety of inflammatory- and stress-related conditions, such as arthritis, and this may expedite trials for CNS indications such as HAD.

In summary, most recent experimental evidence regarding HAD indicates that synergy between excitatory and inflammatory pathways leads to neuronal injury and death. Moreover, these mechanisms may, at least in part, be common to other CNS disorders including stroke, spinal cord injury, and Alzheimer's disease. It seems plausible therefore to argue that the development of new therapeutic strategies for HAD will impact several other neurodegenerative diseases, and possibly vice versa.

Acknowledgments

Marcus Kaul and Stuart A. Lipton are supported by the National Institutes of Health, R01 NS050621 (to M. Kaul), P01 HD029587, R01 EY09024, R01 NS046994, R01 EY05477, R01 NS047973, and R01 NS41207 (to S. A. Lipton). Stuart A. Lipton is a Senior Scholar in Aging Research of the Ellison Medical Foundation, and is or has been a consultant to Allergan, Alcon, Merck, Johnson & Johnson, Forest Laboratories, NeuroMolecular Pharmaceuticals, Inc., and Neurobiological Technologies, Inc. in the field of neuroprotective agents. Stuart Lipton is also the named inventor on patents for memantine in the treatment of neurodegenerative diseases. These patents are assigned to his former institution, Harvard Medical School and its affiliated hospitals, and he participates in a royalty sharing agreement under the rules of those institutions.

References

Adamson, D. C., Wildemann, B., Sasaki, M., Glass, J. D., McArthur, J. C., Christov, V. I. et al. (1996). Immunologic NO synthase: Elevation in severe AIDS dementia and induction by HIV-1 gp41. *Science* **274**, 1917–1921.

Adle-Biassette, H., Chretien, F., Wingertsmann, L., Hery, C., Ereau, T., Scaravilli, F. et al. (1999). Neuronal apoptosis does not correlate with dementia in HIV infection but is related to microglial activation and axonal damage. *Neuropathol Appl Neurobiol* **25**, 123–133.

Adle-Biassette, H., Levy, Y., Colombel, M., Poron, F., Natchev, S., Keohane, C., and Gray, F. (1995). Neuronal apoptosis in HIV infection in adults. *Neuropathol Appl Neurobiol* **21**, 218–227.

Aizenman, E., Stout, A. K., Hartnett, K. A., Dineley, K. E., McLaughlin, B., and Reynolds, I. J. (2000). Induction of neuronal apoptosis by thiol oxidation: Putative role of intracellular Zinc release. *J. Neurochem.* **75**, 1878–1888.

Alkhatib, G., Combadiere, C., Broder, C. C., Feng, Y., Kennedy, P. E., Murphy, P. M., and Berger, E. A. (1996). CC CKR5: A RANTES, MIP-1-alpha, MIP-1-beta receptor as a fusion cofactor for macrophage-tropic HIV-1. *Science* **272**, 1955–1958.

Anderson, E. R., Gendelman, H. E., and Xiong, H. (2004). Memantine protects hippocampal neuronal function in murine human immunodeficiency virus type 1 encephalitis. *J Neurosci* **24**, 7194–7198.

Asensio, V. C. and Campbell, I. L. (1999). Chemokines in the CNS: Plurifunctional mediators in diverse states. *Trends Neurosci* **22**, 504–512.

Batchelor, P. E., Liberatore, G. T., Wong, J. Y., Porritt, M. J., Frerichs, F., Donnan, G. A., and Howells, D. W. (1999). Activated macrophages and microglia induce dopaminergic sprouting in the injured striatum and express brain-derived neurotrophic factor and glial cell line-derived neurotrophic factor. *J Neurosci* **19**, 1708–1716.

Bazan, J. F., Bacon, K. B., Hardiman, G., Wang, W., Soo, K., Rossi, D. et al. (1997). A new class of membrane-bound chemokine with a CX3C motif. *Nature* **385**, 640–644.

Beckman, J. S., Beckman, T. W., Chen, J., Marshall, P. A., and Freeman, B. A. (1990). Apparent hydroxyl radical production by peroxynitrite: Implications for endothelial injury from nitric oxide and superoxide. *Proc Natl Acad Sci U S A* **87**, 1620–1624.

Bezzi, P., Carmignoto, G., Pasti, L., Vesce, S., Rossi, D., Rizzini, B. L. et al. (1998). Prostaglandins stimulate calcium-dependent glutamate release in astrocytes. *Nature* **391**, 281–285.

Bezzi, P., Domercq, M., Brambilla, L., Galli, R., Schols, D., De Clercq, E. et al. (2001). CXCR4-activated astrocyte glutamate release via TNF-alpha: Amplification by microglia triggers neurotoxicity. *Nat Neurosci* **4**, 702–710.

Bigge, C. F. (1999). Ionotropic glutamate receptors. *Curr Opin Chem Biol* **3**, 441–447.

Bleul, C. C., Farzan, M., Choe, H., Parolin, C., Clark-Lewis, I., Sodroski, J., and Springer, T. A. (1996). The lymphocyte chemoattractant SDF-1 is a ligand for LESTR/fusin and blocks HIV-1 entry. *Nature* **382**, 829–833.

Bonfoco, E., Krainc, D., Ankarcrona, M., Nicotera, P., and Lipton, S. A. (1995). Apoptosis and necrosis: Two distinct events induced, respectively, by mild and intense insults with N-methyl-D-aspartate or nitric oxide/superoxide in cortical cell cultures. *Proc Natl Acad Sci U S A* **92**, 7162–7166.

Bormann, J. (1989). Memantine is a potent blocker of N-methyl-D-aspartate (NMDA) receptor channels. *Eur J Pharmacol* **166**, 591–592.

Brack-Werner, R. and Bell, J. E. (1999). Replication of HIV-1 in human astrocytes. *Science Online: NeuroAids* (www. sciencemag. org/NAIDS) **2**, 1–7.

Brauner-Osborne, H., Egebjerg, J., Nielsen, E. O., Madsen, U., and Krogsgaard-Larsen, P. (2000). Ligands for glutamate receptors: Design and therapeutic prospects. *J Med Chem* **43**, 2609–2645.

Brenneman, D. E., Westbrook, G. L., Fitzgerald, S. P., Ennist, D. L., Elkins, K. L., Ruff, M. R., and Pert, C. B. (1988). Neuronal cell killing by the envelope protein of HIV and its prevention by vasoactive intestinal peptide. *Nature* **335**, 639–642.

Brew, B. J., Corbeil, J., Pemberton, L., Evans, L., Saito, K., Penny, R. et al. (1995). Quinolinic acid production is related to macrophage tropic isolates of HIV-1. *J Neurovirol* **1**, 369–374.

Bruno, V., Copani, A., Besong, G., Scoto, G., and Nicoletti, F. (2000). Neuroprotective activity of chemokines against N-methyl-D-aspartate or beta-amyloid-induced toxicity in culture. *Eur J Pharmacol* **399**, 117–121.

Budd, S. L., Tenneti, L., Lishnak, T., and Lipton, S. A. (2000). Mitochondrial and extramitochondrial apoptotic signaling pathways in cerebrocortical neurons. *Proc Natl Acad Sci U S A* **97**, 6161–6166.

Budka, H. (1991). Neuropathology of human immunodeficiency virus infection. *Brain Pathol* **1**, 163–175.

Chakravarty, S. and Herkenham, M. (2005). Toll-like receptor 4 on nonhematopoietic cells sustains CNS inflammation during endotoxemia, independent of systemic cytokines. *J Neurosci* **25**, 1788–1796.

Chao, C. C., Hu, S., Sheng, W. S., and Peterson, P. K. (1995). Tumor necrosis factor-alpha production by human fetal microglial cells: Regulation by other cytokines. *Dev Neurosci* **17**, 97–105.

Chen, H. S. and Lipton, S. A. (1997). Mechanism of memantine block of NMDA-activated channels in rat retinal ganglion cells: Uncompetitive antagonism. *J Physiol* **499 (Pt 1)**, 27–46.

Chen, H. S., Pellegrini, J. W., Aggarwal, S. K., Lei, S. Z., Warach, S., Jensen, F. E., and Lipton, S. A. (1992). Open-channel block of N-methyl-D-aspartate (NMDA) responses by memantine: Therapeutic advantage against NMDA receptor-mediated neurotoxicity. *J Neurosci* **12**, 4427–4436.

Chen, H. S., Wang, Y. F., Rayudu, P. V., Edgecomb, P., Neill, J. C., Segal, M. M. et al. (1998). Neuroprotective concentrations of the N-methyl-D-aspartate open-channel blocker memantine are effective without cytoplasmic vacuolation following post-ischemic administration and do not block maze learning or long-term potentiation. *Neuroscience* **86**, 1121–1132.

Chen, W., Sulcove, J., Frank, I., Jaffer, S., Ozdener, H., and Kolson, D. L. (2002). Development of a human neuronal cell model for human immunodeficiency virus (HIV)-infected macrophage-induced neurotoxicity: Apoptosis induced by HIV type 1 primary isolates and evidence for involvement of the Bcl-2/Bcl-xL-sensitive intrinsic apoptosis pathway. *J Virol* **76**, 9407–9419.

Choi, D. W. (1988). Glutamate neurotoxicity and diseases of the nervous system. *Neuron* **1**, 623–634.

Choi, D. W., Koh, J. Y., and Peters, S. (1988a). Pharmacology of glutamate neurotoxicity in cortical cell culture: Attenuation by NMDA antagonists. *J Neurosci* **8**, 185–196.

Choi, D. W., Yokoyama, M., and Koh, J. (1988b). Zinc neurotoxicity in cortical cell culture. *Neuroscience* **24**, 67–79.

Choi, Y. B., Chen, H. S., and Lipton, S. A. (2001). Three pairs of cysteine residues mediate both redox and Zn^{2+} modulation of the NMDA receptor. *J Neurosci* **21**, 392–400.

Choi, Y. B., Tenneti, L., Le, D. A., Ortiz, J., Bai, G., Chen, H. S., and Lipton, S. A. (2000). Molecular basis of NMDA receptor-coupled ion channel modulation by S-nitrosylation. *Nat Neurosci* **3**, 15–21.

Clifford, D. B. (1999). Central neurologic complications of HIV infection. *Curr Infect Dis Rep* **1**, 187–191.

Cocchi, F., Devico, A. L., Garzino-Demo, A., Arya, S. K., Gallo, R. C., and Lusso, P. (1995). Identification of RANTES, MIP-1 alpha, and MIP-1 beta as the major HIV- suppressive factors produced by CD8+ T cells. *Science* **270**, 1811–1815.

Conant, K., Garzino-Demo, A., Nath, A., McArthur, J. C., Halliday, W., Power, C. et al. (1998). Induction of monocyte chemoattractant protein-1 in HIV-1 Tat- stimulated astrocytes and elevation in AIDS dementia. *Proc Natl Acad Sci U S A* **95**, 3117–3121.

Cunningham, P. H., Smith, D. G., Satchell, C., Cooper, D. A., and Brew, B. (2000). Evidence for independent development of resistance to HIV-1 reverse transcriptase inhibitors in the cerebrospinal fluid. *AIDS* **14**, 1949–1954.

Das, S., Sasaki, Y. F., Rothe, T., Premkumar, L. S., Takasu, M., Crandall, J. E. et al. (1998). Increased NMDA current and spine density in mice lacking the NMDA receptor subunit NR3a. *Nature* **393**, 377–381.

Dawson, V. L., Dawson, T. M., Bartley, D. A., Uhl, G. R., and Snyder, S. H. (1993). Mechanisms of nitric oxide-mediated neurotoxicity in primary brain cultures. *J Neurosci* **13**, 2651–2661.

Dawson, V. L., Dawson, T. M., London, E. D., Bredt, D. S., and Snyder, S. H. (1991). Nitric oxide mediates glutamate neurotoxicity in primary cortical cultures. *Proc Natl Acad Sci U S A* **88**, 6368–6371.

De Clercq, E. (2004). HIV-chemotherapy and -prophylaxis: New drugs, leads and approaches. *Int J Biochem Cell Biol* **36**, 1800–1822.

Digicaylioglu, M., Garden, G., Timberlake, S., Fletcher, L., and Lipton, S. A. (2004a). Acute neuroprotective synergy of erythropoietin and insulin-like growth factor I. *Proc Natl Acad Sci U S A* **101**, 9855–9860.

Digicaylioglu, M., Kaul, M., Fletcher, L., Dowen, R., and Lipton, S. A., (2004b). Erythropoietin protects cerebrocortical neurons from HIV-1/gp120-induced damage. *Neuroreport* **15**, 761–763.

Digicaylioglu, M. and Lipton, S. A. (2001). Erythropoietin-mediated neuroprotection involves cross-talk between Jak2 and NF-kappaB signalling cascades. *Nature* **412**, 641–647.

Doble, A. (1999). The role of excitotoxicity in neurodegenerative disease: Implications for therapy. *Pharmacol Ther* **81**, 163–221.

Dragic, T., Litwin, V., Allaway, G. P., Martin, S. R., Huang, Y., Nagashima, K. A. et al. (1996). HIV-1 entry into CD4+ cells is mediated by the chemokine receptor CC-CKR-5. *Nature* **381**, 667–673.

Dreyer, E. B., Kaiser, P. K., Offermann, J. T., and Lipton, S. A. (1990). HIV-1 coat protein neurotoxicity prevented by calcium channel antagonists. *Science* **248**, 364–367.

Dreyer, E. B. and Lipton, S. A. (1995). The coat protein gp120 of HIV-1 inhibits astrocyte uptake of excitatory amino acids via macrophage arachidonic acid. *Eur J Neurosci* **7**, 2502–2507.

Elkabes, S., DiCicco-Bloom, E. M., and Black, I. B. (1996). Brain microglia/macrophages express neurotrophins that selectively regulate microglial proliferation and function. *J Neurosci* **16**, 2508–2521.

Ellis, R. J., Deutsch, R., Heaton, R. K., Marcotte, T. D., McCutchan, J. A., Nelson, J. A. et al. (1997). Neurocognitive impairment is an independent risk factor for death in HIV infection. San Diego HIV Neurobehavioral Research Center Group. *Arch Neurol* **54**, 416–424.

Erdo, S. L. and Schafer, M. (1991). Memantine is highly potent in protecting cortical cultures against excitotoxic cell death evoked by glutamate and N-methyl-D-aspartate. *Eur J Pharmacol* **198**, 215–217.

Everall, I. P., Heaton, R. K., Marcotte, T. D., Ellis, R. J., McCutchan, J. A., Atkinson, J. H. et al. (1999). Cortical synaptic density is reduced in mild to moderate human immunodeficiency virus neurocognitive disorder. HNRC group. HIV Neurobehavioral Research Center. *Brain Pathol* **9**, 209–217.

Fiala, M., Looney, D. J., Stins, M., Way, D. D., Zhang, L., Gan, X. et al. (1997). TNF-alpha opens a paracellular route for HIV-1 invasion across the blood-brain barrier. *Mol Med* **3**, 553–564.

Fontana, G., Valenti, L., and Raiteri, M. (1997). gp120 can revert antagonism at the glycine site of NMDA receptors mediating GABA release from cultured hippocampal neurons. *J Neurosci Res* **49**, 732–738.

Fox, L., Alford, M., Achim, C., Mallory, M., and Masliah, E. (1997). Neurodegeneration of somatostatin-immunoreactive neurons in HIV encephalitis. *J Neuropathol Exp Neurol* **56**, 360–368.

Galasso, J. M., Harrison, J. K., and Silverstein, F. S. (1998). Excitotoxic brain injury stimulates expression of the chemokine receptor CCR5 in neonatal rats. *Am J Pathol* **153**, 1631–1640.

Grace, E., Caroleo, M. C., Aloe, L., Aquaro, S., Piacentini, M., Costa, N. et al. (1999). Nerve growth factor is an autocrine factor essential for the survival of macrophages infected with HIV. *Proc Natl Acad Sci U S A* **96**, 14013–14018.

Garden, G. A., Guo, W., Jayadev, S., Tun, C., Balcaitis, S., Choi, J. et al. (2004). HIV associated neurodegeneration requires p53 in neurons and microglia. *FASEB J* **18**, 1141–1143.

Gartner, S. (2000). HIV infection and dementia. *Science* **287**, 602–604.

Gelbard, H. A., Dzenko, K. A., DiLoreto, D., del Cerro, C., del Cerro, M., and Epstein, L. G. (1993). Neurotoxic effects of tumor necrosis factor alpha in primary human neuronal cultures are mediated by activation of the glutamate AMPA receptor subtype: Implications for AIDS neuropathogenesis. *Dev Neurosci* **15**, 417–422.

Gelbard, H. A., James, H. J., Sharer, L. R., Perry, S. W., Saito, Y., Kazee, A. M. et al. (1995). Apoptotic neurons in brains from paediatric patients with HIV-1 encephalitis and progressive encephalopathy. *Neuropathol Appl Neurobiol* **21**, 208–217.

Gelbard, H. A., Nottet, H. S., Swindells, S., Jett, M., Dzenko, K. A., Genis, P. et al. (1994). Platelet-activating factor: A candidate human immunodeficiency virus type 1-induced neurotoxin. *J Virol* **68**, 4628–4635.

Gendelman, H. E., Grant, I., Lipton, S. A., Everall, I., and Swindells, S. (2005). *The Neurology of AIDS*. London: Oxford University Press.

Gendelman, H. E., Persidsky, Y., Ghorpade, A., Limoges, J., Stins, M., Fiala, M., and Morrisett, R. (1997). The neuropathogenesis of the AIDS dementia complex. *AIDS* **11 Suppl A**, S35–S45.

Giulian, D., Vaca, K., and Noonan, C. A. (1990). Secretion of neurotoxins by mononuclear phagocytes infected with HIV-1. *Science* **250**, 1593–1596.

Giulian, D., Wendt, E., Vaca, K., and Noonan, C. A. (1993). The envelope glycoprotein of human immunodeficiency virus type 1 stimulates release of neurotoxins from monocytes. *Proc Natl Acad Sci U S A* **90**, 2769–2773.

Giulian, D., Yu, J., Li, X., Tom, D., Li, J., Wendt, E., Lin, S. N., Schwarcz, R., and Noonan, C. (1996). Study of receptor-mediated neurotoxins released by HIV-1 infected mononuclear phagocytes found in human brain. *J Neurosci* **16**, 3139–3153.

Glass, J. D., Fedor, H., Wesselingh, S. L., and McArthur, J. C. (1995). Immunocytochemical quantitation of human immunodeficiency virus in the brain: Correlations with dementia. *Ann Neurol* **38**, 755–762.

Glass, J. D., Wesselingh, S. L., Selnes, O. A., and McArthur, J. C. (1993). Clinical-neuropathologic correlation in HIV-associated dementia. *Neurology* **43**, 2230–2237.

Gleichmann, M., Gillen, C., Czardybon, M., Bosse, F., Greiner-Petter, R., Auer, J., and Muller, H. W. (2000). Cloning and characterization of SDF-1 gamma, a novel SDF-1 chemokine transcript with developmentally regulated expression in the nervous system. *Eur J Neurosci* **12**, 1857–1866.

Gonzalez-Scarano, F. and Martin-Garcia, J. (2005). The neuropathogenesis of AIDS. *Nat Rev Immunol* **5**, 69–81.

Gonzalez-Zulueta, M., Feldman, A. B., Klesse, L. J., Kalb, R. G., Dillman, J. F., Parada, L. F. et al. (2000). Requirement for nitric oxide activation of p21(ras)/extracellular regulated kinase in neuronal ischemic preconditioning. *Proc Natl Acad Sci U S A* **97**, 436–441.

Gonzalez, E., Rovin, B. H., Sen, L., Cooke, G., Dhanda, R., Mummidi, S. et al. (2002). HIV-1 infection and AIDS dementia are influenced by a mutant MCP-1 allele linked to increased monocyte infiltration of tissues and MCP-1 levels. *Proc Natl Acad Sci U S A* **99**, 13795–13800.

Gras, G., Chretien, F., Vallat-Decouvelaere, A. V., Le Pavec, G., Porcheray, F., Bossuet, C. et al. (2003). Regulated expression of sodium-dependent glutamate transporters and synthetase: A neuroprotective role for activated microglia and macrophages in HIV infection? *Brain Pathol* **13**, 211–222.

He, J., Chen, Y., Farzan, M., Choe, H., Ohagen, A., Gartner, S. et al. (1997). CCR3 and CCR5 are co-receptors for HIV-1 infection of microglia. *Nature* **385**, 645–649.

Hesselgesser, J., Taub, D., Baskar, P., Greenberg, M., Hoxie, J., Kolson, D. L., and Horuk, R. (1998). Neuronal apoptosis induced by HIV-1 gp120 and the chemokine SDF-1 alpha is mediated by the chemokine receptor CXCR4. *Curr Biol* **8**, 595–598.

Heyes, M. P., Brew, B. J., Martin, A., Price, R. W., Salazar, A. M., Sidtis, J. J. et al. (1991). Quinolinic acid in cerebrospinal fluid and serum in HIV-1 infection: Relationship to clinical and neurological status. *Ann Neurol* **29**, 202–209.

Ho, D. D., Rota, T. R., Schooley, R. T., Kaplan, J. C., Allan, J. D., Groopman, J. E. et al. (1985). Isolation of HTLV-III from cerebrospinal fluid and neural tissues of patients with neurologic syndromes related to the acquired immunodeficiency syndrome. *N Engl J Med* **313**, 1493–1497.

Jain, K. K. (2000). Evaluation of memantine for neuroprotection in dementia. *Expert Opin Investig Drugs* **9**, 1397–1406.

Jiang, Z. G., Piggee, C., Heyes, M. P., Murphy, C., Quearry, B., Bauer, M. et al. (2001). Glutamate is a mediator of neurotoxicity in secretions of activated HIV- 1-infected macrophages. *J Neuroimmunol* **117**, 97–107.

Johnston, J. B., Jiang, Y., van Marle, G., Mayne, M. B., Ni, W., Holden, J. et al. (2000). Lentivirus infection in the brain induces matrix metalloproteinase expression: Role of envelope diversity. *J Virol* **74**, 7211–7220.

Jones, G. and Power, C. (2006). Regulation of neural cell survival by HIV-1 infection. *Neurobiol Dis* **21**, 1–17.

Kalams, S. A. and Walker, B. D. (1995). Cytotoxic T lymphocytes and HIV-1 related neurologic disorders. *Curr Top Microbiol Immunol* **202**, 79–88.

Kaul, M. (2002). Chemokines and their receptors in HIV-associated dementia. *J Neurovirol* **8[Suppl 1]**, 41–42.

Kaul, M., Garden, G. A., and Lipton, S. A. (2001). Pathways to neuronal injury and apoptosis in HIV-associated dementia. *Nature* **410**, 988–994.

Kaul, M. and Lipton, S. A. (1999). Chemokines and activated macrophages in gp120-induced neuronal apoptosis. *Proc Natl Acad Sci U S A* **96**, 8212–8216.

Kaul, M. and Lipton, S. A. (2001). Knock out of HIV-1 coreceptors attenuates neuronal apoptosis induced by HIV envelope glycoprotein gp120. *Soc Neurosci Abstr* **27**, 678.8.

Kaul, M., Ma, Q., Medders, K. E., Desai, M. K., and Lipton, S. A. (2006). HIV-1 coreceptors CCR5 and CXCR4 both mediate neuronal cell death but CCR5 paradoxically can also contribute to protection. *Cell Death Differ*. doi:10.1038/sj.cdd.4402006 (advance online publication).

Kaul, M., Zheng, J., Okamoto, S., Gendelman, H. E., and Lipton, S. A. (2005). HIV-1 infection and AIDS: Consequences for the central nervous system. *Cell Death Differ* **12** Suppl 1, 878–892.

Keilhoff, G. and Wolf, G. (1992). Memantine prevents quinolinic acid-induced hippocampal damage. *Eur J Pharmacol* **219**, 451–454.

Kerr, S. J., Armati, P. J., Pemberton, L. A., Smythe, G., Tattam, B., and Brew, B. J. (1997). Kynurenine pathway inhibition reduces neurotoxicity of HIV-1- infected macrophages. *Neurology* **49**, 1671–1681.

Kikuchi, M., Tenneti, L., and Lipton, S. A. (2000). Role of p38 mitogen-activated protein kinase in axotomy-induced apoptosis of rat retinal ganglion cells. *J Neurosci* **20**, 5037–5044.

Kim, W. K., Choi, Y. B., Rayudu, P. V., Das, P., Asaad, W., Arnelle, D. R. et al. (1999). Attenuation of NMDA receptor activity and neurotoxicity by nitroxyl anion, NO^- *Neuron* **24**, 461–469.

Koenig, S., Gendelman, H. E., Orenstein, J. M., Dal Canto, M. C., Pezeshkpour, G. H., Yungbluth, M. et al. (1986). Detection of AIDS virus in macrophages in brain tissue from AIDS patients with encephalopathy. *Science* **233**, 1089–1093.

Kohr, G., Eckardt, S., Luddens, H., Monyer, H., and Seeburg, P. H. (1994). NMDA receptor channels: Subunit-specific potentiation by reducing agents. *Neuron* **12**, 1031–1040.

Kramer-Hammerle, S., Rothenaigner, I., Wolff, H., Bell, J. E., and Brack-Werner, R. (2005). Cells of the central nervous system as targets and reservoirs of the human immunodeficiency virus. *Virus Res* **111**, 194–213.

Krathwohl, M. D. and Kaiser, J. L. (2004a). Chemokines promote quiescence and survival of human neural progenitor cells. *Stem Cells* **22**, 109–118.

Krathwohl, M. D. and Kaiser, J. L. (2004b). HIV-1 promotes quiescence in human neural progenitor cells. *J Infect Dis* **190**, 216–226.

Kullander, K., Kylberg, A., and Ebendal, T. (1997). Specificity of neurotrophin-3 determined by loss-of-function mutagenesis. *J Neurosci Res* **50**, 496–503.

Langford, D., Sanders, V. J., Mallory, M., Kaul, M., and Masliah, E. (2002). Expression of stromal cell-derived factor 1-alpha protein in HIV encephalitis. *J Neuroimmunol* **127**, 115–126.

Lannuzel, A., Lledo, P. M., Lamghitnia, H. O., Vincent, J. D., and Tardieu, M. (1995). HIV-1 envelope proteins gp120 and gp160 potentiate NMDA [Ca2+]i increase, alter [Ca2+]i homeostasis and induce neurotoxicity in human embryonic neurons. *Eur J Neurosci* **7**, 2285–2293.

Lapidot, T. and Petit, I. (2002). Current understanding of stem cell mobilization: The roles of chemokines, proteolytic enzymes, adhesion molecules, cytokines, and stromal cells. *Exp Hematol* **30**, 973–981.

Lavi, E., Kolson, D. L., Ulrich, A. M., Fu, L., and Gonzalez-Scarano, F. (1998). Chemokine receptors in the human brain and their relationship to HIV infection. *J Neurovirol* **4**, 301–311.

Lazarov-Spiegler, O., Solomon, A. S., Zeev-Brann, A. B., Hirschberg, D. L., Lavie, V., and Schwartz, M. (1996). Transplantation of activated macrophages overcomes central nervous system regrowth failure. *FASEB J* **10**, 1296–1302.

Le, D., Das, S., Wang, Y. F., Yoshizawa, T., Sasaki, Y. F., Takasu, M. et al. (1997). Enhanced neuronal death from focal ischemia in AMPA-receptor transgenic mice. *Brain Res Mol Brain Res* **52**, 235–241.

Le, D. A. and Lipton, S. A. (2001). Potential and current use of N-methyl-D-aspartate (NMDA) receptor antagonists in diseases of aging. *Drugs Aging* **18**, 717–724.

Lees, K. R. (1997). Cerestat and other NMDA antagonists in ischemic stroke. *Neurology* **49**, S66–S69.

Lei, S. Z., Pan, Z. H., Aggarwal, S. K., Chen, H. S., Hartman, J., Sucher, N. J., and Lipton, S. A. (1992). Effect of nitric oxide production on the redox modulatory site of the NMDA receptor-channel complex. *Neuron* **8**, 1087–1099.

Letendre, S. L., Lanier, E. R., and McCutchan, J. A. (1999). Cerebrospinal fluid beta chemokine concentrations in neurocognitively impaired individuals infected with human immunodeficiency virus type 1. *J Infect Dis* **180**, 310–319.

Lipton, S. A. (1992a). Memantine prevents HIV coat protein-induced neuronal injury in vitro. *Neurology* **42**, 1403–1405.

Lipton, S. A. (1992b). Models of neuronal injury in AIDS: Another role for the NMDA receptor? *Trends Neurosci* **15**, 75–79.

Lipton, S. A. (1992c). Requirement for macrophages in neuronal injury induced by HIV envelope protein gp120. *Neuroreport* **3**, 913–915.

Lipton, S. A. (1993). Prospects for clinically tolerated NMDA antagonists: Open-channel blockers and alternative redox states of nitric oxide. *Trends Neurosci* **16**, 527–532.

Lipton, S. A. (1994). HIV coat protein gp120 induces soluble neurotoxins in culture medium. *Neuroscience Research Communications* **15**, 31–37.

Lipton, S. A. (1997a). Neuropathogenesis of acquired immunodeficiency syndrome dementia. *Curr Opin Neurol* **10**, 247–253.

Lipton, S. A. (1997b). Treating AIDS dementia [letter; comment]. *Science* **276**, 1629–1630.

Lipton, S. A. (1998). Neuronal injury associated with HIV-1: Approaches to treatment. *Annu Rev Pharmacol Toxicol* **38**, 159–177.

Lipton, S. A. (2004). Erythropoietin for neurologic protection and diabetic neuropathy. *N Engl J Med* **350**, 2516–2517.

Lipton, S. A. (2006). Paradigm shift in neuroprotection by NMDA receptor blockade: Memantine and beyond. *Nature Rev Drug Disc* **5**, 160–170.

Lipton, S. A., Choi, Y. B., Pan, Z. H., Lei, S. Z., Chen, H. S., Sucher, N. J. et al. (1993). A redox-based mechanism for the neuroprotective and neurodestructive effects of nitric oxide and related nitroso-compounds. *Nature* **364**, 626–632.

Lipton, S. A., Choi, Y.-B., Takahashi, T., Zhang, D., Li, W., Godzik, A., and Bankston L. A. (2002) Cysteine regulation of protein function—As exemplified by NMDA-receptor modulation. *Trends Neurosci* **25**, 474–480.

Lipton, S. A. and Gendelman, H. E. (1995). Seminars in Medicine of the Beth Israel Hospital, Boston. Dementia associated with the acquired immunodeficiency syndrome. *N Engl J Med* **332**, 934–940.

Lipton, S. A. and Jensen, F. E. (1992). Memantine, a clinically-tolerated NMDA open-channel blocker, prevents HIV coat protein-induced neuronal injury in vitro and in vivo. *Soc Neurosci Abstr* **18**, 757.

Lipton, S. A. and Kieburtz, K. (1998). Development of adjunctive therapies for the neurologic manifestations of AIDS: Dementia and painful neuropathy. In *The Neurology of AIDS*, H. E. Gendelman, S. A. Lipton, L. G. Epstein, and S. Swindells, Eds., 377–381. New York, Chapman and Hall.

Lipton, S. A. and Rosenberg, P. A. (1994). Excitatory amino acids as a final common pathway for neurologic disorders [see comments]. *N Engl J Med* **330**, 613–622.

Lipton, S. A. and Stamler, J. S. (1994). Actions of redox-related congeners of nitric oxide at the NMDA receptor. *Neuropharmacology* **33**, 1229–1233.

Lipton, S. A., Sucher, N. J., Kaiser, P. K., and Dreyer, E. B. (1991). Synergistic effects of HIV coat protein and NMDA receptor-mediated neurotoxicity. *Neuron* **7**, 111–118.

Lipton, S. A. and Wang, Y. F. (1996). NO-related species can protect from focal cerebral ischemia/reperfusion. In *Pharmacology of Cerebral Ischemia*, J. Krieglstein, Ed., 183–191. Stuttgart: Medpharm Scientific Publisher.Liu, R., Paxton, W. A., Choe, S., Ceradini, D., Martin, S. R., Horuk, R. et al. (1996). Homozygous defect in HIV-1 coreceptor accounts for resistance of some multiply-exposed individuals to HIV-1 infection. *Cell* **86**, 367–377.

Liu, Y., Jones, M., Hingtgen, C. M., Bu, G., Laribee, N., Tanzi, R. E. et al. (2000). Uptake of HIV-1 Tat protein mediated by low-density lipoprotein receptor-related protein disrupts the neuronal metabolic balance of the receptor ligands. *Nat Med* **6**, 1380–1387.

Lo, T. M., Fallert, C. J., Piser, T. M., and Thayer, S. A. (1992). HIV-1 envelope protein evokes intracellular calcium oscillations in rat hippocampal neurons. *Brain Res* **594**, 189–196.

Locati, M. and Murphy, P. M. (1999). Chemokines and chemokine receptors: Biology and clinical relevance in inflammation and AIDS. *Annu Rev Med* **50**, 425–440.

Lopalco, L., Barassi, C., Paolucci, C., Breda, D., Brunelli, D., Nguyen, M. et al. (2005). Predictive value of anti-cell and anti-human immunodeficiency virus (HIV) humoral responses in HIV-1-exposed seronegative cohorts of European and Asian origin. *J Gen Virol* **86**, 339–348.

Luo, X., Carlson, K. A., Wojna, V., Mayo, R., Biskup, T. M., Stoner, J. et al. (2003). Macrophage proteomic fingerprinting predicts HIV-1-associated cognitive impairment. *Neurology* **60**, 1931–1937.

Ma, Q., Jones, D., Borghesani, P. R., Segal, R. A., Nagasawa, T., Kishimoto, T. et al. (1998). Impaired B-lymphopoiesis, myelopoiesis, and derailed cerebellar neuron migration in CXCR4- and SDF-1-deficient mice. *Proc Natl Acad Sci U S A* **95**, 9448–9453.

Marshall, D. C., Wyss-Coray, T., and Abraham, C. R. (1998). Induction of matrix metalloproteinase-2 in human immunodeficiency virus-1 glycoprotein 120 transgenic mouse brains. *Neurosci Lett* **254**, 97–100.

Masliah, E., Ge, N., Achim, C. L., Hansen, L. A., and Wiley, C. A. (1992). Selective neuronal vulnerability in HIV encephalitis. *J Neuropathol Exp Neurol* **51**, 585–593.

Masliah, E., Heaton, R. K., Marcotte, T. D., Ellis, R. J., Wiley, C. A., Mallory, M. et al. (1997). Dendritic injury is a pathological substrate for human immunodeficiency virus-related cognitive disorders. HNRC group. The HIV Neurobehavioral Research Center. *Ann Neurol* **42**, 963–972.

Mattson, M. P., Haughey, N. J., and Nath, A. (2005). Cell death in HIV dementia. *Cell Death Differ* **12**, 893–904.

McArthur, J. C., Haughey, N., Gartner, S., Conant, K., Pardo, C., Nath, A., and Sacktor, N. (2003). Human immunodeficiency virus-associated dementia: An evolving disease. *J Neurovirol* **9**, 205–221.

McArthur, J. C., Hoover, D. R., Bacellar, H., Miller, E. N., Cohen, B. A., Becker, J. T. et al. (1993). Dementia in AIDS patients: Incidence and risk factors. Multicenter AIDS Cohort Study. *Neurology* **43**, 2245–2252.

McArthur, J. C., McClernon, D. R., Cronin, M. F., Nance-Sproson, T. E., Saah, A. J., St. Clair, M., and Lanier, E. R. (1997). Relationship between human immunodeficiency virus-associated dementia and viral load in cerebrospinal fluid and brain. *Ann Neurol* **42**, 689–698.

McGrath, K. E., Koniski, A. D., Maltby, K. M., McGann, J. K., and Palis, J. (1999). Embryonic expression and function of the chemokine SDF-1 and its receptor, CXCR4. *Dev Biol* **213**, 442–456.

Melton, S. T., Kirkwood, C. K., and Ghaemi, S. N. (1997). Pharmacotherapy of HIV dementia. *Ann Pharmacother* **31**, 457–473.

Mennicken, F., Maki, R., de Souza, E. B., and Quirion, R. (1999). Chemokines and chemokine receptors in the CNS: A possible role in neuroinflammation and patterning. *Trends Pharmacol Sci* **20**, 73–78.

Meucci, O., Fatatis, A., Simen, A. A., Bushell, T. J., Gray, P. W., and Miller, R. J. (1998). Chemokines regulate hippocampal neuronal signaling and gp120 neurotoxicity. *Proc Natl Acad Sci U S A* **95**, 14500–14505.

Meucci, O. and Miller, R. J. (1996). Gp120-induced neurotoxicity in hippocampal pyramidal neuron cultures: Protective action of TGF-beta1. *J Neurosci* **16**, 4080–4088.

Michael, N. L. and Moore, J. P. (1999). HIV-1 entry inhibitors: Evading the issue [news] [see comments]. *Nat Med* **5**, 740–742.

Miller, R. J. and Meucci, O. (1999). AIDS and the brain: Is there a chemokine connection? *Trends Neurosci* **22**, 471–479.

Milligan, C. E., Cunningham, T. J., and Levitt, P. (1991). Differential immunochemical markers reveal the normal distribution of brain macrophages and microglia in the developing rat brain. *J Comp Neurol* **314**, 125–135.

Minghetti, L. (2005). Role of inflammation in neurodegenerative diseases. *Curr Opin Neurol* **18**, 315–321.

Miwa, T., Furukawa, S., Nakajima, K., Furukawa, Y., and Kohsaka, S. (1997). Lipopolysaccharide enhances synthesis of brain-derived neurotrophic factor in cultured rat microglia. *J Neurosci Res* **50**, 1023–1029.

Mukherjee, P. K., DeCoster, M. A., Campbell, F. Z., Davis, R. J., and Bazan, N. G. (1999). Glutamate receptor signaling interplay modulates stress-sensitive mitogen-activated protein kinases and neuronal cell death. *J Biol Chem* **274**, 6493–6498.

Muller, W. E., Schroder, H. C., Ushijima, H., Dapper, J., and Bormann, J. (1992). Gp120 of HIV-1 induces apoptosis in rat cortical cell cultures: Prevention by memantine. *Eur J Pharmacol* **226**, 209–214.

Nath, A. (1999). Pathobiology of human immunodeficiency virus dementia. *Semin Neurol* **19**, 113–127.

Nath, A., Psooy, K., Martin, C., Knudsen, B., Magnuson, D. S., Haughey, N., and Geiger, J. D. (1996). Identification of a human immunodeficiency virus type 1 Tat epitope that is neuroexcitatory and neurotoxic. *J Virol* **70**, 1475–1480.

Navia, B. A., Dafni, U., Simpson, D., Tucker, T., Singer, E., McArthur, J. C. et al. (1998). A phase I/II trial of nimodipine for HIV-related neurologic complications. *Neurology* **51**, 221–228.

Nicholas, R. S., Stevens, S., Wing, M. G., and Compston, D. A. (2002). Microglia-derived IGF-2 prevents TNF alpha induced death of mature oligodendrocytes in vitro. *J Neuroimmunol* **124**, 36–44.

Nicotera, P., Ankarcrona, M., Bonfoco, E., Orrenius, S., and Lipton, S. A. (1997). Neuronal necrosis and apoptosis: Two distinct events induced by exposure to glutamate or oxidative stress. *Adv Neurol* **72**, 95–101.

Nottet, H. S., Jett, M., Flanagan, C. R., Zhai, Q. H., Persidsky, Y., Rizzino, A. et al. (1995). A regulatory role for astrocytes in HIV-1 encephalitis. An overexpression of eicosanoids, platelet-activating factor, and tumor necrosis factor-alpha by activated HIV-1-infected monocytes is attenuated by primary human astrocytes. *J Immunol* **154**, 3567–3581.

Nottet, H. S., Persidsky, Y., Sasseville, V. G., Nukuna, A. N., Bock, P., Zhai, Q. H. et al. (1996). Mechanisms for the transendothelial migration of HIV-1-infected monocytes into brain. *J Immunol* **156**, 1284–1295.

Oberlin, E., Amara, A., Bachelerie, F., Bessia, C., Virelizier, J. L., Arenzana-Seisdedos, F. et al. (1996). The CXC chemokine SDF-1 is the ligand for LESTR/fusin and prevents infection by T-cell-line-adapted HIV-1. *Nature* **382**, 833–835.

Ohagen, A., Ghosh, S., He, J., Huang, K., Chen, Y., Yuan, M. et al. (1999). Apoptosis induced by infection of primary brain cultures with diverse human immunodeficiency virus type 1 isolates: Evidence for a role of the envelope. *J Virol* **73**, 897–906.

Olney, J. W. (1969). Brain lesions, obesity, and other disturbances in mice treated with monosodium glutamate. *Science* **164**, 719–721.

Olney, J. W. and Sharpe, L. G. (1969). Brain lesions in an infant rhesus monkey treated with monosodium glutamate. *Science* **166**, 386–388.

Osborne, N. N. and Quack, G. (1992). Memantine stimulates inositol phosphates production in neurones and nullifies N-methyl-D-aspartate-induced destruction of retinal neurones. *Neurochem Int* **21**, 329–336.

Parsons, C. G., Danysz, W., and Quack, G. (1999). Memantine is a clinically well tolerated N-methyl-D-aspartate (NMDA) receptor antagonist—A review of preclinical data. *Neuropharmacology* **38**, 735–767.

Paxton, W. A., Liu, R., Kang, S., Wu, L., Gingeras, T. R., Landau, N. R. et al. (1998). Reduced HIV-1 infectability of CD4+ lymphocytes from exposed-uninfected individuals: Association with low expression of CCR5 and high production of beta-chemokines. *Virology* **244**, 66–73.

Paxton, W. A., Martin, S. R., Tse, D., O'Brien, T. R., Skurnick, J., VanDevanter, N. L. et al. (1996). Relative resistance to HIV-1 infection of CD4 lymphocytes from persons who remain uninfected despite multiple high-risk sexual exposure. *Nat Med* **2**, 412–417.

Pellegrini, J. W. and Lipton, S. A. (1993). Delayed administration of memantine prevents N-methyl-D-aspartate receptor-mediated neurotoxicity. *Ann Neurol* **33**, 403–407.

Persidsky, Y., Stins, M., Way, D., Witte, M. H., Weinand, M., Kim, K. S. et al. (1997). A model for monocyte migration through the blood-brain barrier during HIV-1 encephalitis. *J Immunol* **158**, 3499–3510.

Petito, C. K., Cho, E. S., Lemann, W., Navia, B. A., and Price, R. W. (1986). Neuropathology of acquired immunodeficiency syndrome (AIDS): An autopsy review. *J Neuropathol Exp Neurol* **45**, 635–646.

Petito, C. K. and Roberts, B. (1995). Evidence of apoptotic cell death in HIV encephalitis. *Am J Pathol* **146**, 1121–1130.

Pierson, T. C., Doms, R. W., and Pohlmann, S. (2004). Prospects of HIV-1 entry inhibitors as novel therapeutics. *Rev Med Virol* **14**, 255–270.

Pittaluga, A., Pattarini, R., Severi, P., and Raiteri, M. (1996). Human brain N-methyl-D-aspartate receptors regulating noradrenaline release are positively modulated by HIV-1 coat protein gp120. *AIDS* **10**, 463–468.

Power, C., Gill, M. J., and Johnson, R. T. (2002). Progress in clinical neurosciences: The neuropathogenesis of HIV infection: Host-virus interaction and the impact of therapy. *Can J Neurol Sci* **29**, 19–32.

Power, C., McArthur, J. C., Nath, A., Wehrly, K., Mayne, M., Nishio, J. et al. (1998). Neuronal death induced by brain-derived human immunodeficiency virus type 1 envelope genes differs between demented and nondemented AIDS patients. *J Virol* **72**, 9045–9053.

Prospero-Garcia, O., Gold, L. H., Fox, H. S., Polis, I., Koob, G. F., Bloom, F. E., and Henriksen, S. J. (1996). Microglia-passaged simian immunodeficiency virus induces neurophysiological abnormalities in monkeys. *Proc Natl Acad Sci U S A* **93**, 14158–14163.

Rapalino, O., Lazarov-Spiegler, O., Agranov, E., Velan, G. J., Yoles, E., Fraidakis, M. et al. (1998). Implantation of stimulated homologous macrophages results in partial recovery of paraplegic rats. *Nat Med* **4**, 814–821.

Robinson, A. P., White, T. M., and Mason, D. W. (1986). Macrophage heterogeneity in the rat as delineated by two monoclonal antibodies MRC OX-41 and MRC OX-42, the latter recognizing complement receptor type 3. *Immunol* **57**, 239–247.

Rottman, J. B., Ganley, K. P., Williams, K., Wu, L., Mackay, C. R., and Ringler, D. J. (1997). Cellular localization of the chemokine receptor CCR5. Correlation to cellular targets of HIV-1 infection. *Am J Pathol* **151**, 1341–1351.

Ryan, L. A., Zheng, J., Brester, M., Bohac, D., Hahn, F., Anderson, J. et al. (2001). Plasma levels of soluble CD14 and tumor necrosis factor-alpha type II receptor correlate with cognitive dysfunction during human immunodeficiency virus type 1 infection. *J Infect Dis* **184**, 699–706.

Sardar, A. M., Hutson, P. H., and Reynolds, G. P. (1999). Deficits of NMDA receptors and glutamate uptake sites in the frontal cortex in AIDS. *Neuroreport* **10**, 3513–3515.

Sasseville, V. G., Newman, W., Brodie, S. J., Hesterberg, P., Pauley, D., and Ringler, D. J. (1994). Monocyte adhesion to endothelium in simian immunodeficiency virus-induced AIDS encephalitis is mediated by vascular cell adhesion molecule-1/alpha 4 beta 1 integrin interactions. *Am J Pathol* **144**, 27–40.

Sathi, S., Edgecomb, P., Warach, S., Manchester, K., Donaghey, T., Stieg, P. E. et al. (1993). Chronic transdermal nitroglycerin (NTG) is neuroprotective in experimental rodent stroke models. *Soc Neurosci Abstr* **19**, 849.

Sattler, R., Xiong, Z., Lu, W. Y., Hafner, M., MacDonald, J. F., and Tymianski, M. (1999). Specific coupling of NMDA receptor activation to nitric oxide neurotoxicity by PSD-95 protein. *Science* **284**, 1845–1848.

Savio, T. and Levi, G. (1993). Neurotoxicity of HIV coat protein gp120, NMDA receptors, and protein kinase C: A study with rat cerebellar granule cell cultures. *J Neurosci Res* **34**, 265–272.

Scala, E., D'Offizi, G., Rosso, R., Turriziani, O., Ferrara, R., Mazzone, A. M. et al. (1997). C-C chemokines, IL-16, and soluble antiviral factor activity are increased in cloned T cells from subjects with long-term nonprogressive HIV infection. *J Immunol* **158**, 4485–4492.

Schifitto, G., Sacktor, N., Marder, K., McDermott, M. P., McArthur, J. C., Kieburtz, K. et al. (1999). Randomized trial of the platelet-activating factor antagonist lexipafant in HIV-associated cognitive impairment. Neurological AIDS Research Consortium. *Neurology* **53**, 391–396.

Seif el Nasr, M., Peruche, B., Rossberg, C., Mennel, H. D., and Krieglstein, J. (1990). Neuroprotective effect of memantine demonstrated in vivo and in vitro. *Eur J Pharmacol* **185**, 19–24.

Stieg, P. E., Sathi, S., Warach, S., Le, D. A., and Lipton, S. A. (1999). Neuroprotection by the NMDA receptor-associated open-channel blocker memantine in a photothrombotic model of cerebral focal ischemia in neonatal rat. *Eur J Pharmacol* **375**, 115–120.

Stumm, R. K., Rummel, J., Junker, V., Culmsee, C., Pfeiffer, M., Krieglstein, J. et al. (2002). A dual role for the SDF-1/CXCR4 chemokine receptor system in adult brain: Isoform-selective regulation of SDF-1 expression modulates CXCR4- dependent neuronal plasticity and cerebral leukocyte recruitment after focal ischemia. *J Neurosci* **22**, 5865–5878.

Sullivan, J. M., Traynelis, S. F., Chen, H. S., Escobar, W., Heinemann, S. F., and Lipton, S. A. (1994). Identification of two cysteine residues that are required for redox modulation of the NMDA subtype of glutamate receptor. *Neuron* **13**, 929–936.

Susman, E. (2001). Memantine improves function and cognition in advanced Alzheimer's. *Inpharma Weekly* **1292**, 5.

Tachibana, K., Hirota, S., Iizasa, H., Yoshida, H., Kawabata, K., Kataoka, Y. et al. (1998). The chemokine receptor CXCR4 is essential for vascularization of the gastrointestinal tract. *Nature* **393**, 591–594.

Tenneti, L., D'Emilia, D. M., and Lipton, S. A. (1997). Suppression of neuronal apoptosis by S-nitrosylation of caspases. *Neurosci Lett* **236**, 139–142.

Tenneti, L., D'Emilia, D. M., Troy, C. M., and Lipton, S. A. (1998). Role of caspases in N-methyl-D-aspartate-induced apoptosis in cerebrocortical neurons. *J Neurochem* **71**, 946–959.

Toggas, S. M., Masliah, E., and Mucke, L. (1996). Prevention of HIV-1 gp120-induced neuronal damage in the central nervous system of transgenic mice by the NMDA receptor antagonist memantine. *Brain Res* **706**, 303–307.

Toggas, S. M., Masliah, E., Rockenstein, E. M., Rall, G. F., Abraham, C. R., and Mucke, L. (1994). Central nervous system damage produced by expression of the HIV-1 coat protein gp120 in transgenic mice. *Nature* **367**, 188–193.

Tornatore, C., Chandra, R., Berger, J. R., and Major, E. O. (1994). HIV-1 infection of subcortical astrocytes in the pediatric central nervous system. *Neurology* **44**, Pt 1, 481–487.

Tran, P. B. and Miller, R. J. (2003). Chemokine receptors: Signposts to brain development and disease. *Nat Rev Neurosci* **4**, 444–455.

Turchan, J., Sacktor, N., Wojna, V., Conant, K., and Nath, A. (2003). Neuroprotective therapy for HIV dementia. *Curr HIV Res* **1**, 373–383.

Turrin, N. P. and Rivest, S. (2004). Unraveling the molecular details involved in the intimate link between the immune and neuroendocrine systems. *Exp Biol Med (Maywood)* **229**, 996–1006.

UNAIDS. (2005). AIDS epidemic update: December 2005.

Verani, A. and Lusso, P. (2002). Chemokines as natural HIV antagonists. *Curr Mol Med* **2**, 691–702.

Weiss, J. H. and Sensi, S. L. (2000). Ca^{2+}-Zn^{2+} permeable AMPA or kainate receptors: Possible key factors in selective neurodegeneration. *Trends Neurosci* **23**, 365–371.

Welch, K. and Morse, A. (2002). The clinical profile of end-stage AIDS in the era of highly active antiretroviral therapy. *AIDS Patient Care STDS* **16**, 75–81.

Wesselingh, S. L., Takahashi, K., Glass, J. D., McArthur, J. C., Griffin, J. W., and Griffin, D. E. (1997). Cellular localization of tumor necrosis factor mRNA in neurological tissue from HIV-infected patients by combined reverse transcriptase/polymerase chain reaction in situ hybridization and immunohistochemistry. *J Neuroimmunol* **74**, 1–8.

Wiley, C. A., Soontornniyomkij, V., Radhakrishnan, L., Masliah, E., Mellors, J., Hermann, S. A. et al. (1998). Distribution of brain HIV load in AIDS. *Brain Pathol* **8**, 277–284.

Yeh, M. W., Kaul, M., Zheng, J., Nottet, H. S., Thylin, M., Gendelman, H. E., and Lipton, S. A. (2000). Cytokine-stimulated, but not HIV-infected, human monocyte-derived macrophages produce neurotoxic levels of L–cysteine. *J Immunol* **164**, 4265–4270.

Zagury, D., Lachgar, A., Chams, V., Fall, L. S., Bernard, J., Zagury, J. F. et al. (1998). C-C chemokines, pivotal in protection against HIV type 1 infection. *Proc Natl Acad Sci U S A* **95**, 3857–3861.

Zhang, K., McQuibban, G. A., Silva, C., Butler, G. S., Johnston, J. B., Holden, J. et al. (2003). HIV-induced metalloproteinase processing of the chemokine stromal cell derived factor-1 causes neurodegeneration. *Nat Neurosci* **6**, 1064–1071.

Zhang, L., He, T., Talal, A., Wang, G., Frankel, S. S., and Ho, D. D. (1998). In vivo distribution of the human immunodeficiency virus/simian immunodeficiency virus coreceptors: CXCR4, CCR3, and CCR5. *J Virol* **72**, 5035–5045.

Zhao, J., Lopez, A. L., Erichsen, D., Herek, S., Cotter, R. L., Curthoys, N. P., and Zheng, J. (2004a). Mitochondrial glutaminase enhances extracellular glutamate production in HIV-1-infected macrophages: Linkage to HIV-1 associated dementia. *J Neurochem* **88**, 169–180.

Zhao, M. L., Si, Q., and Lee, S. C. (2004b). IL-16 expression in lymphocytes and microglia in HIV-1 encephalitis. *Neuropathol Appl Neurobiol* **30**, 233–242.

Zheng, J., Thylin, M. R., Ghorpade, A., Xiong, H., Persidsky, Y., Cotter, R. et al. (1999). Intracellular CXCR4 signaling, neuronal apoptosis and neuropathogenic mechanisms of HIV-1-associated dementia. *J Neuroimmunol* **98**, 185–200.

Zou, Y. R., Kottmann, A. H., Kuroda, M., Taniuchi, I., and Littman, D. R. (1998). Function of the chemokine receptor CXCR4 in haematopoiesis and in cerebellar development. *Nature* **393**, 595–599.

Index

B

C

D

E

F

G

H

I

J

K

L

M

N

O

P

R

S

T